PRACTICAL CONSERVATION BIOLOGY

DAVID LINDENMAYER & MARK BURGMAN

CSIRO
PUBLISHING

National Library of Australia Cataloguing-in-Publication entry

Lindenmayer, David
Practical conservation biology

Bibliography
Includes index
ISBN 0 643 09089 4

1. Conservation biology – Australia. 2. Nature conservation – Australia.
3. Plant conservation – Australia. 4. Biological diversity
conservation – Australia. 5. Environmental management – Australia.
I. Burgman, Mark A. II. CSIRO Publishing. III. Title

333.95160994

Available from
CSIRO PUBLISHING
150 Oxford Street (PO Box 1139)
Collingwood VIC 3066
Australia

Telephone: +61 3 9662 7666
Local call: 1300 788 000 (Australia only)
Fax: +61 3 9662 7555
Email: publishing.sales@csiro.au
Web site: www.publish.csiro.au

Front cover
Background image: Photo by David Lindenmayer with permission from the people of the Blackstone Community in Western Australia. Other photos from left to right: Mainland Tiger Snake; Remarkable Rocks, Kangaroo Island, South Australia; Sugar Glider (photo by Mike Greer); Market Garden, Dandenong Ranges.
Back cover
Guanacos, Tierra del Fuego, Chile.
All photos by David Lindenmayer unless otherwise noted.

Set in Minion
Cover and text design by James Kelly
Index by Russell Brooks
Typeset by Thomson Press
Printed in Australia by Ligare

Contents

Preface

The title of this book is *Practical Conservation Biology* because part of its focus is 'how to do practical conservation biology'. The inclusion of information on the application of particular methods, we believe, sets this book apart from many other conservation biology texts. Much of the book is aimed at an Australian audience, and this is for good reason - most conservation biology texts are written for the North American market. But students and practitioners of conservation biology in Australia often relate best to examples from this country. Moreover, many aspects of conservation in Australia are different from those elsewhere in the world. Despite the Australian focus, the vast majority of methods and the lessons from the case studies are relevant to conservation elsewhere. We hope that conservation biologists outside Australia will also read this text – the exchange of ideas between scientists in different countries reduces the number of wheels that are reinvented.

Writing this book proved to be a monumental task. There is an enormous body of excellent work on conservation biology. Indeed, the literature is now proliferating so rapidly that it is impossible for any one person or small group of authors to stay up to date with it – any book will be out of date before it is published. More than 120 books published in the last five years have the word 'conservation' in the title. Tim New has brought out an excellent book titled *Conservation Biology. An Introduction for Southern Australia* (2000). There are also updated editions of Northern Hemisphere-focused conservation biology texts by Hunter (2002) and Primack (2002) as well as new books from Europe (Sutherland, 2000; Pullin, 2002). There has been a second round of reporting in the Australian State of the Environment series (published in 2001) as well as the National Land and Water Resources Audit (released in 2001 and 2002). There also have been several major reports on the global environment and the state of biodiversity (e.g. Groombridge and Jenkins, 2002; Millennium Ecosystem Assessment, 2005). Much of the information from those reports has found its way into this book. There is also a profusion of information on the Internet and in several chapters we cite the locations of relevant web sites, although readers need to be aware that the addresses of these can change and that information on web sites is of varying reliability.

We apologise to those people whose excellent work we have overlooked. Inevitably, there had to be a considerable body of material that could not be included – often simply because to include it would have made the book too large and unwieldy.

Based on feedback from reviewers on the first edition of this book (titled *Conservation Biology for the Australian Environment*), two new chapters have been added to *Practical Conservation Biology*: one on fire and another on landscapes and habitat fragmentation. The first was added because fire has major impacts on biota and is an integral part of the Australian environment. The second chapter on landscapes and habitat fragmentation has been included because this topic has become a major part of mainstream conservation biology.

We hope to receive feedback on the deficiencies of this book, so that if we have sufficient energy to produce a new book, it will be better than this one. But most importantly, we hope that the information presented in this book will stimulate further interest in conservation biology and encourage more Australians to make a contribution to the conservation of the country's unique biological resources.

David Lindenmayer and Mark Burgman

May 2005

Acknowledgements

This book would not have been completed without the efficient work of Monica Ruibal and Nikki Munro, who found numerous references and constructed many of the tables and figures. We thank Kate Thompson for making the drawing of the giant insects. We are greatly indebted to Joern Fischer and a number of other students who read parts of the book and made many critical comments that vastly improved earlier versions of the manuscript. We are grateful to Claire Drill, Kuniko Yamada and Rebecca Montague-Drake for their work on the figures and for proof-reading drafts of the document. Catherine Hunt did a magnificent job in editing the text.

The authors have made every effort to contact the owners of copyright for figures used in the text. We thank those people who have given us permission to use their work. In some cases, we were not able to find the appropriate person. We apologise for these and any other omissions or oversights in the use of copyrighted material. Any residual mistakes, omissions, and other errors are entirely the work of the authors. DBL is most grateful to all those who kindly provided photographs for this book.

Nick Alexander championed this work and encouraged it to be completed. DBL is greatly indebted to Harvard University (Harvard Forest), who provided a six-month Bullard Fellowship to facilitate the initial stages of book writing.

Finally, we thank our long-suffering partners and families, who have had to deal with the not inconsiderable stresses associated with our book-writing over the years.

General introduction

Since the mid-1980s, conservation biology has become a major research and teaching discipline throughout the world. This book is intended to provide an introduction for advanced undergraduates who are interested in conserving the Australian environment. We introduce methods that are important for detecting and solving conservation problems. We assume familiarity with Mendelian genetics, linear algebra and probability, but not a working knowledge of calculus or computer programming. We expect the reader to have been exposed to the kind of practical statistics that biologists should be taught and are expected to use in a routine way, such as the calculation of confidence intervals and linear regression.

Stromatilites at Hamelin Pool in Shark Bay, Western Australia – some of the oldest living life forms on the planet. (Photo by David Lindenmayer.)

A focus on the Australian environment

We have taken an Australian perspective for several reasons. First, few texts deal adequately with the conservation of the Australian environment. The second reason is that the Australian continent and its biological resources are unique. Australia is the oldest, flattest, and most isolated continent in the world. It supports some of the oldest living life forms on the planet – the 3.5-billion-year-old blue-green algal stromatolites in Shark Bay in Western Australia. Australia's marine ecosystems support the largest area of coral reef, the longest fringing reef (at Ningaloo in Western Australia), the largest areas of seagrass meadows and the most species-rich marine fish, mangrove and algal assemblages in the world. Australia is the driest inhabited continent, with less than 0.2% of the land surface subject to snowfall, and an even smaller area where snow persists for more than 30 days per year. Australia also has the smallest area of wetland of any continent. The physical environment is also remarkable, because climatic conditions over much of Australia are highly variable over time – more variable than anywhere else (McMahon *et al.*, 1992a). Australia's soils, and its freshwater and marine systems, are nutrient-depleted. All these factors directly influence the ecology and dynamics of the species that inhabit the continent and the marine systems surrounding it.

The terrestrial flora and fauna of Australia, as well as the continent's marine and aquatic ecosystems, differ markedly from those elsewhere on the planet. Most taxonomic groups are species-rich and highly endemic (Table A1). Australia is one of only two of the world's 17 megadiverse countries (i.e. nations that together harbour 60–70% of the earth's species) that have a developed, industrialised economy. Many terrestrial environments in Australia are dominated by eucalypts (*Eucalyptus* spp., with between 650 and 820 species depending on taxonomy) and wattles (*Acacia* spp., with more than 700 species). Many ecosystem processes in Australia are unique; for instance, Australia has a greater proportion of hollow-dependent vertebrates than any

Table A.1. Levels of endemism in Australia and other features of the nations biodiversity. (Data from various sources including Mittermeier *et al.*, 1997; State of the Environment Report, 2001.)

Group	Comment
Marine fish	One of the most diverse fish faunas in the world; >3500 species
Sharks and rays	50% of the world's sharks and rays are confined to Australian inshore waters
Terrestrial vertebrates	1350 endemic terrestrial vertebrates - far higher than the next highest country (Indonesia with 850 species)
Terrestrial mammals	305 species of which 258 (85%) are endemic, more than 50% of the world's marsupial taxa are confined to Australia
Birds	1/6th of the world's parrots occur in Australia; >50 species (second highest level of endemism after Brazil and the same as Colombia)
Reptiles	89% endemic, some groups such as front-fanged snakes (Family Elapidae), pythons, sea-snakes, skinks, goannas and geckos are more diverse than elsewhere in the world. Australian deserts have the world's highest reptile species diversity.
Frogs	93% endemic (highest of any vertebrate group in Australia) >200 species
Marine invertebrates	Southern Australian coastline supports the highest diversity of crustaceans, sea squirts and bryozoans in the world
Vascular plants	85% of flowering plants are endemic, >30 000 species of flowering plants, 12 endemic plant families (the highest in the world), over 50% of the world's mangrove species
Marine algae	Highest known diversity of red and brown algae in the world occurs on the southern coastline; >1150 species
Ectomycorrhizal fungi	95% endemic (22 genera and 3 families endemic)

other nation, even though the processes of cavity formation in trees are slow because the woodpeckers that promote hollow formation in other parts of the world are absent (Gibbons and Lindenmayer, 2002).

The third reason we have focused on Australia is that it has unique environmental problems. For example, in contrast with other continents, the most pressing threats to the Australian biota are not in tropical rainforests or temperate forests, but in temperate lowland woodlands, grasslands, mallee and heath (Hopper, 1997). The magnitude of salinity problems, with potentially up to 17 million hectares affected by 2050, and the approaches used to counter the problems (see Stirzaker *et al.*, 2002) are rare elsewhere in the world. At a finer scale, within fragmented landscapes, fragment-edge processes such as increased bird nest parasitism by cuckoos, which is common in the Northern Hemisphere, are not apparent in Australia. Similarly, large charismatic carnivorous mammals such as wolves and large cats, which are used as flagship and/or umbrella species in the Northern Hemisphere, are absent from Australia. The largest and most widespread mammalian carnivores are exotic. These and many other differences in conservation biology and management set Australia apart from other continents, regions and nations.

Salt-affected land in South Australia. (Photo by David Lindenmayer.)

The fourth factor contributing to the Australian focus of this book is that unique social and economic factors influence biological conservation here. Australia has the longest-known continuous human culture in existence, with Aboriginal inhabitants being present for at least 40 000 years and possibly 70 000 years. Australia is also one of the most urbanised human societies, with one of the lowest population densities, but one of the highest per capita population growth rates (Smith, 1994; Foran and Poldy, 2002). It has a human population density five times lower than the next least densely populated megadiverse nation, New Guinea (State of the

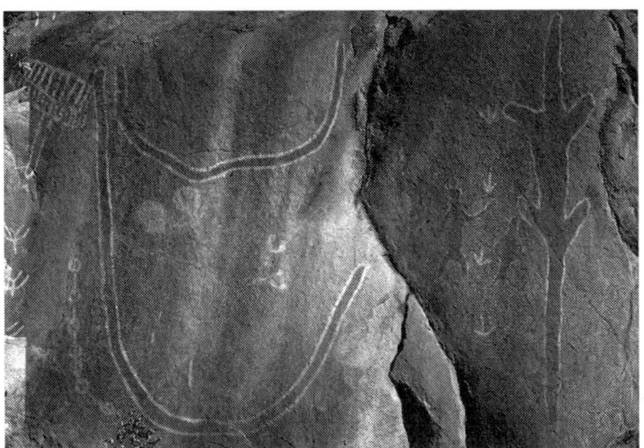

Rock paintings are a key part of indigenous Australia culture – the longest continuous human culture in existence. (Photo by David Lindenmayer with permission from the people of the Blackstone Community in Western Australia.)

Environment, 2001a). Australia's per capita energy and resource use is relatively large, even by the standards of developed economies – 200 tonnes of natural resources are moved to support each Australian each year, which is $2^1/_2$ times the equivalent for a person in the USA and five times that for someone living in Japan (Foran and Poldy, 2002). Thus, Australians have one of the heaviest 'ecological footprints' of any people on earth (Wackernagel *et al.*, 1997).

High levels of consumption translate into high levels of impact on the environment; in fact, Australia has one of the highest rates of land degradation of any nation (Graetz *et al.*, 1995; Chisholm, 1999; State of the Environment, 2001a). Australia also leads the world in recent recorded mammal extinctions (Short and Smith, 1994; although see Balmford, 1996) and has the highest per capita number of threatened species (Table A2). Similarly, Australia has the highest rate of land clearing of any developed nation (Australian Conservation Foundation, 2001). High levels of land clearing in other megadiverse countries such as Brazil, Mexico, Zaire, Madagascar and Indonesia are associated with rapidly expanding human populations, poverty and migration patterns. Such associations do not hold in Australia's low-density, affluent society.

Table A.2. Proportion of groups that are nationally extinct, endangered, or vulnerable. (Data from State of the Environment, 2001.)

Group	Proportion threatened (%)
Higher plants	8
Birds	14
Marsupials	23
Reptiles	8
Frogs	18
Freshwater fish	9

Structure of the book

This book comprises four broad sections. We anticipate that some people will read the book from beginning to end, whereas others will select sections or methods that interest them. The first two sections deal with general principles and are appropriate for introductory subjects in conservation biology. The third section explores technical issues, and is suitable for advanced students and research workers. The fourth section explores some principles of management for conservation. At the end of each chapter we provide a summary of practical implications, details of scientific papers and books for further reading. Key terms and phrases are highlighted in bold throughout the book and are defined in the glossary (Appendix II). A list of taxonomic names for organisms referred to throughout the book is in Appendix I (note also that the names of taxa are spelled with capitals in this book when they refer to a particular genus or species).

Part I deals with the principles of conservation biology. In it, we outline some of the general areas that underpin the discipline of conservation biology, including the basis for conservation, an exploration of what should be conserved, methods for classifying threats, types of protected areas, and *ex situ* conservation.

Part II examines human impacts on the natural world. It encompasses changes in the physical environment, threatening processes, loss of genetic variability, loss of species and populations, range changes, the effects of exotic plants and animals, harvesting of natural populations, habitat loss, habitat fragmentation, and fire. Part II concludes with a review of the demands of the human population.

Part III focuses on a small subset of the many methods of analysis for conservation biology. It is relevant to those using analytical methods to solve problems in conservation. The topics covered include measuring genetic variation, habitat analysis, reserve design, bioclimatic modelling, measuring diversity, monitoring, statistical power, indicators, and risk assessment.

Part IV of this book examines some general themes in conservation biology, particularly as they relate to ecologically sustainable development. In it, we build general arguments for conservation and provide a rationale for the focus on sound scientific practices in conservation biology.

Even in a textbook such as this, it is possible to provide only a cursory treatment of topics associated with conservation biology. We are also acutely aware that a number of important topics, relating both directly and indirectly to conservation biology, have not been tackled. We have omitted (among many other topics) geographic information systems, environmental law, national and international legislation, trade, environmental pollution, epidemiology and toxicology.

Part I: Principles for conservation

The multifaceted nature of conservation biology

Conservation biology is the applied science concerned with the management of biological diversity (or biodiversity). Although biological information is needed to inform decision-making about conservation and natural resource management (Clark, 2002), conservation biology is more than just ecology – much of ecology has no conservation implications. Rather, conservation biology draws on methods and problem-solving skills from many disciplines, including animal behaviour, chemistry, ecology, education, genetics, geography, mathematical demography, medicine, philosophy, policy, political science, sociology and statistics (see Ludwig *et al.*, 2001; Aguirre *et al.*, 2002). In fact, conservation biology could be thought of as a meta-discipline (Brewer, 2001; see Begg and Leung, 2000; Pullin and Knight, 2001; Fazey *et al.*, 2005a).

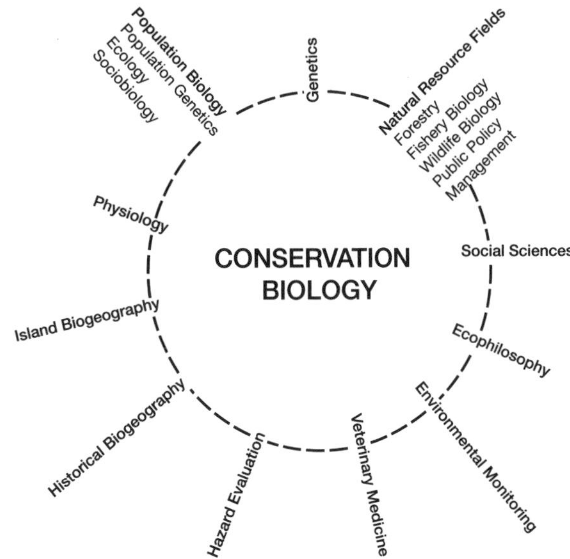

Figure B1. Soulé's (1985) vision of conservation biology as a multidisciplinary field. Some contributing areas are arguably no longer appropriate, such as island biogeography (see Chapter 10). (Redrawn from Soulé, 1985.)

Soulé's conservation biology principles

Two decades ago, Soulé (1985) outlined the broad principles that underpin a multidisciplinary approach to modern conservation biology (Figure B1). In essence, the principles state that biodiversity and potential for evolutionary development are intrinsically valuable, and that conservation biology seeks to reduce rates of species extinction and biodiversity loss. Soulé's (1985) principles influence much of the discussion in this text.

The tension between 'pure' and 'applied' conservation biology

Conservation biology goes beyond the traditional boundaries of the life and physical sciences to include such things as advocacy (New, 2000; Ludwig *et al.*, 2001). It shares this attribute with epidemiology, in which practitioners believe there is an obligation to influence public policy and decision-making, for the sake of public good. There can be tension between the results of 'pure' conservation research and the 'real-world', practical applications of conservation biology. This tension is embodied in the trade-offs between generality, precision and realism illustrated in Figure B2. Whelan *et al.* (2002) describe this as the 'fire triangle', with reference to fire ecology research. The fire triangle is based on earlier ecological work by Harper (1982) and Hairston (1989), and is equally relevant to applied conservation biology.

The philosophy in Figure B2 is important because, although there are some broad generalities in

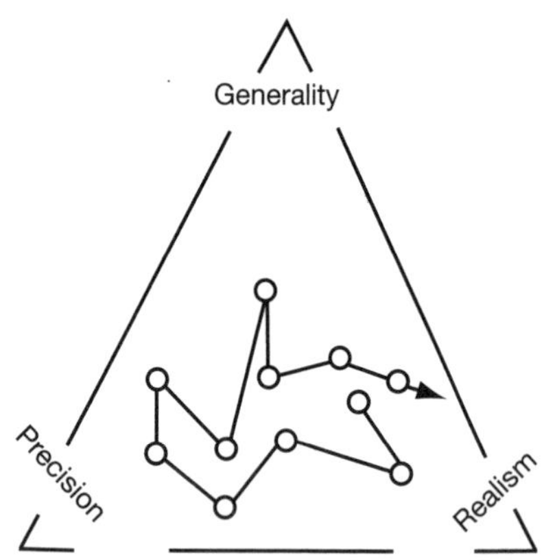

Figure B2. The conservation biology triangle, adapted from fire research (see Whelan *et al.*, 2002) and syntheses of ecological work by Harper (1982) and Hairston (1989).

conservation biology, application details are nearly always specific to a species, a group of species, a site, a landscape or a region. Such complexity is the reality of conservation biology. Krebs (1999) highlighted the disappointment of numerous ecologists who have found that their supposed 'generalities' did not apply to other systems.

At the outset of a book on conservation biology, it is important to establish the boundaries of the discipline and to set the framework within which problems are identified and solved. Part I of this book explores, in general terms, why we wish to conserve biodiversity (Chapter 1), what it is we wish to conserve (Chapters 2 and 3), and some of the mechanisms by which conservation can be achieved (Chapter 4).

Box B1

Real world conservation biology: dealing with `wicked environmental problems'

At the end of almost all sections of this book there are important caveats about the uncritical application of concepts, methods, equations and general conservation tools. Whether the topic is simple rules for reserve design, ratio estimation for species diversity, indicator taxa, species-loss equations, or habitat suitability indices, we outline the reasons why these approaches are often not `magic bullets' and why their uncritical application could actually deliver poor biodiversity conservation outcomes. If anything, these sentiments highlight the challenges posed by practical conservation biology, and emphasise that the science of conservation biology is inexact, young, and still evolving. Much of what is in this book will be found to be wrong in 20 years; but making mistakes and learning how to better manage ecosystems and conserve biodiversity is a natural part of the scientific process and the evolution of conservation biology as a meta-discipline (Redford and Taber, 2000; Berger *et al.*, 2004).

An additional issue associated with the multidisciplinary nature of conservation biology is that the field often deals with what have been termed `wicked environmental problems'. Ludwig *et al.* (2001) noted that such problems are truly complex in that they have: (1) no definitive formulation, (2) no stopping rules to determine when a problem has been appropriately addressed, (3) multiple legitimate human perspectives, (4) radical uncertainty, and (5) no test for a solution – in part because the solution will be unique in each case and it will not represent the final resolution of a problem. The outcome will often depend on how the problem is framed and by whom (Maddox, 2000). Lindenmayer and Franklin (2002) quoted Bunnell *et al.* (2003) as stating that `forest management is not rocket science – it is much harder'. This sentiment is perhaps even more true of conservation biology.

Chapter 1

Why conserve?

This chapter provides a framework for explaining how different attitudes to environmental management develop and coexist. It explores the different values people hold for biodiversity and the natural environment, such as utilitarian, consumptive, productive use, service, cultural, spiritual, experiential, existence, aesthetic, recreational, and tourist values. The chapter also explores the ethical basis for conservation. The topics discussed in this chapter are central to effective conservation practices because species and communities can be viewed either as objects used to serve human welfare, or as entities possessing value *per se*.

1.0 Introduction

Conservation biology attempts to conserve the diversity of living things (often termed biological diversity or **biodiversity**). There are many definitions of biodiversity (Bunnell, 1998; see Box 2.1 in Chapter 2). For the purposes of this book we define it as encompassing genes, individuals, demes, populations, metapopulations, species, communities, ecosystems, and the interactions between these entities.

This definition stresses both the numbers of entities (genes, species, etc.) and the differences within and between those entities (see Gaston and Spicer, 1998). It is similar to the definition of biodiversity proposed by the United Nations Environment Programme (UNCED, 1992):

> *the variability among living organisms from all sources including, inter alia, terrestrial, marine and other aquatic systems and the ecological complexes of which they are a part; this includes diversity within species, between species and of ecosystems.*

Most definitions of biodiversity in the literature consider genetic variation within species, the number of species and their relative abundances, variation in the composition of communities at the level of species and at other taxonomic levels, and the diversity of ecosystems and the processes that drive them (Harper and Hawksworth, 1995; Bunnell, 1998). Further discussion of the concept of biodiversity and its definition can be found in Chapter 2.

Objectives of conservation

The objectives of the United Nations Convention on Biological Diversity (1992, cited in CCST, 1994) include:

> *the conservation of biological diversity, the sustainable use of its components, and the fair and equitable sharing of the benefits arising out of the utilisation of genetic resources.*

Conservation is defined by the World Conservation Strategy (IUCN, 1980) as:

> *The management of human use of the biosphere so that it may yield the greatest sustainable benefit to present generations while maintaining its potential to meet the needs and aspirations of future generations.*

The reasons for conserving biodiversity are influenced by underlying human values and philosophies (see Table 1.1). Philosophies differ markedly between individuals and organisations, even those dedicated to conservation (Redford *et al.*, 2002).

Table 1.1. Summary of values and ethical positions underpinning the conservation of biodiversity. (Each of the values is discussed in subsequent parts of Chapter 1.)

Utilitarian value
Consumptive value
Productive use value
Service value
Scientific and educational value
Cultural, spiritual, experiential and existence value
Aesthetic, recreational and tourist use
Intrinsic value
Ecocentric ethic
Biocentric ethic
Future option value
Future discoveries of utilitarian and/or intrinsic value
Precautionary principle

Temporal changes in philosophies and opinions

Dominant philosophies on the use and conservation of biological resources are not static. For example, the Royal Commission (1931) into resource use and human settlement in northern Queensland reported that:

> *Queensland needs no forestry science for present requirements. There is an abundance and enough of timber for all. Business or common sense management, and not science, is the first requisite. The productive wealth of the country at present suffers from the fact that there are too many, rather than too few trees. That is why ringbarking campaigns are being organised (p. 22).*

This sentiment contrasts with that of Gould (1870), who wrote more than a half a century before the Royal Commission:

> *Australia – a part of the world's surface in maiden dress, but the charms of which will 'ere long be ruffled and their true character no longer be seen! Those charms will not long survive the intrusion of the stockholds, the farmer and the miner, each vying with each other to obliterate that which is pleasing to every naturalist; and fortunate do I consider the circumstances which induce me to visit the country while so much of it remained in its primitive state'.*

Similarly, Aldo Leopold (1949), recommended that we should:

> *Quit thinking about decent land-use as solely an economic problem. Examine each question in terms of what is ethically and aesthetically right as well as what is economically expedient. A thing is right when it tends to preserve the integrity, beauty, and stability of the biotic community. It is wrong when it tends otherwise (p. 262).*

Leopold was a forester, and one of the leading advocates for environmental protection in the USA. Environmental management practices reflect social and ethical attitudes, and in many instances, practices applied in the past would not be acceptable today. For example, in 1963, the Tallangatta District Conservation Officer in the Upper Murray region of Victoria noted in an unpublished report on grazing conditions in the Tatonga Timber Reserve, under the heading of 'vermin', that 55 wombats were killed in one section of the reserve towards the end of 1962, and that the rest of the area would be trapped in 1963. Today, wombats are a protected species.

In the following sections, we outline some current attitudes that motivate conservation. They range from rationalisations of benefits to people, through opportunity costs and ethical considerations of equity, to arguments about the intrinsic worth and moral status of non-human species.

1.1 Utilitarian value

Much of the biodiversity in many countries, including Australia, is on public land or aquatic areas that are subject to government controls. The question of the immediate and long-term use of this land (and its associated biodiversity) is a political and ethical issue. This is evident in the following provocative quote by Jensen (1984):

> *But I am decidedly and emphatically anti-preservationist, in that I do not condone or support the tying up of vast tracts of useful land (agricultural, grazing and others) in wilderness and parks to become nothing more than breeding grounds for all sorts of pests and vermin, into which people are not permitted to enter, except on foot, and which, unmistakably, create a wasteland of underdevelopment, leaving Australia open to criticism, even invasion, from land-hungry, food-starved people outside these shores.*

Given opinions like that of Jensen (1984), there is an imperative to determine the **economic** and **utilitarian** (and numerous other) values of biological

Box 1.1

Valuing ecosystems

Among economists, contingent valuation is perhaps the most popular method for valuing ecosystems and their services (e.g. Tisdell *et al.*, 2005). It uses questionnaires and interviews to elicit preferences and demand functions for environmental goods and services (Garrod and Willis, 1999). It has been applied to a range of environmental issues, including wilderness protection, water quality, and soil erosion. The approach can take into account ownership, access, social context, and perceptual biases (Slovic, 1999). However, the method has several unresolved technical and theoretical problems (Chee, 2004), including the influences of context and framing, and free-riding (in which an individual attains the value of something without outlaying resources or experiencing risk; Garrod and Willis, 1999).

`Market price' and `productivity' methods estimate the economic values of ecosystem products or services that are bought and sold in commercial markets, or that contribute to the production of commercially marketed goods (e.g. `clean' water contributes to agricultural productivity). Other approaches include `replacement cost' and `substitute cost' methods. Hedonic pricing means that the economic value of something is determined by how it influences market transactions. A valuer uses market valuations such as house prices to establish values for environmental attributes (e.g. what is a view `worth'). Stated preference techniques use direct consumer valuations of environmental values. Most methods rely on converting environmental preferences to monetary preferences, and each method has its own peculiarities (Chee, 2004). Approaches such as multicriteria decision analysis can assist diverse stakeholders to reach consensus on preferences and relative values. There is no easy solution to the problem of valuing ecosystems when values have inherently different scales.

Box 1.2

The value of biodiversity

The environment returns an estimated A$44 trillion in goods and services to human society each year (BirdLife International, 2000). Wood products contribute A$530 billion annually to the world market economy (or about 2% of the world's total gross domestic product; World Commission on Forests and Sustainable Development, 1999). Losses of biodiversity and flow-on impacts on ecosystem processes could have substantial social and economic costs (Costanza *et al.*, 1997; Pimentel *et al.*, 1997). For example, in 1992 it was estimated that populations of natural parasites and predators accomplished the equivalent of A$130–265 billion worth of pest control in the USA, compared with the A$25 billion expended on artificial control measures such as spraying (Pimentel *et al.*, 1992).

In an Australian context, crude estimates by Jones and Pittock (1997) valued Australian terrestrial ecosystems at A$325 billion per year. The State of the Environment Report (2001) estimated the annual value of some biodiversity-based industries to the nation's economy. These included: commercial fisheries (A$2.3 billion), woodchips from native forests (A$590 million), honey production (A$300 million), kangaroo harvesting (A$245 million), bushfood production (A$100 million) and wildflower exports (A$30 million). Biodiversity-related tourism, which depends on the nation's unique animals, plants and ecosystems is worth several billion dollars.

resources (Redford *et al.*, 2002; Table 1.2). Calculating the economic value of the products and processes of nature is not a simple task (Chee, 2004), and many attempts have been made (Norton, 1987; McNeely, 1988; Bergstrom, 1990; McNeely *et al.*, 1990; State of the Environment, 2001a; Zedler, 2003).

In the first part of the following section, we outline a classification of the utilitarian value of the natural world. It represents one of several ethical perspectives on the environment, and serves to outline the usefulness of conservation practices from a human perspective. We explore some issues associated with utilitarian values, including tourist impacts and the equitable use of genetic resources.

Consumptive use value

Consumptive use refers to the products of nature that do not pass through a market. The most important direct uses of biodiversity by humans are as food, medicine, fuel and building materials, even though only a very small proportion of the plants and animals that are potentially nutritious are used as food. Consumptive use is usually more diverse, and depends on a much wider spectrum of the available biota than does market-based

Table 1.2. Total estimated economic benefits of biodiversity worldwide (from Pimentel *et al.*, 1997).

Activity	Economic benefit (US$ x 10^9)
Waste disposal	760
Soil formation	25
Nitrogen fixation	90
Bioremediation of chemicals	121
Crop breeding (genetics)	115
Livestock breeding (genetics)	40
Biotechnology	6
Biocontrol of pests (crops)	100
Biocontrol of pests (forests)	60
Host plant resistance (crops)	80
Host plant resistance (forests)	11
Perennial grains (potential)	170
Pollination	200
Fishing	60
Hunting	25
Seafood	82
Other wild foods	180
Wood products	84
Ecotourism	500
Pharmaceuticals from plants	84
Forests sequestering carbon dioxide	135
Total	2928

use. Many people, particularly those who live in traditional ways, for example some Australian Aboriginal and Torres Strait Islander populations, depend directly on the natural environment for live game, firewood, edible plants, medicines, building materials, weapons, transport, cultural and spiritual items, raw materials for other technology, and trade goods.

Consumptive fishing by local Burmese people on Inle Lake in Myanmar. (Photo by David Lindenmayer.)

Consumptive use is a significant part of the livelihood of people in many nations. In Ghana, approximately 75% of the human population depends largely on traditional, natural sources of protein, including fish, insects and snails. Firewood and dung provide more than 90% of the total primary energy needs of Nepal, Tanzania and Malawi (McNeely *et al.*, 1990). As many as 80% of people in developing countries rely on traditional medicines derived from wild plant and animal populations (WCMC, 1992; Population Action International, 2000). Despite their importance, especially to people in developing countries, consumptive use values rarely appear in national income accounts (McNeely *et al.*, 1990; Population Action International, 2000).

Productive use value

Productive use values refer to commercially harvested biological resources. That is, products that pass through the marketplace. Some examples are firewood, timber, fish, animal skins, musk, medicinal plants, honey, beeswax, fibres, gums, resins, oils, construction materials, ornamental plants, animals harvested for game meat, animal fodder, mushrooms, fruits and dyes (Beattie and Ehrlich, 2001; AFFA, 2003a). Biological resources can be taken from natural ecosystems in commercial quantities, and are important in many national economies. For example, drugs derived from rainforest plants had an accumulated net worth of approximately US$150 billion per year in 2000 (Beattie and Ehrlich, 2001). Prescott-Allen and Prescott-Allen (1986, cited in McNeely *et al.*, 1990) estimate that wild species contribute 4.5% to the gross domestic product of the USA. The percentage contribution of wild species to the economies of developing countries is usually higher (Groombridge and Jenkins, 2002).

Wild biological resources also contribute to domesticated production systems (Vavilov, 1949). In particular, native species contribute to pasture, and wild species, especially plants, serve as sources of new domesticates and provide a pool of genetic variation from which new material can be introduced into existing gene pools (Myers, 1990a; Population Action International, 2000). Variety in plant and animal species insures against climate change and catastrophic disease (Myers, 1990a). For example, Wild Rice is the source of resistance to viruses in commercial rice varieties (McNeely *et al.*, 1990), and resistance to rust, a fungal disease common in Wheat, was discovered in the wild relatives of domestic Wheat (Myers, 1993).

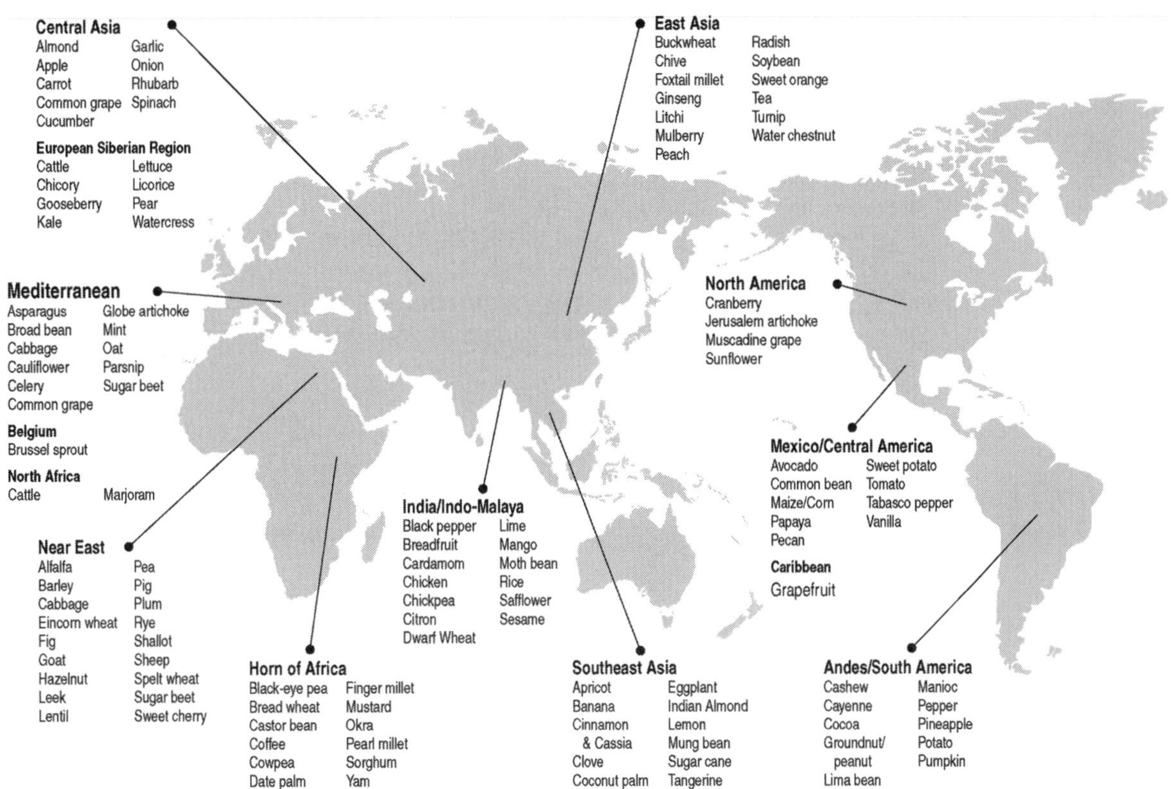

Figure 1.1. Origins of crops and livestock on a global basis. (Redrawn from Myers, 1990a; Population Action International, 2000.)

However, there has been a major loss in the diversity of agricultural genetic resources over the past 50–100 years. Groombridge and Jenkins (2002) reported that only 10% of the Wheat varieties used in 1949 remained 20 years later. Similarly, 80% of the varieties of Apple, Cabbage, Corn, Pea and Tomato were lost between 1804 and 1904.

Just a few species currently provide most of the food for most people on earth. The productivity of natural resources in most countries is dominated by plant and animal species from other countries and regions (Vavilov, 1949; see Figure 1.1). Of the 20 most important food crops and the 20 most important industrial plant species globally, more than 90% of the germ plasm originates from developing countries (Kloppenburg and Kleinman, 1987; Population Action International, 2000). Wild genetic resources from Central America and Mexico serve the needs of maize growers and consumers globally (Myers, 1990a; Population Action International, 2000). Many of the principal Cocoa-growing nations are in West Africa, whereas the genetic resources on which they depend are found in the forests of western Amazonia. More than 98% of the agricultural produce of the USA and Australia is derived from non-native species. Half the crop production of North America is derived from species originating in Asia or Africa, and 30% of Asia's crop production involves species from the Americas or Africa (WCMC, 1992; Population Action International, 2000). Statistics such as these raise issues about the ownership of genetic resources, the rights of access to these resources, and mechanisms for the distribution of wealth that are derived from these resources.

Making use of biodiversity depends on knowledge about available resources. Our knowledge of invertebrates is poor (Chapter 6), and yet their potential economic benefit is substantial. Beattie (1994) and Beattie and Ehrlich (2001) provide examples of products developed from invertebrates, ranging from new adhesives to anticoagulants and antibiotics in human medicine (Table 1.3). Some species of spiders produce a type of web that is adapted to capturing fast-flying insects. These webs have many useful properties: they are light, have considerable mechanical strength, and have an

Table 1.3. Some economic applications of invertebrates (after Beattie and Oliver, 1994; Beattie and Ehrlich, 2001).

Application	Taxonomic groups
Adhesives	Onychophorans (velvet worms), annelids
Biocontrol	Nematodes, mites
Brain research	Sea slugs, nematodes
Ant-repellents	Ants, wasps
Bird-repellents	Leaf hoppers
Termiticides	Ants
Anti-coagulants	Leeches
Biological control of weeds	Leaf hoppers
Cryoprotectants	Collembola, mites
Industrial products (concrete, car parts, ceramics, fibres)	Molluscs, spiders
Control of crystallisation	Molluscs
Development of antibiotics	Ants

ability to absorb large quantities of kinetic energy, thus they have many applications in industry. One of their uses is in the construction of bullet-proof vests, which are packed with spider web (Beattie, 1994; Lewis, 1996; Beattie and Ehrlich, 2001). Invertebrates such as grasshoppers are increasingly being used for food in many countries, particularly in Asia.

Although only a tiny proportion of invertebrate biodiversity is ever likely to have major, direct economic benefits for humans (Lawton, 1991), conserving them will ensure future opportunities to develop such resources (Beattie and Ehrlich, 2001).

Ecosystem service value

Ecosystem services are the processes by which natural ecosystems sustain human life. They include producing goods such as food, fuels and pharmaceuticals, and services such as biodiversity maintenance and waste assimilation (Daily, 1997b, 2000; Hooper *et al.*, 2005). In general, management goals are developed with ecosystem processes or services in mind. Biodiversity has important service values because it supports ecological functions and ecosystem processes (e.g. see Naeem *et al.*, 1994; McGrady-Steed *et al.*, 1997; Naeem, 1998, 2002; Duffy, 2003; Naeem and Wright, 2003) on which consumptive and productive values depend (Commonwealth of Australia, 2002a). Service values include pollination; **gene flow**; predation; **competition**; maintenance of water cycles (see Box 1.3); provision of nurseries for commercial fish species (in mangroves and coral reefs in particular); regulation of climate; soil production and protection; storage and cycling of essential nutrients; and the absorption, breakdown and dispersal of wastes, pesticides and pollutants (Kremen, 2005). Daily (1999) classified these and other ecosystem services into several broad categories:

- *production of goods* such as food, pharmaceuticals, durable materials, energy, industrial products and genetic resources
- *regenerative process* such as cycling and filtration processes
- *stabilisation processes* such as coastal and river bank stability and the control of pest species
- *life-fulfilling processes* such as aesthetic beauty and serenity
- *preservation of future options* such as new goods and services awaiting discovery.

Successful crop cultivation requires several other species in addition to the crop, such as soil

Box 1.3

Ecosystem services: New York's water supply

On average, a 0.06-hectare area of clean uninhabited water catchment is needed to supply each person living in an urban environment with good-quality water (Foran and Poldy, 2002). The more modified a catchment (by roads, for agriculture etc.), the greater the area that is needed to produce a fixed volume of water, the lower the quality of the water, and the higher the treatment costs.

The economic values of ecosystem services are illustrated by the costs of water treatment versus well maintained water catchments for the city of New York. Water treatment plants to supply the city have been estimated to cost A$8–10.5 billion, with annual costs of A$400 million for upkeep and maintenance. In contrast, the 20 000 hectares of land purchased to produce the same amount of water and improve water catchment coverage in the Catskill/Delaware and Croton catchments in north-eastern USA cost much less than a quarter of this – less than A$3 billion (Chichilnisky and Heal, 1998). Notably, similar approaches to set aside forest areas to maintain water quality (and reduce the need for major water treatment works) were taken for the water supply of Boston and neighbouring cities many decades earlier. Entire towns were relocated as part of closing the Quabbin Water Catchment in western Massachusetts to urban and agricultural development.

Large areas of former urban settlements were purchased in the Quabbin Reservoir region to ensure the quality of water catchments that provide water to Boston and surrounding areas. Water production from the forested catchments is an important ecosystem service for many people in the region. (Photo courtesy of Les Campbell and Alfredo's Photographic Gallery, Amherst, Massachusetts.)

micro-organisms and pollinators, and the virtual absence of many others, particularly insect and vertebrate herbivores. The economic value of pollinators can be significant and they can significantly boost agricultural production (De Marco and Coelho, 2004). Coffee production is a useful example: forest-based native pollinators from 1 kilometre or closer can boost yields by as much as 20% (Ricketts *et al.*, 2004). In another example, in the early 1980s, Greathead (1983) estimated that the value of insect-based pollination of commercial palms in Malaysia exceeded US$115 million per year. Nabhan and Buchmann (1997) estimated that the value of native pollinators to the economy of the USA is US$4-6 billion per year.

The growth of trees for timber production often depends on the presence of **symbiotic** fungi (Eldridge *et al.*, 1994). The control of the majority of potential crop pests and disease carriers depends on ecological processes such as predation. Extensive stands of natural vegetation are central components of systems that regulate ozone, oxygen, carbon dioxide and other gases in the atmosphere.

Wetlands have major ecosystem service value. Zedler (2003) estimated that although wetlands cover less than 1.5% of the planet's surface, they generate as much as 40% of the world's renewable ecosystem services. Some of the key services they provide and the associated monetary values of these are presented in Table 1.4. There are equivalent estimates for Australian

Table 1.4. Estimated values of ecosystem services from wetlands*. (Modified from Zedler, 2003, and based on Costanza *et al.*, 1997.)

Renewable ecosystem service	Value (US$ per hectare per year)	Value (US$ billion per year)
Hydrological services		
Water regulation	15–30	
Water supply	3800–7600	
Gas regulation	38–265	
Water quality services		
Nutrient cycling	3677–21 100	
Waste treatment	58–6696	
Biodiversity services		
Biological control	5–78	
Habitat/refugia	8–439	
Food production	47–521	
Raw materials	2–162	
Recreation	82–3008	
Cultural	1–1761	
Disturbance regulation	567–7240	
Global totals		
Coastal wetlands		8286
Inland wetlands		4879
Total for global wetlands		13 165
Total global ecosystem services for entire globe		33 268
Percentage from wetlands		39.6%

*All shallow-water habitats (tidal marshes and mangroves, swamps and floodplains, estuaries, seagrass/algal beds, and coral reefs) are included in the calculations.

aquatic ecosystems. For example, the rivers, wetlands and floodplains of the Murray–Darling Basin in Queensland, the Australian Capital Territory, New South Wales, Victoria and South Australia have been estimated to provide A$187 billion in ecosystem services (Thoms and Seddon, 2000).

The net worth of a proposal for resource development is determined in part by the environmental consequences of the action to undertake the development (Morton *et al.*, 2002). Ecosystem service values are highlighted when ecosystems become degraded to the point that the economics of restoration is measured in dollar terms. Highly degraded ecosystems are not effective providers of goods or services, and they can be very costly to rehabilitate (State of the Environment, 2001a). One of the problems of the ecosystem services concept is that values are often not recognised until they become impaired (McIntyre *et al.*, 2002). The treatment of land degradation in Australia, for example, has current direct costs of hundreds of millions of Australian dollars annually, possibly even billions of dollars (Madden *et al.*, 2000). The annual costs of **secondary salinity** (in terms of lost agricultural production and remedial treatments) within eight subcatchments in the Murray–Darling Basin is up to A$300 million (Wilson, 2000; see Chapter 5).

Ecosystem service arguments are not, on their own, a sufficient reason to conserve all forms of biodiversity: only a small proportion of species actually provide useful services (Hooper *et al.*, 2005). The degree to which a species that provides ecosystem functions can be substituted by different species is related to the concept of **ecological redundancy** (Walker, 1992; see Chapter 17). In fact, a relatively small number of species may be necessary for ecosystems to function successfully. However, given our current limited understanding of ecological processes, it is very difficult to say which species are necessary and which are superfluous (see Ellsworth and McComb, 2003, for a study of the hypothesised impacts resulting from the loss of the Passenger Pigeon in North America). The Thylacine in Tasmania is an example of possible ecological redundancy: to the best of our knowledge, the loss of this charismatic large carnivore has not compromised any ecosystem service. This is not to say this species had no ecological role or that, if it still persisted, attempts to conserve it would be unimportant; rather, no particular service value is apparent, and arguments for its conservation would need to consider other values. For instance, Beattie and Ehrlich (2001) contend that so-called redundant species insure ecosystem services against the loss of important species.

Box 1.4

Ecosystem services and option values (after FAO, 2003b).

Endod, also commonly known as the African Soap Berry, is a perennial plant that has been cultivated for centuries in many parts of Africa, where its berries are traditionally used for laundry soap and shampoo. In 1964, the Ethiopian biologist Aklilu Lemma observed that downstream from where people were washing clothes with Endod berries, dead snails were found floating in the water. Further research revealed that sun-dried and crushed Endod berries are lethal to all major species of snails, but do not harm other animals or people, and are completely biodegradable.

For Africa, where one of the most serious human diseases, schistosomiasis, is transmitted by freshwater snails, discovery of a low-cost and biodegradable lumicide (a chemical that kills snails) represented a major breakthrough. According to the World Health Organization, more than 200 million people are infected with schistosomiasis, and it kills an estimated 200 000 people every year. With support from international donors, Endod is undergoing further toxicological studies to ensure its safety. Dr Lemma views Endod as a product of traditional knowledge that can be developed by and for African communities.

Scientific and educational value

Natural resources are the basis of improved biological knowledge. Many regions support relatively undisturbed areas that can serve as scientific and educational reference areas. Most agencies responsible for managing natural landscapes have a research and education role, but these can only be fulfilled if representative areas are protected (Lindenmayer and Franklin, 2002). Beattie and Ehrlich (2001) provide a detailed discussion of the emerging field of biomimicry, which involves examining the solutions that native species have evolved to solve particular problems, and then using the insights to solve a similar problem faced by humans. For example, when seeking new antibiotics and antifungal agents to treat humans, perhaps the best place to search and learn more about them is in organisms that are likely to encounter large numbers

Box 1.5

Wilderness and biodiversity conservation

The word `wilderness' derives from the old English `wild deer-ness', and referred to uninhabited and uncultivated tracts of land occupied only by wild animals (Shea *et al.*, 1997). Wilderness is a human construct that relates to areas remote from human influence and infrastructure (Mackey *et al.*, 1998; Mittermeier *et al.*, 2003). The concept is usually linked to the spiritual, aesthetic enjoyment, and recreation needs of Western peoples in natural landscapes (Mackey *et al.*, 1999). The views of indigenous peoples (such as those that lived throughout the continents of North America and Australia), who occupied the land before European settlement and who did not have concepts of wilderness, have generally not been taken into consideration by advocates of wilderness (Hammond, 1991).

The idea of wilderness has an insidious implication in Australia and elsewhere. Langton (1998) points out that it is based on a failure to perceive the anthropogenic nature of apparently pristine landscapes, leading to an underestimation of the importance of indigenous knowledge and management systems. Langton calls the concept of wilderness a `science fiction', which is closely linked to the legal fiction of *terra nullius* – the idea that Australia was not occupied at the time of European settlement. The concept denies the imprint of millennia of Aboriginal impacts on, and relationships with, Australian species and ecosystems.

The role of wilderness or remote areas in conserving biodiversity is complex. Some authors assert that large wild areas are essential for conservation, even though such areas may not be species-rich (e.g. Mittermeier *et al.*, 2003). For example, Noss and Cooperrider (1994) state that `unless it contains many millions of acres, no reserve can maintain its biodiversity for long'. (See also initiatives such as the Wildlands Project; Foreman *et al.*, 1992.) There are some places where the only unaltered stream channel and deep-pool architecture in riparian systems occurs in roadless remote areas (McIntosh *et al.*, 2000). The conservation of particular groups (such as fish) can be related to such a lack of human disturbance, highlighting the value of wilderness areas (Baxter *et al.*, 1999).

The type and intensity of human disturbance will always be an important factor influencing biodiversity, but in some cases, measures of the integrity of remote areas may be outweighed by management practices in production landscapes. Lindenmayer *et al.* (2002a) found no correlation between the occurrence of any species of arboreal marsupial in Central Victorian forests and measures of the intensity of human development taken from the National Wilderness Inventory (Lesslie and Maslen, 1995), such as the distance of field survey sites from roads. Better predictors of species occurrence were found to be factors such as the extent of matrix management practices (e.g. levels of tree retention on harvest units). Similarly, at a continental level, there was no relationship between the number of threatened Australian mammal species and wilderness quality (as measured by the National Wilderness Inventory; Figure 1.2), although significant trends were recorded for vascular plants (Mackey *et al.*, 1999).

Some areas will be important for biodiversity conservation precisely because the remainder of the landscape that contains them has been subject to intensive human use. Schwartz and van Mantgem (1997) demonstrated this for the intensively modified landscapes of Illinois in the mid-west of the USA, where small reserves (less than 10–20 hectares) have valuable roles in conserving many plant and animal taxa. Similarly, Kirkpatrick and Gilfedder (1998) highlighted the importance of degraded ecosystems for rare plants in Tasmania. These ecosystems have been extensively transformed by humans, yet the remnants support numerous species that can be conserved by appropriate management.

Remote areas clearly have spiritual, aesthetic, and recreation values. Yet remote areas and biodiversity conservation are not always mutually inclusive (Brown and Hickey, 1990; Lindenmayer and Franklin, 2002). Although large relatively undisturbed areas have considerable conservation value, small reserves that are not remote also have much to contribute to the conservation of biodiversity.

of bacteria and fungi. Fly larvae (maggots) often feed in rotting flesh where bacteria and fungi are common and compete as detrivores. The larvae have evolved an array of chemicals with antibiotic properties that have potential use in human medicine. There are undoubtedly numerous other species, such as worms, ants and other insects that feed in and around rotting vegetation and flesh, that have developed protective

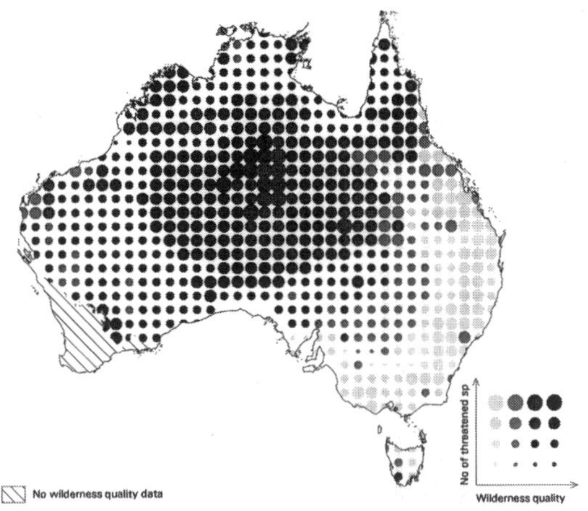

Figure 1.2. The lack of relationship between threatened mammal species and wilderness value. (Based on Mackey *et al.*, 1999.)

chemicals with antibiotic properties (Beattie and Ehrlich, 2001).

Cultural, spiritual, experiential and existence value

Many natural environments in Australia include sites of religious, spiritual and cultural significance, especially for Aboriginal people. Important sites include meeting places, birthplaces, historic sites (such as the location of a battlefield), and sacred sites (such as burial grounds). Natural environments also provide the opportunity for enjoyment and participation in traditional practices. Such values can be associated with large areas of land; for example, in Western Australia the Aboriginal Land Trust alone holds more than 27 million hectares, or approximately 12% of the State (Department of Indigenous Affairs, 2002).

A sense of spirituality in nature is shared by all people, and forms the basis for the view of natural landscapes known as 'deep ecology' (Devall and Sessions, 1985). In its submission to the Resources Assessment Commission (RAC, 1991), the Australian Heritage Commission defined cultural heritage values as:

> *features which are sacred to Aboriginal people, prehistoric archaeological sites going back to as far as 30 000 years, and material remains of historic activity since European settlement. All these places have some value. The value may be either aesthetic, or social, as being of religious, spiritual, symbolic, and/or contemporary importance to groups of people,*

Examples of remote areas in Australia and Australian territories: (top) Kosciuszko National Park in south-eastern New South Wales; (middle) the Australian Antarctic Territory; and (bottom) Cape Gantheaume in South Australia. (Photos by David Lindenmayer and Karen Viggers.)

or it may be historic or scientific, in the information sites may impart about past human activity and behaviour (p. 76).

Experiential and existence values are associated with large, natural, unmodified landscapes, sometimes called **wilderness** (Box 1.5). Remote areas provide services by maintaining natural water-flow rates, sedimentation and turbidity levels, and by providing waters that are free of contamination from fertilisers, weedicides and pesticides (Mackey *et al.*, 1998, 1999). Scenic qualities may enhance the experiential value of a place. Solitude is an important quality of remote areas, and implies limited access by humans. The need to preserve remote areas led to the dedication of national parks such as Yellowstone National Park in the USA and the Royal National Park in New South Wales – the first two national parks in the world. Although remote area values are largely non-consumptive, they can also have scientific, educational, recreational and water resource uses. Vicarious use of remote areas includes access through films, books and photographs, which are often the main sources of public support for the preservation of remote areas.

Box 1.6

Economic benefits of World Heritage Areas

Driml and Common (1995) examined the economic benefits of five major Australian World Heritage Areas and the economic/ecological trade-offs associated with the management of these areas. They studied the Great Barrier Reef and the Wet Tropics Management Area in Queensland, Kakadu and Uluru National Parks in the Northern Territory, and the South-west Tasmanian Wilderness area. These areas receive large numbers of international and domestic tourists (Blamey and Hatch, 1998). Driml and Common (1995) calculated that the total expenditure associated with visits to these five areas by tourists in 1991–92 was approximately A$1.4 billion, although they considered that this was an underestimate because it did not include money spent in travelling to the various World Heritage Areas. More than half of this amount was spent on accommodation and other services purchased in regions adjacent to the World Heritage Areas. Such expenditure supported more than 800 commercial tour operations (Driml and Common, 1995).

Thus, natural areas are extremely important national and regional economic assets for the Australian tourist industry and many regional economies throughout the nation (Australian Bureau of Statistics, 2003). However, the budgets dedicated to managing these natural areas are relatively small (see State of the Environment, 2001a), and typically amount to less than about 5% of the tourist expenditure. Agencies responsible for managing World Heritage Areas are not adequately funded to cope with the rapidly increasing numbers of tourists (Driml and Common, 1995).

Aesthetic, recreational and tourist use

In 2000–01, tourists in Australia consumed more than A$70 billion worth of goods and services, of which 76% was consumed by domestic tourists. About 550 000 people were employed in tourism in 2000–01, or 6% of total employment in Australia (Australian Bureau of Statistics, 2003). During this time, tourism accounted for 4.7% of Australia's gross domestic product (Australian Bureau of Statistics, 2003). Much of this tourism is founded on the unique composition and appearance of Australian animal and plant species, ecosystems and landscapes. For example, the Koala is believed to attract so many international visitors to Australia that its existence is worth more than A$1 billion per year (State of the Environment, 2001a). Because tourism is a significant contributor to the Australian economy, the natural environment also makes a significant direct contribution to the economy (ESDWG, 1991a; Driml and Common, 1995; Australian Bureau of Statistics, 2003). The gross value of tourism

The Wet Tropics Management Area in Queensland is one of the key areas for environment-based tourism in Australia. (Photo by David Lindenmayer.)

Examples of potential conservation problems arising from human impacts associated with tourism and recreation: (top) a major seabird rookery in Antarctica, where human disturbance effects on populations have been examined; (middle) nesting shorebirds on the south coast of New South Wales; and (bottom) vegetation trampling on Kangaroo Island, South Australia. (Photos by Karen Viggers and David Lindenmayer.)

exceeds that of natural resource industries such as agriculture, forestry or fishing (Australian Bureau of Statistics, 2003).

Impacts of tourism on biodiversity

The recreational benefits of natural environments are shared by bushwalkers, climbers, campers, birdwatchers, skiers, naturalist clubs, wildflower enthusiasts, four-wheel drivers, off-road motorcyclists, horse riders, anglers, hunters, cross-country skiers, scuba divers, surfers, white-water enthusiasts and others. However, visitors can represent potential threats to the natural environment (McMillan *et al.*, 2003; Mollner *et al.*, 2004; Table 1.5). Recreational activities can result in the direct disturbance of animals (e.g. Giese, 1996; Berger *et al.*, 2004) or the destruction of habitats (Wipf *et al.*, 2005), such as by divers on the Great Barrier Reef (Rouphael and Inglis, 1997) or by walkers on alpine feldmark in the high country of Kosciuszko National

Box 1.7

Whale-watching and regulating the impacts of whale-watchers

There are approximately 45 species of cetaceans in Australasian waters. Some of these are relatively common and others are rarely sighted (with only a handful of records). Whale-watching has become a major eco-tourism industry and there are guide books to assist people to find the places around the nation where these animals can be observed (Gill and Burke, 1999). In 1994 the global whale-watching industry had become a A$0.7 billion enterprise with more than 5.4 million tourists observing whales in that year (Hoyt, 1995). In Australia, in 1995 whale-watching was estimated to generate A$10 million per year and in 1999 there were more than 150 commercial whale-watching operators nationwide (Gill and Burke, 1999). In 2003, more than 1.5 million Australians watched whales. The industry is estimated to be worth A$270 million and has an annual growth rate of 15%.

However, if eco-tourism enterprises are not carefully managed, tourists can damage the natural resources they come to see (Lasseau, 2003). Australian Federal and State government agencies have developed guidelines to regulate the whale-watching industry and maintain the resource that is the core of the industry. Some of these guidelines are summarised in Table 1.6.

Table 1.5. Impacts of tourist and recreational activities on Australian plant taxa. Note that categories are not independent (after Kelly *et al.* 2003).

Reported threat	No. taxa affected	No. communities affected
Trampling by bushwalkers	20	3
Introduction and spread of Cinnamon Fungus by bushwalkers and vehicles	14	1
Non-specific recreational activities	17	–
Picking or collecting flowers or seed	10	–
Recreational vehicles	7	2
Increased fire frequency as a result of recreational use	4	1
Maintenance and/or construction activities in recreational areas (including road works)		4
Weeds	4	–
Horse riding	3	1
Soil compaction	3	–
Camping	2	–
Erosion by bushwalkers or vehicles	2	–
Tourism	2	–
Trail bikes	2	1
Walking tracks	2	–
Bushwalking	1	–
Degradation of litter accumulation	1	–
Firewood removal	1	–
Habitat degradation	1	–
Mountain-bike riding	1	–
Tourism development	1	–
Track creation (unofficial)	1	–
Track maintenance in recreational area	1	–
Uncontrolled access	1	–
Vegetation damage or degradation	1	–

Note that some taxa and communities are threatened by more than one process.

Table 1.6. Whale-watching regulations in Australia,[1] Australian States[2] and New Zealand. (Modified from Gill and Burke, 1999.)

	New Zealand	Australia	Tasmania	Victoria	New South Wales	Queensland	Western Australia	South Australia
Minimum approach distances (metres)								
Boats	50* (whales only)	100	100	150* (motor vessels)	100	100*	100	100*
Swimmers	100* (whales only)	30	30	30	30	300	30	300*
Aircraft	150	300*	300*	300*	300*	300*	300*	300*
Limit on number of vessels <300 metres	3	3	3	–	–	3	–	–
Feeding permitted	No	No	No	–	–	No*	No*	–

[1]National guidelines (i.e. Commonwealth of Australia).

[2]Whale watching is not a recognised activity in the Northern Territory, and there are no provisions for regulating it in that jurisdiction.

*Special conditions are applied in these cases.

Observers watching seals at Seal Bay on Kangaroo Island. Observers pay to view animals on the beach at Seal Bay, but regulations restrict how closely the public can approach animals. (Photos by David Lindenmayer.)

Park (McDougall and Wright, 2004). Infrastructure for tourists, such as roads, tracks and amenity structures can have detrimental impacts on biodiversity. Kelly *et al.* (2003) estimated that tourism was a threatening process for 72 plant taxa in Australia.

Lonsdale (1999) found that the extent of weed invasions in national parks and reserves increased with the number of human visitors. In these and other cases, tourists and their associated activities need to be managed (see Box 1.7).

In general, people are willing to pay to protect environmental quality. Two of the most important components determining willingness to pay are recreational use and landscape aesthetics (Bergstrom, 1990; Moran, 1994). Elements of public benefit that contribute to perceptions of the economic value of the environment include air quality, water quality, knowledge of the existence of undisturbed landscapes, and the option to access such landscapes in the future (Walsh *et al.*, 1990).

Summary: utilitarian values

Brussard (1994) and Beattie and Ehrlich (2001) summarise the economic value of a natural area as a region's 'natural capital', including both biotic and abiotic resources. These resources consist of natural and semi-natural ecosystems, their genetic resources, and the processes that maintain them (Gaston and Spicer, 1998). Natural resources provide goods in the form of forage, livestock, fish and wildlife for harvest and enjoyment, as well as services such as the regulation of hydrological and nutrient cycles, the detoxification of waste, and the protection of soil.

Many human activities cause degraded ecosystems, which can be defined as ecosystems that are not effective providers of goods or services. Conservation is not necessarily incompatible with other land uses; rather, some compromises may be necessary, and we may have to change some aspects of the way in which we use the natural world to ensure that the benefits we currently enjoy are available for future generations (Holling and Meffe, 1996).

Conservation and economics are inextricably linked – a fact that is reflected in the emergence of ecological economics as a field of study (Grafton and Pezzey, 2005). Conservation actions that ignore the economic and social context are mostly naive and ill-considered, and are therefore likely to fail (Daily and Walker, 2000; Dovers and Wild River, 2003). Sound businesses live off the dividends of investments, and reinvest part of the dividends in their asset base. The asset base of the environment includes the full array of species and ecological processes (Beattie and Ehrlich, 2001), and the environment is poorly managed when the asset base is damaged or diminished. The economy should be managed so that we live off the dividends of our natural resources (Beattie and Ehrlich, 2001). The task of conservation biology can be seen as the provision of the tools and expertise necessary to ensure that the asset base is maintained and enhanced.

1.2 Intrinsic value

Conservation biology is not driven exclusively by economic or utilitarian criteria. Economic criteria for conservation are shifting and opportunistic in their practical application (McNeely *et al.*, 1990; Daily and Walker, 2000). For example, species that are fewest in number, often the ones most likely to become extinct, are least likely to contribute to key ecosystem processes, and are often the ones least likely to have an important perceived economic value (Ehrenfield, 1988).

Ethical considerations contribute to valuation for conservation. Two differing ethical positions, the **ecocentric ethic** and the **biocentric ethic**, are elaborated on in this section. **Intrinsic values** (biocentric and ecocentric values) differ from cultural, spiritual and aesthetic values (which are nevertheless utilitarian) because the former places value on species and communities, independent of people. The latter values are evaluated with respect to the preferences and needs of people.

Ecocentric ethic

The **ecocentric ethic** was conceived by Leopold (1949) and recognises that all species, including humans, are the product of a long evolutionary process and are inter-related in their life processes. The sentiment of the ethic is reflected in Leopold's definition of good environmental management, which was given in the introduction to this chapter. The object of concern for the ecocentric ethic is the biotic community as a whole (Booth, 1992). The ecocentric ethic is served when ecosystem composition and ecological processes are maintained.

The World Charter for Nature (UN, 1982) recognises the ecocentric value of conservation. It states that humankind is part of nature, that every form of life is unique and warrants respect regardless of its worth to human beings. It also recognises that lasting benefits for people depend on the maintenance of essential ecological processes and life-support systems, and upon the diversity of life forms. Both scientific and remote area values, while perhaps having utilitarian motives, are closely related to ecocentric values. Large natural undisturbed areas that are protected from exotic species and modifying processes, and maintained through natural processes, are often valuable for nature conservation.

Biocentric ethic

The **biocentric ethic** argues for the value of all individuals. It proposes that all wild plants and animals are worthy of moral consideration and that humans have an ethical obligation towards them. It requires rules for human behaviour that include a duty not to harm any entity in the natural environment, which, by virtue of its existence, has a good of its own. People are obligated to treat the natural environment with a 'hands-off' approach to whole ecosystems, and to refrain from placing restrictions on individual organisms (Booth, 1992).

The ecocentric and biocentric views differ because if the ecological properties of natural communities are not changed by human use or management, then the ecocentric view would argue that nothing has been lost. The biocentric view regards both individual organisms and the ensemble properties of ecosystems to be inviolable, unless they are needed to satisfy vital needs, such as the provision of food or shelter, or they need to be controlled for reasons such as self-defence.

The ecocentric environmental ethic may prescribe intervention to maintain the characteristics of, for example, an old growth forest if the absence of management would lead to a change in the ecological

Box 1.8

The Koala conundrum

Kangaroo Island lies off south-eastern South Australia. Because the Red Fox was not introduced there, the island supports populations of species that are declining elsewhere in southern Australia, for example the Bush Thick-knee (Ford, 1979). However, population management of the Koala on Kangaroo Island illustrates the potential conflict between conservation and animal welfare. Koalas were deliberately introduced from the mainland in 1923 (St. John, 1997), but the species did not previously occur on Kangaroo Island. The Koala population size is estimated to exceed 30 000 individuals and is so large that it is negatively affecting the vegetation, with potential detrimental effects on other taxa that depend on the same vegetation. Some conservation biologists have called for populations of the Koala to be culled on the island to relieve pressure on the vegetation, including on reserved land that is important for a range of rare and threatened plants (Davies, 1996). However, culling is opposed by organisations such as the Australian Koala Foundation, which is concerned about the welfare of the Koala. Judgements about culling are based on the different value sets of the respective parties.

characteristics of the forest. In contrast, the biocentric ethic may require intervention to protect the life of individual plants, mammals or birds of old growth forest, but would not object to an invasive plant, which would be considered by most people to be a weed. The ecocentric environmental ethic is 'somewhat more friendly to the instrumental use of nature' (Booth, 1991).

Unlike the ecocentric ethic, the biocentric ethic supports both animal welfare and conservation. Conservation biology is not necessarily concerned directly with the welfare of individual animals, although the two issues are often confused. For example, practical conservation biology may require the destruction of exotic pests (such as feral predators) or populations of weedy plants. This would run counter to the philosophies of the biocentric ethic regarding animal welfare.

The Ground Boa, an endemic snake from Round Island (near Mauritius in the Indian Ocean) illustrates the potential for conflict between ethical positions. The Ground Boa was threatened by habitat loss caused by feral Goats and Rabbits. Animal welfare advocates were concerned about the impacts of measures (such as poisons) being used to eradicate feral animals. The delay in implementing the control programs lasted

Manna Gum trees near Kelly Hill Caves on Kangaroo Island, South Australia, have been heavily defoliated by introduced Koalas. (Photos by Esther Beaton and David Lindenmayer.)

Box 1.9

An ethical basis for conservation

The International Union for the Conservation of Nature's Working Group on Ethics and Conservation developed an ethical basis for the conservation of nature (McNeely *et al.*, 1990). It reflects the intrinsic value argument, and combines aspects of biocentrism and ecocentrism. It is composed of the following points:

- The world is an interdependent whole made up of natural and human communities.
- All life depends on the functioning of natural systems to ensure the supply of energy and nutrients.
- The ecological limits within which we must work give direction to how human affairs can sustain the environment.
- All species have an inherent right to exist.
- The ecological processes that maintain the integrity of the biosphere are to be maintained.
- Sustainability is the basic principle of all social and economic development. This principle will enable the utilitarian values of nature to be equitably distributed and sustained for future generations.
- The well being of future generations is the social responsibility of the present generation. The present generation should limit its consumption of non-renewable resources to a level that meets the basic needs of society, and ensure that renewable resources are nurtured for their sustainable productivity.
- All persons should be empowered to exercise responsibility for their own lives and for the life of the earth.
- Diversity of outlooks toward nature is to be encouraged by promoting relationships that respect and enhance the diversity of life.

2 years in the wild – sufficient time for the Ground Boa to become extinct (King, 1988 in Webb, 2002).

Often conservation biologists have a clear social mandate to protect species. However, the World Commission on Environment and Development (WCMC, 1992) argued that there are difficulties in demonstrating that a species (a taxonomic unit that is to some extent a human construct; see Chapter 2) has any greater right to exist than does any one of the individuals that it comprises - any organism that is not genetically identical to another represents a facet of diversity.

Utilitarian and intrinsic value perspectives on conservation have implications for the size of the human population and rates of consumption; that is, environmental impacts would be reduced if the human population was smaller, and if the per capita rate of resource consumption was lower (see Chapter 12). From the perspective of the biocentric ethic, humans have no right to reduce biological richness and diversity except to satisfy vital needs. The ecocentric ethic is also compatible with a substantial reduction in the human population size, and a reduction in human interference in the non-human world (see Devall and Sessions, 1985; Aslin, 1991).

1.3 Anthropocentric bias

The allocation of conservation resources is biased by human attitudes towards the environment. Resources are directed towards those species that are closely related to humans and most like humans in appearance, and those species that have the greatest cultural or economic importance. This bias is reflected in the composition of zoological and botanical gardens, taxonomic and ecological research effort, and recovery planning.

For example, of the 16 Action Plans published by the International Union for the Conservation of Nature (IUCN)/Species Survival Commission between 1949 and 1992, the first published was for African primates, and the second was for Asian primates. There were none for plants and only three for non-mammalian animals (Table 34.2 in WCMC, 1992). The same priorities are evident in the development of recovery plans for Australian species (Table 1.7). In late 2002, more than 170 recovery plans (including plans that covered multiple species) had been approved under the Australian Federal Government's *Environment Protection and Biodiversity Conservation Act* (EPBC Act) of 1999. Of these, very few were for non-vascular plants or invertebrates (Table 1.7). The attention paid to invertebrates, non-vascular plants and fungi is negligible, despite their ecological importance and them being far more numerous and species-rich than groups such as vertebrates (Chapter 2), which receive most attention. We do not suggest that this bias is inherently wrong – we just wish to point out that it exists and that priorities should be based on careful thought.

Despite the existence of ethical positions such as biocentrism, the human population remains primarily preoccupied with itself. However, even within this framework, there are ethical imperatives that lend weight to arguments for the conservation of biodiversity. The notion of responsible care of the environment

Table 1.7. Number of species recovery plans in Australia in 2002 that had been adopted for various groups of organisms. (Based on data supplied by Department of Environment and Heritage, 2005a.)

Group	No. recovery plans adopted[1]	No. critically endangered, endangered, and vulnerable species[2]
Vascular plants	183*	1215
Frogs	29*	27
Mammals	21	85
Birds	22	101
Fish and sharks	10	35
Reptiles	14	50
Other plants	2	Unknown
Invertebrates	3	Unknown
Fungi	0	Unknown

[1] Data do not include recovery plans presently in preparation.
[2] Based on definitions in the Environment Protection and Biodiversity Conservation Act (1999).
* Includes cases where multiple species are included within the one recovery plan.

Leadbeater's Possum – a charismatic species for which a recovery plan has been developed (see Macfarlane *et al.*, 1998). (Photo by David Lindenmayer.)

is part of most societies, and, more recently, explicit government policy and legislation have made a stronger case for considering conservation.

1.4 Custodial responsibility and the precautionary principle

Any judgement made today that has an adverse impact on natural populations, particularly if it involves the extinction of species or communities, is likely to be irrevocable (Beattie and Ehrlich, 2001). **Custodial responsibility**, sometimes called the **principle of inter-generational equity**, underpins both the intrinsic value and the utilitarian cases for conservation. Permanent loss of populations, species or environmental processes reduces the ability of the human population to manage environments (Beattie and Ehrlich, 2001). The consequence of such losses will be a reduction in the ability to cope with unforeseen environmental change, which necessarily creates an onus of custodial responsibility, both for the current generation and for future generations. Indeed, the definition of conservation given at the beginning of this chapter suggests that conservation should aim to maximise the benefits of the environment for both current and future generations.

Other concepts are related to the idea of custodial responsibility. A central tenet of management is to avoid decisions that are irreversible, and other terms have been used to represent essentially the same argument. The **bequest values** of preserved areas are based on the concern that large undisturbed landscapes should be available for future generations. Similarly, **option demand** is the desire by non-users to retain the option to use a resource at some time in the future.

Cautious use of the environment is not a recent phenomenon. For example, Leopold (1953) suggested in relation to environmental management and use that the most important rule of 'intelligent tinkering' is to keep all of the pieces. King Louis XIV's Forest Ordinance of 1669 replaced earlier damaging exploitation practices in French forests; it stated (in Brown, 1883):

> *By the tenet that the woods and forests of a country, not the state forests alone, but all, are national property, it is not understood that the population of a country have one and all of them a right to go into the woods and forests everywhere and cut or fell as it may please them; but that these forests, public and private alike, are the property of the nation in its entirety: not of the individuals composing the nation at any one period, nor of these conjointly; but of the nation in times past, in the passing present and in the times coming - property of which each successive generation has a right...and which it is bound in justice to leave to the succeeding generation in as good condition as it were found, or with an equivalent in national property or national advantage, for any diminution or deterioration which has been occasioned in it.*

This system divided the forest estate into equal areas and limited what could be harvested each year, leading to the concept of forest rotation. The management approach in the ordinance spread throughout Europe over the next 100 years, and was replaced in the late 1700s in England (Saxony) by a system that divided the forest into areas by yield and product type.

The great biogeographer and evolutionary biologist, Alfred Wallace, made a clear statement about custodial responsibility in the mid-1800s (Wallace, 1863):

> *future ages will certainly look back on us as a people so immersed in the pursuit of wealth as to be blind to higher considerations. They will charge us with having culpably allowed the destruction of some...*[species]*...which we had it in our power to preserve; and while professing to regard every living thing,...with a strange inconsistency, seeing many of them perish irrecoverably from the face of the earth, uncared for and unknown (p. 234).*

The **precautionary principle** was first defined and applied in West German environmental legislation in the late 1960s. Another early example of the use of the precautionary principle can be found in the Second International Conference on the Protection of the North Sea, which was held in London, in November 1987 (Gray, 1990b). A ministerial declaration stated that potentially damaging pollution emissions should be reduced 'even when there is no scientific evidence to prove a causal link between emissions and effects' (Peterman and M'Gonigle, 1992).

Box 1.10

Importing exotic plants: a potential application of the precautionary principle

The rules governing the importation of exotic plants into Australia provide an example of where the precautionary principle can be applied, resulting in ecological and economic benefits. Consider the situation in which a customs officer at a port of entry into Australia is presented with a living plant cutting. The individual carrying the plant asks permission to bring it into the country. The instinctive reaction may be to ask: What is it? What is the life form and mode of reproduction? Is it poisonous to stock? Is it poisonous to native animals? What potential does it have to invade extensive areas and damage crops, pasture, urban environments or national parks? Is it a potential competitor to native plants, particularly rare or threatened species?

The precautionary principle would suggest that the onus should be on the importer to provide evidence that the plant is not a threat, and that the default condition under the law should be that plant species cannot be imported unless sufficient cause can be shown that the importation is, on balance, beneficial (by whatever standard of benefit is used).

If the request is accompanied by documentation and scientific evidence to suggest that the plant is not a threat to either production systems or natural communities, the customs officer may be inclined to allow the entry of the plant. However, the decision would depend to some extent on how thoroughly the potential for damage had been evaluated.

This hypothetical situation is an example of an application of the precautionary principle. The relationship between the principle and statistical power is explored further in Chapter 17.

The Intergovernmental Agreement on the Environment (Australian Government Publishing Service, 1992, paragraph. 3.5.1) defined the precautionary principle as:

> *Where there are threats of serious or irreversible environmental damage, lack of full scientific certainty should not be used as a reason for postponing measures to prevent environmental degradation.*

The WCMC (1992) described the precautionary principle as the cautious use of biodiversity; that is, actually or potentially useful resources should not be lost simply because we do not know about them or value them at present. The precautionary principle is incorporated into many international and national treaties and declarations, including the Australian National Forest Policy Statement (NFPS; Commonwealth of Australia, 1992).

The precautionary principle implies a shift in the onus of proof to the developers proposing an action to demonstrate that potential impacts are absent, negligible or worthwhile. Thus, a given action is 'guilty until proven innocent' (Taylor and Gerrodette, 1993). The precautionary principle requires that action is taken to protect the environment in advance of conclusive scientific evidence that harm will occur (Deville and Harding, 1997). Deville and Harding (1997) outline potential actions to implement the precautionary principle, including policy, strategy and planning; legislation and law reform; administrative procedures; organisational management; and plans of management.

The precautionary principle has been enthusiastically endorsed by some people and repudiated by others (see Gray, 1990a, 1990b; Josefson, 1990; Lawrence and Taylor, 1990; Lutter, 1990; Brunton, 1994; Collier, 1998). It has been argued that the precautionary approach is inconsistent with objective scientific methods such as the 'concept of proof and reason' (Collier, 1998), and that it will result inevitably in unnecessarily conservative management of the environment. The principle relies on an assumption that it is possible to establish acceptable levels of risk and to reconcile competing risks. Collier (1998) cites the case of the chemical chlorine, which destroys ozone but also makes water fit to drink; the precautionary principle does not provide guidance on how to reconcile such trade-offs. Others have suggested a direct link between the precautionary principle, the principles behind statistical power, and the sensitivity and reliability of statistical tests. We return to these arguments in Chapter 17.

1.5 Conclusions

There is no single environmental ethic underpinning conservation, and it is possible for utilitarian arguments to be in accord with other ethical values. Different ethical viewpoints are not mutually exclusive, and most people would probably agree with elements of all of the viewpoints outlined so far in this chapter, given appropriate circumstances.

Human society decides on courses of action in the absence of complete knowledge, and conservation decisions cannot be separated from issues of social and political development (McNeely *et al.*, 1990; Collier, 1998; Dovers and Wild River, 2003). Conservation depends on methods that explicitly deal with uncertainty. The precautionary principle may provide a framework for dealing with uncertainty in the decision-making process and provide a motivation to improve knowledge.

1.6 Practical considerations

No simple recipe exists for determining how biological resources can best be conserved, either locally or globally. No single set of prescriptions will determine the appropriate land-use strategies that will best achieve conservation objectives. Changing and inter-related ecological, social, political, ethical, economic and technological factors all play a role. To be effective, practising conservation biologists need to be aware of the social and economic context in which decisions are made, how ecosystems are perceived and valued, and what individual people stand to gain or lose by decisions.

1.7 Further reading

An introduction to the terminology of conservation biology is provided by Fiedler and Jain (1990). A detailed examination of the myriad terms, topics, perspectives and principles in conservation biology can be found in the five-volume *Encyclopedia of Biodiversity* edited by Levin (2001). Gaston and Spicer (1998, 2004) discuss the concept of biodiversity. Low (1988), McNeely *et al.* (1990), the Resource Assessment Commission (RAC, 1991) and WCMC (1992) discuss consumptive use of the natural environment. Daily (1997, 2000) defines and describes ecosystem services. Garrod and Willis (1999) introduce the valuation of ecosystems components and services. Grigg *et al.* (1995) and Webb (2002) make spirited cases and provide many examples in favour of consumptive use values of wildlife in an Australian context. The State of the Environment Report (2001a) gives details of wildlife harvested and consumed in Australia over the past few years.

Ehrlich (1990), Brussard (1994), Daily (1997a,b) and Beattie and Ehrlich (2001) provide detailed discussions of the reasons for protecting biodiversity. Booth (1992) reviews several ethical arguments for the non-utilitarian imperatives for conservation. Singer (2000) outlines the case for animal welfare. Mackey *et al.* (1999) and Mittermeier *et al.* (2003) discuss some of the issues associated with remote areas and biodiversity conservation in Australia and globally.

McNeely *et al.* (1990) provide an overview of the principles and practice of conservation from a global perspective, and Ludwig *et al.* (2001) outline how the scientific method in ecology and conservation has changed over the previous decade. Leopold's (1949) *A Sand County Almanac* informed much of the discussion in this chapter. Redford *et al.* (2002) provide an interesting summary of the different philosophies, values and objectives of some of the world's leading conservation organisations.

McNeely (1988), Bergstrom (1990), Walsh *et al.* (1990) and Chee (2004) describe approaches to the economic valuation of biological resources. Outcomes from a workshop on the economic value of biodiversity are presented in Bennett and Ann (2004). Deville and Harding (1997) discuss the application of the precautionary principle, and Brunton (1994) provides a critique. Mangel *et al.* (1996) and Ludwig *et al.* (2001) provide an overview of principles for conservation.

Chapter 2

What should be conserved?

This chapter describes the foundations of different concepts of biodiversity, particularly the key units of conservation: genetic diversity, species, species diversity and the species concept. It outlines the measurement of some of these entities, for example species richness, endemism and diversity, and provides some examples of community structural diversity, for example rainforest and old growth forest.

2.0 Introduction

What should be conserved depends on why one wants to conserve (see Chapter 1); in fact, many conflicts are the result of different values and ethical positions. Consensus depends in part on reaching agreement regarding appropriate units of conservation (Clark, 2002). The National Population Council (1991) defined national goals for Australia, and the ecological goals included protection of ecological processes, maintenance of natural capital as a production resource and amenity, maintenance of natural systems capable of absorbing wastes, and the preservation of **biodiversity**. Dovers and Norton (1994) claim that these goals may not be mutually compatible, at least in the short term, and that some defy definition. They asked 'Does the preservation of biodiversity imply no loss of species, no loss of vertebrate animals and vascular plants, no loss of genetic diversity within species, no loss of vegetation types, or something else?' It is essential to answer such a question if the role of conservation is to be defined and its consequences measured.

The decision may be taken to protect genes, **populations**, species, **communities**, **ecosystems** or some other component of the natural world. Although these objectives are clearly inter-related, any one may be achieved without necessarily achieving the others. Thus, it is important to have a clear understanding of the structure and composition of biological diversity, and to describe its components in unambiguous terms. This chapter provides a basis for classifying and describing the elements of ecosystems at different organisational and functional levels.

2.1 Units of conservation

The word biodiversity is often used as a synonym for the diversity of life on earth (Dovers and Norton, 1994; Bunnell, 1998; Bunnell *et al.*, 2003). In Chapter 1, we defined biodiversity as encompassing genes, individuals, demes, populations, **metapopulations**, species, communities, ecosystems, and the interactions between these entities. In brief, biodiversity refers to the range of variation or differences among living things.

A unit of conservation is the smallest set of organisms considered in conservation planning. It may be a **genotype**, a group of individuals in a single place (a **deme** or **population**), a set of populations, a taxonomic group such as a species, or groups of species that make up a **community**. At the species level, it is the number of species in a given environment (Bunnell *et al.*, 2003). The units are used to design and manage conservation reserves, to set priorities for allocating resources, and to evaluate the adequacy of protection measures. In this section, we outline different concepts of units of conservation.

Box 2.1

Defining biodiversity

If an objective is to `conserve biodiversity', then it is important to be clear about what is meant by the term `biodiversity'. There is an extraordinary array of definitions of biodiversity – Delong (1996) and Bunnell (1998) reviewed approximately 90 interpretations of the concept! From these reviews, it emerged that many definitions were very abstract, making it difficult to apply them to management practices (see Bunnell *et al.*, 2003). For instance, definitions that include the maintenance of genetic diversity and the maintenance of ecosystem processes or functions are difficult to apply. On this basis, Bunnell *et al.* (2003) argued that the best surrogate for sustaining the diversity of biological entities is the maintenance of species richness. They argued that for practical conservation, concepts to guide ecologically sustainable resource management, such as the maintenance of biodiversity, must be associated with `indicators of success in obtaining that objective'.

Genetic diversity

There are many reasons for conserving genetic diversity (Moritz, 1999; see Chapter 1). From a utilitarian perspective, genetic resources maintain crop production and protect against future contingencies, such as new diseases of food plants (Population Action International, 2000).

Wilson (1992) suggested that genetic information content be used to measure biodiversity, so that conservation value depends on evolutionary distinctiveness, reflected by taxonomy. The International Union for the Conservation of Nature/United Nations Environment Programme/World Wide Fund for Nature (1980, in Vane-Wright, 1992b) noted that:

> *the size of potential genetic loss is related to the taxonomic hierarchy because…different positions in this hierarchy reflect greater or lesser degrees of genetic difference…the current hierarchy provides the only convenient rule of thumb for determining the relative size of a potential loss of genetic material.*

For example, the World Conservation Monitoring Centre (WCMC, 1992) suggested that the two species of Tuatara in New Zealand are relatively important because they are the only extant members of the order Rhynchocephalia. In Australia, the Helmeted Honeyeater (a subspecies of the Yellow-tufted

A Tuatara – a species belonging to an Order with only two members, making conservation very important. (Photo by Phil Treweek and Hilldale Zoo, Hamilton, New Zealand.)

Honeyeater) may be seen as less important than the Plains Wanderer (which has full species status). Similarly, the Western Swamp Tortoise, the only member of its subfamily, would take precedence over other species of tortoises that belong to genera and families with many representatives.

Genetic diversity is the **heritable variation** between individuals within populations, between populations within species, and between species. Within populations, genetic diversity can be caused by different sequences of base pairs on strands of DNA, by different chromosome numbers and arrangements between populations or between species, or by the different genetically based phenotypic traits exhibited by individuals within populations. Genetic variation can be measured in many ways (some of the kinds of genetic data that are useful for answering questions in conservation biology are described in Chapter 13). Variation can be represented by the presence of alternative alleles at a single locus, or by differences among individuals in quantitative traits that are determined by many genes.

The total variation between individuals in a population comprises: (1) variation due to genotype, (2) variation due to the different environmental conditions experienced by individuals (see Box 2.2), and (3) variation in response to genotype that depends on the environment in which the individual occurs (termed the genotype–environment interaction).

Heritable genetic variation is the raw material upon which natural selection acts, enabling populations to adapt to novel environmental conditions. Fisher's fundamental theorem of natural selection states that the rate of

Box 2.2

Variation in environmental conditions and Tiger Snakes on islands

There can be marked variation between individuals within a population and also between populations, even neighbouring ones, because of different environmental conditions. Such differences can, in turn, start the process of speciation, where divergent forms eventually evolve into different taxa (Bush, 2001). A spectacular example of marked variation between neighbouring populations occurs among Tiger Snakes on the islands of southern Australia. These populations are relicts isolated from those on the mainland by a rise in sea levels after the last Pleistocene glacial period. Some islands, such as Chappell Island off the west coast of Flinders Island in Bass Strait, support extremely large-bodied snakes (Schwaner and Sarre, 1988). Snakes on other islands, such as Roxby Island off the Eyre Peninsula in South Australia, are an order of magnitude smaller (Schwaner, 1985). Schwaner (1985) compared Tiger Snake populations on 10 South Australian off-shore islands that had broadly similar climate and seasonal weather patterns. The key driver of the highly significant variations in body form was the availability of prey (Boback, 2003). Unlike mainland areas, where there is a diversity of prey sizes (and less marked variation in body size between populations of Tiger Snakes; Shine, 1987), islands have a limited range of types of prey. Prey size and body size in the Tiger Snake are highly correlated. Small snakes on islands eat smaller prey, such as lizards, whereas the large snakes on other islands eat large prey, such as bandicoots and Muttonbirds or Shearwaters (Schwaner, 1985).

The taxonomy of Tiger Snakes remains controversial (Rawlinson, 1991; Cogger, 2000) and it is possible that the `group' is a single taxon (Shine, 1987). Nevertheless, the differences between populations on islands are marked, and if the populations were to continue to remain isolated, they could eventually evolve into distinct species. Irrespective of the current taxonomic status, such differences in populations highlight the diversity that is an inherent part of biodiversity – and as such, the conservation of each population is therefore warranted. Indeed, conservation efforts to ensure the maintenance of such variation are therefore central to approaches that attempt to maintain the evolutionary potential of biodiversity (e.g. the *Victorian Flora and Fauna Guarantee Act (1988)*; see Chapter 3).

evolution is directly proportional to the amount of heritable genetic variation in a population (Fisher, 1930).

When genetic diversity in a population declines, the potential rate of evolutionary change declines. In addition, there is a positive relationship between the proportion of **heterozygous** loci in an individual and its reproductive fitness (Eldridge *et al.*, 1999; Reed and Frankham, 2003). This relationship between heterozygosity and fitness affects the chances of survival and number of births in a population, which in turn affects the chances of persistence of the population (Frankham, 1995b), particularly in response to factors such as environmental change (e.g. Hughes and Stachowicz, 2004).

Variation in the sizes of adult island Tiger Snakes (see Box 2.2). (Photo by Terry Schwaner.)

Genetic distance (see Chapter 13) measures genetic variation between individuals in a population, between populations within a species, or between species in a higher taxon. Figure 2.1 shows the average genetic distance between species in several animal genera, together with the range of distances in these taxa.

Genetic distances and taxonomy

The decision about whether two populations belong to a single species, separate species, or separate genera is usually somewhat subjective, and is often based on

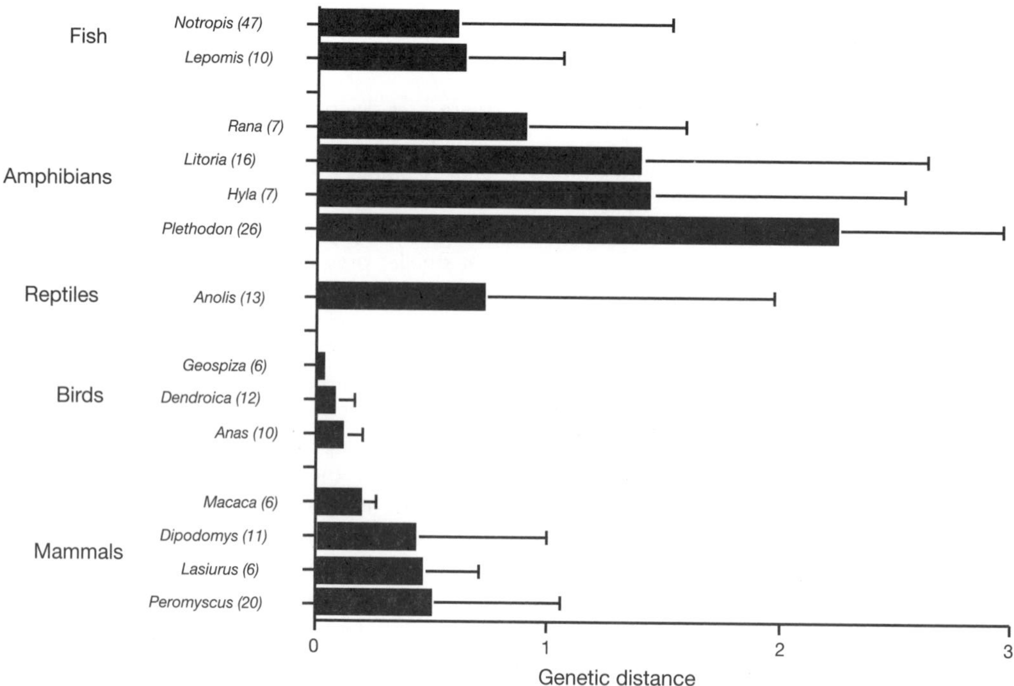

Figure 2.1. Genetic distances among species within 14 vertebrate genera (after Avise and Aguardo, 1982, in Vane-Wright, 1992b). The histogram bars show the mean genetic distance between taxa, and the error bars show the ranges of genetic distances within each taxon. The numbers of species upon which the values are based are shown in parentheses.

morphological evidence alone. In some groups, such as birds, the genetic distances within genera tend to be relatively small. In contrast, there are relatively large distances between species within amphibian genera. Among amphibians, reproductively isolated and genetically distinct but morphologically similar species can be relatively common. Visual cues and associated morphological changes are relatively unimportant mechanisms in their reproductive isolation, so many species are undetected by taxonomists (see the following section on the biological species concept), resulting in large genetic distances within recognised genera and species. Birds, in contrast, rely on visual cues, so morphologically identical species are relatively rare, making a taxonomist's job much easier.

In some cases, species harbour different amounts of genetic variation because they experience different rates of mutation and genetic change. Selection, population bottlenecks and other processes affect the amount of variation in populations and species. Genetic information can be used to monitor population size and trends; identify species, sexes or individuals; evaluate mating systems; reconstruct pedigrees; confirm predictions of population structure; and guide the management of genetic variation in closely managed populations. In Chapters 6 and 13, we explore some pragmatic applications of genetic information, and outline how genetic variation is measured.

One of the difficulties in using genetic criteria to conserve biodiversity is that the requisite information is usually unavailable. Much less than 1% of plant species, and very much less than 1% of animal species could be assayed for genetic variation in the foreseeable future. Conservation planning requires variables that are easily measured and that reliably represent genetic variation. For instance, it is a useful strategy to conserve individuals in spatially separate patches throughout a species' range because genetic differentiation is often a spatial process.

Populations

Genetic diversity is perhaps the most direct measure of biodiversity, as Wilson (1992) argued, but the fate of populations determines the fate of genes. Spatial segregation of populations restricts **gene flow** or isolates populations completely, resulting in **genetic differentiation**. Populations that do not freely exchange genes experience independent **genetic drift** and unique selective pressures. These forces result in genetic differentiation.

Populations are the smallest demographic unit within which the exchange of genetic material is more or less unrestricted. If gene flow between populations is restricted, then populations are the units of evolution (as defined by Nelson, 1989). The advantage of making populations the unit for conservation is that conserving populations or metapopulations (groups of spatially separate populations; see Chapter 17) conserves genetic variation. This level is much less affected by the idiosyncrasies of taxonomy and it is sensitive to the geographic distribution of biological variation.

Sometimes, gene frequencies are distributed in gradients (clines) resulting from genetic drift or natural selection. In other cases, changes in gene frequencies are discontinuous, resulting from isolated breeding populations. **Genetic neighbourhoods** are the largest **panmictic** (completely mixed) unit within a species. **Neighbourhoods** can be defined by the average distance dispersed by reproductive adults between their birth and the birth of their offspring, or by the average dispersal distance of seed or pollen.

Neighbourhoods define the **provenance** of an individual, which is the location or source of genetic material. The term provenance originated in forestry (Eldridge *et al.*, 1994) and is usually employed for botanical samples, but can be applied to any kind of organism. It represents a sample of genetic material from a relatively cohesive genetic unit (a population, subpopulation or neighbourhood). Variation between provenances reflects the distinctiveness and extent of genetic differences between populations (Sydes, 1995; Figure 2.2).

Ecotypes are cohesive genetic units that are adapted to local **microclimatic**, topographic or **edaphic** conditions. For example, Pacific Salmon return to breed in the streams in which they were born. Local populations are essentially genetically and demographically isolated from one another, and thousands of populations make up the salmon fisheries of north-western North America (Hilborn *et al.*, 2003). The physiology of each salmon is adapted to the conditions of particular rivers and streams; for example, salmon have different energy reserves that relate to the distance they migrate to reach spawning grounds. Local adaptation makes it difficult to reintroduce or translocate fish to places where stocks have been depleted or **extirpated** (Withler, 1982; Clark, 1984).

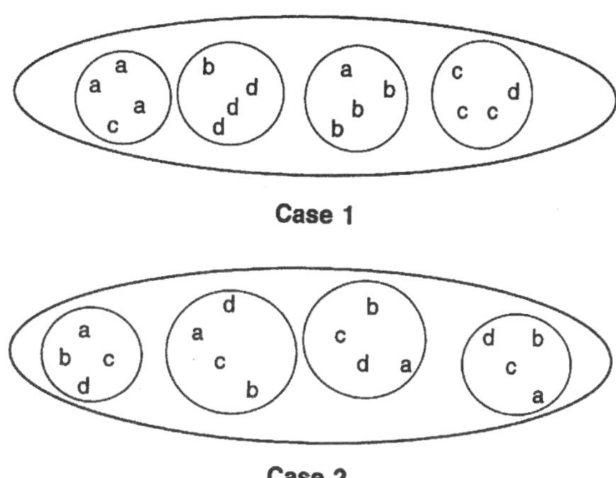

Figure 2.2. Representation of genetic diversity within and between populations (after Sydes, 1995). Case 1: most variation is between populations. Case 2: most variation is within populations.

Often, legislation, regulatory agency policy and industry practice ignore local genetic resources. It may not be sufficient to replant, revegetate or reintroduce individuals of the same species as those that have been eliminated. To maintain ecological function and the full suite of genetic opportunities, it may be important to use genetic material from the same population, provenance or neighbourhood as the original. However,

Salmon at fish markets at Seattle, north-western USA. (Photo by David Lindenmayer.)

using populations as the unit for conservation does not take evolutionary potential into account.

Evolutionarily significant units and management units
Sometimes populations diverge in **allele** frequencies, but without any evidence of associated phylogenetic separation. There is no empirical or theoretical rule for deciding the amount of genetic divergence beyond which a set of populations can be said to be undergoing independent evolution (Moritz, 1994a, 1999). Using appropriate analyses of patterns of genetic diversity among taxa, it is possible to determine which taxa (populations, subspecies, species) are evolving independently (Templeton, 1991; Moritz *et al.*, 1995), termed **evolutionarily significant units** (ESUs; Rojas, 1992; Moritz, 1994b, 1999; Vogler and DeSalle, 1994). Conserving ESUs has the advantage of conserving genetic variation in a form that provides the greatest opportunity for species to adapt to changing environments.

There are, however, some drawbacks to this method. Patkeau (1999) noted that the Polar Bear evolved from the Brown Bear in the Pleistocene, became isolated, and now has a fixed mitochondrial DNA (mtDNA) lineage. In contrast, the Brown Bear has maintained several well-diverged mtDNA lineages. Clearly the Polar Bear is a separate species, but as Patkeau (1999) notes, the species may fail to pass the ESU threshold because it has become fixed for its lineage within the mtDNA clades of the Brown Bear.

Nevertheless, although the ESU approach has some limitations, in some cases it may be the appropriate focus for conservation (Moritz, 1999). ESUs have been recommended for use in marine biodiversity conservation (Moritz *et al.*, 1995), for management of Atlantic Salmon (Dodson *et al.*, 1998), which, like Pacific Salmon, comprise many separate and locally adapted stocks, and for conservation of Brown Bears (Waits *et al.*, 1998).

It is important for both short-term and long-term conservation objectives to recognise the spatial structure of genetic variation and to identify populations for conservation (Moritz, 1994b; Peakall *et al.*, 2003). The appropriate scale for management is where populations show significant divergence, regardless of their phylogenetic histories. **Management units** (MUs) are demographically independent units with distinctive allele frequencies (Moritz, 1994a, 1999).

The most important difficulty with ESUs and MUs is that it is impractical to identify them given current technology for genetic analysis. Considering that there are insufficient resources even to collect and name most species (see Section 2.3), we are unlikely to have sufficient resources to measure and characterise the genetic variation of the populations of which species are composed (Patkeau, 1999). The best rule of thumb in most cases is to aim to conserve viable populations of species throughout their geographic range, thereby using space as a surrogate for genetic variation within species. Conservation should also use subspecific taxonomy, where it exists.

The species concept

Typically, species are the units of conservation. In most countries, there is a social, political and legal mandate for the protection of species, whereas mandates for the protection of other biological entities such as genes, populations and ecological communities are less common (Caughley and Gunn, 1996).

Biologists disagree about how to define species. Taxonomists use morphological, biological and phylogenetic concepts, among others; usually, the choice reflects the philosophy of the person who delimits and classifies the taxon. The main problem is that different definitions result in different classifications.

The biological species concept
The **biological species** concept was proposed by Mayr (1942), who defined a biological species as a group of interbreeding (or potentially interbreeding) natural populations that are reproductively isolated from other populations. Each species forms a separate gene pool. Reproductive isolation is the sole discriminator in this system, although the geographic distribution of genetic variation may be reflected in the identification of subspecies and hybrids (O'Brien and Mayr, 1991).

This concept is difficult to apply in circumstances in which species do not reproduce sexually, such as **clonal** and **parthenogenetic** species. It can also be difficult to apply when populations are separated geographically, or when the degree of reproductive isolation varies over time. Furthermore, in most species, the limits of genetic exchange are only poorly understood and they must be inferred from indirect evidence, such as discontinuities in geographic ranges, observations of behavioural mechanisms that may isolate the species, or unique morphological characteristics.

Many species are morphologically and behaviourally distinct, but do not obey the rule of reproductive isolation. Some species hybridise naturally when their environment is disturbed, and the introgression of genes from one plant species to another is common (Rojas, 1992; Potts *et al.*, 2003). Many of the more than 650 species of the genus *Eucalyptus* are capable of exchanging genetic material directly through hybridisation (Boland *et al.*, 1984; Doughty, 2001; see Box 2.3). For example, Mountain Ash and Messmate often hybridise along the margins of their distributions, as do Mountain Ash and Red Stringybark (Ashton, 1981a; Ashton and Sandiford, 1988). The lack of strict reproductive isolation between eucalypt species has led to **genetic pollution**, an emerging problem in which genes from eucalypt plantation species move by hybridisation into eucalypt species in surrounding native vegetation (Potts *et al.*, 2003; see Box 2.3).

Even though many *Eucalyptus* species exchange genetic information, they have different ecological roles and morphological properties. As new methods for the measurement of gene flow become available, more evidence of genetic exchange between species is found.

Remnant paddock tree within an area being prepared for the establishment of plantation eucalypts in south-western Australia. (Photo by David Lindenmayer.)

Box 2.3

Genetic pollution in eucalypts and the breakdown of the biological species concept

Genes can move from plantation eucalypts to endemic eucalypts by pollen dispersal and subsequent hybridisation (Potts *et al.*, 2001, 2003; Barbour *et al.*, 2003), thus altering natural patterns of genetic variability (Ive and Lambeck, 1997; Strauss, 2001). In eucalypts, species definitions are a poor guide to reproductive isolation, and the potential for gene introgression is extensive (Barbour *et al.*, 2003). The widely used plantation tree Southern (or Tasmanian) Blue Gum is known to hybridise with at least 14 species of other eucalypts (Strauss, 2001). Hybrids can also outcompete naturally occurring eucalypt `species' that inhabit an area. Hybrids have been reported between Shining Gum plantation trees and Swamp Gum in Tasmania (Barbour *et al.*, 2000). Changed genetic composition through gene transfer from plantation trees to local trees could lead to profound ecological impacts, such as host shifting by species-specific herbivorous insects (Strauss, 2001). Potts *et al.* (2001) believe that remnant populations of the main plantation eucalypt tree species in rural areas could be vulnerable to genetic `swamping'.

Wild Bison in Yellowstone National Park, Wyoming, USA, and domesticated Bison in north-western Massachusetts. Bison are known to have bred with domestic Cattle, which are placed in a different genus. (Photos by David Lindenmayer.)

For example, Templeton (1991) found evidence of genetic exchange between Bison and domestic Cattle species, which are placed in separate genera. There is no simple relationship between relatedness in a taxonomic hierarchy and the likelihood that species will exchange genetic material (Vane-Wright, 1992a). The practical application of the biological species concept in the recognition and description of species requires many arbitrary decisions (Bock, 1992).

The morphological species concept

Under the **morphological species concept**, species are defined by a set of shared morphological characters, and they are distinguished from other species by morphological discontinuities (Bock, 1992; Keogh, 1999). This is not always straightforward, because the same species can develop different morphological forms depending on environmental conditions. For example, in reef-building corals, an individual of a species growing in a low wave-action area such as a lagoon may have an erect extensive branching form, whereas another on a high wave-action reef edge may have a more prostrate form (Veron, 2000). The external morphology of marine sponges also varies with wave action. Similarly, New (2000) noted that the morphology of insects can depend on the host-plant species on which they are raised.

In practice, inference of reproductive isolation is based largely on morphology. For example, in many snakes, the reproductive mechanics of mating are associated with the 'lock and key hypothesis'. The male sexual organ (the penis) is the 'key' that will only fit the right 'lock' (i.e. the female organ), ensuring that sex can only occur between members of the same species (Keogh, 1999, 2000). The vast majority of species are known only from their morphology, and limited knowledge of their geographic distributions (Vane-Wright, 1992a). Thus, in pragmatic terms, although the biological species concept may be the theoretical inspiration for many species descriptions, operational procedures rely heavily on the morphological concept (Gaston, 1996).

The phylogenetic species concept

Nixon and Wheeler (1990) define the **phylogenetic species concept** as:

> *the smallest aggregation of populations (sexual reproduction) or lineages (asexual reproduction) diagnosable by a unique combination of character states in comparable individuals.*

This approach defines species on the basis of evolutionary lineages, without recourse to reproductive isolation. The units of conservation become the terminal taxa in a **cladogram**, according more closely with the idea of ESUs than with the morphological or biological species concepts.

The phylogenetic species concept is important because, typically, it results in many more species than are recognised using other species concepts. For example, Cracraft (1992) points out that there are approximately 40 species of birds of paradise (family Paradisaeidae) recognised by using the biological species concept, and approximately 90 by using the phylogenetic species concept.

Species concepts and biodiversity conservation

There is no single definition of what constitutes a species, but congruent taxonomies based on more than one kind of data may generally be more robust (Hillis and Moritz, 1986). In Australia, combinations of data (such as genetic and morphological information) have been used in taxonomic revisions of Western Australian Acacias (Elliott *et al.*, 2002), the Brown Antechinus (*Antechinus stuartii*; Dickman *et al.*, 1998; Sumner and Dickman, 1998), Corroboree Frogs (*Pseudophryne* spp.;

Box 2.4

A new species of Mountain Brushtail Possum

The Mountain Brushtail Possum is a large species of arboreal marsupial that is closely related to the Common Brushtail Possum. Its distribution spans wet forests and rainforests from southern Victoria to central Queensland. Southern populations (from Victoria) have a significantly larger ear conch, longer pes, and shorter tail than northern populations (from New South Wales and Queensland). North–south dimorphism is strongly supported by patterns in genetic data, which show genetic distances of 2.7–3% between the southern and northern populations. Lindenmayer *et al.* (2002d) argued that the combined outcomes of morphological, genetic and phylogenetic analyses suggested the existence of two distinct species. The northern form was renamed the Short-eared Possum (*Trichosurus caninus*), whereas the southern form (which retained the common name of Mountain Brushtail Possum) was given the new Latin name *Trichosurus cunninghami*.

Short-eared Possum and Mountain Brushtail Possum (see Box 2.4). (Photos by Mick Tanton and Esther Beaton.)

Osborne, 1989; Osborne and Norman, 1991), Bentwing Bats (*Miniopterus* spp.; Cardinal and Christidis, 2000) and the Mountain Brushtail Possum (see Box 2.4).

The species concept is important because taxonomy shapes perceptions of biodiversity and the recognition of endangered species (Avise, 1989; Rojas, 1992; see Box 2.5). This affects decisions related to, for example, natural area protection, preservation of taxa in botanical and zoological gardens, priorities for germ plasm banks, legal and economic aspects of *in situ* protection, access rights to genetic resources, and property rights to species. For example, areas of high species richness and endemism (sometimes called **hotspots** – see Section 2.6) targeted for protection (Mittermier *et al.*, 1998; Ovadia, 2003) depend sensitively on species definitions (Crisp *et al.*, 2001).

Taxonomic revision has led to a major alteration of the conservation status of some 'new' species; Australian examples include the Corroboree Frog group (*Pseudophryne* spp.) from southern New South Wales (Osborne, 1989; Osborne and Norman, 1991) and the Mahogany Glider from far north Queensland (Box 2.5).

There may be only weak relationships between genetic divergence and taxonomic status, particularly if the taxonomy rests on the biological or morphological concepts, because **speciation** could have resulted from very few genetic changes (Dowling *et al.*, 1992). Conservation strategies based on all species concepts may fail to protect biodiversity because they ignore genetic divergence within species (Box 2.1; see Schodde and Mason, 1999). Until sufficient genetic information

Box 2.5

The species concept and endangerment: the case of the Mahogany Glider

The Mahogany Glider was first described in 1883. Subsequently, specimens on which the descriptions were based were thought to belong to the closely related (but much more widely distributed) Squirrel Glider. Careful examination of specimens from the Queensland Museum then revealed that gliders from a tiny coastal area of far north Queensland were larger, had a longer tail, and differed in a number of other ways from the Squirrel Glider. As a result, the Mahogany Glider was again established as a separate species in 1993, 110 years after it was first described (Van Dyck, 1993). The Mahogany Glider is rare and highly endangered, and has an extremely restricted distribution, with the smallest range of any glider. Up to 90% of its former habitat has been cleared for Cattle grazing and plantations of Banana, Pineapple, Sugar Cane and Radiata Pine. Less than 20% of the remaining distribution of the Mahogany Glider occurs within nature reserves and national parks. Without adequate protection of the limited remaining areas of habitat, most of which occurs on private land, the Mahogany Glider faces an appreciable risk of extinction within the next few decades (Jackson, 2000). Notably, an economic analysis has indicated that there is a strong willingness among Australians to pay for habitat protection for the conservation of the Mahogany Glider (Tisdell *et al.*, 2005).

Mahogany Glider – a taxon restored to species status in 1993 and now highly endangered. (Photo by Bruce Cowell, Queensland Museum.)

is available, the solutions are to conserve species throughout their geographic ranges, to use higher taxonomic groups (monospecific genera or families) in setting priorities, and to conserve subspecies, races and varieties) when the information exists.

2.2 Number of species

Taxonomists continue to collect and describe species. For instance, in Australia more than 2330 new species were described between 1995 and 1999 (State of the Environment, 2001a). It is difficult to estimate the total number of described species for several reasons:

- Taxonomists work in many different institutions.
- The literature on species descriptions is dispersed because taxonomists publish the results of their work in many different journals, some of which can be difficult to find and the resulting work hard to compile.
- No single institution keeps track of all taxonomic work.
- Many groups contain a substantial proportion of **synonyms** (redundant names resulting from species that have been described more than once, usually by different people).
- Many of the original specimens used to describe a species (e.g. the **holotypes**) are located in the countries where early collectors lived (England, France, Russia etc.) and can be difficult to obtain for comparative purposes such as identifying synonyms.

Figure 2.3. Relative numbers of species in major taxa (after Wilson, 1992 and Hammond, 1992). The area of the drawing of each organism is roughly proportional to the number of species in the taxon it represents. The estimates are based on extrapolations from described species. They include, from left to right, protozoans (1.6%), molluscs (1.6%), algae (1.6%), Lepidoptera (3.2%), other invertebrates (2.8%), other plants (2.4%), Diptera (12.9%), vertebrates (<1%), arachnids (6%), Coleoptera (24.9%), Hymenoptera (19.3%), fungi (8%), nematodes (4%), other insects (4%), viruses (4%) and bacteria (3.2%).

- Taxonomists disagree on what constitutes a species (see Section 2.1).

Given these problems, current estimates of the total number of valid, currently described taxa vary from 1.4 to 1.8 million (Hammond, 1992). The best known groups are birds and mammals; however, these groups make up a tiny proportion of the world's biota (Figure 2.3). Insects make up about 65% of all species (Hammond, 1992; Stork, 1992; Figure 2.4), but the vast majority remain undescribed.

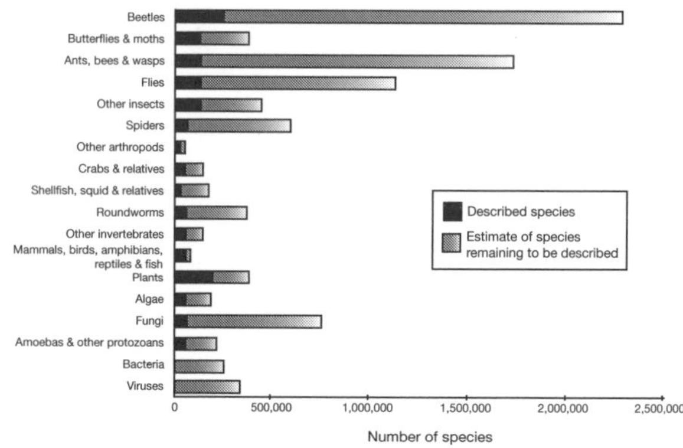

Figure 2.4. The proportion of described and undescribed species. (Redrawn from Stork, 1992.)

Total number of species

Estimates of the total number of species globally remain controversial and highly uncertain. They range from 3 million to more than 100 million. Putting aside the uncertainty associated with species concepts, most uncertainty derives from the fact that the vast majority of species are still undescribed (Figure 2.4). There are no reliable figures for the numbers of undescribed insects, bacteria, viruses, algae, fungi, protozoans or nematodes. For example, in fertile environments, the soil biota can weigh more than 20 tonnes per hectare and contain representatives of most major groups of plants, animals and microbes (Lee, 1996). There may be 3–4 million nematodes alone in a square metre of soil (Bardgett and Cook, 1998). Genetic analyses may further increase the number of species, as has occurred among invertebrates such as velvet worms (see Box 2.6). The total numbers may be revised upwards over time, and published estimates for these groups vary greatly.

Box 2.6

Velvet worms: an example of the discovery of new species

The velvet worms belong to the invertebrate phylum Onychophora. They have soft bodies, lack jointed appendages, and lack a hard exoskeleton. They occur in Australia, New Guinea, south-east Asia, South Africa, South America and Latin America (Monge-Najera, 1995). Until recently, approximately 100 species (including 10 from Australia) had been described worldwide. However, work on Australian onychophorans using allozyme electrophoresis identified more than 110 new species of velvet worms, thus doubling the known world fauna (Tait *et al.*, 1995). This work highlighted the extensive levels of genetic variation among the different species. The outcomes of this work have been validated through the use of other genetic techniques, including chromosome analysis (Rowell *et al.*, 1995), and there appear to also be subtle morphological differences between these taxa (Tait *et al.*, 1995). It is possible that many more species may yet be described, and widespread taxa could actually be composed of several species. The reasons for such high diversity among the Onychophora may be associated with the group's limited dispersal ability and persistence at low population sizes (New, 1995). The conservation of some onychophorans is a major concern (Mesibov, 1990; Taylor, 1990), as many are associated with rotting logs in wet forests (Barclay *et al.*, 2000), and may be susceptible to the effects of fuel reduction burning, harvesting of coarse woody debris, woodchipping, and plantation development (New, 1995; Bonham *et al.*, 2002). In Tasmania, forest management strategies have been developed for some species (Forest Practices Board, 1998).

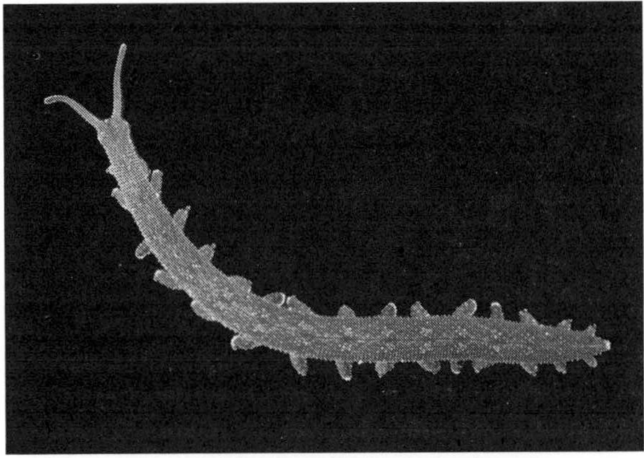

Velvet worm. (Photo by Paul Sannucks and Noel Tait.)

As field surveys, taxonomic work and sampling of previously undersampled habitats (such as tree canopies) progress, estimates of the numbers of species in the most species-rich group, the insects, increase (Majer *et al.*, 1994, cf. Yen and Butcher, 1997; see Box 2.7). Estimates of the total number of species of insects vary from less than 5 million to more than 80 million (Gullan and Cranston, 1994). Some groups of insects appear to be particularly species-rich, especially the beetles (e.g. Major *et al.*, 2003), although this may reflect sampling bias and the preferences of taxonomists. In some places where the insect fauna is relatively well known (e.g. Great Britain), other groups such as flies, bees, wasps and ants are more species-rich than beetles (Gullan and Cranston, 1994). This is also true in some Australian vegetation types that have been sampled for invertebrates, such as the Box–Ironbark forests of northern Victoria (Yen *et al.*, 1999), in which ants and wasps are the dominant invertebrate group (see Figure 2.5).

Number of microbe species

Groups other than invertebrates are also species-rich: possibly only 1% of micro-organisms have been described (Head *et al.*, 1998). For example, an entire previously unknown community of endophytic fungi was recently found with the Northern American tree species Western White Pine (Ganley *et al.*, 2004). In Australia, May and Simpson (1997) calculated that

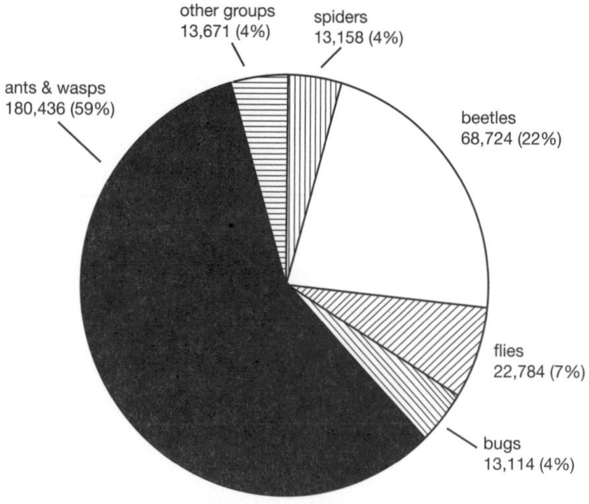

Figure 2.5. Number of specimens and percentages of invertebrates trapped at 80 sites in the Box–Ironbark forests of northern Victoria. (Redrawn from Yen *et al.*, 1999.)

there may be more than 7000 species of mycorrhizal fungi associated with eucalypts alone. In addition, an estimated 75% of Australia's more than 20 000 vascular plants form mycorrhizal associations with fungi (State of the Environment, 2001a). Some of these fungi are critical parts of the diets of many native vertebrates such as potoroos, bandicoots and rodents (Claridge *et al.*, 2000a, 2000b).

The number of species of bacteria may actually be much greater than previously estimated. Soil samples collected from the top 10 centimetres of soil in a Norwegian beech forest contained more than 4000 species of bacteria, the vast majority of which were taxonomically unknown (Torsvik *et al.*, 1990). A similar number of species were recorded from sediment samples off the coast of Norway (Wilson, 1992). Many species cannot be cultured and are recognised only by molecular methods (Crozier *et al.*, 1999).

Box 2.7

Insect species richness in Australian forests

Almost all estimates of global insect biodiversity are based on the number of species found in tropical rainforests. However, work by Davies (1993), Majer *et al.* (1994), Margules *et al.* (1994a; 1995a) and Davies *et al.* (2000) on Australian temperate arthropods indicates that temperate forests could support a significant fraction of the world's species. Majer *et al.* (1994) collected arthropods from the foliage of 80 trees in Western Australia (40 Jarrah, 40 Marri) and a further 80 trees in New South Wales (40 Narrow-leaved Ironbark and 40 Grey Box). Ten trees were sampled in each of several seasons between 1987 and 1988. Insects were killed by applying an insecticide to the canopy of each tree, and the arthropods were sorted and identified (Majer *et al.*, 1994).

More than 67 000 individual arthropods from 23 different orders and 229 families were gathered. There were more than 1500 species of arthropods, of which more than 1300 were insects. A total of 691 species was recorded from the Western Australian trees and 977 species from the New South Wales samples. Few species were common to collections from both sides of the continent. The most diverse groups were the ants, bees and wasps (450 species), beetles (363 species), flies (252 species), spiders (168 species) and leaf-hoppers (150 species).

Estimates of the number of insect species in Australia range between 100 000 and 300 000 species (Yen and Butcher, 1997). If these estimates are accurate, Majer *et al.* (1994) sampled about 1% of the known fauna. They believe that the total insect fauna of Australia is likely to include many more than 200 000 species because:

- they sampled a small fraction of just four of more than 650 species of eucalypts that occur in Australia
- the study did not extend to the extensive suite of vegetation communities not dominated by eucalypt trees
- only canopy foliage animals were collected and other potentially rich arthropod faunas such as those in the litter layer, soil, and bark were ignored
- the vegetation structure and floristic composition of the forested areas where they worked was simple; other areas may support a greater array of species than was recorded in their study.

The areas studied by Majer *et al.* (1994) and others had been disturbed and fragmented by fire, logging, grazing, and urban encroachment. Nevertheless, these areas support extensive and diverse arthropod faunas, and have considerable conservation value.

An entomologist collecting canopy insects. (Photo by Jonathon Majer.)

New habitats and new species

No habitat has ever been comprehensively surveyed for all elements of biodiversity, especially for bacteria and fungi (New, 2000). Even among those taxa that are routinely surveyed, spectacular new species are still being found. In 1994, the Wollemi Pine was discovered by New South Wales Department of Environment and Conservation naturalists in a rainforest gorge in Wollemi National Park (Jones *et al.*, 1995), about 150 kilometres west of Sydney, close to the most densely populated part of the continent. The three known populations consist of few hundred seedlings and fewer than 50 adults, of which the oldest is perhaps 400 years old (Banks, 2002). Mature trees reach 35 metres in height, with a main trunk up to 1 metre in diameter. There are two distinct foliage phases, and the bark is brown and bubbly. It is remarkable that this large and very distinctive species remained undiscovered until 1994.

The discovery brought to light a new genus of conifer (*Wollamia*, Araucariaceae) with a 91-million-year-old Gondwanan history (Benson, 1996). Pollen and fossil records show that the Wollemi Pine was once widespread but its distribution has been reduced by increasing aridity since Gondwanan times (McPhail *et al.*, 1995). The discovery has led to risks of infection by plant pathogens introduced by human visitation (Bullock *et al.*, 2000), increased fire risk, and the illegal removal of seedlings. Notably, Hogbin *et al.* (2000) and Peakall *et al.* (2003) found limited genetic variability in the species, which may make it susceptible to threats such as fungal attack, although the population is capable of sexual reproduction in the wild (Offord *et al.*, 1999).

Wollemi Pine. (Photo by Esther Beaton.)

New habitats that support large numbers of new species (particularly of microbes and invertebrates) continue to be discovered. The discovery of the Wollemi Pine highlights an under-sampled and potentially species-rich habitat: deep terrestrial gorges dominated by cliffs (Larson *et al.*, 2000; Lindenmayer and Franklin, 2002). Beattie and Ehrlich (2001) discuss other new species-rich habitats, including deep-sea vents associated with sea floor spreading, tree canopies, and habitats within the earth's crust.

Sea mounts off southern Australia support species-rich, endemic marine communities, and many of the invertebrates restricted to these areas are undescribed (Koslow and Gowlett-Holmes, 1998). Notably, 60% of the earth's surface comprises the deep-sea abssyal plains (i.e. the sea floor) but 95% of the species described from the marine environment are from the first 20 metres from the surface of the sea. The number of species of nematodes alone from sea floor environments may exceed more than 1 million (Lamshead, 1993; Lambshead and Boucher, 2003). Other newly identified marine environments are likely to support numerous additional undescribed species. For example, deep-water seagrass beds have been discovered recently around the Great Barrier Reef (State of the Environment, 2001b), and a previously unknown extensive coral reef system was found in 2005 off the northern Australian coast between Mornington Island and Groote Eyelandt. In another example, the intestinal tracts of animals may contain numerous unknown species of microbes (Gordon and Fitzgibbon, 1999). Recently, a species representing a new phylum of animals (*Symbion pandora*, phylum Cycliophora) was discovered attached to the mouthparts of the Norwegian Lobster (Funch and Kristensen, 1995).

Sampling methods and new species

New sampling techniques can reveal numerous new species. For example, reducing the mesh size of nets used to trawl marine systems for plankton has revealed a group of new micro-organisms called picoplankton (Beattie and Ehrlich, 2001). The same is true, even for large vertebrates: in the forests of Vietnam and Laos, interviews with hunters and sampling using remote automatic cameras in the mid-1990s revealed several large and spectacular mammals previously unknown to science, including the Soala, the Giant Muntjac Deer (the largest land animal described since 1937), and a species of striped rabbit (Schaller and Rabinowitz, 1995; Schaller and Vrba, 1996; Surridge *et al.*, 1999).

New habitats, new sampling approaches, genetic reassessments of currently described species, and other factors indicate that the number of species may be greater than early estimates of 3–7 million made by earlier workers such as Raven (1985; see also Groombridge and Jenkins, 2002).

Rates of description of new species

The number of new species formally described each year tell us how quickly our knowledge of various taxa is growing. In Figure 2.6, the description rates of two taxa are shown. The vertical lines show the dates by which half of the taxa recognised in 1970 were described. Half of all crustaceans and arachnids documented in 1970 were described between 1960 and 1970, whereas half of the bird species recognised in 1970 were described before 1843. Most of the insects known from the Galapagos Islands were recorded recently (Table 2.1).

Table 2.1. Increase over time in the number of known insect taxa from the Galapagos Archipelago (after Gaston and Williams, 1993).

Year	Orders	Families	Genera	Species
1835–1966	19	129	395	618
1966–1977	22	164	531	883
1977–1990	24	221	790	1339

However, it is difficult to assess how quickly the lists of recognised taxa grow, because most descriptions and revisions involve the detection of synonyms. In some insect groups, for example, species disappear into synonymy at one-quarter to one-third of the rate at which new species are described (Hammond, 1992). Between 1979 and 1988, the annual rate of description of new species around the world was 2000–2700 new species of beetles, 1300–1600 spiders, 900–1300 flies, and 250–400 nematodes (WCMC, 1992; State of the Environment Advisory Council, 1996).

Even for birds, new species are discovered each year (Blackburn and Gaston, 1995). About three new bird species are added each year to the 9000–10 000 known taxa (Diamond, 1985). For example, a new species of Hawk Owl (called the Little Sumba Hawk Owl) was described from the island of Sumba in the Indonesian archipelago in 2002 (Olsen *et al.*, 2002). In contrast, reptiles are a comparatively poorly known group. Over the past 10 years, more than 15 new species have been added to the Australian reptile fauna annually.

Figure 2.6 demonstrates a bias in systematics that extends to the way in which we manage environments, which was termed the **anthropocentric** bias in Chapter 1. Attractive species (such as flowering plants, butterflies and dragonflies), those most closely related to humans (vertebrates, especially mammals, and most especially primates), or those that directly affect our economy (insect pests, plague rodents) or health (some viruses, bacteria and invertebrates) are the favoured objects of study (Hammond, 1992). Further bias is caused by ease of observation and collection. For example, small species, and species that are cryptic or that inhabit remote areas, are relatively unlikely to be described (Yen and Butcher, 1997; New, 2000).

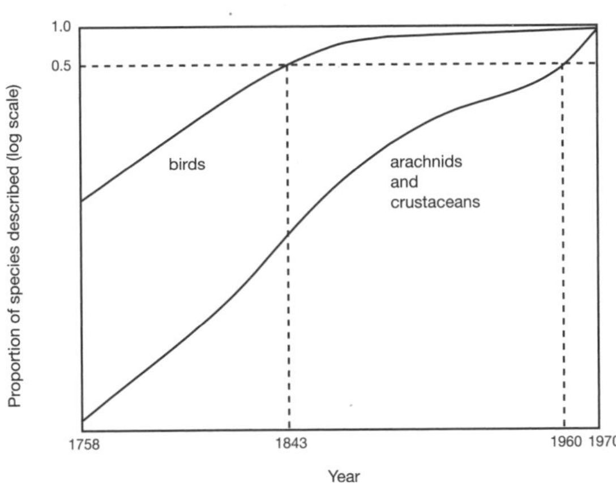

Figure 2.6. Global rate of taxonomic description for birds, arachnids and crustaceans from 1758 (when Linnaeus introduced the binomial naming system) to 1970 (after Simon, 1983, in May, 1990a). Numbers of descriptions are expressed as proportions of the total number of species in each group recognised in 1970. The ordinate uses a log scale.

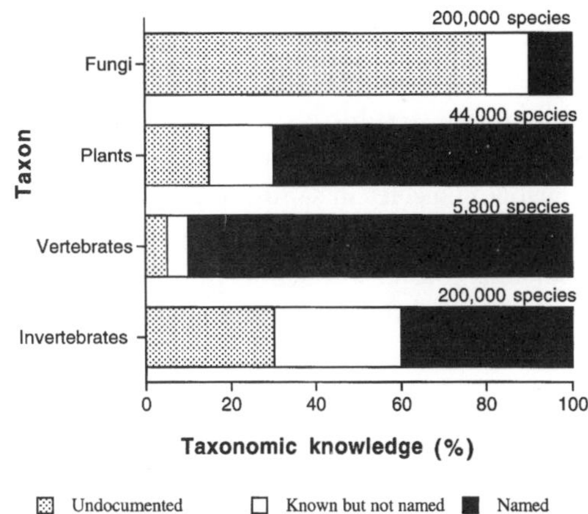

Figure 2.7. The approximate state of taxonomic knowledge in Australia for four taxonomic groups (after OCS, 1992).

The anthropocentric bias common to most scientific effort is evident in the relative taxonomic knowledge of Australian plants and animals. Our knowledge of vertebrates and vascular plants is reasonably good (although it is far from complete), but our knowledge of other groups is poor by comparison (Figure 2.7).

Little Sumba Hawk Owl – described in 2002. (Photo by Jerry Olsen.)

Summary: number of species

The diversity of life on earth is so great that it is beyond our resources to catalogue, study or understand it in the foreseeable future. Unique and striking discoveries such as *Symbion pandora*, the Wollemi Pine and the Little Sumba Hawk Owl have the potential to capture the public imagination, but they are merely symptoms of the fact that science is ignorant of the identity and ecology of most of the earth's biota (Morris, 1995). They highlight the range of life forms that might be lost, possibly before they are known to science (Wilson, 1992; Beattie and Ehrlich, 2001).

2.3 Species richness

The term biodiversity is sometimes used as a synonym for **species richness** (Bunnell *et al.*, 2003), which is the number of different species in a given site, habitat, or defined geographic region. **Species diversity**, in the strict sense, refers to both the number and relative abundance of different species within a site or habitat. Approaches to the measurement of species diversity are treated in detail in Chapter 16.

We can also include the amount of genetic differentiation among species in the definition of species diversity. A species that differs widely from all others (indicated, for example, by its relatively isolated position in a taxonomic hierarchy) contributes more to overall diversity than does a species with many close relatives. This idea underlies the use of genes, populations and ESUs as the units of conservation.

Species richness varies from place to place. Patterns of decreasing species richness are associated with increasing latitude on continents, with increasing elevation (MacArthur, 1972) and with increasing isolation of islands (MacArthur and Wilson, 1967). There are examples of remarkable local diversity; for example, Rice and Westoby (1983) reported many more than 100 vascular plant species in sample areas of 1000 square metres of rainforest in Puerto Rico, grazed woodlands in Israel, and coastal woodlands in New South Wales. Tropical moist forests cover only 7% of the earth's land surface, but contain at least half of the planet's species (McNeely *et al.*, 1990; FAO, 2001). We return to the estimation of species richness in Chapter 14.

Problems with the uncritical use of species richness

In some instances, natural resource managers have aimed to maximise the number of species at a location (see Shields *et al.*, 1992; Attiwill, 1994a, b). This approach has two main problems. First, the scale at which species richness is measured can affect conclusions and management actions. For example, maximum species richness may occur at a local scale following disturbance from invasions by animals and plants that are more typically associated with disturbed environments (Shields and Kavanagh, 1985; Gascon *et al.*, 1999). In this situation, species that depend on undisturbed ecosystems may be eliminated. Species richness at a broader scale (e.g. across an entire landscape) may be reduced as taxa that are sensitive to local disturbance are lost, even though species richness increases at single locations (Noss and Cooperrider, 1994; Lindenmayer and Franklin, 2002; see Chapter 14). The second major problem with the uncritical use of species richness is that all taxa are assigned equal status. For example, the Common Starling, which was introduced to Australia in the mid 1800s and is now a major pest, would be given a weight equal to rare species such as the Sooty Owl, which is threatened by human activities (Milledge *et al.*, 1991; Garnett and Crowley, 2000). Before deciding on management priorities, the identity of the taxa that comprise an assemblage should be appraised (Gilmore, 1990). When species richness is used by itself to set management priorities or when it is applied at an inappropriate spatial scale, it can lead to ineffective conservation strategies (Murphy, 1989).

2.4 Endemism

A species or other taxon is regarded as **endemic** to an area if it occurs uniquely in that area. To consider a taxon as endemic without specifying a geographic range is meaningless. Because the ranges of taxa change over time, time must also be specified, or at least understood (Anderson, 1994). Changes in the ranges of species, either by invasion (range extension) or by range contraction, change the percentage endemism within a given area. Changes in endemism also occur by extinction and speciation, although the rate of speciation is much slower than the rate at which taxa change their ranges. The term endemic is used occasionally to refer to species with restricted geographic distributions.

Causes of endemism

High levels of endemism within an area may be the result of long geographic distances to sources of immigrants (and hence lower immigration rates), large area, long periods since geological formation, or topographic and habitat variety within the area (Mayr, 1965; Anderson, 1994; Rosensweig, 1995). The relationship between area and percentage endemism of mammals is shown in Figure 2.8.

There are usually more species (and more endemic species) on larger land masses, in larger areas of habitat, in ecologically diverse areas, and in areas that have been isolated for long periods (Nelson and Platnick, 1981; Anderson, 1994; Rosenzweig, 1995). For example, approximately 5% of Australia's flora occurs in the Stirling Ranges of south-west Australia (a tiny area) – an ancient and isolated place with unique ecological characteristics.

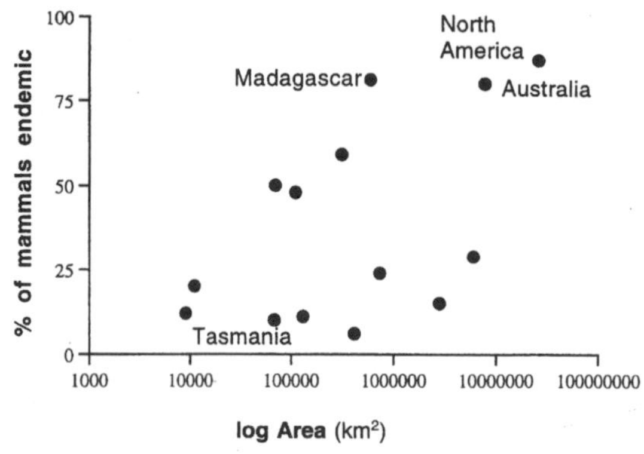

Figure 2.8. Percentage of endemic mammals within several geographic regions (after Anderson, 1994). Values for Australia, Tasmania, North America and Madagascar are shown. The other regions are Cuba, Hispaniola, Puerto Rico, Jamaica, southern Africa, Argentina, Sumatra and the Philippines.

The Stirling Ranges of Western Australia (see Table 2.5). (Photo by David Lindenmayer.)

There can be other reasons for endemism. For example, to explain centres of vascular plant endemism in Australia, Crisp *et al.* (2001) argued that the expansions of desert areas during the Pleistocene prevented range-limited species from persisting in refugia, accounting for the proximity to the coast of all 12 areas of high endemism in Australia (see Figure 2.10).

Megadiverse nations and endemism

The concept of megadiversity is based on the total number of species and the degree of endemism displayed at the species level, and at higher taxonomic levels. The 17 so-called **megadiverse nations** are Australia, Brazil, China, Colombia, the Democratic Republic of the Congo, Ecuador, India, Indonesia, Mexico, Madagascar, Malaysia, Papua New Guinea, Peru, the Philippines, South Africa, the USA and Venezuela. Together, these megadiverse countries support 60–70% of the world's species.

Australia has more known endemic terrestrial vertebrates than any other country (State of the Environment, 1996, 2001a, 2001b; Table 2.2; Figure 2.8): it is relatively rich in mammals, birds and reptiles. The number of endemic amphibians in Australia is equivalent to the numbers recorded in some countries in Central and South America. By world standards, the number of endemic freshwater fish species in Australia is low (Table 2.3), reflecting the small amount of freshwater habitat here and the fact that Australia was once joined to New Guinea, with which many species are shared (Mackey *et al.*, 2001). In contrast, there is a greater number of reef fishes and coral species in Australia than elsewhere (Table 2.4), reflecting the large area of coral reef habitat.

Australia ranks fifth among the 17 megadiverse nations in terms of the number of endemic plants it supports: more than 80% of Australia's plants are endemic (OCS, 1992; State of the Environment, 2001a, 2001b; Figure 2.9). Some other aspects of the endemic status of Australia's biota are detailed in Table A1 in the Introduction to this book and in Box 2.8.

Endemism and global biodiversity hotspots

In a worldwide study, Myers (1988; 1990b) identified 18 relatively small terrestrial areas that are rich in endemic vascular plant species, which are currently

Table 2.2. Distribution of mammal families (and genera in parentheses) among six biogeographic regions studied by Cole *et al.* (1994). The number of families and genera represented in the mammal fauna of each region, the percentage of those families endemic to that region, and the affinities among regions are shown. Definitions of the regions are given in the glossary.

Region	No. families (genera)	Percentage endemic	No. families (genera) in common				
			Nearctic	Neotropic	Palaearctic	Ethiopian	Oriental
Nearctic	17 (184)	5.4 (26.6)					
Neotropic	50 (309)	38.0 (63.1)	30 (131)				
Palearctic	42 (262)	0 (26.7)	17 (32)	12 (10)			
Ethiopian	52 (298)	32.7 (77.5)	13 (13)	11 (18)	28 (61)		
Oriental	50 (312)	6.0 (35.9)	15 (20)	12 (11)	38 (158)	33 (61)	
Australian	28 (135)	42.9 (63.0)	5 (6)	5 (7)	9 (17)	9 (17)	16 (46)

Table 2.3. Numbers of freshwater fish species by continent (after Nelson, 1984 in Banister, 1992).

Continent	No. freshwater fish species
South America	2200
Africa	1800
Asia	1500
North America	950
Central America	354
Europe	250
Australia	170
New Zealand	27

Note: The value for Asia is probably an underestimate because relatively little taxonomic and survey work has been carried out.

experiencing relatively rapid rates of habitat modification or loss (Table 2.5). These are often called biodiversity **hotspots.** These 18 sites contain a total of about 50 000 endemic plant species (20% of the world's total) in about 750 000 square kilometres (0.5% of the earth's surface area). In a more recent study, Mittermeier *et al.* (1998) created a list of terrestrial hotspots, which they estimated support about 50% of the earth's terrestrial biodiversity. These areas cover less than 2% of the earth's land surface (Mittermeier *et al.*, 1998; see Table 2.5). The approach used to identify hotspots has been criticised on a number of grounds (see Brummitt and Lughada, 2003), but from a practical conservation perspective, one of the important outcomes of identifying hotspots of biodiversity is that it can provide a focus for conservation efforts, and attract money for their protection (Myers and Mittermeier, 2003).

Table 2.4. Numbers of reef fishes and coral species in various areas (after Harmelin-Vivien, 1989, in Banister, 1992).

Coral reef	No. fish species	No. coral species
Great Barrier Reef	2000	500
New Caledonia	1000	300
French Polynesia	800	168
Heron Island (Great Barrier Reef)	750	139
Society Islands	633	120
Toliara (Madagascar)	552	147
Aqaba	400	150
Moorea (Society Islands)	280	48
St Gilles (Réunion)	258	120
Tutia Reef (Tanzania)	192	52
Tadjoura (Djibouti)	180	65
Baie Possession (Réunion)	109	54
Kuwait	85	23
Hermitage (Réunion)	81	30

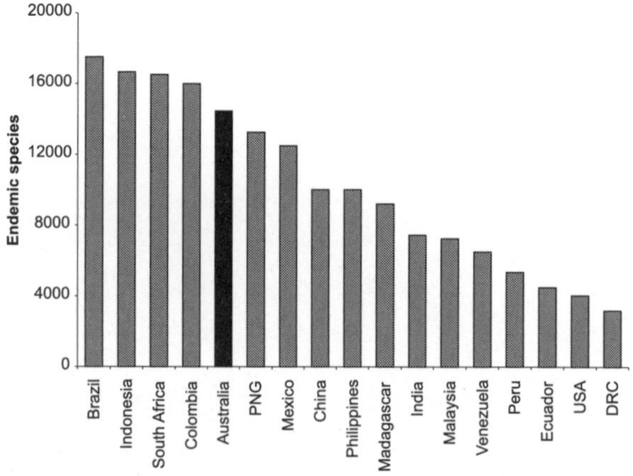

Figure 2.9. Number of endemic vascular plants in each of the 17 megadiverse countries worldwide. (Redrawn from the State of the Environment, 2001a.)

Box 2.8

Species richness and endemism in Australia

Australia is home to about 7% of the earth's species, of which the majority are endemic. The unique nature of the Australian biota places a particular responsibility on the Australian human population to take care of these natural resources, both for future generations and on behalf of other nations. The levels of richness and endemism in some taxonomic groups are as follows:

- 22 000 species of vascular plants (85% endemic)
- 850 species of birds (45% endemic)
- 146 species of marsupials, representing 52% of the world's total (90% endemic)
- 276 species of mammals (84% endemic)
- 174 species of amphibians (93% endemic)
- 700 species of reptiles (89% endemic)
- more than 200 000 species of insects (most endemic)
- 3600 species of fish (most endemic)
- tens of thousands of species of molluscs (most endemic)
- tens of thousands of species of ectomycorrhizal fungi (most endemic)

(Data are from OCS, 1992; Anderson, 1994; State of the Environment Advisory Council, 1996, 2001a.)

Endemism within Australia

Endemism is defined with reference to a specified area. There are areas of endemism within Australia, for

Table 2.5. Plant biodiversity hotspots from around the world, listed in descending order of plant endemism (after Mittermeier *et al.*, 1998).

Biodiversity hotspot	No. endemic plant species
1. Tropical Andes	20 000
2. Mediterranean Basin	13 000
3. Madagascar and Indian Ocean Islands	9700
4. Mesoamerican forests	9000
5. Caribbean Islands	7000
6. Indo-Burma	7000
7. Atlantic Forest Region	6000
8. Philippines	5832
9. Cape Floristic Region of South Africa	5682
10. Eastern Himalayas	5000
11. Sundaland	5000
12. Brazilian Cerrado	4400
13. South-western Australia	3724
14. Polynesia/Micronesia	3334
15. New Caledonia	2551
16. Choco/Darien/West Ecuador	2500
17. Western Ghats/Sri Lanka	2182
18. California Floristic Province	2125
19. Succulent Karoo	1940
20. New Zealand	1865
21. Central Chile	1800
22. Guinean forests of West Africa	1500
23. Wallacea	1500
24. Eastern Arc Mountains and coastal forests	1400
Total no. plant species endemic to hotspots	124 035
Total global plant diversity endemic to hotspots	45.94%

example for birds (Cracraft, 1991; Schodde and Mason, 1999). Crisp *et al.* (2001) recognised 12 centres of endemism for vascular plants around Australia. Three of these: Adelaide–Kangaroo Island, Cape York Peninsula and New England Dorrigo, have been underrated or overlooked in the past (Crisp *et al.*, 2001).

Other important areas of endemism occur in Australia. In the marine environment, for example, the sea mounts off southern Australia support large numbers of endemic fish and invertebrates (Koslow and Gowlett-Holmes, 1998). Work by de Forges *et al.* (2000) highlights the large number of new and endemic species in the south-west Pacific Ocean (Figure 2.10).

2.5 Species diversity: alpha, beta and gamma diversity

Alpha diversity

Alpha diversity accounts for the number of species and their relative abundances within a single location in a

Box 2.9

Other types of hotspots

The concept of biodiversity hotspots has traditionally been linked with parts of particular continents that support extraordinarily high levels of biodiversity, for example south-western Australia (e.g. Myers, 1990b; Mittermeier *et al.*, 1998; Crisp *et al.*, 2001). However, other types of hotspots at smaller scales are important for conservation and management. Within oceans, for example, sea mounts (submarine mountains), and reefs can support very large numbers of species relative to the surrounding area (Worm *et al.*, 2003). In Australia, rocky reef habitats support 50% of the nation's fisheries (State of the Environment, 2001b). On land, cliffs, caves, mound-springs and other features can be small but support many unique species (e.g. Culver *et al.*, 2000; reviewed by Lindenmayer and Franklin, 2002). Many species are restricted to special types of soils and rock formations, for example serpentine soils (Whittaker, 1954a,b; Harrison *et al.*, 2000). Landscapes subject to persistent fog in the tropical forests of South America often support very large numbers of bird species (Laurance *et al.*, 1997). There are numerous other examples.

Cliff-lines are areas that can be species-rich hotspots in many landscapes. (Photo by David Lindenmayer.)

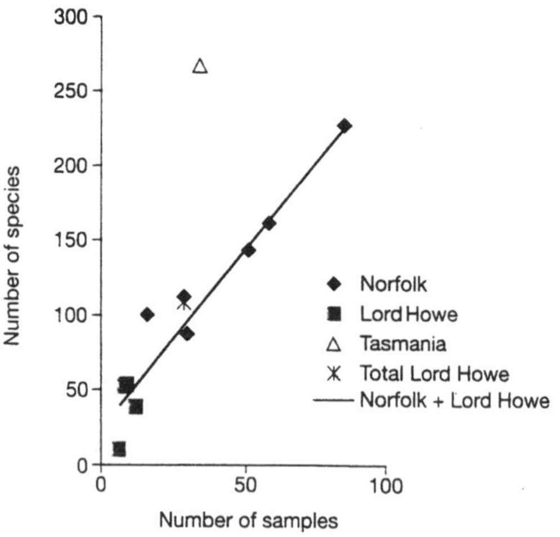

Figure 2.10. Numbers of new species and sampling effort at sea mount sites in the south-west Pacific Ocean. (Redrawn from de Forges *et al.*, 2000.)

Tropical rainforest in north Queensland. (Photo by David Lindenmayer.)

defined geographic region. If a Wheat crop extends throughout a region, as it does in central New South Wales, the environment on a large scale is not very diverse, because most individuals belong to just a few species. In contrast to a tropical rainforest, a monoculture such as a Wheat field is not very diverse: there are relatively few species, most individuals in the community belong to a single species, and only a few weeds add variety. Even if there are many species of weeds, they are not abundant relative to the crop species, so they do not add much to the diversity of the community.

A hay paddock in south-western New South Wales. (Photo by David Lindenmayer.)

Beta diversity

Beta diversity represents the change in species composition that occurs along an ecological gradient within a community (Figure 2.11). The degree of similarity between communities at different points along a gradient provides a guide to beta diversity in a landscape. If the turnover of species is high (that is, if the number of species in common between two places is low), communities at different points on the gradient will be relatively dissimilar (Westoby, 1988). Turnover may also be represented by change in the composition of the biota at a single location over time (Anderson, 1994).

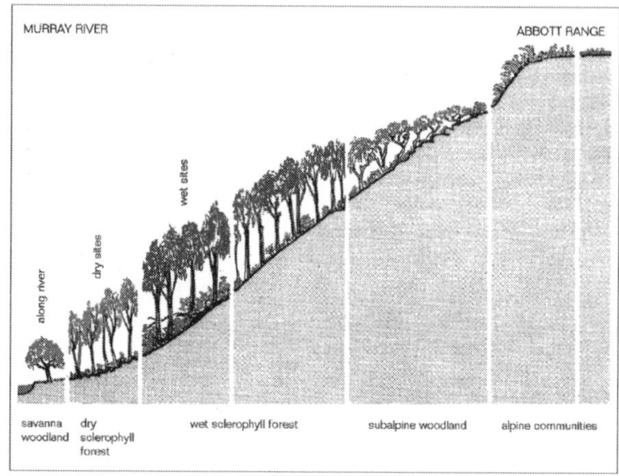

Figure 2.11. Turnover of dominant tree communities with elevation from the Murray River to the top of Mount Kosciuszko in New South Wales (from Costin *et al.*, 1979).

Gamma diversity

The turnover of species between sites within more or less homogeneous habitat is known as gamma diversity. It measures the spatial component of community diversity that is independent of environmental gradients.

Ecosystem diversity

Ecosystem diversity applies at a landscape or global level and refers to the variety of different assemblages and communities of organisms that exist in different places within a landscape. There is no consensus on the meaning of ecosystem diversity, which is reflected in the lack of a consensus on how to measure its properties.

2.6 Vegetation structure as a target for conservation

Variation in the physical structure of vegetation is somewhat independent of species diversity. Variation in size and shape may be related to the life forms of different species, the ages of individuals within species, or the different environments that individuals experience. Communities in which individual plants have similar shapes and sizes, such as commercial crops and plantations, are not structurally diverse. Sometimes, particular structural features associated with vegetation cover can be critical for biodiversity. Mistletoe is a good example: many birds, mammals and other animals use these parasitic plants for food and nesting (Watson, 2001).

Variation within species can be considerable. Some eucalypt species exhibit considerable morphological variation throughout their range, sometimes being difficult to recognise as the same taxon in different places. Manna or Ribbon Gum is a tall, smooth-barked tree in many wetter parts of its range, such as montane environments in the Great Dividing Range; yet it is a much shorter, rough-barked plant in western Victoria, where climatic conditions are warmer and drier (Costermans, 1994).

The regeneration strategies of plants in a community are important determinants of community structure. For example, the life history strategies of different species vary, with some species being relatively sensitive to intense wildfire (Whelan *et al.*, 2002; see Chapter 11). Many species regenerate from seeds (Ashton, 1981b), which often results in dense, relatively evenly aged stands (Attiwill, 1994b). Other species readily survive fire and resprout new growth from epicormic buds (Jacobs, 1955; MacArthur, 1968), resulting in unevenly aged populations.

Projective foliage cover is the proportion of the ground covered by the vertical projection of the vegetation. **Crown cover** is the percentage of the surface of a site that falls within the (projected) crowns of plants, where crowns are defined by polygons tracing their external perimeters. Either crown cover or foliage cover can be used to characterise the distribution and abundance of the biomass of a plant community. For example, the percentage cover and height of foliage is a reasonable indicator of the productivity of a community.

Tropical rainforests are tall and have very large cover values. More than 90% of the surface of a site may be covered by crowns more than 30 metres above the ground. Temperate eucalypt forests can be as tall or taller than their tropical counterparts, but crown cover of the tallest stratum never reaches 90% – values around 30% are more common. Arid zone shrublands dominated by saltbush often have a projective foliage cover of less than 10%, including all height classes.

In 1970, the Australian botanist R.L. Specht designed a table for the classification of Australian vegetation into different life forms and size classes (Specht, 1970; see Table 2.6). The table is based on projective foliage cover and the growth form of plants in the tallest stratum, and has been used, in slightly varying forms, as a basis for vegetation maps of all parts of Australia.

Specht's (1970) classification provides a useful tool for evaluating conservation status impacts at all spatial scales (Specht *et al.*, 1995). When combined with information on species composition and abundance, it is a central element in defining and delineating plant ecological communities (Specht and Specht, 1999). For example, the Atlas of Australian Resources (AUSLIG, 1990) used a slightly modified version of the classification to document changes in the status of Australian vegetation on a continental scale. The use of this classification to assess the adequacy of conservation reserves is outlined in Chapter 4.

There are powerful social mandates to protect rainforest and old growth forest. Neither forest type is specifically identified or named in Specht's classification, but both are usually defined by their structural attributes. There are many kinds of rainforest and old growth forest and it is important to define their characteristics.

Australian rainforest

The Ecological Society of Australia proposed a definition of the structural and ecological characteristics of Australian rainforest (Dale *et al.*, 1980):

Table 2.6. Structural classification of Australian vegetation (see Specht and Specht, 1999).

Growth form of the tallest stratum	Foliage cover of the tallest stratum			
	>70%	30–70%	10–30%	<10%
Tall trees (>30 metres)	Tall closed forest	Tall open forest	Tall woodland	
Medium trees (10–30 metres)	Closed forest	Open forest	Woodland	Open woodland
Low trees (<10 metres)	Low closed forest	Low open forest	Low woodland	Low open woodland
Tall shrubs (>2 metres)	Closed scrub	Open scrub	Tall shrubland	Tall open shrubland
Low shrubs (<2 metres)	Closed heath	Open heath	Low shrubland	Low open shrubland
Hummock grasses			Hummock grassland	
Tufted/tussock grasses	Closed tussock grassland	Tussock grassland	Open tussock grassland	Sparse open grassland
Graminoids	Closed sedgeland	Sedgeland	Open sedgeland	
Other herbaceous species	Dense sown pasture	Sown pasture	Open herbfield	Sparse open herbfield

The rainforests are defined ecologically as closed, broadleaved forest vegetation with a continuous tree canopy of variable height, and with a characteristic diversity of species and life forms. The ecological definition of rainforests includes transitional and seral communities with sclerophyll emergents that are of similar botanical composition to mature rainforests in which sclerophylls are absent.

Typically, rainforests are closed forests composed of trees less than 30 metres tall. They are only one element of the spectrum of structural vegetation types, but are highly valued because of the perception that they harbour many unique species and species associations.

Australia supports many types of rainforest and they extend throughout the country (see Figure 2.12). In addition to the better-known tropical rainforests of northern Queensland, subtropical and warm temperate rainforests extend from Queensland to east Gippsland in eastern Victoria. Monsoonal rainforests occur in the Northern Territory and northern Western Australia (McKenzie and Belbin, 1991; Bowman and Woinarski, 1994). Cool temperate rainforests such as those dominated by Myrtle Beech extend through the wetter, fire-protected parts of Victoria and Tasmania (Howard, 1973; Busby, 1986; Lindenmayer *et al.*, 2000a). There are even so-called 'dry rainforests', which are thickets of dense scrub, in south-eastern Queensland (Bowman, 1999).

Contrasting tussock grassland and tall open forest vegetation in Australia. (Photos by David Lindenmayer.)

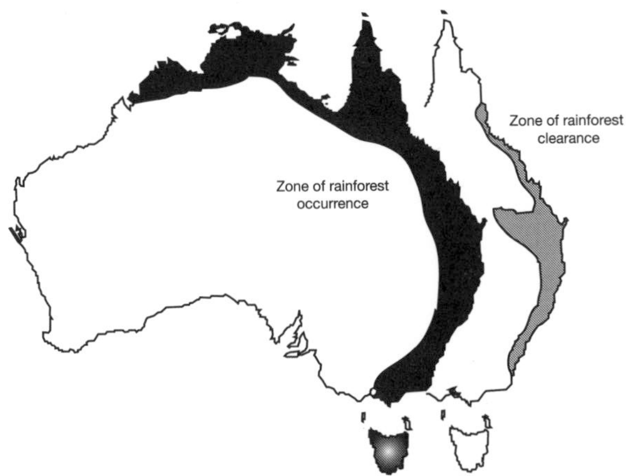

Figure 2.12. Distribution of rainforest and rainforest clearance in Australia. (Redrawn from Bowman, 1999.)

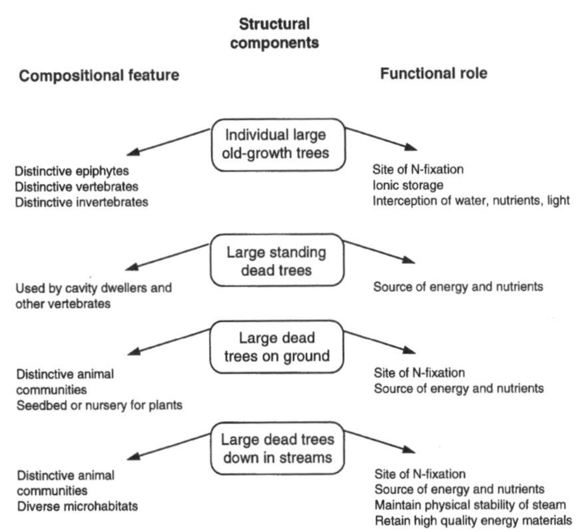

Figure 2.13. Relationships between four structural attributes of Douglas Fir old growth forests and special habitat and ecosystem functions (after Franklin *et al.*, 1981).

In the most detailed review of Australian rainforests to date, Bowman (1999, p. 46) noted that there was 'no consensus on what constitutes rainforest in Australia', and described how rainforest could be defined by (among other things) climatological factors, diagnostic life forms (e.g. epiphytes), canopy measures, the light environment, biogeographically distinctive taxa, and fire susceptibility. Part of the problem of defining (and conserving) rainforest is that rainforests are not static in space or time. In the absence of disturbance, rainforest trees can invade neighbouring wet sclerophyll forest (Harrington and Sanderson, 1994) or become a dominant part of the understorey of eucalypt forest (Lindenmayer *et al.*, 2000a), replacing them over time (Jackson, 1968). The definition of rainforest matters in conservation debates, particularly because large areas of rainforest have been cleared in the past (Bowman, 1999; see Figure 2.13). For example, in both Victoria and New South Wales, State government agencies claim to harvest only non-rainforest stands for timber, so the definition of rainforest in these cases influences what is logged. Wider definitions of rainforest tend to be applied by conservation groups.

Tropical rainforests and high levels of biodiversity

Rainforests in many parts of the world are characterised by high levels of biodiversity. Australian tropical rainforests occupy a relatively small area, but support roughly 25% of the continent's species of plants (Bowman, 1999). A disproportionate number of non-plant species are also found in tropical rainforests, particularly in the forests of northern Queensland. For example, approximately 20% of Australian avifauna species, 60% of bat species, 30% of marsupial taxa, 25% of reptile species, and 30% of frog species are found in north Queensland (Strahan, 2000; Blakers *et al.*, 1984; Reader's Digest, 1990; Cogger, 1995; Churchill, 1998). A significant proportion of species in other groups are also found in north Queensland tropical rainforests (e.g. approximately 60% of butterfly species; Smith, 1994).

Cool temperate rainforests and low levels of vertebrate biodiversity

Not all types of Australian rainforest support diverse vertebrate faunas. Many cool temperate rainforests of south-eastern Australia have ancient biogeographic origins that can be traced to Gondwana, and are dominated by Myrtle Beech and Southern Sassafras. Myrtle Beech forests are cool wet places (Busby, 1986; Lindenmayer *et al.*, 2000a), which often support a rich array of mosses and ferns (Ashton, 1986; Ough and Ross, 1992). In Victoria, although these forests provide important nesting habitats for the Pink Robin (Loyn, 1985), no vertebrate animals appear to be confined solely to Myrtle Beech rainforests. Cool temperate rainforest trees often do not develop cavities (except in the very largest and oldest stems), making them unsuitable for many hollow-dependent animals (Gibbons and Lindenmayer, 2002). Myrtle Beech is a wind-pollinated species, so flower-derived food resources such as nectar, pollen, fruit and seeds, which are important for many

Table 2.7. Patterns of species richness and obligate species use in Tasmanian rainforest (from Read and Brown, 1996).

Biotic group	No. species in Tasmanian rainforest	No. obligate	No. species in Tasmania
Mammals	22	None	35
Amphibians	4	None	10
Reptiles	6	None	19
Birds	21	None	230–256
Invertebrates	?	?	~32 000
Flowering plants	70(–100)	Few	~1500
Conifers	7	None	11
Ferns and allies	48	None	98
Bryophytes	168	?	~650
Lichens	217	?	~700
Fungi, algae	?	?	?

vertebrates and invertebrates, are extremely limited. Myrtle Beech rainforests are also unsuitable habitat for many species of arboreal marsupials that depend on insects and plant gums for food. They are dominated by only a few species of trees, and their relative floristic and structural simplicity means that they may not contain a wide array of foraging substrates for animals. There are many types of rainforest in Australia, and all have important conservation values. However, they are not all rich in unique species of animals and vascular plants (Table 2.7).

Old growth forest

Old growth forests are perceived by the human population to be a valuable structural vegetation type; therefore, there is a need to fully understand what is meant by the term. For instance, the Regional Forest Agreement Process in Australia, underpinned by the National Forest Policy Statement (Commonwealth of Australia, 1992), pledged to 'conserve and manage areas of old growth forest and wilderness as part of the reserve system'. According to the Commonwealth of Australia (2001a), 67% of the old growth forest within the regions where Regional Forest Agreements have been signed is protected in reserves.

Cool temperate rainforest in Tasmania. (Photo by David Lindenmayer.)

Early work on old growth forest

The first major attempt to define the ecological characteristics of old growth forest was by Franklin *et al.* (1981) for Douglas Fir forests in north-western North America. This and subsequent work by the same group (e.g. Spies and Franklin, 1996; Franklin *et al.*, 2002) influenced definitions and procedures for delineating old growth forests throughout the world.

Franklin *et al.* (1981) found that old growth coniferous forests differ substantially from young forests in structure, species composition, rates and paths of energy and nutrient flows, and water cycling (Figure 2.13). They suggested that the distinctive features of old growth forests were related to four structural features: large, live, old growth trees; large stags (i.e. dead trees); large logs on land; and large logs in streams. None of these characteristics uniquely define old growth forest, but together they provide a means to differentiate old growth from other Douglas Fir forest types, because these characteristics play a pivotal ecological role (Figure 2.13).

Subsequent research has revealed more attributes, patterns and processes unique to old growth Douglas Fir forests. Lichens on large standing trees and decaying logs fix important amounts of nitrogen each year, but are absent during the first 100–150 years of the forest's development (Harmon *et al.*, 1986; Marcot, 1997; see accompanying photo). Microclimatic conditions including light, air temperature, soil temperature, wind and moisture profiles, and their diurnal and seasonal patterns, are different in old growth forest (Chen *et al.*, 1991).

Other unique features of old growth forests around the world include the size and shape of understorey and dominant tree species; carbon dioxide exchange patterns; nitrogen transformation pathways; vegetation floristics and structure; the role of tree fall mounds in mixing soils and maintaining stable assemblages of herb species; tree population dynamics; decay and nutrient dynamics of coarse woody debris and leaf litter; and the volume, mass and nutrient content of fallen logs (see Burgman, 1996). There are likely to be some similarities and many differences in the ecological attributes of old growth in different forest types (see Box 2.10).

Few, if any, of the features that characterise old growth forest are unique to it (Bunnell *et al.*, 2003). Rather, the ensemble of species, the structural attributes of the vegetation, and the ecological processes may be unique (Marcot, 1997). Young stands developing after wildfire can have high densities of large dead trees (or stags; Lindenmayer *et al.*, 1991a, 2000b), large logs remaining from the original stands (Lindenmayer *et al.*, 1999a), and varying proportions of recovering trees that predate the fire event (Ashton, 1986). The fruiting bodies of hypogeal fungi are often produced in association with tree roots in decomposing logs and stumps, and provide a food source for insects and mammals (Scotts, 1991; Claridge and May, 1994; Lindenmayer *et al.*, 2002b), although they may also be common in young forests.

Epicormic branch systems in old growth Douglas Fir. (Photo by J. Franklin.)

Australian definitions of old growth forest

In many places in Australia, very little old growth (however it is defined) remains. For example, in a study of 43% of the forested part of south-east Queensland, only 2.7% was classified as 'old growth' and 3.1% was 'likely old growth' (see Kelly, 1998). In the wood production Ash forests of the Central Highlands of Victoria, less than 5% of the forest cover is old growth and the largest single patch is less than 50 hectares (Lindenmayer *et al.*, 2000b). However, the way these two studies defined old growth contrast markedly. Indeed, there are considerable differences among the definitions of old growth forest in Australia suggested by different organisations, reflecting differing ethical, social and ecological perspectives.

The Australian Conservation Foundation (in RAC, 1991) defined old growth as:

> *forest that has not been, or has been minimally, affected by timber harvesting and other exploitative activities by Australia's European colonisers.*

The Resource Assessment Commission (RAC, 1992a) considered old growth forests to have high conservation and intangible values and defined them as forest:

> *both little disturbed and ecologically mature.*

The National Forest Policy Statement (NFPS; Commonwealth of Australia, 1992) defined old growth as:

> *forest that is ecologically mature and has been subjected to negligible unnatural disturbance...in which the upper stratum or overstorey is in the late mature to overmature growth phases.*

In Tasmania, old growth forest is defined as (after Forestry Tasmania, 2004):

> *Forest that is ecologically mature and been subjected to negligible unnatural disturbance such as logging, roading and clearing.*

The latter three definitions make use of the term 'ecologically mature'. The Regional Forest Agreement process in Australia (Commonwealth of Australia and Department of Natural Resources and Environment, 1997) defined old growth forest as:

> *ecologically mature forest where the effects of disturbances are now negligible.*

This definition has some accompanying qualifications, such as old growth having 'characteristics of older growth stages', 'functional qualities expected to characterise an ecologically mature forest ecosystem', and 'negligible disturbance effects' (summarised in Commonwealth of Australia, 2001a).

Woodgate *et al.* (1994) defined old growth forest in East Gippsland as:

> *forest which contains significant amounts of its oldest growth stage in the upper stratum – usually senescing trees – and has been subjected to any disturbance, the effect of which is now negligible.*

Thus, old growth forests are the oldest and least disturbed forests for a given ecological vegetation class, and can include old growth woodlands and old growth mallee. Woodgate *et al.* (1994, 1996) saw their definition as a refinement of the definition in the National Forest Policy Statement (Commonwealth of Australia, 1992), which relied on ecological maturity and the absence of disturbance. They classified forest as ecologically mature if the canopy cover of senescing trees was greater than 10%.

Ecological characteristics of old growth forests

Most definitions of old growth include some measure of time since disturbance. However, most organisms that are most abundant in, or restricted to, old growth do not respond to time itself, but rather to the attributes of a forest that develop or accumulate over time (Bunnell *et al.*, 2003). In this context, Scotts (1991) described the ecological characteristics of old growth forests in south-eastern Australia and related these to forest-dependent vertebrate fauna. There were several 'important' characteristics including:

- a deep multilayered canopy resulting from the presence of more than one tree age class and/or dominant and subdominant members of one age class
- many individual, live, large or old trees
- significant numbers of stags and large logs.

Dyne (1991) argued that all forest types, including rainforest, dry sclerophyll, wet eucalypt forest and cypress forests, can develop 'old growth' forms. Although ecological processes and some structural attributes can be shared by some old growth forests, no single definition is applicable to all forest types (cf. Hunter, 1990). For instance, Alpine Ash trees typically have a much shorter lifespan (250–300 years) than Mountain Ash trees (about 500 years). Tallowwood in northern New South Wales lives longer still (up to 1000 years). In tropical Australia, eucalypts rarely survive beyond 200 years before being destroyed by fire or termites (Ogden, 1981). Thus, structural features such as dead and dying trees accumulate at different rates. Similarly, tree hollow development depends on tree species and the environment (Gibbons and Lindenmayer, 2002). Old growth characteristics develop over different time scales in different forest types, so there cannot be a universal definition of its age or features. However, several attributes are used to measure old growth status in many forest types (Table 2.8).

Old growth forests take time to develop and may senesce, being replaced by other vegetation types, such as cool temperate rainforest (Jackson, 1968; Lindenmayer *et al.*, 2000a), or they are sometimes disturbed naturally by events such as wildfires or windstorms. Therefore, areas of young forest need to be reserved to take the place of existing old forest that will be lost in the future. Management of old growth should embrace an estate of old forests and younger forests destined to replace them, and the estate should be designed to accommodate successional changes,

Table 2.8. Attributes of old growth forests in Australia (after Dyne, 1991).

Attributes
Structural and compositional properties
Relatively large trees and other plants
Relatively old vegetation (mature or over-mature)
Presence of large crown gaps
Characteristic biotic composition
Presence of hollows or fallen logs
Presence of indicator species
Presence of characteristic life forms
Absence of weeds or pathogens
Functional properties
Characteristic levels of gross and net productivity
Nutrient cycles, high litter levels in dynamic equilibrium
Low or negative biomass increment
Low rates of change in composition, structure and function
Reduced transpiration, high soil moisture, dry season stream flows
Permanence of wildlife habitat
Absence of disturbance (fire, logging, grazing etc.)
Ancillary properties[1]
Aesthetics
Wilderness qualities
Size of the forest stand
Context (vegetation mosaic and disturbance)
Public perceptions
Ease of long-term maintenance

[1]Characteristics that do not directly aid scientific evaluation, but which may be important for management or policy.

uncertain environmental events and the spatial processes and dispersal mechanisms of old-growth-dependent flora and fauna (Burgman, 1996; McCarthy and Lindenmayer, 1999).

Species diversity and old growth forests

Scotts (1991) listed 30 vertebrates that find their optimum habitat in the old growth forests of south-eastern Australia (Table 2.9). Tree hollows are one of the most important ecological characteristics of many old growth Australian forests (e.g. Scotts, 1991; State Forests of NSW, 1994; Gibbons and Lindenmayer, 2002). More than 300 species of Australian vertebrates and an unknown but large number of invertebrates depend on them (Gibbons and Lindenmayer, 2002).

The paucity of research on old growth in Australia

There has been limited research into the ecological characteristics of Australian old growth forests. The most detailed empirical work has been conducted in Victorian Ash forests (see Box 2.10 and also Ashton, 1976). Other relatively extensive studies include investigations in East Gippsland in Victoria (Woodgate *et al.*, 1994, 1996) and the Clarence River catchment in northern New South Wales (Clode and Burgman, 1997). Some studies of old growth have been conducted as part of background data gathering and research to support Regional Forest Agreements, such as in south-east Queensland (e.g. Kelly, 1998).

Table 2.9. Groups of species for which old growth eucalypt forests in south-eastern Australia are considered to provide optimum habitat. (Modified from Scotts, 1991.)

Common name	Value
Mammals	
Tiger Quoll	Foraging, nesting
Tuan or Brush-tailed Phascogale	Nesting
Yellow-bellied Glider	Nesting
Squirrel Glider	Nesting
Greater Glider	Nesting
Greater Pipistrelle	Nesting
Birds	
Glossy Black Cockatoo	Nesting
Yellow-tailed Black Cockatoo	Nesting
Gang-gang Cockatoo	Nesting
Superb Parrot	Nesting
Regent Parrot	Nesting
Powerful Owl	Nesting
Sooty Owl	Nesting
Leaden Flycatcher	Foraging
Satin Flycatcher	Foraging
Cicadabird	Foraging
Varied Sitella	Nesting, foraging
Mistletoebird	Foraging
Lewin's Honeyeater	Foraging
Regent Honeyeater	Foraging
Crescent Honeyeater	Foraging
Reptiles	
Diamond Python	Foraging, nesting
Carpet Python	Foraging, nesting
Tree Goanna	Foraging
Spencer's Skink	Foraging
Amphibians	
Southern Barred Frog	Foraging
Spotted Tree Frog	Foraging

The Yellow-bellied Glider – a species strongly associated with old growth Mountain Ash forest in the Central Highlands of Victoria. (Photo by David Lindenmayer.)

There have been few studies of old growth forests in Australia, perhaps because the concept relates most easily to vegetation types where major disturbances (principally wildfires) are stand-replacing (or predominantly stand-replacing). It is difficult to examine the characteristics of old growth stands when many individual trees survive disturbances such as fire (e.g. by resprouting from epicormic buds, as occurs in most eucalypts; see Chapter 11). When trees die, small gaps are filled by new recruits, and the process of gap regeneration creates multi-aged stands, making the definition and subsequent study of old growth complicated.

A lack of basic ecological information hampers attempts to define old growth forests adequately, and delineation of old growth forest in Australia usually ignores its many unique ecological characteristics. Instead, planners and managers use subsets of characters such as the developmental status of dominant tree crowns to discriminate the forest from other ecological types and successional stages. This creates an

Box 2.10

Studying Victorian old growth forests

The Ash-type forests of the Central Highlands of Victoria are perhaps the most thoroughly studied old growth forest type in Australia. Lindenmayer *et al.* (2000b) examined more than 520 sites in 11 age classes ranging from more than 250 years to less than 20 years old. Statistical modelling of the structural characteristics of stands revealed that old forests have: (1) abundant tree ferns and understorey rainforest trees, (2) numerous large living and dead cavity trees that vary considerably in characteristics such as diameter, height, and stage of decay, (3) more vertical layers of vegetation than younger stands, (4) large trees supporting large quantities of bark strips, which are important habitats for arboreal invertebrate communities, and (5) well-developed clumps of mistletoe (and associated vertebrates such as the Mistletoe Bird). Other work showed that the fallen log biomass in old growth Mountain Ash forests is substantial – often exceeding 550 tonnes per hectare (Lindenmayer *et al.*, 1999a).

Most old growth stands contained trees of markedly different ages within the same stand – a result of past fires that burned only some of the trees in the previous stand (Lindenmayer *et al.*, 1999b). Old forests were disturbed by fire whereas younger forests were disturbed by both logging and fire, making it impossible to determine if the recovery would be different following different kinds of disturbance.

Few, if any, vertebrates are confined to old growth Ash stands, although many are strongly associated with old growth structures such as tree hollows that can also occur in regrowth stands. Two species that are closely associated with large continuous areas of old growth Ash stands – the Sooty Owl (Milledge *et al.*, 1991) and the Yellow-bellied Glider (Lindenmayer *et al.*, 1999c) – are uncommon in younger forest. However, the same age-class relationships for these taxa do not hold in forests dominated by different tree species elsewhere in Australia. Populations of other species in Ash forests, for example the Greater Glider, reach their highest abundances in old growth, but they are not uncommon in younger forests provided that old growth attributes such as tree hollows are present (Lindenmayer, 2002a).

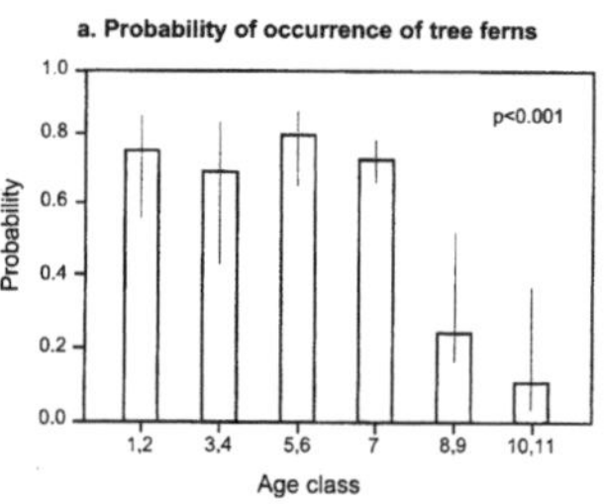

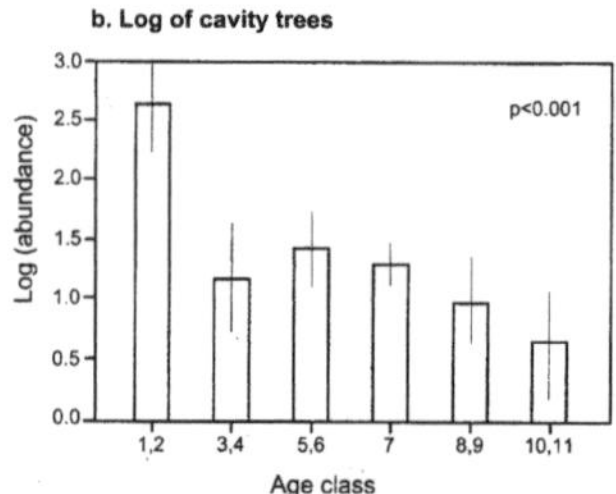

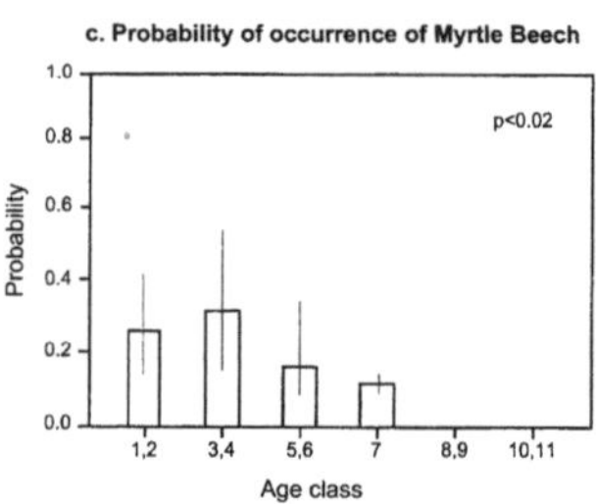

Figure 2.14. Some structural differences between young and old Mountain and Alpine Ash forest. The age classes are 1+2 (old growth stands more than 250 years old) through to age classes 10+11 (young post-logging regrowth forests comprising stands less than 20 years old; from Lindenmayer *et al.*, 2000b). Three features are shown: the occurrence of Tree Ferns (top), the logarithm of the abundance of cavity trees (middle), and the presence of rainforest (Myrtle Beech) in the understorey (bottom).

imperative to demonstrate that the attributes used to define old growth are appropriate surrogates for other ecological characteristics.

Summary: old growth

Old growth forest is not necessarily defined by the presence or absence of old trees, stags or logs. Not all old growth forests have the cathedral-like qualities that are produced by trees of great size and a superficially pristine appearance (e.g. Woodgate *et al.*, 1994). Not all old growth forests necessarily harbour species that depend on them exclusively, and they are not necessarily rich or diverse in species, although some species may be most abundant in them. The concept and definition of old growth only has ecological meaning when it is attached to a particular vegetation type. Each forest type (and other type of vegetation) will probably have a set of characteristics that uniquely defines old growth within it, and that can serve to distinguish old growth from other successional stages.

A stand of old growth Mountain Ash. The figure in the mid-ground indicates the size of the trees in this stand. (Photo by Esther Beaton.)

2.7 Conclusions

Species richness and endemism are important for conservation because they are an index of the biodiversity for which people are responsible. The Australian biota is species-rich and highly endemic, so as a result, the loss of a species in Australia is likely to be a global extinction event.

Biodiversity can be lost through the loss of genes, species or communities. Measures of species richness and diversity ignore the genetic distinctness of species. In addition, most taxa are unknown, with the result that ecologists constrain themselves to small subsets of relatively well-documented species such as vascular plants, mammals and birds. Different species concepts, and variable amounts of data upon which classifications are based, ensure that species are not equivalent entities (Vane-Wright, 1992a). Treating them as such, while ignoring the structure and relationships represented in a taxonomic hierarchy, and ignoring whatever information may be available on the genetic diversity within and among species, does not adequately capture the concept of biodiversity.

Genetic information can assist in the delineation of populations with a unique phylogenetic history. New techniques may allow the rapid delineation of species or other genetic entities and overcome some of the disadvantages of species concepts. However, the conservation of many taxonomic groups in Australia will depend on describing them and mapping their habitat.

Diversity is not a static concept: it changes whenever individuals are born, die, invade or become extinct. Measures of diversity reflect changes in the composition of communities from one place to another, and through time. Diversity can be related to particular types of vegetation, such as rainforest and old growth forest, and although various forms of these forests may have some structural and functional properties in common, the application of such simple concepts belies the complex nature of the ecological processes and biotic assemblages of which they are composed.

Despite problems of definition and the importance of biodiversity represented by genes and populations, species are likely to remain the primary focus of conservation because of the social mandate for species conservation, and because of the impracticality of applying genetic criteria to more than a handful of species. Conservation priorities are also likely to be driven to some extent by perceptions of the importance of a few community types (e.g. rainforest and old growth forest).

2.8 Practical considerations

A lack of taxonomic work hampers conservation in Australia (Hopper, 1997) and worldwide. Although conservation cannot wait for taxonomic information for fear of losing taxa before they are catalogued, it is difficult to know what to protect or how to protect it if we are ignorant of the identity and ecological characteristics of the majority of the biota. Just as importantly, there is no social mandate for the protection of unknown biota. Despite this critical hurdle, teaching, research and regulatory institutions are investing less in taxonomy than in the past. Australia runs the risk of losing significant undiscovered and undescribed components of its biota (Hopper, 1997), particularly among the fungi, algae and insects, as well as vascular flora and some vertebrate groups (such as reptiles). Governments and conservation biologists should invest more heavily in taxonomic research, particularly the description of new species, and encourage scientific institutions and funding organisations to do the same.

Most ecological surveys concentrate on vertebrates and vascular plants, ignoring the vast bulk of biodiversity.

Conservation biologists should strive to broaden the taxonomic focus of surveys, and to explore novel and under-sampled habitats.

Relatively complete taxonomies, where they exist, provide an approximate guide to genetic distinctiveness. However, interpretation should be tempered by knowledge of the species concepts and taxonomic conventions applied in different taxa. Conservation objectives should use subspecific taxa (subspecies, races, varieties) where they exist. The only other reliable surrogate for genetic variation is geographic distribution, because genetic differentiation and speciation are usually spatial processes. In most cases, conserving a species throughout its range will conserve the range of genetic variation within it, and this strategy will make conservation priorities somewhat immune to the vagaries of taxonomic uncertainty.

2.9 Further reading

Detailed discussions of the concept of biodiversity are given by Wilson (1992) and Gaston and Spicer (1998; 2004).

Hopper and Coates (1990) review the conservation of Australia's genetic resources and the information available to form the basis of conservation decisions. The Coordination Committee on Science and Technology (CCST, 1994) discuss access to Australia's genetic resources. Rojas (1992), Moritz (1994b, 1999), and Patkeau (1999) discuss units of conservation. Morell *et al.* (1995) outline methods for distinguishing closely related taxa.

Bock (1992) outlines the species concept. May (1990a) and Gaston *et al.* (1993) review the global number of species, and the State of the Environment (1996, 2001a, 2001b) reports provide information on species richness and diversity in Australia. The *Journal of Biogeography* has a special issue (volume 28, issue 2) on endemism and patterns of species diversity. Mittermeier *et al.* (1997, 1998) provide excellent appraisals of biodiversity hotspots and endemism at a global scale.

Specht *et al.* (1995) and Specht and Specht (1999) give details of a classification of the Australian vegetation. Australia's State of the Forests Report (Bureau of Rural Sciences, 1998) and Commonwealth of Australia (2001a, 2001c) provide an appraisal of forest cover in Australia. The Bureau of Rural Sciences (2003a,b) contains a valuable summary of statistics concerning Australia's forest and woodland cover. The authoritative work on Australian rainforests is Bowman (1999). The seminal work on old growth forest is Franklin *et al.* (1981). Studies by Woodgate *et al.* (1994, 1996) were some of the first comprehensive attempts to classify and map old growth forest in Australia. Burgman (1996) reviews some of the concepts of old growth forests applied in Australia. Lindenmayer *et al.* (2000b) provide a detailed description of the analyses and results of work on contrasting old growth Mountain Ash and Alpine Ash forest with younger age classes in the Central Highlands of Victoria. Keenan and Ryan (2004) provide a brief but up-to-date overview of the status of old growth forests in Australia following the Regional Forest Agreement process.

Chapter 3

Conservation status: classification of threat

This chapter explores the classifications of threats faced by biodiversity and methods for estimating conservation status, particularly for species. Because many conservation programs focus on rare species, we summarise information on rarity, the different forms of rarity and why these forms of rarity can be important for assessing conservation practices. Schemes for assessing threats and the conservation status of species are outlined in the middle section of this chapter, including rule-based and point-scoring procedures. This is followed by a discussion of threatening processes, illustrated with case studies on the sedimentation of streams and the loss of hollow-bearing trees.

3.0 Introduction

Resources for conservation are limited, and there are competing demands on the natural environment. As a result, conservation strategies involve setting priorities. Priorities can take the form of the order in which recovery plans are developed and implemented, or they can be reflected in the importance placed on different areas of land to be included in reserves. Despite the imperatives for protecting biodiversity at all organisational levels, the social mandate embodied in most laws and government policy is to protect species. As a result, formal procedures for the assessment of conservation status are concerned primarily with species. Estimates of the level of threat faced by a species can be combined with estimates of the likelihood of recovery and the cost of recovery to determine which actions are likely to be most valuable.

3.1 Rarity and conservation status

Components of rarity

Rarity is an intuitive concept that is frequently used in conservation, wildlife management, and natural resource planning (Usher, 1986). The term 'rare' generally refers to species that have low abundance, limited distribution, or both. The following terms were defined by Brown (1984) and Rabinowitz *et al.* (1986) to refer to mutually exclusive characteristics of population size and distribution:

- **Abundance:** the density of individuals within a local area. The terms rare and common (or abundant) describe extremes of density.
- **Range:** the spatial distribution of a species. The terms restricted (or localised) and widespread describe extremes of spatial distribution.
- **Specificity:** the range of ecological conditions, both physical and biotic, within which a species can survive. The terms generalised (or wide) and specialised (or narrow) describe extremes of habitat specificity.

Based on combinations of these three criteria, Rabinowitz *et al.* (1986) and others (e.g. Cody, 1986; New, 2000) recognised eight types of species – seven of which they considered had features of distribution, habitat specificity or abundance that made them rare.

For statements about the abundance, range or specificity of a species to make sense, the reference area must be stated, or at least understood. Species that are rare in one region are not necessarily rare everywhere throughout their range, such as rare plants in open eucalypt forests in south-eastern Australia (Murray and Lepshci, 2004) that can be common in other vegetation types such as wet eucalypt forests or temperate woodlands. In another example, the Eastern Barred Bandicoot is rare in Victoria but relatively common and widespread in Tasmania (Watts, 1987; Mallick *et al.*, 1997).

The three terms described above that define rarity summarise the related components of population size, distribution and tolerance recognised in other nomenclatures. For example, Drury (1974; see also Main, 1984) specified three types of geographic distributions for rare species:

- species inhabiting 'stressed sites' (in which there are a few individuals wherever there is habitat)
- widespread but locally infrequent species
- species in large numbers in a few locations.

Under the definition of Rabinowitz *et al.* (1986), the first category includes rare, specialised species, the second includes widespread, rare species, and the third includes locally common species that are restricted.

Rarity and conservation priorities

Rare species are not necessarily threatened, nor threatened species rare (Oredsson, 1997). Nevertheless, small populations are more likely to become extinct than large ones. This idea is recognised by conservation biologists when they write lists of rare species or develop management programs for rare species, which are often

Table 3.1. The eight types of species distributions and patterns of abundance. (Modified from concepts outlined in Rabinowitz *et al.*, 1986; Cody, 1986; and New, 2000.)

Abundance of species within a community (alpha rarity)	Abundance of species across habitat (beta rarity)	Geographic range (gamma rarity)	Description	Examples
Common	Common	Widespread	Widespread, occurs in a wide range of habitats and is abundant in those habitats (and therefore cannot be considered rare)	Noisy Miner, Greater Glider
Common	Common	Restricted/localised	Highly localised distribution but occurs across a range of habitats and is abundant in places where it occurs	—
Rare	Common	Widespread	Widespread and occurs across a range of habitats but is scarce in places where it occurs	New Holland Mouse
Common	Rare/specialised	Widespread	Widespread, but occurs in few habitats, and is common in places where it occurs	Mountain Pygmy Possum
Common	Rare/specialised	Restricted/localised	Highly localised distribution and occurs in few habitats, but is common in places where it occurs	Leadbeater's Possum, Colonial Spider
Rare	Common	Restricted/localised	Highly localised distribution, occurs across a range of habitats but is scarce in places where it occurs	—
Rare	Rare/specialised	Widespread	Widespread, but occurs in few habitats, and is scarce in places where it occurs	Princess Parrot, Hastings River Mouse
Rare	Rare/specialised	Restricted/ localised	Highly localised distribution, occurs in few habitats, and is scarce in places where it occurs	Wollemi Pine, Gilbert's Potoroo

defined as those with small population sizes or restricted geographic ranges.

Classifications of rarity are useful because they assist in setting priorities for allocating scarce conservation resources. Species that have narrow geographic distributions and restricted habitats, but are abundant in at least one place, would benefit from a protected area. Species that have narrow geographic distributions, restricted habitats, and are rare in all locations deserve the most immediate attention, and may benefit from translocation or reintroduction strategies if unoccupied habitat is available. Management strategies may focus on restrictions on trade and consumptive use for those species that have wide geographic distributions, restricted habitat requirements and are rare throughout their distributions. Species that have restricted geographic distributions are likely to be susceptible to habitat loss, irrespective of their abundance or degree of habitat specialisation.

Relationships between components of rarity

Low densities of animals and plants can be related to large size, scarce or dispersed resources, chance adverse environmental conditions, or disease. Restricted ranges can be caused by intolerance to a broad range of environmental conditions, chance events such as the formation of dispersal barriers, or the loss of habitat. Specialised habitat needs may be the result of evolutionary processes leading to physiological adaptation to unique or unusual environmental conditions (Morrison *et al.*, 1992). All three components of rarity may be determined by biotic interactions including predation and competition.

A species' **fundamental niche** is defined as that region of the environment within which a species could persist indefinitely (Hutchinson, 1959; cf. Block and Brennan, 1993). A fundamental niche is determined by the abiotic factors that affect survival and reproduction of a species, and the range of tolerance of a species to each of these factors determines a multidimensional volume, or **multidimensional niche** (Figure 3.1).

Realised niches and rarity

The concept of rarity is based on a species' **realised niche**, that is, the part of the fundamental niche available to a species after biotic effects such as competition and predation reduce its distribution and abundance. Thus, habitat specificity is closely related to environmental tolerance and niche volume, but is theoretically partly independent of population size and geographic range. Tolerance of a species to a range of environmental conditions reflects a lack of habitat specificity and is likely to result in a relatively broad geographic range.

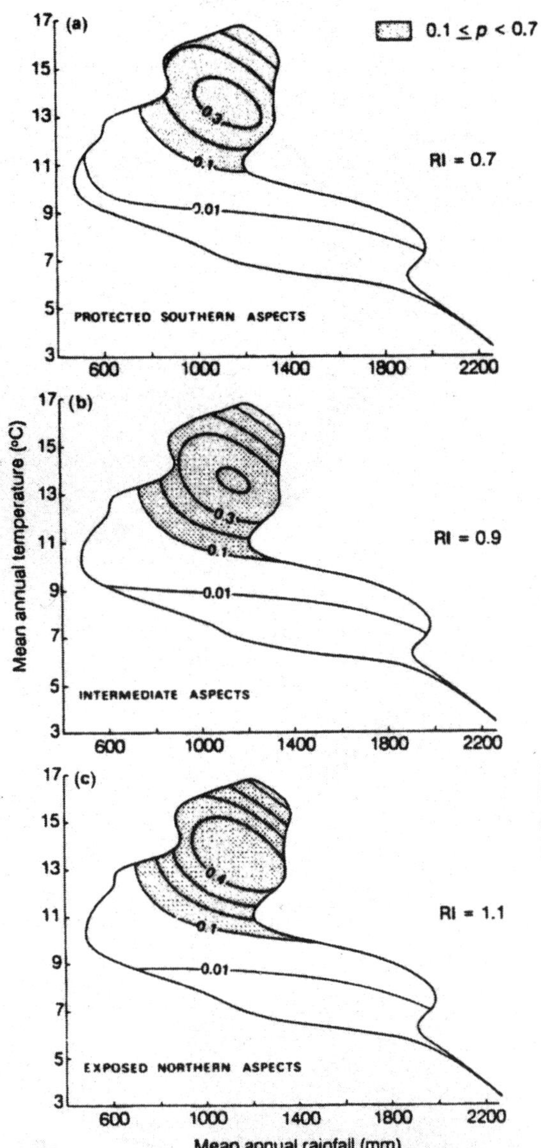

Figure 3.1. Niche space of Silver-top Ash. The series of figures show the quantitative environmental realised niche of the species on sedimentary rocks in relation to annual rainfall, mean annual temperature and radiation. (Redrawn from Austin *et al.*, 1990.)

Species abundance curve

The relationship between the abundance and frequency of species in samples was explored by Preston (1962). Working with insect light-traps, he recorded the number of individuals for each of the species caught. He observed that most species are relatively rare, and that most of the individuals in any higher taxon or guild belonged to just a few very abundant species. Preston (1962) termed this the **species curve**, whereas others call it the species abundance curve (e.g. MacArthur and Wilson, 1967). This pattern has been observed in many different taxonomic groups in many different places around the world, with the same qualitative result (Williams, 1944). The strongly right-skewed distribution of relative abundances of species (Figure 3.2) may be relatively common (although not ubiquitous) in nature (see McGill, 2003).

Explanations for the species curve

Preston's (1962) observations led to several theories to explain the right-skewed distribution using random partitions of resources among species (e.g. Connor and McCoy, 1979; Sugihara, 1980). Hubbell (2001) used a model of random dispersal and recruitment to predict distributions with a heavier left-hand tail (more rare species) than Preston's model (cf. Chisholm and Burgman, 2004).

There is a direct relationship between the relative frequencies of species in samples and their relative abundances (Nachman, 1981; see Chapter 14). The term **ubiquity** refers to the frequency of species in samples (i.e. the proportion of samples in which a species is found), and the terms **scarce** and **ubiquitous** describe the extremes. These categories describe a sampling phenomenon that is a function of a population's abundance, range and specificity (Burgman, 1989).

The relationships among these concepts are important because, in practice, studies of rare species use abundance, geographic range, ubiquity and the number of populations, individually or, less commonly, collectively. For example, Basset and Kitching (1991) classified rare arthropods from north Queensland rainforests as those that were collected just once. Burgman (1989) considered plants in Western Australian mallee and sand heath communities to be rare if they occurred in fewer than

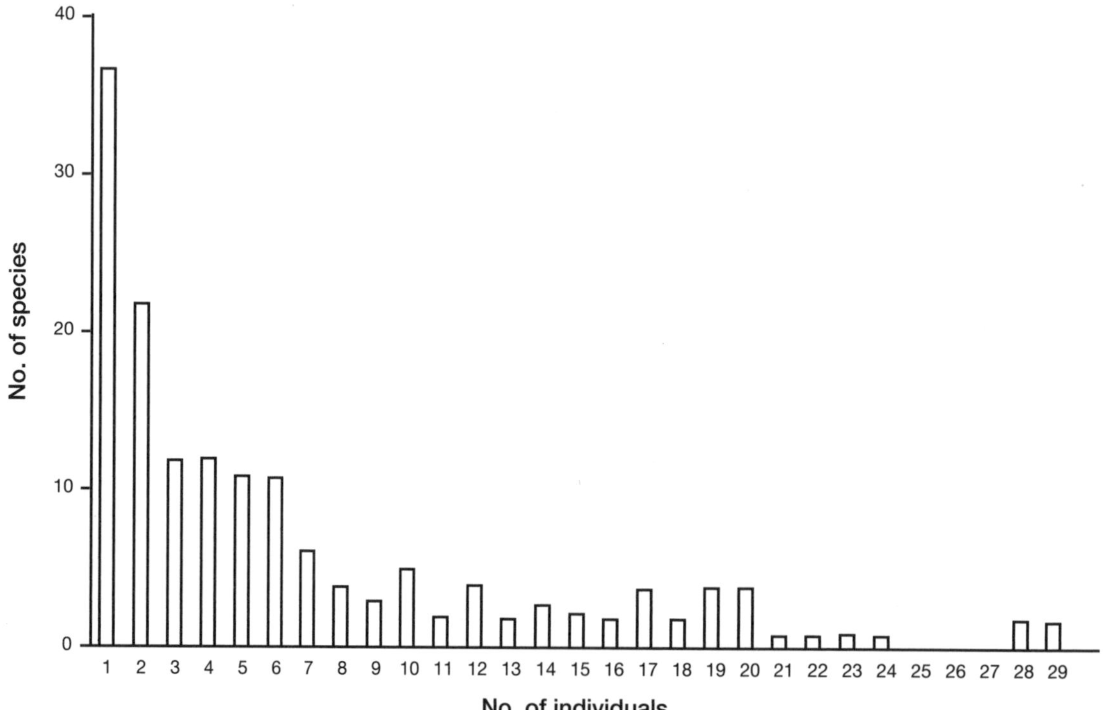

Figure 3.2. Relative abundance of butterflies and moths in a light-trap at Rothamstead, England, in 1935 (after Williams, 1964, in Krebs, 1985). Not all abundant species are represented. One common species (not shown) included 1799 individuals. There was a total of 197 species, with six species making up 50% of the total catch.

3 of 64 sample sites. Gaston (1994) summarised the results of 45 studies that dealt with rare species; of these, a total of 19 studies used only ubiquity to define and identify rare species (Table 3.2). Gaston also found that abundance was used frequently to represent rarity, and geographic range was used once, but habitat specificity was not used in any of the studies.

Species classified as rare using many of the criteria in Table 3.2 might in fact be **vagrant species** – species that are not permanent members of the assemblage or that do not breed and do not have self-sustaining populations within the area of interest (Gaston, 1994). Many rare plant species in New Zealand occur occasionally along the coast after

Table 3.2. Criteria by which rare species have been defined and the attributes that these criteria describe (after Gaston, 1994).

Source	Taxon	Criterion	Attribute
Beebe (1925)	Various	Seldom observed	Ubiquity
Avise (1992)	Various	<10 000 individuals	Abundance
Hall and Mareau (1962)	Birds	Range <250 miles	Range
O'Neil and Pearson (1974)	Birds	Seen in small quantities	Abundance
Ridgely (1976)	Birds	In <25% of trips	Ubiquity
Munves (1975)	Birds	Seen only a few times	Ubiquity
Karr (1977)	Birds	Seen at most a few times	Ubiquity
Pearson (1977)	Birds	<0.04 sightings per hour	Abundance
Thomas (1979)	Birds	Found <5 times	Ubiquity
Ridgely and Gaulin (1980)	Birds	Recorded occasionally	Ubiquity
Roberson (1980)	Birds	Occurred <5 times in any area	Abundance
Stiles (1983)	Birds	Small numbers	Abundance
Haila and Jarvinen (1983)	Birds	Observed <6 times	Ubiquity
DeSante and Pyle (1986)	Birds	Detected in low numbers on <10% of days	Abundance and Ubiquity
Goodman *et al.* (1989)	Birds	1–100 individuals	Abundance
Morgan *et al.* (1991)	Birds	Relative abundance 0.01–1.00	Abundance
Osborne and Tigar (1992)	Birds	1–100 individuals	Abundance
Laurance (1991)	Mammals	<1% of all captures	Abundance
Rands and Myers (1990)	Amphibians and reptiles	Rarely seen	Ubiquity
Werner (1982)	Anurans	Recorded in <25% of samples	Ubiquity
Tonn *et al.* (1990)	Fish	In <6 lakes	Ubiquity
Faith and Norris (1989)	Macro-invertebrates	Abundances <0.5% of total of all taxa	Abundance
Bushnell *et al.* (1987)	Invertebrates	1 individual per field season	Abundance
Goeden and Ricker (1986)	Insects	Collected at <3 sites	Ubiquity
Youtie (1987)	Insects	<10 collected	Abundance
Roubik and Ackerman (1987)	Bees	Ln (mean abundance) <0.6	Abundance
Bassett and Kitching (1991)	Arthropods	One individual collected	Abundance
McGowan and Walker (1979)	Copepods	<100 individuals	Abundance
Buzas and Culver (1991)	Forams	In 1 or 2 localities	Ubiquity
Dony (1953, 1957)	Plants	Local abundance	Abundance
Bowen (1968)	Plants	<1000 in each locality	Abundance
Perring and Walters (1962)	Plants	In <21 vice-counties	Ubiquity
Hubbell and Foster (1986)	Plants	<1 per hectare	Abundance
Usher (1986)	Plants	<10 individuals or <3 clumps	Abundance
Jefferson and Usher (1986)	Plants	In <50 grid squares	Ubiquity
Nilsson *et al.* (1988)	Plants	In <3 sites	Ubiquity
Burgman (1989)	Plants	In <3 sites	Ubiquity
Adsersen (1989)	Plants	Rarely collected	Ubiquity
Verkaar (1990)	Plants	In <30 grid squares	Ubiquity
Deshaye and Morisset (1989)	Plants	In <6 habitat patches	Ubiquity
Dzwonko and Loster (1989)	Plants	In <1/10 localities	Ubiquity
Longton (1992)	Mosses	In <16 grid squares	Ubiquity
Hartshorn and Poveda (1983)	Trees	<0.1 mature trees per hectare	Abundance

being washed up there and germinating (D. Norton, personal communication).

Mass effects is a term used to describe circumstances where species occur in places other than their primary habitat because of the continued flux of propagules from adjacent primary habitat. Shmida and Ellner (1984) identify mass effects as one of the most important biological determinants of species richness. Eliminating vagrants from assessments of rarity can be an important step, depending on the problem at hand.

Quantifying abundance, range and specificity

Measurements of abundance, range and specificity depend on definitions of the limits of a **population.** A common definition of a population is a group of individuals that are sufficiently close geographically that they can find each other and reproduce. In practice, it is any group of individuals of the same species that occur contiguously: the implicit assumption is that if individuals are close enough, genes will flow among generations of individuals. Biologists are often asked to determine the geographic boundaries of a population. The limits of a population depend on the size, social organisation or life form of a species, its mode of reproduction, mode of seed or juvenile dispersal, motility, habitat specificity, and pattern of distribution within its geographic range.

Subpopulations are parts of a population between which gene flow is limited to some degree, but within which it is reasonable to assume individuals are **panmictic.** A **metapopulation** is a set of subpopulations of the same species, usually more or less isolated from one another in discrete patches of habitat, which may exchange individuals through migration (see Chapter 17). All of the factors that make it difficult to define the limits of a population are magnified when trying to determine the limits of subpopulations.

Habitat specificity can be determined by tolerance to temperature, solar radiation, precipitation, soil physical properties, or nutrient status. Ecological theory and empirical observation suggest that many species are most abundant at locations where environmental conditions are ideal (Brown, 1984). There will be many exceptions to this generalisation because local abundance also depends on competition, predation and the chance events that disturb all natural populations (Begon *et al.*, 1996). The conditions that define the niche space of some rare species are unusual (see Gaston, 1994); consequently, environmental tolerance, represented by the range of conditions in which a species is found, may be only weakly related to local abundance (e.g. Austin, 1985; Burgman, 1989; Mac Nally, 1989; Prober and Austin, 1990; McIntyre and Lavoral, 1994a).

Habitat specificity is difficult to quantify because the environmental variables that limit distribution can only be identified confidently by experimental manipulation, and are confounded by biotic interactions (Begon *et al.*, 1996; see Chapter 15). Even identifying the correlates of a species' distribution is more difficult than measuring range or abundance. For these reasons, habitat specificity is often ignored in studies of rare species.

Local population abundances are estimated by many methods (Sutherland, 1996). A complete count (a census) is sometimes possible if the population is small, well circumscribed, and easy to count. Total population size and confidence limits can be estimated from density within fixed areas such as **quadrats.** Subjective visual estimates, crown cover or cover abundance measurements, biomass, scat counts, spotlight or transect encounters, or relative frequency of a species among samples can be used to generate indices of

Eucalyptus paliformis – a rare eucalypt from south-eastern Australia. (Photo by Suzanne Prober.)

relative abundance. If the relationship between the index and population size is known, these indices can be transformed into estimates of population size (Sutherland, 1996). Such approaches are best if the sampling strategy takes into account variations in habitat quality across the study area and samples are **stratified** accordingly.

Population range or **geographic range** can be expressed as a map, an area or a linear distance. Range estimates can depend on the time of year, the inclusion or exclusion of reproducing and non-reproducing individuals, migration pathways, and the exclusion of vagrants (Gaston, 1994). Species distributions are influenced by such processes as successional dynamics and disturbance, particularly in Australian ecosystems (Main, 1984; Attiwill, 1994a,b). Range estimates should account for the disturbance dynamics of the landscape in which the species occurs and the time over which data are recorded.

The **extent of occurrence** is the area containing all observations of a species, that is, its entire geographic range (see Gaston, 1994). The extent of occurrence was defined by the International Union for the Conservation of Nature (IUCN) as the area contained within the shortest contiguous boundary encompassing all known inferred or projected sites of current occurrence of a taxon, excluding cases of vagrancy (IUCN, 1994; Figure 3.3). Most often, the extent of occurrence is represented as a polygon on a map, which is based on the most widely dispersed localities, fitted by eye, and does not indicate any variation in habitat suitability within the limits of the species distribution (Schodde and Mason, 1999). This approach is typical of distribution maps in field guides for different taxonomic groups, including reptiles and amphibians (Cogger, 2000), birds (Pizzey and Knight, 1997), mammals (Strahan, 1995) and plants (Costermans, 1994).

Interpreting the extent of occurrence as a convex polygon does not account for discontinuities in the spatial distribution of a taxon – if the actual range of the species does not have a convex shape, the range will be overestimated (see Colwell and Coddington, 1995). The bias is greater when a species' geographic range is convoluted or when there are many relatively isolated patches or subpopulations. In fact, adding new observations to a convoluted habitat makes range estimates more and more biased (Burgman and Fox, 2003). Furthermore, because range and abundance are not independent, even if two species have the same extent of occurrence, the rarer of the two species will usually be represented by fewer samples and will appear to have a smaller geographic range (Brown, 1984; Burgman, 1989; Wright, 1991).

The IUCN (1994) defined **area of occupancy** as the area within the extent of occurrence that is occupied by a taxon, excluding cases of vagrancy (Figure 3.3). It is a measure of range that disregards areas that may be unsuitable or that are simply unoccupied at present, but measures the smallest area essential at any stage to the survival of a taxon (including, for example, nesting sites and feeding sites). Gaston (1994) pointed out that this concept is sometimes difficult to apply and depends

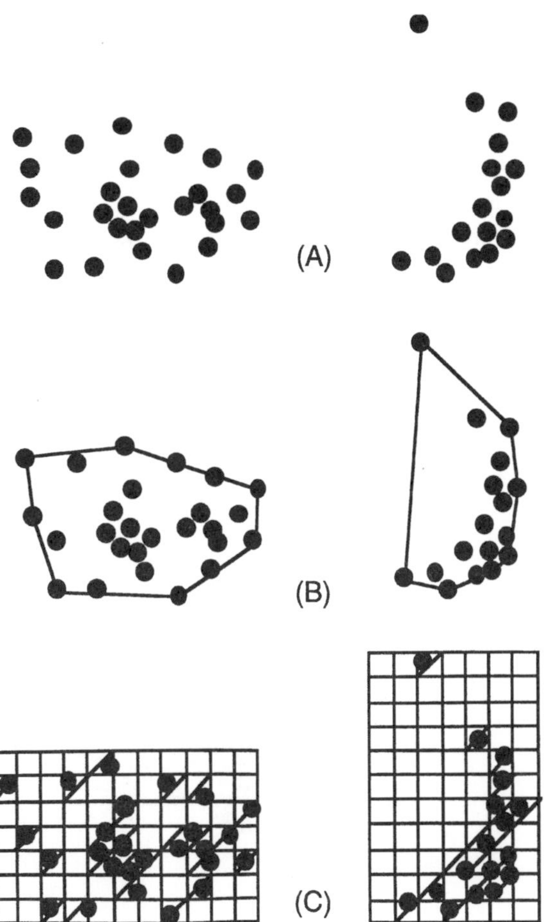

Figure 3.3. Two examples of the distinction between extent of occurrence and area of occupancy (after IUCN, 1994). (A) The spatial distribution inferred from sites of occurrence; (B) one possible boundary to the extent of occurrence; (C) one measure of the area occupied, calculated by summing the occupied grid squares.

critically on the scale of resolution of mapping of the species' distribution. For example, the area of occupancy derived for the example in Figure 3.3 would almost certainly be smaller if a finer grid was applied.

Sampling bias affects abundance and range estimates – rare species remain undetected if sampling effort is insufficient (McArdle, 1990). Yet the effectiveness of surveys is very rarely assessed. Species ranges are seldom mapped with equal reliability across their entire geographic range, and different species are mapped with different levels of reliability (Gaston, 1994). Most existing data have been collected on an opportunistic basis, are coloured by subjective interest, and have little regard for environmental gradients or spatial representation (Margules and Austin, 1994). Such characteristics limit the usefulness of data (see Chapter 15).

The frequency of observations of a species often relate to accessibility rather than patterns of habitat use or quality. For instance, the recorded distribution patterns of plants, snakes (Nix, 1986) and many other taxa in Australia reflect the road system, or are mostly from areas close to cities (Figure 3.4). Roads are not randomly distributed in the landscape, but follow ridges, avoid drainage lines and are associated with human populations (Forman *et al.*, 2002; Kadmon *et al.*, 2004). Vegetation near roads is more likely to be degraded and support exotic species (Wace, 1977).

Range estimates suffer from the implicit assumption that habitat is either suitable or unsuitable (Lindenmayer *et al.*, 2003d). The sharp boundaries represented in most maps of species distributions are abstractions (Gaston, 1994), and habitat suitability is likely to be determined by the response of a species to a set of continuous habitat variables. Different places will support the individuals of a species to varying degrees, depending on the environmental conditions at that place; its proximity to source populations; chance events related to dispersal, survival and reproduction of individuals that determine local population persistence; and the vagaries of environmental change. More sensitive measures of habitat suitability, such as probabilistic habitat models (e.g. Austin, 1985, 1999), use a continuous scale (see Elith, 2002; Fischer and Lindenmayer, 2004). We outline these and other methods for representing range in Chapter 15.

Figure 3.4. Bias in the collection of field data for Australian birds (based on data from Birds Australia, 2005). The lines of records reflect the road system and not the true distribution of taxa.

Ecological correlates of rarity

Several researchers have looked for general factors to explain the distribution and abundance of rare species (e.g. Kirkpatrick and Gilfedder, 1995). Gaston (1994) argued that the search is unlikely to be worthwhile because studies depend on correlations between distribution or abundance and environmental variables, whereas clearly establishing causal relationships requires manipulative experiments. Experiments are uncommon because they are prohibitively expensive and may not be justifiable, particularly if they put a threatened species at greater risk. Correlations are confounded by chance events and biotic processes such as dispersal, competition and predation.

However, the search for generalisations can be valuable where processes at a landscape scale affect many species (Mac Nally and Bennett, 1997). For example, McIntyre and Lavoral (1994b) found general differences in the responses of exotic and rare native plants to disturbance in the New England area of New South Wales. Exotic species were positively associated with water enrichment and soil disturbance, whereas these factors were detrimental to native species. Such studies are a first step in the development of effective management at a landscape level.

It is important to understand what constitutes habitat for rare species, because habitat loss is the most important factor leading to the decline of many of the world's extinct, threatened and vulnerable species (May, 1990a; Primack, 2001; Groombridge and Jenkins, 2002), and is the most important cause of species decline and extinction in Australia (Recher and Lim, 1990; Leigh and Briggs, 1992; Cogger *et al.*, 2003; see Chapter 9). This makes the development of tools for habitat mapping one of the most important objectives of conservation biology (Chapter 15).

Eucalyptus parviflora – a rare eucalypt from south-eastern Australia that is confined to a rare habitat. (Photo by Suzanne Prober.)

Rarity, threat and extinction proneness

Biologists use qualitative categories to communicate the risks faced by different species – termed the conservation status of a species (Department of Environment and Heritage, 2005a,b). The IUCN (1994, 2001) assigns species to various categories, including extinct, extinct in the wild, critically endangered, endangered, vulnerable, and lower risk (see Section 3.2).

Some classifications include the category 'rare' (e.g. Briggs and Leigh, 1988; Briggs and Leigh, 1996), implying a strong relationship between rarity and threat. The inclusion of rarity in classifications of conservation status has translated to the inclusion of the concept in schemes for the protection and reservation of natural areas (Table 3.3; e.g. Kirkpatrick and Brown, 1991).

Table 3.3. Minimum conservation targets for genotypes, species and communities in protected areas (after Kirkpatrick and Brown, 1991).

Genotype or species	Community	Conservation target (minimum percentage protected)
Neither rare nor threatened	Disturbance-resilient, expanding since last glacial period	30
Rare or vulnerable	Contracting since last glacial period	60
Endangered by disturbance or climate change	Endangered by disturbance or climate change	90

Small populations at risk

Abundance is related to conservation status through the levels of population variability that affect all natural populations (Hartley and Kunin, 2003). The amount of random variation in the survival and reproductive success of individuals in a population, the rate of genetic drift and the loss of genetic variation, and the likelihood of a population falling below some critical size, are all inversely related to population size (e.g. Lande, 1993; Burgman *et al.*, 1993; Lacy, 1993a; McCarthy *et al.*, 1994; Possingham *et al.*, 2001; Figure 3.5).

Small populations, populations with limited environmental tolerance, and populations with specific habitat requirements are more likely than large populations to become extinct as a result of these processes, either because they are closer to extinction at the outset, or because they are less able to tolerate extreme environmental conditions (see Chapter 16). Caughley (1994) termed this the 'small population paradigm'. There are numerous empirical studies that support the theoretical conclusion that low abundance increases the likelihood of extinction (e.g. Diamond, 1984; Davies *et al.*, 2000).

Extinction proneness

Other ecological characteristics of species can be correlated with the chances of loss of local populations (e.g. Terborgh, 1974; Pimm *et al.*, 1988; Burbidge and McKenzie, 1989; Laurance, 1991a; Angermeier, 1995; Mac Nally *et al.*, 2000). For example, carnivores at the top of the food chain tend to be large-bodied, are relatively long-lived, occur at low abundances, and have low rates of population growth. These characteristics can make large carnivores vulnerable to extinction (Terborgh, 1974). Other factors correlated with extinction risk include dispersal ability; the degree of habitat specialisation; niche characteristics; tolerance of disturbed environments; susceptibility to hybridisation, competition, predation, disease and catastrophes; and the loss of populations of mutualistic species (Soulé, 1983; Thomas, 1990; Walter, 1990; Karr, 1991; Jenkins, 1992; Tscharntke, 1992; Pimm, 1993; Gaston, 1994; Davies *et al.*, 2000).

Characteristics that predispose a species to extinction as a direct consequence of human impacts include overlap with the habitat preferred by people (e.g. biota of relatively accessible areas with fertile soils and benign climates, and species that inhabit coastlines, major rivers

Box 3.1

Small viable populations

Several factors can place small populations at a greater risk of extinction than large populations (Shaffer, 1981). However, just because a population is small does not mean that it is necessarily doomed to extinction. There are many cases where a species has recovered to substantial population sizes from small numbers of individuals (some examples are cited in Chapter 4). In another example, Walter (1990) outlined the case of the very small but apparently viable population of the Red-tailed Hawk on Socorro Island in the Pacific Ocean. Socorro Island is more than 450 kilometres from the American mainland and the Red-tailed Hawk is one of several endemic taxa. Red-tailed Hawks on Socorro Island do not migrate and are isolated from mainland populations of related subspecies (Walter, 1990). The total population of approximately 30–50 birds appears to have been stable at this size for at least 40 years and possibly considerably longer. Many standard theories in conservation biology would predict the likely demise of such a population due to genetic effects, demographic effects, population dynamics, or environmental variation. As noted by Walter (1990), generalisations based on small population size, demographics and genetics should be made cautiously. Extinction is a random process and the persistence of small isolated populations may simply be the result of good luck.

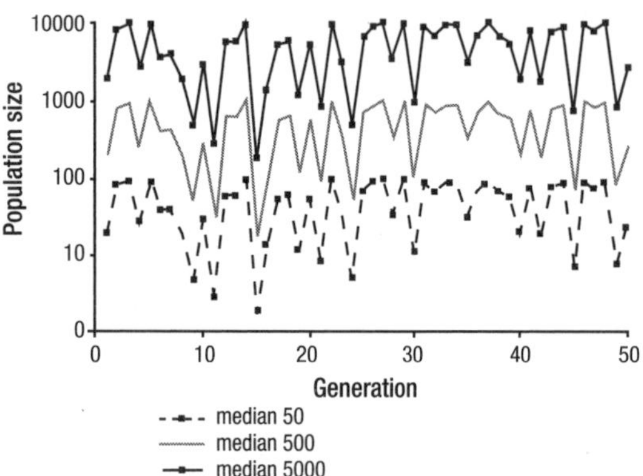

Figure 3.5. Simulated relationships between population size, environmental variability and extinction risk. The smallest population size varies has a median population size of 50 individuals and the largest 5000 individuals. Environmental variability added to the simulated system takes the smallest population closer to zero than the medium and larger sized populations over multiple generations. (Redrawn from Thomas 1990.)

Numbat – a so-called 'critical weight range' species. (Photo by Tony Friend.)

Box 3.2

A reappraisal of the critical weight range concept
The concept of a critical weight range species was recently challenged by Cardillo and Bromham (2001). They argued that because most Australian mammals fall within the critical weight range (35–5500 grams; Figure 3.6), then even if extinction is random with respect to body size, most extinct species would fall within this range; that is, the association is incidental. Their re-analysis showed that the smallest mammals were actually those most resistant to extinction.

and streams), palatability or other consumption values to humans or livestock, and limited adaptability and resilience to environmental disturbance (Diamond, 1989; Saunders, 1994; Margules *et al.*, 1995a). Balmford (1996) suggests that the biota in many places today represents the survivors of direct, intensive human impact. It follows then that species that have coexisted with human populations and modern technology for extended periods are relatively resilient to human impact.

Burbidge and McKenzie (1989) undertook a pioneering study on extinction proneness in Australian mammals. They assessed the fate of a large assemblage of mammals and identified strong correlations between body size and likelihood of extinction or decline. They found that almost all the species that had declined, or had been lost, were non-flying taxa that weighed between 0.035 and 5.5 kilograms – a category Burbidge and McKenzie (1989) named the **critical weight range** (CWR; see Chapter 7). Factors that correlate with the demise of CWR species include diet, habitat requirements and regional patterns in rainfall. However, a more recent study has suggested that the CWR is an artefact, and in fact, most species fall within this weight range (Cardillo and Bromham, 2001; see Box 3.2).

Laurance (1991a) studied extinction proneness among 16 mammal species in fragmented rainforest patches in Queensland. The species represented three guilds: arboreal folivores, predators and omnivorous rodents. Long-lived species with low fecundity, as well as dietary specialists, were the most extinction-prone, and species that were rare or absent in disturbed environments were also most likely to have been lost or to have declined substantially in the rainforest fragments (e.g. Brown Antechinus, Tiger Quoll and Atherton Antechinus). Laurance (1991a) concluded that species persist in rainforest fragments because they can tolerate edge effects and use modified habitats in the surrounding landscape matrix to feed and disperse between fragments (Figure 3.7). Notably, this result has been observed in several other studies of habitat fragmentation (reviewed by Lindenmayer and Franklin, 2002), including studies of invertebrates in forest fragments in south-eastern New South Wales (Davies *et al.*, 2000).

Extinction proneness varies between species assemblages and depends on life history attributes, habitat requirements, disturbance history, and a wide range of other factors. Almost all extinctions result from multiple processes that can act in unexpected ways. The case of the Thylacine on mainland Australia and Tasmania is a good example of this phenomenon (see Box 6.6 in Chapter 6), and the loss of the Red-fronted Parakeet on

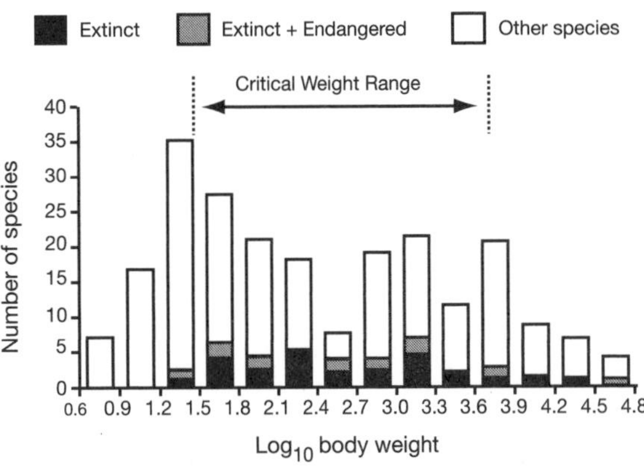

Figure 3.6. Histogram of extinct, endangered and other species of Australian mammals. (Redrawn from Cardillo and Bromham, 2001.)

Rainforest fragments in north Queensland. (Photo by Davo Blair.)

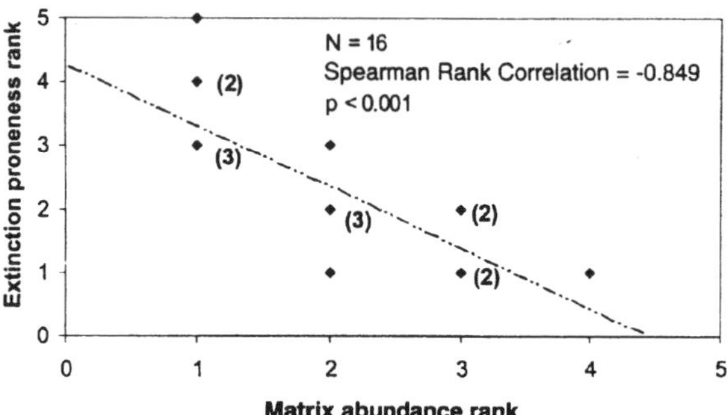

Figure 3.7. Relationships between the occurrence of species in the matrix and their persistence in habitat fragments. (Modified from Laurance, 1991a.) The values in brackets are the number of species with the same plotted values. Extinction proneness is ranked from high (5) to low (1).

Macquarie Island from the interacting impacts of cats and rabbits is another (Taylor, 1979). The Red-fronted Parakeet coexisted with cats for 80 years on the island, until rabbits were introduced in 1879. Over time, the rabbits increased in abundance, as did the numbers of cats that preyed upon them. Altered vegetation cover because of rabbit grazing and increased numbers of cats contributed to the loss of the Red-fronted Parakeet. This example (and many others like it) indicate that species-specific ecological information is required to conserve and manage most species effectively.

3.2 Assessing conservation status

Components of rarity are used as parameters to assign species to different categories of threat (Munton, 1987; Mace and Lande, 1991; IUCN, 1994, 2003). Population size and range, in particular, play important roles in most schemes for assessing conservation status. The purpose of the following section is to outline several methods for assessing conservation status.

Extinct and presumed extinct species

Extinct species are those for which there is no doubt that the last member of the species has died (Groombridge and Jenkins, 2002). **Presumed extinct** species are those that are suspected to have become extinct. Usually, this means that they have not been definitely located in the wild for a considerable period (often more than 50 years), or that intensive searches have failed to locate them. There may be doubt that the species is extinct if, for example, its habitat is remote or inaccessible, its growth form or life history strategies make it difficult to survey, or if there has been a lack of thorough field surveys to establish its absence.

Records over the last 100 years show that it is difficult to know at what point a species should be considered extinct. There are many cases of species that were presumed extinct, only to have been found after some considerable time (Deweerdt, 2003; Table 3.4).

Table 3.4. Examples of species of Australian animals that have been rediscovered in the past 40 years.

Species	Last record before rediscovery	Year of rediscovery
Mountain Pygmy Possum	~15 000 years before present	1966
Parma Wallaby	1932	1966
Leadbeater's Possum	1909	1961
Dibbler	~1884	1967
New Holland Mouse	~1887	1967
Sandhill Dunnart	1894	1969
Bridled Nailtail Wallaby	~1930s	1972
Long-tailed Dunnart	~1940s	1984
Gilbert's Potoroo	1869	1994
Greater Stick-nest Rat	1938	1986
Night Parrot	1912	1979
Noisy Scrub-bird	1889	1961

Table 3.5. Turnover in Australian plants listed as extinct between 1981 and 2001, showing reasons for additions and deletions to lists (after Keith and Burgman 2004). The total in 2001 includes 12 species remaining listed despite rediscovery.

Reasons for change	Additions	Deletions
Presumed extinction between 1981 and 2001	8	
Not previously evaluated	48	
Change in extinction uncertainty	46	1
Taxonomic change	6	25
Change in taxonomic uncertainty	7	15
Rediscovered		89
Correction		34
Change in opinion on native status		3
Total changes in status	115	167
Originally listed in 1981	113	
Remaining listed in 2001		61
Total listings	228	228

Methods that make use of opportunistic observations for estimating the likelihood that a species has become extinct are outlined in Chapter 16. Often, species are considered to be extinct and are later found not to have been a correctly circumscribed species (i.e. there is taxonomic uncertainty). In other cases, there is no change in knowledge or taxonomy, but the people responsible for maintaining lists change their attitude towards uncertainty, with the consequence that species are added or removed from lists of extinct species (Table 3.5).

Diamond (1987) pointed out that extinction rates are greatly underestimated where the flora and fauna are poorly known, because species are considered extinct only after exhaustive field work has failed to locate them (Figure 3.8). He suggested that in these cases, species should be presumed extinct or threatened unless they can be shown to be extant and secure. Such a list may be considerably longer than the current lists of threatened species. For these reasons, it may be appropriate to differentiate between the categories of 'extinct' and 'presumed extinct'.

Types of extinctions

Local extinction, sometimes called **extirpation**, refers to the extinction of a single population in a spatially separate patch of its habitat. **Global extinction** refers to the loss of all members of a species in all of its constituent populations. Ginzburg *et al.* (1982) defined the risk of **quasi-extinction** to be the risk of decline below a specified population size within some specified time frame.

Ecological extinctions (sometimes called functional extinctions) are those in which a given species still occurs in a community, but the numbers of the taxon are so depleted that it 'no longer interacts significantly with other species' (Estes *et al.*, 1989, p. 253; see also Conner, 1988). For example, the Norfolk Island Boobook Owl (*Ninox novaeseelandiae albaria*) was functionally extinct when a single individual (a female) remained (Olsen, 1996; Norman *et al.*, 1998). It could be argued that species are functionally extinct when they have undergone very substantial range contractions or remain only in zoos, botanic gardens or other *ex situ* facilities (see Chapter 4).

Co-extinction occurs when a species is lost because it depends strongly on another species. Species-specific ecto- and endo-parasites of extinct species are

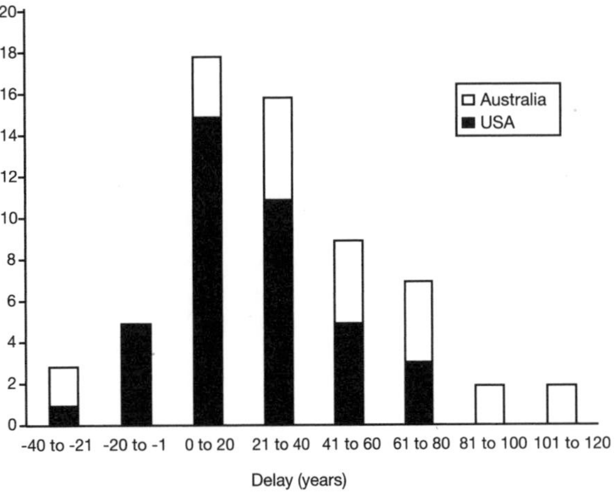

Figure 3.8. Delays between the year of extinction of species, subspecies and evolutionarily significant units (ESUs) and the year in which the extinction event was formally published (after Burgman *et al.*, unpublished data). Negative values mean that the putatively extinct species was extant at the time of publication but became extinct afterwards.

Box 3.3

Rediscovery of Gilbert's Potoroo

Gilbert's Potoroo was first collected in 1840 near Albany in Western Australia. The species was collected again from around the same region between 1866–1869 and 1874–1879 (Start *et al.*, 1995). It was added to the list of extinct Australian mammals because it was not recorded between 1879 and the mid-1990s (e.g. Calaby, 1960). However, in November 1994, a number of animals were captured by scientists working at Two Peoples Bay Nature Reserve in south-western Western Australia (Sinclair *et al.*, 1996). The very dense long-unburned vegetation in that region may have provided suitable habitat (Sinclair *et al.*, 2002). Start *et al.* (1995) suggested that dense vegetation cover reduces the susceptibility of these animals to predation, although field experiments would be required to test the hypothesis. Notably, morphological and genetic studies of Gilbert's Potoroo have indicated that it is a distinct species from the related Long-nosed Potoroo in eastern Australia. It is now considered a separate species (Courtenay *et al.*, 1996; Sinclair *et al.*, 2002). The revised taxonomic status makes Gilbert's Potoroo one of Australia's most critically endangered marsupials (Maxwell *et al.*, 1996; Friend, 2003). A captive breeding program is underway at the Royal Zoological Society of South Australia (Adelaide) using cross-fostering techniques to increase the numbers of this species. Cross-fostering techniques involve taking young from the pouch and transferring them to the pouch of another species of macropod for suckling, leaving the mother to give birth to a new Gilbert's Potoroo.

examples of co-extinction. Because so much biodiversity remains poorly known or undescribed (see Chapter 2), many species will be lost before they are known to science; Wilson (1992) defined these extinction events as **centinelan extinctions**.

Qualitative procedures for assessing threat

Threat and threatened are collective words that refer to taxa that face an appreciable risk of decline or extinction. The conservation status of a species describes the degree of threat it faces, and is related to the species' chances of extinction. Protocols currently in use for assigning conservation status and for setting priorities

Gilbert's Potoroo. (Photo by Tony Friend.)

for conservation action come under three headings: qualitative descriptions, rule sets, and point-scoring procedures. Each has strengths and weaknesses.

When conservation status is assessed, it employs a time scale. A species that has little chance of extinction within 10 years may face certain extinction within a century. The IUCN in the early 1970s defined a set of terms that made reference to both levels of risk and time scales (Scott *et al.*, 1987):

Box 3.4

Ecological extinction of the California Spiny Lobster

The California Spiny Lobster is an important predator of sea urchins, mussels and gastropods in kelp forests. However, extensive harvesting by commercial and recreational fishers has substantially reduced both the average size of animals, as well as the overall abundance of the species. Luxuriant kelp forest occurs in areas supporting healthy populations of large California Spiny Lobsters, and mussels and whelks are important prey items for the species. Conversely, where the California Spiny Lobster has become locally extinct, the marine substrate is dominated by beds of mussels and large populations of whelks (Estes *et al.*, 1989). Attempts to reintroduce the California Spiny Lobster have been unsuccessful because they are eaten by whelks, indicating a reversal of the predator–prey interaction between members of the marine assemblage. Complex spatial and temporal interactions between California Spiny Lobsters, predatory seastars, Sea-otters and other species make it difficult to predict the consequences of hunting and fishing (Pinnegar *et al.*, 2000).

- *Endangered*: species that face a high risk of extinction within one or two decades if present causal factors continue to operate.
- *Vulnerable*: species not currently endangered but at risk over longer periods (usually 50 to 100 years) if factors tending to push the species into decline continue to operate.
- *Rare, restricted and specialised*: species that are not presently vulnerable and may be present in stable populations, but some characteristic of their population sizes or distributions makes them conceivably at risk in the long term.
- *Indeterminate*: species that are known to be endangered, vulnerable or rare, but for which there is insufficient information to say which category is the most appropriate.
- *Insufficiently known*: species for which there is insufficient information on which to base a judgement concerning either their abundance and distribution, or the degree of threat they face.

This classification was applied widely over the following decades and it has been expanded in more recent times to include other categories, such as 'extinct in the wild' and 'lower risk/conservation dependent' (IUCN, 2003; see Table 3.6, Table 3.7). These categories are essentially comparative. Rarely have risk levels or time horizons been specified (i.e. what is

Table 3.6. Summary of the International Union for the Conservation of Nature categories and criteria (after IUCN, 2001).

	Critically endangered	Endangered	Vulnerable
A. Declining population			
Population declining at a rate of ... using either	>80% in 10 years or 3 generations	>50% in 10 years or 3 generations	>20% in 10 years or 3 generations
1. population reduction observed, estimated, inferred or suspected in the past, or			
2. population decline suspected or projected in the future, based on direct observation, an abundance index, decline of habitat, changes in exploitation, competitors, pathogens, etc.			
B. Small distribution and decline or fluctuation			
Either extent of occurrence ...	<100 square kilometres	<5000 square kilometres	<20 000 square kilometres
or			
area of occupancy (AOO)[1] ...	<10 square kilometres	<500 square kilometres	<2000 square kilometres
and two of the following three:			
1. Either severely fragmented or known to exist at a number of locations			
2. Continuing decline in habitat, locations, subpopulations or mature individuals			
3. Fluctuations of more than one order of magnitude in extent, area, locations or mature individuals			
C. Small population size and decline			
Number of mature individuals ...	<250	<2500	<10 000
and one of the following two:			
1. Rapid decline of ...	>25% in 3 years or 1 generation	>20% in 5 years or 2 generations	>10% in 10 years or 3 generations
2. Continuing decline of any rate and either			
Populations fragmented with ...	All subpopulations <50	All subpopulations <250	All subpopulations <1000
or			
All individuals in a single population			
D. Very small or restricted			
Number of mature individuals...	<50	<250	<1000 or AOO <1000 square kilometres or locations <5
E. Quantitative analysis			
Risk of extinction in the wild...	>50% in 10 years or 3 generations	>20% in 20 years or 5 generations	>10% in 100 years

[1]AOO, area of occupancy, usually estimated as the number of grid cells containing a record of the species, multiplied by the area of a grid cell.

Table 3.7. Numbers of species assigned to various International Union for the Conservation of Nature (2003) threat categories (based on data from IUCN, 2003).

Part A: Animals

Class	EX	EW	Subtotal	CR	EN	VU	Subtotal	LR/CD	NT	DD	LC	Total
Mammalia	74	4	**78**	184	337	609	**1130**	66	620	256	2639	**4789**
Aves	129	3	**132**	182	331	681	**1194**	3	731	79	7793	**9932**
Reptilia	21	1	**22**	57	78	158	**293**	3	74	61	20	**473**
Amphibia	7	0	**7**	30	37	90	**157**	2	22	75	138	**401**
Cephalaspidomorphi	1	0	**1**	0	1	2	**3**	0	5	3	1	**13**
Elasmobranchii	0	0	**0**	8	17	32	**57**	1	64	63	74	**259**
Holocephali	0	0	**0**	0	0	0	**0**	0	0	1	3	**4**
Actinopterygii	79	11	**90**	153	126	410	**689**	12	96	270	98	**1255**
Sarcopterygii	0	0	**0**	1	0	0	**1**	0	0	0	0	**1**
Echinoidea	0	0	**0**	0	0	0	**0**	0	1	0	0	**1**
Arachnida	0	0	**0**	0	1	9	**10**	0	1	7	0	**18**
Chilopoda	0	0	**0**	0	0	1	**1**	0	0	0	0	**1**
Crustacea	7	1	**8**	56	73	280	**409**	9	1	32	2	**461**
Insecta	69	1	**70**	46	118	389	**553**	3	76	42	24	**768**
Merostomata	0	0	**0**	0	0	0	**0**	0	1	3	0	**4**
Onychophora	0	0	**0**	3	2	4	**9**	0	1	1	0	**11**
Hirudinoidea	0	0	**0**	0	0	0	**0**	0	1	0	0	**1**
Oligochaeta	1	0	**1**	1	0	4	**5**	0	1	0	0	**7**
Polychaeta	0	0	**0**	1	0	0	**1**	0	0	1	0	**2**
Bivalvia	31	0	**31**	52	28	12	**92**	5	60	7	6	**201**
Gastropoda	260	12	**272**	198	215	462	**875**	14	175	522	39	**1897**
Enopla	0	0	**0**	0	0	2	**2**	0	1	3	0	**6**
Turbellaria	1	0	**1**	0	0	0	**0**	0	0	0	0	**1**
Anthozoa	0	0	**0**	0	0	2	**2**	0	0	1	0	**3**
Total	680	33	**713**	972	1364	3147	**5483**	118	1931	1427	10837	**20509**

Animals: Mammalia (mammals), Aves (birds), Reptilia (reptiles), Amphibia (amphibians), Cephalaspidomorphi (lampreys and hag fish), Elasmobranchii (sharks, skates and rays), Holocephali (chimaeras), Actinopterygii (bony fishes), Sarcopterygii (coelacanth), Echinoidea (sea urchins, starfish, etc.), Arachnida (spiders and scorpions), Chilopoda (centipedes), Crustacea (crustaceans), Insecta (insects), Merostomata (horseshoe crabs), Onychophora (velvet worms), Hirudinoidea (leeches), Oligochaeta (earthworms), Polychaeta (marine bristle worms), Bivalvia (mussels and clams), Gastropoda (snails, etc.), Enopla (nemertine worms), Turbellaria (flatworms), Anthozoa (sea anemones and corals).

Part B: Plants

Class	EX	EW	Subtotal	CR	EN	VU	Subtotal	LR/CD	NT	DD	LC	Total
Bryopsia	2	0	**2**	10	15	11	**36**	0	0	0	1	**39**
Anthocerotopsida	0	0	**0**	0	1	1	**2**	0	0	0	0	**2**
Marchantiopsida	1	0	**1**	12	16	14	**42**	0	0	0	9	**52**
Lycopodiopsida	0	0	**0**	1	2	8	**11**	0	1	0	1	**13**
Sellaginellopsida	0	0	**0**	0	0	1	**1**	0	1	0	0	**2**
Isoetopsida	0	0	**0**	0	0	1	**1**	0	0	0	0	**1**
Polypodiopsida	2	0	**2**	26	22	50	**98**	0	12	45	7	**164**
Coniferopsida	0	0	**0**	16	43	93	**152**	26	53	60	327	**618**
Cycadopsida	0	2	**2**	47	39	65	**151**	0	67	18	50	**288**
Ginkgoopsida	0	0	**0**	0	1	0	**1**	0	0	0	0	**1**
Magnoliopsida	74	21	**95**	1061	1335	3372	**5768**	201	702	354	614	**7734**
Liliopsida	2	2	**4**	103	160	248	**511**	17	84	123	53	**792**
Total	81	25	**106**	1276	1634	3864	**6774**	244	920	600	1062	**9706**

Plants: Bryopsida (true mosses), Anthocerotopsida (hornworts), Marchantiopsida (liverworts), Lycopodiopsida (club mosses), Sellaginellopsida (spike mosses), Isoetopsida (quillworts), Polypodiopsida (true ferns), Coniferopsida (conifers), Cycadopsida (cycads), Ginkgoopsida (ginkgo), Magnoliopsida (dicotyledons), Liliopsida (monocotyledons).

International Union for the Conservation of Nature categories are: EX, extinct; EW, extinct in the wild; CR, critically endangered; EN, endangered; VU, vulnerable; LR/CD, lower risk/conservation dependent; NT, near threatened (includes LR/NT, lower risk/near threatened); DD, data deficient; LC, least concern (includes LR/LC, lower risk/least concern).

meant by 'a high risk of extinction' and over what period is this risk evaluated?). This is largely because species are assigned to different categories without any quantitative analyses; assignments are based on the experience and intuition of biologists.

Australian Federal threat classifications

In Australia, The *Environment Protection and Biodiversity Conservation Act* (1999) (EPBC Act) is the definitive list of extinct, critically endangered, endangered and vulnerable species and subspecies (Department of Environment and Heritage, 2005a,b). These definitions are broadly derived from the IUCN descriptions and are largely qualitative. Definitions for major categories are (as indicated in the EPBC Act):

- *Extinct*: 'there is no reasonable doubt that the last member of the species has died'.
- *Extinct in the wild*: 'known only to survive in cultivation, in captivity or as a naturalised population well outside its past range' or 'it has not been recorded in its known and/or expected habitat, at appropriate seasons, anywhere in its past range, despite exhaustive surveys over a time frame appropriate to its life cycle and form'.
- *Critically endangered*: 'facing an extremely high risk of extinction in the wild in the immediate future, as determined in accordance with the prescribed criteria'.
- *Endangered*: 'facing a very high risk of extinction in the wild in the near future, as determined in accordance with the prescribed criteria'.
- *Vulnerable*: 'facing a high risk of extinction in the wild in the medium-term future, as determined in accordance with the prescribed criteria'.

Lists of fauna and flora assigned to these various categories are maintained and regularly updated by the Department of Environment and Heritage (2005a,b). The EPBC Act (1999) has provisions for public nominations to list populations, subspecies, species and ecological communities. Nominations are assessed by the Threatened Species Scientific Committee – a panel of experts in biodiversity conservation. Some taxa are accepted and assigned to the various threat categories listed here (see Chapter 6). Although some attempt has been made to broaden the lists, in mid-2004, the overwhelming majority of listed taxa were vascular plants and vertebrates – fewer than 2% of the more than 740 listings were for invertebrates, and none were for fungi.

The extinct Paradise Parrot (from the illustration of John Gould and supplied by the Queensland Museum).

Australian State agency threat classifications

Qualitative descriptions of conservation status and threat are used in each Australian State and Territory, with jurisdictional differences in each case. The New South Wales *Threatened Species Conservation Act* (1995) has Schedules 1, 2 and 3, which correspond to (1) extinct and endangered species, populations and ecological communities; (2) vulnerable species; and (3) threatening processes. The taxonomic bias in the listed species is similar to that in the Federal EPBC Act (1999); that is, there are few invertebrates and fungi (NSW National Parks and Wildlife Service, 2003a). Other organisations in New South Wales list species and ecological communities. For example, New South Wales Fisheries (2003) has a list of marine and freshwater species, populations and communities that are extinct, endangered and vulnerable, and it also lists four threatening processes in marine and freshwater environments.

In Tasmania, species are classified as extinct, endangered, threatened or rare (Department of Primary Industries, Water and Environment, 2002). The Northern Territory uses the same categories as the IUCN (IUCN, 2001; Department of Infrastructure, Planning and Environment, 2002).

Strengths and weaknesses of threat classification schemes

The strengths of qualitative schemes include their simplicity, modest data requirements and wide acceptance (Chalson and Keith, 1995). Their principal weakness is their lack of explicit guidelines for assigning taxa to categories of risk (Mace and Lande, 1991). Because of the almost exclusive reliance on expert opinion, consistency among workers using these schemes is difficult to attain, and where conflicting opinions arise, there is no systematic means of resolution (Chalson and Keith, 1995). Problems that derive from expert opinion are not restricted to qualitative descriptions. As an example, qualitative descriptions of plants can favour species with restricted ranges and/or habitat requirements, thus underestimating risks to species with widespread but declining ranges (McIntyre, 1992).

Rule sets

Rule sets for classifying threat grew out of the early IUCN qualitative descriptions. Mace and Lande (1991) suggested that the IUCN categories should be assessed quantitatively, and they proposed the following risk categories (Table 3.6):

- *Critical*: a 50% probability of extinction within 5 years or two generations of the species in question, whichever is longer.
- *Endangered*: a 20% probability of extinction within 20 years or 10 generations, whichever is longer.
- *Vulnerable*: a 10% probability of extinction within 100 years.

Akçakaya (1992) recognised that the classification of risk involves three parameters, namely time, probability of decline, and percentage decline. Threat can then be seen as a combination of the magnitude of the impending decline within some time frame, and the probability that a decline of that magnitude will occur.

After setting down their definitions in the early 1970s, the IUCN continued to develop categories for the conservation status of species. The structure of the various threat categories proposed by the IUCN in 2001 is set out in Figure 3.9. The IUCN (2001) defined the sizes of populations in terms of the number of mature individuals. A mature individual is capable of reproduction; thus, individuals that cannot reproduce because of environment or behaviour are excluded from estimates of the size of a population. Where populations are characterised by extreme fluctuations, the IUCN recommended that the minimum number of individuals should be used to characterise the population. Reproducing individuals within a clone should be counted as individuals, except where such individuals are unable to survive alone (e.g. reef-building corals). In populations with biased sex ratios, it is considered appropriate to use the number of the least abundant sex. The generation time is approximated by the average age of parents in the population.

Given these operational definitions, the IUCN (2001; Figure 3.9) defined categories for the conservation status of taxa at the level of species and below. These categories are supported by decision rules related to range, population size, and population history (see Table 3.6).

- *Extinct* (EX): there is no reasonable doubt that the last individual of the taxon has died.
- *Extinct in the wild* (EW): the taxon is known to survive in cultivation, in captivity or as a naturalised population well outside its past range; exhaustive surveys have failed to record an individual.
- *Critically endangered* (CR): the taxon is facing an extremely high risk of extinction in the wild in the immediate future (Criteria A1 to E1).
- *Endangered* (EN): the taxon is not critically endangered but is facing a very high risk of extinction in the wild in the near future (Criteria A2 to E2).
- *Vulnerable* (VU): the taxon is not endangered but is facing a high risk of extinction in the wild in the medium-term future (Criteria A3 to E3).
- *Conservation dependent* (CD): the taxon is not vulnerable, but is the focus of a continuing taxon-specific or habitat-specific conservation program; the cessation of this program would result in the taxon qualifying for one of the four categories above.

- *Data deficient* (DD): there is inadequate information to make a direct, or indirect, assessment of a taxon's risk of extinction based on its distribution or population status.
- *Low risk* (LR): the taxon has been evaluated and does not qualify for any other category; it may be worthwhile to distinguish between species that are close to qualifying for one of the other categories, and species that are unlikely to face extinction in the foreseeable future.
- *Not evaluated* (NE): the taxon has not yet been assessed against the criteria.

These categories provide an assessment of the likelihood of extinction under current circumstances. Each of the categories in Table 3.6 can be addressed by an appropriate quantitative analysis (Criterion E). The methods for doing so, termed population viability analyses, are addressed in detail in Chapter 16.

Strengths and limitations of rule sets

Rule sets are attractive because of their use of explicit, ecologically meaningful criteria, and their wide applicability and simplicity of use (Mace and Lande, 1991). By necessity, thresholds that delimit categories of risk are arbitrary, effectively ignoring the fact that risk of extinction and the factors related to it vary continuously (Chalson and Keith, 1995). Given and Norton (1993) claim that decision rule sets are often too coarse to assist managers in decision-making, because they do not provide an actual ranking of species from most threatened to least threatened. Quantitative criteria such as extinction probabilities can be difficult or costly to estimate because they involve collection of field information, sound ecological understanding, parameter estimation and predictive modelling (Burgman *et al.*, 1993; see Chapter 16). Even when they are available, they may be too uncertain to be useful

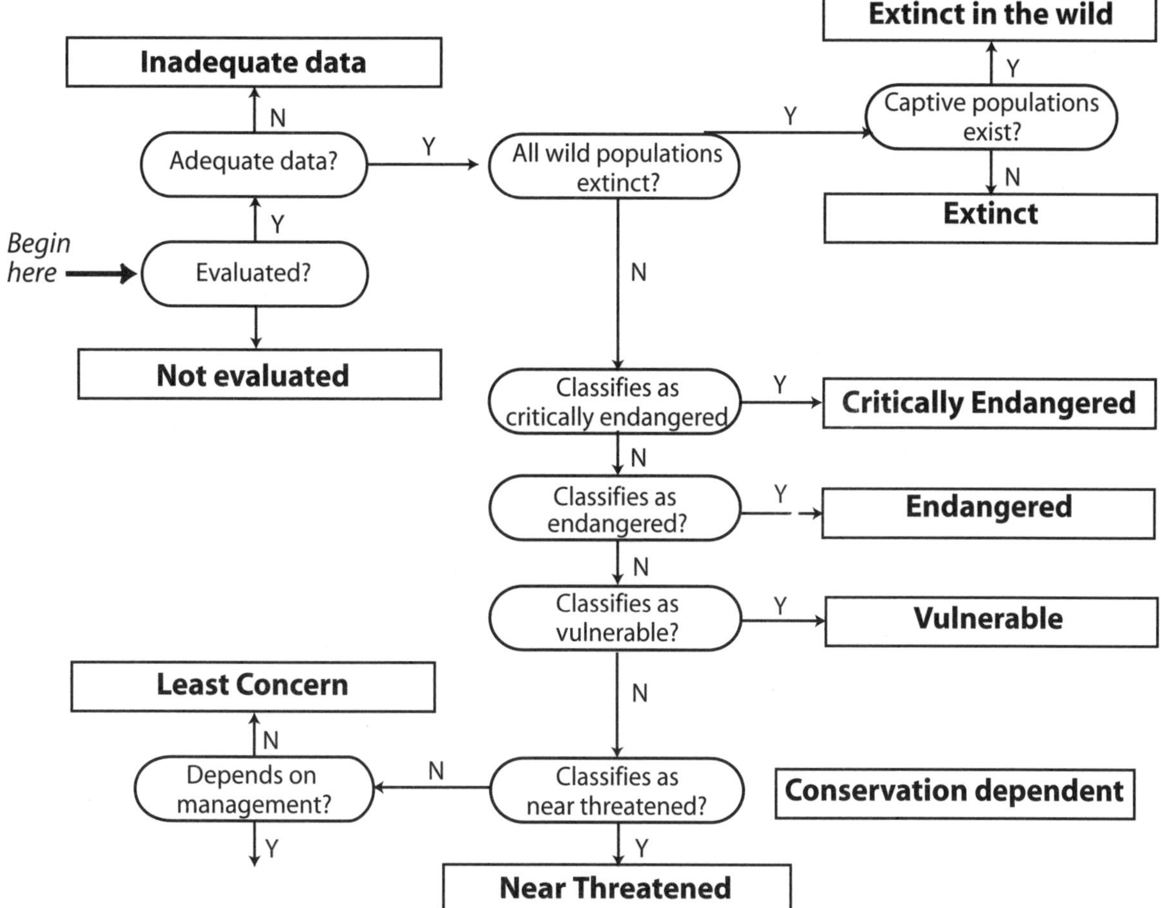

Figure 3.9. Relationships between the IUCN threat categories.

(Taylor, 1995). Usually, biologists use other characteristics of a species to infer the relative likelihood of extinction or decline (Possingham *et al.*, 2001). The problem is a difficult one because trends in abundance and measures of geographic range can also be difficult or costly to estimate. In most applications, direct, reliable estimates of parameters are unavailable.

A related problem is that by allowing membership to a category based on one criterion alone, it is assumed that the criterion is an adequate predictor of the level of threat. For example, the biogeography of Western Australia has resulted in a flora characterised by a high level of endemism and naturally rare and geographically restricted plant distributions (Hopper and Coates, 1990; Hopper, 1997). The application of rules such as those embodied in the IUCN (2001) criteria would inflate the inferred level of threat associated with many plant species in Western Australia. If interpreted unthinkingly, the classifications could result in the misdirection of scarce conservation resources.

Point-scoring procedures

Millsap *et al.* (1990) developed a fundamentally different approach to assessing conservation status based on weighted scores of species attributes. Points are allocated to a taxon for each of a number of demographic, life history and management variables, then the sum of the points determines the rank of the species. Adaptations of the method are used in many jurisdictions, including New South Wales for fauna assessments (Lunney *et al.*, 1996), and Western Australia for both flora and fauna (Table 3.8). A similar system has been applied in New Zealand (Molloy and Davis, 1992).

The scoring system in Table 3.8 was developed by Burbidge and Brown (1993) and is used in conjunction with expert opinion to place taxa in conservation status categories (CALM, 1994). Such schemes are attractive because of their use of explicit and consistent assessment variables, and because they can be used to rank species. They provide many more classes than rule sets or qualitative evaluations, thereby providing a means for allocating priorities among many taxa (Holt, 1987).

Scoring procedures are sensitive to the relative weighting of variables (Chalson and Keith, 1995); but there may be little justification for maintaining consistent weightings across all taxa. In accounting for weights in their system, Lunney *et al.* (1996) explained that biological variables:

> *were weighted to reflect the reliability of available information. Greater weight was given to the ecological attributes of population size and trend, and area of distribution and trend, as these provide estimates of change in populations...The other biological attributes...were given less weight as they do not estimate changes, but are factors that predispose species to vulnerability or inability to recover quickly from sudden widespread declines.*

It is difficult to test the rationale for the weights because calibration data are unavailable.

The arbitrary nature of the total scores in point-scoring systems is influenced by the fact that the variables are not independent of one another. Lunney *et al.* (1996) reported strong correlations between a number of the biological variables, based on an analysis of scores for 883 faunal species. The strength of correlations between variables is taxon-specific, uncontrolled, and may have a substantial influence on the scores assigned to different species, making comparisons between taxa unreliable.

Uncertainty in conservation status assessment

Typically, judgements concerning the threats faced by species are made in the absence of adequate information. Estimates of population size, geographic range, and trends in population parameters are uncertain. Agreement or disagreement with one of the thresholds in a rule set is specified on the basis of a best estimate of the parameter. For example, the IUCN (2001) rules ask if a population consists of fewer than 50 mature individuals. Effectively, only two responses are possible. The structure of the question implies that there will be no error in the answer. Such an approach ignores whatever information is available regarding the reliability of the estimate.

It could be argued that in assessing conservation status, the best strategy is to base judgements on a lower bound. For example, to be 90% certain that no endangered species have been ignored, then the lower 10th percentile for population size would provide a more effective means for classifying the relative risks faced by different species than using the best estimate for population size. Consider the situation in Figure 3.10, in which there are two species, for each of which there is an estimate of population size.

In many applications, neither species A nor species B would be considered critically threatened, at least not on the basis of the total number of mature individuals

Table 3.8. Part of the scoring system for aiding the assessment of threat for research and management of threatened taxa in Western Australia (after CALM, 1994). The system includes categories for effects of fire, competition, reproductive biology and *ex situ* breeding/propagation.

Score	Attribute
	Current geographic distribution
5	Very narrow endemic, total range <50 square kilometres or <20 kilometres (linear)
4	Narrow endemic, total range <500 square kilometres or <100 kilometres (linear)
3	Confined to single phytogeographic district
2	Confined to single phytogeographic region
1	Endemic to Western Australia
0	Not endemic to Western Australia
	No. populations
10	Only one known
8	Only two known
5	3 or 4 known, but thought to be few
3	From 5 to 10 known
0	>10, or unknown but thought to be many
	Effective population size
	Vertebrates and invertebrates
10	<50
8	50–500, or range area <1 hectare (or unknown but thought to be small)
5	500–2000, or range area 1–100 hectares
0	>2000, or range area >100 hectares (or unknown but thought to be large)
	Range decline
10	Occupies <1% of former range area, almost all habitat destroyed or unsuitable
8	Occupies 1–5% of former range area, most habitat destroyed or unsuitable (or range decline unknown but thought to be large)
5	Occupies 5–10% of former range area, apparently suitable habitat remaining
2	Occupies 10–50% of former range area
0	Occupies 50–100% of former range area (or decline unknown but thought to be small)
	Total wild population decline rate
	(if present circumstances do not change)
10	Declining at a rate that threatens survival within 5 years
8	Declining at a rate that threatens survival within 5–20 years (or unknown but thought to be high)
5	Declining at a rate that threatens survival within 20–50 years
2	Declining at a rate that threatens survival within 50–100 years (or unknown but thought to be low)
0	Total wild population stable or increasing
	Protection of habitat
5	No populations known from conservation reserves or State Forest
4	One population known from conservation reserve or State Forest
2	More than one population, mostly small, in conservation reserves or State Forest
0	Several large populations in conservation reserves or State Forest
	Existing habitat loss rate
10	100% of habitat (or breeding habitat) likely to be destroyed of severely modified in <10 years
8	>75% likely to be destroyed of severely modified in <10 years, or unknown but thought to be high
5	50–75% likely to be destroyed of severely modified in <10 years
2	25–50% likely to be destroyed of severely modified in <10 years, or unknown but thought to be low
0	No likely change, or <25% likely to be destroyed or severely modified
	Environmental threats
	Vertebrates and invertebrates
10	Mammal 35–8000 grams, ground-nesting bird in arid or semi-arid area, or reptile >50 grams, or environmental threat (e.g. dieback, hunting) likely to have a severe impact on taxon
8	Mammal 35–8000 grams, ground-nesting bird not in semiarid area, or environmental threats likely to have high impact on taxon
4	Environmental threats likely to have moderate impact on taxon
1	Environmental threats likely to have low impact
0	No environmental threats
	Vascular plants
10	High susceptibility and high risk of infection by *Phytophthora*, most populations already infected, or very high risk of destruction or habitat degradation due to clearing, dieback, salinity, recreation, grazing, etc.
8	High susceptibility and high risk of infection by *Phytophthora*, some populations already infected, or high risk of destruction or habitat degradation
5	Moderate susceptibility and high risk of infection by *Phytophthora* or other fungal diseases, moderate risk of destruction or habitat degradation
3	Moderate susceptibility and risk of infection by *Phytophthora* or other fungal diseases, moderate to low risk of destruction or habitat degradation
1	Low to moderate susceptibility and low risk of infection by fungal diseases, low risk of destruction or habitat degradation
0	No environmental threats

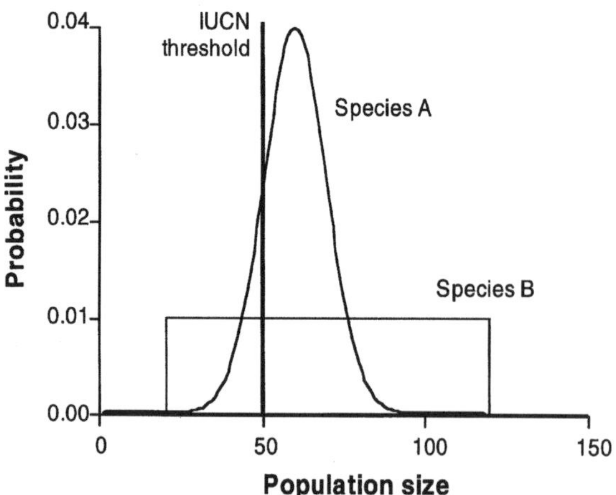

Figure 3.10. Diagrammatic representation of the best estimates for the total number of mature individuals of two hypothetical species (A = 60 and B = 70 individuals), together with the probability distributions that describe the reliability of each estimate. The IUCN (1994) threshold population size below which species must be critically threatened is 50 mature individuals. The distribution of species B is uniform, and it ranges between 20 and 120 individuals. The distribution of species A is normal with a standard deviation of 10.

in the population. However, if population size alone is used as a guide to rank the species, then protection would be afforded to species A before species B.

Uncertainty can be the result of measurement error, year-to-year variation in population size caused by variation in the environment, demographic stochasticity, or taxonomic uncertainty. Species A represents a case in which the estimate for a species is based on a carefully designed survey, from which the form of the distribution and its variance can be estimated. Species B represents a case in which the estimate is based on extrapolation from related species by experts who are able to specify only upper and lower bounds for population size. A significant proportion of the distribution that describes the population size of species B falls below the threshold of 50 mature individuals, whereas very little of the distribution for species A falls below the same threshold. Intuitively, species B can be considered more at risk than species A, simply because there is a greater chance that the true population size for this species is below 50 individuals. If we interpret the lower 5th percentile of each distribution, we can be 95% certain that the population size of species A is greater than 40, and 95% certain that the population size of species B is greater than 25.

Akçakaya *et al.* (2000) designed a system for assessing conservation status that accounts for uncertainty in all the parameters used in the IUCN (2001) protocol. Someone making an assessment specifies best estimates and bounds, or other representations of uncertainty, then the system combines the information to generate a best estimate and bounds for conservation status (Figure 3.11).

When uncertainty is made plain, it provides a new context for decision-making. It can be valuable to know that one species is certain to be critically endangered, whereas another might be critically endangered, but is almost equally likely to be endangered or vulnerable. Our attitude to uncertainty will then contribute to decisions about allocation of resources. For example, we might try to recover critically endangered species rather than species whose threat status is highly uncertain.

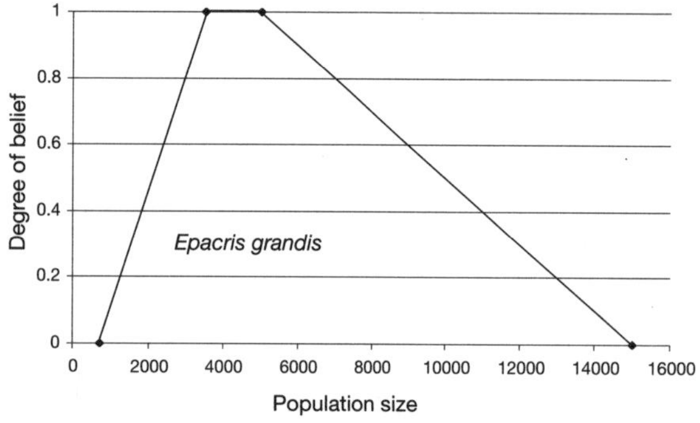

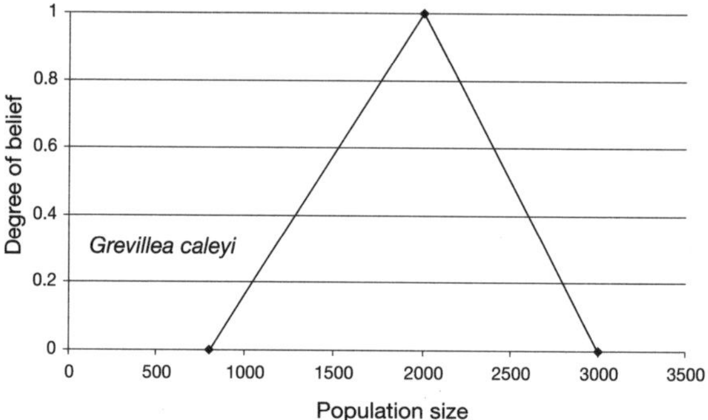

Figure 3.11. Trapezoidal numbers representing uncertainty in the population sizes of two threatened Australian plants: *Epacris grandis* (number of mature individuals ranges between 700 and 15 000, with a best estimate of between 3500 and 5000) and *Grevillea caleyi* (number of mature individuals ranges from 800 to 3000, with a best estimate of 2000; after Akçakaya *et al.*, 2000). The uncertainty in the parameters does not affect the assessment of conservation status for *Epacris grandis* – it is classified as certainly endangered. The classification of *Grevillea caleyi* is affected by uncertainty – it is probably critically endangered, but may be endangered.

Box 3.5

Ecological triage

One approach to the conservation of biodiversity has been termed 'ecological triage', whereby limited resources for conservation funding are targeted at the subset of species for which management success is most likely. The approach stems from the same process used in medicine to set priorities for allocating efforts to treat patients. Possingham (2001) argues that the ecological triage approach is essential because too many resources are directed to species on the brink of extinction that are doomed in the long run, and too few are targeted at declining taxa that are still recoverable. He believes that the concept of extinction debt (see Tilman *et al.*, 1994; McCarthy *et al.*, 1997; Berglund and Jonsson, 2005) means that extinction risk will continue well after inappropriate human actions (e.g. land clearing) have taken place and there will be insufficient funds to deal with the increasing numbers of species under threat.

Some conservation biologists believe that parallels between emergency medicine and conservation biology are inappropriate. For example, the approach may make extinction acceptable and allow decision-makers to get away with allocating insufficient resources to address environmental problems. On this basis, Cameron and Soderquist (2002) argue that nations such as Australia should reject the concept of ecological triage because the country has the knowledge, time and ability to save threatened and endangered species.

These types of debates are important to stimulate discussion among the public, policy-makers and politicians about the long-term trajectory of conservation, and to identify the management actions and expenditure of resources that will provide the maximum benefit.

Summary: assessing threat

Assessments of threat are just one component in the process of assigning priorities for the allocation of resources in conservation. In most cases, conservation status assessments will not, on their own, provide a basis for the allocation of resources. For instance, Possingham (2001) suggested that priorities for recovery should be based on the amount of risk reduction achieved per dollar spent, maximising the overall efficiency of recovery efforts.

Determining priorities is an area of public policy in which numerous factors must be considered in

Table 3.9. Threatening processes listed under the *Environment Protection* and *Biodiversity Conservation Act* (1999). (Data from Department of Environment and Heritage, 2005a.)

Listed key threatening process	Chapter reference in this book
Competition and land degradation by feral Goats	
Competition and land degradation by feral Rabbits	Chapter 8
Dieback caused by the root-rot (Cinnamon) fungus (*Phytophthora cinnamomi*)	Chapter 7
Incidental catch (by-catch) of sea turtles during coastal otter-trawling operations within Australian waters north of 28 degrees south	Chapter 8
Incidental catch (or by-catch) of seabirds during oceanic longline fishing operations	Chapter 8
Infection of amphibians with chytrid fungus resulting in chytridiomycosis	Chapter 7
Injury and fatality to vertebrate marine life caused by ingestion of, or entanglement in, harmful marine debris	
Land clearance	Chapters 8, 9, 10
Loss of biodiversity and ecosystem integrity following invasion by the Yellow Crazy Ant on Christmas Island, Indian Ocean	
Loss of climatic habitat caused by anthropogenic emissions of greenhouse gases	Chapter 5
Predation by feral Cats	Chapter 8
Predation by the Red Fox	Chapter 8
Predation, habitat degradation, competition and disease transmission by feral Pigs	
Psittacine circoviral (beak and feather) disease affecting endangered psittacine species	
The biological effects, including lethal toxic ingestion, caused by Cane Toads	Chapter 8
Reduction in the biodiversity of Australian native fauna and flora due to the Red Imported Fire Ant	Chapter 8

addition to levels of threat; such factors include public sentiment, costs, logistics, chance of success, taxonomic distinctiveness and political motivation, among others. For instance, Shrader-Frechette and McCoy (1994) suggested that:

> *ecological science and environmental policy may require us to move beyond purely scientific rationality and into ethical rationality, into a real recognition of ecological interdependence, not only among all living beings, but also among science and public values.*

Usually, the autecological data and demographic studies necessary to make a direct estimate of the threat faced by a species are unavailable. The procedures described earlier in this section that are based on population size, geographic range, number of populations, kinds of threats and so on are a compromise in which these variables act as surrogates for the risk of extinction. However, even these data are usually not available. In the absence of data, it is not possible to do anything except use value judgements and biological intuition. Because rule sets, point-scoring and qualitative procedures usually rely on expert opinion, there are no guarantees of consistency (Lunney *et al.*, 1996; see Chapter 18). Even where there is consistency, it is weak evidence of reliability. Moreover, sources of bias are difficult to identify. The best that can be done is to be honest about uncertainty, carry it through the chains of logic and

Box 3.6

Loss of hollow-bearing trees in native forests

The example of the loss of hollow-bearing trees illustrates how the deterioration of a structural feature of an ecosystem can have important ramifications for the functioning of that ecosystem, and, conversely, how the management of threatening processes can conserve assemblages of species and their interactions.

Hollows in large old trees provide nesting and/or sheltering sites for more than 300 species of Australian vertebrates (Gibbons and Lindenmayer, 2002). Some animals, for example several species of arboreal marsupials, may spend more than 75% of their lives within a cavity in a large tree (Smith *et al.*, 1982). There is also a large (but unknown) number of invertebrates that require hollows for part or all of their life cycle (Gibbons and Lindenmayer, 1996). It is unlikely that we will know very much about such groups within the foreseeable future. Hollow trees are a major structural element of Australian eucalypt forests (see the section on old growth forests in Chapter 2) and they can take hundreds of years to develop. This is because, unlike many other parts of the world, Australia does not support vertebrates such as woodpeckers that excavate cavities in trees. Instead, hollow development occurs in response to relatively slow-acting decay mechanisms such as termite and microbial activity (Mackowski, 1987).

There has been a substantial decline in the number of hollow trees since the arrival of Europeans in Australia as a result of altered fire regimes, past and present logging practices, vegetation clearing for mining and agriculture, and tree decline in rural areas (e.g. from dieback). The loss of trees with hollows is listed as a threatening process in Victoria (Garnett and Loyn, 1992). Given the dependence of animals on trees with hollows, the decline of these resources could have major implications for a substantial component of wildlife in forests (Gibbons and Lindenmayer, 1996) and woodlands (Traill, 1993). For example, a monitoring program of hollow trees in Central Victorian forests has indicated that the number of hollow trees is declining rapidly and that there will be a prolonged shortage of them for at least the next 75 years (Lindenmayer *et al.*, 1990; Figure 3.12). There will also be a reduction in populations of hollow-using species (Figure 3.13). This could, in turn, have long-term implications for forest ecology and productivity as some of these hollow-dependent animals have roles in pollination (Goldingay and Kavanagh, 1991; Whelan, 1994) and the control of insect populations (Suckling, 1982; Lumsden, 1993).

Attempts to rectify the problems associated with the loss of hollow-bearing trees will require changes in the way forests, woodlands and agricultural areas are managed. In the case of wood production zones, there is a need for more long-term planning in maintaining and/or re-constructing stand structure, and a need to ensure that enough trees are retained in logged areas to meet the requirements of the very wide array of taxa that are dependent on trees with hollows (Gibbons and Lindenmayer, 2002).

calculation that contribute to a conservation status assessment, and make the uncertainty apparent when the results of the assessment are communicated.

Constitutional arrangements between the Australian Commonwealth and State Governments have seen the principal responsibility for the management of rare and threatened species resting with the States. Each State has sought an effective approach to developing priorities for conservation actions within its own political and social constraints and knowledge base. From a national perspective, this fragmented approach has led to two important problems. First, the lack of uniformity between States has made it cumbersome to attempt to describe the status of Australia's biodiversity accurately. Second, the equitable allocation of Commonwealth resources to the States has been difficult in the absence of an easy and transparent means of addressing the differences in each State's methods for setting priorities.

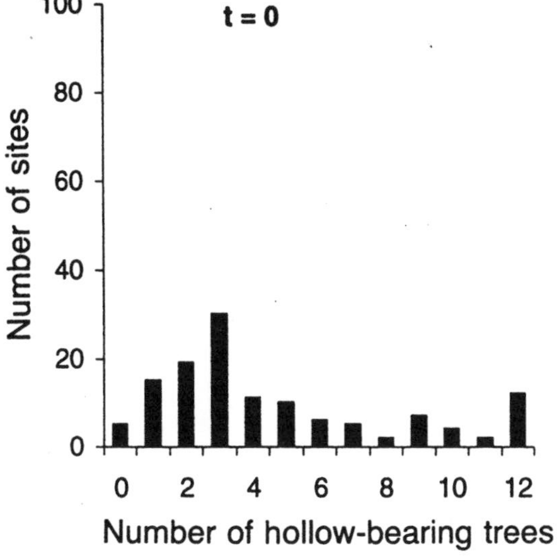

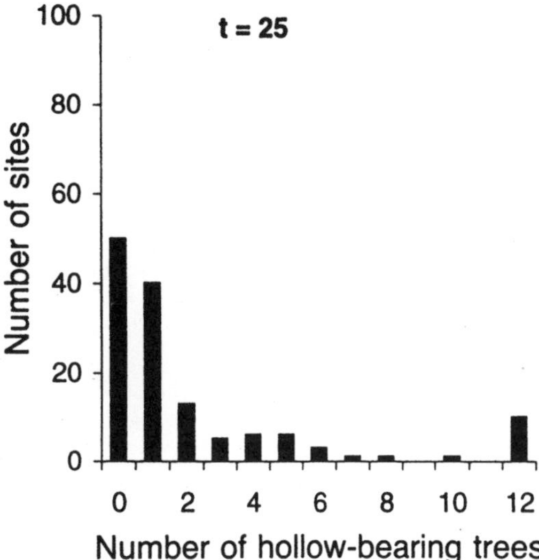

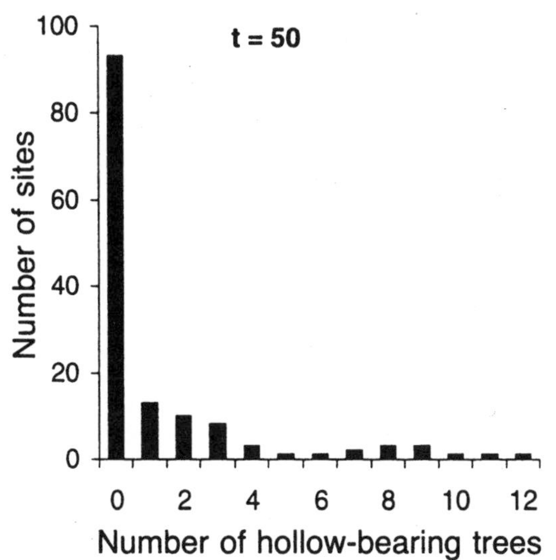

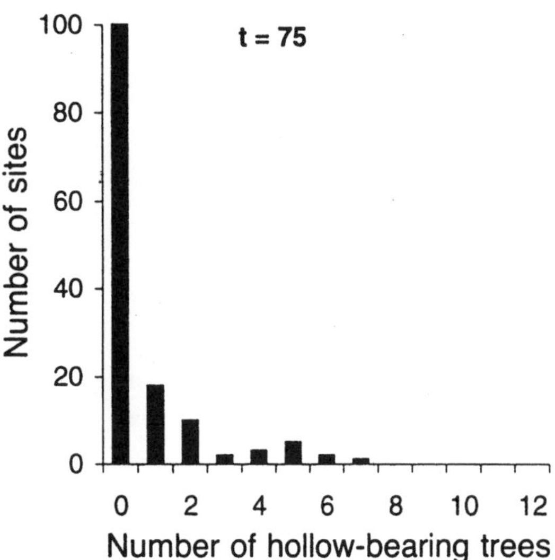

Figure 3.12. Current and expected decline in tree hollow abundance in the montane Ash forests of the Central Highlands of Victoria at 25, 50 and 75 years from now. (Redrawn from Lindenmayer *et al.*, 1997.)

The most important factor in the development of a ranking system is to be clear about its purpose. In the absence of a definitive national approach to setting priorities for conservation action, and because ranking systems have several uses (Possingham *et al.*, 2002), a multiplicity of aims and methods in the States' conservation status assessments is inevitable.

All of the methods applied in Australia have one feature in common. They ignore uncertainty in the data on which they rely. All of them use point estimates, implying implicitly that the data used to make judgements are known exactly. Approaches to assessing status that include uncertainty use more information, are sufficiently flexible to use both field measurements and expert judgement, and provide the means to infer levels of threat and to establish ranks for conservation action that account for the amount and quality of data.

3.3 Threatening processes

Frequently, effective conservation involves the identification of the causes of environmental change, and the implementation of management practices to limit those changes or modify their effects (Caughley and Gunn, 1996). A **threatening process** is a process that detrimentally affects, or may detrimentally affect, the survival, abundance, distribution or potential for evolutionary development of a native species or

Fallen hollow tree in Victorian Ash forest (photo by David Lindenmayer) and a Mountain Brushtail Possum occupying a hollow. (Photos by Esther Beaton.)

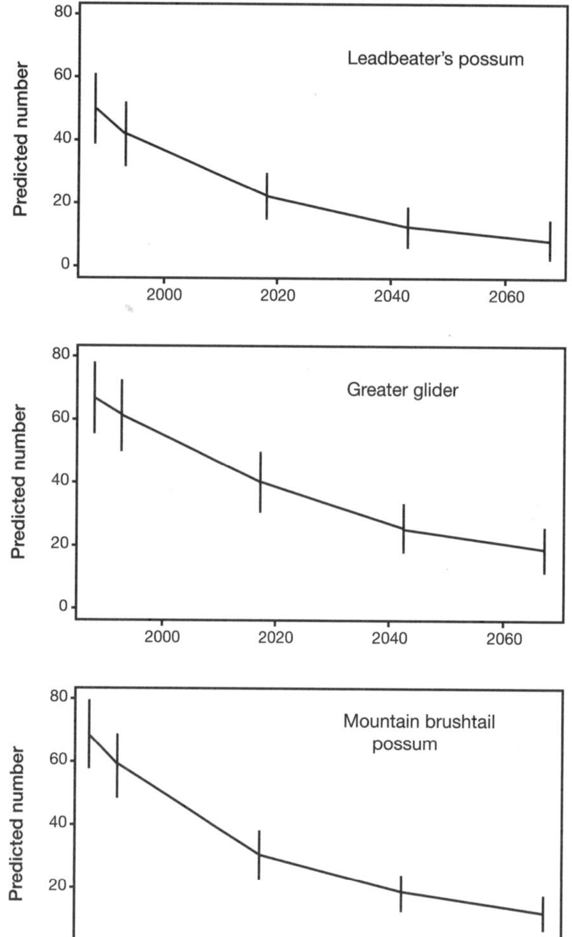

Figure 3.13. Forecast population decline of three species of arboreal marsupials as an outcome of the decline of trees with hollows. (Modified from Lindenmayer *et al.*, 1997.)

ecological community. For example, Falk (1990) classifies some of the threats faced by plants:

- destruction of habitat
- competition by invasive species
- loss of pollinators, dispersal agents, host species or symbionts
- genetic drift, genetic swamping, inbreeding, demographic variation or other consequences of small population size
- destruction of individual plants or populations by disease, foraging or collecting.

Each of these is a threatening process. Falk (1990) suggests that decline in natural populations is usually the result of multiple impacts. In the face of the enormous number of species that require attention, it will be most effective to focus conservation effort on the processes that result in population decline (Caughley and Gunn, 1996). This is analogous to treating population decline as a symptom, and looking for, and correcting, the causes of that symptom.

Threatening processes under the Australian EPBC Act (1999)

The Australian Federal Government's EPBC Act (1999) makes provision for the nomination of threatening processes (see Table 3.9). Once a threatening process has been listed, a Threat Abatement Plan can be implemented.

State-based listings of threatening processes

The State of New South Wales lists threatening processes under the *Threatened Species Conservation Act (1995)*. For example, competition from feral Honeybees is listed (addressed in greater detail in the 'Exotic animals: invertebrates' section in Chapter 7). Another is the collection of bush rock and its impacts on rare reptiles such as the Broad-headed Snake (see Chapter 8). Many other

Box 3.7

Increase in sediment input into rivers and streams due to human activities

Processes that increase the input of sediments into rivers and streams include forestry operations, land clearance for agriculture, grazing and cropping in the riparian zones, road building, construction of dams and weirs, in-stream mining and quarrying, urban run-off, and industrial discharges (Metzeling *et al.*, 1995; Lovett and Price, 1999). Increased sediment deposition can: (1) lead to deterioration in habitat quality by reducing the availability of food for vertebrates such as aquatic mammals, birds and reptiles, (2) affect egg-laying and the behavioural patterns of amphibians, fish and invertebrates by reducing levels of oxygen in the water or covering egg masses with sediment, and (3) adversely influence photosynthetic activity in aquatic plants, and affect the quality of attachment sites (e.g. on logs, rocks and the stream) for some aquatic invertebrates. An advantage of managing threatening processes such as increased stream sedimentation is that it improves the chances of persistence of several species simultaneously, and targets resources such that they rectify the cause of the problem (Caughley and Gunn, 1996). In response, in places such as the Australian Capital Territory, groynes are being constructed in river systems to reduce sediment loads and abate the problems for aquatic ecosystems.

Groynes built into the Tharwa River near Canberra in an effort to reduce problems with sedimentation in aquatic ecosystems. These constructions can also help create deep pools, which provide habitat for crustaceans and fish such as the endangered Trout Cod. (Photos by Mark Lintermans.)

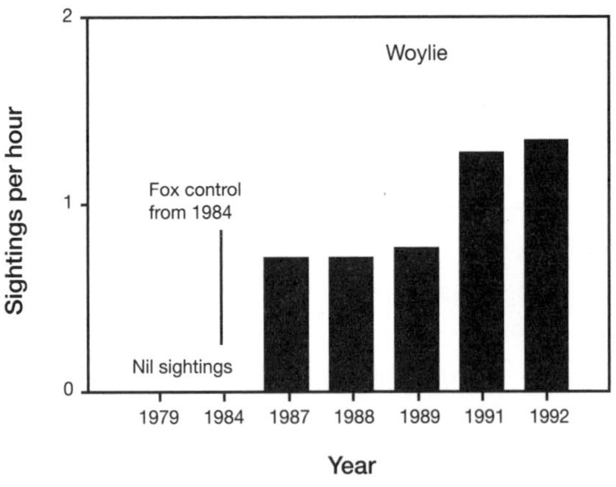

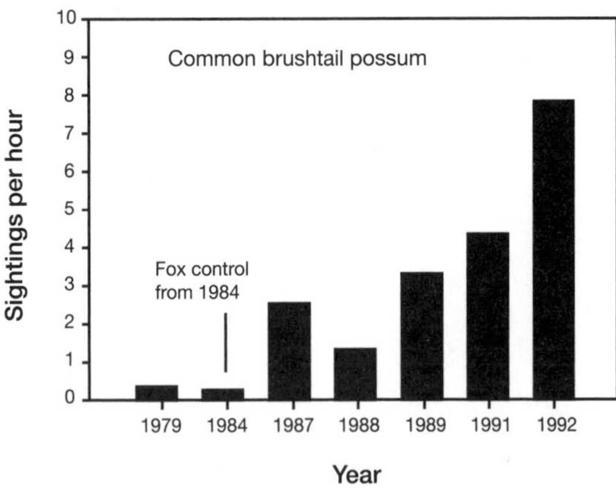

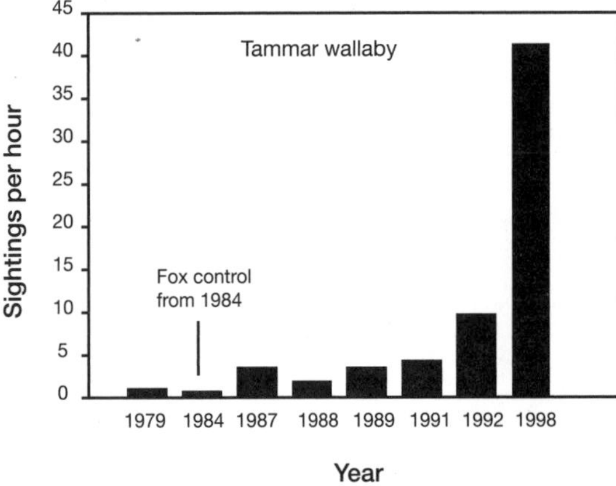

Figure 3.14. Population recoveries of three species of native mammals following fox control in south-western Australia. (Redrawn from Kinnear *et al.*, 2002.)

threatening processes, such as feral Cat and Red Fox predation and anthropogenic climate change are similar to those listed Federally.

In the State of Victoria, the *Flora and Fauna Guarantee Act (1988)* is designed to halt and reverse the decline of species through the identification of threatening processes. The Act defines threatening processes as those that pose 'a significant threat to the survival or evolutionary potential of a range of flora and fauna' (State Government of Victoria, 1988, Section 11, p. 8). Between 1988 and 1993, 12 threatening processes were listed under the *Flora and Fauna Guarantee Act.* By 2003, this had increased to 30 (Department of Sustainability and Environment, 2003) – two of them are reviewed in Box 3.6 and 3.7 respectively. It is notable that despite the best intentions of the *Flora and Fauna Guarantee Act*, there has been a lack of on-ground enforcement of its objectives.

Mitigating the impacts of threatening processes

Threatening processes must be identified (which often requires careful experiments or observational studies; see Caughley and Gunn, 1996), and controlled. An extensive fox-baiting program in Western Australia has been instrumental in the recovery of several mammal taxa, to the extent that the conservation status of some has been downgraded. The status of the Woylie has been changed from endangered to lower risk and that of the Chuditch and Numbat from endangered to vulnerable. The recovery of the Woylie has been such that it was the first species to be removed from Western Australia's endangered species list, and animals are being reintroduced to parts of the species' former range in New South Wales and South Australia. Indeed, the Woylie was the first mammal species to be removed from an international threatened species list (Davidson, 2004). Without the control of the Red Fox, many native animals would be limited to small populations within highly restricted refugia. With control, populations have increased and expanded beyond predation refugia (reviewed by Kinnear *et al.*, 2002; Figure 3.14). The effects of the Red Fox are pronounced in south-western Australia, where a large proportion of mammals have been negatively affected by foxes (Calver and Dell, 1998).

3.4 Conclusions

Lists and maps of the distributions of rare plants and animals have been used to guide conservation priorities. This approach is useful for the conservation of those species that appear on the lists, but it can be a flawed guide to ecosystem conservation, because some threatened groups are under-represented, depending on the distribution and intensity of survey effort, the life histories of the taxa and the reasons for their rarity (McIntyre, 1992).

Relatively little attention has been paid to the way in which classifications of rarity are developed and applied. The quality of the data used to generate classifications of rare species is seldom evaluated. In particular, the inclusion of a species on a rare plant or animal list, and the relative degree of threat implied by the classification, sometimes determine conservation priorities. In these circumstances, it is important to test the sensitivity of the rank order of priorities to uncertainty in the data. Exploratory studies of the reliability, repeatability and biases of systems used to measure and classify the threat faced by species (e.g. McIntyre, 1992; Taylor, 1995; Lunney *et al.*, 1996) are uncommon.

Threatening processes play an important role in determining the future of both species and communities, and are formally recognised by the Australian Federal Government and by several Australian States. Each State in Australia has its own system for setting conservation priorities, which reflects the availability of data, and unique social and biogeographic constraints; however, all systems ignore uncertainty. The systems could be modified to take account of measures of reliability associated with estimates of population size, rates of decline and other quantitative attributes. Such modifications will improve the way in which threats are assessed.

Setting conservation priorities involves ethical decisions: people with exclusively utilitarian motives will set priorities that are very different from those set by people with ecocentric or biocentric motives. The anthropocentric bias in taxonomy and the natural sciences has important consequences for the way in which conservation resources are allocated. Although these factors play a part in decision-making for conservation, rarely are ethical positions stated or rationalised.

3.5 Practical considerations

For statements about the abundance, range or specificity of a species to make sense, the reference area must be stated, or at least understood. Conservation biologists are often responsible for estimating population size, and trends in population size and geographic range, to establish the conservation status of species and the effectiveness of management interventions. This book does not cover estimation methods, but conservation biologists are obliged to make their sampling strategy as efficient as possible, to maximise the chances of detecting important changes.

Measures of range based on convex polygons are likely to be biased, because they overestimate range for convoluted shapes and become increasingly biased as samples are added. Spatial bias in sampling effort and unrealistically sharp boundaries exacerbate the problem. Better estimates of geographic range can be made with other tools; this topic is explored in more detail in Chapter 14.

Conservation status assessments are uncertain. Akçakaya *et al.* (2000) invented a system that takes into account the uncertainties associated with each element of the IUCN (2001, 2003) system. This approach should be used routinely by conservation biologists to make assessments and report uncertainties.

Policy and legislation supporting the listing of ecological communities and threatening processes provides a much broader array of opportunities to develop effective conservation strategies. In most instances, species conservation should begin by managing threatening processes.

3.6 Further reading

Various aspects of the concept of rarity have been discussed by Harper (1981), Margules and Usher (1981), Main (1984), Cody (1986), Rabinowitz *et al.* (1986) and New (2000). An extensive review of the concept of rarity, and the measurement of abundance and range is provided by Gaston (1994).

Apart from the procedures for setting conservation priorities described here, a few multivariate approaches have been suggested by Given and Norton (1993) and Hall (1993). Diamond (1987) introduced the idea of extinct and presumed extinct species.

Mace and Lande (1991) revised the rule-based approach to include quantitative criteria. The IUCN (1994, 2003) provide a detailed account of rule-based procedures and the concepts on which they depend. Akçakaya *et al.* (2001) designed a system for assessing conservation status that accounts for uncertainty in all the parameters used in the IUCN (2001) protocol. Millsap *et al.* (1990) introduced the notion of point-scoring procedures for setting conservation priorities. Mackey (2003) is a colourful and highly readable discussion of the endangered species in various plant and animal groups found throughout the world.

Department of Environment and Heritage (2005a,b) provide extensive details on extinct, threatened and other species of conservation concern as they relate to the EPBC Act (1999). A major listing of extinct species from around the world is given by Groombridge and Jenkins (2002).

Chapter 4

Protected areas, off-reserve conservation and managed populations

This chapter outlines some principles for reserve design, and provides background information on reserve systems in Australia (methods for reserve design are outlined in more detail in Chapter 16). As protected areas will, in isolation, be insufficient to conserve biodiversity, we also discuss the importance of off-reserve conservation. We outline some of the facilities devoted to the preservation of biodiversity, such as seed storage facilities and gene banks, and some of the strategies employed in the direct management of natural populations, such as captive breeding and reintroduction.

4.0 Introduction

In Chapter 3 we examined the systems that can be used for setting conservation priorities and identifying threatening processes. Conservation strategies are often based on setting aside land for the protection of species and ecosystems. The development of reserve systems in most countries has been undertaken on an *ad hoc* basis, targeting areas for which there is an obvious social and political mandate, and areas that are unlikely to provide immediate productive benefits for agriculture, forestry, mining, urban development and human infrastructure. Effective biodiversity conservation should move beyond protected areas to explore different land management strategies to conserve species and communities. This chapter outlines some options.

Protected areas

The idea of setting aside areas for hunting, recreation and timber protection has been established in Europe for many hundreds of years (Nelson, 1991). The original notion of wilderness was of a place where animals lived and could be hunted (Shea *et al.*, 1997; see Box 1.5). Few protected areas were legislated before 1900, but significant examples include Yellowstone National Park in the USA (established in 1872) and Royal National Park in Sydney (established in 1879). The Australian Association for the Advancement of Science lobbied successfully for a national park in Western Australia in 1893, and in 1894 they succeeded in forming a large 65 000-hectare park in the centre of the Jarrah forest in Western Australia, but failed to have the park properly vested. The purpose of the park was reclassified to 'Timber – Government Requirements' in 1911, following pressure from the timber industry (Ride, 1975). Other smaller parks were created around the same time and were more permanent. Most of the northern Jarrah forest was dedicated as State forest in the late 1920s to protect it against uncontrolled clearing and unmanaged timber exploitation. The area of Jarrah forest in reserves was recently expanded under the Regional Forest Agreement process (Commonwealth of Australia, 2001a).

The first reserves in Australia were mostly proclaimed to protect the intrinsic or utilitarian values of a natural resource such as timber or water, to provide recreation opportunities, and to protect and allow for the observation of spectacular natural features. Reserves are still set aside for such reasons, although less commonly than in the past. The Yarra Ranges

Yellowstone National Park. (Photo by David Lindenmayer.)

National Park (in central Victoria) largely encompasses water catchments that have been closed to public access for many decades and are critical for the supply of water to Melbourne.

Most early reserves were in areas of limited economic importance and were acquired opportunistically (Recher and Lim, 1990; RAC, 1992b,c; Margules and Pressey, 2000). The conservation of biodiversity was rarely considered until recently (Chindarsi, 1997).

Why protected areas are important for biodiversity conservation

Lindenmayer and Franklin (2002) listed five reasons why reserves are important for biodiversity conservation:

- Large reserves support some of the best examples of ecosystems, landscapes, stands, habitats, biota, their inter-relationships and opportunities for natural evolutionary processes.
- Many species find optimum conditions within large reserves that become strongholds for these species.
- Some species are intolerant of human intrusions.
- Large reserves provide control areas against which the impacts of human activities in managed landscapes can be compared.
- The effects of human disturbance on biodiversity are poorly known and some impacts may be irreversible. Others, such as synergistic and cumulative effects, can be extremely difficult to quantify or predict. Large reserves are a 'safety net' against unanticipated ecological consequences, because they are relatively free from human disturbance.

Table 4.1. Protected areas in World Parks Congress in the four decades between 1962 and 2003 (from Chape *et al.*, 2003).

Year	Number	Area (millions of square kilometres)
1962	9214	2.4
1972	16 394	4.1
1982	27 794	8.8
1992	48 388	12.3
2003	102 102	18.8

A

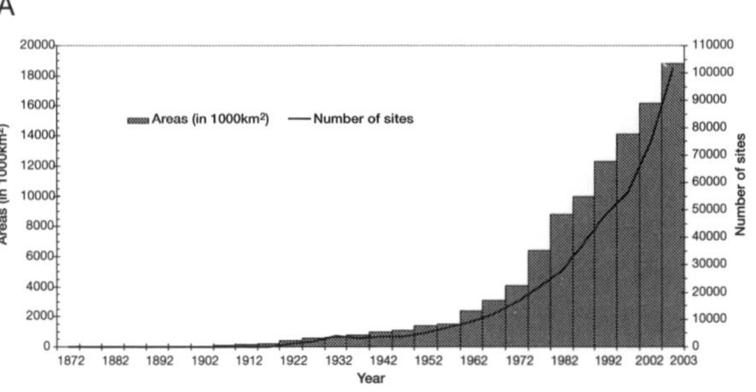

B

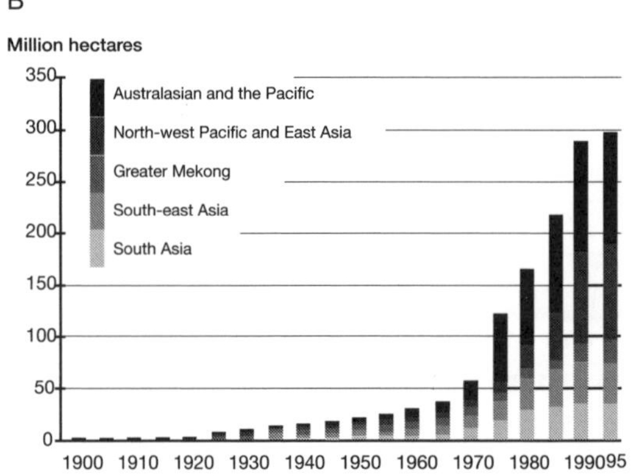

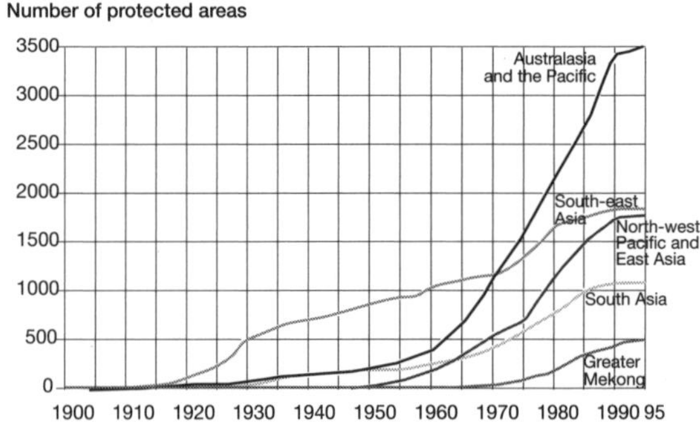

Figure 4.1. (A) Global increase in the number of protected sites and the size of the protected areas (based on data in the United Nations List of Protected Areas, 2003). (B) Increases in the size and number of protected areas in the Asia-Pacific region to 1995 (data for Australia are included in the Australasia category). (Redrawn from WCMC, 1998.)

The allocation of land to reserves is essentially an ecosystem approach to conservation. If representative samples of ecosystems and habitat types are maintained, reserves should maintain component species and ecosystem processes. This approach has the distinct advantage that it does not require detailed knowledge of the taxonomy, distribution or abundance of individual species. However, it is difficult to devise satisfactory habitat and ecosystem classifications, and to ensure the protection of threatened species, particularly when their distributions do not coincide with areas set up to protect ecosystems. Furthermore, the scarcity of resources makes it necessary to develop lists of priorities for land.

The marine reserve system around Heard Island. (Photo by Brett Melbourne.)

Box 4.1

Marine reserves: another part of the protected area system

Most discussions of protected areas focus on reserves on land. However, there has recently been a significant increase in the number of marine reserves set aside in many parts of the world, including Australia. Conservation biologists are also taking an increasing interest in setting aside marine reserves (e.g. Hooker *et al.*, 1999). In many cases the reserves are thought to be valuable in the protection of fish stocks from over-harvesting (Bohnsack, 1998; Ward *et al.*, 2001), which is the case for the Tasmanian Sea Mounts Marine Reserve (Pogonoski *et al.*, 2002).

The Australian Federal Government has 13 marine parks and reserves and a further 6 protected terrestrial areas that have an adjacent protected marine component to them (State of the Environment, 2001; Figure 4.2). These include large marine reserves such as the Great Australian Bight Marine Park (nearly 20 000 square kilometres). The 65 000-square-kilometre reserve around Heard and McDonald Islands, which is managed by the Australian Antarctic Division, is the largest protected marine area in the world (State of the Environment, 2001b; Department of Environment and Heritage, 2005d).

Some State governments have also set aside marine reserves. In June 2002, legislation was passed in the Victorian parliament to establish 13 marine national parks and 11 marine sanctuaries. Fishing will be prohibited or limited in these areas (Parks Victoria, 2002). There are also marine reserves in other States and Territories, including New South Wales, Queensland, Western Australia and the Northern Territory.

Reserve accumulation over time

The legally protected areas of the world increased by 80% between 1970 and 1990 (Groombridge and Jenkins, 2002), and by 100% between 1992 and 2003 (Table 4.2), with approximately two-thirds of the latter increase occurring in the developing world (McNeely *et al.*, 1990). The creation of the Great Barrier Reef Marine Park in 1980, which covers some 34 million hectares, was a significant contribution to the global protection of natural areas, as was the creation of the Greenland National Park. Figure 4.1 shows the increases in the size and number of protected areas in Australasia relative to the rest of the Asia-Pacific region until the mid-1990s (UNEP, 1999).

In mid-2005, the size of the terrestrial protected area estate in Australia was almost 77.5 million hectares, or

Port Campbell area in Southern Victoria – the location of a Marine National Park (administered by Parks Victoria). (Photo by David Lindenmayer.)

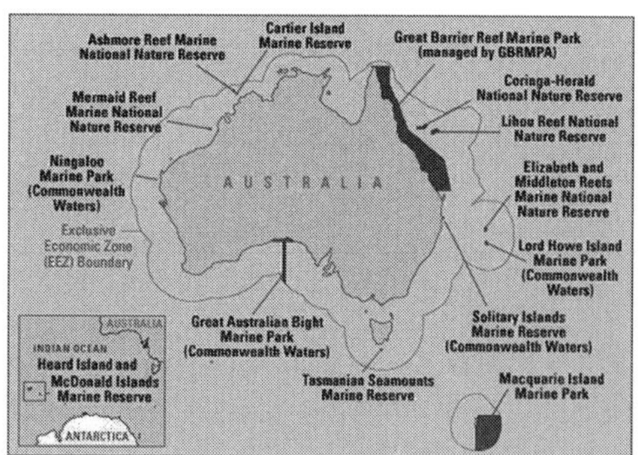

Figure 4.2. Australian marine areas as of mid-2003. (Redrawn from data in Department of Environment and Heritage, 2005f.)

approximately 10.1% of the Australian mainland (including Tasmania; Department of Environment and Heritage, 2005d). This area comprised 6755 individual reserves.

4.1 Categories of protected areas

The International Union for the Conservation of Nature (IUCN, 1984) defined eight categories of protected areas (Table 4.2). Many authors have added an additional three categories to the list of eight, representing areas recognised under international agreements for which conservation is an objective (e.g. Nelson, 1987; Amos *et al.*, 1993; Table 4.1). The nomenclature for protected areas varies in different places. For example, national parks in Canada do not include areas of human habitation, whereas all national parks in Pakistan and the United Kingdom contain human communities (McNeely *et al.*, 1990). The numbers and sizes of terrestrial areas in Australia in the first six IUCN protected area categories are shown in Table 4.3.

Other types of protected areas

Ramsar wetlands

The **Convention on Wetlands of International Importance** was signed in Ramsar (Iran) in 1971 and came into force in December 1975. The **Ramsar Convention** encourages international cooperation to conserve wetland habitats, especially for waterbirds. Wetlands are defined by the convention as areas of marsh, fen, peatland or water, whether natural or artificial, permanent or temporary, with water that is static or flowing, fresh, brackish or salt, including areas of marine waters, the depth of which at low tide does not exceed 6 metres (see Chapter 29 in Johnston, 1992).

As of mid-2005, there were 136 contracting parties to the Ramsar Convention, who have added more than

Table 4.2. Classification of protected areas (after Nelson, 1987; Amos *et al.*, 1993).

	Category	Purpose
I	Scientific reserves/strict nature reserves	To protect nature and maintain natural processes in an undisturbed state.
II	National parks	To protect natural and scenic areas of national or international significance for scientific, educational and recreational use.
III	Natural monuments/ natural landmarks	To protect and preserve nationally significant features because of their special interest or unique characteristics.
IV	Managed nature reserves/ wildlife sanctuaries	To ensure the natural conditions necessary to protect nationally significant species, groups of species, biotic communities or physical features of the environment by direct human management.
V	Protected landscapes	To maintain nationally significant natural landscapes characteristic of the harmonious interaction of humans and land while providing opportunities for recreation and tourism within the normal lifestyle and economic activity of these areas.
VI	Resource reserves	To protect the natural resources of an area for future designation and prevent or contain development activities that could affect the resource.
VII	Natural biotic area/ anthropological reserves	To foster the continuation of the way of life of societies living in harmony with the environment little disturbed by modern technology.
VIII	Multiple-use management area	To provide for the sustained production of water, timber, wildlife, pasture and recreation with the conservation of nature primarily oriented towards the support of economic activities.
IX	Biosphere Reserves	To provide areas for research, monitoring, training, and demonstration, as well as conservation.
X	World Heritage Sites	To conserve areas of natural and cultural value – those of 'outstanding universal value'.
XI	Wetlands of International Importance (Ramsar)	To conserve wetland habitats, especially for waterfowl.

Table 4.3. Number and size of terrestrial areas in Australia in the first six International Union for the Conservation of Nature protected area categories in 2003. (Data from Department of Environment and Heritage, 2005d.)

International Union for the Conservation of Nature category	Number	Area (hectares)
IA	2006	18 103 255
IA (Heritage Agreement Areas)	1193	564 682
IB	32	3 963 356
II	642	28 766 907
III	696	390 948
IV	1527	2 225 208
I–IV subtotal	**6096**	**54 014 356**
V	172	788 779
VI	452	22 635 792
V–VI subtotal	**624**	**23 424 571**
TBA*	35	23 024
Totals	**6755**	**77 461 951**

*Includes areas yet to assigned to particular IUCN categories.

1200 wetland sites to the Ramsar List of Wetlands of International Importance, which includes a total area of 108.7 million hectares (Department of Environment and Heritage, 2005e). Australia is one of the contracting parties and currently has 64 Wetlands of International Importance, which cover approximately 7.37 million hectares (Department of Environment and Heritage, 2005e; Figure 4.3). For example, Kakadu National Park is listed under the Ramsar Convention.

Although the Ramsar Convention is useful, managers may need to consider landscapes and catchments more broadly to maintain wetlands (Farrier and Tucker, 2000). For example, Box 7.2 on shorebirds in the Yellow Sea (see Chapter 7) highlights the problems of developing wetland and intertidal areas, and the over-exploitation of extensive river systems, such as the Yellow and Yangtze rivers, to which such areas are linked (Baxter, 2002). Similarly, the abatement of dryland salinity in Australia, which threatens many wetlands (State of the Environment, 2001f; Commonwealth of Australia, 2002b; see Chapter 5), will involve land management practices that extend well beyond wetland boundaries (Stirzaker *et al.*, 2002).

World Heritage Areas

The Convention Concerning the Protection of the World Cultural and Natural Heritage was adopted in Paris in 1972 and came into force in December 1975. The convention provides for the designation of areas of 'outstanding universal value' as **World Heritage Areas.** In 2003, the 754 properties on the World Heritage list included 582 cultural, 149 natural, and 23 mixed properties (UNESCO, 2003). In Australia there are currently 15 World Heritage-listed sites (Figure 4.4).

For 'natural' areas to qualify as World Heritage Areas, they must (in the words of UNESCO, 2003) meet at least one of the following conditions:

- 'Be outstanding examples representing major stages of earth's history, including the record of life, significant on-going geological processes in the development of landforms, or significant geomorphic or physiographic features'
- 'Be outstanding examples representing significant on-going ecological and biological processes in the evolution and development of terrestrial, fresh water, coastal and marine ecosystems and communities of plants and animals'

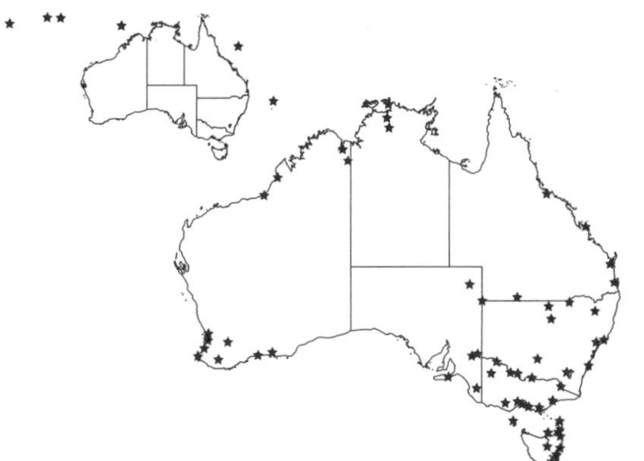

Figure 4.3. Location of Ramsar wetlands in Australia. (Redrawn from data in Department of Environment and Heritage, 2005e.)

Fraser Island, one of 15 World Heritage Areas in Australia. (Photo by David Lindenmayer.)

1 Heard and McDonald Islands
2 Macquarie Island
3 Tasmanian Wilderness
4 Australian Fossil Mammal Sites
  4a Naracoorte
  4b Riversleigh
5 Lord Howe Island Group
6 Central Eastern Rainforest Reserves of Australia
7 Willandra Lakes Region
8 Shark Bay, Western Australia
9 Uluru-Kata Tjuta National Park
10 Kakadu National Park
11 Fraser Island
12 Wet Tropics of Queensland
13 Great Barrier Reef
14 The Greater Blue Mountains Area
15 Purnululu National Park

Figure 4.4. Australia's 15 World Heritage Areas. (Redrawn from Department of Environment and Heritage, 2003).

- 'Contain superlative natural phenomena or areas of exceptional natural beauty and aesthetic importance'
- 'Contain the most important and significant natural habitats for in-situ conservation of biological diversity, including those containing threatened species of outstanding universal value from the point of view of science or conservation'.

Australia's *Environment Protection and Biodiversity Conservation Act (1999)* (EPBC Act) provides protection for areas of environmental significance, such as World Heritage Areas and Ramsar wetlands.

Biosphere reserves

Biosphere Reserves are the product of an international scientific program – the United Nations Educational, Scientific and Cultural Organization (UNESCO) Man and the Biosphere Program. The reserves are designated for research, monitoring, training, demonstration and conservation, and are managed in ways that combine conservation and sustainable use of natural resources. In November 2002, there were 495 Biosphere Reserves in 95 countries worldwide. In mid-2003, Australia had 13 Biosphere Reserves:

- Bookmark Biosphere Reserve (South Australia; see Box 4.2)
- Croajingolong National Park (Victoria)
- Fitzgerald River National Park (Western Australia)

Kata Tjuta, part of the Uluru–Kata Tjuta Biosphere Reserve. (Photo by David Lindenmayer.)

- Hattah-Kulkyne National Park (Victoria)
- Kosciuszko National Park (New South Wales)
- Macquarie Island World Heritage Area (Tasmania)
- Mornington Peninsula and Western Port (Victoria)
- Prince Regent Nature Reserve (Western Australia)
- South-West National Park (Tasmania)
- Uluru–Kata Tjuta National Park (Northern Territory)
- Unnamed Conservation Reserve (South Australia)
- Wilsons Promontory Marine Park and Marine Reserve (Victoria)
- Yathong National Park (New South Wales).

Box 4.2

Bookmark Biosphere Reserve

The Bookmark Biosphere Reserve was officially recognised and listed by the United Nations Educational, Scientific and Cultural Organization (UNESCO) in 1995. UNESCO is the overarching body responsible for Biosphere Reserves. Bookmark occurs in South Australia near the border with Victoria and New South Wales; it is large, and covers an area more than 6000 square kilometres – or three times the size of the Australian Capital Territory. The area supports three major landforms: the riverine environment associated with the Murray River, the mallee environment, and the semi-arid environment. The land tenures encompass private land, grazing leases, conservation reserves, and state forest, and there are accordingly a range of different land uses, including irrigated crops and other forms of horticulture. One of the aims of the Bookmark Biosphere Reserve is to identify approaches to ecologically sustainable development in what is essentially a low productivity landscape with a myriad of major conservation and land management problems (Brunckhorst, 1999). In some cases this aim has corresponded with a shift by some partners in the Biosphere Reserve away from extensive use of chemicals to organically grown crops (such as oranges). In others, it has involved diversifying agricultural industries to reduce land management impacts on landscapes as well as to reduce levels of water use.

Bookmark Biosphere Reserve has experienced problems such as those stemming from partners in the reserve who have differing aims. This has, in part, been tackled by initiatives from organisations such as the Department of Environment and Heritage, who have facilitated meetings to promote communication between partners (Howie, 2002).

4.2 Protecting communities and ecosystems

Many conservation areas have been created because of concerns for a single species or small groups of species (Chapter 2). However, a focus on the conservation of individual species suffers from the anthropocentric bias of taxonomy (see Chapter 2) and is susceptible to the bias of public perception and scientific preferences. More importantly, taxonomic knowledge is so poor for many groups that there is little hope of providing protection for most species in the foreseeable future, simply because we have no idea that they exist.

Alternative approaches to protecting species include protecting entire ecological communities (e.g. Specht *et al.*, 1995; Specht and Specht, 1999) or biogeographic provinces (Thackway and Olsson, 1999), thereby also attempting to protect both the known and unknown biodiversity associated with each community or province. These approaches are based on a hierarchical understanding of different classifications, which, in turn, reflect different spatial scales of biodiversity. Hence, at the continental scale, biomes and biogeographic provinces can be defined (e.g. Thackway and Olsson, 1999; Department of Environment and Heritage, 2005d), and within these, ecological communities are identified (Specht *et al.*, 1995). At finer scales of resolution, floristic or structural associations can be mapped (Specht and Specht, 1999). In the remainder of this section we examine two levels of classification and biodiversity protection within such a hierarchy: ecosystem types or biogeographic bioregions, and ecological communities.

Classifying and protecting ecosystem types: the Interim Biogeographic Regionalisation for Australia

In 1992, the then Prime Minister of Australia made a public commitment to conserve a representative sample of all major ecosystems (Keating, 1992). The National Reserve System Program was established to achieve this, which, in turn, led to the development of the bioregional framework called the **Interim**

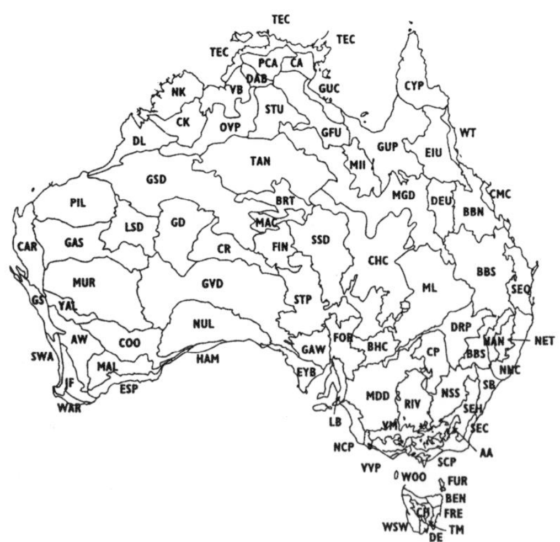

Figure 4.5. Map of Australia showing the various IBRA regions. (Redrawn from Thackway and Olsson, 1999; Department of Environment and Heritage, 2005d.) The key to codes is provided by the Department of Environment and Heritage (2005d).

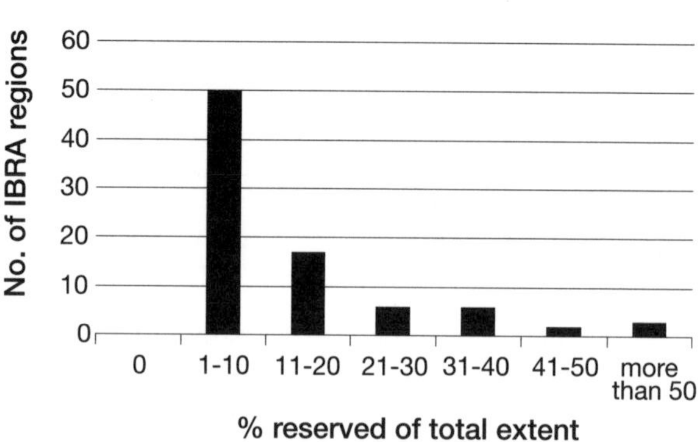

Figure 4.6. Representation of IBRA regions in protected areas. (Based on data from Department of Environment and Heritage, 2005d.)

Biogeographic Regionalisation for Australia (often termed the IBRA regions; Thackway and Creswell, 1995, 1997). The IBRA regions provide a framework within which planners and land managers can identify gaps in the reserve system, acquire new reserves, and implement other conservation-related strategies (e.g. the National Strategy for the Conservation of Australia's Biodiversity; Department of the Environment, Sports and Territories, 1996).

The basis for the bioregionalisation is that: 'key physical processes drive ecological processes, which in turn are responsible for driving the observed patterns of biological productivity and the associated patterns of biodiversity' (Thackway and Cresswell, 1995; Department of Environment and Heritage, 2005d).

Assigning areas to IBRA regions and identifying the boundaries between regions has been based on specialist ecological knowledge, and regional- and continental-scale data on climate, geomorphology, landform, lithology and characteristic flora and fauna. The latest version classifies 85 bioregions (Department of Environment and Heritage, 2005d; Figure 4.5).

The system also provides a subregional classification. In 2003, 354 subregions were recognised (Environment Australia, 2000), which were used in national studies such as the National Land and Water Resources Audit (Commonwealth of Australia, 2001b–e).

Protecting IBRA regions

The representation of IBRA regions in protected areas is highly variable and for many it is extremely poor. Data on IBRA regions in relation to the first six IUCN management categories (see Table 4.3) show that 46 of the 85 regions have protection levels of less than 10% of their total extent (Figure 4.6). Many (33) have less than 5% of their land cover in the six IUCN protected area categories.

The IBRA bioregional classification has a marine counterpart that is used nationally (State of the Environment, 2001a, 2001b) and in States (Parks Victoria, 2002) to guide marine reserve acquisition.

Classifying and protecting ecological communities

An ecological community can be defined as 'an integrated assemblage of native species that inhabits a particular area' (ESSS, 1995). Any assemblage of species that occupies the same place is often assumed to be 'integrated', and 'a particular area' refers to any geographic area defined by geology, landscape features, or geopolitical boundaries.

The EPBC Act (1999) protects threatened ecological communities. As of mid-2005, 30 threatened ecological communities were listed under the Act, of which two were critically endangered and the remainder were endangered (Department of Environment and Heritage, 2005b; see Table 4.4).

The criteria for the various categories of threat are similar to those for subspecies, species and populations as outlined in Chapter 3. That is, an ecological

Table 4.4. Critically endangered and endangered ecological communities in Australia under the *Environment Protection and Biodiversity Conservation Act* (1999). (Based on information in Department of Environment and Heritage, 2005b.)

Community	Status
Aquatic root mat community 1 in caves of the Leeuwin Naturaliste Ridge	Endangered
Aquatic root mat community 2 in caves of the Leeuwin Naturaliste Ridge	Endangered
Aquatic root mat community 3 in caves of the Leeuwin Naturaliste Ridge	Endangered
Aquatic root mat community 4 in caves of the Leeuwin Naturaliste Ridge	Endangered
Aquatic root mat community in caves of the Swan Coastal Plain	Endangered
Assemblages of plants and invertebrate animals of tumulus (organic mound) springs of the Swan Coastal Plain	Endangered
Blue Grass dominant grasslands of the Brigalow Belt bioregions (north and south)	Endangered
Brigalow dominant and co-dominant	Endangered
Buloke woodlands of the Riverina and Murray–Darling Depression bioregions	Endangered
Corymbia calophylla – Kingia australis woodlands on heavy soils of the Swan coastal plain	Endangered
Corymbia calophylla – Xanthorrhoea preissii woodlands and shrublands of the Swan coastal plain	Endangered
Cumberland Plain woodlands	Endangered
Eastern Stirling Range montane heath and thicket	Endangered
Eastern suburbs Banksia scrub of the Sydney region	Endangered
Grassy White Box woodlands	Endangered
Mabi forest (complex notophyll vine forest 5b)	Critically endangered
Natural temperate grassland of the Southern Tablelands of New South Wales and the Australian Capital Territory	Endangered
Perched wetlands of the Wheat belt region with extensive stands of living Sheoak and Paperbark across the lake floor (Toolibin Lake)	Endangered
Sedgelands in Holocene dune swales of the southern Swan coastal plain	Endangered
Semi-evergreen vine thickets of the Brigalow Belt (north and south) and Nandewar bioregions	Endangered
Shale/sandstone transition forest	Endangered
Shrublands and woodlands of the eastern Swan coastal plain	Endangered
Shrublands and woodlands on Muchea Limestone of the Swan coastal plain	Endangered
Shrublands and woodlands on Perth to Gingin ironstone (Perth to Gingin ironstone association) of the Swan coastal plain	Endangered
Shrublands of the southern Swan coastal plain ironstones	Endangered
Silurian limestone Pomaderris shrubland of the South East Corner and Australian Alps Bioregion	Endangered
Swamp Tea-tree forest of south-east Queensland	Critically endangered
Swamps of the Fleurieu Peninsula	Critically endangered
Community of native species dependent on natural discharge of groundwater from the Great Artesian Basin	Endangered
Thrombolite (microbial) community of coastal freshwater lakes of the Swan coastal plain (Lake Richmond)	Endangered

Box 4.3

Box–Gum woodlands – an endangered ecological community in New South Wales

The Box–Gum woodlands of the south-west slopes of New South Wales are mixtures of White Box, Yellow Box and Blakely's Red Gum, along with Apple Box, Grey Box, Candle Bark, Brittle Gum and Red Stringybark (Benson, 1991, 1999; NSW NPWS, 2002). The woodlands are listed under the *New South Wales Threatened Species Conservation Act (1995)* as an endangered ecological community. Since early European settlement (beginning around 1820), Box–Gum woodlands have been highly modified for agriculture and, because they were considered to indicate good soils, were preferentially cleared (Banks, 1997). As a result, Box–Gum woodlands on better soils and on undulating topography are some of the most highly altered vegetation types in Australia (Benson, 1991). More than 85% of the native vegetation has been cleared in the South-Western Slopes Bioregion, making it the most extensively cleared region of New South Wales (Benson, 1999).

community is **critically endangered** if 'at that time, it is facing an extremely high risk of extinction in the wild in the immediate future'. It can be listed under the **endangered** category if 'at that time, it is not critically endangered and is facing a very high risk of extinction in the wild in the near future'. Finally, it can be listed under the **vulnerable** category if 'at that time, it is not critically endangered or endangered, and is facing a high risk of extinction in the wild in the medium-term future'. As in the case of threatened subspecies, species and populations, nominations are forwarded to a panel of experts – the Threatened Species Scientific Committee (Department of Environment and Heritage, 2005b).

State and Territory governments also make provision for the listing and protection of communities. For example, the Victorian *Flora and Fauna Guarantee Act (1988)* and the *Threatened Species Conservation Act (1995)* in New South Wales (NSW National Parks and Wildlife Service, 2003a) list endangered ecological communities. Box 4.3 gives the example of the Box woodlands, an endangered community listed under the EPBC Act (1999; see Table 4.4).

Native grasslands, as found on the Southern Tablelands of New South Wales and the Australian Capital Territory in south-eastern Australia. This vegetation type is listed as an endangered community under the EPBC Act (1999). (Photo by David Lindenmayer.)

It is useful to know the prevalence of all communities in protected areas, not only threatened communities. For example, Specht *et al.* (1995) used their vegetation community classification to assess the adequacy of reservation of plant communities throughout the Australian continent. They based their classification on 650 ecological surveys, which varied in intensity from one to more than 15 surveys per one degree of latitude and longitude. Specht *et al.* (1995) used an ordination analysis to find groups of species that typically co-occur, resulting in 343 floristic groups. This information was used to define a continent-wide set of biogeographic regions. Further analysis resulted in a final classification with 921 floristic groups.

The distribution of each community was then overlaid on a grided map of Australia (at a scale of half a minute of latitude and longitude). Plant communities

Table 4.5. Conservation status of 921 Australian plant communities (after Specht *et al.*, 1995).

Vegetation	Total no. communities	Percentage with very poor or no protection within conservation reserves
Overstorey		
Rainforest	70	6
Monsoon rainforest (Northern Territory)	16	38
Dry scrub	29	16
Eucalypt open forest	237	17
Eucalypt communities (south-west Western Australia)	31	10
Heathland/scrub	83	21
Alpine	24	4
Grassland	23	48
Mallee	62	29
Desert Acacia	104	45
Hummock grassland	10	40
Chenopod shrubland	19	16
Forested wetland	29	17
Freshwater swamp	24	8
Coastal dune vegetation	16	6
Coastal wetlands	32	9
Understorey		
Heathy sclerophyll	21	14
Grassy savanna	39	54
Hummock grass	20	35
Humid and arid wetland	30	30

were assessed in relation to the number and size of conservation reserves in which they occurred, as well as their representation in reserves within each of the different biogeographic regions from which they were known. On this basis, plant communities were classified in reservation status categories including (after Specht *et al.*, 1995; Table 4.5):

- *Adequate*: occurs in several large reserves across all biogeographic regions in which it is known to occur.
- *Reasonable*: occurs in one large reserve or several small reserves in most of the biogeographic regions in which it is known to occur.
- *Poor*: found in a few small reserves and absent from some of the biogeographic regions in which it is known to occur.
- *Very poor*: found in only a few small reserves where human activities occur.
- *Nil*: the plant community is unreserved.

The best-protected ecological regions in Australia include coastal dune vegetation, alpine vegetation, moist rainforests, and some eucalypt communities in south-west Australia, all of which have around 60–70% of their community types conserved. Relatively few of the plant communities from these types are poorly conserved. In contrast, large proportions of the grassy savannah, grassland, hummock grassland, monsoon rainforest, and *Acacia*-dominated desert communities are either very poorly represented or absent from protected areas.

One of the most important results of the work by Specht *et al.* (1995) was to emphasise that protected areas are not uniformly distributed among ecological communities. This is, in part, because past reservation practices have been *ad hoc*, and driven by the economic value of the land (Braithwaite *et al.*, 1993; Pressey, 1995; see Chapter 16). For example, in the case of forests, although considerable work has been undertaken to expand the forest reserve system in Australia under the Regional Forest Agreement process (Commonwealth of Australia, 2001a), protected areas are still biased towards steep terrain on areas of low productivity (Pressey *et al.*, 2000; Lindenmayer and Franklin, 2002; Chapter 16).

Ecosystem types, vegetation communities and the adequacy of protection

It is tempting to assume that if major plant communities are conserved in a reserve system, then most plant species will also be conserved. The implicit assumption here is that communities and ecosystems act as **surrogates** for other attributes such as species occurrence. Chapter 16 discusses doubts about the usefulness of surrogate schemes. The effectiveness of protected areas depends on how they are distributed in space and in relation to the communities and species they are intended to protect. It also depends on the level of threat. Kirkpatrick and Brown (1991) and Amos *et al.* (1993) proposed minimum protection levels for Australian genotypes, species, and communities, which took into account the level of threat they faced, ranging from 30% reservation for disturbance-resilient species and communities, to 90% reservation for communities endangered by, and unlikely to recover from, disturbance or climate change (Table 3.3, Chapter 3). The underlying idea is that the degree of protection afforded to both species and communities should reflect the level of threat they face, and their tolerance of the disturbances generated by the human population.

If plant communities are adequately conserved, it may suggest that the conservation of all biota has been achieved, because animals depend on vegetation. Such assumptions need to be carefully and thoroughly tested in the field. There will be times when a plant community may not be a good surrogate for animal conservation, such as when an animal species requires a seral or successional stage within a plant community (e.g. early post-fire regeneration or old growth; see Chapter 2). Therefore, simply setting aside a proportion of a plant community (irrespective of its developmental stage) may not be sufficient to conserve other species that depend on that community (Lindenmayer and Cunningham, 1996).

The fact that existing reserves may not provide equivalent protection for all landforms or plant community types does not necessarily mean that the biodiversity of each of these elements is not protected. It may be that land-use practices in a particular region are such that there is no need for any strict conservation reserves. In fact, biodiversity can be adequately protected without excluding most human activities. Planning that involves natural, social and economic factors has a much greater chance of success in the long term, if success is counted in terms of the preservation of natural ecosystem processes, populations and communities.

The adequacy of a reserve system depends on the land-use practices that take place around it, the influence these practices have on the ecological composition and processes within reserves, and the ecology and sensitivity of the biota. Cooperation for conservation must extend beyond public boundaries and include protection systems for public and private land, incorporating the idea of both public and private stewardship (Thackway and Olsson, 1999; Maestas *et al.*, 2003). Such a system will allow protected areas to be integrated with other land uses, a fact that emphasises the importance of the following section on off-reserve conservation.

4.3 Off-reserve conservation

Extensive protected area systems cannot conserve all biodiversity (Anonymous, 1995; Dickson *et al.*, 1997; Maestas *et al.*, 2003). In Australia, the area of private land (which is managed primarily for uses other than conservation) is nearly three times larger than the area of public land (Australian Bureau of Statistics, 2001; see Chapter 19). In the case of forests, even if a reserve system includes 15% of the pre-European extent of each forest type in Australia, a large proportion of each vegetation community and its associated biodiversity will still remain outside the reserve system. Even for threatened species known to occur on reserves, many important populations may not be on protected lands. For example, of all the Australian plant species considered to be endangered, vulnerable or rare in 1996, about half were represented on conservation reserves. At that time, only 40% of the endangered plants in Australia were recorded from protected areas, and only eight species of endangered plants were represented by populations of more than 1000 individuals in protected areas. These factors indicate that off-reserve conservation is critical for the conservation of biodiversity. Other reasons for the importance of off-reserve conservation are outlined in the following section.

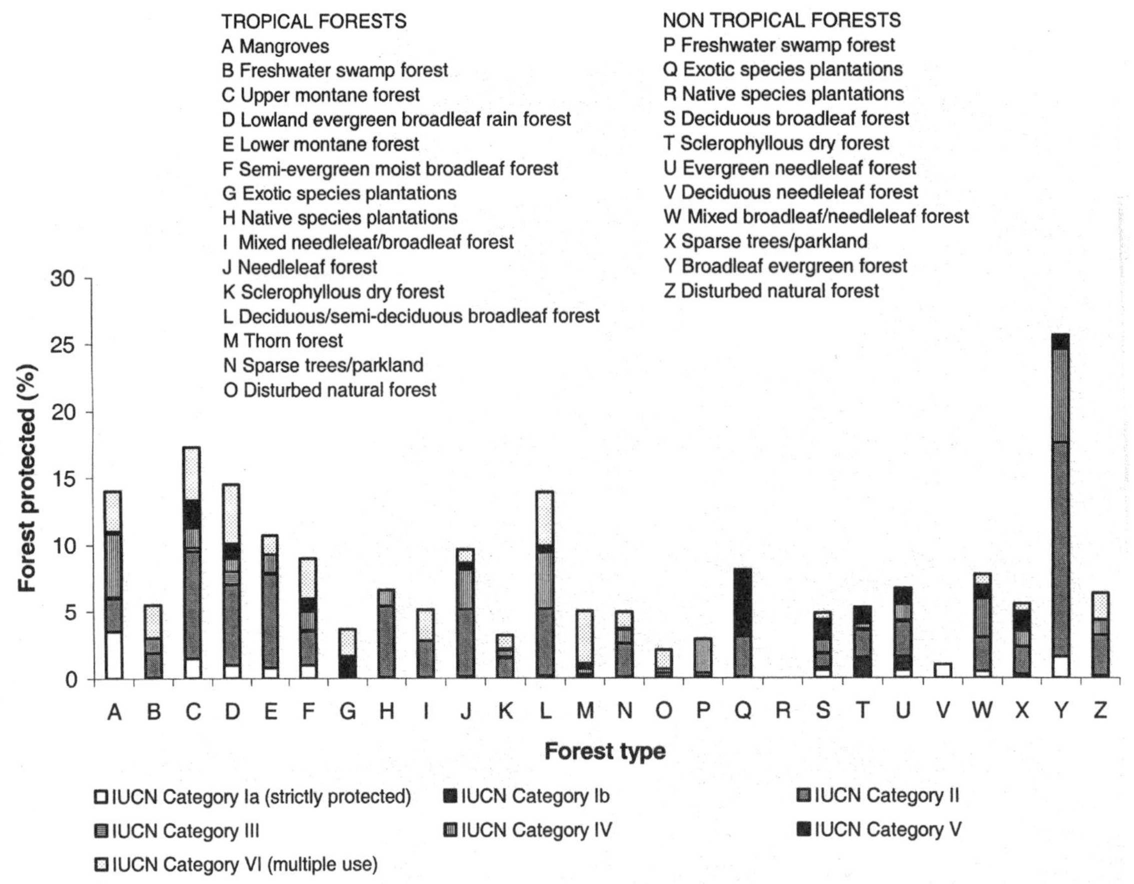

Figure 4.7. Worldwide levels of protection of broad forest types (from Commonwealth of Australia, 1999).

Limitations of a reserve-only focus for biodiversity conservation

Lindenmayer and Franklin (2002) discussed the significant limitations of a reserve-only focus for biodiversity conservation. They include:

- The area available for reserves is limited.
- The size of most reserves is limited.
- Many reserves are on steep terrain or infertile soils, which are not always suitable habitats for many elements of the biota.
- There are social and economic impediments to the expansion and management of reserve systems.
- Mobile taxa such as migratory or nomadic species and species with patchy distributions are not contained in conventional reserves.
- Abiotic and biotic conditions within reserves are unstable.
- Human exploitation in the surrounding off-reserve areas may intensify once reserve systems are established.

Limited area available for reserves

The limited area available for forest reserves is a significant impediment to a reserve-only focus for biodiversity conservation. For example, as of 1998, approximately 38×10^6 square kilometres or about 8% of the world's forest was reserved in strictly protected areas (Commonwealth of Australia, 1999; Figure 4.7).

The IUCN has set reservation targets of 10% for each biome on a global basis (Scientific Advisory Group, 1995; Groombridge and Jenkins, 2002). Targets adopted for reserved areas, such as 10% of the land base are arbitrary – they were not developed using scientific criteria (Scientific Advisory Group, 1995; Soulé and Sanjayan, 1998). In fact, such levels are generally viewed as inadequate, and perhaps as much as 50% of a region and even 100% of some ecosystems may need to be reserved (Noss and Cooperrider, 1994), particularly those that are rare (Mendel and Kirkpatrick, 2002). For example, Kirkpatrick and Fowler (1998) argue that in the case of the now highly relictual alpine communities of Tasmania, which cover <2% of that State, 100% reservation is appropriate. The same would be appropriate for alpine communities on the Australian mainland that are highly restricted and cover only a limited area. For example, the alpine 'windswept feldmark' community covers less than 30 hectares in total. In other cases, such as extremely extensive ecosystems that are relatively species-poor (e.g. polar icefields), 10% protected area may be greater than is needed for biodiversity conservation, although protection may be important for other reasons, such as to maintain experiential and spiritual values.

Windswept alpine feldmark (fore- and mid-ground) in Kosciuszko National Park. (Photo by David Lindenmayer.)

Some species demand ecological attributes that are difficult to provide, even in large reserves. For example, even in the very extensive Yellowstone National Park of central USA, populations of the Bald Eagle appear to depend, in part, on the maintenance of off-reserve populations (Swenson *et al.*, 1986).

Even in the rare cases where reserve targets of 30% or more are achieved for a particular region or vegetation type (such as in Tasmania; see Mendel and Kirkpatrick, 2002), the majority of land and its associated biodiversity will still be unreserved, and reserve levels may still be insufficient for particular taxa. For example, approximately 20% of the montane Ash forests of the Central Highlands of Victoria has been set aside as reserves (Commonwealth of Australia and Department of Natural Resources and Environment, 1997). Despite this level of protection, it is possible that recurrent disturbances such as wildfire will extinguish the endangered Leadbeater's Possum from large ecological reserves (Lindenmayer and Possingham, 1995).

Limited reserve size

Although small and medium-sized reserves have significant conservation value (Zuidema *et al.*, 1996), they may be less likely to support viable populations of some species in the long term (East, 1981; Wilcove, 1989; Armbruster and Lande, 1993; Gurd *et al.*, 2001). Even small species, such as some invertebrates, may be viable only in large numbers (Thomas, 1990; Tscharntke, 1992)

Swift Parrot – a species dependent on off-reserve conservation. (Photo by Chris Tzaros.)

or need large areas for long-term persistence (Økland, 1996; Didham, 1997).

Protecting isolated reserves is not sufficient to maintain resident populations (Baillie *et al.*, 2000) that depend on supplementation from populations in adjacent lands, as is the case for metapopulations of butterflies in the United Kingdom (Thomas *et al.*, 1992) and for the Northern Spotted Owl in north-western USA prior to the adoption of current forest plans (Yaffee, 1994; Lindenmayer and Franklin, 2002).

Small reserves are a problem for mobile species that require resources that vary in time and space (reviewed by Law and Dickman, 1998), and that occur both within and outside reserves (Redford and de Fonseca 1986; Woinarski *et al.*, 1992; Mac Nally and Horrocks, 2000). Examples include Australian forest birds (Woinarski and Tidemann, 1991) such as pigeons and honeyeaters (Keast, 1968; Date *et al.*, 1996), as well as fruit bats (Palmer and Woinarski, 1999). For example, populations of rainforest pigeons in northern New South Wales require different resources in different seasons, and the resources they need occur in spatially separated habitats, both within and outside reserves (Date *et al.*, 1991). Only 2% of the nesting habitat of the Swift Parrot occurs in conservation reserves; the remainder is on private land and publicly owned production forests (Brereton, 1997). The long-term conservation of the species depends almost entirely on management actions outside reserves (Brown and Hickey, 1990; Swift Parrot Recovery Team, 2000).

Bias in environmental and other conditions within reserves

Many reserves are on steep terrain or infertile soils, or occur in high elevation areas, which are not always suitable habitats for many elements of the biota (e.g. Crumpacker *et al.*, 1988; Johnson, 1992; Khan *et al.*, 1997; Scott, 1999; Margules and Pressey, 2000). This is because the reserves were set aside for reasons other than nature conservation (Khan *et al.*, 1997; Lindenmayer and Franklin, 2002), such as:

- low value for commodity production or human settlement (Chindarsi, 1997)
- high value for recreation (Sax, 1980; Pouliquen-Young, 1997)
- high scenic and aesthetic values (Recher, 1996).

Strzelecki National Park on Flinders Island – an example of a steep and rocky 'residual' area that is unsuitable for grazing or other land uses (which dominate the surrounding off-reserve areas). (Photo by David Lindenmayer.)

In the majority of countries, including Australia, most of the productive and often biodiverse land has been converted to agricultural land, urban environments and managed forests. Reserve systems in North America, Europe, Australia, New Zealand and Asia are typically low-productivity areas (Braithwaite *et al.*, 1993; Hunter and Yonzon, 1993; Pressey, 1995; Norton, 1999; Figure 4.8). In north-western North America, highly productive sites were already occupied by private landowners more than a century ago, and so they are poorly represented in reserves. Similar circumstances exist in Scandinavia (Virkkala *et al.*, 1994). In Chile, more than 90% of the protected land is concentrated at latitudes above 43°, and areas of highest biodiversity have virtually no representation in the reserve system (Armesto *et al.*, 1998).

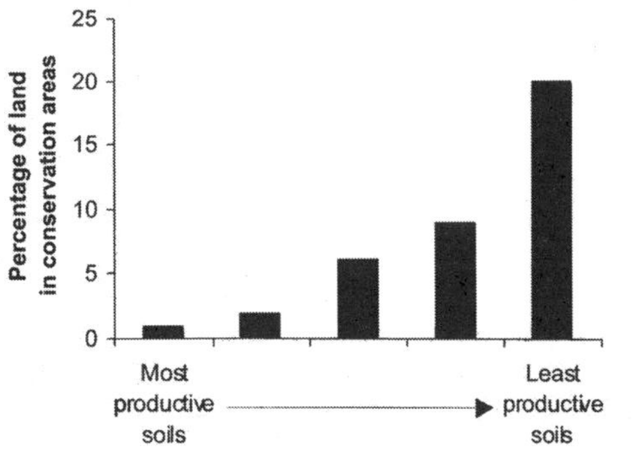

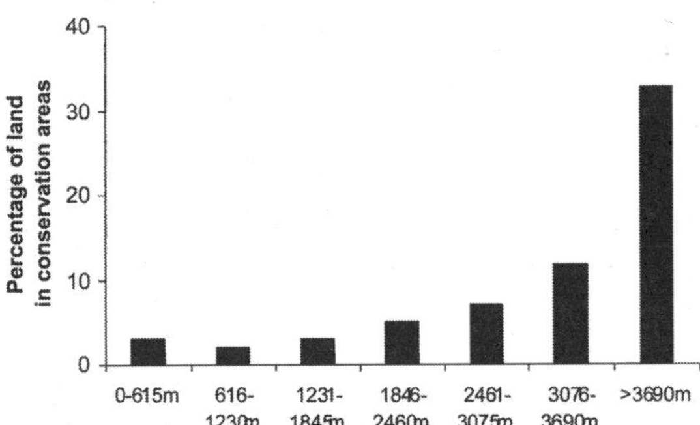

Figure 4.8. Bias in the reserve system in the USA toward areas of high elevation and low-fertility soils. (Redrawn from Scott *et al.*, 2001a.)

The bias towards low productivity environments has implications for conservation. Theoretical and empirical evidence suggest that species diversity is often (although not always) positively associated with productivity (Harris, 1984; Srivastava and Lawton, 1998). Low productivity areas can support marginal-quality habitat for many species (Taylor *et al.*, 1972; Braithwaite, 1984), and several studies have demonstrated that the population sizes of forest-dependent animals are lower in such areas (e.g. Lindenmayer *et al.*, 1991a, 1999c).

Impediments to expanding reserve systems

In many cases there are limited opportunities to expand the size of reserves and diversify the range of environmental conditions captured within them (such as higher productivity areas). Land that is privately owned and is valuable for other uses can be difficult and expensive to add to a reserve system. For example, in the Great Lakes region of Wisconsin, USA, simply absorbing the remaining old growth forest stands within a large ecological reserve was not feasible because of demands for wood production and existing land tenures (Mladenoff *et al.*, 1994). In the monsoon rainforests of northern Australia, it was politically untenable to reserve and manage all representative samples (Price *et al.*, 1995, 1999). In this situation, a combination of strategies was recommended that included both monsoon rainforest reserves and off-reserve management in savannah landscapes surrounding the rainforest patches.

Difficulties in capturing taxa with fine-scale or patchy distribution patterns in reserve systems

Even extensive reserve systems can fail to adequately capture species-rich assemblages with fine-scale or restricted distributions. Local biodiversity hotspots or

Box 4.4

When a formal reserve system won't work: the case of the Superb Parrot

The Superb Parrot is a medium-sized (130–160 grams), fast-flying and highly mobile bird that forages over large areas. There appear to be two populations: (1) a generally resident population that breeds in the riparian River Red Gum woodlands of the Murray–Murrumbidgee valleys, and (2) a migratory population that breeds in the remnant White Box/Yellow Box/Blakely's Red Gum (Box–Gum) woodlands of the south-western slopes of New South Wales (Webster, 1988; Manning, 2004).

The Superb Parrot breeds in modified open woodland habitat in agricultural areas. It nests in tree hollows, and threats to its persistence include clearing and modification of its nesting and feeding habitat (Webster 1988; Webster and Ahern, 1992). However, the species lives on some of the most productive privately-owned land in Australia (Davey, 1997; Manning, 2004) and this, together with the extensive movement patterns exhibited by the species means that formal reservation strategies would be almost impossible for the Superb Parrot. Critical conservation strategies such as the protection of nest trees, and stimulating the natural regeneration of new stands of woodland (much of which is currently subject to dieback and natural mortality) will be critical to the long-term persistence of this species (Manning, 2004).

Superb Parrot – a species dependent on off-reserve conservation management. (Photo by Adrian Manning.)

areas that are critical for species persistence (e.g. over-wintering or calving grounds; Hansen and Rotella, 1999) might be omitted. The conservation of velvet worms (phylum Onychophora; see Box 2.6 in Chapter 2) in southern Australia illustrates this problem. Many onychophoran species have patchy and often highly limited distributions, and there are many **cryptic species** (P. Sannucks, personal communication). Onychophorans have poor dispersal ability and depend on rotting log habitats in wet forests (Barclay *et al.*, 1999), making them susceptible to disturbances such as prescribed burning and timber harvesting (New, 1995). Setting aside reserves for these and many other species-rich assemblages with restricted distributions is problematic, as there are invariably species confined to unreserved land. For onychophorans, forestmanagement strategies within logged stands, such as retaining logs within clumps of intact vegetation and using low-intensity prescribed fires for site preparation after timber harvesting, can have considerable conservation value (Forest Practices Board, 1998).

Instability of conditions in reserves

Natural ecosystems are dynamic – species composition and distribution patterns change over time (e.g. Woinarski *et al.*, 1992; Margules *et al.*, 1994a; Witting and Loeschcke, 1995; Rodrigues *et al.*, 2000). Reserves that assume species distributions are static underestimate conservation requirements. For instance, Margules *et al.* (1994a) discovered that the set of areas needed to conserve all taxa on natural limestone pavements in northern England in 1985 no longer supported all species in 1994, just 9 years later (Table 4.6). If the limestone

Table 4.6. Plant species turnover on limestone pavements in Yorkshire (United Kingdom) over a 9-year period (modified from Margules *et al.*, 1994a).

Change in no. species per pavement	No. pavements
No change	7
Net loss of species	
1–5	21
6–10	10
11–15	10
>15	5
Net gain of species	
1–5	10
6–10	9
11–15	2
>15	0

pavements were treated as reserves and no other populations existed, the network of reserves identified in 1985 would not have conserved several plant species. Studies of birds (Rodrigues *et al.*, 2000) and plants (Virolainen *et al.*, 1999) produced similar results.

To mitigate these problems to some extent, Rodrigues *et al.* (2000) suggested focusing on rare species and selecting the best areas for them. However, it is usually not possible to select the best areas, because they are used by humans for other purposes, for instance agriculture, forestry and urban settlements. Moreover, this approach still ignores the landscape context of reserves and, hence, the contribution of unreserved areas to biodiversity conservation.

Most reserves are disturbed from time to time by events such as wildfires. Thus, if reserves do not exceed the maximum size of disturbed areas, managers should expect species loss (Pickett and Thompson, 1978; Hobbs and Huenneke, 1992). Larger reserves have a greater probability of supporting areas that escape the effects of a single catastrophic event such as a fire (Seagle and Shugart, 1985; Baker, 1992; McCarthy and Lindenmayer, 1999; Norton, 1999).

It is expected that the ranges of many species will change as a result of global climate change (Peters and Lovejoy, 1992; Parmesan, 1996; Thomas *et al.*, 2004). In fact, many may disappear from reserves in the near future (Peters and Darling, 1985; Beaumont and Hughes, 2002; Scott *et al.*, 2002; see Chapter 5). Habitat in unreserved areas will become important as refuges and dispersal pathways.

Intensification of exploitation and downgrading the conservation value of unreserved land

It is possible that an expanded system of protected areas will shift additional harvesting pressures onto the remaining unreserved areas (Davie, 1997). There are signs of such problems in some Australian forests, where there are policies of harvesting intensification in unreserved areas following the establishment of new reserves (Bauhaus, 1999; Lindenmayer *et al.*, 2002b; see Box 4.5).

Small habitat areas within unreserved areas may make valuable contributions to conservation (Schwartz, 1999; see Franklin *et al.*, 1997; Hale and Lamb, 1997; Mac Nally and Horrocks, 2000). For instance, in South American forests, Armesto *et al.* (1998) argued that priority should be given to managing forest remnants in intensively managed landscapes that currently have high levels of unprotected biodiversity. Many other examples highlight the importance of small areas of habitat in unreserved lands (Angelstam and Petersson, 1997; Fries *et al.*, 1998; Semlitsch and Bodie, 1998; McCoy and Mushinsky, 1999), including numerous examples from Australia (Kirkpatrick and Gilfedder, 1995; Prober and Thiele, 1995; Law *et al.*, 1999; Lindenmayer *et al.*, 1999c;

Box 4.5

Intensification of management in the matrix: the land allocation problem in Tasmania

Expanding plantation forestry in Tasmania highlights the trade-offs between reserve systems and managing unreserved land for biodiversity conservation. Here, the conservation movement focused on increasing the size of the reserve system, particularly those forests with high 'wilderness value' (see Box 1.5 in Chapter 1). Under the Regional Forest Agreement for Tasmania, the size of the reserve system increased by approximately 400 000 hectares, and now encompasses 40% of the forest in the State. The trade-off was that up to 85 000 hectares of native eucalypt forest will be cleared for exotic plantations of Radiata Pine and fast-growing eucalypts from mainland Australia, which will be managed on a 15–30-year rotation (Forestry Tasmania, 1999). Between 1996 and 2001, almost 3% of Tasmania's wet eucalypt forests were converted to plantations (Resource Planning and Development Commission, 2002). The percentage of the State's forests managed for plantations will increase from 2.8 to 5.1% in the next 10 years and up to 20% of a given biogeographic province could be converted to plantations. Although the reserve system is now more representative than it was previously, the areas targeted for plantation establishment are mainly in the north-west and north-east of Tasmania, which are areas already fragmented by agriculture and existing plantations.

The focus on expanding the reserve system in Tasmania neglected the importance of unreserved areas, with potential for detrimental impacts on biodiversity values (Smith, 2000), particularly for the 114 forest-associated threatened species in Tasmania. A better conservation outcome may have been achieved by improving the management of unreserved areas and foregoing increases in the reserve system.

Abensperg-Traun and Smith, 2000). The habitat remnants in all these examples are not candidates for inclusion in ecological reserves because of the land tenure status, or because they are too small, too spatially dispersed, or are embedded in production landscapes.

If the conservation values of unreserved areas are considered, it can have important social benefits. For instance, in northern New South Wales, Barrett *et al.* (1994) and McIntyre and Barrett (1992) recommended improving the management of woodland remnants and other small habitat patches within agricultural lands, rather than securing a few large ecological reserves. Focusing only on large ecological reserves would have concentrated efforts on approximately five bird species that depend on large and relatively undisturbed areas of forest. In contrast, conserving remnant areas in unreserved agricultural land helped protect a larger number of species and signalled to private landowners that biodiversity conservation could be enhanced without evicting them from their land and returning it to a natural state (Barrett *et al.*, 1994).

Impacts of external factors on conditions within reserves

Reserves are susceptible to impacts from processes in the surrounding landscape. For example, the changes in hydrological regimes associated with land-use practices can alter patterns of stream flow and groundwater levels (Strizaker *et al.*, 2002), which can, in turn, influence both aquatic and terrestrial ecosystems that are 'protected' within a reserve. Weeds, feral predators and competitors, stock trespass, changed fire regimes, recreational activities, and pathogens have detrimental impacts on the biota of many Australian reserves.

Western Swamp Tortoise. (Photo by Arthur Georges.)

Box 4.6

Conservation strategy for the Western Swamp Tortoise

Simply setting aside a reserve may not guarantee the long-term survival of a species. The Western Swamp Tortoise was discovered in Australia in 1839, but no more specimens were seen until 1953. The species is the only member of its genus (Cogger, 1995) and may possibly be the only living representative of the subfamily Pseudemydurinae (Kuchling *et al.*, 1992). It is one of Australia's most endangered reptiles (Cogger, 1995), and it has a highly restricted distribution (covering about 100–150 square kilometres) near Perth. Two swamps in which the Tortoises were found were reserved, one of 65 hectares and the other 155 hectares. However, the Western Swamp Tortoise disappeared from the larger reserve during the 1980s and by the late 1980s only a few dozen animals remained in the smaller swamp (Kuchling *et al.*, 1992). The species was probably always restricted to temporary swamps in the Swan Valley of Western Australia, and this habitat has been extensively modified for agriculture and human settlement (Kuchling *et al.*, 1992). When swamps dry out, tortoises **aestivate** in holes or under deep leaf litter, where they are vulnerable to Red Fox predation. Populations also have very low reproductive rates.

A captive breeding program was initiated for the Western Swamp Tortoise in 1988, which substantially increased the numbers of animals. In the mid-1980s there were about 30 animals remaining; in 2001, there were 110 animals, of which only 25 were adults. A total of more than 170 tortoises have been successfully reared in captivity (Threatened Species Network, 2003). Other recovery initiatives included feral predator control, reserve expansion, and habitat rehabilitation (Kuchling *et al.*, 1992; Burbidge and Kuchling, 1994).

The species that are most likely to persist in reserved habitats are those for which suitable conditions have been maintained in surrounding areas (e.g. Diamond *et al.*, 1987; Franklin, 1993a,b; Lindenmayer and Franklin, 2002). Management of off-reserve areas for multiple uses (including biodiversity conservation) can strongly influence reserve design and the long-term effectiveness of protected areas (reviewed by Lindenmayer and Franklin, 2002). This also highlights the importance of resource management across different land-use tenures and the need for cooperation between different government agencies as well as between managers of private and public lands (Craig *et al.*, 2000).

Barriers to off-reserve conservation

There are significant social and political barriers to effective off-reserve conservation (Whitten *et al.*, 2002). The attitude of many Australians to private land is that all decisions concerning its use are the business of the owner alone, and, inevitably, most decisions are driven by economics. Planning guidelines provide some constraints, but management of private land for the public good is not a strong sentiment. Management of private land for conservation is even less well developed, although there are important public and private initiatives in this area, including LandCare, Bushcare, Greening Australia, Land for Wildlife, and the Trust for Nature. Box 4.7 describes a scheme in Victoria that appears to have been successful in promoting biodiversity conservation on private land and that has landholder support.

Farmland area with paddock trees. Private lands are a major part of the Australian landscape, and can support important conservation resources, which are sometimes managed poorly. (Photo by David Lindenmayer.)

Box 4.7

A biodiversity credits scheme: the BushTender trial

Biodiversity conservation on private land is a major challenge in all parts of the world (Hilty and Merenlender, 2003). In Victoria, an auction-based system was developed to allocate conservation contracts on private land – the BushTender scheme (Department of Natural Resources and Environment, 2002). This approach pays private landholders to enter into contracts to undertake management to improve the quality or area of native vegetation on their land. Basically, the system works by landholders contacting the government to express interest in the BushTender scheme. A field officer then assesses the significance and quality of their land. Landholders then identify management activities they will undertake, prepare a management plan, and submit a bid that outlines the payment being sought from the Victorian State Government. Bids are then judged on the current conservation value of a site, the amount of service offered by a landholder, and the cost. Successful bidders enter into an agreement with the State Government and receive regular payments on the basis of work completed (Department of Natural Resources and Environment, 2002).

The BushTender Trial commenced in 2001 and spanned two regions: one between Bendigo and Ballarat and the other between Wangaratta and Wodonga. A total of 73 landholders made successful bids and A$400 000 was allocated. Many new populations of rare and threatened plant species were identified (S. Berwick, personal communication) and the scheme was rated highly by participating landholders. The scheme has now been expanded to other regions.

Summary: off-reserve conservation

Substantial conservation benefits can be achieved by taking into account the conservation values of off-reserve areas, including productive forests, agricultural landscapes and urban landscapes. Reserves and off-reserve management should compliment one another. Lindenmayer and Franklin (2002) noted that where reserves and appropriate off-reserve management are established, (1) the overall amount of land for biodiversity conservation will increase, (2) biodiversity conservation will occur across a broader and more representative set of environments, and (3) processes threatening reserve effectiveness and that emanate from off-reserve areas will be reduced. Despite this, few discussions of reserve design take the importance of off-reserve conservation into account. The reserve design algorithms outlined in Chapter 16 do not deal explicitly with the question of how likely it is that the target species will persist outside the reserve, or how susceptible they are likely to be to environmental impacts on adjacent land.

4.4 Botanic gardens and zoos

There are numerous facilities devoted to biodiversity conservation, including zoos, botanic gardens, aquaria, seed storage and microbial storage facilities (IUCN, 1998; Linington and Pritchard, 2002; Lyles, 2002). Although many zoos and botanic gardens have active conservation programs, their primary function is to show animals and plants. Hence, many zoo and garden species are selected on the basis of a strong anthropocentric bias. Motivations aside, *ex situ* conservation provides new ecological knowledge about some species and provides some insurance against inadvertent or unavoidable extinctions. Captive breeding, cultivation and reintroduction are important components of many recovery programs.

Species may be in cultivation or captivity to maintain a living museum of common species, provide for the conservation of threatened species, or protect germ plasm resources. The global distribution of these institutions is related to economic status to some degree: in 1992, there were 532 botanical gardens in Europe, but 82 in Africa and 66 in South America. At that time, 573 zoos and aquaria were in developed countries, representing 65% of all such institutions (WCMC, 1992).

Botanic gardens

There are approximately 1600 botanic gardens worldwide, and they attract more than 15 million visitors annually (Mackey, 2003). In the early 1990s, cultivated material for more than 20 000 plant species, or roughly 8% of the world's vascular species had been collected (WCMC, 1992). However, attempts are underway to maintain much larger proportions of the world's flora in *ex situ* establishments. Pullin (2002) noted that the Kew Gardens in London hopes to collect 24 000 species (more than 10% of the world's seed-bearing plants), including all the species from the United Kingdom.

Approximately half of Australia's threatened plants may be in botanic gardens (Table 4.7). However, because populations in gardens are necessarily small, such a representation is a relatively small sample of the genetic variation present in the wild population. Box 6.2 in Chapter 6 outlines ways to increase levels of genetic variability when collecting seeds from plants – either for *ex situ* conservation or other activities such as replanting schemes.

Zoos

Most zoos have their origins as menageries for public entertainment and education. In the mid-1980s, more than 3000 vertebrate species were being bred in zoos and other captive facilities worldwide, comprising about 540 000 individual animals (Conway, 1986). By 2000, there were more than 181 accredited zoos in North America alone, containing approximately 800 000 individuals (Lyles, 2002). Many zoos make considerable effort to act as mechanisms for conservation (as genetic and demographic reservoirs, and insurance against loss of wild populations; Miller *et al.*, 2004). For example, the zoos of San Diego, Chicago (Brookfield), New York, Washington DC and Frankfurt (among others) have important field research and conservation programs that support *in situ* management (McNeely *et al.*, 1990; Redford *et al.*, 2002; Miller *et al.*, 2004).

Table 4.7. Examples of threatened taxa known in cultivation in botanic gardens (after WCMC, 1992).

Region	No. threatened taxa surveyed	No. known in cultivation in botanic gardens	Percentage
Macronesia	557	419	75
China	338	211	62
New Zealand	230	129	56
South Africa	1051	514	49
Australia	1867	893	48
Europe	1723	558	32
USA	3324	890	27
India	927	105	11
Cuba	874	55	6

Most Australian zoos play an active role in conservation, in addition to heightening awareness about conservation issues. For example, captive-bred animals have been used in many reintroductions; an example of this is the Dibbler in Western Australia (Moro, 2003).

Despite the valuable contributions that zoos make, the vast majority of captive animals are not directly important for conservation purposes. For example, even though more than 600 mammal species are considered to be threatened globally (Groombridge and Jenkins, 2002; IUCN, 2003), in 1990 only 20 500 individuals from 140 threatened mammal species were held in zoos (Olney and Ellis, 1991), representing 10% of the global zoo capacity at that time (Magin *et al.*, 1994). Only nine threatened mammal taxa had captive populations exceeding 500 individuals and only a further 14 had captive populations exceeding 250 individuals. For groups other than mammals, there are much smaller proportions of the threatened biota in captivity (chapter 34 in WCMC, 1992; Pullin, 2002). This is clearly

Box 4.8

Zoos and the rescue of the Lord Howe Island Stick Insect

Lord Howe Island Stick Insects are large, with adults growing to 15 centimetres long. Adults are a deep black colour, but juveniles are green and turn black as they age. Both juveniles and adults are flightless. The species was thought to have become extinct after rats were introduced to Lord Howe Island from a ship that ran aground in 1918.

A scientific expedition by the New South Wales Department of Environment and Conservation in 2001 discovered a tiny population surviving on a *Melaleuca* shrub on Ball's Pyramid, a spectacular 550-metre-high granite plug 23 kilometres from Lord Howe Island. Conservation biologists from Melbourne Zoo and the New South Wales National Parks and Wildlife Service returned in February 2003 and brought back two pairs of animals. One of the females began laying eggs as soon as she arrived.

One of the pairs died, but the pair taken to the Melbourne Zoo survived. Shortly after she arrived in Melbourne, the female became so ill that she was comatose and near death. The zoo's veterinary team and her keeper, Patrick Honan, tried a number of solutions before she revived. The first hatching, of an egg laid in February, was in September 2003. Several more eggs hatched and several juvenile animals were already several centimetres long in November 2003.

The ultimate goal of this program is to breed sufficient numbers to allow the reintroduction of animals to Lord Howe Island. Success will depend on first controlling rats on the island.

Ball's Pyramid near Lord Howe Island (photo by Ian Hutton) and the Lord Howe Island Stick Insect. (Photo from Australian Museum)

because, as outlined in Box 4.9, there are major limits to the *ex situ* capacity of zoos, even large zoo systems such as those in North America.

Studbooks are records used to keep track of all captive individuals of a taxon of conservation concern. Since 1974, the studbook system has been supplemented by the International Species Inventory System, which provides information on the sex and age distribution of captive populations, and the births and deaths within them (Lyles, 2002). Together, these two systems can be used to manage the demographic and genetic composition of captive populations worldwide through the exchange of animals between zoos (e.g. Myroniuk, 1995).

Until recently, most zoos relied on natural populations to provide a source of specimens – an approach that has the potential to deplete wild stocks (Rabinowitz, 1995). Greater reliance is now placed on captive breeding, and in 1992 about 90% of mammals, 71% of birds and the majority of reptiles were captive-bred (WCMC, 1992).

In captivity, characters important for persistence in the wild can be lost, in favour of features that lead to successful persistence in captive conditions (see Chapter 6). For example, captive breeding can select for 'tameness', which is useful in a captive situation but not for a wild existence. Furthermore, captive breeding is expensive (Fischer and Lindenmayer, 2000), and

Box 4.9

The capacity of *ex situ* conservation: North American snake conservation

There are many examples of where animals from a captive population have been successfully restored to the wild (see the review by Griffith *et al.*, 1989; Fischer and Lindenmayer, 2000). However, careful consideration needs to be given to the extent to which the maintenance of captive populations is feasible for providing a 'bank' of individuals from which to attempt to subsequently restore wild populations (Soulé *et al.*, 1986).

Perhaps the first question to ask is: How many species could be maintained in captivity? Quinn and Quinn (1993) examined the ability of the North American zoo system (the largest and most extensive in the world) to support captive breeding programs for endangered snake species under Species Survival Plan (SSP) programs. They surveyed North American zoos to determine the space that could be dedicated to housing snakes in an SSP program. Quinn and Quinn (1993) calculated that, within the zoos participating in SSP programs for snakes, there were more than 3000 snake enclosures. To examine how individuals, populations and species could be partitioned among this resource, Quinn and Quinn (1993) then made some general assumptions concerning captive populations of snakes under SSP programs, including:

- a generation time of about 15 years for each species
- a population growth rate of about 15% per annum
- a total of 26 animals in each founding population
- continuing the breeding program for a period of 100 years
- ensuring that 90% of the original levels of genetic variation were maintained.

Quinn and Quinn (1993) then applied a computer simulation model to predict what was required to maintain a captive population according to these criteria. As with all analyses, their calculations were based on a number of assumptions, including the size of snakes (and thus the dimensions of cages needed to house them), the availability of wild animals to found captive populations, the snakes' ability to breed in captivity, and opportunities to effectively display animals. Their results indicated that, on average, a total of 49 enclosures, each containing two individuals, would be required to maintain a given species. Given this, they concluded that between 13 and 16 species of snakes could be maintained under SSP programs within the current network of zoological institutions in North America.

The choice of which species to include in captive breeding programs is another problem. For example, Quinn and Quinn (1993) considered that the 16 captive populations could comprise one large species (e.g. a species in the boid family) and 15 smaller ones from a wide range of other snake families to maximise taxonomic diversity and thus biodiversity captured within the zoo collections. However, the selection would be subject to many other factors, such as opportunities to capture wild animals and the husbandry of captive animals. Sheppard (1995) completed a parallel study for birds in American zoological institutions; she calculated that there was room for only 141 long-term management programs.

Quinn and Quinn (1993) made recommendations about ways to increase the 'carrying capacity' of the zoological system in North America, but even if these recommendations were followed, they would be unlikely to substantially increase the number of species that could be housed in this way. The results of their work and the study by Sheppard (1995) demonstrate that although captive breeding and associated subsequent reintroductions can be a potentially important strategy in conservation, the ability of such approaches to stem the current losses in biodiversity is limited. Even if it was financially possible to support many captive breeding programs, there is simply insufficient room to house even a tiny proportion of the very large number of the world's declining species.

depends on ongoing policy, government funding and community support.

It is not always possible to breed animals successfully in captivity. Particular environmental, sexual and other cues for reproduction can be absent. In other cases, approaches that were successful in the past can later fail. For example, captive-breeding programs for Leadbeater's Possum have been highly successful in the past, but now, for unknown reasons, the possums rarely reproduce in captivity.

Captive Leadbeater's Possum – a species that formerly bred successfully in captivity but now rarely does so. (Photo by David Lindenmayer.)

4.5 Gene banks and storage facilities

Like zoos and botanic gardens, storage facilities usually have a strong economic mission. Most seed storage facilities, for instance, are dedicated to protecting varieties and wild relatives of agricultural and economically important plants (Linington and Pritchard, 2002). Wild populations are a rich source of novel genetic material that can serve to protect important food crops from new pathogens and other threats (see Chapter 1; Vavilov, 1949; Groombridge and Jenkins, 2002). *Ex situ* storage facilities also provide insurance against inadvertent or unavoidable extinctions (as do zoos and gardens).

Plants can be maintained in seed banks, pollen stores (although this requires female plants to pollinate), and germ plasm collections. Plant storage can be accomplished in several different ways, and which method is most suitable depends on the reasons for storage, the number of species, and the technology and resources available for the task. Storage of genetic material is also possible for animals (at the embryonic or gametophyte stage) but at greater cost. Beyond the involvement of zoos, relatively little effort has been expended on the *ex situ* conservation of animal genetic resources. The Food and Agriculture Organization set up its Animal Genetic Resources Program in 1982, which aims to establish databases, identify endangered breeds, establish gene banks to store semen and embryos, and administer genetic resource management. The program is aimed at endangered breeds and those with economic potential.

Field gene banks

Seeds can be classified as 'orthodox' (those that can be stored for long periods at sub-zero temperatures if their moisture contents are reduced below 10%) or 'recalcitrant' (those that cannot survive low temperatures or dehydration). The seeds of many species cannot survive for even a year; in fact, about 20% of the world's vascular plants have recalcitrant seeds (McNeely *et al.*, 1990). In particular, many tropical species produce seeds with no natural dormancy. Therefore, species such as rubber, palms and many tropical fruits must be maintained in **field gene banks.** Field gene banks are areas of land in which collections of growing plants have been assembled, in populations containing as wide a sample of genetic variation as possible. For example, the Malaysian Agricultural Research and Development Institute maintains plant gene banks in remote areas of the country. A fruit tree gene bank might have as many as 40 varieties of fruit in a field and six trees of each variety. The Food and Agriculture Organization (1996 in Linington and Pritchard, 2002) estimated that at that time there were more than 525 000 accessions in field gene banks around the world. Field gene banks have been established for tropical crop species and for tropical and temperate trees that are important for forestry (e.g. Wood and Burley, 1983). They are particularly appropriate for long-lived perennial trees and shrubs that cannot be adequately conserved in the wild, and which may require decades to produce seed. Field gene banks are also an effective means of conserving species that reproduce vegetatively.

Seed banks

Seed banks are effective for the conservation of sexually reproducing species whose seeds are suitable for long-term storage (Linington and Pritchard, 2002). Seeds are usually small, and each has a different genetic makeup, so a collection can carry a wide range of variation. Cold storage facilities are available in most parts of the world, and they can be used to conserve many valuable species for periods of at least a generation (in terms of the plant species). The method is relatively cheap and space efficient, and because of its extensive use, seed storage provides a mechanism for the exchange between facilities of potentially valuable genetic material (Bonner, 1990). Practical constraints include a dependence on power supply (for cool storage), on continuous monitoring of seed viability, and on regeneration or stock replenishment if seed viability falls. One problem with the method is that genetic changes in the form of chromosomal aberrations and the selective deaths of genotypes can occur during storage (Bonner, 1990).

Orthodox seeds can be conserved for very long periods at sub-zero temperatures, if previously dried to about 5–8% moisture content. The longevity of seeds increases with decreasing storage temperature, from around 5–25 years at 0–5°C, to around 100 years at –10°C to –20°C. About one-third of the world's botanic gardens surveyed by the World Conservation Monitoring Centre had facilities for seed storage and handling, and at least 144 gardens had low-temperature seed storage facilities (WCMC, 1992).

As of July 2003, the Millennium Seed Bank at Kew Gardens in London held 6655 identified species in its cold store, representing 15 613 collections. This represents approximately 2.8% of the world's flora, and 97% of the United Kingdom flora (Royal Botanic Gardens, 2003).

International Agricultural Research Centres have been active in the international coordination of activities concerned with plant resources, particularly gene banks. In 1990, there were 13 such centres worldwide, most of which had specific responsibilities in crop variety development and germ plasm conservation. They covered such species as Rice, Wheat, Potato, Maize, Barley, Cassava, Sorghum, Yam and Soyabean. Crop Genetic Resource Centres and the International Board for Plant Genetic Resources have developed about 60 gene banks with storage facilities, almost all for crop plants. For plants of global economic value such as Wheat, Maize, Oats, Rice, Sorghum, Millet and Potato, more than 90% of the variation among different races has been preserved in *ex situ* collections (McNeely *et al.*, 1990). Of these predominantly seed bank accessions, 95% lack any evaluation data, such as responses to germination tests, and extensive data are held on only 1% of specimens (McNeely *et al.*, 1990).

There are about 6 million plant accessions in more than 1300 gene banks around the world. Of the stored accessions, 60% are in long-term or medium-term facilities, 8% in short-term facilities, and the remainder are in field gene banks. A few are stored using techniques such as cryopreservation. Only 15% of plants in seed banks are wild or weedy plants (FAO, 2003b). The wild relatives of commercially important plants can provide genetic material for maintaining or improving crop production, and for the acquisition of resistance to pathogens and herbivores (Groombridge and Jenkins, 2003). For example, more than 20 wild species have contributed genes to Potatoes. Wild relatives of Wheat, Potato, Tomato and Maize have been extensively collected and preserved in seed banks.

The Australian Tree Seed Centre has extensive seed collections of several Australian genera, including *Eucalyptus*, *Acacia*, *Casuarina* and *Grevillea*. Some of the research at this facility has targeted the maintenance of genetic diversity of particular species (e.g. the endangered Nepean River Gum) and the centre has assisted in the development of ways to conserve a threatened provenance of the River Red Gum (Arnold and Midgley, 1995).

By 2003, the United States Department of Agriculture had recorded nearly half a million accessions from 10 359 species, most with some economic importance, in germ plasm collections (USDA, 2003). They hold collections for Cattle, Sheep, Pigs, Goats, Asses and Horses, as well as invertebrate and microbial culture collections.

In vitro storage

In vitro storage refers to the storage of species in laboratory conditions. For germ plasm storage, *in vitro* plants are usually initiated from meristem tips, buds or stem tips and propagated through division into test tubes. Plants can be stored under a variety of conditions, generally at low temperatures (–3°C to –12°C) to create slow growth conditions and increase the storage period. This method is expensive and labour intensive, and subculturing is usually necessary after 6 months to 2 years. In 1992, about 500 threatened taxa worldwide were stored *in vitro* (chapter 34 in WCMC, 1992). However, the approach can be valuable for protecting key genetic resources if, for example, events such as cyclones or wildfires destroy field gene banks, as occurred on the Caribbean island of St. Lucia (Linington and Pritchard, 2002).

Plants can be stored indefinitely using cryostorage techniques, which involves storing seeds or other plant material in liquid nitrogen at –196°C. The process has been used for many years to store animal semen for breeding. Such techniques have the potential to reduce space and labour requirements over seed storage and other *in vitro* storage techniques. In 1992, eight countries had cryostorage capability in one facility, and at that time only a small number of species had been successfully preserved in this way (chapter 34 in WCMC, 1992). Cryostorage has the potential to extend the storage life of seeds far beyond that of conventional storage, and the likelihood of genetic damage can be reduced. Recalcitrant seeds remain problematic (Bonner, 1990).

The importance of plant conservation is particularly pronounced in Western Australia, where more than 60% of Australia's plant extinctions have occurred and where there are a very large number of threatened taxa (Dixon, 1994; Hobbs and Mooney, 1998). At Kings Park and Botanic Garden in Perth, Western Australia, standard methods such as seed collection and storage, as well as cutting and grafting are used. However, these methods are not suitable for all taxa. Cryostorage is employed for species that consist of very few individuals, for which there is little available seed, and for which there are few immediately promising reintroduction opportunities. The cryostorage technique shows great potential, but procedures for its use are still undergoing development. For example, thawing plant material is not uniformly successful. In some cases, successful propagation of plants held in cryostorage can also involve the storage of other organisms essential to the health of the species, such as symbiotic fungi (Dixon, 1994). Work at Kings Park and Botanic Garden on the recovery of Purdie's Donkey Orchid, a rare and endangered species from south-west Western Australia, involved several steps, including isolation of the symbionts that are essential for germination and seedling development, development of symbiotic *in vitro* propagation, establishment of symbiotic seedlings in pots, transfer of seedlings to sites inoculated with the symbiotic fungus, and monitoring of the success of the reintroductions (Dixon, 1994).

Ex situ conservation of microbial diversity

Collection and management of microbial diversity has been conducted in a largely *ad hoc* fashion (Crozier *et al.*, 1999). Collections of permanently preserved living cultures of micro-organisms are coordinated by the World Federation of Culture Collections. In 1992, there was a total of 345 culture collections distributed among 55 countries. Fifty of these collections were in Australia. The majority maintain a limited number of strains (mostly less than 1000) and are narrowly focused (e.g. only plant pathogenic bacteria, or *Rhizobium*, or human pathogens). However, several institutions have very large collections of bacterial strains. Linington and Pritchard (2002) noted that two based in the USA, one in Illinois and the other in Massachusetts, maintain more than 78 000 and 53 000 microbial strains, respectively.

The United Nations Educational, Scientific and Cultural Organization (UNESCO) sponsors the Microbial Resources Centers (MIRCENs) Network. There are 23 MIRCENs in 19 countries, some of which have their own regional network of collaborating laboratories. The aim of the network is to preserve microbial germ plasm, although the focus is often in the areas of agriculture, biotechnology or informatics (MIRCEN-Stockholm, 2003).

4.6 Reintroduction, translocation and captive breeding

Some populations are closely managed or held in captivity to re-establish natural populations following population crashes and imminent extinction (IUCN, 1998; Lyles, 2002). There are many examples of species that have collapsed to a handful of individuals in the wild, after which the remaining individuals have been taken into captivity. An example is Shevolski's Horse, which was reduced to 12 animals in Prague Zoo in 1912, and then recovered to a current population of about 700. Another is the Arabian Oryx, which was reduced to eight individuals in London Zoo, but recovered to more than 400 individuals, although poaching of re-established wild populations has re-emerged (Ostrowski *et al.*, 1998).

Translocation and reintroduction are being used increasingly as management strategies for the conservation of species, and in 1998 the IUCN updated their set of broad guidelines for such types of conservation efforts. The IUCN (1987) defines reintroductions as the:

> *intentional movement of an organism into part of its native range from which it has disappeared or become extirpated as a result of human activities or natural catastrophe.*

The intention of reintroduction programs is to establish viable populations (IUCN, 1998).

Types of relocations

Many terms are used to describe the deliberate movement of animals and plants between locations (Viggers *et al.*, 1993); however, their use is often not consistent

(Fischer and Lindenmayer, 2000). Konstant and Mittermeier (1982) assigned the methods of transferring populations to three categories: **introduction**, **translocation** and **reintroduction**.

Introduction

Introduction involves the accidental or deliberate movement of animals to previously unoccupied areas. The Common Brushtail Possum was introduced to New Zealand in 1840 (Pracy, 1962; Montague, 2000), where it now causes serious environmental damage (Cowan, 1990; Kerle, 2001). Amphibians are often introduced to areas beyond their natural range in shipments of fruit (O'Dwyer *et al.*, 2000). Chapter 7 discusses examples and consequences of introductions to Australia.

Translocation

The transfer of plants and animals from one part of their range to another is known as translocation (Kleinman, 1989). Populations of the Common Brushtail Possum are sometimes translocated from suburban Melbourne to areas outside the city (Pietsch, 1995; see Box 4.12). Notably, this definition of 'translocation' is somewhat different from that developed by the IUCN (1987).

Reintroduction

Reintroduction is the release of captive-bred animals or herbaria-raised plants and/or wild-caught animals or plants. In the case of animals, Lindburg (1992) recognised two types of reintroduction. The first is

Box 4.10

Captive breeding and the reintroduction of the Mala

The Mala or Rufous Hare-Wallaby is one of a group of hare-wallabies that have undergone dramatic declines since European arrival in Australia. The Mala used to be one of the most abundant and widespread macropods in central and Western Australia (the species' range encompassed almost one-third of the continent) and it was frequently observed by early European explorers in these areas (Johnson, 1988). However, by the early 1980s, the distribution of the Mala was confined to two tiny populations in the Tanami Desert in the Northern Territory. The mainland subspecies is now extinct in the wild after the two populations in the Tanami desert were lost to predation by feral animals, drought and wildfire (Gibson *et al.*, 1995). Two Western Australian islands (Bernier Island and Dorre Island) support separate subspecies of the Mala.

A captive breeding program for the Mala based on wild-caught animals commenced in 1980 and this proved to be extremely successful – there are now captive colonies in the Northern Territory, New South Wales, South Australia and Western Australia. Animals from these programs have been proposed for use in reintroduction programs to re-establish wild populations of the mainland subspecies.

A 'Mala Paddock' was established in the Tanami Desert in 1986. It is on Aboriginal land and, as a result, the recovery program for the species has required close cooperation between conservation biologists and the local indigenous people. Notably, the restoration of Mala populations is considered to be an important cultural goal for the traditional Aboriginal land owners, and more than 300 Aboriginal people have participated in the project (Gibson *et al.*, 1995). There are many other components of recovery efforts for the Mala, including an assessment of potential release sites, habitat management in areas where captive populations occur, clarification of the taxonomy of the species and subspecies, and additional captive breeding.

Recovery efforts for the Mala have been expensive and labour-intensive, and have required considerable long-term commitment of resources (Langford, 1999). However, a truly wild and self-sustaining population of the Mala has yet to be established. Training the Mala to recognise and become cautious of predators such as the feral Cat and the Red Fox could be a valuable component of re-establishment programs (McLean *et al.*, 1996). However, the control of feral animals, particularly the feral Cat and the Red Fox, appears to be an essential element in the long-term success of efforts to establish reintroduced populations of not only the Mala, but possibly other native mammals of similar size in arid Australia (Gibson *et al.*, 1995). This highlights the fundamental importance of addressing the processes that led to a decline of a target species in the first place (Caughley and Gunn, 1996; Fischer and Lindenmayer, 2000).

Mala. (Photo by Jiri Lochman.)

re-establishment reintroduction, in which captive-bred individuals are used to re-establish an extinct wild population. Populations of the Mala or Rufous Hare-wallaby were reintroduced into the Tanami Desert (Northern Territory) using animals bred in captivity at Alice Springs (Gibson *et al.*, 1995; see Box 4.10). Lindburg's second form of reintroduction is **restocking reintroduction**, in which a declining population is supplemented with captive-bred stock. The declining population for the Orange-bellied Parrot in Tasmania (see Box 7.3 in Chapter 7) was restocked by adding captive-bred animals.

Prevalence of translocation and reintroduction

Translocation and reintroduction are often advocated to slow a rapid rate of species decline and extinction, and they are often part of threatened species management (Backhouse *et al.*, 1995; Department of Environment and Heritage, 2005a). For example, the release of captive animals is listed as part of the recovery process of about half the threatened species listed under the Australian Federal Government's EPBC Act (1999; State of the Environment, 2001a). However, translocations and reintroductions can be employed for other reasons. These include: (1) restocking an area with animals for hunting, (2) re-establishing a species that has particular cultural significance, as in the case of the Mala for the Walpiri people of the Tanami Desert (Gibson *et al.*, 1995; Box 4.10), and (3) resolving human–animal conflicts such as shifting aggressive Australian Magpies out of suburban Brisbane (Jones and Finn, 1999; see Fischer and Lindenmayer, 2000; Figure 4.9).

There have been several thousand attempts throughout the world to translocate or reintroduce populations

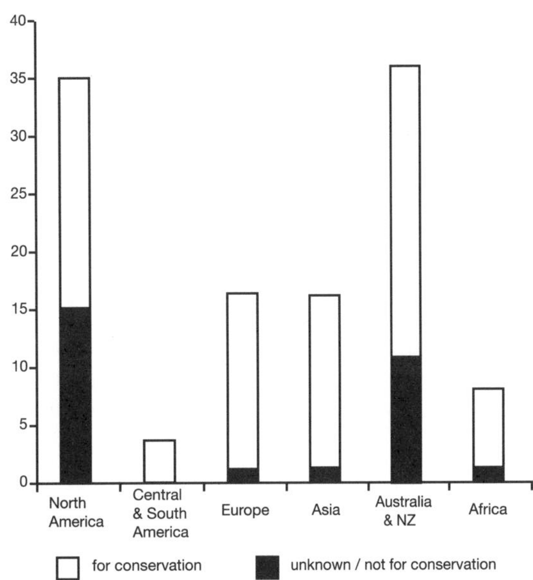

Figure 4.9. Purposes of animal relocations. (Redrawn from Fischer and Lindenmayer, 2000.)

(Wolf *et al.*, 1996; Fischer and Lindenmayer, 2000). Griffith *et al.* (1989) calculated that more than 700 attempts were undertaken annually, mostly in the USA and Canada. Fischer and Lindenmayer (2000) found that birds and mammals accounted for 93% of the animal relocations reported in 12 leading scientific journals between 1979 and 1998, and the remaining 7% were amphibians, reptiles and invertebrates. A provisional survey by the WCMC (1992) indicated that between 1980 and 1990, more than 220 plant reintroduction projects had been conducted worldwide, involving 29 different plant families.

Translocations, reintroductions and former ranges of species

A usual requirement for translocations of vertebrates in Australia is their release into areas of their former range (see Serena, 1995). However, there are examples where endangered animals have been successfully established in areas outside their former known distributions. For example, several species of New Zealand birds and Australian mammals have been established successfully on offshore islands where they previously did not occur (Short *et al.*, 1992; Johnson *et al.*, 2003; Moro, 2003). In other cases, the vegetation of the target area may need to be manipulated to create floristic and/or structural conditions that are suitable for the species being reintroduced or translocated (Vietch, 1995). Translocations beyond the former range of a species need to be carefully considered, as they can sometimes have unintended negative impacts. For example, local populations of freshwater shrimps were deliberately moved between different subcatchments of the same drainage system in south-eastern Queensland, resulting in an inappropriate mixing of genotypes (Hughes *et al.*, 2003).

Effectiveness of reintroduction and translocation strategies

Griffith *et al.* (1989) found that translocation of game species constituted 90% of translocations worldwide, and that the translocations had a success rate of 86%. Translocations of threatened species made up the remainder, and their success rate was much lower, just 44%. In another study, Dodd and Siegel (1991) found that only 19% of translocations of reptiles and amphibians were successful. Fish and plant translocations and reintroductions are also a high risk strategy (Hall, 1987; Henrickson and Brooks, 1991; Maunder, 1992; Fischer and Lindenmayer, 2000); many of the plant reintroduction projects examined by the WCMC (1992) were unsuccessful. In a recent detailed appraisal of 180 published case studies, Fischer and Lindenmayer (2000) noted that few studies provided useful measures of success or failure (Figure 4.10), making it difficult to ascertain the outcome of population re-establishment efforts.

Why reintroductions and translocations fail

There are many reasons why reintroductions and translocations fail. In the case of plants, factors such as poor horticultural practice, a lack of understanding of the ecological requirements of the target species (such as symbiotic dependencies; Dixon, 1994) and limited post-planting maintenance and monitoring can contribute (WCMC, 1992; Pavlik *et al.*, 1993).

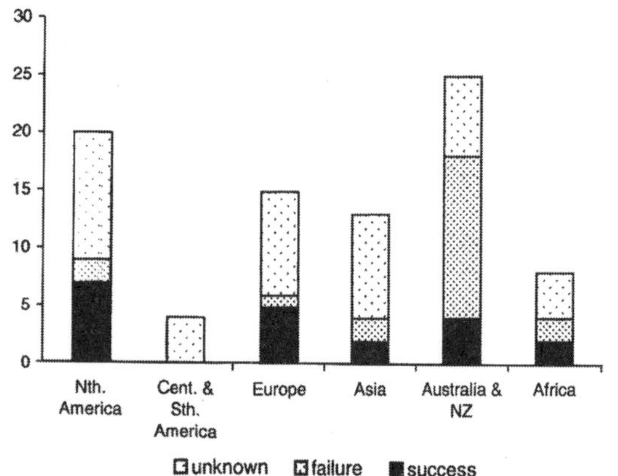

Figure 4.10. Success of translocation and reintroduction programs examined by Fischer and Lindenmayer (2000). Note the large number of cases where the outcome was unknown.

It is critically important for all reintroductions and translocations that the initial reasons for the decline of the population are rectified (Fischer and Lindenmayer, 2000). For example, 31 young captive-bred Malleefowl were released into remnant vegetation in central New South Wales, but all died within 107 days. The release attempt was compromised by predators, particularly the Red Fox (Priddel and Wheeler, 1994). Caughley (1994) recommended the use of experiments involving trial or 'probe' groups of animals to ascertain if the causes of the original population decline have been

Box 4.11

How to improve the success of translocation and reintroduction programs

Translocation and reintroduction programs can be expensive and risky. For example, captive-breeding and reintroduction efforts for the Brush-tailed Rock Wallaby were severely hampered when a wildfire in 2003 in the Australian Capital Territory destroyed animal enclosures and killed all but a few of the individuals they contained. Fischer and Lindenmayer (2000) reviewed translocation and reintroduction efforts and made several recommendations for improving success rates. These were: (1) examine carefully the appropriateness of reintroduction and translocation before a program is instigated, (2) develop rigorous and widely accepted criteria for gauging success or failure, (3) remove the process that caused the species to decline in the first place, (4) in the case of animals, where possible use a wild source of individuals for release, (5) where possible, release a large number of individuals (>100), (6) employ a robust procedure to monitor populations following release, (7) ensure appropriate measures of financial accountability are in place (so that the costs of reintroduction and translocation effort can be assessed and their opportunity costs relative to other recovery measures can be gauged), and (8) report the results of reintroduction and translocation efforts in the scientific literature – both successes and failures – so that other workers benefit from the experience.

Table 4.8. Some general considerations in the design of reintroduction and translocation programs.

Problem	Comments
Disease	Diseases in wild populations can have severe effects on released animals or plants, especially those bred ex situ that have no immunological defences. Similarly, wild populations can be susceptible to diseases carried by released individuals (Viggers *et al.*, 1993; Langdon, 1990).
Genetics	Translocated animals and plants can hybridise with previously isolated subpopulations or populations. Inbred individuals may have impaired survival ability on release (Jimenez *et al.*, 1994; Horowitz, 1995).
Predator recognition	Captive-bred animals may not recognise natural or exotic predators and be relatively susceptible to predation on release (McLean *et al.*, 1995; Priddel and Wheeler, 2004).
Threatening Processes	The original cause of the decline of the population needs to be addressed prior to release; otherwise there is no reason to believe the population will not face the same fate as its predecessor (Short *et al.*, 1992; Gibson *et al.*, 1995; Caughley and Gunn, 1996).
Competition	Conspecifics may kill, drive away, or otherwise exclude released animals (Pietsch, 1995).
Size and Connectivity	Habitat arrangement may influence the size of the released population and/or the ability of released individuals to move between habitat patches (Lindenmayer, 1995).
Climatic Conditions	Climate may influence the physiological tolerance of species (e.g. survival ability of offspring) or susceptibility to environmental factors such as disease (Templeton, 1986).
Demography	The size and composition of the release group should be such that it avoids demographic problems associated with small population size and unusual age or stage structures. Group size and composition may affect factors such as social interactions, foraging, defence, behaviour of pollinators, or Allee effects (Burgman *et al.*, 1995a; McCallum *et al.*, 1995; Sigg et al., 2005).
Habitat	The physical and ecological conditions at the release site should be evaluated to ensure that it provides appropriate substrates, shelter, nest sites, germination conditions or other components of the species niche (Griffith *et al.*, 1989; Pearce *et al.*, 1994; Reading *et al.*, 1996; Fischer and Lindenmayer, 2000).
Impacts on other species	Reintroductions of a given species can have impacts on other taxa, including threatened species (e.g. freshwater fish in North America; Williams *et al.*, 1988, in Noss and Cooperrider, 1994).

mitigated. In some cases, the process leading to the decline of a species may not have been determined, and a pilot release program using a relatively small number of animals may be required to establish it (Armstrong *et al.*, 1995; Danks, 1995).

The wide array of considerations in Table 4.8 highlights the need for careful planning, execution and monitoring of reintroduction and translocation programs. In the case of the endangered Bridled Nailtail Wallaby – a species confined to a small area of central Queensland – McCallum *et al.* (1995) considered that a reintroduction program was more likely to succeed if it involved the release of animals in a single release event, and at least 50 animals were considered necessary. These results were based on estimates of the rate of growth of populations of wallabies, coupled with the potential impacts of predators (McCallum *et al.*, 1995).

There are some cases where reintroduction and translocation programs are inappropriate. One is where the aim is to liberate excess captive stock. Another is where the removal of animals for captive breeding or translocation might contribute to the decline of the source population, as is believed to have happened for the Eastern Barred Bandicoot in Victoria (Todd *et al.*, 2002). Translocations are sometimes used to reduce human–animal conflicts, but Fischer and Lindenmayer (2000) found that these often fail. For example, when aggressive Australian Magpies were removed from suburban Brisbane, new individuals colonised the territory vacated by the translocated animals within a few days (Jones and Finn, 1999). A similar situation for the Common Brushtail Possum is outlined in Box 4.12. Such outcomes highlight the need for a critical risk assessment prior to a translocation being undertaken (Hodder and Bullock, 1997).

Cost and cost-effectiveness of reintroduction strategies

Although *ex situ* strategies can make an important contribution to conservation, the few studies that have published the actual costs of such programs (see Fischer and Lindenmayer, 2000) show that they can be time consuming and expensive. In the USA, the annual cost of the reintroduction program for the Californian Condor has been estimated to be US$1 million (Cohn, 1993). The total cost over an 8-year period for the reintroduction of the Grey Wolf to central Idaho and Yellowstone National Park has been calculated to be US$6.7 million (Bangs and Fritts, 1996). The rehabilitation and reintroduction

Box 4.12

When animals should not be translocated: suburban Common Brushtail Possums

The Common Brushtail Possum lives in Australian cities, particularly in well-established parks and suburban gardens. These animals can create problems in a city environment because they live in the roofs of houses, eat plants in household gardens, and make considerable noise at night (Kerle, 2001). One solution has been to capture animals and translocate them to forested areas outside the urban environment or to bushland within a city area.

Pietsch (1995) followed the fate of 64 nuisance Common Brushtail Possums that were trapped in the metropolitan area and translocated. More than 70% of animals died within a week of being released. Only one remained within the release area for more than 10 weeks. Animals from urban areas were naive about the forest environment and they spent significantly longer periods on the ground than resident Common Brushtail Possums. They tended to select den sites on, or close to, the forest floor, making them susceptible to predation, particularly by feral animals such as the Red Fox. A number of other animals succumbed to post-release stress and trauma. Notably, among the translocated animals studied by Pietsch (1995) that survived, one took up residence in the roof of a house. The results of the study by Pietsch (1995) are consistent with those of other species of possums, for example the Common Ringtail Possum, in which 90% or more of the released animals lived less than 2–12 weeks (e.g. Shaw, 1979; Augee, in Pietsch, 1995).

Furthermore, populations inhabiting areas to which urban animals are translocated may suffer from the introduction of new diseases and parasites (see Viggers *et al.*, 1993). In addition, the removal of animals from their original home ranges in the city environment creates vacant territories available for occupation by other dispersing Common Brushtail Possums. Translocation is not an effective strategy for resolving problems associated with urban populations of native marsupials such as the Common Brushtail Possum.

Common Brushtail Possum. (Photo by Esther Beaton.)

of Sea-otters after the Exxon Valdez oil spill in Alaska has been estimated to cost US$80 000 per animal or US$17 million altogether (Estes, 1998).

High costs are also typical in Australia. In Victoria, the recovery plan for the high-profile threatened Helmeted Honeyeater has included A$232 000 for captive breeding and A$111 000 for reintroduction to new sites (Menkhorst *et al.*, 1997). The cost of translocating captive-bred Dibblers to Escape Island (off Western Australia) was estimated to be A$600 000 (Moro, 2003).

In some cases, the money for reintroduction and translocation can divert scarce conservation dollars from other important projects (Sundquist, 1993), or divert attention from more appropriate and effective conservation strategies (see Box 4.13 on the Sumatran Rhino). For example, in the case of the Orangutan in Borneo, the annual cost associated with the maintenance of facilities for housing wild-caught animals and translocating them to other undisturbed rainforest areas is estimated to be about A$500 000 (MacKinnon and MacKinnon, 1991). The contribution of these captive-housed translocated animals to the persistence of remaining populations of the Orangutan is thought to be negligible, and captive-bred and released animals can even transmit tuberculosis to previously disease-free wild populations (Viggers *et al.*, 1993). Equivalent funding dedicated to the protection of reserves would have major benefits for the protection of the species, as well as many other species inhabiting such areas (which are also considered to be conservation areas of high biodiversity on a global scale; MacKinnon and MacKinnon, 1991).

In other situations, translocation could be cost effective and critically important. The success of mammal recovery efforts in Western Australia (e.g. Richards and Short, 2003) and bird recovery efforts in New Zealand are good examples.

Box 4.13

Captive breeding, reintroduction and conservation of the Sumatran Rhino

There are five species of rhinos around the world, and populations of all of them are in decline as a result of habitat destruction and hunting pressure. Much of their commercial value comes from rhino horn, which has been used in traditional medicine for thousands of years, particularly among Chinese cultures. Rhino horn is still prescribed and/or sold by doctors and pharmacists in countries such as Taiwan (Nowell *et al.*, 1992). Demand for rhino horn products has increased since the turn of the century and between 1970 and 1987 this contributed directly to an 85% reduction in the remaining populations of the various species (Fitzgerald, 1989).

The Sumatran Rhino is the smallest of the rhinoceros species. Its distribution used to encompass countries including Myanmar (Burma), Thailand, China, India, Indonesia and Malaysia. As a result of intensive hunting, habitat clearance and illegal logging, small breeding populations now remain only in Sumatra, Borneo and the Malay Peninsula, although their present status in these places is not known (Rabinowitz, 1995). In the late 1970s, the World Conservation Union set up an Asian Rhino Specialist Group in response to this decline. Management actions recommended by this group, and others that were subsequently created (e.g. the Sumatran Rhino Trust), included field research, surveys and monitoring, setting aside of protected reserves where poaching is prevented, control of trade in rhino horn, and a captive breeding and reintroduction program.

Rabinowitz (1995) traced the history of recovery efforts for the Sumatran Rhino. Despite early and repeated recommendations about the need for *in situ* conservation (protected areas and poaching control), the major focus was on captive breeding using wild-caught animals. This cost millions of dollars and has been spread across a number of separate zoo facilities in Indonesia, Malaysia, the United Kingdom and the USA. The captive breeding program was largely a failure. Many animals died in captivity and there were few successful births. The program has depleted wild populations of animals, although some observers believe that these animals could have died anyway as result of illegal hunting. The captive breeding program diverted attention and resources from more appropriate (and potentially more effective) conservation actions, particularly setting aside forest reserves and protecting them from poaching and illegal timber harvesting (Rabinowitz, 1995). A more recent review (Linklater, 2003) suggested that there has been a major divide between pure and applied conservation research, which has caused the demise of rhino species, and which is despite a major increase in the number of scientific publications on these high profile animals.

More recently, some reserves have been set aside that include habitat for the Sumatran Rhino. One is the Bukit Barisan National Park in Sumatra, which covers over 350 000 hectares, making it the third-largest protected area on the island. To combat poaching, Rhino Protection Units have been established. These units also monitor illegal logging and land clearing (World Wide Fund for Nature, 2002). On reserves that support populations of the Sumatran Rhino in other areas, such as Sarawak, heavy fines and jail sentences are imposed on people for taking protected animals such as the Sumatran Rhino. Despite these important conservation efforts, the Sumatran Rhino is still the most endangered species of rhino (World Wide Fund for Nature, 2002). It is estimated that there are now fewer than 300 individuals worldwide and all occur as small fragmented populations.

4.7 Conclusions

Strictly protected areas are unlikely to cover more than 10% of the earth's surface – in the case of forests the total area is only 8%. The Australian Federal Government implemented the Regional Forest Agreement process in 1995 with the objective of increasing the representation of ecological communities in protected areas to between 10 and 15% of their pre-1780 area (Commonwealth of Australia, 2001a). The question of how much land is enough has been a central issue in conservation biology for several decades. However, it is certain that protection of even 15% of the ecological communities in Australia will be insufficient to protect the majority of the biota for the medium to long term (Lindenmayer and Franklin, 2002).

The underlying causes of environmental problems are often described in terms of insufficient protected

Box 4.14

Reintroductions of Australian macropods

Short *et al.* (1992) reviewed more than 20 reintroduction and translocation efforts for a variety of species of macropods throughout Australia. The release programs included those for the Quokka and Tammar Wallaby in Western Australia, the Brush-tailed Bettong and Black-flanked Rock Wallaby in South Australia, and the Brush-tailed Rock Wallaby and Parma Wallaby in New South Wales. Many of the programs were unsuccessful. The failure rate was particularly high (approximately 90%) in cases where animals were released into areas where control programs for feral predators (especially the Red Fox and the feral Cat) had not been undertaken. Conversely, the successful recovery efforts (9 of the 10 that were considered to have succeeded) were on islands where feral predators were absent, or had been eliminated. Short *et al.* (1992) recommended a range of predator control measures, such as the repeated application of poison baits to assist release programs, particularly those planned for mainland Australia. Often these release programs were on islands not previously part of the known distributions of the target species (actually making the programs 'introductions' rather then 'reintroductions' according to the definition of Konstant and Mittermeier, 1982). Since the review by Short *et al.* (1992), reintroduction attempts have been made for an array of native mammals (including macropods), especially in Western Australia. Many have been successful and in virtually every successful case, the control of feral predators has been an essential feature (Richards *et al.*, 2001; Richards and Short, 2003).

area, poor law enforcement, illegal trade, or land clearance. Thus, quite naturally, solutions such as more protected areas, improved species management and so on are proposed in response. But biodiversity will be conserved only partially by such means. Fundamental problems lie outside the scope of protected areas, arising from sources such as agriculture, mining, pollution, settlement patterns, urban and infrastructure development, capital flows, and the distribution of wealth within and between nations (Chapter 19).

McNeely *et al.* (1990) and Lindenmayer and Franklin (2002) argued that conservation over the coming decades will lie primarily in areas where some human land uses are tolerated or encouraged, or in degraded landscapes that have been restored to productive use for conservation. This view rests on the assumption that protection depends on putting a sufficient economic value on natural resources and biodiversity. It also assumes that changes in ethical conceptions are unlikely, or that, if they do occur, such changes will not effectively translate into the increased allocation of land to protect biodiversity.

Techniques ranging from seed collection to cryostorage provide ways to store large numbers of plant species and a considerable proportion of the genetic diversity of some taxa. These techniques have been important for conserving a range of rare species. Applications of these methods to the conservation of individual species may require an integrated program.

Ex situ conservation strategies are an adjunct to, and not a substitute for, *in situ* management. Maintenance of significant proportions of the world's threatened biota may not be the most effective use of *in situ* conservation resources. The role of botanic gardens and zoos in research and education is at least equal in importance to their role in the cultivation and maintenance of threatened species.

Reintroduction, translocation and captive breeding are often expensive conservation strategies with a high risk of failure, despite considerable human effort and financial investment. The high rate of failure should not be regarded as surprising given the complexity of ecosystems. The need for reintroduction and translocation programs can be considered to be symptomatic of a failure of ecosystem management.

Conversely, reintroduction and translocation programs can be useful in generating public awareness of a particular conservation issue, especially when a high profile or charismatic species is involved. Captive programs provide important opportunities to learn about the ecology of captive species, and in some cases, the instigation of such activities can indicate to society that a particular species is worth conserving (Brereton *et al.*, 1995). Although it is clear that captive breeding, reintroduction and translocation programs can make an important contribution to attempts to conserve biodiversity, such efforts should not distract attention or resources from other conservation strategies (Rabinowitz, 1995). Captive breeding, reintroduction and translocation strategies are not an appropriate substitute for *in situ* ecosystem conservation (Noss and

Murphy, 1995) and all attempts should be made to obviate the need for their use in the first place.

4.8 Practical considerations

Conservation biologists should strive to develop reserve systems in a way that takes into account the comprehensiveness of coverage of different ecological communities, and the nature and imminence of threats. However, even very extensive reserve systems cannot adequately conserve all biodiversity. Effective conservation depends on off-reserve conservation that forms partnerships with private landholders and that encourages land use compatible with conservation objectives.

Small reserves are more likely than large reserves to need close management. Management can include providing some ecological resources artificially, assisting dispersal between patches, mitigating edge effects, managing invasive species, and initiating and controlling disturbance processes. Reserves should exceed the maximum size of potential disturbances, or there should be a source population of susceptible species within dispersal range.

Ex situ conservation, translocation and reintroduction are expensive and high-risk options in most cases. They are likely to provide effective strategies for the conservation of only a very small proportion of the world's threatened biodiversity. On the other hand, they play a critical role in protecting economically, socially and agriculturally important species, and provide an important focus for research and education.

4.9 Further reading

Briggs and Leigh (1996), Specht *et al.* (1995) and Specht and Specht (1999) review the protection of Australian plants and plant communities. The Australian Federal Government approach to protected area conservation is documented in Anonymous (1995), Dickson *et al.* (1997), and Thackway and Olsson (1999). The website maintained by the Department of Environment and Heritage (http://www.deh.gov.au/) provides extensive information on the size of the marine and terrestrial protected area estate in Australia, as well as data on World Heritage Areas, Ramsar sites and Biosphere Reserves. Papers on off-reserve conservation are presented in books edited by Hale and Lamb (1997) and Craig *et al.* (2000). Lindenmayer and Franklin (2002) provide a detailed discussion of ways to undertake off-reserve conservation in forests.

Linington and Pritchard (2002) outline material on the role and importance of gene banks. Lyles (2002) gives a review of zoos and a history of changes in the role of zoos, particularly for biodiversity conservation. A book edited by Serena (1995) contains a number of examples of the use of reintroduction and translocation as a conservation strategy for Australasian fauna. The IUCN (1998) provide a detailed set of guidelines for reintroduction and translocation programs. Assessments of the success of translocations and reintroductions are given by Griffith *et al.* (1989), Wolf *et al.* (1996), and Fischer and Lindenmayer (2000). Jackson (2003) is a key work on the captive breeding and management of Australian mammals.

Part II: Impacts

Human activities impact heavily on the natural world. More than 35% of the planet's land surface area is used for agriculture (Gerard, 1995). Humans directly or indirectly use about 40% of the world's plant growth, 25–35% of the world's marine primary production, and 60% of the planet's readily accessible fresh water (Pimm, 2001). A consequence is that 20–30% of the world's species are threatened (Lawton, 2002).

Part II of this book describes some of the ways humans affect the environment, and outlines the extent of impacts on the Australian environment. Different kinds of impacts are examined: changes in the physical environment (Chapter 5); loss of genetic diversity, species, and populations (Chapter 6); and changes in species distributions and abundance (Chapter 7). Over-harvesting of natural populations is examined in Chapter 8. One major area of study in conservation biology is habitat loss and fragmentation and two chapters in this book (9 and 10) are dedicated to it, in part because loss of habitat is the most important process threatening the persistence of all levels of biodiversity (Fahrig, 2003). Another major factor influencing biodiversity is the alteration of fire regimes (Bradstock *et al.*, 2002). Given its importance in an Australian context – for example, alteration of fire regimes ranks second only to habitat loss as a threatening process for birds – Chapter 11 is dedicated to a discussion of fire.

The final chapter in Part II (Chapter 12) considers the role of the human population as a driver underpinning all of these impacts. Many of the symptoms of threat are closely associated with increases in the demands of the human population. These demands are related to the size of the human population, the per capita consumption of resources, and the kinds of technology employed in acquiring and consuming those resources.

Chapter 5
Changes in the physical environment

This chapter outlines a few of the most important processes that affect the Australian physical environment and which can also have indirect and direct impacts on flora and fauna. There are numerous elements of the physical environment that this book does not cover, including deterioration in soil structure and conditions. Instead, this chapter provides a brief account of water resources, the process of salinisation, and some examples of chemical pollution and climate change in Australia. Some additional detail about the world's population and its demands is provided in Chapter 12.

5.0 Introduction

The physical environment includes soil, air and water, and the dynamic properties and processes that occur in the **abiotic** world. Australia harbours a unique flora and fauna because the continent is dry and flat, the soils are ancient and, for the most part, nutrient-poor, and because it has been isolated for the last 50 million years. These characteristics predispose the physical environment, and the biota that it supports, to be susceptible to particular kinds of change.

Soils, soil water, water and air quality are sometimes affected detrimentally by human activities (see State of the Environment, 1996, 2001a,c). A full exposition of all physical processes and the methods for analysing the environmental problems associated with them are beyond the scope of this book. Instead, we outline a few of the most important.

5.1 Land degradation, water resources and salinisation

Land degradation

Land degradation is the process whereby land deteriorates because of the effects of wind, water, gravity or temperature, and is characterised by changes in soil composition, structure and depth, and water quality (Blyth and Kirby, 1984, in Mercer, 1995; State of the Environment, 2001d). Forms of accelerated degradation associated with human activities include wind and water erosion, dryland salinity, irrigation salinity and waterlogging, soil compaction, vegetation loss, mass movement of soil, chemical contamination of soil and water, and soil acidification (Mercer, 1995; State of the Environment, 2001d; Commonwealth of Australia, 2001d).

The flat and relatively featureless Australian landscape is the consequence of many millions of years of wind and water erosion. However, processes of land degradation associated with human activities are many times faster acting than geological processes. Degradation of potentially productive land is a worldwide problem, which usually occurs in conjunction with agricultural development. In fact, severe land degradation affects nearly 20% of the earth's vegetated land (World Resources Institute, 1992; UNEP, 1999).

Soil degradation is a symptom of land degradation; it can be a consequence of erosion and deposition; mass movement of soil; coastal erosion by marine processes; alteration of soil salinity, structure, fertility, or

Table 5.1. Extent of water, wind, chemical and physical degradation of soils in Australia (based on data in Ghassemi *et al.*, 1995).

Type of soil degradation	Affected area (millions of hectares)	Percentage of total affected area
Water	1094	55.7
Wind	548	27.9
Chemical	239	12.2
Physical	83.3	4.2
Total	1964	100

acidification; changes in water repellency; waterlogging; or soil pollution (Table 5.1). In Australia, water and wind erosion, and changes in soil salinity and structure are important (McTainsh and Boughton, 1993; Stirzaker *et al.*, 2002; Table 5.2). For example, recent data suggest that soil erosion rates in parts of northern Australia may exceed 50 tonnes per hectare per year (Lu *et al.*, 2001), in part because of infrequent major storm events.

Water in the Australian environment

Australia is the driest inhabited continent. The nationwide average annual rainfall is only 465 millimetres and only about 12% of this results in **run-off** – evaporation and transpiration by plants return almost all of the remaining moisture to the atmosphere (Smith, 1998; Commonwealth of Australia, 2002a; Figure 5.1). By comparison, run-off in other places is much higher (57% in South America and 48% in Asia; Ghassemi *et al.*, 1995; Table 5.3).

The variability of annual stream flows is more pronounced in Australia than elsewhere, especially for large water catchments (McMahon *et al.*, 1992a). The **coefficient of variation** in stream flow reflects the variability of flow relative to the average flow. Flows from large catchments in Australia are significantly more variable than those from other continents, and the flows become more variable as catchment size increases (Figure 5.2; Smith, 1998). The largest catchments in Australia are west of the Great Dividing Range; they are significantly drier and their flows are more variable than the smaller catchments along the eastern and southern margins of the continent. Very extensive regions of the central and western parts of the continent experience negligible mean annual run-off (Figure 5.1).

Aridity, variability and the Australian biota

The aridity of the Australian environment has profound effects on the biota. The dynamics of many natural populations, especially those of the arid and semi-arid zones, are intimately linked to rainfall patterns. Floods act as cues for breeding and provide a productive floodplain for subsequent life-history dynamics. For instance, the breeding and life-history strategies of many frogs, freshwater fish and macroinvertebrates take advantage of periodic flood events (Caldwallader, 1986; Smith, 1998; Cogger, 2000). Similarly, many species of birds in arid Australia, for

Table 5.2. Summary of major changes to some Australian soils in the last 200 years and their causes (after Chartres *et al.*, 1992).

Property	Nature of change	Cause
Soil organic matter	Decline of organic carbon in A horizon.	Changes in vegetation and microbiological populations.
Soil structure	Lower stability. Development of massive structured and hard-setting A horizons. Greater incidence of surface crusting.	Declining organic matter from cultivation. Too frequent cultivation. Removal of vegetation.
Soil water	Lower infiltration rates. Lower hydraulic conductivities.	Loss of topsoil structure. Decreasing porosity due to cultivation and blocking of channels. Compaction by stock and traffic.
Soil pH	Acidification. Associated development of aluminium and manganese toxicities. Associated decline in biological activity. Detrimental effects on interactions between plants and bacteria.	Increased nitrate production and leaching. Removal of alkaline products. Use of acidifying fertilisers. Changes in carbon–nitrogen cycle.
Soil waterlogging/salinity	Increased area subject to non-irrigated salinity. Rising watertables. Salinisation of surface water.	Removal of deep-rooted perennial vegetation. Replacement with shallow-rooted annual crops and pastures.

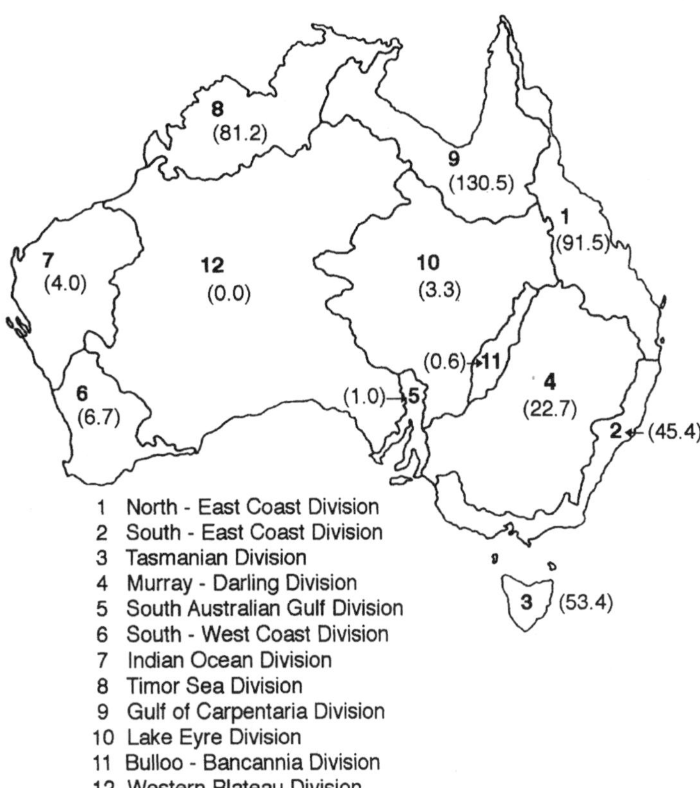

Figure 5.1. Total mean annual run-off in major Australian catchment divisions. (Redrawn from Chartres *et al.*, 1992.)

example the Grey Teal, track the locations of flood events and nest where bodies of water persist long enough to allow successful breeding.

Australia supports a disproportionate number of nomadic bird species that move between different areas in response to rainfall and associated pulses of food (Ford, 1989; Box 5.1). Populations of macropods, small mammals, frogs and reptiles increase substantially following rainfall (Caldwallader, 1986; Rose, 1995; Dawson, 2000).

Floods also recharge wetlands and improve aquatic ecosystem integrity by moving large quantities of sediment, logs and other coarse woody debris (McMahon *et al.*, 1992b; Bayley, 1995; Gregory, 1997; Stirzaker *et al.*, 2002). Damming, diversion and regulation of rivers and streams change river flow and temperature regimes, affecting native flora and fauna (Smith, 1998).

Australian plants have evolved strategies in response to extremely variable and unpredictable water availability. These adaptations include reduction of foliage biomass by shedding branches under conditions of water stress; high stomatal regulation of transpiration; rapid recovery after water stress; and extensive, high-density root systems to capture water.

Table 5.3. Run-off of water in different global regions. (Modified from Ghassemi *et al.*, 1995.)

Area	Annual stream flow (millimetres)	Annual stream flow (cubic kilometres)	Global run-off (%)
Europe	306	3210	7
Asia	332	14 410	31
Africa	151	4750	10
North and Central America	339	8200	17
South America	661	11 760	25
Australia	45	348	1
Oceania	1610	2040	4
Antarctica	160	2230	5

Arid inland Australia, where limited run-off occurs. (Photo by David Lindenmayer.)

The features that characterise Australian moisture availability regimes and, in turn, shape the nature of the biota in the continent, mean that many of the approaches to managing and conserving biodiversity, water, and land developed in the Northern Hemisphere may not be directly applicable to Australian systems. In particular, the variability of rainfall patterns and disturbance regimes will often require novel approaches to problem-solving in Australian conservation biology and natural resource management. Two examples are the approaches used for resolving problems associated with the conservation of natural mound springs and the management of artificial water points in arid and semi-arid Australia (Landsberg *et al.*, 2001; Figure 5.3; see Box 5.2).

Australia's fresh water is controlled by thousands of weirs and hundreds of large dams (>10 metres crest or top height; Kingsford, 2000), and are subject to more than 50 schemes that transfer water within and between

Box 5.1

The Princess Parrot: a 'desert nomad'

The Princess Parrot is a strikingly beautiful native Australian parrot. It occurs in the extensive desert areas of central and west Australia. The species appears to have always been rare and partly nomadic, with birds tracking occasional rainfall events (Forshaw, 2002). There have been a number of recent sightings, particularly around the Canning Stock Route, which is an increasingly popular destination for outback adventure four-wheel driving enthusiasts. In fact, this area may prove to be a core part of the distribution of the Princess Parrot. There are few confirmed breeding records of the species (Baxter and Henderson, 2000), and the Princess Parrot may be more common in captivity than in the wild (Burbidge, 2003). It is possible that the Princess Parrot is vulnerable to the effects of competition from other species of parrots that have benefited from the establishment of artificial water points throughout arid and semi-arid Australia – but there is presently little definitive evidence for this (Forshaw, 2002).

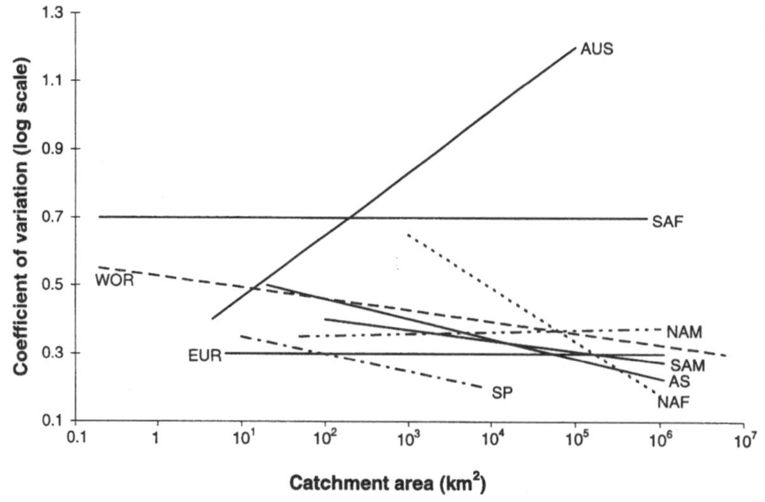

Figure 5.2. Variability of stream flow in Australia (after McMahon *et al.*, 1992a). AUS, Australia; SAF, South Africa; NAM, North America; AS, Asia; NAF, North Africa; SP, South Pacific; EUR, Europe; WOR, world average.

A rare flood event in arid Australia (Congee Lakes in north-eastern South Australia). (Photo by David Lindenmayer.)

catchments (Arthington and Pusey, 2003). The ecological effects of these controlling mechanisms include nutrient loads and **algal blooms**, sediment entrainment, introduction of toxic chemicals, destruction of in-stream habitat, barriers to migration, thermal pollution, stock depletion (over-fishing), and introduction of invasive species (Kingsford, 2000; Todd *et al.*, 2005). About 80% of stored water in Australia is used for irrigation, and Australia has the highest water storage capacity per capita in the world (Arthington and Pusey, 2003). Flood regulation measures have changed flood pulses, water temperatures and longer-term flow regimes.

Thermal pollution by cold water releases from large dams is largely unmanaged and remains one of the most far-reaching changes in Australian fresh waters. Low-level releases from deep impoundments can be 10°C colder than natural temperatures downstream (Acaba *et al.*, 2000). Low temperatures may reduce food availability, favour cold water species, reduce cues for spawning, reduce egg and larval survival, change developmental rates, and reduce ability to capture prey and escape from predators (Koehn *et al.*, 1996; Clarkson and Childs, 2000; Todd *et al.*, 2005). Uncertainties about the ecological effects of changes in flow volumes

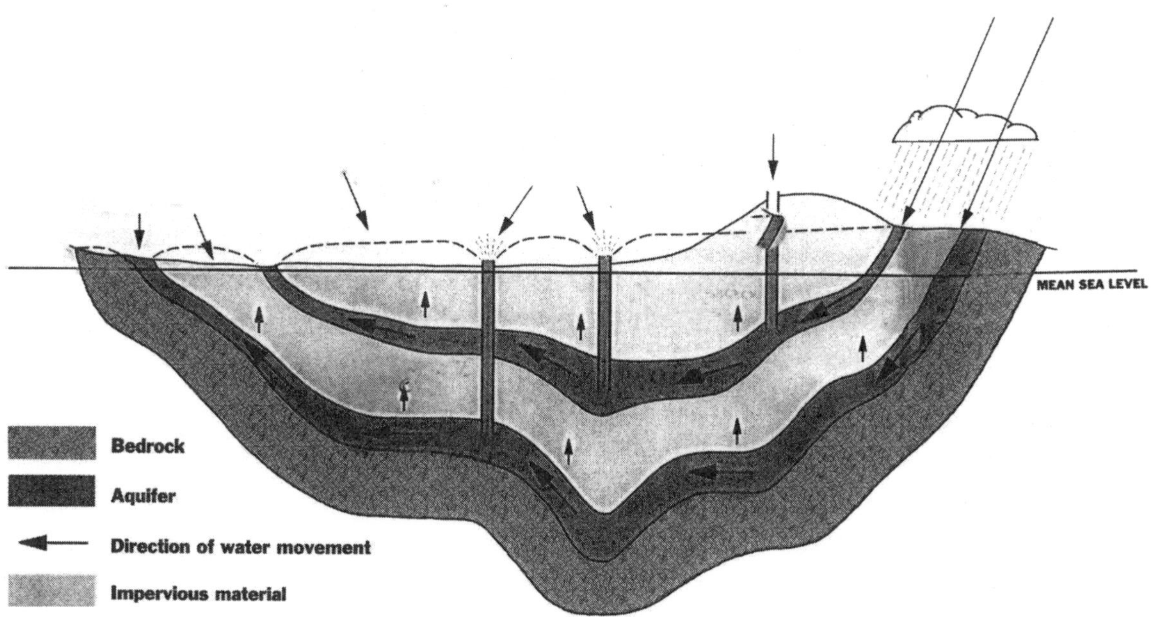

Figure 5.3. Natural and artificial water points in the Great Artesian basin. (Redrawn from van Dugteren, 1999.)

Box 5.2

How to resolve problems associated with artificial water points in inland Australia

Rainfall in inland arid and semi-arid Australia is erratic and limited. The Great Artesian Basin covers approximately 1.7 million square kilometres of Australia. In this area there are 23 000 artificial water points comprising 3000 artesian bores and 20 000 sub-artesian bores. The total flow exceeds 1.5 million litres of water per day, with some individual bores producing more than 10 000 litres daily. Almost all parts of pastoral Australia (roughly 60% of the continent) are within 10 kilometres of an artificial water point. The large volume of water and extensive distribution of artificial water points create new challenges for conservation.

The water supports large numbers of domestic livestock as well as feral herbivores. Landsberg *et al.* (1997) and James *et al.* (1999) studied the changes in species composition of plants, birds, mammals, reptiles and invertebrates at different distances from water points. They estimated that many native species have benefited from the network of bores and increased their abundance and/or range in response (Figure 5.4). Some birds that have benefited from artificial water points include the Galah, White-plumed Honeyeater, and Spiny-cheeked Honeyeater. Conversely, 15–38 % of species in different groups were found to have been disadvantaged, with eight species of birds extinct in many rangeland regions, and 39 species with a lower abundance and/or smaller range that was attributed to the presence of artificial water points and the associated pastoralism. Examples of birds that had declined are Bourke's Parrot, the Grey Honeyeater, and the Chiming Wedgebill.

The simplest way to limit the problem of declining species is to reduce the number of artificial water points by closing some of them. This is particularly relevant in rangeland reserves, and steps to do this have been taken in reserves such as Gundabooka National Park near Bourke in north-western New South Wales (NSW National Parks and Wildlife Service, 2003b). However, there also would be conservation value in maintaining some unwatered areas on pastoral lands. Clearly, like all conservation actions, there would be considerable value in establishing a robust monitoring program to compare the response of biota both before and after water points are closed and compare these areas with areas where water continues to flow.

---

and temperatures make judgments about river management difficult.

Several approaches have been developed to evaluate **environmental flow** requirements. The 'holistic' approach assumes that if the essential features of natural flow regimes are identified and incorporated into managed flows, then functional integrity and ecological communities will be preserved. Under this approach, ecological requirements take precedence over other demands (see Arthington, 1998). The 'habitat analysis' approach relies on technical workshops and hydrological models to identify habitat elements in a catchment, determine their flow requirements, and develop management and monitoring systems to meet these requirements and measure their effectiveness (Burgess and Vanderbyl, 1996). The objective with this approach is to manage the river ecosystem rather than individual species, and impacts are assessed by stakeholders. Both approaches have been used in Australia to guide catchment management (Tharme, 2003).

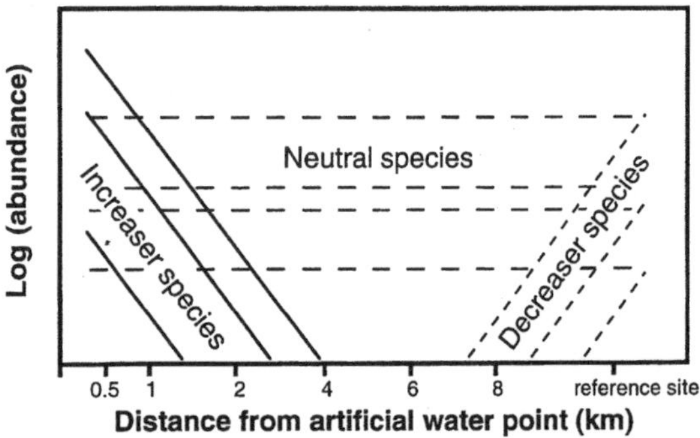

Figure 5.4. 'Increaser' and 'decreaser' species in relation to artificial water points in inland Australia. (Modified from Landsberg *et al.*, 1997.)

Recent legislation and polices in many jurisdictions indicate that it is now legitimate to use water for river and wetland ecology, but there is little guidance on how much water should be allocated to different uses. Arthington and Pusey (2003) suggested that 80–90% of natural mean annual flows may be needed in some northern Australian rivers to maintain a low risk of environmental degradation.

A natural mound spring in South Australia, where artesian water comes naturally to the surface. (Photo by Nikki Munro.)

Salinisation

Salinisation is one of the most important consequences of clearing native vegetation in Australia (Stirzaker *et al.*, 2002). It threatens productive agricultural activities, water quality, and the persistence of remnant vegetation in agricultural landscapes.

Ghassemi *et al.* (1995) defined salinisation as 'the process whereby the concentration of total dissolved solids in water and soil is increased due to natural or human-induced processes'.

Salinisation can result from changing climatic patterns, irrigation, vegetation clearing, the replacement of trees by shallow-rooted plants such as crops and pastures, and the discharge of saline agricultural or industrial water (Ghassemi *et al.*, 1995).

Soils and salt

All soil contains some salt. The salt comes from three key sources: (1) weathering of soil and rock, (2) marine deposition in earlier geological periods (termed connate salt), and (3) salt from the atmosphere over the ocean that is deposited to the ground via rainfall (termed cyclic salt) (Chartres *et al.*, 1992). Most salt in Australian soils is probably cyclic (Stirzaker *et al.*, 2002). In some places, large quantities of salt have accumulated in the soil over long periods of time; for instance, in inland Western Australia, up to 10 000 tonnes occur per hectare (Stirzaker *et al.*, 2002).

A soil is considered **saline** when it contains such a quantity of salt that plant growth is significantly reduced (Peck, 1993; Stirzaker *et al.*, 2002). Salt in available soil water results in water stress and reduced osmotic pressure in plants, as well as reduced growth. Excessive concentrations of some solutes, including chloride, harm plants. An imbalance of sodium relative to calcium and magnesium results in a soil condition known as **sodicity** (Peck, 1993; Commonwealth of Australia, 2001d).

Soils believed to have been saline for thousands of years as the result of natural landscape processes, such as saltpans and claypans in Australia's arid interior, exhibit **primary salinity** (Figure 5.5). About 5% of the Australian continent is affected by primary salinity. Areas of previously productive land, which are now salt-affected to the point where plant growth is inhib-

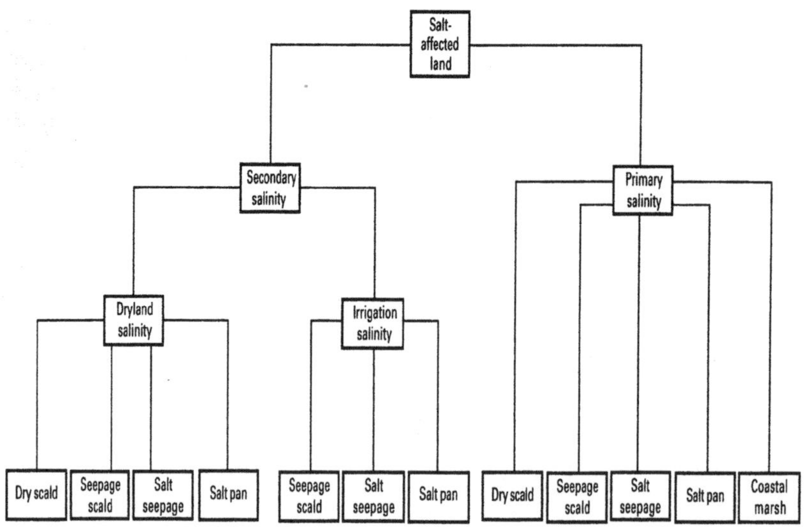

Figure 5.5. Hierarchical classification of primary and secondary salinity. (Redrawn from Smith, 1998.)

ited, exhibit **secondary salinity**, usually resulting from irrigation and/or excessive clearing. Areas with the highest risk of secondary salinity are those with an annual rainfall of 700–1100 millimetres. In places with higher rainfall than this, salt is leached from the soil profile by rainfall, and in places with less rainfall, there is insufficient water to recharge ground water reserves (McIntyre *et al.*, 2002).

Salinity and vegetation removal

Land clearing affects the level of the watertable, and the processes of ground water discharge, recharge and soil salinity (Commonwealth of Australia, 2002a). Changes in ground water dynamics in turn have consequences both for agricultural production and for the persistence of remnant vegetation.

Island Lagoon: a salt lake in central Australia and an example of primary or natural salinity. (Photo by David Lindenmayer.)

Hydrological cycles can be characterised by the expression (after McMahon *et al.*, 1992a):

Rainfall – Evapotranspiration = Run-off + Ground water recharge.

Stirzaker *et al.* (2002) noted that before European settlement of Australia, in catchments receiving less than 1000 millimetres rainfall per year there was a balance between woody vegetation growth and precipitation. The vegetation used most of the incoming rainfall and watertables remained deep underground. **Dryland salinisation** occurs when salt rises towards the soil surface following the removal of native vegetation. Where trees are removed and replaced by plants with lower rates of transpiration (e.g. crop and pasture plants), water is discharged from the ground water system and moves more or less vertically upwards through salt-laden sublayers of the soil (Figure 5.6). These processes can result in salt-scalded areas or saline 'seeps' (Ghassemi *et al.*, 1995). The 1999 salinity audit published by the Murray–Darling Basin Commission estimated that most of the salt in river systems comes from dryland salinity.

Prior to clearing in Australia, shallow watertables (less than two metres from the soil surface) occurred only in the bottoms of river valleys or near springs. Clearing increased ground water recharge rates with the result that some areas, such as parts of Western Australia, have watertables that have risen by as much as 25 metres since clearing (George *et al.*, 1996). In the Queensland part of the Murray-Darling Basin, it is possible that 633 000 hectares or 2.4% of the region will have watertables within two metres of the surface by 2020 (Murray Darling Basin Commission, 1999).

Irrigation water, which is often salty, can also mobilise salt stored in the soil or ground water. It can raise the level of the watertable, giving rise to **irrigation salinity**. Irrigation salinity occurs through evaporation of irrigation water, leaving the salts at, or close to, the soil surface, or by the water moving through the soil profile to enter a watercourse (George *et al.*, 1996; Stirzaker *et al.*, 2002).

Scale and magnitude of salinity problems in Australia

There were many early warnings about the potential impacts of clearing and associated salinity in Australia. For example, in 1917, scientific evidence highlighting the risks of salinity in Western Australia was rejected by a Royal Commission on the Mallee and Esperance Lands. Similar evidence was provided by scientists in the 1920s and 1940s (State of the Environment, 2001d). Nevertheless, widespread clearing continued to take place in Western Australia and elsewhere around the continent (see Chapter 9).

By 2000, more than 2.5 million hectares of cultivated land in Australia was affected by secondary salinity (Commonwealth of Australia, 2000; Table 5.4). The National Land and Water Resource Audit (Commonwealth of Australia, 2001b, 2002a) subsequently concluded that more than 5.7 million hectares of land was at risk or was already threatened by dryland salinity, and that given the rate of increase, the area affected could increase to 17 million hectares by 2050. Furthermore, the National Land and Water Resource Audit calculated that dryland salinity threatens more than 41 000 kilometres of rivers, 20 000 kilometres of major roads, 1600 kilometres of railways, 68 towns, and 130 wetlands.

Figure 5.7 shows that salinity problems are widespread in Australia: they occur in every State and Territory, and more than 20 major catchments require attention (State of the Environment, 2001a,d). Ongoing high rates of clearing in tropical and subtropical woodlands in northern Australia may create a salinity problem in areas well beyond those currently affected (Williams *et al.*, 1997).

Secondary salinity is predicted to dramatically reduce the quality of drinking water in some regions. For example, in south-west Western Australia, 52% of the divertible water is no longer drinkable, or is of only marginal quality. There are similar trends in other parts of the country (Figure 5.8). In Adelaide, drinking water is forecast to fall below World Health Organization standards within a few decades (Stirzaker *et al.*, 2002).

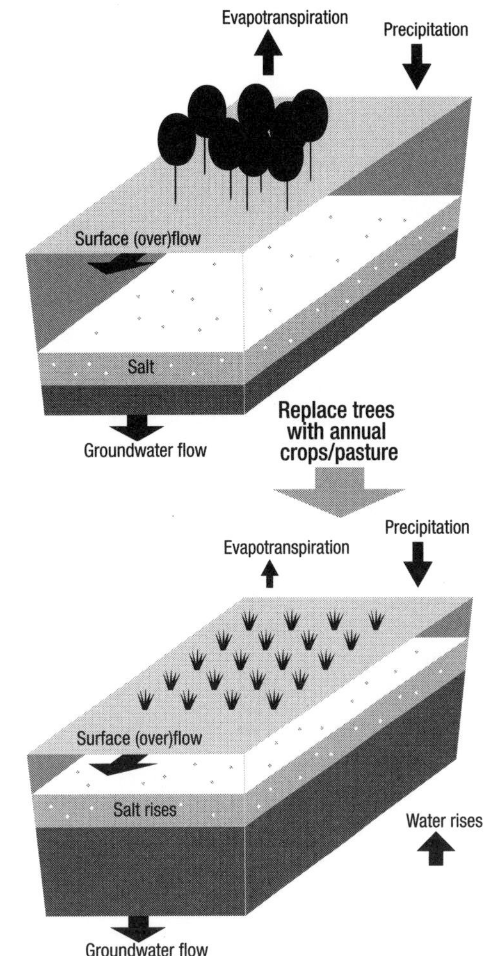

Figure 5.6. The bathtub analogy: relationships between land clearing and secondary salinity. (Redrawn from Stirzaker *et al.*, 2002.)

Figure 5.7. Location of areas at high risk of salinity in Australia (from Commonwealth of Australia, 2001c and the Australian Dryland Salinity Assessment, 2000).

Table 5.4. Australian agricultural, water, infrastructure and biodiversity assets at high risk from shallow watertables or with a high salinity hazard now and in 20 and 50 years time (from Commonwealth of Australia, 2001c; Australian Dryland Salinity Assessment, 2000).

Asset	2000	2020	2050
Agricultural land (hectares)	4 650 000	6 371 000	13 660 000
Remnant and planted perennial vegetation (hectares)	631 000	777 000	2 020 000
Length of streams and lake perimeter (kilometres)	11 800	20 000	41 300
Rail (kilometres)	1600	2060	5100
Roads (kilometres)	19 900	26 600	67 400
Towns (no.)	69	125	219
Important wetlands (no.)	80	81	130

Until a new equilibrium is reached between the rate of water input and output, salinity will continue to develop in extent and severity throughout Australia. For example, in agricultural districts in Western Australia, the area affected by secondary salinity is expected to increase by 2% annually (Peck, 1993). George *et al.* (1996) estimated that if ground water input rates continue, up to 25% of many agricultural landscapes in Western Australia, and as much as 40% of some regions, will become salt-affected within the next century.

Costs of salinity

The loss of 5% of the cultivated land in Australia (2.5 million hectares) to secondary salinity represents a significant direct economic cost. The State of the Environment Report (2001a) estimated that every 5 000 hectares of land affected by salinity costs A$1 million annually. By 2050, the annual cost of secondary salinity could be A$1 billion.

Costs of up to A$5 billion (in 1990 dollars) were estimated for a proposed salinity mitigation scheme for the Murray-Darling Basin, which involved the development of an outfall system to discharge salt-affected water (Ghassemi *et al.*, 1995). The annual costs of secondary salinity in eight catchments within the Murray–Darling Basin, in terms of lost production and other economic impacts, are between A$200 and A$300 million (Table 5.5). Such cost estimates ignore species decline and other ecosystem changes that are difficult to value in economic terms.

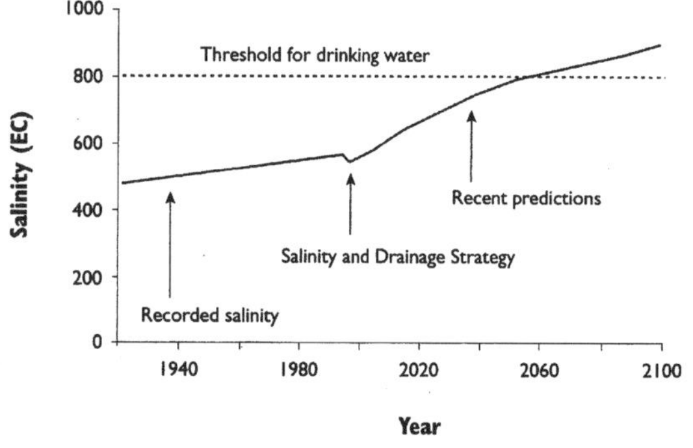

Figure 5.8. Relationships between past and predicted levels of salinity in the Murray River at Morgan in South Australia and threshold levels for drinking water. (Redrawn from Stirzaker *et al.*, 2002.)

Salinity and biodiversity

Large areas of native vegetation (currently more than 630 000 hectares) are at risk from salinity, and this could increase fourfold by 2050 (Commonwealth of Australia, 2001c). Many river systems, wetlands and terrestrial habitats are potentially threatened by salinity (see Table 5.6), although to date limited research has been conducted on salinity impacts on biodiversity, particularly in vegetation communities such as woodlands that are already heavily modified by human activities such as clearing and domestic livestock grazing (Briggs and Taws, 2003).

In the Western Australian wheat belt and other parts of the south-west of that State, many areas of remnant native vegetation are at risk from rising watertables (George *et al.*, 1996; Wallace *et al.*, 2003). Small and large reserves, and remnants on farms, lakes, wetlands, river systems and estuarine systems are threatened. In the same area, the Department of Conservation and Land Management (2000) estimated that more than 450 endemic plants and 200 species of aquatic invertebrates were at risk of extinction from salinity. Other work has indicated that elevated salinities in aquatic ecosystems in south-west Western Australia are leading to substantial changes in biological communities (Halse *et al.*, 2003). Little of the 2.6 million hectares of remnant native vegetation on private land in the agricultural region of Western Australia is fenced, and most areas of remnant vegetation experience Sheep or Cattle grazing, nutrient enrichment, or weed invasion, which

Salt-affected land on Kangaroo Island in South Australia (photo by David Lindenmayer) and in the wheat belt of Western Australia (photo by Richard Hobbs).

compound the effects of salinisation. As noted in Chapter 3, areas adjacent to and surrounded by agricultural land in the south-west of Western Australia support the greatest floral diversity in Australia, including more than 8000 species of flowering plants, most of which are endemic to the area.

Table 5.5. Annual costs of dryland salinity in eight catchments within the Murray–Darling Basin. (Based on data from Commonweath of Australia, 2001c; Australian Dryland Salinity Assessment, 2000.)

	Lower estimate (A$ million per year)	Upper estimate (A$ million per year)	Best estimate (A$ million per year)
Local government	–	–	14.69
Households	41.03	139.23	90.13
Businesses	8.45	8.96	8.71
State government agencies and utilities	–	–	16.31
Environment	?	?	?
Agricultural producers	–	–	121.80
Total	**202.38**	**300.99**	**251.64**

Table 5.6. Current and predicted areas of native vegetation at risk from the impacts of salinity (from Commonwealth of Australia, 2001c; Australian Dryland Salinity Assessment, 2000).

State	Current	2020	2050
New South Wales	7000	32 700	81 000
Victoria	6000	11 800	24 300
Queensland	N/A	N/A	92 000
South Australia	18 000	22 000	25 000
Western Australia	600 000	710 000	1 800 000
Total	**631 000**	**776 500**	**2 022 300**

N/A, not available.

Increased salinity modifies wetland environments, resulting in losses of food sources for some vertebrates (e.g. ducks and other waterbirds), the death of salt-intolerant plants and animals, and an overall reduction in plant and animal species diversity. McKenzie *et al.* (2003) found that as areas of woodland vegetation contract because of increases in salinity, communities that depend on such habitats, such as small ground-dwelling animals, are also reduced. In contrast, Saunders and Ingram (1995) noted that some water-bird species have increased in abundance in the Western Australian wheat belt, in part because water bodies have become saline, killing trees that cavity-nesting birds then use. However, the longer term prospects are not good for the many taxa that depend on hollow-bearing trees in these environments (Saunders and Ingram, 1995), particularly large parrots and owls (Mawson and Long, 1994; Gibbons and Lindenmayer, 2002). The trees killed by salinity will collapse eventually and there will be no recruitment of new stems under saline conditions.

In Victoria, almost 100 species of plants, terrestrial vertebrates and fish are thought to be susceptible to increasing salinity (Salinity Planning Working Group, 1992, in Ghassemi *et al.*, 1995). Many invertebrates and aquatic plants (such as algae) are also likely to be susceptible (Brock, 1981; Hart *et al.*, 1991).

In summary, considering the conservation significance of remnant vegetation in agricultural landscapes (including riparian areas), predicted increases in salinity over the next 50 years suggest that this process is one of the most significant current threats to the conservation of the Australian biota (State of the Environment, 2001a; Stirzaker *et al.*, 2002).

Tackling salinity

The information given in the previous sections highlights the scale, magnitude, costs and biodiversity impacts of salinity in Australia. Serious efforts have commenced to tackle the problem (see, for example, the 2000 National Salinity Action Plan, Commonwealth of Australia, 2000). The State of the Environment Report (2001a) outlined strategies to mitigate the impacts of salinity on biodiversity in Australia including (among others): (1) mapping salinity hazards, (2) improving the condition of native vegetation and stream water quality, and (3) protecting and rehabilitating waterways and wetlands and associated vegetation, including improved environmental flows. Of course, these actions must sit within broader social and economic actions that develop more ecologically sustainable farming systems and water management practices, maintain rural and regional human communities, and a raft of interrelated issues.

Salinity and revegetation

Salinisation can be reversed by revegetating large areas of land. However, revegetation requires care because, as noted by, Stirzaker *et al.* (2002):

- The extent of revegetation needed in some areas may be greater than can be reasonably achieved.

Replanted area for salinity control in the Riverina region of southern New South Wales. (Photo by David Lindenmayer.)

- The areas most visibly affected by salt may not be the best places to plant trees; planting trees in source areas higher in a catchment may be more effective.
- Plantings in some areas may be killed by salt water flows from elsewhere in a catchment.
- In high-rainfall catchments it may be unwise to plant large areas of trees because these areas provide large quantities of fresh water that dilute saltier water lower in the same catchments.
- Tree planting will be influenced by factors such as catchment sizes, flow rates through catchments, soil types, salt storage capacities and ground water discharges – factors that vary markedly between locations.

Given the extent of the salinity problem in Australia, resolving it will require significantly greater levels of resources than are currently being expended (Stirzaker *et al.*, 2002). Success may also depend on the financial markets for products from revegetated land, such as timber and pulp from farm forestry (Salt *et al.*, 2004).

5.2 Chemical pollution

Chemical pollution influences environmental conditions, human health and biodiversity in urban, rural and marine environments, and in developed and relatively natural areas. For example, particulate matter in the air is responsible for up to 2400 human deaths annually in Australia (State of the Environment, 2001c). The State of the Environment (2001e) estimated that at that time there were more than 80 000 contaminated terrestrial sites in Australia, with 7000 sites in New South Wales alone estimated to cost in excess of A$2 billion to clean up. The following section focuses on just a small number of cases of chemical pollution in the Australian environment: agricultural

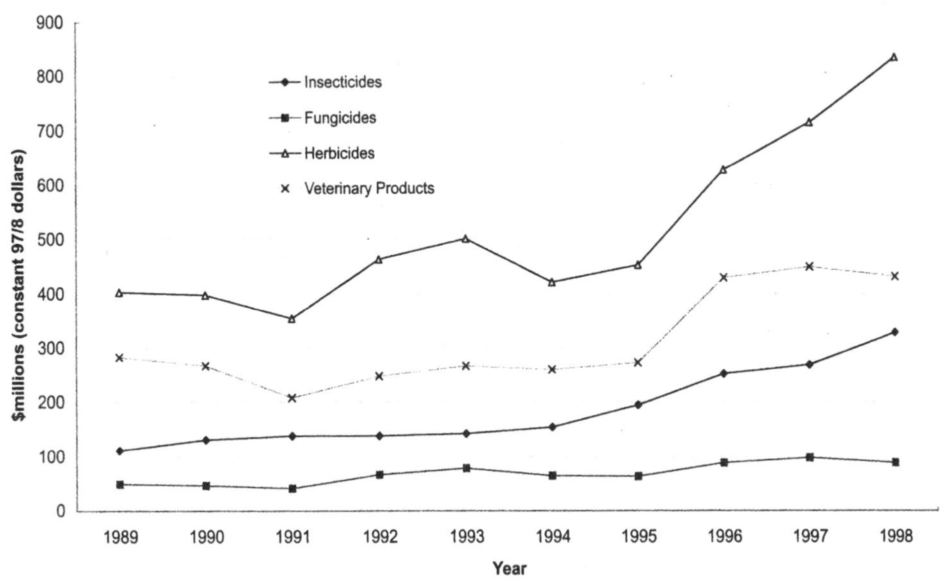

Figure 5.9. Increasing use of fungicides, insecticides and herbicides in Australia between 1989 and 1998. (Redrawn from State of the Environment, 2001d.)

chemicals, excess nutrients in ecosystems, and chlorofluorocarbons (CFCs).

Agricultural and other chemicals

The Australian Bureau of Statistics (1999a) reported that approximately 2500 chemicals and 2000 animal health products are used on farms in Australia. The extensive and increasing use of fungicides, insecticides, herbicides and other chemicals (Figure 5.9) has the potential to contribute to losses of natural resources, including biodiversity.

Many organic chemicals are fat-soluble and accumulate in the fatty tissues of animals. Such chemicals concentrate in predators, increasing the chemicals' potential for harm. The lethal and sublethal effects of pesticides, for example reproductive failure in natural populations, are well established (Carson, 1962). Such impacts were identified in Australia by the report of the Senate Select Committee on Agricultural and Veterinary Chemicals (SSCAVA, 1990), which included links to the deaths of birds and fish. For example, eggshell thinning associated with the bioaccumulation of pesticides is known in several species of Australian birds, including pelicans and birds of prey (Olsen *et al.*, 1993; Olsen, 1995). Bird diversity is higher on farms where spraying is conducted at intervals longer than five years (Birds Australia, 2000). Herbicides and insecticides leach from farms into streams and rivers; however, there is limited information about the effects of herbicides and pesticides on Australian biota and ecosystem functions.

Peregrine Falcon – a species susceptible to eggshell thinning as a consequence of pesticide use both in Australia and overseas. (Photo by Esther Beaton.)

Excessive inputs of nutrients into ecosystems

Nutrient-rich sediments, agricultural fertilisers, domestic sewage and some industrial effluents are responsible for **eutrophication**, **algal blooms** (Figure 5.10) and subsequent reduction of oxygen levels in inland waters, resulting in intolerable conditions for much of the freshwater biota. Such conditions often occur in the Murray-Darling catchment (Smith, 1998; Land and Water Resources Research and Development Corporation, 1999; State of the Environment, 2001f; see Chapter 9) and are common elsewhere. For instance, 115 algal blooms were reported in New South Wales between 1993 and 1994.

Increased nutrients can have detrimental impacts in marine environments and influence phytoplankton and macroalgal productivity, biomass, and nutrient concentrations (Brown *et al.*, 1990; Pogonoski *et al.*, 2002). For example, increased nutrient and sediment loads from agriculture and human wastes in north Queensland have substantial impacts on the Great Barrier Reef (State of the Environment, 2001b). Nearly 12 million tonnes of sediment are transported from freshwater systems to the reef every year (State of the Environment, 2001b).

Nitrogen-based fertiliser consumption in Australia more than tripled between 1980–81 and 1998–99 (Australian Bureau of Statistics, 2001), and

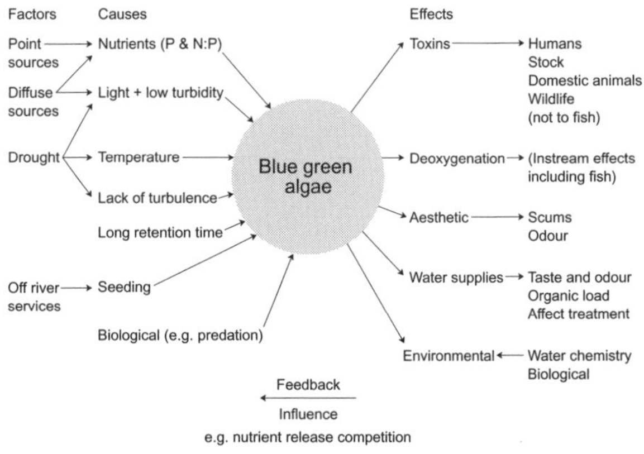

Figure 5.10. Causes and effects of algal blooms. (Redrawn from Smith, 1998.)

Algal bloom conditions in the Murray–Darling Basin. (Photo by David Lindenmayer.)

the use of phosphate-based fertilisers increased by 25% during the same period. This trend builds on previous increases in fertiliser use in Australia and mirrors patterns elsewhere around the world. Global fertiliser application has risen exponentially since the 1950s and in some places, for example south-eastern USA, nitrogen inputs are 20 times higher than natural levels. The United Nations Environment Programme noted that humans contribute more to the global supply of fixed nitrogen than do natural processes (UNEP, 1999). Agriculture accounts for more than 85% of human-generated nitrogen and this has other major impacts, such as soil acidification (see Box 5.3).

CFC-induced ozone depletion

One major environmental 'surprise', unforseen by scientists, was the impact of CFC-induced ozone depletion. Stratospheric ozone absorbs all ultraviolet-C and most ultraviolet-B radiation from solar radiation. Depletion of stratospheric ozone over Antarctica in spring has been linked with CFCs, chloroform, methane and halons released to the atmosphere in aerosol propellants, refrigerants, plastic foams and so on. Several of these compounds, including CFCs, are entirely created by humans (McNeely *et al.*, 1990). Elevated ultraviolet radiation at the earth's surface may have consequences ranging from lower algal production in the ocean's surface water to increased levels of cataracts and skin cancer among humans (Amos *et al.*, 1993; State of the Environment, 2001c).

Box 5.3

Fertilisers and soil acidification

The use of fertiliser has increased dramatically worldwide. In Australia, fertiliser use tripled in the 1980s and 1990s, and now more than five million tonnes of fertilisers are used annually (Commonwealth of Australia, 2001a). Although fertilisers have extended the range of farming and increased the productivity of agricultural land, such benefits come at an ecological cost. The rates of soil acidification in Australia have increased dramatically as a consequence of increased fertiliser use, and now threaten productivity on 25% of the nation's agricultural land (Commonwealth of Australia, 2002a). The National Land and Water Resources Audit (Commonwealth of Australia, 2001d) estimated that between 29–66 million hectares of land will reach limiting soil pH values within the next decade. Up to 66 million tonnes of lime will be needed to adjust soil pH levels, with an additional annual 3–12 million tonnes needed to maintain soil pH within a satisfactory range. These findings suggest a need for the better informed management of fertiliser application in Australian agriculture, which would both reduce the costs of farm production and reduce the ecological impacts on streams, coastal environments and their associated biota (Commonwealth of Australia, 2002a).

Other chemical pollutants

Sulfur and nitrogen oxides produced by fossil fuel combustion and production processes are washed out of the atmosphere by rain as sulfates and nitrates. Lakes, rivers and soils have become acidified by this process, particularly in Europe and North America. Natural sources of these gases are substantial, but environmental impacts can result from localised elevated concentrations of anthropogenic emissions. The principal sources in Australia are smelting, power generation and oil refining (Amos *et al.*, 1993; State of the Environment, 2001c). To date, impacts of this kind in Australia have been relatively minor.

Oil spills from tankers and rigs, and chemicals used in clean up operations damage marine environments. Oil fouls plants, animals and coastal habitat, and contains hydrocarbons that can have lethal, sublethal or carcinogenic effects on marine mammals, birds, fish and invertebrates (Amos *et al.*, 1993; Estes, 1998). Although oil spills are comparatively rare in Australia, a substantial spill in Bass Strait in the mid-1990s affected seabird populations and other forms of marine life (State of the Environment, 2001a). In more recent times, the Australian Maritime Safety Authority has developed a rapid response initiative to reduce the environmental impacts of spills (State of the Environment, 2001b).

Limiting chemical pollution

Industry-based regulations have the potential to improve environmental practices and accountability. For example, the Australian Plastics and Chemicals Industries Association launched a Code of Practice in 1989. The code is intended to promote transparency and the acceptance of responsibility in dealing with the environment. It provides for public comment through an advisory panel that includes non-industry representatives. The code stipulates company and external audits to assess compliance with environmental standards and to measure environmental performance. Audit protocols, particularly external independent audits, also have the advantage that they foster credibility for the industry in the eyes of the community.

Box 5.4

Impacts of tributyltin

Amos *et al.* (1993) reviewed the impacts of tributyltin (TBT), an organo-metallic compound that is the principal biocide in many anti-fouling paints used on boats. Toxic lethal and sublethal effects of low concentrations of TBT have been described for many groups, including molluscs, crustaceans, whelks and salamanders. The compound is also believed to lead to deafness in marine mammals. TBT has been found to concentrate in the tissues of the Sydney Rock Oyster and is also associated with shell deformities and reduced weight in that species (Batley *et al.*, 1989; I. White, personal communication). Concentrations of TBT are likely to be highest in enclosed waters such as bays and estuaries.

Concerns about the effects of TBT have led to it being banned for use on small boats in some Australian States. However, the use of TBT is not banned on large ships, or on boats from other areas that use the waters of Australian ports, estuaries and marinas. The New Zealand Navy banned the use of tributyltin paints on its vessels and in dry dock in 1989. The New Zealand Government had already banned the use of the paints on boats less than 25 metres in length. The two actions together effectively eliminated the toxin from New Zealand waters, except for leaching from foreign vessels (Oryx, 1990; Amos *et al.*, 1993). In Australia in late 2000, a Malaysian-registered container ship became grounded on the Greater Barrier Reef and released large quantities of TBT. During the clean-up operation, divers removed TBT paint flakes from 1500 square metres of reef (State of the Environment, 2001b). Under the Australian Federal Government's Oceans Policy, there are plans to ban TBT by 2006, unless the International Maritime Organization bans it sooner (State of the Environment, 2001b).

5.3 Climate change

The human basis for climate change

Global warming, or the enhanced **greenhouse effect**, is a direct consequence of human activities. It has the potential to disrupt or permanently change global ecological processes. There has been a steady rise in global temperatures over the past 100 or more years (World Meteorological Office, in Australian Bureau of Statistics, 2001; Figure 5.11). Temperatures during the 1990s were the highest since records began in the 1860s.

Global warming occurs because carbon dioxide, methane, nitrous oxide and CFCs accumulate in the atmosphere (see Figure 5.12). Together these gases influence temperatures by increasing the reflective capacity of the atmosphere, retaining heat on the earth. Major increases in the levels of atmospheric carbon dioxide have come from the use of fossil fuels, fires, leaks from oil and gas wells, land clearing, ruminant animals, coal mining, rice paddies and land fills.

Studies in 1990 and 2002 showed that on a per capita basis Australia releases greater volumes of

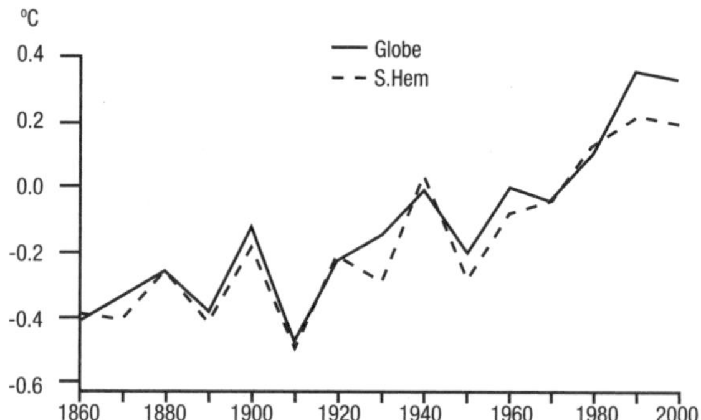

Figure 5.11. Trend in temperatures (expressed as a departure from the 1961–1990 mean) globally and for the Southern Hemisphere. (Redrawn from the Australian Bureau of Statistics, 2001.)

Box 5.5

Global warming and the Northwest Passage

Many mariners in the 15th, 16th and 17th centuries searched in vain for the fabled Northwest Passage, a link between Europe and the Far East. One of the more unexpected potential outcomes of a much warmer planet, resulting from the enhanced greenhouse effect, is the opening of a Northwest Passage between the Atlantic and Pacific Oceans. Studies by the United Nations have indicated that by 2080 (and perhaps as much as 50–60 years earlier) the shrinking Arctic ice cap could connect the Atlantic and Pacific Oceans with a northern sea route. Although this would be a boon to shipping companies (cutting thousands of nautical miles off transport between countries such as Venezuela and Japan), it could substantially change the ecology of regions inhabited by indigenous human communities living in the far north, as well as many elements of biodiversity, including the Caribou and Polar Bear.

greenhouse gases into the atmosphere than any other country (Figure 5.13; see also Chapter 12). A striking feature of the profile of emissions from Australia is the relatively large contribution made by forestry and agriculture (Table 5.7). Nearly 10% of the continent's total greenhouse gas emissions result from the permanent clearing of native vegetation for agriculture (Australian Greenhouse Office, 2002; see Chapter 9). In most cases, no productive use is made of this vegetation – it is stacked into windrows and burnt.

Despite increasing evidence of conservation problems arising from the enhanced greenhouse effect (see reviews by Hughes, 2000; McCarty, 2001), net national greenhouse emissions from Australia continue

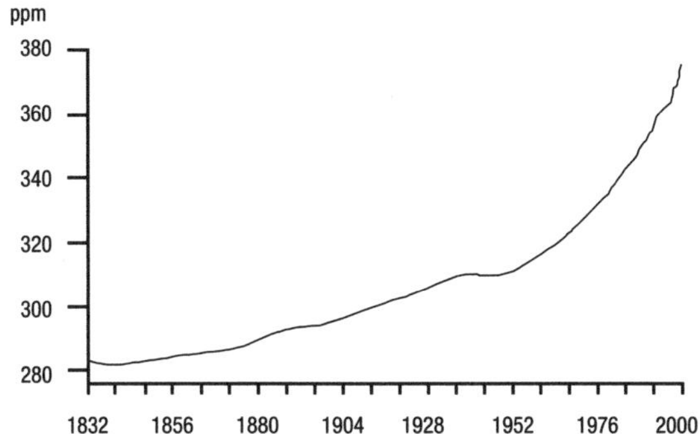

Figure 5.12. Increasing concentrations of carbon dioxide in the atmosphere. (Redrawn from Australian Bureau of Statistics, 2001.)

Land clearing in Queensland – the Australian State where this conservation problem is most pronounced. Before it was cleared, this area supported a population of the Mahogany Glider. (Photo by Bruce Cowell.)

to increase. Data from the Australian Greenhouse Office (2002) indicated that net emissions increased by 2.1% between 1999 and 2000 and by 6.3% in the 10 years between 1990 and 2000. Foran and Poldy (2002) predicted that by 2050 emissions could rise by as much as 300% above 1990 levels.

Predicting future climate change

Climate change predictions are uncertain, but almost all predict substantial global temperature increases in the future. The enhanced greenhouse effect may lead to an increase of 2°C in the mean world temperature and an increase in mean sea levels of 30–50 centimetres over the next 40 years (Pearman, 1988; State of the Environment, 2001c). The Intergovernmental Panel on Climate Change (2001a,b) suggested that changes of almost 6°C in temperature are possible by the year 3000. There may also be positive feedbacks between climate and vegetation such that climate change will be accelerated by accompanying vegetation change (Foley *et al.*, 2003). There may be marked changes in local rainfall and temperature regimes (Pittock and Nix, 1986). In fact, there is strong evidence that changes are already taking place (Hughes, 2000, 2003; Thomas *et al.*, 2004). For example, temperatures and rainfall patterns in many parts of Australia have changed over the last 50 years (Bureau of Meteorology, 2002; Figure 5.14), for example in south-west Western Australia, where there has been a 10–20% decrease in annual rainfall since the mid-1970s (Timbal, 2004),

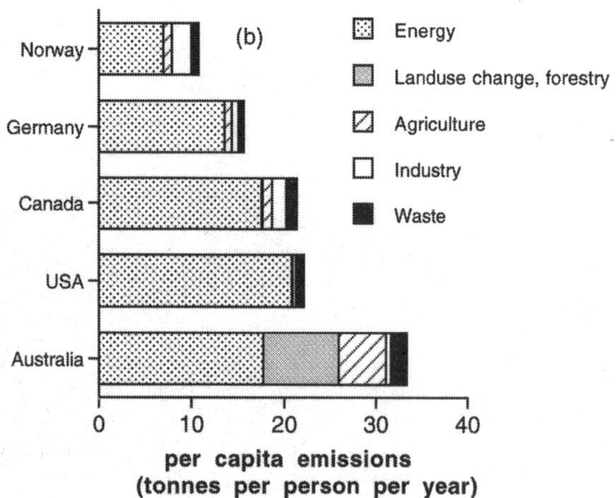

Figure 5.13. Greenhouse gas emissions in Australia and four other countries per capita. (Based on data from the National Greenhouse Gas Inventory, 1988–1998.)

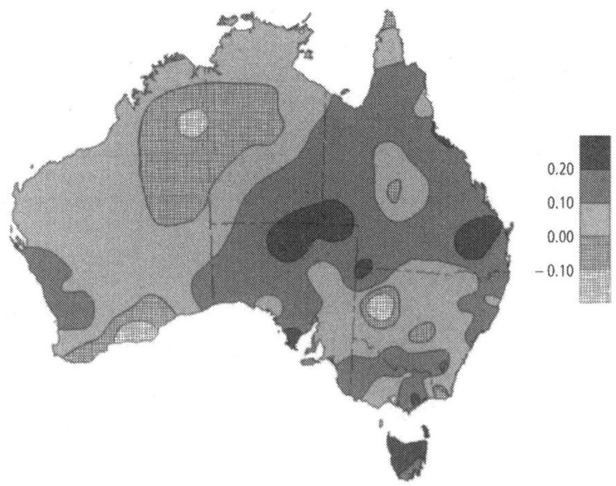

Figure 5.14. Trends in maximum daily temperature between 1950 and 2001. (Based on data from the Bureau of Meteorology, 2002.)

although these may be cyclical changes rather than long-term trends (Mackey *et al.*, 2002).

Impact of climate change on species distribution patterns

Changes in climate threaten biological resources because the predicted rates of change lie outside the range of variation to which living organisms have been exposed over the last several hundred thousand years (Pearman, 1988; Hughes *et al.*, 1996; Hannah *et al.*, 2002). That is, although climatic conditions have been as warm and wet in the past as they are predicted to become, the *rate of change* is unprecedented – it is too fast to allow an evolutionary response. Furthermore, many past mass extinction events (see Chapter 2) have been directly or indirectly associated with climate change. In the present, many species will need to deal with both an extraordinarily rapid rate of climate change, increased variability in climatic conditions, and the added impacts of human activities such as land clearing (Novacek and Cleland, 2001).

Climate affects the broad-scale distributions of plants and animals, influencing the availability of water, the nature of temperature regimes and the frequency

Table 5.7. Australian net greenhouse gas emissions by sector in 2000 (from the Australian Greenhouse Office, 2000).

Sector and subsector	CO_2		CH_4		N_2O		CO_2-e	
	Amount (megatonnes)	Percentage of total	Amount (megatonnes)	Percentage of total	Amount (megatonnes)	Percentage of total	Amount (megatonnes)	Percentage of total
All energy (combustion and fugitive)	339.2	89.3	1.31	22.7	0.017	16.3	371.8	69.5
Stationary energy	261.0	68.7	0.09	1.6	0.003	3.2	264.0	49.3
Transport	71.7	18.9	0.02	0.4	0.013	13.0	76.3	14.3
Fugitive emissions from fuel	6.5	1.7	1.19	20.7	0.000	0.1	31.5	5.9
Industrial processes	7.8	2.0	0.00	0.1	C	C	10.3	1.9
Solvent and other product use	N/A	N/A	N/A	N/A	N/A	N/A	N/A	N/A
Agriculture	N/A	N/A	3.47	60.3	0.082	79.9	98.4	18.4
Land use change and forestry	32.8	8.6	0.21	3.7	0.002	2.2	38.0	7.1
Waste	0.0	0	0.77	13.3	0.002	1.7	16.7	3.1
Total net emissions	**379.9**	**100.0**	**5.8**	**100.0**	**0.103**	**100.0**	**535.3**	**100.0**

N/A, not available.

and severity of extreme events (Nix, 1986; Woodward, 1987; Mackey, 1994). MacArthur (1972) suggested that a change of 3°C may lead to a shift in habitat of roughly 250 kilometres in latitude or 500 metres in altitude. Hence, global warming may result in shifts in the distributions of many species (Figure 5.15; Table 5.8; Hughes *et al.*, 1996; Hannah *et al.*, 2002; Thomas *et al.*, 2004) and prevent breeding in others (e.g. seabirds; Gjerdrum *et al.*, 2003).

Computer models have been used to predict the distribution patterns of Australian plants and animals by predicting the locations of suitable bioclimatic conditions (Busby, 1988; Bennett *et al.*, 1991; Lindenmayer *et al.*, 1991b; Brereton *et al.*, 1995; Beaumont and Hughes, 2002; see Chapter 15 for details). To account for uncertainty in climate change predictions, Brereton *et al.* (1995) modelled a range of possible climate change scenarios. They used many different values for winter and summer rain, and maximum and minimum temperatures, to predict changes in the distribution patterns of suitable bioclimatic conditions for 39 vertebrates and two invertebrates, most of which were 'threatened'. Most areas of suitable bioclimatic conditions were predicted to contract (Table 5.8). Changes in the range of suitable conditions for the Swamp Antechinus were typical of the results for many of the taxa (Brereton *et al.*, 1995; Figure 5.16).

One of the consequences of increased temperatures may be a shift by species to higher elevations (Brereton *et al.*, 1995). In the case of the Mountain Pygmy Possum, which is presently confined to areas characterised by winter snow (Broome and Mansergh, 1995), suitable climatic conditions may be totally eliminated (Brereton *et al.*, 1995). Other organisms that are restricted to high elevation areas, such as the plants in the windswept and snowpatch feldmark communities, may also become extinct if temperatures rise.

In a study of climate change impacts on butterflies, Beaumont and Hughes (2002) suggested that species that can tolerate a narrow set of climatic conditions, that have limited geographic distributions

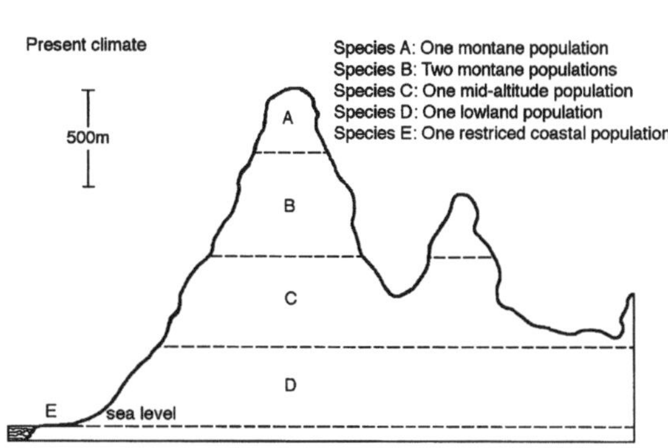

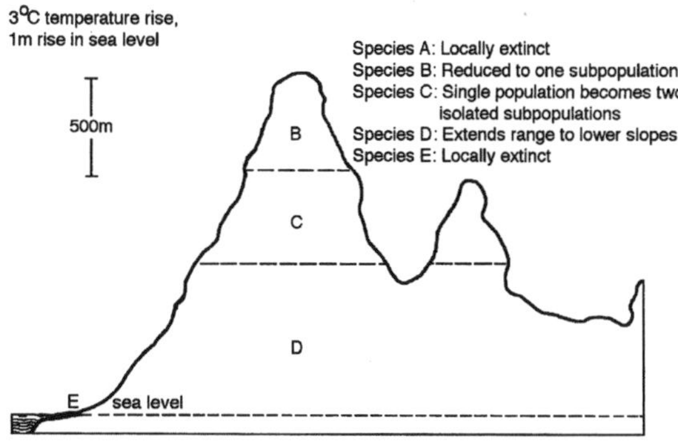

Figure 5.15. Idealised representations of changes in the distribution of species in response to changes in mean temperature (after Mansergh and Bennett, 1989).

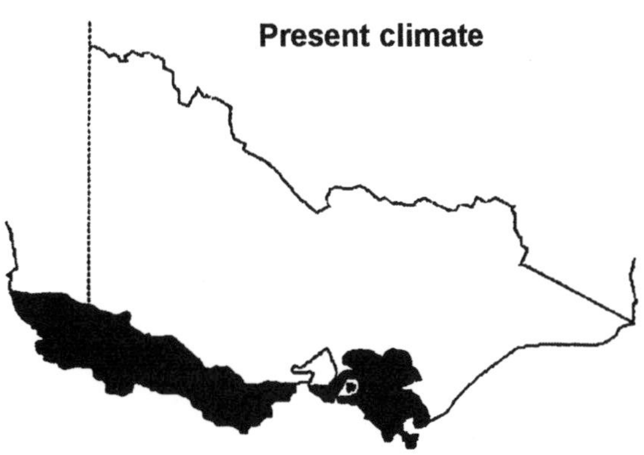

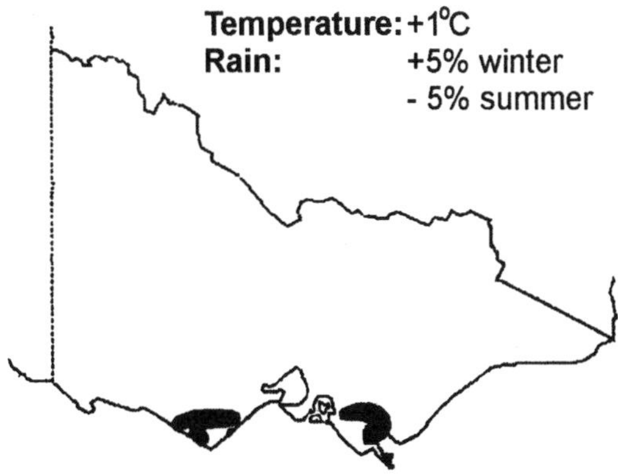

Figure 5.16. Predicted changes in the distribution of the Swamp Antechinus (after Brereton *et al.*, 1995).

Table 5.8. Changes in the predicted distributions of suitable bioclimatic conditions for a range of species in south-eastern Australia in response to a range of scenarios for the enhanced greenhouse effect (after Brereton *et al.*, 1995).

Species	Present	Distribution (x 100 square kilometres)					
		A	B	C	D	E	F
Swamp Antechinus	353	44	1	1	0	1	1
Southern Brown Bandicoot	1752	1519	1034	520	461	535	574
Eastern Barred Bandicoot	267	111	9	0	0	0	0
Leadbeater's Possum	224	264	119	50	96	109	113
Little Pygmy Possum	434	127	10	0	0	0	0
Mountain Pygmy Possum	9	0	0	0	0	0	0
Long-footed Potoroo	43	25	18	12	24	15	4
Greater Pipistrelle	1873	1469	702	397	418	427	406
Broad-toothed Rat	477	306	222	119	163	161	145
New Holland Mouse	732	735	660	596	447	439	389
Smoky Mouse	520	300	171	69	101	129	128
Heath Mouse	280	17	3	0	0	2	4
Mitchell's Hopping Mouse	514	34	6	0	0	0	0
Malleefowl	3281	2406	166	252	133	231	511
Plains Wanderer	4369	4241	3606	2664	2193	2376	2544
Sooty Owl	972	853	579	347	367	356	304
Red-tailed Black Cockatoo	356	42	3	0	0	4	7
Gang-gang Cockatoo	2619	2377	2082	1506	1521	1521	1482
Regent Parrot	747	433	122	5	1	1	2
Orange-bellied Parrot	237	279	110	3	3	6	11
Swift Parrot	3285	3133	2417	1517	1472	1585	1628
Ground Parrot	656	380	135	48	59	58	59
Pink Robin	1566	377	806	466	443	461	463
Red-lored Whistler	723	358	28	4	0	1	3
Western Whipbird	168	2	0	0	0	0	0
Eastern Bristlebird	337	408	353	351	331	242	158
Rufous Bristlebird	271	93	7	0	2	2	5
Mallee Emu-wren	383	8	3	0	0	0	0
Regent Honeyeater	3004	3232	3094	2377	2208	2164	2061
Helmeted Honeyeater	3	7	0	0	0	0	0
Striped Legless Lizard	1052	627	222	15	8	35	112
Swamp Skink	303	274	76	6	6	12	20
Spencer's Skink	1182	780	444	256	256	258	234
She-oak Skink	414	271	164	101	133	102	71
Brook's Striped Skink	526	265	18	0	0	0	1
Glossy Grass Skink	1099	710	348	201	234	249	265
Spotted Tree Frog	69	39	3	0	2	2	13
Growling Grass Frog	4602	3467	1978	946	1184	1272	1321
Alpine Tree Frog	365	265	160	83	117	124	112
Giant Gippsland Earthworm	26	4	0	0	0	0	0
Altona Skipper Butterfly	127	134	4	0	0	0	0

Climate change scenarios:

A: +1°C; +5% rainfall in summer; −5% rainfall in winter

B: +2°C; +7.5% rainfall in summer; −7.5% rainfall in winter

C: +3°C; ±10% rainfall in summer; −10% rainfall in winter

D: +3°C; +10% rainfall in all months

E: +3°C; same rainfall as present

F: +3°C; −10% rainfall in all months

Swamp Antechinus – a species predicted to suffer a major contraction in distribution as a result of climate change (see Figure 5.16). (Photo by Jason van Weenen.)

or poor dispersal capabilities, might be particularly vulnerable to climate change. For example, the Mahogany Glider in far north Queensland has a highly restricted distribution that may be lost if the climate changes substantially. Suitable habitat for the Yellow-bellied Glider is interspersed with extensive areas of unsuitable habitat, which may be difficult to traverse (Lindenmayer, 2002b). The translation of the climatic envelope of a species to a new location will not result in its demise if all of its requirements are met in the new location and all other species upon which it is ecologically dependent also move in concert. Given these kinds of contingencies, it is very difficult to predict the consequences of climate change for any species (Hampe and Petit, 2005).

Parmesan (1996) surveyed populations of Edith's Checkerspot Butterfly in Mexico, USA and Canada, and provided some of the first pieces of evidence of changes in species distribution patterns in response to global warming. Populations inhabiting warmer areas, that is, those in the south (for example Mexico) and at lower elevations (less than 2400 metres above sea level), were more likely to become extinct than those inhabiting cooler areas (in Canada and in locations above 2400 metres; Parmesan, 1996). The study accounted for the influences of population isolation and land-use practices. Other studies, such as the work by Pounds *et al.* (1999), have documented extinctions that have occurred already as a result of climate change.

Climate change and reserve design

The study by Brereton *et al.* (1995) and others like it (e.g. Hughes *et al.* 1996; Beaumont and Hughes, 2002) have a number of important implications for nature conservation. Places presently suitable for a species may not support suitable habitat in the future (Tellez-Valdes and Davilla-Aranda, 2003). Hence, reserves set aside for the conservation of a particular species may not be effective in the long-term, including those in marine systems (MacLeod *et al.*, 2005). Reserves that protect biodiversity hotspots, where large numbers of species occur, may be particularly vulnerable (Hannah *et al.*, 2002). Boundary fences around some reserves may limit the ability of some species in protected areas to shift their ranges in response to climate change (Ogutu and Owen-Smith, 2003). Current strategies for selecting reserves do not take climate change into account (Hannah *et al.*, 2002), although it is possible to do so (Hampe and Petit, 2005).

One strategy is to ensure that reserves capture full environmental gradients, such as complete ranges of topography and altitude. If this was the case, some plants and animals would have the opportunity to shift to higher elevations in response to changed climatic conditions. Other species would need to be carefully managed in areas outside parks and reserves so that they can make gradual shifts in distribution (Lindenmayer and Franklin, 2002).

Other impacts of climate change on biodiversity

Many vertebrate species may cope with change by moving. Invertebrates, such as soil-borne fauna, may

Box 5.6

Climate change and potential impacts on folivorous animals

Changes in climate may have subtle but nevertheless important impacts on biota – sometimes in unexpected ways. Kanowski (2001) speculated that elevated levels of carbon dioxide in the atmosphere might not only alter the growing patterns of plants, but could also change aspects of leaf chemistry, such as the amount of nitrogen and other plant compounds. This could, in turn, change the palatability and nutritional value of food plants for many folivorous animals, ranging from the Greater Glider to an array of leaf-eating invertebrates. Reductions in populations of animals such as the Greater Glider could influence other species associated with them, for example the Powerful Owl, which preys on large numbers of gliders in parts of its distribution (Kavanagh, 1988).

be incapable of moving fast enough to keep pace with suitable bioclimatic conditions. In the case of long-lived plants such as trees, adult plants may tolerate change, but new seedlings may not germinate or survive after the parent plants die. The movement of whole communities of species in response to shifts in bioclimatic conditions is unlikely. Species persistence will depend on the magnitude of changes, and the presence of sufficient genetic variability so that at least some individuals survive novel environmental conditions.

Research on the predicted changes in plant and animal distributions needs to be interpreted with caution. Climate is only one of several processes that influence the distribution of organisms; competition and predation (see Chapter 15), the dispersal abilities of species and those they depend on, and the degree of habitat fragmentation can also be important. Broad-scale changes in environmental conditions may produce unexpected responses (e.g. Schindler *et al.*, 2005) and create new assemblages of taxa. This may lead to new interactions between species (Stachowicz *et al.*, 2002), including pest species (Logan *et al.*, 2003). Other possible impacts of climate changes on biota, some subtle and not readily predictable, are outlined in Table 5.9.

The amount of carbon dioxide in the atmosphere generally limits plant growth. Increased levels will probably result in faster plant growth in many natural, semi-natural and agricultural systems. However, as Possingham (1993) noted, the persistence of natural populations depends on survival and reproduction, not on growth rates. Competitive interactions between plants and animals may be affected by elevated carbon dioxide in unpredictable ways. For instance, the toxicity and unpalatability of leaves may increase, affecting the survival of fauna populations (Box 5.6). Other complex predator-prey inter-relationships may develop as a result of climate change (Schindler *et al.*, 2005).

The consequences of sea level rises may be very substantial, particularly when we consider the huge proportion of the world's population and its associated infrastructure that is located on or close to the coast (Chapter 12). In addition to the disruption of coral reefs, seagrass beds, **mangroves** and other highly productive coastal and shallow water ecosystems, disruption to and changes in the distribution of the human population because of sea level rises may have consequences for the environment that far outweigh any direct influences of climate change (see Chapter 12).

5.4 Conclusions

Some changes in the physical environment can be seen as magnifications of natural processes, such as climate change due to the enhanced greenhouse effect. Others, such as the input of novel chemicals to the environment, are entirely new phenomena. Many of the threats and impacts outlined in this chapter are inferred, descriptive, or are theories based on an understanding of ecological processes. Because of limited survey and monitoring data at a continental scale, it is difficult to estimate the absolute effects of many potential or actual threats and impacts. The absence of information concerning the occurrence and magnitude of environmental changes does not

Table 5.9. Potential impacts on biodiversity resulting from rapid changes in climate associated with global warming.

Potential changes	Reference
Contractions or expansions in the ranges of native species	Parmesan (1996)
Altered breeding times (e.g. flowering patterns, bird migration, frog spawning)	McCarty (2001), Corn (2003), Cotton (2003)
Altered breeding outcomes (e.g. sex of reptile offspring)	
Altered growth patterns (e.g. growing seasons for plants) and problems with germination of seedlings	Hughes *et al.* (1996)
Extensive coral bleaching	
Changed palatability of plant-based food for herbivorous animals	Kanowski, 2001
Increased exotic invasions and rate of growth of invading species	Stachowicz *et al.* (2002)
Fragmentation of a continuous range of a species into a patchy and disjunct distribution	Peters and Darling (1985)
Complete extirpation of species	Mansergh and Bennett (1989)
Altered food webs and inter-species trophic interactions	Petchey *et al.* (1999)
Altered fire regimes	Cary (2002), Mackey *et al.* (2002)

Greater Glider – a currently widespread folivorous arboreal marsupial that could be influenced by climate change. (Photo by David Lindenmayer.)

imply that the impacts are absent. If we take the view that we should act to protect the environment only when damage is demonstrable, or when we have conclusive evidence that the impact is severe, we create a disincentive to monitor and report environmental impacts. The precautionary principle outlined in Chapter 1 seeks to reverse this pattern. Methods for exploring the likelihood of impact are outlined in Chapter 17.

McNeely *et al.* (1990) argued that the severity of changes in the physical environment can be explained by the distribution of the costs and benefits of exploitation and conservation. One solution to these problems may be to ensure that those people who exploit natural resources pay the full cost of conservation.

5.5 Practical considerations

Reliable measures of the ecological impacts of chemicals, nutrients and salt are important. Managers should monitor human activities that are likely to directly affect these variables and develop systems in which proponents and users of environmental services pay the full cost of environmental degradation.

Reserves should be designed to include as full a range of continuously distributed ecological gradients as possible. The gradients may provide pathways for plants and animals to track climate changes and relocate to places with suitable climatic conditions. The dispersal of species with narrow climatic tolerances, geographic ranges and limited dispersal capabilities may have to be managed directly.

5.6 Further reading

McTainsh and Boughton (1993), Ghassemi *et al.* (1995), Graetz *et al.* (1995), and Barson *et al.* (2000) review aspects of land degradation in Australia. Detailed assessments of the extent of land degradation are provided in the National Land and Water Resource Audit (Commonwealth of Australia, 2001b–e).

Chartres *et al.* (1992), McMahon *et al.* (1992a), Smith (1998) and Stirzaker *et al.* (2002) provide readable accounts of the human influences on land degradation and hydrological cycles in Australia. Kingsford (2000) contains an excellent review of the wide range of ecological impacts associated with dams, water diversions and other aspects of river management. In addition, there is an extensive set of major government documents on salinity in Australia – perhaps reflecting the magnitude and seriousness of the problem. These include (among many others) the Australian Dryland Salinity Assessment (2000) (part of the National Land and Water Resource Audit, 2001), and the Salinity Audit of the Murray–Darling Basin (Murray–Darling Basin Commission, 1999). Volume 56, issue 6 of the *Australian Journal of Botany* contains a series of articles on the impacts of salinity on biodiversity. Stirzaker *et al.* (2002) is an excellent source of information on revegetation strategies to reduce salinity impacts.

Amos *et al.* (1993) and State of the Environment (2001a,b,d) provide a detailed description of the threats to biological resources in Australia. Williams (1987), Hart *et al.* (1991) and Metzeling *et al.* (1995) describe a wide range of taxa that are potentially at risk from salinisation, ranging from frogs and aquatic plants, to macro- and microinvertebrates.

The seminal work on the sublethal effects of pesticides on wildlife populations is by Carson (1962).

Nix and Kalma (1972), Nix (1978, 1986), Austin *et al.* (1990), Mackey (1993, 1994) and Mackey and Lindenmayer (2001) discuss the influence of moisture regimes and other environmental attributes (e.g. temperature) on species distribution and abundance.

Caldwallader (1986), Blakers *et al.* (1984), Carstairs (1976), Caughley (1978), Mansergh and Bennett (1989), Rose (1995), Beaumont and Hughes (2002) and Chambers *et al.* (2005) outline the impact of climatic fluctuations on the Australian biota. Kinrade (1995) and Foran and Poldy (2002) review Australia's greenhouse gas emissions.

The State of the Environment reports (State of the Environment, 1996, 2001a–d) and the National Land and Water Audit (Commonwealth of Australia, 2001b–e, 2002a,b) contain considerable additional information on the Australian physical environment.

Chapter 6

Loss of genetic diversity, populations and species

This chapter outlines the ways in which human activities affect genetic diversity, the abundance of populations, and the distributions of species. It also examines the past, present and future magnitude of these changes through an exploration of past background extinction rates and mass extinction events. If processes that cause biodiversity loss continue for a few hundred years, the present biodiversity extinction crisis is likely to at least parallel past mass extinction events that have occurred periodically during the geological history of the earth.

6.0 Introduction

The impacts of human activities on biodiversity translate directly into losses of genes, populations and species. Chapter 5 described some of the impacts of human activities on the physical environment, and their potential consequences. These activities affect individuals and populations and hence may affect the amount or spatial distribution of genetic variation. Human impacts also result in local and global extinctions of species.

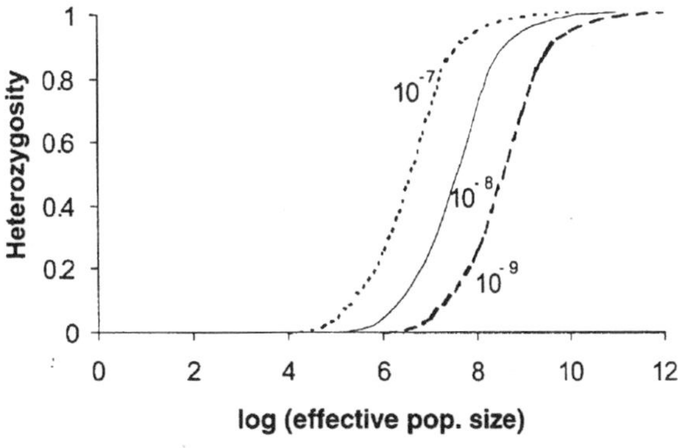

Figure 6.1. Predicted relationship between heterozygosity (H) and logarithm of effective population size (log N) for three different neutral mutation rates. (Redrawn from Frankham, 1996.)

6.1 Loss of genetic variation

In general, larger populations have more genetic diversity than small ones (Billington, 1991; Lacy, 1993a; Frankham, 1996; Young *et al.*, 1996; Figure 6.1). Genetic diversity protects species against unpredictable environmental change, and contributes to the reproductive effectiveness of individuals and the viability of their offspring. The impacts of reduced or modified genetic variation are summarised in the following section.

Inbreeding

Inbreeding is breeding between genetically related individuals and it can result in elevated levels of **homozygosity** in a population. When two relatives breed, there is an increased chance that **alleles** at each gene **locus** will be homozygous because the same allele is passed to an offspring from a common ancestor, via the parents. Consider an example in which there are two alleles, C and c, at a locus. The c allele is **recessive.** The frequency of the C allele in the population is 0.99. Thus, the chance of a random individual being

Box 6.1

Losses of genetic variability in the Black-footed Rock Wallaby

Eldridge *et al.* (1999) compared patterns of genetic variability in an isolated population of the Black-footed Rock Wallaby on Barrow Island off north-western Australia with populations on mainland Australia. The Barrow Island population had extremely low levels of genetic variability and suffered from symptoms of inbreeding depression such as reduced fecundity of females and a skewed (highly female-biased) sex ratio. In addition, whereas animals on the mainland showed normal patterns of bilateral symmetry in their limbs and other parts of their anatomy, animals in the isolated Barrow Island population were asymmetrical.

Inbreeding depression has reduced the fitness of the Barrow Island population of the Black-footed Rock Wallaby, and is likely to contribute to its extinction proneness. Other small and isolated populations of mammals on the islands of Western Australia are likely to suffer from the same genetic pressure (Eldridge *et al.*, 1999). This is an important issue given that island populations are both a refuge and a reservoir for an array of mammal species that are threatened on the Australian mainland (Richards *et al.*, 2001).

homozygous for the c allele is 1 in 1000. If two individuals breed, one of which is heterozygous for the c allele, 50% of their offspring will be Cc and 50% will be CC. If these offspring mate with each other, 1 in 16 of their progeny in the next generation will have the cc genotype.

Black-footed Rock Wallaby. (Photo by Jeff Cole.)

Inbreeding depression is the reduction in survival or fecundity that results from elevated homozygosity for **deleterious alleles** (e.g. the cc case above, if c is a deleterious allele). For example, Saccheri *et al.* (1998) demonstrated that low levels of genetic variation and subsequent inbreeding depression significantly increased the risk of extinction in fragmented populations of the Glanville Fritillary Butterfly in Finland. Another example is the Black-footed Rock Wallaby (Eldridge *et al.*, 1999).

Inbreeding depression can result from the loss of **heterosis** (the vigour associated with heterozygous loci), or it can result from the expression of rare recessive deleterious alleles in homozygous form. All

Box 6.2

How to maintain high levels of genetic variation in tree populations

Tree planting is an increasingly important activity in rural and regional Australia, particularly in areas where previous woodland cover has been significantly reduced and only small, isolated remnants remain (see Gibbons and Boak, 2002). A key problem when sampling seeds from local populations of trees is how to maintain levels of genetic variability among the species targeted for planting. Mlot (1989) recommended collecting seed from 50–100 trees. However, this may be impossible in many cases where remaining woodland stands are limited in size (and tree numbers). Others have recommended that seed be collected from only 10 trees, but that it should come from trees some distance apart, for example several canopy widths apart, so that closely related plants are not sampled (Boland *et al.*, 1980; McIntyre *et al.*, 2002). Prober and Brown (1994) recommended that seed for reconstructed White Box woodlands should be collected from populations with more than 500 individuals. Andrew Young (personal communication) recently suggested that smaller amounts of seed should be gathered from larger numbers of trees (in larger patches) rather than vice versa. This collection strategy maximises the genetic diversity of the restored population. In addition, seed should be collected during droughts and over several years to ensure that planted trees can persist over a range of climatic conditions. All of these rules of thumb may also be useful when collecting seed for the establishment of *ex situ* populations of trees and other plants (see Chapter 4).

populations carry a **genetic load** – a proportion of deleterious alleles that is maintained by the balance between mutation and selection. The genetic load in a population is expressed when inbreeding results in the expression of these alleles.

Both these mechanisms can reduce traits directly associated with fitness, such as the number of offspring per litter, the number of viable offspring, the birth-weight of offspring, the survivorship of lactating mothers, and so on. Ralls *et al.* (1988) reviewed 50 species in which naturally outbreeding, wild-caught populations experienced inbreeding in zoos. Of these, 40 species exhibited symptoms of inbreeding depression in fitness-related traits, such as reduced survival and reproduction. Detrimental effects of inbreeding have been found in many wild populations, for example butterflies, fish, snails, snakes, Lions, shrews, Deer Mice, birds and plants (Orr, 1994; Frankham, 1995a; Newman and Pilson, 1997; Eldridge *et al.*, 1999; Madsen *et al.*, 1999; Schwartz and Mills, 2005).

Inbreeding does not always lead to inbreeding depression, and even when it occurs, it may not be easy to detect. Many plants are adapted to breeding with themselves or with adjacent plants that are very likely to be relatives. For example, Gungurru occurs naturally in isolated stands on granite outcrops in south-western Australia. These populations may well have been isolated from one another for more than a thousand years, and most populations have low levels of variability remaining, probably due to inbreeding and drift. There are no known detrimental effects of inbreeding in this species, perhaps because deleterious recessive alleles were eliminated by selection in past generations (Moran and Hopper, 1983; Hopper and Coates, 1990).

Worthington-Wilmer *et al.* (1993) examined the relationship between levels of inbreeding and growth rates of individuals in a captive Koala population. They failed to find an important effect (Figure 6.2) and speculated that Koalas may be somewhat immune to inbreeding depression because low historical population sizes, inbreeding and **genetic drift** (the random

Box 6.3

An experimental study of inbreeding depression

A North American study examined the inter-relationships between inbreeding and survival among populations of White-footed Mice in a natural habitat. Jimenez *et al.* (1994) collected wild White-footed Mice from an area of mixed deciduous forest at Brookfield Zoo in Chicago, USA, and used them to establish captive populations. A breeding program was then instigated in which some animals were deliberately inbred. Inbred animals, together with non-inbred descendants of the wild mice, were then released back to the field site on the grounds of Brookfield Zoo.

The researchers found that the weekly survivorship of inbred mice was 56% lower than that of the non-inbred conspecifics. Inbred male White-footed Mice declined in body weight throughout the study. Conversely, males that were not inbred rapidly regained the body weight lost after the first few days of release. The results of the study suggested that there were significant impacts of inbreeding depression on the survival of the released White-footed Mice.

There are a number of other important aspects to the work by Jimenez *et al.* (1994). For example, the mortality rates among inbred mice were low in the captive populations held in the laboratory. Therefore, the negative effects of inbreeding depression were far more pronounced in the natural environment than under laboratory conditions. Furthermore, the inbreeding program for the mice produced animals that were only moderately inbred compared with other rodents and mammals typically held in captivity, which highlights both the consequences of inbreeding that can arise from captive management, and the potential for such problems to compromise strategies such as species reintroduction (Miller, 1994; see the section on reintroductions in Chapter 4).

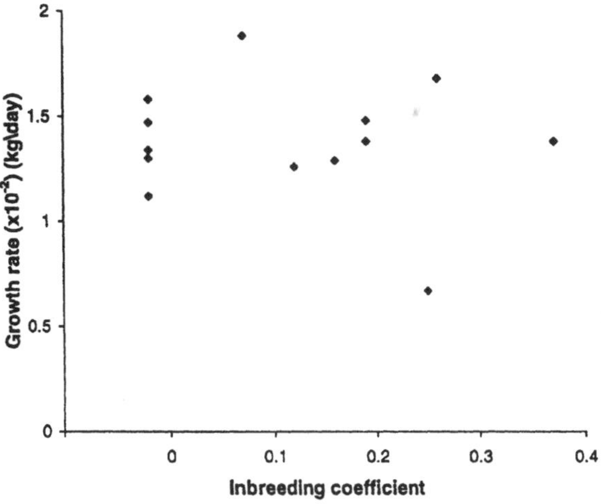

Figure 6.2. Relationships between levels of inbreeding and juvenile growth rates in a captive Koala population (after Worthington-Wilmer *et al.*, 1993).

White-footed Mouse. (Photo by Bob Lacy.)

selection of genes from one generation to the next) have led to the expression and subsequent elimination of rare, recessive and deleterious alleles. The negative impacts of inbreeding are hard to detect in mammals that are mobile and have comparatively long generation times (Lacy, 1993a).

Captive populations do not experience extreme climatic conditions, predation or disease (Miller, 1994; McPhee, 2003; see Chapter 4). Little is known about the relationships between inbreeding and survivorship in wild organisms where climate, predation and disease influence population dynamics. Although inbreeding depression can affect traits such as age-specific fecundity or survivorship, or even the variability in these traits, the effects can be complex and usually there is little or no information relating population dynamics to levels of inbreeding.

Outbreeding

Outbreeding is breeding between individuals that have been isolated from one another in more or less separate gene pools (Goldberg *et al.*, 2005). Different populations are likely to be fixed (i.e. only a single allele exists in the population, so that all individuals are homo-zygous) for at least some alleles, so that any crosses will be heterozygous at these loci. Outbreeding can then result in enhanced vigour and improved performance in fitness-related traits. This is known as hybrid vigour or heterosis.

Outbreeding can also lead to **outbreeding depression**, which is a reduction in performance in fitness-related traits such as resistance to infectious disease (Goldberg *et al.*, 2005). This can occur, for example, because local populations may be adapted to local conditions, and individuals from remote populations may introduce genes that are poorly adapted to local conditions. Another mechanism for outbreeding depression is the breakdown of **intrinsic coadaptation**. Genetic complexes evolve under the influence of other genes, and the result is an integrated genome that can disintegrate if novel alleles from remote populations are introduced into the population. Rock Isotome is a small perennial shrub restricted to granite outcrops in Western Australia, and like the Gungurru, populations are isolated from one another by sandplain heathlands and eucalypt woodland, and it is likely that they have been isolated from one another for a long time. James (1982) found that as many as 90% of the ovules produced by the plants under natural conditions fail to develop because of the expression of recessive lethal genes. He suggested that isolated populations are adapted to local environmental conditions, and that the local gene pools are co-adapted for their lethal

Box 6.4

The Koala and the maintenance of genetic variation

There have been many studies of patterns of genetic variability in the Koala (reviewed by Sherwin *et al.*, 2000). The species is widely distributed and exhibits considerable morphological and genetic variation among populations across its range (Martin and Handasyde, 1999), and there appears to have been considerable local adaptation to local conditions. Locally adapted gene pools may be disrupted by the mixing of genotypes from highly divergent populations (Sherwin *et al.*, 2000).

The importance of maintaining the integrity of gene pools is at odds with: (1) the past history of extensive translocations of the species (Martin and Handasyde, 1999), and (2) animal welfare advocates wanting to 'solve' the problems of overbrowsing feral Koala populations in places such as Kangaroo Island by transporting many animals to the mainland (see Box 1.8).

Rock Isotome. (Photo by David Poyner.)

recessive alleles. Different gene pools may be differently co-adapted, and if different local populations hybridise, it may break down the co-adapted gene structures, exposing recessive lethal genes and thereby reducing the fitness of the populations.

Bottlenecks

A population **bottleneck** occurs when a population is reduced to an extremely small size (e.g. Sinclair *et al.*, 2002). One of the consequences is a rapid loss of genetic variation by genetic drift. The rate of loss of variation is approximately proportional to $1/2N$ per generation, where N is the number of breeding adults. For example, if a population is reduced to two reproductive individuals, 25% of the variation will be lost in the first generation. If the population recovers rapidly, little more variation will be lost. Furthermore, most losses would be of alleles that are rare in the parent population. Typically, these alleles are neutral or deleterious. Because of heterosis, survivors are likely to be heterozygous at more loci than the average individual in a population, implying perhaps even smaller losses than anticipated. If, however, the population remains small for several generations, then very appreciable quantities of genetic variation will be lost, irrespective of the condition or composition of the initial population (Lande, 1988; Carson, 1991; Briskie and MacKintosh, 2004).

Coates (1992) studied genetic variation in isolated populations of a rare and geographically restricted Triggerplant in Western Australia. One population went through a prolonged decline, leading to a bottleneck of just three plants, and yet there was no measurable reduction in genetic diversity in the population. Coates (1992) attributed the maintenance of heterozygosity to the rapid recovery of the population following the bottleneck, the elimination of inbred plants during seed development, and selection favouring heterozygotes as plants developed towards maturity. In support of these findings, there was a significant decrease in the level of inbreeding from younger to older plants. Coates (1992) suggested

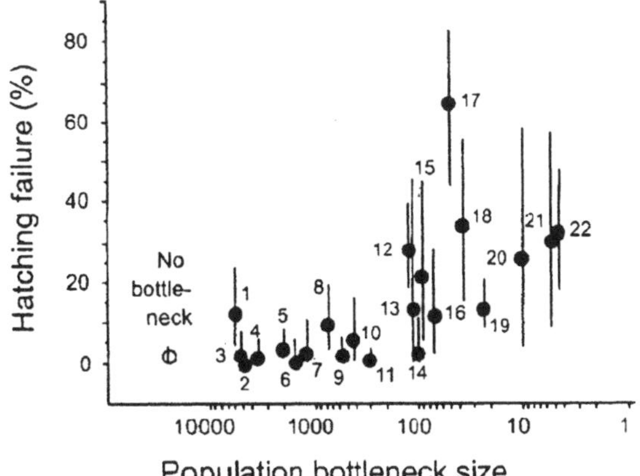

Figure 6.3. Increase in hatching failure with increasing severity of population bottleneck in native New Zealand birds. Circles are means 95% ± confidence intervals. Open circle shows mean hatching failure in 15 species that did not pass through a bottleneck. Species are: 1, *Eudyptes pachyrhynchus*; 2, *Megadyptes antipodes*; 3, *Anarhynchus frontalis*; 4, *Procellaria parkinsoni*; 5, *Haematopus unicolor*; 6, *Coenocorypha pusilla*; 7, *Charadrius obscurus aquilonius*; 8, *Petroica macrocephala chathamensis*; 9, *Anas aucklandica*; 10, *Philesturnus rufusater*; 11, *Pterodroma axillaris*; 12, *Porphyrio hochstetteri*; 13, *Ha. chathamensis*; 14, *Pt. magentae*; 15, *Thinornis novaeseelandiae*; 16, *Anas nesiotis*; 17, *Strigops habroptilus*; 18, *Ph. carunculatus*; 19, *Himantopus novaezelandiae*; 20, *Sterna nereis*; 21, *Pe. australis*; and 22, *Pe. traversi*.

that the processes of population decline and recovery may be commonplace because the environment where the plant occurs is subject to frequent disturbance, which has resulted in considerable divergence in genetic composition between subpopulations of the species.

In other cases, there have been severe impacts caused by population bottlenecks. For example, Briskie and Mackintosh (2004) studied the effects of bottlenecks on the New Zealand avifauna (Figure 6.3). They found that problems such as hatching failure were common and widespread among species of birds that had been through past population bottlenecks. Moreover, they documented evidence of bottleneck effects at population sizes well above those used to establish new populations of endangered bird species – a finding that has implications for reintroduction and translocation programs (see Chapter 4).

Hybridisation and swamping

Disturbance can increase the likelihood of hybridisation between previously isolated species (Knopf, 1992; Potts *et al.*, 2003). Disturbance sometimes breaks down the ecological, behavioural or biogeographical barriers that isolate populations or species, leading to hybridisation and the **introgression** of genes. If the number of hybrids is high compared with the size of the population, if the new genes confer a selective advantage, or if random drift is an important process (most likely in small populations), then the genes of one species can spread through the genome of the other species, a process termed **genetic swamping**.

Templeton (1991) described the effects of domestic Cattle on Javanese Banteng. Banteng were widespread in Java in 1900, but by 1960, their forest habitat was fragmented and greatly reduced in area, and only 1000 Banteng remained. Following conservation efforts, the population was stabilised in the early 1990s at about 700 individuals. However, many populations have been exposed to domestic Cattle and hybridisation has occurred. For example, in 1922, three female domestic Cattle were introduced to a nature reserve that now supports a population of 60 Banteng. The exact extent of gene introgression is unknown, but hybridisation is considered to be the most important current threat to Banteng. In another example, Riley *et al.* (2003) described how the declining California Tiger Salamander is hybridising with the exotic Tiger Salamander, which was introduced to central California as fish bait. Hybrids are common and continued interbreeding is threatening the 'purity' of the native species (Riley *et al.*, 2003).

There are many examples of genetic swamping in Australia. Breeding between domestic Dogs and Dingoes may result in the demise of the latter (Corbett, 2001; Daniels and Corbett, 2003). Closely related but formerly isolated populations of *Acacia* spp. trees have been brought geographically closer by human movements of plants. There is increasing evidence of gene flow between introduced plantation eucalypts and native eucalypt species (e.g. Barbour *et al.*, 2003).

Rare species are particularly at risk from hybridisation because of their small population size and limited gene pool (Prober *et al.*, 1990; Sampson *et al.*, 1995). The rare Buxton Gum is restricted to a handful of localities north-east of Melbourne, and is threatened by swamping from the far more common and widespread Swamp Gum (A. Jelinek, personal communication).

Hybridisation can have significant impacts on an entire biota. Of the 2834 plant species in the British Isles, 1264 are aliens. There are also 715 hybrids, 70 of which are the products of hybridisation between native and alien species; in fact, approximately 4% of the native plants of the British Isles have hybridised with alien species (Abbott, 1992). The extent of genetic introgression and the implications for the conservation of genes are unknown.

Hybridisation can occur naturally on the boundary between the ranges of two species (e.g. Ashton, 1981b); however, **hybrid zones** may not disrupt the genetic

The rare Buxton Gum, which is threatened by genetic swamping from the related Swamp Gum. (Photo by Ann Jelinek.)

integrity or the composition of either species (O'Brien and Mayr, 1991). Witham *et al.* (1991) emphasised that about 70% of flowering plant species arose from hybrids.

Hybrid zones can be centres of diversity for species associated with the hybridising species. For example, Witham *et al.* (1991) recorded insects on adjacent populations of Risdon Peppermint (an endangered species), Black Peppermint (widespread) and Risdon Peppermint × Black Peppermint hybrids. Of the 40 insect species sampled, 5 were restricted to the hybrid zone, 29 were more abundant in the hybrid zone and 20 coexisted only in the hybrid zone. O'Brien and Mayr (1991) urged the implementation of management policies that discourage hybridisation between species. However, natural hybrids may be harmful *or* beneficial to the conservation of biodiversity and they require careful study before management decisions are made (Hopper, 1997).

Mutational meltdown

Eventually, random genetic drift within a population at any **polymorphic locus** will ensure that only one allele remains. Large fluctuations in gene frequencies will result in more homozygotes and fewer heterozygotes because more genes will become fixed by chance. The expected frequencies of the three phenotypes of a two-allele (A and a) locus in a population are given by (Falconer, 1989):

$$\mathrm{AA} = p^2 + \sigma^2$$
$$\mathrm{Aa} = 2pq - 2\sigma$$
$$\mathrm{Aa} = q^2 + \sigma^2$$

where p and q are the frequencies of the alleles A and a, respectively, and σ^2 is the variance in gene frequency.

The fluctuations in gene frequencies are larger in small populations because sampling variation from generation to generation is inversely related to the size of the population (see Falconer, 1989; Maynard-Smith, 1989).

Each population carries a load of deleterious recessive genes determined by the rate at which new mutations occur, and the rate at which they are lost by random drift and natural selection. In traits determined by a single locus, variations in the trait caused by a mutation occur about 10^{-8} to 10^{-5} times per locus per year. In traits determined by many loci (heritable quantitative traits), variations in the trait caused by a mutation occur about 10^{-3} to 10^{-2} times per generation (Lynch, 1996). Most mutations that affect phenotype have a deleterious effect. If individuals in a population accumulate enough mutations, eventually the mutations will become lethal. The mean fitness of individuals in a population is determined by the mutation rate per genome per generation (Lynch, 1996).

In small populations, the dominant genetic process is drift. If the size of the breeding population is very small, then random drift can overwhelm natural selection and a population can accumulate and become fixed for quite deleterious mutations (Lynch, 1996). If the decline in fitness that results from the accumulation of new mutations reduces fecundity and survival to the extent that the population declines, feedback between random genetic drift and mutation is set in motion. As the population size decreases, random genetic drift becomes a more significant force and the rate of fixation of deleterious mutations increases, further reducing population size (Lynch, 1996).

In captive populations, mutations that are significantly deleterious in nature can be rendered almost neutral, and their effects may go undetected until the population is reintroduced into the wild (Jiminez *et al.*, 1994). The consequences of deleterious mutations are magnified in harsh and novel environments (see Box 6.1), explaining the poor performance of many captive-bred individuals following reintroduction (see Chapter 4).

The threat faced by small populations from the accumulation of mutations is argued by Lynch *et al.* (1995) and Lynch (1996) on theoretical grounds. However, Gilligan *et al.* (2005) failed to find evidence of greater genetic loads in relatively small Fruit Fly populations. It remains to be seen if the predictions of mutational meltdown are substantiated in other empirical laboratory studies and in natural populations.

Summary: loss of genetic variation

The ability of species to adapt to future change depends on their reserves of heritable genetic variability. The consequences of human impacts on natural populations often result in reduced population sizes and the loss of genetic information. Loss of heritable variation and the accumulation and expression of deleterious mutations affects the chances of persistence of populations. This loss may occur through selection and drift, or it may occur through the extirpation of local populations (when there is significant variation between populations). Other issues, such as fragmentation and migration, estimation of effective population size, adaptation to captivity, reintroduction, and identification of the units of conservation, are fundamental issues for conservation genetics. Many of these topics are treated in more detail in Part III of this book. The loss of genetic variation invariably results from the loss of species and populations, which is the topic of the following section.

6.2 Background extinction rates

Extinction is a natural process and most species (perhaps as many as 95–99%) that have ever existed are now extinct (Levinton, 2001), although human impacts can certainly accelerate the process. The extent of human impacts can be measured by comparing the rate of current extinctions with the natural (background) rate (Primack, 2001). The period since the Early Cambrian (600 million years ago) marks the beginning of the diversification of multicellular organisms, but terrestrial diversification did not commence until about 450 million years ago. Most analyses of extinction in the fossil record rely on changes in the composition of genera and families; the fossil record at the species level is incomplete and highly biased. Fossils come predominantly from areas where sedimentary material accumulates (swamps, river deltas and lake beds; Clarkson, 1979), and animals and plants from other environments, and animals lacking relatively large and heavy bones are not well represented. The fossil record for marine invertebrates with shells is relatively complete and reliable because marine sedimentation is continuous, and shelled invertebrates are abundant, well-studied, and fossilise well. Relatively comprehensive data also exist for some terrestrial vertebrate, insect and plant groups.

Extinctions in the fossil record are generally typified as falling within one of two regimes: background periods and **mass extinction events.** Mass extinctions are substantial biodiversity losses that are global in extent, taxonomically broad, and rapid relative to the average duration of the taxa involved. The **background rate of extinction** is the average rate of loss of species globally in geological time, outside the periods of mass extinction.

The average lifetime of a species varies considerably but is usually of the order of a few million years. For example, the species lifetimes of marine invertebrates averages 5–10 million years, and mammals average 1 million years (May *et al.*, 1995). Times to extinction after the time of initial radiation are usually strongly skewed (Figure 6.4), with most species persisting for times less than the average within any one taxon, and a few species persisting for much longer periods. In this kind of statistical distribution, the median time to extinction (the time by which half the species within a taxon have become extinct) is less than the average time to extinction.

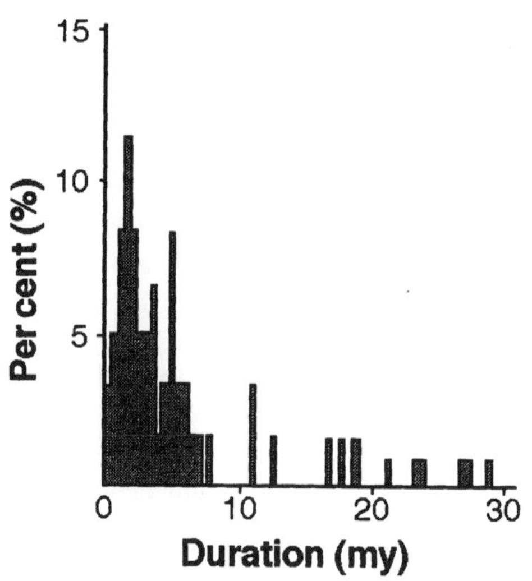

Figure 6.4. Time to extinction of species in a family of Foraminifera (after Levinton and Ginzburg, 1984).

The total number of species ever to have evolved has been estimated to be 5 to 50 billion (Raup, 1991). Today's biotic diversity constitutes perhaps between 1 and 10% of the total that has ever lived. However, many species were not eliminated by extinction in the sense that we use the term today. Rather, natural selection and mutation transformed many species, giving the appearance of losses and gains in the fossil record. These have sometimes been termed **pseudoextinctions** (Levinton, 2001). Other species underwent change following geographic subdivision of an ancestral population, or following hybridisation between different species after disturbance or dispersal. A large proportion of what we consider to be extinctions in the past may have actually been the by-product of speciation (Brooks *et al.*, 1992). The average (background) rate of loss of species globally (including both species extinction in the usual sense of the word, and the 'loss' of species by phyletic evolution) in geological time has been of the order of one or two species per year (Lyell, 1832; van Valen, 1985; Raup, 1986).

Even during background extinction periods, the rate of extinction has varied considerably among continents and at different times. Many groups have undergone periods of unusually high extinction, after which adaptive radiation has increased diversity (Erwin *et al.*, 1998; Levinton, 2001).

6.3 Mass extinction events

The background rate of extinction over the last 600 million years has been punctuated by mass extinction events, each associated with relatively severe environmental change (Erwin, 1998; Pfefferkorn, 2004; Figure 6.3). Six major mass extinction events and a major biotic extinction crisis (that was less pronounced than the mass extinction events) have been identified for marine invertebrates. Extinction patterns among terrestrial animals and plants correlate with mass extinctions detected in the oceans. For example, a major dieback of woody vegetation appears to have occurred during the mass extinction event at the end of the Permian (Visscher *et al.*, 2004). The precise magnitude and timing of land-based and marine mass extinctions is still controversial. Five causes of mass extinction events have been proposed (after Levinton, 2001):

- impacts of extraterrestrial objects such as meteorites
- volcanic activity
- climate change
- falls in sea levels, reducing the amount of habitat for marine taxa
- spread of anoxic deep water onto continental shelves.

The most severe extinction event was 245 million years ago (during the late Permian period; Figure 6.5), at which time about 50% of the families of marine animals in the fossil record were lost (the number of genera declined by about 80% and the number of species by 95%). At the same time, the number of fish families declined by 44% and the number of tetrapod families by 58%. The late Permian event is generally accepted to have occurred rapidly (over 5–8 million years), and losses were experienced globally (Table 6.1). The losses appear to have been associated with global physical changes including climate change and volcanic activity (Visscher *et al.*, 2004).

The most recent mass extinction, which marks the boundary between the Cretaceous and Tertiary periods 65 million years ago, is the best documented, and may have been associated with the impact of a meteorite on the earth. This late Cretaceous event resulted in a decline of about 15% in marine families in the fossil record. A total of 40% of tetrapod families and perhaps 75% of plant species in the fossil record were lost at that

Table 6.1. Summary of the extinction and recovery patterns for the six major mass extinctions and one smaller biotic crisis (after Erwin *et al.*, 1998).

Extinction and/or recovery event	Total duration of extinction (million years)	Percentage loss of marine taxa	Duration of survival interval (million years)	Duration of recovery (million years)
Early Cambrian (512 mya)	1	G, 50	?	5
End-Ordovician (439 mya)	1–2	G, 60 F, 26	<1	7?
Frasnian-Famennian (376 mya)	2	G, 57 F, 22	?	3?
End-Permian (251 mya)	11	G, 82 F, 51 vF, 19	1–3	3–9?
End-Triassic (206 mya)	<0.5	G, 53 F, 22 vF,12	?	3
Cenomanian-Turonian (93.5 mya)	0.52	S, 70	0.21	1
Cretaceous-Tertiary (65 mya)	<1	G, 47 F, 16 vF, 18	0.2	2

G, genera; F, families; vF, vertebrate families; S, North American species of molluscs; mya, millions of years ago.

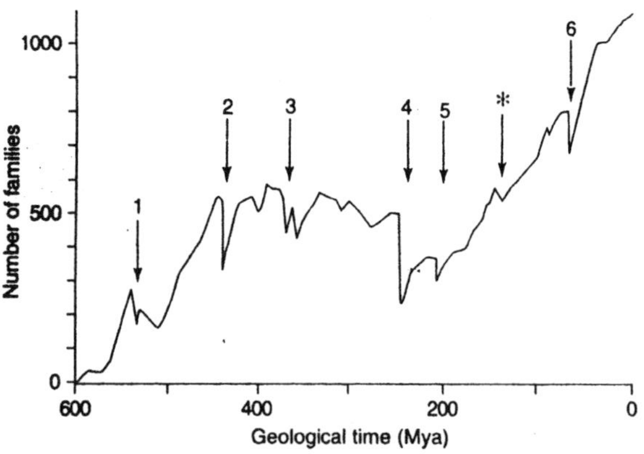

Figure 6.5. Global diversity of marine families during the past 600 million years showing the six mass extinction events and one biotic crisis (marked with a star). (Based on Erwin *et al.*, 1998.)

time (Jenkins, 1992; Erwin *et al.*, 1998). The loss of plant species was unusually high compared with other mass extinction events.

A number of other important extinction events have been recorded, but they do not have the severity, geographic scale or taxonomic breadth of the six large events (Levinton, 2001). Typically there has been a recovery of species diversity after mass extinction events so that biodiversity has increased steadily since life began on earth about 4 billion years ago (see Figure 6.5). Recovery of global diversity to pre-mass extinction levels has taken between 10 and 100 million years (Erwin *et al.*, 1998; Table 6.1).

The rules that govern the likelihood of extinction of species during background times, such as population size and range, appear to be unimportant in determining the likelihood of extinction during mass extinction events (Erwin *et al.*, 1998). Despite the severity and extent of mass extinction events, it is likely that more than 90% of all extinctions occurred in background times.

6.4 Extinction rates in recent history

There have been about 700 animal extinctions (see Groombridge and Jenkins, 2002) and 600 plant extinctions recorded globally since 1600. The number of reported extinctions increased dramatically in the 100 years up to 1960, and the current extinction rate is about 0.25 to 0.50% per decade. The apparent decline in extinctions in the 30 years to 1990 in Figure 6.6 is at least partly due to the time lag between extinction events and their detection and recording. In addition, conservation efforts over the last 40 years have slowed the rate of extinction.

Table 6.2. Number of recorded animal extinctions since 1600 listed by the International Union for the Conservation of Nature (after Groombridge and Jenkins, 2002).

Group	No. extinct species
Mammals	83
Birds	128
Reptiles	21
Amphibians	5
Fishes	81
Insects	72
Molluscs	291
Other invertebrates	12

The numbers in Figure 6.6 and Table 6.2 are clearly an underestimate, because values for groups such as amphibians are higher for Australia alone than are given by Groombridge and Jenkins (2002). In addition, the data focus only on named species that are recorded as extinct, but a species is presumed to be extinct only when: (1) a specific search has not located it, (2) it has not been recorded for several decades, or (3) expert opinion suggests it has been eliminated (see Chapter 3, on extinct and presumed extinct species, and Chapter 16, on methods for estimating the likelihood that a species has become extinct). Most taxa have not been described or named and there is no reason to expect that undescribed groups are less susceptible to extinction than groups that are better described. The vast majority of described species are not monitored and species

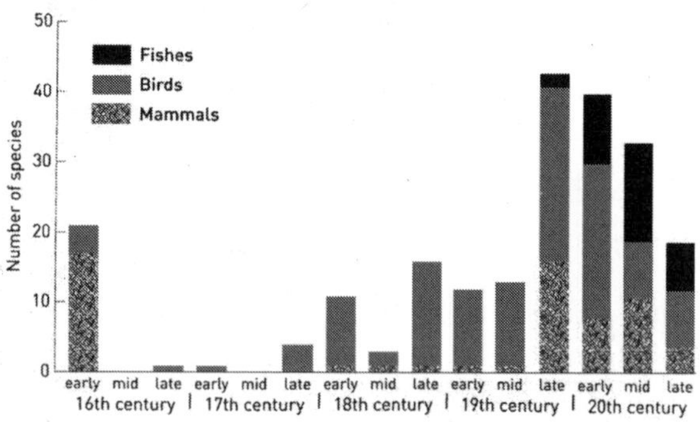

Figure 6.6. Number of extinct fishes, birds and mammals since 1500. (Redrawn from Groombridge and Jenkins, 2002.)

Table 6.3. Number of species listed as endangered in Australia, USA and China (each of these countries has relatively large numbers of endemic species) compared with the number of species listed as endangered in the United Kingdom and the number thought to exist globally. (Data from Groombridge, 1994; Anonymous, 1995b; Burgman, 2002.)

Taxon	Estimated total no. species in the world[1]	Number listed as 'Threatened'			
		Australia	USA	China	United Kingdom[2]
Fungi	500 000	0	0	0	46
Non-vascular plants	100 000	0	0	0	70
Vascular plants	250 000	1597	1845	343	61
Invertebrates	3 000 000	372	860	13	171
Fish	40 000	54	174	16	7
Amphibians	4000	20	16	0	3
Reptiles	6000	42	23	8	6
Birds	9500	51	46	86	25
Mammals	4500	43	22	42	18

[1]Estimates of the global numbers of species in each taxon were estimated crudely from numbers of currently described species and expert judgments of the proportion remaining to be described (see May *et al.*, 1995; Burgman, 2002).

[2]Numbers from the United Kingdom represent all those species in the United Kingdom's Biodiversity Action Plan for which conservation action plans were written (Anonymous, 1995b).

could be locally or globally eliminated without our knowledge.

Table 6.3 provides recent estimates of the proportion of species in various taxonomic groups that are known to have become extinct. It suggests that extinction rates have been highest among birds and mammals, intermediate in reptiles, fish, amphibians, molluscs and plants, and low elsewhere. Unfortunately, this is more a popularity poll than an index of relative rates of extinction – it reflects the interest and taxonomic effort expended on different groups over the last several decades (May *et al.*, 1995).

Several lines of reasoning suggest that extinction rates over the last 400 years have been substantially higher than those experienced globally over the last 50 million years. There are about 5300 species of mammals on earth, and if the recent extinction rate of mammals was approximately the same as the background rate of extinction in geological time (between one and five species globally per year), if extinctions occur with more or less equal frequency among different taxa, and if we assume that there are a total of 10 million extant species, we would expect there to have been fewer than one mammal extinction within the last 400 years. In fact, more than 80 mammalian extinctions in the last 400 years have been documented (Primack, 2001; Groombridge and Jenkins, 2002). If we compare the observed and expected current extinction rates in three taxa (mammals, birds and molluscs), it is evident that all three are substantially above background levels (Table 6.4), even if we make the very conservative assumption that there have been no unobserved extinctions of mammals, birds or molluscs in the last 400 years (Regan *et al.*, 2002).

Evidence for elevated extinction rates was examined by Smith *et al.* (1993) and May *et al.* (1995). They observed that any poorly studied taxonomic group or geographic region appears to be in good health because recorded extinctions are limited. For example, only a small proportion of recorded extinction events since

Table 6.4. Observed number of extinctions globally in the last 400 years, and the predicted number of extinctions in the same period assuming current rates are equal to background rates in geological time.

Taxon	No. species	Expected number of extinctions if background rate is one per year	Expected number of extinctions if background rate is five per year	Observed number of extinctions
Mammals	5300	<1	<1	69
Birds	9600	<1	2	115
Molluscs	200 000	8	40	191

Box 6.5

The status of the Night Parrot
The Night Parrot is one of the most enigmatic and mysterious species of Australian birds. Its status illustrates the difficulties in knowing whether a species is extinct and when extinction has occurred. Very little is known of the biology of the Night Parrot, although it is thought to be nocturnal, extremely secretive and an inhabitant of the remote parts of arid Australia (Forshaw, 2002). The Night Parrot may always have been irruptive (experiencing occasional rapid increases in population, like many desert animals), nomadic and extremely rare. There are 23 known specimens, including 16 from one collector, and most specimens were collected in the late 1800s. Another specimen was obtained in 1912, but it rotted away because of poor preservation (Forshaw, 2002). Because of the paucity of subsequent records and the lack of confirmed reports, in the early 1990s it was concluded that the Night Parrot was probably extinct. However, the discovery of a road-killed animal in 1990 indicated that the Night Parrot was extant at that time, although the persistence of the species remains highly uncertain. The species was not recorded during the last Atlas of Australian Birds conducted between 1998 and 2002 (Barrett *et al.*, 2003).

Night Parrot. (Image from William T. Cooper, courtesy of the National Library of Australia.)

1600 are from continental tropical forest ecosystems, yet these ecosystems are now subject to extremely high rates of clearing and modification (see Chapter 9). Only one bird species was listed in international databases as extinct on the Solomon Islands, but Diamond and colleagues reported that 12 species had no definite records since 1953, and residents of the islands reported that several of these had been eliminated by Cats (Diamond and May, 1976). Similarly, there were no recorded extinctions of fish on the Malay Peninsula, but only 122 out of a total of 266 previously described species of freshwater fish were found over a 4-year survey period. Almost all the vascular plant extinctions recorded in Africa have been recorded in the relatively affluent Republic of South Africa. However, in this latter case, it is difficult to know how much of the difference between countries is because of different taxonomic and monitoring effort, and how much is due to the impact of developed economies on their environment.

Extinctions in Australia

Extinctions and indigenous people

Nearly all species of very large mammals and reptiles, and nearly half of the large flightless birds were lost in the Pleistocene (Lövei, 2001). The activities of Aboriginal people may have contributed to the extinction of megafauna in Australia in the late Pleistocene (see Figure 6.7), perhaps as a result of hunting

Figure 6.7. Silhouettes of extinct late Pleistocene vertebrates – body sizes are scaled relative to the human on the left side of the figure. (Redrawn from Lovei, 2001.)

(Merrilees, 1968; Miller *et al.*, 1999), altered fire regimes (Bowman, 2003) or a range of other factors (Brook and Bowman, 2004). In many parts of Australia, Aboriginal people were the top predators in the food chain for tens of thousands of years. North and South America lost 73 and 80%, respectively, of their genera of large mammals at the time of the arrival of humans, and similar impacts were experienced in Madagascar and New Zealand, where an entire avian megafauna was eliminated after the arrival of humans (Martin and Klein, 1984; Jenkins, 1992). However, mainland Africa did not experience such losses.

Severe climate changes (ice ages and glacial maxima that led to much drier and colder climatic conditions) during the Pleistocene (between 10 000 and 100 000 years ago) may also have contributed to the elimination of megafauna. In Australia, extinctions of megafauna occurred primarily during the mid-Pleistocene, but megafauna extinctions occurred elsewhere in world at different times: 500–800 years ago in New Zealand, 3500 years ago in New Caledonia, 11 000 years ago in North and South America, and 12 000–14 000 years ago in northern Europe. These times correlate with the arrival of humans in each place, but they are inconsistent with widespread climate change. It is also possible that loss of the megafauna led to environmental change. For example, Kirkpatrick (1994) hypothesised that dispersal agents for plants such as Cycads were lost when the megafauna became extinct in Australia.

Recent extinctions

Australia has one of the worst records of species loss in the world. Of the species extinctions recorded globally, about 10% have occurred in Australia. Of the nearly 80 mammal extinctions recorded globally since 1600, about 20 have been in Australia and all of these occurred during the last 200 years (Jenkins, 1992; State of the Environment, 1996, 2001a). Currently, 85 of Australia's mammal species and subspecies, and 103 species of birds are listed as critically endangered, endangered or vulnerable (Table 6.5). Sixty-one species of vascular plants are listed as extinct in Australia, compared with 27 species in Europe and 74 species in the USA (see Keith and Burgman, 2004). The loss of species has not occurred equally across different habitat types in Australia; in desert areas, for example, roughly one-third of mammals that inhabited the region at the time of European settlement are now extinct.

Table 6.5. Numbers of extinct, critically endangered, endangered and vulnerable species, subspecies and populations listed for Australia and its island territories in 2003. (Based on data from Department of Environment and Heritage, 2005b.)

Group	Extinct	Critically endangered	Endangered	Vulnerable
Plants	61	54	509	676
Mammals	27	12	34	52
Birds	23	5	37	63
Retiles	0	1	11	38
Frogs	4	0	15	12
Fish	0	2	15	20
Invertebrates	0	4	5	6

The fact that Australia harbours about 7% of the world's species and is one of only two of the 17 **megadiverse** nations with a developed economy (Mittermeier *et al.*, 1997) creates a special imperative to maintain existing levels of biodiversity. The richness of the flora and the magnitude of the threat are reflected by the number of higher taxa (genera and families) that are threatened (Briggs and Leigh, 1996), and the number of species that are threatened compared with the total number in Australia and globally (Table 6.6; Figure 6.8). The figures for threatened Australian plants typically ignore fungi and non-vascular plants such as bryophytes and liverworts.

Extinction and documented changes to the Australian environment

In Australia, visits by botanists and zoologists preceded or coincided with permanent European settlement, so that science has been able to document many extinctions and range contractions that may otherwise have been unobserved. In many other places, the impacts occurred much earlier, in times when both lack of interest and the level of scientific sophistication meant that impacts were not recorded. Balmford (1996) suggested that intensive human activity is an 'extinction filter', quickly eliminating sensitive species as new technologies emerge. Unrecorded human-induced extinctions would have been most extensive in places that first experienced the impacts of modern technology, and in places that have been most densely populated. Therefore, the biotas that remain on most continents are collections of

Box 6.6

Extinction of the Thylacine

The Thylacine (or Tasmanian Tiger or Tasmania Wolf) was a large carnivorous marsupial, weighing up to 35 kilograms, with stripes on its back and rump. It was the only representative of its family, although fossil evidence indicates that there may have been other species in this group. The Thylacine was described scientifically in 1808 but was known to Aboriginal Australians long before then (Guiler, 1985). The species was confined to Tasmania at the time of European settlement, but its past distribution encompassed large parts of Papua New Guinea and mainland Australia. The Thylacine was probably displaced from the mainland by the Dingo (Corbett, 2001), although humans may have also played a role in the species' demise on the mainland (Paddle, 2000), from which it disappeared approximately 1000–2000 years ago. The Thylacine persisted in Tasmania, where it was isolated from the mainland and from Dingoes. Its major prey included Kangaroos and Wallabies (Rounsevell, 1988). Sheep accompanying European settlers to Tasmania created an additional prey item, although stock losses attributed to the Thylacine were probably exaggerated (Guiler, 1985). Bounties for scalps were offered by the Van Diemen's Land Company as early as the 1830s, and it was only 10 years later that some observers forecast that the Thylacine could become extinct. Official bounties from the Tasmanian State Government were not offered until the late 1880s, and in the following two decades approximately 2060 bounties were paid (Guiler, 1985). The last year the Thylacine appeared to be relatively common was 1908, and by 1915 there were insufficient animals to maintain the export trade to zoos around the world. The last known living Thylacine died in captivity in 1936.

The extinction of the Thylacine illustrates typical extinction processes; that is, several processes contributed to the extinction. First, Aboriginal Australians contributed to the extinction of the Thylacine by importing the Dingo, which outcompeted it on the mainland (and therefore eliminated the species from much of its range). Second, within Tasmania, bounty hunting probably had a major impact on populations of the Thylacine. Third, prey items for the Thylacine in Tasmania such as the Emu and Eastern Grey Kangaroo were reduced after European settlement, and this may have pushed animals into more marginal habitats (Paddle, 2000). Other factors also contributed to the extinction of the species. Population decline was not actually greatest in areas where human hunting pressure was greatest, or where it first started. There was an abrupt crash in the number of scalps presented for bounty payment toward the end of the first decade of the twentieth century (Smith, 1982), when there were many anecdotal reports of Thylacines suffering from a 'distemper-like' disease (Guiler, 1985). Remains of animals from the National Zoo in Washington show a form of haemorrhagic enteritis (Booth, 1994). Population crashes of large dasyurid marsupials such as Quolls (Godsell, 1983; Lunney and Leary, 1988) and the Tasmanian Devil (Guiler, 1983) have been widely recorded, possibly as a response to disease. Other factors that may have influenced the dynamics of populations of the Thylacine include competition with and predation by Dogs introduced by European settlers, poison baiting programs undertaken by fur-trappers to protect the pelts of other animals caught in traps, clearing of land for agriculture, and the fencing of pastures.

Thylacine. (Photo from David Fleay Trustees.)

species that have survived modern technology and increasing human population densities. The surviving species may be relatively resilient to current, anthropogenic sources of impact.

The magnitude of impact of any single threatening process is essentially a product of the intensity of the effect of the process on the vegetation, and the area over which the process operates (Chapters 9 and 10). Leigh and Briggs (1992) and Briggs and Leigh (1996) evaluated the principal causes of vascular plant extinctions and threats in Australia. They found that many of the principal causes of threat are directly related to agricul-

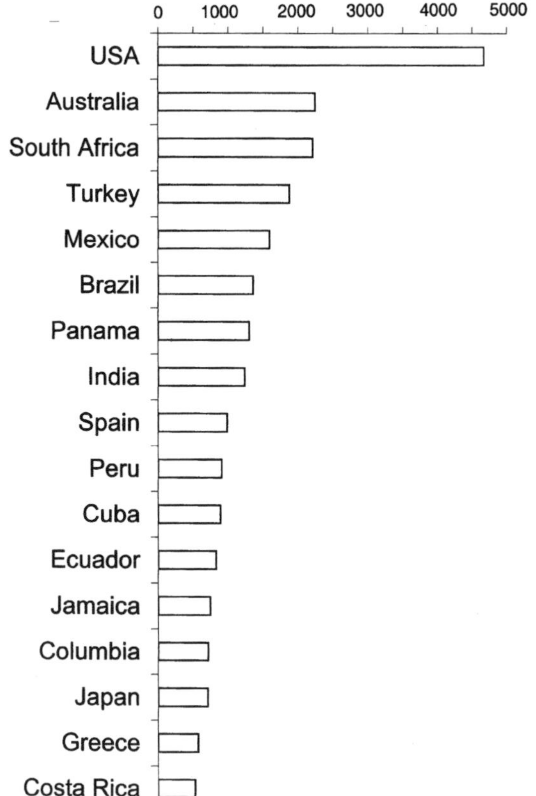

Figure 6.8. Number of threatened plant species in countries with more than 500 threatened plant species (after IUCN, 1997). The data for the USA include Hawaii.

ture and urban development. The vast majority of recorded extinctions of plants are directly attributable to clearing land for agriculture. Similarly, agriculture threatens far more species of birds than any other process (Garnett and Crowley, 2000; State of the Environment, 2001a). Table 6.8 shows an assessment of the current and future threats facing Australian vascular plants. Most threats relate to the fact that many species exist in small, restricted populations. In most cases, limited distributions and small populations arose because plant habitat was eliminated by agricultural and urban expansion. For instance, plants are at risk in rail and road verges but in most cases they persist in these environments because they are the only fragments of habitat remaining in regions converted to agriculture. Industrial, urban and agricultural activities remain important sources of threat, but new processes such as disease, salinity and weeds will become increasingly important in the future.

Mining and forestry are more stringently regulated than the agricultural sector, yet they play a relatively minor role. That is not to say that the impacts of these activities are overemphasised. Rather, it suggests that the priorities for management of the Australian environment may not be determined by the relative impacts of different activities. The effects of urban development and agriculture have received scant attention relative to their impacts. Social and political factors play a substantial part in determining how different activities are regulated. Agriculture is largely on private and

Table 6.6. Australian rare or threatened plant species in a world context (after Court, unpublished; Briggs and Leigh, 1996).

Plant group	Total no. species known in the world	Total no. species known in Australia	Rare or threatened species in Australia	New Australian species yet to be named and described[1]
Vascular				
Flowering plants	295 000	15 500	3236	3–5000
Cycads and conifers	620	60	12	10
Fern and fern allies	13 500	410	81	25–30
Total	309 120	15 970	5031	3035–5050
Non-vascular				
Mosses	14 000	1200	Unknown	500
Liverworts	8000	800	Unknown	500
Algae	20 000	3000	Unknown	1–2000
Fungi	150 000	20 000	Unknown	20 000
Lichens	30 000	3000	Unknown	500
Total	222 000	28 000	Unknown	22 500–23 500
Total	531 120	43 970	Unknown	25 535–28 540

[1]All counts of new Australian species yet to be named and described are approximate.

Box 6.7

Extinctions on islands: the example of Lord Howe Island

A disproportionately large number of the recorded animal and plant extinctions in the last 400 years have been from oceanic islands. For example, 75% of recorded animal extinctions since 1600 have been of species inhabiting islands, even though islands support a small fraction of the number of animal species found on continents.

Extinctions on Lord Howe Island are a typical example of the process. The island is 25 square kilometres in area and it is located approximately 550 kilometres east of northern New South Wales. The Lord Howe Island Woodhen is a flightless species of rail that almost became extinct in the late 1970s (Hutton, 1986; 1990) and recovered only after the eradication of feral Pigs (Miller and Mullette, 1985). The Lord Howe Island Woodhen is probably similar to the numerous species of rails (more than 2000 species) that formerly occurred on islands throughout the Pacific and that became extinct as a result of human settlement in the region.

Lord Howe Island was one of the few islands that remained undiscovered by Pacific peoples, and it did not experience the major extinction events that characterised faunal assemblages on numerous other islands (Flannery, 1994; Caughley and Gunn, 1996; Simberloff, 2000). The first contact between humans and the fauna of Lord Howe Island did not occur until 1788 when crew members from the HMS *Supply* landed while on their way to establish a settlement on Norfolk Island (Hutton, 1990). The animals were not afraid of humans and several of the bird species were flightless. Hunting by settlers and crews from passing ships resulted in the extinction of several endemic bird species by 1870 (Whitely, 1974). In addition, populations of other animals such as seabirds, turtles and fish were devastated. The invasion of the Black Rat in 1918 from a ship that ran aground off the island led to the rapid extinction of a host of other species of endemic birds (see Table 6.7). Nine of the 15 endemic species of birds on Lord Howe Island are now extinct (Hutton, 1990). Thus, the direct consequences of European settlement (over-hunting) together with the accidental introduction of pest vertebrates have resulted in at least two waves of extinction and markedly changed the faunal assemblage that now occurs on Lord Howe Island (Hutton, 1990).

Table 6.7. Extinct birds from Lord Howe Island (based on descriptions in Hutton, 1990).

Common Name	Reasons for demise	Year of extinction
Tasman Booby	Hunting, egg removal	?
White Gallinule	Hunting	~1834?
White-throated Pigeon	Hunting	1853
Red-fronted Parakeet	Eradicated as a pest	1869
Lord Howe Boobook	Competition from introduced owl	1950s?
Vinous-tinted Blackbird	Predation by introduced rats	1920s?
Lord Howe Starling	Predation by introduced rats	1920s?
Lord Howe Fantail	Predation by introduced rats	1920s?
Robust White Eye	Predation by introduced rats	1920s?
Lord Howe Gerygone[1]	Perhaps predation by rats	Post-1930s

[1]The taxonomy of the Lord Howe Gerygone has not been resolved and two or more closely related taxa may have formerly inhabited the island.

Lord Howe Island and the Lord Howe Island Currawong. (Photos by Karen Viggers.)

Table 6.8. The causes of current and future threats to vascular plants in Australia. The number of species listed as critically endangered or endangered in Australia in November 2004 by the responsible Federal agency (Department of Environment and Heritage) that are affected by various categories of threat (after Burgman *et al.*, unpublished data). The sources of threat are relatively important causes of extinction risk in the next 20 years (for instance, they ignore threats that affect only one of several populations or whose effects are not imminent).

Categories/causes of threat	No. species
Demographic factors	
Few populations	226
Small range	205
Low numbers	200
Narrow habitat	56
Industry and urban development	
Road/rail verge environments	39
Urban/coastal development	27
Mining	1
Forestry	0
Agriculture	
Land clearing	30
Domestic grazing	18
Landscape factors	
Disease	28
Weeds	21
Fire/changed disturbance regimes	11
Fragmentation	4
Salinity/hydrology	3
Feral and native grazing	
Feral grazing	11
Native grazing	3
Other human activities	
Collecting	5
Trampling	1
Other causes	
(herbicides, rubbish dumping, recreation, extreme environmental conditions, road construction, fire wood collection, pollution, lack of supportive habitat, flooding, dam construction, mowing)	2

Note: many species are affected by more than one threat and only the major ones are listed here.

leasehold land and is more difficult to control than activities such as forestry, much of which takes place on public land.

6.5 Future extinction rates

Attempts to estimate future extinction rates are based on extrapolation of past trends assuming exponential increases in human impacts, extrapolations for land clearance and forest loss combined with species–area curves, and estimates of future losses of habitat in species-rich tropical rainforests and other areas with large numbers of endemic species (e.g. Thomas *et al.*, 2004). Estimates are uncertain for numerous reasons (Heywood and Stuart, 1992; Pimm *et al.*, 1995; Jablonski, 1995; May *et al.*, 1995; Newman *et al.*, 2003). Estimates of the percentage global loss of species per decade over the next 50 years range between about 0.25 and 11% (Table 6.9; Pimm *et al.*, 1995; Pimm, 2001; although see Thomas *et al.*, 2004). One of the tasks of conservation biology is to validate these predictions and improve the information on which they are based (May *et al.*, 1995).

The sources of bias and uncertainty concerning future extinction rates are reflected in the state of our knowledge concerning recent extinctions (May *et al.*, 1995; Primack, 2001). Estimates of past, current and future extinction rates such as those shown in Table 6.9 should account for uncertainty and provide confidence limits or other measures of reliability. Extrapolations are difficult because global rates over the last 50 years of 0.25–0.5% per decade may abate if susceptible species are lost quickly. On the other hand, if the human population and its consumption of land and resources continue to increase (Chapter 12), the rate may accelerate, particularly if the predicted effects of climate change on biodiversity are realised (Pounds *et al.*, 1999). Thomas *et al.* (2004) explored future extinction rates in different regions based on species–area relationships, habitat loss effects, and altered distribution patterns resulting from climate change. They estimated that between 15 and 37% of species in their range of study regions might be lost in the next 50 years. For example, based on species–area relationships, losses from habitat destruction alone are forecast to approach 25–30% for temperate deciduous forests and shrublands (see Table 6.10).

The work by Thomas *et al.* (2004) and others indicates that extinction rates in the near future will greatly exceed those of the last 50 million years, and will probably exceed those of the last few hundred years. If current rates persist for only a few hundred years, then we are in the midst of a mass extinction event matched in magnitude and extent only by the mass extinction events of the geological past (Lövei, 2001).

Table 6.9. Estimates of future global rates of extinction (after Reid, 1992).

Estimate	Percentage global loss per decade[1]	Method of estimation	Reference
One million species between 1975 and 2000	4	Extrapolation of past trends, assuming exponential increase	Myers (1979)
15–20% of species between 1980 and 2000	8–11	Species–area curve and forest loss projections	Lovejoy (1980)
12% of neotropical birds, 15% of bird species in Amazon basin	–	Species–area curve (z = 0.25)	Simberloff (1986)
2000 plant species per year in tropics and subtropics	8	Assuming loss of half of species in areas likely to be deforested by 2015	Raven (1987)
25% of species between 1985 and 2015	9	As above	Raven (1988a,b)
At least 7% of plant species	7	Half of species lost in 10 `hotspots' covering 3.5% of forest area	Myers (1988)
0.2–0.3% per year	2–3	Half of rainforest species assumed to be local endemics and assumed lost following clearing	Wilson (1988, 1989)
2–13% loss between 1990 and 2015	1–5	Species–area curve ($0.15 < z < 0.35$), and assuming current rate of loss of f orest, and 50% increase in rate	Reid (1992)

[1]Assumes a total species number of 10 million. Reid (1992) provides additional details.

Table 6.10. Predicted extinction levels based on habitat loss (based on Thomas *et al.*, 2004).

Biome	Percentage of world surface area			Percentage of species expected to go extinct by using the species–area approach (z = 0.25)
	Undisturbed directly by humans	1990	Area lost	
Cropland	0.0	10.9	0.0	0.0
Pasture	0.0	23.1	0.0	0.0
Ice	1.7	1.7	0.0	0.0
Tundra	4.8	4.6	0.2	1.0
Wooded tundra	2.0	1.9	0.1	1.1
Boreal forest	13.0	12.5	0.5	0.9
Cool conifer forest	2.7	2.1	0.6	6.1
Temperate mixed forest	5.2	2.2	3.0	19.2
Temperate deciduous forest	4.5	1.5	3.0	24.2
Warm mixed forest	4.7	1.9	2.8	20.3
Grassland/steppe	13.7	6.9	6.8	15.7
Hot desert	14.9	11.8	3.1	5.6
Scrubland	7.3	1.9	5.4	28.9
Savannah	11.9	6.2	5.7	15.1
Tropical woodland	6.1	4.4	1.7	8.0
Tropical forest	7.6	6.4	1.1	4.0

6.6 Conclusions

Only a small proportion of recorded extinction events since 1600 are from continental tropical forest ecosystems. Most recent projections for future species loss take into account the expected loss of tropical forest. Estimates are uncertain mostly because the necessary information on numbers of species, population sizes, distribution, and kinds of impacts are themselves uncertain. However, even given conservative assumptions concerning current and future extinction rates, simple calculations suggest that the earth is experiencing a loss of populations and species at a rate that has been matched only five or six times in the geological past. Human-caused mass extinction is an event that has been underway for at least a thousand years, and is likely to continue beyond the next century.

6.7 Practical considerations

Genetic variation provides the resources for species to adapt to human impacts and environmental change. Conservation strategies should aim to protect the full range of species genetic variation.

One of the central tasks of conservation biology is to make reliable assessments of current extinction rates. Lists of threatened species are uncertain and biased and do not communicate the full extent of impact or threat. For example, greater effort should be expended on assessing the conservation status of insects and fungi, which make up the bulk of biodiversity. New methods are required urgently to estimate extinction rates reliably.

6.8 Further reading

Daniels and Corbett (2003) outline the problems of genetic swamping in wildcats and the Dingo. Hammond (1992) provides detailed information on the state of the global species inventory, including the number, distribution, taxonomic status and conservation status of species in Australia and around the world. Gaston (1991) and Erwin (1991) discuss the estimation of insect species numbers. May (1990b) discusses the strengths and weaknesses of various methods for the estimation of total numbers of species.

Jenkins (1992), the World Conservation Monitoring Centre (WCMC, 1992), Lövei (2001) and Groombridge and Jenkins (2002) provide detailed discussions of extinction events in recent history. Data on extinct and other categories of threatened species in Australia are provided on the web site maintained by the Department of Environment and Heritage (http://www.deh.gov.au/). Primack (2001) is a very readable account of extinction processes. Pimm (1995), May (1994) and May *et al.* (1995) review knowledge about extinction rates. Reid (1992) describes the methods and limitations of estimating future extinction rates. Brooks *et al.* (1992) discuss the relationship between phylogeny and extinction. Lande (1988) and Frankham *et al.* (2002) review conservation genetics. Maynard-Smith (1989) provides an overview of evolutionary genetics and Lynch (1996) provides an introduction to some genetic perspectives on conservation biology. Jablonski (1995), Erwin *et al.* (1998) and Levinton (2001) review extinctions in the fossil record. Attempts to estimate future extinction rates have been made by Lovejoy (1980), Myers (1988), Wilson (1988), Reid (1992) and Thomas *et al.* (2004).

Chapter 7

Changes in species distributions and abundances

This chapter explores some of the characteristics of changes in the range and abundance of Australian native species, examining taxa that have declined and those that have increased. This chapter also examines the impacts and control of exotic animals, plants and pathogens, including the use of quarantine programs. Lastly, it examines the impacts of genetically modified organisms.

7.0 Introduction

Extinction is a coarse indicator of impact (Lindenmayer and Gibbons, 1998). An exclusive focus on the number of extinctions can mask other problems that are better reflected by changes in range and abundance (Hobbs and Mooney 1998). Thus, although many native species persist (and are likely to do so for the foreseeable future), they may have been eliminated from part or most of their former range (Hobbs, and Mooney, 1998). Part of the role of conservation biology is to predict which species are likely to become threatened (e.g. Oredsson, 1997; Murray *et al.*, 2002).

Rather than becoming extinct and declining, some native and introduced taxa have greatly expanded their ranges in Australia following introduction or in response to changed landscapes and ecological processes (Hobbs and Mooney, 1998). Introductions to Australia cover the entire spectrum of species from large vertebrates to fungi and viruses, and plant and animal breeders have modified many species so that they make more efficient use of Australian conditions. Furthermore, genetic engineering provides new possibilities for the modification of useful species. Changes in species' ecological performances can have consequences for the natural environment through the **introgression** of genes into wild populations, and the escape and naturalisation of domesticated exotic populations.

7.1 Range contraction and depletion

Range contractions have been inferred from major changes in the reporting rates of many species of Australian birds over the past two decades, including some migratory shorebirds and woodland taxa (Olsen *et al.*, 2003). The ranges of many Australian mammals have contracted and many species of Australian frogs have undergone major declines. The following sections examine such changes for mammals and frogs in some detail.

Mammals

There have been substantial contractions in the ranges of a large number of Australian mammal species (Figure 7.1, Table 7.1). For example, the Greater Stick-nest Rat once occurred in parts of Victoria, New South Wales, South Australia and Western Australia, but by the 1980s, the species was confined to Franklin Island off Ceduna in South Australia (Copley, 1988). It has since been successfully reintroduced to a number of other islands (Copley, 1995; Richards *et al.*, 2001).

Many arid and semi-arid species of Australian mammals have become extinct or have undergone

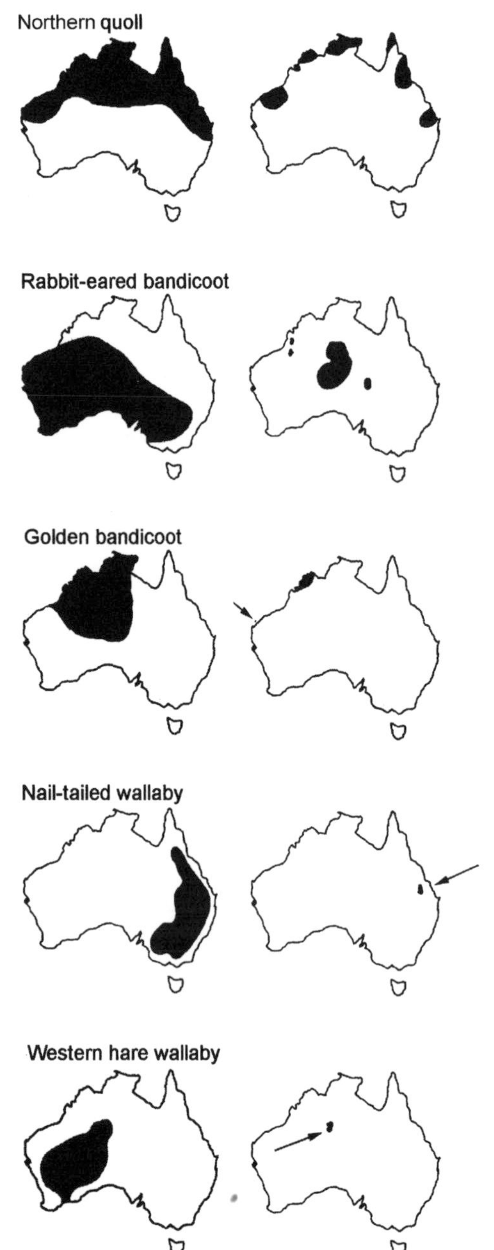

Figure 7.1. Range changes of some Australian mammals (after Short and Turner, 1994; Lomolino and Channell, 1995).

substantial reductions (see Burbidge and McKenzie, 1989; Maxwell *et al.*, 1996). Changes have been caused by land clearance, competition with introduced herbivores (e.g. Rabbits), soil compaction and grazing by domestic livestock, feral predators, changed fire regimes, and modifications to vegetation patterns resulting from the dislocation of Aboriginal people

Long-tailed Dunnart – a species that has disappeared from all but 1% of its former (pre-European) range. (Photo by Jiri Lochman.)

(Morton, 1990; Short and Turner, 1994; see also Chapter 11).

Proximate causes of local and global extinctions vary from species to species and from region to region, and some or all of the factors listed here may play a part in affecting each species. The common threat and perhaps the single ultimate cause of all these extinctions and range contractions is the arrival of Europeans, their land-management systems, and the exotic species that accompanied them.

Amphibians

Range contractions and population declines can be confounded or masked by environmental changes and sampling errors. Populations of amphibians in many areas around the world appear to have undergone substantial declines in the past few decades (reviewed by Alford and Richards, 1999). This observation has received considerable attention (Richards *et al.*, 1993; Laurance, 1996; Stallard, 2001) and is controversial (e.g. Alford *et al.*, 2001 *vs* Houlahan *et al.*, 2001; Corn, 2003 *vs* Blaustein *et al.*, 2003; see also volume 15, issue 4 of the journal *Conservation Biology*).

In Australia, of the 208 described frog species, 20 are classified as endangered, of which eight may be extinct (Campbell, 1999). However, it is difficult to distinguish between changes that result from human activities, and those that are part of the natural variation of populations. Indeed, not long after reductions in amphibian populations first began to be identified in 1989, some 'declines' were described instead as natural fluctuations in response to environmental variation (Pechmann *et al.*, 1991). However, more recent analyses suggest that many species (more than 500 taxa worldwide;

Table 7.1. Estimates of the extent of range collapse of 18 species of Australian mammals. (Modified from Lomolino and Channell, 1995.)

Common name	Historic range[1] (square kilometres)	Current range (square kilometres)	Collapse (%)
Long-tailed Dunnart	1 175 319	15 489	99
Dibbler	99 006	10 326	90
Numbat	1 924 243	58 918	97
Red-tailed Phascogale	176 450	28 852	84
Burrowing Bettong	4 371 154	607	100
Tasmanian Bettong	512 342	47 681	91
Brush-tailed Bettong	1 771 786	53 451	97
Leadbeater's Possum	43 733	5163	88
Rufous Hare-wallaby	1 961 902	1215	100
Banded Hare-wallaby	489 868	607	100
Northern Hairy-nosed Wombat	105 991	1519	99
Bridled Nailtail Wallaby	1 097 876	10 022	99
Bilby	5 295 921	946 026	82
Dusky Hopping Mouse	902 900	42 518	95
Smoky Mouse	150 939	12 755	92
Hastings River Rat	269 078	7593	97
Heath Rat	235 975	14 881	94
Greater Stick-nest Rat	1 325 043	607	100

[1]Historic range at the time of European settlement.

Alford and Richards, 1999) are in decline (Houlahan *et al.*, 2001; Blaustein and Kiesecker, 2002).

Separating the effects of environmental **stochastic** processes from human impacts and establishing that a population is not declining requires sampling over a prolonged period. This is particularly important in amphibians that undergo large annual changes in population size (Reed and Blaustein, 1995). Thus, it is critical to distinguish between a situation where there is no decline and a situation where there is a lack of sufficient statistical power to identify a decline that is actually taking place (i.e. a **type II** error – see Chapter 17). Attempts to determine population declines and range contractions in amphibians have often been limited by the short intervals between resurveys (see Skelly *et al.*, 2003).

Amphibians may be declining because of habitat degradation and destruction, land-use practices (e.g. logging), stream sedimentation, direct human impacts such as road kills and over-collection, acid rain, pollution from chemicals such as pesticides and herbicides, climate change, unusual weather conditions, disease (see Box 7.1), feral animals (e.g. trout, Mosquito Fish, Cane Toad, Red Fox), altered fire regimes, or increased levels of ultraviolet radiation (after Alford and Richards, 1999). Both Richards *et al.* (1993) and Laurance (1996) emphasised the potential impacts of diseases on frog populations in Queensland (see Box 7.1). Increased levels of ultraviolet radiation are unlikely to be important in tropical and subtropical forests where limited light penetrates to the forest floor (Richards *et al.*, 1993); however, this threat may be more important in open, high elevation areas.

The causes of amphibian declines vary between places and over time (Blaustein and Kiesecker, 2002). Factors act at local scales (e.g. habitat loss or change), at regional scales (e.g. chemical pollution) and at global scales (climate change and increased ultraviolet radiation).

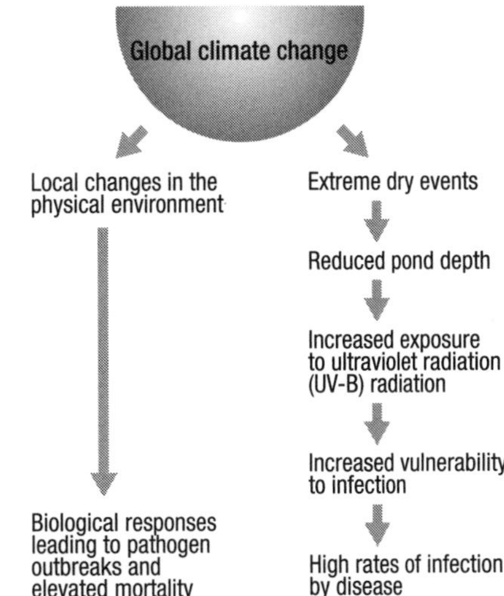

Figure 7.2. Interaction pathways underpinning the decline of some species of amphibians. (Redrawn from Pounds, 2001.)

Box 7.1

Disease and frog declines

There are a number of amphibian diseases that have emerged recently, which harm many populations in this group. A fungal disease, chytridiomycosis, has been identified as one of the factors contributing to the decline of a number of frog populations in Australia and Central America (Daszak *et al.*, 2000). Clinical signs of amphibian chytridiomycosis include abnormal posture, lethargy and the loss of the righting reflex (i.e. the ability to revert to a normal upright position if knocked over). There can also be skin lesions and abnormal sloughing of skin. Notably, retrospective histological surveys of museum specimens from Australian and Central America collected up to 10 years previously showed no signs of chytridiomycosis, indicating that the disease may be: (1) recently introduced, (2) endemic but only recently pathogenic, or (3) endemic but amphibian declines were only recently discovered (Daszak *et al.*, 2000). However, the rapid decline of some populations and substantial rates of population loss strongly suggest a recently introduced and highly virulent pathogen.

In addition to chytridiomycosis, there are other potentially important diseases of amphibians, such as the ranaviruses. These have been considered for use in genetically modified pathogens as a method to control the rapidly expanding populations of the introduced Cane Toad in Australia, although impacts on native amphibians would need to be carefully considered. Finally, it appears that infections of parasites such as the nematode worm *Ribeiroia ondatrae* are causing limb deformities in amphibians in areas such as North America.

Although there seems to be strong evidence of the effects of disease on amphibians in parts of Australia, it is by no means the sole factor implicated in population declines. Trout have been implicated in the decline of some species of frogs (Gillespie, 1995). The introduced Mosquitofish has excluded the Green and Gold Bell Frog from habitats in south-eastern Australia (Hamer *et al.*, 2002). In other cases, habitat loss and habitat degradation through, for example, draining wetlands and overgrazing riparian areas, as well as predation by feral animals appear likely to have influenced the populations of some species (Hazell et al., 2001; Lindenmayer *et al.*, 2003a). Wildfires in high altitude bogs occupied by the Corroboree Frog appear to have influenced populations of that species (K. Green, personal communication). In summary, as with the declines of almost all species and species assemblages, there are multiple contributing factors, and the magnitude of their relative effects varies depending on the species, the location and other taxa in the assemblages.

There could be substantial cascading effects of the decline of amphibians in some ecosystems. For example, in Central America, amphibians comprise a substantial proportion of the vertebrate biomass. In Australia, some species of snakes depend on amphibians. In both places, the tadpole stages of amphibians consume aquatic plant material and a reduction in their numbers could lead to changes in algal growth.

Some factors interact, triggering other processes such as the outbreak of disease. Pounds (2001) illustrated the potential pathways associated with amphibian decline (see Figure 7.2); the combination of factors at different spatial and time scales indicates the difficulty in predicting which species in which areas might be at risk.

Range contraction and natural distribution and abundance patterns

Attributes of animal populations such as density and stability tend to be higher at the geographic centre of their distributions (e.g. Brown, 1984; Brussard, 1984). Range contractions may occur initially at the edge of a range and move inward (Lawton *et al.*, 1984), and this pattern can be expected particularly in circumstances where an impact affects all of the range simultaneously.

Lomolino and Channell (1995) tested this hypothesis for 18 species of Australian mammals by comparing their original and current distributions, but in only one case (Leadbeater's Possum) did a species' range contract inward from the periphery to the core. In all of the other Australian examples, the remaining populations were at the edge of the original distributions (see Figure 7.1). Some threats, such as increases in populations of feral vertebrates or competition with large grazing animals, began first at the centre or at one edge of the distribution. In other cases, peripheral populations (such as those on the islands off Western

Green and Gold Bell Frog – a species that has undergone major declines in eastern Australia over the past decade. (Photo by Esther Beaton.)

Australia) were quarantined from the threats. The work of Lomolino and Channell (1995) has a number of important implications. First, isolated populations at the margins of a species' original distribution can be important for population persistence. Second, programs to re-establish populations of animals should not necessarily be targeted at areas at the core of the original distribution. Finally, surveys to detect new populations of taxa that have undergone range contractions should examine both the core of the original distribution and its margins.

Migratory species: a special case of range conservation

Frith (1990) defined **migration** as: 'the regular, annual movement of animals between two places, each of which is occupied for part of the year'. The conservation of many species depends on maintaining sufficient suitable habitat in geographically distant locations. Migratory species require conservation strategies in disjunct parts of their range, and possibly along the route travelled between different places. There are large numbers of migratory species, ranging from the largest animals on the planet – whales – to invertebrates such as the Bogong Moth, which congregate in the high country of south-eastern Australia between September and January (Yen and Butcher, 1997; Figure 7.3).

Birds in many parts of the world migrate and there are three broad types of migration:

- **Latitudinal migration**, in which birds fly from low to high latitudes in the spring and summer and then return to low latitudes in the autumn and winter. In Australia, birds from southern areas travel northward in autumn to warmer environments such as those in northern New South Wales, Queensland, Papua New Guinea and the islands to the north of the continent. Australian species in this category include Whistlers, Pigeons, Cuckoos, Nightjars, Kingfishers, Bee-eaters, Honeyeaters, Flycatchers, and Cuckoo-shrikes.
- **Altitudinal migration**, in which birds move to lower elevations during the cooler winter months of the year (Green and Osborne, 1994). In the Australian Alps of southern Australia, some of the species that undergo such migration patterns include the Gang-gang Cockatoo, Flame Robin and Olive Whistler (Green and Osborne, 1994).

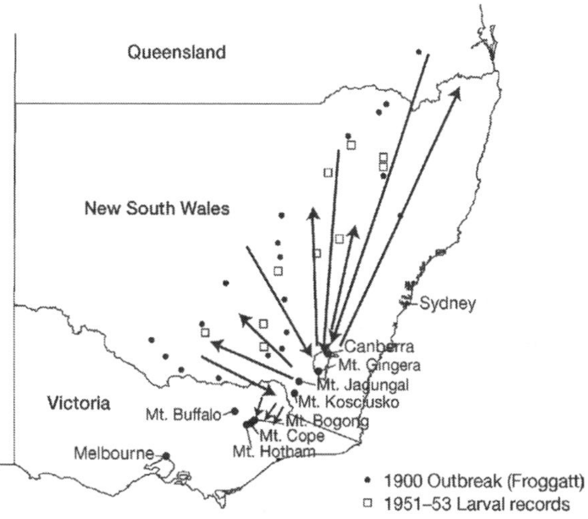

Figure 7.3. Movements of the Bogong Moth. (Redrawn from Yen and Butcher, 1997.)

- **Trans-equatorial migration**, in which birds cross the equator to breed. In Australia, waders and shorebirds fly northward in autumn to breed in north Asia (see Box 7.2).

In addition to migratory taxa, nomadic species move between areas in response to resources such as flowers and nectar (which are important for Lorikeets), fruit (which provides food for rainforest pigeons and doves; Date *et al.*, 1996), and water (which stimulates responses from ducks, pelicans and teals). For example,

Box 7.2

Trans-continental migration: waders and conservation of the Yellow Sea in north Asia

A large number of waders and similar shorebirds migrate each year between Australia and north Asia along a route termed the east Asian–Australasian flyway. Research carried out in the Yellow Sea around the shorelines of China and Korea highlights the importance of the massive intertidal zones and near-coastal wetlands in the region for waders and shorebirds. Baxter (2002) estimated that nearly two million birds use the flyway on their northward migration and one million on the southward migration. More than 35 species of shorebirds occur in internationally important numbers in the area.

The Yellow Sea region is experiencing massive and rapid economic growth, and huge areas are subject to land reclamation. A total of 43% of the intertidal area that existed in the Chinese part of the Yellow Sea in 1950 had been reclaimed by 2002, with plans existing to reclaim a further 45%. Similar levels of development are planned for the South Korean part of the Yellow Sea shoreline. In addition, water flows (and associated sediment inputs) from major feeder rivers such as the Yellow and Yangtze Rivers have been altered dramatically (Baxter, 2002). The changes will clearly impact on shorebirds, both in China and Korea and, because of migratory behaviour, also Australia.

The rate of development and the substantial human populations that depend heavily on the Yellow Sea environment makes migratory shorebird habitat conservation particularly challenging. Traditional reservation is not possible. Baxter (2002) believes that the best way will be to adopt management strategies based on ecologically sustainable management, including coordinated conservation and human development between North and South Korea and China.

Box 7.3

Management of the migratory Orange-bellied Parrot

The Orange-bellied Parrot is an endangered migratory bird. It breeds during October and November on the west coast of Tasmania and migrates northwards in March and April across Bass Strait to salt marsh habitats in southern Victoria and South Australia (Brown *et al.*, 1995). In winter, it feeds on the seeds of shrubs in coastal heath communities. Both the range and abundance of the Orange-bellied Parrot have undergone considerable declines in the past century (Menkhorst *et al.*, 1990; Brown *et al.*, 1995). During the past few decades there have been a number of management issues associated with the conservation of the Orange-bellied Parrot (see Menkhorst *et al.*, 1990; Brown *et al.*, 1995). In the mainland component of the species range, these have included:

- livestock grazing in salt marsh habitats
- destruction of suitable over-wintering habitat
- maintenance of the floristic composition and nutrient status of the winter vegetation
- the nutritional status of introduced weed species (the birds feed on the seeds of weed species that have become established in part of their feeding area, but the nutritional value of these seeds may be less than native food sources and could affect the ability of the birds to survive and migrate back to Tasmania).

In the breeding grounds in Tasmania, some of the conservation issues have included:

- the need to employ appropriate fire regimes (possibly to emulate past Aboriginal burning patterns) to ensure a continuing supply of food resources in sedge and Buttongrass habitats diseases among captive birds that developed during breeding trials
- the limited extent of suitable habitat in a major breeding area
- pressures associated with eco-tourism.

Any one of these problems makes the small remaining populations of the Orange-bellied Parrot vulnerable to extinction. Managing the risks involves close cooperation among State government authorities and researchers in three States. Fortunately, many management actions have been undertaken, including the reservation of habitat and the implementation of an integrated research program that includes projects investigating the birds' feeding ecology, population dynamics and breeding behaviour (Menkhorst *et al.*, 1990; Brown *et al.*, 1995).

Orange-bellied Parrot. (Photo by Davo Blair.)

Woinarski and Tidemann (1991) found that the bird populations in tropical woodlands in northern Australia undergo substantial regional movements. The woodlands are refuges from drought in the arid interior, a transit route for regular migratory species, and they provide a range of fruits, flowers, seeds and insects for a variety of resident and nomadic populations. The survival of many bird species depends on access through, and to, a range of habitats and geographic locations.

Many aspects of the migratory behaviour of birds are poorly understood. For example, not all individuals of a 'migratory species' migrate. In a few populations of the Golden Whistler, some males migrate, whereas some females are year-round residents. In other cases, entire populations are sedentary, whereas others are completely migratory.

7.2 Range expansion

The ranges of many native Australian animals have increased since European settlement. For example, the distributions of some of the large macropods (e.g. Eastern Grey Kangaroo) expanded in response to the creation of extensive pastures and watering places for stock (Viggers and Hearn, 2005). Flannery (1994) noted that early European explorers in Australia observed large Kangaroos infrequently and considered their rare sightings to be notable. John Gould (in Flannery, 1994) expressed concern in the late 1800s about the long-term persistence of the Red Kangaroo and its rapid extirpation. Currently, there may be as

Box 7.4

Trends in Australian parrot populations

Australia supports a substantial proportion of the world's parrots, and they are a species-rich and highly diverse assemblage, including representatives of markedly different size, dietary requirements and other features (reviewed in Higgins, 2001; Forshaw, 2002). Since European settlement, many species have declined and are considered to be at risk of extinction (Garnett and Cowley, 2000), whereas others have increased dramatically in population size and geographic range (e.g. the Sulphur-crested Cockatoo, Galah and Long-billed Corella) and have even been assigned `pest' status in some jurisdictions (Bomford and Sinclair, 2002). An interesting question arises from considering the status of these species: What ecological characteristics differentiate species that have expanded their ranges and/or increased in population size from those that have declined? This question is not unlike the one of assessing extinction proneness (see Chapter 10), but is posed in reverse. Two extensive continent-wide samples for all Australian birds were completed between 1977 and 1981 (Blakers *et al.*, 1984) and 1998 and 2002 (Garnett and Crowley, 2000; Barrett *et al.*, 2003). Cunningham and Lindenmayer (unpublished data) integrated the data for parrots with information on life history attributes and found that the species of parrots that had increased were habitat generalists with vegetarian diets, which used large hollows (Figure 7.4).

Little Corella – a native Australian species that has increased dramatically in range and has attained pest status in some areas. (Photo by Esther Beaton.)

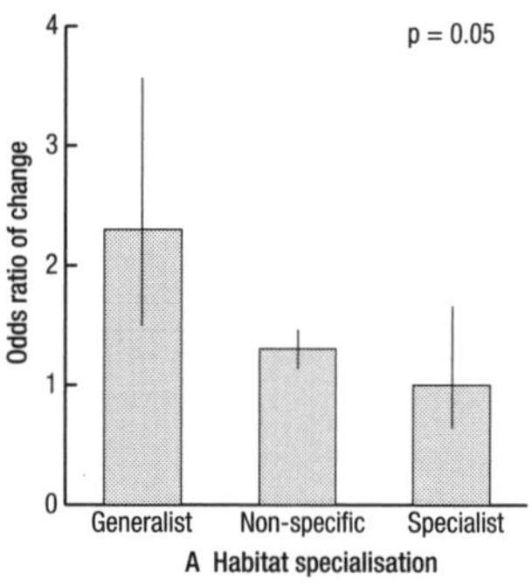

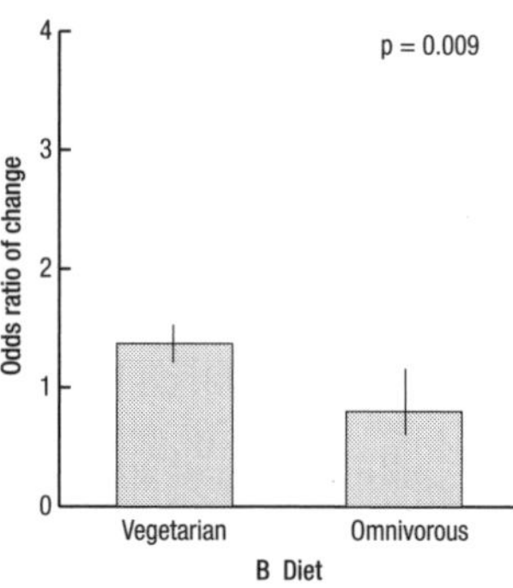

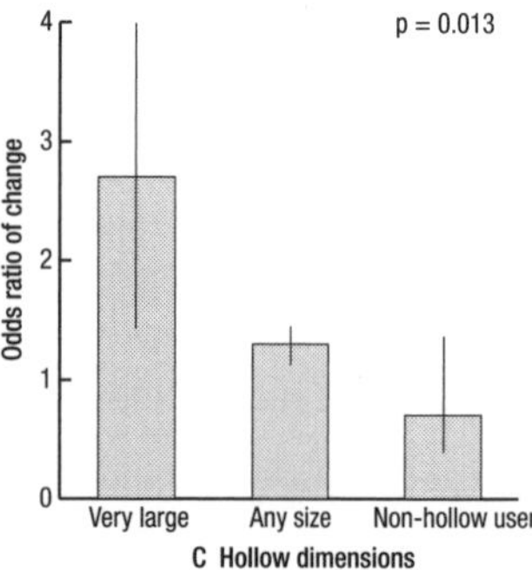

Figure 7.4. Relationships between life history attributes of parrots and the chances of them becoming a pest.

many as 10 million individuals. However, information on range expansions of most native wildlife in Australia is typically very scarce, apart from data on a few species (e.g. a few large parrots; Olsen *et al.*, 2003). It remains an interesting question why some species have increased in range while others have declined (Forshaw, 2002; see Box 7.4).

Detailed studies of the avifauna in the Western Australian wheat belt have documented range expansions (Saunders and Ingram, 1995; see also Box 7.5 on the Galah). More than 90% of the native vegetation in this area has been cleared or highly modified in the last 200 years (Hobbs and Mooney, 1998; Hobbs and Yates, 2000), and vegetation remnants are small, highly fragmented and often badly degraded (Saunders *et al.*, 1993; Wallace *et al.*, 2003; see Chapter 5). Saunders and Ingram (1995) recorded 195 species of birds (excluding vagrants) in the area. Of these, almost half (95 species) had declined both in range and abundance in the last 200 years (see Table 7.2). The distribution and abundance of a further 66 species (34%) remained unchanged. A total of 34 species (17%) had increased in range and/or abundance (Saunders and Ingram, 1995).

Table 7.2. Change in the status of birds in the Wheat belt of south-west Western Australia. (Modified from Saunders and Ingram, 1995.)

Category	Declining	Unchanged	Increasing	Total
Non-passerines				
Resident	28	12	17	57
Nomad	5	26	4	35
Migrant	4	12	0	16
Total	37	50	21	108
Passerines				
Resident	50	7	12	69
Nomad	5	6	1	12
Migrant	3	3	0	6
Total	58	16	13	87

Human activities have created suitable foraging and nesting habitat for some species, typically as a result of native vegetation being cleared, the creation of dams and troughs to water domestic stock, or the establishment of buildings, gravel pits, extensive areas of exotic grassland, or crops. For example, populations of the Black-shouldered Kite have increased markedly, possibly in response to an increase in the abundance of prey such as the House Mouse and insects such as Grasshoppers.

Only three of the species studied by Saunders and Ingram (1995) that had increased their distribution or abundance were introduced taxa. Other species that were previously rare or absent have benefited from human activities. For example, farm dams and rising watertables (resulting in saline water bodies) create suitable habitat for the Mountain Duck and the Wood Duck (Saunders and Ingram, 1995). However, a relatively rapid change in the status of a species may not reflect long-term trends in population dynamics. In the case of the Mountain Duck and Wood Duck, the persistence of populations depends on the availability of suitable nest sites in hollow-bearing trees. The limited recruitment of these types of trees throughout large parts of the agricultural regions of Western Australia (see Mawson and Long, 1994) may ultimately result in a decline (Saunders and Ingram, 1995; Gibbons and Lindenmayer, 2002).

Box 7.5

Range expansion of the Galah

The Galah has benefited from changes in the environment brought about by human activities (Rowley, 1990; Saunders and Ingram, 1995; Forshaw, 2002). The Galah's diet is predominantly seeds, especially those from cereal crops and agricultural weeds. It is probably the most abundant species of Australian cockatoo, and flocks exceeding 1000 birds are not uncommon, particularly in places where supplies of food are abundant (Forshaw, 2002).

The distribution of the Galah now spans most of the continent (Blakers *et al.*, 1984; Forshaw, 2002; Barrett *et al.*, 2003), but prior to 1750 it was typically associated with watercourses in arid and semi-arid areas (Saunders *et al.*, 1985). The expansion of the Galah's distribution followed the clearing of native vegetation for agriculture, the provision of dams as watering places for stock (and inadvertently for the Galah), and the establishment of pasture and cereal crops that provided a major food source for the species. The species is able to make extensive use of trees that are placed within remnant native vegetation and that contain cavities and provide places for birds to nest and breed.

Saunders and Ingram (1995) described the expansion of the Galah into south-west Western Australia during the past 90 years. Their atlas of bird distribution records shows that the Galah was rare in the region at the turn of the century. By the 1940s the species had begun to invade, and now the Galah is common throughout the area (Figure 7.5), and its distribution continues to expand. There are similar examples of the range expansion of the Galah in other parts of Australia, for example, on the New England Tablelands of northern New South Wales (Blakers *et al.*, 1984).

The expansion of the distribution of the Galah has had negative consequences for other species of cockatoos. For example, there is competition between the Galah and other hole-nesting birds such as the rare Carnaby's Cockatoo in south-west Western Australia (Saunders and Ingram, 1987). Moreover, the Galah is known to ringbark and kill trees in which it is nesting (Rowley, 1990; Gibbons and Lindenmayer, 2002), and this is contributing to the deterioration of already highly degraded remnant vegetation in the region (Merriam and Saunders, 1993).

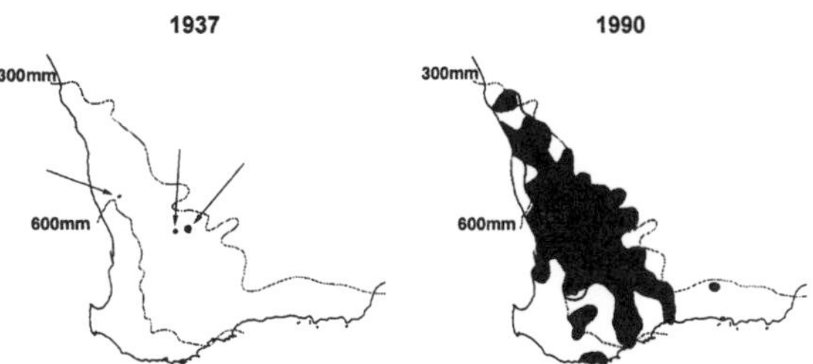

Figure 7.5. Expansion of the range of the Galah in south-west Western Australia (after Saunders and Ingram, 1995). The maps show the 300- and 600-millimetre isohyets.

Galah. (Photo by Esther Beaton.)

7.3 Exotic animals

Exotic species are a major and increasing threat to the conservation of biodiversity worldwide (Primack, 2001). For example, although many processes impact on the 1880 endangered species of plants and animals in the USA, invasive species are a significant threat to the persistence of more than 50% of them (The Nature Conservancy, 2003). Only habitat loss affects more endangered native taxa in that country. The biota of Australia has also been altered rapidly and substantially by exotic animals and plants. Introduced organisms do well in new environments (Lampo and De Leo, 1998; Low, 1999) because they can utilise resources not exploited by native organisms and because reproductive output, offspring success and population densities are not limited by the parasites, competitors and predators typical of their original habitat. In the following sections, we discuss examples of introduced vertebrates, invertebrates and environmental weeds in Australia.

Exotic vertebrates

Within the confines of this book, it is not possible to explore the ecological impacts of all animal species

Table 7.3. Factors linked with extinctions or threatened extinctions in major vertebrate groups around the world. Introduced species feature prominently in the Table. (Modified from Primack, 2001.)

Group	Percentage due to each cause[1]					
	Habitat loss	Over-exploitation[2]	Species introduction	Predators	Other	Unknown
Extinctions						
Mammals	19	23	20	1	1	36
Birds	20	11	22	0	2	37
Reptiles	5	32	42	0	0	21
Fish	35	4	30	0	4	48
Threatened extinctions[3]						
Mammals	68	54	6	8	12	–
Birds	58	30	28	2	1	–
Reptiles	53	63	17	3	6	–
Amphibians	77	29	14	–	3	–
Fish	78	12	28	–	2	–

[1]These values represent the percentage of species that are influenced by the given factor. Some species may be influenced by more than one factor, thus the rows may exceed 100%.

[2]Over-exploitation included commercial, sport, and subsistence hunting, as well as live animal capture for any purpose.

[3]Threatened species and subspecies include those given in the critically endangered, endangered and vulnerable International Union for the Conservation of Nature categories.

introduced to Australia. Instead, we concentrate on a few vertebrates (the Camel, European Rabbit, Cane Toad, and feral Cat) and a small set of invertebrates (the Honeybee, Bumblebee and Red Imported Fire Ant).

In Australia, **exotic species** (also termed **alien** or **introduced** species) are defined as those introduced accidentally or deliberately since European settlement. Species become **naturalised** when they are able to persist and reproduce in the wild. Naturalised animals are often termed **feral** animals. Exotic vertebrates have been a primary cause of many extinctions in Australia and worldwide (Caughley, 1994; Primack, 2001; Table 7.3).

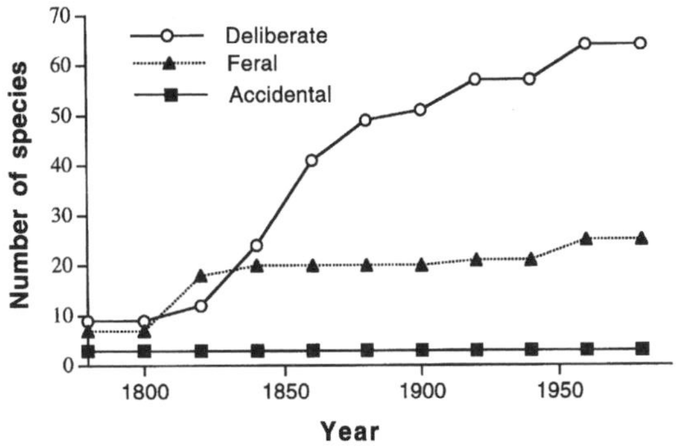

Figure 7.6. Cumulative number of mammal introductions into Australia since the arrival of Europeans. Feral animals include domesticated species that have established viable wild populations as well as species that have been deliberately released to become part of the wild biota (after Rolls, 1969; McKay, 1984; Dyne and Walton, 1987).

There has been a steady stream of deliberate introductions of mammals to Australia since the arrival of Europeans, so that today, introduced species comprise approximately 10% of Australia's terrestrial mammal fauna (Figure 7.6). Many of these introductions were deliberate, and were often actively encouraged by acclimatisation societies during the 1800s, who viewed Australia's wildlife as inadequate (Paddle, 1996; Low, 1999). The Dingo does not meet our definition of an exotic species because it was brought to Australia not by Europeans, but by Aboriginal people. However, since its introduction several thousand years ago (Corbett, 1995), it has been implicated in the demise of mainland populations of the Thylacine and Tasmanian Devil.

Despite the widespread environmental problems created by exotic vertebrates, deliberate introductions continue today and are even advocated by some scientists. For example, Flannery (1994) suggested that the giant varanid lizard, the Komodo Dragon, should be introduced to Australia to control over-abundant feral herbivores and replace the native carnivorous megafauna that were lost tens of thousands of years ago. However, the problems caused by feral animals could be exacerbated, not solved, by the release of yet another exotic species in Australia. Although there are numerous examples of increases in the populations of exotic species coincident with decreases in populations of native species (see, for example, Jones, 1986; Fox, 1990; Grigg, 2000), there is little direct evidence of competition or displacement.

Before the arrival of Europeans, Australia lacked mammals with hooves, but Sheep, Cattle, Buffalo, Pigs, Horses, Donkeys, Camels and Goats have all now established substantial feral populations in Australia (Wilson *et al.*, 1992). The effects of trampling and grazing by these animals are novel in the Australian environment, and their impacts can be difficult to quantify. However, they are known to affect the species composition of communities in ephemeral and permanent water bodies in arid and northern Australia (James *et al.*, 1999), and to alter the structure and composition of vegetation and soils (see Chapter 8).

Numbers of exotic vertebrate species in Australia

In a detailed exploration of invasive animals and plants in Australia, Low (1999) listed exotic species that have become established in Australia and its offshore islands. The list includes 24 species of mammals (see Table 7.4), 26 species of birds, 6 species of reptiles, 1 species of amphibian, and 31 species of fish.

The list in Table 7.4 does not include species of Australian mammals that have been moved to parts of the continent where they did not occur previously.

Table 7.4. Some exotic mammals established in the wild in Australia. (Modified from Wilson *et al.*, 1992.)

Common name	Year introduced
Rabbit	1858
Brown Hare	1862
Feral Horse	1788
Feral Donkey	1866
Feral Buffalo	1826
Ferret	?[1]
Banteng	~1840
Feral Goat	1788
Feral Camel	~1840s
Feral Pig	1788
Red Fox	~1870s
Feral Cat	~15th–17th century[2]
Feral Sheep	>1788
House Mouse	~15th–17th century[2]
Black Rat	~15th–17th century[2]
Brown Rat	~15th–17th century[2]
Indian Palm Squirrel	1898
Fallow Deer	~1840s
Chital Deer	1803
Red Deer	1914
Hog Deer	Mid-19th century
Sambar	~1850s
Rusa	1907

[1]A wild population is known south of Launceston, Tasmania (Wilson *et al.*, 1992).

[2]With shipwrecks and European seafarers.

Examples for birds include the introduction of the Emu to Kangaroo Island (Ford, 1979), the introduction of the Kookaburra to Western Australia and other parts of Australia (such as Kangaroo Island and Tasmania), and the introduction of the Lyrebird to Tasmania. The impacts of these introduced species are not well known, although in Tasmania anecdotal records suggest that the Lyrebird is impairing the regeneration of some plants following timber harvesting (J. Hickey, personal communication), and the removal of leaf litter by the Lyrebird is thought to be a key factor threatening the highly endangered Myrtle Elbow Orchid in Tasmania. Anecdotal observations of populations of the Kookaburra in Tasmania suggest that the introduction of this species may be having an impact on populations of small reptiles such as skinks (S. Lloyd, personal communication).

Camels

Camels were first brought to Australia in the early 1840s and over the next 80 years they were used for transport and exploration in arid areas (Wilson *et al.*, 1992). Feral populations now occur in central and north-western Australia (Menkhorst and Knight, 2001), and are concentrated in spinifex grassland and semi-arid dunes (Wilson *et al.*, 1992; Figure 7.7).

Populations of Camels may increase by as much as 15% in good years, but the populations may also experience high mortality during droughts (Dorges and Hoecke, 1989, in Wilson *et al.*, 1992). No wild populations of Camels occur outside Australia (Newman,

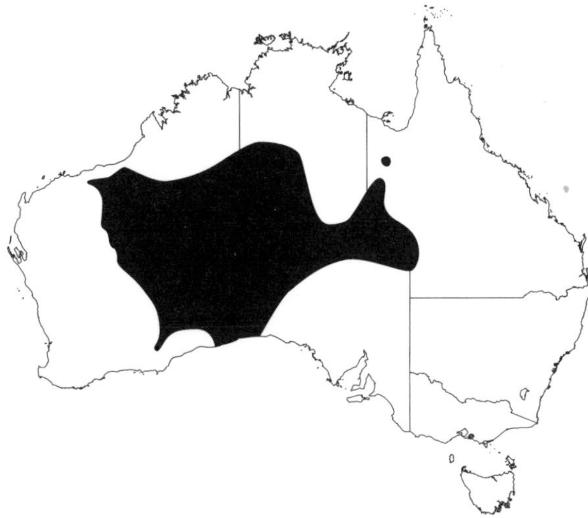

Figure 7.7. Distribution of the Camel. (Redrawn from Menkhorst and Knight, 2001.)

1983). In the early to mid-1990s there were an estimated 50 000–100 000 animals in Australia, but populations appear to be increasing (Walsh, 1992; Dorges and Huecke, 1995; State of the Environment, 1996). Edwards *et al.* (2004) estimated that populations of Camels in the southern part of the Northern Territory increased by approximately 10% per annum between 1993 and 2001, which is equivalent to a doubling of the population every 8 years (Edwards *et al.*, 2004). They calculated that in 2001 there were probably a minimum of 300 000 Camels in Australia.

The impacts of Camels on the Australian environment are poorly understood. Feral Camels eat a wide range of plants, particularly shrubs and trees, including those not usually consumed by other herbivores. Newman (1983) and Wilson *et al.* (1992) speculated that they damage native vegetation (but see Dorges and Huecke, 1995), especially during drought periods. They may reduce vegetation cover for small mammals and therefore increase predation risks, and damage fences and watering points. To reduce these impacts, fencing programs for culturally and environmentally important waterholes and other areas have been established. In addition, aerial culling takes place on some pastoral properties. Presently, there are no systematic control programs for Camels. Conversely, feral Camels are increasingly being used in the tourism industry. Camel meat is served in a limited number of specialist restaurants, and live animals are exported (O'Brien, 1990). Edwards *et al.* (2004) believed that the large and increasing size of the Camel population is having a similar effect on the environment as populations of feral Goats – the impacts of which are listed as key threatening process under the *Environment Protection and Biodiversity Conservation Act (1999)* (see Table 3.9). However, the impacts caused by Camels have yet to be listed as a threatening process and a threat abatement plan has not yet been prepared.

Figure 7.8. Distribution of the European Rabbit. (Redrawn from Menkhorst and Knight, 2001.)

Rabbits

Rabbits accompanied the first European settlers to Australia in 1788 (Myers, 1986), then more than 30 additional introductions took place after that. Within 60 years, the species had spread to almost all parts of Australia except the tropical forests and high-elevation alpine areas (Griffin and Friedel, 1985; Stodart and Parer, 1988; Figure 7.8). Their spread, particularly into arid Australia, was coincident with relatively high rainfall years when food was abundant. Rabbit populations were recognised as a major problem by 1888 and in many respects, this species is probably the most damaging herbivore introduced to Australia.

Rabbits prevent shrub and tree regeneration, eliminate herbaceous species and their burrows cause soil destabilisation and erosion. For example, Cooke (1987) demonstrated that two tree species, Drooping Sheoak and Salt Paperbark, regenerated poorly in the Coorong National Park unless Rabbits were excluded or

Sign at the port of Lady Barron on Flinders Island warning of the dangers of Rabbit introductions. (Photo by David Lindenmayer.)

controlled to low densities. Seedlings are quickly consumed by Rabbits, particularly if alternative food sources are scarce. Rabbits and domestic livestock are known to affect stands of Mulga trees in semi-arid and arid Australia. Mature trees are too tall to be eaten, but even a very low density of Rabbits is sufficient to preclude regeneration. Without seedling recruitment, the structure and composition of the vegetation of large parts of inland Australia could change.

Box 7.6

Immunocontraception and rabbit populations control

Most approaches to pest animal population management involve increasing the mortality of the target species, typically through traditional measures such as poisoning, shooting, and trapping. More recent approaches involve disruption to reproduction via immunocontraception. Essentially, the method involves identifying an antigen (or vaccine) that will block reproduction, and a vector to carry the antigen, such as a virus or a bait that will be consumed by the target species (Rutberg *et al.*, 2004). Figure 7.9 shows how such a reproductive blocking agent would work. The technology has been developed successfully for use in feral Horse (Kirkpatrick *et al.*, 1997) and deer populations (Rutberg *et al.*, 2004) and it is being investigated for possible application in Rabbit, Red Fox and House Mouse population control in Australia (Pech *et al.*, 1997; Twigg *et al.*, 2000; Hinds *et al.*, 2003). It is also part of a research program seeking to control the Common Brushtail Possum in New Zealand (Ji *et al.*, 2000; Ramsey, 2005; see Box 7.13).

Population control via immunocontraception is a potentially humane and permanent method of feral pest control, but has risks (Carruthers, 2004; Hoddle, 2004). First, it is not known if natural selection will diminish the efficacy of such control measures over time. Second, once an agent to instigate immunocontraception is released it cannot be retrieved, which means that safety and other issues should be fully explored before release (Tyndale-Biscoe, 1995; Marvier and van Acker, 2005). For example, although studies of immunocontraception attempt to identify a species-specific reproductive blocking mechanism, the potential for impacts on non-target species need to be carefully assessed. Finally, the spatial scale of application of immunocontraception needs to be carefully considered. This is because immigration of unsterilised animals from non-treatment areas may negate the effects of fertility control in a given target area (Ramsey, 2005).

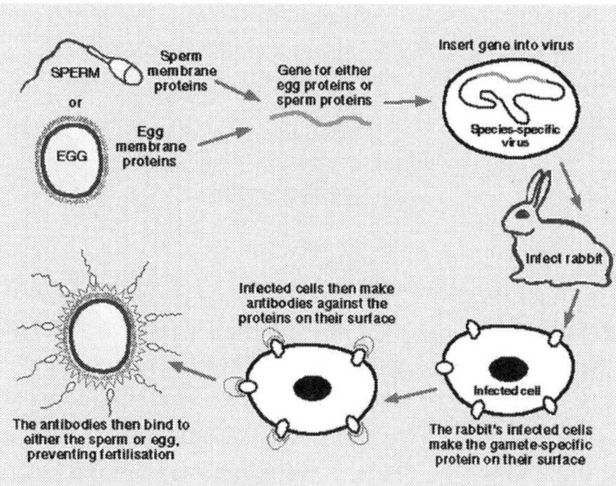

Figure 7.9. Reproductive interference via immunocontraception. (Redrawn from Hinds *et al.*, 2003.)

Rabbits compete directly for food and burrows with some medium-sized native mammals, including the Bilby, which now has a very restricted distribution in arid Australia. Rabbits are an important component of the diet of introduced Cats and Foxes – generalist predators that have important impacts on medium-sized native mammals, especially in the arid zone (Burbidge and McKenzie, 1989; Robley *et al.*, 2002). The number of feral Cats on Macquarie Island (a sub-Antarctic island south of Tasmania) increased following the introduction of Rabbits, possibly leading to the extinction of the Red-fronted Parakeet. Prior to the introduction of Rabbits, feral Cats and Red-fronted Parakeets co-existed for almost a century (Taylor, 1979).

Myxomatosis, a mosquito-borne pathogenic virus of Rabbits, was introduced to Australia by the CSIRO in 1950–51. The virus was partially successful in controlling the Rabbit population for 30 years and initially resulted in more than 90% mortality in areas that supported sandfly and mosquito vectors. However, the combined effects of the evolution of resistance in Rabbits, and the evolution of less virulent strains of the virus, have led to a decline in its effectiveness over the last decade.

Rabbit Calicivirus was released accidentally from a trial facility on Kangaroo Island in South Australia to the Australian mainland in 1996. This pathogen dramatically reduced Rabbit populations, particularly in arid areas (State of the Environment, 2001d), although the impacts of the disease have been patchy (see Williams *et al.*, 1995).

It has been estimated that Rabbits cause annual losses in excess of US$370 million to Australia's agricultural sector (McNeely *et al.*, 2003). Partly because of such economic costs, new biological control measures are currently under evaluation, for example immunocontraception – a form of genetic engineering (see Box 7.6).

Cane Toads

The Cane Toad was introduced to Queensland from South America in 1935 to control the beetle pests of sugar cane, for example the Grey Back Cane Beetle. The Cane Toad is an example of a biological control program that went badly wrong. The background to the release of the Cane Toad is testament to the arrogance of the scientists involved; for example, at the time of the release of the Cane Toad, it was stated that:

> *The introduction into Queensland was made only after a careful analysis of the pros and cons, and, according to the behaviour of the toad up to the present, there appears to be no reason for the assumption that we have made an error in our judgement* (Mungomery, 1936).

Low (1999) noted that the introduction of the Cane Toad was not actually accompanied by *any* research! The Cane Toad failed to control the Grey Back Cane Beetle, instead relying on a range of other prey, including native invertebrates and small vertebrates. It now competes for food with predators such as snakes and lizards. The Cane Toad has also been found to be a predator of bird's eggs and nestlings, such as those of the tunnel-nesting Rainbow Bee-eater (Boland, 2004). Some species of native frogs that share water bodies with the Cane Toad may experience reduced growth and survival rates (Williamson, 1999). The Cane Toad has toxic skin secretions that can kill vertebrate predators including goannas, skinks, quolls and snakes; Phillips *et al.* (2003) speculate that up to 30% of Australia's terrestrial snake species may be at risk from the Cane Toad, and other reptile predators such as goannas may also be at risk. Laurance (1991a,b) suggested that the toad may be responsible for the rarity of the Spotted-tailed Quoll on the Atherton Tableland, which has declined in parts of Cape York following the arrival of the Cane Toad (Grigg, 2000). In addition, the early life history stages of the Cane Toad (eggs, tadpoles, hatchings) are toxic to some aquatic predators (Crossland and Alford, 1998; Crossland, 2001).

From the initial release sites near Cairns, the Cane Toad has now become established in large areas of Queensland, northern New South Wales and large areas of the Northern Territory (Van Dam *et al.*, 2002), and the species is likely to invade the whole of tropical northern Australia (Freeland, 1986; Amos *et al.*, 1993; Figure 7.10). The rate of spread is about 25 kilometres per year, but it is possible that it might reach other areas faster because of accidental translocations in shipments of food such as bananas (O'Dwyer *et al.*, 2000).

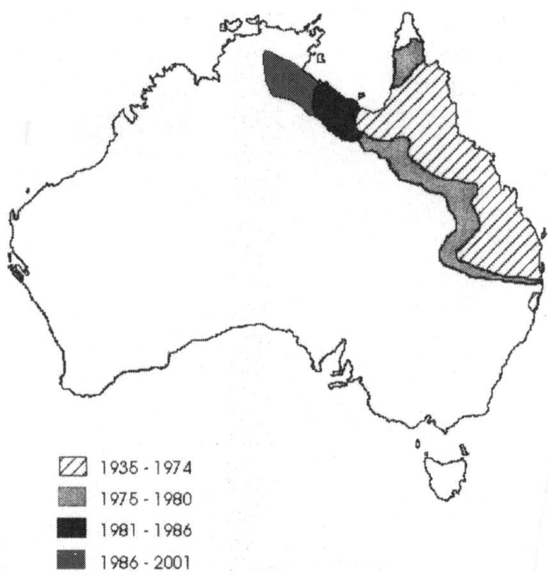

Figure 7.10. Expansion in the range of the Cane Toad. (Redrawn from the State of the Environment Report, 2001.)

Some of the impacts of the Cane Toad on the Australian fauna have been rapid and quite remarkable. For example, Phillips and Shine (2004) reported morphological changes in two species of snakes vulnerable to toxic poisoning after ingesting Cane Toads. In

Cane Toad. (Photo by Esther Beaton.)

Box 7.7

The Cane Toad in Kakadu

In 2001 the Cane Toad arrived in Kakadu National Park (Department of Environment and Heritage, 2001). It was recognised well before this that it would be impossible to limit its spread (e.g. Grigg, 2000). Prior to the arrival of the Cane Toad, earlier studies assessed the risk it posed to biodiversity in the park. More than 150 species of predators and a range of prey species (such as beetles, ants and termites) were identified as being at risk. Predators were ranked in terms of risk priority: the Northern Quoll was judged to be at greatest risk, followed by several species of varanid lizards (goannas), three elapid snakes and the Dingo (Department of Environment and Heritage, 2001). Monitoring programs were established to assess the impacts on the fauna following its arrival.

Detecting changes in faunal assemblages in far northern Australia is not always straightforward. Comparing biota before and after Cane Toad arrival is confounded by highly seasonal and variable environmental conditions that strongly influence the abundance and detectability of animal populations (Catling *et al.*, 1999). For example, some frog species call only after heavy rain. In an attempt to overcome this problem, some unique forms of monitoring have been implemented to track the expansion of the Cane Toad and the changes in native biota. For example, Grigg (2000) described a series of remote recording stations where computers using voice recognition software identify particular frog and toad calls over prolonged periods of changeable weather. These recording stations were in place a few years before the Cane Toad arrived in Kakadu National Park, providing a useful basis on which to evaluate information on the subsequent impacts on biodiversity.

the 80 years since the Cane Toads have started spreading in Queensland, the body size of the Red-bellied Black Snake and the Common Tree Snake has increased. Phillips and Shine (2004) suggest that such rapid morphological changes have occurred because the probability of eating a toad large enough to be fatal decreases with an increase in body size. Snake species with a low risk from Cane Toads have not exhibited changes in morphological traits during the past 80 years.

Australia is not the only place where the Cane Toad has successfully colonised: populations have become established in 32 of 40 countries worldwide where it has been released (Lever, 2001).

Cats

Feral Cats have been in Australia since at least 1788, and they may have even been in the country earlier as a result of European contact (perhaps from shipwrecks). However, recent genetic and morphological work suggests that they have expanded their range rapidly from initial releases at the time of English settlement in 1788. The spread of the species was aided by land owners looking for ways to control plagues of the introduced Rabbit (Rolls, 1969), as well as other animals such as the House Mouse. Feral Cats are now well established throughout Australia (Figure 7.11).

Dickman (1993) estimated that at that time there were between 5 and 10 million feral Cats in Australia, although densities of animals vary widely (Wilson *et al.*, 1992; Commonwealth of Australia, 2001c). Feral Cats weigh up to 2 kilograms more than domestic Cats, although the reasons for such weight differences are not known. Based on known energetic requirements, adult breeding feral Cats consume food weighing approximately 20% of their body weight each day (Jones, 1977). They prey on birds, mammals, reptiles and invertebrates. Animals as large as the Common Brushtail Possum and the Tiger Snake can be eaten, although in many parts of Australia, Rabbits are the major prey item (May, 2001).

Feral Cats carry diseases such as toxoplasmosis (Obendorf and Munday, 1983; Whitely, 1989), which can be passed to native animals (especially marsupials),

Figure 7.11. The continent-wide distribution of the Cat in Australia (from State of the Environment Report, 1996).

domestic stock and even humans. Toxoplasmosis may have contributed to the decline of the Eastern Quoll (Dickman, 1993), as well as other species of carnivorous marsupials (Braithwaite and Griffiths, 1994; Corbett, 2001).

Feral Cats kill many million native animals annually. Almost 50% of the volume of the guts of 80 feral Cats in Victoria sampled by Brunner and Coman (1972) consisted of the remains of native species, and the balance was mostly Rabbits and House Mice. Predation by feral Cats appears to have contributed to the decline of small mammals such as the Eastern Barred Bandicoot in western Victoria (Seebeck *et al.*, 1990). The presence of feral Cats has confounded attempts to re-establish populations of a wide range of species (Smith and Quin, 1996), including the Rufous Hare Wallaby and the Mala in the Northern Territory (Gibson *et al.*, 1995; see Chapter 4), and the Golden Bandicoot in Western Australia (Christensen and Burrows, 1995).

Domestic Cats also eat large numbers of native animals (Lepczyk *et al.*, 2003). Paton (1991) studied well-cared-for household Cats in South Australia and found that between 50 and 60% of them ate birds, mammals and reptiles. On average, each animal killed 30 vertebrates annually (Table 7.5). Tens of millions of Australian native vertebrates are killed by domestic (and feral) Cats each year, and similar problems are known in other parts of the world, including North America (Lepczyk *et al.*, 2003). Around 60 million domestic Cats and 30–40 million feral Cats in the USA kill in excess of one billion small mammals and 200 million birds each year (Coleman *et al.*, 1997, in Population Action International, 2000).

The feral Cat is difficult to control by poison baiting, although there are cases on islands where the method has worked (e.g. Brothers, 1982; Twyford *et al.*, 2000). Cat populations do not vary much in response to environmental conditions such as drought, perhaps because of their ability to capture reptiles. Moreover, populations of

Table 7.5. Consumption of vertebrate prey by domestic Cats in South Australia (after Paton, 1991).

Type of study area	No. Cats monitored	Captures of prey per Cat per year		
		Birds	Mammals	Reptiles
Urban	335	5.4	11.5	5.2
Country town	178	9.7	17.2	9.8
Rural	99	13.8	27.4	13.1
Totals	612	8.0	15.7	7.8

Sign highlighting a baiting program for the Red Fox. (Photo by David Lindenmayer.)

the Feral Cat may increase markedly in response to baiting programs that control other predators such as the Red Fox and the Dingo (Pettigrew, 1993; Christensen and Burrows, 1995; Risbey *et al.*, 2000).

Co-abundance of exotic animals is a major problem for native mammal conservation. When large populations of Rabbits occur in an area, large populations of Cats and Foxes can also be sustained. This has impacts on populations of native mammals, and is perhaps one of the reasons why conilurine rodents (native rodents in the tribe Conilurini, including tree rats, stick-nest rats, native mice and hopping mice) have suffered so many significant declines and/or extinctions in Australia (Smith and Quin, 1996). It may be necessary to coordinate the control of introduced species so that feral predators and their exotic prey are both controlled (Smith and Quin, 1996). Impacts of control programs such as poison-baiting on non-target animals like the Spotted-tailed Quoll also need to be considered as part of efforts to reduce populations of feral predators (Murray and Poore, 2004).

Predicting which vertebrates will become pests

One of the tasks of conservation biology is to determine which exotic species are most likely to become serious pests if introduced to Australia. Reliable predictions would support, for example, the assessment of risks associated with proposals for new imports for the Australian pet industry. Several studies have examined this issue for exotic birds (e.g. Newsome and Noble, 1986; Duncan *et al.*, 2001; Bomford and Sinclair, 2002). Of the 55 species of birds introduced to Australia, 19 have become naturalised (Long, 1981). An exotic bird is most likely to become naturalised and develop into a serious pest species if (after Bomford and Sinclair, 2002):

- Many individuals are released at each of several release sites.
- There is strong congruence between the climate domain (see Chapter 14) of the species in its natural distribution and the climatic conditions present in Australia.
- The introduced species is widely distributed outside Australia.
- The species produces multiple broods annually.
- The species is a generalist, survives in disturbed environments, and has a broad diet.

The release of large numbers of animals reduces extinction risk due to demographic stochasticity, and multiple release sites increase the chance that at least one introduced population will survive (Duncan *et al.*, 2001). Given that climate strongly affects the distribution patterns of plants and animals (Nix, 1978; Woodward, 1987; Mackey, 1994; Mackey and Lindenmayer, 2001), similarities between source and destination areas should improve the chances of establishment success.

Exotic invertebrates

Many exotic invertebrates have become established in Australia and some are now serious pests (Low, 1999), for example the *Sirex* wood wasp, which can damage Radiata Pine plantations. Exotic invertebrates are common in some vegetation types (Strauss, 2001). Most of the 40 species of pest invertebrates in Australian Radiata Pine plantations are introduced (Neumann, 1979). Loch and Floyd (2001) and Hobbs *et al.* (2002) found large numbers of newly introduced pest invertebrates in Blue Gum plantations in Western Australia, and in adjacent remnant native vegetation. Almost nothing is known about the impacts of the approximately 500 introduced invertebrates in Australia, except for those few species that are economically important (Amos *et al.*, 1993; Low, 1999). The Honeybee, a controversial exotic invertebrate in Australia, is the focus of the following section.

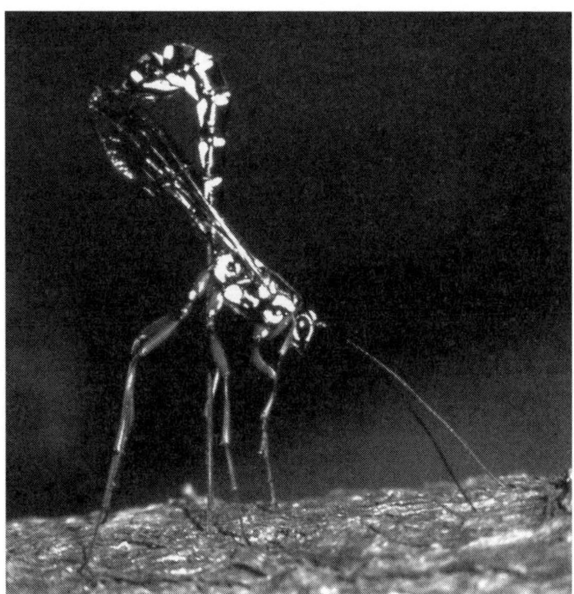

Sirex wood wasp. (Photo by David Lindenmayer.)

Honeybees

Honeybees were introduced to Australia in 1826 and have played an important role in the honey and beeswax industries since that time (Oldroyd *et al.*, 1994). The honey industry is now estimated to be worth A$300 million annually. However, colonies of Honeybees have not remained confined to hives managed by apiarists. Feral populations have become established throughout most of Australia (Paton, 1993, 2000) and are probably absent only from some desert areas where there is insufficient water to maintain them, and from high altitude landscapes that provide limited winter food supplies (Pyke, 1990). Feral Honeybee densities in parts of the Mallee National Parks in western Victoria are higher than those recorded anywhere else in the world (C. Miller, personal communication).

Populations of feral Honeybees may have a range of impacts on Australian ecosystems (although evidence of impacts is controversial (Manning, 1997)). Colonies of feral Honeybees use hollows in trees (Ambrose, 1982; Gibbons and Lindenmayer, 2002), competing with a large number of other cavity-dependent vertebrates and invertebrates in woodlands and forests. For example, feral Honeybees compete with the Forest Red-tailed Black Cockatoo for cavities in the forests of south-western Australia. When Suckling and Goldstraw (1989) erected artificial cavities to assist in the reintroduction of the Sugar Glider, they found that 30 of the 59 nest boxes they installed became occupied by Honeybees. Oldroyd *et al.* (1994) investigated competition for tree hollows between feral Honeybees and the endangered Regent Parrot in western Victoria. They found overlap in the types of cavities used by Honeybees and parrots. Despite this, Oldroyd *et al.* (1994) believed that competition was unlikely to be important because of the large number of trees with hollows in the River Red Gum and Black Box forests used by the Regent Parrot. However, their work indicates that competition between feral Honeybees and hollow-dependent animals could be a problem in forests where cavities are a limiting resource (Oldroyd *et al.*, 1994). The problem was reviewed by Gibbons and Lindenmayer (2002) and they concluded that, in general, infestations of cavities by Honeybees appears to be greatest in open habitats such as woodlands as well as in disturbed landscapes close to human habitation.

Nectar and pollen from plants are important food resources for thousands of native Australian animals, including birds, arboreal marsupials, and many invertebrates, including the more than 2000 species of native bees, which comprise one of the richest and most distinctive bee faunas in the world (Michener, 1965; Cardale, 1993). Honeybees are relatively aggressive and colonial, and they can out-compete and displace smaller, typically solitary, Australian native bees, especially for food (Paton, 1993, 1997; Paini and Roberts, 2005). Honeybees are extremely efficient at gathering pollen, nectar and water, and can remove up to 30–90% of the nectar and pollen produced by some plants (Paton, 1985; Paton, 1993; Table 7.6). Studies of Honeybees in Kosciuszko National Park demonstrated that they reduced the foraging efficiency of native bees on wildflowers (Pyke and Balzer, 1985). In Western Australia, the number of nests produced by the Banksia Bee (a native solitary bee) was almost 25% lower in areas where commercial hives were maintained than in places where hives were absent (Paini and Roberts, 2005). Paton (1993) found that most of the pollen from the native flowers of some plant species was removed by feral Honeybees before native bees began foraging. As a consequence, a reduction in native bee populations can occur when there is an increase in the abundance of feral Honeybees (Paton, 1993). However, some other studies of the overall effects of competition between indigenous and exotic bees have been inconclusive (Sugden and Pyke, 1991). In the Ngarkat Conservation Park in South Australia, Paton (1999) showed that although bees removed large quantities of pollen and nectar, there was still a sufficient quantity remaining to meet the needs of other animals.

The resource requirements and foraging efficiency of Honeybees can lead to competition with a range of vertebrates with similar food requirements, for example birds (particularly honeyeaters). Paton (1993) examined the activity patterns of Honeybees and honeyeaters on Scarlet Bottlebrush flowers near Goolwa in south-eastern South Australia. He found that the number of times honeyeaters attended individual flowers dropped from almost ten to three visits per day when Honeybees were present. In response, birds tended to favour flowers generally unavailable to Honeybees (such as those deep within the canopy of the plant). Moreover, when Honeybees were present, the dominant birds increased both the number of flowers they defended and the size of their territories – activities that displaced juvenile and

Table 7.6. Quantities of pollen and nectar removed by Honeybees and native animals visiting several plant species. (Modified from Paton, 1993.)

Plant species	Month and year	Percentage of resource removed		
		Honeybee	Bird	Native bee
Cup Gum	Early August 1987	14.1	85.8	0.1
Cup Gum	Late August 1987	29.9	70.0	0.1
Mount Taylor Mallee	January 1989	16.1	83.7	0.2
Scarlet Bottlebrush	December 1988	92.1	7.9	
Gland Flower	August 1987	–	100	
Gland Flower	January 1989	97.2	2.8	
			Pollen	
Common Correa	May 1987	38.7		
Common Correa	July 1987	24.1		
Common Correa	August 1987	6.9		
Mount Taylor Mallee	January 1989	12.0	<0.01	
Gland Flower	August 1987	100.0		
Gland Flower	January 1989	1.0		

Feral Honeybees. (Photo by Esther Beaton.)

female birds (Paton, 1993; Table 7.7). However, Paton (1993) believes that honeyeaters would not be completely displaced because they can harvest some nectar early in the morning before Honeybees become active.

Differences in the foraging behaviour of Honeybees and native pollinators (bees and birds) can influence plant reproduction and disrupt pollination (Pyke, 1990; Paton, 1993, 2000; Vaughton, 1996; Celebrezze and Paton, 2004). Honeybees pollinate weeds in Australia (Parker, 1997; Barthell *et al.*, 2001), and they also pollinate numerous species of Australian plants, although they may not be as effective as indigenous pollinators that have co-evolved with native plants. For example, Paton (1993) showed that Honeybees concentrated their foraging on the flowers of a single Scarlet Bottlebrush plant before returning to a hive, often failing to strike the anthers to pick up pollen. Limited cross-pollination of Scarlet Bottlebrush by Honeybees led to reduced seed and fruit production. In contrast, honeyeaters were more effective pollinators, moving often between different plants and frequently striking the stigma (Paton, 1993).

There are no firm indications of the effects of commercial or feral Honeybees on native Australian plants and animals (reviewed by Paini, 2004). However, Honeybees can occupy nest sites in cavities, compete for food resources, and provide ineffective pollination. Their impacts on native bees can be difficult to determine because native bee diversity and abundance varies naturally both in space and time (Paini, 2004; but see Paini and Roberts, 2005). It is also possible that Honeybees play a beneficial role in Australian ecosystems, pollinating plants that have lost their native pollinators (e.g. as a consequence of fire or clearing; Paton, 1993).

Although it is now impossible to control feral populations of Honeybees, it may be beneficial to regulate the hives kept by apiarists, especially within and adjacent to national parks and nature reserves (Gross and Mackay, 1998). Paini (2004) noted that even though commercial apiary practice involves leaving boxes in an area for a relatively short time (1–3 months), large numbers of hives, sometimes more than 100, can be deployed. These bees will be additional to feral Honeybee populations established in that area. Considerable further work is required to establish the effects of strategies such as placing hives only in areas where nectar is abundant. Paton (1993) recommended that the size of hives of bees be reduced rather than them being eliminated from national parks, with the aim of maintaining the role that Honeybees play in pollination, while retaining nectar and pollen for native

Table 7.7. Patterns of use of Bottlebrush flowers by New Holland Honeyeaters in the presence of Honeybees. (Modified from Paton, 1993.)

Position of flower or inflorescence	Frequency of use by New Holland Honeyeaters	
	Honeybees absent	Honeybees present
Inflorescences (visits per inflorescence per hour)		
Exposed	3.44	1.26
Partially covered	3.68	2.62
Fully covered	3.00	4.06
Flowers (percentage of probes)		
Proximal third	34.2	28.2
Middle third	35.8	47.9
Distal third	30.0	23.9

animals. There are no straightforward solutions to the Honeybee problem; management will need to balance the potential ecological and economic costs and benefits of having Honeybee hives in natural areas (Paton, 2000).

Other invertebrate invaders: Red Imported Fire Ants and Bumblebees

Low (1999) documented exotic invertebrates that are known to have become established in Australia, some of which have created problems in many other parts of the world. The Red Imported Fire Ant can become very abundant in disturbed areas and has invaded many areas of conservation significance (e.g. the Galapagos Islands). Its invasion and impacts are listed as a threatening process under the Australian Federal Government's *Environmental Protection and Biodiversity Conservation Act (1999)* (see Table 3.8 in Chapter 3) and the *Threatened Species Conservation Act (1995)* in New South Wales.

Fire Ants have powerful venom that can cause painful skin reactions in humans, and they are also capable of badly damaging or destroying human infrastructure and agricultural crops. Fire Ants are native to South America and have colonised large parts of North America, where their range is expanding (Wojcik *et al.*, 2001). Tens of millions of dollars are spent annually trying to control them in the USA. The species was discovered in mainland China in late 2004 and has now spread to Hong Kong. Red Imported Fire Ants were first discovered in Australia in the Brisbane region in 1999. The species is now well established in many of the suburbs of Brisbane (Nattrass and Vanderwoude, 2001) and a major control program is underway, costing in excess of A$120 million. Red Imported Fire Ants prey on the young of many native species and are believed to have contributed to the decline of several North American taxa (Wojcik *et al.*, 2001). They could have similar impacts in Australia, and climate analyses indicate that they have the potential to spread beyond Brisbane (Low, 2002). The Red Imported Fire Ant is one of several species of exotic ants that have accidentally been introduced to Australia (and offshore islands such as Christmas Island; e.g. the Crazy Ant; O'Dowd *et al.*, 2003), most probably as a result of international trade (Low, 1999). Genetic evidence indicates that the Red Imported Fire Ant has been introduced to Brisbane on at least two separate occasions in the past few years.

The Bumblebee is a large, efficient foraging insect that could have impacts that far exceed those of Honeybees. The Bumblebee was deliberately introduced into Tasmania (without government approval) and was first discovered in Hobart in 1992 (Semmens *et al.*, 1993). It seems likely that it was released by tomato growers to increase pollination rates in greenhouse crops. An extensive survey almost a decade after its initial discovery in Tasmania revealed that the Bumblebee is now widespread throughout the State, including remote areas (Hingston *et al.*, 2002). It is possible that the species could invade the mainland using the Bass Strait islands as stepping stones. The Bumblebee can remain active when other insects cannot, such as during periods of cold temperature. The species could displace other insect pollinators (including native bees), reduce pollination rates in some native plants, and promote the spread of weeds (Hingston *et al.*, 2002).

Exotic marine organisms and ballast water

Numerous species of marine animals and plants have been introduced accidentally to Australian waters in discharged ballast water (Joint SCC/SCFA National Taskforce, 2000) or by hull fouling (Pogonoski *et al.*, 2002). Port Phillip Bay in Victoria is believed to support between 300 and 400 introduced species (Hewitt *et al.*, 1999), the populations of some of which are substantial. In 1998, the number of exotic seastars in Port Phillip Bay was estimated to exceed 15 million individuals.

Fish introduced from Chinese and Japanese ports include the Yellowfin Goby, which could eliminate native gobies and affect stocks of Whiting, and a Japanese Sea Bass, which could affect stocks of commercial native fish (RAC, 1993). The Northern Pacific Starfish from Japan and Alaska is a threat to the Spotted Handfish in the Derwent River in Hobart (Newton, 1999). Economically valuable shellfish industries may be vulnerable to introduced organisms such as the Mediterranean Fanworm, European Shore Crab, Black-striped Mussel, Asian Mussel, European Clam, New Zealand Screwshell and Pacific Oyster. Introduced toxic dinoflagellates and the introduced seaweed Wakame may also impact on the shellfish industry (RAC, 1993; Joint SCC/SCFA National Taskforce, 2000). Finally, organisms such as Cholera can be translocated in ballast water, which has direct implications for human health.

The problems of translocations of marine organisms are substantial because of the large number of commercial ships that visit Australia, and the variety of places these ships were previously berthed (RAC, 1993). More than 120 million tonnes of ballast water are discharged into Australian waters each year (State of the Environment, 1996), from 10 000 vessels from more than 600 international ports (Low, 1999).

Some strategies used to mitigate these problems include discharging ballast in deep ocean areas, preventing ships from taking ballast water in locations where there are infestations of exotic organisms, and regulating ship movements so that oceanic conditions in discharge areas are as dissimilar as possible from the places where the ballast water was obtained (RAC, 1993). Other methods are being investigated, including heating ballast water in the ship's engine, water filtration, treatment with ultraviolet radiation (State of the Environment, 2001b), and deoxygenating the ballast water (Hendricks, 2004).

As of 2003, the Australian Federal and State Governments agreed to a National System for Prevention and Management of Marine Pest Incursions, which includes a capability for emergency response (S. Barry, personal communication). This is a useful step; however, the issues associated with marine pests from ballast water and hull fouling are likely to become more problematic in the future, especially with increased global trade, ship-based bulk transport and climate change (see Stachowicz *et al.*, 2002).

7.4 Exotic plants

Types of weeds

Many species are considered undesirable, depending on perceptions and the context in which the species occur. Undesirable plants are usually termed **weeds. Agrestal weeds** (of agricultural land) and **ruderal weeds** (of waste areas or disturbed places) have been the primary focus of weed scientists and weed management, particularly in agricultural zones. These are often termed **noxious weeds. Environmental weeds** are plant species that invade natural vegetation, usually adversely affecting the survival of the native flora. Such species can be Australian native plants growing in new locations within Australia, or they can be introduced species.

Weeds in Australian plant communities

The Australian flora comprises about 25 000 species, including at least 2200 naturalised species (Hnatiuk, 1990) and possibly as many as 3000 (Low, 1999; Weeds

Table 7.8. Origin and current and potential distributions of the top 20 weeds of `national significance' in Australia in 1999. (Redrawn from Australian Bureau of Statistics, 2001.)

Common name	Origin of weed	Current distribution (×1000 square kilometres)	Potential distribution (×1000 square kilometres)
Alligator weed	Argentina	30	500
Athel pine	North Africa, Arabia, Iran and India	80	3646
Bitou bush/Boneseed	South Africa	231	1258
Blackberry	Europe	691	1425
Bridal creeper	South Africa	385	1244
Cabomba	USA	35	181
Chilean needle grass	South America	14	242
Gorse	Europe	233	870
Hymenachne	Central America	73	415
Lantana	Central America	389	1052
Mesquite	Central America	410	5110
Mimosa	Tropical America	73	434
Parkinsonia	Central America	950	5302
Parthenium	Caribbean	427	2007
Pond apple	North, Central and South America and West Africa	27	181
Prickly acacia	Africa	173	2249
Rubber vine	Madagascar	592	2850
Salvinia	Brazil	383	1376
Serrated tussock	South America	171	538
Willows	Europe, America and Asia	63	135

Australia, 2000; Australian Bureau of Statistics, 2001). Hence, exotic weeds make up least 10% of the flora of Australia (Table 7.8). They come from all parts of the world (Figure 7.12) and are established in the plant communities of all regions.

About half of the naturalised exotic plants in Australia are environmental weeds (Williams and West, 2000). Some environmental weeds are plants native to Australia that have been translocated to habitats where they previously did not occur. For the State of Victoria, Carr *et al.* (1992) listed 534 taxa from overseas, 28 Australian species from outside Victoria and 22 species outside their pre-European distribution within Victoria (such as the Cootamundra Wattle) that threaten natural plant communities in Victoria.

Different biomes and plant communities are differently affected by weed invasion (Lonsdale, 1999). The extent of invasion is reflected in the proportion of the flora that is composed of exotic species. Savannas, deserts and hummock grasslands in Australia are relatively little affected (Table 7.9), perhaps reflecting the relatively limited extent and severity of human impacts in these areas relative to the impacts in temperate and coastal ecosystems.

No vegetation type is immune from weed invasion, and the degree of invasion depends on edaphic and microclimatic conditions, vegetation structure, the type and frequency of natural and artificial disturbance (such as grazing, see Box 7.8), and the proximity of source populations of weed species (Fox and Fox, 1986; Carr *et al.*, 1992; Lonsdale, 1999). In a wide-ranging study of plant invasions worldwide, Lonsdale (1999) found other important variables related to the proportion of exotic species in a vegetation community.

Table 7.9. Proportion of the local flora that is exotic in several Australian biomes (after Humphries *et al.*, 1991; Lonsdale, 1992).

Biome	Percentage exotic
Islands	33
Savanna	7
Temperate (urban/agricultural)	29
Desert	7
Montane	16
Mediterranean	13
Hummock grassland	5

Box 7.8

Weed invasions in Tasmania

Introductions of weeds in Australia have often been associated with introductions of other organisms, particularly domestic stock (Kirkpatrick, 1994). This was highlighted by a study of plant diversity and abundance on islands in Bass Strait by Kirkpatrick *et al.* (1974), in which they compared the floras of Rodondo and Hogan Islands. The first is fringed by steep cliffs, is rarely visited by humans, and only 7% (4 of 56 species) of plant species were exotic. In contrast, on Hogan Island, which has been grazed for almost a century, more than 35% of the plants were exotic (Kirkpatrick *et al.*, 1974).

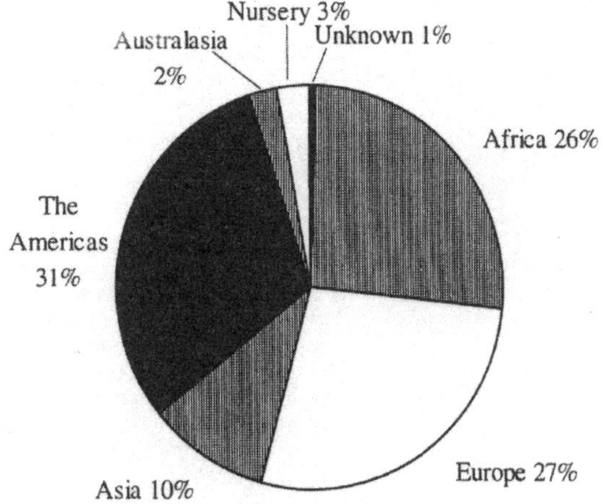

Figure 7.12. Origin of plant species that became naturalised in Australia between 1971 and 1995. (Redrawn from Groves, 1997.)

Extensive infestations of Bitou Bush at Jervis Bay on the New South Wales coast, where a major control program is underway. (Photo by David Lindenmayer.)

Box 7.9

Two weed problems: Bitou Bush and the Giant Sensitive Plant

Bitou Bush is a shrub that was introduced to Australia from South Africa in 1852 to stabilise coastal sand dunes (Carolin and Clarke, 1991). It grows in coastal ecosystems containing grasslands, dune scrub, heaths and woodland. Seeds from the subspecies *rotundata* were introduced to Stockton, New South Wales, in ballast from a South African ship in 1908. The species was sown extensively by the Soil Conservation Service between 1946 and 1968 to control sand dune drift, and by sand mining companies for revegetation (Humphries *et al.*, 1991; Low, 1999). By 1982, the species occupied 60% of the entire New South Wales coastline and extended to Rockhampton in Queensland (Figure 7.13). It was the dominant species for more than 220 kilometres (Love, 1985), was present on 80% of headlands, and occurred in many national parks and reserves. Since then, Bitou Bush has spread rapidly (through seed dispersal by birds), particularly in coastal New South Wales, and it could expand inland (State of the Environment, 2001b). In places where it dominates, Bitou Bush forms impenetrable thickets and excludes native shrubs, including Banksias, Wattles, Tea-trees, grasses and heaths (see Bell, 1987; Amos *et al.*, 1993; Matarczyk *et al.*, 2002).

Biological control seemed to be a most promising management option, and several insect species were released in the late 1980s and early 1990s for that purpose (Scott and Adair, 1991). Of these, the Bitou Tip Moth has been the most successful, and had been released at more than 60 sites in New South Wales by 1995. It is having a significant impact on flowering and seed production at several locations (Holtkamp, 1996). In other areas, such as Booderee National Park on the south coast of New South Wales, a battery of measures including poisoning, burning and ripping are used. However, these approaches need to be carefully applied because the application of herbicide treatments can have unintended impacts on non-target species, including some endangered plant taxa (Matarczyk *et al.*, 2002).

The Giant Sensitive Plant is a prickly tropical shrub that grows in thickets up to 6 metres tall. It was introduced to the Darwin Botanical Gardens in the 1870s from Brazil, probably as an ornamental garden plant. Despite its long history in the tropical north of Australia, it has only become a serious environmental weed since the 1970s. It has spread rapidly since about 1970, probably in conjunction with floods and disturbance caused by Water Buffalo. In the 1980s, it was distributed over 45 000 hectares of land between Darwin and north-west Arnhem Land. Infestations spread rapidly, and threaten extensive areas of Kakadu National Park. In 2001, the species was naturalised in the high rainfall areas of coastal north Queensland (Land Protection, 2001). The Giant Sensitive Plant is an opportunist and invades places where vegetation has been cleared or disturbed. It prefers Paperbark swamps and open sedgelands associated with riparian environments, and excludes tree seedlings, reduces the diversity of the shrub and herb layers, and reduces or excludes some species of reptiles, amphibians and mammals (Lonsdale, 1992). Management is targeting individual plants as they are discovered in Kakadu, and five potential biological control agents had been released by 1990 (Humphries *et al.*, 1991). Introduced psyllids are effective in some areas, and other control measures include herbicides, slashing, fire and grazing by domestic livestock (Land Protection, 2001).

Although the number of exotic plants varies widely between different parts of the world, the proportion is higher outside nature reserves, within plant communities that are species-rich, and on islands. For example, Hawaii has more introduced plants than native ones (860 *vs* 850) (Eldredge and Miller, 1995).

The most widespread plants in southern Australia are possibly *Hypochoeris glabra* and *Hypochoeris radicata*, which are both known as Flatweed. Flatweed occurs in almost all vegetation formations in all southern States, although usually in low numbers and with low cover. There are numerous weeds around Australia and therefore it would be impossible to examine them in detail in this book, but two examples of widespread weed problems are briefly outlined in Box 7.9. Table 7.8 shows the origins and current distribution of the top 20 weeds of 'national significance' in Australia. This table also shows the potential distribution of the same species, highlighting the increasing magnitude of weed problems in Australia.

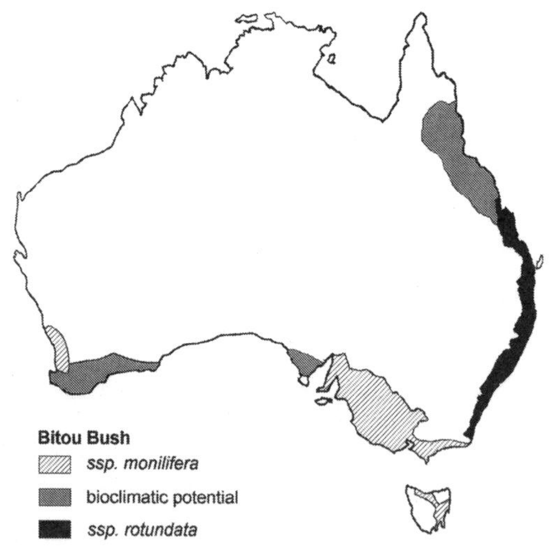

Figure 7.13. Current and potential distribution of Bitou Bush (after Howden, 1985; Humphries *et al.*, 1991).

Rates of naturalisation

Early data from Specht (1981) suggested that the rate at which exotic species become naturalised in the Australian environment has been more or less constant at about 5–10 new species per year between 1860 and 1980. However, Groves (1997) found that between 1971 and 1995 there was a trend toward an increasing number of naturalisations (Figure 7.14).

Mechanisms of introduction

Most exotic plants in Australia were introduced deliberately, particularly for use as ornamental plants in urban gardens and to improve agricultural pasture (Groves, 1997; see Figure 7.15). For example, of the environmental weeds listed in Victoria by Carr (1993), up to 70% were deliberately introduced to areas where they are now a problem. However, there were also accidental introductions with livestock, as contaminants in grain, and in ballast and soils (Parsons and Cuthbertson, 1992; Kirkpatrick, 1994; Low, 1999).

Almost half of the 200 species of exotic plants now classified as noxious weeds in Australia were deliberately introduced (Panetta, 1993; Groves, 1997). Examples include the importation of 450 herbaceous species for trials on land affected by salt in Western Australia (Malcolm, 1971, in Carr *et al.*, 1992) and the importation into Western Australia of 331 grass species that were potentially useful for fodder (Rogers *et al.*, 1979, in Carr, 1992).

Hundreds of environmental weed species are available commercially for land improvement, pasture improvement, and ornamentation, many through the horticultural industry (see Box 7.10). For example, Tall Wheat-grass and Cord-grasses, which were imported for land reclamation and pasture improvement, are serious invaders of salt marsh in Victoria (Carr *et al.*, 1992). Exotic weeds are often recommended for soil stabilisation. For example, Hill *et al.* (1985) recommended 10 species for use in soil stabilisation work in coastal situations, of which most pose serious threats as environmental weeds (Carr *et al.*, 1992). Similarly, Zallar (1980) suggested 153 plant species for erosion control in Victoria, of which more than 50 are environmental weeds (Carr *et al.*, 1992).

Weeds and pasture productivity for grazing

A very large number of plants have been introduced to enhance the productivity of grazing lands for the

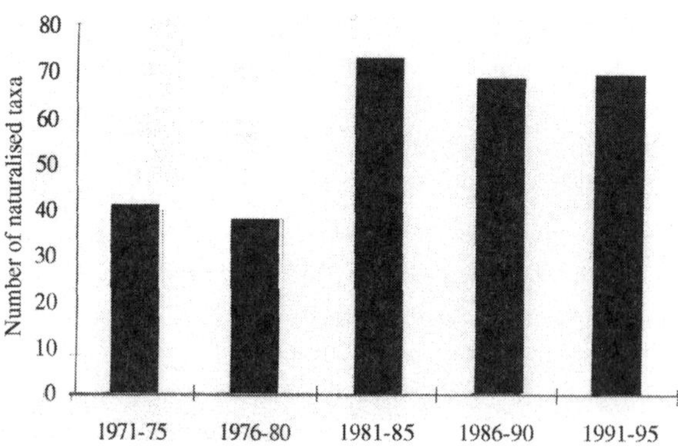

Figure 7.14. Numbers of plant naturalisations between 1971 and 1995. (Redrawn from Groves, 1997.)

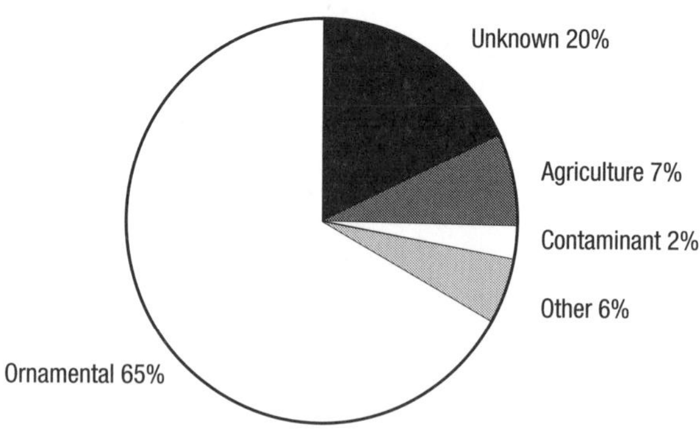

Figure 7.15. Mechanism of introduction of plants that became naturalised in Australia between 1971 and 1995. (Redrawn from Groves, 1997.)

Box 7.10

The Olive industry: a horticultural time-bomb

Many weed species in Australia were deliberately imported for pasture improvement or horticulture. Serious weed infestations are now created by wild-growing Olive trees. Olives infest the foothills of the Adelaide Hills of South Australia and now form a distinct `Olive zone' that was formerly occupied by native eucalypt woodlands.

Olives were first introduced to South Australia in 1836, and the species rapidly became naturalised there probably because of the similarity of conditions there to those in parts of the Mediterranean (Animal and Plant Control Commission, 1999). Many Olive orchards were abandoned in the late 1800s when harvesting became uneconomic because of low prices. The species then became a major weed problem. Sheep grazing checked seedling development, but when it ceased in the foothills there was an increase in wild Olive populations. The seeds of Olives are spread by the Red Fox as well as by a range of birds (Spenneman and Allen, 2000).

Olives not only threaten the integrity of woodlands in parts of South Australia (Weed Management Society of South Australia, 2003), but also populations of several significant plant species (Crossman, 1999), including the endangered orchid *Pterostylis cucullata* (Davies, 1995). The control of wild Olives has been estimated to exceed A$10 000–15 000 per hectare, and the problem is so widespread that reduction of infestations is unlikely in the near future (Animal and Plant Control Commission, 1999). To limit the further spread of wild Olives from existing orchards, the Animal and Plant Control Commission (1999) has recommended that all orchards be netted – at a cost that is far less than that of weed control.

The story of infestations of Olives in South Australia is a sobering one, particularly because the Olive industry is undergoing rapid expansion in other parts of Australia. The suitable `Mediterranean' climate and soil conditions that exist in many other parts of Australia suggest that the weed problem in the Adelaide Hills could easily be repeated in many other Australian woodland communities, adding yet another threatening process to vegetation communities that have been heavily modified in the past 200 years (Hobbs and Yates, 2000). For instance, areas of farmland adjacent to Jarrah forest in Western Australia are currently being planted with Olives without any detailed plans for weed monitoring and control.

An Olive plantation on the southern Highlands of New South Wales. (Photo by David Lindenmayer.)

pastoral industry in northern Australia. More than 200 exotic grasses (mostly from Africa) and 263 legumes (predominantly of Central or South American origin) were introduced to the region between 1947 and 1985 (Winter *et al.*, 1985), but the introductions were rarely successful (Lonsdale, 1994; Table 7.10). Only 21 of the 463 introduced species were useful, and all but 4 of the 21 useful species have also become weeds (Lonsdale, 1994). Thus, more than 95% of the introduced plant species served no useful purpose. Very few of the introduced plants were subjected to field trials (such as palatability testing for domestic stock) before release. Furthermore, 60 of the various introduced species had negative impacts on cropping activities and/or conservation reserves (Table 7.10). Of these, 13 species became major environmental weeds requiring extensive weed control (Lonsdale, 1994).

Apart from ecological and ethical considerations, the question of the relative economic costs and benefits of pasture plant introductions in northern Australia depends on the benefits of the 21 useful grasses and legumes outweighing the costs of the release of 60 new weed species. This is a complex question because assessing the profitability of the pastoral industry is difficult and the extent of use of exotic plants for pasture improvement is not readily estimated (although, for example, in Queensland less than 3% of the area used for Cattle grazing supports improved pasture). Lonsdale (1994) calculated that the annual production of the pastoral industry in northern Australia was about A$1.50 per hectare. Conversely, the cost per hectare per year of weed

Table 7.10. Impacts of exotic pasture species introduced into northern Australia. (Modified from Lonsdale, 1994.)

	Grasses (no. species)	Legumes (no. species)	Total
Fates of the various introduced taxa according to type			
Solely useful	3	1	4
Solely weedy	24	19	43
Useful and weedy	8	9	17
Total useful	11	10	21
Total weedy	32	28	60
Total species introduced	186	227	463
	Useful (no. species)	**Non-useful (no. species)**	**Total**
Impacts of reintroductions on various sectors			
Solely cropping	2	19	21
Solely conservation	6	14	20
Cropping and conservation	9	10	19

control was A$30–120 (depending upon whether an annual or woody weed was being controlled).

Weed dispersal and the impacts of motor vehicles

Weeds disperse by wind, water, animal movements, and the transport of soils and agricultural products. Motor vehicles are a significant form of weed dispersal within Australia. Between 1973 and 1975, Wace (1977) collected mud and sludge washed from vehicles cleaned in a commercial car washing facility in the centre of Canberra. The sludge was sorted and the seeds it contained were germinated in a glasshouse, grown to seedling stage, and then identified. Wace (1977) germinated more than 15 600 seedlings representing almost 225 species from the sludge samples. More than 50% of the total number of seedlings were ruderal weeds. This group contained 82 taxa of which only nine were native, and many of these species were common in disturbed roadside environments.

The actual number of plant species that comprise 'car-borne' flora is considerably greater than is reflected by the information collected by Wace (1977). At least some species may not have been detected because their seed would not have survived the oils, detergents and immersion experienced at the car wash. Furthermore, Wace (1977) confined his analyses to seeds from the outside of vehicles. He found that samples vacuumed from the inside of motor cars contained similar quantities of seeds to those in the mud and sludge. Weed control is likely to be very difficult in areas intersected by roads (Forman *et al.*, 2002). Wace's (1977) results suggest a need to maintain at least some conservation areas without roads.

Environmental impacts of weeds

There are many potential environmental effects of weeds, including (after Williams and West, 2000):

- competition with indigenous plants for space, light, nutrients and other resources
- replacement of indigenous plant communities
- restriction of natural regeneration
- altered movement of water through the soil and in watercourses
- altered water quality
- altered microtopography of the landscape
- addition of toxins into the soil and water
- provision of food and shelter for pests and indigenous animals
- introduction of new genes into native populations of plants
- altered fire regimes (e.g. in northern Australia).

In the following section we briefly examine just one of the impacts of environmental weeds – their inter-relationships with pests and indigenous animals.

Weeds and animal populations

Weed invasions can have significant consequences for animal populations. The Greater Stick-nest Rat is a threatened Australian native rodent (Moseby and Bice, 2004). Two islands in the Nuyts Archipelago off Ceduna in South Australia contain critical habitat for the species. A South African annual, the Ice Plant, now covers a substantial part of the rat's habitat. The annual is a food item for the rat; however, the Ice Plant accumulates salt and inhibits the regeneration of indigenous perennial species, partly because of salt accumulation. It also contributes to soil erosion because it undergoes annual summer die-off (Copley, 1988). Thus, the Ice Plant is part of the diet of the rat, but it degrades the environment of associated vascular plant species.

In other cases, weeds provide important habitat for native animals, such as cover for the Eastern Barred Bandicoot in Tasmania (Mallick *et al.*, 1997). Bitou Bush in coastal woodland at Jervis Bay on the New South Wales coast provides habitat for species such as the Long-nosed Bandicoot and Eastern Bristlebird and gives them cover from feral predators such as the Red Fox. In the Tumut region of southern New South Wales, Blackberry often provides the only understorey cover for birds and small mammals (Lindenmayer and McCarthy, 2001). Species at Tumut such as the Yellow-faced Honeyeater and the threatened Olive Whistler are found most commonly in close association with dense thickets of Blackberry (Lindenmayer *et al.*, 2002c; see also Suckling and Heislers, 1978). The uncommon Long-nosed Potoroo inhabits understorey thickets of Lantana in Hoop Pine plantations in central Queensland (Lindenmayer and Viggers, 1994). Lantana is also an important winter food source for many birds such as rainforest pigeons (Date *et al.*, 1996). In another example, the weed Spanish Heath on South Mount Midway on Bruny Island in Tasmania is being parasitised by a threatened subspecies of the native plant Eyebright. Uncontrolled spraying of Spanish Heath would also kill the threatened parasitic plant (Potts, 2000). Thus, although it is important not to encourage the further spread of weeds such as Blackberry, Lantana and Spanish Heath, these weeds are valuable in some respects. In the case of Blackberry and Lantana, they appear to act as structural mimics of native understorey plants that have been eliminated by plantation management.

Another example of relationships between weeds and animal populations is that of Paterson's Curse and the Honeybee, itself an introduced species. The flowers of Paterson's Curse are important for apiarists, so much so that in southern New South Wales they have renamed it the 'Riverina Bluebell' and opposed control methods, despite it being one of Australia's 20 most noxious environmental weeds.

Weed control

In addition to the array of environmental impacts of weeds listed earlier, the direct and indirect economic costs of weeds in Australia and other nations are enormous. The 50 000 established non-indigenous animals and plants in the USA cost the economy US$137 billion annually (Pimentel *et al.*, 2000). The overall total economic impact of weeds in Australia in terms of lost production and costs of control is more than A$3.3 billion (Williamson *et al.*, 2000, in McNeely *et al.*, 2003; Australian Bureau of Statistics, 2001).

Blackberry thickets at Tumut in southern New South Wales. This area provides habitat for the Olive Whistler and is probably acting as a structural mimic of *Leptospermum* spp. thickets that were key habitat for the species prior to clearing for plantation establishment. (Photo by David Lindenmayer.)

There are two consequences of the environmental and economic costs of weeds. First, the most effective strategy is almost certainly to prevent exotic species from arriving (Low, 1999; Myers *et al.*, 2000). Second, when an invasive and damaging species is first discovered, early action can eliminate the problem before it becomes logistically and financially impractical to do so (McNeely *et al.*, 2003; see Figure 7.16). These recommendations are relevant to all invasive organisms (State of the Environment, 2001a).

Prevention

Quarantine aims to prevent the importation of environmental weeds. For most of the last 100 years, the only restriction on the importation of seeds and plants

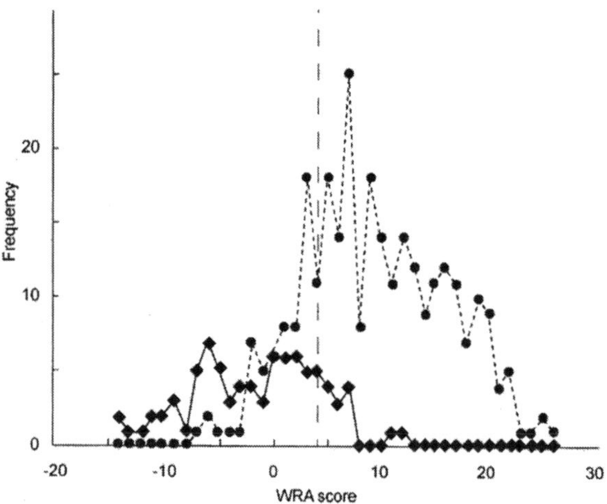

Figure 7.16. Frequency distribution of weed risk assessment scores plotted separately for weeds (circles) and non-weeds (diamonds) (after Hughes and Madden, 2003). The status of species (weeds or non-weeds) was determined retrospectively from data on invasive behaviour after introduction. The vertical dashed line represents the threshold score of 4 used to discriminate weeds from non-weeds before introduction.

into Australia was the *Federal Quarantine Act (1908)*. This act prohibited a small number of plant species, and required that the remainder be fumigated to remove potential animal pests. More recently, a weed risk assessment procedure was established (Steinke and Walton, 1999) to detect plants of concern (State of the Environment, 2001a).

Quarantine services are too under-resourced to adequately perform the functions they are assigned, not only in Australia (Tyndale-Biscoe, 1997), but in all other countries around the world (Simberloff *et al.*, 2005). There are also some glaring inconsistencies in quarantine policies, such as a provision that allows millions of aquarium fish into Australia. Low (1999) provides examples of the failure of the quarantine process in Australia. These issues are exacerbated by increased trade associated with the integration and globalisation of the economies of trading nations (see Padilla and Williams, 2004, for a review of problems caused by the aquarium trade). It is impossible to assess

Table 7.11. Predicted and actual ecological status (weed and non-weed) for 980 plant species introduced onto the Australian continent (after Smith *et al.*, 1999).

True outcome	Predicted outcome	
	Weed	Non-weed
Weed	17	3
Non-weed	147	833

Box 7.11

Assessing the risk posed by an imported plant

Lonsdale (1994) examined the fate of introduced pasture species in northern Australia. He noted that any useful exotic pasture plant can become a weed in some circumstances. However, it is difficult to predict which species are most likely to become weeds, partly because of the changes that can occur when an exotic species is introduced into a new environment, and is free from its natural predators, competitors and pathogens (Smith *et al.*, 1999). Lonsdale (1994) suggested a process to maximise the value of such introductions (given that so many appeared not to be useful) and to minimise the risk of introduced species becoming major environmental weeds. These steps were (after Lonsdale, 1994):

- Screen out the species of no known value.
- Examine the range of options available for the control of the species (e.g. biological or chemical agents).
- Complete an assessment of the potential costs and benefits of the introduction.
- Field trial the species under rigorous quarantine restrictions across a range of geographical areas in the proposed release region, with follow-up treatments to eliminate plants from the experimental sites. Such trials would include examinations of factors such as the palatability of plants for stock.
- Examine the potential impacts of release on other land management plans, then undertake further cost–benefit analyses that examine both ecological and economic criteria.
- Release exotic plant species if steps 1–4 have been satisfied.

Such steps would be costly, and this cost could be borne by those who propose to introduce a plant (Humphries *et al.*, 1991; Lonsdale, 1994). A consequence of this approach would be to ensure that fewer species are imported and more extensive evaluations would be completed for those taxa already naturalised in Australia. The onus would be on the proponent to demonstrate minimal risk (see also Ruesink *et al.*, 1995).

Notably, the recently implemented weed risk assessment procedure (Pheloung et al., 1999; Steinke and Walton, 1999) for detecting plants of concern seems likely to limit the number of new pasture grasses that will be brought into Australia.

traded goods adequately for all potential invasive species (Everett, 2000). Loch and Floyd (2001) called for tighter quarantine-style restrictions within Australia to limit the movement of plantation eucalypt seedlings between areas in an attempt to limit the spread of unwanted organisms between regions.

As Figure 7.15 shows, ornamental plants make up a large proportion of plants that have become naturalised in Australia and/or that have become serious environmental weeds (State of the Environment Report, 2001a; see Table 7.8).

Given the large number of weeds originating in gardens (see Low, 1999, for further details), risks can be reduced by educating nurseries, gardeners and high-profile hosts of television and radio gardening programs about the sorts of plants that should not be sold or grown, especially in places close to native bushland.

Extensive areas of weed infestations. (A) Paterson's Curse in south-eastern New South Wales dominating almost the entire ground layer; (B) Cape Weed in Canberra. (Photos by David Lindenmayer.)

Methods of weed control

There are numerous weed control methods, but it is beyond the scope of this book to delve into them. Some are (after Wright, 1991) herbicide application, fire (e.g. burning after flowering and before seed set), removal of seedlings by hand, ringbarking, cutting at ground level, and the removal of stems and roots. The appropriate method will depend on the extent of the problem, available resources, the biology and habitat of the target species, and the potential impacts on non-target species. As shown in Figure 7.17, eradication success is typically highest in the early stages of infestation when the distribution of the weed species is limited.

Exotic species and their expansion into Interim Biogeographic Regionalisation for Australia (IBRA) regions (see Chapter 4) are listed as indicators under the State of the Environment reporting process (State of the Environment, 2001a). Similarly, the extent of weed distributions and steps taken to control them are indicators of sustainability under international environmental agreements for ecologically sustainable forest management, such as the Montreal Protocol (Commonwealth of Australia, 1997).

Biological control

Biological control is the use of a predatory or parasitic organism to control a pest plant or animal (Harley and Forno, 1992). An example is the control of Prickly Pear, which was introduced into Australia as an ornamental plant, and is now spread over hundreds of thousands of hectares of grazing land and woodland in eastern

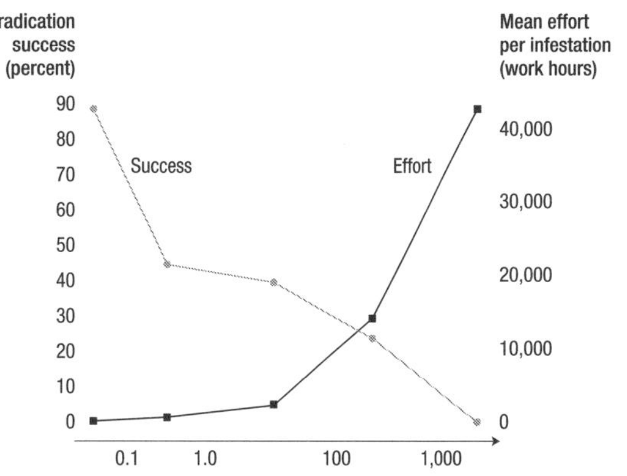

Figure 7.17. Weed control relationship in California showing: (A) eradication success and effort expended, and (B) eradication effort in relation to the extent of weed infestation. (Redrawn from McNeely *et al.*, 2003.)

Australia. It was controlled following the deliberate introduction of the Pyralid Moth, a predator of Prickly Pear in its natural habitat. However, in some cases, the introduction of biological control agents has led to substantial environmental damage; the Cane Toad is a notable example (see earlier in this chapter).

Few environmental weeds have been successfully managed with biological control agents and some, such as those with large seed reserves in the soil, may not be suited to such control measures (Briese, 2000). Blackberry is a serious problem in southern temperate forests (Amor and Richardson, 1980; Lindenmayer and McCarthy, 2001) and Blackberry Leaf Rust was released in 1991 in an attempt to control it. A less virulent strain was also released illegally in the mid-1980s (Parsons and Cuthbertson, 1992). The consequences of these releases have not been fully documented.

7.5 Australian exports

Many Australian species have had significant environmental impacts after being introduced overseas. For example, tree species from the genera *Eucalyptus*, *Acacia*, *Casuarina* and *Melaleuca* are serious weed species in South Africa, Mexico, California, Florida, and elsewhere. Australian Tree Ferns are serious pest plants in Hawaii.

> **Box 7.12**
>
> **The other side of the Prickly Pear story**
> The control of Prickly Pear by *Cactoblastis* caterpillars and moths is one of the most celebrated examples of biological control and it is still widely discussed. There is another side of the story though, which is still unfolding. Prickly Pear control was so effective in Australia that eggs from *Cactoblastis* were sent to other places where there were problems with cacti – including Hawaii, South Africa and the Caribbean Islands. Unfortunately, the species eventually invaded mainland USA and began decimating populations of native prickly pear species, several of which are rare or endangered (Low, 1999). *Cactoblastis* is spreading though Florida and will move into Texas and Mexico, harming the cacti flora and affecting human populations that depend on these plants for food and fodder for their stock (Low, 1999). Therefore, an agent of effective biological control in one area cannot necessarily be successfully exported to other areas.

Many Australian invertebrates have become established overseas, and several cause substantial economic and environmental damage. Low (1999) describes how the Acorn Barnacle invaded estuaries in Europe and has become a pest species in oyster and mussel beds. Another example is the Carpet Beetle, which destroys carpets in Europe and New Zealand (Low, 1999).

At least 24 species of Australian vertebrates have established feral populations in other parts of the world, mostly in New Zealand and the Pacific islands (Low, 1999; Table 7.12). For example, there may be more than 70 million Common Brushtail Possums in New Zealand, which disrupt ecological processes and threaten native fauna by predating on bird nests and competing for resources (Kerle, 2001; Montague, 2000). Ecological research on some of these feral animals has greatly increased scientific understanding of their biology and ecology. For example, far more research is conducted on the enormous feral populations of the Common Brushtail Possum in New Zealand than takes place in Australia (see Montague, 2000).

Some Australian animals that have become established overseas are declining, rare or threatened in Australia. Examples from the list in Table 7.12 include the Brush-tailed Rock Wallaby and the Green and Gold Bell Frog. Such populations could provide sources of animals for future reintroductions in Australia.

7.6 Genetically engineered species

Agricultural science has long sought to improve species by breeding and selection. For example, exotic species in Australia have been 'improved' to increase their

Stands of Australian *Casuarina* and *Melaleuca spp.* in Florida that have established as major environmental weeds. (Photo by Dan Simberloff.)

Table 7.12. Australian vertebrates that have established populations overseas. (Based on data in Low, 1999.)

Species	Locations of population establishment
Mammals	
Common Brushtail Possum	New Zealand
Tammar Wallaby	New Zealand
Parma Wallaby	New Zealand
Red-necked Wallaby	England, New Zealand, Scotland
Brush-tailed Rock Wallaby	New Zealand
Birds	
Brown Quail	Fiji, New Zealand
Black Swan	New Zealand
Peaceful Dove	St. Helena Island
Sulphur-crested Cockatoo	New Zealand
Rainbow Lorikeet	New Zealand
Crimson Rosella	New Zealan
Eastern Rosella	New Zealand
Budgerigar	New Zealand
Laughing Kookaburra	New Zealand
Noisy Miner	Solomon Islands
Australian Magpie	Fiji, New Zealand, Solomon Islands
Red-browed Finch	Tahiti
Chestnut-breasted Mannikin	New Caledonia, Society Islands
Reptiles	
Eastern Grass Skink	New Zealand, Hawaii
Amphibians	
Green and Gold Bell Frog	New Caledonia, New Zealand, Vanuatu
Brown Tree Frog	New Zealand
Dwarf Tree Frog	Guam
Green Swamp Frog	New Zealand

ecological tolerance to such processes as drought, even when such species are environmental weeds (e.g. Toowoomba Canary-grass; Oram and Freebairn, 1984). Plant and animal breeders are not asked to predict the consequences of their work for the natural environment in which the modified species are to be released. Such consequences would be very difficult to predict without extensive ecological research, field experiments, and mathematical modelling including risk assessment (see Godfree *et al.*, 2004).

Transgenic varieties and genetically modified organisms

A **transgenic variety** is one that contains a gene or genes from an unrelated organism (such as a bacterial gene in a plant) that could not be transferred to the species using traditional breeding techniques (Shaner, 1996; Marvier and van Acker, 2005). The result of such processes is called a **genetically modified organism**. One of the most active areas of research in molecular biology, and one with potential for immediate economic gain, is the development of transgenic crop plants that are resistant to predation by insects and infection by fungal and viral pathogens (Wilkinson *et al.*, 2003; Marvier and van Acker, 2005; Snow *et al.*, 2005). Targets for research also include the development of plants with traits for pesticide resistance, viral resistance, nitrogen fixation, tolerance to heavy metals for mine site rehabilitation, salt tolerance, and tolerance to low nutrient status (Table 7.13).

The use of transgenic crops has begun with the introduction of herbicide- and insect-resistant varieties in several countries, including the United Kingdom and the USA (Wilkinson *et al.*, 2003; Snow *et al.*, 2005). Globally, in 2002 there were nearly 60 million hectares of genetically modified crops (James, 2002). In the USA, since field trials first began, more than 3000 varieties of genetically engineered plants have been developed and as of 2003 more than 42 million hectares of genetically modified crops were being grown there. In Australia, there have been more than 50 deliberate releases of genetically modified organisms (State of the Environment, 2001a).

Table 7.13. A sample of current and planned genetically modified organisms in the USA (based on Snow *et al.*, 2005)

Host organism	Function and production of introduced gene	Intended use
Microbes		
Pseudomonas syringae, P. fluorescens	Deletion of ice-nucleating cell membrane protein	'Ice minus' bacteria sprays on crops to protect from frost
Pseudomonas fluorescens	Several genes for hydrocarbon degradation and light production	Detect and degrade pollutant (polycyclic aromatic hydrocarbons); fluorescent marker
Pseudomonas putida	4-ethylbenzoate-degrading enzyme	Degradation of pollutant (benzene derivatives)
Clavibacter xyli	Bt crystal protein toxin	Colonise plant vascular tissue to protect plant from insect pests
Baculoviruses	Scorpion neurotoxin; proteases from Rat, Human, and Flesh Fly	Biological control for specific insects
Plants		
Corn, Cotton, Potato	Insect-specific toxin	Kill or deter target insects eating plant tissues (many lines deregulated)
Soybean, Cotton, Corn, Canola, Wheat	Glyphosate resistance	Ability to withstand application of glyphosate herbicide (many lines deregulated)
Corn	Pharmaceutical and industrial compounds (e.g. avidin and many others)	Purify as inputs into other commercial products
Rice	Provitamin A	Provide vitamin A precursor for better nutrition
Rice	Ferritin	Increase iron content of Rice to reduce anaemia
Tomato	Anti-sense polygalacturonase	Delay ripening when red, allowing more time on-vine (Flavr Savr brand was deregulated, but is no longer produced commercially)
Poplar	Modified lignin	Enhanced paper-making qualities; less pollution during milling
Animals		
Pink Bollworm	Marker and sterility genes	Research method for tracking dispersal of adult moths; reduce moth population by suppressing mating
Mouse	Virus-neutralising monoclonal antibody	Model system for testing protection against viral encephalitis
Atlantic Salmon	Growth hormone	Accelerated growth rate, improved feed conversion efficiency (now under regulatory review)
Pig	Insulin-like growth factor I	Accelerated growth rate, improved feed conversion efficiency, leaner carcass composition
Pig	Phytase	Ability to utilise phytate in plant-derived feeds, decreasing phosphorus in wastes
Pig	Human factor VIII	Secrete blood-clotting factor in milk, to be administered to haemophiliacs
Goat	Human tissue plasminogen activator	Production of anti-clotting agent
Sheep	Human α 1 anti-trypsin	Production of agent for treatment of asthma and emphysema

Potential benefits of genetically modified organisms

Many of the results of transferring genes between species can be environmentally beneficial. For example, a special form of genetic manipulation involves developing immunocontraception for pest species such as Rabbits, Foxes and even European Carp (this topic is explored in greater detail in Box 7.6 earlier in this chapter). Development of innate pest resistance could decrease the dependence of agriculture on pesticides (Barnes, 2000). Genetic engineering also provides the prospect of biological control of weeds.

Risks of genetically modified organisms

Genetically engineered organisms pose a range of ecological risks (Snow *et al.*, 2005) through five main avenues:

- the establishment and spread of transgenic crop plants into natural areas (Marvier and van Acker, 2005)
- the transfer, by hybridisation and introgression, of transgenes from crops and domestic animals to native or feral species (Amos *et al.*, 1993; Abbott, 1994; Shaner, 1996; see Wilkinson *et al.*, 2003)
- the development of more drought- and salt-tolerant commercial species that extend the boundaries of agricultural areas beyond those currently considered too marginal, exacerbating agricultural land clearance
- the development of pesticide-resistant crops that could have detrimental effects on non-target species
- unintended impacts on non-target species that are of conservation significance (see Box 7.13).

In the case of weeds, the likelihood that a transgenic plant will become an environmental weed will depend on, among other things, the ability of the transgenic to compete with native species, the density of the released plants, and the habitat into which they are introduced (Hatchwell, 1989; Abbott, 1994; Shaner, 1996; also see the section earlier in this chapter on environmental weeds). Hybrids could form in areas where transgenic plants are raised together with conspecific plants or related species. Data from the United Kingdom show that hybridisation does occur (Wilkinson *et al.*, 2003).

Many of the species targeted for genetic engineering in Australia have related native species (Brown and Brubaker, 2000). For instance, genes could move by hybridisation from plantation eucalypts to adjacent wild populations (Potts and Wiltshire, 1997). The prospect of introgression of genetically modified genes to wild populations of the same or related species highlights an issue that exists but has escaped scrutiny in conventional breeding and production programs (see Wilkinson *et al.*, 2003).

If a transgene is expressed in a non-target species, it is possible that it will confer a competitive advantage upon it (Snow *et al.*, 2005). To find out whether this is the case, manipulative experiments of the demographic attributes of the species will probably be required (Abbott, 1994), as well as detailed modelling (Hails, 2000). The fate of a new gene is likely to be determined largely by genetic drift, and is therefore unlikely to be entirely predictable (State of the Environment, 2001a).

Pesticide effects on non-target species may be an important issue. Pollen from maize that had been genetically engineered to include pesticide genes affected non-target butterfly populations in north America (Pullin, 2002). Non-target species that play beneficial roles in the environment (e.g. in pollination

Box 7.13

Unintended effects of genetically modified organisms: Australian marsupials and New Zealand possum control efforts

A problem with genetically modified organisms is the potential for major unintended negative effects on biota (Gilna *et al.*, 2005). An important example is that of attempts to perfect immunocontraception techniques for pest populations of the Common Brushtail Possum in New Zealand (Ji *et al.*, 2000; Ramsey, 2005). The parasitic nematode *Parastrongyloides trichosuri* is the vector that is used for disseminating genetically-derived immunosterility (Henzell, 2002). This parasite also occurs in Brushtail Possums in Australia (Viggers and Spratt, 1995; Viggers and Lindenmayer, 2005). If effective methods are developed for the immunosterility of possums in New Zealand, and if the virus was to inadvertently cross the Tasman Sea, it could have substantial impacts on the same or closely-related species of (phalangerid) possums in Australia (Henzell, 2002). The parasite lays eggs that are passed out in possum faeces and can then remain viable in the soil for up to 2 months. This means that the modified parasite could be readily transferred to Australia by visitors from New Zealand, or Australians returning home from New Zealand (Gilna *et al.*, 2005).

The conservation challenge posed by the use of genetically modified organisms to control the Common Brushtail Possum in New Zealand is representative of a growing tension where the need to eradicate an introduced species in one nation may be at odds with the conservation needs of that species in another (Cooke *et al.*, 2004; Gilna *et al.*, 2005). Other similar examples include: (1) the potential for Australian Rabbit immunocontraceptive viruses to be distributed in Europe, where the target species is endemic and threatened, and (2) for Spanish genetically modified viral vaccines to disable Australian Rabbit biocontrol (Angulo, 2001; Angulo and Cooke, 2002).

and seed dispersal) may be affected more strongly than by conventional pesticide applications because exposure to modified crop plants is continuous.

Assessing the effects of genetically modified organisms

The impacts of new introductions and genetic transfer are very difficult to predict (Barnes, 2000). Current methods of assessing the effects of the released organisms are crude (State of the Environment, 2001a), although some interesting new approaches to monitoring programs for tracking the environmental costs associated with transgenic crops are being developed in the USA (Andow and Ives, 2002).

Shaner (1996) suggested several procedures that could minimise the risks from transgenic crops. Field testing of transgenic plants could provide information on seed production, dormancy, dispersal, pollination, germination, growth, environmental tolerance and resistance to pathogens. Such information would provide a guide to the potential, for example, of a transgenic crop to develop into an environmental weed. This information could be used to build explicit models to assess the ecological risks posed by transgenic crops. Other steps include avoiding traits that would increase weed potential in high-risk crop types, developing genetic barriers to protect against hybridisation and gene introgression, and the spatial isolation of transgenic crops. Programs of evaluation, quantitative risk assessment, monitoring and remediation should accompany the release of any new transgenic species.

In Australia, the Gene Technology Bill (2000) established the Office of the Gene Technology Regulator. All proposals to develop genetically modified organisms are assessed by Genetic Manipulation Advisory Committee, part of the Office of the Gene Technology Regulator. Part of the assessment process is the preparation of a risk assessment plan for the release of genetically modified organisms into the environment. The plan is prepared in consultation with experts and stakeholders (Henzell, 2002; Australian Office of the Gene Technology Regulator, 2003). The assessment is essentially subjective.

As of 2001, nearly 5000 projects involving genetically modified organisms had been assessed by the Genetic Manipulation Advisory Committee. Almost all of these proposals were deemed acceptable. Only two were rejected, and both of these involved the development of transgenic gut bacteria for ruminants that would have given them resistance to naturally occurring poisons in native Australian plants (State of the Environment, 2001a). The concern in these cases was that the escape of such genes into feral populations could make poison-baiting strategies ineffective (State of the Environment, 2001a).

A United Nations multilateral accord, the Conference of Parties to the Cartagena Protocol on Biosafety, came into force in February 2004, and aims to protect biodiversity from the potential threats of biotechnology by ensuring that nations are provided with sufficient information before allowing genetically modified organisms into their countries.

Despite the efforts of the Genetic Manipulation Advisory Committee and agreements such as the Cartagena Protocol to regulate the development and release of genetically modified organisms, it remains to be seen how effective the risk assessment and regulation process will be.

Box 7.14

`Ice-minus'

The development and release of genetically engineered organisms is increasing significantly. Amos *et al.* (1993) relate the history of the development of `ice-minus', a genetically engineered strain of the bacterium *Pseudomonas syringae*, which reduces the temperature at which frost forms on the surfaces of leaves. The natural form of *P. syringae* facilitates frost formation, causing damage to crops and other vegetation. It was intended that the `ice-minus' strain would be applied to crops to replace the wild strain. The intent to release the new strain caused concern in the USA for several reasons, including concerns about the bacterium's potential to alter the competitive status of species in natural communities in frost-prone areas. Furthermore, *P. syringae* cells are nuclei around which ice forms in clouds, and their displacement in the atmosphere has the potential to alter weather patterns. The release of the `ice-minus' strain was prevented by an American court after environmental groups claimed that it should be the subject of an environmental impact statement. The company who developed the genetically engineered strain of *P. syringae* nevertheless proceeded with an illegal rooftop release for which they were subsequently fined.

7.7 Pathogens

Diseases of plants and animals are part of the natural environment. Human impacts on the environment interact with diseases in many ways; for example, land clearing in the Amazon has been associated with an increased prevalence of malaria (Pearson, 2003). Human migration to the Australian continent has brought many viruses and fungi that have detrimental impacts on natural populations (Low, 1999), and which we classify as diseases. Humans have spread some endemic, geographically isolated, diseases to new locations within the continent, and have changed ecological processes to enhance or modify disease dynamics.

Cinnamon Fungus

Cinnamon Fungus is a soil and soil-water borne pathogenic root fungus that is widespread in Australasia, and which probably has tropical origins (Weste and Marks, 1987). It moves between plants in soil-water and by root-to-root contact. It is lethal to plants in many different taxa, including species that are characteristic of rainforests, sandplain heaths and sclerophyll forests. Cinnamon Fungus was originally called 'Jarrah dieback' because it killed extensive stands of Jarrah in Western Australia; 10% of Jarrah Forest was affected by the disease in 2001 (State of the Environment, 2001a,d). However, it also occurs widely in Victoria and Tasmania (e.g. Reiter *et al.*, 2004). Outbreaks in North America (Rhoades *et al.*, 2003) are currently devastating a range of tree species populations, including Port Orford Cedar (Zobel *et al.*, 1985) and American Chestnut (Rhoades *et al.*, 2003). The disease has been listed as a threatening process under the *Environment Protection and Biodiversity Conservation Act* (EPBC Act; see Chapter 4).

Different taxonomic groups have different susceptibilities to the fungus (Reiter *et al.*, 2004; Shearer *et al.*, 2004); Jarrah trees are relatively resistant, except on poorly drained sites. Podger and Brown (1989) isolated Cinnamon Fungus from 39 species of indigenous cool temperate rainforest flora in Tasmania. They rated 30% of the rainforest flora as highly susceptible to the disease and 5% as highly resistant. Families such as the Proteaceae and Epacridaceae (which are responsible for much of the species richness of the highly endemic flora of south-west Australia) are highly susceptible to the disease. Indeed, more than 25% of the 8000 plant species that are endemic to south-western Australia are at risk from Cinnamon Fungus. The disease has the potential to completely eliminate susceptible plant species from sites where it becomes established, and to change the composition and structure of entire vegetation communities and their dependent fauna (Shearer *et al.*, 2004). For instance, the Honey Possum would suffer from the loss of Proteaceae species from a community.

Cinnamon Fungus can be spread to uninfected areas in mud on vehicles, heavy machinery and the boots of bushwalkers. Activities such as mining, forestry and recreation have been implicated in the spread of the disease in Western Australia. Cinnamon Fungus has also been identified in ornamental nursery plants.

The distribution of Cinnamon Fungus in Tasmania is symptomatic of recent invasion, although it may have been present for a very long period (Burbidge, 1960). In a State-wide survey, Podger *et al.* (1990) noted that the distribution of Cinnamon Fungus in remote areas of Tasmania was associated with mining exploration, human settlement and walking tracks. Samples taken more than 2 metres uphill of sharp boundaries between healthy and diseased vegetation tested negative, suggesting a localised point infection and subsequent spread in soil water. Podger and Brown (1989) found the disease to be patchily distributed along exposed road and track edges, but they did not recover it from healthy roadside vegetation or from undisturbed rainforest. The disease responds positively to elevated soil temperature and moisture. The analysis of Podger *et al.* (1990) of its distribution in Tasmania in relation to climatic variables suggested that the ecological limits of the disease appear to be where annual rainfall is less than 600 millimetres and the mean annual temperature is less than 7.5°C. They concluded that it is unlikely that practical measures for control could be devised for areas where the disease is widely scattered.

In Victoria, Cinnamon Fungus is widespread in local populations (Marks and Smith, 1991), although there is potential for it to spread (Wilson *et al.*, 2003). Although the disease causes extensive damage to plant communities in southern Western Australia, it is not easy to predict its effects in south-eastern Australia (Wilson *et al.*, 2003). Weste and Vithanage (1978) suggested that high organic matter content in soils might stimulate competitive and antagonistic micro-organisms. As a result, the fungus could be less damaging in areas with nutrient-rich soils than in areas such as Western Australia that typically have nutrient-depleted soils.

Cinanmon Fungus-affected-areas (lighter-coloured dying vegetation patches) in Strzelecki National Park on Flinders Island. (Photo by David Lindenmayer.)

Cinnamon Fungus control area on Kangaroo Island. (Photos by David Lindenmayer.)

However, recent studies of small mammals in Victoria have shown that populations are reduced and habitat-use patterns are altered in areas that are subject to high levels of infection (Wilson *et al.*, 2002). Even in areas where the effects of Cinnamon Fungus are likely to be relatively benign, the disease has the potential to cause significant damage to individual elements of the biota, including Tall Astelia, one of Victoria's most endangered forest-dependent plants.

Other diseases

There are numerous other exotic and endemic pathogens that have important effects on the Australian biota, although most are generally less well studied, and their biology and symptoms are less thoroughly documented than those of Cinnamon Fungus. The impacts of the fungal disease Chytridiomycosis on frog populations (see Daszak *et al.*, 2000) were discussed in Box 7.1. The protozoan *Toxoplasma gondii*, a parasite of the introduced Cat, also causes disease and mortality in marsupials. The parasite may have contributed to the decline of the mainland population of the Eastern-Barred Bandicoot, in addition to the impacts of predation by the Cats themselves (Lenghaus *et al.*, 1990).

7.8 Conclusions

Changes in the abundance and distribution of both native and introduced species in Australia are largely a consequence of the spread of European land management practices, combined with increased rates at which species are transported into Australia from outside, and from region to region within Australia. Although changes in the distribution and abundance of species is a natural process, the rate at which these changes are occurring in the Australian landscape is without precedent. The ecological consequences of plant and animal invasions include impacts on soil erosion, fire regimes, and the recruitment and persistence of native species.

Management of the consequences of these changes may involve the development of strategies to limit, reduce, or eliminate those species with the most severe impacts. We can avoid problems in the future by putting in place regulations that govern the way in which species are introduced or modified.

7.9 Practical considerations

The best way to mitigate the detrimental impacts of invasive plant and animal species is to prevent their arrival. We recommend, in addition to full cost–benefit and risk analyses, that contingency funds, biological control agents (if they exist) and rapid response systems be put in place *before* any new releases of exotic species or genetically modified organisms are contemplated. In all such practices, the burden of proof, and the cost of testing, should be borne by those proposing to introduce a species.

If risk assessments of genetically modified organisms are to protect the environment effectively, then they must consider the implications of novel characteristics for the spread of agricultural systems and exacerbation of land clearing, as well as examining the potential for impacts on non-target organisms and gene flow to related species.

When an invasive and damaging species is first discovered, early action can eliminate the problem before it becomes logistically and financially impractical to do so. Early detection will depend on extensive and reliable monitoring systems, the cost of which should be borne by the proponent of any new introduction or release. Risks could be reduced by educating nurseries, gardeners, pet aquarium stores, school children, and high-profile hosts of television and radio gardening programs about the sorts of plants and animals that should not be sold or grown, especially in places close to native bushland, waterways or wetlands.

7.10 Further reading

The paper by Lomolino and Channell (1995) is an interesting discussion of range changes in Australian mammals. Low (1999) gives an extraordinary account of the plants and animals that have been introduced to Australia since European settlement as well as the array of organisms that have been exported to other nations. Research into particular sorts of exotic species is provided by Brunner and Coman (1972), Cooke (1987), Howden (1985), Scott and Adair (1991), and Wilson *et al.* (1992). Lever (2001) provides a valuable account of the history of introductions of the Cane Toad to various countries around the world, including Australia. Other examples of work on individual exotic species include compendia of scientific material on introduced rodents and their management (Singleton *et al.*, 1999), and the large exotic populations of the Common Brushtail Possum in New Zealand (Montague, 2000). A summary of the impacts of introduced species in a global context is provided by McNeely *et al.* (2003). General treatments of invasions are provided by Dyne and Walton (1987), Humphries *et al.* (1991), Lonsdale (1994, 1999) and Myers *et al.* (2000). An excellent book on exotic plants is Cronk and Fuller (1995), and there have been useful issues on invasive plants in the journals *Austral Ecology* in 2000 (volume 25, issue 5) and *Conservation Biology* in 2003 (volume 17, issue 1). Simberloff *et al.* (2005) provide some valuable insights into the policy, management and research needs for invasive species. The journal *Plant Protection Quarterly* examines issues associated with weeds and weed control in Australia.

The journal *Molecular Ecology* published a special issue on the ecological risks of transgenic crops in 1993 (volume 3, issue 1) and there is now a newsletter called *Biocontrol News and Information* that canvasses a range of issues associated with genetically modified organisms. The State of the Environment Report (2001a,d) contains information on the number and types of genetically modified organisms in the Australian environment. Andow and Ives (2002) provide an interesting discussion on a monitoring program for tracking the environmental costs associated with transgenic crops. Wilkinson *et al.*, (2003) is a valuabe quantitative study of hybridisation between transgenic and other crops in the United Kingdom. Godfree *et al.* (2004) and Marvier and van Acker (2005) provide some valuable insights into ecological and risk assessment issues associated with transgenic plants. A major review of the topic is presented by Snow *et al.* (2005).

Chapter 8
Harvesting natural populations

This chapter focuses on harvesting natural populations, particularly as it relates to potential impacts on biodiversity. It includes discussions of the potential impacts on biodiversity of native forest harvesting and the conversion of native forests to plantations in Australia. The harvesting of natural populations of marine organisms (particularly fish stocks) and Kangaroos are also discussed.

8.0 Introduction

Hunting, fishing, forestry, and collecting for trade are important activities in many parts of Australia and elsewhere around the world. These activities have the potential to cause local and global extinctions (Primack, 2001), particularly in conjunction with habitat loss, competition or disease. They can also change ecological interactions between populations of different taxa (e.g. Estes *et al.*, 1998; Springer *et al.*, 2003).

Birds, reptiles, amphibians and fish are collected illegally for the local and overseas pet markets. Populations of declining taxa such as the Carpet Python, Diamond Firetail, and numerous species of parrots are vulnerable to collection. More than 100 000 marine fish are collected for aquaria each year in Queensland alone (Pogonoski *et al.*, 2002). The impacts on wild stocks are unknown but studies elsewhere in the world suggest that they may be significant (Kolm and Berglund, 2003; Tissot and Hallacher, 2003). Similarly, the development of a native terrestrial mammal pet industry in Australia may cause damaging levels of illegal harvesting (e.g. Mooney, 2002; Viggers and Lindenmayer, 2002).

Harvesting can change the structure and composition of the associated plant and animal communities. The Broad-headed Snake is a spectacularly coloured venomous snake that is largely confined to the Hawkesbury Sandstone rock formation, an area encompassing about 250 kilometres around Sydney (Cogger, 1995). The species requires weathered rocky sandstone outcrops where it shelters during the day (Shine and Fitzgerald, 1989; Shine *et al.*, 1998), but people harvest bush rock from its habitat to decorate suburban gardens (Shine *et al.*, 1998). These habitats are important for many other species, including a diverse invertebrate fauna (Goldsborough *et al.*, 2003).

Carpet Python – a native species of reptile threatened by several factors including illegal collection for the pet trade. (Photo by Esther Beaton.)

The decline of populations of the Broad-headed Snake as a result of habitat destruction was recognised more than 130 years ago (Krefft, 1869, in Shine and Fitzgerald, 1989). Conservation of the species depends on protecting rocky areas within national parks around Sydney, and changes in public attitudes so that people value snakes and decorate their gardens with quarried rocks rather than ones collected from weathered sandstone outcrops (Shine and Fitzgerald, 1989).

Some exploitative industries can operate sustainably. There is a large legal Australian wildflower industry that harvests stems from natural stands. In 1999 nearly 31 million stems were harvested (State of the Environment, 2001a). The wildflower industry is regulated by government agencies, and over-exploitation is rare; only illegal collections of rare or threatened plants pose a real threat.

Broad-headed Snake. (Photo by Damian Michael.)

Sandstone cliff environments in Kangaroo Valley, where the Broad-headed Snake has been found but bush rock collection is known to be a problem. (Photo by David Lindenmayer.)

Heavily exploited species can recover provided harvesting pressure is reduced and suitable habitat is available. For instance, populations of the Saltwater Crocodile in northern Australia were over-harvested for many decades but have recovered rapidly (see Box 8.1).

Aquarium fish, wildflowers and crocodiles are examples of the direct and indirect effects of harvesting. The remainder of this chapter examines three major industries. Two of these, forestry and fisheries, have global and Australian perspectives. A third, Kangaroo harvesting, is uniquely Australian but has similarities with large herbivore harvesting elsewhere around the world.

Box 8.1

Crocodiles in northern Australia: a population recovery following over-harvesting

The Saltwater Crocodile occurs in northern Australia, eastern India, throughout South-East Asia and on Pacific islands including Vanuatu (Cogger, 2000). The skin of the Saltwater Crocodile is the most valuable of all species of crocodiles, and following World War II, a major skin industry developed in northern Australia based on hunting wild populations. By the late 1960s populations were significantly depleted and the Saltwater Crocodile was rarely seen. Some biologists were concerned that the species was in danger of extinction. State legislation to protect the species was introduced in Western Australia, the Northern Territory and Queensland, after which populations began to recover. For example, in the Northern Territory, the population rebounded from less than 5000 in 1971 to stabilise at approximately 75 000 in 1999 (Richardson *et al.*, 2002). Although the size of the population is now stable, the average size of individuals continues to increase. Cannibalism and exclusion prevents increases in the numbers of small crocodiles, with survival rates of less 1% for animals less than 8 years of age. The field data for the Saltwater Crocodile highlight its ability to recover, although like almost all species, the capacity to do this depends on the maintenance of suitable habitat (Richardson *et al.*, 2002). The species is now farmed extensively in northern Australia (Webb, 2002), and considerable numbers of skins and quantities of meat products are harvested each year (State of the Environment, 2001).

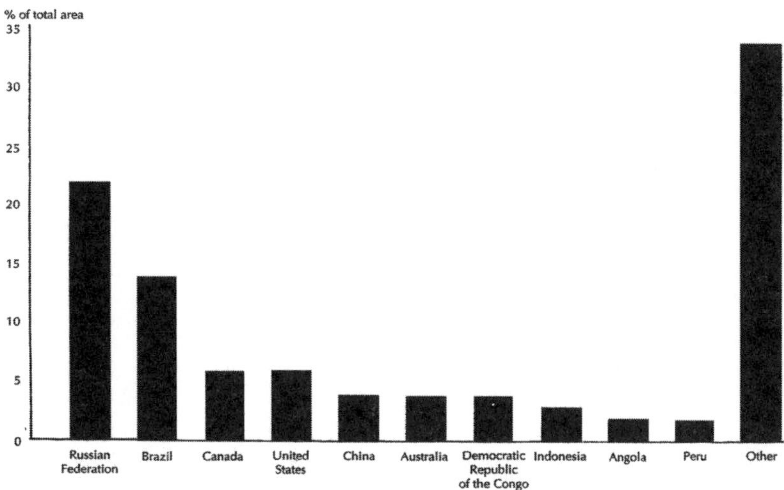

Figure 8.1. Proportion of the total world area of forest found in Australia and other nations. (Redrawn from FAO, 2000.)

8.1 Native forest harvesting

Definition of forest cover

In the late 1990s, Federal Government agencies (e.g. AFFA, 2003a; in Australia) and global organisations (e.g. FAO, 2000) redefined forest as 'any treed vegetation with canopy closure exceeding 10%'. Canopy closure is the proportion of an area of ground that is covered by the vertical projection of tree crowns (Philip, 1994). This definition encompasses a range of cover types, from closed forest through to open forest and woodland (Figure 8.2). This definition substantially changes estimates of forest cover in places such as Russia and Australia. Under previous definitions, Australia's forest cover was about 41 million hectares (National Forest Inventory, 2003). Under the new definition, forests in Australia cover almost 165 million hectares or approximately 20% of the continent's land surface, constituting about 4% of the world's forests (FAO, 2000; Figure 8.1; Table 8.1). Of the 165 million hectares, only 1% is plantation. The vast majority of forest in Australia is leasehold or privately owned. Reserves and multiple-use forests comprise a relatively small proportion (Figure 8.3).

A broad range of species can be found in Australian forests: from Mountain Ash, the tallest flowering plant in the world, to short multi-stemmed **mallee**-form eucalypts. The extent of different forest types is shown in Table 8.1.

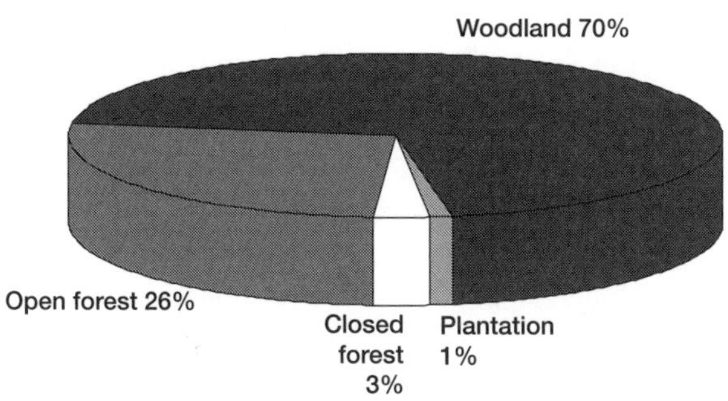

Figure 8.2. Types of forest cover by vegetation type. Woodland = 20–50% canopy cover; open forest = 50–80% canopy cover; and rainforest = >80% canopy cover. (Redrawn from AFFA, 2003a, and based on data in the National Forest Inventory.)

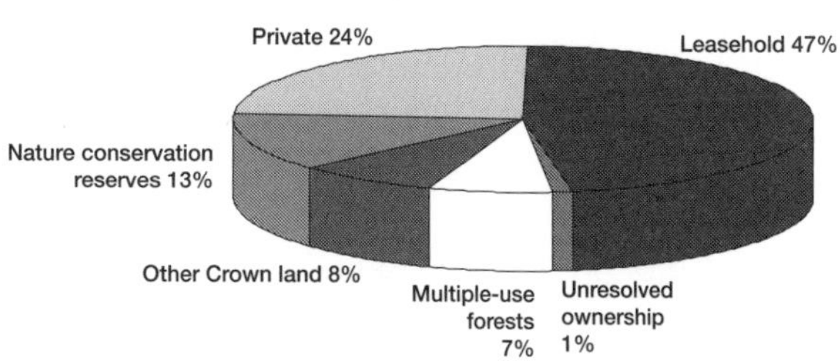

Figure 8.3. Forest cover tenure. (Redrawn from AFFA, 2003a, and based on data in the National Forest Inventory.)

Table 8.1. Extent of different forest types in Australia (from AFFA, 2003a, and based on data in the National Forest Inventory).

	Area (×1000 hectares)	Percentage of area
Native forest		
Acacia	16 488	10.0
Callitris	2330	1.4
Casuarina	2039	1.2
Eucalypt	127 025	77.3
Mangrove	750	0.5
Melaleuca	7037	4.3
Other	2799	1.7
Rainforest	4214	2.6
Total native forest	162 681	
Plantation		
Hardwood plantation	588	0.4
Softwood plantation	980	0.6
Unknown plantation	1	0.0
Total plantation	1569	1.0
Total forest	164 250	

Two forest industries are discussed in our treatment of forestry: native forest harvesting and plantation harvesting. A third, the firewood industry, is also outlined briefly (see Box 8.3).

Early vegetation clearing and the establishment of State forests

In all Australian States, initial European settlement was followed by almost a century of largely uncontrolled clearing (Kellas and Hateley, 1991). Red Cedar, discovered on the banks of the Hawkesbury River in 1790, was the first export product from the new colony of New South Wales. Hand tools and animal haulage made the selection and extraction of single trees an almost universal harvesting technique up until the mid-1900s. By 1842, Red Cedar cutters had penetrated to northern New South Wales and the species had been extensively depleted in the wild (Boland *et al.*, 1984). Similarly, Bog Onion was once an important rainforest component in the Richmond–Tweed Heads region of New South Wales. It is a prized timber species and because of selective logging, it is now scarce in the wild (Floyd, 1990). In Western Australia, there was no legislative control of timber cutting until the 1890s. In Victoria, forests were heavily exploited to provide timber and fuel for gold mines and towns.

The Conservator of Forests in Victoria reviewed the condition of the forests in 1896 and emphasised the 'forest vandalism' practiced by miners, timber splitters and sleeper hewers who produced more waste than timber. He concluded that it was:

> *for the want of proper supervision and expert knowledge of the conservation of forests in the past 40 years which has brought the forests of Victoria to their present condition of dismal ruin and decay (Kellas and Hateley, 1991).*

Forest Commissions were established in most Australian States to halt the overuse and degradation of forests. Formal and relatively effective control of forest exploitation was in place by the early 1900s. By then, large areas of forest had been cleared for agriculture and many remaining forests were degraded, particularly in areas close to population centres. The first timber reserves in New South Wales were set aside in 1871 to protect forest resources from uncontrolled exploitation. Unauthorised logging of Red Cedar was forbidden by the Governor of New South Wales in 1902. By 1920, harvesting in most of the forest estate in New South Wales was regulated. In Tasmania there was no effective legislative control over forest use until 1920.

Native forest harvesting

Native forests provide multiple values for human societies: timber, pulp, firewood, water, honey, wildlife and recreation. Timber production forests have protected a vast spectrum of the Australian biota that would have been lost had these forests been converted to agricultural land. The economic incentive for retaining native forests has both commercial and conservation benefits. Timber production can modify the environment directly through the construction of roads and tracks,

Clearfell logging operations in Victorian forests. (Photos by David Lindenmayer.)

felling and removing mature trees, burning to reduce the likelihood and intensity of wildfire and to encourage regeneration, the control of plant and animal pests and diseases, the inadvertent introduction of weeds, the thinning and culling of less desirable tree species, compaction of the soil, siltation of streams from roads and cutting areas, and the application of fertilisers (RAC, 1991; Lindenmayer and Franklin, 2002). Roads and railways constructed for harvesting, even in well-managed forests, provide access to previously remote regions, encouraging the expansion of human populations and associated agricultural practices (Harcourt, 1992; Putz *et al.*, 2000). Roads and railways also facilitate the movement of native predators (e.g. Quolls in Tasmania; Taylor *et al.*, 1985) and feral predators (e.g. the Red Fox and feral Cat; May and Norton, 1996; May, 2001). Harvesting and regeneration aimed at maximising timber productivity can lead to long-term changes in forest composition and structure (Kellas and Hateley, 1991; Mueck *et al.*, 1996; Ough, 2001;

Clearfelling is the localised removal of most or all trees followed by burning of debris. The soil tends to be exposed to erosion, and regrowth is often less diverse, both in terms of species and the age of trees.

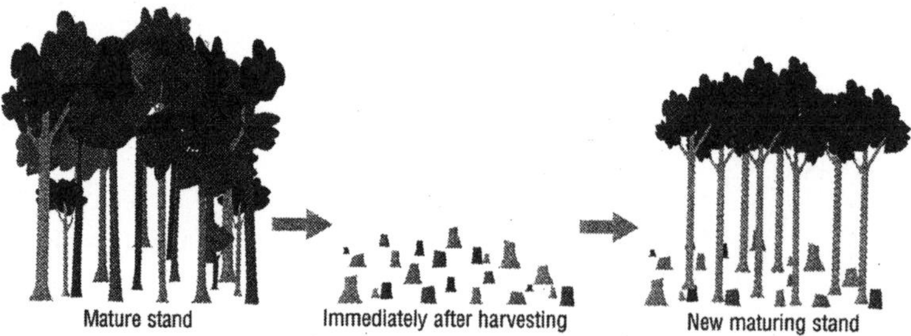

Modified clearfelling keeps some trees for conservation purposes such as mammal habitat and to allow further growth of immature trees. The resulting forest retains a greater diversity of species and age classes.

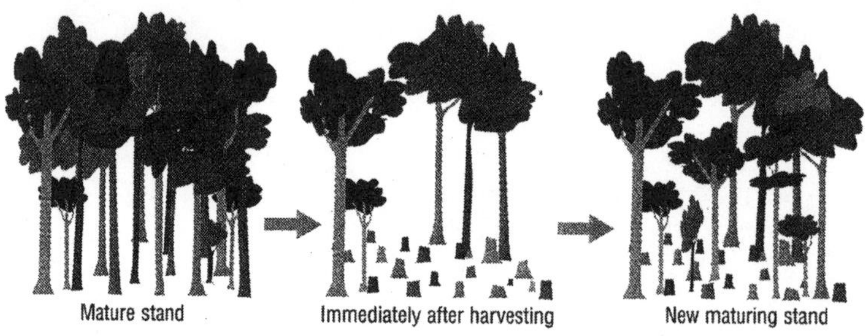

Shelterwood logging in highland forests minimises snow or frost damage to seedlings. Shelter trees are retained and then felled once some regrowth is established.

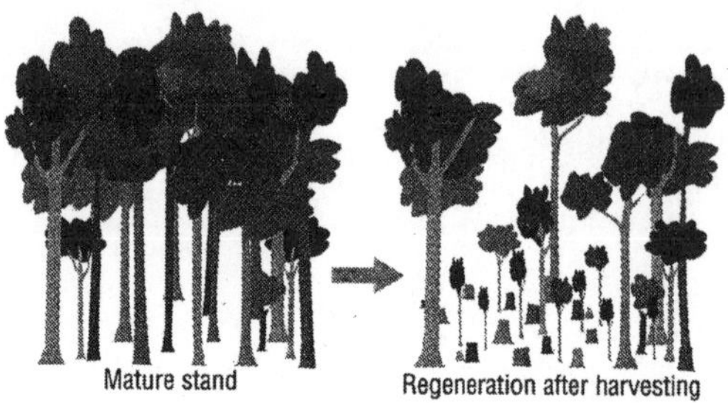

Selective logging removes individuals or patches of trees at relatively short intervals. There are many variations including focus on a particular species or thinning of small trees. The general aim is to retain diversity of species, sizes and ages.

Figure 8.4. Types of logging. (Redrawn from RAC, 1992c.)

Furusawa *et al.*, 2004), including the invasion of weed species (e.g. Brown and Gurevitch, 2004).

Harvesting affects animal populations that depend on particular forest attributes such as hollow-bearing trees, fallen logs and decaying wood (Gibbons and Lindenmayer, 2002; Lindenmayer *et al.*, 2002a). Changes in the spatial pattern of forest age classes can also affect large forest owls (Milledge *et al.*, 1991) and some arboreal marsupials, such as the Yellow-bellied Glider (Lindenmayer *et al.*, 1999c).

One of the objectives of conservation within managed forest landscapes has been the protection of old growth forest. Old forests are a diminishing resource in the Australian forest estate (RAC, 1992a,b; Burgman, 1996) and they are in some respects unique (see Chapter 2).

In 1992, the Resource Assessment Commission (RAC, 1992a,b) estimated that 42% of the remaining total area of forest in Australia had not been disturbed by logging. Further, 27% of the original area of rainforest and eucalypt forest had never been logged, including 10% in reserve systems. These estimates account only for logging history and exclude other disturbance processes, such as grazing and weed invasion. In Queensland and New South Wales, for example, many forests are used for Cattle grazing (Smith *et al.*, 1992; McAlpine *et al.*, 2002).

Disturbance impacts vary greatly among forest types. Logged forests have not lost all old growth or other conservation values. The single-tree selection method (see Figure 8.4), in particular, leaves behind many of the biological characteristics of old growth forests. Clear-felling has been in place in some Australian wet eucalypt forest types since the mid-1960s and most forests before that time were logged by single-tree selection. Many of the ecological attributes of old-growth forest can be regenerated in the long term, although the values associated with some intangibles such as the 'pristine qualities' of an unlogged stand will not be replaced (RAC, 1992a,b; see Chapter 2; Burgman, 1996; Lindenmayer and Franklin, 2002).

Since the mid-1970s there has been protracted debate over native forest harvesting in Australia (Dargarvel, 1995; Lindenmayer and Franklin, 2003). There have been more than 75 enquiries into the industry since 1945 – more than one per year! There have been many attempts to resolve conflict, including the Resource Assessment Commission (1992a–c) and the National Forest Policy Statement (Commonwealth of Australia, 1992). In the following paragraphs we outline three recent major policy developments in Australia: the Regional Forest Agreement process, forest industry certification, and sustainability benchmarking through criteria and indicators.

Box 8.2

Land allocation and harvesting intensification

The Regional Forest Agreement (RFA) process was designed to deliver certainty to resource users (i.e. the timber industry) and conservationists aiming to secure increased reserve areas. It is essentially a land allocation process. An inherent danger in the allocation approach to land management is that it can trigger more intensive utilisation of unreserved areas (Lindenmayer and Franklin, 2002; see Chapter 14). In 2002, a proposal emerged to harvest large quantities of timber from native forests in coastal southern New South Wales to make charcoal for silicon smelting (Environmental Resources Management Australia, 2001). This would have greatly intensified logging with the removal of so-called 'waste wood', including standing defective trees and fallen timber on the forest floor.

Species conservation in wood production forests in the medium to long term is determined, in part, by the intensity of harvesting operations (Lindenmayer and Franklin, 2002). More intensive logging simplifies the structural diversity of stands and reduces the number and type of structural features such as tree hollows, and large decaying logs. 'Waste wood' is habitat for many elements of forest biodiversity (Lindenmayer *et al.*, 2003b). Intensification of harvesting could, in an extreme form, blur the distinction between native forest management and plantation forestry. Ecologically sustainable forest management, a key element of the RFA process, depends on maintaining biota and ecosystem processes in wood production forests outside the reserve system. Nevertheless, with changes in markets for forest products (e.g. charcoal production) and pressure to increase the diversity of wood products derived from native forests, it seems likely that there will be continued pressure to intensify harvesting.

Regional Forest Agreement process

The Regional Forest Agreements (RFA) are legal agreements between the Federal and State Governments (e.g. Commonwealth of Australia and Department of

Natural Resources and Environment, 1997) that aim to balance commercial timber and pulpwood production with other values of forests including biodiversity conservation, water production, and recreational values (Slee, 2001). Many such agreements have been signed in various Australian States, including Tasmania, Victoria, and New South Wales. The RFA process was typically preceded by a Comprehensive Regional Assessment – essentially a survey of forest resources and values, as well as discussions with stakeholders. A core theme in the RFA process has been the development of a comprehensive, adequate and representative (CAR) reserve system (Dickson *et al.*, 1997; see Chapter 16 on reserve design for further details of this approach to protected area selection).

The RFA process has been relatively successful in some cases and problematic in others. Many new reserves were created and timber industries have better assurances about resource supply (Australian Bureau of Statistics, 2001). Some conservation scientists have questioned the credibility of the underpinning ecological science (e.g. Horowitz and Calver, 1998; Kirkpatrick, 1998) and pointed out that some non-market values were overlooked (Slee, 2001). In some cases, the intentions of the RFA process appear not to have been translated to the actual outcomes, leading to increased conflict (Slee, 2001).

Criteria and indicators of sustainability

Governments and forest management agencies around the world have sought to develop criteria and indicators of sustainability. Arborvitae (1995) listed organisations developing forest sustainability criteria including the Convention for Sustainable Development Intergovernmental Panel on Forests, United Nations Food and Agriculture Organization, Helsinki Process, Montreal Process, International Standards Organization, International Tropical Timber Organization, Amazon Process Guidelines, and the World Commission on Forests and Sustainable Development.

The Montreal Process was adopted for Australian forests (Commonwealth of Australia, 1998). It uses 33 indicators of sustainability including soil conditions, long-term site productivity, water quality, and the maintenance of biodiversity (Commonwealth of Australia, 1998). Some criteria are practical, such as the area of forest types and the areas of forest in different age classes. Others seem less practical (Tickle *et al.*, 1998), and many have not been tested or validated. For example, one of the key criteria (Criterion 1.1e) is the development of landscape measures of fragmentation, a legacy of concerns about the potentially negative effects of forest fragmentation on biota (see Chapter 10). However, Smith (2000) and Lindenmayer *et al.* (2002a) found that landscape metrics have little general applicability for the conservation of species.

Forest industry certification

The World Commission on Forests and Sustainable Development (1999) estimated that the international trade in wood products exceeds US$100 billion annually, which is about 3% of world merchandise trade. Forest certification systems aim to enhance the prospects for ecological sustainability and biodiversity in forests (Wallis *et al.*, 1997; Daily and Walker, 2000).

The concept of certification arose, in part, from the 1992 Earth Summit in Rio de Janeiro. It was agreed then that decisions regarding the exploitation of natural resources would link both producers and consumers. Further, decisions should not impinge negatively on residents of another country or on future generations (Viana *et al.*, 1996). Certification involves landowners (including governments) entering into an agreement with a certification authority. This agreement requires forest managers to make specific undertakings with respect to forest management. The organisation certifying a forest owner is responsible for compliance, for example, by conducting random checks of forest practices. Following certification, the products derived from a certified forest can be marked with a logo that signals to potential consumers that timber products have been produced in an ecologically sustainable way.

By mid-2000, more than 18 million hectares of forest worldwide had been certified by the Forest Stewardship Council (Cauley *et al.*, 2001) and certification and consumer concern for the environment had influenced some forestry operations. Despite the potential benefits of the certification, problems remain, including (after Lindenmayer and Franklin, 2002):

- Significant parts of the production forest estate are yet to be certified and some nations do not want to join global certification agreements (e.g. Japan and other Asian countries; Jenkins and Smith, 1999) or want only very minimum standards (Viana *et al.*, 1996).
- International certification agreements may not be appropriate in countries where much of the

timber consumption is domestic (Ghazoul, 2001).

- There are difficulties in developing common sustainability criteria that can be applied uniformly across all countries, in part because of substantial inherent differences between forest ecosystems (e.g. for biodiversity and structural attributes).
- Secondary impacts of logging on biodiversity sometimes receive limited attention. For example, roads associated with logging in many tropical countries lead to increases in hunting for game animals or bush meat (Redford, 1992; Bennett, 2000).

Some countries and organisations, including those in Australia and Europe, have developed local versions of international certification systems (e.g. the Australian Forestry Standard, 2003).

In February, 2004, Hancock Victorian Plantations, Australia's largest private forest plantation owner, was certified by the Forest Stewardship Council on the basis that it manages its forest operations in a socially and environmentally sustainable manner. It was the first large plantation owner in the country to be certified. In 2003, the company owned and managed 245 000 hectares of forest plantations across Victoria (T. Cadman, personal communication).

Certification could be important for ecologically sustainable forest management (Putz and Romero, 2001) if a believable, secure, incorruptible and equitable certification system and a robust set of common protocols can be developed (Viana *et al.*, 1996; Wallis *et al.*, 1997), which also reflect local, regional, national and continental differences. By 2005, very few Australian forests had been certified under any system.

Box 8.3

Firewood production: the other logging industry

Considerable debate has been associated with native forest logging (e.g. Brand, 1997; Lindenmayer and Franklin, 2003) and the effects of plantation forestry (Ray *et al.*, 1983; Lindenmayer and Hobbs, 2004). However, there is a third industry that has largely escaped public attention – the firewood industry. More than 4.5 million tonnes of firewood are cut annually for domestic consumption in Australia (Driscoll *et al.*, 2000; Wall, 2000); this is approximately two-thirds of the volume of woodchips exported each year (AFFA, 2003a). The State of the Environment Report (2001) noted that firewood is the third largest source of energy in Australia after gas and coal.

The firewood industry is based on cutting dead standing and fallen trees, particularly on private land and on roadside reserves. These trees and the vegetation they are removed from can have significant value for many elements of biodiversity – including some rare and threatened species, for example the Superb Parrot (Webster and Ahern, 1992; Manning, 2004). Moreover, much of the timber is sourced from temperate woodlands that have been heavily cleared and are subject to other threatening processes, such as overgrazing of domestic livestock, rural dieback and salinity (Hobbs and Yates, 2000). On this basis, firewood cutting could be considered an additional process threatening the integrity of many areas of temperate woodland, and will have enormous impacts on the many elements of biodiversity that are strongly associated with dead standing trees and fallen timber (Driscoll *et al.*, 2000). Indeed, Garnett and Crowley (2000) identified 21 species of birds at risk from firewood harvesting. In fact, more than 50 species of vertebrates are threatened by firewood harvesting (Driscoll *et al.*, 2000).

The firewood industry is presently unregulated and cannot in any way be construed as being ecologically sustainable (Wall, 2000). Major reform is needed, including better regulation and licensing, appropriate environmental standards, and the development of more sustainable alternative sources of timber, such as purpose-grown woodlots (Wall, 2000).

In New South Wales, dead trees have limited legal protection because they are not considered native vegetation in the State Environmental Planning Policy No. 46/*New South Wales Native Vegetation Conservation Act (1997)*. Dead trees can, however, have protection under the *New South Wales Threatened Species Conservation Act (1995)* if they are critical habitat for endangered species. Local councils can also apply tree preservation orders on dead trees (Farrier *et al.*, 1999). Fortunately, the removal of dead wood, dead trees and logs has recently been proposed as a key threatening process under the *New South Wales Threatened Species Conservation Act (1995)*.

Firewood cutting. (Photo by Mason Crane.)

A nest tree known to be used repeatedly by the Superb Parrot – a native species that nests in large dead and partially dead trees that are often targeted for firewood collection. (Photo by David Lindenmayer.)

Summary: native forest harvesting

Native forest harvesting remains an important industry in Australia. Australian Agriculture, Fisheries and Forestry (AFFA, 2003a) calculated that the forestry industry was the second-largest manufacturing industry in the country, with nearly 80 000 employees. Despite processes such as the RFA and attempts to develop sustainability measures, native forest harvesting is beset by conflict and social division (Dargarvel, 1995; Lindenmayer and Franklin, 2003). Much of this conflict is underpinned by the different values held by protagonists (see Chapter 1).

8.2 Plantation forestry

Plantations are:

> *Stands of trees of native or exotic species, created by the regular placement of cuttings, seedlings or seed selected for their wood-producing properties and managed intensively for the purposes of future timber harvesting* (Australian Forestry Standard, 2003).

In 1996, the combined area of plantations worldwide was estimated to exceed 130 million hectares (Cubbage *et al.*, 1996) and in 2001 it was 187 million hectares (FAO, 2001). Plantations are established to produce wood and fibre (e.g. for timber and paper production) (Doughty, 2001; Wood *et al.*, 2001). There is a global trend toward a greater reliance on wood sourced from plantations (FAO, 2001; Lindenmayer and Franklin, 2003).

Plantations of exotic and native trees have long been established in Australia. In some States, including South Australia, production from plantations far exceeds native forest production. Similar trends are forecast for New South Wales (Figure 8.5), where prolonged conflicts over native forest harvesting and its potential environmental effects has led to the expansion of plantations.

Australian plantations

There are two broad types of Australian plantation forests: those dominated by stands of exotic softwood Pines and those dominated by eucalypts such as Blue Gum and Shining Gum (Eldridge *et al.*, 1994; Wood *et al.*, 2001; Salt *et al.*, 2004). Softwood plantations predominate in Australia (Wood *et al.*, 2001), but the area of eucalypt plantations in Australia has been increasing steadily (Figure 8.6). For example, the area increased from 19 000 hectares in 1995 to 84 000 hectares in 2000 (Wood and Allison, 2000). Blue Gums are often planted as part of the extensive and rapidly

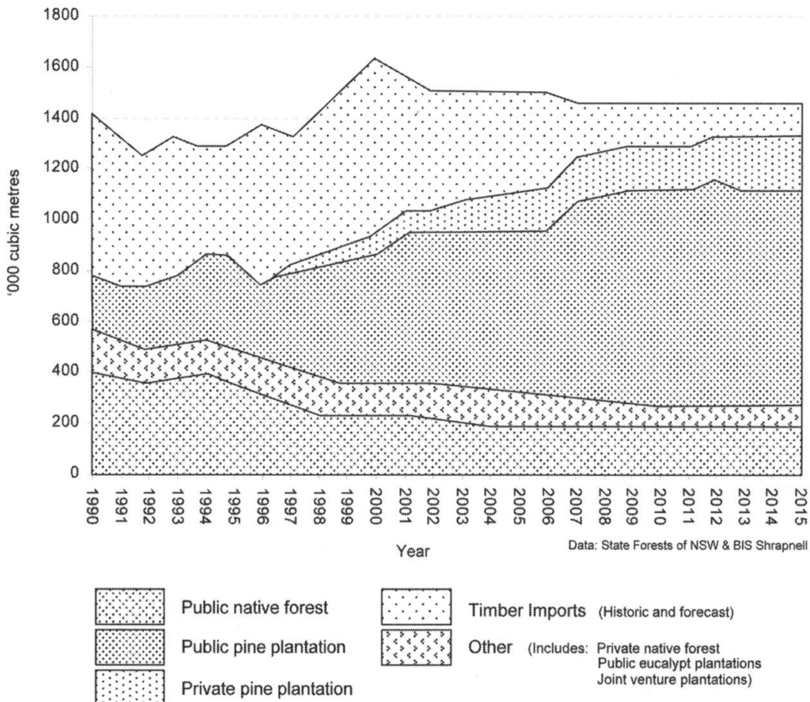

Figure 8.5. Predicted forest product trends in New South Wales. (Redrawn from State Forests of NSW, 2000.)

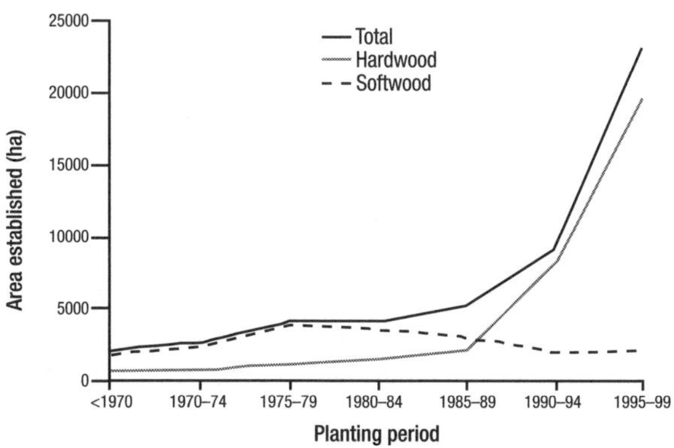

Figure 8.6. Increases in eucalypt and softwood plantation areas (from Wood *et al.*, 2001.)

developing farm forestry industry (Wood *et al.*, 2001; Hobbs *et al.*, 2002; Salt *et al.*, 2004).

Why biodiversity conservation within plantations is important

Intensive or simple plantation forestry (*sensu* Kanowski, 1997) has a narrow and intensive focus on producing a forest crop, but it may not meet social demands for a range of outputs from plantations in addition to wood and pulp. Lindenmayer and Hobbs (2004) argued that biodiversity conservation is an important part of plantation management and should be part of the design of new plantations. Balancing values such as wildlife conservation with other uses, such as wood and pulp production, become relevant when extensive areas of land are involved (Kanowski *et al.*, 2005). Computer modelling suggests that up to 19 million hectares of semi-cleared agricultural land in Australia may be suitable for plantation establishment (Burns *et al.*, 1999). Given that the plantation estate in Australia is likely to increase significantly in the coming decades, the location and management of new plantations will impact on existing biota. Maintaining some elements of the biota within plantations could have benefits for ecosystem processes such as pest control, and is an element of the certification of the ecologically sustainable management of plantations in many parts of the world.

Plantation establishment and biodiversity

In the past, plantation establishment typically occurred by clearing native forest. Clearing to establish plantations has largely ceased in Australia, although it is still occurring in Tasmania, and to a lesser extent in Victoria and New South Wales. Clearing of native vegetation to establish plantations has significant negative impacts on biodiversity (Lindenmayer and Hobbs, 2004; Kanowski *et al.*, 2005). When areas of eucalypt forest in the Tumut region of southern New South Wales were cleared and replaced with Radiata Pine trees (Tyndale-Biscoe and Smith, 1969), there was heavy mortality among populations of the Greater Glider. Almost all animals remained faithful to their home ranges and did not move to uncleared neighbouring areas. Animals simply died *in situ* or were caught by predators, including diurnal ones such as the Wedge-tailed Eagle, which would otherwise rarely take the Greater Glider (Tyndale-Biscoe and Smith, 1969). Many native species cannot survive in softwood plantations because they do not support resources such as trees with hollows or food from saps, gums or eucalypt leaves (Disney and Stokes, 1976; Lindenmayer and Hobbs, 2004).

Patches of forest and woodland were sometimes retained within the plantation boundaries, particularly along watercourses and on hilltops. The size and location of these patches appears to be important. Lindenmayer *et al.* (1999d, 2002c) found that patches 1 hectare or larger, and particularly those exceeding 10 hectares, supported populations of birds, mammals, reptiles and frogs. Other studies have similarly shown the importance for biodiversity of areas of remnant native vegetation within plantations (e.g. Suckling *et al.*, 1976; Recher *et al.*, 1987; Fisher and Goldney, 1998; Hobbs *et al.*, 2002).

Plantations are now being established on semi-cleared grazing lands that sometimes support relatively small areas of the original vegetation. Patches of remnant forest and woodland can be retained (rather than cleared) during establishment. This provides an opportunity for plantation systems to become patchy landscapes comprising stands of exotic trees interspersed with remnants of the original forest and woodland.

Patches of remnant vegetation within a softwood plantation at Tumut in southern New South Wales. (Photo by David Lindenmayer.)

Plantations of Blue Gum, Shining Gum and other eucalypt species continue to be established (Salt *et al.*, 2004). The few studies on their ecological effects have shown that, like conifer plantations, some species of native animals and plants are largely absent (Lindenmayer and Hobbs, 2004). Vertebrate and invertebrate assemblages are less diverse than those of native vegetation, largely because of the structural simplicity of the plantations (Bonham *et al.*, 2002; Hobbs *et al.*, 2002; Cunningham *et al.*, 2005). For example, hollow-using arboreal marsupials are virtually absent from all types of plantations. Birds that use tree hollows are sometimes found in plantations, but they do not nest there. When artificial hollows are added to plantations, they are used by some cavity-dependent fauna including arboreal marsupials and bats (Smith and Agnew, 2002). Nevertheless, as for Pine plantations, Blue Gum plantations are not 'biological deserts', but provide habitat or resources for a range of species, including a selection of bird species considered to be at risk.

Box 8.4

Invertebrates in Tasmanian plantations

Studies in Tasmania have compared invertebrate species diversity and composition in pine and eucalypt plantations as well as nearby native eucalypt forests (Bonham *et al.*, 2002). Native land snails and millipedes were more diverse in native forests than in either sort of plantation. Introduced snails were most abundant in plantations. However, plantations did contain many species of native invertebrates, including a number of rare taxa, including some types of velvet worms (see Box 2.6). It was found that some types of management practices were likely to increase levels of native invertebrate species richness in plantations. These included retaining windrows of rotting logs that remained after the original native forest was cleared, mound ploughing to create furrows of leaf and twig-litter microhabitat, and leaving pruned branches on the forest floor (Salt *et al.*, 2004).

A hardwood plantation being established in Western Australia. (Photo by David Lindenmayer.)

8.3 Kangaroo harvesting

In 1990, there were about 10 million Red Kangaroos in Australia (Fletcher *et al.*, 1990). These populations and those of a small number of other species are the basis of a major Kangaroo harvesting industry. The first three species in Table 8.2 comprise more than 95% of the total number of harvested animals (Pople and Grigg, undated).

Indigenous Australians maintained fire regimes to stimulate the seasonal growth of vegetation and hunted the Kangaroos attracted by it (Stanbury, 1987; see Chapter 11). It has been proposed that hunting may have maintained some Kangaroo populations at considerably lower levels than are presently observed (Flannery, 1994). Large population increases in the past 150 years may have been caused by reductions in predators such as the Dingo, the clearance of trees and shrubs for grazing and crops, the cultivation of improved pasture, and the construction of numerous permanent watering points such as farm dams.

Table 8.2. Commercially harvested species of kangaroos and wallabies. (Based on information in Pople and Grigg, undated.)

Harvested species	Location of harvest
Red Kangaroo	Queensland, New South Wales, South Australia, Western Australia
Eastern Grey Kangaroo	Queensland, New South Wales
Western Grey Kangaroo	New South Wales, South Australia, Western Australia
Euro or Wallaroo	Queensland, New South Wales, South Australia, Western Australia
Whiptail Wallaby	Queensland
Bennett's Wallaby	Tasmania
Tasmanian Pademelon	Tasmania

History of Kangaroo harvesting in Australia

The history of the Kangaroo harvesting industry has been reviewed by the Australian Nature Conservation Agency (ANCA, 1995a) and Pople and Grigg (undated). Large populations of Kangaroos can have negative impacts on agriculture, including providing competition for water and pasture for domestic stock, accelerating land degradation, destroying crops by eating or trampling, and damaging fences. As a result, Kangaroos are often regarded as pests. In some States (e.g. New South Wales and Queensland), they were classified as noxious pests. A government bounty system led, in turn, to the development of a major Kangaroo harvesting industry that included markets for skins and which, in the 1950s, began to provide meat for the pet food industry (when populations of the introduced Rabbit crashed). During this time the Kangaroo industry was not controlled or regulated.

In 1974, the Red, Eastern Grey and Western Grey Kangaroos were placed on the endangered species list in the USA. This resulted in a ban on Kangaroo export

Eastern Grey Kangaroo – a key species in the Kangaroo harvesting industry. (Photo by Esther Beaton.)

to that country, because of the perception that the harvested species were threatened. The Australian Federal Government set about assisting the various States to develop management programs and quota systems to facilitate the export of Kangaroo products. This was achieved in the mid-1970s, and the trade ban imposed by the USA was lifted in 1981. Since that time there have been a number of changes to the way various State and Federal bureaucracies plan and administer the Kangaroo harvesting industry. These plans require the conservation of each species throughout its known range, the regulation of licensed shooting, the monitoring of population sizes, and limitations to be placed on the extent of illegal shooting (ANCA, 1995a; Pople and Grigg, 1994).

Data input to guide Kangaroo harvesting

Information on macropods, such as life history data (e.g. fecundity, response to environmental variability etc.) and population models assist management. Population data are gathered regularly, both through aerial survey (Fletcher *et al.*, 1990; Pople and Grigg, undated) and harvest returns (ANCA, 1995a). These data are critical because populations of Kangaroos fluctuate markedly in response to droughts and floods (e.g. Grigg, 1982; Caughley *et al.*, 1982; Robertson, 1986; Hume, 1999). In years of high rainfall, food is relatively abundant and Kangaroo numbers increase through increased birth rates and reduced mortalities (Newsome, 1966; Caughley, 1987; Dawson, 1995; Figure 8.7).

There are large areas where little or no Kangaroo harvesting takes place, such as national parks. These populations can seed recovery elsewhere if an error in quota setting (or other unforeseen problems) should arise. However, adjacent landholders may experience the effects of animals moving from neighbouring protected areas (Viggers and Hearn, 2005).

Setting quotas for Kangaroo harvesting

Setting quotas typically involves six types of input: (1) current population trends, (2) previous harvest levels, (3) current and past climatic conditions, (4) the level of non-commercial harvest, (5) the size of the population exempt from harvesting, and (6) other forms of mortality (apart from harvesting) (Department of Environment and Heritage, 2005c). For many years the actual cull of each commercial species has been 30–50% less than the quota (ANCA, 1995a;

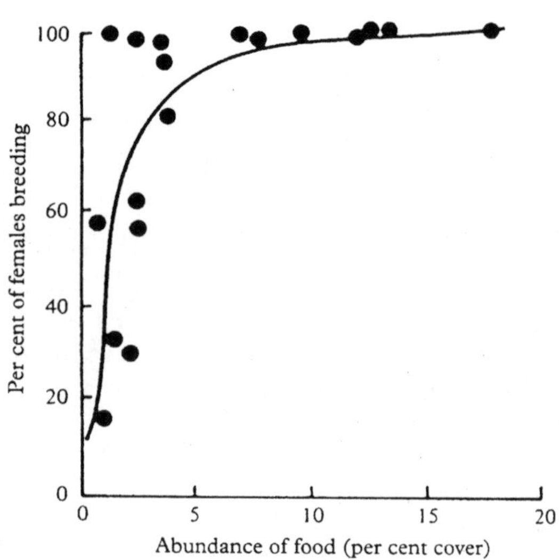

Figure 8.7. Relationship between the abundance of food and the proportion of adult Red Kangaroos breeding. (Redrawn from Newsome, 1966, in Hume, 1999.)

Department of Environment and Heritage, 2005c). For example, in 2003, the Kangaroo quota was just over 6.5 million animals, almost 370 000 less than 2002 and less than 15% of the total estimated population size of the major commercial species (Department of Environment and Heritage, 2003c). Population sizes declined as a result of the drought in 2001–03 and quotas were reduced.

Environmental changes brought about by European settlement in Australia have been beneficial for some taxa, for example several species of large Kangaroos, so much so that they have become both an economic resource and a problem for land management (Viggers and Hearn, 2005). Most importantly, the relevant management authorities have recognised the importance of reliable information on the life history attributes and dynamics of populations of Kangaroos for calculating harvest levels that will be sustainable in the long term (Pople and Grigg, undated; Department of Environment and Heritage, 2005c). There may be detrimental economic effects of large populations of Kangaroos (Gibson and Young, 1987) as well as problems associated with land-use management (e.g. land degradation; Viggers and Hearn, 2005). These species also represent a potentially important economic resource and an additional, or alternative, source of income for land-holders in rural Australia (Grigg, 1991; Grigg *et al.*, 1995; Hale, 1995). Notably, the State of the

Environment Report (2001a) estimated that the Kangaroo harvesting industry was worth almost A$250 million annually to the Australian economy. Grigg (1995) suggested that Kangaroo farming may be a more ecologically sustainable way to manage rangeland environments than grazing introduced domestic livestock. The option could be economically feasible if prices per kilogram of Kangaroo meat were similar to those for beef and lamb (Young and Wilson, 1995).

Ethical positions and perspectives on Kangaroo harvesting

The debate over the management of Kangaroo populations has been controversial. The biocentric ethic (Chapter 1) prioritises the protection of individual animals, and therefore the use of Kangaroos for purposes not vital to basic human requirements is seen as inappropriate and unnecessary (Rawlinson, 1988). Using Kangaroos for pet food and leather may be construed as satisfying superfluous human demands. The utilitarian view would permit harvesting to satisfy maximum sustainable yield, irrespective of the uses to which the resource is put. Kangaroos inhabit extensive semi-arid landscapes along with Sheep and other domestic stock. The ecocentric view of Kangaroo harvesting insists that it be carried out within a management framework that considers all of the ecological implications of various management options (Grigg, 1988; Pople and Grigg, undated). These different viewpoints lead to fundamental differences of opinion concerning whether harvesting should take place, how it should be carried out, how many animals should be taken, and how the process should affect other human activities in the same landscape, such as the grazing of domestic stock.

8.4 Fisheries

The complexity of fisheries management

Fish are hard to observe and population sizes fluctuate naturally, sometimes in response to phenomena that are difficult to predict (such as El Niño episodes). Adults can be highly fecund and the mortality rates of eggs, larvae and juveniles vary markedly. These factors can make it extremely difficult to predict the size and recruitment capabilities of fish stocks (Kearney, 1995) and to determine why a stock has declined or collapsed (Myers *et al.*, 1997; Kurlansky, 1999). Because individual females can carry millions of eggs, recruitment can still be substantial when there are few adults in a population or, conversely, it can be limited when there are large numbers of adults because of competition (Kearney, 1995). In addition, stocks of particular fish species may in fact be an amalgam of a range of diverse local populations and the overall behaviour of the fishery may be due to differences in the diverse substocks that comprise it (Hilborn *et al.*, 2003). The Sockeye Salmon fishery in Alaska comprises several hundred discrete and diverse spawning populations (Hilborn *et al.*, 2003). A final factor that adds to the complexity of fisheries management is that in addition to pressures on fish stocks from commercial fishers (Rosenberg, 2003), recreational fishing affects many species, with recreational catch sizes exceeding commercial harvests in some cases (see McPhee *et al.*, 2002).

In 2000, it was estimated that there were 5 million recreational fishers in Australia and their annual catch was crudely estimated to be 30 000 tonnes (Fisheries Research and Development Corporation, 2000). However, recreational harvest levels are often difficult to document – there can be cumulative effects of large numbers of individuals fishers, and their impacts are usually ignored in stock management (McPhee *et al.*, 2002).

Fisheries management is further complicated by complex inter-relationships between fish and other components of marine food chains (e.g. phytoplankton and zooplankton) and prey-predator relationships (Springer *et al.*, 2003). Jackson *et al.* (2001) highlighted the long-term impacts of over-fishing and the subsequent collapse of coastal and other systems brought about by altered food webs and interactions between species (Figure 8.8). They noted time lags of decades to centuries in the decline and collapse of fisheries and the start of over-fishing (by indigenous people or early European colonists to an area) and more recent intensive (industrial) and global exploitation of fish stocks (Jackson *et al.*, 2001). They attributed the time lags to factors such as other non-target species at the same level in the food chain assuming the roles of the over-fished species until the stocks of the non-target species were also depleted. Given the prolonged nature of some of the time lags, Jackson *et al.* (2001) suggested that many more fisheries were likely to collapse in the future and that associated marine food webs were likely to be simplified.

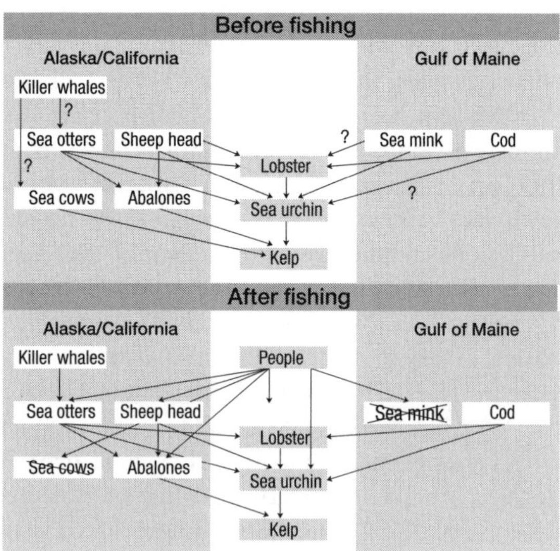

Figure 8.8. Simplification of food webs as a result of over-fishing. (Redrawn from Jackson *et al.*, 2001.)

Box 8.5

A top-down fisheries collapse

Many fish and marine mammal stocks have been subject to significant mismanagement. Although poor management practices may have largely stopped, the impacts of these may still be ongoing in some marine ecosystems. Recent evidence suggests that there has been a 'top-down' collapse in the marine food-web structure of the North Pacific Ocean. Springer *et al.* (2003) argued that collapses of populations of large marine mammals such as seals, sea-lions and Sea-otters in the North Pacific Ocean and southern Bering Sea may be a result of industrial whaling that focused on stocks of large whales after World War II. This removed key prey items for the major predator of large whales, the Killer Whale, which then prey-switched to smaller marine mammals such as seals, sea-lions and Sea-otters. Springer *et al.* (2003) termed the collapse of populations of smaller marine mammals as a 'fishing-down' effect of the marine food-web. The work by Springer *et al.* (2003) not only illustrates the complex array of effects that can result from 'fisheries', but also highlights the fact that the impacts of resource harvesting can continue to occur well after such activities have ceased.

Stages of fisheries collapse

Many management strategies aim for a steady yield from a fishery in perpetuity (Ludwig *et al.*, 1993), which is termed 'maximum sustained yield' (see Chapter 17). For example, the US *Fishery Management and Conservation Act (1977)* specified that fishery management should be based on optimum yield, which is defined as maximum sustainable yield adjusted for ecological and socioeconomic factors (Dennis *et al.*, 1985). Larkin (1977; see also Ludwig, 1993) pointed out that fishery managers have rarely been able to control the amount, distribution and technique of fishing effort, even though such controls are necessary to achieve sustainable yields. Rosenberg *et al.* (1993) and Rosenberg (2003) reported numerous examples in which fishery managers consistently allowed higher catch levels than indicated by the consensus of scientific advice.

Uncertainty in estimating the available resources often leads to over-exploitation, because in the absence of reliable information, there is a tendency for resource managers to become overly optimistic. Over-exploitation of populations can lead to the collapse of the resource. Some fisheries can recover after an initial collapse, but they will re-establish at levels considerably lower than virgin stocks, and/or at levels that are somewhat below maximum productivity of the target species (Kearney, 1995; Rosenberg, 2003).

The collapse of fisheries throughout the world often follows a familiar pathway, which includes a number of well-recognised stages (after Talbot, 1993):

- *Stage 1.* A new fishery, or a new method to harvest an existing stock, is discovered.
- *Stage 2.* The new resource is rapidly developed with little or no regulation.
- *Stage 3.* Major fishing effort results in over-capitalisation of the equipment used to harvest the resource.
- *Stage 4.* The fishing capacity rapidly outstrips the potential of the fishery to sustain levels of harvesting.
- *Stage 5.* The fishery is depleted and the level of harvest begins to decline.
- *Stage 6.* Fishing effort is intensified to offset the decline in the harvest.
- *Stage 7.* The intensive fishing effort continues so as to service investments made on over-capitalised equipment.
- *Stage 8.* The fishery is depleted to levels below which it is uneconomic to harvest.

In some cases, attempts to manage the fishery occur at Stages 6 and 7 – approaches such as putting in place quotas and economic subsidies, or reducing the fishing capacity of the fleets are employed. However, stock conservation and management efforts at Stages 6 and 7 are often belated or ineffective, particularly given the uncertainty about the resource, the lack of information on the biology and ecology of the target fishery, and the fact that an industry with vested interests will lobby hard to protect capital investments (Talbot, 1993; Losos *et al.*, 1995; Rosenberg, 2003). We further explore some of the motivations for resource over-use in Chapter 17.

Following the collapse of a given fish stock (at Stage 8 above), fishing effort will often be switched to another species and the series of steps in the depletion of the new stock recommences. The Spiny Dogfish was the much-loathed by-catch (see following section) of the now collapsed Atlantic Cod fishery in north-eastern USA and eastern Canada. Fishers sometimes called them the 'cockroach of the sea'. With the collapse of Atlantic Cod stocks (see Myers *et al.*, 1997), the fish was renamed 'Cape Shark', and the Spiny Dogfish fishery turned into a US$11 million per year industry in the mid 1990s. Regulations reducing the catch were implemented in 2000 (5 years after calls by fisheries scientists to do so) and the size of the industry has been reduced significantly. However, the fishery is now heavily exploited and Federal Government regulators are attempting to further limit the size of the catch, while State and Province governments in the USA and Canada have ignored Federal Government scientists and lobbied hard to increase the quota. In essence, the problems resulting from the decline of Atlantic Cod stocks have been repeated for the Spiny Dogfish. The types of feedback between the status of a fish stock, warnings by scientists and levels of political resistance to change are illustrated in Figure 8.9 and discussed in detail by Rosenberg (2003).

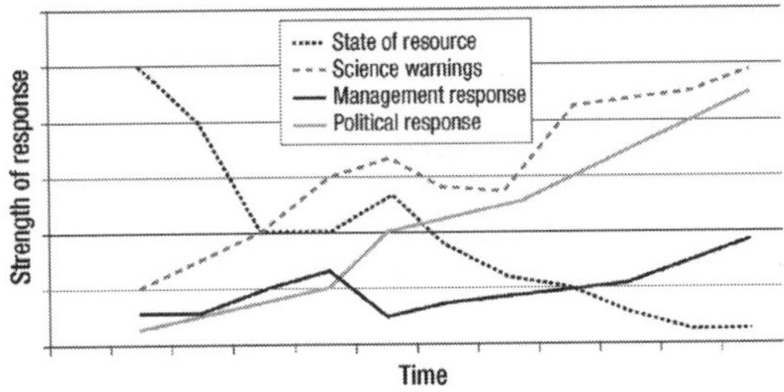

Figure 8.9. Management, scientific and political responses to over-fishing. (Redrawn from Rosenberg, 2003.)

By-catch impacts

By-catch are the fish and other animals caught as part of fishing operations targeted at other species (Larkin, 1977). The fish in the by-catch often have limited commercial value and are returned dead or dying to the sea (Barratt *et al.*, 2001). For example, sea-snakes are commonly caught in prawn trawls. Those animals that survive being caught almost certainly die soon after release (Barratt *et al.*, 2001). Almost 40 000 tonnes of up to 500 non-target species are caught annually as part of harvesting prawns in northern Australia. The by-catch is 4–6 times the tonnage of the prawns taken (Pender *et al.*, 1992) and 95% of the total weight landed (State of the Environment, 2001a).

By-catch can threaten some species (see Box 8.6) unless specific methods are used to protect them. For example, marine turtle exclusion devices are now mandatory in some Australian fisheries, including the northern prawn fishery (State of the Environment, 2001b).

Seabirds are caught by hundreds of millions of hooks set in longline fisheries in the Southern Ocean

Box 8.6

Endangered by-catch: the Green Sawfish

Australia supports more half of the world's known species of sawfishes, which are spectacular marine animals with a flattened, blade-like snout with rows of lateral 'teeth' (Last and Stevens, 1994). The Green Sawfish is listed as endangered by the World Conservation Union and also in New South Wales under the *Fisheries Management Act* (Pogonoski *et al.*, 2002). It is also listed as a threatened species under the Commonwealth's *Environment Protection and Biodiversity Conservation Act (1999)*. The species used to occur as far south as Sydney, where it was occasionally caught in nets that were used to exclude sharks from swimming beaches. Pogonoski (2002) noted that the last confirmed record in New South Wales was in 1970. These animals are observed mostly when entangled in fishing nets, and their capture as by-catch in other fisheries is considered to be the greatest threat to their long-term persistence (Pogonoski, 2002). An important future conservation strategy will be to reduce the amount of by-catch of Green Sawfish in commercial trawling activities in northern Australian waters (Pogonoski *et al.*, 2002).

(Tuck *et al.*, 2003). In 1995 nearly 45 000 albatross may have been killed by fishing hooks set for Southern Bluefin Tuna in the Southern Ocean. In the Pacific Ocean, longlining for Southern Bluefin Tuna includes an estimated bycatch of 60 000 turtles, 20 000 dolphins, and 3.3 million sharks. It is difficult to estimate the effects of by-catch mortality on even a well-studied species (Larkin, 1977; see Myers *et al.*, 1997) and managers may need to be innovative. For example, Robertson (2000) measured the diving depths of various albatrosses in relation to the sinking rate of longlines set in the Patagonian Toothfish fishery and estimated the chances that birds might be caught and subsequently die (Figure 8.10). This work showed that it is possible to set longlines in ways that minimise the capture rates of seabirds.

Cascading impacts of over-fishing

Despite the pressure exerted by fishing industries on fish stocks, there are no documented cases of the extinction of any marine fish as a direct consequence of over-harvesting (Kearney, 1995), although there are many documented cases where ecosystems have changed significantly because of over-fishing (Jackson *et al.*, 2001). Changes in the abundance of harvested species can change community dynamics and composition (Bascompte *et al.*, 2005). For example, the removal of large predators (such as Southern Blue-Fin Tuna) in marine food webs can lead to realignments of, or compensatory changes in, populations of other predators (see also Box 8.5). In other cases, localised stocks have been eliminated or depleted (Kearney, 1995), leading to loss of genetic diversity and with it the loss of localised adaptations to spatial variations in environmental conditions (and possibly variations in their ability to support harvesting; Larkin, 1977).

The economic importance of fisheries has led to a recognition of the importance of habitats and nursery grounds for fish, including estuaries, seagrass beds and mangrove forests – places that can be important for aquaculture. Terrestrial activities can also affect fisheries; for example, land clearing, Cattle grazing, urban development and pollution have all had impacts on the Sydney Rock Oyster (White, 2001; see Chapter 12).

Australian fishing industry

The early development of the fishing industry in Australia was underpinned by a belief that the seas that surround the country supported inexhaustible fish stocks. In reality, Australia's seas are highly nutrient-depleted. Limited geological activity (such as volcanism and glaciation) on the continent, low levels of nutrients in our deeply weathered 'fossil' soils, and limited stream run-off (see Chapter 5) mean that relatively few nutrients flow through our river systems to the ocean. This, coupled with a paucity of deep, nutrient-rich ocean upwellings,

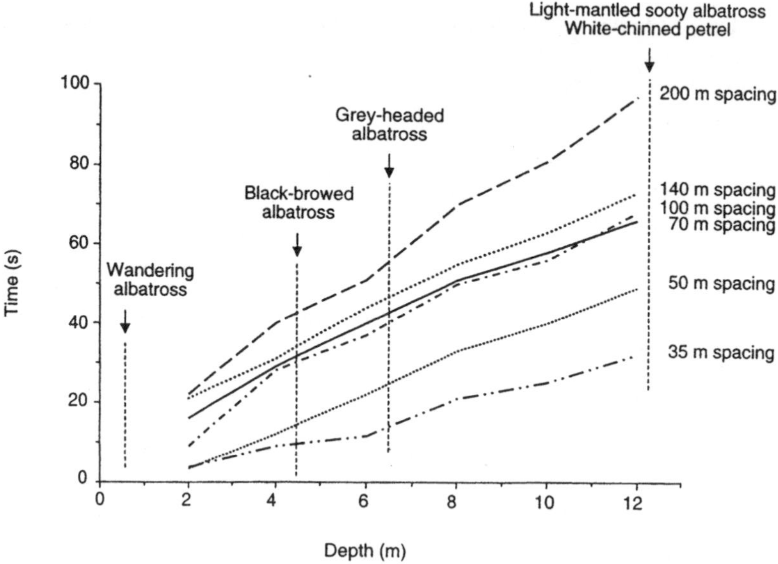

Figure 8.10. Relationships between weight-spacing on longlines and the time taken to reach various depths. The vertical lines show the maximum recorded diving depths of various species of albatross – seabirds that are highly susceptible to being caught on longlines. (Redrawn from Robertson, 2000.)

New South Wales	Northern Territory	Queensland	South Australia
Sydney rock oyster Prawns Abalone Crabs Mullet Eastern rock lobster Bream Snapper	Mackerel Barramundi Snapper Mud crab Pearls Shark	Prawns Scallops Mud, sand, spanner crabs Reef fish Mackerel Whiting Barramundi Mullet Shark	Tuna ranching Southern rock lobster Western king prawn Abalone King George whiting Pacific oyster Crabs Snapper
Tasmania	**Victoria**	**Western Australia**	
Abalone Pacific oyster Scallops Wrasse Banded morwong Atlantic salmon culture Southern rock lobster	Abalone Southern rock lobster Trout culture Bream Mussels King George whiting Pilchard	Western rock lobster Pearls Prawns Abalone Scallops Snapper Shark Mackerel Crabs Pilchard	

Figure 8.11. Major Australian fish stocks. (Redrawn from AFFA, 2003b.)

means that the Australian seas are relatively infertile (Kailola *et al.*, 1993; State of the Environment, 2001b).

Australia maintains an economic fishing zone of about 9 million square kilometres (RAC, 1993), which is the third largest in the world (after France, because of its external colonies, and the USA), and 16% larger than the size of the continent's landmass. However, the annual fish catch (in terms of tonnes landed) ranks 50th globally (Australian Bureau of Statistics, 1997). Of the estimated 4000–4500 species of fish in Australia, 200 are caught and sold commercially. In addition, more than 60 species of crustaceans, 30 species of molluscs, and some echinoderms are harvested (Australian Bureau of Statistics, 1997). The gross value of commercial fisheries in Australia approached A$2.5 billion in 2001–02, of which A$2 billion was export income (AFFA, 2003a). The nation's major fish stocks are shown in Figure 8.11. Although the export value of fishery products has increased by close to 50% in the past 5 years (AFFA, 2003b), this sector comprises only a small part of Australia's gross domestic product (Australian Bureau of Statistics, 1997).

Status of Australian fishery stocks

Australia is no exception to the general pattern of over-exploitation and collapse of many fisheries (RAC, 1993; State of the Environment, 2001b; Edgar and Samson, 2004; Table 8.3). There are now no commercial fisheries of Flathead in New South Wales, although up to 60 boats worked this resource before World War II. The Orange Roughy is an extremely long-lived and slow-breeding fish – individuals can live for more than 100 years, and can take 20–30 years to reach sexual maturity. Some workers have even termed the harvesting of the species as 'old-growth fishing'. The fish lives off the continental shelf of Australia and New Zealand, particularly near the South Tasman Rise (a large submerged plateau south of Tasmania). Nearly 40 other species of fish are associated with important marine topographic features such as sea-mounts and submarine plateaux (Koslow and Gowlett-Holmes, 1998). The Orange Roughy has low rates of fecundity and forms spawning stocks in winter, making it extremely vulnerable to over-fishing (Kearney, 1990). The catch of Orange Roughy (and co-occurring fish such as Smooth Oreo) has declined dramatically from peaks in the late 1980s of about 40 000 tonnes per annum. New stocks of Orange Roughy were discovered in 1997, but the stocks were quickly depleted, despite scientific information about the vulnerability of this fishery! Australian Agriculture, Fisheries and Forestry (AFFA, 2003b) reported that in the 2000–01 fishing year, only 830 tonnes of Orange Roughy and 290 tonnes of Oreo were harvested, despite considerable searching and fishing effort on the South Tasman Rise. Hence, it appears that fishers from Australia, New Zealand and other countries have depleted the stock (AFFA, 2003b).

Table 8.3 shows the change in status of commercial fisheries in Australia from 1992 to 2000–01. These

Table 8.3. Number of Australian fisheries stocks with a given harvest status between 1992 and 2000–01. (Redrawn from AFFA, 2003b.)

Status	Year of fishery status report							
	1992	1993	1994	1996	1997	1998	1999	2000–01
Under-fished	1	11	6	8	2	1	1	0
Fully fished	15	14	18	16	15	14	13	11
Over-fished	5	5	3	3	4	6	7	11
Uncertain	9	15	17	26	35	35	36	35
Status not classified	37	22	23	14	11	11	10	10

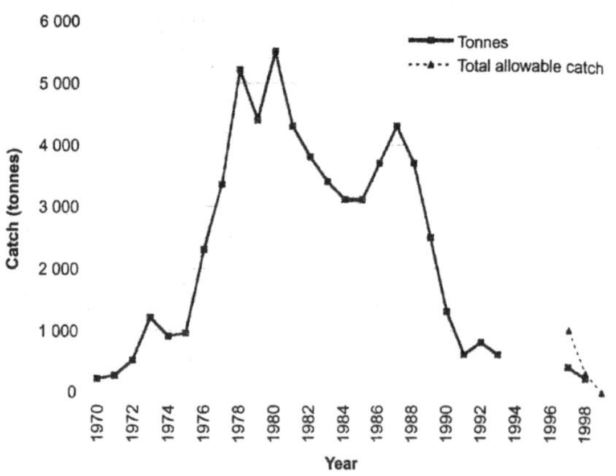

Figure 8.12. Collapse of the Gemfish in southern New South Wales. (Redrawn from State of the Environment, 2001b.)

Lobster boats at Lady Barron Port, Flinders Island, Tasmania. (Photos by David Lindenmayer.)

categories are defined by Australian Agriculture, Fisheries and Forestry (AFFA, 2003b) as follows, and they are sobering in the context of the nation's fish resources. **Under-fished** fisheries are those that can sustain higher catches than currently taken. There were no fisheries in this category in 2000–01. **Fully fished** refers to stocks in which fishing pressure is close to optimum. **Over-fished** fisheries are those in which there are too few fish left or there is too much fishing (AFFA, 2003b; see Figure 8.12). The largest category in Table 8.4 is 'uncertain', highlighting the large number of fisheries for which there is insufficient information to assess their status. In 2001, less than half of the nation's 144 managed marine and estuarine fisheries had management plans (State of the Environment, 2001a).

Example of a sustainable fishery

The Western Rock Lobster Industry in Western Australia (see Box 8.7; Figures 8.13, 8.14) is an exception to the general trend of discovery, expansion, over-exploitation and collapse.

Future issues and approaches to sustainability

Unfortunately, the Western Rock Lobster industry is one of the few exceptions in a natural resource sector plagued with stock collapses (see Table 8.3). These problems are mirrored almost everywhere else around the world (Rosenberg, 2003; but see Hilborn *et al.*, 2003 for a rare exception of the Sockeye Salmon fishery in Alaska). Estimation of sustainable yields is hampered by poor data and lack of understanding about the basic ecology of the resource species and other species in its ecosystem.

Given the fluctuations in fish populations and the lack of ecological understanding, sustainable yield calculations may never be particularly robust (see Kearney, 1990; Ludwig *et al.*, 1993). Moreover, the history of fisheries and their over-exploitation highlights the fact that appropriate conservation and management has just as much to do with the management of human values and perceptions as it does with the management of the fish stocks (Ludwig *et al.*, 1993; Rosenberg, 2003).

In most instances, to be sustainable, fishing industries must fish less (Pauly and MacLean, 2003;

Box 8.7

The Western Australian Rock Lobster industry

In 2002, more than 55% of the value of exports of Australia's fishery products was from high-value crustaceans and molluscs (AFFA, 2003a). Various types of rock lobsters accounted for 20% (>A$500 million) of the value of the nation's fish production sector in 2001–02 (AFFA, 2003a). The Western Rock Lobster fishery is managed by the State of Western Australia and is valued at A$300 million annually, making it the most lucrative single-species fishery in Australia (AFFA, 2003b). Most of the catch is exported, and the USA takes more than 60% of the catch (AFFA, 2003b).

The Western Rock Lobster occurs in a single stock in the seas off the Western Australian coast where it lives on reefs, rock shelves and ledges from just north of Shark Bay to south-east of Cape Lewin (Kailola *et al.*, 1993). Adult Western Rock Lobsters are captured in pots baited with fish or other organic matter (Kailola *et al.*, 1993). Pots are usually retrieved 1–5 days after being set.

The Western Rock Lobster industry is one of the few fisheries in Australia where there is a substantial body of scientific knowledge on the life history, mortality, egg production, recruitment, and habitat requirements of the species, including information on the breeding success of the species and how it is correlated with changes in sea levels and the movements of ocean currents (e.g. Caputi *et al.*, 1995a,b; Wallace *et al.*, 1997; Hall and Chubb, 2001). Mathematical models predict likely catch sizes up to 4 years ahead as well as the relationships between fishing intensity and stock levels (e.g. Brown *et al.*, 1995). On this basis, the annual catch over the past 20 years has remained stable at between approximately 10 000 and 12 000 tonnes (Kailola *et al.*, 1993; Rock Lobster Industry Advisory Committee, 2002; Figure 8.14).

The Western Rock Lobster industry is regulated by a body with statutory authority (the Rock Lobster Industry Advisory Committee), which reports to the Western Australian Government (Rock Lobster Industry Advisory Committee, 2002). The industry underwent a substantial expansion in the 1950s and 1960s, during which the addition of new boats and extra lobster pots increased the pressure on the Western Rock Lobster resource. In response, the Western Australian Government prevented new entrants to the industry and limited the number of pots used by existing fishers in 1963. In addition, harvesting is limited to particular areas and seasons, and undersized animals and females with eggs cannot be taken. Contravention of fishing regulations results in penalties such as the loss of equipment. There is also a strong management response to new data that suggests over-harvesting is taking place. For example, in the early 1990s it appeared that the adult population was declining, so restrictions were introduced that enabled it to recover (State of the Environment, 2001).

The case of the Western Rock Lobster highlights the importance of good scientific information and sound management practices for a natural resource. Research is continuing to further assist the management of the industry, and new simulation models are being developed to better forecast future catches and to explore the consequences of changed management regimes (Hall, 1994; Hall and Chubb, 2001). However, the success of harvesting operations depends on management of the human side of the industry, through regulation of the number of fishers and the kinds of pots used. This approach to the management of the Western Rock Lobster means that it is among the best-managed fisheries in the world (RAC, 1993; AFFA, 2003b). In March 2000, the Western Rock Lobster industry became the first fishery in the world to be certified by the Marine Stewardship Council (Rock Lobster Industry Advisory Committee, 2002). Part of the certification involved completion of an ecological risk analysis. Four hazards were considered to pose a moderate threat: mortality of sea lion pups in lobster pots, direct damage to coral by pots, entanglement of Leatherback Turtles in fishing ropes, and dumping of domestic waste on islands (IRC, 2002).

Rosenberg, 2003). However, other strategies can assist in the transition of fisheries towards sustainability. One of these is setting aside marine reserves (Neubert, 2003; Russ and Alcala, 2004; Williamson *et al.*, 2004) with the potential to protect populations from depletion and assist recovery elsewhere (Ward *et al.*, 2001; Neubert,

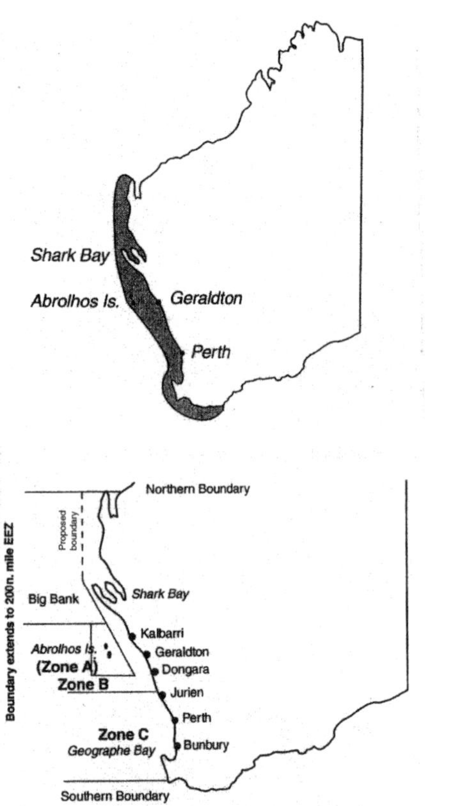

Figure 8.13. Distribution of the Western Rock Lobster and Western Rock Lobster fishing zones. (Redrawn from Rock Lobster Industry Advisory Committee, 2002.)

2003; Grafton *et al.*, 2005). Unfortunately, there has been strong community opposition to the setting aside of marine reserves in several parts of Australia (e.g. in Victoria, New South Wales and Western Australia), particularly by recreational fishers (see McPhee *et al.*, 2002). This has occurred even when the evidence for over-fishing-related stock collapse has been unequivocal.

Fishing industry certification may be valuable. To this end, the Marine Stewardship Council was established in 1997 through a partnership between the transnational corporation Unilever and the World Wide Fund for Nature. The aim was to develop an environmental standard for fisheries management (Marine Stewardship Council, 2003). Certification results in being able to apply product labels that indicate to environmentally- conscious consumers that the fishery is well-managed. The certification process requires strong evidence of environmentally responsible fisheries management, an explicit ecological risk assessment, monitoring, and demonstrated continued improvement of management practices. The system provides opportunities for new, high-value markets. As outlined in Box 8.7, the Western Rock Lobster fishery was the first in the world to be certified by the Marine Stewardship Council (Rock Lobster Industry Advisory Committee, 2002).

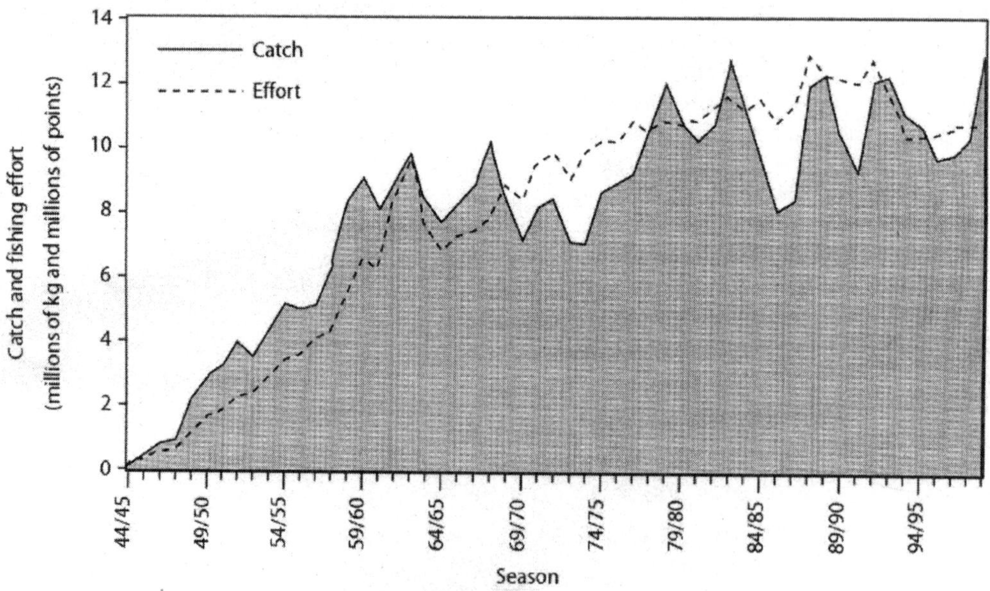

Figure 8.14. Catch and index of effort in the Western Rock Lobster fishery since it began in the 1940s (from IRC, 2002).

8.5 Conclusions

There are many similarities among the long-term trajectories of industries that harvest natural populations. Forestry and fisheries, in particular, have many problems in common, such as poor resource inventories, a lack of monitoring and impact assessment, high levels of human conflict over resource availability, and the unsuccessful application of economic instruments and regulatory controls such as quotas and subsidies. Indeed, as noted by Overton Price (1902, in Gregory, 1997): 'it is the history of all great industries...that the necessity for modification is not seen until the harm has been done and results have been felt'.

Forestry and fishing industries around the world have, with only rare exceptions, been characterised by ecologically unsustainable levels of over-harvesting that not only rapidly depleted the target resource but have had major flow-on impacts on other biotic elements and ecosystem processes in their respective ecosystems. In Australia, forest industries have converted old growth forest to younger stands and higher-volume, lower-value industries such as export woodchipping that have become prominent at the expense of sawlog production. In fisheries, former by-catch stocks have become important as species that were initially targeted have become depleted.

In both the fishing and forestry industries, the problems of ecological unsustainability are difficult to resolve because there are many intertwined human social and economic issues. Better regulation of harvests, sustainability and certification criteria, and better-informed natural resource management outcomes based on better scientific data (that is implemented on the ground and embraced by all stakeholders) could assist transitions to sustainability in both sectors.

It is possible that with continued sound management, the Kangaroo-harvesting industry will not suffer the same problems that are apparent in the fishing and forestry sectors of natural resource management. Harvesting levels should be ecologically and economically sustainable in the long-term if they are set: (1) with high-quality data based on extensive and repeated field modelling, (2) with well-informed population modelling, and (3) by taking into account factors such as considerable year-to-year variations in environmental conditions (e.g. drought) and other forms of uncertainty.

8.6 Practical considerations

In developing approaches for sustainable harvesting, stakeholders should be involved in all aspects of planning and management. Models and reliable monitoring data are most effective when they are used to cross-examine the ideas of stakeholders and to find strategies that are robust to uncertainty. Experience in natural resource management suggests that sustainable practices depend on better regulation (including self-regulation), industry-wide environmental ethics and codes of practice, international accreditation systems and, ultimately, economic incentives for sustainable innovations.

8.7 Further reading

Dargarvel (1995), Ferguson (1996) and Lindenmayer and Franklin (2002, 2003) discuss forestry and forest biodiversity conservation. Ray *et al.* (1983), Wood *et al.* (2001), and Lindenmayer and Hobbs (2004) review plantation forestry. Salt *et al.* (2004) discuss the emerging field of farm forestry. The most detailed discussion of the firewood industry is by Driscoll *et al.* (2000), although another useful examination is presented by Wall (2000). The most up-to-date statistics on Australia's forests and forestry industry is the *State of the Forests Report 2003* (Bureau of Rural Sciences, 2003a,b).

Grigg *et al.* (1995) contains a wide-ranging set of papers on various issues associated with the sustainable use of wildlife (including Kangaroo harvesting). The May 2004 issue of *Australian Mammalogy* explored a wide range of issues associated with Kangaroo biology and the Kangaroo harvesting industry.

Pogonoski *et al.* (2002) give a detailed account of Australian marine systems and the conservation status of marine biota. Edgar (2001) provides an excellent overview of marine habitats in temperate waters.

Chapter 9

Vegetation loss and degradation

This chapter reviews land clearance and habitat modification, both globally and in Australia. Land clearance and habitat modification are major problems in conservation biology because of both their direct impacts on biodiversity loss and flow-on effects leading to habitat deterioration and degradation, which also have negative implications for biota. In this chapter we also examine the potential impacts of landscape change that can result from different land-use practices, including mining, urbanisation and pastoralism.

9.0 Introduction

Land clearing, primarily for agriculture, is perhaps the single most important cause of environmental degradation, loss of species, and depletion of ecological communities, both in Australia and worldwide (Clark *et al.*, 1990; Schur, 1990; Sivertsen, 1994; Possingham *et al.*, 1995; Population Action International, 2000; State of the Environment, 2001a). About 80% of the world's original cover of forest has been cleared, fragmented, degraded or converted to plantations (World Resources Institute, 1997), of which about 50% has been totally cleared. Almost 80% of mammals and about 60% of birds listed by the International Union for the Conservation of Nature (see Chapter 3 for details of classifications of threat) have declined primarily as a result of habitat loss (Groombridge, 1992; IUCN, 2003).

Forest loss and degradation are particular problems for conservation biology because forests are some of the most species-rich environments on the planet (Lindenmayer and Franklin, 2002), particularly for birds (Gill, 1995) and invertebrates (Erwin, 1982; Majer *et al.*, 1994). The United Nations Food and Agriculture Organization reviewed global forest clearance, and found that between 1990 and 2000, the average rate of forest loss was 16.1 million hectares per year (FAO, 2000). Of this, 14.6 million hectares was lost through deforestation and 1.5 million hectares was lost through conversion to plantations each year. Although there were some net gains in forest area through plantation establishment and the natural expansion of forest, these were small in comparison with the overall rate of loss, particularly in tropical countries (see Table 9.1). These estimates do not include the loss of the millions of hectares of forest that are burned annually (see Chapter 11).

Rates of vegetation clearance are likely to be associated with the size of the human population, developments in technology, and per capita consumption of resources. Shifting cultivation causes forest loss in all tropical regions, and accounts for 70% of deforestation in Africa, 50% in Asia and 35% in the Americas. Social and political factors underlie the movement of people between different places in significant numbers (e.g. transmigration programs in Brazil and Indonesia). Such movements to relatively pristine environments usually result in forest modification and loss.

Factors that influence vegetation clearing in many nations, such as shifting cultivation, increases in the human population, and transmigration, do not apply

Table 9.1. Annual gross and net changes in forest area, 1990–2000 (from FAO, 2000).

Domain	Deforestation ($\times 10^6$ hectares)	Increase in forest area[1] ($\times 10^6$ hectares)	Net change in forest area ($\times 10^6$ hectares)
Tropics	− 14.2	+ 1.9	− 12.3
Non-tropics	− 0.4	+ 3.3	+ 2.9
World	− 14.6	+ 5.2	− 9.4

[1]Increase in forest area represents the sum of natural expansion of forest and afforestation.

Box 9.1

Vegetation loss in the Cerrado, Brazil
A large amount of the native vegetation that is cleared around the world each year is in Brazil, and much of it in Amazonia where, in 2003, almost 15 000 square kilometres of forest was cleared. However, the extent of land clearing may be even greater in other parts of Brazil. The second-largest ecoregion in Brazil is the Cerrado, which is a mosaic of grassland, palm stands and forests. This ecoregion supports more than 10 000 species of endemic plants and an array of threatened vertebrates. The amount of land clearing occurring in the Cerrado appears to be as high or perhaps higher than in the Amazon region, and is taking place to establish crops ranging from Soyabeans to Rice, Wheat and Maize.

in Australia. There are different underlying causes, as outlined in the following sections.

The Cerrado in Brazil. (Photo by Darius Tubelius.)

9.1 Vegetation clearing and habitat loss in Australia

Australia's contribution to global levels of land clearing and vegetation loss

Over the decade 1990–2000, Australia had the sixth highest annual rate of land clearing in the world (Figure 9.1). Notably, Australia is the only country in the top 20 land-clearing nations with a developed first world economy! More than 550 000 hectares of native vegetation are cleared in Australia each year (Australian Conservation Foundation, 2001). Even this staggeringly high figure may be an underestimate. In 1999, permits to clear more than 730 000 hectares of vegetation were issued by the governments of New South Wales and Queensland (State of the Environment, 2001a). The figure of 550 000 hectares per year equates to clearing an area equivalent to 10 average suburban house blocks per minute, every minute of the year.

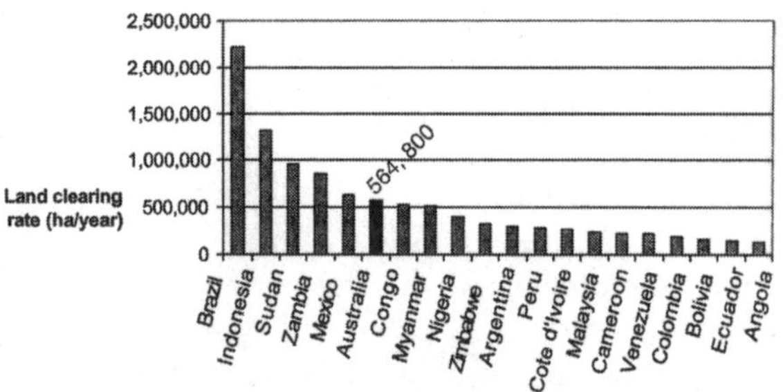

Figure 9.1. Annual rates of vegetation clearing for the top 20 land clearing nations worldwide. (Redrawn from the Australian Conservation Foundation, 2001.)

Land clearing in southern New South Wales. (Photo by David Lindenmayer.)

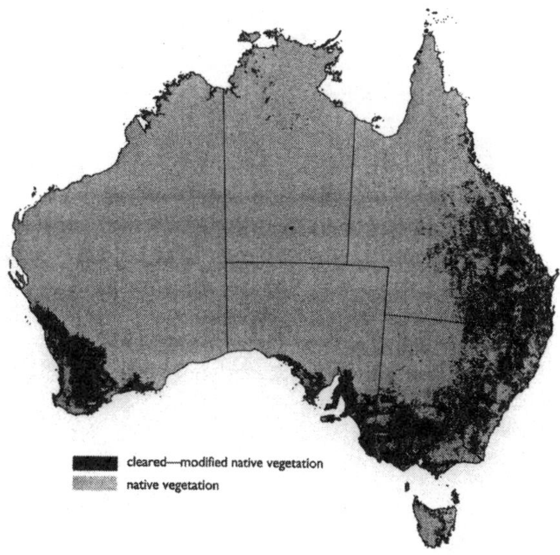

Figure 9.2. Broad distribution pattern of cleared land in Australia (from Commonwealth of Australia, 2001c).

Extensive areas of ringbarked woodland near Coolac in southern New South Wales. (Photo by David Lindenmayer.)

Past land-clearing patterns in Australia

The first Europeans to arrive in Australia cleared land in order to establish farms. Land clearance continued throughout the 19th and 20th centuries as the population expanded from the areas that were first settled. Patterns of growth of the human population after Europeans arrived in 1788 reflect the places that were settled earliest and the growth of towns and cities (Figure 9.2). The vegetation and fauna of the places unlucky enough to be closest to early European settlements suffered the greatest impacts. This general trend

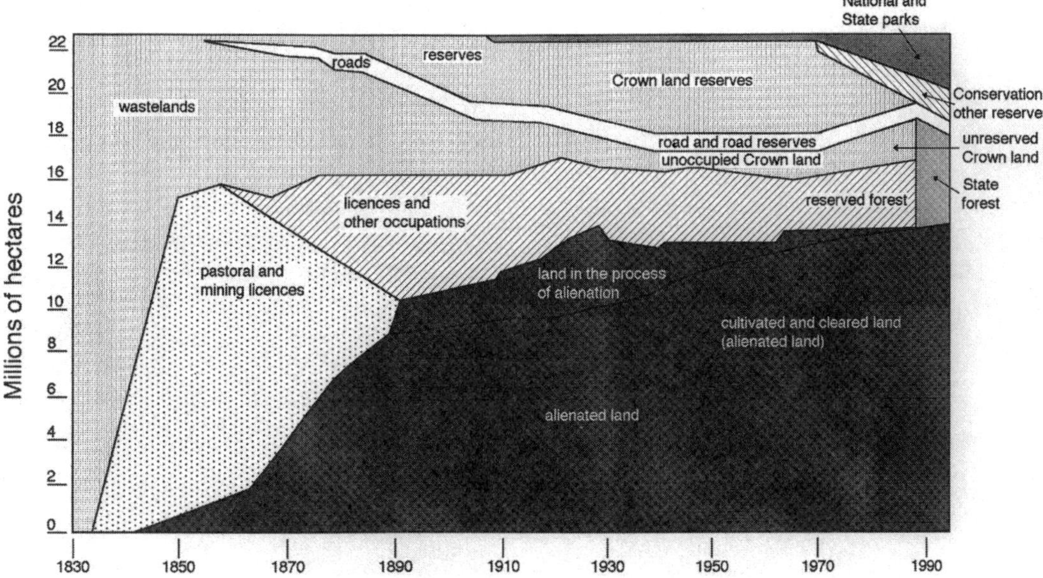

Figure 9.3. Vegetation changes in Victoria (after State of the Environment, 1996).

is still evident in the broad distribution patterns of cleared vegetation in Australia (Commonwealth of Australia, 2001c).

In contrast to current knowledge (see Chapter 5), early land-clearing efforts were considered to 'improve' the land, a fact that is illustrated in the following quote: 'ring-barking the trees, a process guaranteed to sweeten the soil, improve the quality and quantity of native grasses, and build up the soaks and springs' (Durack, 1938, in McIntyre *et al.*, 2002).

Areas of arable land were cleared first and were most extensively targeted. Farming systems tended to reflect the practices in Europe where rainfall was predictable, and salinisation and wildfires were rare. Over time, vegetation clearing and habitat loss continued, as seen in the patterns in Victoria, Australia's most heavily cleared state (Figure 9.3).

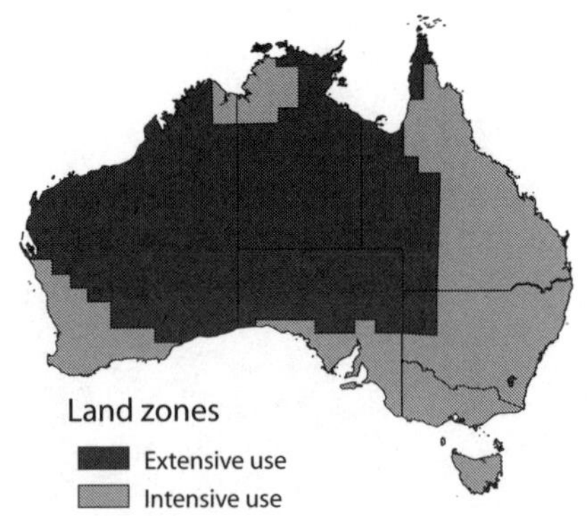

Figure 9.4. Location of the extensive and intensive land use zones (*sensu* Graetz *et al.*, 1995). (Redrawn from Commonwealth of Australia, 2001c.)

Clearing by land-use zone

Graetz *et al.* (1995) used satellite images taken in the early 1990s to examine the amount of disturbance in Australia since 1770. As part of their work, they used a continent-wide data set representing natural vegetation cover at the time of European settlement (AUSLIG, 1990). Graetz *et al.* (1995) reduced 230 original categories to 34 by grouping areas with similar overstorey structure and floristics. Satellite information on present vegetation cover was then compared with pre-European vegetation. To complete this task, they partitioned the continent into an **intensive land-use zone** and an **extensive land-use zone**; Figure 9.4). These zones equate broadly to the classes used by Hobbs and Hopkins (1990), namely, 'complete vegetation removal and replacement', and 'vegetation utilisation', respectively. The intensive land-use zone encompasses eastern Australia, an area west of Darwin in the Northern Territory, and south-west Western Australia; it covers almost 3 million square kilometres or about 39% of the continent, and includes areas of national parks and other protected areas, and remnant vegetation on private land. Areas within the zone were assigned to one of several categories: 'uncleared', 'thinned', 'cleared', 'indeterminate' and 'other' (Table 9.2).

Graetz *et al.* (1995) calculated that more than 1 million square kilometres or almost 52% of the intensive land-use zone (about 20% of the Australian continent) had been cleared or thinned (Table 9.2). On average, about 47% of each vegetation type had been cleared, with the values ranging from 0 to 99%. The most extensive forms of disturbance were clearing for cropping, clearing for forestry plantations, and the establishment of pastures. Other activities, such as urban development and mining, had impacts on smaller scales. Graetz *et al.* (1995) suggested that they probably underestimated the extent of thinning and clearing in the intensive land-use zone, because they were unable to distinguish areas where grazing had occurred in the understorey, or where forests had been logged and then regenerated; these areas were counted as undisturbed natural vegetation.

Table 9.2. Extent of disturbance in the area of Australia classified as the intensive land-use zone by Graetz *et al.* (1995).

Cover class	Area (square kilometres)	Percentage that is intensive land-use zone	Percentage of continent
Uncleared	1 059 741	35.5	13.8
Thinned	518 223	17.4	6.7
Cleared	1 029 640	34.5	13.4
Indeterminate	365 350	12.2	4.7
Other	10 953	0.4	0.1
Total	2 983 908	100	38.8

A more recent study by Barson *et al.* (2000) suggested that between 1990 and 1995, more than 1.2 million hectares of woody vegetation in the intensive land-use zone had been cleared – primarily for agriculture and grazing, but also for urbanisation. Both the study by Graetz *et al.* (1995) and the later one by Barson *et al.* (2000) indicated that existing remnants within agricultural and pastoral areas are now highly fragmented and continue to be cleared.

Clearing rates and land tenure

Rates of land clearance depend to a large extent on land tenure. Land in private hands is typically cleared more quickly than land owned by Federal or State governments. A substantial proportion of the remaining forests and woodlands in Australia are privately owned or are leased for agricultural production (Figure 9.5).

Land clearing has continued relatively unabated on private land in some States, but has slowed in others at different times, depending on the introduction of relevant laws. In Victoria, relatively stringent controls were introduced in 1987, but in the 15 years before that time, land in private hands was cleared at a rate of about 1% per year (Figure 9.6). In parts of western New South Wales, land clearing actually increased after restrictions were introduced (see Scully, 2003). In the south-west of Western Australia, 54% of the land cleared for agriculture was cleared between 1945 and 1982 (Hobbs and Hopkins, 1990).

The information in Figure 9.6 is thought to be conservative and does not include illegal clearing or the clearing of vegetation types such as heathland and grassland.

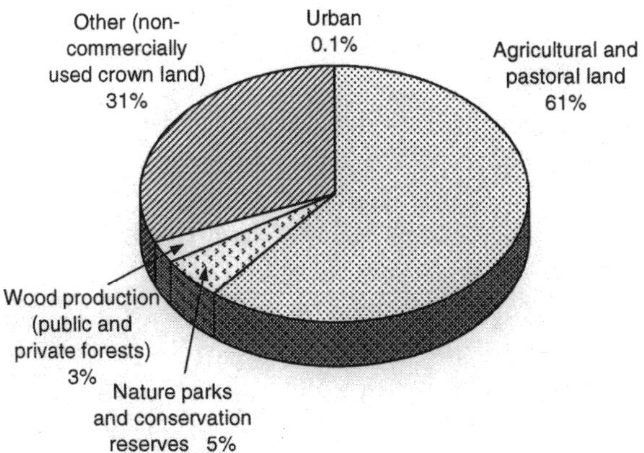

Figure 9.5. Land use in Australia (after RAC, 1992a,b).

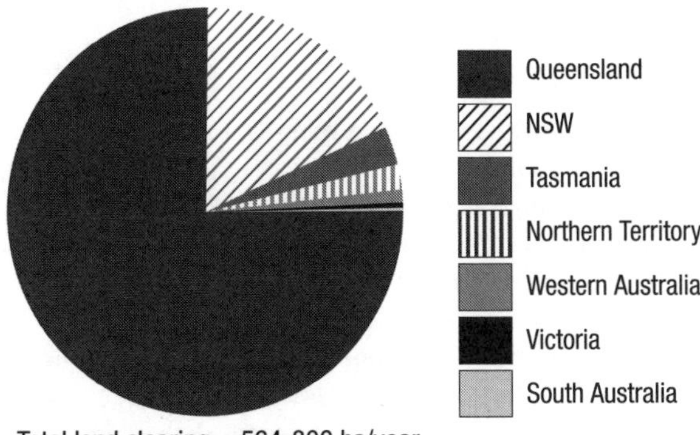

Figure 9.6. Estimated annual rates of land clearing in Australian States and Territories (based on data from various sources).

The special case of Queensland

Even though land clearing is listed as a threatening process under the Federal Government's *Environmental Protection and Biodiversity Conservation Act (1999)* (EPBC Act), the Commonwealth has no jurisdiction over State actions. Nowhere is this more starkly illustrated than in Queensland, where the extent of recent land clearing is more than 425 000 hectares per year (Figure 9.6). Between September 2001 and August 2003, approximately 1 051 000 hectares of woody vegetation was cleared (Government of Queensland, 2005). If Queensland were a country, it would rank 9th worst in the world in terms of land clearing. The rapid rate of vegetation loss in Queensland is not new; in fact, it has been ongoing for several decades. For instance, permits were issued in 1994 for the clearance of a total of 1.1 million hectares of land (Alexandra, 1995). Importantly, laws were passed in Queensland in 2004 to phase out broad-scale clearing by the end of 2006.

Smith *et al.* (1994) examined clearing in a 7.5-million-hectare area between Rockhampton, Emerald, Taroom, Roma, and Charleville. In the 100-year period from the commencement of the rapid increase in European settlement (between 1850 and 1870) until the early 1970s, approximately 3.4 million hectares of land (or almost half the region) was cleared (Smith *et al.*, 1994). The rate of clearing during this time was 22 000 hectares per year. However, in the following two decades, more than 1 million hectares were cleared for cropping and pastoral production. This corresponds to a rate of clearing of 43 000 hectares per annum, almost twice that of the preceding 100 years (Smith *et al.*, 1994).

Box 9.2

Calculating the costs and benefits of not clearing in Queensland

Queensland has one of the highest rates of land clearing in the world, and the environmental impacts of this are increasingly being discussed. Some organisations, for example farm lobby groups, present arguments about the economic *costs* of stopping clearing. However, Morton *et al.* (2002) presented an alternative view based on the economic *benefits* of bringing a halt to vegetation clearing in Queensland. They noted that there are about 2.3 million hectares of native vegetation in the 'of conservation concern' category. These are in regional ecosystems with less than 30% cover remaining or where the remaining vegetation remnants are small. The Queensland Government has estimated that the compensation costs of halting land clearing would be A$200 million or about A$88 per hectare. Morton *et al.* (2002) calculated that there would be massive benefits of not clearing – mainly in the form of carbon credits, reduced salinity and erosion control. Carbon credits at A$1400 per hectare sum to A$3178 million for the land in question. Morton *et al.* (2002) assumed that one-third of the area was susceptible to salinity, and that its prevention would be valued at A$830 million over a 10-year period. Hence, the collateral benefit of not clearing was more than A$4 billion – or 20 times more than the A$200 million needed for farmer compensation. Of course, these calculations are crude, but they illustrate the long-run economic benefits of altering clearing practices. These calculations also do not take into account the amount of biodiversity conserved, which Morton *et al.* (2002) estimated would be substantial (in excess of 5300 species).

The trends in extensive and rapid land clearing outlined here continue in Queensland today (Accad *et al.*, 2001; State of the Environment, 2001a). Some of the highest rates of clearing are in particular vegetation types, for example the Brigalow belt (Cogger *et al.*, 2003). The reasons for rapid rates of tree clearing in Queensland are complex, but some clearing is linked to the concerns of landholders, who are worried that legislation limiting clearing will restrict future development opportunities (McIntyre *et al.*, 2002), compelling them to clear now. Notably, a number of these areas may be susceptible to secondary salinity (Queensland Department of Natural Resources, 1997; see Chapter 5), causing a risk that extensive and costly rehabilitation work will be required in the future to address the cascading impacts of inappropriate actions taking place now.

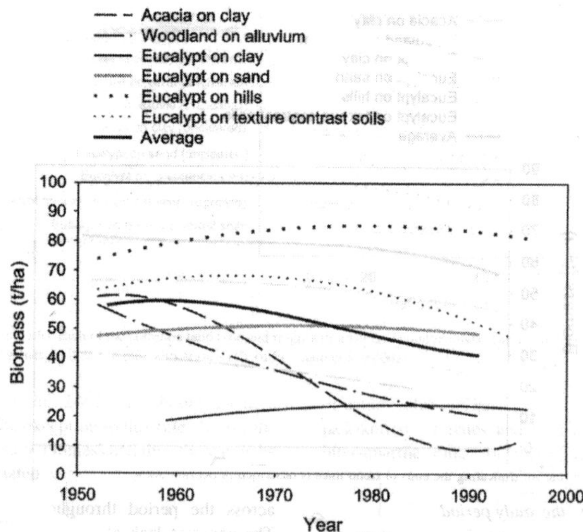

Figure 9.7. Vegetation thickening in Queensland. (Redrawn from Fensham *et al.*, 2003.)

Another issue in Queensland is the case of vegetation 'thickening' – the term given to regenerating and growing trees and understorey shrubs that increase the vegetation cover and reduce the potential domestic stocking rates. Research by Fensham *et al.* (2003) indicates that vegetation thickening is indeed taking place in Queensland (Figure 9.7), and landholders are clearing land in an attempt to control it.

Vegetation types that have been cleared

Morgan (2001) documented the uneven pattern of vegetation clearing in the intensive land-use zone. Of the 166 subregions of the Interim Biogeographic Regionalisation for Australia (IBRA) classification (see Chapter 4) in the intensive land-use zone, 97 had less than 50% of the original vegetation remaining, and 57 had less than 30% of the original vegetation remaining (Morgan, 2001).

Most vegetation losses have occurred in higher rainfall zones and on relatively fertile and arable soils, mostly because of cropping and pasture improvement. The changes in various ecosystem types since 1788 can

Table 9.3. Pre- and post-European vegetation cover by vegetation type in Australia (in thousand hectares) (from Commonwealth of Australia, 2001c).

Major vegetation group	Present	Pre-European	Percentage remaining
Rainforest and vine thickets	30 231	43 493	70
Eucalypt tall open forests	30 129	44 817	67
Eucalypt open forests	240 484	340 968	71
Eucalypt low open forests	12 922	15 066	86
Eucalypt woodlands	693 449	1 012 047	69
Acacia forests and woodlands	560 649	657 582	85
Callitris forest and woodlands	27 724	30 963	90
Casuarina forests and woodlands	60 848	73 356	83
Melaleuca forests and woodlands	90 513	93 501	97
Other forests and woodlands	119 384	125 328	95
Eucalypt open woodlands	384 310	513 943	75
Tropical eucalypt woodlands/grasslands	254 228	256 434	99
Acacia open woodlands	114 755	117 993	97
Mallee woodlands and shrublands	250 420	383 399	65
Low closed forests and closed shrublands	8749	15 864	55
Acacia shrublands	654 279	670 737	98
Other shrublands	98 947	115 824	85
Heath	25 861	47 158	55
Tussock grasslands	528 998	589 212	90
Hummock grasslands	1 756 104	1 756 962	100
Other grasslands, herblands, sedgelands and rushlands	98 523	100 504	98
Chenopod shrubs, samphire shrubs and forblands	552 394	563 389	98
Mangroves, tidal mudflats, samphires and bare areas, claypans, sand, rock, salt lakes, lagoons, lakes	106 999	112 063	96

be summarised as follows (after Hopper, 1997; Commonwealth of Australia, 2001c):

- open forests: 45% cleared
- woodland forests: 32% cleared
- **mallee** forests: 30% cleared
- rainforests: 25% cleared
- coastal wetlands in southern Australia: >60% destroyed
- temperate lowland grasslands: >99% cleared.

Table 9.3 gives a detailed breakdown of the extent of land clearing by vegetation type throughout Australia.

Uneven patterns of land clearing have resulted in different effects in different biological regions. The 'big scrub' of northern New South Wales once covered 750 000 hectares, and was the largest area of subtropical rainforest in Australia. Only about 0.07% remains, mostly as scattered remnants that are smaller than 15 hectares in size (Floyd, 1990). It was once possible to walk from Melbourne to Sydney through almost continuous woodland cover, but now much of it is gone and the remaining patches are small and highly disturbed. For example, about 0.01% of grassy White Box woodlands remain in eastern Australia, although this vegetation type once covered several million hectares (Prober and Thiele, 1995). In the Wheat-producing Avon Botanical District of Western Australia, only about 7% of the natural vegetation of this district remains, mostly as scattered remnants (Hobbs and Hopkins, 1990). At a broader scale, 65% of the vegetation cover of south-western Western Australia (a region that comprises 12.3% of that State) has been cleared, with only 0–2% of some vegetation types remaining (Hobbs and Mooney, 1998).

The structure, and floral and faunal composition of large areas of many vegetation types, including

Box woodland patches in grazing lands in eastern Australia. (Photo by David Lindenmayer.)

Box 9.3

The Brigalow and land clearance

Brigalow refers both to the tree species *Acacia harpophylla* and to the vegetation communities it dominates. Brigalow communities contain a wide array of other species of plants (Johnson, 1984), and in a mature state, stands of Brigalow contain 10–25-metre-tall trees and shrubs.

Brigalow has a number of distinctive features. It is reasonably salt tolerant and the lateral root system is extensive. If the root system or the trunk is damaged, then extensive suckering can take place, producing what is often termed 'whipstick', that is, stands of vertical stems. These aerial suckers provide valuable fodder for stock during periods of drought. However, such a regeneration strategy created problems for those land holders who employed methods such as bulldozing, spraying and ringbarking to clear stands of Brigalow (Nix, 1994). As a result, although regions supporting Brigalow were settled during the 1840s–1860s, it was not until the 1880s that Brigalow communities began to be extensively cleared, when trees and shrubs were cut with axes or brush-hooks. The vegetation was then allowed to dry before it was burnt in a high-intensity fire. The land was then cropped for a period before being sown to pasture.

Some of the most extensive clearing in Australia has taken place in the Brigalow belt (Smith *et al.*, 1994; Sattler and Webster, 1984; Alexandra, 1995; Accad *et al.*, 2001; Table 9.4). Brigalow originally extended from Collinsville in northern Queensland to Narrabri in north-western New South Wales (Nix, 1994; Figure 9.8), but much of its distribution encompassed areas of undulating country that had considerable value for land uses such as cropping and grazing. The pace and extent of clearing of Brigalow increased dramatically following the end of World War II and the advent of new technologies to clear vegetation, for example the use of large steel balls that were pulled between bulldozers. Various forms of government assistance also promoted the rate of clearing.

Clearing in many parts of the Brigalow continues, and in some places (e.g. central Queensland), the rate of clearing is now significantly faster than in earlier times (Smith *et al.*, 1994; Cogger *et al.*, 2003). There are some small remnants of the community left outside parks and reserves, including those along roadsides and in stock reserves. In 1994, of the 2.1 million hectares of Brigalow present at the time Europeans arrived, only 300 000 hectares (14%) remained uncleared (Smith *et al.*, 1994). In 2003 it was estimated that more than 60 million trees in the Brigalow were being cleared annually (Cogger *et al.*, 2003).

There have been some major impacts from clearing, cropping and grazing in the areas formerly occupied by Brigalow, including soil compaction, dryland salinity, modified hydrological regimes, erosion and weed invasion. Extensive clearing of this vegetation community has resulted in losses of biodiversity (Gordon, 1984; Cogger *et al.*, 2003; see Box 9.4). For example, the southern Brigalow belt of Queensland supports almost 330 species of birds – 50% of Australia's avian species diversity (Cogger *et al.*, 2003).

It is estimated that more than 90% of the original cover of Brigalow has been cleared and, as a result, the Threatened Species Scientific Committee has recommended that it be listed as an endangered community under the *Environment Protection and Biodiversity Conservation Act (1999).*

woodland, open woodland and mallee, have been modified. In particular, the densities of stems in many of the extensive vegetation types have declined. Much of the area of tall shrubland has been replaced by tall open shrubland, and low shrubland has been replaced by low open shrubland. Much of this change reflects the impacts of grazing in the semi-arid and arid zones.

Land clearing impacts on biodiversity

Millions of animals and plants are killed directly when native vegetation is cleared (Cogger *et al.*, 2003; see Box 9.4). Permanent or semipermanent vegetation clearance affects species through loss of habitat, including critical habitat and refuge areas. For example, wetlands were commonly drained to provide agricultural or urban land. The disappearance of Magpie Geese from southern Australia is probably related, at least in part, to the loss of wetland habitat in agricultural regions (Frith and Davies, 1961).

The ecosystems that remain after land clearing are usually fragmented and modified. They are subject to new disturbance regimes, invasive species, disease, increased nutrient loads, and changes in physical edge effects, including changes in wind, temperature, light

Table 9.4. Rate of tree-clearing in various Queensland bioregions (from Cogger *et al.*, 2003).

Bioregion	Estimated no. trees cleared (millions per year)	Proportion of total trees cleared (%)
Northwest Highlands	2.00	1.0
Gulf Plains	9.19	4.8
Cape York Peninsula	0.0	0.0
Mitchell Grass Downs	9.51	5.0
Channel Country	0.31	0.2
Mulga Lands	34.89	18.3
Wet Tropics	0.77	0.4
Central Queensland Coast	1.38	0.7
Einasleigh Uplands	3.66	1.9
Desert Uplands	13.00	6.8
Brigalow Belt	112.32	58.8
South-east Queensland	3.27	1.7
New England Tableland	0.66	0.3

and humidity. The impacts of habitat loss and fragmentation on biodiversity have dominated much of conservation biology research since the early 1990s. We explore these issues in detail in Chapter 10.

Land clearing affects ecosystem processes and services; the removal of perennial vegetation contributes to soil erosion by wind and water, leading to declines in water quality (see Chapter 5). The removal of deep-rooted perennial vegetation and its replacement by shallow-rooted pasture and crop species reduces evapotranspiration, thereby allowing ground water to rise and ground and surface water flows to increase (Stirzaker *et al.*, 2002). Salts in the soil are dissolved and brought to the surface by the rising ground water, where they are concentrated by evaporation. Given the current rates of land clearing and land degradation, the Commonwealth of Australia (2001c) has predicted

Brigalow. (Photo by Eleanor Collins.)

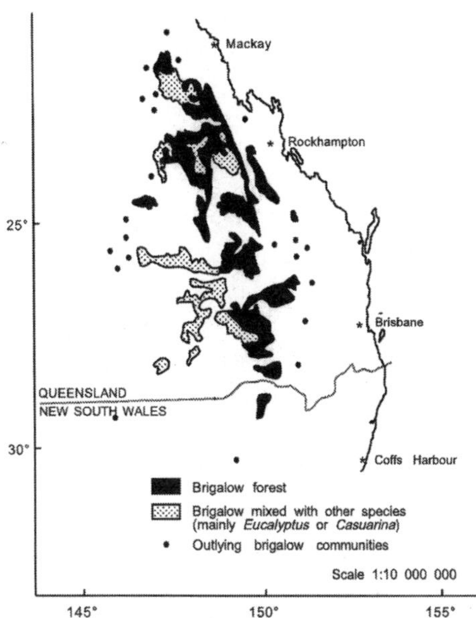

Figure 9.8. Distribution of Brigalow forests and scrubs (with *Acacia harpophylla* as the dominant or co-dominant species) prior to 1840 (after AUSLIG, 1990; Nix, 1994).

Box 9.4

Direct and indirect impacts of land clearing on the biota

Many animals maintain territories or strict home ranges that they defend against conspecifics. They are unable to move to other areas when their habitat is cleared and therefore often perish (see Tyndale-Biscoe and Smith, 1969). Animals that succeed in moving to an adjacent area are often unable to compete with the individuals that have already established a territory there (Pietsch *et al.*, 1995), so these animals also die. A recent study by Cogger *et al.* (2003) estimated the direct impacts of clearing on several groups of vertebrates. They focused their work on Queensland – the State with by far the highest rate of land clearing in Australia and one of the highest rates anywhere in the world. Cogger *et al.* (2003) calculated that about 2 million mammals, 8.5 million birds and 89 million reptiles die each year as a direct result of land clearing. They are killed at the time of vegetation removal or die soon after from predation or starvation. Notably, these estimates do not include impacts that have a longer lag time, such as the so-called extinction debt (Tilman *et al.*, 1994; McCarthy *et al.*, 1997; Berglund and Jonsson, 2005), whereby small non-viable isolated populations in remaining habitat fragments go extinct long after clearing has ceased (see Chapter 10).

that more than 17 million hectares of Australian farmland will be severely salt-affected by 2050 (see Chapter 5).

Dieback

Altered physical and biological processes can degrade remnant vegetation. Since the mid- to late 1960s, the premature decline and death of trees in rural landscapes has increased markedly (Wylie and Landsberg, 1987; Landsberg and Wylie, 1988; Reid and Landsberg, 2000). This phenomenon is known in Australia as rural **dieback**. Dieback has been reported from all Australian States and affects a wide range of forest and woodland types (Old *et al.*, 1981; Reid and Landsberg, 2000; McIntyre *et al.*, 2002; Rice *et al.*, 2004). The problem has become particularly pronounced in some areas; for example, Platt (1995) reported that approximately 28% of the trees in a 3300-hectare area of pastoral land in north-eastern Victoria died in a 22-year period between 1971 and 1993. If this trend continued, then almost all remaining trees in the area would be dead in about 80 years. Dieback is also becoming increasingly common in south-eastern Queensland (McIntyre *et al.*, 2002).

Some tree species are relatively susceptible to rural dieback. In the New England Tablelands in New South Wales, Lowman and Heatwole (1992) found that Blakely's Red Gum in woodland lost only 7.5% of leaf area per annum to herbivores. In contrast, trees in pasture lost between 16 and 88% of their leaf area per annum. Fox and Morrow (1983) found that Blakely's Red Gum in disturbed woodland and forests lost 29% of its leaf area compared with 16% for the next most affected species. Journet (1981) found that Blakely's Red Gum trees on farms in the Australian Capital Territory lost up to 70% of their leaves to insect herbivory per annum. Similarly, in the tropical savannas of north-eastern Australia, ironbark eucalypts are particularly prone to dieback, whereas bloodwoods are not (Rice *et al.*, 2004).

McIntyre *et al.* (2000) found that dieback is most serious where less than 30% of native vegetation cover remains, but there is no single cause of dieback (Landsberg and Wylie, 1983). Some of the contributing factors include (after Landsberg and Wylie, 1991; Reid and Landsberg, 2000):

- defoliation by native invertebrates (such as stick insects) and introduced insects
- salinity

(A) Eucalypt dieback in Western Australia (photo by David Lindenmayer), (B) southern New South Wales (photo by Mason Crane) and (C) southern Victoria (photo by David Lindenmayer).

Box 9.5

The Noisy Miner and eucalypt dieback

The native Noisy Miner is extremely abundant in many fragmented woodlands in eastern Australia, particularly in areas without understorey trees and shrubs. The social organisation of the species can be complex, and populations consist of strongly territorial groups of 6–30 individuals. The Noisy Miner scares away most other small insect-eating birds. Woodlands supporting Noisy Miner populations have fewer other small insect-eating birds, leaving eucalypt trees susceptible to dieback when they are being overgrazed by insects (MacDonald and Kirkpatrick, 2003). Grey *et al.* (1998) performed several removal experiments and found that when Noisy Miners are removed, smaller birds quickly return. This makes a strong case for Noisy Miner control on many farms to protect remnant vegetation and limit rural dieback. Noisy Miner populations are markedly less likely to occur in remnant vegetation with a well-developed understorey layer, so encouraging understorey regeneration is useful. As Noisy Miners prefer to live at the edges of patches of remnant vegetation, conserving larger areas of remnant vegetation can also be useful.

Noisy Miner. (Photo by Esther Beaton.)

- application of nutrients to improve pastures, which can lead to changes in soil fertility, and to changes in insect grazing pressure
- impacts of plant pathogens such as fungi
- soil acidification
- drought and fire
- deterioration of the structure of the soil
- mechanical damage caused by livestock and farm machinery
- increased numbers of parasitic mistletoes
- airborne salt
- decreased abundance of wildlife such as the Sugar Glider, which can consume large quantities of tree-defoliating insects, and the Echidna, which eats insect larvae
- increased abundance of aggressive native birds such as the Noisy Miner (see Box 9.4) and the Bell Miner, which displace other species, reducing predation pressure on defoliating insects
- increased browsing of trees by mammals such as the Koala and the Common Brushtail Possum.

Causes of rural dieback vary between regions. For example, in parts of south-eastern Queensland, New South Wales and Western Australia, dieback appears to have resulted from secondary salinisation, whereas Offor (1990, in Landsberg and Wylie, 1991) found that dieback among trees in coastal Victoria resulted from the effects of airborne salt. In Western Australia, Cinnamon Fungus has killed trees and shrubs throughout the relatively moist south-west region (see Chapter 7). In north-eastern Australia, drought stress (e.g. from extreme El Niño events) appears to be an important factor contributing to dieback among eucalypts on tropical savannas (Rice *et al.*, 2004).

In many instances, dieback is unlikely to result from just one cause (Landsberg and Wylie, 1983). Many large old trees in paddocks, which were retained to provide shelter for stock, began growing before the arrival of European settlers. They are now reaching the ends of their lives and are beginning to senesce. The extra nutrients left by Sheep and Cattle that camp under them can make insect damage more likely (Anonymous, 2001).

There are many positive feedback systems that may reinforce and perpetuate dieback (Landsberg and Wylie, 1991). For example, climatic extremes or fires may produce stress in trees. In response, the branches of stressed trees may produce epicormic shoots. These new shoots contain reserves of nutrients that are, in turn, a highly attractive food resource for foliage-feeding insects. A prolonged defoliation–refoliation cycle may therefore result (McIntyre *et al.*, 2002). Fungal agents may then invade weakened trees and further debilitate them, thereby exacerbating dieback (Landsberg and Wylie, 1991; Reid and Landsberg, 2000).

Although there has been some important research on rural dieback in the past few decades, many of the processes leading to this problem remain poorly understood. New forms of dieback are still appearing: in South Australia, Paton *et al.* (2000) described a new form of dieback among roadside River Red Gums and other woodland trees in which trees developed characteristic yellow foliage before dying.

Vegetation remnants in rural landscapes are important for biodiversity conservation (Saunders and Hobbs, 1991; Gibbons and Boak, 2002) and for maintaining farm productivity (e.g. Chartres *et al.*, 1992). It is important to understand the underlying causes of dieback before designing planting strategies for trees in rural Australia, so that rehabilitated vegetation remains viable (Landsberg and Wylie, 1991; Lindenmayer *et al.*, 2003a).

9.2 Mining and urbanisation

In general, the consequences for conservation of different forms of land use are related to the intensity of the impact and the area over which the activities are conducted. Table 9.5 shows some of the most important land-use activities in Australia. Mining and urbanisation are intensive but influence a small area relative to most other land-use activities. Relatively few plant species, for example, are threatened by mining activities. This does not mean that mining impacts are unimportant; rather, it implies that the impacts of other activities, such as cropping and grazing, should be managed at least as carefully as those of the mining industry.

Mining

In the last 25 years, mining-related exports have accounted for, on average, approximately 35–40% of Australia's total export income. In 1998–99 they represented about 4% of gross domestic product (Australian Bureau of Statistics, 2001). Australia is a major mineral producer and is the world's largest exporter of mineral sands (including ilimenite, rutile and zircon), bauxite, diamonds and lead (Australian Government Survey Organisation, 1999).

Table 9.5. Impacts on Australian terrestrial vegetation (estimated for the year 1989) by various forms of land use (after Hobbs and Hopkins, 1990).

Vegetation modification	Area affected (% of total)
Complete vegetation removal	
Mining	< 0.1
Urbanisation and transport	1.3
Vegetation replacement	
Crops and pastures	6.1
Utilisation of native vegetation	
Non-arid grazing	17.4
Arid grazing	43.7
Forestry	2.0
No deliberate modification	
Unused or traditional use (mainly desert)	26.0
Nature conservation	3.5

Impacts of mining

Mining directly affects less than 1% of the land and marine area of the Australian continent (see Hancock, 1993). Despite such relatively small areas, the local impacts of mining can be intense. Mines and mining operations have the potential to spread weeds, disturb and destroy habitat, damage cultural heritage, and pollute air, streams, ground water and soils, especially with heavy metals and strongly acidic and alkaline materials. **Spoil heaps** erode and **tailings dams** sometimes leak, and mining can also have indirect impacts on the environment. For example, processing bauxite for aluminium production requires major quantities of electricity, which has flow-on impacts for greenhouse gas production (Foran and Poldy, 2002). Blue Gum trees from eastern Australia were planted in the Jarrah forest of Western Australia as part of a revegetation program on bauxite mine sites in the forest, changing the forest composition. Until recently, Jarrah was difficult to germinate and it is also susceptible to the soil-borne pathogen Cinnamon Fungus (see Chapter 7).

A few of the impacts of mining are briefly explored in this chapter, and others are touched on in the case studies in the following section. Many mining operations produce toxic substances as the by-products of extraction or processing, and wastes in solution are stored in tailings dams designed for long-term containment. If dams leak or overflow, or if tailings leach and contaminate ground waters, then the contaminants can pollute surrounding environments. This risk has been identified in mines within Kakadu National Park in the Northern Territory (Wasson *et al.*, 1999). Tailings

(A) Opal mining and (B) open-cut coal mining in South Australia. (Photos by David Lindenmayer.)

retained on site may also pose a toxic risk to some wildlife species. In addition, local site conditions, the chemical composition of the ore-bearing rock, and on-site processing of the ore can produce tailings and other wastes that are very difficult to deal with. Once operations have ceased, remediation of sites to a point where natural communities of plants and animals are once again established can be a long, difficult and expensive process.

Potential impacts vary from project to project, depending on the mineral in question, the type and size of mining and treatment operations, and the location of the mine in the landscape. Two examples of mining impacts are described below, from Rum Jungle in the Northern Territory and Queenstown in Tasmania. Environmental standards and regulations have improved substantially since the times of these early mines. For contrast, we outline the codes of practice for gas exploration off the north-west shelf of Western Australia.

Rum Jungle

The Rum Jungle mine is located about 65 kilometres south of Darwin in the Northern Territory. The mine was a series of five open cuts that extracted uranium, copper and lead. It operated between 1949 and 1971, and the total area of the mine site was about 200 hectares. Ore was treated at a plant near one of the larger open-cut mines (Richards *et al.*, 1996). When mining first commenced, solid waste was left near the treatment plant, and liquid effluent drained directly into a tributary of the Finniss River. Conway *et al.* (1975, in Richards *et al.*, 1996) estimated that up until 1984, when rehabilitation work was well advanced, 4.7 tonnes of copper, 3.5 tonnes of manganese, 320 tonnes of sulfate and unknown amounts of lead and acid water were released annually into the Finniss River. The water immediately below the operations and tailings site was undrinkable and supported few living organisms, and much of the **riparian** vegetation died (Farrell and Kratzing, 1996).

The leachates and effluent produced by mining, and the radiation associated with the extraction of uranium ore created human health risks (Farrell and Kratzing, 1996; Richards *et al.*, 1996). At the time the mine was first built, there were few laws to regulate the potential impacts of mining and there was only very limited public concern for the environment (Farrell and Kratzing, 1996).

An initial clean-up of the Rum Jungle mine site was undertaken in 1977, and more extensive rehabilitation commenced in 1982. Rehabilitation involved reshaping and revegetating waste dumps (some were relocated), constructing subsoil drains and coarse stone barriers to reduce erosion risks, and treating waste water (after Richards *et al.*, 1996). The rehabilitation program was completed by mid-1986 and cost more than A$18 million (Richards *et al.*, 1996), in addition to the cost of ongoing repairs.

Native and exotic grasses and *Acacia* spp. trees have now established on the waste rock dumps, and discharge of heavy metals into the Finniss River has fallen by 50–70%. However, high concentrations of heavy metals continue to leach into the ground water. Environmental monitoring at the Rum Jungle mine site continues and Richards *et al.* (1996, p. 553) noted that it is: 'one of the most detailed case studies available to help with the development of strategies for the management of mine site pollution'.

Queenstown

One of the most striking examples of the effects of mining occurred around Queenstown in western Tasmania (Hancock, 1993). The Mount Lyell mine was opened in Queenstown in 1893 and large copper smelters came into operation 3 years later. Major environmental impacts have been associated with mining at Queenstown for more than a century.

The effects of mining and processing ore extend well beyond the operations site: the area is characterised by spectacularly bare hills that support almost no native vegetation. Sulfur dioxide emissions from copper smelters that operated for more than 70 years, together with high rainfall in the area, resulted in acid rain. In addition, extensive erosion followed clearing of the vegetation on the hills surrounding Queenstown, which was exacerbated by the high rainfall (Farrell and Kratzing, 1996). Timber cut from these areas was used in the mining and smelting processes.

Waste from the mine and the smelter contained high levels of heavy metals, which were discharged directly into the Queen River, which flows through Queenstown and eventually into Macquarie Harbour on the west coast of Tasmania. This has resulted in the pollution and siltation of watercourses. Tailings have also been transported from Queenstown and dumped in Macquarie Harbour. Heavy metals from tailings deposited in the marine environment and river sediments threaten stocks of marine fish (Farrell and Kratzing, 1996).

Rehabilitation efforts in Queenstown include a revegetation program, treatment of waste water, flooding of underground mines to reduce levels of oxidation to restrict the development of acid water, and the development of enhanced drainage systems to reduce run-off from waste dumps (Hancock, 1993).

There are many other examples of major mine-related environmental impacts throughout Australia (see Farrell and Kratzing, 1996, for an extensive review).

North West Shelf Gas Project

The North West Shelf Gas Project is located near Karratha on the central coast of Western Australia. Natural gas and liquid condensate is extracted from under-sea reservoirs by drilling platforms located 130 kilometres off the coast. The gas is then piped to treatment plants near the towns of Karratha and Dampier (Wright and Nunn, 1996). The project began with an environmental impact assessment in 1978, although some environmental studies had been instigated some years earlier, then construction started in 1980. Since then, approximately A$20 billion has been expended on the development of the project, making it the single largest resource development project in Australia (Wright and Nunn, 1996).

There are a number of important environmental considerations associated with the North West Shelf Project. The region supports coral reefs, mangrove forests, and large areas of intact native vegetation. In addition, the islands of the Dampier Archipelago (off the coast from Karratha and Dampier) have important conservation value. Notably, the first employee of the project was an on-site environmental officer. Impact detection and mitigation include (after Wright and Nunn, 1996):

- surveys of flora, fauna and Aboriginal sites prior to the construction of the onshore processing plant (this included documentation of all Aboriginal sites and artefacts as well as the development of a herbarium of native plants from the region where the treatment plant is located)
- control of feral animals around the treatment plant
- re-establishment of native vegetation on and around areas such as roads, quarries, and stockpile areas, which were modified in the process of the onshore infrastructure construction
- treatment of effluents and run-off from the gas processing plant before they are discharged into the ocean
- long-term monitoring of heavy metals in the marine environment and emissions to the atmosphere surrounding the treatment plant
- surveys of seabed condition and sediment drift around the off-shore drilling platforms
- ongoing training of staff in environmental management.

As of 1996, environmental studies associated with the project had cost about A$20 million. Continued expenditure supports monitoring of the effects of marine pollution and the success of revegetation efforts (Wright and Nunn, 1996). However, the project *has* actually caused environmental problems; for example, an oil spill damaged an area of mangroves close to the onshore gas treatment plant. Project managers do, however, attempt to anticipate and minimise impacts (Wright and Nunn, 1996). Although there are no

Marine organisms growing on the surface of the North West Shelf Project gas pipeline. (Photo by Clay Bryce, Western Australian Museum.)

Extensive area of mined sand dunes in south-eastern Queensland. (Photo by David Lindenmayer.)

absolute guarantees that serious environmental impacts will not occur, the project has gone considerably further than the Rum Jungle and Queenstown projects to mitigate environmental impacts.

Legislative and other controls on mining

Community attitudes to mining have changed since operations such as Rum Jungle and Mount Lyell in Queenstown (Hancock, 1993; Bradfield *et al.*, 1996). Moreover, whereas mining has been subject to only limited regulation in the past, there are now many acts, regulations, conventions and protocols associated with the industry. For example, Bradfield *et al.* (1996) compiled a list of 15 different Commonwealth acts relating to the mineral and energy industries and there are many other pieces of relevant State and Territory legislation (Bradfield *et al.*, 1996).

Non-binding guidelines also assist the industry. For example, the Environmental Protection Agency (1996) published a series of documents on 'best practice environmental management in mining'. In Western Australia, where there are more mines than elsewhere on the continent (Australian Bureau of Statistics, 2001), the State Government has guidelines for the preparation of detailed standardised environmental reports that should be completed each year for each mine (Government of Western Australia, 1996). In addition to legislative controls, the mining industry has adopted many non-binding, self-imposed guidelines. The Australian Minerals Industry Code for Environmental Management has been established and many major mining companies in Australia are signatories (Minerals Council of Australia, 2002).

Larger and more profitable companies are more likely to have the resources and management structures to anticipate and deal effectively with environmental problems. For example, in the past, sand mining operations were highly controversial because they took place in ecologically and visually sensitive environments such as beaches and associated dune systems. BHP-Billiton, the largest mining conglomerate in the world, is an important miner of mineral sands, but its operations no longer target beach environments. Rather, the company seeks inland locations, where beach and dune systems developed several thousand years ago, when sea levels were higher.

Concluding remarks: mining impacts in Australia

Of course not all mining operations or extractive industries are as well managed or monitored as the North West Shelf Gas Project. Buckley (1991b) and Burton *et al.* (1994) pointed out that there are no clear scientific or legal criteria that define successful rehabilitation. Given the vagaries of the dynamics of natural populations and the long time scales over which community succession takes place, it is not always possible to specify what constitutes successful rehabilitation, and at what point it has been achieved. Long-term management and monitoring practices may be deficient, particularly after mining operations have ceased (Mercer, 1995). These problems are not debilitating, but neither can they be ignored. Environmental management in the mining industry will improve if companies commit to long-term sustainable management of the ecosystems in which they operate (Chapter 17).

Impacts of urbanisation

The impacts of urbanisation in Australia are usually considered minor compared with other land-use activities. Although the area of the impacts is relatively small (less than 1.3% of the area of the continent), the magnitude of local impacts (and other impacts, see Box 9.5) is relatively large, mainly because they involve more or less complete and permanent removal of habitat and fundamental changes to the physical environment, and because they are in relatively productive and fertile environments. Indeed, many more vascular plant species have been eliminated by urban development than have been eliminated by mining and forestry combined.

Recher and Lim (1990) reviewed changes in the distribution and abundance of the vertebrate fauna of Australia since European settlement. They concluded that extinction rates and declines in abundance and range have been highest in regions where settlement first occurred, and are attributable to the use of land for human purposes, habitat loss and fragmentation, overexploitation, and the spread of exotic herbivores, predators and disease. In contrast, some native animals are now more abundant in urban areas than in the wild (e.g. the Common Brushtail Possum and some species of crows).

The physical and biotic environment in cities is very different from the pre-urban state (Parris and Hazell, 2005; Tait *et al.*, 2005). Urban remnants of native vegetation are subject to human activities, some of which are unique to urban environments; these include:

- invasion by weeds from suburban gardens
- planned and *ad hoc* mowing programs
- removal of dead and dying trees
- incursion of exotic animals
- trampling by walkers, bicycles, horses, stock etc.
- dumping of household and industrial waste
- dumping of soils and garden cuttings
- picking of flowers, fruit and foliage
- malicious damage to trees and wildlife
- alteration to ground water and surface water flows
- alteration of fire regimes
- chemical pollution of the air and water
- drift of pesticides and herbicides
- leaching of fertilisers from lawns and gardens.

Vegetation remnants in urban environments are often linear or **dendritic** in shape, and are therefore relatively susceptible to **edge effects** (see Chapter 10). Many weed problems start at the urban–bushland interface with escapees from gardens or with garden waste dumped in nearby bushland (Low, 1999). National parks such as those in the Blue Mountains and others in and around Sydney have active 'friends and support' groups who spend large amounts of time removing weeds (e.g. Pallin, 2000). The potential magnitude of the exotic plant problem in Australian cities is highlighted by experiences elsewhere. For example, work in Buenos Aires (Argentina) demonstrated a strong relationship between prevalence of exotic species and urbanisation, with virtually all plants being non-native by the time housing density exceeds 12 houses per square kilometre (Rapaport, 1993; Figure 9.9).

The management (and attempted control) of fire in bushland adjacent to urban developments (Whelan, 1995, 2002; see Chapter 11) can have consequences for conservation. Cities such as Sydney, Melbourne, Hobart

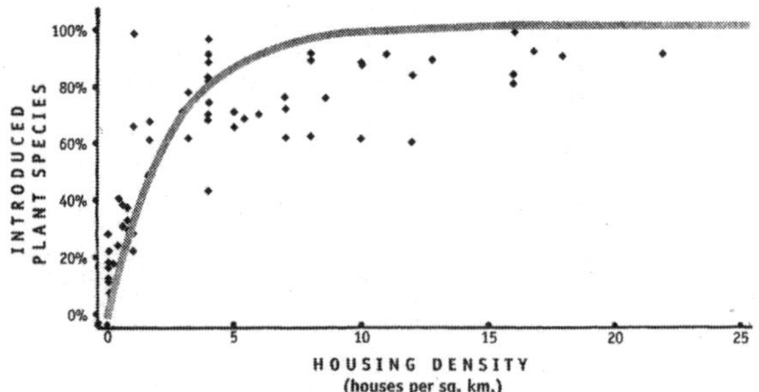

Figure 9.9. Relationship between non-native species and housing density in Buenos Aires. (Redrawn from Rapaport, 1993.)

Box 9.6

'Ecological footprints' of cities

Many people consider that the impacts of cities are generally confined to area they cover, which in Australia is tiny relative to the size of the continent. However, some workers have calculated the 'ecological footprint' of cities based on the quantity of food and forest products that are consumed and the extent of the area needed to assimilate waste products such as carbon dioxide emissions (Wackernagel and Rees, 1996). For example, the ecological footprint for the city of London is 125 times larger than the size of the city (IIED, 1995). Such calculations, although difficult, provide a crude indication of the extent and magnitude of human urban-derived impacts on the environment. This is important because by the late 1990s, there were 326 cities worldwide, with populations exceeding one million people – an increase of more than 50 cities from the start of the same decade (UNEP, 1999).

Box 9.7

Urban effects on native grasslands

Habitat loss within urban areas is ongoing. Native grasslands are a highly endangered ecological community in Victoria, and have been reduced to less than 1% of their original extent. Williams *et al.* (2004) mapped remnant native grassland in urban Melbourne, using spatial data from 1985 and 2000 (Figure 9.11). Despite the status of grasslands, Williams *et al.* (2004) found that of the 7238 hectares of native grassland in the area in 1985, only 4071 hectares remained in 2000. A total of 1690 hectares had been cleared, mostly for urban development and roads, and 1465 hectares was substantially degraded by invasive species. Patch loss could be explained by proximity to the city centre and to roads (Figure 9.12). Patch loss was less likely close to watercourses because of planning provisions for open space.

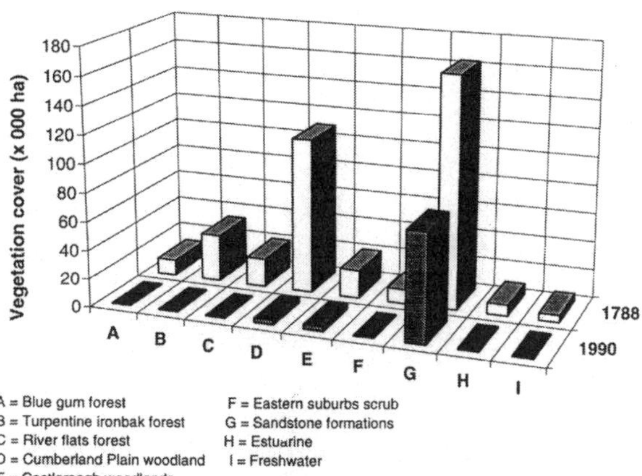

Figure 9.10. Vegetation cover in the County of Cumberland in Sydney's western suburbs in 1788 and 1990 (after Benson and Howell, 1990).

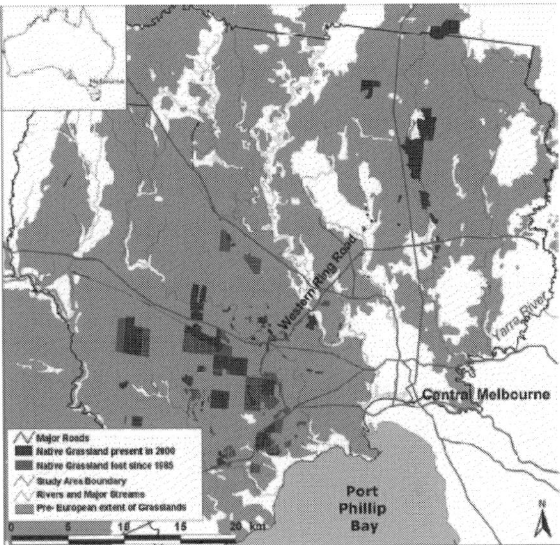

Figure 9.11. Map of part of urban Melbourne showing the patches of native grassland present in 1985 and those remaining in 2000. The pre-European extent of native grasslands, major roads and streams in the Melbourne metropolitan region are also illustrated.

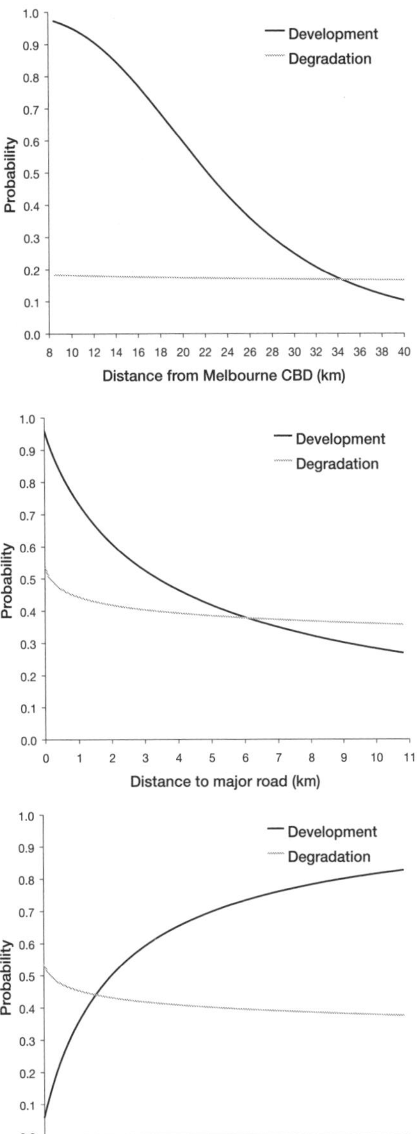

Figure 9.12. Probability of a patch being destroyed by development or substantially degraded with (A) distance to the Melbourne central business district, (B) distance to a major road and (C) distance to a third-order or higher stream (after Williams *et al.*, 2004).

and Canberra (among others) experience large fires that destroy houses at the urban–bushland interface (Whelan *et al.*, 2002; Cary *et al.*, 2003). This issue is addressed in further detail in Chapter 11.

Biodiversity in cities

Human impacts can be substantial in cities. For example, Tait *et al.* (2005) completed a detailed study of the biota of the city of Adelaide. They found that between 1836 and 2002, 132 species of native plants and animals had become extinct, 50% of native mammals had been lost, and almost 650 species of exotic plants had been introduced. Urban environments in Australia often support many fragments of original bush – some with important conservation value (Box 9.7). Road verges, railway reserves and stream reserves provide the sole habitat for many Australian plant communities. For example, the plant communities of western Sydney persist entirely within an urban landscape (Figure 9.10).

Conservation of the Eltham Copper Butterfly is a useful case study for the impacts of urban development on a remnant ecosystem (Box 9.9). The Eastern Quoll is another example. The Eastern Quoll now appears to be extinct in Victoria, and some of the last records of the

Box 9.8

Conservation of the Eltham Copper Butterfly

The Eltham Copper Butterfly was first recorded in 1938 from Eltham, an outer suburb of the city of Melbourne. A large number of animals were collected from Eltham in the 1940s and 1950s (Braby *et al.*, 1992), but the populations at this location declined markedly after about 1950. The Eltham Copper Butterfly was recorded from eight localities in central and western Victoria, including Eltham, but a survey of sites in the late 1980s revealed that populations no longer existed at five of these sites (Braby *et al.*, 1992).

A colony was found near Eltham in 1987 on land destined for urban development, so the land was subsequently purchased and protected, with cooperation from the developer, the local community, the Eltham Shire Council and the State Government (New, 1991). Today, several colonies are known from around Eltham, and two populations occur in country Victoria (Braby *et al.*, 1992, 1999). The butterfly is rare and persists in a few disjunct isolated populations (New, 1991, 2000).

The butterfly occurs in dry open eucalypt woodland in which the understorey is composed of a diverse array of native shrubs, herbs and grasses. It is very localised even within the remaining patches of habitat, preferring dry, open microhabitats that contain scattered patches of the shrub Sweet Bursaria. Larvae of the Eltham Copper Butterfly feed only on *Bursaria spinosa* var. *spinosa*. Caterpillars occur only on stunted plants associated with nest chambers of the ant *Notoncus*, into which the caterpillars retreat during the day (New, 1991, 2000). Thus, persistence of the species depends on the persistence of its habitat and on two species with which it shares a dependent ecological relationship.

The species has experienced extensive contraction of its range in Victoria. Prior to 1750, the Eltham Copper Butterfly was probably widespread, but with restricted niche requirements, so that it may have always been localised, with populations restricted to the eucalypt woodlands of the foothills and plains of the Great Dividing Range. Its range overlaps the extensive Wheat-growing regions of Victoria that are now largely cleared. It is also likely that gold-mining activities and the associated soil disturbance would have impacted on host plants, larvae and the associated ant colonies (Braby *et al.*, 1992). The decline of the populations at Eltham are attributable to urbanisation, with the associated destruction of native vegetation and habitat alienation (New, 1991).

The Eltham Copper Butterfly remains at risk from weeds, urban development and land clearance. Associated threats include trampling of vegetation, garbage dumping, invasion by garden plants, and the activities of exotic animals (Vaughan, 1988). Weeds inhibit the regeneration of Sweet Bursaria and are likely to render the habitat unsuitable for the butterfly's host ant species (Vaughan, 1987). Regular prescribed burning has been applied in an effort to control weeds and promote host plant growth (New, 2000). Part of the subdivision supporting the large suburban butterfly population at Eltham has been reserved, but it is protected only by a small buffer (New, 1991). Intrusion of housing on the remainder of the subdivision poses a significant threat, although a large sum of money was gathered to purchase land to improve the protection of Eltham Copper Butterfly habitat (Braby *et al.*, 1999).

Eltham Copper Butterfly. (Photo by Alan Yen.)

species were from the suburbs of Melbourne (Seebeck, 1977). It is possible that small remaining populations around the city were isolated from diseases that reduced numbers of the Eastern Quoll elsewhere in Victoria (Menkhorst, 1995a). Finally, there have been problems stemming from increasing populations of Flying Foxes in a number of east coast Australian cities (Parris and Hazell, 2005), and a good understanding of the biology of these animals has been employed to address them (see Box 9.9).

Future urban expansion

Continued growth in the earth's human population (see Chapter 12) has been coincident with a worldwide demographic shift toward an increasing proportion of people living in cities (UNEP, 1999). This has occurred through increases in urban populations, the migration of people from rural to urban areas, and the transformation of villages and towns into cities. Australia is no exception to this general trend (Foran and Poldy, 2002). In the first half of the 21st century, the Australian human population will exert increased pressure on the coastal and subcoastal landscapes that support the majority of Australia's cities (see Chapter 12).

In most Australian cities, population growth is negative in the core and inner suburbs; most growth takes place in the outer fringing suburbs, where land subdivision still takes place. Almost invariably, expanding suburbs destroy native vegetation, and the destruction of natural vegetation in most cases is so complete that it is difficult or impossible to reconstruct the original cover (Bridgeman *et al.*, 1995). The severity of these impacts is reflected in the disproportionate number of extinct and threatened Australian plants in urban environments. Industrial and urban development, road works, railways and rubbish dumping together affect only about 1.3% of the landscape, but account for 19% of present and future threats (see Chapter 6). Many suburban vegetation remnants are restricted to creeks or steep slopes, in places that are unsuitable or uneconomic for housing. Careful planning and the rehabilitation of existing remnants will be required if future impacts are not to exceed those of the past (Rawling, 1996).

Box 9.9

Urban environments and Flying Foxes

Several Australian cities, including Melbourne, Sydney and Brisbane, support large camps of Flying Foxes, for example the Grey-headed Flying Fox, which is a nationally threatened species (Pavey, 1995; Pallin, 2000). These camps are, in turn, important as a prey source for large predators such as the Powerful Owl (e.g. Pavey, 1995). Large camps of Flying Foxes can damage and eventually kill vegetation where the camps are located. In addition, the spatial movements of animals foraging away from camps can affect surrounding commercial orchards (Tidemann, 2003). Droppings from large numbers of Flying Foxes can soil the roofs of homes, cars and washing, leading to complaints from local residents. The animals can also carry viruses that pose a human health risk.

Hall and Richards (2000) described the management of Flying Fox camps in the city of Ipswich west of Brisbane. Early attempts to relocate animals were futile, so an alternative approach based on an understanding of Flying Fox biology was initiated, which involved planting fast-growing trees in an appropriate location. Within 5 years, campsites were relocated away from suburban backyards (Hall and Richards, 2000).

In another example, the Grey-headed Flying Fox arrived in Melbourne for the first time in the 1980s – possibly because urbanisation has led to the city becoming warmer and experiencing fewer frosts, making the climate suitable for the Flying Fox (see Parris and Hazell, 2005). The population of the species has grown exponentially since that time, and numbers had increased to more than 20 000 in the Royal Botanic Gardens of Melbourne by 2002. Ecologists encouraged the animals to establish roost sites elsewhere in the city by using a mixture of sounds to make roost sites in the Botanic Gardens less attractive. The population relocated in 2003 to a new roost that is more tolerable to both bats and people.

Table 9.6. Extent of land cover disturbance as a result of grazing and recurrent fires in the area of Australia classified as the extensive land-use zone by Graetz *et al.* (1995).

	Disturbance class				
	Slight	**Substantial**	**Significant**	**Indeterminate**	**Total**
Area (square kilometres)	2 773 596	655 700	1 152 127	6189	4 708 092
Percentage extensive land use zone	58.9	13.9	24.5	0.1	100
Percentage continent	36.1	8.5	15.0	0.1	61.2

9.3 Traditional Aboriginal use and pastoralism

According to Graetz *et al.* (1995), the extensive land-use zone covers about 61% of Australia (more than 4.7 million square kilometres). The dominant forms of disturbance in these areas are grazing and recurrent burning (Table 9.6).

The extensive land-use zone is host to three major human influences: harvesting of natural populations of animals and plants (see Chapter 8), traditional Aboriginal land use, and pastoralism. In contrast with the intensive land-use zone, about 60% of the extensive land-use zone has been only slightly disturbed (Table 9.6). The largest areas in the 'significantly disturbed' class are in Western Australia (482 000 square kilometres) and Queensland (271 000 square kilometres). However, the States with the largest proportion of total area of significantly disturbed land are New South Wales and Queensland. Places with the highest primary productivity are more likely to have been cleared and/or significantly disturbed. The overall extent of vegetation disturbance estimated by Graetz *et al.* (1995), based on the sum of information from the intensive and extensive land-use zones, shows that the Australian continent is less heavily disturbed than Europe, Asia or North America, but has similar levels of disturbance to South America (Figure 9.13).

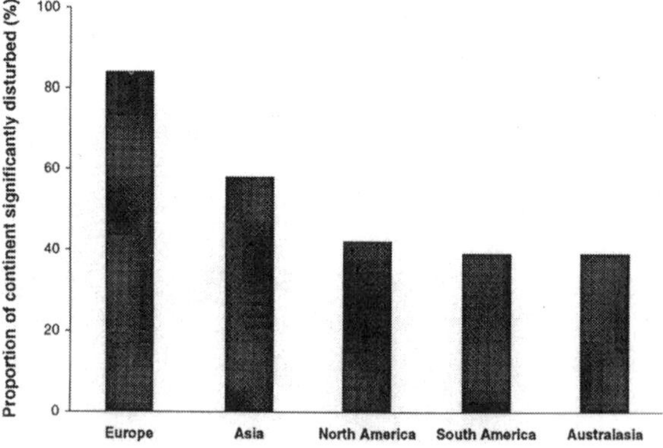

Figure 9.13. Extent of disturbance on different continents (after Graetz *et al.*, 1995).

Traditional Aboriginal land use

The Australian environment developed partly in response to extensive Aboriginal influence (Bowman, 1998, 1999, 2003; Langton, 1998). It is in the context of Aboriginal management that the impact of European occupation is generally viewed as having an 'unnatural' impact on the environment (Woodgate *et al.*, 1994). The arrival of the first humans in Australia may well have been more than 50 000 years ago (Kirkpatrick, 1994), and if that is the case, then Europeans have occupied the Australian continent for less than 0.5% of the time that indigenous people have inhabited it.

A key element of Aboriginal land use was the use of fire. Burning practices often involved (and still involve) burning of limited areas in a mosaic pattern, depending on seasonal conditions (Langton, 1998; see Chapter 11). Fire is used to create fire breaks, to manage fuel loads, and for signalling, hunting, 'cleaning' country, and promoting food plants (Bowman, 1998). As an example, fire was used by the Martu people of the Sandy Desert to broaden the plant resource base, thus stimulating elevated species diversity in a range of landscapes (Walsh, 1990; see Chapter 11). Other management practices included scattering the seeds of food plants, introducing species (e.g. the Dingo; see Box 9.9), and increasing the productivity of preferred species (particularly those close to water resources and resource-rich areas; Walsh, 1990).

The colonisation of Australia by Aboriginal people almost certainly had a substantial impact on wildlife populations and probably led to the extinction of some plant and animal species (Bowman, 2003; Brook and Bowman, 2004). Some species, such as the Koala, were relatively uncommon at the time of European settlement because of Aboriginal hunting (Martin and Handasyde, 1999). The effective management of many ecosystems may well depend on continuing Aboriginal

Box 9.10

The Dingo – the indirect impacts of Aboriginal Australians?

The Dingo accompanied Aboriginal Australians to continental Australia approximately 5000 years ago (Savolainen *et al.*, 2004) and it then became established in all parts of the country except Tasmania (Corbett, 2001). Two vertebrates, the Thylacine and the Tasmanian Native Hen became extinct on the mainland at the time that the Dingo arrived, but survived in Tasmania where the Dingo was absent. A third species, the Tasmanian Devil, became much reduced in distribution and abundance, but appeared to survive in small numbers on the mainland after the Dingo arrived (until the time of European settlement; see Paddle, 2000).

The Dingo is widely believed to have outcompeted the Thylacine, which became extinct on the mainland about 3000 years ago (e.g. Corbett, 2001). Most records of mainland populations of the Tasmanian Native Hen date from the Late Pleistocene, and there is only one record from the Holocene (4500 years ago).

The notion that the Dingo was solely responsible for the extinction of a number of mainland vertebrates has been questioned by several authors. For example, Paddle (2000) does not accept that the Thylacine was outcompeted, because of significant niche differences. Johnson and Wroe (2003) believed that changes in the characteristics of human populations in mainland Australia could have contributed to extinctions. For example, about 4000 years ago, human populations increased substantially. Hunting technology advanced, and there was an intensification of resource use (including increased hunting pressure on large vertebrates such as those that were prey for the Thylacine and the Tasmanian Devil). It is likely that, in each case, extinction was the result of several unique and interacting factors, of which changes in human technology was just one (Johnson and Wroe, 2003).

(A) The Dingo (photo by Gary Steer) and (B) poisoned Dingoes near Rocky Plain in southern New South Wales (photo by David Lindenmayer).

management, including burning regimes (Langton, 1998; see Chapter 11). Aboriginal land use has been integrated successfully with conservation management in several national parks (see Birckhead *et al.*, 1992), for example Uluru and Kakadu in the Northern Territory (Fraser *et al.*, 2003; Liddle, 2003). Other opportunities for continuing Aboriginal management exist, particularly in areas such as Western Australia, where 12% of the State is held by the Aboriginal Land Trust (Department of Indigenous Affairs, 2002). However, a detailed treatment of these topics is beyond the scope of this book (see Cary *et al.*, 2003, for examples).

There may also be situations in which the objectives of conservation are not consistent with modern Aboriginal land and resource management practices; pastoralism and hunting declining populations are examples of such situations. The Dugong was once abundant in northern Australia around seagrass beds from Moreton Bay in southern Queensland through northern Australian waters to Shark Bay in Western Australia. Populations of Dugong have been hunted by indigenous people for many thousands of years (Smith and Marsh, 1990; Marsh *et al.*, 1997). On Cape York Peninsula in northern Australia, harvesting rates of the Dugong appear to be well above sustainable levels (Heinsohn *et al.*, 2004). Reducing harvest rates to appropriate levels will depend on finding suitable alternatives for isolated indigenous communities, who are socially and economically disadvantaged. Sustainability is critical for permanent Aboriginal residents; pastoralism, resource development and tourism create economic options that are essential for the survival of Aboriginal communities (Langton, 1998).

Pastoralism

Grazing by Sheep and Cattle on natural or semi-natural pastures is the major land use for 4.5 million square kilometres (60%) of Australia's land surface (Commonwealth of Australia, 2002a; Figure 9.14). Products from pastoralism contribute about 3% of the nation's annual average gross domestic product, particularly through exports (AFFA, 2003a).

Griffin and Friedel (1985) noted three major changes to land-use practices following European settlement and the development of pastoralism in the rangelands and semi-arid parts of Australia: (1) domestic livestock were introduced to better watered country close to markets,

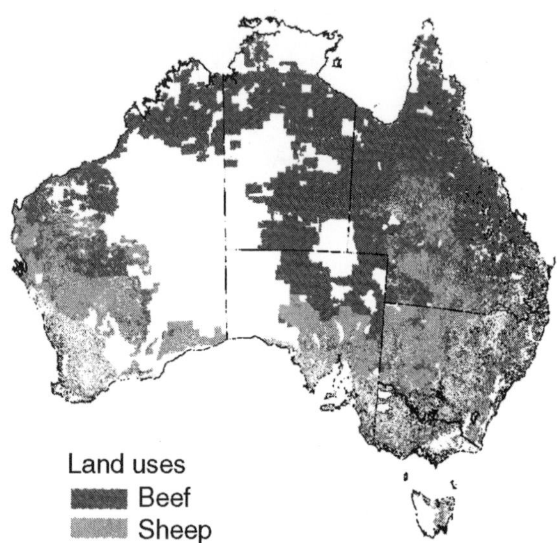

Figure 9.14. Major land uses in Australia, showing the extensive areas of the continent where grazing by Sheep and Cattle are prevalent (from the Commonwealth of Australia, 2002a).

(2) the European Rabbit and other exotic animals and plants invaded or were introduced, and (3) Aboriginal burning practices were suppressed, effectively ceasing in many areas of Central Australia by 1920.

Pastoralism and biodiversity loss

Introduced **ungulates** are more dependent on permanent water than are native mammals. They congregate near permanent water, especially during drought (Landsberg *et al.*, 1997; James *et al.*, 1999). They eat and trample native vegetation, compact soil, alter soil structure and disrupt the soil crust (Hobbs and Hopkins, 1990; see Chapter 7). Landsberg *et al.* (1997) and James *et al.* (1999) measured the vegetation in several Australian rangelands along 10-kilometre gradients between water points and areas remote from water. They found that grazing intensity increased in areas that were closer to water (see Box 5.2 in Chapter 5). About 20% of the native species at each location were abundant only near water where grazing intensity was high, and about 30% were abundant only at lightly grazed sites. Spatial analysis of the distribution of water points in the rangelands showed that potential habitat has become rare for species that are adversely affected by grazing.

Dickman *et al.* (1993) examined the conservation status of vertebrates in the Western Division of New South Wales. More than one-third of the 71 mammal species have been lost from the region since European settlement and a further 11 have declined substantially (Dickman *et al.*, 1993; Smith and Quin, 1996). Similarly, one-fifth of the more than 300 bird species in the area are threatened, more than 100 have decreased in abundance, and 6 species no longer occur in the region (Smith *et al.*, 1994).

Pastoralism, over-grazing and fire

In general, pastoralists protect pastures from fire, although they sometimes use fire to control inedible woody species (see Chapter 11). Changes in fire regimes alter the vegetation composition: the absence of low-intensity fire has increased perennial shrubs and small trees in some areas (Griffin and Friedel, 1985). Fire and grazing have modified ground cover over extensive areas of Australia, and, together with the impacts of feral predators, are probably the most important factors determining the decline and loss of arid and semi-arid bird and mammal species (Burbidge and McKenzie, 1989; Morton, 1990; Woinarski, 1999; Garnett and Crowley, 2000; see Chapter 11).

Box 9.11

Impacts of pastoralism in north Queensland
Gardener *et al.* (1990) described the pastoral use of the open woodlands of northern Queensland, which have been grazed by domesticated Cattle for about 120 years. Vegetation changes did occur after settlement, but there was little land degradation during the first 100 years. The main limitation to Cattle production at the time of settlement was poor quality herbaceous fodder. Mortality among British Cattle breeds was high during droughts, and increases in herd size were limited by poor breeding performance. Feed supplements and the introduction of better adapted Zebu Cattle resulted in greatly improved survival and growth rates, with an accompanying increase in vegetation consumption and associated grazing pressure. Some pastures, including those in the Burdekin Catchment, are now suffering the effects of over-grazing, including loss of herbaceous cover, shrub invasion and soil erosion. Gardener *et al.* (1990) concluded that only a reduction in grazing pressure would prevent a rapid expansion of this problem.

Grazing by Cattle and Rabbits can have relatively severe ecological impacts in drought years (McIntyre *et al.*, 2002). Over-grazing leads to desertification, and up to 4 million hectares of pastoral inland Australia could become permanent desert in the near future (Recher and Lim, 1990).

Pastoralism and vegetation response
At the time of early pastoralism, many graziers believed that enormously high stocking rates were possible. This is illustrated by the following quote from the 1850s, before settlement in the Western Division of New South Wales (by Captain William Randell in Main, 2000):

> *Sheep carrying qualities can scarcely be overrated, the grass for miles and miles... being so thick and long that it can only be walked through with difficulty.*

However, the major impacts of pastoralism on vegetation communities were realised by many graziers not long after European settlement. Cain (1962) reported that:

> *By 1895 the Western Division presented a very different aspect from that which met the eyes of settlers in 1880; many of the grasses and much of the edible scrub had gone.*

Similarly, the Pastoral Review (1945) noted that in the Western Division:

> *most properties were bare of feed and denuded of stock.*

Native vegetation communities within Australia's pastoral regions have been modified to different extents. Shrublands dominated by Mulga are much more degraded than grasslands such as those dominated by Mitchell Grass (Figure 9.15). A total of 85% of Australian grasslands show minor or no degradation, whereas only 35% of arid zone shrublands are relatively undisturbed. None of the Mitchell grasslands are severely disturbed, but about 25% of the shrublands are severely disturbed (Christie, 1993).

Area subject to pastoralism near the Dingo fence, Lake Frome, South Australia. (Photo by David Lindenmayer.)

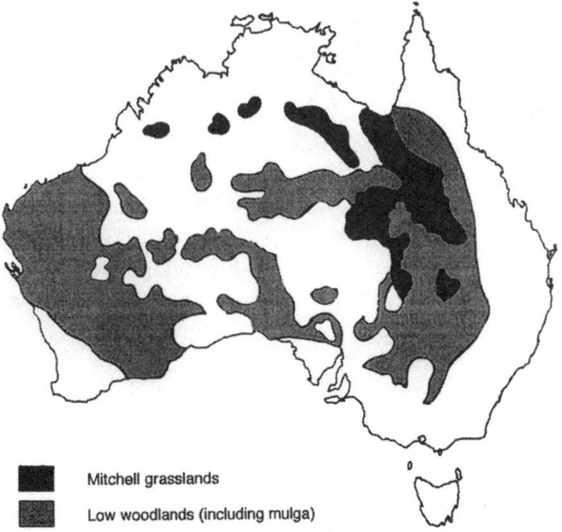

Figure 9.15. Distribution of Mulga shrublands and Mitchell grasslands in Australia (after Christie, 1993; AUSLIG, 1990).

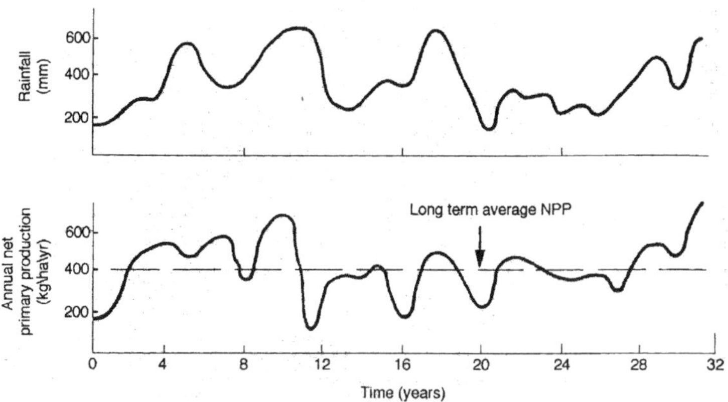

Figure 9.16. Rainfall and annual net primary productivity for a natural Mulga community in the Charleville area of Queensland (after Christie, 1993).

Box 9.12

Grazing in temperate woodlands

Grazing by domestic livestock is a major land use in areas that used to be dominated by temperate woodland vegetation types. Temperate woodlands are now some of the most extensively cleared, heavily modified, and highly degraded vegetation types in Australia (Er, 1997; Goldney and Wakefield, 1997; Hobbs and Yates, 2000; McIntyre *et al.*, 2002). For instance, less than 3% of the original cover of woodland dominated by White Box and Yellow Box remains, and most patches are smaller than 1 hectare (Prober and Thiele, 1995; Gibbons and Boak, 2002). Almost one-third of the threatened ecological communities listed under the *Environment Protection and Biodiversity Conservation Act (1999)* are woodlands (Department of Environment and Heritage, 2005b). Many of the species in these communities are threatened or endangered (Hobbs and Yates, 2000).

There are many impacts of grazing in temperate woodlands (Lindenmayer *et al.*, 2005a; Maron and Lill, 2005; Martin and Possingham, 2005) and it is well beyond the scope of this book to discuss them in detail. One of the most significant impacts is that vegetation clearing to promote grass growth for grazing removes key plant species and modifies or destroys habitat for the array of organisms that depend on that vegetation. As grazing pressure increases, native plant diversity is reduced and exotic plant species diversity and cover increases (Yates *et al.*, 2000). There tends to be limited natural regeneration of native plants (Spooner *et al.*, 2002). Grazing by domestic livestock can also cause soil compaction (Bezkorowajny *et al.*, 1993), add nutrients to the soil, and damage remnant native trees (e.g. by rubbing; McIntyre *et al.*, 2002). Livestock often concentrate in more productive parts of paddocks and affect areas such as watercourses (and associated riparian vegetation; Jansen and Robertson, 2001).

Most temperate grassy woodlands occur in 700–1100-millimetre rainfall zones, meaning that they are in areas that make them potentially susceptible to secondary salinity (see Chapter 5; Freudenberger, 2001). In addition, practices commonly associated with grazing, such as chemical spraying, ploughing, the removal of woody debris and a range of other management activities, can negatively affect biodiversity (see Lindenmayer *et al.*, 2003a, 2005a).

Many types of temperate woodland have declined dramatically as a direct or indirect consequence of grazing (reviewed in Hobbs and Yates, 2000). Other vegetation types have suffered similar fates; the formerly extensive areas of temperate lowland native grasslands are now substantially altered (Kirkpatrick *et al.*, 1995; Eddy, 2002). Indeed, no intact areas of some grassland vegetation types remain, for example those from parts of north-eastern Victoria (McDougall and Kirkpatrick, 1994). In 2002, the National Land and Water Resources Audit (Commonwealth of Australia, 2002a,b) identified the temperate grasslands on the New South Wales/Australian Capital Territory border as the most threatened ecosystem in Australia.

Grazed temperate woodlands in south-eastern Australia showing limited understorey and natural regeneration. (Photo by David Lindenmayer.)

The resilience of different plant communities is reflected in the differences in the extent of damage in Mitchell grasslands and Mulga communities. High-rainfall years and conservative stocking rates in the grasslands favour Blue Grasses and Three-awned Spear Grasses. Over-grazing and drought lead to an increase in Mitchell Grass, which has superior drought resistance and lower palatability than its competitors. The composition of the grasslands at any point depends on stocking rates, rainfall (Figure 9.16), the time of year in which rain falls, and the composition of the seed bank.

Patterns of floristic change in Mulga shrublands do not display the same kind of resilience. The structure and composition of the shrublands is very different from the vegetation at the time of European settlement. Management practices have included the thinning and clearing of Mulga to promote grass cover, and retention of some Mulga as a drought fodder reserve. Thinning and over-grazing have promoted small woody weeds and relatively unpalatable grasses. The soils of the Mulga are phosphorus-deficient, which restricts seedling growth. Poor seedling growth and low and erratic rainfall mean that Mulga shrublands do not regenerate quickly enough to keep pace with the impacts of over-grazing and drought in Australia (Christie, 1993).

Sustainable grazing regimes in the pastoral zone

In Chapters 1 and 5, we noted that the Australian environment is characterised by large amounts of spatial and temporal variability in annual rainfall (McMahon *et al.*, 1992a; Smith, 1998), which results in large variations in annual net primary production (Figure 9.16). Current pastoral management practices cannot change livestock numbers quickly enough to track changes in productivity. Consequently, in drought years, livestock perish, but not before the natural vegetation on which they depend has been damaged. As noted by Stafford Smith *et al.* (2000), land managers in many pastoral areas are experiencing pressures that mean they will disregard the future viability of their land as a result of immediate productive and financial gain.

Land managers need a method for assessing primary production and appropriate livestock numbers in order to regulate grazing to match land condition (Christie, 1993). Unfortunately, the pastoral industry is still a long way from matching activities to land capability and ecology (see Australian and New Zealand Environment and Conservation Council, 2001). In the absence of a code of practice or reliable monitoring programs, there is no imperative to manage sustainably except for local, short-term self-interest. Notably, the National Land and Water Resources Audit (Commonwealth of Australia, 2001e) recommended the adoption of the Australian Collaborative Rangeland Information System to help monitor, assess and improve pastoral management in Australian rangelands. Despite this innovation, and although many pastoralists manage land in a sustainable way, a very small number of people operating in an irresponsible fashion can do substantial damage to very large areas of the country. As long as economic factors at the scale of the individual farm are the dominant imperatives for environmental management, serious and long-term environmental impacts will accrue in Australian pastoral landscapes.

There have been few studies of land rehabilitation in Australian rangelands (Noble *et al.*, 1997). The prospects for land rehabilitation are limited because rehabilitation is costly and economic returns from grazing enterprises are poor (McIntyre *et al.*, 2002). Thus, the effects of feral animals, weeds, soil erosion, salinity and altered fire regimes will persist, even in the absence of pastoralism.

9.4 Conclusions

Habitat modification and loss, species extinctions and range contractions, depletion of ecological communities, land and water degradation, salinisation, accelerated soil erosion, and nutrient loss have been experienced most severely in the agricultural and grazing sectors of the country (ESDWG, 1991a;

Commonwealth of Australia, 2001d, 2002a). Pre-European settlement patterns of vegetation cover cannot be retrieved, and in many cases may be irrelevant because of changed disturbance regimes, the intrusion of exotic weeds and pests, and changes in vegetation structure and composition.

The agricultural sector lacks many of the key environmental considerations that characterise other resource management sectors, including management practices for multiple land uses and the environmental prescriptions that apply to timber harvesting (see Chapter 8), the process of environmental impact assessment (see Chapter 16) that applies to forestry and mining (see Chapter 8), and the procedures for monitoring that apply to forestry, fishing and mining. Therefore, by any conceivable measure, agriculture in Australia has been far more detrimental in its effects on the natural environment and on biodiversity than the forest harvesting, mining or fishing industries. On this basis, it may be beneficial for the agricultural sector to borrow approaches from other natural resource management industries in an attempt to make a transition to improved sustainability. For example, a process similar to the Regional Forest Agreements could produce interesting outcomes in non-forest lands, as might the certification of agricultural products as being sustainably produced or harvested, as in the case of 'dolphin-friendly' tuna or certified wood products. Conversely, attempts to intensify agricultural production may produce negative environmental and biodiversity conservation outcomes, as occurs in forest ecosystems (Lindenmayer and Franklin, 2002).

Hopper (1997) commented that Australia is unlike other continents in that the most pressing threats to the biota are in temperate lowland woodlands, grasslands, mallee and heath, rather than in tropical rainforests or temperate forests. He notes that this fact is at odds with the perceptions of many people in the community, and contrasts with the attention paid by the media to the conservation of Australian forests relative to the attention paid to the conservation of woodlands, mallee and heath in agricultural landscapes.

9.5 Practical considerations

The most pressing imperative of conservation and land management in Australia, and in many other parts of the world, is to halt land clearing and vegetation degradation, because it is a major contributor to biodiversity loss worldwide. This includes not only land clearing and habitat deterioration for agricultural purposes, but also for other land uses, including mining operations, urbanisation and pastoralism. In the case of pastoralism, there is an urgent need for the development of grazing methods and associated monitoring and rehabilitation protocols that can assist with a more rapid and lasting transition to ecological sustainability in this sector. The sustainability of agricultural systems will be enhanced by the widespread adoption of practices from other resource industries, including voluntary codes of practice, environmental risk assessments, long-term sustainability goals, environmental accreditation schemes, impact assessments and independent auditing.

9.6 Further reading

The State of the Environment (State of the Environment, 2001a–d) and the National Land and Water Resources Audit (Commonwealth of Australia, 2000b, 2001b–d, 2002a,) provide extensive overviews of environmental conditions in Australia.

Landsberg and Wylie (1991) and Reid and Landsberg (2000) review various aspects of eucalypt dieback. The book by McIntyre *et al.* (2002) is an excellent discussion of the relationships between grazing management and woodland conservation. Hobbs and Yates (2000) contains many useful chapters on the demise of woodlands in Australia – primarily as a consequence of clearing, grazing and other impacts.

Hopper (1997) and New (2000) provide an overview of directions for conservation in Australia. Stafford Smith *et al.* (2000) provide a valuable appraisal about pathways to sustainability in the Australian pastoral industry. Volume 29, issue 1 of Austral Ecology (2004) contains a series of valuable papers on biodiversity and monitoring in Australian rangelands.

Chapter 10

Landscapes and habitat fragmentation

This chapter reviews landscapes, the ways in which they can be altered by human activities, the various models that are used to describe landscape characteristics, and biotic responses to landscape change. The key processes associated with landscape change are examined, in particular habitat loss, habitat subdivision and patch isolation, and edge effects. The use of true experiments, 'natural experiments', observational studies and modelling in studies of habitat loss and habitat fragmentation are briefly outlined. The final part of the chapter discusses ways in which the impacts of habitat loss and habitat fragmentation might be mitigated, including limiting and reversing habitat loss, increasing connectivity and buffering sensitive habitats.

10.0 Introduction

In Chapter 9 we paid considerable attention to habitat loss because it is a major factor influencing the loss of biodiversity in Australia and worldwide (Groombridge, 1992; Hobbs and Yates, 2003), and is thought to be contributing significantly to current extinctions (Wilcox and Murphy, 1984; Wilcove *et al.*, 1986; Saunders and Hobbs, 1991; Craig *et al.*, 2000; Fahrig, 1999). This pattern is clear in the USA (Figure 10.1) and is thought to be repeated in many parts of the world (Fahrig, 2003).

In this chapter, we expand our discussion of habitat loss by coupling it with an exploration of habitat fragmentation effects (Bunnell, 1999a; Fahrig, 2003). Habitat fragmentation often takes place within human-modified landscapes, so much of our discussion is embedded within a wider treatment of landscape ecology and its intersection with conservation biology.

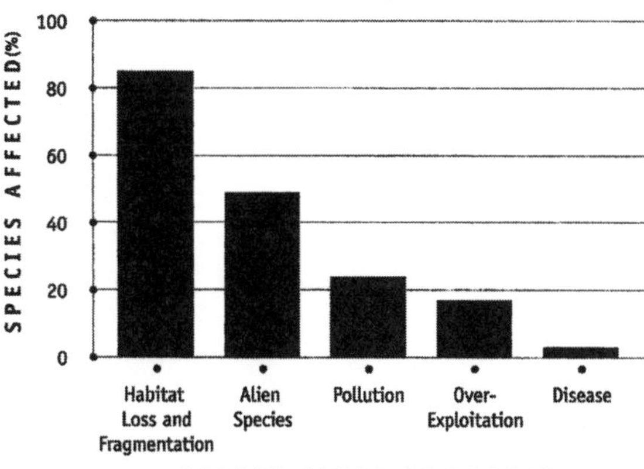

Figure 10.1. Relative impacts of different threatening processes on species in the USA. (Redrawn from Population Action International, 2000, and based on data in Wilcove *et al.*, 1998.) Note that percentages do not add up to 100% because many species are influenced by several threatening processes.

First, we briefly examine some of the ways in which landscapes can be altered. We outline how vegetation patterns change when landscapes are altered and habitat is lost. This leads to an exploration of different conceptual models of landscape cover: the island model, the landscape mosaic or patch-matrix-corridor model, the habitat variegation model, and the landscape contour model. The next major section focuses on ecological processes and species responses to habitat loss and fragmentation, including threshold effects, cascading fragmentation effects, edge effects, and landscape supplementation. This is followed by a discussion of how to predict which species are most likely to be vulnerable to fragmentation. Given the risks that habitat loss and habitat fragmentation pose to biodiversity, we also outline ways to gain a better understanding of the problem, including experiments, observational studies and modelling. Finally, we discuss some strategies to mitigate the effects of habitat fragmentation.

Themes associated with landscape change have become a major focus of conservation biology and landscape ecology (MacGarigal and Cushman, 2002; Fahrig, 2003; Hobbs and Yates, 2003). Indeed, the body of work is becoming so extensive and growing so rapidly that it is almost impossible for any review to do it justice. For example, in a review of all the papers published in 2001 in the journals *Conservation Biology, Biological Conservation and Biodiversity and Conservation*, Fazey *et al.* (2005a) found that 13% focused on habitat fragmentation and habitat loss. This was by far the largest subject area in these journals in that year. Similarly, a search of online journal articles prior to writing this chapter produced more than 2500 published papers where the abstract or key words contained the phrase 'habitat loss' and/or 'habitat fragmentation'. Hence, the treatment given to the subject in this chapter could only ever be somewhat cursory. Therefore, as for other chapters in this book, at the end of the chapter we provide a list of books and articles that give a point of entry into the topics summarised below.

The study of landscapes dates back many decades (von Uexküll, 1926; Troll, 1939). More recently, interest in landscape ecology was sparked by Forman and Godron (1986) and Forman (1995). They emphasised that conservation issues should be considered in a landscape context because:

- The persistence of viable populations of many species depends on networks of habitat patches (Hanski, 1999a,b).
- Many species use resources from spatially separate areas and move through landscapes to persist (Law and Dickman, 1998). For example, frugivorous birds inhabiting the monsoonal rainforests of the Northern Territory make use of patch networks (Price *et al.*, 1999).
- Many phenomena in a given patch (e.g. edge effects, population dynamics and viability, species movement patterns) are related to a combination of patch conditions, conditions in neighbouring patches, and conditions in the surrounding landscape (Forman, 1995).
- The spatial arrangement of patch types affects ecological processes such as dispersal, foraging patterns, the spread of natural disturbances and physical flows (e.g. of water and wind; Bunnell *et al.*, 2003).

Box 10.1

Landscape matters: people, reserves and biodiversity in Africa and the USA

Harcourt *et al.* (2001), Parks and Harcourt (2002), and Harcourt and Parks (2003) examined the effects of landscape context on species persistence within reserves in Africa and the USA. The work showed that small reserves in both Africa and the USA were in areas where the surrounding population density was significantly higher than in areas surrounding large reserves. In Africa, local human population density correlated significantly with human-caused mortality of carnivores (Harcourt *et al.*, 2001). In the USA, extinction rates of large mammals correlated significantly with local human population density, but not with reserve size – a result that contrasts with other studies that have shown strong relationships between reserve size and mammal extinctions in North America (e.g. Newmark, 1987). Parks and Harcourt (2002) concluded that landscape context is extremely important for small reserves, because wildlife populations within them are susceptible to small population size (as a function of small reserve size), and high human population density in the surrounding area, which can affect biodiversity within the reserves.

Natural processes such as volcanic eruptions (Croizat, 1960; Franklin *et al.*, 1985), long-term climatic change (Cunningham and Moritz, 1998) and wildfires (Williams and Gill, 1995; Agee, 1999) can destroy habitat and fragment landscapes. However, landscape modification by humans is by far the most important cause of habitat loss and fragmentation (Saunders *et al.*, 1987; see Chapter 9), and causes damage to biodiversity worldwide.

10.1 Ways that landscapes can be altered

Forman (1995) outlined five ways in which humans can alter landscapes spatially, resulting in changes in

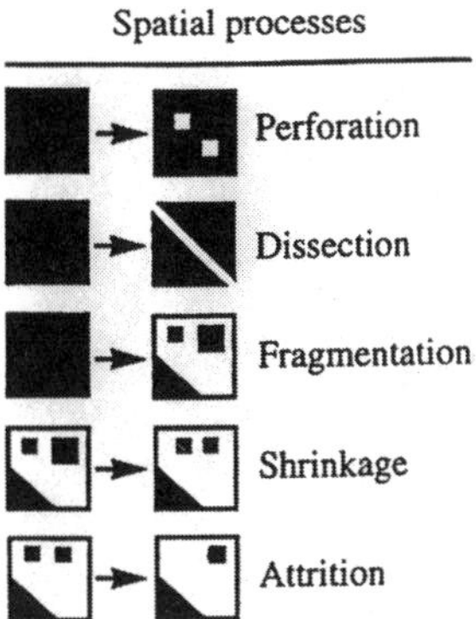

Figure 10.2. Ways in which landscapes can be altered by human activities.

ecological processes and the distributions of plants and animals (Figure 10.2). Some examples of the ways in which landscape change might occur are given in Table 10.1.

Table 10.1. Ways in which Forman's (1995) forms of landscape change might manifest.

Type of landscape change	Example
Perforation	Mine sites in remote areas (e.g. Irian Jaya)
Dissection	Roading in remote areas (e.g. Pan-American Highway)
Fragmentation	Residual areas in grazing lands (poor quality country left)
Shrinkage	Patch size reduction due to paddock realignment
Attrition	Poorest quality patches cleared last in a heavily cleared landscape

Vegetation cover patterns that arise from habitat loss and habitat fragmentation

Loss of habitat often results in discontinuities in what remains – a spatial process that is termed 'fragmentation'

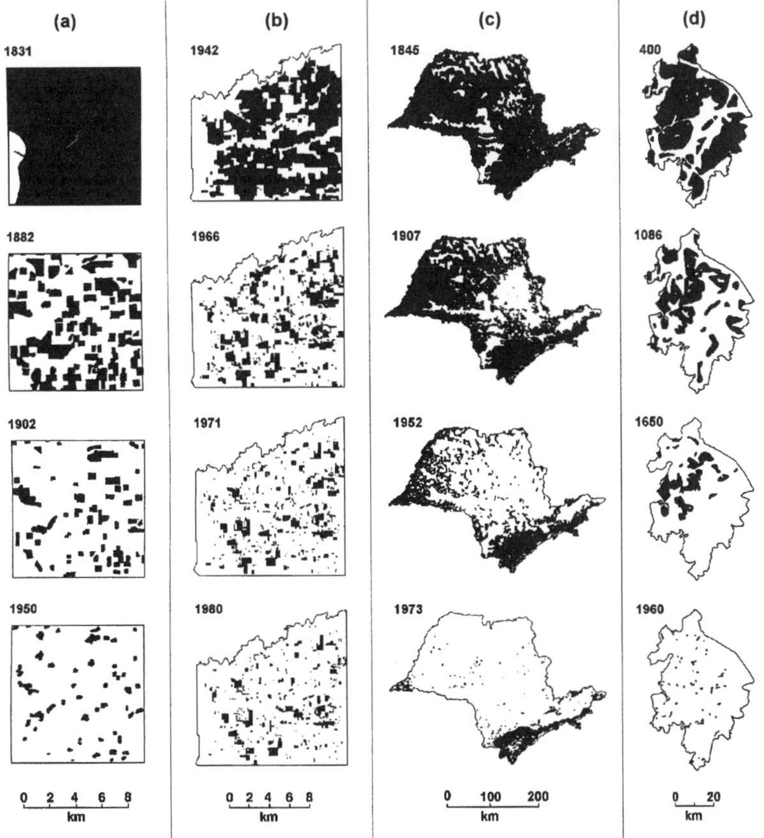

Figure 10.3. Land clearance patterns and habitat fragmentation in: (A) Cadiz township, Green County, Wisconsin, USA, (B) Naringal, Victoria, Australia, (C) Sàn Paulo, Brazil, and (D) Warwickshire, England (after Bennett, 1990b; Shafer, 1990; Primack, 2002; Smith and Hellmund, 1993).

(see Figure 10.2). Fragmentation typically results in a decrease in the average size of habitat patches, an increase in the average distance between them, a decrease in connectivity between patches, and in an increase in the edge:size ratio (Figure 10.2). Such changes can be exacerbated by changes in fluxes of wind, water, radiation and nutrients (Saunders *et al.*, 1991; Laurance *et al.*, 1997).

Landscape fragmentation is not a random process. In many cases, most of the most productive land in an area is extensively modified, leaving remnants of vegetation that tend to be unsuitable for other land uses. Some vegetation and habitat types survive relatively unscathed, whereas others may be eliminated completely. These factors have some important implications. First, the spatial distribution of vegetation types reflects spatial heterogeneity in landscape productivity. Furthermore, remnant vegetation is not a representative sample of the unmodified landscape (e.g. Norton *et al.*, 1995). Usually, different faunal assemblages associate with different vegetation types, so previously abundant species that were characteristic of productive vegetation may be rare or locally extinct.

Remnant vegetation associated with roadsides: (A) South Australia (photo by David Lindenmayer) and (B) Western Australia (photo by Richard Hobbs).

A unique feature of the Australian environment is the *rate* at which habitat loss and fragmentation have occurred. Processes that took several thousand years in Europe and several hundred years in North America are happening in Australia in decades. The changes documented in Figure 10.3 that occurred in Naringal, Victoria, took less than 150 years, and most of the damage was done in a few decades. Land clearance elsewhere in Australia has occurred at about the same rate or even faster (see Chapter 9).

The trends in vegetation cover shown in Figure 10.3 include those studied at Naringal in south-western Victoria by Bennett (1990a,b). The region was originally covered by 'thick forest' and was first settled by Europeans in the late 1830s. The forests were used to graze domestic stock, and extensive clearing began some 30–40 years after first settlement. By 1947, approximately half the native vegetation in the Naringal region had been cleared (Figure 10.3), although the rate of forest removal accelerated substantially during the following decades. By 1966 a total of 19% of the original vegetation cover remained, and in 1980 only 8.5% of the original vegetation cover remained. Not surprisingly, there was a coincident marked change in the size of remnant blocks of vegetation. In 1947, about 90% of remnant vegetation occurred as patches exceeding 100 hectares in size. In 1980, 92% of remnants measured less than 20 hectares and none was larger than 100 hectares (Bennett, 1990b; Figure 10.3). The area now supports numerous small dairy farms, and the small patches of native vegetation are either surrounded by pasture or occur as remnant riparian strips or roadside reserves. The changes in vegetation cover at Naringal have had profound impacts on the vertebrates of the region, which we describe in more detail in the section on observational studies at the end of this chapter.

Dynamism in the patterns of vegetation cover

Many perspectives on landscape alteration are unidirectional, with alteration continuing to reduce both the size of individual patches ('shrinkage' in Figure 10.2) and the overall number of patches ('attrition' in Figure 10.2). There are many cases where trends in landscape change are reversed. For example, in some wood production forests, cover patterns change rapidly

Forest cover in western Massachusetts on areas formerly cleared for agricultural production. (Photos by David Lindenmayer.)

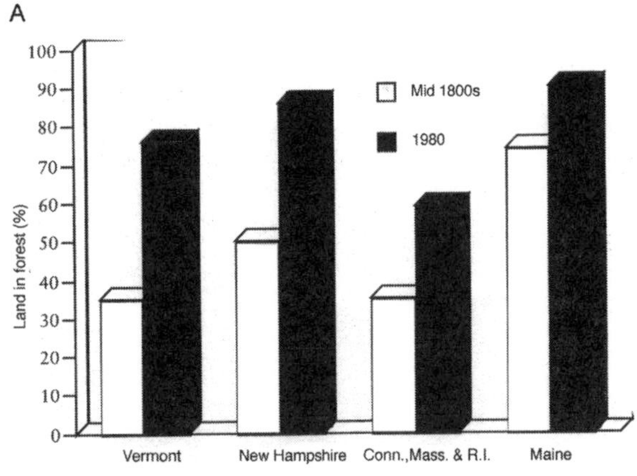

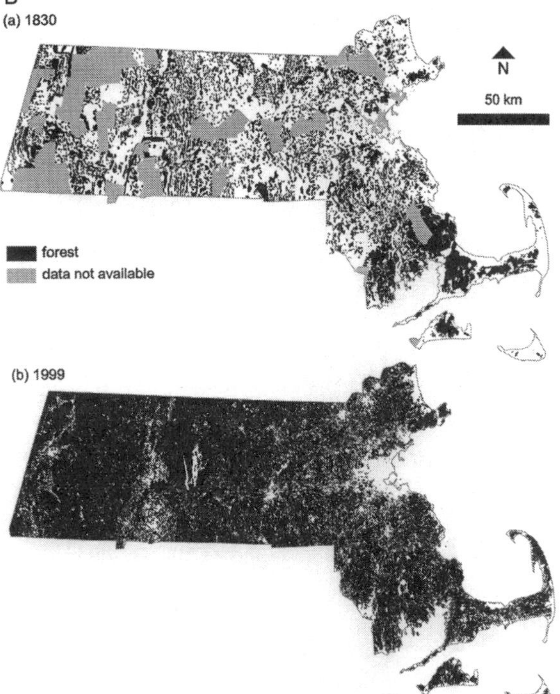

Figure 10.4. (A) Trends in forest cover across several States in north-eastern USA in the mid-1800s and in 1980. (Redrawn from MacCleery, 1996.) (B) Forest cover patterns in 1830 and 1999 in the State of Massachusetts, north-eastern USA. (Provided courtesy of D. Foster.)

because of regeneration in logged areas surrounding uncut stands (Fahrig, 1992). In the forests of north-eastern USA, extensive forest clearing for agricultural development accompanied European settlement. In Massachusetts, less than 30% of the original forest cover existed by the mid-1800s, and the amount of old growth forest was less than 500 hectares, occurring primarily in 1.5-hectare or smaller patches (D. Foster, personal communication). However, following the industrial revolution, many workers left agricultural employment and began working in factories and service industries. As a result, abandoned agricultural fields regenerated and forest cover increased rapidly (Figure 10.4). Extensive areas of Massachusetts are now covered in forest, although pressure on the forest has returned, this time from urban expansion.

There are many other examples of native vegetation regeneration following clearing. For example, some woodlands in southern New South Wales have been cleared at least three times in the last 150 years and have partially recovered each time. Trends in landscape cover and associated biodiversity patterns can be reversed. Mather (1990) summarised the notion in a model of forest change and associated human perceptions based on historical trends in Great Britain (Table 10.2).

Table 10.2. Simple sequential model of forest change (modified from Mather, 1990).

Stage	Description	Trend in forest area	Public perception of trend
1	Unlimited resource	Contraction	Positive or neutral
2	Depleting resource	Contraction	Negative
↓			
3	Expanding resource	Restoration/ expansion	Neutral/negative stability (?)
4	Equilibrium (?)		Positive/neutral/ negative (?)

10.2 Models of landscape cover

The example from north-eastern USA in the previous section assumed that clearing creates two conditions: remaining uncleared forest (habitat for biota) and cleared land (non-habitat). This perspective is, in essence, a simple model that relates intact and cleared vegetation to biotic responses. Models characterise landscape patterns and provide a basis to quantify the response of biodiversity to landscape change. There are many alternative models, and four are treated in some detail in the following section. However, this area of landscape ecology and conservation biology is controversial (Haila, 2002; Manning *et al.*, 2004). Ultimately, the choice of the best model depends on the objective of the study at hand, the characteristics of the landscape, and the habitat requirements and ecology of the species or group targeted for study.

Island model

The underlying premise of the **island model** is that fragments of original vegetation surrounded by cleared or highly modified land are analogous to oceanic islands in an 'inhospitable sea' of unsuitable habitat

A Black Noddy nesting on Heron Island, off the coast of central Queensland. The presence of birds on the island is strongly associated with access to marine-derived food resources in the surrounding Coral Sea, which highlights the interaction between island biotas and the surrounding marine landscape. (Photo by David Lindenmayer.)

Table 10.3. Some limitations of the island model for studies of biodiversity response to landscape cover (based on Manning, 2004).

Issue	Description
Edge effects	Edge effects can either increase or decrease species diversity or the likelihood of persistence within habitat fragments. They can also mean that the physical size of a patch and the functional size of a patch are different.
Surrounding landscapes can have habitat value	In some landscapes, particular species may not be confined to habitat fragments, but rather can find suitable habitat in the surrounding cleared or semi-cleared area.
Surrounding landscapes can be a complete barrier to movement	The landscape surrounding a fragment can be so hostile that no movement occurs between fragments. This can result in extinctions within patches, or population sizes and density can be elevated well above 'typical' carrying capacities (e.g. `fence effects'; Wolff *et al.*, 1997)
Uniformity of pre-fragmentation conditions	The island model assumes that landscape conditions prior to fragmentation were uniform. This is a gross simplification and ignores the reality of natural landscape heterogeneity and associated variations in conditions that can markedly influence species distribution patterns (e.g. Austin, 1999).
Differences in human versus organism perspectives of landscapes	Human perspectives of a landscape (e.g. island vs non-island) will often be markedly different from those of the organism/s of interest. For example, functional patch isolation for a species may not be congruent with human - calculated measures such as inter-patch distance.
Historical factors influence species response	Species may be lost from a landscape not because of present levels of fragmentation, but because of past historical factors that are no longer visible.

Box 10.2

The value of islands for conservation

Islands support a wide range of endemic fauna and flora and are valuable reservoirs of biodiversity. For example, Bruny Island supports populations of all 11 species of Tasmanian endemic birds as well as important populations of threatened invertebrates and plants (Cochran, 2003). Maxwell *et al.* (1996) noted that the 140 islands that occur around the Australian continent support 67 species of marsupials. However, the evolutionary reasons for endemicity (usually the isolation of populations) mean that islands can be valuable in other ways for conservation. For example, islands can sometimes be quarantined from processes that threaten plants and animals on continents. In 1996, seven species of marsupials that had become extinct on the Australian mainland persisted on islands. Thus islands can be valuable places for reintroduction and translocation programs (e.g. Short *et al.*, 1992; see Chapter 4).

(reviewed by Haila, 2002). There are three broad assumptions made under the island model. They are: (1) that islands (and habitat patches) can be defined in a meaningful way for all species of concern, (2) that clear patch boundaries can be defined that distinguish the islands from the surrounding landscape, and (3) that environmental, habitat and other conditions are relatively homogenous within an island or patch.

Much empirical work has shown that large areas support more species than smaller ones and that the number of species can be predicted from **species–area curves** (e.g. Arrhenius, 1921; Preston, 1962; Rosenzweig, 1995). The **theory of island biogeography** (Macarthur and Wilson, 1963, 1967) was developed to explain species–area phenomena for island biotas. Part of this theory considers the aggregate species richness on islands of varying size and isolation from a mainland source of colonists (Shafer, 1990). The balance between extinction and colonisation produces an equilibrium number of species, which is a function of island size and isolation (Macarthur and Wilson, 1967).

The literature on the theory of island biogeography is immense and it is well beyond the scope of this chapter to review it (e.g. see Simberloff, 1988; Shafer, 1990). Much of it deals with the adaptation of the island biogeography theory to reserve design – a topic we explore in detail in Chapter 16. As a model of landscape dynamics, the theory often fails because vegetation remnants and the landscapes surrounding them interact (see Table 10.3), and can significantly influence patterns of species occurrence, abundance and diversity. This highlights the limitations of the oceanic island analogy.

Even true oceanic islands do not exist in isolation from the surrounding marine environment. Oceans are a source of dispersalists, provide food (e.g. for seabirds), and generate edge effects (e.g. windblown salt driven onshore).

Nested subset theory

The **nested subset theory** (Patterson and Atmar, 1986) extends the species–area relationship by tracking both the numbers of species and their identities on islands. The premise is that species-poor small islands should support assemblages that are subsets of larger, species-rich islands (Patterson and Atmar, 1986; Patterson, 1987; Cutler, 1991; Figure 10.5). The concept has been extended to human-modified landscapes, assuming that habitat patches are unaffected by conditions in the surrounding landscape (Doak and Mills, 1994).

Computer programs are available to examine patterns of nestedness (Fischer and Lindenmayer, 2002a). Significant nestedness has been identified in almost all studies where it has been examined (Wright *et al.*, 1998). Simberloff and Martin (1991) suggested that nestedness may be a widespread phenomenon because habitats are nested. Nestedness may also be a consequence of species-specific differences in immigration and extinction in modified landscapes.

The nested subset theory predicts that all the species in a system should occur in the largest patch. The patch

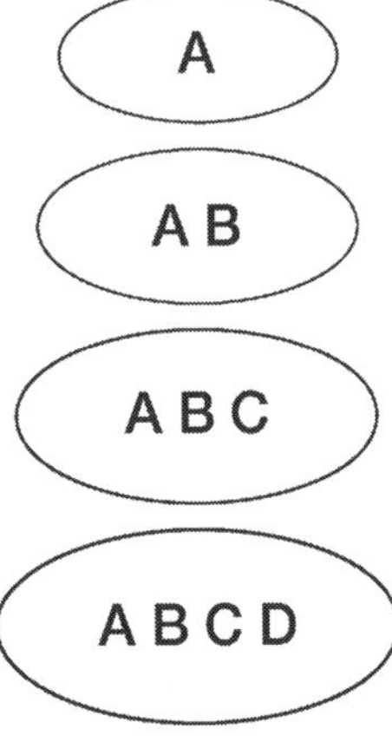

Figure 10.5. Simplified patterns of nestedness in four patches of increasing size. In each case a smaller patch has a subset of the species (as represented by letters) inhabiting the next largest patch.

containing the most area-sensitive species will have all of the less area-sensitive taxa nested within it (Patterson, 1987). The conservation implications are clear – the largest patch should be targeted for reservation. However, nestedness is usually not a perfect model, and the largest patch often does not capture the full array of taxa in landscape (J. Fischer, unpublished data). Part of the problem is that, like the island biogeography theory, the nested subset theory assumes that habitat fragments are 'islands' located in a sea of 'unsuitable lands' (Doak and Mills, 1994). These constructs are too pessimistic in their predictions of species loss and they fail to recognise the potential value of conservation strategies in the matrix surrounding habitat fragments. This led Wiens (1994) to conclude that:

> *Fragments of habitat are often viewed as islands and are managed as such; however, habitat fragmentation includes a wide range of spatial patterns…Fragments exist in a complex landscape mosaic and dynamics within a fragment are affected by external factors that vary as the mosaic structure changes. The simple analogy of fragments to islands, therefore, is unsatisfactory.*

Finally, the nested subset theory and the island biogeography theory focus largely on collective measures (assemblage composition and species diversity, respectively), ignoring the need for species-specific information to support sound conservation decisions (Doak and Mills, 1994).

Patch-matrix-corridor model

Forman (1995) developed a patch-corridor-matrix model in which landscapes are conceived as mosaics of three components: patches, corridors and the matrix (Figure 10.6), defined as follows:

- **Patches** are relatively homogeneous non-linear areas that differ from their surroundings.
- **Corridors** are strips of a particular patch type that differ from the adjacent land on both sides.
- The **matrix** is the dominant and most extensive patch type in a landscape (Forman, 1995). It is characterised by extensive cover, high connectivity, and/or a major control over dynamics.

In the patch-corridor-matrix, the matrix is intersected by corridors or perforated by smaller patches, and patches and corridors are readily distinguished from the background matrix (Forman and Godron, 1986; Kotliar and Wiens, 1990, Figure 10.6).

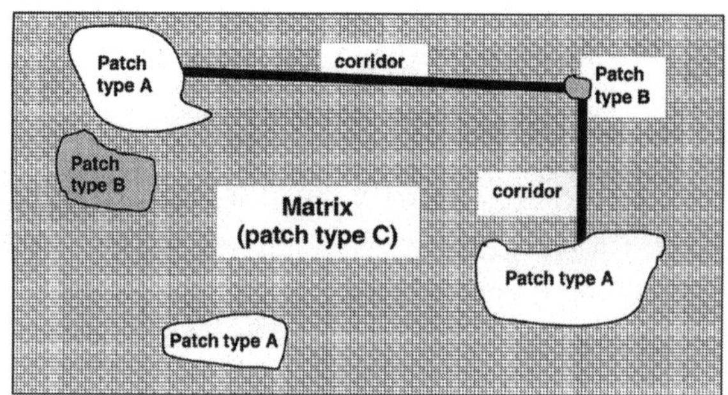

Figure 10.6. A landscape perspective based on the patch-matrix-corridor model (*sensu* Forman, 1995). The model has a number of patch types, including the most extensive type, the background matrix (patch type C).

Forman (1995) noted that every point in a landscape is either within a patch, a corridor, or the background matrix, and that the matrix can be extensive to limited, continuous to perforated, and variegated to nearly homogenous.

The patch-matrix-corridor model has been widely adopted in landscape conservation biology. It helps land managers and researchers to translate their ideas into a spatial context, and it builds on the island theory, which simplifies landscapes into areas of habitat and non-habitat. The model assumes that a simple classification of the landscape can work for all species. However, many organisms do not perceive landscapes in the same way as humans (Bunnell, 1999a,b). Patterns of landscape cover seen from a human perspective may not provide a useful framework for interpreting biotic responses to landscape conditions (Ingham and Samways, 1996; Bunnell, 1999a; Manning *et al.*, 2004). The patch-matrix-corridor model does not generally deal with spatial continua (apart from edge effects; see Laurance *et al.*, 1997). Furthermore, planning at a single scale with a single landscape concept may not provide strategies to conserve all species and communities.

Habitat-variegation or landscape continuum model

Harrison (1991) objected to the sharp boundaries and discrete classes prevalent in most landscape concepts. McIntyre and Hobbs (1999) used ideas about ecological processes to develop the **habitat-variegation** or **landscape continuum model.** In many landscapes, the boundaries between patch types are diffuse, and differentiating them from the background matrix may not be straightforward. The landscape continuum model was developed, in part, to incorporate gradual spatial changes or gradients in habitat quality (McIntyre, 1994; McIntyre and Hobbs, 1999; Figure 10.7).

The landscape continuum model was proposed originally for semi-cleared grazing and cropping landscapes in rural eastern Australia. These landscapes are characterised by small patches of woodland and relatively isolated native trees scattered throughout grazing lands (McIntyre and Barrett, 1992; McIntyre *et al.*, 1996). In such landscapes, from a human perspective, 'patches' and 'corridors' are difficult to identify among the loosely organised and spatially dispersed trees and ecological communities (see Fischer and Lindenmayer, 2002b). For instance, numerous trees scattered across a landscape collectively provide habitat for some species (e.g. for some woodland birds; Barrett *et al.*, 1994). The landscape continuum model takes account of small habitat elements that might otherwise be classified as 'unsuitable habitat' in the background matrix (Tickle *et al.*, 1998).

McIntyre and Hobbs (1999) extended their model to include tropical and temperate forests, and recognised four broad cover classes: intact landscapes and relictual landscapes are at opposite ends of the continuum, and variegated landscapes and fragmented landscapes are intermediate (Figure 10.7). The concept of variegated landscapes was a new idea in landscape ecology and conservation biology, and McIntyre and Hobbs (1999) outlined what they meant by the idea:

> *In variegated landscapes, habitat* [not necessarily an unsuitable environment] *still forms the matrix, whereas in fragmented landscapes, the matrix comprises 'destroyed habitat' (McIntyre and Hobbs, 1999).*

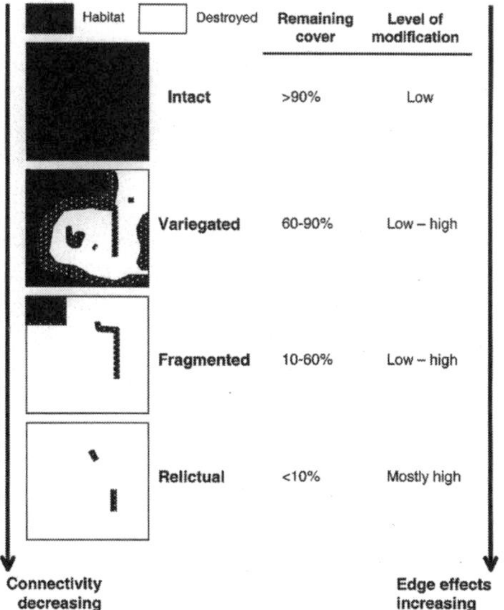

Figure 10.7. States of landscape condition in the landscape continuum model (from McIntyre and Hobbs, 1999).

Variation in vegetation cover across landscapes. (A) A uniform matrix; (B) a patchy landscape; (C) a variegated landscape (*sensu* McIntyre and Barrett, 1992) of scattered paddock trees in south-eastern Australia. (Photos for A and B by J. Franklin. Photo for C by David Lindenmayer.)

McIntyre and Hobbs (1999; see also Wiens, 1994; Pearson *et al.*, 1996) believed that landscapes were more complex than could be described by the patch-matrix-corridor model, because it assumes that the matrix is non-habitat. Also, human disturbances give rise to a range of landscape conditions. McIntyre and Hobbs (1999) equated the patch-matrix-corridor model to their 'fragmentation' stage of landscape condition (see Figure 10.7). McIntyre and Hobbs (1999) believed that by recognising different landscape conditions, conservation strategies could be better focused. In their view, for numerous elements of the biota, many forest landscapes will conform to an intact or variegated condition if the matrix is managed appropriately. If appropriate management practices are not embraced in the matrix, the system may degenerate from a variegated to a fragmented or even relictual one (*sensu* McIntyre and Hobbs, 1999; McIntyre *et al.*, 1996). Appropriate matrix management can 're-variegate' fragmented landscapes, restoring connectivity and halting degrading processes (Hobbs *et al.*, 1993; McIntyre and Hobbs, 1999).

Some of the assertions about the advantages of the landscape continuum model ignore the details of Forman's (1995) original patch-matrix-corridor model. The interpretation of the patch-matrix-corridor model by McIntyre and Hobbs (1999) is a landscape comprising patches and corridors of suitable habitat embedded within a hostile (non-habitat) matrix. However, Forman (1995) stated that the matrix was often the largest and most dominant landscape component 'characterised by extensive cover, high connectivity, and/or a major control over dynamics' (see definitions given in the section on the patch-matrix-corridor model). Therefore, the patch-matrix-corridor model does not predict the response of a given species to particular landscape components (Forman, 1995).

Congruence between the patch-matrix-corridor and continuum landscape models

The patch-matrix-corridor model and landscape continuum model are congruent on several levels. Both recognise a gradient in the complexity of landscape conditions and a continuum of landscape 'naturalness'. The matrix is usually the dominant landscape component.

However, the two models are different in several ways. The patch-matrix-corridor model focuses on the *pattern* or *form* of different landscape units (e.g. the size and shape of different patches and patch types), whereas the emphasis of the landscape continuum model is on the specific *function* of a given landscape across a gradient of vegetation cover, requiring that the organism or assemblage of interest and the processes influencing species' responses are specified.

Landscape modification can lead to both altered patterns of vegetation cover (e.g. habitat subdivision) and altered ecological processes (e.g. changed population dynamics; Wiens, 1994; Cale, 1999). Looking at the attributes of both models may encourage landscape ecologists, conservation biologists and land managers to think about the relationships between form (or structure) and function.

Limitations in the application of the landscape models

Irrespective of the original intentions of the architects of the patch-matrix-corridor and continuum models, both are often simplified by biologists who represent landscapes as patches contrasting markedly with remaining areas of non-habitat. Mapping tools such as geographic information systems (GIS) are used to define patches, under the assumption that species perceive patches in the same way and at the same scale as humans (Bunnell, 1999a), instead of in a way based on the habitat requirements, movement patterns, and other attributes of the particular species itself.

Habitat quality for a particular species is often patchy or a gradient of conditions that are suitable to varying degrees (Whittaker *et al.*, 1973; Block and Brennan, 1993; Lindenmayer and Cunningham, 1996; Austin, 1999), or comprises elements that are used at different stages of an organism's life history. This is not unlike the source–sink concept of Pulliam *et al.* (1992) and others (e.g. Howe *et al.*, 1991). Habitat within patches is heterogeneous (Fox and Fox, 2000; Knight and Fox, 2000), which influences population dynamics (Harrison and Taylor, 1997).

Different species use areas between patches to different extents (Andrén *et al.*, 1997). Species respond uniquely to the same landscape (Kotliar and Wiens, 1990; Villard *et al.*, 1999), depending on the scales at which they use the environment (von Uexküll, 1926; Wiens, 1994; Andrén, 1996; Manning et al., 2004), where suitable habitat occurs, and how animals move within and disperse between areas (Tickle *et al.*, 1998).

Some workers have attempted to overcome the problem of species-specific responses to landscape conditions by classifying species according to their use of modified landscape. Such classes of species include 'forest-interior species' (Tang and Gustafson, 1997; Villard, 1998), 'edge species' (Howe, 1984; Bender *et al.*, 1998; Euskirchen *et al.*, 2001) and 'generalist species' (Andrén, 1994; Williams and Hero, 2001). Others classify patches according to the extent to which species use them or by their propensity to provide dispersers (the source–sink concept; Pulliam *et al.*, 1992). Some concepts discriminate between intact and relictual habitat (McIntyre and Hobbs, 1999; Figure 10.7), or the core and the remainder of the bioclimatic domain (e.g. Nix, 1986; see Chapter 15), or describe the matrix in more detail (Lindenmayer and Franklin, 2002). All of these approaches to classify species or landscapes describe extreme points on continua, which are segmented to reduce complexity.

Landscape contour approach

Although refinements in labelling species or matrix types will sometimes be sufficient to explain ecological patterns in modified landscapes, it may be useful to consider a more flexible landscape model. Fischer *et al.* (2004) developed the **landscape contour model** to incorporate multiple species and their unique habitat requirements, and to characterise gradual habitat change across multiple spatial scales. The conceptual foundation comes from Wiens (1995), who suggested viewing landscapes as 'cost–benefit contours', and Lindenmayer *et al.* (1995b), who pointed out how spatially explicit habitat models are similar to contour maps (see Fischer *et al.*, 1995b).

Topographic maps provide a familiar graphical representation of complex spatial information. Similarly, the landscape contour model represents a landscape as a map of habitat suitability contours overlain for different species (Figure 10.8). The properties of the contour-based landscape model can be described as follows (after Fischer *et al.*, 2004):

- Habitat is a species-specific concept (see Chapter 15; Figure 10.8A).
- The spatial grain at which species respond to their environment varies between species (see Kotliar and Wiens, 1990). This is translated onto a contour map through different contour spacing. Species may have densely spaced contours with many peaks and troughs (fine spatial grain) or widely spaced contours with few peaks and troughs (coarse spatial grain; Figure 10.8B).
- Different resolutions can be used to represent responses at different spatial scales. For example, at the continental scale, the interval between contours can be coarse.
- A contour model can be simplified to correspond to either the patch-matrix-corridor model or the variegation model. The corridor-patch-matrix model translates into a contour model if contours are spaced widely within a given patch (continuously high suitability), if contours undergo a rapid transition from high suitability to low suitability at the edge of a patch, and if contours are widely spaced and indicate low suitability within the landscape matrix (Figure 10.8C).
- A habitat contour map for a single species emerges from many ecological processes operating at many spatial scales. The map may not reflect the processes causing the pattern of distribution or abundance.
- A contour map does not have a temporal component. Multiple 'snapshots' of contour maps at different times could reflect temporal dynamics such as changing habitat requirements at different stages of a life cycle or changes in suitability

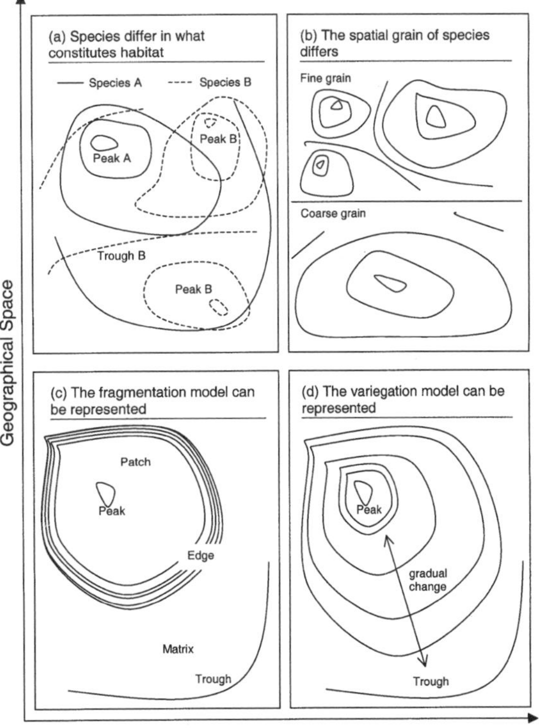

Figure 10.8. Graphical presentation of a contour-based conceptual landscape model. Key model features are that it: (A) allows for species to differ in what constitutes suitable habitat, (B) recognises differences in the spatial grain of species, (C) contains the fragmentation model, and (D) contains the variegation model. (Redrawn from Fischer *et al.*, 2004.)

with ecological succession (e.g. Palomares *et al.*, 2000; Blab and Vogel, 1996).

Fischer *et al.* (2004) combined data on the habitat requirements of the Greater Glider in the montane Ash forests of the Central Highlands of Victoria (including tree age and hollow tree abundance) with mapped information on these variables. The resulting 'contour map' of the species' predicted distribution appears in Box 15.2 in Chapter 15. The approach depends on being able to quantify the habitat requirements of the species and to map the habitat features.

The model we use to represent landscapes affects the way we perceive and study landscapes, and subsequently interpret the effects of changes (such as habitat loss and fragmentation). Landscape models are, by definition, a simplified abstraction. As with other kinds of models, it is important to know their assumptions and limitations.

Box 10.3

Bunnell's panchreston

Bunnell (1999a) argued that habitat fragmentation can sometimes be a panchreston, which he defined as:

a proposed explanation intended to address a complex problem by trying to account for all possible contingencies, but typically proving to be too broadly conceived and therefore oversimplified to be of any practical use.

Bunnell noted that fragmentation means 'to break into pieces'. But the use of the term in conservation biology is often ambiguous, vague, underspecified, and context-dependent (Regan *et al.*, 2002), limiting its practical use. An array of different effects have been collected together under the term `fragmentation', thereby making a panchreston and obscuring different processes. Bunnell argued that a solution to the problems created by the fragmentation panchreston was to acknowledge that multiple processes take place and to work in ways to separate them so that consequences can be better assigned to particular processes. This would, in turn, help guide management. For example, although habitat loss and subdivision often go together, the distinction between them is important because dealing with habitat loss will often require quite different conservation strategies than those that limit the effects of habitat subdivision (Lindenmayer and Franklin, 2002).

10.3 Ecological processes and species responses to habitat loss and fragmentation

Context for habitat loss and habitat fragmentation

A key issue in studies of landscape change is that habitat fragmentation has been regarded by some workers as a catch-all problem; that is, it has been considered to be *the* reason for how and where species and populations come to be where they are in a landscape, and *the* reason for the loss or decline of species. However, we believe that habitat fragmentation needs to be put in context as one of several factors that influence patterns of species richness and the distribution and abundance of particular species in a landscape. In some cases, factors unrelated to fragmentation will be more important (e.g. Jellinek *et al.*, 2004). For example, the ecological literature is replete with examples of relationships between high latitude or high elevation and low levels of species richness in largely natural landscapes (reviewed by Gaston and Spicer, 2004). In intact forests in parts of Central America, many species of vertebrates have been lost because of intensive hunting pressure from humans (Redford, 1992). Similarly, in some largely unaltered Australian landscapes, species loss and population decline has been substantial – as a result of the impacts of introduced pests such as the Cane Toad and feral predators such as the Red Fox (see Chapter 7). The decline of forest birds on the island of Guam is another classic example where habitat loss and habitat fragmentation have been largely irrelevant to species loss (see Box 10.12).

Problems with the term `habitat fragmentation'

The term 'habitat fragmentation' is vague and ambiguous (Haila, 2002; Regan *et al.*, 2002). Frequently, the term encompasses the myriad processes and changes that accompany landscape alteration (Bunnell, 1999a ; see Box 10.3). It may be useful to decompose the concept into its subcomponents and focus on a subset of the most important factors for the problem at hand (whether it be the conservation of a particular species or assemblage, or the control of a threatening process; see Zuidema *et al.*, 1996; Hobbs and Yates, 2003; Figure 10.9).

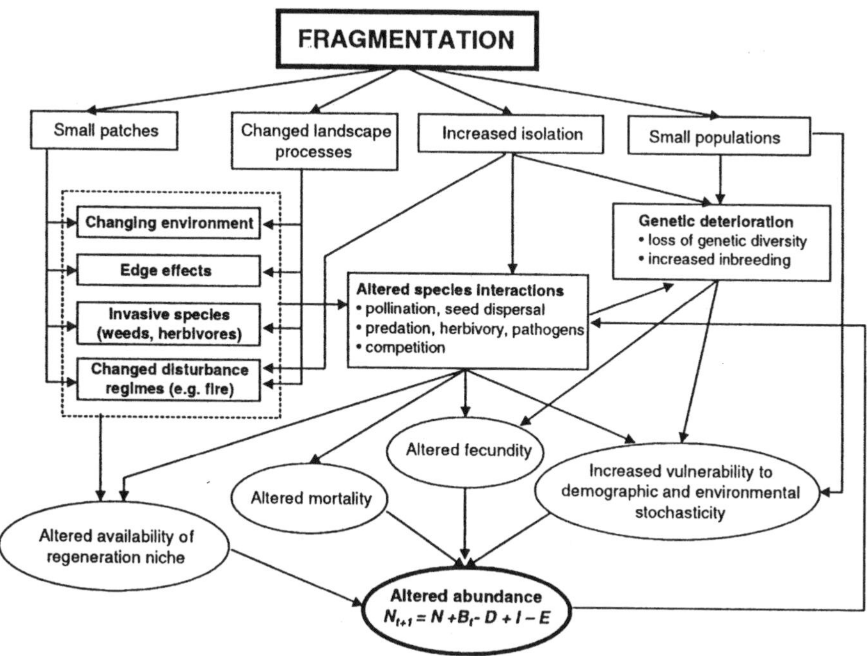

Figure 10.9. A conceptual habitat fragmentation model for plants. N is population size, B is number of births, D is number of deaths, I is immigrants and E is emigrants. (Redrawn from Hobbs and Yates, 2003.)

Five processes associated with landscape change

Landscape change can alter the dynamics of populations and ecological processes through five interrelated processes. These are:

- *Habitat loss.* Remnant habitat patches are considered 'samples' of the original (pre-fragmentation) habitat (Connor and McCoy, 1979; Zuidema *et al.*, 1996). The total amount of habitat determines the fate of many species (Delin and Andrén, 1999; Fahrig, 1999; Trzcinski *et al.*, 1999; Villard *et al.*, 1999). For some generalist species that can use disturbed environments, habitat may increase as the area between fragments expands (Saunders and Ingram, 1995; Lindenmayer *et al.*, 2002c).
- *Habitat degradation.* Habitat suitability is reduced and may eventually lead to an area becoming uninhabitable for a particular species (Hobbs, 2001). Habitat degradation is a gradual process leading ultimately to habitat loss (New, 2000).
- *Habitat subdivision.* Habitat loss can lead to increased subdivision of remaining habitat, resulting in more smaller patches. Fahrig (2003) considered fragmentation to be a separate process from habitat loss (see also Fahrig, 1997; Bender *et al.*, 1998). Habitat subdivision may result in non-viable populations of some species in small remaining fragments (Shaffer, 1981; Shaffer and Samson, 1985; Armbruster and Lande, 1993). It may also have a range of other ecological effects; for example altering the structure of food webs (Holyoak, 2000) and disrupting ecological processes such as the decomposition of wastes (Klein, 1989).
- *Patch isolation.* Habitat loss and habitat fragmentation lead to larger distances between patches and greater patch isolation (Fritz, 1979; Smith, 1980; Hanski, 1994a), thus reducing inter-fragment movement (Powell and Powell, 1987; Sarre *et al.*, 1995; Desrochers and Hannon, 1997). A modelling study by Franklin and Forman (1987) found that patches of old growth forest can become fragmented even when approximately 70% of the landscape cover remains. Below a particular level of habitat loss, distances between habitat patches can increase exponentially (Gardner *et al.*, 1987; Gustafson and Parker, 1992).
- *Edge effects.* The ratio of patch perimeter to patch interior area is higher in fragmented landscapes.

This can result in edge effects within remaining patches, for example weed invasion (Brothers and Spingarn, 1992; Burdon and Chilvers, 1994; Lindenmayer and McCarthy, 2001), altered micro-climatic conditions (e.g. modification of wind speeds, light fluxes and temperature regimes; Saunders *et al.*, 1991; Chen *et al.*, 1991; Matlack and Litviatus, 1999), increased nest predation and brood parasitism (Paton, 1994; Robinson *et al.*, 1995), and lowered rates of fledging success among birds with territories located at habitat edges (Hutha *et al.*, 1998).

Because each of these processes is a major research topic in conservation biology, they are discussed in some detail in the following sections.

Habitat loss

Effects of habitat loss

Habitat loss leads to (after Fahrig, 2003; see also Chapter 9):

- increased species loss (Temple, 1986; Zimmerman and Bierregaard, 1986; Klein, 1989; Thomas and Morris, 1995; Gurd *et al.*, 2001; Schmiegelow and Monkonnen, 2002)
- reduced abundance and distribution of species (Hinsley *et al.*, 1995; Best *et al.*, 2001)
- reduced breeding success (Robinson *et al.*, 1995; Kurki *et al.*, 2000)
- reduced dispersal success (Pither and Taylor, 1998)
- reduced genetic diversity (Gibbs, 2001)
- altered species interactions (Taylor and Merriam, 1995).

Habitat loss is species-specific (Morrison *et al.*, 1992; see Section 10.3; Chapter 15). However, habitat loss is often equated simply with loss of vegetation cover (e.g. Fahrig, 2003). This may be misleading because, as outlined in Section 10.2, landscape modification is a not a random process; the native vegetation remaining in most areas is of low productivity and has little value for other uses such as agriculture or forestry (Braithwaite *et al.*, 1993; Scott *et al.*, 2001a,b).

Simple maps of vegetation versus cleared land, or habitat versus non-habitat, ignore gradients and the variation in distribution and abundance patterns. Such variation is essential for understanding why species respond to landscape change in the way they do (Brooker and Brooker, 2001, 2002; see Box 10.12).

Habitat loss versus habitat subdivision

Fahrig (2003) noted that the impacts of habitat subdivision are often less important than those of habitat loss. Notably, this is despite the fact that few studies in conservation biology focus specifically on habitat loss and many examine habitat subdivision. Fazey *et al.* (2005a) found that of the 13% of published articles in the journals *Conservation Biology, Biological Conservation and Biodiversity and Conservation* that

Box 10.4

Why distinguishing threatening processes is important in studies of habitat fragmentation: the case of the Brown Treecreeper

Identifying which of the processes associated with landscape change, habitat loss and habitat fragmentation is the one that is threatening a given species of conservation concern is critical for the development of efficient and effective conservation strategies. This is illustrated by the case of the Brown Treecreeper – a declining bird in many areas of woodland in parts of south-eastern Australia (Cooper and Walters, 2002). The species is absent from many fragments of apparently suitable woodland and in Mulligan's Flat Nature Reserve near Canberra, populations have declined in recent years to a single individual in mid-2005 (J. Bounds, personal communication). Developing strategies to conserve the species will depend on identifying the underlying reasons for decline and low levels of patch occupancy.

Work on the Brown Treecreeper indicates that a lack of connectivity and hence disrupted dispersal is the most likely reason for the decline of the species in landscapes that are subject to major human modification (Walters *et al.*, 1999). If a habitat patch is isolated and the landscape between it and another patch does not support suitable habitat, then Brown Treecreepers may be absent, even if that patch contains all the essential habitat components that it needs (fallen logs, tree hollows, etc.; Cooper *et al.*, 2002). Therefore, if a small population in a patch is lost (e.g. through predation by a Red Fox), sources of dispersing animals may be too remote to reverse the local extinction. Conservation of the species will depend on finding ways to promote connectivity, and these may include revegetation programs to link remnants and promote natural dispersal, or 'artificial dispersal' through translocation programs of birds to empty patches or patches with a few remaining birds.

examined landscape change impacts, only 2% focused on habitat loss.

Attempts to quantify the effects of subdivision are often not made independently of habitat loss. In studies where habitat subdivision was examined, holding constant the effects of habitat quantity, the effects on biota were equivocal (Fahrig, 2003). In some cases, positive effects were due to factors such as increased opportunities for predators and their prey and competitors to coexist in more fragmented landscapes (Fahrig, 2003). Negative effects were due to edge effects and the non-viability of populations occupying small fragments. These topics are revisited later in this chapter.

In most landscapes, habitat loss and subdivision occur simultaneously, and it is difficult to separate their effects (Andrén, 1997; Harrison and Bruna, 2000). Hence, the two processes are nearly always confounded. Nevertheless, determining the relative importance of habitat loss versus habitat subdivision is important. If habitat loss is the primary factor influencing the persistence of species, then the focus of landscape management could be to maximise the area of suitable habitat (Lindenmayer and Franklin, 2002). Conversely, if habitat subdivision is the key problem, then enhancing connectivity (e.g. through the provision of wildlife corridors) might be appropriate. Practical methods for addressing vegetation loss and subdivision are examined later in this chapter.

Vegetation loss, threshold effects and species loss

In many landscapes, especially those with relatively high levels of original vegetation cover, the proportion of native vegetation lost is thought to be the best predictor of species loss (Fahrig 1998). However, it has been hypothesised that below certain threshold levels of vegetation cover, species loss will accelerate (Andrén, 1994; With and Crist, 1995; With, 1997; With and King, 1999; Figure 10.10).

In a review of fragmentation studies on birds and mammals, Andrén (1994) calculated that populations declined more rapidly than predicted by vegetation loss alone when less than 10–30% of the vegetation cover remained. Rapid decline below a threshold may be a consequence of 'cascading effects' (see the section later in this chapter on cascading effects). For example, populations of the Capercaillie Grouse in Scandinavia declined faster than predicted by habitat loss alone when vegetation cover fell below 30% (Rolstad and Wegge, 1987; see Enoksson *et al.*, 1995; Jansson and Angelstam, 1999). McIntyre *et al.* (2000) identified a threshold for Australian native woodland cover of >30% to limit the effects of tree dieback.

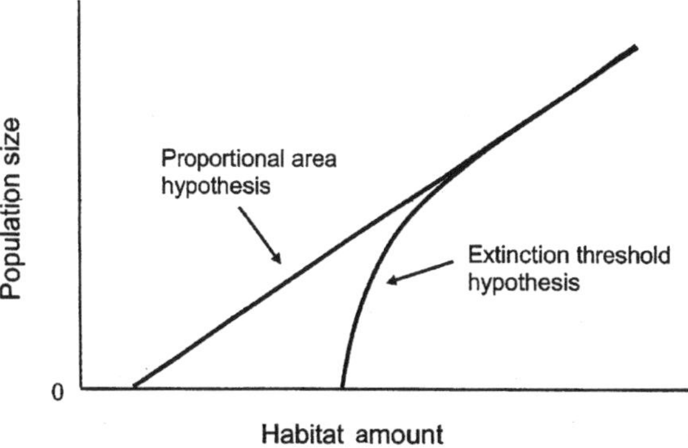

Figure 10.10. Comparison of rates of species loss proportional to the remaining amount of habitat and the threshold hypotheses. (Redrawn from Fahrig, 2003.)

Caveats associated with the threshold theory

Although threshold effects have been identified in several studies (e.g. see Figure 10.11), other investigations have failed to find them in work on invertebrates (Parker and Mac Nally, 2002), and birds and reptiles (Lindenmayer *et al.*, 2005b; Figure 10.12). In another example, small and intermediate-sized mammals can be lost from virtually intact forest ecosystems because of the spread of feral predators (e.g. the Red Fox and feral Cat) along roads (May, 2001).

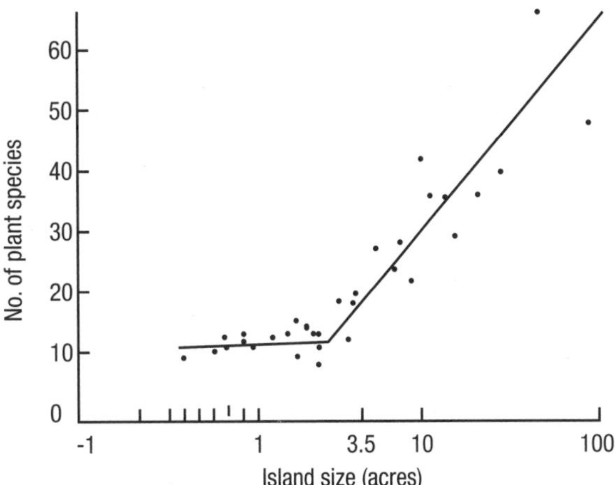

Figure 10.11. Empirical evidence of a threshold relationship between island size and the number of species of higher plants on the islands of the Kapingamarangi Atoll, Micronesia, unlike that predicted by Figure 10.10. (Redrawn from Macarthur and Wilson, 1967.)

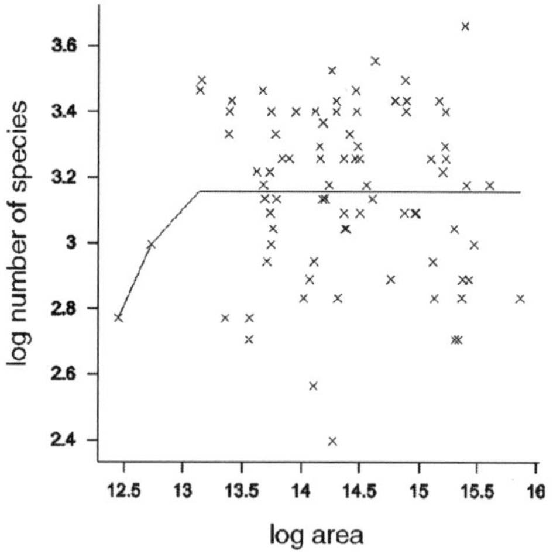

Figure 10.12. Relationship between the diversity of bird species at Tumut in south-eastern Australia and the amount of native vegetation cover in the surrounding landscape. The solid line shows a hypothetical critical change point at the 30% cover level, which would be expected if a threshold effect existed. The scatter of the measured data indicate the poor fit of the data to a hypothesised threshold.

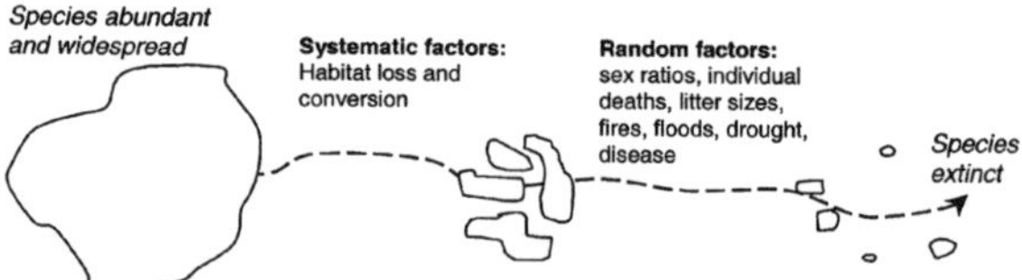

Figure 10.13. Inter-relationships between the loss of a species and vegetation loss (as a systematic threatening process), and subsequently cascading fragmentation effects that operate in addition to vegetation loss at low levels of vegetation cover. (Modified from Clark *et al.*, 1990.)

The terminology of the threshold theory is imprecise: for some, a threshold means any abrupt change in response, whereas for others it refers to changes at a precise value (e.g. exactly 30% cover), but the 'threshold' response isn't associated with an abrupt critical change, as depicted in Figure 10.10. Moreover, the perception that a curve such as that in Figure 10.10 is a threshold is scale-dependent; it may be altered if the scales on the *x* and *y* axes are changed.

The theoretical underpinnings of the threshold concept are problematic (Table 10.4) and it seems highly unlikely that there will be generic rules for threshold levels of native vegetation cover (e.g. 10, 30 or 70%; see Fahrig, 2003). Where threshold relationships exist, they will depend on the characteristics of the landscape, the species, and the ecological processes occurring in the landscape.

Cascading fragmentation effects

The preceding section outlined observations and speculations that species loss is influenced by

Table 10.4. Caveats and potential problems associated with threshold relationships for species loss and habitat loss relationships.

Issue	Explanation	Reference
Other plausible responses	For some (and perhaps most) taxa, there may not be distinct thresholds. Responses might be linear, curvilinear etc.	Bunnell *et al.* (2003)
Matrix permeability	Matrix conditions affect movement, colonisation of fragments and threshold relationships.	Taylor *et al.* (1993), Andrén (1999)
Landscape context	Threshold relationships can vary depending on whether areas surrounding habitat fragments are cropped areas, or managed forests.	Mönkönnen and Reunanen (1999)
Binomial treatment of landscape cover	Threshold theory treats landscape cover as either habitat or non-habitat. However, not all fragments are alike and internal differences between them significantly affects species occurrence.	
Non-random habitat loss	Threshold theory assumes that modification results in a given level of habitat cover of uniform quality. But fragmentation is non-random, with the most productive parts of landscapes modified first. Species loss and persistence of individual species can be strongly affected by the quality of *what* remains as well as *how much* remains.	Lindenmayer and Franklin (2002)
Species-specific responses	Threshold levels for habitat loss will vary for each species and depend on the scale at which it interacts with the patch mosaic in a landscape.	Mönkönnen and Reunanen (1999), With (1999)

Table 10.5. Potential 'cascading' impacts of habitat loss and habitat fragmentation.

Impact and examples	Citation
Destabilised population dynamics (e.g. disrupted sex ratios and Allee effects)	McCarthy *et al.* (1994)
Altered movement patterns (e.g. modified home ranges)	Barbour and Litvaitis (1993), Pope *et al.* (2004)
Altered social systems	Ims *et al.* (1993)
Food shortages in fragments	Robinson (1998), Zanette *et al.* (2000)
Increased disease prevalence	Allan *et al.* (2003)
Altered breeding success (e.g. smaller young)	Hinsley *et al.* (1999)
Increased short-term population densities in habitat fragments	Darveau *et al.* (1995)
Altered vulnerability of populations to catastrophic events	Burgman *et al.* (1993)
Increased human hunting pressure	Redford (1992)
Altered gene frequencies in populations	Lacy (1993a), Sarre (1995)
Altered interactions between species (e.g. inter-specific aggression) or weed/parasite invasions	Catterall *et al.* (1991), Lavorel *et al.* (1999)
Disrupted ecological processes (e.g. pollination)	Kapos (1989), Klein (1989)
Altered evolutionary processes (e.g. morphological changes in organisms)	Hill *et al.* (1999), Weishampel *et al.* (1997)

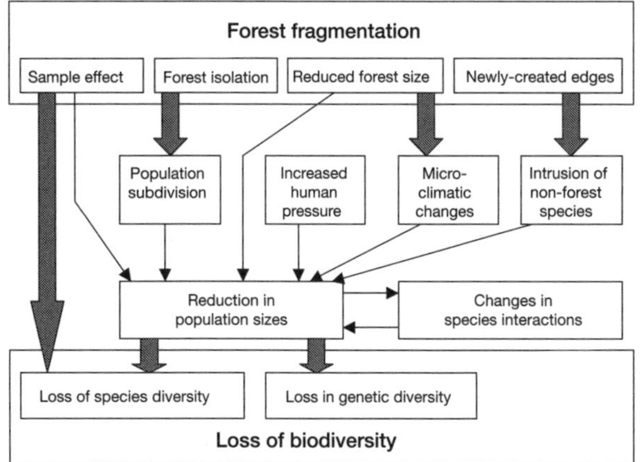

Figure 10.14. Graphical summary of the interacting effects of vegetation loss and habitat subdivision. (Modified from Zuidema *et al.*, 1996.)

vegetation loss. The causes of loss, sometimes called 'cascading fragmentation effects' (Table 10.5; Figure 10.13), include demographic stochasticity, genetic stochasticity, environmental variability, altered ecosystem processes (e.g. pollination; Cunningham, 2000), disrupted species interactions (e.g. aggressive inter-specific behaviour), and altered intra-specific dynamics (e.g. Allee effects; Allee *et al.*, 1949) (Figure 10.14). For example, Ims *et al.* (1993) found that with increasing vegetation loss and habitat subdivision, the social system of the Capercaillie Grouse changed from a **lekking system** to one of solitary displaying males. Other types of Allee effects include problems with finding mates and a reduced ability to engage in group defence against predators.

Vegetation subdivision, patch isolation and dispersal

Patch subdivision creates several smaller patches from a larger one, increasing the distances between remaining patches, and potentially harming species and ecological processes (Holyoak, 2000). Smaller patches support fewer species because populations tend to be smaller and therefore at greater risk of local extinction (MacArthur and Wilson, 1967; MacNally and Bennett, 1997), although not all species respond in this way (see the review of fragmentation experiments by Debinski and Holt, 2000).

Box 10.5

Why smaller habitat fragments can be worse

Many studies have found that the sizes of populations decrease with decreasing patch size. Ecological theory (Hutchinson, 1958; Macarthur and Wilson, 1967) predicts this relationship from simple generalisations of per capita consumption of resources and the finite nature of those resources. Zanette *et al.* (2000) examined this phenomenon in a study of the Eastern Yellow Robin in forest fragments in northern New South Wales. They found that the total biomass of insects was a function of patch size. As a result, in comparison with large habitat fragments, nesting females of the Eastern Yellow Robin in small habitat fragments: (1) left their nests more often to find food, (2) had a shorter breeding season, (3) laid eggs that were lighter in weight, and, (4) reared smaller nestlings. Zanette *et al.* (2000) concluded that per capita food limitations explained why the Eastern Yellow Robin was less abundant in small fragments.

Eastern Yellow Robin. (Photo by Esther Beaton.)

In many fragmented ecosystems, local extinctions are reversed by dispersal and recolonisation (Brown and Kodric-Brown, 1977; Hanski, 1994a, 1999a,b). Subdivision and isolation can impair inter-patch dispersal and recolonisation, making data on dispersal and recolonisation important for sustainable landscape management (Lamberson *et al.*, 1994), designing reserve networks (Burkey, 1989), establishing wildlife corridors (Andreassen *et al.*, 1996; Rosenberg *et al.*, 1997) and determining the ability of populations to recover from disturbance (e.g. Whelan, 1995; see Chapter 11). Despite this need, knowledge about dispersal and recolonisation in fragmented landscapes is limited (Chepko-Sade and Halpin, 1987; Dieckmann *et al.*, 1999). For example, few empirical studies have examined how patches are recolonised (but see, for example, Middleton and Merriam, 1981; Johnson and Gaines, 1985) and there are few tests of the prediction that recolonisation probability depends on distance from a source population (Fritz, 1979; Johnson and Gaines, 1990; Hanski, 1994a).

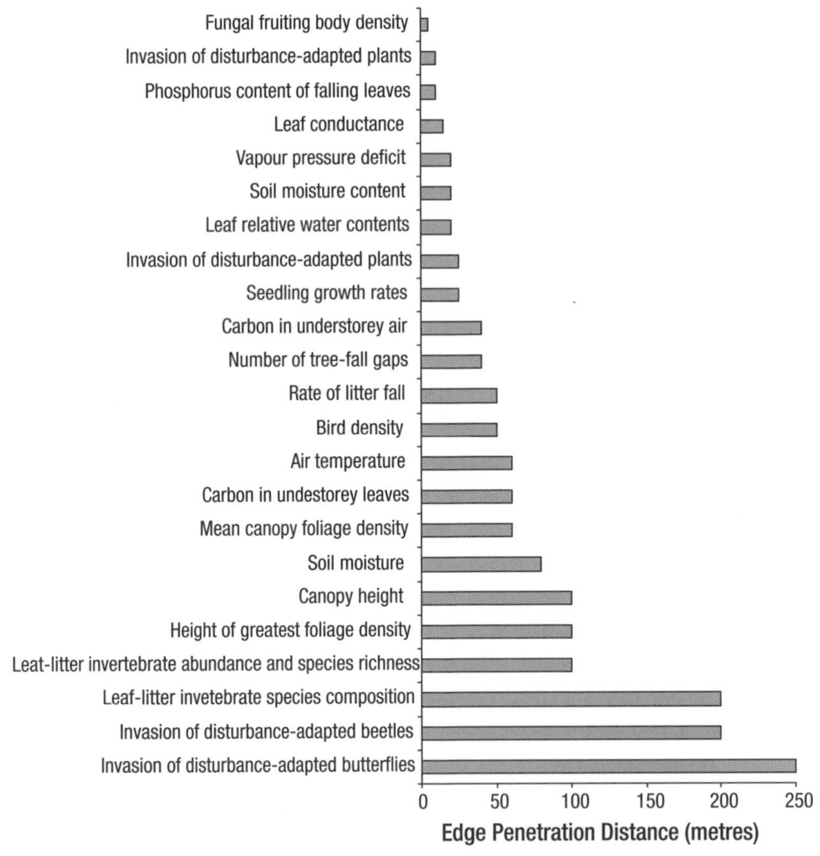

Figure 10.15. Variation in edge penetration for a range of measures in the Biological Dynamics of Forest Fragments Project in Brazil. (Modified and redrawn from Laurance *et al.*, 1997.)

The paucity of data on dispersal and recolonisation arises because these processes are difficult to study using standard field techniques such as colour banding of birds or **radio-tracking** (Oakleaf *et al.*, 1993; Baptista and Gaunt, 1997). Data from genetic markers combined with knowledge of movement patterns should improve our understanding of dispersal and recolonisation (e.g. Peakall *et al.*, 2003; see Chapter 13).

Edge effects

An edge can be viewed as a marginal zone of altered microclimatic and ecological conditions that contrasts with the interior of a patch (Matlack, 1993). Edge effects refer to the changes in biological and physical conditions that occur at an ecosystem boundary and within adjacent ecosystems (Wilcove, 1985; Temple and Cary, 1988; Kremsater and Bunnell, 1999). The term 'edge effect' is also used to describe increases in species diversity in places where two habitats meet. The ecological edge that results from a disturbance is the result of interactions between the type and intensity of the disturbance event and the ecological dynamics within the adjacent undisturbed environment. The affected measure can be an environmental variable (e.g. air temperature), an ecological process (e.g. rate of organic matter decomposition) or a community interaction (e.g. predation).

The intensity of edge effects and the area of a patch subject to significant edge influence depend on the parameter of interest. Each type of affected measure has a unique response pattern (see Figure 10.15). Two broad types of edge effects are recognised: biotic edge effects and abiotic edge effects.

Abiotic edge effects

Edges can experience microclimatic changes such as increased temperature and light and decreased humidity that extend tens or hundreds of metres from an edge, depending upon the environmental variable, the physical nature of the edge, and weather conditions (Chen *et al.*, 1990, 1991; see the review by Saunders *et al.*, 1991). These changes affect fire ignition probabilities (e.g. Kauffman and Uhl, 1990), precipitation, frost and fire behaviour (e.g. Roberts, 1973), insect activity (e.g. Simandl, 1992), nutrients (e.g. Yanai, 1991), and the composition of soil-borne bacterial and fungal populations (e.g. Jha *et al.*, 1992; reviewed by Bradshaw, 1992). In some cases, edge effects can lead to the degradation of

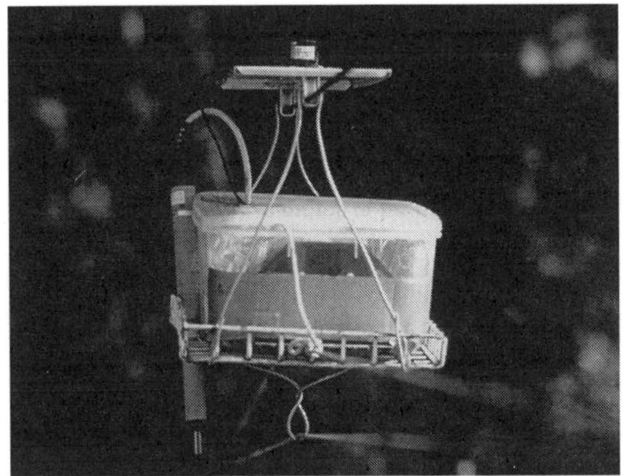

(Top) Equipment for measuring abiotic edge effects in Victorian Mountain Ash forests (photos by Brooke Parry). (Bottom) Accelerated tree fall at edges of clear-felled areas in Victorian Mountain Ash forests (photo by David Lindenmayer).

habitat in protected areas such as wildlife corridors (Lindenmayer *et al.*, 1997), and disrupt connectivity between larger reserved areas (Lovejoy *et al.*, 1986).

Biotic edge effects

Biological parameters that affect ecological communities across a boundary, such as diseases, weeds and predators, can penetrate hundreds of metres into retained areas (e.g. Wilcove *et al.*, 1986; Andren and Angelstam, 1988; Laurance, 1997b). Edge environments can affect reproduction, growth, and mortality in plants (Hobbs and Yates, 2003). Chen *et al.* (1991) observed increased reproduction and growth of surviving mature trees in old growth forests bordering recently clear-felled areas. Differences in invertebrate community composition have been observed between edge and interior environments (e.g. Bellinger *et al.*, 1989; Hill, 1995). For instance, amphipods responded negatively to changes in wind, moisture and temperature at the edges of forest fragments in south-eastern New South Wales (Margules *et al.*, 1994a).

In the Northern Hemisphere, some game species have strong preferences for edge environments where foraging and cover are close together (Patton, 1974; Matlack and Litvaitus, 1999). Large populations of White-tailed Deer graze food plants heavily at forest edges (Johnson *et al.*, 1995), including endangered taxa (Miller *et al.*, 1992), which can limit stand regeneration (Tilghman, 1989) and alter patterns of species diversity (McShea and Rappole, 2000).

Box 10.6

Abiotic edge effects in wet eucalypt forests

Parry (1997) measured environmental parameters at different distances from edges between clear-felled areas and adjacent unlogged stands in the Mountain Ash forests of Victoria. There were strong gradients in air temperature, vapour pressure and radiation regimes from logged/unlogged harvested area (or coupe) boundaries into unlogged forest (Parry, 1997). The measures depended on time of day, aspect, cloud cover, and forest type.

Davies *et al.* (2001b) reported that solar radiation was highest at the edges of fragments in the Wog Wog Fragmentation Experiment (see Box 10.9). Such changes may influence some elements of the invertebrate biota, for example amphipods (Margules *et al.*, 1994a).

Box 10.7

Biotic edge effects in the South Australian Mallee region

Luck *et al.* (1999a,b) studied two types of edges in the Murray Mallee region of south-east South Australia: (1) induced or abrupt human-created edges between eucalypt vegetation and human infrastructure, and (2) inherent natural edges that were gradual transitions between eucalypt vegetation and adjacent shrubland. Luck *et al.* (1999a) found that open country birds, for example the Australian Magpie and Little Raven, were often found at human-created edges but were rare in patch interiors and at natural edges. Other species, for example the Spotted Pardalote and Southern Scrub-robin, avoided human-created edges. Predation rates on artificial nests were higher at human-created edges than in interior areas, whereas there were no important edge–interior differences for natural edges (Luck *et al.*, 1999b).

These results make it possible to infer which species of birds will be susceptible to alterations of landscapes that fragment the remaining Murray Mallee habitats in ways that increase the amount of human-created edge.

Some vertebrates avoid edges (e.g. Fletcher, 2005) and are classified as 'interior' species (Gates and Gysel, 1978; Robinson *et al.*, 1995); many bird species in tropical forests fall into this category (Terborgh, 1989; Frumhoff, 1995), in part because arthropod densities are lower at the edges (Burke and Nol, 1998) or because foraging efficiency is impaired by edge conditions (McCollin, 1998).

Elevated nest predation is commonly but not universally associated with edges (Berg *et al.*, 1992; Hanski *et al.*, 1996; reviewed by Lahti, 2001). Elevated nest predation is often observed in agricultural landscapes where there are strong contrasts between vegetation remnants and the surrounding environment (e.g. Andrén, 1992; Bayne and Hobson, 1997, 1998; Hannon and Cotterill, 1998). Conversely, nest predation is less pronounced at edges in landscapes where contrasts are weaker (Schmiegelow *et al.*, 1997), such as continuous eucalypt forest dissected by minor bush tracks, or native forest abutting softwood plantations (Sargent *et al.*, 1998; Lindenmayer *et al.*, 1999g; Piper *et al.*, 2002).

Biotic responses to edges are inconsistent. Forest edges on the boundary of clear-cut areas in Sweden have lower bird species diversity (Hansson, 1983), but no such pattern occurs in clear-cut forest edges in north-eastern USA (Rudnicky and Hunter, 1993). Similarly, patterns of brood parasitism that are characteristic of edge environments in many Northern Hemisphere landscapes appear to be rare in Australia. Even within North America, the increased nest predation and brood parasitism that is observed on the eastern side of the continent is less common in the west (Kremsater and Bunnell, 1999; Marzluff and Restani, 1999).

Factors influencing edge effects

As noted for birds, the magnitude of many types of edge effects is related to the strength of the contrast between the matrix and other landscape units: where the contrast is strong, there will often be more intense interactions and edge effects (Laurance and Yensen, 1991; Mesquita *et al.*, 1999). The extent of the area supporting high contrast conditions influences the magnitude of edge effects. For example, microclimate edge effects in forests may be greater where a large clear-felled area abuts a retained patch than where a logged area or coupe is small (Lindenmayer *et al.*, 1997). Buffers can play an important role in mitigating edge effects.

Artificial nest and quail egg used in a nest predation experiment at Tumut in southern New South Wales. (Photo by Matthew Pope.)

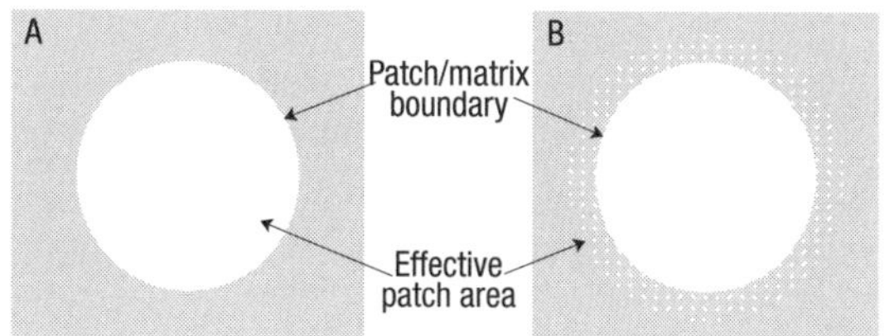

Figure 10.16. Negative and positive edge effects. (A) Processes from the matrix reduce functional patch size for the species of interest. (B) Suitable conditions in the matrix enable the species to expand its functional patch size.

'Reverse' edge effects

Edge effects are not unidirectional processes: fragments exert edge influences on adjacent matrix areas (Figure 10.16), and retained trees affect light and other abiotic factors in clear-cut areas. Birds are more likely to occur in stands of Radiata Pine in south-eastern Australia when the stands are adjacent to patches of native eucalypt forest (Tubelius *et al.*, 2004). Pest species move from remnant native forest patches into adjacent planted stands and browse and defoliate trees (Bulinski, 1999; McArthur, 2000).

Summary: edge effects

For many species, edge effects amplify the harmful impacts of fragmentation (see, for example, Lovejoy *et al.*, 1986). Changes in the environmental conditions at the edge of a patch (e.g. temperature, humidity, wind and light, and disturbances such as fire, grazing and weed invasion) decrease the effective size of a patch for those species that inhabit the original habitat. Some species favour edge habitat; regenerating vegetation and patch edges experience a seed rain of environmental weeds and other introduced plants that are frequently better adapted to exposed and disturbed environments than native plants (Janzen, 1983).

10.4 Studying habitat loss and fragmentation

Despite the importance of studying the response of species to habitat loss and fragmentation, investigations are difficult for several reasons. Some of these include:

- The way that species perceive and respond to a landscape may be very different from the way humans perceive that same landscape (see Section 10.3). For example, edge effects might mean that a 60-hectare area of native forest contains only 30 hectares of suitable habitat for a species that avoids edge environments. Thus, the boundary between two habitat types may be much more gradual than the sharp edges defined by mapping tools such as GIS (see Dangerfield *et al.*, 2003).

- No two parts of any landscape are the same, making true replicates impossible.
- Productive areas often support more species (Braithwaite, 1984) and are often more extensively cleared and subdivided than less productive ones (Armesto *et al.*, 1998), confounding comparisons between fragmented and unfragmented areas (Lindenmayer and Franklin, 2002).
- Many processes within patches are linked to processes in the surrounding landscape. Thus the **landscape context** (the conditions surrounding a patch) causes otherwise similar habitat patches to support markedly different biota (Harris, 1984; Lindenmayer *et al.*, 1999c). For example, a patch of native forest surrounded by extensive plantation can support markedly different bird, reptile and mammal assemblages when compared with a similar patch of native forest embedded within a larger area of native forest (Lindenmayer *et al.*, 2002c).
- Many species respond to several threatening processes that have impacts at several spatial scales. These can be difficult to tease apart because they are confounded, cumulative or interact with each other.
- Suitable habitat (see Chapter 15) and responses to habitat loss and fragmentation are species-specific (Davies *et al.*, 2000; Lindenmayer *et al.*, 2002c), making it difficult to extrapolate results from one species to another, or even the same species in different landscapes (Freudenberger, 2001; Lindenmayer *et al.*, 2001c).
- The life history of a species in an unfragmented landscape (e.g. home range size, patterns of fecundity and sex ratios) may be markedly different from that of the same species in habitat fragments (Barbour and Litvaitis, 1993). This means that species can respond to altered landscapes in sometimes quite unexpected ways (Lindenmayer *et al.*, 2003d).
- Landscape history can affect species distributions. Species may continue to respond to landscape processes that took place many years previously (see Lindborg and Eriksson, 2004). Successional change and disturbance responses may mask the effects of fragmentation.
- Habitat fragmentation is just one of many factors that can influence the distribution and abundance of a particular species, but it may not be the most important one (Jellinek *et al.*, 2004).

Box 10.8

Extinction debts and species loss

The effects of landscape history can sometimes be related to concepts such as 'extinction debts' and 'species richness relaxation', which are well known in the theoretical conservation biology literature (Macarthur and Wilson, 1967; McCarthy *et al.*, 1997; Brooks *et al.*, 1999) and from empirical data (e.g. Loyn, 1987; Robinson, 1999). These concepts suggest that populations go extinct well after landscape change (Tilman *et al.*, 1994; McCarthy *et al.*, 1997), and that a species can respond to circumstances that are no longer directly visible to researchers. Long-lived species can be particularly slow to respond to change. For instance, trapdoor spiders in the Wheat belt of Western Australia can live for more than 20 years (Main, 1999). The substantial changes in landscape cover that have occurred in that region (Saunders *et al.*, 1987) may take quite some time to be reflected in population dynamics. Thus, the current distribution and abundance of a species may not be a good indicator of its future persistence (Lindenmayer, 1995). Hence, we need to avoid over-confidence in our ability to ensure the conservation of some taxa in the long term (Niemelä *et al.*, 1993; Noss and Cooperrider, 1994; Lindenmayer and Franklin, 2002; Berglund and Jonsson, 2005).

As outlined in section 10.4, many studies have not treated habitat fragmentation and habitat loss as separate processes and subsequently have not been able to distinguish their effects (Fahrig, 2003). Indeed, habitat fragmentation sometimes refers to all processes that affect species in human-modified landscapes (Haila, 2002; Villard, 2002; see Box 10.3).

The difficulties associated with investigations of landscapes make it impossible to design ideal studies of habitat loss and fragmentation (Box 10.9). The major approaches to studying habitat loss and fragmentation – experiments, 'natural experiments', observational

Box 10.9

The difficulties of studying forest fragmentation in Australia

Many studies of habitat fragmentation in Australia have focused on agricultural ecosystems, where there is a strong contrast between the vegetation in the fragments (e.g. woodland) and that of the surrounding landscape, which is typically extensively cleared cropping and/or grazing land (e.g. see Saunders *et al.*, 1987; Bennett, 1990a; Saunders and Hobbs, 1991; Dunstan and Fox, 1996; Barrett *et al.*, 1994; Zanette *et al.*, 2000). There have been comparatively fewer investigations in Australian native forests where landscapes have been altered by timber harvesting. The contrasts between logged and unlogged forests are often not as marked as between cropped areas and remnant vegetation in agricultural zones, particularly if logging impacts are mitigated by significant large tree and log retention at the time of harvesting. Both cut and uncut areas can eventually provide suitable habitat for many (although not all) species as stand regeneration occurs (Lindenmayer and Franklin, 2002). Determining what constitutes a fragment, and whether a landscape is fragmented can be difficult in such forests.

Boundaries between distinct forest types are rarely sharply defined in many forests. Rather, most boundaries are gradual changes in associations of tree species across multiple and interacting environmental gradients of elevation, aspect, soil type and fertility, and climate (Nix and Gillison, 1985; Austin, 1999). This makes it hard to determine whether species assemblages have changed because of environmental conditions or because of human-derived habitat fragmentation.

Human and natural disturbance patterns vary between forest types and also in response to different environmental conditions in those forest types. For example, some forest types are more likely to be logged than others because the timber is more valuable. Other forms of disturbances, for example fires, will be more frequent and/or more intense in some forest types or in some parts of landscapes (e.g. on north- and west-facing slopes close to ridges). Overlaying human and natural disturbances on natural patterns of environmental heterogeneity in forests complicates attempts to determine if habitat fragmentation effects are important or if other processes have a more significant influence on species presence and abundance.

An Australian forest landscape characterised by marked gradients in tree species composition. (Photo by David Lindenmayer.)

studies and modelling – are outlined below, because investigations of fragmentation have become an important part of the conservation biology literature.

Experiments

'True' experiments in field ecology and conservation (see Oksanen, 2001) typically include (1) replicate sites of the particular treatment in question (e.g. replicates of a specified patch size), (2) control sites in which a treatment is not applied and which can be compared against sites where a treatment is applied, and (3) adequate description of pre-treatment conditions.

It is well beyond the scope of this book to discuss experimental design fully. The first rule for all conservation biologists is first to consult with a statistician. Robust data and outcomes can only be generated by experiments with adequate replication of treatments, adequate sample sizes (and adequate statistical power; see Chapter 17), quantification of before-treatment conditions in each replicate, and replicated 'control' sites matched to the treatments.

Experiments are comparatively rare in studies of landscape change and habitat fragmentation, principally because it is difficult, time-consuming, and expensive to manipulate large areas. It can also be difficult to find sufficient replicates, particularly for large patches. In a detailed literature search Debinski and Holt (2000) identified only 20 fragmentation experiments in terrestrial ecosystems: 6 in forests and 14 in grasslands or old fields. Most of these experiments were conducted in North America or Europe, but one of the most extensive and longest running was conducted in south-eastern Australia (see Box 10.9).

Box 10.10

The Wog Wog fragmentation experiment

The Wog Wog fragmentation experiment was established near Bombala in south-eastern New South Wales (Margules, 1992). The experiment was set up to examine the fragmentation responses in remnant eucalypt patches surrounded by stands of exotic Radiata Pine forest. The experimental design of the Wog Wog fragmentation experiment includes six replicates of fragment sizes (Figure 10.17). Each replicate contains three plots: one small (0.25 hectares), one medium (0.875 hectares), and one large (3.062 hectares). Four of the six replicates became fragments when the eucalypt forest surrounding them was cleared in 1987 and planted to Radiata Pine in 1988 (Margules, 1992). Within each plot are eight sites stratified by location (edge or centre) and topography (slope and drainage line). This gave 144 sites within the 18 plots. A further 44 sites were located within the Radiata Pine stands once the plantation was established. One of the strengths of the Wog Wog fragmentation experiment is that the data gathered include 2 years of pre-fragmentation information.

Much of the work at Wog Wog has focused on the response of invertebrates, particularly millipedes (Margules *et al.*, 1994a) and beetles (Davies and Margules, 1998). Edge effects resulted in increased species richness at fragment edges and increased numbers of fungivorous and detritivorous beetles (Davies *et al.*, 2001a). Contrary to expectations from conservation biology theory, there was no evidence of a decline in species richness or increased extinction rates for beetles within the fragments (see Davies *et al.*, 2000, 2001a).

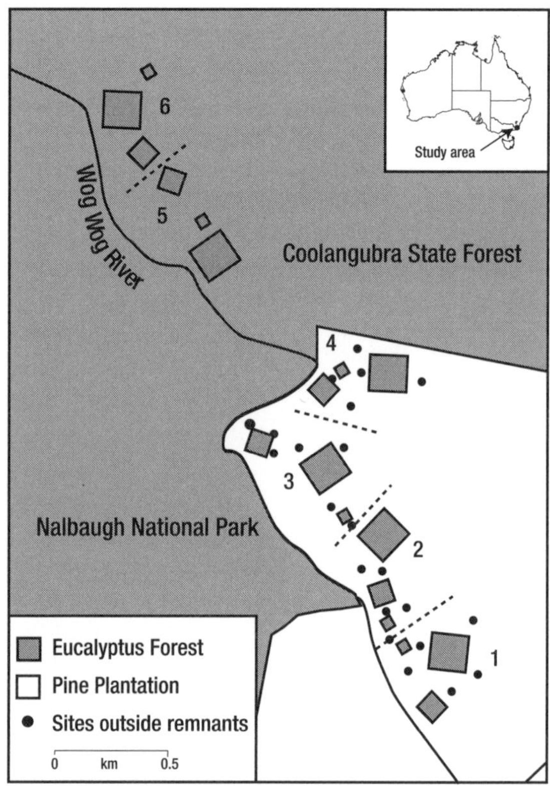

Figure 10.17. Location and design of the Wog Wog fragmentation experiment. (Redrawn from Margules, 1992.)

Extrapolating the results of small-scale experiments to large-scale ecosystems is error-prone because most spatial processes are scale-dependent (Carpenter *et al.*, 1995; Thrush *et al.*, 2000). Many factors are controlled in true experiments, making extrapolation of the results to other species, ecosystems or ecological phenomena unreliable (Davies *et al.*, 2001b; Oksanen, 2001).

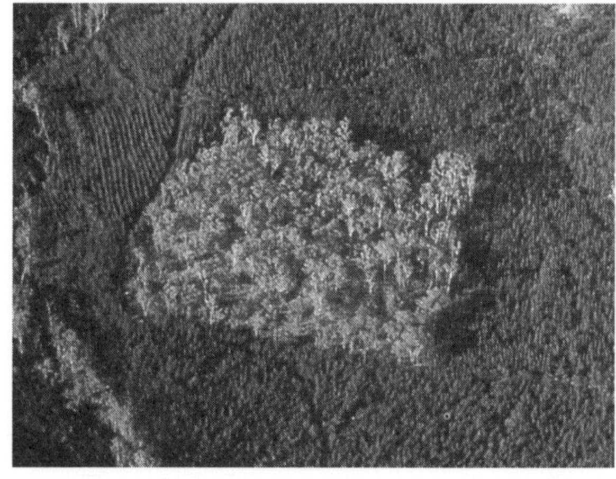

A eucalypt fragment in the Wog Wog fragmentation experiment. (Photo courtesy of CSIRO.)

Fragmentation experiments have examined two important hypotheses (Debinski and Holt, 2000): (1) species richness and abundance increase with increasing patch area, and (2) movement and species richness increase with increased connectivity between patches. Debinski and Holt (2000) found many species-specific responses with many inconsistent results across different studies. They found that arthropods showed the best fit with theoretical expectations; for example, larger fragments supported greater numbers of species. However, many other groups did not respond in ways that would be 'expected' from theory, including birds, mammals, early successional plants, mobile generalists and long-lived species (Debinski and Holt, 2000).

Box 10.11

The Tumut fragmentation study

The experimental design for the natural experiment at Tumut covers 166 sites in three broad classes: (1) 86 eucalypt fragments stratified across four patch size classes, five forest types and two ages of surrounding pine forest, (2) 40 sites located in large areas of continuous native forest, and (3) 40 sites dominated by stands of Radiata Pine, which form the matrix surrounding the eucalypt fragments. The design features include:

- stratifying the eucalypt remnants to ensure the full environmental space of the study region was represented
- randomly selecting the 86 eucalypt remnants within strata to minimise the chance of bias, enabling averaging over random factors
- locating sites in large areas of continuous eucalypt forest to provide `control' areas, which are large enough to support viable populations of the array of species that once occurred in the study region
- using climate, forest type and geology data to ensure the range of environmental and other conditions were matched across the three broad classes of sites
- selecting sites dominated by Radiata Pine stands so that the potential habitat value of the landscape matrix surrounding the fragments was not ignored (see Wiens, 1994; Beier and Noss, 1998).

From the work at Tumut it was found that small- to intermediate-sized patches (0.5–3 hectares in size) supported considerably higher levels of vertebrate biodiversity than anticipated, highlighting the importance of remnants as small as 0.5 hectares. Many species of native animals occurred in stands of Radiata Pine, but their presence was often related, in part, to nearby remnant eucalypt forest. Thus, a mosaic of remnant native vegetation and Pine stands will have significantly higher biodiversity value than a Radiata Pine monoculture (Lindenmayer *et al.*, 2002c).

At Tumut, as in most natural experiments, there is no information on species occurrence patterns prior to the beginning of landscape alteration in the 1930s.

`Natural' experiments

Natural experiments overlay an experimental design on an ecosystem where change or active manipulation has occurred or is planned (Carpenter *et al.*, 1995). Natural experiments are similar to 'true' experiments (Diamond, 1986), but the changes are not controlled by the researcher.

A patch of remnant eucalypt vegetation surrounded by stands of Radiata Pine at Tumut in southern New South Wales. (Photo by David Lindenmayer.)

Usually, they occur at larger scales than true experiments. For example, like the Wog Wog fragmentation experiment (see Box 10.10), the Tumut fragmentation Study in New South Wales examines patterns of species richness and abundance in fragments of eucalypt forest embedded in an extensive Radiata Pine plantation (Lindenmayer, 2000). The work is taking place over a large area (>100 000 hectares) and the primary focus is on birds and mammals (Box 10.11).

Observational studies

Observational studies rely on passive sampling to infer ecosystem responses to fragmentation. The number of observational studies in conservation biology is increasing rapidly (McGarigal and Cushman, 2002; Fahrig, 2003). Bennett (1990b) compared the historical and current status of mammals in the Naringal area of south-western Victoria using historical and anecdotal records, collections of road kills, museum archives and fauna surveys in 39 patches of 0.3–92 hectares in size. At the time of European settlement there were 33 species of native mammals in the region. Of these, the Dingo, Tiger Quoll, Eastern Quoll, Common Wombat, Koala and Eastern Pygmy Possum are now regionally extinct. The first four were hunted extensively in pest control programs. Six introduced taxa have invaded since European Settlement: the House Mouse, Black Rat, Red Fox, feral Cat, Rabbit, and Brown Hare (Table 10.6). Although the particular taxa present have changed dramatically, the total number of mammal species is unchanged.

Bennett (1990a) showed that larger taxa are more vulnerable to fragmentation: the Southern Brown

Table 10.6. Frequency of occurrence of mammals in remnant vegetation in south-western Victoria (after Bennett, 1990b).

Patch size (hectares)	<2	3–7	8–15	16–40	41–100
No. patches	8	8	8	8	7
Occurrence of species (%)					
[*]Rabbit	**63**	**100**	**100**	**100**	**100**
Bush Rat	38	**75**	**100**	**100**	**100**
Common Ringtail Possum	25	**88**	**100**	**88**	**100**
[*]Red Fox	0	**63**	**100**	**100**	**100**
Echidna	13	50	**100**	**100**	**100**
Brown Antechinus	13	50	**100**	**100**	**86**
Swamp Wallaby	13	13	**75**	**63**	**86**
Long-nosed Potoroo	0	13	50	**63**	**100**
Eastern Grey Kangaroo	13	50	13	**63**	**57**
[*]House Mouse	25	25	38	38	**57**
[*]Cat	25	38	50	38	43
Swamp Rat	0	13	13	25	29
Long-nosed Bandicoot	0	13	13	0	43
Red-necked Wallaby	0	0	0	38	29
Sugar Glider	0	0	13	25	29
Southern Brown Bandicoot	0	0	13	13	14
Common Brushtail Possum	0	0	13	0	14
[*]Black Rat	13	25	0	0	0
[*]Brown Hare	13	0	0	13	0

[*]Exotic species
Frequencies greater than 50% are in bold face.

Box 10.12

Temporal dynamics in fragmentation studies: the Swift Parrot

Many fragmentation studies are single snapshots of species composition, ignoring year-to-year variations in conditions and population responses. A striking example of the risks of ignoring temporal processes comes from work on the endangered Swift Parrot in north-eastern Victoria. This species is migratory and birds overwinter in areas such as the Box–Ironbark forests of north-eastern Victoria. Mac Nally and Horrocks (2000) found that the Swift Parrot used the landscape in different ways in successive years. Their surveys in 1996 showed that the majority of observations were in large tracts of woodland and only 16% of records were from small remnants. However, in the following year, 40% of observations were from small fragments. Birds were strongly associated with flowering Golden Wattles. If conservation strategies had been based on data from 1996, the value of small remnants would have been discounted and their importance for the Swift Parrot overlooked. From a statistical perspective, the differences between years remain important only if the sample sizes are large and the observation effort is scaled to patch size in the same way for large and small patches, respectively.

Bandicoot, Sugar Glider and Red-necked Wallaby were absent from smaller patches of habitat at Naringal.

Bennett (1990a) concluded that it is important to avoid further clearing and to maintain connectivity and the suitability of habitat within existing patches (Bennett, 1990a,b). Even if managers achieve these objectives, species might still be lost to isolation (the extinction debt; see Box 10.8). Bennett (personal communication) has begun revisiting his sites in the Naringal region to survey vertebrates and monitor the long-term impacts of habitat loss and habitat fragmentation.

Often, spatial patterns of vegetation cover created by landscape alteration are linked to the responses of particular elements of biodiversity. The metrics used to characterise such patterns are termed **landscape indices** and they are used widely in fragmentation research, particularly in observational studies (Haines-Chopping and Young, 1996; McAlpine *et al.*, 2002). We examine the utility of landscape metrics in Chapter 16.

Modelling

In computer modelling studies, a model of a landscape and the organism or organisms inhabiting it is created and ecological responses are simulated. Spatially explicit models sometimes include patch occupancy (patches are either vacant or occupied; abundance is ignored) and dispersal (Hanski, 1999a,b). More complex models include age or developmental stage structure and density dependence, usually assuming that landscapes are made up of discrete, sharply defined habitat patches (Akçakaya, 2002). These abstractions can be useful but suffer from the same limitations as patch-matrix-corridor models (see Section 10.3). Alternatively, the landscape can be divided into discrete cells (cellular automata), each with its own characteristics. At the most detailed level, models can track individuals and use rules to govern their movements and interactions with continuous environmental conditions.

Spatially explicit models have been criticised because they lack data and cannot make reliable predictions (Doak and Mills, 1994). Even so, they can be valuable in several ways. For example, the models organise information for subsequent empirical tests (Walters, 1986). They also identify knowledge gaps (Burgman *et al.*, 1993), highlight problems for which pre-emptive action may be beneficial (Tilman *et al.*, 1994), and help to identify complex ecological processes that might otherwise be overlooked in fieldwork (Gilpin and Soulé, 1986; Temple and Cary, 1988; Bender *et al.*, 2003).

The best approach to modelling is for the architects of models to work closely with field-based conservation biologists to ensure that the modelling process is an open one, where the strengths and limitations of both the model and the data used in it are clear. The results of modelling are often inaccurate, but the primary roles of models are to organise information, assist thinking and ensure internally consistent arguments (Walters, 1986; Starfield and Bleloch, 1992; Burgman *et al.*, 1993).

Problems in the way fragmentation is studied

Many fragmentation studies focus almost exclusively on the use of fragments or corridors, ignoring the role of the surrounding landscape matrix (see critiques by Simberloff *et al.*, 1992; Laurance and Bierregaard, 1997). However, matrix areas make substantial contributions to population persistence (Gascon *et al.*, 1999). On this basis, Wiens (1989) noted that:

> *A focus exclusively on fragmentation of habitats misses the point that it is often the structure of an entire landscape mosaic rather than the size or shape of individual patches [that matters]... The likelihood that dispersal can occur between fragments and forestall the extinction of sensitive species on a regional scale is influenced by the configuration of the fragments and the landscape mosaic in which they are embedded.*

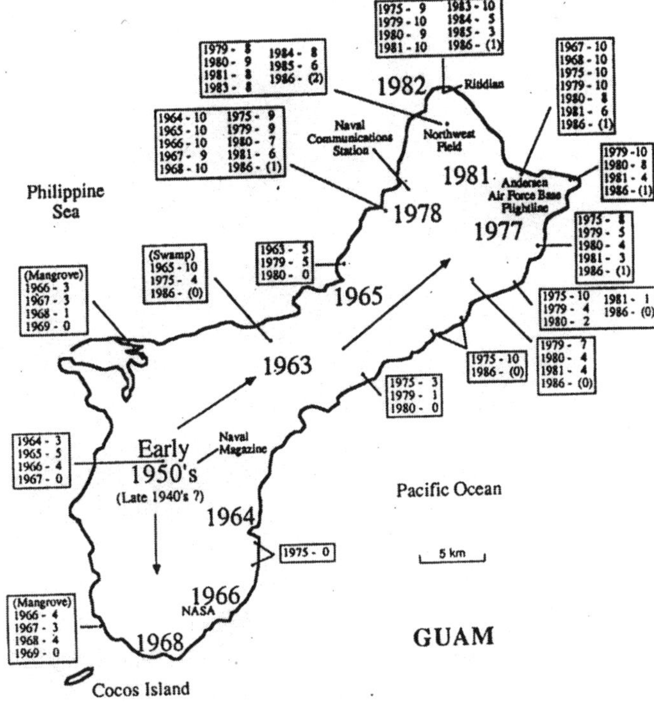

Figure 10.18. Relationship between the number of extant native bird species and time taken to respond to the northward range expansion of the Brown Tree Snake on Guam. (Redrawn from Savidge, 1987.)

Box 10.13

Diagnosing a landscape decline: forest birds and the introduction of the Brown Tree Snake on Guam

Island faunas are particularly susceptible to introduced species (Simberloff, 1995). For example, the bird assemblages on the island of Guam in the North Pacific Ocean experienced a massive collapse during the second half of the 20th century (Wiles *et al.*, 2003; Figure 10.18). The reasons for the initial decline are unclear, but some of the candidate reasons were: hunting by humans; pesticides (and their effects on eggshell thinning); competition with exotic birds; habitat loss; diseases; predation by introduced Cats, Dogs and Rats; and predation by the introduced Brown Tree Snake.

Savidge (1987) described an elegant diagnostic approach to reveal the underlying cause of the decline by examining the relationships between the symptoms associated with a threatening process. Savidge (1987) first ruled out pesticides as the cause of avifaunal decline because the pesticide levels on Guam were lower than were known to lead to problems such as eggshell thinning. Human hunting was ruled out because some declines occurred where hunting had not taken place. Competition with exotic birds was also considered unimportant because there was no evidence of resource overlap between native and introduced species and because introduced and exotic taxa co-occurred in some places. Habitat loss was excluded because native bird species were lost from many areas where suitable habitat remained. There was also no evidence that bird diseases had significantly affected populations, despite field pathology and other assays for the presence of bacterial or viral diseases. Finally, predation by Rats, Cats and Dogs was ruled out because bird declines had taken place in many areas where these animals were rare or absent.

After eliminating other possible threatening processes, the most likely cause of bird decline appeared to be predation by the introduced Brown Tree Snake (Savidge, 1987). The species occurs in Australia, New Guinea, Indonesia, the Solomon Islands, and other nearby islands (Cogger, 2000). It is believed to have been accidentally introduced to Guam in the 1940s or 1950s with cargo transported to the large military base on the island.

A number of lines of evidence supported the hypothesis that predation by the Brown Tree Snake was responsible for the crash in native bird populations. First, bird populations on nearby islands without snakes were reasonably stable. Furthermore, following the introduction of the species to Guam, the snake's range expansion coincided with a contraction in the range of various bird species (see Figure 10.18). The timing of expansions of the Brown Tree Snake's range (in the early 1960s) correlated closely with dramatic declines in the range of ten species of forest-dependent birds on Guam. The northern area of the island was the last known habitat supporting all ten species and it was the last place to be invaded by tree snakes (Savidge, 1987). Predation rates of birds were higher where birds were declining than where populations of birds were stable. Other species that were prey items for the Brown Tree Snake were also in decline, such as small mammal and lizard populations. Experiments subsequently confirmed the role of the Brown Tree Snake in the decline of bird populations on Guam. Bird lures in artificial cages (where native birds were extinct) were subject to heavy predation by snakes, and artificial nests suffered massive egg loss to snakes. The Brown Tree Snake is a particularly effective predator (Savidge, 1987) because it has a broad, non-specific diet. It consumes large numbers of small lizards as an alternative prey source, thereby maintaining high numbers that, in turn, devastate bird populations (which breed more slowly than lizards). Furthermore, there are few predators or competitors to limit population increases of the Brown Tree Snake.

The Brown Tree Snake invasion of Guam highlights the speed with which extinctions and range contractions can occur following the establishment of an exotic predator. Of the 18 native species of birds on Guam, populations of all but one have been either extirpated or severely reduced (Wiles *et al.*, 2003). Many other species of mammals and reptiles have also been impacted. Snake control measures are underway, including aerial baiting with mice injected with the human painkiller acetaminophen, which is toxic to snakes. However, the loss of so many bird species means that the prospects for restoration are limited (Wiles *et al.*, 2003). Hence, a key additional conservation strategy is to develop predator-proof fences to limit accidental introductions of snakes to other islands via shipments of cargo from ports and airports on Guam (Engeman and Linnell, 1998; Engeman *et al.*, 1999). In addition, specially trained Dogs are used to detect snakes in air cargo.

We believe that a full understanding of fragmentation effects requires an understanding of how biota use all landscape components, including habitat fragments, corridors and the matrix in which they are embedded (Äberg *et al.*, 1995; Flather and Sauer, 1996; Lindenmayer and Franklin, 2002).

The scale of many studies is not well defined. The majority of 'landscape studies' examine patches within a single landscape (Fahrig, 2003). Studies of whole landscapes include those that have examined the relationships between species diversity and landscape patterns (e.g. Bennett and Ford, 1997), the occurrence of individual species (e.g. the Superb Parrot) and the occurrence of large-scale habitat cover (Manning, 2003).

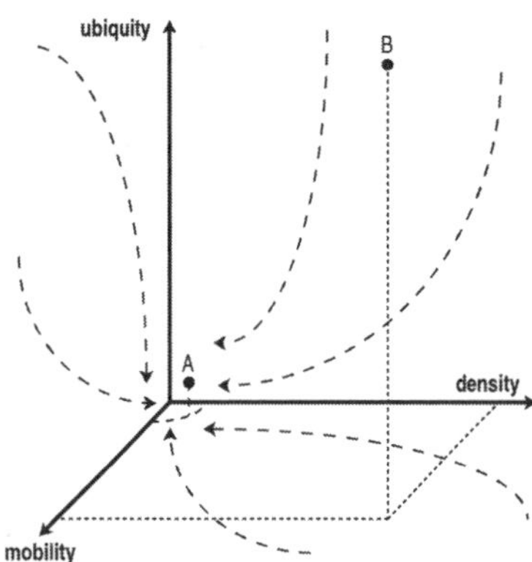

Figure 10.19. Mac Nally and Bennett's (1997) extinction proneness model based on population density, mobility and habitat specificity. The arrows show paths of increasing habitat specificity in which species B would not be at risk but species A would be highly extinction prone. (Redrawn from Mac Nally and Bennett, 1997.)

10.5 Forecasting fragmentation effects

The species least susceptible to fragmentation (and hence most likely to persist in habitat fragments) are those for which the surrounding matrix is suitable or partially suitable (Szaro and Jakle, 1985; Renjifo, 2001). The modified landscape is less fragmented for these species (Davies and Margules, 1998; Ås, 1999) because it is suitable for foraging (Yahner, 1983; McAlpine *et al.*, 1999), extends home ranges, or supports breeding (McCarthy *et al.*, 2000).

Other work on extinction proneness has focused on factors that make species susceptible to altered landscapes. Three factors emerged:

- *Mobility*. The extinction proneness of species is related to migration and dispersal (Bright, 1993). For example, empty patches are more likely to be recolonised by more mobile species (Mac Nally and Bennett, 1997).
- *Population density*. Common, numerically abundant species have a greater chance of persisting than rarer ones (Lacy, 1987; Fagan *et al.*, 2002; see Chapter 17).
- *Habitat specificity*. Species with broad habitat requirements are more likely to persist in fragmented landscapes because they can occupy or disperse across a wide range of conditions in a landscape and they can occupy larger proportions of habitat fragments (Mac Nally and Bennett, 1997).

Mac Nally and Bennett (1997) developed a generalised model for extinction proneness based on these three factors (Figure 10.19) and calculated an index of expected extinction proneness for 43 species of birds in the heavily cleared and extensively fragmented Box–Ironbark forests of north-eastern Victoria.

Mac Nally *et al.* (2000) tested their model and found it had virtually no predictive power. They suggested that the effects of mobility, density and habitat specificity were clouded by habitat suitability and inter-specific interactions between fragments and unfragmented reference areas. For instance, fragments were characterised by large populations of the Noisy Miner, an aggressive native honeyeater that drives away otherwise common mobile bird species (see Box 9.4 in Chapter 9).

Predictive ability, generality and meta-analyses

The complexity and diverse array of fragmentation effects limit the ability of conservation biologists to forecast the responses of biota to landscape change (Mac Nally and Bennett, 1997; Mac Nally *et al.*, 2000). Many studies yield results that are specific to a particular landscape, and are not readily transportable to others. Krebs (1999) noted that many researchers are disappointed when their 'general findings' do not extrapolate to other systems. A major challenge remains for conservation biologists to develop the study of habitat loss and fragmentation beyond case-specific outcomes (Mac Nally and Bennett, 1997). Predictive capabilities may improve if more studies link patterns of species occurrence with underlying ecological processes (Bunnell, 1999a).

General principles to guide landscape management might also come from syntheses of results of studies of different sets of taxa in the same landscape (e.g. Robinson *et al.*, 1992; Gascon *et al.*, 1999) and studies of a range of landscapes subject to different disturbance regimes (Lindenmayer and Franklin, 2002). **Meta-analysis** (e.g. Murtaugh, 2002; Lajeunesse and Forbes, 2003; Bengtsson *et al.*, 2005) has been used widely to synthesise studies in medicine and agriculture, and there is considerable potential for its application to fragmentation studies (Rosenberg *et al.*, 2000; Peek *et al.*, 2003), for example nest predation and brood parasitism at patch boundaries (see Rudnicky and Hunter, 1993; Kremsater and Bunnell, 1999; Lahti, 2001).

Syntheses of fragmentation studies (e.g. McGarigal and Cushman, 2002; Debinski and Holt, 2000) and model tests (e.g. Mac Nally and Bennett, 1997; Mac Nally *et al.*, 2000) have highlighted inconsistencies, the difficulties of making accurate predictions, and of determining *a priori* which species will be those most vulnerable to landscape change. Few general principles that are applicable to many landscapes and groups have emerged.

10.6 Limiting the effects of habitat loss and fragmentation

As outlined previously in this chapter, at least five key inter-related processes are associated with habitat fragmentation: loss of habitat, habitat degradation, patch subdivision and size reduction, increased patch isolation, and edge effects. Managing the impacts of fragmentation involves managing these processes (Figure 10.20).

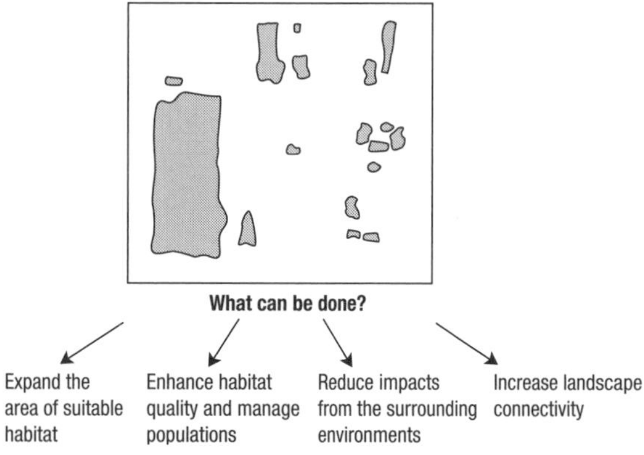

Figure 10.20. Ways to address key problems associated with habitat loss and habitat fragmentation. (Redrawn from Bennett, 2003.)

Limiting and reversing habitat loss

Slowing habitat loss has important political and social dimensions. Policy instruments to reduce the rate of habitat loss and its effects on biodiversity can include new environmental legislation, incentives to retain and manage native vegetation (e.g. the BushTender Scheme in Victoria; see Box 4.7 in Chapter 4), disincentives (e.g. fines and other penalties), compensation packages for landowners who elect to conserve vegetation, and purchasing land for conservation. Some of these approaches are discussed in more detail in Chapter 8, but a detailed exploration of them is beyond the scope of this book (see Whitten *et al.*, 2002).

Increasing the area of habitat in a landscape should increase species richness and the sizes of populations. In woodlands, for example, reduced grazing by domestic livestock can improve tree and shrub regeneration. In a forest example, McCarthy and Lindenmayer (1999) showed that the most effective way to increase the area of habitat for Greater Gliders was to expand the size of old growth patches by maintaining a harvest-free zone around them.

Active replanting and revegetation can be a very long-term process to reverse habitat loss, especially if important habitat features take a long time to develop. Vegetation restoration is a large and complex topic; detailed discussions are provided by Saunders *et al.* (1993), Bennett *et al.* (2000), Hobbs and Yates (2000), and Salt *et al.* (2004).

Natural regeneration around a paddock tree at Table Top near Albury, New South Wales. (Photo by David Lindenmayer.)

Natural regeneration may have biodiversity value but may also be economically detrimental for landowners. 'Vegetation thickening' of regrowth woodland in Queensland concerns private landowners and is one of the reasons why broad-scale clearing has occurred in that state (see Chapter 9). Apart from the **focal species approach** (Lambeck, 1997; Watson *et al.*, 2001; Box 10.14), there are few scientifically defensible principles to guide the extent and location of habitat restoration efforts for biodiversity conservation (Bennett *et al.*, 2000; Salt *et al.*, 2004).

Box 10.14

The focal species approach and vegetation restoration

The focal species approach involves the identification of a suite of species to guide the management of threatening processes and vegetation restoration. Together, the set of focal species have '*requirements for persistence [which] define the attributes that must be present if [the landscape] is to meet the needs of the remaining biota*' (Lambeck, 1999). Brooker (2002) applied the focal species approach using bird assemblages in fragmented landscapes of Western Australia. The study identified the most patch-area-sensitive and patch-isolation-sensitive bird species and presented maps showing where strategically important revegetation efforts might be focused to best conserve those species. This method assumes that the strategies will provide the resources that are necessary to conserve other taxa in the same landscape (Brooker, 2002).

Lindenmayer and Fischer (2003) examined the work of Brooker (2002) and raised some concerns about the method. For example, the responses of birds, which are the sole group for which the focal species approach has been applied, may not be good species-based indicators of the responses of other vertebrates (for example reptiles, amphibians and mammals), for which the selected focal species is often purported to be a surrogate. Perhaps the main reservation is that, in the absence of detailed population models supported by extensive field data, it can be extremely difficult to identify those species that are most sensitive to given threatening processes (Lindenmayer *et al.*, 2002e). The approach would be more robust if the assumptions about species sensitivities were supported by independent evidence and were subsequently tested with new field data (Lindenmayer and Fischer, 2003).

Maintaining habitat quality

An effective strategy to limit the impacts of habitat fragmentation is to maintain suitable habitat at multiple spatial scales (Lindenmayer and Franklin, 2002). As outlined earlier in this chapter and discussed in detail in Chapter 15, habitat is species-specific, but some habitat attributes can be critical for many species. In forest environments, these include large living trees, large and dead living trees with hollows, fallen trees, and logs and thickets of understorey trees and shrubs. Retaining these structures may require the management of activities that could otherwise threaten them, such as firewood harvesting, frequent prescribed burning, weed invasion, and high-intensity grazing.

Increasing connectivity

'Natural' landscapes – those relatively little disturbed by humans – have high levels of **connectivity**. Noss (1991) defined connectivity as: 'linkages of habitats…communities and ecological processes at multiple spatial and temporal scales'. If landscapes are connected it implies that species persist, recolonise empty habitat patches, and exchange individuals and genes among subpopulations (see Chapter 16).

Landscapes that retain more connections between patches of otherwise isolated habitat are assumed to

Box 10.15

Problems with a lack of connectivity: Fairy-wrens in Western Australia

Habitat subdivision, increased patch isolation and decreased connectivity are considered to be important processes influencing plant and animal populations. However, very few field studies have empirically demonstrated the relationships between connectivity and population persistence. One of the best studies to date is on the Blue-breasted Fairy-wren in the Wheat belt of Western Australia (Brooker and Brooker, 2001, 2002). Brooker and Brooker studied coloured banded birds in woodland and heath remnants of different sizes over 6 years (1993–98), and were able to demonstrate that low connectivity reduced recruitment to unoccupied patches below the levels required to replace mortality. Subsequent calculations by Smith and Hellman (2002) using data from Brooker and Brooker (2002) showed that annual population growth rates of +5% characterised well-connected networks of patches.

Blue-breasted Fairy-Wren. (Photo by Graeme Chapman.)

be more likely to maintain populations of the species that inhabited the original landscape (Brown and Kodric-Brown, 1977; Haddad and Baum, 1999). Angermeier (1995) showed that a lack of connectivity contributed to extinction proneness in fish. Dispersal pathways maintain or increase demographic and genetic population size (Leung *et al.*, 1993; Mills and Allendorf, 1996), thereby enhancing the chances of long-term persistence (Scotts, 1994).

For plants, connectivity may include movements of individuals and propagules such as spores, pollen and seeds. For animals, connectivity involves five broad types of movement (modified from Hunter, 1994):

- Day-to-day movements such as those within home ranges or territories. These can be small for species such as adult frogs or large in the case of wide-ranging animals such as bats (e.g. Lumsden *et al.*, 1994, 2002) or large vertebrates such as the Black Bear in North America (Klenner and Kroeker, 1990).
- Dispersal between the **natal territory** and suitable habitat patches (Wolfenbarger, 1946). These movements are typically made by juvenile or subadult animals attempting to establish new territories (Stenseth and Lidicker, 1992).
- Annual patterns of long-distance migration, which can span continents or hemispheres (Keast, 1968; Flather and Sauer, 1996).
- Nomadic movements made in response to temporal and spatial variability of important resources (e.g. food for wide-ranging rainforest pigeons; Price *et al.*, 1999).
- Large shifts in distribution in response to climate change. These changes have typically been slow in the past (Keast, 1981), but more rapid and extreme changes are expected in response to global climate change (Peters and Lovejoy, 1992; Hughes, 2004; see Chapter 5).

Connectivity is species-specific: it is an outcome of dispersal behaviour, mode of movement (e.g. whether the main form of movement is flying or crawling), and how these interact with landscape patterns. For these reasons, a map of vegetation cover may not reveal whether a landscape is connected or fragmented for a given species (Ingham and Samways, 1996; Wiens *et al.*, 1997; With, 1999; Fischer *et al.*, 2004).

Wildlife corridors as a way to maintain connectivity

Wildlife corridors have been proposed to maintain connectivity and limit the effects of habitat fragmentation. They are believed to:

- facilitate the movement of animals through suboptimal habitat
- provide habitat for resident populations
- provide access to unexploited habitat
- enhance dispersal success by reducing mortality during dispersal (Beier, 1993)
- prevent and reverse local extinctions by allowing empty patches to be recolonised
- promote the exchange of genes between subpopulations, thus increasing the effective population size and reducing genetic drift and inbreeding depression.

Some species benefit from wildlife corridors that link suitable habitat (Gilbert *et al.*, 1998; Haddad and Baum, 1999), for example those species that avoid open areas (Martin and Karr, 1986; Desrochers and Hannon, 1997) and those that disperse only through suitable habitat (e.g. Baur and Baur, 1992; Nelson, 1993).

Corridors do, however, have potential disadvantages. For example, they can facilitate the spread of genes that break up co-adapted gene complexes in naturally isolated populations. They can exacerbate the spread of weeds, pest animals, diseases and fires. Corridors can connect high-quality habitat patches to areas of poor-quality habitat that act as population 'sinks'. If individuals disperse when populations are below the carrying capacity of the local environment, sinks may be detrimental to metapopulation

Table 10.7. Factors influencing wildlife corridor use (based on Lindenmayer, 1994, 1998).

Factor	Example
Target species characteristics	Foraging strategy/colonial versus solitary social system (Recher *et al.*, 1987)
Biotic interactions	Aggressive inter-specific behaviour (Catterall *et al.*, 1991)
Abiotic edge effects	Microclimatic conditions (e.g. light regimes) reduce habitat suitability (Hill, 1995)
Dispersal behaviour	Random dispersal versus movement along habitat gradients (Murphy and Noon, 1992)
Habitat suitability within corridor	Structural features influence movement (Lindenmayer *et al.*, 1994; Bowne *et al.*, 1999; Merritt and Wallis, 2004)
Corridor characteristics	Corridor width and length (Andreassen *et al.*, 1996)
Topographic location and variation	Confined to a gully or capturing multiple topographic positions (Claridge and Lindenmayer, 1994)
Vegetation gaps	Roads and tracks through corridors pose barriers to movement (Lindenmayer *et al.*, 1994)
Size of areas connected	Small connected patches provide few dispersalists to move through corridor (Wilson and Lindenmayer, 1996)
No. other corridors	Influences chances of corridors being found by an animal when its moving (Forman, 1995)
Matrix condition	Clear-cut versus selectively logged adjacent forest

persistence. If vegetation loss continues in the matrix surrounding a corridor, the corridor may make only a limited contribution to biodiversity conservation (Rosenberg *et al.*, 1997; Harrison and Bruna, 2000).

Not all species use corridors, and their usefulness may depend on the ecology of the species in question, habitat suitability, attributes of the corridors such as their width and length, the location of corridors in the landscape (e.g. the range of landscape features captured within a corridor such as gullies and ridges), the type and pattern of land-use patterns in the matrix surrounding the corridors, and the size and value for fauna of habitat connected by corridors (Table 10.7).

It is important to identify explicitly both the species targeted for conservation and the principal objectives of establishing a network of wildlife corridors within a given landscape (Hobbs, 1992; Lindenmayer, 1998). On this basis, key questions in corridor design and establishment include:

- Which species move between habitat patches without corridors and which species depend on corridors and to what degree (Beier and Noss, 1998)?
- How is corridor use influenced by the conditions and human activities in the surrounding landscape (Rosenberg *et al.*, 1997)?
- Is a corridor to function solely as a conduit for movement or also to provide suitable habitat?

Strip of retained native vegetation within stands of plantation Pine at Tumut in southern New South Wales. These areas act as wildlife corridors for some species of small mammals. (Photo by David Lindenmayer.)

Specifically built grassed overpass in Banff National Park, Alberta (Canada), to reduce collisions between large mammals and motor vehicles and improve connectivity for wildlife populations. (Photo by David Lindenmayer.)

- How suitable are areas being connected by the corridor for the species of interest?

When considering the advantages and disadvantages of maintaining and/or establishing networks of wildlife corridors, the value of the corridor strategy depends on the ecology of the species in question and the types of change corridors are intended to mitigate. It may not be possible to design corridor systems that conserve all species that are vulnerable to the effects of habitat fragmentation (Box 10.16). Perhaps most importantly, the reservation, rehabilitation or maintenance of corridors may be relatively expensive. An uncritical insistence on corridors may divert financial resources from other potentially more appropriate or less costly conservation strategies, such as setting aside large consolidated protected areas (Simberloff *et al.*, 1992).

Box 10.16

Wildlife corridors and arboreal marsupials

Wildlife corridors, or strips of retained habitat, are used to mitigate the impacts of timber harvesting on forest fauna in wood production forests in many States of Australia. Typically, the corridors are unlogged strips in riparian areas (streamside reserves) that maintain water quality as well as play a role in the conservation of wildlife that may be sensitive to the impacts of forestry operations.

Lindenmayer *et al.* (1993a) surveyed arboreal marsupials in 49 retained linear strips that varied from 40 to 250 metres in width. Each strip was surrounded by young recently cut forest that was unsuitable for animals. The aim of the study was to determine if the strips were being used as habitat by possums and gliders. Two questions were examined:

1. Is the number of corridors occupied by different species of arboreal marsupials similar to that expected, based on the suitability of the forest in the strips for these animals?
2. Are there any features of the strips (for example their width or length) that make them more likely to be used by arboreal marsupials?

The habitat requirements of several species of arboreal marsupials within intact forest were known from earlier studies (Lindenmayer *et al.*, 1991a), providing the basis for estimating the suitability of the habitat for arboreal marsupials in each of the retained strips. Four out of five species, including the two most common ones, were observed with a frequency that did not differ substantially from expectations, based on the quality of the habitat. In the case of Leadbeater's Possum, 17 sites supported suitable habitat, but the species was recorded from just one. Therefore, even though suitable forest occurred in many sites, other factors appeared to preclude the use of the retained linear strips by Leadbeater's Possum.

Animals with a complex social system (e.g. those that live in colonies) and that consume widely-dispersed food (e.g. large flightless insects), were relatively rare in the retained strips. It seems likely that the narrow linear habitat strips make it difficult for these species to harvest food efficiently and undertake some aspects of group social behaviour. The species that were most commonly observed in wildlife corridors have a diet comprising readily available food (e.g. leaves) and have a relatively simple social system; that is, they are solitary or live in pairs. These findings indicate that although setting aside networks of wildlife corridors may be a valuable strategy for some arboreal marsupials, they may not be effective for the conservation of all species.

Some features of the strips were related to their use by arboreal marsupials (question 2). Significantly more animals occurred in strips that supported numerous trees with hollows, probably because these trees provide nest sites for arboreal marsupials. In addition, the position of the strips in a forest landscape influenced their use. More animals were found in corridors that spanned forests on different parts of the topographic sequence (e.g. linked gullies to ridges) than sites confined to only a gully or a mid-slope. The reasons for this result may be related to the need for some animals to move through different parts of the forest to harvest a range of types of food (Claridge and Lindenmayer, 1994). The results highlight the importance of a good understanding of the biology and ecology of those species that it is thought can be managed by establishing networks of wildlife corridors.

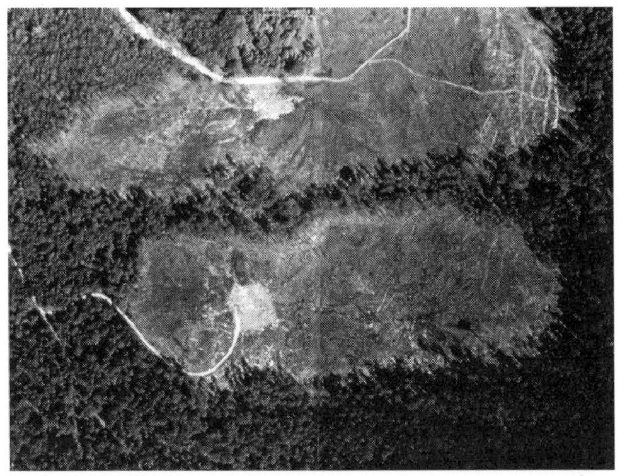

One of the wildlife corridors targeted for the study of arboreal marsupials in the Central Highlands of Victoria (see Box 10.16). (Photo by David Lindenmayer.)

Other approaches to enhancing connectivity

Connectivity can be achieved by means other than corridors. Strategic management, for example retaining trees during the logging of forest landscapes, or modifying the landscape surrounding habitat remnants, can facilitate animal and plant dispersal (Forest Ecosystem Management Assessment Team, 1993). If species can disperse through the matrix, it may lessen the need for corridors (Rosenberg *et al.*, 1997) or reduce the corridor widths required (Forman, 1995; Lindenmayer, 1998). Few dispersal events may then be required to 'rescue' populations in habitat remnants (Stacey and Taper, 1992; Mills and Allendorf, 1996). Conversely, if conditions in the matrix are hostile, then larger corridors linking larger retained patches will be required to maintain connectivity (Taylor *et al.*, 1993; Sarre *et al.*, 1995).

Management of the matrix may provide better connectivity than wildlife corridors for species that disperse randomly in landscapes (e.g. Eastern Screech-Owl; Belthoff and Ritchison, 1989). In Australia, suitable matrix habitat supports migratory and nomadic species such as native pigeons and honeyeaters (Date *et al.*, 1996; Price *et al.*, 1999). Connectivity via stepping stones or dispersed 'islands' of potentially suitable habitat may be the best strategy for these and other mobile species (e.g. butterflies and bats; Lumsden *et al.*, 1994; Schultz, 1998; Law *et al.*, 1999).

Most existing knowledge about connectivity, corridors and the matrix comes from theory (e.g. Murphy and Noon, 1992) and simulation modelling (e.g. Burkey, 1989; Boone and Hunter, 1996). Because empirical data linking matrix conditions and connectivity are limited (Nicholls and Margules, 1991; but see Brooker and Brooker, 2002), carefully designed field studies of the relationships between matrix conditions and connectivity are urgently required.

Reducing edge effects: buffer systems

It is possible that habitat fragments may not maintain their species composition because of edge effects; in these situations, **buffers**, which are natural areas that surround and protect sensitive habitats, can be helpful. Buffers may need to have different attributes for different purposes; for example, the width of buffer strips to protect riparian areas from pesticides and to reduce in-stream invertebrate mortality (more than 50 metres in Australian eucalypt plantations; Barton and Davies, 1993) may be quite different from that required to mitigate changed wind patterns, which may otherwise penetrate several hundred metres from the boundaries of logged forests (Harris, 1984; Saunders *et al.*, 1991).

The aim of a buffer may be to limit the impacts of a disturbance regime or maximise species diversity within a protected area (Baker, 1992). Kelly and Rotenberry (1993) formalised a general approach for buffer zone design through a set of inter-related questions:

- What external forces or processes are likely to have an impact on the protected entity (species, community or resource)?
- To what extent are the external forces likely to penetrate the boundary of the protected area or sensitive site and result in negative impacts?
- Can these forces be ranked in terms of their impacts to support a priority list of buffering requirements?
- Are the potentially negative forces amenable to hypothesis testing?
- How can data be gathered to test these hypotheses?
- How can the external forces be mitigated?

Burgman and Ferguson (1995) considered several of these issues in assessing threats to cool temperate rainforest fragments in Victoria. They recommended:

- improvements to the planning and mapping of forest landscapes
- adoption of a system of buffers
- establishment of special protection zones around rainforest areas

Cool temperate rainforest in central Victoria – a vegetation type that requires buffering from disturbances such as logging of adjacent wet eucalypt forest. (Photo by David Lindenmayer.)

- modification of timber harvesting practices in eucalypt stands adjacent to rainforest
- planning road construction to avoid sensitive areas
- undertaking new research to fill knowledge gaps and assess hypotheses regarding the sensitivity of rainforest communities to human activities.

Noss and Harris (1986) outlined a strategy termed the 'multiple-use-modules' (MUMs), whereby concentric management zones in the matrix buffer a core reserved area. Mladenoff *et al.* (1994) described 100-metre restoration zones around remnant old growth patches in second-growth matrix lands designed to buffer and reduce edge effects. Unlogged or selectively harvested stands can have a positive buffering effect for adjacent sensitive areas (Recher *et al.*, 1987; Macfarlane and Seebeck, 1991). Similarly, planning of the spatial alignment of harvest units can mitigate the impacts of abiotic effects such as wind damage (Lindenmayer *et al.*, 1997; Perry, 1994). Aquatic zones can be buffered by staggering logging operations to ensure that both sides of a riparian buffer are not logged at the same time (Recher *et al.*, 1987). All of these approaches have 'adjacency rules', which specify waiting periods before an area neighbouring a regenerating harvest unit can be cut.

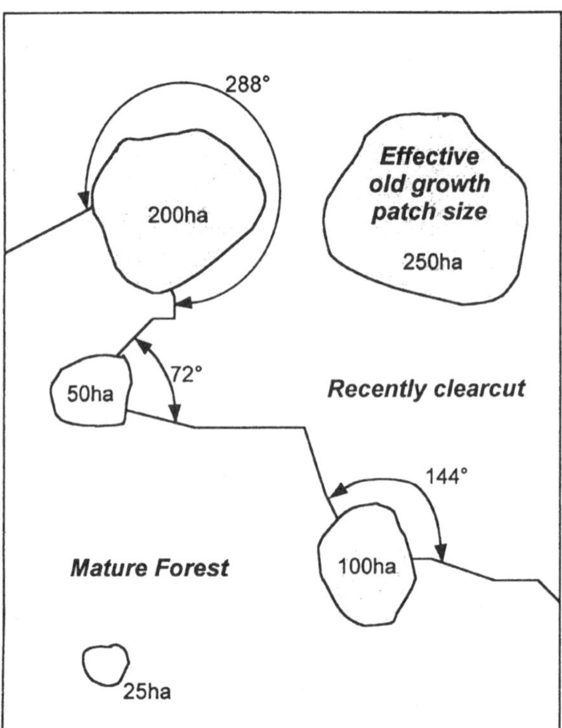

Figure 10.21. Buffering effect of surrounding matrix conditions on stands of old growth forest. The diagram shows the size of an area of old growth forest needed to maintain interior conditions in a matrix dominated by recently cut forests (250 hectares in size) contrasted with one surrounded by mature forest (25 hectares in size). (Redrawn from Harris, 1984.)

To illustrate the potential value of buffering, Harris (1984) showed how an old growth patch bounded by a recently clear-felled area may need to be an order of magnitude larger than one surrounded by mature forest to achieve the same area of interior forest habitat (Figure 10.21). In the case of fire risks and edge effects, mature forest buffers may reduce the chance of a fire in an old growth patch (Harris, 1984) because the probability of ignition and spread declines with increasing age in some forest types (Agee and Huff, 1987).

General principles for landscape management to mitigate habitat loss and fragmentation

Most of Section 10.7 focused on ways in which the effects of habitat loss and fragmentation on biodiversity can be reduced. Surprisingly, in comparison with studies of fragmentation effects, much less work in conservation biology has focused on how these types of mitigation strategies might be further developed and integrated within a whole-of-landscape approach. There are no generic guidelines for landscape design; rather, the specific outcomes of a landscape conservation strategy depend on the objectives of management, differences in the biota between different vegetation types and regions, and variations in the responses of the biota to disturbance. Therefore, the best strategies will vary according to the conservation objectives. Nevertheless, Lindenmayer and Franklin (2002) provided some broad principles to guide the conservation of biodiversity in any landscape, in which the key principle was to maintain habitat suitability across a range of spatial scales. They proposed five broad strategies to help achieve the goal of habitat maintenance:

- maintaining connectivity
- maintaining landscape heterogeneity
- maintaining structural complexity (the range and spatial arrangement of features of stands of trees, shrubs or other forms of vegetation)
- maintaining hydrological and geomorphological processes
- using knowledge about natural disturbances to guide more ecologically sustainable human disturbance regimes and develop improved resource management practices.

Suitable habitat, connectivity, stand complexity, landscape heterogeneity and ecosystem integrity are defined on a species-specific basis. Because defining these variables for many species is essentially impossible, a practical response is to create a range of conditions, thereby providing habitat for different species in at least some parts of a landscape. Strategies at three scales are available (Lindenmayer and Franklin, 2002): (1) very large-spatial-scale strategies (large ecological reserves), (2) mid-spatial-scale or landscape-level strategies (e.g. protected areas within production lands), and (3) stand-level approaches (e.g. the retention of individual structural features within patches.

10.7 Conclusions

Landscape change and associated habitat loss and fragmentation are massive topics in conservation biology, and encompass many diverse and inter-related processes (see the panchreston problem discussed in Box 10.3). As noted by Bissonette and Storch (2002): '*straightforward predictions do not capture the multi-causal nature of organism response to fragmentation*'. Despite this sobering conclusion, there do appear to be some broad and pragmatic generalisations that can be made about the field of landscape change and habitat fragmentation. These represent a platform from which additional insights might come and from which improved landscape management might evolve. The generalisations are as follows:

- No single model of habitat will work best in all circumstances. The most appropriate will depend on the objective of a study, the landscape, and attributes of the species or group targeted for study.
- Models of landscape cover that define patches on the basis of human perceptions and characterise patches with sharp and well-defined boundaries will often be inappropriate for many species. In many cases there will not be complete congruence between the amount of suitable habitat and the amount of native vegetation that remains in an altered landscape.
- Habitat loss and habitat fragmentation are often confounded, but their effects should be evaluated separately. In most cases, habitat loss will have significantly greater negative impacts on biodiversity than will habitat subdivision.
- Although species diversity and species occurrence patterns have been well documented, better links between pattern and process should improve predictions and the chances of identifying general principles.
- Habitat is species-specific, as are responses to edge effects, habitat loss and habitat fragmentation. Responses (even for the same species) may not be able to be reliably extrapolated to different landscapes.
- A range of strategies can be employed to address the impacts of landscape change. These need to be aimed at countering the processes that accompany landscape alteration, including habitat loss, habitat fragmentation, patch isolation, and edge effects.

10.8 Practical considerations

Models are unlikely to be useful for precisely predicting the consequences of habitat loss and fragmentation.

They can, however, be useful for ordering ideas, exploring data requirements and perhaps ranking management options for mitigating the effects of habitat loss. Conceptual landscape models will be most useful if they account for the potential effects of conditions in the matrix, include diffuse boundaries between habitat and non-habitat, allow definitions of habitat that are species-specific, and provide a framework for dynamic changes in landscapes.

There is a growing collection of attempts to study ecosystem processes that have been unsuccessful. Confounding factors and lack of replication undermine many of these studies. We recommend that statisticians play a central role in designing spatial analyses and fragmentation experiments, from true experiments to observational studies.

10.9 Further reading

As outlined at the start of this chapter, the literature on habitat loss and habitat fragmentation is vast and sometimes complex. Some of the citations below provide a lead into some of the key topics.

Shafer (1990) gives a detailed explanation of island biogeography, although it is not a particularly critical appraisal. Burgman *et al.* (1988), Doak and Mills (1994) and Haila (2002) are shorter but more critical assessments of the generic value of the island theory. Despite the clear problems associated with the island biogeography theory that have emerged in recent years, the original text on the subject by Macarthur and Wilson (1967) is well written and worth reading. It also examines many more topics in ecology than simply the island biogeography theory.

Forman (1995) is a seminal book on landscape mosaics and contains an extensive set of background references on a wide range of topics covered in this chapter. The habitat variegation model as an alternative model to the corridor-patch-matrix model of landscape cover is described by McIntyre and Barrett (1992) and McIntyre and Hobbs (1999). Fischer *et al.* (2004) discuss an integrated model that attempts to conjointly extend the best features of the corridor-patch-matrix and habitat variegation models. Burgman *et al.* (2005) is a useful paper describing landscape modelling.

Numerous books and articles examine the many subtopics associated with habitat loss and habitat fragmentation. An outstanding set of studies in the highly fragmented wheat belt of Western Australia has been published in the series of books by Saunders *et al.* (1987, 1993) and Saunders and Hobbs (1991). A valuable set of papers on fragmented tropical systems appear in the edited volume by Laurance *et al.* (1997); the volume contains several valuable reviews of habitat loss and habitat fragmentation, which provide a useful entry into this massive and complex topic. The following papers give quite different but informative perspectives: Davies *et al.* (2001), Haila (2002), McGarigal and Cushman (2002), Bennett (2003), Fahrig (2003) and Hobbs and Yates (2003). Debinski and Holt (2000) provide a useful review of the limited number of fragmentation experiments that have been conducted to date.

There are useful reviews of several of the major topics within the broad field of habitat fragmentation. For example, Saunders *et al.* (1991), Laurance *et al.* (1997), Kremsater and Bunnell (1999), and Lindenmayer and Franklin (2002) give syntheses of material on edge effects. Rudnicky and Hunter (1993), Paton (1994) and Lahti (2001) provide more detailed examinations of nest predation and parasitism. A range of aspects of corridor biology have been reviewed (see Saunders and Hobbs, 1991; Hobbs, 1992; Lindenmayer, 1994, 1998; Beier and Noss, 1998; Bennett, 1998), but their value for the conservation of biodiversity remains controversial (Rosenberg *et al.*, 1997). This issue has been debated vigorously in the conservation biology literature (e.g. Noss, 1987; Simberloff *et al.*, 1992). Part of the debate is caused by the fact that there is remarkably little information available by which to rigorously assess the effectiveness of networks of wildlife corridors (Hobbs, 1992; Simberloff *et al.*, 1992; Lindenmayer and Franklin, 2002), although an increasing number of studies are showing that they will be useful for some species (e.g. see Saunders and de Rebeira, 1991; Bennett, 1998; Haddad and Baum, 1999).

Kelly and Rotenberry (1993), Burgman and Ferguson (1995), and Semlitsch and Bodie (2003) discuss some practical approaches for the design of buffers to better protect particular elements of biodiversity from the effects of habitat fragmentation such as edge effects.

Chapter 11

Fire and biodiversity

This chapter examines a range of aspects of the impacts of fire on biodiversity in Australia. Two broad types of fires are discussed – wildfires and prescribed fires. Their impacts on plants and animals are discussed. Some species that are vulnerable to the effects of fire and ways to mitigate fire impacts are outlined. Other areas that are briefly explored in this chapter include fire and reserve design, the importance of biological legacies in ecosystem and species recovery after fire, relationships between fire and logging, and the relevance of the intermediate disturbance hypothesis.

11.0 Introduction

Fire has global social, economic, health, and ecological consequences (UNEP, 1999; Gill, 2001; Bradstock *et al.*, 2002). Each year, huge areas of vegetation are burned by wildfires. For example, in 1997–98, the Food and Agriculture Organization (FAO, 2000) estimated that nearly 10 million hectares in Indonesia burned, costing up to US$10 billion and influencing the health of 75 million people. During that time there were major fires in Russia (7 million hectares), Mongolia (2.7 million hectares), Brazil (4 million hectares), and Mexico (800 000 hectares). In Australia, there were major fires in the south-east in 2002 and 2003 (Cary *et al.*, 2003), and extensive areas of northern Australia are burned every year (Commonwealth of Australia, 2001e; Andersen *et al.*, 2003).

Fire influences the majority of Australian terrestrial ecosystems, and many endemic Australian species are threatened by inappropriate fire regimes (State of the Environment, 2001a; Bradstock *et al.*, 2002; Brook *et al.*, 2002b; Keith *et al.*, 2002a). For groups such as birds, only vegetation clearing threatens more species. Attempts to control or prevent high-intensity fire attract public interest; there are conflicts between interests, in part because there is an ongoing expansion of human settlements into areas of flammable and fire-prone native vegetation (Whelan, 2002).

Fires burn differently in different vegetation types, even when they have the same fuel load and are adjacent to one another (Whelan, 1995); apparently similar fires can have very different effects. A **fire regime** is the sequence of fires typical of a given area. A fire regime has four key components (after Gill, 1975): (1) fire intensity, (2) fire type (e.g. crown or ground fire), (3) between-fire interval (or frequency), and (4) season. Organisms respond to all attributes of a fire, including post-fire conditions such as the amount of precipitation and the features that remain unburned after the fire.

This chapter is limited to a relatively brief appraisal of the relationships between fire and the conservation of biodiversity. Much of this chapter is based on seminal material on fire and biodiversity in Gill *et al.* (1981, 1999), Whelan (1995), and Bradstock *et al.* (2002).

11.1 Brief history of fire in Australia

The presence of charcoal in pollen samples collected from swamps around Australia shows that fire has been a prominent part of the Australian landscape for millions of years (Kershaw *et al.*, 2002). Long-term trends toward a drier climate (since the late Tertiary) coupled with periodic droughts and flammable vegetation probably meant that before the arrival of humans, fire periodically burned large areas of vegetation (Woinarski, 1999). The historical record shows an increase in fire activity about 40 000 years ago, which appears to correspond with increased Aboriginal activity (Kershaw *et al.*, 2002). Fire was used extensively (and still is used) by Aboriginal people (Bowman, 1998, 2003). From an evolutionary perspective, Bowman (2003) argued that: '*one of the great triumphs of the Pleistocene Australians was the taming of wildfires through the development of igniculture*'.

Aboriginal people use fire for many reasons:

- to smoke out animals such as possums and gliders from trees (Kerle, 2001; Lindenmayer, 2002b) or to trap smoke-tracking species such as some types of raptors (Boekel, 1980)
- to encourage the development of grasses for herbivores such as Kangaroos (Johnson *et al.*, 1989)
- to stimulate the growth of particular food plants ('firestick farming', sensu Jones, 1969)
- as part of warfare (Martin and Handasyde, 1999)
- to communicate (by signalling; Jones, 1969)
- to clear paths for travel
- to manage fuel loads and landscape mosaics of woodlands and rainforests (Langton, 1998).

Given the widespread use of fire (Bowman, 1998), it is probably not surprising that James Cook referred to Australia as a 'continent of smoke' during his exploration of the east coast in 1770 (Martin and Handasyde, 1999).

There were strong inter-relationships between Aboriginal people, fire, and food supply, as noted by the explorer Thomas Mitchell:

> *Fire, grass, kangaroos and human inhabitants, seem all dependent on each other for existence in Australia; for any one of these being wanting, the others could no longer continue (Mitchell, 1848, p. 412).*

Langton (1998) points out that the environmental concerns of Aboriginal people go beyond consumption and the exchange of goods. Fire has spiritual meaning; for instance it is used to 'clean the country' after the death of landowners.

Fensham and Fairfax (1996) reported grasslands reverting to rainforest in the absence of Aboriginal burning in the Bunya Mountains of south-eastern Queensland. Regular burning maintained Cypress Pine stands in western New South Wales as open woodlands with an extensive grassy ground cover, but these areas reverted to denser forest when European settlers discouraged burning (Jeans, 1972). Similar changes have been reported for the forests near Sydney (Mitchell, 1848) and in the valley floors of eastern Victoria (Howitt, 1890). Such patterns of fire management and the vegetation cover that resulted might have benefited species that are vulnerable to high-intensity fire (see Box 11.1), but disadvantaged others that prefer habitats with a dense ground cover or understorey.

Bowman (2003) speculated that indigenous fire regimes would have had negative effects on organisms dependent on long-unburned fire refugia, such as several

Box 11.1

The vulnerability of the Koala to fire

The Koala is an Australian 'icon' species that is well known to all of the nation's inhabitants and probably the vast majority of people overseas as well. As an arboreal and slow-moving animal that does not use hollows where it can shelter, the Koala is particularly vulnerable to the impacts of wildfires that burn the tree canopy (where the species feeds and roosts). It is notable that even in the early part of the 20th century, leading naturalists such as Fred Lewis, the Chief Inspector of Wildlife in Victoria, regarded wildfire (and not land clearing and shooting for the fur industry) as the key process threatening populations of the Koala (Martin and Handasyde, 1999). It is possible that frequent low-intensity burning of the ground and understorey layers of forests and woodlands by Aboriginal people (e.g. see Howitt, 1890) limited the number of high-intensity conflagrations prior to European settlement. However, populations of Aboriginal people were substantially reduced following European settlement and many of their approaches to land management were no longer practiced. Several extensive wildfires (such as those in 1851, 1898 and 1939) probably contributed significantly to the decline of Koala populations in the forests and woodlands of Victoria. Although prescribed burning does now occur in some parts of the forest estate, frequent major wildfires are still common in eastern Australia (such as those in 1983, 1994 and 2003) and these are likely to harm Koala populations (Martin and Handasyde, 1999).

Cape Barren Island in Bass Strait, where European settlers used fire extensively as part of hunting wallabies. (Photo by David Lindenmayer.)

now-extinct genera of leaf-eating kangaroos (e.g. *Stenurus*, *Simosthenurus* and *Procoptodon*), and probably contributed to their extinction. He also noted that Aboriginal fire regimes maintain some species, for example the Cypress Pine (*Callitris intratropica*) in northern monsoonal Australia (Bowman, 1998). Irrespective of such speculations, the Australian environment of 200 years ago was the result of millennia of Aboriginal environmental management, and fire was one of their most important land management tools (Bowman, 2003).

Aboriginal fire regimes are tailored to specific locations and conditions to achieve particular objectives (Langton, 1998). In some landscapes, for example the tall montane forests of Victoria, fire does not appear to have been employed routinely. These are cold and wet environments that do not burn readily. In addition, the dense vegetation, tall trees and large quantities of fallen debris characteristic of these areas could have made hunting inefficient relative to the more benign locations nearby.

European settlers, like the Aboriginal people they partially or totally displaced in many parts of Australia, also often used fire in land management, primarily to assist with land clearing. In landscapes such as those on

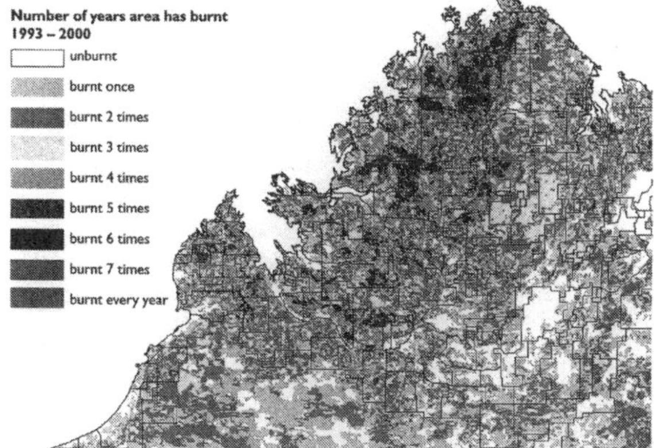

Figure 11.1. Satellite data showing the extent and frequency of burning in the Kimberley. (Sourced from the Commonwealth of Australia, 2001e.)

Box 11.2

Fires in northern Australia

In more recent times, there has been increasing attention focused on fire regimes in northern Australia. Organisations such as the Tropical Savannas Management Cooperative Research Centre are researching many aspects of fire impacts on ecosystems and biodiversity. In addition, satellite data are being increasingly used for examining the timing and extent of burning (Bowman *et al.*, 2004; Figure 11.1) – particularly according to Interim Biogeographic Regionalisation for Australia region (as defined in Chapter 4; State of the Environment, 2001a). For example, data from the National Land and Water Resources Audit show that extensive areas of places such as the Kimberley in north-western Australia burn as frequently as three or more times in 7 years (Commonwealth of Australia, 2001e). Moreover, the amount of land burned in northern Australia is increasing annually and in 1998 it was five times greater than in the 4 years previously (25 million hectares versus 5 million hectares; Commonwealth of Australia, 2001e). Other work shows that there are major differences between current fire regimes in areas such as the savanna landscapes and those used by indigenous Australians (State of the Environment, 2001a; Andersen *et al.*, 2005). The timing of these fires are different – fires often now occur much later in the dry season than previously occurred and are extremely intense, which can negatively affect many elements of biodiversity (Pardon *et al.*, 2003), including small mammals (Andersen *et al.*, 2005). The fires that occur now are also non-systematic, extensive and frequent, and result from uncontrolled fires that began elsewhere. This is in marked contrast to the patchy and heterogenous fire regimes used by indigenous people (State of the Environment, 2001a; Bowman *et al.*, 2004). A major management challenge in ecosystems such as the tropical savannas of northern Australia will be to maintain long unburned areas (Andersen *et al.*, 2005).

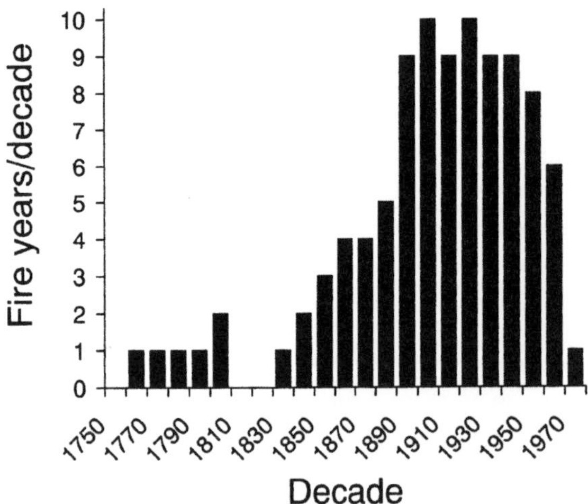

Figure 11.2. Temporal trends in fires per decade in stands of Snow Gum woodland. (Redrawn from Banks, 1982, and Cary, 2002.)

Cape Barren Island in Bass Strait, fire was also used to improve hunting. European settlers burnt vegetation to stimulate grass growth and then captured the wallabies that grazed there (Edgecombe, 1999).

Once fences and buildings became features of 'settled land', the use of fire by Aboriginal people and European settlers was actively discouraged by the European settlers. Hence, fire regimes in Australia have changed considerably since European settlement (King, 1963; Williams and Gill, 1995; Bowman, 1998; Ward *et al.*, 2001). This is illustrated by the change in the frequency of fires in Snow Gum woodland since 1770 (Banks, 1982; Figure 11.2).

Bowman (2003) speculated that the 'tamed fire regimes' developed as part of igniculture then became 'feral' following the cessation of Aboriginal fire management. Managing fire is now a challenge for land managers given the extent of infrastructure that now characterises many landscapes, including bushland –urban interfaces (Whelan, 2002). It seems likely that solutions will be found in the approaches developed over thousands of years by indigenous Australians (Bowman, 1998; Whelan, 2003; see Box 11.3). Indigenous fire management varied between vegetation types, landscapes and regions; so, accordingly, there is no single 'recipe' for all places (Baker, 2003). Note, however, that reimposing past regimes may not achieve particular conservation goals, for example maintaining rare plant species that have little food or other value for indigenous people (Keith *et al.*, 2002a).

Box 11.3

Indigenous fires and better informed fire management in Australia

Indigenous Australians now manage almost 20% of the continent. Fire was and still is a major management tool used by Aboriginal people (Bowman, 2003; Liddle, 2003). Reintroducing fire into some landscapes by drawing on indigenous knowledge will not prevent wildfires, but it could reduce their impact. However, changes fire management practices need to be made relative to the range of other objectives set for a given area. Moreover, it is also important to note that landscapes are now substantially modified from those managed by indigenous Australians in the past. Factors such as weed invasion can dramatically alter fire regimes and fire impacts on landscapes. Nevertheless, there are important opportunities for exchanges of indigenous and scientific knowledge about fire and land management (Baker, 2003). A potential problem is that although knowledge about indigenous fire regimes is available for many parts of Australia, in other places the knowledge base has been (and continues to be) eroded, thus creating some urgency to keep the remaining information 'alive'. Land managers can learn many important lessons from indigenous people simply from spending time together in the bush. However, this is related to other key issues, including those of: (1) ensuring that indigenous knowledge is passed on by the 'right people to other right people', and (2) how knowledge is stored and who has access to it. There is considerable sensitivity about these issues – Aboriginal people can be protective of their knowledge because in the past it has been exploited by Europeans (Hill, 2003). Part of an approach to address some of these knowledge-management issues has been taken in communities in north-east Australia, where physical examples of fire management practices are applied on the ground to demonstrate to traditional owners and the wider community how 'country' can be managed (Davis, 2003).

11.2 Types of fire

There are two broad types of fire: **wildfire** and **prescribed** fire. Although they are treated separately in this chapter, they are not always mutually exclusive, as there are many documented cases where prescribed fires have developed into wildfires.

Wildfire

Wildfires are unplanned. They vary substantially in their timing (e.g. the time of the year or time of day), frequency (or return interval; Banks, 1982), intensity (McCarthy *et al.*, 1999a), size (Gill, 1981; Figure 11.3) and heterogeneity (i.e. variation in intensity and impact within the limits of the total area affected; Mackey *et al.*, 2002).

Fire intensity is the amount of heat released at a point on a fire edge, and it is a function of the heat yield of the fuel, the amount of fuel per unit area, and the rate of spread of the fire (Byram, 1959). Gill and Catling (2002) provide a crude scale of intensities: (1) low (<350 kilowatts per metre of fire front), (2) high (350–3500 kilowatts per metre), (3) very high (3500–35 000 kilowatts per metre), and (4) extreme (>35 000 kilowatts per metre). The maximum intensity for burns in Australian forests has been estimated to be 100 000 kilowatts per metre (Gill and Moore, 1994).

Fire timing, frequency, intensity, size and heterogeneity interact. For example, few landscapes experience frequent high-intensity wildfires because fuel loads do not accumulate to high levels. In contrast, many landscapes and vegetation types are characterised by recurrent low-intensity disturbances. Similarly, timing and intensity can be strongly correlated. In the tropical savannas of northern Australia, early dry-season fires are less intense than those late in the same season (Williams *et al.*, 2002). Climate and topography further influence how fire behaves in particular landscapes and vegetation types.

Biological legacies are the structural and ecological elements that remain after fire, including recovering vegetation, fallen and standing logs, ash, soil and canopy-stored seed, and surviving animal populations (Foster *et al.*, 1998; Franklin *et al.*, 2000; see Section 11.10). The numbers, types and spatial arrangements of biological legacies influence the successional dynamics of fire-disturbed plant and animal communities (Lindenmayer and Franklin, 2002; see Section 11.10).

Fires that burn at the wrong time of year can kill plants before seed is set, or interrupt the breeding activities of mammals and invertebrates (Greenslade, 1996; Woinarski and Recher, 1997; Keith *et al.*, 2002b; Whelan *et al.*, 2002). Wildfires that occur too frequently or infrequently can make an area uninhabitable for some species (Gill *et al.*, 1999; Bradstock *et al.*, 2002). We return to the interactions between biota and fire characteristics later in this chapter.

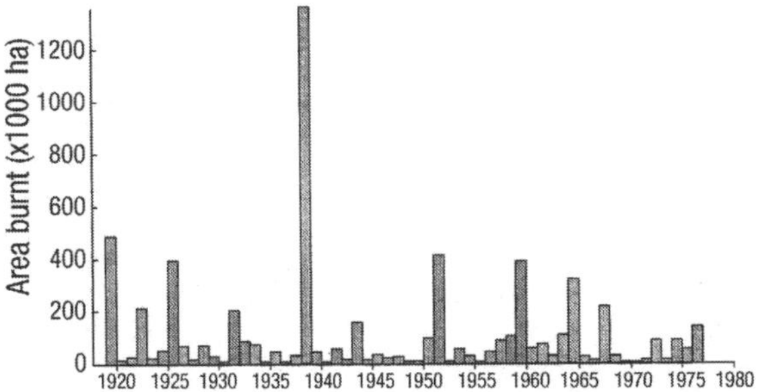

Figure 11.3. Mean sizes of wildfires in Victoria between 1920 and 1980 showing the major spike at 1939. (Modified from Gill, 1981.)

Prescribed fires

Prescribed fires (or 'hazard reduction burns') usually involve the routine use of low- to moderate-intensity controlled fires (Morley *et al.*, 2004). These are lit to reduce fuel loads to reduce the risk of extensive, high-intensity fire that can damage property and kill people (P. Cheney, personal communication). Controlling prescribed fires depends on topography, access, wind speed, temperature, humidity, fuel loads and fuel moisture content (Whelan, 1995). Wildfires in recent decades have been suppressed and fuel loads have accumulated in some forest areas to create unacceptable risks to people and infrastructure (e.g. Hurditch and Hurditch, 1994). For example, the Jarrah forests of south-western Australia are burned extensively and frequently to reduce fuel loads (Christensen and Abbott, 1989).

Prescribed burning in the Jane State Forest in Western Australia. (Photo by David Lindenmayer.)

Regeneration burn in the Central Highlands of Victoria. (Photo by David Lindenmayer.)

Prescribed fires are sometimes used to manage habitat for wildlife, control weeds (Hodgkinson and Harrington, 1985), and promote plant species for animal survival (as in the case of the Eltham Copper Butterfly; see Box 9.6 in Chapter 9; New, 2000). Prescribed fire is used in montane ash forests and other wet eucalypt forest types in southern Australia. Regeneration burns (or slash burns) are burns that are carried out after timber harvesting to create an ash bed to promote stand regeneration (Campbell, 1984).

Prescribed burning is controversial in Australia. Some vegetation types and the biota they support are susceptible to any fire. Despite the widespread application of prescribed fires in some jurisdictions, in some extreme fire weather situations, prescribed fire may not prevent wildfire. Gill and Bradstock (2003) noted that '*the extent to which fire regimes can be controlled or imposed by people is largely unknown*'.

Frequent burning can select for fire-prone species, making some vegetation types more flammable than they otherwise might have been. In some vegetation types there can be a rapid accumulation of fuel following prescribed fire, so that fuel management can require repeated prescribed fire. For example, Park (1975, in Florence, 1994) notes that:

> *Where regrowth develops rapidly following perturbation, the forest floor biomass builds up rapidly to a point of peak fuel energy storage during the forest's rapid early growth stage.*

Decomposition of the litter layer may make fuel reduction burns unnecessary (Crockford and Richardson, 1998a,b, 2002). Conversely, frequent burning can destroy the organisms that decompose litter, resulting in a rapid accumulation of new litter requiring further burning.

Equilibria between decomposition and accumulation are not characteristic of all fuel types in all vegetation types. In the Mountain Ash forests of Victoria, the volume of large logs increases with stand age (Lindenmayer *et al.*, 1999a). However, bryophyte cover on logs, log decay levels, and the moisture content of logs also increases with age (Ashton, 1986; Lindenmayer *et al.*, 1999a). There are no simple, linear relationships between fuel accumulation, fuel levels, fuel flammability and time since fire.

Prescribed burning is expensive and can compete for resources with other management activities. The costs of maintaining a prescribed burning program over large areas of the national park and State forest estates would be prohibitive, even if it was feasible (Whelan, 2002).

Prescribed fire involves other trade-offs. Perhaps most importantly, fire affects the amount and quality of water produced by catchments. Smoke can have impacts on natural assets, such as discolouring cave formations (A. Spate, personal communication). Smoke can influence air quality and can pose a human health risk, particularly for people with respiratory conditions such as asthma (Johnston *et al.*, 2002).

Some areas do not burn readily, making management of fuel with prescribed fires difficult. In the montane ash forests of the Central Highlands of Victoria, for example, the high moisture content of the large quantities of woody debris on the forest floor (see Ashton, 1986) make low-intensity prescribed fires almost impossible. Thus, most wildfires in landscapes dominated by these forest types are high-intensity ones (Smith and Woodgate, 1985; Attiwill, 1994a,b; McCarthy and Lindenmayer, 1998). If forest biotas are adapted to infrequent, high-intensity fires, low-intensity fuel-reduction fires can produce unwanted ecological outcomes (Lindenmayer and Franklin, 2002). Repeated low-intensity prescribed fire can affect the biota negatively or positively (Gill *et al.*, 1999). We explore this topic in the following section.

11.3 Response of biodiversity to wildfire

Wildfire and Australian animals

Fires can kill many individual animals at the time of the conflagration (e.g. birds in the Nadgee Nature Reserve

Table 11.1. Numbers of birds killed directly in Nadgee Nature Reserve after the 1972 wildlife. (Modified from Fox, 1978; Keith *et al.*, 2002b.)

Species	No. carcasses
New Holland Honeyeater	266
Little Wattlebird	120
Eastern Yellow Robin	49
Brown Thornbill	25
Rufous Whistler	21
Eastern Spinebill	12
Striated Thornbill	11
White-throated Treecreeper	10
Crimson Rosella	7
Blackbird	5
Ground Parrot	5
Yellow-faced Honeyeater	5
Brush Bronzewing	4
Superb Fairy Wren	3
Beautiful Firetail	3
Red-browed Finch	3
Crested Shrike Tit	2
Grey Butcherbird	2
Owlet Nightjar	2
Pied Currawong	2
Stubble Quail	2
Varied Sitella	2
Bell Miner	1
Boobook Owl	1
Barn Owl	1
Eastern Whipbird	1
Fan-tailed Cuckoo	1
Grey Shrike Thrush	1
King Parrot	1
Leaden Flycatcher	1
Olive-backed Oriole	1
Rainbow Lorikeet	1
Satin Bowerbird	1
Scarlet Robin	1
Southern Emu Wren	1
Spotted Pardalote	1
Tawny Frogmouth	1
White-browed Scrub Wren	1
White-naped Honeyeater	1

on the far south coast of New South Wales in 1972; Fox, 1978; Keith *et al.*, 2002b; see Table 11.1). Survival of individual animals depends on mobility (to escape fire) and the insulating potential of fur and feathers (Whelan, 1995). Additionally, many animals that survive fire subsequently perish because of limited food and shelter, or because of increased predation due to a lack of vegetation cover (Christensen, 1980; Russell *et al.*, 2003).

Species have a wide array of responses to a single fire (Letnic, 2003). There can be different responses within a given vegetation type, between vegetation types, and following successive fires at the same site (Fox, 1982; reviewed by Whelan, 1995; Woinarski, 1999; Whelan *et al.*, 2002). Figure 11.4 shows the changes in occurrence or abundance with time since fire of a small subset of birds and small mammals. Some species are attracted to burning areas and others, for example some raptors, can even attempt to promote the spread of fire by dropping smouldering sticks in unburned areas. The Black Kite and the Pied Butcherbird find prey more easily in burned areas (Woinarski and Recher, 1997). Fire-damaged areas provide suitable habitat for some species, for example several species of beetles (Schmitz and Bleckmann, 1998), including jewel beetles (Beattie and Ehrlich, 2001). A range of invertebrates forage on burned logs and fire-damaged standing trees (Grove *et al.*, 2002).

Wildfires promote habitat development for some animal species after a given interval following disturbance (Catling *et al.*, 2001) through stimulating the germination and subsequent growth of particular food plants (e.g. Meredith *et al.*, 1984). Wildfires can produce significant quantities of dead wood (including logs) and stimulate the development of cavities in trees (Inions *et al.*, 1989), which are essential habitat elements for many animal species, and which can be depleted by forestry operations (Grove, 2001; Gibbons and Lindenmayer, 2002) and prolonged high-intensity grazing by domestic livestock (Gibbons and Boak, 2002).

Wildfire suppression can also affect the biota. Woinarski *et al.* (2004) found a range of species that were strongly associated with long unburned areas in tropical open forests in northern Australia. In other cases, back-burning is used to check the spread of wildfires, but it can close off the escape options for animals and incinerate them (Baker, 1997). Roads built to give access to fire-fighters can cause damage, providing conduits for the spread of weeds (Wace, 1977), causing soil erosion and stream sedimentation (O'Shaughnessy and Jayasuriya, 1991), and facilitating the movement of feral predators (May and Norton, 1996). After wildfires in Victoria in 2003, aerial mapping showed that several thousand kilometres of roads had been constructed to fight the fires.

Fire-damaged forests are sometimes logged (termed salvage logging). Post-fire salvage logging in Victorian forests following the 1939 wildfires continued for nearly 20 years (Noble, 1977). The logging significantly reduced the abundance of large trees with hollows and probably had corresponding negative effects on an

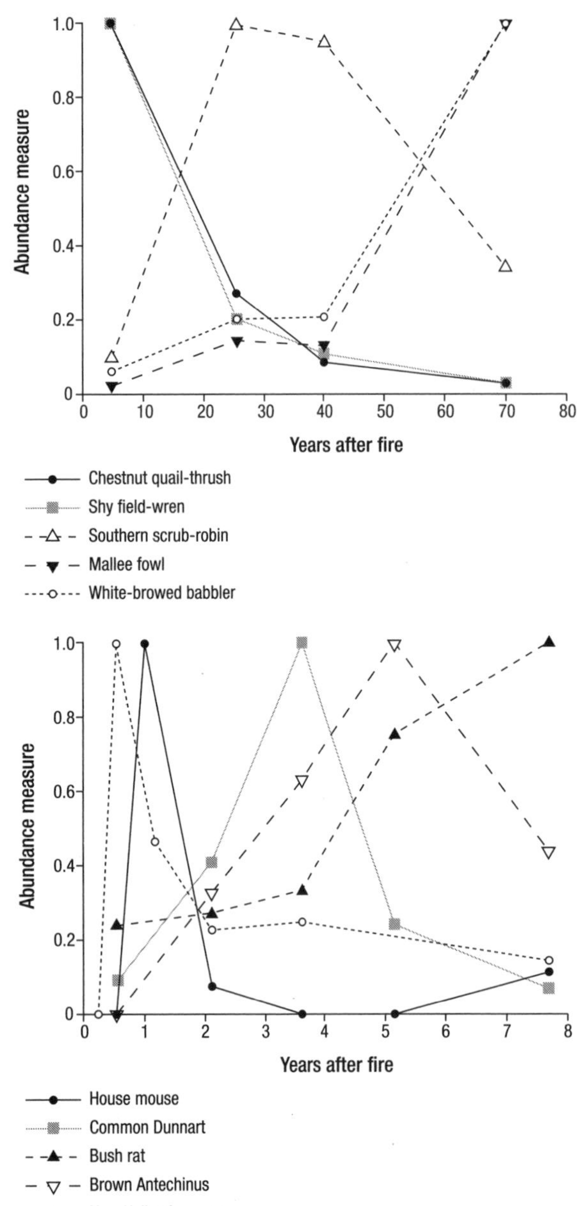

Figure 11.4. Temporal changes in the abundance of (1) selected birds following fire in Mallee–Broombush vegetation in north-eastern Victoria (modified from Woinarski, 1989), and (2) an array of small mammals at Myall Lakes in New South Wales (redrawn from Fox and McKay, 1981; Whelan, 1995).

array of cavity-dependent species (Lindenmayer *et al.*, 2004b). Salvage logging also has detrimental impacts on biota in many other forest ecosystems (e.g. Hutto, 1995; Foster *et al.*, 1997; Saab and Dudley, 1998; McIver and Starr, 2001).

Box 11.4

The impact of salvage harvesting in forests

Many ecologists view natural disturbances such as wildfires as key ecosystem processes rather than ecological disasters that require human repair (e.g. Turner *et al.*, 2003). Indeed, major disturbances can sometimes aid ecosystem restoration by recreating some of the structural complexity and landscape heterogeneity lost through previous intense management (e.g. intensive forestry). For example, major wildfires can generate significant volumes of dead trees and logs that provide important habitat for many organisms, but which are depleted by traditional forestry practices. However, salvage harvesting operations in forests can have significant negative impacts on ecosystems and undermine many of the benefits of major disturbances. The negative ecological impacts of salvage operations can manifest in at least three ways.

First, there can be major impacts on ecosystem processes following salvage operations. Extensive salvage harvesting after the 1938 New England hurricane produced a long-lasting shift in hydrological regimes that were manifested at a regional scale (Foster *et al.*, 1997). Second, the removal of large quantities of biological legacies threatens the persistence of some taxa. For example, salvage harvesting of burned trees removes critical habitat for many early successional species (e.g. Hutto, 1995). Third, there can be compounding, cumulative or magnified effects on ecosystem processes and elements of the biota if an intense natural disturbance event is soon followed by an intensive (and often prolonged) human disturbance. That is, some taxa may be maladapted to the interactive effects of two disturbance events in rapid succession. For example, in South-East Asia, salvage logging of burned rainforests led to significant forest deterioration and loss with major negative impacts on the regenerative potential of stands, as well as a range of other undesirable effects (van Niewenstadt *et al.*, 2001). The impacts of salvage harvesting indicate a need for large areas to be exempt from such activities, including national parks. In addition, carefully formulated policies are needed wherever salvage harvesting continues in order to ensure the retention of biological legacies such as dead trees, live trees, logs, and islands of undisturbed or partially disturbed vegetation (Lindenmayer *et al.*, 2004b).

Salvage logging following the 2003 wildfires near Canberra (middle). Stream-bank damage and in-stream sedimentation can be seen (bottom), resulting from burned Radiata Pine trees having been removed from the riparian zone. (Photos by David Lindenmayer.)

Wildfire and Australian plants

Some plants survive fire and then resprout from **epicormic buds**, roots or lignotubers. Others are killed by fire, but germinate from fire-resistant seeds in the canopy or soil. The seeds of some species (e.g. various species of Acacia) remain viable for many hundreds of years.

The timing of fire can be critical. Dry season burning in northern Australia can have no apparent negative effects on some grass species, but a wet season fire will eliminate populations because there is no seed bank at that time of year (Stocker and Onwin, 1986).

Fire frequency can also have a major impact. In the case of Mountain Ash forest in Victoria, trees do not reach sexual maturity until they are at least 20 years old

> **Box 11.5**
>
> **Epicormic buds in eucalypts**
>
> Some species of eucalypts, for example Mountain Ash, are quite sensitive to wildfire; this is particularly the case for young trees, which have thin bark and are usually killed by wildfire. However, a large proportion of eucalypt species have the ability to resprout even when most of the outer layers of bark are burned by fire. Strands of epicormic tissue run from the bark surface to the woody part of the tree and form buds along their length. Some species are particularly potent resprouters. In a classic experimental study, Chattaway (1958) found that certain species of Victorian eucalypts could have their leaves clipped off and then resprout more than 25 times before finally dying. Many other species of plants are also capable of resprouting (for example, many Banskia species), but they recover fewer times before finally dying (Zammit, 1988).

Epicormic buds regrowing on eucalypts following fires in the Canberra region in 2003. (Photo by David Lindenmayer.)

and seeds do not remain viable in the soil for a prolonged period like those of many other plant species. Stand-replacing fires at intervals shorter than 20 years (e.g. those in 1926 and 1939) lead to replacement by stands of Acacia spp. trees. Reinvasion of Mountain Ash trees then needs to take place from seeds dispersed from adjacent stands (Mackey *et al.*, 2002). If the fire interval exceeds the lifespan of Mountain Ash (>350–500 years), then eucalypt stands will eventually be replaced by Myrtle Beech in parts of landscapes suitable for the development of cool temperate rainforest (see Lindenmayer *et al.*, 2000a,b).

There can be large differences in the sensitivity of individual plants at different times in their life cycle. For example, large old eucalypts can be more fire resistant (because of thick bark) than younger seedlings or saplings (West, 1979). Conversely, very large and decayed trees can be very fire-prone because of the accumulation of flammable rotting material and the limited amount of water in the conducting tissues of the stem (Gibbons and Lindenmayer, 2002).

Plants respond in different ways to fire: some species grow more rapidly after fire, others undergo massive flowering (e.g. Grass Trees), and yet others germinate seedlings en masse (e.g. Mountain Ash and Alpine Ash). Responses are influenced by variables such as (among others) post-fire climate conditions (Whelan, 1995), the life history attributes of the species (Noble and Slatyer, 1980), and the quality (productivity) of the sites (Ashton and Martin, 1996).

Grass Trees on the floor of a coastal woodland two weeks after a fire, that are beginning to resprout. (Photo by David Lindenmayer.)

Wildfire and identifying patterns of species responses

The array of species responses to fire gives rise to what Bowman (2003) termed a 'blizzard of ecological details'. In an effort to simplify the problem, several authors (e.g. Whelan, 1995; Gill, 1999) have developed response curves to explain changes in biota with time since fire using measures such as species diversity, abundances of particular taxa, and rank abundance (of plants). Some examples are shown in Figure 11.5, and they highlight the array of possible species responses to fire: some species remain relatively immune or

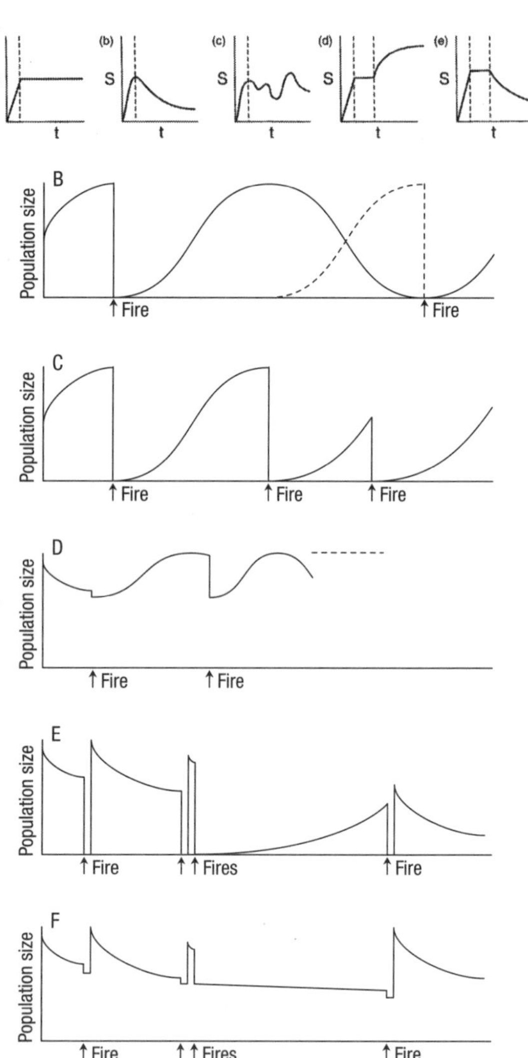

Figure 11.5. (A) Hypothetical plant species diversity response curves with time since fire (after Gill, 1999). (B–F) Hypothetical individual population size response curves with time since fire (redrawn from Whelan *et al.*, 2002).

benefit considerably from repeated fire, whereas others are impacted negatively.

General fire response patterns, such as those based on particular life history attributes (e.g. animal dispersal ability, or whether a plant is a **resprouting** or an **obligate seeding plant**) can assist managers to predict the responses of species assemblages to particular management regimes (such as fire exclusion or frequent prescribed fire). Noble and Slatyer (1980) proposed a model for plant responses based on 'vital attributes'. Three groups of attributes were proposed in their model:

- method of arrival and/or persistence of a given species during and following disturbance
- establishment and growth of a plant species in a plant community
- time required for the species to reach key life history stages (e.g. to produce viable seed).

A given vital attribute can occur as a result of different biological mechanisms (Noble and Slatyer, 1980).

Response categories are only approximate guides to ecological responses (Whelan, 1995; Whelan *et al.*, 2002). For example, a species may or may not resprout, depending on fire intensity. Myrtle Beech can resprout after low-intensity fires, but is killed outright by high-intensity conflagrations. Whelan *et al.* (2002) concluded that for animals, attempts to identify 'clear associations between fire responses and fire regimes' have not been particularly fruitful. Outcomes are often site-specific, and there is unlikely to be a set of 'vital attributes' that determines responses for animals as there is for plants. In fact, fire responses vary within and among sites even for the same species (e.g. the Ground Parrot discussed in Section 11.7; Woinarski, 1999; Keith *et al.*, 2000b). Whelan *et al.* (2002) argued that more (and better designed) experiments are required to identify the underlying ecological processes that give rise to the emergent fire-response patterns.

A fire regime that benefits a particular species or set of species may not suit others. Keith *et al.* (2002a) noted that '*the full spectrum of biodiversity cannot be maintained if fire management only addresses a single species in isolation*'. This injects considerable complexity into the inter-related topics of wildfire, fire management and biodiversity conservation. The objectives of managment need to be carefully articulated (as examined later in this chapter).

11.4 Response of biodiversity to prescribed fire

The impact of prescribed burning on biodiversity is controversial and research results are equivocal (reviewed by Williams and Gill, 1995; York, 1999; see also Collett, 2003 among many other studies).

King (1985) found that log cover was reduced by prescribed burning and concluded that there could be cumulative impacts on litter invertebrates and their small mammal predators. Similarly, in a study by State Forests of New South Wales Northern Region (1996) it is noted that frequent low-intensity prescribed fires accelerated the decay of large logs. Smith *et al.* (1992) and Hannah *et al.* (1998) found that both logs and many vertebrates were less common in frequently burned forests (see also Singh *et al.*, 2002). In one of the longest running studies completed to date, York (1996, in Gill *et al.*, 1999) showed how prescribed burning changed the extent of charring and desiccation of the outer surfaces of logs, leading to a change in the composition of invertebrate communities to include more species that were more tolerant of drier and more open forest environments (York, 1999, 2000).

Fire is used on farms to reduce fire hazards around paddocks and farm buildings. As in the case of forests and woodlands, such practices can affect biota that are dependent on logs, fallen timber and leaf litter. Lindenmayer *et al.* (2003a) recommended that ecologically-sensitive areas on farms such as wetlands (especially those with peat) be quarantined from fire; elsewhere, burning should be concentrated on paddock boundaries, leaving the interior parts unburned or less intensively burned.

Burning off in a low-lying swampy area – an activity that can cause long-term damage to these habitats. (Photo by David Lindenmayer.)

Mosaic or patchy burning can maintain suitable habitat for some ground-dwelling organisms and provide escape routes for others. The interval between burning should be sufficient to regenerate native vegetation and allow fire-sensitive species to recover. For example, Lambert and Elix (undated) and McIntyre *et al.* (2002) recommended that the intervals between fires in grassy woodlands should be at least 5–10 years. Fire can facilitate weed establishment and promote weed growth (Hobbs and Atkins, 1991), especially if fires are frequent (Hobbs and Atkins, 1990).

Prescribed burning is designed to reduce fuel loads and, in turn, reduce the risk of a high-intensity fire. In some ecosystems, frequent prescribed fires will be necessary to reduce fuel. However, fuel is also habitat for animals and plants (e.g. logs that provide nursery sites for plant germination; Howard, 1973; McKenny and Kirkpatrick; 1999); therefore, changes to fuel such as logs and leaf litter can impact on the elements of the biota that depend on these features (York, 1999; Lindenmayer *et al.*, 2002b).

Box 11.6

Fire as a threatening process for the Golden-shouldered Parrot

The case of the Golden-shouldered Parrot is a useful illustration of the way in which altered fire regimes can become a process that threatens the persistence of a species (Keith *et al.*, 2002a). The species occurs on Cape York Peninsula and its range has been contracting because the lowland grassland habitats that it requires are being replaced by Paperbark (*Melaleuca* spp.) thickets and woodlands (Crowley and Garnett, 1998). These changes have occurred because of the loss of Aboriginal burning patterns and the expansion of pastoralism. More frequent burning late in the dry season and early in the wet season will be need to be undertaken (Garnett and Crowley, 1995) – not only to halt and reverse broad-scale vegetation change but also to stimulate the availability of the seeds of herbs and grasses that are eaten by the Golden-shouldered Parrot. Importantly, the notion of returning to past burning regimes is in the interests of pastoralists, as emerging woodland habitats are not efficient grazing areas for livestock. By spelling pastures from grazing and allowing fuel to accumulate, subsequent high-intensity fires can assist the transition of vegetation cover back toward grassland-dominated systems that are more suitable for the Golden-shouldered Parrot (Garnett and Crowley, 1997).

11.5 Species vulnerability to fire

Animal and plant groups threatened by altered fire regimes

Inappropriate fire regimes threaten biodiversity in Australia (State of the Environment, 2001a). In fact, they have already contributed to the extinction of two species and three subspecies of Australian birds (Woinarski, 1999). For example, the Kangaroo Island Emu – a dwarf species in comparison to the mainland form – is thought to have become extinct because of wildfires in the 19th century (Ford, 1979). Inappropriate fire regimes threaten more than 50 extant bird species, they are second only to land clearing as a threat to avifauna in Australia (Garnett, 1992; Maxwell *et al.* 1996; Australian Bureau of Statistics, 1999b). For example, the Malleefowl is most abundant in long-unburned mallee habitats (Bradstock and Cohn, 2002), probably because of its reliance on litter for nest building (although it is also vulnerable to other factors, such as predation by the Red Fox).

Approximately 45% of south-western Australia's mammals and birds may be negatively influenced by altered fire regimes (Calver and Dell, 1998) (e.g. the White-bellied and Orange-bellied Frogs; Wardell-Johnson and Roberts, 1993).

Leigh and Briggs (1992) list almost 20 plant species that are threatened with extinction because of inappropriate fire regimes. There are also many examples of local plant extinctions (Leigh *et al.*, 1984; Gill and Bradstock, 1995) and changes in the composition of vegetation communities as a result of inappropriate fire regimes (e.g. Niewenhuis, 1987). The decline of fire-sensitive Mulga stands in arid Australia may be due to infrequent large fires in spinifex ecosystems (Gill, 2000). Previously, these communities experienced more frequent, smaller fires as part of Aboriginal management (Allan and Southgate, 2002).

Vegetation communities sensitive to fire

Some vegetation communities in the Australian environment appear to be particularly sensitive to the impacts of fire, suggesting that they have evolved largely in the absence of fire. An endemic Tasmanian conifer,

King Billy Pine forest burned in 1914. (Photo by J. Hickey.)

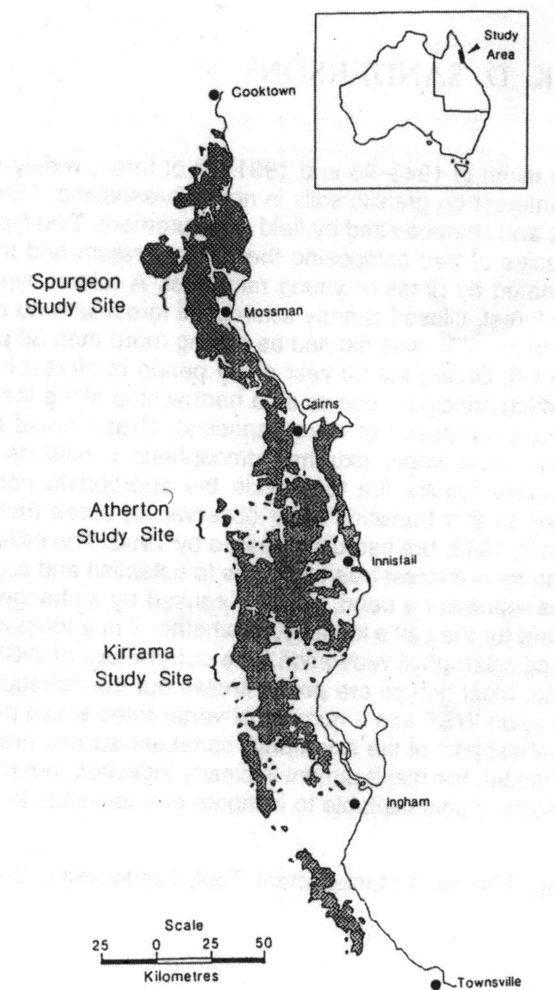

Figure 11.6. Map showing the band of wet sclerophyll forest in north Queensland. (Redrawn from Harrington and Sanderson, 1994.)

King Billy Pine, is highly sensitive to fire (Brown, 1988). Living trees of this species can exceed 1000 years of age. Stands of the species in western Tasmania were damaged by a fire in 1914, and there has been only minimal regeneration since then.

Some wetland habitats, such as those that contain peat, require protective management because they are readily and badly damaged by fire, and take a long time to recover. Fires in peat bogs can dramatically reduce the quality of water within adjacent creeks and destroy drought refuge areas for native animals (Lindenmayer *et al.*, 2003a). Fire can damage moist high-elevation vegetation types such as endemic plant-rich **feldmark** (which often supports long-lasting snow patches) and sphagnum bogs. Species that are closely associated with sphagnum bogs, for example the Corroboree Frog, may be particularly sensitive to wildfires (K. Green, personal communication).

Rainforest is also sensitive to fire (Bowman, 1999; Fensham *et al.*, 2003). In the prolonged absence of fire, cool-temperate rainforests (e.g. those in Tasmania; Jackson, 1968), warm-temperate rainforests, tropical rainforests (Harrington and Sanderson, 1994; Mullen, 1995), and monsoonal rainforests (Russell-Smith and Bowman, 1992) invade adjacent vegetation (such as wet sclerophyll forest).

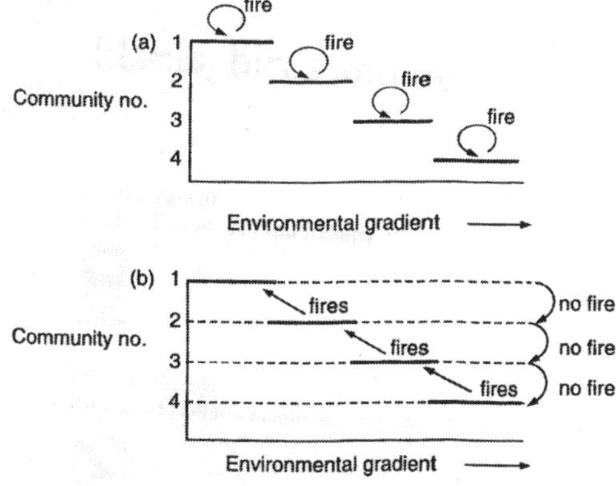

Figure 11.7. A model for the invasion of a given vegetation type by another (e.g. rainforest) in the absence of fire (after Gill, 1999).

Wet sclerophyll forests in northern Queensland occur as long narrow belts adjacent to tropical rainforest that is protected from logging and fire (Figure 11.6). Wet sclerophyll forest supports many species that are found virtually nowhere else, for example the northern subspecies of the Yellow-bellied Glider and the Northern Bettong (Laurance, 1997a,b; see Table 11.2). Gill (1999)

Table 11.2. Mammal and bird species virtually restricted to wet sclerophyll forest adjacent to tropical rainforest in North Queensland or for which wet sclerophyll forest provides primary habitat (from Harrington and Sanderson, 1994).

Restricted to wet sclerophyll forest
Yellow-bellied Glider (northern subspecies)
Northern Bettong
Eastern Yellow Robin
Yellow Thornbill
Buff-rumped Thornbill
Wet sclerophyll forest as major habitat
Greater Glider (small northern subspecies)
Squirrel Glider
White-naped Honeyeater
Yellow-faced Honeyeater
Crested Shrike Tit

describes a model of how sclerophyll forest can be invaded by rainforest (see Figure 11.7), a process that could have a significant impact on the persistence of many species (Table 11.2). The conservation of biodiversity closely allied with wet sclerophyll forests together with that of adjacent rainforest will require management of different disturbance regimes in close proximity.

11.6 Spatial variability in fire behaviour: fire refugia, landscape mosaics, and Aboriginal burning patterns

Wildfires and prescribed fires rarely burn all areas with equal intensity, and this spatial variability can be important. For example, patches that escape fire provide refugia for fire-sensitive taxa (Mackey *et al.*, 2002). Animals and plants survive fire in microhabitat features such as rock piles, soil cracks and tree hollows (Clarke, 2002; Whelan *et al.*, 2002). Large logs can be fire refuges for plants and animals because of their diameter and length, the moisture they contain, and the moisture levels in the adjacent litter (Campbell and Tanton, 1981; Andrew *et al.*, 2000; reviewed by Lindenmayer *et al.*, 2002b). For example, Christensen (unpublished data) in Christensen *et al.* (1981) recorded large numbers of skinks surviving under logs after moderate- to high-intensity wildfires.

At a landscape scale, streams, lakes and rocky clifflines are natural firebreaks. Differences in flammability between different vegetation types mean that, for example, sedgelands burn more readily than neighbouring woodlands, thus reinforcing heterogeneity in landscape vegetation patterns (Burrough *et al.*, 1977). Within widespread areas of the same vegetation type, microclimate, terrain and vegetation structure create fire refugia. In the ash-type forests in the Central Highlands of Victoria, steep slopes and areas with low levels of radiation experience lower fire intensities. In these areas, more trees survive and multi-aged stands are relatively common (Mackey *et al.*, 2002). Multi-aged forests support the highest diversity of species of arboreal marsupials (McCarthy and Lindenmayer, 1998).

Patchy burned area on Flinders Island, Tasmania. (Photo by David Lindenmayer.)

Fine-scale vegetation mosaics and Aboriginal burning

Aboriginal people burned the spinifex-dominated landscapes in dry inland parts of Australia, producing a fine-scale vegetation mosaic that provided suitable habitat for medium-sized native mammals (Bolton and Latz, 1978; Burbidge *et al.*, 1988). These landscapes now experience large fires (Burrows and Christensen, 1990) and suitable habitat for medium-sized native mammals may have been lost (see Short and Turner, 1994; Allan and Southgate, 2002).

In the examples of the montane ash forest and the spinifex landscapes, fire patterns and the resulting heterogeneity in vegetation at local and landscape scales have important implications for biodiversity (Lindenmayer and Franklin, 2002; Allan and Southgate, 2002). A range of structural and floristic conditions support more taxa than one or only a few vegetation conditions. Fire management is an important part of conservation management, which is the topic of the following section.

11.7 Fire management and biodiversity conservation

The most appropriate fire regime for a system will depend on the objectives of management and the characteristics of the system. Management may involve both the suppression of unwanted fires (including wildfires) and the ignition of prescribed fires (Gill, 1999). Objectives for management will vary depending on the proximity of people and property and the relative importance of conservation and water production. The situation is complicated by the fact that few (if any) areas have just one economic or ecological value (Keith *et al.*, 2002a).

Even if biodiversity conservation is the primary management objective, differences between vegetation communities and individual elements of the biota in their response to fire means that there are no simple recipes. Given such complexity, what can be done? One approach is to vary fire regimes between and within landscapes, creating a range of conditions. Therefore, if unsuitable habitats are created in one area, there will be other places where a species can survive. This is termed 'risk-spreading' (den Boer, 1981; see also Lindenmayer and Franklin, 2002). Large parts of some landscapes, for example the edges of national parks bordering grazing properties or urban developments, might be subject to frequent low-intensity prescribed burns (as is now done by the New South Wales Department of Environment and Conservation; Whelan, 2002). Other landscapes that are more remote from human infrastructure may be burned less frequently and/or less of the area may be burned, and others still may remain unburned by prescribed fire (although wildfire may still occur).

There are two key variables in this simple example: the interval between prescribed fires (fire frequency) and the proportion of a landscape burned (fire extent). Unfortunately, very little is known about how frequent and extensive prescribed fire must be to reduce the risk of wildfire (see Cary and Bradstock, 2003), or about the impacts on biodiversity of varying prescribed fire frequencies and sizes. Such uncertainty highlights the need to record actions, monitor the response of the biota, then feed these observations back to managers so that practices could be improved. This process is called **adaptive management** or learning by experiment and monitoring (Walters and Holling, 1990; Bunnell *et al.*, 2003; see Chapter 19). Despite its potential to improve biodiversity outcomes, the method has rarely been used by resource management agencies (Dovers and Lindenmayer, 1997; Beese *et al.*, 2003; Bunnell *et al.*, 2003).

The prescribed burning example given in this section is highly simplified. Landscapes are not homogenous – some areas, such as the wet gullies in valley floors, are less likely to burn than other places. Conducting patchy burns of varying intensity is not easy in practice, and the timing and intensity of burning add further complexity. Such variation may provide opportunities for natural experiments (see Box 10.11) and a feedback process to inform management.

In summary, perhaps the best informed perspective on fire management comes from Gill (1999, p. 47), who believes that the management of fire for biodiversity conservation should:

> *aim at achieving suitable proportions of landscape with a variety of times-since-fire stages within appropriate intensity levels at appropriate times of the year and within appropriate frequency range.*

This is an important objective, but it is not easy to achieve, because using fire as a management tool is influenced by many practical constraints, including in many cases a limited time window during which prescribed fires can be applied safely (Whelan, 1995), and limited financial and human resources for fire management. Where prescribed burning is difficult, it may be better to use other forms of management (Baker, 1997). For threatened grassland lizards such as the Striped Legless Lizard and the Grassland Earless Dragon, low-intensity grazing may be more appropriate than frequent prescribed burning (Robertson and Cooper, 1997; Robertson, 1999).

Fire management and conservation of the Eastern Bristlebird and the Ground Parrot

Active management is sometimes required to maintain or restore important ecological values or conserve particular taxa. Two high-profile cases are the Eastern Bristlebird and Ground Parrot, which are birds that are associated with heathlands (Garnett, 1992). In general, field data suggest that many areas of heathland have been burned more frequently than is desirable for some threatened bird species (Woinarski, 1999).

The Ground Parrot is a strong flier that quickly recolonises burned areas and then occupies them for many years (see Woinarski, 1999; Keith *et al.*, 2002b). Early work suggested that burning at intervals of 4–5 years was appropriate for this species. More recently, recommendations have extended the time between fires to 8–10 years in Queensland and south-west Western Australia and 10–25 years in Victoria (Woinarski, 1999). In Victoria, animals were least common in heaths that were less than 3 and more than 18 years old (Meredith *et al.*, 1984). However, Baker and Whelan (1994) found that Ground Parrot populations did not decline with heathland age and recommended fire exclusion. The most appropriate fire management strategy for the Ground Parrot may vary between areas and in relation to food resources such as seed production in particular habitats.

The Eastern Bristlebird requires multi-layered heathland vegetation (Baker, 1998). It disperses poorly and appears to be vulnerable to the effects of large-scale fires that do not leave unburned refugia (Pyke *et al.*, 1995; Baker, 1997). Fire exclusion is the most appropriate conservation strategy, but this may not be possible given the proximity of human infrastructure to its habitat, such as at Jervis Bay Village on the south coast of New South Wales. Because of the need for fuel management in these areas, Baker (1997) recommended the strategic slashing of vegetation. In other cases where prescribed burning is essential, the direction of fire fronts should be planned to provide escape routes for animals,

Ground Parrot habitat that was burned to promote habitat suitability at Wilsons Promontory, Victoria. (Photos by David Lindenmayer.)

Burned habitat formerly occupied by the Eastern Bristlebird at Jervis Bay prior to fires in December 2003. (Photos by David Lindenmayer.)

given their limited movement ability. Many of these recommendations have been embraced as part of management of Eastern Bristlebird habitat, such as that in the Booderee National Park in New South Wales (M. Fortescue, personal communication). Nevertheless, in late 2003, a major wildfire burned much of the complex multi-layered heathland habitat occupied by the Eastern Bristlebird.

The Eastern Bristlebird and the Ground Parrot co-occur in some areas of heathland, including those that may be periodically burned. Baker (1997) believed that the management of the Eastern Bristlebird should have priority in these circumstances (primarily achieved through fire exclusion) because its distribution is more restricted, it has poorer recolonisation ability, and its populations are smaller and more sensitive to frequent fire. This situation emphasises the importance of setting explicit objectives for ecological management, and in so doing making assessments of the potential trade-offs involved.

11.8 Studies to examine the effects of fire

One of the consistent themes emerging from the literature on fire is reinforced by statements in the most recent State of the Environment Report (2001, p. 8): 'the effects of various fire regimes on the conservation of biodiversity remain uncertain'. There is general consensus about the need for more detailed work on the impacts of fire on the Australian biota. But what are the best ways to study fire? There are several ways improving our knowledge of fire impacts, including experiments, observational studies and modelling. Each is briefly outlined below.

Experiments

Fire experiments are critical because they identify the underlying ecological processes that give rise to patterns of species response (Whelan *et al.*, 2002). However, the experimental design of many fire studies is poor, limiting the validity of inferences. For example, many studies take place only after a fire has occurred (Whelan, 1995). A key part of experiments is that many factors need to be controlled to test a small number of variables of interest. Experiments can enable robust conclusions to be reached about, for example, the effects of the timing of a fire on small mammal population dynamics if conducted with adequate replication of treatments, quantification of before-treatment conditions in each replicate, and replicated 'control' sites that are matched to the treatments (see Whelan, 1995).

The interpretation of fire effects from experiments (and other types of studies) needs to be made in

Table 11.3. Experimental design issues associated with fire studies. (Modified from Whelan, 1995.) The table outlines various approaches that have been used to make inferences about the impact of a single fire on population size. The table also indicates the appropriateness of the different approaches for each level of question asked, and points to possible shortcomings.

	Level of question asked			
	Indication of population change	**Separation of fire effects from temporal change**	**Separation of fire effects from site effects**	**Inference of fire effect on population**
Record of dead organisms after fire	No – dead individuals may not have contributed to population dynamics even without the fire	No	No	No
Census before and census after fire	Yes, but choice of time for post-fire sample may mean a change is missed	No	No	No
Census before and after fire in a burned and a control site	Yes	Yes, but there will be a confounding interaction of season and fire effects if timing of samples not carefully planned	No	No
Census before and after fires in randomly allocated replicate burned and control sites	Yes	Yes	Yes	Yes, but care must be taken not to treat consecutive samples as replicates, as they are not independent

conjunction with an understanding of the biology of the species in question. For example, if long-lived species survive after fire, their persistence may mask impacts on other processes such as reproduction. In other cases, more captures of animals after fire may be due to increased movements to find food rather than greater abundance (Whelan, 1995). In the case of plants, many taxa that appear after fire were present before the fire in cryptic forms such as dormant seeds or ground tubers (Gill, 1999).

Despite the considerable value of experiments, it is surprising how few have been conducted in Australia (and elsewhere around the world; Whelan, 1995). Studies such as the Kapalga Fire Experiment in Kakadu National Park in the Northern Territory (see Braithwaite, 1995; Brook *et al.*, 2002b; Andersen *et al.*, 2005 among numerous other publications) are among the few Australian exceptions. Perhaps part of the reason for the lack of experiments is that experimental fires have to be of low to moderate intensity and constrained to small areas, otherwise they can become uncontrolled wildfires. Hence, they cannot represent the wide range of intensities that occur in nature (Gill, 1977).

Because experiments often take place on a small spatial scale, they can be of limited value for examining species that move over larger scales, such as birds and large mammals. By carefully controlling some conditions in experiments, initially unforeseen biases can be added, which confound the results and the interpretation of effects (Huston, 1997). Complex interacting factors that can have an important influence on species persistence, such as cumulative effects (sensu McComb *et al.*, 1991; Burris and Canter, 1997) in which two factors in isolation have limited impacts but are substantial in combination, often cannot be readily examined with field experiments. Finally, experiments can be expensive to establish and maintain, but changes in populations (or other measures of interest) can take a long time to appear.

Observational studies

The vast majority of studies of fire in Australia have been observational investigations that took advantage of unplanned fires. Typically, they lack the replication of sites and the controls that characterise well-designed experiments (e.g. Thompson *et al.*, 1989; Loyn, 1997; Singh *et al.*, 2002; see Table 11.3). Despite the inherent problems of observational studies, they can nevertheless produce important insights into the impacts of fires on biota (Gill *et al.*, 1999), including the effects of very high-intensity fires (e.g. Recher *et al.*, 1975, 1985). Indeed, much of what is currently known about fire impacts on biota comes from observational studies (reviewed by Whelan, 1995; Woinarski, 1999).

Modelling

The value of modelling in studies of fire is controversial, with some authors arguing that modelling has provided few valuable insights, whereas others contend the opposite (see debates in Cary *et al.*, 2003). However, we believe that modelling can make a useful contribution to a better understanding of fire and its impacts on the biota, particularly when it is underpinned by sound ecological data (Burgman *et al.*, 1993).

Modelling can be a vehicle for synthesising what is known about particular phenomena (Burgman *et al.*, 1993). It can help form hypotheses for testing with field-based empirical studies, such as in adaptive management programs (e.g. Walters, 1986). It can provide a mechanism to compare the relative effectiveness of management options (e.g. the risk of extinction of a given species under several conservation strategies; Possingham *et al.*, 1993). With modelling, it is possible to describe the possible long-term consequences of actions taken now. For example, McCarthy and Burgman (1995) demonstrated that fire and logging led to quite different amounts of forest in different age classes in a landscape, even if the average fire-return interval was the same as the logging rotation time (Figure 11.8). This has implications for forest management and the conservation of species associated with

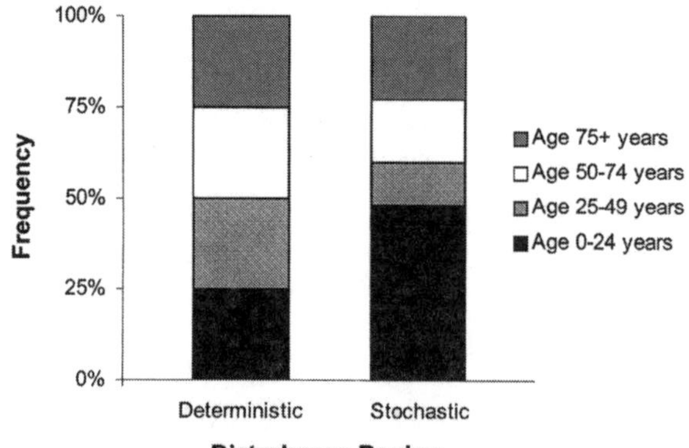

Figure 11.8. Differences in age class frequency between deterministic (logging) and stochastic (fire) disturbance regimes in an Australian eucalypt forest. (Redrawn from McCarthy and Burgman, 1995.)

Box 11.7

How to assess burning strategies for a reserve: the use of decision theory

A major problem facing park managers is determining the most appropriate fire regime to apply. Richards *et al.* (1999) used decision theory as a way to identify an optimal prescribed burning regime. Seven key steps were involved (after Richards *et al.*, 1999; Possingham *et al.*, 2001):

1. *Specify the management objectives.* For example, the objective may be a balance of three post-fire successional states within the reserve or a minimum area of each stage.
2. *List the management options.* The park manager may have the options of doing nothing, attempting to stop all wildfires, or prescribing burns in specific areas of the park.
3. *Specify the variables that describe the state of the system.* In this example it might be the area of the park in the three successional stages.
4. *Develop equations to capture the dynamics of the system.* For example, these might be the transition probabilities of an area moving from one successional stage to another.
5. *Specify constraints for the variables describing the state and dynamics of the system.* In this case, it may be the impossibility of complete suppression of a wildfire in all areas of the park.
6. *Specify the level of uncertainty for various parameters.* This might include (among others) the relationships between successional stage and the risk of ignition, or the level of resources available to undertake prescribed burning or suppress a wildfire.
7. *Find a solution to the problem.* Richards *et al.* (1999) and Possingham *et al.* (2001) used stochastic dynamic programming to find the optimal solution.

This is a useful way to make optimal decisions for practical conservation, and Richards *et al.* (1999) applied it to fire management issues in the Ngarkat Conservation Park in South Australia. The optimal management strategy turned out to be a function of the current state of the park. Presently, the key management strategy is to suppress wildfires when they ignite. However, Richards *et al.* (1999) noted that if fire-free conditions remained for a prolonged period, then prescribed burning would be required to recruit an increased area of early successional vegetation to the park's vegetation cover patterns.

particular age classes, and would not have been identified from long-term field observations. Similarly, phenomena such as the sensitivity of fire regimes to future climate change cannot be studied empirically (Cary, 2002). Despite the value of modelling, model-users should be clear about the limitations of models.

Another important role for modelling is that of risk assessment, in which different management options can be compared and ranked in terms of the likelihood of an unfavourable conservation outcome (e.g. species extinction). Such management options can include the application of fire-related actions or possible chance occurrences of wildfires. There are some useful examples of such applications of risk assessment modelling to the conservation of plants and animals, including the Heath Banskia (Bradstock *et al.*, 1997), Splendid Fairy Wren (Brooker and Brooker, 1994), Carpentarian Rock Rat (Brook *et al.*, 2002b), and Leadbeater's Possum (Lindenmayer and Possingham, 1995; McCarthy and Lindenmayer, 2000). Richards *et al.* (1999) used a decision theory framework to explore fire management options within reserves (Box 11.7).

In summary, all three types of investigations – experiments, observational studies and simulation modelling – are useful for achieving a better understanding of the relationships between fire and biota. The value of links between various sorts of studies was neatly summarised by Dooley and Bowers (1998, p. 969) who argued that experiments were: '*an important intermediary between the inherent abstraction of simulation modelling and what is observed in the real world*'.

11.9 Ecological theories, fire disturbance and biodiversity conservation

Several areas of ecological thinking are relevant to fire management and attempts to conserve biodiversity, and three of these are examined here: (1) the biological legacies concept, (2) creating greater congruence between human and natural disturbance regimes, and (3) the intermediate disturbance hypothesis.

The biological legacies concept and biodiversity

Media descriptions of fires often report that 'x hectares of land have been destroyed'. The reality is that forests, woodlands and other vegetation types are rarely, if ever, completely destroyed by fire. Patches of vegetation actually remain intact within the broad boundaries of a fire (Delong and Kessler, 2000; Mackey *et al.*, 2002). Within burned areas, fires leave behind considerable biological legacies: organisms, organically-derived structures, and organically-produced patterns that persist from the pre-disturbance ecosystem (Franklin *et al.*, 2000; see Section 11.2). The presence, type and number of biological legacies remaining after fire affects the persistence of populations and the trajectory of succession and recovery (Turner *et al.*, 1998). The concept of 'destroyed' forest and woodland is particularly inappropriate in Australia, because many species have evolved strategies for surviving fire (see Box 11.5).

Biological legacies have a range of important ecological functions. They can:

- survive, persist and regenerate after disturbance and be incorporated as part of the recovering stand
- assist other species in persisting in a disturbed area through a variety of mechanisms (often termed a 'life-boating' function)

Fire in Victorian Messmate forests and vegetation recovery following fire at the same site. (Photo by David Lindenmayer.)

- provide habitat for species that eventually recolonise a disturbed site – a phenomenon referred to as 'structural enrichment' of the post-disturbance area (Lindenmayer and Franklin, 2002)
- influence patterns of recolonisation in the disturbed area; for example, biological legacies within a disturbed area can provide foci that facilitate population recovery; that is, recovery can occur not only via colonisation from neighbouring disturbed areas, but also from organisms and structures persisting within a disturbed area (Spies and Turner, 1999)
- provide a source of energy and nutrients for other organisms; this function is particularly important as it relates to maintaining a flow of energy into the soil to maintain soil organisms, including mycorrhizal fungi (Hooper *et al.*, 2000)
- modify or stabilise environmental conditions in the recovering stand (Perry, 1994).

The floristic composition of an area is strongly influenced by pre-disturbance vegetation and the form of individuals, seeds and other propagules (Franklin *et al.*, 2000). The abilities of many animal species to persist and recolonise are influenced by biological legacies (Whelan *et al.*, 2002). This has implications for decisions about the intensity, timing and frequency of prescribed burning.

Congruence between human disturbance and natural disturbance: values and limitations

There is a widely held view that impacts of human disturbances on biodiversity are less when these disturbances resemble natural ecological disturbances (Attiwill, 1994a,b; Hunter, 1994). The underlying premise is that organisms are best adapted to the disturbance regimes under which they have evolved (Bergeron *et al.*, 1999; Hobson and Schieck, 1999). Conversely, organisms may be poorly adapted to novel ecosystem disturbances, including those involving different disturbance agents, different frequencies or intensities of disturbance, or new combinations of disturbances (Paine *et al.*, 1998; Lindenmayer and McCarthy, 2002).

Congruence between human disturbance and natural disturbance has been extensively explored in forest management (Hunter, 1993; Lindenmayer and McCarthy, 2002), where similarities and differences between logging and natural disturbance (particularly fire) are thought to be important (Korpilahti and Kuuluvainen, 2002). Patch types, sizes, shapes and the internal complexity of patches (i.e. biological legacies) in unmanaged landscapes can guide the size, location, spatial arrangement and rotation period of logged areas or coupes and the structures to be retained in wood production landscapes (Franklin, 1993c; Mladenoff *et al.*, 1993; Franklin *et al.*, 2000). In Sweden, forest management systems are based on silvicultural systems that aim to emulate natural disturbance regimes (Figure 11.9). For example, in wetland forest, ravines, and small islands in lakes there is no forestry activity and fire is excluded. Forest on watercourses and flat moist areas are burned on average every 200 years, and selective harvesting is recommended for these forests. Forest in most moist or wet areas, for example, the majority of **boreal** forest in Sweden, is burned about every 100 years, the same as the recommended rotation length. Logged forest is burnt and seed trees are retained on site. All dry forest land (e.g. Pine forest on sedimentary soils on flat terrain) is burned every 50 years. These areas support trees that survive recurrent low-intensity fires, so controlled burning of the forest is used. Because the fire regimes result in multi-aged stands, the final felling operation includes the retention of seed trees.

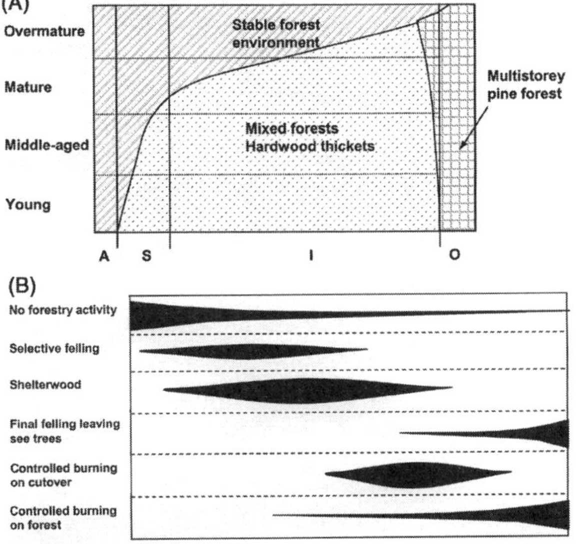

Figure 11.9. (A) Variability in fire regimes in Swedish boreal forests and associated variations in harvesting regimes. (Modified from Rülcker *et al.*, 1994.) The acronym ASIO on the *x*-axis highlights the frequency of fires: A, almost never, S, seldom, I, infrequent, O, often. (B) The black polygons indicate the extent of natural disturbance (fire on the *x*-axis) in relation to human disturbance (on the *y*-axis).

Limitations of the human–natural disturbance congruence approach

In Sweden, prescribed burning is linked with the frequency of natural fires in the same system. The same ideas could be useful in planning prescribed burning regimes in Australian landscapes. However, natural fire patterns are complicated because fire frequencies and fire sizes are variable (Chou *et al.*, 1993; Gill and McCarthy, 1998). Historically, landscapes were disturbed by indigenous people (King, 1963; Bowman, 1998), making it difficult to determine 'natural' patterns (Hunter, 1996; Keith *et al.*, 2002a). Disturbance regimes also change following long-term climate change (Hiura, 1995). Bergeron *et al.* (1998) noted that large changes in the fire frequency in the boreal forests of Canada during the Holocene meant that there was no single characteristic fire regime for this system. Finally, few disturbances in modern landscapes are 'natural'. For example, fires are often extinguished to limit threats to human life and property. Despite these problems, natural disturbance regimes are a useful model for management practices.

Fire and logging

From time to time, the suggestion appears that logging results in acceptable environmental impacts when the timing and extent of operations mimics natural disturbance regimes (Attiwill, 1994a,b). This notion has been used together with assumptions that logging reduces the likelihood of wildfire to recommend timber harvesting in national parks (Tuckey, 2000).

In fact, logged native forests are no less fire prone than unlogged native forests. Many major fires have occurred in areas that were previously heavily logged, including the 1983 Ash Wednesday fires in Victoria (Smith and Woodgate, 1985) and others from south-eastern New South Wales (e.g. Recher *et al.*, 1985) and north-eastern Victoria (Loyn, 1993). Logging in some wet forests in East Gippsland appears to have shifted the vegetation community composition toward one more characteristic of drier forest (Mueck and Peacock, 1992), which may be more fire-prone. Similarly, Whelan (1995) noted that clear-felling operations led to dense regrowth of saplings, which created more available fuel than if the forest was not clear-felled. Logging does not protect forests from fire, and national parks do not need to be logged to prevent it. This does not mean, however, that reserves should not be managed; this need is widely recognised and well accepted (Gill *et al.*, 1999; Woinarski, 1999).

There is little evidence to support the assertion that logging has the same effects on forests as fire (Lindenmayer *et al.*, 1990; Ough, 2001). There are major differences between logged forests and those burned by wildfire (Lindenmayer and McCarthy, 2002). For example, large trees with hollows are depleted in Australian logged forests, which potentially affects the more than 300 species of animals that require hollows (Gibbons and Lindenmayer, 2002). In contrast, many Australian plants and animals have evolved adaptations to recover from fire (Whelan, 1995), although, as outlined above, changes in the fire regimes in the past few centuries threaten many of them (Leigh *et al.*, 1984; State of the Environment, 2001a; Bradstock *et al.*, 2002).

Intermediate disturbance hypothesis

The intermediate disturbance hypothesis predicts highest species diversity at intermediate rates and intensities of disturbance (Connell, 1978; Shea *et al.*, 2004; Figure 11.10). The concept was developed to explain species richness in coral reefs and rainforests (e.g. Rogers, 1993; Aronson and Precht, 1995; Molino and Sabatier, 2001), and it has also been applied in studies of other systems, such as plankton communities (e.g. Wilson, 1990), temperate forests, and prairies (Collins *et al.*, 1995). The idea appears intuitively logical, because relatively few species can survive frequent high-intensity disturbances and most seem likely to be better adapted either to frequent low-intensity or infrequent high-intensity perturbation. However, support for the intermediate disturbance hypothesis is inconsistent (e.g. Collins, 1992; Schwilk *et al.*, 1997; Bascompte and Rodriguez, 2000; Beckage and Stout, 2000), possibly because it is unclear what actually defines a 'disturbance' and what are appropriate spatial and temporal scales for testing (reviewed by Shiel and Burslem, 2003). Intermediate rates and intensities of disturbance will vary between vegetation types: high-intensity fires 20–30 years apart may correspond to long inter-fire intervals in heathland but in forests they would be sufficiently frequent to eliminate fire-sensitive trees.

Even if appropriately defined and scaled, the intermediate disturbance hypothesis has limited utility for

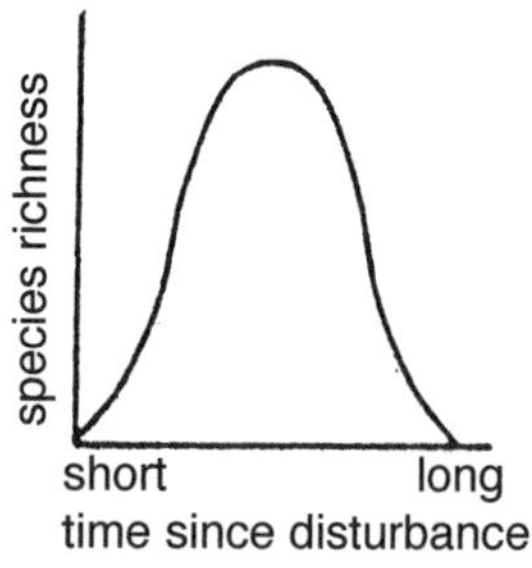

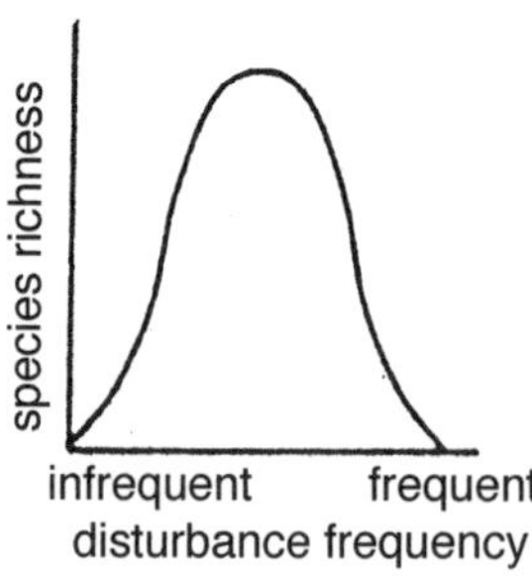

Figure 11.10. Relationships between species diversity and disturbance regimes under the intermediate disturbance hypothesis. (Redrawn from Wilson, 1994.)

practical conservation biology. Species diversity is only one measure of species response and often not the one of greatest interest or relevance. Other factors, such as the conservation of individual threatened species or species likely to respond (positively or negatively) in a particular way to a given fire regime, will be of greater concern. The largest and most important effects of fire are often in terms of the relative abundances of taxa (Recher *et al.*, 1985).

As outlined earlier, responses to fire are varied and highly complex (as shown in the curves derived by Whelan, 1995; Gill, 1999; see Figure 11.5). Many species will not be present and/or most abundant at intermediate rates of disturbance. Some species will be most likely to occur only after a prolonged period without disturbance (e.g. old growth forest specialists; see Chapter 2). In heterogenous landscapes composed of several vegetation types and where a management objective is to maintain biodiversity at a landscape level (such as in national parks), managing for intermediate fire frequencies and intensities may not be appropriate. In this situation, an intermediate disturbance regime would benefit some taxa but disadvantage others. A better approach would be to follow the reasoning of Gill (1999) and apply burning regimes that vary between vegetation types and within a given vegetation type, so that each type is represented by a range of different seral or successional stages, thereby enabling species associated with those stages to persist.

11.10 Cumulative effects of fire and other disturbance processes

Organisms in many landscapes often have to contend with the combined effects of two or more types of disturbance. The impacts of two types of disturbance may be independently unimportant but together create significant challenges for biodiversity conservation (Taylor, 1979; Caughley and Gunn, 1996; Paine *et al.*, 1998). Grazing–burning interactions are more detrimental to plant mortality then either alone (Leigh and Holgate, 1979). For example, in the Kimberleys of north-western Australia, domestic Cattle grazing results in the transport of grass seeds into rainforest remnants that then grow into fuel for fires, thereby changing fire and rainforest patch dynamics (McKenzie and Belbin, 1991).

Fire, grazing, or logging acting individually may not have significant detrimental impacts in the forests of south-eastern Queensland. However, when combined, the recruitment of new trees to replace harvested stems is impaired – young seedlings are either burned by subsequent fires or eaten by domestic livestock (Smith *et al.*, 1992).

The following circumstances illustrate some of the pressures on animal populations, particularly those living in wood production forests. In dry years, populations must contend with drought and high temperatures, which create problems for some species (How *et al.*, 1984; Rubsamen *et al.*, 1984; Smith, 1984b). These conditions can then be followed by wildfire (which can further reduce populations; Keith *et al.*, 2002b). Fire-damaged forests may then be subject to salvage logging, also with possible impacts on forest biota (Lindenmayer and Franklin, 2002; Lindenmayer *et al.*, 2004b). Although studies of such potentially compounding processes are difficult and therefore rare, it seems likely that recovery from the combination of the processes will be slow.

In tropical forests in South America, logging and fire are inextricably linked because logging opens the canopy, creates additional coarse and fine fuels, dries the understorey, and promotes the development of fire-prone grasses (Holdsworth and Uhl, 1997). Fire frequency can increase and preclude forest recovery. In addition, fires are not constrained to logged ignition points; Putz *et al.* (2000) noted that in Bolivia in 1999, the area of logged forest was 10% of the total area burned by fire.

Where introduced weeds displace native flora, fire regimes can change because of the flammability or biomass of the invasive species (Cochrane, 1963), leading to additional effects on native plants (Wardell-Johnson and Nichols, 1991). This phenomenon is of increasing importance in northern Australia (Whitehead and Dawson, 2000; D. Bowman, personal communication), where many exotic pasture grasses have been introduced (Lonsdale, 1994; Low, 1999; see Chapter 7). In particular, some of the grasses introduced from Africa have substantial biomass late in the dry season, and the additional fuel changes fire regimes. Burning in the late dry season may create novel problems for a number of native mammals (Pardon *et al.*, 2003).

As noted in Section 11.6, fire refugia, whether they are individual rocky areas (Whelan *et al.*, 2002) or entire patches of unburned vegetation (Delong and Kessler, 2000; Mackey *et al.*, 2002), can be critical for species survival and post-fire recovery. Populations depending on post-fire refugia may be particularly susceptible to additional impacts, for example bush rock collection or post-fire salvage harvesting (Mackey *et al.*, 2002).

11.11 Fire and reserve design

Three guiding principles – comprehensiveness, adequacy, and representativeness (CAR; Dickson *et al.*, 1997; see Chapters 4 and 16) – underpin reserve design. Lindenmayer and Franklin (2002) argued for a fourth principle: replication. This refers to the need for a reserve system to contain multiple protected areas of a given vegetation type, forest community or species (see Chapter 16). Replication limits the risk that all reserved examples of a vegetation type, population or community will be affected by a single catastrophic event such as a wildfire (Pickett and Thompson, 1978; Lindenmayer and Possingham, 1994).

The need for replication is influenced by the size of a single reserve, especially whether it is large enough to exceed the maximum size of single disturbance event. Large reserves have a better chance of supporting populations that survive natural disturbances such as wildfires (Pickett and Thompson, 1978; Baker, 1992). Large fires are more likely to leave unburned patches within their boundaries and contain a mosaic of post-fire recovery patterns that are needed for a range of biodiversity (Keith *et al.*, 2002a). If some unaffected area remains after a disturbance, propagules or offspring can recolonise disturbed areas.

A recurring theme after major fires is that government agencies are blamed because there has been insufficient fuel management and that, as a result, there should be no additional reserves until they can be 'managed properly'. In fact, more fires burn into national parks than out of them (Barnett, 1994; B. Gilligan, personal communication). National parks are set aside for many reasons other than to manage fuels (see Chapter 16). The ability to reduce fuel over vast parts of the protected area network (even if it were desirable) is clearly not feasible (Whelan, 2002). In response, park managers are now increasingly focusing prescribed burning at the boundaries of parks adjacent to human infrastructure (I. Pulsford, personal communication; S. Troy, personal communication).

11.12 The future

Climate shapes the broad-scale distribution patterns of biota, including vegetation types (Woodward, 1987), and it is likely to alter future fire regimes through changes in temperature, rainfall, evaporation and the frequency of lightning strikes (Cary, 2002). Goldammer and Price (1998) reported that lightning frequency could double over much of continental Australia if the current levels of CO_2 in the atmosphere were to double. However, predicting the impacts of future climate change on fire regimes will be difficult, particularly for rare catastrophic fire events. In addition, the predictions of global climate models (GCMs) for rainfall, temperature and other meteorological patterns are very uncertain. Regional climate models (RCMs) are even more uncertain. For example, Mackey *et al.* (2002) were unable to obtain appropriate input data or suitable regional climate models to simulate future fire regimes in the forests of central Victoria.

There is general consensus about the need for more detailed field-based empirical work on the impacts of fire on the Australian biota. For example, Whelan (1995, p. 307) wrote that:

> *long-term experimental studies [of fire] are badly needed, but are rare.*

In the case of particular groups (such as birds), Woinarski (1999, p. 57 and 59) noted:

> *there are few data on the long-term impacts of a sustained regime of control burning*

and

> *The search for general pattern in response is further hampered by the very variable and limited research effort, with few long-term studies and little experimentation with a range of fire treatments.*

Keith *et al.* (2002a) and Whelan *et al.* (2002) highlighted the critical need for monitoring and predicting the impacts of fire regimes on biodiversity. The problem of a lack of monitoring pervades all areas of conservation biology. Nevertheless, these statements, together with the number of species threatened by fire in Australia, indicate that a major research and management challenge exists to identify ecologically appropriate fire regimes for different vegetation types and their associated biota (Gill, 1999; Cary *et al.*, 2003).

11.13 Conclusions

Fire has a complex relationship with the Australian biota. The impacts of fire vary between vegetation types and species. The response of biota is influenced by the intensity of fires; the interval between the spatial pattern of fires (including the size, location and condition of unburned areas within the fire boundary); the number, quantity and types of biological legacies left within a disturbed area (Franklin *et al.*, 2000); and the life history attributes of the biota in the area subject to fire (Whelan, 1995; Woinarski, 1999). Other factors make the situation even more complex, including weather patterns, changing climate conditions, altered past burning regimes, reduction of the total overall area of forest and woodland due to land clearing, and the need to protect human lives and assets.

There are no simple or generic fire management guidelines that can be applied uncritically to all species and all landscapes. Some species will remain relatively immune or benefit considerably from more frequent fire, whereas others will be negatively affected (Whelan, 1995; Gill, 1999). Designing fire management strategies that are sympathetic to the conservation of biota requires careful planning, setting objectives and subsequent monitoring. This is important particularly because inappropriate fire regimes have contributed to the extinction of several species of plants and animals and threaten the long-term persistence of many others.

It is perhaps unsurprising that the impacts of fire on the vast majority of Australian organisms remain poorly known. A major and sustained research effort is required to better understand such a key environmental process, particularly given the number of species threatened by fire. Sustaining the research effort is critical because public concern about fires and fire impacts is usually short-lived following major fire years in southern Australia. The lack of relevant monitoring data and well-designed field experiments hampers efforts to improve our understanding of fire and its impacts. The most obvious shortcoming is the lack of a national fire mapping system (Gill and Bradstock, 2003). As concluded by Cary *et al.* (2003):

> *The crucial task of identifying ecologically sustainable fire management policies and practices that are consistent with community safety will not be appropriately tackled until significantly increased levels of research, funding and public interest are maintained in the long term.*

11.14 Practical considerations

Fire management for biodiversity conservation is extremely complex – responses to fire will be species-specific, vegetation-type-specific, fire-event specific and even landscape-specific. The best approach is to ensure that the same fire management regime is not applied uniformly everywhere. Variation in the timing, intensity, frequency and spatial location of fires within and between landscapes should reduce the risks of extensive species loss from inappropriate fire regimes (Gill, 1999; Turner *et al.*, 2003). Nevertheless, targeted fire management prescriptions will often be needed to meet the requirements of particular ecological communities. Fire management for conservation must include monitoring of the values that fires are believed to offer.

11.15 Further reading

There are several excellent reviews of fire and the Australian biota. They include Gill *et al.* (1981, 1999), Whelan (1995), and Bradstock *et al.* (2002). All of these except Whelan (1995) are edited books, and the chapters in each one give excellent entry points into fire ecology, control, management, and biotic responses. The edited volume by Cary *et al.* (2003) provides a useful set of papers related to integrated fire, environmental, and policy dimensions in an Australian context. Some of the key findings of the Kapalga Fire Experiment in the savannas of northern Australia – the world's largest fire experiment – are discussed in the edited book by Andersen *et al.* (2003). Bowman (1998, 2003) gives an excellent summary of fire in an indigenous context. Richards *et al.* (1999) gives an elegant example of decision theory modelling to manage fire in a reserve. The State of the Environment Report (2001a) outlines a range of impacts of fire in the Australian environment. Volume 1, issue 5 of the journal Frontiers in Ecology and Environment (June 2005) contains some informed insights into fire regimes and related issues, such as biodiversity conservation by some of the world's leading fire scientists. Shea *et al.* (2004) is a valuable review of the intermediate disturbance hypothesis. A major report by the State Government of Victoria (2003) provides a detailed and wide-ranging account of many facets of fire management, control and other issues (including conservation ones) following the inquiry into the 2002–03 bushfires in Victoria. A wide-ranging discussion of fire and fire management and its impacts on Australian birds has been compiled by Olsen and Weston (2005).

Chapter 12

Demands of the human population

This chapter examines issues associated with the world population and its demands, including potential population sizes, growth rates and per capita consumption. We then examine the demands of the Australian population, particularly its possible size by 2050 under different growth and immigration scenarios. Rates of Australian per capita consumption and the impacts of the Australian population on biodiversity are discussed. The final part of this chapter outlines two areas where the impacts of the Australian population have been pronounced – the coastal environment and the Murray–Darling Basin.

12.0 Introduction

There is often (but not always) a direct relationship between the size of the human population and its impact on biodiversity. For example, the rate at which forest cover is lost, both within most countries and within broad biogeographic regions, is often proportional to the number of people that inhabit the region in question – although this is not the case in Australia, where rapid vegetation loss is also occurring in relatively sparsely populated areas (see Chapter 9).

In many developing countries, forest loss is caused largely by conversion of forest to farmland and by the collection of firewood and charcoal for domestic and industrial use, particularly in savanna woodlands, shrublands and, increasingly, in tropical moist forests (Harcourt, 1992; FAO, 2000). However, the relationship between habitat loss and population size is not simple, and the inequitable distribution of land and other resources frequently plays an important part in the rate of loss of forest cover (see Harcourt, 1992; Population Action International, 2000). For example, in circumstances in which an average-sized farm could easily support a family, many societies tolerate a situation whereby most of the land is owned by a few people, thereby creating a need for many people to develop new farms, simply to subsist. Human population growth raises several issues associated with the absolute size and rate of increase of the population, and the increase in the per capita consumption of biological resources.

Human activities impact heavily on the natural world. Humans directly or indirectly use 40% of the world's plant growth, 25–35% of the world's marine primary production, and 60% of the planet's readily accessible fresh water (Pimm, 2001). The consequence is that human populations now contribute more to the global fixed nitrogen budget than do natural processes (UNEP, 1999). Humans have reduced terrestrial net primary production by about 13% (Vitousek *et al.*, 1986) and have produced conditions that threaten the persistence of 20–30% of the world's species (Lawton, 2002).

Human population control, immigration, the sustainable carrying capacity of the human population, limits to per capita consumption, and the questioning of perpetual economic growth have frequently been taboo subjects. They are symbolic of an underlying contradiction in human attitudes toward natural resource conservation (Ludwig, 1993). However, there is some evidence

Box 12.1

The 'frontier ethic'

Beattie and Ehrlich (2001) cited three major elements of the natural resource problem created by humans: over-population, over-development and over-consumption. A major part of these problems stems from what Chiras (1994) has termed the 'frontier ethic' with respect to the use of natural resources which, in turn, has three interrelated components (after Chiras, 1994; Pullin, 2002):

- There will always be more resources and they are all for human use.
- Humans are not a part of nature and human societies do not have to conform to natural environmental relationships.
- The control of the natural environment is a critical part of human success.

Many of these themes actually have religious and nationalistic origins, and they continue to drive many aspects of economic policies, for example the ideals of constant economic growth. They run counter to notions of ecologically sustainable development (see Chapter 18).

that things are changing. Foran and Poldy (2002) examined human population size, resource use and environmental impact in Australia, and some findings of that report figure prominently in this chapter.

12.1 The world population

The global population now exceeds 6 billion people; this is 2.5 times the global population in 1930. It is presently growing at about 80 million people annually (UNEP, 1999). The United Nations Long-Range Projections (UN, 1998) estimate that the global population in 2100 will be between 6 billion and 18 billion people, although it is estimated to stablise at around 9 billion people by 2050. The difference in the lower and upper bounds in Figure 12.1 is related to whether fertility rates drop to 1.6 children per woman (as has occurred in most European countries) or whether fertility rates level off at 2.6 children per woman as has occurred in some of the more affluent South American nations (Population Action International, 2000).

It is likely that the rate of growth of the human population will decrease within the next 100 years. Projections that assume a constant rate of increase result in unreasonable forecasts. For example, if the

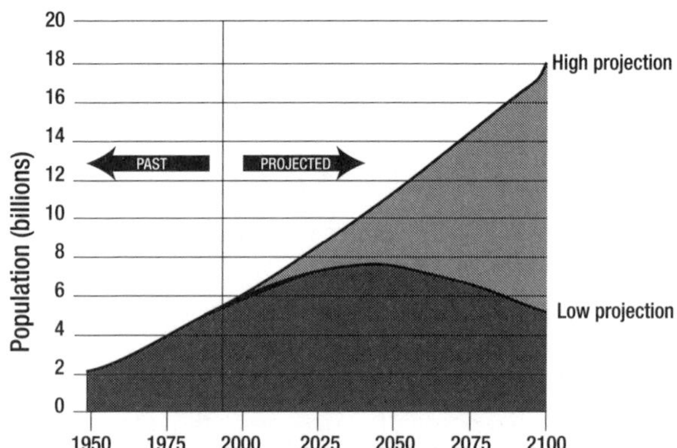

Figure 12.1. Range of variation in forecast population size. (Redrawn from United Nations Long-Range Projections, UN, 1998; Population Action International, 2000.)

current growth rate of 1.0188 is sustained until the year 2100, there would be 41 billion people on the planet. The planet simply cannot sustain this many people, even given the most benign assumptions about the interactions between people and the environment.

The birth rate in many nations may decline to equal the death rate over the next few decades (Hardin, 1993). Cultural factors and beliefs result in large family sizes, which were developed originally to compensate for child mortality. However, modern medicine has substantially reduced mortality rates, particularly those of children, so the propensity towards large families may erode over one or two generations as people realise that most children that are born will survive. Nevertheless, the age structure of a population can impart inertia to population growth. For example, Myers (1993) noted that the long-term status of Kenya's population was related not only to its present growth rate of 3.7%, but also to its population profile, which now contains many young people. As a result, even if each couple produced just two children (an unlikely outcome), the population would still double within two generations. The same profile is true of all countries that have had high population growth rates over the last few decades.

Per capita consumption

There are direct impacts of the size of the human population on the planet's resources, and there are also impacts of per capita consumption. Average per capita consumption of energy is increasing, reflecting growing affluence and the spread of modern technology. Energy use per person has consistently risen over the

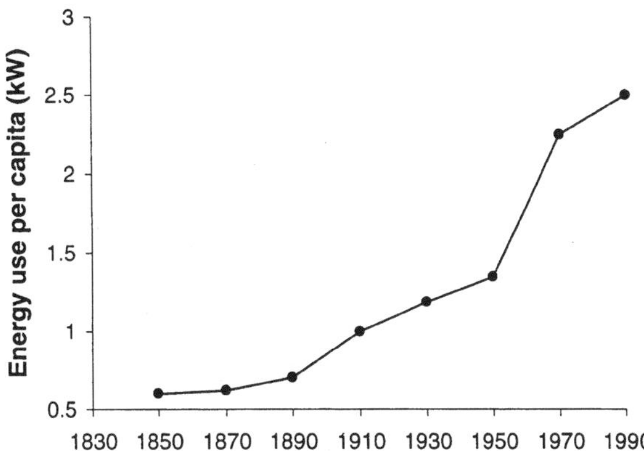

Figure 12.2. Global average energy use per person between 1850 and 1990 (after Holdren, 1991; Vitousek, 1994).

Table 12.1. Comparisons of 'ecological footprints' for the top and bottom five consuming nations. (Based on data in Australian Bureau of Statistics, 2001.)

Nation	Ecological footprint without sea products (hectares per capita)	Ecological footprint with sea products (hectares per capita)
Top five consuming nations of the world		
Iceland	4.2	10.1
New Zealand	8.6	9.8
United States	7.4	8.4
Australia	**7.4**	**8.1**
Canada	6.0	7.0
Bottom five consuming nations of the world		
Egypt	0.8	1.2
Ethiopia	1.0	1.0
Pakistan	0.8	0.8
India	0.7	0.8
Bangladesh	0.4	0.7
World average	**1.8**	**2.3**

last 150 years and more than doubled since 1930 (Figure 12.2). Even in countries with little or no population growth, per capita consumption can grow exponentially. There have been parallel increases in the associated problems of waste management, destruction of natural habitats, loss of biodiversity and degradation of productive landscapes.

There are gross inequities between nations in the consumption of energy and other resources on a per capita basis. The 20% of the world's population that live in high-income countries account for 60% of the world's commercial energy use (UNEP, 1999). Each person in Japan (a high-income country) consumes 40 tonnes of natural resources annually. The equivalent figure for the USA is 80 tonnes, and in Australia it is 200 tonnes. That is, to supply the goods and services for our lifestyle, and the exports to pay for our imports, more than 200 tonnes of resources need to be moved per Australian per year (Foran and Poldy, 2002).

Another way of examining the impacts of humans in high- and low-income countries is to compare per-capita **ecological footprints** (*sensu* Wackernagel *et al.*, 1997) – the amount of land required to provide the energy and natural resources to support a person. Australians rank in the top five in the world by this measure, with levels of per capita resource consumption 2–3 times the world average and 5–10 times higher than a person from one of the bottom five consuming nations (see Table 12.1).

Correcting the distortions in per capita consumption between nations will require massive changes in lifestyles. To release adequate resources for the economic growth of people living in developing countries, there would need to be a tenfold reduction in resource consumption by people inhabiting industrialised nations (UNEP, 1999).

Ehrlich and Daily (1993) and Beattie and Ehrlich (2001) stated that the fundamental threats to biological resources are human population growth and per capita consumption, global warming, and the depletion of biodiversity. It is an inescapable fact that increasing human populations will eventually make insupportable demands on natural resources. This is illustrated by forecasts by the United Nations Environment Programme (UNEP, 1999), which estimated that doubling the human population size coupled with rising levels of consumption by increasing numbers of affluent people will result in a fourfold increase in food consumption and a sixfold increase in energy consumption. The number of motor vehicles on the world's roads increased from 40 million at the end of

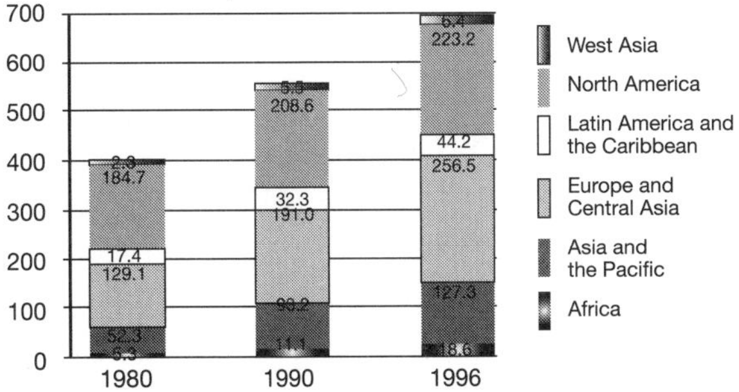

Figure 12.3. Increase in the number of motor vehicles between 1980 and 1996. (Redrawn from UNEP, 1999.) Australia is included as part of the Asia and Pacific region.

World War II (1945) to nearly 700 million in the late 1990s (International Road Federation, 1997; Figure 12.3). At current rates of expansion the number will exceed 1 billion by 2025 (UNEP, 1999).

Impacts of the human population on the environment

The size of the human population and its energy use are relatively easy to measure, but the impact per unit of consumption is more difficult. Even if the environmental impact per unit of consumption remains constant, changes in environmental impact are measured by the product of increasing affluence and increasing population size. Impacts of the human population on the environment are a product of the absolute size of the population (P), the per capita consumption of biological resources (called affluence, A), and the environmental damage caused by the technologies employed in supplying each unit of consumption (T). In general terms, the impact (I) of the human population on the environment can be expressed as (see Ehrlich and Ehrlich, 1990; Hardin, 1993; see also Cocks 1996):

$$I = P \times A \times T$$

Levels of consumption are tied more closely to the number of households than the number of people (Hugo, 1999). Social factors such as increasing divorce rates and smaller numbers of people in a larger number of households can influence environmental effects such as greenhouse gas production (Lutz, 1999; Foran and Poldy, 2002).

Human populations and biodiversity loss

Naeem *et al.* (1999) published a hypothesised relationship between human population size and species loss. It is likely that there are important inter-relationships between the size of the human population, per capita consumption and the conservation of biodiversity (e.g. Andelman and Willig, 2003). Nevertheless, few studies exist (see Myers, 1993; Population Action International, 2000; McKee *et al.*, 2003).

There is not necessarily a simple cause and effect relationship between human population size and species loss. In Australia, many species of native mammals have been lost from arid areas where humans are scarce. Feral predators and changing fire regimes may be the primary causes in these cases (see Chapter 7). However, in general, there does appear to be general relationship between the number of bird and mammal extinctions and the size of the human population (WCMC, 1992; McKee *et al.*, 2003; Figure 12.4).

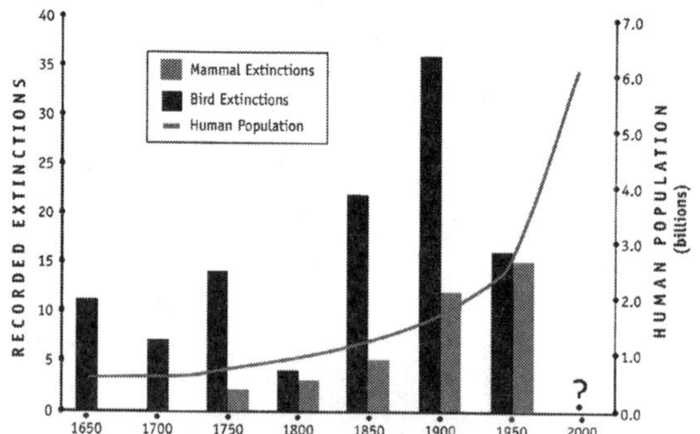

Figure 12.4. Relationships between bird and mammal extinctions and the size of the human population. (Redrawn from WCMC, 1992.)

12.2 Demands of the Australian population

The current rate of population increase in Australia is relatively high (Smith, 1994; Foran and Poldy, 2002); in fact it is one of the highest in the developed world (UNEP, 1999). The maintenance of the present growth rates could result in an Australian population of 50 million by 2100. Given the size, demands and potential impacts of the human population in Australia, a major study examined various scenarios for population size and resource demands (Foran and Poldy, 2002).

Future size of Australia's population

The future size of Australia's population will depend upon the net rate of immigration and birth rates. Table 12.2 shows the projected Australian population size under three immigration scenarios.

Table 12.2. Predicted size of the Australian human population (in millions) under three different immigration scenarios (from Foran and Poldy, 2002).

Year	Zero net immigration per year	70 000 people net immigration per year	0.67% of population net immigration per year
2050	20.6	25.1	32.5
2010	16.7	25.5	50.6

Foran and Poldy (2002) argued that there are various advantages and risks associated with each of the three population scenarios. The 'zero scenario' was believed to be likely to lead to the smallest increases in energy use and waste production. However, economic growth due to population increase might not be replaced by another mechanism of growth, and it could lead to rural decline and longer working hours for an ageing work force.

The intermediate population growth scenario simulated by Foran and Poldy (2002) was forecast to best make a transition to what they termed 'a sustainable population size', although a risk was thought to be the maintenance of inadequate environmental management practices. The high growth scenario was considered to be likely to lead to a high growth economy, but to cause high levels of energy use, greenhouse gas emissions, and deteriorating environmental conditions including increased salinity, land degradation, and biodiversity loss.

Box 12.2

Tax reform

Taxation can be a powerful tool for changing behaviour. Hamilton *et al.* (2000) reviewed taxation in Australia and made a number of recommendations based on the idea that restructuring tax systems in revenue-neutral ways can be used to promote employment and discourage environmental damage. They pointed out that less than 50% of the external and infrastructure costs of transport are reflected in the prices users pay for roads and rail systems. In the electricity sector in Australia, prices capture less than one-third of the costs of the environmental damage caused by its production, and in most countries the price covers none of the environmental damage (Hamilton *et al.*, 2000). Instead, companies offering environmentally friendly and efficient technologies, processes, products and services could be encouraged at the expense of damaging industries.

Tax reform has been adopted by several countries, including the Netherlands, Germany, France and Belgium. In Australia, the revenue from a carbon tax or a system of emissions trading could amount to A$7 billion (Hamilton *et al.*, 2000). The German Finance Minister, Eichel (2000), stated that the revenue from ecological tax reform in that country was being used to reduce pension insurance contributions, and that economic initiatives had led to 'the development of energy-saving technologies and production processes...more than 200 000 new jobs have been created'. Such changes are possible in Germany, in part, because private property law embodies both rights and custodial responsibilities (Raff, 2003). Other economic instruments can achieve conservation goals, including regulation (penalties), voluntary agreements (e.g. the Land for Wildlife scheme in Victoria), conservation covenants (legally binding agreements on land title), land purchase, conservation contracts (compensated agreements between landholders and government), government-assisted community programs (e.g. Birds Australia owns and manages land for conservation), and biodiversity credit trading systems (see Stoneham *et al.* 2000; Agius, 2001).

Energy demands and greenhouse gas production of Australia's human population

In Chapter 5, we noted that Australians produce more greenhouse gases on a per capita basis than any other nation. Australians also consume more energy on a per capita basis than most people in developed countries. Table 12.3 shows data for 1995. In 1998 the figure for Australians had increased to approximately 260 gigajoules per person (Australian Bureau of Statistics, 2001; Foran and Poldy, 2002). Only per capita energy consumption in Canada (325 gigajoules per person) and the USA (360 gigajoules per person) was higher (World Resources Institute, 1998).

Foran and Poldy (2002) recognised that per capita energy consumption was far from uniformly distributed across the Australian population: an individual

Urban development in Brisbane, Queensland. (Photo by David Lindenmayer.)

Table 12.3. Per capita energy consumption in 1995 in various regions around the world. (Based on data in UNEP, 1999.)

Region	Per capita energy consumption (gigajoules)
West Asia	8.31
Africa	12.15
Asia and the Pacific	28.56
Latin America and the Caribbean	41.76
North America	101.68
Europe and Central Asia	114.14
Australia	260

Data for Australia are for 1998 and were calculated by Foran and Poldy (2002).

living in a high-income household used almost twice as much energy as someone in a low-income household.

Under the various population growth scenarios modelled by Foran and Poldy (2002), Australian greenhouse gas emissions could increase massively in the next 50 years. For example, Australia's greenhouse emissions derived from transport alone increased by 24% between 1990 and 2000 to 76 million tonnes (or about 4 tonnes per capita; P. Laird, unpublished data). If improved technologies are implemented, then emissions could rise 170–230% higher than those in 1990, but without improved technologies, they could exceed 300% by 2050 (Foran and Poldy, 2002).

Some of the present causes of high greenhouse gas emissions in Australia are to some extent unique: nowhere else is there such a high proportion of emissions derived from agriculture and burning vegetation following land clearance (Figure 12.5). Land clearance rates in Australia throughout the 1980s, 1990s and early 2000s are as high or higher than those in the vast majority of developing tropical countries.

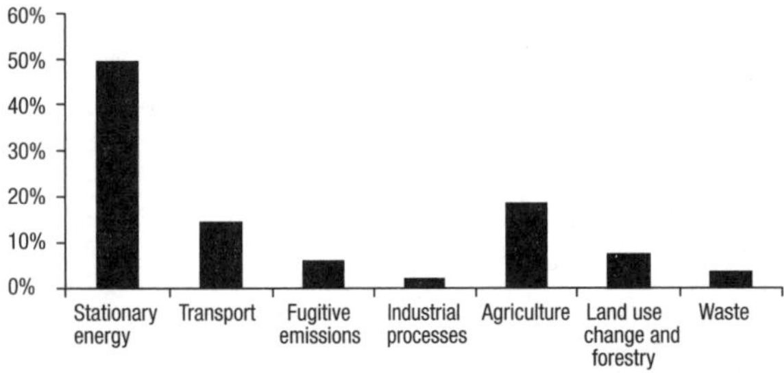

Figure 12.5. Contribution to total net CO_2-equivalent emissions by sector in 2000. (Redrawn from data provided by the Australian Greenhouse Office, 2005.)

Future Australian populations and future resource use

Under the highest growth scenario modelled by Foran and Poldy (2002), the population equivalent of another 90 cities the size of Canberra (~300 000 people in 2005) would need to be accommodated in Australia by 2100. These would occupy a large area of land that presently has other values, including conservation. Cities such as Sydney and Melbourne (~3.5 million and ~4 million people, respectively, in 2005) could exceed 10 million people and have the associated environmental problems that are common in very large cities around the world.

Under this scenario, water extraction would need to almost double from present levels to meet increased agricultural production. This is a critical consideration because Australia relies on irrigated water for the production of many agricultural commodities, for example cotton, wine and dairy products. Human water use in urban areas would also increase substantially with population increases (Figure 12.6). The total annual water use in Australia more than doubled between 1985 and 1996–97. In addition, more than 25% of the nation's surface water is close to, or has exceeded, sustainable extraction limits (State of the Environment, 2001f). Given the current overcommitment of water supplies in southern Australia (see Smith, 1998), most of the increase in water extraction would have to occur in northern Australia. In some locations, increased water allocation for agricultural production would compete with the need to maintain the integrity of aquatic ecosystems, as well as demands for urban water supply (which will increase with human population size; Foran and Poldy, 2002).

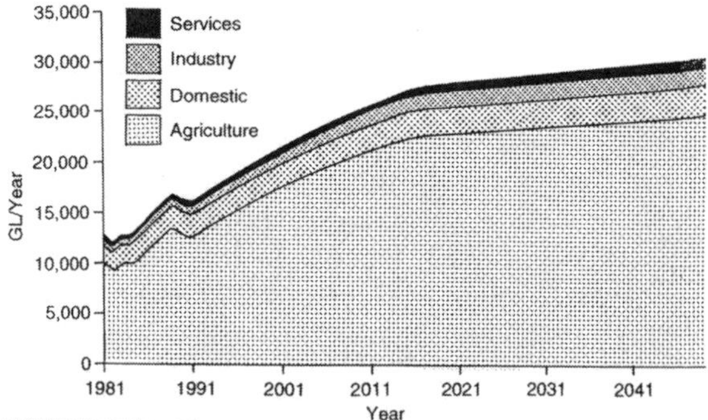

Figure 12.6. Forecast increase in human water requirements in Australian urban areas under various population growth scenarios. (Redrawn from Foran and Poldy, 2002.)

Other aspects of natural resource use will also change. Domestic grain, meat and fish consumption would double under the highest growth scenario. Consumption of other manufactured goods, such as houses and motor vehicles, could also increase considerably. For example, the annual requirement for new cars could be between 400 000 and 700 000 depending on population growth rates (Foran and Poldy, 2002), and the extent of the road network on which to drive them could increase by 3000–10 000 kilometres.

The work by Foran and Poldy (2002) emphasises the importance of both the size of the human population and rates of consumption of natural and other resources. It also highlights the inter-relationships between disciplines such as economics, human demography and environmental management, and the implications they have for conservation biology.

Australia's carrying capacity

> *When I affirmed that Australia was capable of supporting 400 millions of people, I did not mean Australia as we now have it, but as it might be, and probably will be, when its water is carefully conserved and its soil scientifically irrigated and cultivated (Cole, 1947).*

This quote is symptomatic of the ignorance about the impacts of the human population on the Australian environment over the past 200 years. Given the demands of an increasing future Australian human population, a wise course of action would be to determine an appropriate carrying capacity for the continent – a controversial undertaking. Part of the controversy lies with the fact that Australia's carrying capacity is a function of the rates of resource consumption and waste production of the human population that the inhabits the continent. In a political context, it has proved impossible to decouple the human population size debate from issues associated with the intake of refugees, racism and other topics. As early as the 1920s, Taylor (in Flannery, 1994; Cocks, 1996) suggested that Australia's carrying capacity might be approximately 20 million people. Estimates of the optimum human carrying capacity of Australia vary from 6 to 50 million people (see Smith, 1994). For example, Flannery (1994) calculated a value of 6–12 million based on an analysis of the environmental history of Australia (as well as other areas in the Southern Hemisphere), the dynamics of indigenous human populations prior to European settlement, and the continent's renewable and non-renewable natural resources. Regardless of the validity of these figures, it is critical that better informed estimates of the long-term carrying capacity of Australia are made and that ecological criteria are central to such calculations (Foran and Poldy, 2002).

Australian population and biodiversity loss

As noted in many places throughout this book, the human population in Australia has had major direct

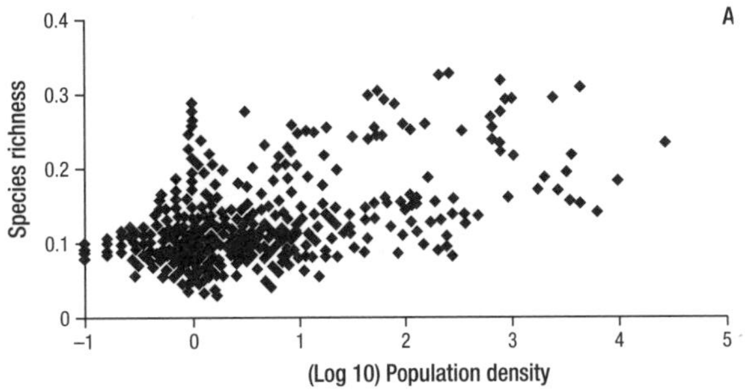

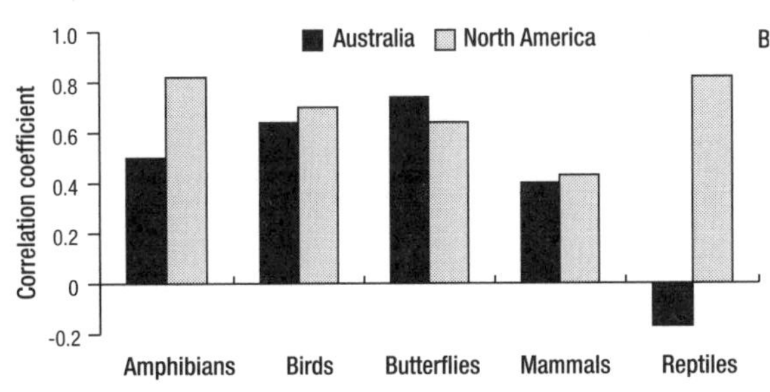

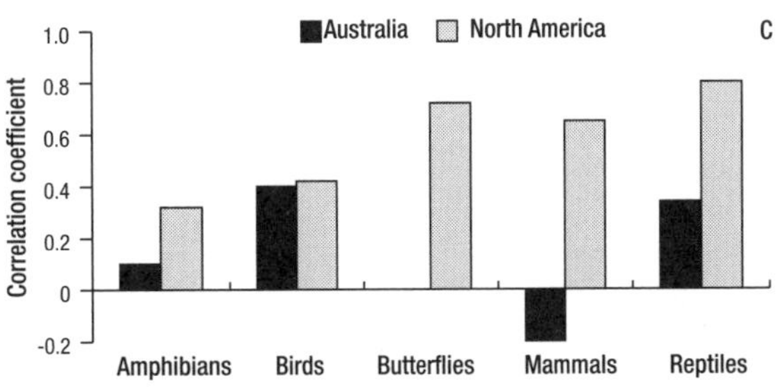

Figure 12.7. Correlations between human population density and the richness of amphibians, birds, butterflies, mammals and reptiles in Australia and North America. (A) Data for all species in Australia. (B) Percentage of threatened species in each group. (C) Percentage of range-restricted species in each group. Reptiles in Australia show a negative trend in both cases, which is possibly due to sparsely populated arid areas supporting most species of reptiles. (Redrawn from Luck *et al.*, 2004.)

and indirect impacts on the continent's biota. Luck *et al.* (2004) showed that there are strong positive correlations between human population density and species richness for birds, mammals, amphibians and butterflies in Australia (Figure 12.7), suggesting that there is considerable potential for future spatial 'conflict' between the expansion of the nation's population and the loss of biodiversity. This highlights the importance of the concept of stewardship, which also embodies custodial responsibility (Chapter 1). Australians are responsible for about 7% of the world's biota. The degree to which a country meets its responsibility can be reflected in the impacts per head of population.

By any measure, Australians have a per capita conservation record that is the worst on earth. There have been more recorded mammal extinctions in Australia over the last 200 years than on any other continent or in any other country (Chapter 3). The proportion of threatened vascular plants and animals in Australia is equivalent to or exceeds the proportions found in other megadiverse countries, and, on a per capita basis, there are more threatened species in Australia than anywhere else (Figure 12.8). Of course, some of Australia's appalling conservation record is due to the fact that the continent was only recently subjected to a developed economy (Balmford, 1996). Historical changes in other continents happened earlier and may not be as visible.

The Australian population has had a major impact on many parts of the Australian environment, two of which are considered below: the coastal zone and the Murray–Darling Basin.

12.3 Coastal zone

Uniqueness of the Australian coastal zone

Australia has a unique coastal environment. The marine zone covers approximately 8.9 million square kilometres and the coastline is also one of the world's longest – almost 37 000 kilometres (State of the Environment, 2001b). When areas such as the

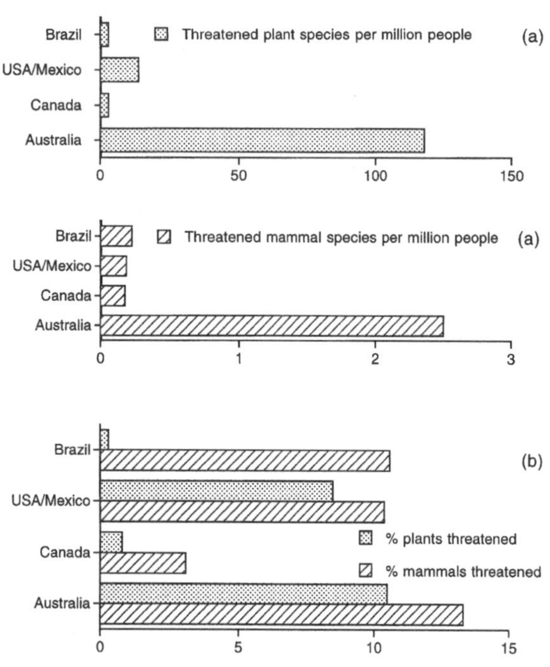

Figure 12.8. Number of threatened species per million people and percentage of threatened plant and mammal species in four countries (after OCS, 1992).

Images of the Australian coastline: (A) Kangaroo Island (South Australia) and (B) Wilsons Promontory (Victoria). (Photos by David Lindenmayer.)

Australian-claimed parts of Antarctica, Heard Island and other offshore islands are included, together with the 200-mile (322-kilometre) exclusive economic zone, the nation's total marine area encompasses more than 16 million square kilometres – one of the largest in the world (State of the Environment, 2001b).

Australia's marine and coastal ecosystems contain environments that vary markedly in geology, topography, and bioclimate. They include both marine systems and associated elements of the terrestrial environment, such as dunes, rivers, mudflats, wetlands, estuaries, coastal plains and mountains. Australia's marine environments support the largest areas of coral reef on the planet, the largest and most species-rich areas of tropical and temperate seagrasses, and the third-largest area of mangroves (Zann and Kailola, 1995; State of the Environment, 2001b). The most species-rich mangrove flora in the world occurs in Australia, and the southern coastline supports the highest known levels of diversity of red and brown algae, lace coral, crustaceans and sea squirts (State of the Environment, 2001b). Australia's marine environments also support many endemic and endangered species (Pogonoski *et al.*, 2002).

Coastal zone and the human population

The problems associated with the over-exploitation and mismanagement of commercial and recreational marine fisheries are outlined in Chapter 8. However, there are many other difficulties linked closely to the size of the Australian human population and the rate at which it uses resources. The Australian coastal environment experiences both intensive and extensive land use practices and extensive urban development (Figure 12.9). About 75% of Australia's population lives within 3 kilometres of the coast, and about 90% of all building activity in Australia between 1983 and 1991 took place in this area (RAC, 1993). About 60 000 Australians are moving to the coast every year. There were very rapid rates of population growth in the coastal zone in the two decades between 1971 and 1991, particularly in parts of Queensland, New South Wales, Victoria and Western Australia (RAC, 1993), and this trend has continued since then (State of the Environment, 2001b; Foran and Poldy, 2002). These rates of growth were particularly pronounced in non-metropolitan regions of the coastal zones outside of the existing large cities and industrial centres of Australia.

Urban sprawl in the coastal zone of eastern Australia means that the landscape between the south coast of New South Wales and the central Queensland coast could become a largely continuous 'urbanised ribbon' over the next few decades (RAC, 1993). The problems of human population and urban sprawl will be further exacerbated by the expansion of the Australian tourist industry (Foran and Poldy, 2002; Australian Bureau of Statistics, 2003), which will focus much of its activities on the coastal fringe of the continent (State of the Environment, 2001b).

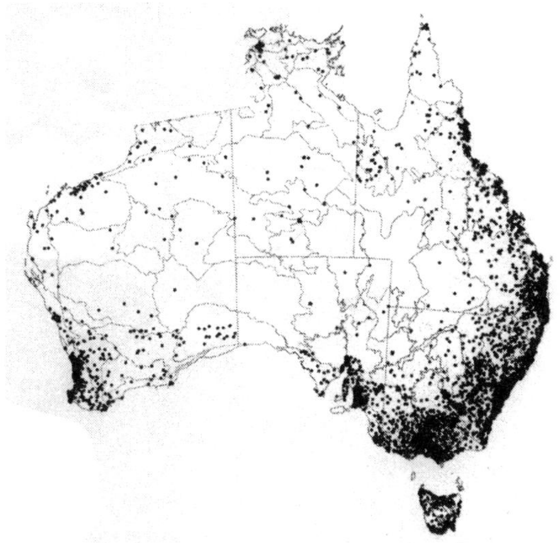

Figure 12.9. Human population distribution patterns for Australia showing a concentration of settlements around the coastal fringe of the continent. (From the Australian Bureau of Statistics, 2003.)

The expansion of the human population in the Australian coastal zone will result in competing interests between users of marine environments. There are potential conflicts of interest, for example, between commercial and recreational fishers, industrial and residential developers (RAC, 1993), and mariculturists (e.g. oyster growers; see Box 12.3). Given the sensitivity of mariculture to pollution, problems may also be created by the elevated levels of human waste that are invariably associated with urban sprawl in coastal environments (State of the Environment, 2001b; White, 2001).

There are other management problems in Australia's coastal zone. The issues of environmental weeds such as Bitou Bush in the coastal zone and invasive species in ballast water were outlined in Chapter 7. There are numerous other problems caused by human populations, for example accelerated erosion and increased amounts of beach litter. Acid sulfate soils and

Box 12.3

The demise of the Sydney Rock Oyster

The Sydney Rock Oyster grows in estuaries along the east coast of Australia from Moreton Bay in Queensland to Flinders Island in Bass Strait. The species has been farmed in New South Wales for more than 130 years (Nell, 1993). Production levels of the Sydney Rock Oyster peaked in the 1970s and have been declining substantially since then – despite improved cultivation methods. Extrapolation of current trends suggests that oyster production will fall to zero by 2019 (White, 2001; see Figure 12.10). The number of lease permits, the size of leases, and the oyster yield per lease are all declining. Leases are being relinquished and production yields reduced because of the failure to adequately protect oyster-growing areas and maintain estuaries in a condition suitable for production.

A complex set of interacting factors has contributed to the decline of the industry and is threatening its long-term viability. Historical expansion and over-harvesting soon after European settlement were important (Kirby, 2004), and more recently, coastal and upstream catchment development, drainage and run-off from acid sulfate soils, increased risks of human and animal waste contamination, and competition from introduced oyster populations have contributed to the problems. Threats have not been countered by coherent land development and management policies by the relevant regulating authorities (White, 2001).

acid run-off into coastal catchments are also increasingly serious problems in Australia (State of the Environment, 2001b).

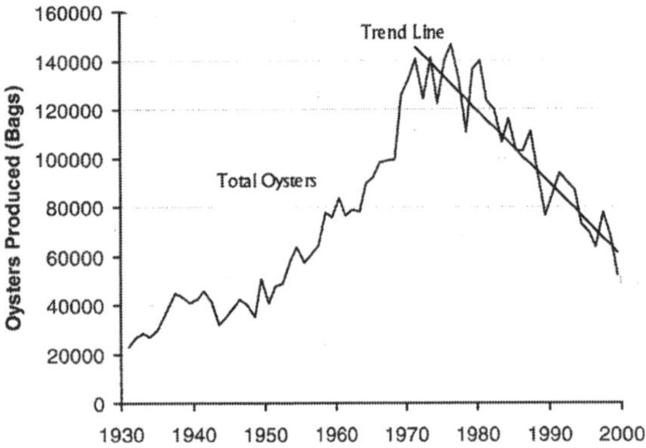

Figure 12.10. Trend lines for production levels of the Sydney Rock Oyster. (Redrawn from White, 2001.)

Oyster leases in the Clyde River on the south coast of New South Wales, where there are major pressures from nearby expanding urban settlements such as Batemans Bay and Nelligen. (Photo by David Lindenmayer.)

In 2002, the National Land and Water Resources Audit concluded that half of Australia's 1000 estuaries were degraded (Commonwealth of Australia, 2002b). In the case of the Hawkesbury–Nepean system in central New South Wales, there have been major impacts from agricultural and urban run-off as well as waste water flows from sewage treatment. The levels of nitrogen in rivers in this system can be up to four times higher than Australian standards (RAC, 1993; Commonwealth of Australia, 2002b). The flow patterns of the Hawkesbury–Nepean rivers have been greatly altered (through the construction and dams and weirs) and the ability of the system to flush itself

Estuary system near Kalamburu on the north-western coast of Western Australia. (Photo by David Lindenmayer.)

Box 12.4

The challenge of coastal management and rapid urban expansion – Moreton Bay in Queensland

Moreton Bay is a large embayment in south-eastern Queensland, which borders the city of Brisbane and other urban areas that have population growth rates that are faster than anywhere else in Australia (10–13% per annum; Skinner *et al.*, 1998). The population of the area is 1.6 million, and it is growing at an estimated rate of 1000 additional people each week, making it one of the fastest growing regions in the world (Skinner *et al.*, 1998). Moreton Bay is a significant marine environment because it supports many significant ecosystems, extensive mangrove forests and seagrass meadows, as well as important marine mammal and other vertebrate populations (e.g. Chilvers *et al.*, 2005).

The management of Moreton Bay represents a substantial challenge not only because of its environmental values and large and rapidly increasing human population, but also because: (1) it supports large commercial fisheries and many recreational fishers, (2) it is a major shipping lane and there are many associated industrial developments, (3) there is a high catchment to estuary ratio (14:1), with substantial discharges from numerous subcatchments, particularly during large flood events, and (4) there can be long residence times for pollutants in some parts of the bay (~200 days).

In an effort to deal with the myriad of environmental challenges associated with the management of Moreton Bay, the Moreton Bay Waterways and Catchments Partnership was established. Under the auspices of the partnership, the condition of more than 250 estuarine and marine sites is monitored monthly, and more than 100 freshwater sites are sampled twice per year (Moreton Bay Waterways and Catchments Partnership, 2003). Environmental management is based on rigorous science and an adaptive management approach. Annual 'report cards' are produced for the numerous catchments, estuaries and marine areas that comprise the Moreton Bay region. Major problems that need to be addressed (for example stream bank stabilisation or the need for sewage or industrial waste treatment) are identified and remediation programs are implemented in response. Significant improvements have been made, for example the reduction of the sewage plume at the mouth of the Brisbane River, and major reductions in the levels of nitrogen in the waterways.

The approach to managing Moreton Bay has been complex because of the range of issues to be tackled, but it is a model approach that should be adopted more widely in Australia and overseas.

Box 12.5

The demise of Broulee harbour

The demise of the deep-water harbour at Broulee on the south coast of New South Wales highlights the lost opportunities that can arise from fragmented and short-sighted decision-making. During the mid-1800s, Broulee supported a sheltered deep-water harbour (a relatively rare geographic feature on the New South Wales south coast) with the potential to be a shipping hub for the surrounding region. However, land developers sold land on the headland near the town and cleared the trees on the sand spit that connected it to the mainland. The spit eroded and the sand deposited at the bottom of the bay. Today the landscape at Broulee is markedly different from that of 150 years ago and the opportunity to develop a deep-water harbour has been lost (modified from RAC, 1993).

has been lost. These types of problems are repeated in many other river systems around Australia (Commonwealth of Australia, 2002b). For example, the levels of heavy metals in the Derwent River in

The beach at Broulee on the New South Wales coast. (Photo by David Lindenmayer.)

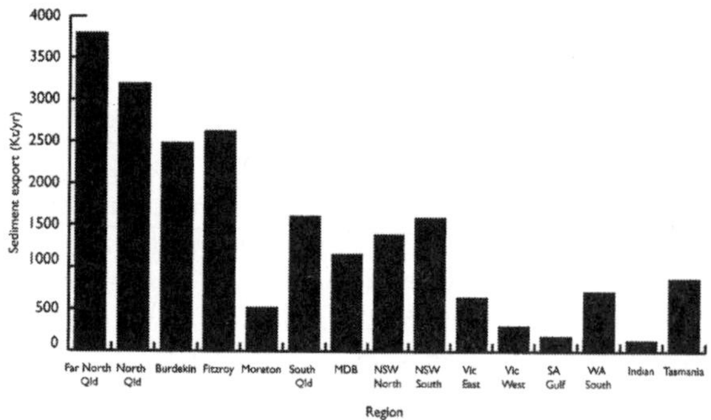

Figure 12.11. Sediment loads exported to the coast in rivers around Australia. (Redrawn from Commonwealth of Australia, 2001d.)

Hobart (e.g. mercury, zinc, cadmium and lead) are extremely high (RAC, 1993).

Agricultural land management practices have had a major impact on rivers that flow to the Australian coastline (Figure 12.11). Each year, nearly 141 000 tonnes of nitrogen and 19 000 tonnes of total phosphorus from areas subject to intensive agriculture are transported down rivers to the coast (Commonwealth of Australia, 2001d). In coastal Queensland (Bell and Elemtri, 1995), more than 14 million tonnes of sediment are being discharged from that State's coastal catchments annually (Moss *et al.*, 1992; Commonwealth of Australia, 2001d). In the region of the Great Barrier Reef in north Queensland, where there are major agricultural and grazing industries, the total amount of sediment run-off is now 3–4 times higher than it was before the arrival of Europeans (State of the Environment, 2001a). In some catchments, such as those of the Burdekin and Fitzroy rivers, sediment loads can be 20 times higher than those prior to European settlement (Commonwealth of Australia, 2001d).

Policy problems and solutions in coastal management

One of the major problems underlying the conservation and informed management of coastal environments in Australia has been the complexity of the management process itself (Wescott, 2001). Coastal management encompasses all levels of government – Local, State and Federal, each of which often has different aims and aspirations. In the early 1990s, the Resource Assessment Commission (RAC, 1993) identified in excess of 1500 different organisations with some type of responsibility for the direct or indirect management and regulation of land and/or building in Australia's coastal zone. In addition, there were more than 900 different regulatory systems in the zone (RAC, 1993).

The complexity of management of the Australian coastline is elegantly summarised by Toyne (1994), who described the array of different jurisdictions and boundaries that would be crossed by a Green Turtle making its way to nest on an island in the Great Barrier Reef in northern Queensland. In this case, the animal would cross (after Toyne, 1994):

- Australia's exclusive economic zone (about 200 miles (322 kilometres) off the coast)
- Australian National waters (12 miles (19 kilometres) off the coast)
- the Great Barrier Reef Marine Park (the waters of the park are administered by the Commonwealth except if they are within 3 miles (5 kilometres) of the coastline, where they become the responsibility of the Queensland State Government)
- land administered by the Queensland State Government (once it moved onto the beach to nest).

As a result of the complexity in coastal management, decision-making has been fragmented and typically narrowly focused (Wescott, 2001). A great number of small decisions sometimes create very large environmental problems that transcend the physical and jurisdictional boundaries affected by any one decision. Hence, there has been a general recognition that there must be a national approach to solving problems in Australia's coastal zone (Wescott, 2001). This requires greater coordination between different levels of government, and different arms of government at the same level (for example, various land management agencies at the State level). This has been termed 'integrated coastal management' (State of the Environment, 2001b) and many of the reforms recommended by the Resource Assessment Commission (RAC, 1993) to achieve it have been implemented in some Australian States and Territories.

In response to the problems that affect Australia's coastal areas, the Commonwealth Government developed a new policy on the coastal environment – Australia's Oceans Policy – which was released in 1998. Nevertheless, given the growing population pressures and associated human-derived environmental problems in the coastal zone, a challenge remains to adequately develop the institutional structures and arrangements needed to fully implement the aims of the Oceans Policy and achieve integrated and ecologically sustainable coastal zone management (State of the Environment, 2001b; see Boxes 12.4, 12.5).

12.4 Murray–Darling Basin

The Murray–Darling Basin (see Figure 12.12) covers about 1 million square kilometres or about one-seventh of the Australian continent (Ghassemi *et al.*, 1995; Smith, 1998). The two rivers that drain the basin (the Murray and the Darling) together cover a length of about 4500 kilometres, making them among the longest rivers in the world. Together with the Murrumbidgee River, the Murray and Darling Rivers are the longest three rivers in Australia. The aquatic ecosystems of the basin include more than 30 000 wetlands and are highly varied.

Annual stream flows in the Murray–Darling Basin are comparatively low: only 0.15 megalitres per second (the equivalent figure for the Amazon River is 180 megalitres per second; Smith, 1998). There are also huge year-to-year variations in annual stream flows, setting the Murray and Darling Rivers apart from other major river systems elsewhere in the world (Ghassemi *et al.*, 1995; Smith, 1998).

Thoms and Sheldon (2000) recently estimated that the rivers, wetlands and floodplains of the Murray–Darling Basin provide A$187 billion in ecosystem services. Indeed, the Murray–Darling Basin is one of Australia's most important areas for agriculture and grazing: it produces more than 40% of the nation's total output from the rural sector (Australian Bureau of Statistics, 2001; Commonwealth of Australia, 2001d, 2002a). As much as 80% of the water in the Murray–Darling Basin has been developed (e.g. diverted) for human use through dams and irrigation systems (State of the Environment Advisory Council, 1996). Water flows have been heavily reduced, requiring two dredges working full-time to keep the mouth of the Murray River open to the sea at an estimated annual cost of A$7 million.

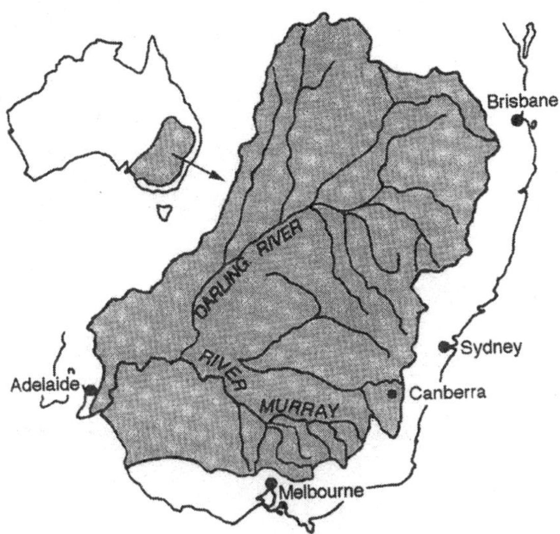

Figure 12.12 Murray–Darling Basin. (Redrawn from Ghassemi *et al.*, 1995.)

Degradation in the Murray–Darling Basin

The Murray–Darling Basin was settled by Europeans in the early 1800s and since that time there has been massive tree clearing in the region. Walker *et al.* (1993) estimated that 12–15 × 10^9 of an original 20–23 × 10^9 trees have been cleared from the Murray–Darling Basin in the past two centuries. About 95% of the river length assessed by Norris *et al.* (2001) was in an environmentally degraded condition.

The development of the Murray–Darling Basin has resulted in a number of major impacts, including the loss of biodiversity, dryland salinity, stream salinity, drowning of large areas of forest, loss of wetland habitats (and thus breeding areas for many animals), degradation of in-stream and riparian habitat for wildlife, modifications to river geomorphology (e.g. changes in, and erosion of, bank conditions), and **eutrophication**

Murrumbidgee River in western New South Wales. (Photo by David Lindenmayer.)

Extensive areas of drowned forest near Corryong. The trees were killed during the establishment of a large dam. (Photo by David Lindenmayer.)

(with associated human health problems; Smith, 1998; State of the Environment, 2001f). For example, a toxic blue-green algal bloom caused by eutrophication in the Darling River in 1991 stretched over a distance of 1000 kilometres.

Other aspects of the 'health' of the Murray–Darling Basin are continuing to deteriorate. Salinity levels are forecast to continue to increase over the next two centuries – a problem that will be extremely expensive to resolve (see Chapter 5). Ghassemi *et al.* (1995) estimated that the construction of an outfall system to carry saline water to the ocean from the lower reaches of the Murray River would cost up to A$5 billion.

Biodiversity in the Murray–Darling Basin

The Murray–Darling Basin supports a rich and diverse biota, including many vulnerable and endangered species (35 birds, 16 mammals, and 35 native fish). Changes to the condition of the Murray and Darling Rivers and the Murray–Darling Basin have had a major effect on biodiversity and have resulted in substantially modified habitat conditions for an array of aquatic native plants and animals (Commonwealth of Australia, 2001c). About 40% of the river length assessed by Norris *et al.* (2001) had severely depleted biota and 10% of the river lacked 50% of the aquatic invertebrates that should have been there. The loss of large trees in streams has altered stream flow patterns (Gippel *et al.*, 1996) and eliminated places for fish to breed and hide from predators (Koehn, 1993). Changed flooding regimes and the highly regulated colder water influxes resulting from the construction of large dams have led to the decline of most species of native fish in the Murray–Darling Basin (e.g. Trout Cod; Cadwallader, 1986; McKinnon, 1993; Smith, 1998; Thoms *et al.*, 2000; Norris *et al.*, 2001; Department of Environment and Heritage, 2005b). The changes have also created suitable habitat for exotic species such as the introduced Common Carp.

Solutions to problems in the Murray–Darling Basin

Environmental challenges in the Murray–Darling Basin are serious and will require substantial efforts to stabilise and eventually reverse. The problems are multi-scaled, multi-disciplinary and transcend political boundaries (e.g. State governments and the domains of influence of management agencies). For example, water quality problems toward the mouth of the river in South Australia are a legacy of management actions (or inaction in some cases) not only in that State but also Queensland, New South Wales, the Australian Capital Territory and Victoria. A compounding factor has been the major over-allocation of water across all jurisdictions (Goss, 2003).

The Murray–Darling Basin Commission and its predecessors have often been held up as an example of excellent practice in multi-government catchment management (Powell, 1993). However, Smith (1998, p. 343) expressed an alternative view that:

> *its performance over some eighty years has led to over-commitment of resources, scant regard for the environment, and has led to a future soil salinisation time bomb that could cripple the agricultural industry.*

Irrespective of the accuracy of Smith's (1998) arguments, there is now increasing intergovernmental cooperation in the Murray–Darling Basin, as well as community group support (e.g. Landcare and Greening Australia) that attempts to address the problems that characterise the area.

12.5 Conclusions

Australia's coastal zone is an important, varied, and biodiversity-rich environment that is presently under considerable human, economic and ecological pressure. The problems associated with the conservation and management of the coastal zone highlight the complexity that has become common in natural resource issues in Australia. Underlying all of these considerations is the inertia of the human population growth rate. The human population continues to grow and individuals continue to make increasing demands on the environment (Foran and Poldy, 2002). Without consideration of the plateaus to which both the size of the population and its ability to consume might rise, all management structures will eventually fail.

It is easy to become pessimistic about the way in which humans use natural resources. Greed and short-sightedness seem to govern most resource-use decisions (Ludwig, 1993; Ludwig *et al.*, 1993, 2001). As Holling (1993) pointed out, at the minimum, such views are an antidote to those who hold cornucopian beliefs in the capacity of humans to learn, adapt and substitute (e.g. Simon and Kahn, 1984) and believe in the ultimately benign nature of human development and interactions with the environment (Brunton, 1994; Lomborg, 2000). The solutions to many problems lie in managing resource consumption. Ludwig (1993) recommends that the rate of human population increase be halted, and that the per capita consumption of resources by humans be reduced. In the case of non-renewable resources, net consumption will eventually have to be reduced to zero. The success with which natural resources are managed to meet present as well as future needs will depend critically on the flexibility and adaptability of management. We will return to these ideas in Chapter 19.

12.6 Practical considerations

High per capita consumption of energy and resources drives environmental impacts in Australia. The greatest environmental advances may eventually be achieved by changes in attitude, brought about by education. Ecological considerations must play a central role in discussions about Australia's population growth.

12.7 Further reading

Important treatments of the subject of human population growth and the environment include Ehrlich and Ehrlich (1990), Hardin (1993), Pulliam and Haddad (1994), United Nation's Environment Programme (UNEP, 1999) and Population Action International (2000). McKee *et al.* (2003) provide an interesting study of relationships between human population growth and threats to global biodiversity. A seminal work on future Australian population size, resource use, and environmental impacts was conducted by Foran and Poldy (2002).

The Resource Assessment Commission (RAC, 1993), the State of the Environment (1996, 2001a–d) reports and Foran and Poldy (2002) provide overviews of environmental impacts in Australia and the effects of the human population.

Part III: Methods of analysis

Methods for the conservation of genes, species and ecosystems include:

- field and laboratory procedures for survey and data collection
- analytical and statistical procedures for data analysis
- mathematical models that predict the distribution and abundance of biodiversity
- methods for reserve design
- tools to assist conservation and resource management decision-making (i.e. decision support tools)
- data-management tools
- legislation, bureaucratic regulations and management strategies.

There is far more, even in this incomplete list, than could be described adequately in any single textbook. All of the skills and techniques available to biologists, ecologists, or wildlife managers might be applied to meet the challenges they present. Our intention is to focus on just some of the methods for data analysis that can be used to inform decision-making, and to focus on methods that are particularly pertinent to the types of problems faced by conservation biologists working in the Australian environment.

In Part I of this book we outlined some of the principles and themes of conservation biology. Part II concentrated on describing the types and extent of impacts of the human population on the Australian environment. In Part III we maintain our predominantly Australian focus, moving to address some of the problems that often confront conservation biologists. We have chosen those methods of analysis that are relevant to many of the most pressing conservation issues in the Australian environment, are useful for the management of species in fragmented environments, and are useful for identifying and protecting the most important habitat in the face of widespread clearance. The themes addressed in Parts I and II provide a background against which we examine analytical methods in conservation biology.

We will *not* address several related topics, mainly because each deserves treatment that is beyond the scope of this book. The topics that we omit include weed and feral animal control, pollution control and remediation, habitat rehabilitation, restoration ecology, rapid biodiversity surveys, farm planning and management, and the laws regarding protection, quarantine and export control of plants and animals. All of these topics have more complete treatments in the scientific literature and in specialist texts, to which we will refer. Notably, we have not completed a detailed examination of the 'human dimension' of conservation and use of natural resources, that is, topics such as legislation, policy, economic instruments, and education to foster change in public attitudes.

We assume a working knowledge of Mendelian genetics and introductory statistics. We address some uses of genetic variation, methods for measuring biodiversity, techniques for estimating the parameters of species' habitats, reserve design, the principles of monitoring and impact assessment, and some principles of risk assessment. We intend Part III to be most useful to advanced conservation biology students who are interested in solving environmental problems relevant to biological conservation in the Australian environment.

Chapter 13
Measuring, managing and using genetic variation

This chapter describes different ways of measuring genetic variation, and how this information can be used to understand the social structure, spatial structure and dynamics of plant and animal populations. It presents a brief summary of the types of genetic data that can be used, gives an introduction to molecular ecology, and discusses aspects of gene conservation.

13.0 Introduction

Genes and genetic diversity can be conserved by maintaining genetic structure within and among subpopulations (Chapter 2). This strategy enhances the ability of populations to adapt to novel environmental conditions and avoids the potentially harmful effects of inbreeding (Frankham, 1995a; Reed and Frankham, 2003). However, there is a much broader role for genetics in conservation biology (Moritz, 1994a, 1999). Genetic information can be used to guide the management of captive populations, to set conservation priorities, and to provide information on population dynamics and mating systems that would otherwise be unobtainable.

13.1 Types of data

Vast sections of the genome of many species appear to be composed of DNA that serves no obviously useful purpose, does not comprise genes and thus does not code for proteins; these regions are described as 'non-coding'. The proportion of the genome that is non-coding differs greatly between species (Crozier, 1990). Furthermore, different parts of the genome have different characteristics, which are useful for solving different problems. Some DNA regions evolve very quickly (e.g. the non-coding regions, which are not subject to selective pressure), so these regions can be used to identify individuals uniquely and trace genealogies. Other regions are conserved within reproductive populations more than would be expected by chance (given that background rates of mutation cause divergence), providing a means of identifying population subdivisions. The purpose of this section of the book is to describe some of the types of genetic data that can be gathered and the uses to which they can be put (see also Crozier, 1990; Amos and Hoelzel, 1992; Mallet, 1996; Sunnucks, 2000; Frankham *et al.*, 2002).

DNA and electrophoresis

DNA consists of four different base nucleotides: adenine (A), guanine (G), cytosine (C) and thymine (T). Perhaps 5% of the genome is involved in coding for proteins. The remainder includes very short sequences of nucleotides that are 'tandemly' repeated thousands of times, in apparently functionless sequences. As mentioned previously, different regions and different kinds of DNA have different characteristic rates and modes of evolution, providing means for analysing patterns ranging from individuals in a population (using fast-evolving regions) to patterns among genera and higher taxonomic units (using slow-evolving regions).

Measurement of genetic variation within a population involves scoring the presence or absence of attributes (e.g. genes or genetically determined morphological features) in individuals, then extrapolating to arrive at estimates of the relative frequency of the attributes in the population. **Electrophoresis** is a method used to separate molecules (proteins or DNA fragments) that have different sizes, charges or configurations. In electrophoresis, the molecules to be separated are dissolved in solution, samples of which are then placed in shallow wells on one end of a gel (usually made of agarose, polyacrylamide, or starch). Then an electric current is passed through the gel. In response to the current, the molecules migrate through the gel at a rate that depends on their charge, weight and configuration. When the electric current is switched off, the different types of molecules will have travelled different distances. Finally, the gel is treated by 'blotting' to make the patterns formed by the migrating molecules visible. These patterns are recorded by counting the genetic types.

In the following sections we outline a few of the more common methods for measuring genetic variation (see Mallet, 1996; Frankham *et al.*, 2002; Wayne and Motin, 2004 for more complete introductions). Electrophoresis is an essential technique in many of these methods of genetic analysis. The cost and resolution of the different methods varies considerably. This field is undergoing very rapid change, but these methods are some of the most commonly employed at present.

Restriction fragment length polymorphism

Restriction endonucleases are enzymes that cut DNA at specific sequences of nucleotides, which are called 'restriction sites'. There are many restriction endonucleases, which cut at different restriction sites; for example, the restriction endonuclease called *Eco*RI cuts DNA wherever it comes across a particular six-nucleotide sequence. When DNA is mixed (or 'digested') with a particular restriction endonuclease, a collection of fragments of different sizes is produced, the size of which depends on the locations of the restriction sites on the DNA. Because each individual has characteristic sequences of nucleotides in their DNA, each will have restriction sites in different places, producing differences between individuals in the arrays of fragments produced by digestion with a restriction endonuclease. These variations in the size of fragments between different individuals are called **restriction fragment length polymorphisms** (RFLPs). RFLPs are visualised by using electrophoresis to separate the different fragments of DNA.

A DNA fragment of a particular size that is produced by a particular restriction endonuclease can be used as an attribute that is characteristic of an individual. Thus, it is possible to derive gene frequencies in a population by sampling a large number of individuals and recording the frequencies of these unique fragments. Analysis of RFLP data can be based on the presence or absence of particular fragments, or on the presence or absence of the restriction sites that gave rise to them (see Frankham *et al.*, 2002). RFLPs in rapidly evolving regions such as **introns** can be used to examine genetic diversity within species (i.e. between individuals). Slower evolving regions are useful for examining genetic diversity between populations or species.

RFLP analysis requires relatively large amounts of DNA. RFLPs tend not to be as variable as **microsatellites** (which are discussed later) and depend on the availability of known locus **probes**. Probes are short sequences of DNA of known composition that are labelled radioactively and then used to hybridise (i.e. join) with the target fragments via complementary base pairing (see Frankham *et al.*, 2002).

DNA sequencing

DNA sequencing is the process of determining the sequence of the four base nucleotides in a segment of DNA. To sequence DNA, it is first isolated and then cut with a restriction enzyme. The DNA is then run in an electrophoretic gel to separate fragments by size. As with RFLPs, specific fragments are identified by hybridising the DNA in the gel to radioactively labelled probes with a known sequence. Hybridisation only occurs with probes that are complementary to particular DNA fragments.

Traditionally, sequencing involved isolating the DNA, then **cloning** it repeatedly until there was sufficient material for analysis, making the process time-consuming and expensive (Frankham *et al.*, 2002). However, the **polymerase chain reaction** (PCR) is now often used to 'amplify' (generate multiple copies of) DNA. PCR can be used on very small samples of DNA (as small as a few molecules) to produce amounts of DNA that are suitable for sequencing. Using PCR, sequence data can be obtained relatively quickly and inexpensively. Automated methods and sequencing machines are developing quickly. Sequence data are useful for analysing genetic variation within genera, species or subpopulations, depending on the region of the genome sampled.

Single nucleotide polymorphism

Single nucleotide polymorphisms (SNP) are the DNA sequence variations that occur when a single base pair is altered at one position in the genome in more than 1% of individuals (Brookes, 1999). SNPs make up the bulk of DNA polymorphisms within species (because DNA from individuals of the same species tends to be very similar) and can be detected by DNA sequencing. SNPs can provide **markers** for virtually every gene in a genome, and can also be associated with **phenotypes** such as congenital diseases. The potential of SNPs to assist conservation biology is untested (Frankham *et al.*, 2002).

Randomly amplified polymorphic DNA

PCR can be used to amplify random, variable sequences of the genome, which are termed 'randomly amplified polymorphic DNA' (RAPD). The PCR products are separated on an electrophoretic gel and then stained, producing a pattern of bands that is characteristic for each individual (given a particular set of PCR **primers**). Each band corresponds to DNA fragments of a certain size and can be scored as present or absent for each individual. RAPDs sample many loci without the need to identify appropriate primers or sequence the genome. The method is fast and relatively cheap, but because it provides only information on presence/absence rather than a full sequence, it provides limited information on social structures, mating systems, and levels of inbreeding in the organism being studied.

Minisatellite and microsatellite analysis

Minisatellites and **microsatellites** are tandemly repeated sequences of DNA that are identified by specific probes. They include sequences of DNA that are highly conserved (i.e. very similar) in virtually all **eukaryotes**. These highly conserved sequences are found at several places in the genome, and at each location, clusters of the repeated sequences occur with variable numbers of repeats. These clusters experience high rates of gain and loss of repeat sequences. Minisatellites can occur clustered in as many as one hundred sites in the genome. The patterns of bands are visualised by digesting samples with a restriction endonuclease followed by electrophoresis. Individual bands are inherited codominantly (i.e. bands from both parents are expressed and can be detected) in strict Mendelian fashion. Microsatellites have relatively short repeated

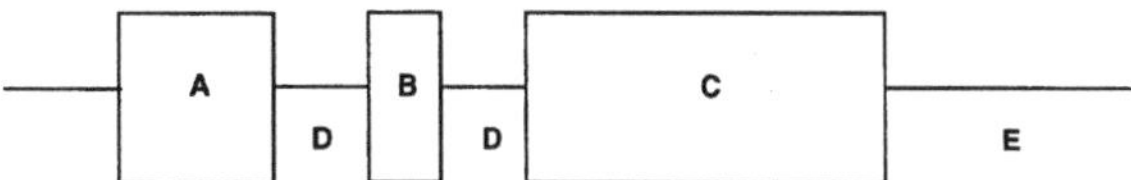

Figure 13.1. rDNA structure (after Wirgin and Waldman, 1994).

sequences compared with minisatellites. Microsatellites are more variable than protein analysis and therefore provide more information about the variations in DNA sequence between individuals.

Multilocus minisatellites are used for DNA fingerprinting to identify individuals and assign parentage. For example, analysts could look for a match between the novel minisatellite DNA fragments (i.e. novel bands on electrophoretic gels) in DNA samples from a child and from potential fathers to determine paternity (see Section 13.2).

Ribosomal DNA analysis

During a cell's lifetime, its DNA is continually 'read' (transcribed into RNA) and repaired. Through this imperfect repair process, sections of DNA can be duplicated or lost. In fact, the cells' repair mechanisms can actually spread mutations throughout the genome. Repeated sequences tend to mutate as units within reproductive populations, so that the reproductive population can then be identified by characteristic sequences of repetitive DNA. Ribosomal DNA (rDNA) is the region of the DNA that contains the genes that code for rRNA. Variability in the rDNA gene family within reproductive populations is less than would be expected by chance, given average levels of mutation over the genome. rDNA has three coding regions (Figure 13.1) that are separated by longer, faster-evolving spacer regions. The spacer regions are potentially rich sources of information on reproductive isolation between populations, although the levels of variability within and between populations are inconsistent.

Mitochondrial DNA analysis

Mitochondrial DNA (mtDNA) resides in the mitochondria of all animals and most plants. It is maternally inherited (i.e. each individual has only one copy of each gene, inherited from the female parent), and thus inheritance is clonal (i.e. essentially **haploid**). Different genes in mtDNA evolve at different rates, and one of an appropriate rate can be found to assist in solving almost any phylogenetic problem (Crozier, 1990). The higher mutation rate of mtDNA relative to nuclear DNA, at least in vertebrates, may make it a more useful tool in assessing population structure than would an analysis of nuclear DNA. In the past, mtDNA data were examined using RFLPs, but sequencing following DNA amplification using PCR tends to be used now, and has the advantage that museum and other preserved specimens can be analysed (because very small original amounts of DNA can be used with PCR).

mtDNA is a particularly useful tool in conservation genetics because it is sensitive to genetic drift and is relatively easily isolated, manipulated and amplified (Moritz, 1994a). mtDNA can be used to infer the impact of a population **bottleneck** (Beheregaray *et al.*, 2003; see Box 13.2). For example, if a **diploid** organism is reduced to one breeding pair, the population retains four copies of each gene on the nuclear genome, but only one transmissible copy of each mtDNA gene. Because regions of the mtDNA evolve faster than nuclear DNA, reduced variation in the mtDNA in a population can imply that the population has been through a bottleneck. Given estimates of average divergence rates, the timing of the bottleneck can be estimated (Amos and Hoelzel, 1992). The main uses of mtDNA have been in taxonomy, defining management units and understanding aspects of species biology (Frankham *et al.*, 2002).

Box 13.1

mtDNA and the origins of the Dingo
The origin and timing of the arrival of the Dingo into Australia has been controversial and was only recently resolved using mtDNA (Savolainen *et al.*, 2004). Using mtDNA extracted from samples from Eurasian Wolves, Dogs from all continents, and Dingoes from all Australian States, Savolainen *et al.* (2004) were able to show that the origins of the Dingo lie with domestic Dogs in East Asia (rather than India as speculated by some earlier workers). Further, the Dingo became established in Australia about 5000 years ago. Analysis of mtDNA also revealed that the Dingo population was founded from a small number of animals and subsequently remained effectively isolated from other Dog populations on the Australian mainland until European colonisation.

Box 13.2

mtDNA and a population bottleneck in an island population of tortoises
mtDNA was used in a study of Giant Tortoises on the island of Isabela, one of the Galapagos Islands (Beheregaray *et al.*, 2003). Isabela supports five subspecies of tortoises, each occurring around the five major volcanoes on the island. One of these, the Alcedo taxon (*Geochelone nigra vandenburghii*), occurs in vegetated areas around the Alcedo volcano, where there was a massive eruption in prehistoric times. The Alcedo taxon has the largest population of the five subspecies. Beheregaray *et al.* (2003) carried out analyses of the genetic variation of 10 nuclear microsatellite loci and 700 base pairs of the mtDNA control region; they showed that the Alcedo subspecies had undergone a major population bottleneck that coincided with the timing of the Alecedo volcanic activity (between 74 000 and 120 000 years ago). This work highlights the impacts of historical events on patterns of genetic variability.

Alcedo taxon of the Giant Tortoise from the island of Isabela in the Galapagos Islands group. (Photo by Luciano Beheregaray.)

Chloroplast DNA (cpDNA) analysis

cpDNA is a circular molecule of DNA that is unique to the chloroplasts of plants. cpDNA is maternally inherited in angiosperms, but is paternally inherited in many conifers. Mutations in cpDNA are relatively rare and there is no sexual recombination. As a result, intraspecific variability is typically very low, so it is potentially a powerful tool for **hybridisation** and **intro-**

gression studies, where genetic material from another species is involved. For example, introgression can be detected by the presence of cpDNA from another species.

Allozyme analysis

Allozymes (isozymes) are different forms of the same enzyme that are coded by different alleles of the same gene. Allozymes are present in the tissues of all plants and animals. Different forms can be separated by using electrophoresis, after which the enzymes are detected by immersing the electrophoretic gel in a bath containing enzyme-specific substrates. The enzymes catalyse a reaction involving the substrates and the products of the reaction are stained to produce bands on the electrophoretic gel. Allozymes are usually codominantly expressed, and gene frequencies are scored by counting the relative frequencies of different bands in a group of individuals from a population.

This approach detects variation in the amino acid sequences of functional proteins. It is suited to the measurement of genetic variation within and between subpopulations, and between species. It is cheap, fast and relatively easy to do. There is a vast literature on levels of enzyme polymorphism that can be used for comparisons.

Unlike variations in non-coding regions, variations in enzyme loci are expressed phenotypically, allowing investigations into the effects of natural selection on gene frequencies at those loci (because different forms of an enzyme may be more or less conducive to reproductive success). However, the levels of the enzymes, and thus the pattern of electrophoretic bands, can be affected by the tissues sampled, the season in which they were collected, and the conditions in which they were stored prior to analysis. Isozymes are less variable than molecular markers such as microsatellites. For instance, it is possible to sample 30 individuals from each of two populations, evaluate 30 different isozyme loci for each of the 60 individuals, and detect no variation either within or between the populations. An absence of variation is not evidence that there has been no genetic divergence unless the number of individuals and the number of loci studied are very large.

Quantitative characters

Quantitative characters, for example the lengths of bones, are typically the result of complex interactions among many loci, and interactions between the environment and the genetic composition of the individuals. Quantitative characters are rarely used in conservation genetics, even though such traits represent the material on which natural selection acts, and are therefore arguably the most relevant form of genetic variation (see Box 13.3). There is no guarantee that neutral genetic variation (either variation in non-coding regions, which isn't expressed, or selectively neutral single locus variation) measured by RAPDs or satellites, for instance, will act as a useful surrogate for important heritable genetic variation that is associated with traits that determine individual fitness or the ability of a species to adapt to change and recover from disturbance.

Narrow-sense heritability is the proportion of the variation in a character that is genetically influenced and can be modified by natural selection. Narrow-sense heritability can be estimated, for example, by regression

Box. 13.3

Genetic variation in Monkey Puzzle trees

The Monkey Puzzle is a very long-lived and vulnerable tree species that is endemic to southern South America (Figure 13.2). The species is susceptible to drought, fire, seed harvesting, timber harvesting and clearance for agricultural development. The remaining populations are fragmented. Bekessy *et al.* (2003) used RAPDs and quantitative genetic techniques to measure genetic variation within and among populations throughout the natural range of the species. Both methods detected significant differentiation among populations. However, the amount and spatial pattern of the variation identified by the two approaches were essentially unrelated. Quantitative trait variation was strongly associated with climate, whereas RAPD variation was not. Most importantly, the neutral markers failed to detect an important quantitative genetic divergence across the Andean Range that was related to drought tolerance. Had recommendations for conserving genetic variation been based solely on neutral markers, this important feature would have been overlooked. This result suggests that neutral markers may not be an appropriate tool for evaluating a population's adaptive potential, or for estimating local adaptation or population divergence. They may, however, provide useful information on mating systems, gene flow and evolutionary history.

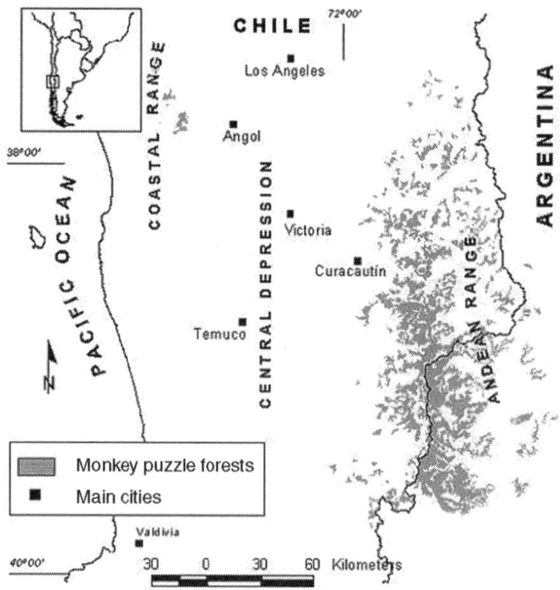

Figure 13.2. Distribution of Monkey Puzzle trees in South America (from Bekessy *et al.*, 2003).

between the values of a trait in offspring and their parents (Falconer, 1989). One serious practical difficulty in determining the heritability of traits in plant and animal populations is that few pedigrees are known, particularly in natural populations and captive populations that live and breed communally. However, DNA fingerprinting can be used to reconstruct genealogies, circumventing the problem (Lynch and Ritland, 1999).

13.2 Molecular ecology

Moritz (1994a) suggested two roles for conservation genetics: to identify and manage genetic diversity (called 'gene conservation'), and to study population dynamics (called 'molecular ecology'). The types of data described in Section 13.1 can be used for both purposes (see Sunnucks, 2000). Genetic research that discriminates between morphologically similar species (Chapter 2) or clarifies options for translocation or reservation by identifying evolutionarily distinct populations is part of gene conservation (dealt with in more detail in Section 13.4).

This section describes some of the uses of genetic research for molecular ecology. The aim is to describe some of the ways in which measures of genetic diversity can help us to understand the ways populations behave, respond to disturbance and move through the landscape.

Understanding social structure

Strategies to conserve species depend on an understanding of their social systems. Management options may be limited by the behavioural rules that govern reproduction when population densities are very high or very low. Mating systems influence the rate at which genetic variation is lost from a population and how species react when habitat is lost and fragmented. Although detailed observational studies can be used to monitor breeding, the breeding system and genetic structure may be different from that suggested by behavioural data (Poldmaa *et al.*, 1995). Molecular genetic techniques can be used to examine patterns of maternity and paternity in a population, thereby conclusively determining the actual mating system (e.g. Mulder *et al.*, 1994; Austin and Parkin, 1995; Poldmaa *et al.*, 1995; Clinchy *et al.*, 2004).

The Superb Fairy Wren is a small (~10 grams), common species of bird from eastern mainland Australia and Tasmania (Blakers *et al.*, 1984; Barrett *et al.*, 2003). The birds live in groups of 6–12 animals, and females raise several broods each season. Some adult males are characterised by striking iridescent blue plumage, whereas others are brown. This observation led to the speculation that groups comprised an adult male and a 'harem' of females. Subsequently, it was learned that some of the brown animals are males that moult into female-like plumage at the end of summer (Reader's Digest, 1990). Groups of birds include a dominant male, an adult female and often a number of her sons (or 'helpers'), who assist with nest defence and the care of the young (Rowley, 1965; Pruett-Jones and Lewis, 1990).

Superb Fairy Wren. (Photo by Esther Beaton.)

Evolutionary theory suggests that such care should be directed predominantly to relatives. Mulder *et al.* (1994) coupled extensive field observations with DNA fingerprinting of blood samples from adult and juvenile animals. They showed that for groups of wrens with extra helpers, most of the offspring (76%) were not fathered by the dominant male, but by extra-group males (males outside the offspring's social group) that contributed no parental care. The proportion of young in a group fathered by the dominant male declined with increasing group size. All but two of 40 groups contained young sired by extra-group fathers – the highest recorded incidence of cuckoldry amongst cooperative breeding birds (Mulder *et al.*, 1994).

The biology of the Noisy Miner is strikingly different. The Noisy Miner is an aggressive honeyeater that is often found in edge habitats in fragmented woodland environments (Loyn, 1987; see Box 9.4 in Chapter 9). It is a common species, and occupies a large part of eastern Australia (Blakers *et al.*, 1984; Barrett *et al.*, 2003). Social organisation is based around territorial groups of 6–30 animals that comprise a loose colony of up to several hundred birds (Reader's Digest, 1990). Behavioural studies indicated a social system in which populations are characterised by a 3:1 male-biased sex ratio (Dow, 1979). Males move freely among small groups throughout a colony. Conversely, females are very aggressive, and their home ranges generally do not overlap. Some males dedicate most of their time to a single nest, whereas others distribute their efforts among many nests (Dow, 1978; Poldmaa *et al.*, 1995). Field observations indicated that the species was highly promiscuous, with males mating with many females and vice-versa (Dow, 1978) – a behavioural pattern described as 'nest-orientated promiscuity' (Brown, 1987, in Poldmaa *et al.*, 1995).

Poldmaa *et al.* (1995) used DNA fingerprinting to examine the parentage of 85 nestlings from 35 different broods. They found that multiple paternity was rare (with only one instance from 31 nests examined). It was found that 80 of the 85 nestlings were sired by the males that assisted most in the feeding of those young.

Thus, extra-group paternity was rare (less than 6% of nestlings) in Noisy Miners, and the male that achieved paternity was also responsible for most of the feeding of the young, although up to 20 other males contributed to the provisioning of the offspring. These males may have been related to the father of the offspring. Apparently promiscuous behaviour did not lead to fertilisation. Thus, complex extended family relationships could be an important part of the social biology of the Noisy Miner (Poldmaa *et al.*, 1995). In contrast to assumptions made on the basis of field observations of its patterns of social behaviour, the Noisy Miner is not polyandrous or highly promiscuous (see Dow and Whitmore, 1990).

DNA fingerprinting studies of the mating systems of the Superb Fairy Wren and the Noisy Miner revealed important new insights into the biology of both species. This type of information can be important for understanding, managing and modelling the population dynamics of animals. Neither the Superb Fairy Wren nor the Noisy Miner is threatened; in fact, both species are common and widespread. However, such studies highlight the value of integrating field observations with genetic analysis.

Estimating effective population size

Genetic variation in small isolated populations is lost by drift and gained by mutation. **Effective population size** (N_e) determines the rate of loss of genetic variation. The rate of loss of genetic variation may be a concern for gene conservation in itself, and it may also tell us a great deal about mating systems, reproductive demography, and natural and human-induced fluctuations in population size, which are important for the management of both captive and wild populations.

The effective population size is the pool of individuals that are each equally likely to contribute genes to the next generation. More exactly, it is the number of individuals in an ideal, randomly mating population that would give the same rate of random genetic drift as occurs in the actual population. When N_e is small, the rate of loss of genes by chance is high, and the likelihood is high that individuals in the population will have copies of genes that are identical by descent.

There are different ways of defining effective population size. The 'variance effective population size' accounts for changes in gene frequency by genetic drift. The 'inbreeding effective population size' accounts for homozygosity by common ancestry (see Crow and Dennison, 1988). Under a set of common (although not entirely plausible) assumptions, including diploidy, discrete generations and constant population size, the two measures of effective population size are the same (Burgman *et al.*, 1993).

In most populations, reproduction is restricted to relatively few individuals. Even among reproductively mature individuals, a few may be inordinately successful. Species with limited dispersal abilities tend to mate with related individuals. Populations that fluctuate in size or that pass through a bottleneck usually show reduced genetic variation. In all these cases, the effective population size is smaller than the census population size (the actual number of individuals in the population). Generational overlap, correlations between the fertility of parents and the fertility of their offspring, lifetime family size and the sex ratio in the population all influence effective population size (Sherwin *et al.*, 1989).

Direct observations of populations can be used to adjust the census population size to provide an estimate of the effective population size. Crow and Kimura (1970), Sherwin *et al.* (1989), Sherwin and Brown (1990), and Frankham *et al.* (2002) outline some observations of social behaviour, breeding systems, demography and population dynamics that provide some indication of these factors and describe derivations of the following equations.

If population size varies between generations, and if N_e was estimated at regular intervals, the **harmonic mean** of values from each time will give an overall measure of N_e. Some members of a population may be far more successful than others; effective population size (N_e) accounting for differential reproductive success is given by:

$$N_e = \frac{(4N-2)}{(2+\sigma_k^2)} \qquad (13.1)$$

where σ_k^2 is the variance in the number of progeny produced per generation per individual, and N is the census number of individuals in the population.

Effective population size can be larger than the census size if there is no variation in reproductive success. In real terms, this can be useful when managers of captive populations equalise the reproductive contributions by founders, thereby reducing the rate at which genetic variation is lost (see Section 13.6).

If the sex ratio in a population is not 1:1, then the preceding calculations can be performed separately for males and females and the values combined using:

$$\frac{4}{N_e} = \frac{1}{N_f} + \frac{1}{N_m} \qquad (13.2)$$

where N_f and N_m are the estimates of the census numbers of females and males.

Effective population size can be estimated by using genetic rather than demographic or ecological data. It is possible to estimate current N_e by measuring recent fluctuations in allele frequencies between generations. The approach makes use of the observation that random drift and inbreeding will increase fluctuations between generations. If heterozygosity has been measured over several generations, then:

$$H_t = H_0\left(1-\frac{1}{2N_e}\right)^t \qquad (13.3)$$

where H_0 is initial heterozygosity, and H_t is heterozygosity at generation t.

Alternatively, levels of genetic **disequilibrium** at multiple loci can be used to estimate current N_e because drift increases **linkage disequilibrium**. Assuming genetic equilibrium, then

$$H = \frac{4N_{e\mu}}{(1+4N_e\mu)} \qquad (13.4)$$

where H is heterozygosity and μ is the mutation rate (assumed to be approximately 2×10^{-6} for single locus variation; Crow and Kimura, 1970). However, this equilibrium value is approached only very slowly after disturbance, and so it is unlikely to be useful in many circumstances.

Genetic methods for estimating effective population size depend on having available gene frequency data from large numbers of polymorphic loci. All methods assume selective neutrality, random mating, and no migration between genetically differentiated populations.

Table 13.1 shows large disparities between the present-day population size and the long-term effective

Eastern Barred Bandicoot. (Photo by Kay Aldridge.)

Box 13.4

Effective population size of the Eastern Barred Bandicoot

The Eastern Barred Bandicoot is widespread and abundant in Tasmania, and there were relatively large populations in the far eastern part of southern Australia at the time of European settlement. A population of the species near Hamilton in south-eastern Australia was isolated from other populations between 1937 and 1960, after which time the remaining Australian mainland populations became extinct. In 1990, Sherwin and Brown (1990, see Burgman *et al.*, 1993) estimated the size of the remaining mainland population to be 633 ± 24.

Sherwin and Brown (1990) then used demographic and reproductive data to estimate the effective size of the population. Eastern Barred Bandicoots are difficult to age, so Sherwin and Brown (1990) approximated the correction for overlapping generations using the number of new recruits per generation; they estimated that there were 153 male and 340 female new weanlings per generation.

Marsupials remain attached to their mothers from birth until they are sufficiently developed to fend for themselves, so it was relatively easy to count the number of offspring born to different females in the isolated population. Two females in the sample of trapped animals shared more than 75% of the successful reproductive output, whereas 11 females had no offspring at all (Figure 13.2). From the data on female reproduction in the trapped population, they estimated that on average each individual successfully reared one offspring in its lifetime (Figure 13.3) and that the variance in the number of progeny produced per generation per individual was 11.6. These values gave a corrected value for lifetime reproductive success of 24 for males and 54 for females.

Lastly, Sherwin and Brown (1990) accounted for the unequal sex ratio in the population using Equation 13.2, giving an effective population size of 67.

Thus, the effective size of the Eastern Barred Bandicoot population is nearly one-tenth the estimated census population size. The rate of loss of genetic diversity from the population will be ten times higher than the rate of loss from an ideal randomly mating population. This rate of loss is not unusual – many other natural populations have effective population sizes less than one-tenth the census population size (Frankham, 1995a; Frankham *et al.*, 2002).

The population at Hamilton studied by Sherwin and Brown (1990) is close to extinction (there may be a few animals still in the wild near Hamilton), except for captive populations and populations created from deliberate release programs. There was no genetic differentiation between Victorian and Tasmanian populations (based on 27 protein loci), and no genetic variation within either population. Heterozygosity in the populations is low, probably below the average for mammals. It is likely that because heterozygosity is low, the study design (30 individuals in each population, 27 loci studied) was insufficient to detect variation. Reintroductions or translocations from Tasmania to Victoria should occur only after experimentation with interstate crosses to determine if outbreeding depression will occur. The population at Hamilton may have been approximately constant at about 600 individuals between 1960 and 1990 (22 generations) and perhaps since 1937 (43 generations). Over this period, the loss of heterozygosity may have been 15–28%.

---

population size in a range of species, which are larger than the discrepancy between census size and effective size noted for the Eastern Barred Bandicoot (Box 13.4; Figure 13.3). Effective population size is almost always less than the number of adults of reproductive age. Effective population sizes in a variety of species average about 0.1 of the actual population size. In some cases, effective population size can be even smaller, compared to the number of adults (Frankham, 1995c). The most important reasons are usually unequal numbers of males and females, variation in population numbers over time,

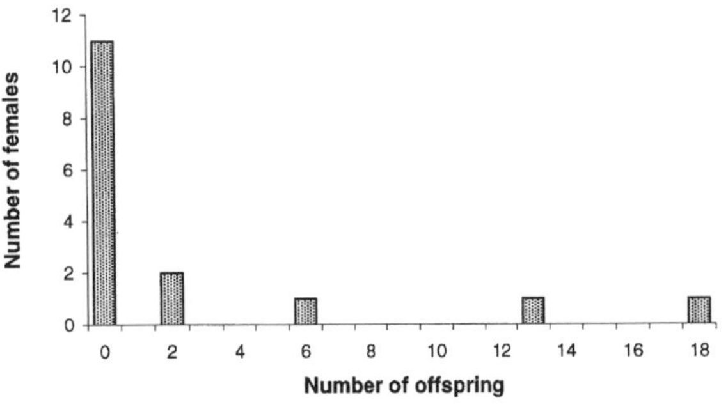

Figure 13.3. Lifetime reproductive success of a population of the Eastern Barred Bandicoot (Sherwin and Brown, 1990).

Table 13.1. Effective population size and census population size of three species (after Avise *et al.*, 1988).

Species	Current female population size	Long-term effective size	Discrepancy
American Eel	5 000 000	5500	909-fold
Hardhead Catfish	10 000 000	45 000	222-fold
Redwing Blackbird	20 000 000	36 700	545-fold

extinction and recolonisation of populations, and relatively large variability in the number of offspring per parent (Avise *et al.*, 1988; Crow and Dennison, 1988; Frankham, 1995a,c; Frankham *et al.*, 2002).

Social insects pose difficulties for estimating effective population size. Ants and termites have an array of life-history patterns, including annual or perennial colonies, single or multiple nests, and single or multiple queens. N_e can be influenced by the longevity of colonies, or the dispersal ability and tolerance of disturbance of the species in question. Effective population size can be small for apparently abundant species (Pamilo and Crozier, 1997). In **eusocial** Hymenoptera (species that care for young cooperatively and divide labour), many species show **inbreeding depression**, mainly in the production of sterile males caused by homozygosity at the sex locus. Such characteristics make many social insects particularly susceptible to the effects of inbreeding (Pamilo and Crozier, 1997).

Detecting migration

Migration between habitat patches determines the persistence of species, particularly in fragmented habitat (Chapter 10). Standard field techniques for observing movement patterns in animals, such as tagging and **radio-tracking**, are expensive and time-consuming, and do not tell us anything about the reproductive success of migrants (Oakleaf *et al.*, 1993). The movement of plant gametes between patches by pollinators cannot be detected reliably by physical observations.

Genetic data can indicate population structure and the number of genetically effective individuals exchanged between populations per generation (N_m; Crozier, 1990). Estimates of N_m rely on the variance in allele frequencies or the frequency of private alleles (those that occur in only one population; Slatkin, 1985). Slatkin (1985) proposed that the mean frequency of private alleles ($\bar{p}$) was related to N_m by:

$$\ln(\bar{p}) = -c(\ln(N_m)) - K \qquad (13.5)$$

where the parameters c and K depend on the model chosen to represent the biology of the population (see also Slatkin and Barton, 1989; Baur and Schmid, 1996).

A second method that uses the genetic variation among populations (Slatkin, 1987) is:

$$N_m = \frac{\left(\frac{1}{G_{ST}} - 1\right)}{4} \qquad (13.6)$$

where G_{ST} is defined in Table 13.3.

Prober *et al.* (1990) used these models to estimate migration rates among populations of Small-leaved Gum. They assumed genetic and demographic equilibrium, neutral mutation, an island model of population structure, and discrete, non-overlapping generations (Table 13.2). The study concluded that the rates of migration for the Small-leaved Gum populations were too high given the present discontinuous distribution of the populations. The values may represent a more continuous distribution and more effective gene flow prior to European settlement.

Table 13.2. Estimates of gene flow among populations of two eucalypt species, two marsupial species and two species of North American rodents. Estimates were based on isozyme variation among populations (after Prober *et al.*, 1990; Sherwin and Murray, 1990).

	Wadbilliga Ash	Small-leaved Gum	Virginia Opossum	Bennett's Wallaby	California Mouse	Rock Squirrel
No. loci	25	17	14	9	16	21
F_{ST} method						
F_{ST}	0.039	0.054	0.013	0.122	0.157	0.284
N_m	6.26	4.38	3.44	2.30	1.64	0.69
Private allele method						
No. private alleles	0	10	10	4	14	12
N_m	–	4.71	0.43	1.75	2.19	0.86

Box 13.5

Testing mammal dispersal between forest fragments

A major change in landscape cover has taken place in the Tumut region of southern New South Wales, with extensive stands of Radiata Pine plantation replacing the original native eucalypt forest (Figure 13.4). Patches of remnant eucalypt forest remain within the plantation, and areas of State forest and national park are at the boundary of the plantation. Populations of the Greater Glider reside with the remnant patches as well as in the large continuous areas of native forest. Several key questions arise from this situation, and Lindenmayer *et al.* (1999f) posed three questions prior to the commencement of integrated demographic and genetic work on the Greater Glider in the region. These questions and the approaches designed to address them are set out below.

Question 1. Do populations in continuous eucalypt forest show genetic differentiation related to geographic distance?

To answer this question, maps of genetic differentiation will be constructed and the degree of differentiation will be plotted against different geographic distances.

Question 2. Are the populations in the fragments in the Radiata Pine plantation isolated from each other and/or from continuous eucalypt forest?

The geographic versus genetic distance plots from continuous forest will be compared with a similar plot generated for subpopulations of the Greater Glider in the remnant eucalypt patches surrounded by extensive stands of Radiata Pine. This is intended to determine if: (1) populations in the patches are genetically isolated from those in the continuous forest, (2) populations in the patches are genetically isolated from one another, and (3) populations in the patches are genetically isolated from both continuous forest and from one another. If the Radiata Pine matrix influences the dispersal of the Greater Glider, then genetic differentiation among patches, and/or between the patch system and continuous forest, should be greater than expected on the basis of geographic proximity.

Even if dispersal between patches is frequent, it is possible that immigration into the patch system from the continuous forest is negligible. Animals dispersing from continuous forest may not move into the adjacent Radiata Pine matrix. If there is inter-patch movement, but the populations in the patches are isolated from those in continuous forest, then the expected genetic outcomes are:

- limited overall genetic differentiation among patches (with reference to the calibration curve for expected genetic differentiation based on distance alone)
- greater similarity between current patch and historic samples than between patch and continuous forest samples (again with the latter comparison weighted by geographic distance)
- no correlation or only a weak correlation between patch size and genetic variation. If inter-patch dispersal is rare (perhaps because animals can persist for a long time in the remnants), with the only movements being immigration from the continuous forest into the patch system, the expected genetic outcomes are:
- greater similarity between patch and continuous forest populations than among patches
- greater similarity between historic and continuous forest samples than between historic and patch populations, possibly because of genetic drift in small patches
- genetic variation within patches will be a random subsample of that from continuous forest, depending on: (1) distance from the source (patches closer to source areas in continuous forest may be more likely to be colonised than distant ones; Hanski, 1994b), and (2) patch size (smaller patches may be less likely to receive immigrants), although there may have been confounding effects between (1) and (2).

If there had been complete isolation of the Greater Glider in the patches then:

- patches may contain unique, or 'private' alleles
- genetic variation within the patches will be: (1) lower than that over a similar spatial scale in continuous forest, (2) lower than that in the historic samples, and/or (3) positively correlated with patch size
- patch populations will contain a random subset of the original genetic variation (i.e. the genetic variation in the historic samples).

Question 3. What are the genetic origins of animals in the patches?

The outcomes of questions 1 and 2 should make it possible to establish the genetic origins of animals in the patches. There may have been dispersal from other patches, from continuous forest or both types of potential sources. Alternatively, the subpopulations may represent 'relict' populations that survived as complete isolates since fragmentation 35 years ago. If this second scenario arose, then it may be possible to explore founder effects and other phenomena such as inbreeding. The work on the Greater Glider is continuing.

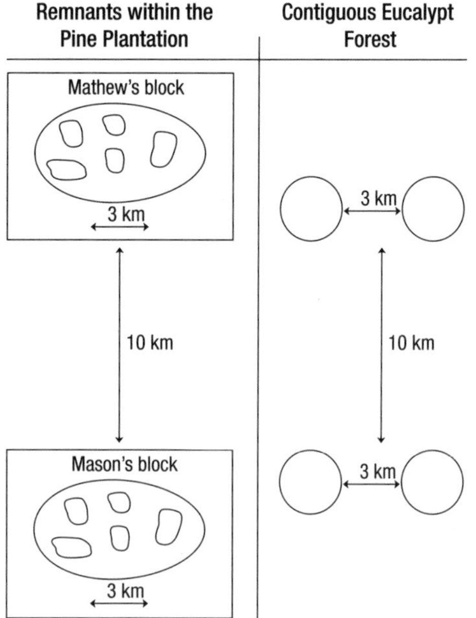

Figure 13.4. Diagrammatic representation of the location of sampling or proposed sampling areas at Tumut within eucalypt remnants surrounded by Radiata Pine and within large areas of continuous eucalypt forest. (Redrawn from Lindenmayer *et al.*, 1999f.)

The gene flows estimated by the two methods outlined here are generally consistent (Table 13.2). The levels of gene flow measured in the examples given earlier are probably sufficient to prevent substantial genetic differentiation between populations by drift alone. Values for the number of migrants per generation as low as 0.10 have been found for sedentary salamanders, and as high as 249.8 have been found in fish populations (Sherwin and Murray, 1990).

Several other approaches to detecting migration depend on genetic information. Slatkin and Maddison (1990) and Takahata and Slatkin (1990) outline the allele phylogeny approach. Some more recent measures use microsatellites (Goldstein *et al.*, 1995; Slatkin, 1995; Sunnucks, 2000). Patkeau *et al.* (1995) describe the use of **assignment tests** to determine the place of origin of a dispersalist. Wayne and Motin (2004) describe 'non-invasive monitoring' techniques to count and identify individuals, detect movement patterns and infer parentage. This involves new molecular techniques for extracting DNA from animal and plant remains, and animal faeces, feathers, hair and bone (see for example, Banks *et al.*, 2003).

Effects of genetic change on demographic parameters

Inbreeding depression refers to the expression of deleterious recessive alleles, resulting in reduced fitness (Chapter 6; Reed and Frankham, 2003; Briskie and Mackintosh, 2004). Reduced fitness can be expressed in reductions in live birth weight, disease resistance, seed viability, or tolerance to competition or environmental stress. Such changes in the performance of individuals translate into changes in demographic parameters such as age-specific fecundities and survivorships.

Models used to predict the viability of populations (Chapter 16) should make assumptions about the feedback between levels of genetic variation, effective population size, and demographic attributes. Unfortunately, many such models ignore the effects of genetic change simply because the data are rarely available on levels of inbreeding or the response to inbreeding. However, ignoring the potential for an effect makes the implicit assumption that there is no effect, and empirical observations show clearly that most habitually outbred species experience inbreeding depression when population size is reduced.

Inbreeding depression that results from the loss of **heterosis** can be modelled by

$$s' = se^{-bF} \tag{13.7}$$

where s' is survivorship under the influence of inbreeding, s is survival in the absence of inbreeding, b is the number of lethal equivalents per haploid genome, and F is the individual's **inbreeding coefficient**, which is essentially the proportion of an individual's genes that are identical because they are derived from an ancestor common to its parents (e.g. Lacy, 1993b). Frankham (1995c) explored the form of the relationship between inbreeding and extinction and found elevated extinction risks at intermediate levels of inbreeding, suggesting a threshold response. Models that ignore genetic uncertainty underestimate extinction risk.

13.3 Gene conservation

The objective of conservation biology can be seen as the maintenance of the full array of genetic variation that currently exists (Chapter 2). The protection of genetic variation retains options for future use. To conserve genetic variation, we must first identify it and then manage it effectively. The purpose of this section is to outline some of the quantitative tools for interpreting genetic data for gene conservation.

Spatial structure

Genetic diversity exists both as the differences among individuals in a population, and the differences among populations of a species. Appropriate management units for conservation depend on the evolutionary history and the spatial distribution of genetic variability of the species in question (Chapter 2; Box 13.3). Standard measures of genetic diversity are based on traits determined by a single locus, for example allozyme and mtDNA band frequencies (Table 13.3). They can be used to measure genetic variation within species, within populations, and between populations.

Average heterozygosity within Australian marsupial populations is not substantially lower than that in other mammal populations; however, between-population variation is somewhat lower than the average value for mammals (Table 13.4). The genetic diversity within and among plant species is most strongly influenced by the breeding system used (Table 13.4). Selfing plants exhibit a much greater degree of population differentiation than outcrossing species.

Average values of genetic diversity across taxa identify general trends, but reliable generalisations cannot be made from such measurements. The average values ignore the considerable variability that exists within each of the categories in Table 13.4. Data from five threatened plant species from Western Australia serve to illustrate the point. All five are restricted to local or regional populations with ranges that cover a few hundred kilometres at most. Yet there are substantial differences among species in the total amount of genetic variation (H_T, which ranges from 0.139 to 0.356) and its distribution within and among populations (G_{ST}, the amount of genetic variation distributed between populations, ranges from 12% in the Laterite Mallee to more than 60% in Gungurru).

The distribution of genetic variation within and among populations determines the appropriate conservation strategy for a species. In Corrigin Grevillea, only 13% of the variation is between populations, suggesting that there has been relatively little genetic differentiation. This may have been because of strong gene flow by pollinators or seed dispersal, or it may indicate that populations fragmented recently from a more or less continuous system. The reintroduction program will therefore mix plants from all available populations (Rossetto *et al.*, 1995).

In contrast, Gungurru appears to have low levels of variation overall, and most of this variation is between separate populations. The species lives on isolated granite outcrops in naturally fragmented habitat. The populations may have been isolated from one another for thousands of years. Genetic drift may have been responsible for the loss of genetic variation and population differentiation. The optimal strategy in this case is to protect as many as possible of the small populations *in situ* on isolated granite outcrops (Moran and Hopper, 1983).

Box 13.6

Spatial structure and gene flow in the Bush Rat

Peakall *et al.* (2003) merged detailed information on the demography of populations of the native Bush Rat at Tumut in south-eastern Australia with information on patterns of genetic variability in the population. The work was a fine-scaled set of microsatellite-based genetic analyses that was aimed at improving our understanding of dispersal patterns and social organisation and spacing in the Bush Rat. Spatial autocorrelation analysis demonstrated that animals occurred in high-density clusters (approximately 200 metres across) interspersed with gaps in the populations or areas of lower population density. Peakall *et al.* also found strong positive spatial genetic autocorrelation patterns and showed that rat genotypes were not randomly distributed, but proximate animals were more genetically alike than distant ones. They concluded that gene flow (via dispersal) was sufficiently limited to create the fine-scale patterns of positive spatial structure observed.

Setting priorities for conservation

Taxonomy shapes perceptions of biodiversity and the recognition of endangered species (Chapter 2). Wilson (1992) suggested that biodiversity can be measured ultimately by genetic information content, and that sequence diversity is the best measure of diversity. It may be more effective to set priorities for conservation on evolutionary grounds, using procedures that tend to retain genetic diversity (see Chapter 2). If information is available to construct phylogenies or estimate genetic distances among populations and species, it can be used to set priorities for conservation. The intention is to develop priorities that will maximise the protection of genetic diversity.

Table 13.3. Measures of genetic diversity (after Nei and Li Wi, 1979; Burgman *et al.*, 1993; Peakall *et al.*, 1994; Hamrick and Godt, 1996).

Measure	Description
	Diversity within populations or species
P	Percentage of polymorphic loci.
AP	Mean number of alleles per polymorphic locus.
A	Mean number of alleles per locus.
A_e	Effective number of alleles per locus, calculated as $1/\Sigma x_i^2$, where x_i is the frequency of the *i*th allele. It measures the evenness of allelic frequencies, and is known as the effective number of alleles, expected panmictic heterozygosity, polymorphic index, and gene diversity (see Brown and Weir, 1983).
H_o	Observed proportion of loci heterozygous per individual.
H_S	Expected proportion of loci heterozygous per individual, calculated as $$H_s = 1 - \sum_{i=1}^{k} x_i^2$$ where x_i is the mean frequency of the ith allele. H_s is averaged across loci.
H_T	The total genetic variation within any species may be calculated from $$H_T = 1 - \sum_{i=1}^{k} \bar{x}_i^2$$ where H_T is total genetic diversity and $\bar{x}_i$ is the mean frequency of the ith allele at a polymorphic locus.
	Diversity among populations
G_{ST}	Between-population differentiation, the genetic diversity between populations, is given by $$G_{ST} = \frac{H_T - \bar{H}_S}{H_T}$$ where $\bar{H}_S$ is within population genetic diversity averaged over all populations.
F_{ST}	Standardised interpopulational variance in allele frequencies, calculated as $$F_{ST} = \frac{S_p^2}{x_i(1-x_i)}$$ where s_p^2 is the variance in allele frequencies among populations and $\bar{x}_i$ is the mean frequency of the ith allele. FST values are calculated for individual loci and alleles.
D_{ST}	Sometimes an alternative measure of between population diversity is used $$D_{ST} = H_T - \bar{H}_S$$
I	Genetic identity or genetic similarity, calculated as $$I = \sqrt{\frac{\sum x_i y_i}{\sum x_i^2 \sum y_i^2}}$$ where x_i and y_i are the frequencies of the ith alleles in populations *x* and y. Nei's (1972) genetic distance is calculated as $-\ln(I)$.
$1 - F$	For haploid or dominant markers, genetic distance between individuals (or populations) is given by $1-F = 1-[2n_{xy}/(n_x+n_y)]$ where F is the ratio of shared bands between *x* and *y*, $2n_{xy}$ is the number of shared bands, and n_x and n_y are the number of bands observed in individuals (or populations) *x* and y, respectively.

When faced with a choice between two species, one of which has a close relative that is not endangered, it is intuitively reasonable to protect the species that has no close relative. For example, it could be argued that Leadbeater's Possum, the sole member of its genus, should be given priority over the highly endangered Mahogany Glider, which belongs to a genus with three other members in Australia and at least two in New Guinea. Several people have attempted to generalise this notion to a set of rules for setting priorities for conservation. Vane-Wright *et al.* (1991), Williams *et al.* (1991), and May (1990b) developed a method for setting priorities that depends on the topological structure of a cladogram (a graphical representation of the

Table 13.4. Distribution of genetic variation within and among populations of some representative taxa (after Moran and Hopper, 1987; Sherwin and Murray, 1990; Sydes, 1995). Diversity values are averages calculated across all polymorphic loci.

Taxon	HS	HT	GST
Marsupials[1]	0.039	0.052	0.250
All mammals[2]	0.046	0.104	0.558
All plants[3]	0.178	0.230	0.224
Breeding system			
Selfing	0.164	0.334	0.510
Mixed	0.238	0.304	0.216
Outcrossing	0.257	0.302	0.149
Range			
Local	0.212	0.281	0.245
Regional	0.241	0.308	0.216
Widespread	0.274	0.347	0.210
Grevillea scapigera[4]	0.310	0.356	0.130
Gungurru[5]	0.068	0.176	0.614
Eucalyptus lateritica[5]	0.278	0.318	0.126
Johnson's Mallee[5]	0.084	0.139	0.396
Eremaea purpurea[6]	0.256	0.317	0.194

H_T is total genetic diversity, H_S is mean genetic diversity within populations, and G_{ST} is the mean diversity between populations, given by $(H_T - H_S)/H_T$.

[1]Average of three studies, including 18 species reported by Sherwin and Murray (1990).

[2]Average of two summaries reported by Sherwin and Murray (1990).

[3]Summarised by Hamrick and Godt (1989, 1996) from 400 studies.

[4]Rossetto *et al.* (1995). The only study reported here that used RAPD markers.

[5]Moran and Hopper (1987).

[6]Coates and Hnatiuk (1990).

evolutionary relationships among taxa) for a set of species (Figure 13.5). For each taxon, a count is made of the number of branches from each node to the root. The priority for conservation is set by the inverse of the count, rescaled by dividing each by the lowest number recorded among the taxa. The taxon with the highest value is the highest priority for conservation.

This method tends to give priority to species that are taxonomically isolated but it ignores **patristic distances** (the branch lengths of the cladogram). The branch lengths reflect accumulated genetic change within a lineage. However, the method is sensitive to the position of the root and therefore to taxonomic revisions that affect tree topology (Crozier, 1992; Williams and Humphries, 1993). The scale for the method has no upper bound, and species are given more weight if they come from a tree that has many species. Hence, priorities are not comparable between taxonomic groups.

Box 13.7

Management units for *Lambertia orbifolia*

Lambertia orbifolia is a large, bird-pollinated, woody shrub that is known from seven populations in south-western Australia. The species is distributed in two disjunct regions about 150 kilometres apart. It is possible that birds move regularly between the disjunct populations, resulting in little genetic differentiation. Coates and Hamley (1999) measured the genetic divergence and gene flow between the populations. It was found that the genetic differentiation was very high (D = 0.252) and gene flow was low (Nm = 0.32). The biogeography of the area suggested that the current population structure is the result of the intervening populations becoming extinct following the Pleistocene climate change (Coates, 2000). The spatial patterns of genetic variation were reflected in both allozyme and cpDNA data. The populations in the two regions will be conserved as separate evolutionary lineages.

A subset of species on a phylogenetic tree make a subtree. Their total phylogenetic diversity is the total length of the branches. An alternative approach to setting priorities is to conserve the subtree with the greatest length, thereby retaining maximum phylogenetic diversity (see Crozier, 1992; Faith, 1992, 1994). The value of a species is given by how much it adds to the total branch length spanned by the protected set.

Crozier (1992) suggested that genetic distances between taxa may improve estimates of their relative value (see Table 13.3 for the formula for Nei's (1972)

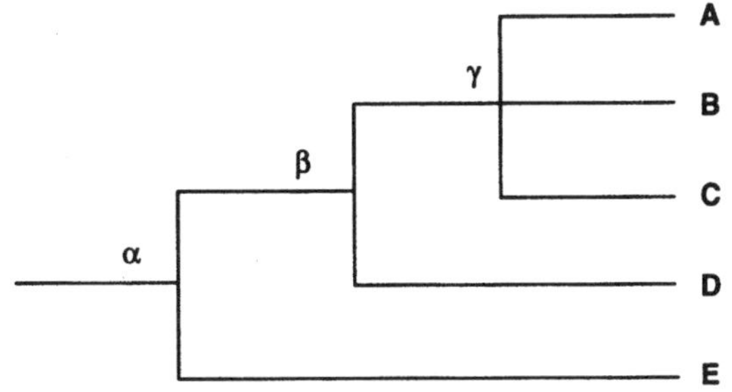

Figure 13.5. A phylogenetic tree for five taxa and estimates of priority (after May, 1990a). The nodes (labelled) from the root to the tip have two, two and three branches, respectively. In moving from the root to the tip, we count seven branches for taxa A, B and C, four for D and two for E. Taking the inverse and rescaling by the smallest value gives the priority scores.

genetic distance measure). Priorities based on genetic distinctiveness will tend to minimise the loss of genetic information. The value of a taxon is defined as the proportional decrease in diversity that occurs when the taxon is removed. Values of localities can be assessed by removing taxa from the tree that are unique to the locality. For this measure, a genetic distance measure is required that gives the probability that one allele will be replaced by another, rather than the more usual measure that gives the probable number of substitutions, although the differences between the two will be small and probably negligible (Crozier and Kusmierski, 1994). Different genes exhibit different rates of divergence, so any comparison of distance between populations or species should sample the same genes. The number of branch points is unimportant, so the inclusion of different numbers of taxa will not affect the analysis.

Genetic distances rely on the availability of genetic data. Phylogenetic distances rely on the availability of a cladogram based on a quantitative analysis of attribute data. In many cases, these types of information are unavailable. The advantage of the method developed by May (1990b) is that it can be applied in any circumstances in which the relationships among taxa are described by a cladogram or any other hierarchy, including traditional taxonomies.

Managing captive populations

For captive populations, we can attempt to maximise genetic variation, keep founder representation equal, select to reduce the frequency of introgressed genes, or maximise the effective population size, thereby minimising the rate at which genetic variation is lost. Selection acts to maximise fitness; in the captive environment, inadvertent selection for tameness and general adaptation to captivity is inevitable (McPhee, 2003). In the long term, functions essential to survival in the wild (flight from predators, resistance to disease and parasites, ability to capture live prey, ability to digest toxic plants) may be lost unless they are retained by appropriate selection (Frankham *et al.*, 1986; see also Box 6.3 in Chapter 6; Jimenez *et al.*, 1994).

It may not be ideal in all captive populations to maximise genetic variation, because the fitness costs of inbreeding vary. Many species have breeding systems that reduce heterozygosity (e.g. self-fertilisation, mating between siblings, or short dispersal distances) and mating dissimilar individuals may disrupt favourable gene combinations, resulting in outbreeding depression.

Pedigrees document which individuals are related. With pedigree analysis, it is possible to set and achieve goals such as equal founder representation and to monitor the level of inbreeding in the population. Calculations for determining the level of inbreeding depend on the relatedness of parents. The calculations measure the proportion of genes that are homozygous in an individual because they are derived from the same ancestor, through both parents (i.e. identical by descent). The inbreeding coefficient for individual X ($F(Z)$) is given by:

$$F(Z) = R(X,Y) = R(\text{father of } Z, \text{mother of } Z) \qquad (13.8)$$

where $R(X,Y)$ equals the relatedness, or the coefficient of kinship between individuals X and Y. Effectively, it is the proportion of genes shared by two individuals because of common ancestors.

The coefficient of kinship can be estimated from:

$$R(X,Y) = 0.5(R(X, \text{father of } Y) + R(X, \text{mother of } Y)) \qquad (13.9)$$

and extended back through the pedigree. The expression for $R(X,Y)$ can be substituted into the expression for $F(Z)$ to give the inbreeding coefficient for an individual.

Studbooks record parentage in a captive population, keeping track of inbreeding. For example, they are used to manage the captive population of Long-footed Potoroos at Healesville Sanctuary (see Table 13.5).

The first five animals in this population are founders and are assumed to be unrelated to one another (i.e. they have a kinship of zero). The degree of kinship among extant animals and the representation of each founder in the current population can be calculated by using the recursive formula above. To perform the calculations, it is easiest to write down the relationships among the animals' ancestors in the form of a tree (Figure 13.6). Note that the same ancestor, animal no. 5, appears in two places in the tree. The inbreeding coefficient for animal no. 10 measures the proportion of genes that are homozygous because both copies are derived from the same ancestor, animal no. 5.

Half of the genes for animal no. 10 are derived from its mother, animal no. 5, and half from its father, animal no. 4. However, animal no. 5 is also the grandmother of animal no. 4. Thus, 0.25 of the genes in animal no. 4 are

Table 13.5. Fragment of the studbook for the captive population of Long Footed Potoroos at Healesville Sanctuary, Victoria (D. McDonald, personal communication).

Number and gender	Male parent	Female parent	Inbreeding Coefficient
1. Female	Wild	Wild	0
2. Male	Wild	5	0
3. Female	Wild	Wild	0
4. Male	2	3	0
5. Female	Wild	Wild	0
6. Male	2	1	0
7. Female	2	5	0.2500
8. Male	4	3	0.2500
9. Male	2	1	0
10. Female	4	5	0.1250
11. Female	6	7	0.1875
12. Male	2	3	0

derived from animal no. 5. Applying the above equations:

$$\begin{aligned} F(\text{no. } 10) &= R(\text{mother of no. } 10, \text{father of no. } 10) \\ &= 0.5(R(\text{no. } 5, \text{father of no. } 4) + \\ &\qquad R(\text{no. } 5, \text{mother of no. } 4)) \\ &= 0.5((0.25) + (0)) \\ &= 0.125. \end{aligned}$$

The information in Table 13.5 can be used to calculate the degree of kinship between all animals and to calculate the contribution of each founder to the current population. One management strategy may be to manage reproductive opportunities such that each founder contributes about the same proportion of genes to the current population. In this way, the effective size of the population is maximised, minimising the rate of loss of the original complement of genetic

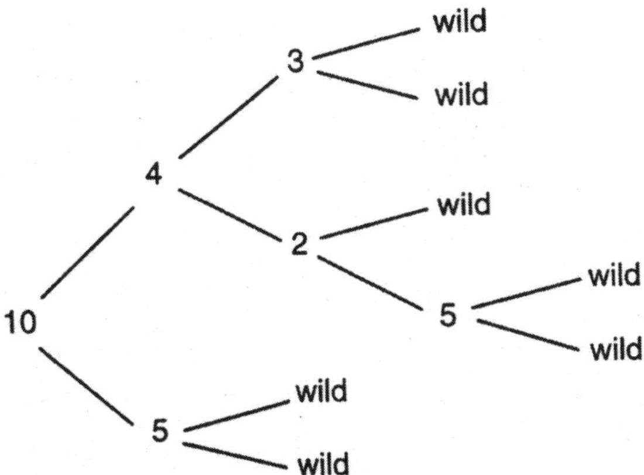

Figure 13.6. Lineage of Long-footed Potoroo no. 10 at Healesville Sanctuary (from D. McDonald, personal communication).

Box 13.8

Breeding systems of Red-necked Wallabies

There can be interactions between the breeding system of a species and the conditions under which a species is held captive, leading inexorably to changes in the genetic composition of a species. Watson *et al.* (1992) noted that female Red-necked Wallabies tend to exhibit cryptic copulation strategies to mate with subordinate males. These males contribute a significant proportion of genes to the next generation in wild populations. In captivity, it can be more difficult for females to elude their consorts, simply because of their proximity. Mating with subordinates can thus be less frequent, decreasing the effective population size and increasing the rate of loss of genetic variation compared with wild populations of the same size and composition.

Red-necked Wallaby. (Photo by Esther Beaton.)

variation. A different strategy would be to minimise the level of inbreeding in the population, but achieving this goal may mean some compromise in achieving the target of equal founder representation.

The rate of change in inbreeding coefficients is determined by the effective population size. The inbreeding coefficient between two individuals in generation t, the probability that genes x and y are identical by descent, can be calculated from the coefficient in the preceding generation by:

$$F_t = \frac{1}{2N_e} + \left(1 - \frac{1}{2N_e}\right)F_{t-1}. \qquad (13.10)$$

This expression has limited value for small populations. In most circumstances where genetic management is important, pedigrees from studbooks or DNA fingerprinting are the only effective tools.

Frankham *et al.* (1986) suggested that it is important not to select for inbreeding tolerance. Rather, we should seek to maintain genetic diversity by maximising effective population size and equalising founder representation. The best way to avoid inbreeding and outbreeding depression is to minimise changes to the genetic structure and composition of the population. Templeton (1991) suggested that management should focus on either avoiding changes to the genetic composition of the population, or on adapting the population to the alterations. If a natural population is highly structured, and if the intention is to maintain the genetic environment as it was prior to reintroduction, captive management may be most effective if it keeps individuals from known subpopulations separate and releases them to the region from which they were obtained (see Chapter 4 on reintroductions). Such a strategy presupposes sufficient information to detect population subdivisions. Management objectives are constrained by the source population size, the availability of individuals for release, and the continuing presence of threatening processes in release sites (see Section 4.6).

13.4 Conclusions

Frankham (1995b) listed several issues for conservation genetics, including inbreeding depression, accumulation and loss of mutations, loss of genetic variation, adaptation to captivity and its effects on reintroduction success, outbreeding depression, fragmentation and migration, and taxonomic uncertainties and introgression. Of course, not all of the issues in this list are independent. Inbreeding depression, adaptation to captivity, and the accumulation of deleterious recessive alleles are closely related. Moritz (1994a) pointed out that conservation genetics has a role to play in guiding the future of closely managed populations, improving the predictions of population viability models, contributing to setting priorities for species protection and reserve design, and providing information on population demography, breeding and migration.

13.5 Practical considerations

The best way to conserve genetic variation is to conserve all individuals from all populations. Peakall and Sydes (1996) describe examples where genetic studies may be relatively unimportant from a conservation perspective, such as when a species is known from only a single population, or when no choice of populations for reservation is possible. Genetic analysis is useful only when it contributes to understanding population dynamics, or when it assists with setting priorities and identifying strategies.

PCR, RAPDs, minisatellites, microsatellites and cpDNA provide new opportunities for understanding population dynamics and reliably identifying the spatial distribution of genetic variation. Wayne and Motin (2004) noted that there are exciting new possibilities for better understanding population structure, dynamics, and movement through the non-invasive analysis of materials such as faeces and hair. However, for the foreseeable future, genetic data will be available for only a small handful of species. Therefore, most decisions and planning strategies will need to be guided by surrogates such as geographic range and biogeographic history.

Hogbin *et al.* (2000) argued that genetic studies have had few substantial practical outcomes. In many cases, the recommendations based on genetic studies are redundant, irrelevant or ignored. For example, studies of population differentiation may only serve to reinforce decisions not to mix populations. Even if a genetic sample failed to detect differences between two populations, the differences may exist for traits not specifically studied, particularly quantitative traits that affect fitness. In the interests of efficiency and cost-effectiveness, Hogbin *et al.* (2000) suggest that genetic research should be encouraged in cases where potential research outcomes could lead to operationally and socially feasible management options. For this to work, procedures must be in place that incorporate genetic results in reserve design or operational planning (Hogbin *et al.* 2000).

13.6 Further reading

Reviews of the use of genetic information for conservation are provided by Amos and Hoelzel (1992), Crozier (1990), Moritz (1994a), Frankham (1995a,b), Sunnucks (2000), Frankham *et al.* (2002) and Reed and Frankham (2003). Sherwin *et al.* (1991) and Frankham (1995c) provide pragmatic advice on estimating effective population size. Crow and Dennison (1988) provide technical details on genetic studies. May (1990b), Crozier (1992), and Faith (1992; 1994) discuss approaches to setting conservation priorities using taxonomic and genetic information. Maynard-Smith (1989) and Falconer (1989) provide general introductions to genetics. Frankham *et al.* (2002) provide a general introduction to conservation genetics, including many of the fundamentals of population genetics. Avise and Hamrick (1996) produced a series of papers on conservation genetics. Mallet (1996) and Sunnucks (2000) summarise the molecular and genetic markers used in conservation biology, their mode of operation and their most useful applications.

Chapter 14

Measuring diversity

This chapter outlines measures of species diversity and community diversity, including alpha, beta and gamma diversity. It discusses ways to detect rare species and the strengths and limitations of ways to assess landscape diversity.

14.0 Introduction

Much of conservation biology is concerned with how humans affect the environment. Detecting and solving environmental problems depends on measurements that reflect human impacts and rehabilitation efforts. Because definitions vary and biodiversity has many elements (Bunnell, 1998; see Chapter 2), many different quantitative tools are used to measure change. In Chapter 13, we outlined ways to measure genetic diversity. The purpose of this chapter is to outline measures of species and community diversity.

14.1 Estimating species richness

Species richness (sometimes called **species diversity**) – one of the most intuitive and frequently used measures of biodiversity – is the total number of species in a defined sample. It is used to compare the conservation value of different areas, to analyse the effects of variables such as area, latitude or altitude on community and population dynamics, and as an index of environmental change following the isolation of a vegetation remnant or exposure of a community to a hazardous chemical.

Even with a statistic as simple as the number of species, there are many factors that can make measurement in a single sample or comparisons between samples difficult or unreliable. Estimates of the number of species can be made at different times, or within different sized areas. Usually, such estimates involve a subset of taxa. For example, studies may focus on the number of species of insects inhabiting a rainforest canopy, the number of plant species in 1000-square-metre quadrats, or the number of benthic species at a number of points affected by sewage discharge. The sampling effort, field methods, and sampling strategies for each circumstance will be different.

Local species richness can be measured with a complete **census**, but this is only practical if the number of taxa and the size of the area are both small. The number of different kinds of species, the range of techniques necessary to sample them, and taxonomic difficulties make complete inventories possible usually only for vascular plants and some vertebrates, and then only within small geographic areas. Most discussions deal with **relative richness**, although few researchers make it clear that they are working with this measure. Comparisons of relative richness attempt to standardise sample areas, sizes, effort and timing to minimise arbitrary biases (Gaston, 1996).

Consider a circumstance in which you wish to count the total number of plant species in a sample site. The problem seems straightforward. However, the following questions arise:

- Is the sample site to be visited once, or more than once?
- What time of year will the sample site be visited?
- How will annual plants be recorded, particularly if the site is to be visited only once?

- Will micro-organisms, algae, bryophytes and other non-vascular plants be included?
- Will the soil seed bank be considered?
- Will vagrant species be included?
- How big should the sample site be?
- What shape should the sample site be?
- How much time should be spent searching?

Even these questions ignore taxonomic difficulties and changes in the composition of the biota from one year to the next. Estimates of species richness in a fixed area, or in a survey with fixed time or sample effort are only useful for comparisons of relative numbers of species.

There are many instances in which it is impossible to arrive at a complete count of the number of species, even for a specific set of taxa and given a specific sample effort (e.g. a fixed area quadrat or a fixed amount of time in a bird survey; Cunningham *et al.*, 1999). In these cases, the various estimators outlined in the following sections can be useful. However, little is known about the reliability of species richness estimators (Gaston, 1996); if they are used at all, they should be used cautiously.

Species accumulation indices

Accumulation curves estimate the total number of species by extrapolation. Usually, they are plots of the cumulative number of species on the *y*-axis versus sampling or collection effort on the *x*-axis. Effort can be represented by time, the number of samples or individuals examined, or the area searched. Effort is partitioned into uniform units such as quadrats, volume samples, fixed duration observations or trap-nights (Sutherland, 2000).

Random or *ad hoc* searches tend to focus on previously uncollected species and are usually more efficient than other more uniform approaches for detecting rare species (Magurran, 2003). However, *ad hoc* searches introduce a bias toward uncommon or rare species if the data are used to estimate the total number of species by extrapolation (Colwell and Coddington, 1995).

The order in which species are added to a species accumulation curve can affect its shape. Furthermore, a bias can arise from sampling error and heterogeneity among the sample units. To address this problem, the sample order can be randomised to eliminate the bias (see Colwell and Coddington, 1995).

When samples are unequal (e.g. fixed-area quadrats of different sizes), a technique called **rarefaction** can be used, in which the number of species expected ($E(S)$) in a sample of size n, is estimated by

$$E(S) = \sum\left\{1 - \left[\binom{N-N_i}{n} \Big/ \binom{N}{n}\right]\right\} \tag{14.1}$$

where $E(S)$ is the expected number of species, n is the standardised sample size, N is the total number of individuals recorded and N_i is the number of individuals of the ith species recorded (see Hurlbert, 1972; and discussion in Magurran, 2003). This is useful because it allows us to compare samples from different areas in which sampling effort has been different. The term $\binom{n}{r}$ is a binomial coefficient calculated using factorials in the following way:

$$\binom{n}{r} = \frac{n!}{r!(n-r)!}. \tag{14.2}$$

Several different species accumulation curves can be used to estimate the total number of species of a taxon or guild within a prescribed area. Three of these equations are described in detail here, although there are alternatives (Keating *et al.*, 1988; Colwell and Coddington, 1995; Magurran, 2003). A widely used expression is:

$$S_A = \frac{S_{max}A}{b+A} \tag{14.3}$$

where S_A is the number of species recorded in an area of size A, and S_{max} and b are fitted constants. S_{max} is the expected total number of species. This equation is the same as the Michaelis–Menten equation in chemistry and there are numerous methods for estimating S_{max} and b (see Soberon and Llorente, 1993).

Species–area curves are commonplace in the ecological literature and derive from the work by Arrhenius (1921), Preston (1962) and MacArthur and Wilson (1967) (see also Rosenzweig, 1995). There are several forms of these curves, including

$$\log S_A = c + z \log A \tag{14.4}$$

$$S_A = c + z \log A \tag{14.5}$$

where c and z are fitted constants (see Chapter 17). The expected total number of species in a sample area is estimated by substituting the value for the area into the equation, once the parameters have been estimated. Confidence intervals and prediction intervals for the estimated number of species can be generated using ordinary regression.

It is difficult to choose the most appropriate method for estimating the number of species. Different methods make different mathematical assumptions, such as the existence of a total (maximum) number of species (Equation 14.3), or a theoretically unlimited number of species (Equations 14.4 and 14.5). The equations also make different assumptions about the ways in which species are assembled into communities. Perhaps the best way to choose among them is to compare them to data, although extrapolations beyond the limits of data are error-prone. These models have produced substantial overestimates when tested (Colwell and Coddington, 1995), but they can be used more reliably to estimate the proportional change in the number of species following a change in area (Chapter 18).

Many methods for estimating species richness rely only on presence/absence information. Methods that make use of the relative abundance of species include the Chao statistic (S_C, Colwell and Coddington, 1995):

$$S_C = S_{obs} + \frac{a^2}{2b} \tag{14.6}$$

where S_{obs} is the number of observed species in a sample, a is the number of observed species that are represented by a single observation, and b is the number of observed species represented by exactly two individuals in the sample.

If the data are in the form of presence/absence information, then S can be calculated as shown in Equation 14.6, but a will be the number of species that occur in only one sample and b will be the number of species that occur in exactly two samples. The variance of this statistic is given by:

$$\mathrm{var}(S_C) = b\left[\left(\frac{a/b}{4}\right)^4 + (a/b)^3 + \left(\frac{a/b}{2}\right)^2\right]. \tag{14.7}$$

The Chao statistic is likely to underestimate true species richness if the sample is too sparse (Colwell and Coddington, 1995). However, it performs well when most of the information in the sample is in the lower frequency classes, that is, in circumstances in which there is a preponderance of rare species, as is often the case (see Chapter 2).

Perhaps the most reliable information from accumulation curves relates to the predicted increase in species richness given additional samples. In all of the expressions here, the area A can be substituted by an index that represents the time or effort spent searching. All methods assume that the data are in the form of sightings per unit effort, for which area is just one index.

There is no guarantee that any of these methods will provide reliable estimates of the total number of species in an area. Few empirical studies have attempted to validate their predictions. Their effectiveness will be determined by how closely the implicit assumptions of a method match the ecological and historical processes that have determined the assemblage.

Ratio estimation

An alternative to accumulation curves is to make extrapolations based on a knowledge of the ecology and taxonomy of different groups. Essentially, **ratio estimation** assumes that a fixed number of species exist in a taxon for every species in another taxon. Usually, one of the taxa is relatively easy to count (e.g. trees). The number of species is then multiplied by a factor to provide an estimate of the number of species in a taxon that is hard to count (e.g. beetles that live on the trees). If we make the assumption that the ratio of beetle species to tree species is fixed, it is relatively simple to estimate beetle numbers anywhere in the landscape, simply by counting the number of tree species. If we know the number of insect species associated with each tree species, the number of tree species, and the level of host-specificity within the insect groups, these data can be used to estimate the total number of insects.

Hawksworth (1992) found that there have been 12 000 fungus species recorded from the British Isles. There are just over 2000 vascular plant species in the same area, giving a ratio of fungus species to vascular plants of 6:1. A conservative estimate of the number of vascular plant species globally is 270 000, which yields a global total for fungi of about 1.6 million species. Using the number of beetle taxa occurring on a tropical tree species, the number of rainforest tree species throughout the world, and the ratio between the number of beetles in rainforest canopies compared with the forest floor, Erwin (1982) estimated that there were about 30 million species of arthropods worldwide. However, other workers have questioned the basis for the extrapolations, for example the complete separation between ground and arboreal assemblages and the extent of host plant–beetle species specificity; they considered that 30 million species was an overestimate (e.g. May, 1990a; Gaston, 1991; but see Gullan and Cranston, 1994; Chapter 2).

Ratio estimation is not restricted to taxon:taxon ratios; extrapolations can be made from a focal group to a more inclusive group, from site to site, from sample to inventory, and across spatial scales (Hammond, 1995).

Williams and Gaston (1994) used a taxon ratio to examine the relationships between richness at a family level and species level for a range of groups. They found that family-level richness was a good predictor of species-level richness for Australian **passerine** birds, British butterflies and ferns and bats from North and Central America. In the case of Australian passerines, Williams and Gaston (1994) examined data on records of birds from 5- and 10-degree grid squares (Blakers *et al.*, 1984). They generated species versus family plots (see Figure 14.1) and found a strong correlation between the two measures. Williams and Gaston (1994) concluded that family-level taxonomy was a useful surrogate for species richness. However, they acknowledged a number of potential problems including that:

- recording all families (or some other taxonomic level) within a given area may require considerable effort
- the nature of the predictive relationship is likely to change with the taxonomic 'distance' between levels chosen
- some regions may be atypical and support many species that belong to only a few families (e.g. the elapid or front-fanged family of snakes that dominate the Australian snake fauna).

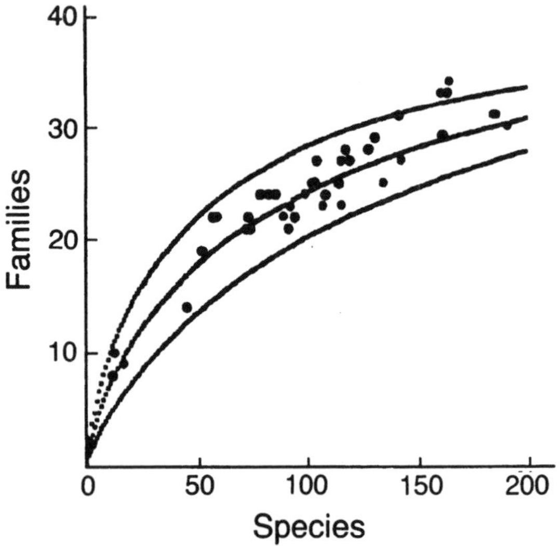

Figure 14.1. Richness of families and species of Australian passerine birds (after Williams and Gaston, 1994).

There is limited empirical evidence to support the notion that taxon ratios can safely be assumed to be constant (Colwell and Coddington, 1995). Ratio estimation is very sensitive to errors in either of the quantities used to calculate the ratio. Researchers should at least endeavour to have several independent estimates. However, when applied with care, ratios may provide reasonably reliable estimates of species numbers in some circumstances.

14.2 Detecting rare species

Most communities have many rare species, which is reflected in the canonical log-normal distribution of species abundances (Preston, 1962; MacArthur and Wilson, 1967; Hubbell, 2001; see Chapter 3). Rare species are the focus of many conservation biology studies (e.g. Hartley and Kunin, 2003) and are scarce in survey data (Welsh *et al.*, 1996). It is often useful to estimate how reliably we have surveyed and mapped the distributions of rare species.

McArdle (1990) defined f to be the probability that a species would appear in a single randomly selected sampling unit. The confidence (α) with which the species will be detected in a sample of size N is:

$$\alpha = 1-(1-f)^N. \tag{14.8}$$

This relationship can be used to estimate the probability of detecting a scarce species, given resources for a certain number of samples. For example, if a species on average turns up in one sample in 50 (f = 0.02), the chance that it would be found at least once in 20 samples is only 0.33. If it is necessary to determine what level of scarcity could be detected with a given level of confidence, then we could use:

$$f = 1-(1-\alpha)^{\frac{1}{N}}. \tag{14.9}$$

For example, a total of 20 samples will detect a species of scarcity (f) of 0.1 with a probability (α) of 0.9.

For the design of surveys, it is often useful to know the number of samples required to achieve a given level of effectiveness. To estimate the number of samples (N) required to detect a species with a level of scarcity f, with a level of confidence α:

$$N = \frac{\log(1-\alpha)}{\log(1-f)}. \tag{14.10}$$

Thus, to detect species of scarcity $f = 0.02$ with a probability of 0.9 would require more than 100 samples. McArdle (1990) pointed out that these formulae depend on two assumptions: that the value of f is constant in all sampling units (i.e. that it is true for any random sample), and that the sampling units are independent.

Environmental conditions, social structure or local dispersal may create clusters of sightings. Whenever the probability of co-occurrence of a species at two points depends on the distance between them, the sample sites are not independent and are said to be **spatially autocorrelated.** Positive spatial autocorrelation will result in overestimates of the probability of detecting a species. A solution is to ensure that the scale is sufficiently large so that most samples are uncorrelated (McArdle, 1990).

Scarcity, the frequency of a species among samples, is a statistical property of species that depends on abundance, range and habitat specificity (Chapter 3). There is a direct relationship between the relative frequency of species and their relative abundance given by:

$$\log(-\ln p_0) = \log\beta_1 + \beta_2 \log\bar{x} \tag{14.11}$$

where p_0 is the proportion of unoccupied samples, $\bar{x}$ is the mean density of the species in the samples, and β_1 and β_2 are fitted constants (Nachman, 1981). Relative frequency is sometimes a good indicator of local abundance. If the samples span the geographic range of the species, then relative frequency will integrate local abundance across this range.

14.3 Species diversity

Alpha diversity

The concept of species diversity (**alpha diversity**) was introduced in Chapter 2. It reflects both the numbers of different species in a community and their relative abundances. To measure changes in species diversity, it is necessary to formulate clear ideas of what is meant by the concept, and to use measures that represent the concept unambiguously. Abundance can be measured in several ways, and relative abundance and the number of species may be combined in an index in several ways. Thus, for any sample, a range of different indices is possible, each of which represents something different about a community. A common index that accounts for the proportional abundance of species is the Shannon–Wiener index (D_{SW}):

$$D_{SW} = -\sum_{i=1}^{S} p_i \log p_i \tag{14.12}$$

where p_i is the proportion of individuals of species i in the community, and S is the number of species present. The log can be to any base, although originally calculations were made by taking logarithms to base 2. It is generally easier to use natural logarithms. The Shannon–Wiener index ranges from zero in communities comprising a single species, to a maximum of log S.

Another diversity index that has been applied widely over the last few decades is Simpson's index (D_S):

$$D_S = 1 - \sum_{i=1}^{S} p_i^2 \tag{14.13}$$

where the parameters are the same as described for the Shannon–Wiener index. This index is equivalent to calculating the probability of detecting two individuals at random that are different species. The value of the index ranges from zero in communities made up of single species, to almost one in communities made up of many species, each present in equal numbers.

These and other indices of species diversity are special cases of a general expression that relates the number and relative abundances of species:

$$D_\beta = \left(\sum p_i^\beta\right)^{\frac{1}{1-\beta}} \tag{14.14}$$

where D_β is the 'β' order of diversity, and p_i is the proportional abundance of the ith species. Where $\beta = 0$, D is species richness, when $\beta = 1$, D is the exponential of the Shannon–Wiener index and when $\beta = 2$, D is the reciprocal of Simpson's index (Hill, 1973; Gove *et al.*,

1994; Magurran, 2003). Different values of the parameter β emphasise either species richness (weighting towards rare species) or dominance (weighting towards common species).

Measures of species diversity need to be applied appropriately. More species may not necessarily be better; species identity is also important (Murphy, 1989; Doak and Mills, 1994). For example, invasions of exotics may increase species numbers but reduce the conservation value of an area, particularly if they result in the loss of native taxa (Hunter, 1996, 2002).

Consider the case of a plant community composed of 10 species, two of which are shrubs. The proportions represented by the variable p in the equations above can be measured using either counts of the number of individuals per species in the sample area, or biomass represented by, for example, crown cover. In this hypothetical case, we may wish to measure the immediate impact of cattle grazing on species diversity. We will assume that grazing removes almost all individuals of the two palatable shrub species, removes two understorey species susceptible to soil compaction, and introduces two weed species (Table 14.1).

On the basis of counts of the number of individuals, and applying Simpson's index, we have:

$$D_S \text{ (before impact)} = 0.593$$
$$D_S \text{ (after impact)} = 0.592.$$

Grazing, in this hypothetical example, has very little effect on species richness. The number of species is identical before and after impact. The composition of the community is different, but these indices are blind to the types of species that make up a community; they reflect only the number of species.

On the basis of crown cover, and again applying Simpson's Index, we have:

$$D_S \text{ (before impact)} = 0.572$$
$$D_S \text{ (after impact)} = 0.791.$$

The results for crown cover suggest that species diversity has increased. It may seem counterintuitive that ecological impacts can increase species diversity, but beliefs that human impacts are always negative or that increases in biodiversity are always beneficial are naive (Noss and Cooperrider, 1994). In this case, removing the shrubs has made the contribution of each species to the total biomass much more even. Measures of diversity reflect both the number and relative abundance of species. The 'equability' of species crown cover is much higher following grazing.

A change in a diversity index is, in itself, neither 'good' nor 'bad'. In the example above, counts and crown cover give different results, but there is no general rule that says one or the other of these is correct. The choice of a measure should be sensitive to the kinds of changes it is important to detect.

Obviously, indices of species diversity are only a small part of the full picture of biodiversity. For instance, weeds can cause further changes in the abundances of other species (Low, 1999; see Chapter 7); the elimination of two species may be important, particularly if they are rare or threatened; and the impacts of cattle on the soil may preclude recolonisation of the site by the threatened species.

Table 14.1. A hypothetical example of changes in a plant community following cattle grazing.

Species	Before impact		After impact	
	Count	Crown cover (%)	Count	Crown cover (%)
Shrub 1	8	10	1	1
Shrub 2	14	60	1	1
Understorey 1	5	1	–	–
Understorey 2	13	1	–	–
Understorey 3	2	1	2	1
Understorey 4	44	9	44	9
Understorey 5	407	3	407	3
Understorey 6	300	8	300	8
Understorey 7	1	1	1	1
Understorey 8	2	1	2	1
Weed 1	–	–	2	1
Weed 2	–	–	36	2

Results such as these reveal little about changes at larger spatial scales, or over time. Shrubland may recover, in terms of species composition, frequency and crown cover, at least within the limits of natural variation. The loss of one or two common species from a particular location may be seen as a tolerable change, particularly if these species are common elsewhere or are well represented within national parks or other protected areas. Weeds may be acceptable if the species do not change the composition of the community by competing for space, light, water or nutrients. A fuller understanding of impacts must be explored over broader temporal and spatial scales than is possible with measures of α diversity alone.

Beta and gamma diversity

Beta diversity is the turnover of species along an environmental gradient (Chapter 2). The smaller the ecological tolerance of the species, the smaller will be their range on the gradient, the greater will be the turnover and the greater will be the beta diversity on that gradient. Beta diversity is often represented by how different two sites are in terms of species composition and relative abundance, but strictly it refers to the difference between sites when they are equally spaced with reference to an ecological gradient.

There are numerous indices for measuring the similarity between communities in different locations. Whittaker (1960) suggested the index:

$$\beta_w = \frac{S}{\alpha} - 1 \qquad (14.15)$$

where S is the total number of species recorded in all samples, and α is the average species richness per sample.

Other indices use the concept of overall similarity between species lists. For example, beta diversity can be represented by the proportion of species shared by two sites. Jaccard's coefficient (β_J) gives the shared proportion, relative to the total number in both sites. Lists of species recorded from different locations in a community can be arranged in a table in which the rows (or columns) are sites, and the columns (or rows) are species. Jaccard's coefficient is calculated as:

$$\beta_j = \frac{a}{a+b+c} \qquad (14.16)$$

where a is the number of species present in both sites, b is the number of species present in the community at site 2 that do not occur in the community at site 1, and c is the number of species present in the community at site 1 that do not occur in the community at site 2.

These indices are particularly useful in studies where there are many rows and columns of information. They can assist decisions about which areas to include in a conservation park to ensure that the park includes representative samples of the variety of communities.

There are numerous other measures of beta diversity, some of which account for relative abundances and other attributes of the species (see Legendre and Legendre, 1998). When there are many samples, average measures of diversity can be computed (Colwell and Coddington, 1995).

Although they are usually applied to species data, any of the indices of species diversity can also be applied to data of any taxonomic rank. Thus, for example, the diversity of families could be studied, taking into account the number of families and their relative abundances. Harper and Hawksworth (1995) call this **organismal diversity** rather than species diversity, to emphasise the fact that these measures are not constrained to any taxonomic level.

Whittaker (1960) defined the concept of **gamma diversity** as the turnover of species that occurs between spatially separate sites, within a more or less homogeneous vegetation type or habitat (see Chapter 2). It measures the rate at which the composition of a community changes as a function of geographical distance, rather than ecological distance. The same statistics that are used to measure beta diversity are also used to measure gamma diversity.

A test for change in community structure

Natural variation and sampling error cloud comparisons between communities, for example between control and impact sites. Solow (1993b) described a randomisation test for the difference in diversity between two community samples. If two lists of individuals, m_1 and m_2, are sampled from a community at times 1 and 2 respectively, there will be a difference (δ) between the index of diversity computed from these two lists. The test operates as follows:

1. Merge the two lists into a single list of $m_1 + m_2$ labelled individuals.

2. Partition the list at random into two populations (list 1 consists of m_1 and list 2 consists of m_2 individuals).
3. Calculate the value of δ (the difference between the indices) based on the random partition.
4. Repeat the partition of species and the computation of δ a large number of times (~10 000 times).
5. Estimate the *P*-value for a two-tailed test by the proportion of partitions with values of $|\delta|$ greater than the observed value of $|\delta|$.

The test can be modified in an obvious way for one-tailed tests. If the total number of individuals is small (less than about 10), then it would be better to calculate δ for all partitions, rather than for a random sample of such partitions.

14.4 Landscape diversity

Spatial processes can be quantified by measuring the spatial patterns they create (Wegner, 1994; Forman, 1995; Smith, 2000; Lindenmayer *et al.*, 2002a; McAlpine *et al.*, 2002). Measures of landscape diversity describe aspects of the 'patchiness' of landscapes, particularly with respect to the size, shape, composition, juxtaposition and arrangement of landscape units (e.g. habitat patches; see Figure 14.2). They also quantify the spatial and temporal relationships between these units (Cale and Hobbs, 1994; Wegner, 1994; Nicholls, 1994; Haines-Young and Chopping, 1996; McAlpine *et al.*, 2002).

The spatial structure of landscapes is related to ecological processes such as fire, the flow of water, erosion, and the supply of nutrients (e.g. eutrophication; Swanson *et al.*, 1988; Mackey, 1993; Mackey, 1994; McAlpine *et al.*, 2002) – processes that influence species distribution patterns (Wardell-Johnson and Roberts, 1993; Lindenmayer *et al.*, 1999c; see Chapter 10).

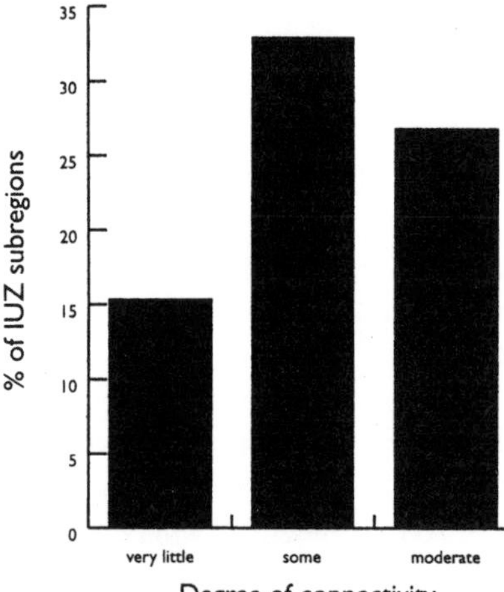

Figure 14.2. Degree of connectivity for native vegetation in the intensive land-use zone (sensu Graetz *et al.*, 1995; see Chapter 9). (From Morgan, 2001.)

There have been some suggestions that human disturbance regimes should mimic the landscape patterns created by natural disturbance (Welsh and Healy, 1993; Attiwill, 1994a; McNeely, 1994a; Korpilahti and Kuuluvainen, 2002; see Chapter 11). To implement this idea, managers need measures of spatial patterns to guide them (Cale and Hobbs, 1994; Lindenmayer and Franklin, 2002). For example, indices of forest fragmentation are now a part of the Montreal Process criteria and indicators of sustainable forest management (Commonwealth of Australia, 1998; Tickle *et al.*, 1998).

Descriptions of the spatial structure of landscapes typically include three elements:

- *composition*: the identity and characteristics of the different types of patches
- *configuration*: the spatial arrangement of the patches
- *connectivity*: the ease with which organisms can move between particular landscape elements.

There are many landscape heterogeneity indices (Turner, 1989; Wegner, 1994; reviewed by Haines-Young and Chopping, 1996). For instance, the Shannon–Wiener Index (Equation 14.2) is used to quantify landscape diversity (e.g. Mladenoff *et al.*, 1993), where p_i is the proportion of the landscape in patch type *i*, and *S* is the number of patches. Short and Turner (1994) used the index to examine factors influencing the abundance of macropods and possums on Barrow Island in Western Australia. They found no significant ecologically meaningful relationships between landscape indices and any species of mammals they studied.

O'Neill *et al.* (1988) devised a 'spatial contagion' index (D_2) to measure the extent and adjacency of *K* landscape units or patches;

$$D_2 = 2K \ln(K-1) \sum_{i=1} \sum_{j=1}^{K} p_{ij} \ln p_{ij} \tag{14.17}$$

where p_{ij} is the probability of land-use type *i* being found adjacent to land-use type *j*. Landscapes with smaller and more dispersed patches have lower values.

Knaapen *et al.* (1992) interpret connectivity as a measure of the probability that an organism can successfully traverse a landscape. Connectivity should be specified relative to the organism of interest, although this is rarely done (Figure 14.2). If patches are considered to be discrete and more or less isolated, then connectivity can be defined as the number of connections between these patches, relative to the maximum number of potential connections. If there are direct links between all N patches in a landscape, then there will be $N(N-1)/2$ such links. The connectivity (C) of a landscape composed of patches and links is

$$C = \frac{2L}{N(N-1)} \tag{14.18}$$

where L is the number of links and N is the number of patches. A fully connected landscape has a value of 1, and an unconnected landscape has a value of 0. This index can be applied to structural links (e.g. corridors between patches of natural vegetation) or it can reflect the ecological links between populations that depend on the biology and dispersal abilities of a species (Chapter 10).

The indices outlined in this section are a small sample of all those available (e.g. see O'Neill *et al.*, 1988; Taylor *et al.*, 1993; Haines-Young and Chopping, 1996). Their use in Australian studies is increasing (Nicholls, 1994; Short and Turner, 1994; Incoll *et al.*, 2001; Lindenmayer *et al.*, 2002a; Morgan, 2002). Cale and Hobbs (1994) and Lindenmayer *et al.* (2002a) critically appraised landscape indices. In particular, they noted that:

- The methods used to develop indices are often not provided.
- It is not always clear what to do with indices when they are generated and they are often not well linked to appropriate aspects of land management.
- Few measures are used consistently in different investigations.
- Values for indices are often scale-dependent, making it difficult to compare results from different landscapes and spatial scales.

Landscape indices also fail to account for factors such as the vertical heterogeneity of vegetation, which is known to be important for bird species richness and abundance (e.g. Gilmore, 1985). Many indices provide sophisticated ways of highlighting intuitively obvious landscape patterns and, as such, they have led to few new insights. Many landscape indices are highly correlated (and therefore provide the same sorts of information about the landscape).

Each species responds differently to the same spatial scale of landscape change and human disturbance (e.g. Davies and Margules, 1998; Villard *et al.*, 1999). Hence, no single measure adequately reflects change for all biota. In addition, indices will fail when species responses to and perceptions of a landscape are different from the way humans characterise and map that landscape.

Most indices provide an instantaneous and static measure, whereas temporal dynamics, for example the length of isolation of habitat remnants, may be important (Loyn, 1987; Bennett, 1990b; Gascon *et al.*, 1999; Fahrig, 1992). Cale and Hobbs (1994) further highlighted the need to match the scale at which measurements are taken to the ecology of the species in question (see also Smith, 2000; McAlpine *et al.*, 2002) and they stated that:

> *Indices of diversity can indicate that differences in heterogeneity exist when no actual change in habitat has occurred from the organism's point of view, or can fail to detect important changes in habitat. The principal reason for this is the problem of matching the scale of measurement with the scale at which organisms perceive the environment.*

For example, the landscape spatial scales relevant to the congeneric Yellow-bellied Glider and Sugar Glider are markedly different. The home range of the latter is 2–4 hectares, which is 15–30 times smaller than that of the former (Lindenmayer, 1997, 2002b).

Landscape indices can be useful to help characterise, for example, patterns of vegetation cover in a landscape and establish whether a pattern has changed. Landscape indices can also link patterns of landscape cover to species responses. When using landscape indices to describe landscapes and guide management actions, it is important to be clear about the reason for using them. Different indices have different strengths and limitations, depending on the context. The units and the context (e.g. the size of a 'landscape') should be clearly defined. The system for application should establish a framework for evaluating the predictions and decisions that flow from application of the indices (Lindenmayer *et al.*, 2002a).

14.5 Conclusions

Estimating the total number of species within a taxon or within a prescribed area is conceptually simple, but practically difficult. Various approaches that have been developed to circumvent problems of varying sampling effort and incomplete lists have not been evaluated empirically, and so should be used with care. Each measure of diversity captures a different aspect of the composition and structure of a community, landscape or set of communities. The appropriate measure depends on context and the availability of the necessary data. The application of a measure must be underpinned by a clear understanding of the question it will be used to address, and the ecology of the organism, community, or landscape. Thus, there is a need to couple diversity indices with those aspects of species, community and landscape heterogeneity that are functionally important and appropriate.

14.6 Practical considerations

Surveys are limited, by necessity, to small samples of an area and a few taxonomic groups. The validity of comparisons of relative species richness and diversity depends on how carefully sampling effort is controlled to eliminate arbitrary bias. Extrapolations are error-prone and should be validated. Landscape indices are useful descriptors of some ecologically important attributes, some species and some spatial scales. The justification for their use should be based on clear links between biological attributes and quantitative measures.

14.7 Further reading

Colwell and Coddington (1995) review methods for estimating the number of species. Manly (1991) provides a complete treatment of randomisation tests for biology. Magurran (2003) provides a complete introduction to estimating diversity. Legendre and Legendre (1998) provide a relatively complete catalogue of the different kinds of statistics for measuring beta and gamma diversity. Cale and Hobbs (1994) and McAlpine *et al.* (2002) review the application of heterogeneity measures. Noss (1991), Taylor *et al.* (1993) and Bennett (1998) describe different concepts of landscape connectivity. Gaston (1994) provides a complete treatment of rarity (see also Chapter 3).

Chapter 15
Identifying habitat

This chapter focuses on ways to quantify and study habitat. It includes overviews of qualitative methods (particularly habitat suitability indices), quantitative statistical approaches such as generalised linear modelling (e.g. logistic and Poisson regression), and envelope methods such as bioclimatic modelling.

15.0 Introduction

Habitat loss is the most important cause of decline and loss of species in Australia and elsewhere around the world (Primack, 2001; Chapters 6, 9 and 10). Managing habitat loss depends on understanding what constitutes habitat for a species. Plant and animal species are distributed discontinuously (Elton, 1927; Begon *et al.*, 1996), and many are clearly linked to particular features, for example swamps, fast-flowing streams, deep sands or hollows in trees. With knowledge of the factors that affect species distributions and abundances, it may be possible to predict and mitigate human impacts.

Many factors influence the distributions of species (see Table 15.1 for an example of Australian marsupial gliders), including:

- history (e.g. biogeographic barriers or rare past events such as a flood that dispersed and germinated River Red Gum seeds), competitors, predators, parasites and pathogens
- natural and human-induced disturbance
- population dynamics and social behaviour
- size, shape and spatial distribution of remnant patches
- habitat suitability, including climatic conditions and their relationships with the physiological tolerances of species.

In this chapter, we focus on measuring and mapping habitat suitability.

Defining habitat

Block and Brennan (1993) and Hall *et al.* (1997) defined habitat as the '*subset of physical environmental factors that permit an animal (or plant) to survive and reproduce*'. We assume that **habitat** is associated with a place – a geographic location (Krebs, 1985). Differences in the habitat requirements of different species allow them to coexist in the same area (Lindenmayer, 1997). In contrast, a niche is not tied to a geographic location. The **fundamental niche** of a species is the set of physical limits within which a species can live and reproduce. The **realised niche** is the environment to which a species is limited both by tolerance of physical variables, and by historical events, and social and biotic interactions, including predation and competition. Related concepts include 'microhabitat', 'critical habitat', 'habitat use', 'habitat selection', 'habitat suitability', 'habitat requirements', 'habitat preference' and 'habitat association' (Block and Brennan, 1993). All of these terms indicate some type of relationship between an organism and its environment (Morrison *et al.*, 1992).

It is impossible to survey all parts of the landscape and record the location of all individuals in a population (Margules and Stein, 1989; Chapman and Busby, 1994). Even if it were feasible, the information would

Table 15.1. Factors influencing the distribution and abundance of Australian marsupial gliders at different temporal and spatial scales. (Modified from Lindenmayer, 2002b.)

Factor	Example
Short-term to long-term temporal effects	
Prevalence of predators	Impacts of the Powerful Owl preying upon local populations of the Greater Glider.
Natural disturbance	Impacts of an absence of wildfires on the persistence of the north Queensland form of the Yellow-bellied Glider.
Biogeographic history	Absence of all gliding possums, except the Sugar Glider, from Tasmania
Small- to large-scale spatial effects	
Social behaviour	Enforced dispersal of subadult males by adult males in the Greater Glider.
Characteristics of individual nest trees	Selection of particular trees with hollows as nest sites by the Sugar Glider and the Squirrel Glider in the Box–Ironbark forests of north-eastern Victoria and the Yellow-bellied Glider in south-eastern Queensland.
Structure of forest stands	Increased probability of occurrence of the Greater Glider and Sugar Glider in areas of Victorian Mountain Ash forest supporting numerous trees with hollows. The occurrence of the Mahogany Glider and Sugar Glider in particular parts of woodland environments in north Queensland.
Spatial pattern of habitat patches	Absence of the Yellow-bellied Glider from Mountain Ash landscapes supporting small patches of old growth forest.
Broad-scale patterns of forest types	Concentration of possums and gliders in high productivity forest types on fertile soils in south-eastern New South Wales.
Broad-scale climate conditions	Relationships between temperature and rainfall regimes and the distribution of the Squirrel Glider and Mahogany Glider, and patterns of heat intolerance in the Greater Glider

erode as the condition of the area changed. Fire or the invasion of a competitor may change the suitability of the environment for a species, perhaps permanently.

In this chapter, we outline methods for understanding a species' habitat and how to use them to predict a species' distribution and abundance. Through this process, it is possible to predict changes in habitat suitability as a consequence of disturbance. This provides a basis for predicting the spatial distribution of the species and temporal changes in their distributions – a foundation for impact assessment, reserve design, management planning, and policy formulation for biodiversity conservation (Noss and Murphy, 1995; Lindenmayer and Franklin, 2002).

15.1 Methods for identifying habitat requirements

All methods for identifying habitat are based on detecting associations between environmental attributes and the presence or abundance of species. Typically, habitat measures relate to the biology of the species: attributes such as tree hollows are potentially important for an arboreal marsupial such as the Yellow-bellied Glider, whereas forest skinks and invertebrates may depend on coarse woody debris on a forest floor (Brown and Nelson, 1992; Goldsborough *et al.*, 2003; Singh *et al.*, 2002). The scale at which variables are measured should be appropriate for the species in question. For example, the Sooty Owl has a home range of hundreds of hectares, so measurements of variables from a 2-hectare site would not be useful (Milledge *et al.*, 1991; Kavanagh, 1998).

A variety of methods can be used to combine variables into an estimate of habitat suitability. Qualitative methods link species and habitat variables using functions created subjectively by people with an ecological knowledge of the species. Envelope methods bound the spatial distribution of a species using location data or climatic envelopes inferred from digital terrain models. Statistical approaches use explicit statistical models to link the occurrence of a species with habitat parameters.

The purpose of this chapter is to describe qualitative habitat models, two envelope analyses and statistical habitat methods. These methods introduce the spectrum of approaches to developing an understanding of species' habitats, some of which have been applied widely in Australia.

15.2 Qualitative habitat models

Habitat can be characterised by a description of the features of the environment that determine the distribution and abundance of a species, which are derived from field experience and unquantified human perceptions (e.g. Short, 1984). Potentially important habitat attributes may be assigned weights that reflect expert beliefs about the relative importance of different components (e.g. Laymon *et al.*, 1985; Minta and Clark, 1989).

Habitat suitability indices (HSIs) are based on the **habitat evaluation procedure** (HEP) developed by the United States Fish and Wildlife Service in the early 1980s (United States Fish and Wildlife Service, 1980, 1981). The approach has been applied extensively in North America (see Van Horne and Wiens, 1991, for a review) and indices have been constructed as part of the assessment of the habitat requirements of many vertebrate groups. HSIs have been used as predictive tools to evaluate the quality of habitat for wildlife (O'Neill *et al.*, 1988; Rand and Newman, 1998) and to predict the effects of logging, vegetation clearance and dam construction on a given species or set of taxa (Gray *et al.*, 1996). HSIs have been used rarely in Australia (but see Box 15.1).

HSIs are a descriptive synthesis of information on the biology and life history of a species. Information is often provided by scientists who have experience working with the species. Key elements of the method are:

- selecting a limited number of variables considered to be the most important for the survival and reproduction of the species
- creating a suitability index (SI) on a scale between zero and one for each of the variables
- estimating the form (shape) of the suitability response curve for values of each critical variable
- assigning weights to each of the variables
- combining the SI values into a single HSI.

Various mathematical operations can be used to combine and rescale the habitat variables into an HSI. For example, the various suitability indices could simply be multiplied together and rescaled using the geometric mean of the SI values:

$$HSI = \sqrt[n]{\prod SI_n} \qquad (15.1)$$

where Π is the product of the n numbers. This formula implies that all elements are equally important and that if one is missing, the habitat is unsuitable.

Alternatively, SIs could be added instead of multiplied, implying that one element can substitute for another. For example, if a reduction in the quality of a foraging area is offset by an increase in the availability of places to make a nest, indices for foraging area and nest sites should be added. The choice of operations should reflect the understanding of the way in which a particular species depends on its habitat (Van Horne, 1983). Suitability values can be multiplied by coefficients to represent their relative contribution to overall suitability and the degree to which one factor can substitute for another (Van Horne and Wiens, 1991).

Potential release area for the Eastern Barred Bandicoot recommended on the basis of HSI work (see Box 15.1). (Photo by Mandy Watson.)

Potential limitations of the HSI approach

HSI models have some problems. They do not account explicitly for correlations among the variables. Combining variables and assigning weights may account for non-independence, but the procedures for deciding on the model structure and the weights are not explicit (Minta and Clark, 1989; Reading *et al.*, 1996). Similarly, the procedure for specifying the function relating habitat suitability to the value of the environmental variable is not usually explicit.

Suitability index curves are often assumed to be linear or sigmoidal functions. However, threshold responses are possible, below which a species simply cannot exist and above which an increase in a given measure will make little or no difference in habitat suitability. Predictions of HSI models that involve a threshold can be sensitive to small changes in the value assigned to a threshold (Van Horne and Wiens, 1991).

There are no means of representing the goodness-of-fit of the model to the observations used to construct it. The only way of exploring the predictive power of an HSI is to compare the predictions with new field observations – a process known as **validation** (O'Neill *et al.*, 1988).

Box 15.1

A habitat suitability index for the Eastern Barred Bandicoot

Mainland populations of the Eastern Barred Bandicoot have undergone major declines in the past few decades (Clark and Seebeck, 1990; Menkhorst, 1995b; Menkhorst and Knight, 2001). The species is now functionally extinct in the wild in Victoria (Seebeck in Todd *et al.*, 2002). Reading *et al.* (1996) developed a HSI model to examine potential reintroduction sites for the Eastern Barred Bandicoot, and this case remains one of the few Australian applications of the technique. The HSI model comprised five variables: size (V_1), habitat structure (V_2), predation (V_3), shape (V_4), and security (V_5). These five variables were then combined to give a value for the HSI for various reintroduction sites:

$$HSI = (V_1 \times V_2 \times V_3 \times V_4 \times V_5)^{1/5}$$

Size (V_1) was an index of the size of the release site and, in turn, a surrogate for its carrying capacity. Reading *et al.* (1996) assigned a value of 0.1 to sites of 50 hectares, 0.9 for 1000-hectare sites and 0.99 for 2000-hectare sites. The equation for the relationship was:

$$V_1 = 1.1e^{-0.0023x}$$

where *x* is site size in hectares.

Habitat structure (V_2) reflected aspects of vegetation cover as well as the importance of habitat heterogeneity. The ideal habitat for this species is composed of a mosaic of open pastures for grazing and areas of vegetation cover for shelter. Given this, V_2 comprised two components: an index of percentage vegetation cover (*CV*) and an index termed distribution cover (*DC*), which represented the percentage of land more than 50 metres from cover. Sites without vegetation cover were assigned a *CV* of 0, those with 2% cover were 0.1, those with 20% cover were 0.66 and those with 100% cover were 1.0. Sites with 70% of their area more than 50 metres from cover were assigned a value of 0.1 or less, whereas sites with all areas closer than 50 metres from cover were assigned a value of 1.0 (the maximum possible value). The specific equations for these relationships are provided by Reading *et al.* (1996). The values for *DC* and *CV* were combined:

$$V_2 = (DC \times CV)^{1/2}.$$

The predator index (V_3) reflected the impacts of feral predators at a release site. This index comprised two parameters: one reflecting the ability of managers to control feral animals (*PC*) and the other the distance of the release site from a major town (*DT*). It was assumed that densities of feral predators (especially Cats) would decline with increasing distance from human settlements (Reading et al., 1996). *PC* was given twice the weight of *DT*. The values for *PC* were ratings ranging from 1 for limited control ability to 5 for sites with the greatest opportunity to control feral predators. The rating scores for *DT* were: >10 kilometres = 5, 5–10 kilometres = 4, 2–5 kilometres = 3, 1–2 kilometres = 2, <1 kilometres = 1. In this case, an additive model was used and *PC* and *DT* were combined:

$$V_3 = (2 \times (PC/5)) + (DT/5))/3$$

V_4 was an index of the shape of a release site. Linear sites were considered to be more susceptible to increased predation by introduced carnivores. V_4 comprised a combination of two parameters: *AR* (axis ratio) was derived by dividing the greatest width by the greatest length, and *SS* (site shape) was an index of the area to perimeter ratio. V_4 was calculated as:

$$V_4 = (AR \times SS)^{1/2}$$

Finally, the security index (V_5) included other attributes that Reading *et al.* (1996) considered to be important for reintroduction sites, including risk of fire, land tenure status, level of support from the local community, on-site support from the land management agency, and availability of the site for reintroduction. Each of these attributes was rated on a scale of 1 (worst) to 5 (best).

Reading *et al.* (1996) then applied the HSI to three sites where reintroduction programs for the Eastern Barred Bandicoot had been undertaken: Gellibrand Hill Park, Hamilton Community Parklands and Mooramong Nature Reserve. Values for the subindices and final HSI for these three sites are shown in Table 15.2. As with all HSI exercises, the HSI for Eastern Barred Bandicoots was largely a synthesis of `subjective guesses and expert opinion' (Reading *et al.*, 1996). Reading *et al.* (1996) argued that the method is a relatively objective way of assessing the potential of sites for reintroduction. There are opportunities in Tasmania, where wild populations of the species are still extant (Watts, 1987; Mallick *et al.*, 1997) to independently test some of the hypothesised relationships between various subindices in the HSI model, for example the habitat structure (V_2) and predation (V_3) indices.

Table 15.2. Values for parameters and final habitat suitability indices for three reintroduction sites for the Eastern Barred Bandicoot in Victoria. (Modified from Reading *et al.*, 1996.)

Site	Gellibrand Hill Park	Hamilton Community Parklands	Mooramong Nature Reserve[1]
Size index (V_1)	0.60	0.20	0.32
Percentage cover (*CV*)	0.99	0.99	0.98
Distribution of cover (DC)	0.99	0.99	0.99
Habitat structure (V_2)	0.99	0.99	0.99
Predator control (*PC*)	5.0	5.0	4.0
Distance to town (*DT*)	1.0	1.0	4.0
Predator index (V_3)	0.73	0.73	0.80
Shape of site (*SS*)	0.56	0.67	0.61
Axis ratio (*AR*)	0.70	0.84	0.46
Shape index (V_4)	0.63	0.75	0.53
Fire potential	5.0	4.5	4.0
Land tenure	5.0	5.0	1.0
Local support	5.0	4.5	5.0
On-site personnel	5.0	3.5	4.0
On-site equipment	5.0	4.0	4.5
On-site support	5.0	3.75	4.25
Availability	5.0	5.0	5.0
Security index (V_5)	1.0	0.91	0.77
HSI value	0.77	0.63	0.63

[1]Does not include areas to be added to the nature reserve in the future (which would change the size (V_1) and shape (V_4) indices; see Reading *et al.*, 1996).

The process of testing and validating models has often been neglected (Block and Brennan, 1993).

Burgman *et al.* (2001) noted that because the HSI method is based on expert opinion, there is rarely empirical evidence linking habitat suitability to the demographic success of the species. Therefore, indices can be highly uncertain. Burgman *et al.* (2001) used fuzzy numbers to calculate bounds on habitat suitability. Explicit representation of uncertainty can take into account the shape of SI functions, their weights, and alternative ways of combining functions, thus improving the quality of decision-making (Burgman *et al.*, 2001).

Advantages of the HSI approach

On the positive side, HSI models are simple to construct and they make use of whatever knowledge is available. Experience and expert judgement are difficult to incorporate into more explicit statistical procedures, and often they are all that is available. HSI models assist scientists and resource managers to draw together the available information on a species. This can be done relatively quickly using HSI models, and the process can help to identify gaps in knowledge. Thus, the method can assist in the development of hypotheses associated with species–habitat relationships and the impacts of environmental changes, which can subsequently be tested more rigorously (Minta and Clark, 1989; Van Horne and Wiens, 1991). In other cases, an HSI-like approach can be used to generate variables for input into statistical habitat models. Catling and Burt (1995) and Catling *et al.* (2000) have constructed a structural complexity index for use in models of the distribution and abundance of mammals in south-eastern Australia. Lassau and Hochuli (2005) used an approach based on that of Catling *et al.* (2000) to demonstrate that wasp species diversity and abundance was higher in forests of high habitat complexity than where habitat complexity was low. This illustrates the valuable links between qualitative habitat approaches and the more quantitative statistical methods outlined in the following section.

15.3 Statistical habitat models

There are numerous statistical methods for examining habitat relationships. Here, we explore the use of two types of regression analysis that are particularly well suited to habitat studies: **logistic regression** and **Poisson regression**. They are part of a set of procedures called **generalised linear models** (McCullagh and Nelder, 1988; Crawley, 1993). These methods are effective for habitat analysis and they have been widely applied (e.g., Austin *et al.*, 1990; Nicholls, 1989, 1991; Lindenmayer *et al.*, 1999c).

Logistic regression

Often a species is recorded as either present or absent, with no indication of abundance. Vegetation quadrats,

Box 15.2

The advantages of representing uncertainty in HSI models

In HSI models, each variable is linked to a single suitability index by additive, multiplicative or logical functions, the forms of which are determined by expert judgment. Thus, HSI maps provide a means of integrating expert knowledge and creating habitat maps that synthesise subjective interpretations of biological processes. But they are susceptible to the full range of idiosyncrasies, and contextual and motivational biases that characterise expert opinions, as well as the natural variation that makes habitat maps uncertain (Burgman, 2004). However, experts can also provide estimates of the uncertainty surrounding the functional forms and numerical values. Burgman *et al.* (2001) elicited functions and uncertainty bounds for variables describing the habitat of the Florida Scrub Jay (e.g. Figure 15.1). They used the intervals implied by the bounds to calculate plausible bounds on sites that were considered for development within the species range (Figure 15.2). The advantage of representing the bounds in the calculations is that subjective uncertainties about ecological relationships are explicit. This process also provides decision-makers with additional information, so that they can account for their attitude to risk in reaching a decision. For instance, consider the circumstance in which you are asked to protect a single site, one of sites 4, 5, 6 or 7, and that an HSI value of less than 0.4 is unacceptable. You may then be inclined to be risk-averse and select site 5, because the lower plausible bound is greater than 0.4, even though sites 6 and 7 provide slightly higher expected values. Without uncertainty bounds, risk-based decisions are impossible.

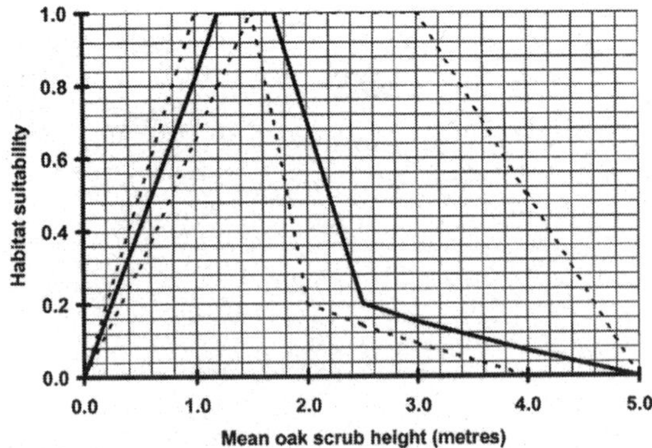

Figure 15.1. Uncertainty bounds for the suitability function linking Florida Scrub Jays to the mean height of Oak scrub. The heavy line is the suitability function; the dashed lines represent expert plausible limits.

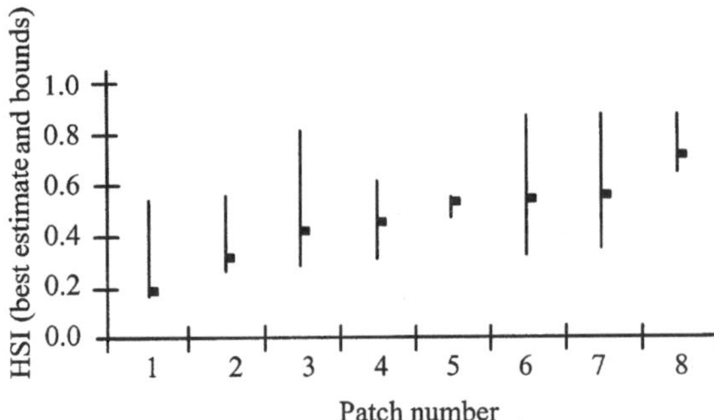

Figure 15.2. Habitat suitability and plausible bounds for eight sites considered for development within the range of the Florida Scrub Jay (after Burgman *et al.*, 2001).

mammal hair tubes, scat samples, spotlight surveys, frog and bird call recordings, and bird sighting transects are all methods that result in binary records (Sutherland, 1996). These data can be used together with habitat attributes to build a model of the relationship between a species and its environment. Because the dependent variable in the relationship is binary (the species is either present or not), logistic regression is used in place of ordinary regression (Collett, 1991).

Logistic regression models predict the probability of an event (the presence of a species) rather than the event itself (0 or 1). If we plot a continuous independent variable (for example altitude) against the response variable (presence/absence of the species), all of the observations fall at the extremes of the *y*-axis. For example, the presence or absence of the bryophyte *Riccardia cochleata* was recorded at 100 locations in the wet forests and rainforests of Central Victoria. Several environmental variables, including altitude, slope, aspect, soil type, and vegetation cover, were recorded at these same locations. In the example in Figure 15.3, the probability of presence of the species is plotted against altitude.

The regions of presence and absence of the plant on the *x*-axis overlap. There is no clear threshold for altitude below which the plant cannot survive. Its presence at any location can only be predicted imperfectly because of sampling error and because, in addition to environmental preferences, the chance events of births, deaths and dispersal of individuals determine where a species is found. Logistic regression fits an s-shaped (logistic) curve to the

Box 15.3

The variable use of the term `model'

The term `model' is employed widely in conservation biology and can have a range of quite different and sometimes confusing meanings. Cockburn (1991) described a useful `taxonomy' of types of models that are characterised by markedly different limitations, assumptions and knowledge.

Type 1: Statistical models

These types of models are equations or relationships derived from analyses of data that typically include a **response variable** (e.g. the presence or absence of a Helmeted Honeyeater at a site – see Box 15.4) and **explanatory variables** (e.g. measures of the vegetation at sites surveyed for Helmeted Honeyeaters).

Type 2: Data-free or theoretical models

Cockburn (1991) defines these types of models as ones in which biological and ecological phenomena are explored using mathematics. These models can entirely lack data and/or only be loosely underpinned by biological reality. However, they can provide new insights into the possible behaviour of ecological systems and open new avenues for other studies (Cockburn, 1991). A good example is the early and highly simplified theoretical model of metapopulation dynamics developed by Levins (1970), in which all patches were equidistant and were of identical size (see the section on metapopulations in Chapter 18). This theoretical model was a progenitor of more detailed and realistic models used in studies of metapopulations and investigations of the fate of populations in fragmented landscapes (e.g. Hanski and Gilpin, 1991; Hanski, 1999a,b).

Type 3: Data-rich models

In some circumstances, empirical data are available for particular ecological phenomena, and models are developed on the basis of a perceived understanding of these relationships. These types of models can be useful for a range of reasons (Starfield and Bleloch, 1992; Burgman *et al.*, 1993; Tyre *et al.*, 2001), including: (1) identifying the extent of congruence between existing knowledge and theory, (2) determining if there are additional insights that can be gained from existing knowledge (for example, such models may add value to field data that have been gathered; McCarthy *et al.*, 2000), and (3) examining the outcomes of interactions between many factors that may influence particular ecological phenomena but which would otherwise be very difficult to track (e.g. see Mackey *et al.*, 2002).

A model of the dynamics of the endangered remaining population of the Helmeted Honeyeater is an elegant example of a data-rich model (see McCarthy *et al.*, 1994). Here, the model was based on a very detailed understanding of the biology and ecology of the species, including its population size, breeding system, and habitat requirements (see Box 15.4). The model demonstrated the potential vulnerability of the population to demographic stochasticity. Furthermore, it provided new insights into factors that place the species at risk and highlighted ones that had previously been overlooked (McCarthy *et al.*, 1994).

Type 4: Empirical models

Empirical models are those that can be used to examine fundamental empirical questions. For example, Cockburn (1991) notes the value of the Fruit Fly as a useful empirical model from which to explore key issues in genetics.

Type 5: Conceptual models

We add conceptual models to Cockburn's (1991) list. These models are representations of ideas about how a system works. Conceptual models can be presented in natural language or as pictures, flow diagrams or logic trees. They form the starting point for developing statistical and theoretical models, and they should capture what a person believes to be true about the system in question, at a scale that suits the context and level of understanding.

---

observations (McCullagh and Nelder, 1988; Austin *et al.*, 1994b).

Logistic regression is a form of generalised linear model in which the binomial distribution is used to model the errors for presence–absence data. The 'link function' relates the (independent explanatory) linear predictors to the expected value (the predicted probability of occurrence). It transforms the mean probability into a linear combination of explanatory variables (Collett, 1991). The link

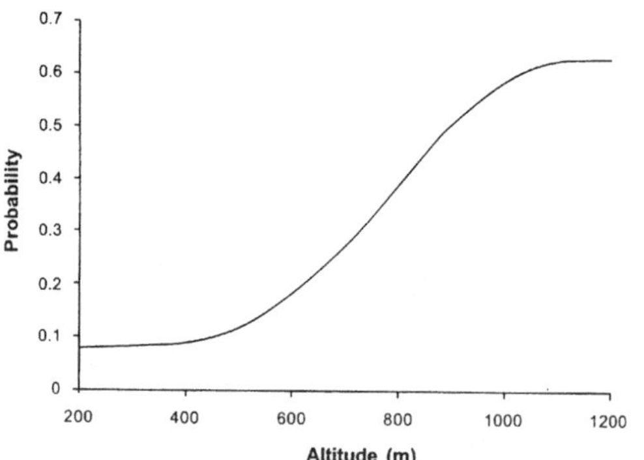

Figure 15.3. Relationship between altitude and the probability of presence of the bryophyte *Riccardia cochleata* (after Blanks, 1996).

function in Figure 15.3 is a logistic curve and it has the form:

$$p = \frac{e^z}{1+e^z} \tag{15.2}$$

where p is the probability that the species will be present, and z is a linear combination of the independent variable(s). The relationship between p and the independent variables takes the form:

$$z = \text{logit}(p) = \log\left(\frac{p}{1-p}\right) = b_0 + b_1x_1 + \ldots + b_nx_n. \tag{15.3}$$

where the x's are the observed, independent variables that may explain p. That is:

logit [expected probability (p) of event] = constant + effects of each variable + possible interactions between variables.

In the example in Figure 15.3 the function is:

$$z = 0.00358 \times \text{altitude} - 3.674$$

where altitude is measured in metres above sea level.

Explanatory variables can be continuous (e.g. daily temperature) or discrete categories (sometimes called

Table 15.3. Classification table for the logistic regression model in Figure 15.3.

	Predicted present	Predicted absent
Observed present	True positive 0.368	False negative 0.633
Observed absent	False positive 0.236	True negative 0.764

Box 15.4

Habitat requirements of the Helmeted Honeyeater

The Helmeted Honeyeater is a highly endangered species (Robinson, 1993) that is confined to an area of about 550 hectares of swamp forest at Yellingbo State Nature Reserve near Melbourne, Victoria (Menkhorst and Middleton, 1991; Pearce and Minchin, 2001). Adult Helmeted Honeyeaters weigh approximately 35 grams (Blakers *et al*., 1984). The Helmeted Honeyeater had a widespread distribution that formerly included parts of western Gippsland in Victoria, but its range contracted markedly as a result of the clearing of its habitat for agriculture and housing. A wildfire in 1983 eliminated one of two populations remaining at that time (Backhouse, 1987). The small (but increasing) remaining population at Yellingbo numbered about 100 birds in 2003 (Parks Victoria, personal communication). The species appears to suffer from competition with the Bell Miner (Woinarski and Sykes, 1983) and is at risk from processes that predispose small populations to decline (Menkhorst *et al*., 1997), including demographic stochasticity (McCarthy *et al*., 1994).

Pearce *et al*. (1994) completed a study of the structural habitat attributes influencing the distribution of the Helmeted Honeyeater within Swamp Gum forests where the species is now confined, and Ribbon Gum forests where it used to occur. They collected information on more than 20 attributes of habitat structure and floristics at 629 sites. The types of attributes measured represented features of the forest environment that are considered potentially important in the biology and ecology of the Helmeted Honeyeater.

Pearce *et al*. (1994) used logistic regression to model the habitat requirements of the Helmeted Honeyeater. The species was more likely to occur in places where there was standing water, a high density of Swamp Gum trees, and a large amount of loose bark hanging on the branches and trunks of trees. The last two of these variables reflected the suitability of the foraging substrate for the Helmeted Honeyeater, for example whether it contained insect microhabitat and surfaces from which to glean sugar-rich plant exudates such as manna. The work confirmed that controlling the Bell Miner could provide vacant habitat that would be suitable for colonisation by the Helmeted Honeyeater (Menkhorst *et al*., 1997).

factors; e.g. classes of rock types). The *b* values in Equations 15.2 and 15.3 are the coefficients (the weights) for the explanatory variables. Thus, the equations estimate the probability (p) of an event (e.g. the occurrence of a species) given estimates of explanatory variables.

Helmeted Honeyeater. (Photo by Davo Blair.)

Reliability measures for statistical models

If there is more than one potential explanatory variable, a researcher can add them in turn to the model, or exclude terms if they are not statistically significant – processes that are known as **stepwise regression** (Sokal and Rohlf, 1995). Alternatively, researchers can explore alternative models with a measure of fit, for example Akaike's information criterion, and choose the model that fits the data best. Strategies for building statistical models are a mixture of statistical expertise and biological understanding. A general rule, however, is that the number of potential explanatory variables should be substantially less than the number of sites from which data are gathered; this is to limit problems associated with over-fitting models and some predictors being significant simply by chance (see Burnham and Anderson, 1998).

In most circumstances, the most important feature of a good habitat model is that it can predict the quality of habitat reliably. A model can be evaluated by predicting the individual observations used to build the model. This method is easy but prone to be very optimistic. It is better to hold back some of the data from model-building and then use the model to predict these observations (called cross-validation). However, the best method of all is where a model based on all the data is used to make predictions at a scale relevant for management planning, and then new field data are collected to assess how frequently and in what circumstances the model falsely predicts presences and absences (Elith 2000). The proportion of observations for which a model correctly predicts the presence of a species is termed 'model sensitivity'. The proportion of observations for which a model correctly predicts the absence of a species is termed 'model specificity' (see Crawley, 1993; Elith and Burgman, 2003).

A classification table summarises predictions and observations. Each element of the table represents the number of correct and incorrect predictions (Table 15.3). The classification table is generated by summing the number of true positive, true negative, false positive

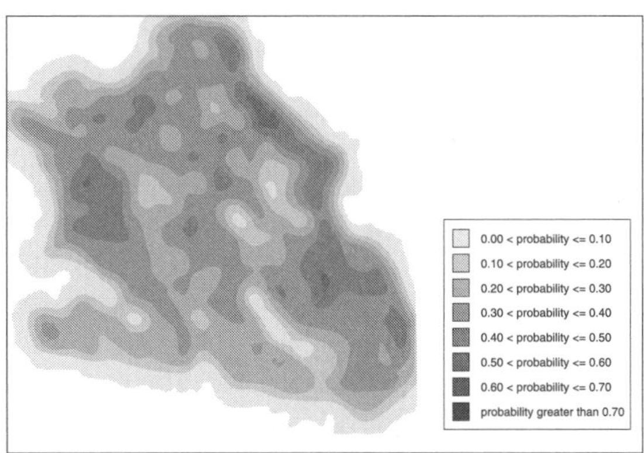

Figure 15.4. Map of the predicted probability of occurrence of the Greater Glider in the Ada Forest Block in the Central Highlands of Victoria (after Lindenmayer *et al.*, 1995b; note that the original maps included 95% confidence intervals around mean predictions). A spatial smoothing procedure was used to incorporate the spatial dependence properties of the data. The map is a product of this smoothing procedure (see Box 15.5).

Making a spatial prediction of potentially suitable habitat

It is possible to predict the form and magnitude of the response of a species to disturbance from the way in which management is likely to affect the independent variables of the model. With a map of the independent variables, it is possible to map the potential distribution and response of the species (e.g. Box 15.5; see also Gibson *et al.*, 2004).

Box 15.5

Spatial predictions for the Greater Glider

The Greater Glider is an arboreal marsupial that reaches a maximum weight of about 1400 grams and feeds almost exclusively on eucalypt leaves (Kavanagh and Lambert, 1990). The species is nocturnal and during the day it shelters within large trees with hollows. Adult animals are fairly sedentary and have a home range of 1–3 hectares. The range of the Greater Glider spans a wide range of eucalypt-dominated forest types from Victoria to northern Queensland (Winter, 1979; McKay, 1983). Lindenmayer *et al.* (1995b) developed a habitat model for the species and used it to predict its spatial distribution. Predictions were made for a small area in the southern part of the species' range: the 6700-hectare Ada Forest Block in the Central Highlands of Victoria.

A logistic regression model of the habitat requirements of the species was developed from the presence or absence of Greater Gliders at many different 3-hectare sites throughout the Central Highlands of Victoria, and the structure, plant composition and environmental conditions at the same sites. The fauna surveys and associated vegetation measures were confined to stands of montane Ash forest dominated by Mountain Ash, Alpine Ash or Shining Gum. Thus, the habitat model is valid only for these forest types. Of 30 different vegetation measures, only two were found to be significant: the age of the forest, and the abundance of trees with hollows. The equation for this habitat relationship was:

$$\text{logit}(p) = -0.993 + 1.106\ \text{AGE} + 0.554\ \text{HBT}$$

where AGE is a categorical variable that takes a value of 1 if the age of the forest stand is 90 years or older, and 0 otherwise. HBT is the abundance of trees with hollows on the survey site. The Greater Glider is more likely to occur on sites with numerous trees with hollows and where the forest is relatively mature (i.e. is more than 90 years old).

Given information on both the age of the forest and the abundance of hollow-bearing trees, it is possible to predict the probability of occurrence of the Greater Glider (p), together with an associated measure of uncertainty (i.e. the upper and lower values for 95% confidence interval; Figure 15.4). Probability values were calculated for each of the approximately 2000 3-hectare grid squares in the Ada Forest Block that were dominated by stands of Mountain Ash forest.

Predictions are affected by the quality of the habitat of the area surrounding a given site (Sokal and Oden, 1978; Ferrier, 1991), the presence of competitors or predators in these areas (Kavanagh, 1988), and metapopulation processes such as extinction–recolonisation dynamics. For example, animals may not persist in small and/or isolated habitat patches (Possingham *et al.*, 1994). The confidence intervals provide an estimate of the reliability of a prediction for a point in the landscape. Such predictions will be much more reliable when aggregated over larger areas. For example, we can be more certain of accurately predicting the number of animals in 100 cells than we can be of predicting the abundance of animals in any single one of those 100 cells.

Lindenmayer *et al.* (1994) tested the reliability of their model by making predictions for an area of habitat, and then recording the presence and absence of the species in that area with additional field work. They predicted that the species should occur at an average of 4.55 sites in the 20 they visited, with a 95% confidence interval of 2–7 sites. When they carried out the measurements, they found the animal at 4 sites.

---

and false negative predictions over the spectrum of predictions (for instance, the region over which management actions are planned and validation samples were collected). A model provides a perfect fit when there are no false predictions. It has no predictive value when there are approximately equal numbers of true and false predictions.

Poisson regression

Some species are ubiquitous, so that it is not possible to discriminate between sites where the taxon is present and where it absent. In these circumstances, the abundance of the species might be a more useful indicator of habitat suitability. Poisson regression is a form of generalised linear model in which the response variable is a count of the number of 'incidents' – for example the abundance of animals (Vincent and Haworth, 1983; Lindenmayer *et al.*, 1991a). A special form of Poisson regression, zero-inflated Poisson regression, deals with rare species that are absent from many sites (see Welsh *et al.*, 1996, 2000).

Ideally, habitat analyses measure habitat quality by survival and breeding success (Van Horne, 1983).

Greater Glider. (Photo by David Lindenmayer.)

Ada Forest Block, Central Victoria. (Photo by David Lindenmayer.)

Poisson regression habitat models assume that there is a relationship between habitat quality and the abundance of the species. The abundance of animals at a site may reflect social and demographic aspects of a species' life history (e.g. colonial behaviour or territoriality, which precludes other animals from inhabiting high quality areas; Van Horne, 1983). For example, the Sulphur-crested Cockatoo exhibits long-term site fidelity even when habitat quality has changed (Lamm and Calaby, 1970).

Density can be determined by non-breeding individuals congregating in suboptimal habitat (as is the case of the Australian Magpie; Carrick, 1963). Large populations of the predominantly male-dominated Kentucky Warbler are a consequence of heavy nest parasitism in forest fragments in the USA (Gibbs and Faaborg, 1990). Finally, the abundance of a species may reflect predators or competitors. The Lord Howe Island Woodhen only occurred in a single location where the impacts of feral Pigs were limited. Other places on the island were otherwise more favourable for the species (Miller and Mullette, 1985). Biological understanding should underpin all habitat models (Table 15.4).

Other forms of statistical models have different properties. Generalised linear mixed models (Schall, 1991; Searle *et al.*, 1992) and generalised additive models (Hastie and Tibshirani, 1990) have been useful in habitat modelling (e.g., Elith, 2000; Manning, 2004), but a detailed exploration of these methods is beyond

Sulphur-crested Cockatoo. (Photo by Andrew Claridge.)

Lord Howe Island Woodhen. (Photo by Karen Viggers.)

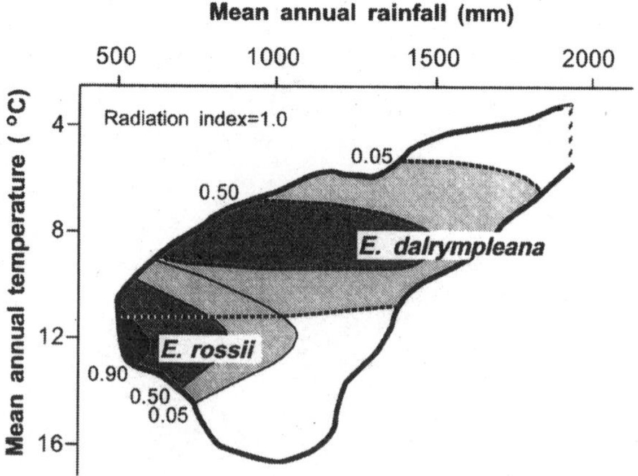

Figure 15.5. Environmental niche (expressed as a probability of occurrence in a site) for two Eucalyptus species in relation to mean annual temperature and mean rainfall (after Austin, 1985; Austin *et al.*, 1990). The outline represents the distribution of 1286 sample sites in relation to these two variables. Radiation was also measured, but in this example the index was held constant at 1.

the scope of this book. Statistical methods use presence–absence or count data that are not always available. Instead, often there are presence-only data. In some instances, the best solution may be to generate pseudo-absence data by sampling random locations (perhaps constrained to be within some distance of presence records; see Zaniewski *et al.*, 2002). Other methods, for example bioclimatic modelling (outlined in Section 15.4) and multivariate factor analysis (beyond the scope of this book, but see Carpenter *et al.*, 1993) are designed specifically to work with presence-only data (see Elith, 2000; Steinbauer *et al.*, 2002, for some examples).

Quantitative habitat analyses result in hypotheses that can subsequently be tested in the field (Block and Brennan, 1993). However, relationships between a dependent (response) variable and independent (explanatory) variables do *not* imply causality. There is no guarantee that the results of statistical analyses will correctly identify the important underlying ecological factors (Van Horne, 1983). Hence, as noted by Whelan *et al.* (2002; see Chapter 11), ecological studies can subsequently identify the specific ecological processes that give rise to the emergent ecological patterns of species distribution and abundance patterns (see Figure 15.5).

Finally, wildlife–habitat relationships do not provide information about how much habitat is needed to support a population in the long-term. Other tools such as population viability analysis (PVA) help with this type of question (see Chapter 18). Habitat information is essential for spatially-explicit forms of such models (e.g. McCarthy *et al.*, 2000; Lindenmayer *et al.*, 2003c).

Table 15.4. Some factors that need to be taken into consideration in the collection of field survey data designed for the construction of statistical models of the habitat requirements of animals.

Wide range of sites surveyed
Will enable a range of environmental variation to be captured.
Can help reduce the effects of other factors resulting in species absence (e.g. predation events).
Replication of survey sites
Assists in estimating random variation.
Randomisation
Essential to avoid potential bias.
Size of survey sites
Survey site dimensions should be matched to the scale of the animal movement/home range.
Selection of measured attributes
Appropriate variables should be measured for the target species.

Box 15.6

General guidelines for model building and some limitations

It is important to ensure that the assumptions associated with a method are satisfied. Assumptions are rarely tested, even though in many cases tests are available (e.g. Nicholls, 1989). In logistic regression, it is important to test for co-linearity (linear dependence between independent variables), interactions between variables, and the form of the relationship between the explanatory and response variables (e.g. whether it is linear or some other form; see Austin 1994a,b). Confidence limits on predictions should be reported routinely, and extrapolations of predictions beyond the extremes of the independent variables should be avoided. Attempts should be made not to overfit models by testing large numbers of potential explanatory variables (Manning, 2003). Statistical models can generate spurious relationships between dependent and independent variables. For example, a relationship can arise simply because a species occurs in places characterised by particular features (e.g. rocky outcrops), but the features may not necessarily be needed as a part of the habitat of that species.

A model with good explanatory power will not necessarily predict reliably. For example, in a study of the Rufous Treecreeper in Western Australia, Luck (2002b) found that a model of nest site use could not make robust predictions because of the large number of unused (but apparently suitable) nest sites. The best way to evaluate a model's predictive power is to test it with independent data (Lindenmayer *et al.*, 1994; Elith, 2000; Pearce and Ferrier, 2000; Luck, 2002b; Figure 15.6). A statistical model is more likely to have good predictive power if both statistical and biological understanding are part of model formulation.

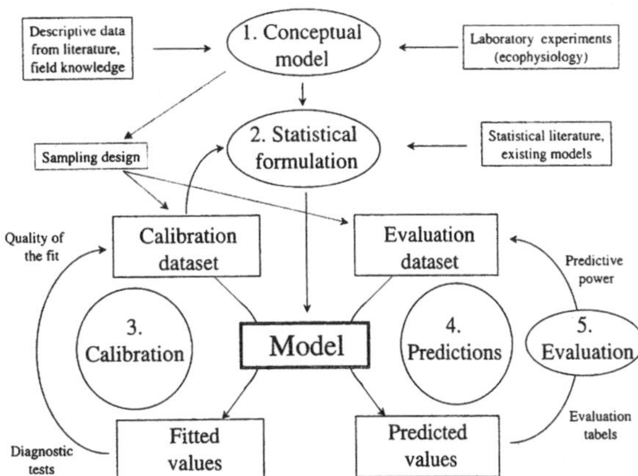

Figure 15.6. Model-building and testing in habitat requirements analyses. (Redrawn from Guisan and Zimmerman, 2000.)

Summary: statistical habitat modelling

Deriving an understanding of the habitat requirements of a species is a key part of conservation biology (Clark and Shutler, 1999). Statistical habitat modelling can involve links with qualitative approaches, for example HSI models, which can act as conceptual 'models' for variable selection and for generating testable hypotheses to explain occupancy patterns (Guisan and Zimmerman, 2000; Pearce and Ferrier, 2000; Tyre *et al.*, 2001; see Figure 15.6). Both generalised linear models and generalised additive models can result in reliable habitat models, but both require competent statistical modellers.

15.4 Envelopes and bioclimatic modelling

Many data sets comprise only presence-only information (Austin *et al.*, 1994a). For example, herbarium and museum records provide an indication of where a species was present, but no indication of places from which it was absent.

In the absence of any other information, the simplest way to map habitat is to draw a line around known occurrences. The **minimum convex polygon** (IUCN, 2001) encloses all observations so that no internal angle exceeds 180 degrees. However, this approach overestimates habitat even without spatial error or sampling bias (Burgman and Fox, 2003).

Alternatively, 'alpha hulls' can be constructed. To construct an alpha hull, triangulate edges (lines) between all known locations in such a way that no edges cross (known as a Delauney tessellation; Burgman and Fox, 2003), then break all edges that are greater than, for example, twice the average edge length. The remaining triangles provide a crude but unbiased approximation of the distribution of a species' habitat.

The broad limits of the distribution of most taxa are constrained by climatic factors such as temperature, rainfall and radiation (Jarvis and McNaughton, 1986; Woodward, 1987). Bioclimatic modelling predicts the bioclimatic limits of a species (Nix, 1986). The approach emphasises environmental regimes (e.g. temperature and water) that drive physical processes and biological responses (Nix, 1986; Mackey *et al.*, 1988; Lindenmayer *et al.*, 1996). The following section outlines how **bioclimatic modelling** works.

BIOCLIM and bioclimatic modelling

Bioclimatic modelling can be applied when presence-only location (latitude, longitude) and elevation data are available. The computer program BIOCLIM derives continent-wide estimates of a range of climatic parameters in the form of climatic 'surfaces'. They include long-term mean monthly precipitation and minimum and maximum temperature, and levels of variation in these attributes interpolated from a network of meteorological stations (Hutchinson, 1995, 1999). Climate variables are calculated using a spatially-referenced, regular grid of elevation points called a 'digital elevation model' (DEM; Hutchinson and Dowling, 1989; Hutchinson, 1989, 1995, 1999).

Bioclimatic models are sensitive to errors in latitude, longitude and elevation (Nix, 1986). BIOCLIM uses these data to generate a summary of the climatic conditions that characterise the locations where the species has been recorded, called a 'bioclimatic profile' (e.g. Mountain Brushtail Possum; Table 15.5).

The next step is to identify areas that experience similar climatic conditions to those of the bioclimatic profile (termed **homoclime** matching; Booth *et al.*, 1987). Any set of values in the bioclimatic profile can be used in this matching process (e.g. the 5–95% or minimum–

Table 15.5. Bioclimatic profile of the Mountain Brushtail Possum. (Redrawn from Fischer *et al.*, 2001.) The 5, 10, 50, 90, and 95 percentile values for each bioclimatic variable are given in the bioclimatic profile.

	Parameter	Mean	SD	Minimum	5%	10%	50%	90%	95%	Maximum
1	Annual mean temperature	11	2.62	5.1	7.3	8.2	10.6	14.2	17	20.3
2	Mean diurnal range (mean (monthly maximum – minimum))	10.6	1.32	7.1	8.3	8.7	10.8	12.1	12.5	14.5
3	Isothermality (2/7)	0.48	0.03	0.36	0.42	0.43	0.48	0.51	0.52	0.54
4	Temperature seasonality (CV)	1.5	0.13	1.07	1.28	1.34	1.5	1.7	1.76	1.89
5	Maximum temperature of warmest month	23.3	2.29	16.7	19.8	20.4	23.3	26.4	27.6	31
6	Minimum temperature of coldest month	1.2	2.22	−3.6	−2	−1.6	1	4.1	5.5	10.4
7	Temperature annual range (5 – 6)	22.1	1.79	16.3	19.2	19.8	22.2	24.6	25.3	28.8
8	Mean temperature of wettest quarter	8.7	5.46	−0.7	2.3	2.9	7	18.2	21.3	24.5
9	Mean temperature of driest quarter	14.8	2.48	3.6	10.2	12.1	15	17.8	18.3	20.5
10	Mean temperature of warmest quarter	16.2	2.41	11	12.8	13.4	15.9	19.1	21.6	24.5
11	Mean temperature of coldest quarter	5.7	2.72	−0.7	2.3	2.8	5.3	9.3	11.8	15.6
12	Annual precipitation	1324	299.89	596	847	906	1305	1707	1831	2838
13	Precipitation of wettest month	154	46.17	63	80	97	150	211	225	451
14	Precipitation of driest month	67	11.99	29	44	50	68	79	87	112
15	Precipitation seasonality (CV)	25	10.49	10	11	12	25	39	44	62
16	Precipitation of wettest quarter	432	130.4	182	221	260	420	595	621	1271
17	Precipitation of driest quarter	222	40.84	108	149	167	224	271	301	350
18	Precipitation of warmest quarter	264	106.38	138	149	160	247	385	532	1067
19	Precipitation of coldest quarter	382	130.58	121	184	214	368	571	599	839
20	Annual mean radiation	16	0.91	14.9	15	15.1	15.7	17.4	18.4	20.2
21	Highest month radiation	25.2	0.5	23.3	24.5	24.7	25.2	26.1	26.3	26.7
22	Lowest month radiation	7	1.44	6.1	6.1	6.2	6.5	9.2	11.2	13.5
23	Radiation seasonality (CV)	43	5.85	21	26	31	46	46	46	46
24	Radiation of wettest quarter	13.4	5.46	6.9	7.2	7.5	10.8	21.9	22.5	24.5
25	Radiation of driest quarter	20.8	3.04	7.2	13.8	15.9	21.2	23.1	23.5	25.6
26	Radiation of warmest quarter	22.8	1.29	20.7	21.2	21.3	22.7	24.6	25.1	25.7
27	Radiation of coldest quarter	8.3	1.58	7.3	7.4	7.4	7.8	11.2	12.8	15.3
28	Annual mean moisture index	0.9	0.08	0.39	0.75	0.79	0.92	0.99	0.99	1
29	Highest month moisture index	0.99	0.04	0.58	0.96	0.96	0.98	1	1	1
30	Lowest month moisture index	0.65	0.17	0.23	0.36	0.41	0.66	0.88	0.91	1
31	Moisture index seasonality (CV)	14	8.65	0	2	4	13	27	30	42
32	Mean moisture index of high quarter MI	0.99	0.04	0.55	0.95	0.96	0.98	1	1	1
33	Mean moisture index of low quarter MI	0.71	0.16	0.29	0.44	0.48	0.73	0.91	0.93	1
34	Mean moisture index of warm quarter MI	0.79	0.13	0.31	0.54	0.6	0.82	0.93	0.96	1
35	Mean moisture index of cold quarter MI	0.98	0.05	0.55	0.91	0.95	0.97	1	1	1

SD, standard deviation; CV, coefficient of variation; MI, moisture index. Temperature is measured in °C; precipitation in millimetres; radiation in megajoules per metre per year; moisture index in millimetres.

maximum values in Table 15.5). Maps of matched areas or homoclimes represent a species' predicted potential distribution, which is sometimes called its predicted bioclimatic domain (e.g. Figure 15.7) and represents the likely broad distributional limits of the species.

Within the bioclimatic domain, other factors, for example soil nutrients, competition and disturbance history, may be important. Thus, only part of the predicted bioclimatic domain may be occupied. For example, the actual distribution of the Mountain Pygmy Possum covers only a few square kilometres in the alpine regions of mainland south-eastern Australia (Broome and Mansergh, 1995; Heinze and Williams, 1998). However, using BIOCLIM, the predicted distribution is almost 100 times this area (Brereton *et al.*, 1995). Other factors, such as the distribution of suitable boulder fields, limit its distribution.

Many bioclimatic indices are not independent (Nix *et al.*, 1992; see Table 15.5). Bias in the range of points used to generate a bioclimatic prediction may bias the resulting envelope. This is particularly important if parts of the target species' range are not represented in distribution records (Sumner and Dickman, 1998; Claridge, 2002). In general, because of the way in which BIOCLIM is structured, analyses will usually result in a species being predicted to occur when it is in fact absent rather than vice versa (Austin *et al.*, 1994a).

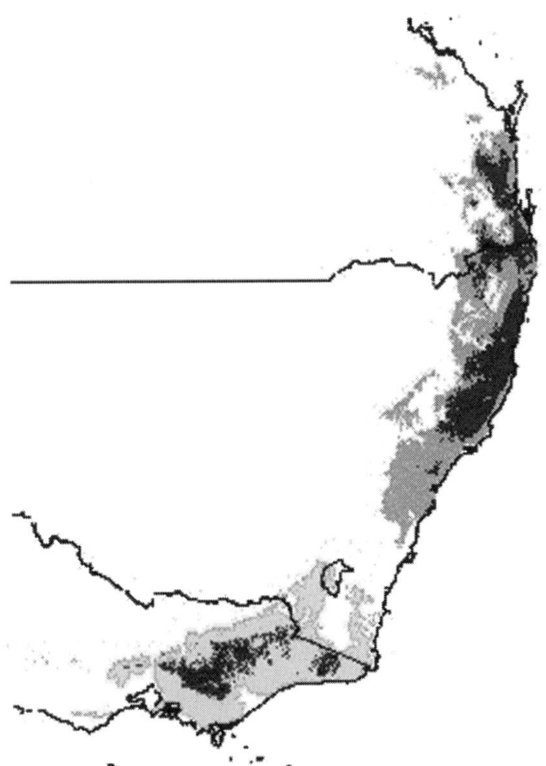

Figure 15.7. Predicted distribution of the Mountain Brushtail Possum. (Redrawn from Fischer *et al.*, 2001.) The separation of the bioclimatic domains highlighted the possible existence of two separate taxa, which was later confirmed by genetic and morphological analyses (see Box 2.4 in Chapter 2).

Applications of bioclimatic analyses

BIOCLIM has been used widely in Australia in studies of the distributions of vertebrates, invertebrates and plants (see Table 15.6). For example, Nix (1986) modelled the distributions of nearly 80 species of snakes, and Nix and Switzer (1991) modelled 60 rainforest vertebrates in north Queensland. Beaumont and Hughes (2002) completed the bioclimatic modelling of almost 80 species of butterflies. Bioclimatic modelling has been used to predict changes in distribution patterns in response to global warming (see Chapter 5; Busby, 1988; Bennett *et al.*, 1991; Brereton *et al.*, 1995; Beaumont and Hughes, 2002; Pearson and Dawson, 2003).

There are a number of examples of good concordance between the predicted and actual distributions of a species (e.g., Williams, 1991; Law, 1994). However, predictions of the potential bioclimatic domain of the endangered Long-footed Potoroo (Busby, 1988) did not include areas near Mount Buffalo in north-central Victoria, where the species is now known to occur (Menkhorst and Knight, 2001). The species was long considered to be confined to East Gippsland in north-eastern Victoria and possibly also in south-eastern New South Wales. Re-analyses of the distribution of this species using BIOCLIM have been used to better direct survey effort (Claridge, 2002).

Climate estimates from BIOCLIM have been used to develop bio-environmental classifications to support reserve design and environmental management (e.g. Austin *et al.*, 1990; Nix *et al.*, 2000; Mackey *et al.*, 2001), termed **environmental domain** analysis (see Chapter 16 for further details). Other attributes of the physical environment, for example the substrate (the geology and accompanying soils) can be incorporated in such analyses. For example, Richards *et al.* (1990) identified 12 environmental provinces in south-eastern New South Wales and north-eastern Victoria based on climate, terrain and geology. These classifications can be used to stratify the environment for subsequent field

Table 15.6. Australian examples of bioclimatic modelling.

Examples	Reference
Plants	
Various eucalypts	Booth (1985), Booth *et al.* (1987), Booth *et al.* (1988), Booth and Pyror (1991), Austin *et al.* (1990), Austin *et al.* (1994b), Lindenmayer *et al.* (1996)
C3 and C4 grasses	Prendergast and Hattersley (1985)
Myrtle Beech	Busby (1986), Busby (1988)
Alpine vegetation	Busby (1988)
Skeleton Weed	Panetta and Dodd (1987)
Temperate rainforest	Hill *et al.* (1988)
Wadbillga Ash	Prober and Austin (1990)
Snow Gum	Williams (1991)
Various rainforest plants	Mackey (1994)
Swamp Mahogany	Chapman and Busby (1994)
Species of orchids	Peakall *et al.* (2002)
Fungi	
Various species	Steinbauer *et al.* (2002)
Invertebrates	
Caledia captiva (grasshopper)	Kohlmann *et al.* (1988)
Giant Gippsland Earthworm	Brereton *et al.* (1995)
Altona Skipper Butterfly	Brereton *et al.* (1995)
Various species of 'true bugs' (Heteroptera)	Steinbauer *et al.* (2002)
Various species of butterflies	Beaumont and Hughes (2002), Thomas *et al.* (2004)
Fish	
Roman-nosed Goby	Nix and Switzer (1991)
Cairns Rainbowfish	Nix and Switzer (1991)
Lake Eacham Rainbowfish	Nix and Switzer (1991)
Reptiles and amphibians	
Bynoe's Gecko	Kearney *et al.* (2003)
Roseate Frog species complex	Wardell-Johnson and Roberts (1993)
Various elapid snakes	Nix (1986)
Various lizards and frogs	Nix and Switzer (1991), Brereton *et al.* (1995)
Birds	
Helmeted Honeyeater	Pearce and Lindenmayer (1998)
Malleefowl	Chapman and Busby (1994)
Common Myna	Martin (1996)
Various species	Nix and Switzer (1991), Brereton *et al.* (1995)
Mammals	
Antechinus spp.	Sumner and Dickman (1998)
Long-footed Potoroo	Busby (1988), Claridge (2002)
Antilopine Wallaroo	Busby (1988)
Leadbeater's Possum	Lindenmayer *et al.* (1991b)
Mountain Brushtail Possum	Fischer *et al.* (2001)
Squirrel Glider	Menkhorst *et al.* (1988)
Yellow-bellied Glider	Norton and Nix (1991)
Mahogany Glider	Van Dyck (1993), Jackson and Claridge (1999)
Common Blossom Bat	Law (1994)
Golden-tipped Bat	Walton *et al.* (1992)
Various species	Bennett *et al.* (1991), Nix and Switzer (1991), Brereton *et al.* (1995)

surveys (e.g. Nix and Gillison, 1985; Austin and Heyligers, 1989), or to determine the representativeness of areas assigned to different land tenures and/or land uses (e.g. reserves; Nix, 1997; Mackey *et al.*, 2001, see Chapter 16).

Mackey *et al.* (2002) used bioclimatic variables to model the height of Mountain Ash, Alpine Ash and Shining Gum trees in the Central Highlands of Victoria (Figure 15.8). Bioclimatic variables have been used to identify factors influencing the integrity of bushland remnants in Tasmania (Gilfedder and Kirkpatrick, 1999) and to explain the separation of genetically different races of Bynoe's Gecko (Kearney *et al.*, 2003).

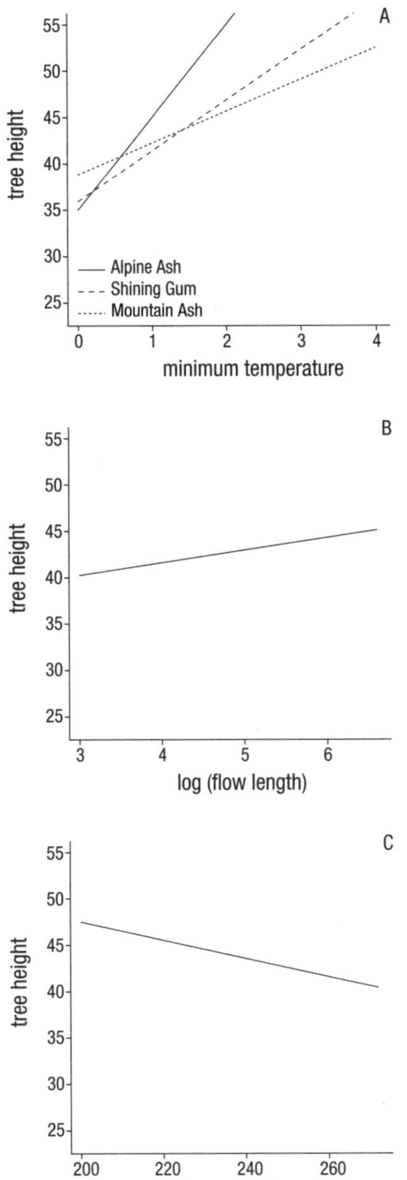

Figure 15.8. Relationships between tree height for Mountain Ash, Alpine Ash and Shining Gum and attributes derived from BIOCLIM. (A) Differences between tree species in relation in minimum temperature. (B,C) Mean tree height value (averaged for Mountain Ash, Alpine Ash and Shining Gum) in relation to the logarithm of stream flow (a measure of topographic wetness) and rainfall in the wettest quarter (winter).

15.5 Conclusions

Many techniques are available to examine the distributions of plants and animals and explore their habitat requirements. Each has strengths, limitations and

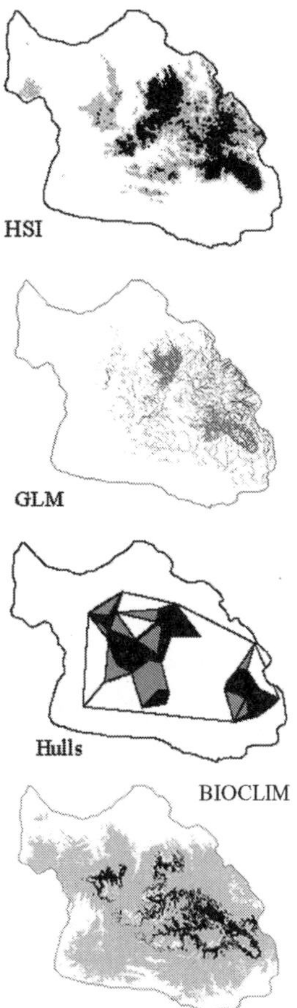

Figure 15.9. Habitat models for *Leptospermum* spp. in Victoria, based on HSI, logistic regression, convex hull/alpha hull and a BIOCLIM model (after Elith, 2000). Dark and light grey represent two levels of resolution for each model.

assumptions. No approach is better than all others in all circumstances. In many circumstances, different methods will generate different predictions (e.g. Figure 15.9). The usefulness of an approach will be related to factors such as the questions being addressed, the nature of the available data, and the availability of technical skills to implement them. Lack of data and appropriate skills limit the application of habitat modelling techniques for most species.

At best, rigorous statistical models will be available for only a handful of species. Qualitative models can be valuable in circumstances in which there is little more than expert judgement. For circumstances in which it is important to look at broadscale pattern, bioclimatic modelling may be the most appropriate tool.

Typically, models are developed from field data and are then used to predict the abundance and distribution of species in new locations. Such uses include the identification of release sites for translocation programs, and the prediction of the spatial distribution of habitat for conservation reserve design. In practice, it is often assumed that the modelled relationships are important causal determinants of species abundances, without testing the performance of the model, which is actually one of the most important steps in the modelling process.

15.6 Practical considerations

Variables that are directly related to the habitat requirements of species will predict future occurrences better than will variables that are indirectly related (Austin, 1994a). Different methods for modelling habitat are not mutually exclusive; they can be used together to improve the understanding of habitat relationships. For example, qualitative methods (e.g. habitat suitability models) can be used to generate hypotheses to stratify the environment to guide field sampling. Presence data can be used for models but lack of absence records limits their usefulness. As data accumulate and conceptual models for a species habitat improve, there may be sufficient presence–absence data to construct reliable statistical models. At each step in the evolution of models, the best solution will be achieved by combining ecological understanding with the methods for which technical skills are available and data are appropriate. In all cases, the predictive value of the habitat model can be assessed most reliably with new field data, guided by the model's predictions and collected at a scale that makes sense for the management context at hand.

15.7 Further reading

The seminal early work on habitat requirements was by Thomas (1979). A useful reference on habitat is Morrison *et al.* (1992). Guisan and Zimmerman (2000) and Rushton *et al.* (2004) have reviewed statistical methods for habitat studies.

A useful reference that contains a critical appraisal of the HSI model-building procedure is Van Horne and Wiens (1991). An excellent application of the technique for identifying potential reintroduction sites for the highly endangered Black Footed Ferret in Montana, USA, is provided by Minta and Clark (1989). Burgman *et al.* (2000) outline the use of a fuzzy numbers approach for improving decision-making based on HSI models. Burgman and Fox (2003) describe how to construct alpha hulls. Elith (2000) outlines many methods and provides worked examples of model-building and validation for Australian plants.

Reference texts on generalised linear models have been written by McCullagh and Nelder (1988), Collett (1991), and Crawley (1993), and they provide a detailed account of this class of statistical model. Worked examples where the approach is well explained include Austin *et al.* (1990) and Nicholls (1989, 1991). Austin *et al.* (1994a) and Guisan and Zimmerman (2000) provide an outline and appraisal of some of the methods described in this chapter. Nicholls (1991) contains a valuable demonstration of the use of generalised linear models in modelling plant and animal distributions. Burnham and Anderson (1998) is an excellent discussion of statistical modelling, model selection and inference. Luck (2002a,b) provides an elegant set of studies of modelling the habitat of the Rufous Treecreeper and a subsequent test of model performance. Receiver operating characteristic (ROC) curves provide a useful and convenient way of summarising the predictive performance of habitat models (see Elith, 2000; Swets *et al.* 2000; Mackey *et al.*, 2002).

There is a rapidly developing and expanding group of techniques that can be used to examine relationships between species' distributions and environmental factors. They encompass approaches such as decision trees (Stockwell *et al.*, 1990), neural networks (Skidmore *et al.*, 1994), and genetic algorithms (Lees, 1994). These approaches are still in the relatively early developmental stage and doubts remain about the assumptions and limitations that underpin them, as well as how robust and ecologically appropriate they are for examining species distributions. Guisan and Zimmerman (2000) examine some of these (and many other) methods for use in habitat-related applications. In volume 41, issue 2 of the Journal of Applied Ecology, some new approaches to statistical modelling of habitat and species distribution patterns are discussed (e.g. Gibson *et al.*, 2004; Rushton *et al.*, 2004).

Detailed discussions of the application of BIOCLIM are presented in Nix (1986), Lindenmayer *et al.* (1996) and Mackey *et al.* (2002). The background to the mathematical approaches for deriving the climate surfaces that underpin BIOCLIM is given by Hutchinson (1995, 1999).

Chapter 16
Reserve design

This chapter describes some of the strengths and weaknesses of approaches to designing nature reserves, including some of the ecological and environmental parameters that can be used as surrogates for biodiversity. The limitations of the island biogeography theory for reserve design are also discussed because it has been such a strong focus for much of conservation biology in the past.

16.0 Introduction

Reserves are a critical component of all credible conservation strategies (Lindenmayer and Franklin, 2002). They are important particularly because of the pressures exerted by the human population on the natural environment (Soulé, 1991; McNeely, 1994b; Margules and Pressey, 2000; Struhsaker *et al.*, 2005). Reserve design research focuses on questions such as: How big should a reserve be and what is the best shape? How should reserves be spatially arranged and connected? Where should reserves be located? How should the surrounding landscape be managed to best ensure the persistence of species or communities?

16.1 *Ad hoc* developments

Reserve systems have developed in an *ad hoc* way in Australia (Pressey, 1995) and elsewhere (Terborgh, 1992; Noss and Cooperrider, 1994; Khan *et al.* 1997; Margules and Pressey, 2000). Often areas became reserves because they were not valued for other types of land use, for example agriculture, grazing, forestry or urban development, or because they supported places of spectacular scenic beauty and aesthetic interest (Pressey, 1995; Mendel and Kirkpatrick, 2002; see Chapter 4). An *ad hoc* approach to adding protected areas to a reserve system results in some species, communities and ecosystems remaining unprotected (see Pressey, 1994a; Pressey and Tully, 1994). New additions to fill gaps may be relatively expensive (Chapter 4). The political nature of land-use allocation means that future attempts to add new areas (of high conservation value) to the reserve system may be compromised by past acquisitions of less suitable land or land that was already well represented in the reserve system. Even when reserve design strategies seek to represent all major communities, ecosystems, and rare or threatened plants and animals, the proportion and total amount of each species or community within protected areas is often determined on an *ad hoc* basis. Priorities should be determined by considering:

- how much of a community remains in a relatively undisturbed state
- the resilience of a species or community to disturbances it is likely to face outside the reserve system
- how much area or how large a population is available for inclusion in a reserve
- the range of non-target communities that would be conserved within an area
- the long-term viability of populations within prospective areas.

For instance, larger percentages of conservation reserves should be allocated to communities that are rare, vulnerable to disturbance, or those with high biodiversity values (Kirkpatrick and Brown, 1991). Analytical methods provide an alternative to *ad hoc* reservation.

16.2 CAR reserve system design principles

If conservation biology were the only factor influencing land allocation, reserve design would depend solely on ecological considerations such as:

- whether the aim is to conserve a particular species, the maximum number of species possible, or to protect complete or representative sets of taxa associated with a given vegetation community
- the viability of populations in any reserve or set of reserves
- threats to the integrity of reserves resulting from activities in surrounding unreserved areas.

In most cases, the objectives of a reserve system are broader than the conservation of an individual taxon or assemblage (Austin and Margules, 1986). Comprehensive, adequate, and representative (CAR) reserves (Dickson *et al.*, 1997) attempt to protect the full array of biodiversity of a region. Comprehensiveness refers to the need to include the complete array of biodiversity, ranging from species (and their associated genetic variation) to communities and ecosystems. Adequacy relates to the need to support populations that are viable in the long term. Representativeness means that a reserve system should sample species, forest types, communities, and ecosystems from throughout their geographic ranges (Pressey and Tully, 1994; Anon., 1995a; Burgman and Lindenmayer, 1998).

Lindenmayer and Franklin (2002) argue that 'replication' could be added to the CAR principles to make reserves comprehensive, adequate, representative and replicated (CARR). Replication refers to the need to protect multiple areas of a given vegetation type, forest community or species, to limit the risk that all reserved values could be affected by a single catastrophic event, such as a wildfire (Pickett and Thompson, 1978; Lindenmayer and Possingham, 1994).

The need for replication is influenced by the size of reserves and whether they exceed the maximum size of a single disturbance event. The International Union for the Conservation of Nature (IUCN, 1994, 2001) recognised the relationship between protected areas and catastrophic disturbance when they defined a 'location' as an area that is likely to be affected by a single disturbance. The number of locations is one of the measures that contribute to species conservation status assessments. Replication may be unnecessary if biodiversity values or their surrogates (e.g. a forest type or age cohort) are sufficiently well represented such that there is a high probability that part of their range will be unaffected after a disturbance, so that propagules or offspring can recolonise disturbed areas.

Replication may better protect taxa that might otherwise be lost from a single reserve because of species turnover (localised extinction; Margules *et al.*, 1994a) following climate change or community succession (Peters and Lovejoy, 1992; see Chapter 5).

16.3 Reserve design and biodiversity surrogate schemes

Types of surrogates

The full spectrum of biodiversity will never be fully documented. There are too many species, too few of them are described, and too little is known about their distributions, abundances and ecological dependencies. A few opportunistic taxonomic collections, or data on climate, terrain and topography, landforms, soils, or vegetation may be all that is available to guide reserve design. Biological surveys cannot be used to assess all kinds and levels of biodiversity because taxonomy is uncertain (see Chapter 2), surveys are expensive, and ecosystems change over time.

CARR reserve design will often depend **on biodiversity surrogates** that are thought to indicate the distribution and abundance of unmeasured species, assemblages, or other elements of biodiversity (Hunter, 1994). Surrogates can include:

- records of taxa from biological surveys and opportunistic records in museums and herbaria (Nix *et al.*, 2000)
- plant communities (Brown and Hickey, 1990; Specht *et al.*, 1995; Mendel and Kirkpatrick, 2002)
- climate parameters (e.g. rainfall)

- eco-regional maps (Sims *et al.*, 1995), or environmental or climatic domains (Mackey *et al.*, 1988, 2001; Richards *et al.*, 1990; Chapter 15).

Biodiversity surrogates in a landscape can be classified and mapped, providing a spatial data set on which reserve selection procedures can be applied. Box 16.1 provides an example.

Surrogates have been used routinely to predict the distribution of the Australian biota. Braithwaite (1984) used vegetation and site geology to predict the density of arboreal mammals. Nix (1986) used bioclimatic indices to predict the distribution of snakes (Chapter 15). Margules and Stein (1989) used rainfall, temperature and substrate to predict the likelihood of occurrence of 32 tree species. Cork *et al.* (1990) used rainfall, temperature, substrate and nutritional requirements to predict the distribution and abundance of the Koala. The following section outlines some common surrogates.

Box 16.1

Reserve selection and historical levels of vegetation cover

One aspect of a comprehensive, adequate, representative and replicated (CARR) reserve system involves comparisons between current levels of biodiversity or biodiversity surrogates and those at some time in the past. For example, reservation targets of 10–15% for each biome (Scientific Advisory Group, 1995) need to be based on the original extent of those biomes. In the case of Australian forest reserve planning, this has entailed comparisons with the modelled pre-1750 distribution of particular forest types (Anonymous, 1995a; Dickson *et al.*, 1997; Austin *et al.*, 2000). This period was immediately prior to the arrival of Europeans in Australia and the concomitant onset of widespread vegetation clearing and landscape change (Walker *et al.*, 1993). A similar approach was used in New Zealand (Awimbo *et al.*, 1996).

Environmental domains

When information is scarce, a simple strategy is to conserve areas that incorporate a range of environmental variation – a useful technique when direct information about the distribution and abundance of species is lacking (Faith and Walker, 1993; Nix, 1997). **Environmental domains** are geographic regions that enclose a continuous range of physical environmental parameters that are expected to be important in determining the distributions of species (see the section on bioclimatic analysis in Chapter 15). They can be constructed from a digital elevation model of the landscape, and soils, slope, aspect, radiation indices, and precipitation can also be included. For example, Mackey *et al.* (1988, 1989) classified the World Heritage Wet Tropics area in north Queensland and examined how completely the various environments were sampled in proposed World Heritage areas and reserves. Environmental domains were also used to examine potential areas for reserves on the south coast of New South Wales (Richards *et al.*, 1990) and for identifying priority areas for biodiversity conservation in Papua New Guinea (Nix *et al.*, 2000).

Vegetation maps

Vegetation maps are usually part of reserve design procedures (e.g. Mendel and Kirkpatrick, 2002). Much of the vegetation of the Australian continent has been classified and mapped (AUSLIG, 1990; Specht *et al.*, 1995; Specht and Specht, 1999). In most cases, vegetation mapping involves interpretation of aerial photographs, stratification of the landscape into relatively uniform units, ground sampling for vegetation structure and floristic composition (usually for vascular plants), classification of ground samples, description of map units, and, finally, detailed mapping of each of the vegetation types (see Chapter 4).

Some classifications rely on the structural composition of the tallest stratum (Specht, 1970; Table 2.6), whereas others use the floristic composition or the dominant tree species. Forest type maps have been developed for most of the productive forests of Australia; these are based on the distribution of commercial tree species, together with their growth stage and condition (Jacobs, 1955; see Chapter 2). Vegetation maps may combine floristic units into larger associations that reflect landscape morphology, vegetation structural attributes and/or disturbance regimes (e.g. the Ecological Vegetation Classes of Victoria; Woodgate *et al.*, 1994; Mac Nally *et al.*, 2002).

Centres of diversity

Often, priority for protection is given to areas that are species rich for a particular taxon – so-called 'hotspots'. The International Union for the Conservation of

Nature has defined 'Centres of Plant Diversity' (see WCMC, 1992; Chapter 18) as places that are particularly rich in plant life, which would, if protected, safeguard the majority of wild plants in the world. The intention is to include geographically defined regions or vegetation types with high species diversity and/or endemism in the scheme. Sites can be included if they harbour an important gene pool that is valuable or potentially useful to humans, a diverse range of habitat types, or a significant proportion of species adapted to special edaphic conditions. Sites under imminent threat of large-scale destruction are given priority.

The south-west botanical province of Western Australia is a Centre of Plant Biodiversity, even though the total number of species is not known precisely (Crisp *et al.*, 2001). The region is species-rich, the flora is highly endemic, and many species are highly specialised and restricted to unique edaphic conditions or geomorphological features (Hopper and Wardell Johnson, 2004).

Potential limitations of surrogates

Investigations into the reliability of surrogates as proxies for biodiversity values have been equivocal. Kerr (1997) found that a reserve system established on the basis of mammal diversity would protect invertebrates poorly. Kirkpatrick and Brown (1994) found good correspondence between areas selected on the basis of environmental domains and plant communities in Tasmania, although many rare species missed selection.

Vegetation types are perhaps the most commonly employed biodiversity surrogate; however, the strengths of association between faunal distributions and vegetation patterns are inconsistent. Forest type was a good surrogate for bryophytes in New South Wales (Pharo *et al.*, 2000). Mac Nally *et al.* (2002) found that the distribution and abundance of reptiles and invertebrates were independent of vegetation type. Some widespread dominant tree species occur across a wide range of environments, and species assemblages vary between these different environments (Hunter, 1991). Some plants, invertebrates and small mammals are closely associated with edaphic variables or understorey plants (Lyons *et al.*, 1974; Gullan and Robinson, 1980; Woinarski and Cullen, 1984). In these cases, simply relying on a vegetation map to identify places for protection may miss important areas. The Australian environment is characterised by variable environments and disturbance regimes (Attiwill, 1994b; Bradstock *et al.*, 2002). Some species depend on particular successional stages within a vegetation community (e.g. the old growth stage of a particular type of forest; see Chapter 2). The biota is temporally and spatially dynamic (Whelan *et al.*, 2002), and any conservation strategy must take into account the contingencies of the existing disturbance regime.

Relationships between elements of the biota may not be symmetrical: Pharo *et al.* (2000) cite the case where reserves selected for vascular plants capture large numbers of bryophyte species, but sites important for bryophytes may be relatively unimportant for vascular plants.

The strategy of reserving hotspots protects relatively large numbers of potentially threatened taxa efficiently (Kati *et al.*, 2004). However, it is not a panacea, because threatened species occur outside species-rich areas. Sites of endemism and richness for different taxa do not usually co-occur (Prendergast *et al.*, 1993). For example, areas of tropical rainforests in northern Queensland are renowned for their extraordinary plant and bird diversity, and these areas now have World Heritage status and are well protected. However, areas of wet sclerophyll forest that border the margins of tropical rainforest are not protected, but they support endemic bat and ant faunas, and are probably critical for the long-term conservation of a range of species, including the northern subspecies of the Yellow-bellied Glider (Harrington and Sanderson, 1994; see Chapter 11).

It is possible that apparent hotspots may simply reflect sampling effort (Connor and McCoy, 1979). Estimates of the number of species at a given location are sensitive to the type and extent of sampling (Chapter 14), so any consideration of hotspots must account for variation in sampling intensity.

The need to test surrogates

For most applications of surrogates in ecology and conservation biology, the reliability of one measure as a surrogate for another is unknown (Pressey, 1994b; Wardell-Johnson and Horowitz, 1996; Gustafsson *et al.*, 1999; Lindenmayer *et al.*, 2000c). The validity of protected-area conservation effort in Australia depends on the veracity of assumptions about surrogates, which creates an imperative to validate these assumptions. Unfortunately, few such validation exercises are undertaken. A notable exception is the work of Mac Nally *et al.* (2002), who found that ecological vegetation classes within the Box–Ironbark forests accommodated birds, mammals and trees reasonably well, but were poor surrogates for reptiles and invertebrates. In the absence of validation of predictions, we don't know how well reserves achieve conservation goals.

Given that the validity of biodiversity surrogate schemes remains largely untested, some authors (e.g. Nix *et al.*, 2000, p. 11) have recommended that: '*A sound strategy is to adopt as many...surrogates as possible to maximise the likelihood of representing more of the biodiversity in priority areas*'. Nix *et al.* (2000) used four biodiversity surrogate schemes to set conservation priorities in Papua New Guinea, including environmental domains, vegetation types, species distribution models based on bioclimatic profiles, and records of rare and threatened species.

16.4 Reserve selection

A traditional goal of reserve planning is to represent the full array of biodiversity of a region. Existing protected area networks typically contain an arbitrarily biased sample of biodiversity (Pressey, 1995; see Chapter 4). Reserve selection algorithms (Kirkpatrick, 1983; Margules *et al.*, 1995b; Pressey, 1997) and gap analysis (Scott *et al.*, 1993; Burke, 2000) can be used to resolve a lack of representativeness (e.g. Williams *et al.*, 1996; Pressey, 1997).

Gap analysis integrates information on biodiversity, land tenure and management regimes to identify biodiversity surrogates that are poorly represented in the existing reserve system (e.g. Burke 2000). Reserve selection algorithms search for candidates for inclusion in a reserve system to optimise conservation goals. Reserve selection procedures are typically implemented in a stepwise fashion (after Margules and Pressey, 2000):

1. Compile data on biodiversity in the region.
2. Identify conservation goals for the region.
3. Review the existing reserve system.
4. Identify additional reserves (using reserve selection or gap analysis techniques) that are required.
5. Implement conservation actions.
6. Maintain (manage) the specified values on the reserves.

Gap analysis and reserve selection algorithms can also be used in combination (Pressey and Cowling, 2001). Three principles govern most reserve selection protocols: complementarity, flexibility and irreplaceability (Pressey *et al.*, 1993).

Complementarity is the degree to which an area adds previously under-represented features (e.g. species, land units, ecosystem types, or environmental climatic domains) to a reserve system. Most algorithms operate by identifying new reserves to be added to a set of protected areas until all species or units are

Box 16.2

An application of a reserve selection algorithm

Nicholls and Margules (1993) and Margules *et al.* (1995b) used a reserve selection algorithm to plan conservation objectives for the forests of southern New South Wales (Figure 16.1, Figure 16.2). The area was approximately 3800 square kilometres in size and the study aimed to identify the minimum area required to sample 10% of the 26 environmental domains that occur in the area as well as 10% of the area occupied by each of 31 communities of forest trees. A grid of 1-square-kilometre pixels was overlaid on the study area and an algorithm was used to achieve the reservation goal of sampling 10% of the area of the various environmental domains and forest communities. The algorithm was to:

1. Identify a subset of grid cells that sample 10% of the area of each of the 36 environmental domains.
2. Determine the extent to which the 31 different forest communities have been sampled by the procedure in Step 1.
3. Add more grid cells until 10% of the area of each forest community has been sampled.

A total of 34 grid cells achieved the 10% sampling target for the environmental domains. A further three cells were required to sample the tree communities to the same level. The 37 grid cells were equivalent to 8.7% of the entire study area, and some of the grid cells could be substituted for others in the region without compromising the reservation goals.

Margules *et al.* (1995b) compared the distribution of the 37 grid cells with current land tenure. They found that many of the grid cells (and thus particular tree species and environmental conditions) were not represented in the national park system as it existed at that time.

In this example, representation corresponded to the presence or absence of a unit within a map cell. Additional considerations may include the amount of land or habitat required to support viable populations, the habitat quality of selected areas, the location of selected cells, and the management implications of various options.

represented, or are represented a number of times (Kirkpatrick, 1983; Margules *et al.*, 1988).

It may be possible to develop a representative network of reserves from different combinations of areas – the principle of flexibility. Different areas can be substituted in a reserve system if they contribute the same conservation values (Pressey *et al.*, 1993).

The concept of irreplaceability has two meanings in reserve selection: (1) the degree to which an area is essential to achieving a completely representative reserve system, and (2) the contribution a given area makes to representativeness (Pressey *et al.*, 1993). Irreplaceability provides a means to explore planning options.

Efficiency is a measure of the area (or cost) required to achieve conservation goals. An efficient solution is one that achieves conservation goals (e.g. adequate populations of all threatened species) in a relatively small area.

The principles of complementarity, flexibility, irreplaceability and efficiency can be applied to any type of biodiversity surrogate: species, species assemblages, vegetation classes, ecosystem types, land units, or environmental classes (Margules *et al.*, 1995b).

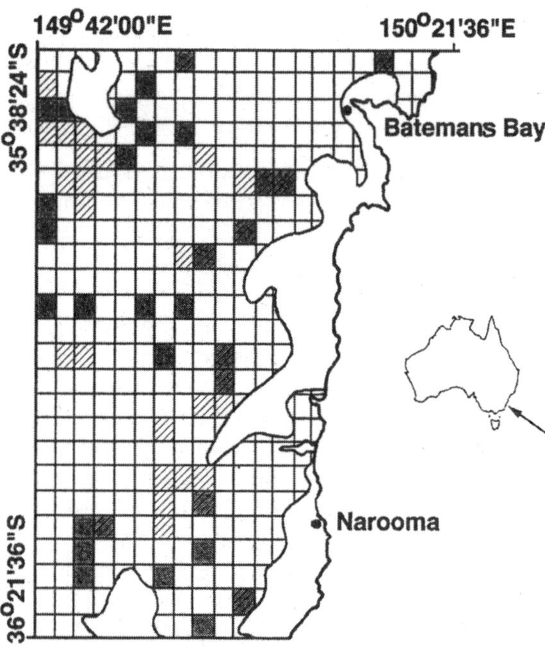

Figure 16.1. Area of south-eastern New South Wales that was the focus of reserve design by Nicholls and Margules (1993), and was intended to include 10% of forest types and environmental domains. Step 4 of their algorithm specified the selection of sites nearest in space to a site already selected.

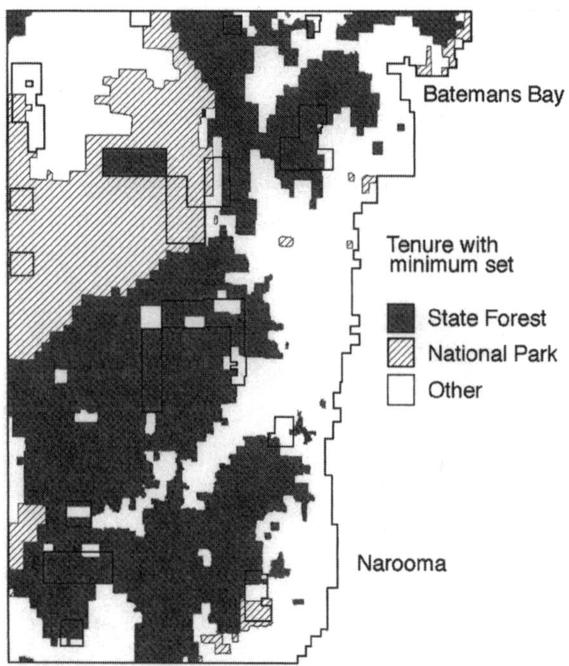

Figure 16.2. Land tenure lines representing one alternative design for conservation reserves in the area of south-eastern New South Wales that was the focus of reserve design by Nicholls and Margules (1993). Note that much of this area is now in reserves following a Regional Forest Agreement.

Approaches like the one described in Box 16.2 treat all environmental classes equally and assume that there are sufficient resources to acquire at least one example of each class for the reserve system. If there is some hierarchical relationship among environmental classes, and if resources are insufficient to acquire an example

View of the area targeted in the reserve selection algorithm in Box 16.2. (Photo by David Lindenmayer.)

of each type, then a method that recognises the degree of similarity among environmental classes will be more efficient (e.g. Woinarski *et al.*, 1996). The usefulness of different approaches can be judged by comparing their efficiency and flexibility (Pressey *et al.*, 1994), or by other relevant pragmatic criteria.

There are many applications of reserve design (see Margules *et al.*, 1995b; Margules and Pressey, 2000 for reviews) and a vast (and increasing) array of alternative algorithms and associated strategic methods (e.g. see Csuti *et al.*, 1997; Pressey *et al.*, 1997). Examples in Australian studies include those of remnant mallee in South Australia (e.g. Margules and Nicholls, 1987), forests in southern New South Wales (Bedward *et al.*, 1992; Belbin, 1992; Nicholls and Margules, 1993; Pressey *et al.*, 1996), vertebrates in subtropical north-western Australia (Woinarski, 1992), and semi-arid lands in western New South Wales (Pressey and Nicholls, 1989; Pressey and Tully, 1994).

Potential limitations of reserve selection methods

There are some potential limitations in using reserve selection algorithms. These are:

- Extensive baseline data are often required to provide adequate spatial coverage of surrogates. These data are unavailable for many areas, particularly in developing nations (Norton, 1999). Pressey and Cowling (2001) argue that this should not preclude the application of reserve design algorithms; rather, users need to be aware of data limitations.
- As outlined in the previous section, the units used as a basis for reserve selection may not be reliable surrogates for other forms of biodiversity (Lindenmayer *et al.*, 2000c).
- Different selection algorithms can produce quite different optimal reserve design outcomes, even for the same data (e.g. Csuti *et al.*, 1997; Pressey *et al.*, 1997).
- Changing the scale to which biodiversity units are subdivided (e.g. the segregation of land units) can significantly affect reserve selection outcomes (Pressey and Logan, 1994).
- Reserve design algorithms often attempt to identify a minimum set of areas to achieve representativeness (i.e. an efficient solution; see Box 16.2). This minimalist approach to reserve design (Crome, 1994) can mean that protected areas are vulnerable to species loss (Virolainen *et al.*, 1999; Rodrigues *et al.*, 2000). The problem of reserve adequacy is discussed in the next section.
- The genetic component of biodiversity is rarely an explicit part of reserve design algorithms. Genetic data are usually unavailable.

Only some of these problems are relevant to gap analysis, where the existing reserve system is the beginning of the reserve design procedure, except when the aim is to minimise the amount of land in protected areas (Pressey, 1994a). Despite these shortcomings, reserve selection algorithms and gap analysis are better than an *ad hoc* approach to reserve design (Pressey *et al.*, 1996; Margules and Pressey, 2000; Pressey and Cowling, 2001). Thorough examination of reservation options will highlight the weaknesses in existing systems and will identify an array of alternatives, any of which may improve the conservation status of protected areas.

Reserve adequacy

Often targets for reserves are set arbitrarily (e.g. 10–15% of a given forest type). Reserve selection algorithms often fail to consider *adequacy*. They do not, for example, reveal *how much* land needs to be reserved. Typically, they also ignore the potential for environmental changes, for example those that are forecast to occur with the enhanced greenhouse effect (Scott *et al.*, 2002), and ecosystem and population dynamics (Cabeza and Moilanen, 2001). For example, most reserve selection algorithms do not provide an indication of the viability of populations that occur initially in a reserve system (Witting and Loeschcke, 1995; but see Cabeza, 2003).

Extinction risk assessment tools such as population viability analysis are useful for exploring issues associated with the viability of populations in large ecological reserves (Armbruster and Lande, 1993; see Chapter 18). Moilanen and Cabeza (2002) optimise reserve selection, accounting for the likely persistence of target species in a reserve network. Long-term empirical studies of the dynamics of populations of different sizes (Berger, 1990) and the presence of species within reserves of different sizes (Newmark, 1985, 1987; Gurd and Nudds, 1999; Gurd *et al.*, 2001) can also provide an indication of the chances of persistence (and hence the viability) of populations in reserves.

The adequacy of a reserve system is also underpinned by such considerations as the size, number and shape of individual protected areas, the connectivity between them (Burkey, 1989), and the temporal and spatial characteristics of disturbance regimes (Pickett and Thompson, 1978). The importance of these factors depends on the objectives of a reserve system and the biodiversity values targeted for conservation.

Generally, many adequacy issues can be addressed by making reserves as large as possible (Soulé and Simberloff, 1986; Hunter, 1994). This is because large reserves:

- usually support more species (Preston, 1962; Connor and McCoy, 1979; Rosenzweig, 1995)
- support larger and more genetically variable (and thus less extinction-prone) populations (Armbruster and Lande, 1993; Billington, 1991; Lacy, 1987; Saccheri *et al.*, 1998)
- have a better chance of incorporating natural disturbance regimes (Pickett and Thompson, 1978; Baker, 1992; see the CARR principles outlined earlier in this chapter)
- contain a greater area of 'interior' habitat buffered from edge effects (Janzen, 1983; e.g. Richardson *et al.*, 1994).

However, reserve adequacy involves more than just the size of protected areas. Processes in adjacent unreserved areas can affect reserve integrity (Nelson, 1991; reviewed by Lindenmayer and Franklin, 2002). Disturbance regimes do not recognise reserve boundaries (Norton, 1999), and pollutants and pests flow into reserves via river systems and wetlands (Calhoun, 1999). For example, many World Heritage sites suffer from threats originating outside their boundaries (World Resources Institute, 1992). Janzen (1983) suggested that adequacy can be assessed by the extent to which threatening processes originating in surrounding unreserved areas can be controlled or mitigated.

16.5 Reserve design and selection in the real world

The first task of reserve planning is to define the objective(s) of setting aside a reserve. Is the reserve intended to capture a representative sample of all biota or a subset of biota, to conserve a particular threatened species, or to protect a scenic feature? Issues such as representativeness, adequacy, replication, reserve size, and reserve shape follow logically from the objectives, and are influenced by context, including the type and size of natural disturbance regimes, conditions in the surrounding unreserved areas, and the potential for management within reserves.

Despite the potential value of reserve selection methods, their use to date has been largely hypothetical (Ehrlich, 1997; Kremen *et al.*, 1999; Prendergast *et al.*, 1999; but see Pressey *et al.*, 2000; Mendel and

Box 16.3

The historical development of reserve acquisition in Tasmania

Mendel and Kirkpatrick (2002) traced the development of the reserve system in Tasmania. Prior to the early 1970s, the reserve system was biased toward areas of unproductive land of high scenic and aesthetic value but low economic value – a trend typical of reserve systems in Australia and worldwide (Margules and Pressey, 2000). Over the following decades there was a substantial increase in the area of the reserve system, with an improvement in the level of representation of many vegetation communities, including economically valuable ones (Mendel and Kirkpatrick, 2002). For example, by 1992 the reserve system supported 15% of the pre-European land cover of one-third of Tasmania's plant communities. Some of the gaps in the representation of the reserve system have been filled since then (see Table 16.1), particularly during the Regional Forest Agreement (RFA) process in 1997. Mendel and Kirkpatrick (2002) found that the RFA identified gaps in the reserve system and attempted to achieve representation levels of forest reservation according to set criteria of 15% of pre-European area (Dickson *et al.*, 1997; JANIS, 1997). After the end of the RFA, there was an increase from 16 to 27 forest communities (out of 50) that had 15% reservation levels or higher.

The reserve system in Tasmania evolved gradually and now the network is more representative than virtually anywhere else in the world. However, despite the outstanding conservation gains in Tasmania, there are still some vegetation communities that are poorly represented in the reserve system (see Table 16.1), especially ones that have experienced large rates of loss since European settlement. In addition, there are some major remaining conservation problems, such as land clearing for plantation establishment (see Box 8.2), and within-reserve management issues, including altered fire regimes and the spread of pathogens such as Cinnamon Fungus.

Table 16.1. Representation of vegetation communities in the reserve system in Tasmania over time (from Mendel and Kirkpatrick, 2002).

Plant community	Pre-European area (hectares)	Pre-European area (%)	Pre-European area reserved (%)				Pre-European area remaining (%)
			1937	1970	1992	2000	
Dry eucalypt forests							
Coastal *Eucalyptus amygdalina*	361 300	5.0	0.4	0.4	5.6	15.8	53
E. amygdalina on dolerite	286 900	3.9	0.2	0.3	2.5	9.9	62
E. amygdalina inland	65 700	0.9	0.0	0.0	1.1	3.3	33
E. amygdalina on sandstone	65 800	0.9	0.0	0.0	1.4	8.7	50
E. viminalis – *E. ovata* – *E. amygdalina* – *E. obliqua* damp	80 000	1.1	0.0	0.0	5.1	14.7	50
grassy *E. globulus*	23 700	0.3	5.0	5.0	13.8	26.7	57
E. pulchella – *E. globulus* – *E. viminalis* grassy/shrubby forest	227 200	3.1	0.2	0.3	3.2	18.4	67
E. viminalis grassy	223 900	3.1	0.1	0.1	0.2	1.2	49
E. viminalis + or – *E. globulus* coastal	7600	0.1	0.0	0.0	3.4	3.7	17
E. tenuiramis on granite	3000	0.0	28.1	28.2	43.7	92.1	100
E. tenuiramis on dolerite	10 000	0.1	0.0	0.0	21.4	59.2	83
Inland *E. tenuiramis*	164 700	2.3	0.0	0.1	1.0	4.1	35
E. sieberi on granite	20 600	0.3	0.0	1.0	2.8	25.0	86
E. sieberi on other substrates	61 300	0.8	0.0	0.0	4.8	17.9	75
E. obliqua dry	288 200	4.0	0.9	0.9	5.9	17.7	56
E. nitida dry[1]	260 800	3.6	1.3	7.9	36.1	52.5	61
E. delegatensis dry	353 200	4.8	2.2	6.9	14.8	25.8	83
E. pauciflora on dolerite[1]	27 400	0.4	0.0	0.0	1.4	14.1	69
E. pauciflora on sediments	41 200	0.6	4.8	4.3	6.9	11.1	44
Furneaux *E. nitida*[1]	77 100	1.1	0.0	2.1	5.4	8.1	39
Furneaux *E. viminalis*	33 800	0.5	0.0	0.0	0.0	0.4	<1
Shrubby *E. ovata*	188 800	2 .6	0.0	0.0	0.0	0.2	4
E. rodwayi dry	18 100	0 .2	0.0	0.0	0.1	4.3	38
E. risdonii dry	560	0.0	0.0	0.0	7.1	30.2	100
E. morrisbyi dry[1]	20	0.0	0.0	0.0	0.0	0.0	100
Subalpine eucalypt forests							
E. coccifera sub-alpine[1]	58 400	0.8	15.6	30.8	49.8	70.2	93
E. subcrenulata subalpine[1]	11 000	0.2	9.5	9.5	69.4	78.2	93
Wet eucalypt forests							
E. obliqua wet	655 500	9.0	0.1	1.4	5.6	16.3	64
E. regnans wet	95 200	1.3	2.6	2.7	6.1	17.1	79
E. nitida wet[1]	118 400	1.6	0.8	9.1	42.5	56.6	63
E. delegatensis wet	355 500	4.9	3.5	7.0	15.9	24.4	80
E. brookerana wet	15 700	0.2	0.0	0.0	0.3	6.1	29
King Island *E. globulus* – *E. brookerana* – *E. viminalis* wet	152 800	2.1	0.0	0.0	0.1	0.5	2
E. viminalis wet	49 000	0.7	0.0	0.0	0.2	1.4	7
Non-eucalypt dry forests							
Allocasuarina verticillata forest[1]	6000	0.1	4.3	4.8	7.2	10.2	22
Notolea ligustrina and/or *Pomaderris apetala* closed forest[1]	430	0.0	0.0	0.0	44.2	44.9	67
Callitris rhomboidea forest	930	0.0	0.0	0.0	21.5	39.8	85
Banksia serrata woodland[1]	80	0.0	0.0	0.0	100.0	100.0	100

Table 16.1. (Continued)

Plant community	Pre-European area (hectares)	Pre-European area (%)	Pre-European area reserved (%)				Pre-European area remaining (%)
			1937	1970	1992	2000	
Non-eucalypt wet forests							
Pencil Pine – deciduous beech wet[1]	4500	0.1	3.1	6.3	8.0	4.2	85
Pencil Pine wet	5600	0.1	1.4	4.0	3.4	5.9	37
King Billy Pine wet	20 000	0.3	4.7	9.9	50.8	75.3	91
King Billy Pine deciduous beech wet[1]	800	0.0	18.8	22.5	25.0	96.3	97
Huon Pine wet	13 100	0.2	0.0	2.2	47.7	56.5	81
Tall rainforest	252 000	3.5	1.4	5.0	18.3	40.8	76
Short rainforest[1]	507 400	7.0	3.1	9.9	30.8	55.1	73
Acacia dealbata wet	55 700	0.8	1.1	2.3	7.2	22.4	97
Acacia melanoxylon on flats	14 900	0.2	0.0	0.0	4.0	15.3	60
A. melanoxylon on rises	26 200	0.4	0.0	0.0	2.1	14.0	51
Leptospermum lanigerum – Melaleuca squarrosa swamp forest	32 900	0.5	0.2	8.8	23.0	30.6	57
Melaleuca ericifolia coastal swamp forest	38 200	0.5	0.0	0.0	0.5	0.6	2
Non-forest vegetation							
Buttongrass moorland/scrub[1]	1 150 000	15.8	1.9	11.3	47.4		99
Coastal heathland	425 800	5.8	1	1.8	8.8		53
Alpine/subalpine[1]	215 300	3.0	22.7	58	81.6		94
Grassland	85 000	1.2	0.0	0.0	1.4		60
Wetland	24 500	0.3	0.0	0.0	6.8		67
Saltmarsh[1]	4000	0.1	0.0	0.0	6.8		82
Total	7 285 720	100					

[1]Not an economically valuable plant community.

Kirkpatrick, 2002; Box 16.3). In some cases, minimum selection algorithms may generate results that are unacceptable for political, social or management reasons – although this is not a problem with the methods *per se.* Price *et al.* (1995) applied a minimum selection algorithm to the monsoon rainforests of the Northern Territory, in which 6 of 16 forest types and 18% of plant species were not represented in existing reserves. The algorithm selected too many additional areas from a political perspective, their management was beyond the budget of all Northern Territory parks, and the solution did not adequately consider the spatial arrangement of patches. These authors found it impossible to design an efficient solution in which the patches were clustered together, and recommended management of vulnerable species and conservation planning for the both the monsoon rainforests and the savanna landscape in which they are embedded.

Kirkpatrick *et al.* (1980) were the first to apply a reserve algorithm to conservation planning. They identified areas to protect plants on the east coast of Tasmania, and all of these areas are now in reserves (Mendel and Kirkpatrick, 2002). The CARR criteria reserve design algorithms have been applied (in part) to some parts of the forest estate (e.g. in New South Wales) during the Regional Forest Agreement process (e.g. Pressey *et al.*, 1996, 2000). The reserve acquisition process under the Regional Forest Agreement in Tasmania was strongly influenced by reserve design science (see Box 16.3). Reserve selection in the real world is rarely a single process in which suitable areas are identified and the allocation task is completed quickly. Rather, reserve allocation is usually gradual, and constrained by historical development and social realities (Mendel and Kirkpatrick, 2002).

Differences in the land base and competing demands for land

No two parts of a landscape are identical: they are better seen as an environmental continuum (Austin 1999;

Cradle Mountain National Park – an early part of the reserve system in Tasmania. (Photo by Esther Beaton.)

Manning *et al.*, 2004). Competing human demands and the extent of past impacts are almost always different in different places. Hence, there is *never* a set of identical candidate reserves from which to select (Prendergast *et al.*, 1999).

Political, economic and social factors *always* take precedence over ecological goals when land is considered for reserve allocation (Hunter, 1994; Pressey *et al.*, 1996; Lawton, 1997; Margules and Pressey, 2000). The tension between economic, social and ecological objectives makes reserve selection highly idiosyncratic. What is appropriate in one jurisdiction will be entirely unsuitable in others (Struhsaker *et al.*, 2005). For example, in the United Kingdom, reserve design is influenced by the fact that many reserves are acquired by different institutions, including government and voluntary bodies (Prendergast *et al.*, 1999). In contrast, in many less developed nations, setting aside reserves without considering the interests, behaviour and attitudes of the local human population typically leads to conflicts and ultimately to degradation of the reserve (e.g. Western and Gichohi, 1993; Smith *et al.*, 1997; Harcourt and Parks, 2003; Struhsaker *et al.*, 2005). For example, Khirthar National Park is Pakistan's first and one of its most important national parks. It is has several threatened species and some spectacular natural features, and it is also the home of almost 100 000 people who grow crops and graze animals there. These people are a critical element in all conservation strategies (e.g. Yamada *et al.*, 2004).

Successful reserve design is neatly illustrated by the example of Masoala National Park in Madagascar, where the design was based on a blend of ecological and socioeconomic criteria, including strong local community and high-level political support (Kremen *et al.*, 1999) – the paper describing the design acknowledged the support of the US Ambassador and the President of Madagascar! Such idiosyncrasies are perhaps best summarised by Prendergast *et al.* (1999), who noted that: '*no single procedure for identifying areas of conservation interest is likely to be universally appropriate*'.

16.6 Island biogeography and the design of nature reserves

As noted previously in this chapter, large areas usually support more species than smaller ones (e.g. Arrhenius, 1921; Preston, 1962; Rosenzweig, 1995; see also Chapters 5 and 14). Species–area relationships are a pervasive feature of ecology (e.g. Newton, 1998; Fahrig, 2003). The **theory of island biogeography** (Macarthur and Wilson, 1963, 1967) was developed, in part, to explain the species–area phenomenon, particularly for island biotas. This theory was then extended by Diamond (1975) and many others (e.g. Terborgh, 1974; Diamond and May, 1976; Shafer, 1990), who likened oceanic islands to reserves and used the theory of island biogeography to develop generic design principles for protected areas. These principles were incorporated in the International Union for the Conservation of Nature's 1980 World Conservation Strategy (IUCN, 1980), and were recommended for use in guiding the management of wildlife populations in wood production forests (Davey, 1989b).

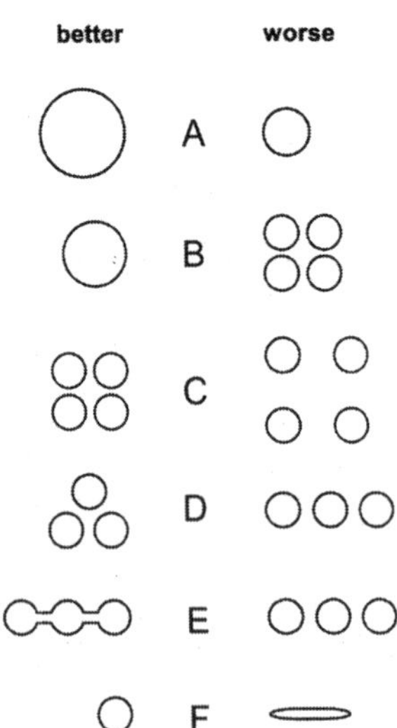

Figure 16.3. Principles for reserve design based on island biogeography theory (after Diamond, 1975).

Six general reserve design principles were derived largely from island biogeography theory (shown graphically in Figure 16.3):

- *Principle 1.* Large reserves are better than small reserves.
- *Principle 2.* A single large reserve is better than a group of small ones of equivalent total area (the basis for the so-called Single Large Or Several Small (SLOSS) debate).
- *Principle 3.* Reserves close together are better than reserves a long way apart.
- *Principle 4.* A compact cluster of reserves is better than a line of reserves.
- *Principle 5.* Circular reserves are better than long thin reserves.
- *Principle 6.* Reserves connected by a corridor are better than reserves not connected by a corridor.

Problems with the 'generic reserve design principles' derived from the island biogeography theory

The relationships between island biogeography theory and reserve design have been a highly controversial topic in conservation biology for several decades (see reviews by Gilbert, 1980; Burgman *et al.*, 1988; Simberloff, 1988; Shafer, 1990; Doak and Mills, 1994; Haila, 2002). There are many circumstances in which they do not hold, or in fact, their adoption could have an adverse effect on conservation values (Simberloff, 1988).

Principle 1. Large reserves are better than smaller ones

In general, larger reserves harbour more species, and all other things being equal, larger reserves are more effective for conservation than smaller ones. However, uncritical application of Principle 1 has led to excessive emphasis on large reserves (Lindenmayer and Franklin, 2002). Although large reserves are unquestionably important, small remnants can also have considerable value for biodiversity (Gascon, 1993; Powell and Björk, 1995; Turner, 1996; Palmer and Woinarski, 1999; McCoy and Mushinsky, 1999; Abensperg-Traun and Smith, 2000; Mac Nally and Horrocks, 2000). For example, Kirkpatrick and Gilfedder (1995) showed that small reserves, even those in poor condition, contained important populations of rare plants in otherwise extensively cleared parts of Tasmania. In other situations, larger reserves may not be appropriate. Studies of terrestrial gastropods in patches of remnant native vegetation in New Zealand showed that they were virtually confined to smaller habitat patches; smaller patches did not support the populations of feral Pigs that were a major predator of the snails (Ogle, 1987).

Principle 2. A single large reserve is better than several smaller ones

This principle has been discussed more than any other in the conservation biology literature over the past 30 years. The benefit of a large reserve compared with several smaller ones depends on:

- the typical size of disturbance events in the landscape
- spatial contagion in disturbance regimes (or the spatial extent of areas typically affected by the same catastrophic event, for example a high-intensity wildfire)
- spatial autocorrelation in year-to-year environmental fluctuations
- the dispersal capabilities of taxa targeted for conservation, that is, their ability to recolonise disturbed areas
- the demographics of populations in reserves.

If we ignore persistence and aim to maximise the number of species in reserves at the outset, then several reserves are likely to encompass a *greater* diversity of habitats and are therefore likely to harbour a greater number of species (see Kirkpatrick, 1994; Honnay *et al.*, 1999, for examples). A single reserve may be more susceptible to a single catastrophic event than a set of smaller, spatially separated ones (see Box 16.4). If a

Box 16.4

Reserve design for Leadbeater's Possum

Wood production activities and wildfires have important impacts on the distribution and abundance of Leadbeater's Possum (Macfarlane and Seebeck, 1991; Lindenmayer, 2000). The aim of management strategies is to conserve populations throughout the range of the species. How then should reserves be designed? Computer simulations were conducted for populations of animals in a nominal reserve area of 300 hectares, set aside as a single 300-hectare area, 2 × 50-hectare reserves, 3 × 100-hectare reserves and so on (Lindenmayer and Possingham, 1995). The probability of extinction was lowest for the set of intermediate-sized reserves (8 × 50-hectare or 6 × 75-hectare patches; Figure 16.4). In scenarios where a single 300-hectare reserve was modelled, there was a high probability that the entire population would be eliminated in a single fire. Such risks were lower when the reserve system comprised several patches. These patches were sufficiently close that recolonisation of empty habitat was likely. Conversely, if the patches set aside were too small (e.g. 25 or 50 hectares), factors such as demographic stochasticity increased extinction risk. Thus, in the case of the design of reserves for Leadbeater's Possum, there is an important trade-off between the impacts of processes that influence very small populations at one extreme (demographic stochasticity) and the influence of fire regimes at the other. More recent work corroborated the need for several spatially separate reserves, but indicated that the size of individual reserves for Leadbeater's Possum should be larger than 75–100 hectares (McCarthy and Lindenmayer, 2000).

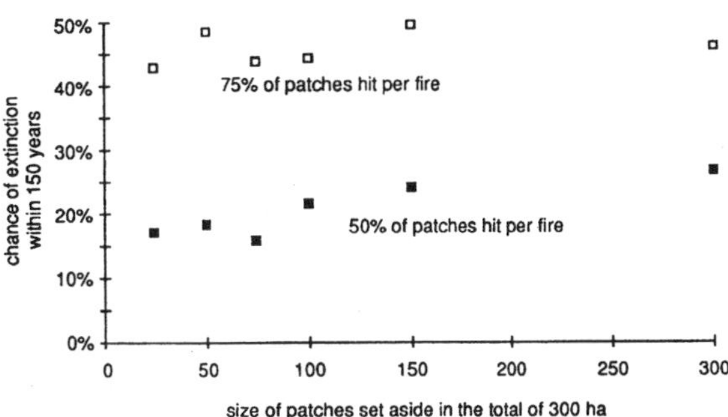

Figure 16.4. Reserve size, reserve number and extinction of Leadbeater's Possum.

species has poor movement capabilities, then it is relatively unlikely to recolonise patches. In these cases, fewer larger reserves or a number of reserves located close together may be required, but at the cost of increased risk from correlated disturbance events and environmental variation.

The trade-off between large and small reserves only applies where the total area of a single large reserve is more or less equal to the total area of small reserves. A fragmented habitat that has several small patches contains a smaller (and a more extinction prone) total population than the original, non-fragmented habitat because the total area of habitat is reduced and the movement of individuals (migration, dispersal) may be restricted (see Chapter 10). Furthermore, the resulting patches are generally no more independent of each other than they were before fragmentation (because they are at the same location in the environment), although fires or disease may spread more slowly, making fragments more independent than parts of a single large habitat.

Principles 3 and 4. Reserves close together are better than reserves far apart and a compact cluster of reserves is better than a line of reserves

Assertions that reserves close together are better than ones a long way apart, and that a compact cluster of reserves is better than a line of reserves (Figure 16.3) suffer from the same limitations as the issues outlined in the previous section. A lack of environmental correlation between habitat patches may mean that organisms can recolonise locally extinct patches more easily. Berger (1990), Murphy *et al.* (1990) and Stacey and Taper (1992) provide examples where a lack of environmental correlation between habitat patches or reserves benefited species persistence.

The best spatial arrangement of areas set aside for nature conservation will depend on the biology of the species at hand, the objectives of the reserve system and the management of the habitat around the reserves. It will be determined by the inter-play between correlations in environmental variation, disturbance regimes or catastrophic events, dispersal capability and the size and number of reserves.

Principle 5. Circular reserves are better than long, thin ones

The shape of reserves rarely receives much attention (Blouin and Connor, 1985). Circular reserves are considered to be superior to linear reserves because of

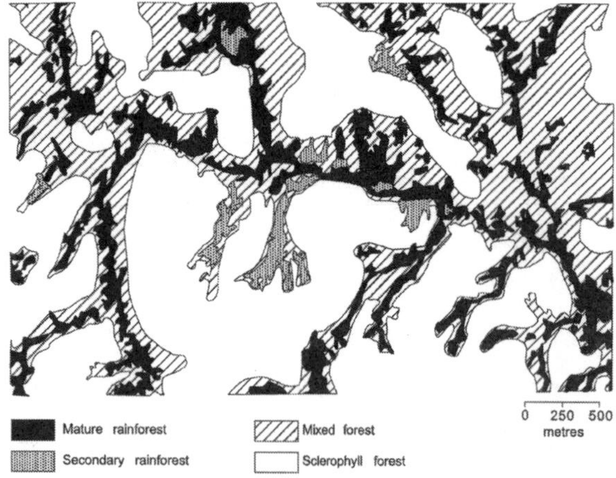

Figure 16.5. An example of the linear spatial arrangement of cool temperate rainforest patches in the O'Shannassy Catchment, Victoria (after Burgman and Ferguson, 1995).

the potential problems created by biotic and abiotic edge effects in linear reserves (see Chapter 10 on edge effects). For example, Reading *et al.* (1996) considered that round reserves would be better reintroduction sites than linear reserves for the Eastern Barred Bandicoot.

Linear riparian areas of cool temperate rainforest in the Central Highlands of Victoria. (Photo by David Lindenmayer.)

As with the other 'general reserve design principles', the issue should be considered carefully. Local context is likely to be more important than any generalisation. Linear reserves may encompass a larger range of habitat types and potentially more species. Configurations that conform with natural boundaries such as water catchments may be easier to manage, and landscape elements such as riparian vegetation may not be captured effectively by a circle. For example, cool temperate rainforests in Victoria (e.g. forests dominated by Myrtle Beech and Southern Sassafras) are largely associated with cool, moist riparian areas (Busby, 1986; Lindenmayer *et al.*, 2000a), and are characterised by linear spatial distribution patterns within larger areas of wet sclerophyll eucalypt forest (Figure 16.5). Similarly, relatively narrow linear reserves would effectively capture areas of wet sclerophyll forest adjacent to tropical rainforest in northern Queensland.

Circular reserves may also be more prone to degradation than linear reserves when the threat spreads from within a patch. For example, many vegetation types are associated with riparian environments, and the streams that run through their centre are a mode of transport for pathogens and weeds. In such cases, the larger the edge:area ratio, the slower will be the rate of loss of unaffected habitat.

Principle 6. Reserves connected by a corridor are better than reserves not connected by a corridor

There has been considerable debate about the conservation value of wildlife corridors (Simberloff and Cox, 1987; Noss, 1987; Hobbs, 1992; Simberloff *et al.*, 1992; Bennett, 1998). Corridors may enhance the value of connected reserves in some instances. Chapter 10 discusses the advantages and disadvantages of corridors and highlights the fact that connected reserves may not always be better because the connections play a role in dispersal, and they affect the correlation of variation between environments in different patches (reviewed in Lindenmayer, 1998). Connections may be beneficial or detrimental to the persistence of species, depending on the species' biology and interactions with the environment (Lindenmayer and Franklin, 2002).

Why island biogeography theory has limited applicability to reserve design

The island biogeography theory treats reserves (and habitat patches) as 'isolated oceanic islands' and does not account for conditions and processes in surrounding unreserved areas (Haila, 2002; Manning *et al.*, 2004). The theory ignores disturbance regimes, the magnitude and type of edge effects at reserve boundaries, and the suitability of surrounding unreserved areas for habitat and dispersal pathways, or as a source of invading species (Aldrich and Hamrick, 1998; Saurez *et al.*, 1998; Ås, 1999; Richetts, 2001). Therefore, the theory may fail where the surrounding area provides even temporarily suitable habitat (Zimmerman and Bierregaard, 1986), or when the importance of reserve size is outweighed by other factors, for example disturbance regimes, habitat conditions within and outside reserves (e.g. Fitzgibbon, 1997; Fox and Fox, 2000), heterogeneity and connectivity across landscapes (Brereton, 1997; Metzger, 1997), and the influx of organisms displaced from the adjacent unreserved areas (Bierregaard and Stouffer, 1997; Darveau *et al.*, 1995).

Island biogeography theory focuses on the numbers of species; however, in many reserve design cases, species diversity *per se* is not important. The number of species may increase (see Box 16.5) but large changes in species composition may be detrimental (Bennett, 1990b; Hutchings, 1991, in Gascon and Lovejoy, 1998). Rather than the number of species, reserve design is concerned with biotic composition, the identity of taxa in reserve systems (Murphy, 1989; Doak and Mills, 1994), the viability of populations (Grumbine, 1990; Lamberson *et al.*, 1994; Gurd *et al.*, 2001), and the need to optimise representativeness or comprehensiveness (Mackey *et al.*, 1988; Scott *et al.*, 1993, 2001).

Box 16.5

Empirical examples of the shortcomings of island biogeography theory

One of the general rules of reserve design derived from island biogeography theory is that smaller patches created by habitat fragmentation support fewer species. Natural resource management and conservation require simple rules to guide decisions. However, simple and general `rules' are nearly always wrong in some circumstances. The Biological Dynamics of Forest Fragments Project in Brazil found that edge effects and matrix conditions were key factors and that their impacts were substantially more important than fragment size for some animals (Brown and Hutchings, 1997). In that study, there was an *increase* in frog, small mammal, and butterfly species richness following the isolation of rainforest patches, which was contrary to predictions based on island biogeography theory (Gascon and Lovejoy, 1998). Species richness was elevated by an influx of taxa capable of using the changed matrix. Estades and Temple (1999) similarly recorded increased bird species richness in small rather than large Southern Beech (*Nothofagus* spp.) forest remnants embedded within an exotic Radiata Pine plantation in Chile.

Summary: Island biogeography theory and reserve design

Zimmerman and Bierregaard (1986), Burgman *et al.* (1988) and Doak and Mills (1994) showed why island biogeography theory often has little practical value in reserve design. Reserves are not isolated oceanic islands (Saunders *et al.*, 1991; Haila, 2002). Attributes of and processes in surrounding unreserved land are important (Gascon *et al.*, 1999), and species diversity alone is not sufficient for reserve selection. Rarely are there two or more ecologically similar areas from which to choose (Simberloff, 1988; Larsen, 1994). On this basis, Lindenmayer and Franklin (2002) argued that it is better to design reserves to satisfy specific objectives rather than apply general design principles from island biogeography theory.

Despite the clear limitations of island biogeography theory, it has had a major impact on the evolution of conservation biology, because of the considerable effort dedicated to testing its predictions (see for example, Shafer, 1990). Testing and falsifying theory is a critical part of the scientific process. Indeed, in the preface of their seminal book, *The Theory of Island Biogeography*, Macarthur and Wilson (1967, p. v) noted that:

> *We do not seriously believe that the particular formulations advanced in the chapters to follow will fit for very long the exacting results of future empirical investigation. We hope instead that they will contribute to the stimulation of new forms of theoretical and empirical studies, which will lead to stronger general theory.*

16.7 Conclusions

Protected areas serve the needs of both species and ecosystem conservation, and they are a core component of conservation strategies worldwide. It is tempting to think that a reserve system will guarantee the persistence of the biodiversity of a region, but this notion is naive. Reserves are one of the most important approaches to conservation, but in isolation they will surely fail (Craig *et al.*, 2000; Lindenmayer and Franklin, 2002). The fact that not all species will be effectively conserved in a network of protected areas highlights the importance of off-reserve conservation. The management of areas outside reserves for multiple uses including conservation is essential in ensuring the conservation of biodiversity (Department of the Environment, Sports and Territories, 1995, 1996; Lindenmayer and Franklin, 2002).

Reserve design algorithms focus on issues such as representativeness and capturing the maximum number of species, and rarely consider the viability of the populations initially captured in the reserve system (Witting and Loeschcke, 1995; Cabeza, 2003). Other tools are required to explore questions of population viability, and these are treated in detail in Chapter 18.

The importance of off-reserve conservation is paralleled by the need for management within reserved areas. Challenges in reserve management include recreation, weed control, fire management, disease management in animal and plant populations, and mitigation of the effects of the surrounding environment. The planning and management of conservation values, both on and off reserves, is essential if the goals of conservation reserves are to be met. The management of the human population and the way in which it uses the environment within, adjacent to, and remote from, reserves is a major component of any attempt to conserve biodiversity.

16.8 Practical considerations

Strong correlations between surrogates and the targets of conservation are usually assumed but are not tested. The adequacy of reserve design procedures depends on the strength of these relationships. The process of validating reserve design options and measuring their success in capturing important elements of the biota must be an essential part of all reserve design protocols.

Reserve designs need to account for social and political constraints, and for the resources required to manage them effectively. They need to account for spatial context, the extent and frequency of disturbance, and the types of human activities that are likely to take place in the surrounding matrix. Reserve design algorithms that seek CARR reserve systems are a qualitative improvement over those based on the theory of island biogeography and *ad hoc* systems. Although realities are such that reserve planners take what they can get, CARR principles will provide guidance on decisions whenever options arise. The CARR principles should be used routinely to guide planning in the light of social and other constraints.

16.9 Further reading

Reserve design methods are discussed by Pressey *et al.* (1993), Pressey and Tully (1994), Margules and Pressey (2002), and Williams *et al.* (2004). Burke (2000) presents a diverse set of recent examples of the application of gap analysis. Volume 112, issues 1 and 2, of the journal *Biological Conservation* (2001) contains a series of interesting articles on reserve design. Nix *et al.* (2000) provide a valuable example of the application of biodiversity surrogates and biodiversity priority-setting tools in a major case study in Papua New Guinea.

The Theory of Island Biogeography was written by MacArthur and Wilson (1963; 1967). Reviews of the theory are presented by Gilbert (1980), Margules *et al.* (1982), Burgman *et al.* (1988), Simberloff (1988), Shafer (1990) and Doak and Mills (1994).

Discussions of the importance of 'spreading the risk' and avoiding potential problems associated with the spatial correlation of catastrophes between reserved areas can be found in Gilpin (1987), Simberloff and Cox (1987), Quinn *et al.* (1989) and Lindenmayer and Possingham (1995).

Hale and Lamb (1997), and Craig *et al.* (2000) are edited books that contain many examples of the role and importance of off-reserve areas for biodiversity conservation. Lindenmayer and Franklin (2002) provide a detailed discussion of the combination of reserves and off-reserve areas for forest biodiversity conservation.

Chapter 17

Monitoring, assessment and indicators

This chapter gives a broad introduction to monitoring and environmental assessment for the conservation of natural populations. It includes a brief outline of statistical power, management goals and endpoints, and the predictive ability of impact assessment procedures. A discussion of the use of indicators and biodiversity surrogates in monitoring and assessment is provided because of their widespread use in monitoring and the assessment of environment impacts, and also because of the increasing focus on them in conservation biology.

17.0 Introduction

An **environmental impact assessment** (or effects assessment) is a document describing the existing environment, predicting the environmental effects of a proposal, and making recommendations to mitigate its effects. Government officers and public interest groups then assess the accuracy and completeness of the document and recommend whether and how the proposal should proceed (Thomas, 1996). Thus, assessment is the process of predicting the type, risks and magnitude of the likely effects on the environment of a particular proposal. Monitoring is defined as sampling and analysis designed to 'ascertain the extent of compliance with a predetermined standard or the degree of deviation from an expected norm' (Hellawell, 1991). Monitoring may be undertaken to gauge the effectiveness of policy or legislation, to audit compliance with guidelines, or to detect changes from a reference value.

The *Australian Environment Protection (Impact of Proposals) Act (1974)* created a legal mandate for environmental protection in 'matters affecting the environment to a significant extent' (section 5A). The Australian *Environment Protection and Biodiversity Conservation Act (1999)* protects endangered and vulnerable species and endangered ecological communities. It applies where Commonwealth actions or decisions are required. There is equivalent legislation in some Australian States, for example New South Wales (*NSW Threatened Species Act, 1995*).

Where there is a legal requirement to assess the impacts of a project on the environment or its component ecosystems or species, it is necessary to be able to detect important changes in the environment, usually with reference to a standard or control. Assessment and monitoring programs fail when they are inappropriately designed (Spellerberg, 1994; Franklin *et al.*, 1999). This chapter provides an introduction to concepts relevant to the conservation of natural populations, and outlines some of the fundamentals of effective monitoring and assessment methods. However, it is beyond the scope of this book to provide a complete outline of statistical methods and experimental design protocols for environmental impact assessment (see Quinn and Keough, 2002).

Box 17.1

The tragedy of monitoring in conservation biology

Monitoring generates the empirical data that measures the degree to which a management program is achieving its objectives. Monitoring can be required as a part of government plans, permits, or agreements (e.g. habitat conservation plans), as part of a certification process, or to fulfil other legal obligations. For example, monitoring is one of the key elements for forestry certification by the Forest Stewardship Council (1996). It is also a part of many endangered species recovery programs (see Clarke *et al.*, 2003)

Relatively few credible long-term forest monitoring programs have been implemented anywhere in the world. Furthermore, many monitoring programs fail. For example, almost half of the monitoring programs initiated in New Zealand were unreported or not completed (Norton, 1996). Lindenmayer and Franklin (2002) discussed the many reasons why adequate monitoring programs are difficult to develop, implement and maintain. They include:

- Meaningful (credible) and practicable monitoring programs are difficult to design and expensive to implement.
- Many monitoring programs are poorly designed and, consequently, their outcomes have limited value for conservation management. For example, field methods often fail to address useful questions (Norton, 1996). Roberts (1991) noted that too often monitoring has been `planned backwards on the collect now (data), think-later (of a useful question) principle'.
- Objectives of monitoring programs are often poorly defined, because explicit statements regarding the expected outcomes are lacking (see Roberts, 1991; Macdonald and Smart, 1993).
- Monitoring is often considered a routine activity that does not merit significant management effort (i.e. an activity that does not contribute to achieving immediate goals) or scientific input.
- Management of long-term environmental data sets is challenging and requires substantial technical expertise and financial support (Michener and Brunt, 2000).
- Adequate and sustained funding for monitoring programs is rarely available. Even if adequate funds are provided, most programs have subsequently been starved of financial and logistical resources. Most government agencies and corporations have annual budget cycles that make monitoring programs, which are necessarily long term, vulnerable to budget cuts as short-term shifts occur in organisational priorities and objectives. Often meaningful monitoring results cannot be obtained for some species within, for example, the 5-year timeframe of an endangered species recovery program (Clarke *et al.*, 2003).
- When an individual or a small group provides the initial momentum for its establishment, failure of an organisation to institutionalise a monitoring program can cause it to fail when the dedicated individual or cadre leaves.

Some of these impediments to monitoring programs are overcome when monitoring becomes a required activity, for example to fulfil legal and market-based requirements for credible assessments of the effects of resource management activities on biodiversity and other environmental variables. Nevertheless, rigorous monitoring programs are very rarely conducted, particularly over prolonged periods.

17.1 Statistical power and the precautionary principle

The detection of deviations from standards of compliance for an environmental action or development is based on statistical inference. The first step in formulating a statistical test is to specify two competing hypotheses. The convention is that a **null hypothesis** (usually denoted H_0) is described, in which there is no difference between the various groups in the experiment. These groups may have been subjected to a set of fixed treatments under the control of the experimenter, or subjected to a set of uncontrolled and, to some extent, random effects. The statistical test evaluates the veracity of the null hypothesis against the **alternative hypothesis** (usually denoted H_1) that there is a difference among the groups in the study (Spellerberg, 1994).

Box 17.2

Shifting baselines in monitoring

Monitoring will often be required to examine changes in the environment with reference to a standard or control. Determining what is an appropriate control can sometimes be complex. This is illustrated by the concept of 'shifting baselines' that has begun to emerge in monitoring and other studies, where management history has created modified environmental conditions that strongly influence human perceptions of what is 'natural'. The concept is used increasingly in marine research, for example to highlight historical population sizes of pelagic sharks and the extent of recent declines (Baum and Myers, 2004), and the magnitude of recent novel changes in the cover of coral reefs (Greenstein *et al.*, 1998). The concept of shifting baselines is also relevant in terrestrial conservation biology. The forests of Sweden have been extensively modified over the past 300 or more years and forest managers in that country find it hard to believe that unmanaged forests naturally support comparatively high levels of deciduous trees and fallen timber (Angelstam *et al.*, 2004). Similarly, in Victorian montane Ash forests, recurrent logging, frequent high-intensity wildfires and post-fire salvage logging has led to widespread even-aged regrowth stands. Forest managers failed to recognise that multi-age stands were widespread because evidence of them had largely been lost from wood production zones. Recently, work in unlogged water catchments has highlighted that the prevalence of multi-aged stands is much greater than previously recognised (Lindenmayer and McCarthy, 2002; Mackey *et al.*, 2002).

For example, in an experiment to test a treatment that is designed to reduce the impact of a chemical on a fish population, one possibility is that the treatment has no effect, whereas the alternative is that the treatment is effective. To distinguish between the two hypotheses, the possibility that the treatment has no effect is the null hypothesis (H_0), whereas the other possibility, that the treatment is effective, is the alternative hypothesis (H_1; Table 17.1).

Irrespective of whether there are impacts or not, the composition of two independent samples will be different. For example, surveys carried out at different times in the same place will almost certainly result in different lists of species (Mac Nally, 1994b; Cunningham *et al.*, 1999). Measurement error and natural variation will ensure that no two samples are identical. If we know the magnitude of natural environmental variation and we have specified the magnitude of change it is important to detect, if we then detect a change that is 'unusually' large in a statistical sense, we can conclude that there *is* an environmental impact.

Table 17.1. Null and alternative hypotheses for a standard statistical test.

Hypothesis	Symbol	Description
Null hypothesis	H_0	The treatment has no effect
Alternative hypothesis	H_1	The treatment is effective

There are two possible conclusions of environmental impact assessments: one is that an impact exists, the other is that there is no significant impact. Independently, it is also possible for these conclusions to be either correct or incorrect. If an effect is thought to exist, the likely magnitude of the effect must be evaluated to determine if it is acceptable. Furthermore, any conclusion may be incorrect. The likelihood of an error depends on how intensive the study is, the magnitude of the change, and the level of natural variation (Table 17.2). The probabilities of errors are central to the concepts of sensitivity and reliability in scientific testing, and are the backbone of sound environmental management.

A **type I error** occurs when a true null hypothesis is rejected, and the incorrect conclusion that there is a difference among the groups is reached. In the context of environmental impact assessment, a type I error would involve the study leading to a conclusion that an impact has occurred when in fact it has not. For this reason, it is sometimes termed a false alarm. Such mistakes can lead to expensive, inappropriate and unnecessary economic opportunity costs or remediation costs to industry and society.

The probability of a type I error can be set directly by those conducting the hypothesis test. By convention, the type I error rate is set at 0.05. This level has no particular scientific import; it simply represents a 5%

Table 17.2. Statistical outcomes in relation to detecting environmental impacts through hypothesis tests.

Actual state of the environment	Conclusion of the study	
	Impact	No impact
Impact has occurred	Correct	Type II error
No impact has occurred	Type I error	Correct

probability that the null hypothesis will be incorrectly rejected. Anyone using a probability of 0.05 in a hypothesis test is agreeing implicitly that this probability of making a mistake is acceptable.

It is also possible to set other levels for the type I error rate, such as 0.10 or 0.01. Using a higher value (such as 0.10) means that it is more likely that natural variation in the environment will be misinterpreted as being due to some human action. Using a lower level for the type I error rate (such as 0.01) will decrease this likelihood. Conversely, it will result in the study concluding that a significant impact has occurred only when there have been relatively large changes in the environment.

A type II error is made when a false null hypothesis is accepted, and the incorrect conclusion that there is no difference among the groups is reached. A type II error in environmental impact assessment would lead to a conclusion that there was no impact when in fact an impact occurred. This is sometimes termed a false sense of security. Although type I and type II errors are two sides of a statistical coin, the attitudes of the scientific community and of society in general to the two types of errors is different. Conventionally, a type I error rate of 0.05 is acceptable. If scientific protocol, convention or regulatory guidelines were to specify that a particular type II error rate was acceptable, it would be necessary to plan for this before a study was undertaken. Usually, the probability of a type II error is arbitrary and depends upon how a study was designed. Type II error rates are rarely calculated, and regulatory authorities and scientific convention do not suggest acceptable thresholds. For example, Fairweather (1991) did not find a single estimate of type II error rates among more than 40 environmental impact statements. Mapstone (1995) noted that impact assessment has inherited a preoccupation with type I error rates. This preoccupation pervades ecological research, even though type II errors are at least as important in impact assessment. There is no real statistical or scientific justification for this.

Statistical power

Statistical power is a measure of the confidence with which a particular effect can be detected, if one exists (Keough and Mapstone, 1995; Hatch, 2003). It is defined as one minus the type II error rate. A monitoring program that has low power has a high type II error rate, that is, a large chance of making a type II error. Such a program is relatively unlikely to detect impacts, even when they do occur. The statistical power of a study depends on sample size used in the test, the effect size to be detected, and the variability inherent in the data. These parameters are related by the expression:

$$power \propto \frac{E\sqrt{n\alpha}}{\sigma} \tag{17.1}$$

where E is effect size, n is sample size, α is the type I error rate and σ is the standard deviation of the variability of the data.

The **effect size** is the magnitude of change that we wish to detect confidently, if there is an impact. The effect size must be specified prior to any analysis of the power or relative power of a monitoring or assessment study (Fairweather, 1991; Mapstone, 1995; Hatch, 2003).

Power analyses are an essential feature in the design of sensitive tests and in the correct interpretation of their results. Low power for a particular test may mean that the test is insensitive and will produce inconclusive results (Fairweather, 1991). For example, Williams *et al.* (2001) noted that even the intensive sampling regime they employed in studies of logging effects on birds in Western Australia were insufficient to detect declines in abundance of less than 80–90% for most taxa.

The methods used for power analyses are not always straightforward, and they can require computer simulations or the use of equations derived specifically for the problem at hand (e.g. Gerrodette, 1987). For example, in a study on marine mammals by Taylor and Gerrodette (1993), as population size decreased, then so did the likelihood of detecting a population decrease. The minimum detectable rate of decline also increased.

Power and the precautionary principle

Recall from Chapter 1 that the **precautionary principle** is defined as: 'Where there are threats of serious or irreversible damage, lack of full scientific certainty should not be used as a reason for postponing measures to prevent environmental degradation'.

There has been considerable debate on whether the precautionary principle is scientific (e.g. Deville and Harding, 1997, *vs* Ollier, 1998). For instance, in debates concerning marine pollution (see Gray,

Box 17.3

Power and the detection of environmental effects

Consider circumstances in which it is necessary to detect the effects of rural tree dieback on a bird population. The bird depends on large hollow-bearing trees for nest sites, and the number of hollows limits the size of the population. The trees within the area are subject to dieback. Although many large hollow-bearing trees are retained on farms, no provision has been made for the recruitment of new hollow-bearing trees as the older trees decay and collapse. Thus, the number of hollow-bearing trees is decreasing exponentially. Field observations and aerial photo interpretation establish that the rate of decay of nest trees is about 2% per year. The suggestion is that the population is declining at the same rate, determined largely by the collapse of nest trees.

Surveys are conducted in two consecutive years, with the following results.

Year	Bird population size	Standard error
1995	303	41
1996	291	37

These data represent a change in population size of about 4%, which is somewhat higher than that predicted by the hypothesis that a decline in population size is due to nest tree collapse. However, a standard statistical test of the difference between the means in 1995 and 1996 shows that there is no significant difference between these values, so we cannot reject the null hypothesis. Convention stipulates that we accept that there is no effect of the loss of nest trees on the size of the bird population.

Before reaching any final conclusions, we should ask how likely it is that the sampling program could have detected an impact, if one actually existed. Given that the means of the two populations must be about two standard deviation units apart before they can be considered different, we would have to observe a decline to about 220 individuals before the result would be statistically significant, representing a decline of about 25%, more than 10 times greater than the expected decline. Therefore, even without detailed or exact power calculations, it is possible to see that there were too few observations (or too much variation in the samples) to test the hypothesis of a 2% decline per year. In this situation the sampling design was inadequate. The options available to us are to increase the sample size, accept a higher type I error rate, or reduce the measurement error by using more reliable counting methods.

1990a,b; Josefson, 1990; Lawrence and Taylor, 1990; Lutter, 1990) it has been argued that the precautionary approach is inconsistent with traditional objective scientific methods. Brunton (1994) suggested that obtaining proof that proposals would cause no damage is logically impossible, and that 'even the most ardent regulatory authority would be unlikely to make such a totally unattainable demand on developers'.

Brunton (1994) argued that the principle will be used to provide justification for non-existent links between human actions and environmental effects where these could be invoked to further the interests of environmental groups, and that it will encourage notions of causation based on spurious correlations and jeopardise the ability to develop systems of management that are resilient to change, and that it ignores the potential for beneficial impacts of development on the environment.

The debate concerning the impacts of marine pollution and the precautionary principle centres on when regulatory action should take place (e.g. penalising polluters or enforcing remedial actions). In the past this has occurred only when the null hypothesis (that there is no significant impact) has been rejected. With the introduction of the precautionary principle, even though statistically significant changes in the environment have not been detected, there may still be reasons to believe environmental degradation is occurring or will occur in the future, especially if sample sizes are small or natural variation is large.

Crawley *et al.* (2001) conducted a 10-year study, planting four different genetically modified (GM) herbicide-resistant or pest-resistant crops (Canola, Maize, Sugar Beet and Potato) and their conventional counterparts in 12 different habitats. Within 4 years, most plots had died out naturally. One plot of potatoes survived the tenth year, but that was a conventional potato plot.

Goklany (2001, p. 44) interpreted this result as follows:

> *GM plants were no more invasive or persistent in the wild than their conventional counterparts... The study confirms that such GM plants do not have a competitive advantage in a natural system unless that system is treated with the herbicide in question.*

Goklany (2001) was convinced that there is no risk because no difference was observed between the treatment and the control. This interpretation is wrong. There may be a substantial difference between them, but this study failed to detect it. We have no idea how likely it is that the study would have detected an important difference (see Burgman, 2005).

The debate should not degenerate into whether scientific methods should be disregarded in favour of a totally precautionary approach to environmental protection. Statistical power calculations can help to identify those cases in which precautionary regulatory actions may be justified and those cases where they are not. The following example taken from Peterman and M'Gonigle (1992) illustrates this point. If analysis of data from a monitoring program fails to reject the null hypothesis, and if power is high (say at least 0.8) for some biologically or economically important effect, then we can say that there was at least an 80% chance of rejecting the null hypothesis if such an effect had been present (and less than a 20% chance of committing a type II error). In such a case, the enforcement of regulatory action on the grounds of uncertainty would probably not be justified (assuming a willingness to accept a 20% probability of making a type II error). However, if power is low for an effect size considered to be important, the monitoring program may not be a reliable source of information because it is likely (with a probability of '1-power') to have failed to detect such an effect, even if it was present. In that case, little confidence should be placed in results from the monitoring program and a precautionary approach to regulation may be justified until an experiment or monitoring program with acceptably high power is carried out.

17.2 Management goals, assessment endpoints and measurement endpoints

Once the objectives of a monitoring study are set, the size, kind, frequency, distribution and number of samples must be specified. Monitoring programs are frequently designed in circumstances in which time and money are insufficient for a comprehensive study. Effectiveness rests on achieving a design such that the data are sensitive to the kinds of changes that they are intended to detect. The specification of an effect size is a decision that may involve considerations of biology, chemistry, physics, aesthetics, politics, ethics or economics (Mapstone, 1995; Franklin *et al.*, 1999). The choice of the variables to monitor is determined in part by anticipated ecological responses and in part by the social and political context of the study.

Environmental impact assessment and monitoring involve measurement of some index of environmental condition. Differences in measurements before and after the human action are compared, considering both type I and type II error rates. This process recognises natural fluctuations by placing limits on changes in the environment that could reasonably be attributed to natural variation. If a larger change is detected, it is concluded that a significant impact has occurred. If no such change is detected, and we are reasonably certain that such a change would have been detected had it occurred, we conclude that there has been no significant impact. A critical step in this process is the choice of indicators with which we are to assess environmental change. Suter (1993) described a process for choosing indices of environmental change that depends on the identification of **management goals**, and the subsequent specification of **assessment** and **measurement endpoints**.

Management goals represent the broad aspirations of managers, including such things as clean water and healthy ecosystems. For instance, one of the management goals of the *Victorian Flora and Fauna Guarantee Act (1988)* is to maintain species in sufficient numbers such that they can survive and flourish in the wild. These goals are socially, ecologically and politically relevant. Because of their general nature, they tend to be too broad for a scientific program to measure, and usually the protocols for measurement are unspecified. For example, there are no legally or scientifically established means for assessing whether adequate conservation of a gene pool, species, community or ecosystem has been achieved.

Assessment endpoints are formal expressions of the environmental values to be protected (Suter, 1993). They provide a means by which management goals can be identified, measured, audited and evaluated. Generally, they should be relevant to society, biologically important, unambiguously defined, measurable, and susceptible to the human impacts under consideration. If the intention of management is to provide clean water, then an assessment endpoint may be the maintenance of natural populations of freshwater biota (Suter, 1993) or an absence of algal blooms. An assessment endpoint for the maintenance of natural populations of flora and fauna may be represented by the persistence of all species within the full extent of their geographic ranges.

Measurement endpoints (and **test endpoints**) are measured in the field or the laboratory, and act as surrogates for other components of an ecosystem. In the case of water management, it would not be possible to record the freshwater biota at all places in the system continuously. Instead, the composition of the freshwater biota could be represented by a list of species and their relative abundances, which could be measured weekly at each of a set of sample stations. Evidence that species are maintained within their ranges could be provided by field surveys conducted annually at each of several locations, targeting those species considered most likely to be affected by land-use management practices.

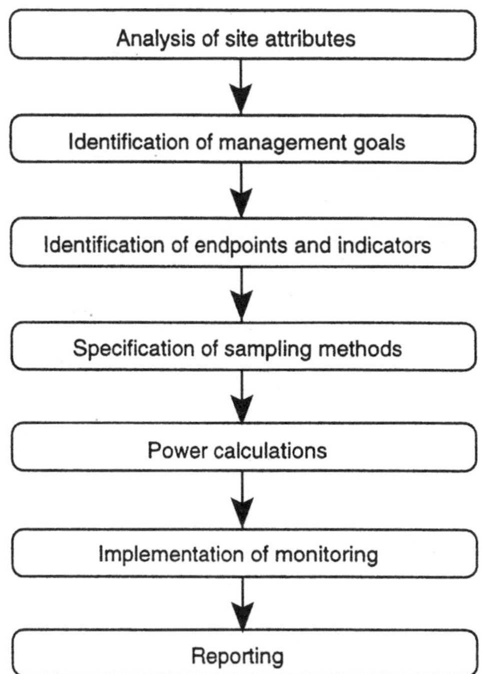

Figure 17.1. Schematic representation of the development of monitoring strategies (after Hellawell, 1991).

The development of a monitoring program is an *iterative* process, of which power calculations are a central part (Figure 17.1). Without them, there can be no guarantee that the program will provide the information necessary to answer appropriate questions. If power calculations find the program lacking, then either the sampling strategy must be redesigned, or the choice of endpoints and indicators must be revised. A monitoring program that does not include power analyses is blind to potential type II errors.

Assessment and measurement endpoints should be sensitive to anticipated impacts. To some extent, the characteristics of the expected impacts will determine the most appropriate sampling strategy. For example, it would be inappropriate to specify the use of sample quadrats only from the centre of the range of a species if the impact is likely to result in a range contraction. Such quadrats would be best suited to circumstances in which changes in local abundance were likely (Figure 17.2).

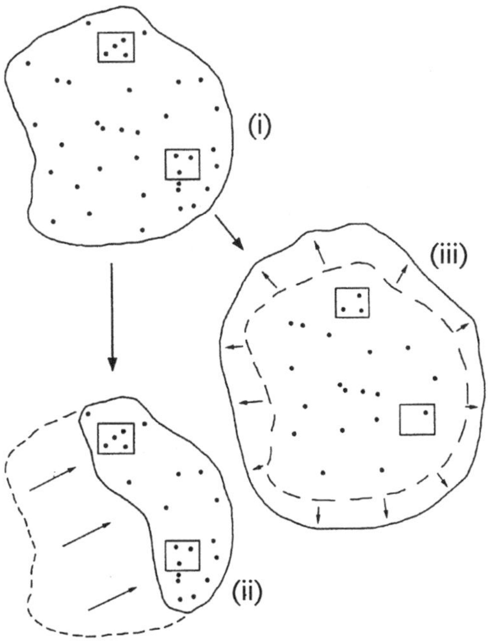

Figure 17.2. The use of fixed quadrats to monitor changes in population size. (i) The average density of the population based on two sample quadrats is 3.5 individuals per quadrat; these data are used as a baseline against which changes in population size are identified. (ii) The sampling strategy results in a type II error: an impact on the range of the species reduced population size without affecting the density in the sample quadrats. (iii) The sampling strategy results in a type I error: there is no impact on population size, although an increase in range reduces population density in the sample quadrats (after Hellawell, 1991).

Figure 17.3 shows two kinds of responses to a chemical spill. A continuous recording device would detect the chemical concentration. By using an index of biotic diversity, it may be possible to detect impacts on longer time scales, and to evaluate the ecological consequences of the impact.

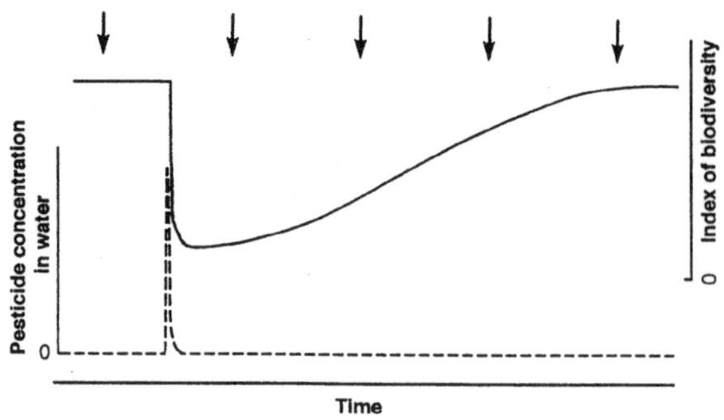

Figure 17.3. Representation of a chemical spill, and two responses (after Usher, 1991).

Monitoring programs are essential because they validate planning decisions (Franklin *et al.*, 1999). Without them, there can be no feedback between planning and its consequences (this concept of a feedback between planning and monitoring will be revisited in Chapter 18). Indicators are used extensively in marine ecology and pollution control to monitor environmental impacts (e.g. Warwick, 1993; see Section 17.3). Efficient statistical designs can be implemented at a range of spatial and temporal scales to detect changes in a variety of environmental parameters (Buckley, 1991a; Eberhardt and Thomas, 1991; Margules and Austin, 1991; Green, 1993).

Box 17.4

How many samples?

One of the first questions asked in the design of any sampling program is: How many samples should we take? The answer to this question lies in knowing the variability of the population that will be sampled, the magnitude of the change that we wish to detect, the reliability with which we wish to detect that change, and the characteristics of the change. Overlain on these strictly statistical considerations are questions of cost and time. Any sampling program that is expensive and time consuming will not be acceptable, and will be rejected on the grounds that it is not sufficiently pragmatic.

The question of the number of samples that are necessary is related to the concepts of type I and type II error rates. If we wish to estimate a parameter of a population, it is necessary to specify the error in the estimate that we are willing to tolerate and the reliability we require of the estimate:

$$n = \frac{\left(\frac{s}{\bar{x}}\right)^2 z^2_{[\alpha]}}{e^2}$$

where s is the standard deviation of the sample population, $\bar{x}$ is the mean (and the ratio $s/\bar{x}$ is the coefficient of variation of the parameter of interest), $z_{[\alpha]}$ is the appropriate value from a normal distribution for a sample size of n and a reliability of $1 - \alpha$, and e is the error that we are prepared to tolerate, expressed as a proportion of the mean.

For example, as part of a study on water quality in a catchment, we wish to estimate the average turbidity of the water in a stream. To plan the monitoring program, we need to estimate the number of samples required to estimate the parameter reliably. The context of the study and the potential for impact from turbidity suggest that it would be acceptable to estimate mean turbidity to within 10% of the true mean. That is, if the value of the estimate was to be either 10% higher or 10% lower than the true mean, the program would still be effective. However, we can never be entirely certain that an estimate will be within a fixed proportion of the true value. Rather, we must specify that we wish to be, for example, 95% certain that the estimate will be within 10% of the true mean. An appropriate value from the z-distribution represents this level of reliability. The coefficient of variation of turbidity in streams in this area is known from other studies to be about 15%. Given these data, it is possible to solve for n:

$$n = \frac{(15)^2 1.96^2}{10^2} \approx 9$$

If it turns out that taking nine samples is prohibitively expensive, then we have several alternatives: we could reconsider the tolerance for reliability and error; we could attempt to reduce the coefficient of variation by stratifying the samples or by improving the quality of the instruments for measuring turbidity; or, similarly, we could reduce variability by reducing measurement error through training programs for field staff. Other approaches to the estimation of required sample sizes can be found in some standard statistical texts (e.g. Sokal and Rohlf, 1995).

Box 17.5

Non-standard statistical designs for monitoring programs

The vast majority of monitoring programs for animals and plants involve resurveying the same sites from year to year. If designed appropriately, these surveys can produce precise estimates of change, as there will be no component of sampling variance. Such programs are conceptually simple and the statistical analysis is usually straightforward. However, such designs typically have low coverage of the broader population of the monitored entity because of logistical and financial constraints on the number of sites. As outlined in Box 17.4, precision and accuracy can depend substantially on sample size. Monitoring programs that use repeated measures may have restricted generality; that is, the outcomes may not be relevant to the full range of environmental conditions in the area of concern.

A novel approach to the design of monitoring programs was developed for studies of seabird nesting in the Coral Sea (see Cunningham *et al.*, 1995; Cunningham and Welsh, 1996; Welsh *et al.*, 2000). It involves the use of overlapping and rotating samples, making it possible to deal with a much larger population of sites than could be used if the study measured all sites each year. In the initial year of a monitoring program, a complete census of all sites is undertaken. In subsequent years, a proportion of sites (e.g. 50%) are selected for resampling. A subset of these is common to the previous year's sample, and the remainder are selected from the full pool of monitoring sites. Sites can be selected to measure animal abundance or the occurrence of a rare species.

In addition to work on seabirds, the method has been applied to monitor long-term trends in arboreal marsupials in the montane Ash forests of the Central Highlands of Victoria (Figure 17.4). There, the survey design was based on overlapping and rotating samples from 161 sites, each 1 hectare in size. The complete census of all sites was completed in 1997–98 and approximately half of these were resampled each year, with the site selection criteria being governed in part by the occurrence of Leadbeater's Possum (the key species of management concern) at a site in the previous year (Lindenmayer *et al.*, 2003b). Notably, the criteria for subsampling from the full pool of sites can be changed if necessary, giving rise to an `adaptive monitoring' approach (*sensu* Ringold *et al.*, 1996).

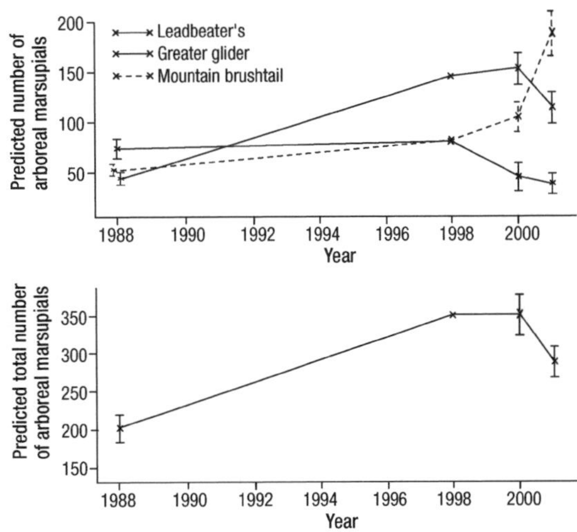

Figure 17.4. Population changes in three species of arboreal marsupials (top) and total number of arboreal marsupials (bottom) identified as part of a monitoring program in the forests of the Central Highlands of Victoria. (Modified from Lindenmayer *et al.*, 2003b.)

Such methods can involve parametric and non-parametric analyses (Margules and Austin, 1991; Clarke, 1993; Green, 1993; Underwood, 1993; Warwick, 1993). In addition, new non-standard statistical approaches for monitoring programs are being developed (see Welsh *et al.*, 2000; Lindenmayer *et al.*, 2003b; Burgman, 2005 on statistical process control; Box 16.3).

Very few environment impact statements (EISs) engage in long-term monitoring, which is a serious flaw in the EIS procedure (see Box 17.6).

Few development proposals make testable predictions, and fewer still are audited. Those that are audited are typically imprecise and inaccurate (Buckley, 1989). Ecological impacts are difficult to predict because ecological experiments take a long time; replication, control and randomisation are hard to achieve; and ecological systems change over time (Hilborn and Ludwig, 1993; Holling, 1993; Taylor *et al.*, 1997; Bunnell *et al.*, 2003). Controls are necessary to ensure that the effects of a process are not confounded by some other change acting on the system. In ecology, no two systems are identical and the larger the sample size, the more heterogeneous the controls will be. Natural and unplanned experiments such as wildfires and oil spills rarely come complete with controls (Whelan, 1995), so that it is frequently impossible to distinguish between the results of some putative impact, and other

Box 17.6

Structural and other problems in EISs

Many individuals and organisations have expressed concern over the standard of EISs in Australia (e.g. Pyke, 1995; Warnken and Buckley, 1998). There have been strong cases made for more rigorous peer review of EIS documents, better designed studies, and better referencing within EIS documents to support assertions (see Ecological Society of Australia, 2003). Lindenmayer and Gibbons (2004) further argued that there are some underlying problems with the current EIS process *per se*. Most importantly, the EIS process is often managed by the proponent. That is, the proponent of a particular project either completes the assessment 'in-house' or contracts a consulting company to write the EIS. The proponent-driven process tends to deliver the outcomes desired by the proponent and can fail to critically assess a proposal.

Lindenmayer and Gibbons (2004) proposed a new approach to EISs, involving an independent body (that is not at the mercy of vested interests both within and outside government) to commission independent environmental analyses. They argued that in taking the concept of an 'environmental impact statement' in a literal sense, where the environmental impacts could be substantial but have yet to be adequately quantified, there should be a mandatory requirement for long-term impact research (which would be independently assessed). Their comments were made in reference to a proposal (and associated EIS) for a large-scale charcoal-making plant on the south-east coast of New South Wales (Environmental Resources Management Australia, 2001). Lindenmayer and Gibbons (2004) contended that forestry operations associated with the proposal and their potential impacts on biodiversity needed to be quantified over relevant time frames (at least several decades). They believed that this analysis could lead to a valuable EIS, in which the impacts on the environment of the development were quantified and changes after the operation commenced were monitored and evaluated independently.

differences between sites that are the result of natural processes.

Issues of sustainability (see Chapter 19) often relate to large systems and long time scales – circumstances in which replication is not feasible. It is important, therefore, to design research programs that learn about a system faster than the system changes (see Hilborn and Ludwig, 1993; Taylor *et al.*, 1997; Bunnell *et al.*, 2003).

As described, measurement endpoints must be both socially and ecologically relevant. In the following section, we will explore the selection of indicators that attempt to serve the purposes of ecology and conservation biology.

17.3 Indicators

Indicators are biological entities whose interactions with the ecosystem make them especially informative about the quality of habitat, communities and ecosystem processes (Soulé and Simberloff, 1986; Kavanagh, 1991; Lindenmayer *et al.*, 2000c). Their purpose is to reflect changes in ecosystem dynamics at many spatial, temporal and organisational scales. The role of an indicator can be played by gene frequencies, populations, species, sets of interacting species, or whole assemblages of communities on a landscape scale (Conner, 1988; Landres *et al.*, 1988; Doak, 1989; Kelly and Harwell, 1990; Noss, 1990; McKenney *et al.*, 1994; Weaver, 1995; Lawton and Gaston, 2001). Indicators are tools to gauge the extent of ecological change, and they act as surrogates for other forms of biodiversity and other components of the ecosystem.

Species as indicators

Very often, species are chosen to indicate the condition of other species or ecosystems. Conservation effort has been directed towards many hundreds of species worldwide, but this is a very small part of the problem. In some instances, species are assumed to act as **umbrella species** in the sense that conserving them will result in the protection of numerous other species (Noss, 1990; Lambeck, 1997, 1999). However, such claims are rarely verified (Lindenmayer *et al.*, 2002e; Lindenmayer and Fischer, 2003). For example, Caro (2001) showed that protected areas set aside for large mammals in East Africa did not conserve small mammals. Similarly, Majer *et al.* (1994) showed that highly degraded remnant woodland areas in Australia can be extremely valuable for invertebrate conservation, even though native vertebrates (particularly mammals) have all but disappeared from them.

Landres *et al.* (1988, p. 317) defined an indicator species as:

> *an organism whose characteristics (e.g. presence or absence, population density, dispersion, reproductive success) are used as an index of attributes too difficult, inconvenient, or expensive to measure for other species or environmental conditions of interest.*

Indicator species are necessary because there are insufficient resources to manage all species explicitly. Some species are much more sensitive than others to management regimes and, for practical reasons, managers can only use data for a limited range of species (e.g. Brinkman, 1990a,b; Lambeck, 1997, 1999).

The term 'indicator species' has several meanings (Spellerberg, 1994; Hilty and Merenlender, 2000). Some of these are (after Lindenmayer *et al.*, 2000c):

- **Bio-indicator species** respond to environmental changes such as global warming (Parsons, 1991) or long-term shifts in fire regimes (Wolseley and Aguirre-Hudson, 1991). In North America, Kirtland's Warbler has been proposed as a biological indicator of global warming because it nests only in Jack Pine trees at the southern edge of their range (Botkin *et al.*, 1991). Plankton community composition (particularly diatoms and flagellates) appears to be a reliable indicator of eutrophication and impaired growth of shallow water corals on the Great Barrier Reef (Bell and Elemtri, 1995). Recher *et al.* (1980, 1987) suggested that Mountain Grey Gum may indicate the presence of high-quality habitat for wildlife in the forests of south-eastern New South Wales.
- **Site-type indicator** species are those that indicate environmental conditions, such as certain soil or rock types. For example, Klinka *et al.* (1989) provide an extensive summary of plants that reflect various soil conditions in coastal British Columbia.
- **Keystone species** (*sensu* Terborgh, 1986) are species whose addition to, or loss from, an ecosystem leads to large changes in abundance or occurrence of at least one other species (e.g. Mills *et al.*, 1993).
- **Dominant species** are species that provide much of the biomass in or that numerically dominate an area.
- **Recovery indicator species** reflect the extent of recovery of an ecosystem following disturbance. Recovery can be gauged by the presence or abundance of given species. For example, ant communities indicate the effectiveness of mine site rehabilitation (e.g. Majer *et al.*, 1984; Andersen, 1990, 1993; see Box 17.7). Birds are used to indicate wetland recovery (Weller, 1995).
- **Management indicator species** reflect the effects of a disturbance regime and/or environmental management efforts (Lehmkuhl, 1984; Conner, 1988; Milledge *et al.*, 1991; Sydes, 1994). For example, Grover and Slater (1994) considered that the composition of bird populations in remnant patches near Brisbane in south-eastern Queensland would indicate 'ecosystem health'. Many species have been proposed to reflect the consequences of forest management (Dyne, 1991), including bats (Richards, 1991), owls (Milledge *et al.*, 1991), and arboreal marsupials (Davey, 1984, 1989a,b; Cork *et al.*, 1990; Goldingay and Kavanagh, 1991; Kavanagh, 1991; Milledge *et al.*, 1991).
- **Pollution indicator species** reflect the effects of pollution on biota (e.g. Scharenberg, 1991; Dickson, 1993; Plenet, 1995). Spellerberg (1994) recognised five broad categories:
  - **Sentinel species** are highly susceptible to pollution and provide an early warning of a pollutant (the 'canary in the coal mine' effect). For example, Lake (1986) considered that species such as the Yabbie (a large crustacean) could be a useful sentinel species for heavy metal pollution in southern Australia.
  - **Detector species** are those that exhibit changes in behaviour, mortality patterns or age-class structure in response to environmental changes (Freedman, 1989).
  - **Exploiter species** are abundant in polluted areas because competing species have failed to survive, or because they make use of the polluted environment or the pollutants themselves.
  - **Accumulator species** concentrate pollutants in their body tissues or in the environment immediately around them (Rosenberg and Resh, 1993). Many lichens are accumulator species.
  - **Bioassay organisms** can be used in laboratory conditions to determine the relative toxicity of concentrations of pollutants.

Box 17.7

Ants as indicators

Ants have been used extensively and, in some cases, successfully to monitor the recovery of disturbed environments (Hoffman and Andersen, 2003; Andersen and Majer, 2004). Much of this work has been done in relation to mine site rehabilitation (e.g., Majer, 1978; Andersen *et al.*, 2003), but applications have been made to the evaluation of land-use practices in Western Australian forests (Majer *et al.*, 1984) and to habitat evaluation in South Australia (Greenslade and Yeatman, 1980). Ants are considered useful indicators of environmental impact and recovery, particularly of land degradation and mine site rehabilitation (Rosenberg *et al.*, 1986; Andersen, 1990; Kremen, 1994; Weaver, 1995; Andersen and Majer, 2004). This is because they:

- are a dominant faunal group
- are small and easily sampled
- are abundant and diverse
- occupy many trophic levels
- respond rapidly and are sensitive to different kinds of environmental stress
- are closely linked to a diverse variety of ecosystem processes, including nutrient cycling, the development of soil structure, and predation.

Two methods can be used to examine communities of ants and their response to environmental change. One is to monitor ant species succession and temporal changes in species' richness (e.g. Majer, 1985). The other is to track changes in the `functional group' composition of ant communities (Greenslade and Yeatman, 1980; Andersen, 1990). `Functional groups' include suites of taxa categorised into units such as `hot climate specialists', `cryptic species', `opportunists', and `large solitary foragers' (see Andersen, 1993; reviewed by Hoffman and Andersen, 2003). Such uses presuppose a sound understanding of the ecology of the species attributed to each functional group (Landres *et al.*, 1988; Andersen, 1991).

Andersen (1993) used assemblages of ant species as indicators of success of the rehabilitation of the Ranger open-cut uranium mine 250 kilometres east of Darwin in the Northern Territory. This work highlighted some strong relationships between the structure of the regenerating vegetation and the composition of the associated ant fauna (particularly the prevalence of some types of functional groups). The predominance of *Acacia* trees in regrowth appeared to lead to a halt in both vegetation and ant community succession, and prescribed fires were required to recommence these processes.

In a detailed review, Hoffman and Andersen (2003) found that the value of ants as indicators in Australia varied between ecosystems, being most useful in mesic habitats and least useful in arid environments. Although the work by Andersen (1990, 1993) and Andersen *et al.* (2003) highlights the potential value of using ants as recovery indicators, this approach nevertheless faces a number of difficulties, in common with many other types of indicator species. In particular, the taxonomy and life histories of ants, like virtually all groups of invertebrates, are very poorly known (Cranston, 1990). In many ecosystems, the environmental and biological relationships between the status of indicator species, land-use disturbance and ecosystem recovery are poorly understood (Lindenmayer *et al.*, 2000c).

These are not mutually exclusive categories, and many other classifications are possible. The particular characteristics researchers look for in choosing indicator species will depend on what it is that they want to detect by monitoring them.

Milledge *et al.* (1991) suggested that, ideally, management indicator species should be sedentary specialised species, or those at the top of food chains, which have large territories or home ranges. Species with the former characteristics can be useful in indicating local changes, whereas the latter can be useful in detecting ecological changes at a landscape scale. An ideal indicator may have many different roles apart from just these two. The usefulness of a species as an indicator is affected by its disturbance–response characteristics, its ecological function, its spatial and temporal scales of response as a function of the spatial and temporal scales of disturbance, and its ability to reflect changes in ecosystem processes at several organisational levels.

Keystone species and indicator species

We know very little about the functional relationships among the different elements of ecosystems. We have

Box 17.8

The Southern Cassowary as a keystone species

The Southern Cassowary is a ratite – a group that includes the Emu, Ostrich, the extinct elephant birds (from Madagascar), and the Kiwis and Moas (from New Zealand; Cracraft, 1973). The Southern Cassowary is confined to rainforest in north Queensland and Papua New Guinea. It weighs up to 80 kilograms and is 1.5–2 metres tall (Crome, 1976; Reader's Digest, 1990).

The Southern Cassowary may be a keystone species in the tropical rainforests of north Queensland. Although it occasionally eats foods such as carrion, snails and fungi, it is a specialist frugivore, consuming fruits from several hundred rainforest plants (Crome, 1976, 1990). It excretes the seeds of fruits intact, often with parts of the original fleshy covering adhering to the seed, making it a particularly effective dispersal vector for rainforest plants. The Cassowary is the largest vertebrate in the north Queensland rainforest, and is the only disperser of more than 100 plant species that have very large fruits (Crome, 1994). It is also the major dispersal agent for a very large number of other rainforest plants (Crome and Bentrupperbaumer, 1993). The role of the Southern Cassowary is filled by an array of other vertebrates (mammals, birds and reptiles) in rainforest communities elsewhere in the world (Jones and Crome, 1990). Without the Southern Cassowary, the dynamics of rainforest communities in north Queensland would be changed markedly and the status of many plants (and the animals that are, in turn, associated with them) would change. Reductions in numbers of the Cassowary could have important consequences for ecosystem function.

The conservation of the Southern Cassowary has become an issue of considerable concern. Its long-term persistence is threatened by a wide range of factors, including habitat destruction, hunting, competition with animals such as feral Pigs, predation by domestic Dogs, collisions with motor vehicles, and disease (Crome and Moore, 1990; Crome and Bentrupperbaumer, 1993).

Southern Cassowary. (Photo by Esther Beaton.)

little understanding of how far biodiversity can be reduced before ecosystem function is compromised or irreversibly damaged (DiCatstri and Younes, 1989). The United States Forest Service defines indicator species as having an important functional role in the working of an ecosystem (Landres *et al.*, 1988). If removed, keystone species precipitate important changes in ecosystem structure and function (Moore, 1962). When the Red Land Crab on Christmas Island was extirpated by the exotic Crazy Ant, it precipitated catastrophic and cascading changes in rainforest ecosystem processes, including seedling recruitment, leaf litter decomposition, invertebrate species composition, and tree canopy dieback (O'Dowd *et al.*, 2003). Mistletoes have a disproportionately large impact on the structure and function of some faunal assemblages, because they provide food and nesting resources, particularly during extreme droughts (Watson, 2001, 2002).

Some authors argue that keystone species will be useful indicator taxa because their status could reflect the integrity of the rest of the ecosystem, and the loss of such species could result in considerable ecosystem degradation (Soulé and Kohm, 1989).

Some keystone species play such a dominant role that they function in ways that could be considered detrimental in an ecosystem; in fact, many species may not be able to persist in their presence. The Noisy Miner is an aggressive native honeyeater that drives away many other smaller forest and woodland birds (Grey *et al.*, 1998; MacDonald and Kirkpatrick, 2003;

Mistletoes, seen here as darker clumps on the mid- and other parts of the canopy of Red Ironbark trees, have keystone roles in some ecosystems, for example woodlands. (Photo by David Lindenmayer.)

see Box 9.4). Removal studies have shown that many bird species return to areas when populations of the Noisy Miner are reduced (Grey *et al.*, 1997).

Ecological redundancy

Walker (1992) discriminates between species that are determinants ('drivers') of community structure and function and others that are 'passengers'. He suggested that it may be more important to manage habitat to enhance those species that play a critical role in ecosystem function. **Ecologically redundant species** are analogous to Walker's (1992) 'passengers'. The loss of these species results in no important changes in the composition or function of an ecosystem. For example, the loss of the Thylacine from Tasmania precipitated no noticeable changes in any other component of the biota.

Guilds as indicators

A guild is a group of species that exploit the same class of environmental resources in a similar way and between which competition can be expected (Root, 1967). The concept has been used extensively to examine the structure and dynamics of communities, particularly birds (Simberloff and Dayan, 1991).

The guild concept led to the idea that a given impact may affect all members of a guild in the same way (e.g. Severinghaus, 1981, in Simberloff and Dayan, 1991). This, in turn, became linked with the notion of indicator species; that is, the response of one member of a guild to environmental change would indicate the response of the remaining members.

Box 17.9

Foraging guilds of birds in south-eastern Australia

Mac Nally (1994a) examined the foraging patterns of birds in the forests and woodlands of Victoria and assigned 66 species to ten different foraging guilds. Ten species were identified as granivores and nine as ground carnivores. Other guilds contained a smaller number of species (e.g. only two species were assigned to the 'sweeper' guild). There were marked differences in the composition of the guilds between some of the different forest and woodland types. However, guild composition was similar among replicate survey sites within a given forest type, indicating a systematic basis for the assemblage of guilds related to habitat structure (Mac Nally, 1994a).

Very often, indicator species are defined as species that are sensitive to a particular land-use activity and whose presence is likely to indicate habitat quality for other species or communities (Landres *et al.*, 1988), implying membership of a guild and a degree of redundancy in ecological roles (Walker, 1992). For example, Milledge *et al.* (1991) suggested that Sooty Owls and Yellow-bellied Gliders in the forests of south-eastern Australia are likely to be lost from forest clear-felled on rotations of less than 80 years because they depend on habitat found in old growth forests. Thus, these species could indicate the status of the guild of hollow-dependent vertebrates.

Sooty Owl – a species proposed as a management indicator species in the wet forests of south-eastern Australia. (Photo by Esther Beaton.)

Verner (1984) and Szaro (1986) added a condition to Root's (1967) definition of guilds for the purposes of wildlife management: that guild members should respond similarly to perturbations in habitat conditions. Walker (1992) suggested that the appropriate way to define functional types (guilds) is by the way that they regulate ecosystem processes. However, guild membership is often decided on the basis of taxonomic relatedness, body form, foraging behaviour, life history, and habitat characteristics.

Problems with indicator species and related concepts

Problems with the indicator species concept have been reviewed by Landres *et al.* (1988), Temple and Wiens (1989), Niemi *et al.* (1997), Simberloff (1998) and Lindenmayer *et al.* (2000c). An extensive appraisal of these problems is beyond the scope of this book, but briefly, the chief problems are that: (1) members of the same guild respond differently to change, (2) some species are insensitive to change, and (3) there is a general lack of understanding of causal relationships between indicator species and the entities they are assumed to indicate (Lindenmayer and Franklin, 2002). Each of these categories is further explored below.

Species respond differently to change

Taxonomically or ecologically similar species respond differently to disturbance (Simberloff and Dayan, 1991; Morrison *et al.*, 1992; Thiollay, 1992; Kruess and

Box 17.10

Selection of indicator species in the montane forests of Victoria

The selection of an indicator species was explored for the montane Ash forests of the Central Highlands of Victoria, based on a choice from seven species of arboreal marsupials that occur in these forests: Leadbeater's Possum, Greater Glider, Sugar Glider, Mountain Brushtail Possum, Feathertail Glider, Eastern Pygmy Possum and Yellow-bellied Glider. The notion of a management indicator species was explored in the context of identifying a species that reflected the co-occurrence of a maximum number of other members of the arboreal marsupial guild.

The data on arboreal marsupials that were analysed included counts of animals on more than 150 sites scattered throughout the Central Highlands. The sites varied in slope, aspect, disturbance history and the management regimes that had been employed in the past. Each area typically supported 1–2 species. However, the average number of taxa co-occurring with each species ranged from 0.7 additional species for the Common Ringtail Possum to 1.8 additional species for the Feathertail Glider (Figure 17.5). Further analysis showed that there were few consistent patterns among groups of co-occurring species.

None of the species could be regarded as a particularly reliable or consistent (and therefore useful) surrogate for the presence or abundance of any other members of the arboreal marsupial guild in the montane Ash forests. Hence, the suitability of a given area of forest for one species does not imply that it is appropriate for another, even among this relatively closely related group. The reasons for this may be related to resource partitioning among the different members of the arboreal marsupial guild. There are large differences between each species in diet, body size, nest tree use, mating system and other life history characteristics. Given this, it is perhaps not surprising that attempts to identify any one (or even two or more) species whose presence reflects the occurrence of other taxa are unlikely to be successful. The chances that a species from this assemblage could be a useful indicator of the response of taxa in more distantly related groups (for example birds or small mammals) is even more remote and is not likely to be based on ecologically robust relationships.

One important habitat component for virtually all of the species of arboreal marsupials inhabiting montane ash forests is the availability of hollow trees as nest and den sites. The abundance of these trees is a factor limiting populations of these animals in large areas of montane ash forest (Lindenmayer, 1994, 2000). As a result, rather than seek a management indicator species in montane ash forests, the prevalence of hollow trees could serve as a useful structural indicator, reflecting the response of many species to forest management actions. However, even then there would be numerous non-hollow-dependent taxa that would not benefit from such an approach.

Tscharntke, 2002). Superficially similar species have different habitat requirements and respond differently to change (see Kotliar and Wiens, 1990; Robinson *et al.*, 1992; Berg *et al.*, 1994; Villard *et al.*, 1999; Debinski and Holt, 2000). Even the same species can respond in different ways in different landscapes (Dooley and Bowers, 1998; Lindenmayer *et al.*, 2001c). Therefore, the possibility that there are valid indicator species seems remote, whether it be for invertebrates (Cranston and Trueman, 1997), birds (Temple and Wiens, 1989), mammals (Caro, 2001) or any other group. In an experiment conducted at the US Brookhaven National Laboratory, a second growth natural forest of Pitch Pine was subjected to ionising radiation from cesium-137 for 15 years. Pitch Pines were much more sensitive to this impact than the other dominant tree species: radiation levels of 10 roentgens per day killed 100% of the Pitch Pine but had no effects on tree species such as Scarlet Oak (Woodwell and Rebuck, 1967). Scarlet Oak would be a poor indicator species for this impact.

Guild members rarely respond in the same way to environmental change, irrespective of the way in which species are partitioned into guilds (Landres, 1983; Verner, 1984; Landres *et al.*, 1988), raising doubts about the validity of extrapolation of predictions from a single species to other guild members (Block *et al.*, 1986; Szaro, 1986; Simberloff and Dayan, 1991; Morrison *et al.*, 1992). For example, Bayer and Porter (1988) used habitat models for two bird species to predict the distribution of other guild members. They assumed that variation in abundance was related to local habitat conditions, that censuses were not subject to appreciable measurement error, that the model

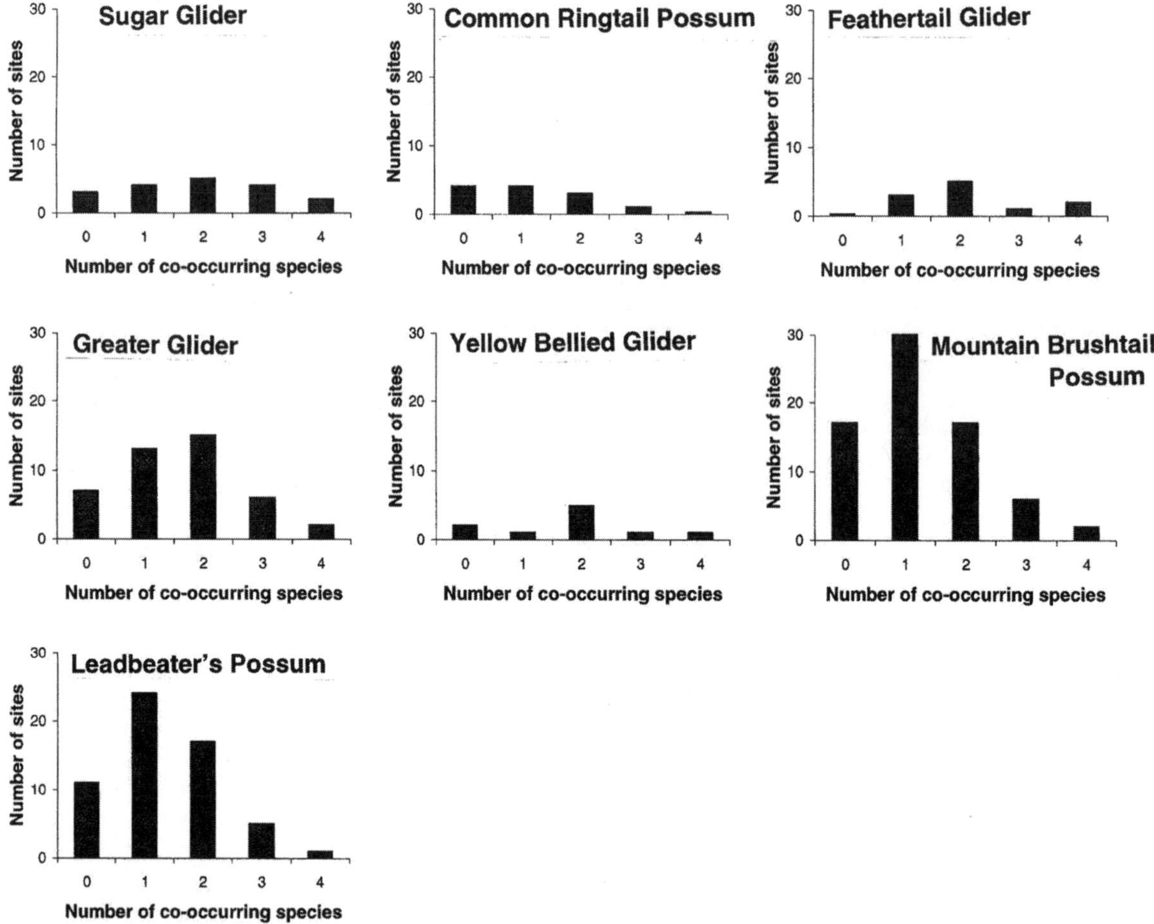

Figure 17.5. Numbers of co-occurring species of arboreal marsupials in the montane ash forests of the Central Highlands of Victoria (after Lindenmayer and Cunningham, 1997).

included all critical habitat variables, and that guild members were appropriately classified and responded in essentially the same manner to habitat conditions. Their models were poor predictors of habitat quality for both ubiquitous and uncommon species. Similarly, Reader (1988) tested the assumption that guild members respond similarly to disturbance by measuring the responses of six members of a guild of deciduous forest herbs to selective tree harvesting. Mechanical disturbance from harvesting operations and subsequent microclimatic changes affected different species differently, and species exhibited different responses over time to altered site conditions.

Lack of sensitivity to change

It is difficult to monitor indicator species with enough statistical power to identify important environmental changes reliably. Many so-called indicator species are insensitive to change, resulting in some environmental problems being well advanced (and difficult to reverse) before they are detected (Lindenmayer *et al.*, 2000c).

There is often a time lag between an impact and its expression in species responses (McIntyre *et al.*, 2002). A species or set of species can respond to disturbances that took place some time ago. The **extinction debt** concept (Tilman *et al.*, 1994; McCarthy *et al.*, 1997; see Chapter 10) reflects the time lag between habitat loss and the gradual, stochastic loss of species that ensues over decades or centuries.

Box 17.11

Are frogs good environmental indicators?

Many people assume that frogs are good indicators of environmental health. Some species of frogs are sensitive to toxins because they are exposed to them both on land and in the water and because they absorb moisture through the skin. However, frogs are not always good indicators of a healthy environment: some species readily use mud puddles on the roadside and others exist in quite polluted areas. An example is the Common Eastern Froglet, which inhabits virtually any type of water body throughout its distribution in south-eastern Australia. More useful information may be gained by asking questions such as: How many frog species are present? Which species are present? How many species have been lost from the area since European settlement? Are they breeding successfully? Do they show signs of environmental stress such as extra, missing or misshapen limbs (see Lindenmayer *et al.*, 2003a)?

Common Eastern Froglet – a widespread species of amphibian in south-eastern Australia and a clear example of a species that has limited sensitivity as an indicator of environmental change. (Photo by Andrew Claridge.)

The term 'management indicator species' been used somewhat ambiguously for species that are the focus of management activities. However, if a species is the focus of management activities, then it can have no role as an indicator of any other component of the ecosystem (Landres *et al.*, 1988). The survival of a target species may be achieved by habitat management. However, such actions confound its usefulness in reflecting the impacts of disturbance on other, perhaps similar, species.

Lack of understanding of causal relationships

Ryan (2000) noted that there is often insufficient knowledge to guide the selection of a given indicator species. The causal relationships between disturbance (or other forms of perturbation) and species response have not been established for *any* taxon recommended as an indicator species (Lindenmayer *et al.*, 2000c). Although many workers have contended that particular taxa are indicator species (e.g. Davey, 1989b; Johnson, 1994; Hill, 1995), the entities they are supposed to indicate are often not explicitly stated (Lindenmayer and Cunningham, 1997). Even where the indicator species concept is well developed, as in the context of pollution indicator species (Spellerberg, 1994), it can be difficult to identify indicator species (Hilty and Merenlender, 2000). For example, early research on the bivalve mollusc *Velesunio ambiguus* in Australian rivers suggested that the species was an indicator of heavy metal pollution (Walker 1981). Subsequent work found that the uptake of heavy metals by *V. ambiguus* did not reflect the extent of pollution in the surrounding river system, making the mollusc an unreliable and unsuitable indicator species (Millington and Walker 1983; Maher and Norris 1990).

Walker (1992) suggested that to reduce declines in biodiversity, it may be effective to focus first on aspects of biodiversity that maintain the resilience of the ecosystems. Species can be weighted by the importance of their functional role in much the same way that they can be weighted by their taxonomic distinctness. Evaluation of the viability of a species should consider interactions of that species with others, with the objective of maintaining ecological processes (Conner, 1988). Populations should be conserved in sufficient abundance to remain ecologically functional in their community, thereby maintaining ecosystem function (Conner, 1988; Ehrlich, 1990; Beattie and Ehrlich, 2001). Such considerations are severely hampered by a lack of ecological understanding. Furthermore, the loss of species that appear to be 'redundant' is nevertheless still a loss of species, and is the antithesis of what many consider to be a general aim of conservation biology – to maintain biodiversity.

Inferences from indicators are further limited when relationships between indicators and other elements of biodiversity vary in space or time (see Lindenmayer *et al.*, 2001c; Gégout and Krizova, 2003), when indicators respond differently to field sampling or collection methods, and when taxonomy is unstable (New, 2000).

Indicators create a potential for circularity; for example, Spencer (1991) noted that '*those woods that contain ancient woodland indicators are by definition ancient woodland*'. Similarly, in the Australian Regional Forest Agreements process, old growth forests were defined as those with more than 30% cover of senescent crowns. This is just one of many features of old growth forest (see Chapter 2), but it was used exclusively because it could be interpreted on aerial photographs and mapped. When indicators (for example the percentage of senescent crowns) define the entity they represent, the definition is trivial unless the indicator is unique and perfectly associated.

Many biotic groups have been claimed to be biodiversity surrogates at some stage or other (Table 17.3); but few ecological relationships between surrogates and the ecological properties they reflect are understood, few postulated relationships have been validated empirically, and many replace the entities they were intended originally to represent.

Summary: indicator species

Conservation biologists can send misleading messages when they use terms differently (Caro and O'Doherty, 1999). For example, the term 'focal species' was defined by Lambeck (1997, 1999) to represent umbrella species for threatening processes, in contrast to its widespread

Table 17.3. Some examples of groups suggested to be indicators or biodiversity surrogates.

Birds	Barrett and Davidson (2000), Brooker (2002), Suter (2002)
Owls	Milledge *et al.* (1991)
Arboreal marsupials	Cork *et al.* (1990), Kavanagh (1991), Milledge *et al.* (1991)
Large terrestrial mammalian carnivores	Carroll *et al.* (2001)
Small terrestrial mammals	Pearce and Venier (2005)
Bats	Richards (1991)
Whales	Hooker *et al.* (1999)
Amphibians	Blaustein (1994), Blaustein and Johnson (2003)
Ants	Andersen *et al.* (2003), Hoffman and Andersen (2003)
Bees (as pollinators)	Kevan (1999)
Butterflies	Kerr *et al.* (2000), Mac Nally and Fleischman (2002), Maes and van Dyck (2005)
Moths	Summerville *et al.* (2004)
Freshwater fish	Harris (1995), Commonwealth of Australia (2002a)
Diatoms	Reid *et al.* (1995)
Aquatic crustaceans	Lake (1986)
Molluscs	Gladstone (2002)
Vascular plants	Klinka *et al.* (1989), Kremen *et al.* (1998), Pharo *et al.* (1999), Lunt *et al.* (2005)
Bryophytes	Edwards (1986)
Fungi	Bredesen *et al.* (1997)
Lichens	Loppi *et al.* (1998)

usage in North America where it is simply the species on which conservation efforts are focused.

Indicator species, and related concepts such as guilds, umbrella species and focal species, are used as surrogates for biodiversity (e.g. Lambeck, 1997, 1999). There is a growing literature on the array of problems associated with such approaches (Temple and Wiens, 1989; Niemi *et al.*, 1997; New, 2000; Lindenmayer *et al.*, 2002e; Rolstad *et al.*, 2002). It is sobering to consider the results of a study of biodiversity surrogates by Andelman and Fagan (2000). They examined the efficacy of biodiversity surrogates, including indicator species, flagship species, and umbrella species, and found that none captured more species or better protected habitat areas than organisms selected at random from the large databases they assembled to conduct their tests! Similarly, Caro *et al.* (2004) found that small reserves in Belize selected on the basis of conserving flagship species were not effective at conserving other elements of the biota, and were no better than reserves selected using non-flagship taxa.

These outcomes and the arguments outlined in the previous section highlight some of the problems associated with relying on a small number of indicators from a relatively small subset of taxa, and from a single level of organisation (Niemi *et al.*, 1997). Different taxa and organisational levels will almost certainly respond differently to disturbance. A monitoring program that uses vertebrates exclusively is unlikely to be useful for predicting the response of other elements of the biota to environmental impacts (Cranston and Trueman, 1997; Prendergast and Eversham, 1997; Pärt and Söderstrom, 1999). For example, claims that birds or carnivores are good indicator species (Barrett and Davidson, 2000; Carroll *et al.*, 2001) are contradicted by empirical observations (e.g. Linnell *et al.*, 2000; Read *et al.*, 2000; see also the review by Temple and Wiens, 1989) and would not result in the adequate conservation of other elements of biodiversity (e.g. plants or invertebrates; Hansson, 1998; Oliver *et al.*, 1998). Birds rarely use habitat resources such as rock outcrops and soil cracks and would therefore be poor surrogates for taxa that depend on such structures, for example reptiles (Lindenmayer *et al.*, 2003a).

Although the notion of indicator species is a potentially useful one, much more work needs to be done to confirm the relationships between species chosen as indicators and environmental change (Simberloff, 1999). In Box 17.12 we discuss a general approach that might assist in testing the validity of a hypothesis that a measure acts as a surrogate for another entity.

Box 17.12

Assessing the validity of an indicator

Given the enormous and rapidly expanding literature on indicators (there are even entire journals now dedicated to their study), how should we determine whether a given indicator is a valid or 'useful' tool? Ideas from clinical trails in medicine provide two principles for evaluating the utility of surrogate endpoints. They are (after Begg and Leung, 2000):

- The best attainable inference must arise from direct measures of the target response.
- The validity of a surrogate (indicator) measure should be judged by the probability that results based on the surrogates alone are 'concordant' with the results that would be obtained from the direct measures.

Thus, the general principle is that an analysis based on the direct measure represents the highest standard. The definition of 'concordance' is somewhat arbitrary, but one interpretation is that management decisions arising from an analysis of the surrogate will be the same as those based on the direct measure. For example, the results of tests of significance on both the direct measure and the surrogate should lead to the same conclusions (Begg and Leung, 2000). Clearly, it is possible that biological processes might affect the surrogate but not the direct measure, and so conclusions will be different, even though the correlation between the two measures is high. However, in general, for the principles outlined here, the smaller the association between the two measures, the greater the possibility for misleading inference.

17.4 Selecting indicators

Lindenmayer and Franklin (2002) argued that the indicator species concept oversimplifies conservation situations, and obscures the need for careful approaches to conservation and management practices at a range of spatial and temporal scales for *all* components of ecosystems (Franklin, 1993a; Rolstad *et al.*, 2002). One possible approach is outlined in Box 17.13 and it includes managing a wide range of entities, including habitat elements such as large living and dead trees and

Box 17.13

Structure-based indicators in forest management
Lindenmayer *et al.* (2000c) explored a range of issues associated with the identification of robust indicators of biodiversity for application in ecologically sustainable forest management. Given problems with the indicator species approach (and related biodiversity surrogate schemes) they recommended the use of forest management practices that sustain (or recreate) characteristics of forest ecosystems, termed 'structure-based indicators'. These are stand- and landscape-level features of forests, such as stand structural complexity, plant species composition, connectivity, and heterogeneity. However, Lindenmayer *et al.* (2000c) acknowledged that information was presently lacking to determine whether such stand- and landscape-level features of forests will serve as successful indices of (and help conserve) biodiversity. On this basis, Lindenmayer *et al.* (2000c) recommended four strategies to enhance biodiversity conservation in forests:

- Establish biodiversity priority areas (e.g. reserves) managed primarily for biodiversity conservation.
- Within forests managed for paper and timber production, apply structure-based indicators.
- Employ a risk-spreading approach in wood production forests using multiple conservation strategies at multiple spatial scales.
- Adopt an adaptive management approach to test the validity of structure-based indices of biodiversity by treating management practices as experiments to provide new knowledge to managers and improve the effectiveness of management strategies.

fallen logs, species guilds, plant associations, stages of ecological succession, landscape attributes such as connectedness of patches, the degree of fragmentation, and measures of spatial pattern together with strategies to conserve particular species (Franklin, 1988; Lindenmayer and Franklin, 2002).

Indicator species are very unlikely to reflect a full range of possible responses to ecological perturbations at all temporal, spatial and organisational levels (Landres *et al.*, 1988; Probst, 1991; Wood, 1991; Lindenmayer *et al.*, 2000c). Careless use of indicators ignores the uncertainty involved in extrapolating the responses of one species to predict the responses of another, and may create the false impression that biodiversity is adequately conserved (Block *et al.*, 1986; Temple and Wiens, 1989; Simberloff and Dayan, 1991; Lindenmayer *et al.*, 2000c). The use of indicator species could be improved by selecting indicators from a broader taxonomic spectrum and from a broader range of spatial, temporal and organisational scales than is currently employed in most studies.

Salwasser (1990) argued that indicators should include aspects of genetic resources, species, biological communities, ecosystems and landscape attributes. Ehrlich (1992) suggested using representatives from vascular plants, amphibians, birds and herbivorous insects to monitor ecological status and function. Ehrlich's rationalisation for these choices is that plants play a fundamental role in ecosystem dynamics, and that insects and terrestrial arthropods make up the vast majority of organic diversity. The poorly developed taxonomy of most insects makes their use as indicators difficult (Majer, 1987; Gullan and Cranston, 1994), but the taxonomy of butterflies is relatively well developed (New, 2000). Butterflies are likely to respond to a variety of microclimatic conditions and interact as larvae and adults with different host plants with which they may have close evolutionary relationships (New, 1991; Ehrlich, 1992; Kremen, 1992). The taxonomy of birds is more or less complete, and bird abundances are closely related to vegetation structure and floristics. Identification of birds is relatively easy, and counting techniques for birds are well developed (Ehrlich, 1992).

Quantitative methods can be used to select potential indicators (e.g. McGeoch, 1998; Fleishman *et al.*, 2000; Mac Nally and Fleishman, 2002). Multivariate methods have been used extensively to assign species to guilds (e.g. Inger and Colwell, 1977; Landres and McMahon, 1980; Mac Nally, 1994a). Kremen (1992) used the ordination of species abundances to evaluate the response of butterflies to topographic/moisture and disturbance gradients. Meilleur *et al.* (1992) used multivariate analyses of species assemblages to identify groups of plant species that associate with particular drainage classes, topographic positions, and geomorphological characteristics of the substrate, as well as those taking advantage of disturbances of different kinds. Ordination of species on environmental or disturbance gradients will at least provide a sample of the range of ecological roles, and may form the basis for the selection of and a means of evaluating taxa that could act as indicators of ecosystem processes.

Examples of the selection of suites of indicators

Physical variables such as water quality or soil condition may reflect the status of ecosystems reliably (Noss, 1999). Some are targets for monitoring and management under agreements such as the Montreal Process for ecological sustainability (e.g. Santiago Declaration, 1995; Commonwealth of Australia, 1998).

The World Conservation Monitoring Centre (WCMC, 1992) recommended indicators for national environmental monitoring programs (Table 17.4), which are intended fulfil the basic information needs of national and international policy makers. Many of these indicators may provide useful feedback at a national scale, but they may be less useful for environmental assessment at the scale of individual projects. The list illustrates the need to choose indicators relevant to the questions and problems at hand, and at appropriate spatial and temporal scales.

The indicators used in the Australian State of the Environment report (State of the Environment, 2001a) highlight the need for social and political relevance (Table 17.5). Saunders *et al.* (1998) recommended 65 biodiversity indicators, none of which were species. Rather, they related to pressures on biodiversity (14 indicators), the condition of biodiversity (17 indicators), and responses to the loss or perceived threat to biodiversity (34 indicators). The indicators range from numbers of pests and endemic species, to the amount of funding for taxonomy and Australia's international role in conservation (State of the Environment Report, 2001a).

As Suter (1993) emphasised, conservation is only one of several factors contributing to environmental decision-making. If indicators are to serve the purpose of providing evidence that management goals have been satisfied, then they must also satisfy the expectations of policy makers, politicians, and others with a stake in the use of the environment. As a result, the selection of indicators has a social dimension. This sentiment is clear in the goals of the indicators in the Montreal Criteria adopted for forest management by the Australian Government (Commonwealth of Australia, 1998): '*indicators provide measures of change in criteria which describe broad forest values that society wishes to maintain*'.

Table 17.4. A recommended set of indicators for monitoring biodiversity at a national level (after WCMC, 1992).

Indicator	Genetic diversity	Species diversity	Community diversity
Wild species and genetic diversity			
Species richness	x	x	
Species threatened by global extinction	x	x	
Species threatened by local extinction	x	x	
Endemic species	x	x	
Endemic species threatened by extinction	x	x	
Species risk index	x	x	
Species with stable or increasing populations	x	x	
Species with decreasing populations	x	x	
Threatened species in protected areas	x	x	
Endemic species in protected areas	x	x	
Threatened species in *ex situ* collections	x	x	
Threatened species with viable *ex situ* populations	x	x	
Species used by local residents	x	x	
Community diversity			
Area dominated by non-domesticated species		x	x
Change from non-domesticated to domesticated species		x	x
Area dominated by non-domesticated species in patches greater than 1000 square kilometres		x	x
Area strictly protected		x	x
Domesticated species			
Crops and livestock in *ex situ* storage	x		
Accessions regenerated in the past decade	x		
Crops and livestock grown as a percentage of the number 30 years before	x		
Number of varieties as a percentage of the number 30 years before	x		
Coefficient of kinship or parentage of crop	x		

Table 17.5. Biodiversity indicators (BD in column 1 of the table) for State of the Environment reporting (from State of the Environment, 2001a).

No.	Title
BD 1.1	Human population distribution and density
BD 1.2	Change in human population density
BD 2.1	Extent and rate of clearing or major modification of natural vegetation or marine habitat
BD 2.2	Location and configuration or fragmentation of remnant vegetation and marine habitat
BD 3.1	Rate of extension of exotic species into IBRA region
BD 3.2	Pest numbers
BD 4.1	Distribution and abundance of genetically modified organisms
BD 5	Pollution
BD 6	Areal extent of altered fire regimes
BD 7	Human-induced climate change
BD 8.1	Lists and numbers of organisms being trafficked and legally exported
BD 8.2	No. permits requested and issued for legal collecting or harvesting by venture
BD 8.3	Proportion of numbers collected over size of reproducing population
BD 8.4	Ratio of by-catch to target species
BD 9.1	No. subspecific taxa
BD 9.2	Population size, numbers and physical isolation
BD 9.3	Environmental amplitude of populations
BD 9.4	Genetic diversity at marker loci
BD 10.1	No. species
BD 10.2	Estimated no. species
BD 10.3	No. species formally described
BD 10.4	Percentage of number of species described
BD 10.5	No. subspecies as a percentage of species
BD 10.6	No. endemic species
BD 10.7	Conservation status of species
BD 10.8	Economic importance of species
BD 10.9	Percentage of species changing in distribution
BD 10.10	Number, distribution and abundance of migratory species
BD 10.11	Demographic characteristics of target taxa
BD 11.1	Ecosystem diversity
BD 11.2	Number and extent of ecological communities of high conservation potential
BD 12	Integrated bioregional planning
BD 13.1	Extent of each vegetation type and marine habitat type in protected areas
BD 13.2	No. protected areas with management plans
BD 13.3	No. interest groups involved in protected area planning
BD 13.4	Resources committed to protected areas
BD 14	Proportion of bioregions covered by biological surveys
BD 15.1	No. recovery plans
BD 15.2	Amount of funding for recovery plans
BD 16.1	No. *ex situ* research programs
BD 16.2	No. releases to the wild from *ex situ* breeding
BD 17.1	No. management plans for ecologically sustainable harvesting
BD 17.2	Effectiveness of by-catch controls
BD 18.1	Area of clearing officially permitted
BD 18.2	Ratio of area cleared to area revegetated
BD 18.3	No. lending institutions considering biodiversity
BD 19.1	No. management plans for exotic/alien/genetically modified organisms
BD 19.2	No. research programs for exotic/alien/ genetically modified organisms
BD 19.3	Funding for research and control of exotic/alien/ genetically modified organisms
BD 20	Control over the impacts of pollution
BD 21	Reducing the impacts of altered fire regimes
BD 22	Minimising the potential impacts of human-induced climate change on biodiversity
BD 23.1	No. local governments with management plans for biodiversity
BD 23.2	No. companies with management plans for biodiversity
BD 24.1	No. species described per reporting cycle
BD 24.2	No. taxonomists involved per reporting cycle
BD 24.3	Amount of funding for taxonomy
BD 24.4	No. research programs into surrogates
BD 24.5	No. research programs into the role of biodiversity in ecological processes
BD 24.6	No. long-term ecological monitoring sites
BD 24.7	Percentage of budgets spent on conservation
BD 24.8	Amount of indigenous ethnobiological knowledge
BD 25.1	Local government management of biodiversity
BD 25.2	Involvement of community groups in conservation
BD 26	Australia's international role in conservation

17.5 Conclusions

Given data from carefully designed surveys and experiments, it is theoretically possible to control the twin statistical errors of accepting the hypothesis when it is false, and rejecting it when it is true. However, the question remains of how to specify the sizes of the environmental mistakes society is willing to tolerate, and therefore how to set the standards of proof (the effect size) by which the statistical tests are to be evaluated. Setting a standard of proof requires an assessment of the seriousness of the risk in question. Thus, people are generally more willing to risk environmental error that causes the deaths or damage of natural populations than to risk human lives (Fischhoff *et al.*, 1982). A standard of proof cannot reasonably be set without an appreciation of its consequences for the perceived fairness of the decision. Such judgements require legal, technical, social and ethical considerations.

In classical statistical inference, the type I error rate is controlled to be equal to or less than 0.05. In contrast, the type II error rate is not controlled. Effective decision-making involves a balance between type I and type II errors (Mapstone, 1995). This means that the trade-offs implicit in a decision need to be balanced. For example: How bad would it be to ban a harmless pesticide? There may be substantial economic costs or famine in some localities. How bad would it be to license a dangerous pesticide? There is a possibility that there would be a greater incidence of cancers than would otherwise be the case, and there may be sublethal impacts on natural populations of vertebrates that live in freshwater streams. Are these effects worth the benefits?

Environmental decisions should be based on formal statements of the costs and benefits of alternatives. Conservation biology is particularly concerned with costs and benefits to natural populations, but weighing public sentiment and political mandates against scientific rigour is a difficult balancing act. Charismatic species play an important part in raising public awareness and are often the focus of management plans and relatively costly recovery efforts.

Conservation biology has attempted to identify species that can act as indicators for other species or attributes that are too difficult and/or expensive to measure. Assumed causal links between the dynamics of those taxa and the entities for which they are purported to indicate have rarely, if ever, been tested. Solutions to this problem will lie in selecting a broader organisational and taxonomic spectrum of indicators, including physical variables such as those used in the State of the Environment report (Saunders *et al.*, 1998). Social and political values influence choices about indicators and the appropriate trade-off between environmental and other values in conservation decision-making.

17.6 Practical considerations

The design and execution of monitoring studies should encompass considerations of statistical power, the risks of type I and II errors and the consequences of such errors. Conservation biologists have a special responsibility to be aware of type II error rates and to estimate sample sizes before a study commences.

The use of some types of indicators is essential for any form of monitoring of assessment of environmental effects or impacts. Whether such indicators are species, physical structures or other entities, it is essential to be aware of their limitations and, on the basis of these acknowledgements, to make explicit statements about the uses for which they are intended.

Monitoring programs should be constructed around the idea of validating predictions and testing presumed relationships between supposed impacts and ecological responses to them.

17.7 Further reading

Mapstone (1995) outlines strategies for accommodating both type I and type II errors in ecological assessment and monitoring programs. Goldsmith (1991) provides an overview of methods of monitoring for conservation. Values of type II error rates or power are given for specific tests, sample sizes, and measures of effects or variability in several books and computer programs. Fairweather (1991) provides some examples along with an extensive list of references, including 'how to' guides for calculating power. Hatch (2003) discusses statistical power with respect to seabird monitoring and cites some computer packages to assist with such analyses.

Thomas (1996) provides a general introduction to the context and structure of environmental impact assessment in Australia. The Ecological Society of Australia (2003) has a position statement on the conduct of environmental impact assessments in Australia and steps needed to improve the process. Quinn and Keough (2002) introduce aspects of experimental design and statistical analysis relevant to ecological monitoring programs. Welsh *et al.* (2000) outline the statistical basis for overlapping and rotating designs for monitoring programs, and a worked example of non-standard monitoring is given in Lindenmayer *et al.* (2003b).

Spellerberg (1994) provides a discussion of environmental change, monitoring, and the selection of indicators for different circumstances. Landres *et al.* (1988), Simberloff (1998), Lindenmayer *et al.* (2000c) and Rolstad *et al.* (2002) provide detailed critiques of the indicator species approach and outline strategies that may assist with the selection of alternative types of indicators, particularly in a forest biodiversity conservation context. Roberge and Angelstam (2004) give a useful overview of the umbrella species concept. Burgman (2005) outlines statistical process control and related approaches to monitoring. Saunders *et al.* (1998) provide a list of detailed ecological, social and political indicators for measuring the state of biodiversity in Australia, and the State of the Environment Report (2001a) sets out comprehensive documentation against those indicators.

Chapter 18

Risk assessment

This chapter deals with methods for estimating extinction rates among plants and animals, the likelihood that a single species has become extinct based on observational records, extinction proneness, and methods for conceptualising and estimating conservation risks.

18.0 Introduction

Risk is the probability that an adverse event with specific consequences will occur in a given time frame, and **risk assessment** is the process of estimating risks. Risk assessment involves building conceptual, qualitative or quantitative representations of the way the natural world works. Risk assessment may focus on a process, such as the diffusion of a chemical from an industrial source, or the dynamics of a natural population and the way in which it interacts with landscape changes. Models are necessary because people can be bad judges of risk. We carry with us a set of psychological disabilities that can make it next to impossible for us to visualise and communicate risks reliably. The realisation that we judge as badly as we do is relatively recent. In the 1970s, psychologists began doing experiments on the ways in which people perceive and react to risks. Their results were strikingly counter-intuitive and led to exciting generalisations. Cognitive biases are heightened by the politically charged and value-laden contexts of most risk assessments, particularly those in conservation biology (Burgman, 2005).

18.1 Estimating extinction rates

At the level of communities within regions, the **theory of island biogeography** (MacArthur and Wilson, 1967) is sometimes used to predict numbers of extinctions. At equilibrium between immigration and extinction, the number of species (S) on an island (or in a patch of habitat) is estimated by:

$$S = cA^z \qquad (18.1)$$

where A is area and c and z are fitted constants (see Chapter 14 on estimating the number of species). The proportional change in the number of species (ΔS) is given by (Preston, 1962; Koopowitz *et al.*, 1994; Dial, 1995):

$$\Delta S = 1 - \left(\frac{S_1}{S_0}\right) = 1 - \left(\frac{A_1}{A_0}\right)^z \qquad (18.2)$$

where S_0 and A_0 are the number of species and the area, respectively, before fragmentation, and S_1 and A_1 are the number of species and the area, respectively, after fragmentation. Normally, the value of z is taken to be about 0.25 (Preston, 1962), with a range of about 0.15–0.40 (Connor and McCoy, 1979; Dial, 1995). The equation predicts that the loss of 90% of the habitat in an area will result in a loss of between 30 and 60% of the species present (depending on the value of z).

Other expressions have been proposed to predict the relationship between area and number of species (at equilibrium; see Chapter 4). No single model fits all data best (Connor and McCoy, 1979). Typically, confidence intervals are wide, particularly outside the domain of the data from which the parameters were estimated.

Box 18.1

Faunal relaxation and hotspots

Information in the text gives some equations for calculating species loss as a function of area. Of course, any area is spatially heterogeneous. Calculations of species loss with area change make unrealistic assumptions about the homogeneity of the landscape. In fact, small areas of a landscape can be hotspots for particular species. For example, serpentine soils support many endemic and highly restricted plants (Whittaker, 1954a,b). Seasonal and permanent rock pools on the tops of granite outcrops such as inselbergs (e.g. Uluru in the Northern Territory) support highly specialised floras (Hopper, 2000). Spawning grounds for fish and over-wintering grounds for butterflies may be species-rich but very localised. Caves often support numerous endemic species in a relatively small area (Hose, 2003). Culver *et al.* (2000) estimated that there were 927 species in the 48 contiguous states of the USA that were limited entirely to caves and associated subterranean habitats. Many of these species are highly restricted; almost half (44%) of the cave-dependent aquatic species and subspecies have ranges limited to a single county (Culver *et al.*, 2000). Even if a large area of a landscape was destroyed, highly species-rich cave assemblages could remain intact. Likewise, many species might be lost if, for example, localised mining operations targeted rock deposits around caves.

Serpentine rocks are unusual geological formations characterised by soils with extremely low levels of nutrients. They can sometimes support unusual plant assemblages, including stands of large Grass Trees. (Photo by David Lindenmayer.)

There is no theory that predicts the rate at which the new equilibrium will be approached (Simberloff, 1992; Doak and Mills, 1994). **Faunal relaxation** is the decline in species richness following isolation of a habitat patch, and **extinction debt** is the difference between current species richness and the number of species expected at equilibrium (Chapter 10). The situation is complicated by extinctions resulting from processes other than habitat loss, for example the impacts of feral predators (see also Box 10.7).

18.2 Estimating the likelihood of extinction from collections

In conservation biology, it is often necessary to make inferences about the status of species in circumstances in which data are scarce. The only information for many species is from museum and herbarium collections. Estimates of risk rely heavily on the opinions of taxonomic experts or simple **heuristics**. For example, it is conventional to conclude that a species is extinct if it has not been recorded for 50 years (Smith *et al.*, 1993). However, the period that we are prepared to wait before we conclude that a species has become extinct should be conditioned by the frequency with which it was seen before the last observation. If a species was often seen, or was easy to see, the time we are prepared to wait should be less than if the species was difficult to observe, cryptic or inhabited a remote area.

Scientific collections are time series composed of dates and locations, sometimes with additional information such as vegetation descriptions or habitat information. The period of time between the first record of a species and the present time can be conveniently partitioned into equally sized units of time or effort (McCarthy, 1998).

If we assume that collections are random, so that the observations are distributed randomly among the time intervals, and the average chance of collecting a species does not change over time, then the probability (p) that the species has not become extinct is given by (Solow and Roberts, 2003; see also Solow, 1993a; Burgman *et al.*, 1995b):

$$p = \frac{T_n - T_{n-1}}{T - T_{n-1}} \qquad (18.3)$$

The period ($T_n - T_{n-1}$) is the interval between the most recent sighting (T_n) and the second most recent sighting (T_{n-1}). ($T - T_{n-1}$) is the interval between the second most recent sighting (T_{n-1}) and the end of the observation period (T). The value of p is the probability of observing a ratio as large or larger, by chance alone, given the null hypothesis that the sighting rate is constant (i.e. the species is not extinct).

The equation results in relatively large p values (i.e. small probabilities of extinction) if a species was observed relatively recently. If the probability is small, we can conclude that the assumptions are wrong and infer that the population may have declined in range or abundance. For the equation to be of use, the chance of collection should reflect the relative abundance of the species.

Table 18.1. Observation data for *Acacia aprica* (B. Maslin, personal communication) and two species of Stick-nest Rats (Copley, 1988).

	Blunt wattle	Greater Stick-nest Rat	Lesser Stick-nest Rat
	1957	1835	1835
	1962	1844	1853
	1967	1857	1857
	1970	1879	1872
	1973	1896	1873
	1976	1897	1903
	1980	1907	1929
		1920	1933
		1921	
		1930	
		1933	
		1938	
		(1986)	
Probability species is extant	0.235	0.094	0.062

Each species was observed in the years listed here. The end of the observation period for both rats was 1988, and for the Acacia it was 1993. The probabilities that the species still exist were calculated using the expression by Solow and Roberts (2003). The 1986 observation of the Greater Stick-nest Rat is the result of a specific effort to find the species and is ignored in the calculations.

Table 18.1 shows records of collections of the Greater and Lesser Stick-nest Rats, which are rare and endangered species from arid central Australia (Copley, 1988; Moseby and Bice, 2004), and Blunt Wattle – a rare shrub from Western Australia. Most collections were made opportunistically because both species inhabit relatively remote places.

The approach outlined in the preceding paragraphs suggests non-random patterns in the collection records for both rats, especially the Lesser Stick-nest Rat. The Greater Stick-nest Rat was not observed between 1938 and 1986. However, this species survived on an offshore island, and there is a program to reintroduce the species to mainland Australia (Moseby and Bice, 2004) and other Australian islands (Richards *et al.*, 2001). The Lesser Stick-nest Rat is considered extinct. Blunt Wattle

Franklin Island off Ceduna in South Australia, where populations of the Greater Stick-nest Rat have been researched for several decades. (Photos by David Lindenmayer.)

occurs in two populations: one on a road verge and one in a small reserve. The pattern of observations for the species suggests that there may have been a change in its status after 1980, but there is no compelling evidence that it is extinct.

The definition of time intervals makes a difference to the probabilities (Grimson *et al.*, 1992). Generally, the finer the scale of resolution (the smaller the time interval), the more sensitive the tests will be. The interval should be decided by the resolution in the available data unless there is a reason the data should be aggregated. For example, sightings made on a monthly basis might reflect seasonal changes in abundance or survey effort, and under these circumstances it would be best to aggregate the data into years.

Equation 18.3 assumes a process in which collections are opportunistic. The most serious problem is likely to be that collection frequencies will reflect changing trends in museum and herbarium collections. For example, efforts may focus on rare species at the expense of common species. Such a bias will tend to mask real decreases in population abundance.

Equation 18.3 should be used together with other information to infer relative threats to different taxa, where such information exists. For example, if the equation suggests a non-random pattern in the collections for a species, it may do so because the taxon has been the focus of an ecological study, because its taxonomy has been confounded with other species, or because a special expedition was mounted to a remote region.

Given the biases in scientific collections, a probability generated by Equation 18.3 is unlikely to be convincing about the fate of any single species. Rather, it is a potentially useful screening tool for large numbers of taxa, especially in circumstances where expert experience is lacking and most collections have been opportunistic.

The equation examines only one dimension of conservation status: past trends in distribution and abundance. Species are also considered threatened when their populations are small or geographically restricted, or if there are foreseeable threats that will result in population decline (Chapter 3). Equation 18.3 should be used together with whatever other information is available.

18.3 Population management and risk

If the long-term growth rate of a population is negative, it will become extinct, no matter how variable the environment. These populations are the victims of **systematic pressure** and their extinction results from deterministic causes (Shaffer, 1981; Caughley and Gunn, 1996). Populations that would persist indefinitely in a constant environment nevertheless face some risk of extinction through variation in average birth and death rates or because of demographic accidents or chance catastrophes. These populations are the victims of **stochastic** factors (see Chapter 3). Over a given time period, there is a chance that any population will become extinct, or fall below a specified small population size. This is so, even if the population would persist indefinitely in a uniform environment. This is the **background risk of extinction** or decline.

Risk analysis uses probabilistic language to describe impacts on populations. 'Added risk' is the increase in risk that results from some impact; it provides a means to compare the impacts of different management practices (Burgman *et al.*, 1993). Ecological risk analysis uses models to understand ecological systems and estimate risks. The models summarise the way a system works, at a scale and level of detail that reflect data and the problems we need to solve.

Diagrams are the simplest form of a conceptual model. They communicate the scale of the problem, the level of detail, conceptual compartments (populations, age classes, habitat fragments), and discontinuities such as boundaries in habitat mosaics. Figure 18.1 is a conceptual model of a population of Ibex in Khirthar National Park, Pakistan (Yamada *et al.*, 2004). The figure summarises decisions by biologists to recognise three kinds of animals (juveniles, yearlings and adults), two processes (survival, *s*, and reproduction, *f*) and an annual time step (because reproduction is pulsed,

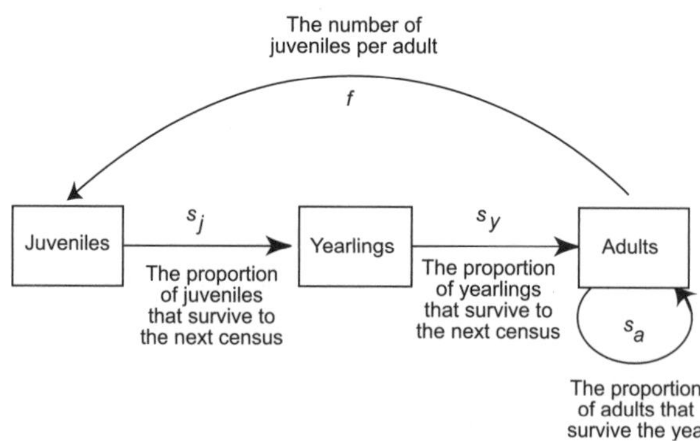

Figure 18.1. Conceptual model for a single population of Ibex in Khirthar National Park, Pakistan (after Yamada *et al.*, 2004).

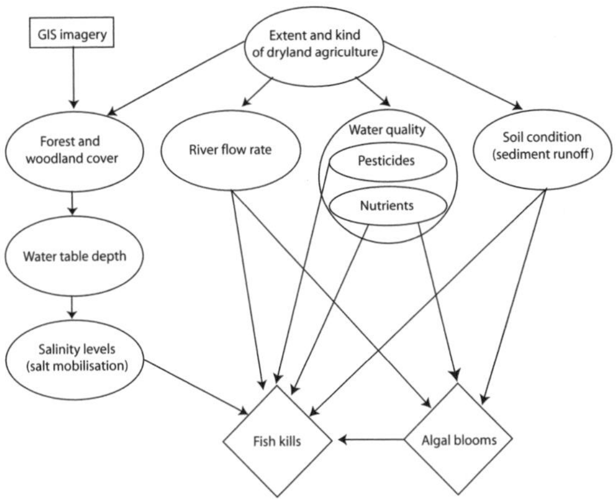

Figure 18.2. Influence diagram showing conceptual relationships among system components in a freshwater catchment (after Hart *et al.*, unpublished data).

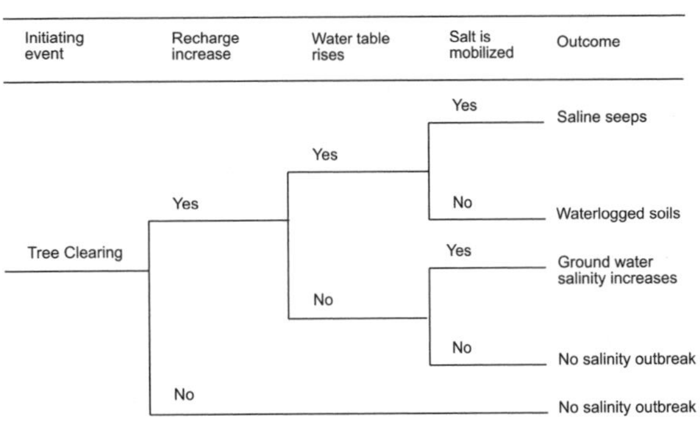

Figure 18.3. Event tree analysis for assessing risk of salinity effects after tree clearing (from Bui, 2000).

reflecting annual monsoonal rains). Conceptual models such as this are the basis for population viability analysis (outlined in Section 18.5).

Influence diagrams represent the functional components and dependencies of a system. Shapes (ellipses, rectangles) represent variables, data and parameters; arrows link the elements, specifying causal relations and dependencies. Figure 18.2 was part of a modelling exercise designed to predict the conditions under which fish kills might occur in the Murray–Darling Basin, including those that might affect populations of the endangered Murray Cod.

Logic trees are diagrams that link processes and events that could lead to a hazard (a situation that can cause harm), or could flow from it, if it occurs. **Fault trees** link chains of events to the outcome. **Event trees** take a triggering event and describe its consequences. These methods are closely related to other techniques that are not discussed in this book, for example decision trees, decision tables, classification and regression trees, and Bayesian networks (see Burgman, 2005, for an introduction).

Figure 18.3 shows an event tree used to understand the ecological consequences of land clearing in Queensland. This example illustrates some of the assumptions in a logic tree. Its discrete nature is awkward when applied to continuous processes. For example, recharge is the volume of water that flows into an underground aquifer. If trees are cleared, the amount of water reaching the aquifer may increase because the transpiration rate of the vegetation is reduced. But how much change qualifies as an increase? The volume of water flowing into the aquifer is a continuous variable, so the question is whether the recharge rate has increased substantially compared with any increase that may have been expected in the presence of the trees (Burgman, 2005).

The International Union for the Conservation of Nature (IUCN, 1994, 2001) developed a protocol for assessing the risk of extinction of species (see Chapter 3; Figure 3.9; Table 3.6). The intent is to classify each species as belonging to one of the following categories: extinct, critically endangered, endangered, vulnerable, near threatened, or safe. The system is usually represented as a table (see Table 3.6 in Chapter 3). The overall structure of decisions about the status of a species is given in Figure 18.4. Species can be assessed as extinct, near threatened, conservation dependent or low risk. Thus, the table can be represented as a decision tree that ends in a classification of species. Each branch in the decision tree asks a question. The answer to the question (yes/no) is determined by the conditions outlined in the table. The logic in this table can be reduced to a few rules. For example, a species is regarded as being critically endangered:

IF Decline of ≥ 80% in 10 years or three generations
OR Range < 100 square kilometres or occupied habitat < 10 square kilometres
AND at least two of the following three conditions are met:
1) severely fragmented or in one subpopulation
2) continuing to decline
3) fluctuations > 1 order of magnitude

OR <250 mature individuals
AND at least one of the following two conditions are met:
1) ≥25% decline in 3 years or one generation
2) continuing decline and one subpopulation or ≤50 per subpopulation
OR <50 mature individuals
OR ≥50% risk of extinction in 10 years or three generations.

This logic can in turn be represented as a fault tree (Figure 18.5) and the condition of being 'critically endangered' and it occurs if the various conditions are

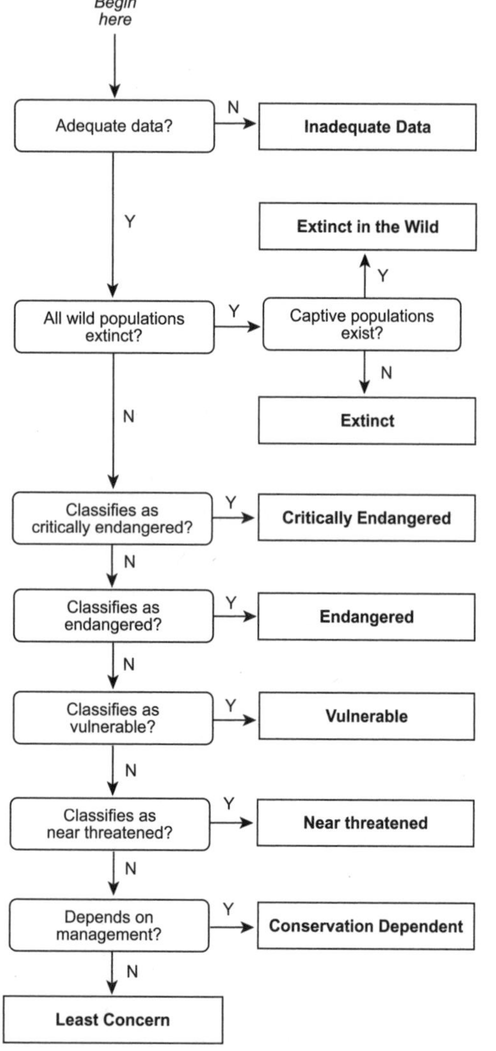

Figure 18.4. Structure of International Union for the Conservation of Nature classification decisions (after Akçakaya and Colyvan, in Burgman and Lindenmayer, 1998). See Figure 3.9.

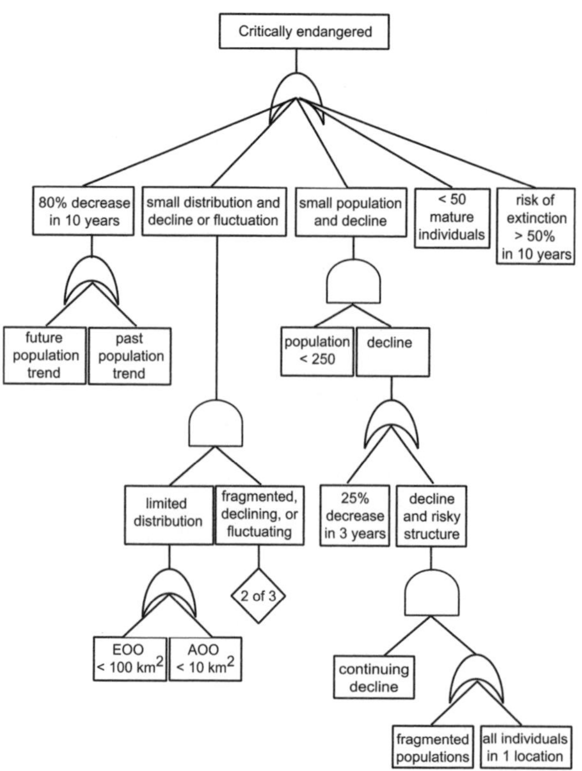

Figure 18.5. Logical structure of the International Union for the Conservation of Nature (IUCN, 2001) fault rule set (resulting in the classification of a species as either critically endangered or not critically endangered; after Burgman, 2005).

met. If the fault is true, the species is classified at that level (Figure 18.5).

The logic tree for classifying conservation status imposes sharp boundaries such as 80%, 10 years and three generations on vague criteria. The logical structures and levels of thresholds affect decisions. As with other trees, the best we can do is to accumulate outcomes by monitoring the consequences of decisions, and updating the thresholds and structures iteratively (Burgman, 2005).

Statements about the risks faced by natural populations have two important components: a time frame and a risk level (Chapter 3). Statements are expressed in terms such as 'the population has a 95% chance of surviving for at least 50 years'.

Often, assessing the risks faced by a population (or a species) involves estimating the chance that it will fall below some specified size. This population size can be thought of as an unacceptably small, critical threshold. It could be the population size at which demographic accidents, behavioural constraints, or genetic factors such as **inbreeding depression** begin to play an important role in the survival of a species.

Ginzburg *et al.* (1982) suggested the use of the term **quasi-extinction risk**, which is the chance of crossing some small population threshold – a lower bound that may be unacceptable for conservation, management, economic or aesthetic purposes. The question is usually phrased as 'what is the chance the population will fall below the critical population threshold size (N_C) at least once in the next *t* years, given an initial population size of N_O?'.

In many models developed for conservation biology (e.g. Leigh, 1981; Goodman, 1987a,b; Quinn and Hastings, 1987; Pimm *et al.*, 1988; Jackson, 2000), time to extinction is used to express risk. Because the longevity of a species cannot be predicted exactly, the shape of the distribution of the predicted time to extinction is important. Expected persistence times are not distributed in a regular fashion where the mean, median and mode have the same value. Shaffer and Samson (1985), for example, point out that although a Yellowstone Grizzly Bear population has an expected persistence time of 114 years, there is a 56% chance that the population will be extinct in less than 114 years.

McCarthy and Thompson (2001) suggested measuring risk by the **minimum expected population** (MEP) – the smallest size to which a population is expected to fall in the next *t* years. This method provides a more complete summary of risk than time to extinction or the risk of quasi-extinction, and a more reliable way to compare options.

Biological intuition is useful for constructing models but it is less useful for assigning species to risk categories such as those outlined in Chapter 3. The functional relationships determining the dynamics of a population are mechanistic. However, people are poor judges of risks, even though judgement of risk is commonplace in human society. When estimating risks subjectively, people (including scientists) are often irrational and emotive (Zeckhauser and Viscusi, 1990), overestimating the likelihood of low probability events (e.g. death by cyclone) and underestimating higher risk levels (e.g. death by heart disease). Risk perception is coloured by such things as the visibility of the risk and the appearance of control of the circumstances. Such biases account for many emotional public responses and will eventually affect decisions made by regulatory organisations, politicians and wildlife managers (Burgman *et al.*, 1993).

Types of uncertainty

The events that lead to extinction may appear in retrospect to be deterministic. But when faced with judging the future of a population, predictions are uncertain. All population are engaged in a random walk in which birth and death rates are buffeted by the environment and other chance factors that make rates of population growth or decline variable. This variability (sometimes called stochasticity) has different sources, and different classifications of variability have been used (see Burgman *et al.*, 1993; Regan *et al.*, 2002). Phenotypic, demographic, environmental and spatial variation, and catastrophes are used conventionally to describe the variation that affects population growth (Akçakaya *et al.*, 1997).

Phenotypic variation is the variation between individuals within the same population. These individuals experience the same environmental and demographic conditions, and differences are due to genetic and developmental differences among individuals.

Demographic variation is the variation in the average chances of survivorship and reproduction that occur because populations are discrete, structured, and often quite small (May, 1973; McCarthy *et al.*, 1994). Consider an orchid population made up of 100 plants and with a growth rate of 1.1. Over the next 4 years, we predict that the population sizes will be 110, 121, 133.1 and 146.41 plants; however, there is no such thing as 0.1 or 0.41 of an orchid. The growth rate we specified is an average based on observations. This result says that in some 4-year periods, the population will increase to 146 plants, in others it will increase to 147. On average, over a 4-year period, there will be between 46 and 47 more germinations than deaths, but we cannot be sure exactly how many. Many other elements of population structure contribute to this kind of uncertainty. For example, in species with separate sexes, uneven sex ratios can arise by chance and affect population increases.

Extinction risk is inversely related to population size through phenotypic and demographic uncertainty, just as the variance of the binomial distribution is inversely related to sample size (Akçakaya, 1990). Large populations are inherently less variable (Thomas, 1990) because the chance events related to births and deaths tend to cancel one another out.

Environmental variation is unpredictable change in the environment through time. For example, even in circumstances where we know precisely the average annual rainfall based on records going back centuries, it is impossible to say with certainty if next year will be

Box 18.2

A Koala population crash in south-western Queensland

In the summer of 1979–80, approximately 63% of the Koala population along Mungallala Creek in south-western Queensland died. The drought and heat wave conditions of that summer caused extensive leaf fall and browning of foliage in food trees. The Koalas died from malnutrition and dehydration, and individuals showed evidence of poor condition, including anaemia and high levels of tick infestation. The habitat was heterogeneous, and in areas near permanent water, mortality was low. Mortality was highest among young animals, which may have been excluded from optimal habitat by older, dominant animals. The population was not threatened because animals could survive in areas close to permanent water, but the drought continued in subsequent years and prevented the immediate recovery of the population. The population crash was considered to be a rare event precipitated by unusual climatic conditions (Gordon *et al.*, 1988).

relatively wet or dry. Environmental variation results in fluctuations in population levels because environmental variables affect the number of survivors and the number of offspring in a population. If water is scarce, for example, more juveniles and adults may die than if water is plentiful (e.g. Box 18.2).

Spatial variation is the variation in environmental conditions at different places in a landscape. Most species consist of populations in more or less discrete patches of habitat (see Chapters 10 and 15), linked by dispersal. In most real circumstances, patches experience some environmental changes in common (e.g. winter) and other changes that are unique (e.g. the local water hole drying out). The patterns of **local extinction** and **recolonisation** of populations have profound effects on the risks of extinction of **metapopulations**.

Catastrophes include natural events such as floods, fires and droughts. They differ from environmental variation because they have a relatively large effect on the chances of persistence of a population – much greater than those environmental changes that cause the normal year-to-year fluctuations in population size. The category is useful because some ecological processes are driven by relatively frequent catastrophic events. In Australia, fire is one of the most important ecological processes determining the structure and composition of most terrestrial communities. Modelling fire explicitly is often useful because most species are adapted to its effects (see Chapter 11).

Hilborn (1987) classified rare and unexpected events and termed them, appropriately, 'surprises'. That does not mean that surprise itself is rare, only that each event is essentially unexpected. The terms of reference of Hilborn's definition are a little broader than the definition of catastrophes. Surprises include anything we do not expect – anything that is unaccounted for by our model or by intuition (Burgman *et al.*, 1993).

The terms **variability** and **incertitude** make a simple and useful taxonomy of uncertainty. Variability is naturally occurring unpredictable change – differences in parameters attributable to 'true' heterogeneity or diversity in a population. Incertitude is lack of knowledge about parameters or models. Incertitude can usually be reduced by collecting more and better data. Variability is better understood and more reliably estimated, but is not reduced, by additional data. These terms have been useful in routine applications of risk assessment (Hoffman and Hammonds, 1994; Kelly and Campbell, 2000). The distinction between variability and incertitude can help structure quantitative investigations of risk (Burgman, 2005) and influence priorities for collecting data. Regan *et al.* (2002) provide a more detailed taxonomy.

18.4 Expert judgement

Conservation assessments assign species to risk categories on the basis of such factors as population size, trends and geographic distribution (see Chapter 3). Decisions to list particular species or ecological communities under threatened species legislation are usually based on recommendations by scientific experts. For example, Table 18.2 gives the numbers of species of plants and animals thought to be at risk in several countries. Such lists are published and used routinely by governments for a number of purposes (Possingham *et al.*, 2002). However, the lists reflect the funding opportunities and personal interests of scientists and others who make recommendations (May *et al.*, 1995). For instance, the numbers of threatened insects, non-vascular plants and fungi are low because of scientific and popular lack of interest.

There is uncertainty associated with lists of threatened species, which is reflected in the turnover of

Table 18.2. Number of species listed as endangered in Australia, USA and China (each of these countries has relatively large numbers of endemic species) compared to the number thought to exist globally (data from Groombridge 1994; May *et al.*, 1995; Burgman, 2002; see Burgman, 2004). See also Table 6.3.

Taxon	Estimated total number of species in the world[1]	Number listed as 'threatened'		
		Australia	USA	China
Fungi	1 000 000	0	0	0
Non-vascular plants	100 000	0	0	0
Vascular plants	250 000	1597	1845	343
Invertebrates	5 000 000	372	860	13
Fish	40 000	54	174	16
Amphibians	4000	20	16	0
Reptiles	6000	42	23	8
Birds	9500	51	46	86
Mammals	4500	43	22	42

[1]The global numbers of species in each taxon were estimated crudely from numbers of currently described species and expert judgments of the proportion remaining to be described (see May *et al.*, 1995; Burgman, 2002).

species in the lists through time. For instance, Briggs and Leigh (1996) created a list of 'threatened' Australian plants, listing 4955 species of vascular plants 'at risk', that is, 'endangered, vulnerable, rare or poorly known and thought to be threatened'. Briggs and Leigh's list was modified as new information came to hand, as people and agencies responsible for the list changed, and as taxonomists revised species descriptions. Five years later, only 65% of the species in the earlier list remained on the official government list (Burgman, 2002).

Some of the reasons for changes to endangered species lists are illustrated in Table 18.3. Changes in attitudes to extinction and taxonomic uncertainty caused many changes in lists of plants. In some instances, the listing authority became more convinced of a taxon's extinct status after reconsidering the literature or specimens. Expert judgment weighed heavily on these outcomes.

Table 18.3. Reasons for changes to lists of extinct Australian vascular plants since 1981 (after Keith and Burgman, 2004). See also Table 3.5.

Reason for change in listing status	Additions	Deletions
Presumed extinction during 1981–2001	8	
Change in attitude to extinction uncertainty	45	1
Taxonomic change	6	25
Change in taxonomic uncertainty	7	12
Not previously evaluated	46	
Rediscovered		86
Correction		35
Change in opinion on native status		3
No change despite rediscovery		13

However, the considerable uncertainty in the status of listed species is not communicated. The consequence of failing to acknowledge and communicate uncertainty is that people are misled into believing that the lists are reliable and unbiased.

Scientific experts and groups are expected to make reliable, transparent and consistent judgments on behalf of the broader community. However, the reliability of expert judgments depends on how the experts are interrogated. Methods for elicitation can include correspondence, personal interviews, traditional meetings, structured group meetings aimed at achieving consensus, and meetings that combine consensus with numerical aggregation.

Context can corrupt the quality of expert judgement. Experts sometimes make judgments just to be helpful, or to retain the semblance of expert respectability (because the experts have been coerced into providing an answer, or won't admit that they can't provide an answer). Many conceptual models are complex, and for ecological systems in particular, there may be more than one plausible opinion. Time is limited, the process of providing expert judgement is demanding, and people tire of estimating parameters and assessing outcomes (Burgman, 2005).

Experts are susceptible to a range of psychological frailties (Kahneman and Tversky, 1979; Fischhoff, 1995; Slovic *et al.* 2000; see Burgman, 2002). Elicitation is particularly error-prone when the topic involves very

Table 18.4. Positions on use and uncertainty, and institutional affiliation of 36 marine turtle experts (after Campbell, 2002).

Positions	Non-governmental organisation	University	Government	Total
Consumption is supported, learn through use	3	5	2	10
Support limited consumptive use	2	2	0	4
Uncertainty dictates caution, consumptive use is not supported	5	7	7	19
Consumptive use is unacceptable	0	2	1	3

low probability events such as extinction. Conservation listing decisions often involve novel circumstances where data and prior experience are limited. Conservation risk assessments are invariably subject to distorting influences, perhaps more so than other types of scientific analysis, because of the politically charged and value-laden nature of many of the problems. Typically, there is considerable pressure to produce results to diffuse social tension. Unfortunately, the experts are themselves susceptible to the same set of pressures. They cannot occupy the independent, objective ground that politicians and policy makers wish them to (Campbell, 2002).

People acting as champions for a project are likely to emphasise the potential benefits and under-state the potential costs – a consequence of enthusiasm and ambition, both of which breed optimism. Motivational biases are tied to ethical positions, and experts are often unaware of the value-laden nature of their opinions. Campbell (2002) interviewed experts interested in the conservation and management of marine turtles. The experts were asked about the sustainable consumptive use of turtles and their eggs. Campbell's results revealed four 'positions' on use, all of which were defensible on scientific grounds (Table 18.4).

A total of 19 experts thought that uncertainty dictates caution, so that managers should err on the side of non-use. Three experts were implacably opposed to commercial use, irrespective of uncertainty. Most experts saw opposing views as influenced by 'emotions', at the same time claiming dispassionate scientific objectivity for their own views, irrespective of their positions. Participants attributed emotional involvement to conservationists without biological training, as though scientific training protects people against emotion. Few propositions are so plainly self-deluded (Burgman, 2005).

Routinely reliable estimates of parameters or the chances of specific outcomes based on expert judgements have only ever been demonstrated in people who make frequent, repeated, easily verified, unambiguous predictions so that they learn from feedback, for example weather forecasters (Figure 18.6).

Expert engineers underestimated the lifetimes of long-lived components and over-estimated the lifetime of short-lived components. Army doctors' subjective probabilities that patients had pneumonia were a poor predictor of the outcomes of more reliable diagnoses based on radiography. The performances of experts in business and military intelligence are likewise mixed and unimpressive (Morgan and Henrion, 1990).

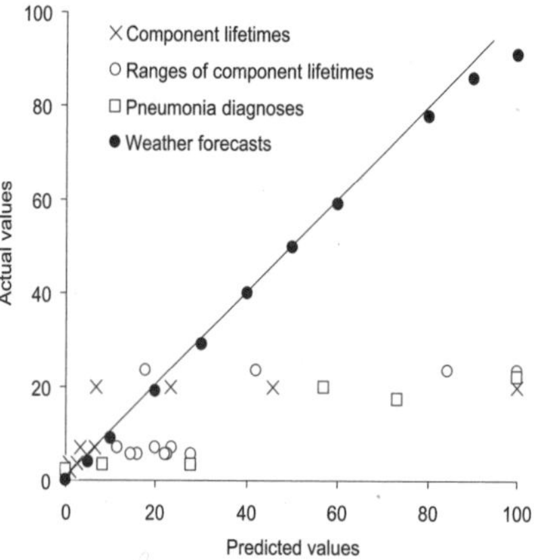

Figure 18.6. Calibration curves: expert predictions plotted against actual outcomes (after Burgman, 2005). Crosses are estimates by engineers of the mean lifetime of components in nuclear power systems, versus measured lifetimes. Open circles are estimates of the ranges for the mean lifetimes of the same components, versus measured ranges. Ranges are expressed as the maximum divided by the minimum. The squares are army doctors' subjective probabilities that patients have pneumonia, versus more reliable diagnoses based on radiography. Solid circles are meteorologists' predictions for the probability of rain on the following day, against the observed relative frequencies. The diagonal line provides the line of correct estimation for all sets of observations. Values are scaled so that the maximum value in each set is 100. (Data from Christensen-Szalanski and Bushyhead, 1981; Cooke, 1991; Plous, 1993.)

Group assessments may have advantages over those made by individuals. In general, if we want to estimate an underlying fact, if participants act rationally (from the perspective of classical probabilities) and are provided with information that reflects utilities and expectations in an unbiased way, group (consensus) judgments can be expected to be better than individual ones. Multiple experts with different backgrounds increase the extent of knowledge and experience contributing to an assessment (Clemen and Winkler, 1999).

However, the larger and more diverse the set of experts, the greater will be the possibility for disjoint opinions. Cognitive psychology theory predicts that conflict between experts can be minimised if the people share common values, experiences, professional norms, context, cultural background, and stand to gain or lose in the same way from outcomes. However, stratifying a sample of experts to include a range of demographic and social attributes increases the chances of disagreement, and reduces the chances of bias. Selecting experts with a narrow set of social attributes will tend to underestimate uncertainty. In groups, strong personalities influence outcomes, participants promote decisions before examining the problem, views are anchored, change is resisted, people hold covert opinions, and there is substantial pressure for conformity.

For example, salmon have great economic, social and cultural value in the Pacific north-west of the USA. Ruckelshaus *et al.* (2002) described how substantial declines in wild salmon populations have been attributed to habitat degradation, dams, harvesting, fish hatcheries, predation and invasion of exotic organisms. There is extensive disagreement among experts over the causes of decline, and the data on the problem are often unavailable or equivocal. For instance, some salmon populations have recovered following reductions of harvests, but others have not. Recovery teams represent a range of stakeholders, but all members have technical backgrounds. Ruckelshaus *et al.* (2002, p. 691) noted that '*major technical disagreements stemming from philosophical differences that seem to run as deep as religious beliefs are commonplace in such technical teams*'.

Regan *et al.* (2003) documented uncertainty among experts in conservation assessments. They asked 18 conservation biologists to assess the status of 13 species from a broad range of taxonomic groups. The assessors were provided with standard information sets and interpreted them using the International Union for the Conservation of Nature (IUCN, 2001) logical rules

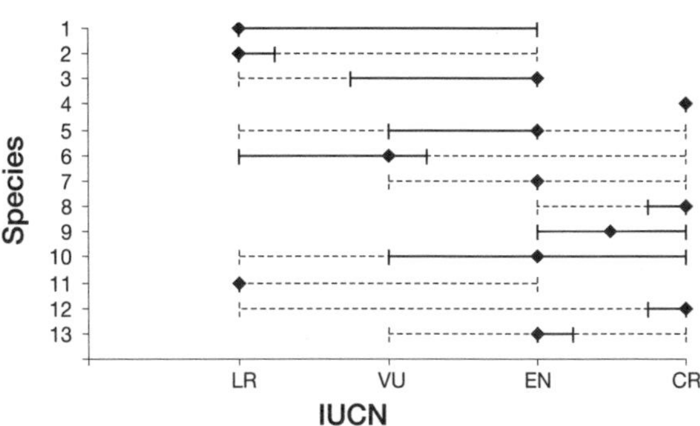

Figure 18.7. Assessments of the conservation status of 13 species made by 18 assessors (Regan *et al.*, 2003). The diamonds represent the median assessment. 50% of assessments are contained within the solid bars. The dashed lines show the full range of assessments.

(Figure 18.7). Different attitudes and interpretations of language in the guidelines and the information sets led some participants to be unwilling to make inferences. Some parameter estimates spanned thresholds in the rule set. Variation in qualitative judgments, inconsistent logic and mistakes entering data all played a role. The lesson from the study of Regan *et al.* (2003) is that expert judgments can differ, even when the only source of uncertainty is the subjective interpretation of language and data, which is a largely unavoidable element in almost all conservation risk assessments. It is always useful to eliminate arbitrary disagreement based on language or differences in the interpretation of terminology. Differences of opinion about scientific detail, as well as about substantial ethical issues, can be made transparent without compromising the ability of a group to reach a decision.

Akçakaya *et al.* (2000) used **non-probabilistic uncertainty methods** to combine parameters in the International Union for the Conservation of Nature (IUCN, 2001) criteria. The result of the analysis is a best estimate and bounds. The bounds encapsulate uncertainty about parameters, and uncertainty about dependencies between the elements of the rule sets (Figure 18.8).

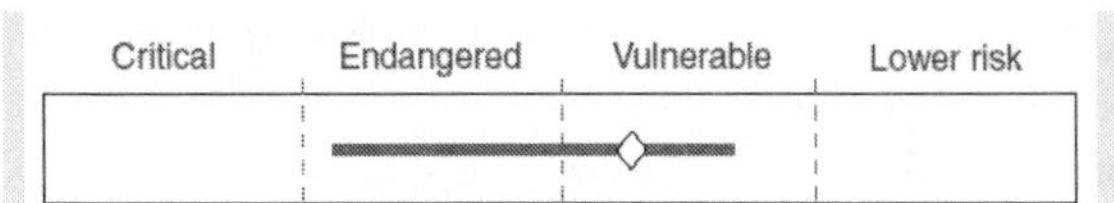

Figure 18.8. Example output from the method developed by Akçakaya *et al.* (2000). The species is most likely to be 'vulnerable', but it is possible that it is 'endangered'.

Almost all experts have personal, value-laden opinions about the outcomes of environmental decisions. The work of Campbell (2002), which was outlined earlier in this section, makes such issues plain. It is naive to think that scientists are anything but advocates of scientific positions; they also advocate personal value systems.

Making decisions to list species or communities on threat lists is a form of risk assessment, which is mediated largely by expert judgement. Unless experts are open to critical questioning, their opinions appear to be infallible (Walton, 1997). If we see scientists as advocates, valid questions from any source should be considered, and anyone with a stake in the outcome should be able to criticise an expert's opinion. Honest differences of opinion among experts – those that remain after linguistic uncertainties are eliminated – should be communicated to stakeholders with interests in the outcome. In many respects, the application of mathematical models in the form of population viability analysis is an attempt to alleviate the subjectivities inherent in expert judgement, and to ensure that arguments are internally rational and consistent with data and theory.

18.5 Population viability analysis

Natural populations live in uncertain environments. We cannot exactly predict the consequences of management options for any population. The best that can be done is to estimate the chances of particular outcomes based on the variations observed in the past and any mechanistic understanding of the population. If we ignore random fluctuations in population size by using measures of the asymptotic (very long-term) behaviour of the population, we miss the fact that the behaviour of the population in the short term may be important, perhaps deciding the fate of the population before the long-term behaviour becomes established.

This section describes the application of **population viability analysis** (PVA) for the measurement of impacts for wildlife management and conservation in terms of risks. Possingham *et al.* (1993, 2001) identified five main tasks for wildlife managers once a species has been identified as threatened:

1. Collate existing information.
2. List and cost management options.
3. Rank management options.
4. Use sensitivity analysis to test ranks and guide future research.
5. Implement the best option(s) together with monitoring and repeated re-evaluation of options.

PVA may be useful in steps 3 and 4. The objective of PVA is to provide insight into how management can influence the probability of extinction (Boyce, 1992; Burgman, 2000; Possingham *et al.*, 2001). PVA provides a basis to evaluate data and to anticipate the likelihood that a population will persist (Boyce, 1992; Burgman *et al.*, 1993; Possingham *et al.*, 2001). It may be seen more generally as any systematic attempt to understand the processes that make a population vulnerable to decline or extinction (Gilpin and Soulé, 1986; Shaffer, 1990; Possingham *et al.*, 2001). PVA has been used widely in Australia and there are many examples of its use in studies of a range of species (Table 18.5).

Models for PVA

In practice, PVA usually refers to building computer-based models of the likely fate of a population (Boyce, 1992). Probabilities are estimated by **Monte Carlo simulation.** The most appropriate model structure depends on the availability of data, the essential features of the ecology of the species and the kinds of questions that managers need to answer. PVA starts when people identify the factors affecting a species, and this information forms the basis of conceptual models of a species and its environment. Usually, mathematical models are then used to combine information into predictions about the persistence of species under different environmental conditions and management options.

Risk assessment is introduced into stochastic models very easily, at least conceptually. A deterministic model that represents our ecological understanding is the starting point. It may include age or stage structure, behavioural ecology, predation, competition, density dependence, or any other ecological mechanism influencing the future of the population. The result is a projection showing the expected future of the population – a single prediction without uncertainty.

Once the deterministic form of the model is established, elements of stochasticity are added to represent kinds of uncertainty. For **demographic variation**, survival and the number of offspring per pair are sampled from a statistical distribution. **Environmental variation** is modelled by sampling time-dependent survivorships and fecundities (Burgman *et al.*, 1993;

Table 18.5. Some Australian examples of the use of PVA or related forms of population modelling (updated from Lindenmayer and Possingham, 1995).

Species	Geographic range	Primary risks	References
Plants			
Matchstick Banksia	Western Australia	Altered fire regimes Agriculture, habitat loss, changed fire regime	McCarthy *et al.* (2001c) Burgman and Lamont (1992)
Mammals			
Eastern Barred Bandicoot	Victoria	Road kills, Cat predation	Lacy and Clark (1990), Maguire *et al.* (1990)
Eastern Barred Bandicoot	South-eastern Australia	Habitat loss, source population depletion as part of translocation efforts	Todd *et al.* (2002)
Southern Brown Bandicoot	South Australia	Habitat loss	Possingham and Gepp (1993)
Leadbeater's Possum	Victoria	Habitat loss, logging	Lindenmayer *et al.* (1993c), Burgman *et al.* (1995a), Lindenmayer and Possingham (1995), Lindenmayer and Lacy (1995)
Greater Glider	South-eastern Australia	Habitat loss, logging	Possingham and Noble (1991), Possingham *et al.* (1994), Norton and Possingham (1992)
Mountain Brushtail Possum	Eastern Australia	Habitat loss	Lindenmayer and Lacy (1993), Lacy and Lindenmayer (1995)
Yellow-bellied Glider	South-eastern Australia	Habitat loss, fragmentation	Goldingay and Possingham (1995)
Mahogany Glider	North-eastern Australia	Habitat loss, fragmentation	Jackson (1999)
Mountain Pygmy Possum	South-eastern Australia	Natural population fluctuations, habitat loss	McCarthy and Broome (2000)
Various species of arboreal marsupials[1]	South-eastern Australia	Habitat loss, fragmentation	Lindenmayer *et al.* (1999, 2000d, 2003c)
Bush Rat and Agile Antechinus[1]	South-eastern Australia	Habitat loss, fragmentation	Lindenmayer and Lacy (2002), Ball *et al.* (2003)
Carpentarian Rock Rat	Northern Australia	Altered fire regimes, grazing by domestic livestock and feral herbivores, weed invasion	Brook *et al.* (2002b)
Greater Bilby	North-central Australia	Habitat loss, predation, grazing competition	Southgate and Possingham (1995)
Birds			
Powerful Owl	South-eastern Australia	Habitat loss, logging	Possingham and Noble (1991), Kavanagh (1997), McCarthy *et al.* (1999b)
Masked Owl and Sooty Owl	South-eastern Australia	Habitat loss, logging	Kavanagh (1997)
Red-browed Treecreeper and White-throated Treecreeper*	South-eastern Australia	Habitat loss, logging	McCarthy *et al.* (2000)
Sacred Kingfisher and Laughing Kookaburra*	South-eastern Australia	Habitat loss, logging	Lindenmayer *et al.* (2001b)
Helmeted Honeyeater	Central Victoria	Habitat loss, competition, release strategies for reintroduction	McCarthy *et al.* (1994, 2004), Akçakaya *et al.* (1995), McCarthy (1996b)
Capricorn Silvereye	Island populations off north-eastern Australia	Cyclones, diseases, habitat loss, inbreeding depression	Brook and Kikkawa (1998), McCallum *et al.* (2000)
Splendid Fairy Wren	Western Australia	Fire, fragmentation	Brooker and Brooker (1994)

[1]Part of assessing the predictive ability of various PVA models in the fragmented forest landscape at Tumut, southern New South Wales.

Akçakaya *et al.*, 1997). The result is a cloud of possibilities for the future of the population. Thus, even though a deterministic model may tell us that the population will increase, and the stochastic model for the same population may tell us that the population will probably increase, there is a measurable risk that the population will not increase.

Management may target the expected (average) population size. Some options may have no effect on average population size, but greatly affect variability, thus increasing the chance of population decline. The task of wildlife managers then is to implement plans to minimise risks and perhaps to maintain populations within specified limits.

The probabilities generated by stochastic models allow us to pose different kinds of questions. For instance: If things go as badly as possible, what will the population size be? How likely is it that we will have less than half the current population a year from now? We might like to know which parameter is most important; for example, if it turns out that adult survival influences the persistence of the population more strongly than any other parameter, it may be best to concentrate resources on estimating that parameter.

A model for Matchstick Banksia

The best way to explain the details of building a model is to provide an example. The purpose of this section is to describe a model for Matchstick Banksia, a threatened Western Australian plant. This is a somewhat simplified treatment of a model by Burgman and Lamont (1992; see also Burgman *et al.*, 1993). In very general terms, the dynamics of any population can be described by the following equation:

$$N(t+1) = N(t)s + N(t)m + I(t) - N(t)e \quad (18.4)$$

where $N(t)$ is the population size at time t, s is survivorship (the probability that an individual survives from time t to time $t + 1$, m is fecundity (the average number of offspring at time $t + 1$ per individual alive at time t), e is the emigration rate and I is the number of immigrants to the population.

Matchstick Banksia grows in six separate populations, but has no obvious means of long-distance dispersal. Therefore, we can assume that immigration and emigration are zero in each of these six populations. It will be sufficient to construct a model for a single population, rather than for the entire metapopulation. If $s = 0.5$ and $m = 0.5$, the number of births (germinations) and deaths in the population will be the same each year and the population should remain stable (Figure 18.9). If $s = 0.95$ and $m = 1.4$ (values

Stands of Matchstick Banksia along a road that passes through a salt lake in Western Australia. (Photo by Richard Hobbs.)

that could be expected in a real population of Matchstick Banksia), then using Equation 18.4, we expect that the population will increase exponentially (Figure 18.9). The parameters do not vary in this model, so it is termed a deterministic model: it provides a single forecast of population size, but no indication of how uncertain the forecast may be.

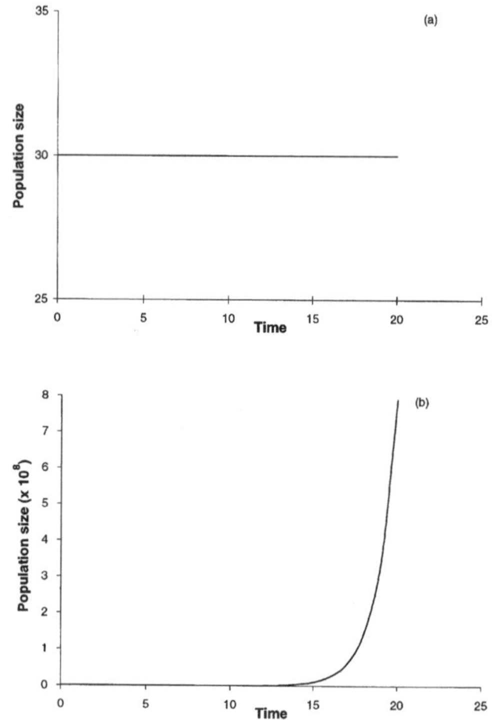

Figure 18.9. Deterministic expectations for a Matchstick Banskia population over 20 years in which the initial population size (N) is 20 and (A) $s = 0.5$ and $m = 0.5$, and (A) $s = 0.95$ and $m = 1.4$.

Rather than multiplying the whole population by survival and fecundity values of 0.5, we could add an element of realism by following the fate of each individual. At each time step, we could generate uniform random numbers, scaled between 0 and 1 (i.e. each number in that range has an equal chance of being sampled). If a random number is greater than the fecundity value (0.5 in this example), the individual produces an offspring. If another random number is greater than the survival value (0.5 in this case), then the individual dies. Otherwise, the individual lives. We ask these questions for each individual in the population, using a different random number each time. Thus, if there are 10 individuals in the population, there is no guarantee that five will survive, although it is the most likely outcome. There is some smaller chance that four or six will survive and some smaller chance still that three or seven will survive. This represents chance births and deaths and is what we mean when we talk of demographic variation. The results for the model in Figure 18.9A are represented by the scenario in Figure 18.10.

Populations can increase or decrease by chance alone, even though deterministically they remain constant. In statistical terms, the above procedure is called a set of **Bernoulli trials**, which is equivalent to sampling the binomial distribution (Akçakaya, 1990). Births come in integers (for example, no plant will produce 1.4 offspring; rather, some will produce none, some 1, some 2 and so on). This can be modelled by sampling the **Poisson distribution** (see Akçakaya, 1990; Burgman *et al.*, 1993).

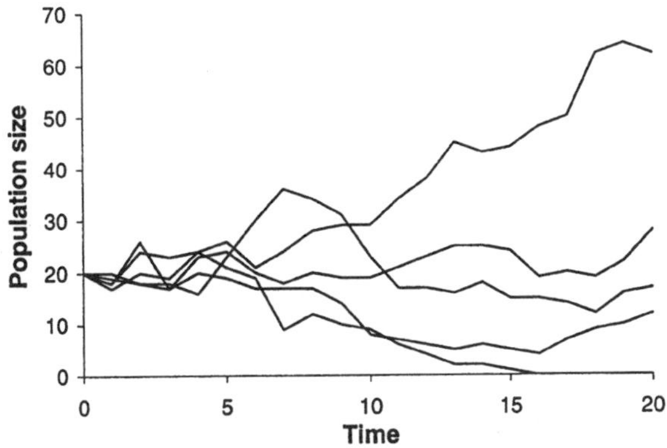

Figure 18.10. The first five projections of a model for Matchstick Banksia in which deaths are subject to binomial uncertainty, $s = 0.5$, $m = 0.5$, and the initial population size (N) is 20.

Fires kill the plants and stimulate the release of canopy-stored seed. Successful regeneration from seed depends on good winter rains following a fire during summer. Thus, to develop a more realistic picture of the species, we need to modify the model to accommodate rainfall and fire. Production of viable seeds does not take place until plants are about 5 years old. Thus, we should distinguish between juvenile (J) and adult (A) plants. We can write the equations for the population as follows:

$$J(t+1) = J(t)s_i + A(t)m$$
$$A(t+1) = A(t)s_a + J(t)s_d \qquad (18.5)$$

In rewriting the equations, we have changed the meaning of survivorship. The term s_j represents the proportion of juvenile plants alive at time t that survives to time $t + 1$ and remains as juveniles. The term s_d represents the proportion of juvenile plants alive at time t that survive and develop sufficiently to be counted as adult plants at time $t + 1$. The term s_a is the survival probability of adults. Thus, s_j and s_d combined represent the fate of juvenile plants (they either stay as juveniles with probability s_j, mature into adults with probability s_d, or they die with probability of $1 - s_j - s_d$.

In the absence of fire, new seedlings germinate and juvenile plants die rarely – the associated probabilities are assumed to be zero. Once plants reach about 5 years of age, they begin to produce seeds. Adult plants die naturally, mainly through brittleness and branch splitting. The average probability of survival of an adult plant is 0.95. When fires occur, all adults and juveniles die, but the adults release seed and an average of 1.4 new plants are produced per adult.

Thus, we can specify two sets of parameters. In years in which a fire occurs, $s_a = s_j = s_d = 0$, and $m = 1.4$. In years in which there is no fire, $s_j = 0.8$, $s_d = 0.2$, $s_a = 0.95$, and $m = 0$. The parameter s_d is set to 0.2 because, on average, it takes 5 years for a plant to mature (therefore, the chance of a plant doing so in any year is 1/5). The parameter s_j is 0.8 because in the absence of fire there are no juvenile deaths and $s_j + s_d = 1$.

We assume that the populations are only subject to wildfires, so the timing of fires is unpredictable. We can treat this element of environmental uncertainty in much the same way that we treated deaths in the population. We assume that there are, on average, four fires every 100 years, giving a chance per year of a fire of 0.04. Therefore, we select a uniform random number between 0 and 1; if the random number is greater than 0.04, there is no fire in the current year, but otherwise the population burns. We apply the appropriate set of population parameters, depending on the outcome of this trial.

Lastly, rainfall following a fire determines regeneration. We can represent this by making fecundity depend on rainfall. If an average rainfall year follows a fire, adults produce 1.4 new juvenile plants. The mean rainfall in areas where Matchstick Banksias grow is close to 350 millimetres annually. The data suggest that when rainfall falls to 300 millimetres, the average number of new recruits in the population falls to about 0.8 juveniles per adult. We can replace the fecundity value for years in which there is a fire with the expression:

$$m = 0.02R - 5 \tag{18.6}$$

where R is the rainfall received in that year.

In good years, more new plants will be produced than in bad years. In each year, we select a random number to represent R with an average of 350 millimetres and a variation that reflects that observed in the natural rainfall regime. There are many ways of doing this. For the sake of example, we will choose a uniform random number scaled between 250 and 450. Rainfall values outside this range are unlikely, although using a uniform random number overemphasises the likelihood of very good and very bad rainfall years. At the extremes, the average fecundity of the Matchstick Banksia populations can be as little as zero or as large as four recruits per adult.

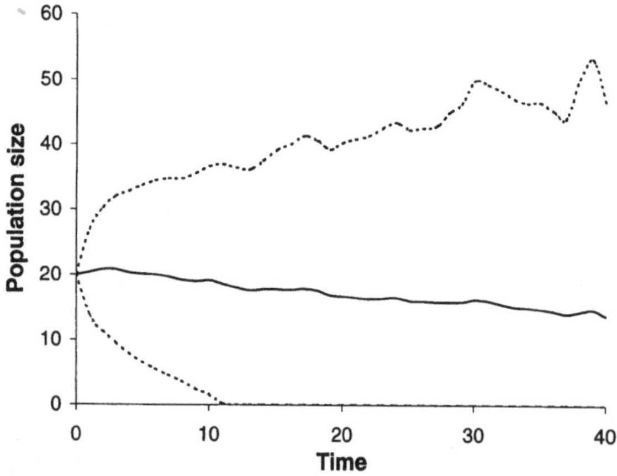

Figure 18.11. Expected population size for the Matchstick Banksia over 20 years (solid line) and 95% confidence intervals, given demographic uncertainty, wildfires on average every 10 years, and rainfall variation.

The predictions of this model (Figure 18.11) are qualitatively different from those predicted by our first model (Figure 18.9). The management practices that can be implemented to minimise the risks of population decline (e.g. reducing the effective variability of rainfall) might be very different from those that maximise expected population size (e.g. improving the germination success following fire).

The model allows us to calculate the risks of extinction of the population, and to explore the sensitivity of the results to different parameter estimates, for example the relationship between fire frequency, expected population size and the chance of extinction of the population (Figure 18.12). This is termed **sensitivity analysis.** By exploring the response of a model to variations in parameters, it is possible to develop a 'feel' for the parameters on which the results depend most sensitively. It is these parameters that are most important in determining the likelihood of decline or the size of the future population.

We can use the model to assess the consequences of management options. A more complete and realistic model for the Matchstick Banksia was developed by Burgman and Lamont (1992). They compared management options including human intervention by watering seedlings in dry years after fires, and explored assumptions about inbreeding depression, survivorship, rainfall, and recruitment.

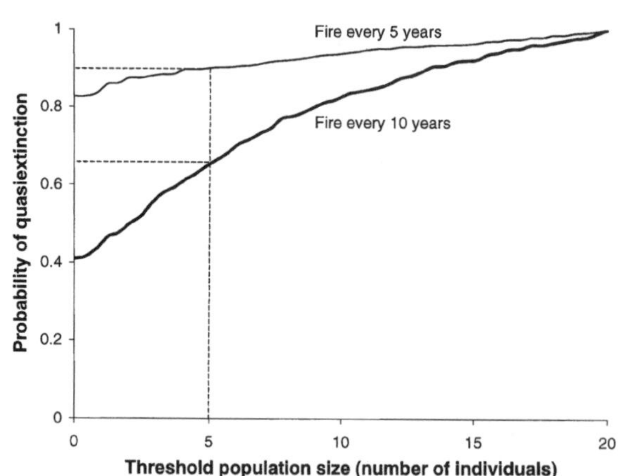

Figure 18.12. Probability of quasi-extinction of the Matchstick Banksia population within 20 years, given fires at frequencies of (A) one in 10 years, and (B) one in 5 years. For example, the chance of the population falling below five individuals within 20 years is about 0.65, given a fire every 10 years. The chance of it falling below five individuals within 20 years is about 0.9, given a fire every 5 years.

Models are useful only insofar as they help to order ideas and suggest relationships that may not be obvious otherwise. They point out where there is a lack of data, but they rarely provide trustworthy predictions of future events. For instance, the model described here is too simple and there is too little empirical information for us to be convinced that the risks of extinction it predicts are reliable. However, the model is sufficiently robust to guide data collection and rank management options.

Metapopulations

For many species, the environment varies in space (Chapter 10) and habitat is patchy (see Hastings and Harrison, 1994; Harrison and Taylor, 1997; Niemien and Hanski, 1998; Hayward *et al.*, 2003). Physical variables (e.g. soil conditions and elevation), extreme events (e.g. droughts and fires), diseases (e.g. Stapp *et al.*, 2004), and social interactions affect populations. How an individual or a population fares depends on where it happens to be. For example, fires burn in mosaics that depend on fuel loads, moisture conditions, natural barriers and prevailing winds (Catchpole, 2002). Different parts of the habitat burn at different intensities and with different frequencies (Whelan, 1995; Bradstock *et al.*, 2002; see Chapter 11).

A **metapopulation** is defined as 'a set of local populations which interact via individuals moving between local populations' (Hanski and Gilpin, 1991, p. 7; Hanski, 1999a,b). Two kinds of habitat structure characterise the extremes of metapopulation dynamics (Hanski and Gyllenberg, 1993; Chesson, 2001). The **mainland–island structure** includes a large 'mainland' area in which populations are secure and very rarely (if ever) become extinct, and an array of small patches in which extinctions are common. The mainland provides a source to recolonise the patches and reverse localised extinctions (see Figure 18.13; e.g. Harrison *et al.*, 1988; Hanski and Thomas, 1994; Hanski, 1999a,b). Several butterflies are thought to have a mainland–island metapopulation structure (Hanski and Thomas, 1994; Harrison *et al.*, 1988). The **Levins structure** takes its name from Levins (1970) and the model assumes that patches are approximately the same size. In this model, local extinctions are frequent and empty patches are recolonised by migration (Hanski and Gyllenberg, 1993; Figure 18.13).

Most metapopulations have patch sizes intermediate between the mainland–island and Levins models.

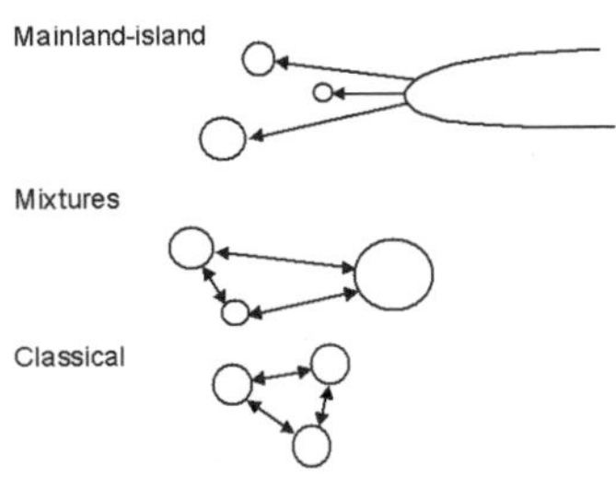

Figure 18.13. Different kinds of metapopulation structure (after Hanski and Gyllenberg, 1993): (A) mainland–island population, (B) a metapopulation composed of habitat patches of different sizes, and (C) a metapopulation with equal patch sizes (a 'classical' Levins structure).

Arnold *et al.* (1993) examined Euro populations in a fragmented landscape in the Wheat belt of Western Australia. They identified a metapopulation (Figure 18.14), including several populations comprising just a few individuals. Animals moved between and occupied remnants in the fragmented landscape. Adults in the largest patch (1196 hectares) were sedentary, but their offspring dispersed (see also Hill *et al.*, 1996; Carlson and Edenhamn, 2000 for other examples of metapopulations).

Figure 18.14. Spatial structure in a metapopulation of the Euro (after Arnold *et al.*, 1993). The map shows remnants of native vegetation in an agricultural landscape. The dashed line encloses patches of habitat in which Euros were captured and patches postulated to be part of the metapopulation.

Euro. (Photo by Marie Lochman.)

Other criteria for defining metapopulations
The term 'metapopulation' is sometimes inappropriately applied to all species distributed in habitat fragments (Arnold *et al.*, 1993; Hanski and Thomas, 1994). However, true metapopulations are defined by dynamic properties including the following (after Hanski, 1997, 1999a,b):

- Suitable habitat must be restricted to particular patches that can be differentiated from the surrounding unsuitable matrix.
- All patches must have a significant risk of extinction at some stage.
- There must be inter-patch dispersal and colonisation.
- The local dynamics of populations within a patch should not be synchronous with the dynamics of populations in other patches.

Patchily distributed populations will not conform to a metapopulation structure where the matrix is so hostile that it precludes movement and populations are confined to isolated habitat patches (Sarre *et al.*, 1995), the matrix is sufficiently suitable to facilitate frequent movements between habitat patches (Price *et al.*, 1999; Fraser and Stutchbury, 2004), or matrix conditions allow organisms to forage or live there (Tubelius *et al.*, 2004).

Hanski and Simberloff (1997) contended that the metapopulation paradigm is most useful when successful inter-patch dispersal is infrequent and migration distances are limited. Hastings (1993) suggested from simulation modelling that populations could be considered independent where less than 10% of resident populations disperse.

Some species do not disperse between patches. For example, like the Matchstick Banksia, geckos in Western Australia exist in isolated populations that are confined to woodland patches with no dispersal between them (Sarre *et al.*, 1995). Such populations are not a true metapopulation.

Some species appear to be patchily distributed but are not true metapopulations because inter-patch dispersal is very frequent. If species have a metapopulation structure, then there should be a characteristic spatial distribution pattern in which patches close to occupied patches are more likely to be occupied than more distant ones (see Hanski, 1994; Smith, 1994; Koenig, 1998). There was no evidence of the expected spatial pattern in 76 species of birds from eucalypt patches surrounded by Radiata Pine plantation in south-eastern Australia (Lindenmayer *et al.*, 2002c). Many species either moved among many different patches to harvest food resources, or regularly foraged or were breeding in the matrix of Radiata Pine surrounding the eucalypt forest.

Metapopulations in a PVA framework

The effects of environmental and demographic stochasticity become more pronounced when populations decrease in size, leading to more frequent local extinctions and recolonisations. This may prevent each population from reaching its equilibrium size, although the metapopulation as a whole may persist much longer than any individual population. The turnover of populations produces a temporal pattern referred to as a **shifting mosaic** – a pattern of occupancy of local populations in a metapopulation.

Differences in productivity can result in populations that receive migrants but seldom produce emigrants. These habitat patches are **sink populations**, which absorb migrants from other patches. Other subpopulations act as sources, producing a net number of emigrants because of their size and the quality of their habitat. Source–sink dynamics (see Howe *et al.*, 1991; Pulliam *et al.*, 1992) are characterised by subpopulations maintained by immigration from other subpopulations – a process known as the **rescue effect** (Brown and Kodric-Brown, 1977; Harrison, 1991). A metapopulation can be maintained in a state of dynamic equilibrium that depends sensitively on migration among patches.

Spatial structure and metapopulation dynamics are important for conservation. Conservation biology is concerned with the protection of species. If individuals move between patches, the risk of extinction of an entire species cannot be estimated based on the risk of extinction of its separate populations because of the interdependency of, and the interactions among, the populations.

Metapopulation dynamics are especially important for endangered species, many of which exist in small, relatively isolated populations that have resulted from habitat loss and fragmentation (Chapter 10). Multiple populations introduce new susceptibilities. In addition to impacts that decrease the mean survivorship or fecundity in a single population, metapopulations are sensitive to impacts that affect the movement of organisms and increase the isolation of populations, for example road building, construction of dams, and agriculture.

When wildlife managers deliberately translocate individuals and reintroduce populations to empty patches, they effectively increase migration rates. These activities involve new kinds of decisions, such as how to formulate the schedule of translocations, or whether to reintroduce a large number of individuals to a single patch, or smaller numbers of individuals to several patches.

An important aspect of reserve design involves selecting patches of habitat for the protection of a variety of species. Which combination of nature reserves gives an endangered species the highest chance of survival? Such questions can only be assessed by an analysis of metapopulation dynamics on a case-by-case basis.

Caveats for metapopulation modelling

Wiens (1997a) noted that most metapopulation models assume that '*the matrix separating subpopulations is homogeneous and featureless*'. They ignore spatial and temporal changes in the suitability of the matrix surrounding habitat patches (see Pope *et al.*, 2000), and assume that the matrix does not influence the success of inter-patch dispersal and that dispersal is a random process (Cale, 1999). Metapopulation models predict poorly when this assumption is violated. Hanski's (1994b) incidence metapopulation model predicted poorly the distributions of two species of arboreal marsupials that persisted in the matrix. McCarthy *et al.* (2000) found the fit between model-predicted and actual values for patch occupancy was poor for birds that were able to forage both in habitat fragments and the surrounding matrix.

Metapopulation models generally do not accommodate the changes in home range sizes that follow habitat fragmentation (e.g. Barbour and Litvaitis, 1993), changed landscape mosaics (Milne *et al.*, 1992), or frequent movements between many patches (e.g. to gather spatially separated food resources; Boone and Hunter, 1996). Model estimates of patch occupancy for an Australian kingfisher, the Laughing Kookaburra, were inaccurate because birds alter home range movements in response to landscape change and move between many different patches to gather food (Lindenmayer *et al.*, 2001b).

Most metapopulation models assume that the suitability of habitat patches and the surrounding matrix does not change. In wood production forests, habitats become more suitable over time for some species. Spatially and temporally explicit models such as those developed by Possingham and Davies (1995) and Akçakaya (2002) can be useful in modelling such changes. In these models, different habitat quality values can be assigned to patches, and the habitat quality values can vary in time depending on, for example, ecological succession in logged forests.

Models simplify the systems they portray (Burgman *et al.*, 1993), and metapopulation models are no exception (Hanski, 1999a); models such as Hanski's (1994b) incidence function model consistently favour simplicity over complexity (Ludwig, 1999). For many species, temporal and spatial variation in habitat and matrix conditions is important. It is vital to check model assumptions before models are applied to real landscapes and real conservation problems (Doak and Mills, 1994; Fahrig and Merriam, 1994; Wiens, 1994; Pope *et al.*, 2000). Decisions that rely on models should be taken only if model assumptions are validated and outcomes are monitored.

Minimum viable populations

Often, circumstances demand that we answer the question: How many individuals are enough? A **minimum viable population** (MVP) is the smallest population that has a reasonable probability of surviving for a specified time (Soulé, 1987). In some applications, the role of PVA has been to identify the MVP for a species. MVPs depend on estimates of absolute extinction risks, a different focus for PVA than that outlined earlier in this chapter, which is to synthesise data, explore the consequences of ideas, and set the rank order of management options.

Initially, MVPs were defined by the size of a population large enough to maintain genetic variability and evolutionary potential. Franklin (1980) and Soulé

(1980) calculated that if inbreeding is less than 1%, a population can avoid inbreeding depression. This translates to an effective population size (N_e) of 50 individuals, although more recent work suggests that 50 individuals are too few (Reed and Bryant, 2000). For longer-term evolutionary potential, Franklin (1980) speculated that the effective population size should be at least 500. For example, in Australia, Mackowski (1986) calculated that a reserve of 15 000 hectares was needed to conserve 500 Yellow-bellied Gliders.

However, estimates of N_e of 500 animals assume life history strategies characterised by a balanced sex ratio, no overlapping generations and other assumptions (Chapter 14). There have been several attempts to account for these factors. For example, Davey (1989a) applied methods developed by Lehmkuhl (1984) to estimate an N_e for the Greater Glider of 2300–5000 individuals.

The main problem with MVPs is that the earliest estimates were based on the rate of loss of genetic variation that was reflected in the dynamics of bristle numbers in Fruit Flies. There is no direct link between a quantitative trait in Fruit Fly and the risks of extinction of other taxa. **Genetic variation** is only one of several factors influencing the dynamics of plant and animal populations. The purposes of PVA – to evaluate the extent of our knowledge, to integrate that knowledge, to test the efficacy of competing management options, and to provide a vehicle to guide future research and decision-making – become clouded when MVPs are considered. Nevertheless, the notions of MVPs persist in the notion of 'adequate' reserves (Chapter 16) that support 'viable' populations (i.e. those that have an acceptably low probability of extinction).

The limits of population viability analysis

Shaffer (1981) distinguished between **systematic pressure** and **stochastic perturbation** as causes of population extinction. Caughley (1994) also differentiated between the **small-population paradigm**, which deals with stochastic influences on small populations, and the **declining-population paradigm**, which deals with the (largely deterministic) ecological causes and cures of population decline. He suggested that the principal contribution of the small-population paradigm is the theoretical underpinning it provides to conservation biology, even though the theory does little to aid species in trouble. He also claimed that determining why a given species is declining (the declining-population paradigm) lends itself directly to scientific testing, whereas the small-population paradigm does not.

Our view is that this dichotomy is artificial. Boyce (1992) noted that all ecological processes are stochastic processes with deterministic elements. The omission of an important process such as loss of habitat, competition or predation from introduced species, impacts of disease or parasites, or the impacts of rare catastrophic events, can substantially affect expectations. The ecology of species and the role of management should be, in the words of Boyce (1992), '*the nuts and bolts of modelling exercises*'.

Risk analysis is also not the end of PVA, which often also includes implementation of conservation options, as well as long-term monitoring and evaluation. Obviously PVA models can and do incorporate deterministic decline (e.g. Lahaye *et al.*, 1994), human-caused factors (Burgman *et al.*, 1993), external influences (Burgman and Gerard, 1990), and the effects of habitat loss (Lindenmayer and Lacy, 1993, 2002; Akçakaya *et al.*, 1995). If an ecologist has ideas concerning the forces that govern the chances of persistence of a population, then these ideas can and should be included in the PVA model, irrespective of their origin, and irrespective of their deterministic or stochastic nature (Possingham *et al.*, 2001).

Predictive ability of PVA

Modellers aim to represent the important aspects of the dynamics of a system (Burgman *et al.* 1993; Fieberg and Ellner, 2001). PVAs are critical in conservation because the assessment of species extinction risk lies at the heart of conservation biology (Burgman *et al.*, 1993; Fagan *et al.*, 2001; Fieberg and Ellner, 2001; Beissenger and McCullough, 2002). However, PVAs have been criticised for failing to predict extinction probabilities accurately (Ludwig, 1999; Ellner *et al.*, 2002 vs Brook *et al.*, 2002a). Although there are several examples where predictions from models and actual dynamics of populations have compared favourably (e.g. Puerto Rican Parrot, Whooping Crane, and Lord Howe Island Woodhen; Lacy *et al.*, 1989; Mirande *et al.*, 1991; Frankham, 1995a), in general, PVAs have produced very variable predictions (Coulson *et al.*, 2001; Ellner *et al.*, 2002).

Since Brook *et al.* (2000) assessed the predictive accuracy of PVA for 21 populations (8 birds, 11 mammals, 1 reptile and 1 fish; see Figure 18.15), discussion of the predictive accuracy of PVAs has

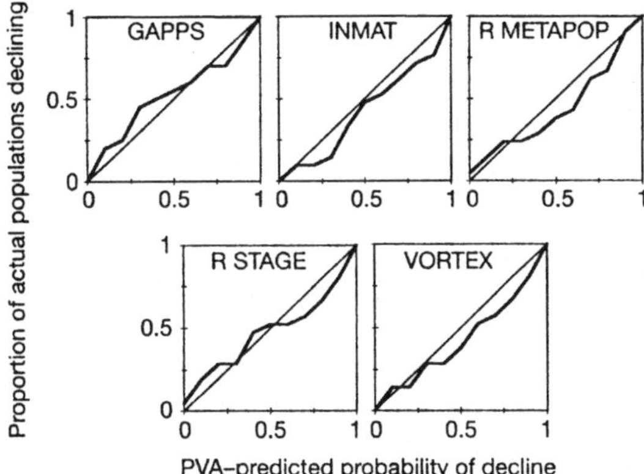

Figure 18.15. PVA-predicted probability of decline versus the actual proportion of 21 real populations that were modelled that declined below the corresponding threshold size. The headings on the diagrams represent different computer packages that were employed in PVA. (Redrawn from Brook *et al.*, 2000.)

increased (e.g. Brook, 2000; Burgman, 2000; Fagan *et al.*, 2001; Fieberg and Ellner, 2001; Beissenger and McCullogh, 2002; Brook *et al.*, 2002a). McCarthy *et al.* (2001a, 2003) outlined methods for testing the accuracy of forecasts from population viability analysis and the reliability of relative predictions. Perhaps most importantly, in the words of McCarthy *et al.* (2001a):

> *the role of model testing is not to prove the truth of a model, which is impossible because models are never a perfect description of reality. Rather, testing should help identify the weakest aspects of models so they can be improved.*

Lindenmayer *et al.* (2003c) compared predictions of population size and patch occupancy with field observations for several vertebrates and a range of models for a fragmented forest landscape at Tumut in south-eastern New South Wales (Lindenmayer *et al.*, 2003c). The predictive abilities of the models differed markedly between species. Predictions for the Greater Glider were accurate, whereas for two species of small terrestrial mammals (the Bush Rat and Agile Antechinus), the predictions from all the models were poor. These are some of the best-studied animals in Australia, and there are abundant, high-quality life history data available for them. Factors other than good life history information appeared to have an important impact on the accuracy of modelling, for example species-specific relationships between patch size, quality and animal abundance, complex spatial responses to the patch system that could not be readily captured by simple measures of patch size and patch isolation, and life histories of species in fragmented settings that were very different from those in continuous forests. The kinds of populations and life history characteristics for which model predictions are likely to be accurate were identified by extensive model testing (see Table 18.6).

Data limits the predictive accuracy of most population viability analyses (Boyce, 1992; Burgman *et al.*, 1993; Caughley, 1994; Taylor, 1995; Ellner *et al.*, 2002). Even the simplest models require more parameters than are usually available. Nevertheless, PVA models are useful if they summarise the available data consistently and transparently (Burgman, 2000; Brook *et al.*, 2002a). Meaningful results are possible even if information is incomplete, because there is value in building a model for its own sake (Starfield and Bleloch, 1992; Burgman *et al.*, 1993; Burgman, 2000). Models clarify assumptions, integrate knowledge from all available sources, and force biologists to be explicit and rigorous in their reasoning. They allow us to identify, through

Table 18.6. Contrasts between characteristics of populations whose fates are likely to be more accurately predicted by PVA than those likely to be less accurately predicted.

More accurate population prediction	Less accurate population prediction
Single population	Metapopulation
Closed population	Open population
Discrete habitat boundaries	Diffuse habitat boundaries
Uniform habitat conditions	Heterogeneous habitat conditions
Constancy of life history attributes across habitat types and landscape conditions	Variation in life history attributes between habitat types and landscape conditions (e.g. fragmented vs unfragmented landscapes)
Constancy of species interactions across habitat types and landscape conditions	Variation in species interactions between habitat types and landscape conditions
Distance-related dispersal patterns	Habitat-related dispersal patterns
Simple social systems	Complex social systems

sensitivity analyses, which model structures and parameters matter, and which do not (Possingham *et al.*, 2001). Models result in a set of logical statements that are internally consistent, and they allow us to explore the consequences of ideas and options, even in the absence of relevant, complete data (Burgman *et al.*, 1993; Burgman, 2000).

Advantages and caveats for model application

People who use a model (whether computer-implemented or not) should be aware of its assumptions and limitations. All too often, the temptation to use a computer model as a 'black box' without understanding its underpinning assumptions and limitations is too great, and inappropriate models are developed to guide conservation and management (Boyce, 1992).

There are many pitfalls in writing even the simplest model, and using one that has been tested and debugged extensively has obvious advantages. The possibility of misuse should be a cause for caution and education, but not discouragement. There are many ways to misuse statistics for example, but we hope that the benefits of accessible statistical analysis outweigh the costs of the unthinking use of statistical software packages.

Last, it is easy to be unaware of the limitations and assumptions of one's own model, both because it is hard to question one's own biases and because most 'easily available' computer-implemented PVA models come with extensive documentation that includes lists of assumptions and limitations (e.g. Lacy, 1993b, 2000; Akçakaya, 2002). Biologists should work closely with modellers or learn to build their own models (with or without software) rather than creating separate niches for 'understanding' and 'modelling', by segregating the essential steps of PVA by allowing them to be carried out by different sets of professionals working in isolation.

Judgement and acceptable risks

Judgement as to what is and what is not an acceptable risk involves a judgement that is, by definition, anthropocentric (Burgman *et al.*, 1993): the judgement will depend on the economic and cultural value to humans of the species under consideration. It is essential, therefore, to distinguish the tool (risk analysis) from the objective (the conservation and protection of species). The tool, risk analysis, can apply equally well to plagues and population explosions of unwanted species such as locusts and rats. Risk analysis permits us to make decisions that modify our actions to reduce the chances that a population will become extinct. It contributes to the setting of conservation priorities by facilitating the comparison of the relative impacts and benefits of different management options (Possingham *et al.*, 1993, 2001; Beissenger and Westphal, 1998; McCarthy *et al.*, 2003).

The use of PVA to rank and compare management options is clearly a different approach from the application of the tool to generate an accurate prediction of absolute extinction risk or population numbers *per se* (McCarthy *et al.*, 2003). Notably, an important advantage of such an approach is that the outcomes of modelling can be useful even where data are limited (Burgman *et al.*, 1993; Possingham *et al.*, 1993, 2001; Lindenmayer and Possingham, 1995; McCallum *et al.*, 1995; Starfield *et al.*, 1995; Burgman, 2000).

18.6 Conclusions

Threatened species management has two broad objectives. The short-term objective is to minimise the risk of extinction. The long-term objective is to promote conditions in which species retain their potential for evolutionary change without intensive management. Within this context, PVA can be used to address three aspects of threatened species management (Possingham *et al.*, 1993, 2001):

- *Planning research and data collection.* PVA may reveal that the viability of a population is insensitive to particular parameters. Research can then be guided by targeting factors that may have an important impact on extinction probabilities or on the rank order of management options.
- *Listing of threatened species.* PVA is an objective mechanism for assessing vulnerability. Together with cultural priorities, economic imperatives and taxonomic uniqueness, it can be used to set priorities for allocating scarce conservation resources.
- *Ranking management options.* PVA can be used to predict the most likely response of species to reintroduction, captive breeding, prescribed burning, weed control, habitat rehabilitation, or different designs for nature reserves or corridor networks.

There is time and information to conduct robust PVAs for only a relatively small number of species. A crucial step is to identify those species to which PVA is best suited (Possingham *et al.*, 1993, 2001). These may include well-studied species, **keystone** species, threatened species, or species of special cultural or economic value.

The relevance of models for environmental decision-making is in the mind of the policy maker. A biologist who builds a model provides a service, a skill, and the end product is a set of recommendations that are bounded by assumptions and uncertainties. It is as important, if not more important, for biologists/modellers to communicate those uncertainties and assumptions as it is for them to communicate predictions.

Analyses are coloured by what the modeller believes to be important. The fact that models of the same natural system may generate different expectations is not surprising to modellers, but it is a source of frustration to decision-makers. In such circumstances, the modellers may have failed in terms of communicating their findings, because frustration implies that the sensitivities, limitations and assumptions of the models have not been explained.

The object of environmental models should be to improve communication and understanding. In such circumstances, the results they produce will be integrated quite naturally with value judgements and political constraints, to produce better decisions than could be made in the absence of models. To achieve these ends, models must be carefully and thoroughly documented, and limitations, sensitivities and assumptions must be stated explicitly. The people affected by decisions should understand the model and trust that it summarises knowledge reasonably and makes sensible assumptions. Modellers must be sensitive to the needs and limitations of those people who intend to use the models.

Risk assessment is essential if we are to allocate scarce resources to conservation and wildlife management as efficiently as possible, thereby minimising the number of species that will become extinct within the next few years (Burgman, 2000).

It is important when dealing with natural populations to recognise that there is never zero risk. Although we may define and consider limits that we set arbitrarily close to zero, we can never, whatever we do, guarantee the survival of a species for any period of time. This point is particularly important because people have a bias towards situations in which they believe there is no risk (Zeckhauser and Viscusi, 1990). Models are one of the few antidotes to the frailties of human perceptions of risk.

18.7 Practical implications

Estimates of species loss can be based on species-area functions or observation data from museums and herbariums. In both cases, the estimates are uncertain and sensitive to sampling bias. They should be used cautiously, and used together with other evidence for decline and loss.

Expert opinion is often used in conservation biology to make decisions. Expert opinion is susceptible to individual biases, but can be difficult to question. A very broad range of qualitative and quantitative techniques may help conservation biologists to make decisions, and PVA is one of these. PVA provides a means for synthesising knowledge about a species and dealing explicitly with uncertainty. PVAs are best suited to evaluating the ranks of management options, and to assisting decision-makers to understand the ecology of the system and the importance of missing data. Typically, extinction risk estimates from PVAs will be too uncertain to provide reliable estimates. The full benefit of modelling will be gained if it can be done in such a way that it is accessible and transparent to all those affected by decisions influenced by the PVA.

18.8 Further reading

Solow and Roberts (2003) provide details on estimating extinction probabilities from observations. Hanski and Simberloff (1997), Hanksi (1999a,b) and Chesson (2001) provide detailed discussions on metapopulations.

Burgman (2005) provides an introduction to a broad range of tools for risk analysis. Burgman *et al.* (1993) provide a complete introduction to PVA, upon which much of the discussion in this chapter is based. Akçakaya *et al.* (1997) provide a gentle and complete introduction to the discipline of stochastic models for natural populations. PVA has been reviewed and discussed by Lande (1988), Simberloff (1988), Boyce (1992) and Caughley (1994), Caughley and Gunn (1996), Beissenger and Westphal (1998), Possingham *et al.* (2001) and Reed *et al.* (2002). Possingham *et al.* (1993), Burgman (2000) and Ellner and Fieberg (2003) provide straightforward and sensible introductions to the use of PVA for threatened species management and resource management given uncertainty. Beissenger and McCullough (2002) is a valuable edited volume with chapters on various aspects and applications of PVA. Lindenmayer *et al.* (1995a) compare different software packages for PVA. McCarthy *et al.* (2003) discuss approaches for testing the accuracy of predictions from PVA models and Brook *et al.* (2000) and Lindenmayer *et al.* (2003b) outline empirical studies where model forecasts have been tested against field data.

Part IV: Management principles for conservation

Different groups of people view landscapes differently. Perceptions depend on experience, education, social background, knowledge of an area, and future expectations. The first European settlers viewed the Australian landscape and its resources as virtually limitless, and clearing and 'taming' the environment were paramount (see Chapters 9 and 11). Different concepts and ethical positions can lead to divergent interpretations of the best way to manage an ecosystem. Management intended to enhance one facet of biodiversity will not necessarily maintain another. Management for conservation is an inherently social and political activity in which the best management decisions are informed by scientific understanding.

Many of the concepts outlined in Part I, together with many of the tools described in Part III, have particular applications to solving problems relevant to individual species; however, this does not mean that we advocate a focus on single species. The social and political practicalities of conservation dictate that single species are frequently the target of management activities, but most good management plans and research activities have a wider focus. The subject of a conservation effort may be a gene, individual, population, geographic variant, species, community or ecosystem. A complete management strategy will address conservation at all levels of ecological organisation (Falk, 1990; Lindenmayer and Franklin, 2002).

Chapter 19

Sustainability and management

This chapter focuses on a range of topics associated with ecologically sustainable natural resource management. It explores the origins and limitations of sustained yield harvesting of populations, species and communities and the more recent concept of sustainable development. Ways to attempt to improve natural resource management, for example adaptive management and ecosystem management, are outlined.

19.0 Introduction

The notion of **sustainability** is closely linked with **custodial responsibility** (Chapter 1), based on **intra-generational equity** in the use of natural resources. There are several interpretations of sustainability in environmental management, each of which reflects different perspectives on and motivations for conservation. The purpose of the final chapter of this book is to outline some concepts relevant to the management of populations, species and communities, to provide definitions of terms for management, and to describe some of the mechanisms by which management objectives can be achieved.

19.1 Sustainability

Use of the natural environment usually has a strong economic motivation. In many cases, we are faced with a choice between using an entire resource immediately, or using less of the resource now, in the hope that it will remain available in the future or indefinitely. If we foreclose a given option now, we can reinvest in other options that may be more valuable. This perspective is common to most of us; when we borrow, we are prepared to pay interest to gain access to resources now. Likewise, when we put money in the bank, we expect interest in return for delaying consumption.

The utilitarian value of a resource decreases as a function of the time between the present and the time we make use of it. **Discounting** is the process by which expected future values can be translated into present-day values. The most common form of discounting is given by Henderson and Sutherland (1996):

$$V_p = \frac{V_t}{(1+r)^t} \qquad (19.1)$$

where V_p is present value, V_t is future value, t is the time delay before receipt of the future value, and r is the discount rate (usually assumed to be somewhere between 0.01 and 0.10). There is some debate about whether discount rates should be exponential (as above) or hyperbolic (Henderson and Sutherland, 1996). Irrespective of the form of the discount rates, beyond about 100 years into the future, values are negligible when converted to current values. This means that purely economic incentives to conserve natural resources are weak. Yet many decisions regarding the management and conservation of the natural world have considerably longer time horizons.

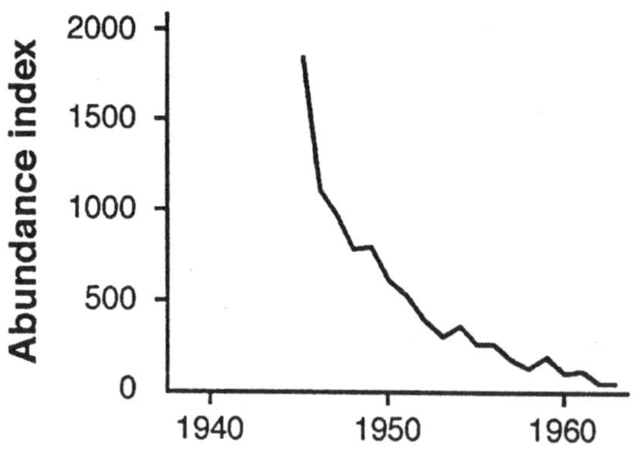

Figure 19.1. An index of abundance of Blue Whales in the Antarctic (estimated as the number of whales caught per catcher-ton-day) between 1945 and 1963 (after Gulland, 1971).

Infrastructure associated with whale harvesting and processing at the now closed whaling station at Albany in south-western Australia. (Photos by David Lindenmayer.)

In decisions about whether to take the entire population now (assuming reinvestment elsewhere for a higher rate of return), or harvest on the basis of sustainable yield, the economic incentive to act immediately can be overwhelming (Caughley, 1994). May (1976) identified the economic incentive as the reason why the whaling industry did not operate on a sustainable basis (Figure 19.1). Hence, 'rational economics' gave rise to 'irrational natural resource use'. This mechanism is believed to operate most forcefully in free and fully informed markets, particularly where the resources are privately owned (Caughley, 1994).

Discounting makes it seem sensible to clear-fell natural forests or harvest fisheries as quickly as possible, disregarding any attempt to plan a continuous supply over time (Ferguson, 1996; Henderson and Sutherland, 1996). This motivation is made stronger when individuals compete for resources, and individual benefits are maximised by maximising individual access within the shortest possible time, leading to what is sometimes called 'the tragedy of the commons' (see Box 19.1).

Box 19.1

The `tragedy of the commons'

Human behaviour is one of the major challenges to the sustainable use of any natural resource. As noted by Ludwig *et al.* (1993), `*Resource problems are, after all, human problems that are generated through economic and political systems which humans design*'.

One of these human/resource problems is called the `tragedy of the commons', which acts as follows. When a natural resource is made available for people to exploit, there is competition between people for it. Rather than share that resource equitably for mutual benefit, an individual will take as much of that resource as possible before anyone else can. Unless the size of the take is heavily regulated (e.g. see the case of the Western Rock Lobster in Chapter 8), the resource is then quickly overexploited. No individual will then wish to reduce harvest levels to benefit the wider community or the recovery of the resource. The result is that resources held in common frequently deteriorate.

Maximum sustainable yield

Maximum sustainable yield is the amount of a renewable natural resource that can be taken while ensuring the indefinite availability of that resource. The term 'sustainable forestry' dates back to at least 1713 in Germany, when von Carlowitz discussed issues of sustainable wood production for the local timber industry in an early textbook on forestry (Schuler, 1998). Similar developments occurred in France in the 1660s (Brown, 1883; see Chapter 1). Faustmann (1849, cited by Ludwig, 1993) used the concept of sustainable forestry to calculate a forest rotation period that would maximise economic benefits. **Optimum catch**, later called maximum sustainable yield, was estimated for a Scandinavian fishery in 1933 (Rosenberg *et al.*, 1993). The underlying ecological principle of the optimum catch concept is density-dependent population growth. As harvesting reduces a population, per capita net population growth increases until the point is reached at which the population cannot compensate for additional mortality caused by harvesting. The harvesting rate that equals net population growth will, in theory, provide the maximum yield in perpetuity.

Forests

The 'regulated' forest or 'normal' forest has long been a focus of forest management (Oliver and Larson, 1996; Davis *et al.*, 2001). In the simple case of a single commercial tree species, uniform site conditions and a single silvicultural system, the prescription can be reduced to (after Davis *et al.*, 2001):

$$\text{Total area/rotation age} = \text{area in each age class}$$

resulting in roughly equal areas in each age class (Figure 19.2; e.g. Department of Natural Resources and Environment, 1996). This management strategy led to maximum economic benefit and output of forest products that could theoretically be sustained over time (e.g. Leuschner, 1990). It takes into account practical perspectives such as the development of the transportation network, stand priorities for harvesting and sizes of harvested areas (or logged coupes) based on logging methods and harvest economics. The ultimate objective was the perpetual, even flow of

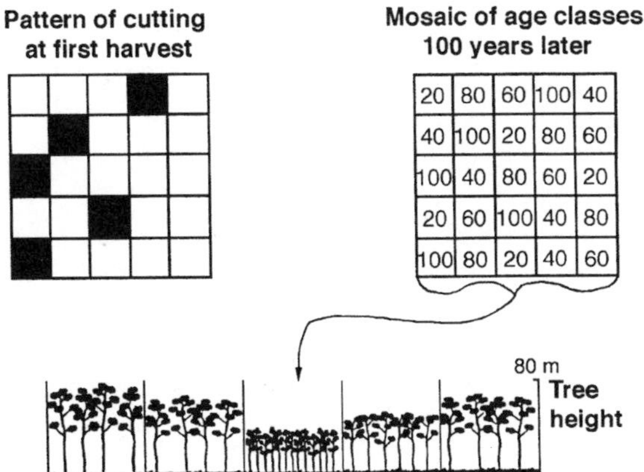

Figure 19.2. The `normal' forest. An illustration of the way an even-aged eucalypt forest might be converted to an uneven-aged forest over a period of 100 years, with harvesting at 20-year intervals (after Squire, 1987).

Box 19.2

Perceptions of sustainability and resource use

The principle of maximum sustainable yield in forestry is focused on timber supply. Non-wood forest values remain a secondary consideration for most forest managers (Lindenmayer and Franklin, 2002, 2003). The concept of maximum sustainable yield has only recently become part of forest management in Australia. For example, for the first half of the 20th century, White Cypress Pine forests were regarded as a resource that could be used immediately, with no intention to provide a continuing supply (NSW Forestry Commission, undated). White Cypress Pine, a relatively small tree with a scented, naturally termite-resistant timber, was the dominant species of the most widespread forest type in New South Wales. This once immense forest belt is now reduced to small fragments in a matrix of Wheat and pasture, and even the remnants are still subject to grazing and timber production. The species does not recruit continually, and thus was not amenable to development of a `normal' forest structure. Rather, it responds to infrequent recruitment windows – periods of conducive environmental conditions – resulting in pulses of even-aged regeneration. Almost all existing stands were produced by two such recruitment events, one in the 1890s and the other around 1950 (NSW Forestry Commission, undated). Thus, because the biology of the species did not lend itself to standard forest management practices, it was classified as an expendable resource, reflecting the utilitarian perspective of management agencies of the time.

Stands of White Cypress Pine in western New South Wales. (Photo by Bree Galbraith.)

wood products for an industry (ESDWG, 1991a; Ferguson, 1996; Lindenmayer and Franklin, 2002).

Management for conservation or other values imposes additional constraints (Davis *et al.*, 2001; Lindenmayer and Franklin, 2002). For instance, rotation length (the time between harvest events) determines the proportion of a landscape in older, more developed forest and in dense, young, regenerating stands (e.g. Franklin *et al.*, 2000). Rotation periods determined primarily by investment economics (i.e. the discounted present net worth of a forest stand) vary from 20 to 60 years, depending upon site productivity and species (Lindenmayer and Franklin, 2002). Biologically-based rotations usually maximise productivity and are typically substantially longer than economic rotations for the same tree species and site conditions.

Maximising productivity usually means harvesting forests when they are mature. For example, rotation periods can be 90 to 150 years in the Douglas Fir forests of north-western North America and the Mountain Ash forests of the Central Highlands of Victoria. However, further problems arise if a goal is to provide, for example, old growth conditions required by some elements of biodiversity or to produce large volumes of good quality water (Franklin *et al.*, 1981; O'Shaughnessy and Jayasuriya, 1991; Smith, 1998; see Chapter 2). Rotations that maximise productivity are at least a century short of allowing the development of structural conditions typical of old growth stands of Douglas Fir forest (Franklin *et al.*, 2000), and 30–50 years short of allowing old growth stand conditions in Mountain Ash forest.

Fisheries

Calculations in fisheries management that rely on catch per unit effort are based on density-dependent compensation in a growing population. Total catch and catch per unit effort increase, up until a point where population growth balances harvesting effort. If the number of fishing boats reflects harvesting effort, the catch per boat can be measured. Boats can be added until the catch per unit effort plateaus, at which point a fishery will be running at maximum efficiency, without compromising future returns from the resource (Figure 19.3; see the critique by Larkin, 1977).

Maximum sustainable yield and uncertainty

The maximum sustainable yield concept ignores **uncertainty** (Kearney, 1995). Uncertainty can arise from measurement error, natural variation that affects the distribution and abundance of the resource, or a lack of understanding of the ecology of

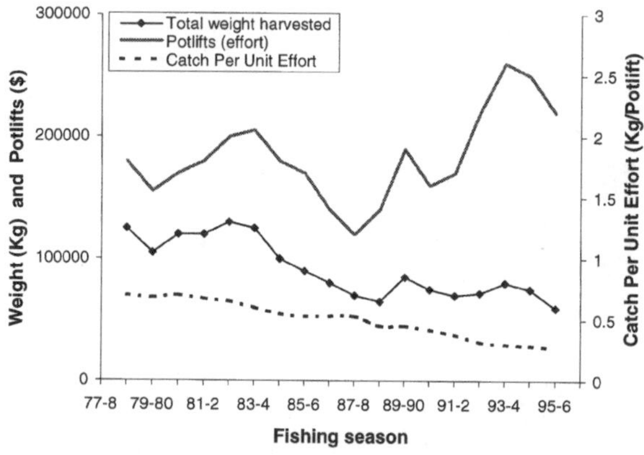

Figure 19.3. Catch per unit effort for the Rock Lobster in the Eastern Fishing Zone of Victoria, showing a long-term decline in catch and catch per unit effort (from D. Hobday, Victorian Marine and Freshwater Institute, personal communication).

the species. If disturbances and unpredictable environmental variation such as bushfires, diseases and random variation in juvenile survival are not accounted for in forecasts, it is likely that the resource will be over-exploited. Such considerations are further complicated by uncertainties in the volumes and values of products of different kinds.

It is unrealistic to expect managed natural forest in Australia to have a balanced distribution of size and quality classes producing an even flow of different forest products, because fires are virtually certain within the rotation times of most forest types (McCarthy *et al.*, 1999a). Similarly, a fixed return per unit effort from a fixed number of fishing boats will never be achievable, simply because fish populations fluctuate naturally and fishing success is inherently variable (Larkin, 1977; Rosenberg, 2003).

An alternative to a maximum sustainable yield strategy is to track the availability of the resource through time. For example, fisheries managers may regulate the harvest of a fixed proportion of the population, rather than a fixed number of fish, or harvest only when the population exceeds a specified size.

Ludwig (1993) made the point that because many of the utilitarian values of natural systems are limited, a conflict between human objectives and conservation of resources is inevitable, unless human use rates are also limited. During periods of relatively stable environmental conditions and with few changes in technology, harvesting rates tend to stabilise at positions determined by bio-economic systems that presume a steady state. Exploiting irregular or fluctuating resources is subject to the **ratchet effect** (Ludwig, 1993), whereby a sequence of good years may encourage investment in infrastructure and capital. In sequences of poor years, an industry will appeal to government or the general population for help (see Chapter 8), because investments and jobs are at stake. The appeals then elicit direct or indirect subsidies. The ratchet effect is caused by the lack of inhibition on investment during good periods, and strong pressure not to disinvest during poor periods. The long-term outcome is a heavily subsidised industry that over-harvests the biological resource on which it depends.

Sustainable development

Despite its operational flaws, the concept of maximum sustainable yield has as its goal the provision of maximum benefit from the environment, without compromising future management options. This philosophy was recognised in the definition of sustainable development in the United Nations Commission Brundtland Report (WCED, 1987):

> *development which meets the needs of the present without compromising the ability of future generations to meet their own needs. …sustainable development is not a fixed state of harmony, but rather a process of change in which the exploitation of resources, the direction of investments, the orientation of technological development, and institutional change are made constant with future as well as present needs.*

This definition focuses on human needs. The Ecologically Sustainable Development Working Groups established by the Australian Government in early 1990 (ESDWG, 1991a,b) defined the ecologically sustainable use of natural resources as:

> *the optimisation of tangible and intangible social and economic benefits which ecosystems can provide to the community, with the goals of maintaining the functional basis of the ecosystems, biodiversity, and the options available for future generations.*

Ecologically sustainable development extends the notion of sustainable development. It does not imply the maintenance of a given standard of living, environmental amenity, or stock of natural resources. It encompasses the changing processes of human needs through time, developments in knowledge and technology, and the notion of protection and enhancement of ecosystems (Dovers and Norton, 1994; Dale *et al.*, 2000; McIntyre *et al.*, 2002; Dovers, 2002). Many hundreds of Australian government policies and more than 120 statutes set out the goals, objectives and principles of ecologically sustainable development (S. Dovers, personal communication).

Ecologically sustainable development rests on the principles of inter- and intra-generational equity (see Chapter 1). The difference between maximum sustainable yield, even one that accounts for uncertainty, and ecologically sustainable development, is that the latter concept puts resource use into an ecological context. This perspective was put succinctly by Callicott and Mumford (1997) who suggested that ecological sustainability means '*meeting human demands without compromising the health of ecosystems*'.

Most commentaries on sustainable development argue for strategies that rely on resilience, learning from current practices, and making choices that retain options (e.g. Aslin, 1991; Dovers, 1990, 1994; Brunton, 1994; Dale *et al.*, 2000; Dovers and Wild River, 2003). This approach implies harvesting renewable resources on a sustainable-yield basis; using non-renewable resources in a way that does not preclude easy access to them by future generations; depleting non-renewable energy sources slowly enough to ensure an orderly transition to renewable energy sources; and maintaining biological productivity and diversity (Goodland and Ledec, 1987, cited by Dovers, 1990). The economy should be managed so that we live off the dividends of our resources, and maintain and improve the asset base (Dovers, 1990; McNeely *et al.*, 1990; Amos *et al.*, 1993; Beattie and Ehrlich, 2001), without damaging ecological processes or communities (Beattie and Ehrlich, 2001; Dovers, 2002). However, the concept of ecologically sustainable development is far from ubiquitous across resource management sectors. For example, in the Australian pastoral industry, Stafford-Smith *et al.* (2000) noted that because many land managers are facing a wide range of economic pressures they are 'discounting' the future viability of their land for immediate productive gain.

Both utilitarian and ecological perspectives on sustainable development require cautious use of natural resources. For example, in the case of fisheries management, the Food and Agriculture Organization (FAO Fisheries Department, 1994) suggested that the 'benefit of the doubt' should be given to the status of a resource, and that preventive action should be taken in fisheries to prevent stock collapses. In contrast, the pursuit of maximum sustainable yield does not embody the precautionary principle (Chapter 1).

Scientific certainty, were it to exist, could not guarantee ecologically sustainable resource management because many social, economic and political factors contribute to decisions. Even in cases where there is abundant evidence and many prior examples of ultimately destructive practices, scientific advice may be ignored. Ludwig *et al.* (1993) cited the example of Sumer, an ancient civilisation that existed in what is now Iraq. Approximately 3000 years ago, Wheat, which had been highly productive in the past, was replaced by Barley. The reason for this change was because Barley was more tolerant of the soil salinity that had been caused by irrigation. In California, developers were warned in 1899 that the consequences of planned irrigation would be similar to that which occurred in Sumer (Ghassemi *et al.*, 1995). The warnings were not heeded, despite knowledge of the Sumerian and other examples, and a much improved understanding of the processes of soil salinisation (Ghassemi *et al.*, 1995). Similarly, in Western Australia there has been clear evidence since the 1920s that clearing native vegetation leads to salinity (State of the Environment, 2001d; Stirzaker *et al.*, 2002). Large-scale dryland salinity is now common in the Wheat belt of that State (Hobbs and Mooney, 1998; Department of Conservation and Land Management, 2000; Chapter 5), yet major attempts to rehabilitate the landscape are only recent (Lambeck, 1999; Brooker, 2002).

In response to current land management problems and the failure to learn from history, in a report to the Ecological Society of America, Dale *et al.* (2000, p. 639) outlined eight ecological guidelines for managing land:

- Examine the impacts of local decisions in a regional context.
- Plan for long-term change and unexpected events.
- Conserve rare landscape elements and associated taxa.
- Avoid land uses that deplete natural resources.
- Retain large contiguous or connected areas that contain critical habitats.
- Minimise the introduction and spread of non-native species.
- Avoid or compensate for the effects of development on ecological processes.
- Implement land-use and management practices compatible with the natural potential of an area.

Many of the examples in this book indicate that such principles are difficult to implement (see also Ludwig *et al.*, 2001). The Australian Federal Government's *Environment Protection and Biodiversity Conservation Act (1999)* (EPBC Act) encourages ecologically sustainable development by requiring sustainability principles to be considered in the approval of a project. Although this is useful in theory, the States and Territories in Australia have jurisdiction over land and resource management and the EPBC Act failed to curtail damaging land management practices such as vegetation clearing (see Chapter 9).

Orr (2002) argued that four approaches are needed to assist human populations to make a transition to sustainability, including:

- The development of more accurate models and measures of human endeavour and their effects on the biosphere.
- Improvement in the 'art' of citizenship and governance.
- Improved education to better highlight issues of ecological sustainability.
- Improved ways to recognise and solve divergent problems.

These are as much social requirements as they are ecological. Part of the challenge of implementing such recommendations is that sustainability is not easily defined and is not an endpoint. Rather, sustainability is a direction – an attempt to balance conservation and resource management that reflects evolution in societal perspectives and scientific knowledge (Lindenmayer and Franklin, 2003; see McIntyre *et al.* 2002; Figure 19.4). In moving toward sustainability, there will be many transitions and not all of the movement will be forward!

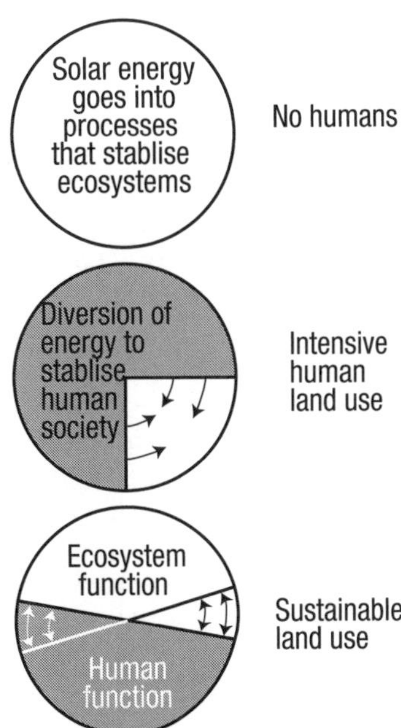

Figure 19.4. The balancing act that attempts to identify sustainable land use practices. (Redrawn from McIntyre *et al.*, 2002.)

International conventions on sustainability

In 1992, the United Nations Conference on Environment and Development (UNCED, 1992) outlined a framework to enhance the productive, protective, environmental and social roles played by forests throughout the world. The conference recommended formulation of '*scientifically sound criteria and indicators for the management, conservation and sustainable development of all forest types*'.

The objective of these criteria and indicators was to provide an internationally accepted basis for measuring sustainable forest management. Subsequently, international agreements were reached involving almost all countries with significant areas of natural forest:

- The International Tropical Timber Organisation (ITTO, 1990, 1992, 1993), representing most countries that export timbers from tropical forests, created guidelines for the sustainable management of natural tropical forests.
- The Second Ministerial Conference on the Protection of Forests in Europe, held in Helsinki in 1993, resulted in the Helsinki Process (1995). All European countries, including Russia and Turkey, agreed to guidelines and criteria for the sustainable management and conservation of European forests.
- The Seminar of Experts on Sustainable Development of Temperate and Boreal Forests held in Montreal in 1993 resulted in the Montreal Process (1995). Ten participating countries including Australia agreed on a set of criteria and indicators of sustainable forest management. These were incorporated into a document known as the Santiago Declaration.
- The Tarapoto Proposals (Tratado de Cooperacion Amazonica, 1995) are an outcome of the Tratado de Cooperacion Amazonica, which linked eight countries of the Amazonian region to a common set of criteria and indicators for sustainable forest management at both national and global levels.
- The Food and Agriculture Organization (FAO, 1995) met in collaboration with the United Nations Environment Program in Nairobi to

develop criteria and indicators for the sustainable management of dry and semi-dry African forests (that is, those not represented by the International Tropical Timber Organisation).

The agreements identified many different criteria and indicators for providing an internationally accepted basis for measuring sustainable forest management. The Helsinki Process uses, among others, the extent of forest resources, forest ecosystem health, biological diversity, and the political and legal capacity to implement sustainable forest management as criteria for the assessment of sustainable forest use. The Montreal Process specifies seven criteria, each of which has a number of indicators associated with it (Commonwealth of Australia, 1998). The first criterion is the conservation of biodiversity. The indicators of species diversity are:

- extent of area by forest type
- extent of area by forest type and by age class or successional stage
- extent of area by forest type in protected area categories
- extent of area by forest type in protected areas by age class or successional stage
- fragmentation of forest types.

Although these indicators must be defined and quantified, their clear intention is to describe status and record temporal trends of forest ecosystems (McCarthy, 1996a; Tickle *et al.*, 1998).

These agreements are a non-binding commitment to implement the sustainable use of natural forests. It remains to be seen how legislative and regulatory processes in Australia cope with the undertakings embodied in the Santiago Declaration. In a broader context, Ferguson (1996) concludes that it is unlikely that many developing countries will be able to find the funding for the necessary improvements in government policies and forest management practices in the foreseeable future. The agreements outlined here apply only to forested landscapes; there is no analogous international process for marine environments or other terrestrial landscapes.

Globalisation, sustainability and biodiversity conservation

Many nations are associated through formal and informal trade links, which are a part of increasingly globalised trade and commerce. International trade can have both positive and negative outcomes for biodiversity conservation. For example, many of the problems associated with invasive species are linked to increased trade, for example the exotic species transported between countries in the ballast water and cargo of ships (Tyndale-Biscoe, 1997; Low, 1999).

Box 19.3 shows the unexpected problems of patterns of resource use in one country (wine in Australia) on forest cover in another (cork trees in Spain and Portugal). There are numerous parallel situations. Approaches such as certification standards in the wood products and fisheries industries (see Chapter 8) attempt to reduce the effects of resource consumption in one country on the environment of another. However, there are problems with such approaches across sectors (e.g. agriculture and forestry). Only a few industries have embraced the concept of certification; the certification of products from the agricultural sector could have substantial conservation benefits, given the extent of environmental damage from that sector in Australia and worldwide. Indeed, there appears to have been some success with the application of certification protocols in the production of coffee. In this situation, more environmentally sensitive growing methods not only better conserve biodiversity (Has and Dietsch, 2004; but see Philpott and Dietsch, 2003), but also promote the delivery of key ecosystem services such as improved pollination of coffee plants (De Marco and Coelho, 2004).

Box 19.3

Thinking locally *and* globally

In conservation biology, seemingly simple issues can often require complex responses to reduce impacts on biodiversity. The use of `plastic corks' and metal screw-tops aim, in part, to reduce levels of wine spoiling in wine production. However, this practice may have serious indirect impacts on populations of some vertebrates. Much of the cork used in the Australian wine industry comes from Spain and Portugal. Cork forests in these countries provide habitat for many species, including threatened ones such as the Iberian Lynx, one the most endangered members of he cat family (Ferreras, 2001). New types of stoppers will reduce demand for cork and reduce the economic value of cork forests, leading to habitat loss when they are cleared for other (more valuable) land uses.

A cork plantation in Portugal, where an Iberian Lynx was seen several weeks previously (see Box 19.3). (Photos by David Lindenmayer.)

19.2 Adaptive management

Ecosystem management depends on a rudimentary understanding of its major components, the processes that underpin its function, and its response to disturbance. Presently, the ecology of virtually every Australian ecosystem is poorly understood. For example, all of the 75 inquiries into the timber industry in Australia since the end of World War II have emphasised the urgent need for more fundamental information on ecosystem processes (RAC, 1992a; Lindenmayer and Franklin, 2003).

Adaptive management can underpin the development of ecologically sustainable practices (as defined by Holling, 1978; Walters and Holling, 1990; Beese *et al.*, 2003) in ecosystem management. Adaptive management involves gaining knowledge and using it to modify practices to achieve management goals.

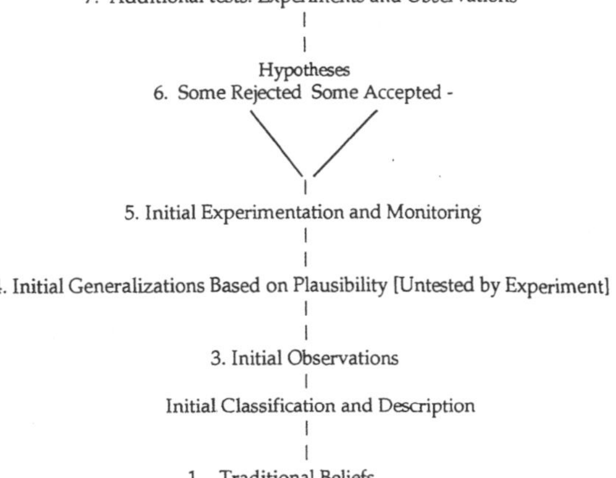

Figure 19.5. 'Knowledge hierarchy' for learning about ecological systems through experimentation, monitoring and testing (from D. Botkin, personal communication).

Botkin (personal communication) has formulated this as a 'knowledge hierarchy' (see Figure 19.5).

Adaptive management requires an active, planned, and systematic effort to acquire information from management experience, monitoring and research. It also demands comprehensive and well-documented management and research records (Walters, 1986, 1997; Taylor *et al.*, 1998; Bunnell *et al.*, 2003).

Adaptive management deals explicitly with uncertainty. Davis *et al.* (2001) note that:

> *The process of adaptive management includes highlighting uncertainties, developing hypotheses around a set of desired system outcomes, and structuring actions to evaluate or test these ideas. Although learning occurs regardless of the management approach, adaptive management attempts to make that learning more efficient.*

Management activities are viewed as experimental tests of hypotheses about ecosystem or species responses, involving active testing of alternative proposals, and passive observation and data collection. Four elements of adaptive management are identified by Davis *et al.* (2001):

- *'an acknowledgement of the uncertainties of proposed management policies'*
- *'The description of key management policies as testable hypotheses'*

- *'The search for, and use of, information that will enable testing the hypothesis or hypotheses...[which] can range from informal observations of foresters and other specialists...to formal replicated experimental design, but does require a conscious attempt to assess the validity of the hypothesis or hypotheses in question'*
- *'An institutional mechanism that ensures that the hypotheses will undergo periodic, fair-minded review and that management policies can change as a result of the review'.*

Davis *et al.* (2001) noted that true adaptive management maintains a *'ruthless hold on uncertainty'*. However, this admission of uncertainty can make stakeholders nervous. Accepting adaptive management means accepting impermanent decisions about management and resource use, which can be disconcerting to stakeholders who would like assurances about timber harvest levels, fisheries quotas or ecological reserve boundaries. The struggle to maintain the flexibility needed for adaptive policies conflicts with stakeholders' desires for assured outcomes (Davis *et al.*, 2001).

A formalised approach to adaptive management

Holling (1973), Walters (1986, 1997), and Walters and Holling (1990) proposed a formal integrated approach to adaptive management involving research, monitoring and management designed to assess and improve the effectiveness of resource management prescriptions (Shaw *et al.*, 1993). An adaptive management system is defined by Holling (1973) as one that *'can absorb and accommodate future events in whatever unexpected form they may take'*.

In implementing a formalised adaptive management program, there are a series of linked steps (Figure 19.6):

- *Step 1.* All available information about a system is gathered. Based on that information, alternative models are created regarding management of that system and policies are clarified on approaches that will meet management goals, possibly by using simulation models (Walters, 1986).
- *Step 2.* A small set of testable hypotheses for different management options is created (Taylor *et al.*, 1998). Sometimes, this step involves consideration of entirely new management approaches that are outside existing procedures and policies.
- *Step 3.* An experimental design and monitoring program is developed. The design must specify which system components are to be measured to assess the success of different management options. A pilot study may be required (Silsbee and Peterson, 1993; Urban, 2000).
- *Step 4.* Management changes are implemented, for example altered harvesting strategies, reserve designs or silvicultural systems, based on alternative models or the results of experiments. Modified management strategies are monitored, coupling field research with result-driven management.
- *Step 5.* The adaptive management experiments, results and monitoring outcomes are fully documented (Taylor *et al.*, 1998).

Experimentation is essential to improve understanding. The goal of experiments is to learn as much as possible from both successes and failures (Taylor *et al.*, 1998). The adaptive management framework uses natural disturbances and human activities as 'experimental' opportunities (e.g. Lindenmayer and Franklin, 1997, 2002). The approach increases the likelihood that new knowledge will be generated and embraced by managers (Taylor *et al.*, 1998).

Unfortunately, monitoring and validation usually take a back seat, and are the first processes to be jettisoned given budgetary constraints and funding cuts. In particular, long-term monitoring studies typically receive limited funding and are often treated as non-essential. However, without monitoring and

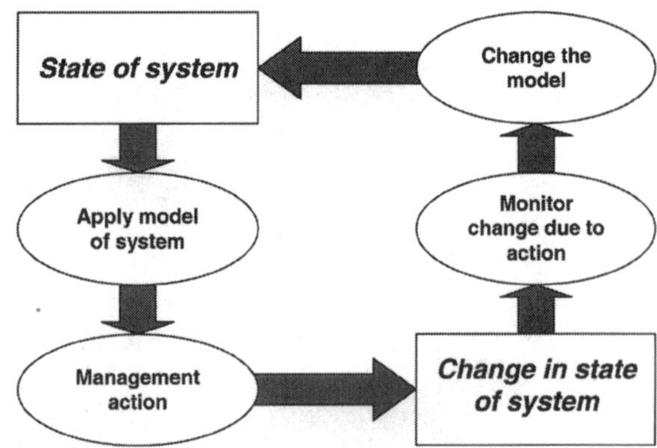

Figure 19.6. The connecting linkages in an adaptive management framework. (Redrawn from Walker, 1998.)

validation, the quality of planning decisions cannot be assessed, and the planners and managers cannot learn from their experiences.

Successful adaptive management retains as many future management options as possible. Experiments and monitoring systems need to be powerful enough to detect important change before it is too late to alter management regimes without serious impacts on the environment (Chapter 17). Tight linkages and feedbacks between scientists, managers and policy makers are essential so that scientific research and monitoring results inform decisions (Dovers and Norton, 1994; Taylor *et al.*, 1998). For example, Underwood (1995) outlined a case where the results of research on invertebrates inhabiting rocky shorelines of New South Wales indicated that the animals were being over-harvested both for human consumption and for use as bait. The research findings were largely ignored by the relevant State Government management agency, and inappropriate and ineffective control measures were implemented (Underwood, 1995). As noted by Hilborn (1992): '*If you cannot respond to what you have learned, you really have not learned at all*'.

Box 19.4

How to implement an adaptive management natural 'experiment' in a production forest

If the same method of logging is used throughout a given forest type, there is little or no variation in silvicultural systems to examine. For example, more than 98% of Victorian Alpine Ash forest is clear-cut (Lutze *et al.*, 1999) and alternative methods of harvest are very rarely contemplated. Instead, after progressing through the initial steps outlined in the text and defining the entities for measurement (e.g. particular biota), a range of logging regimes can be identified. Some areas might be clear-felled, others selectively harvested, and varying levels of stand retention employed in various other places subject to logging. Treatments would need to be matched with 'control' areas with no timber harvesting (e.g. those in large reserves).

Variation in logging regimes provides a basis to test different silvicultural systems. Although variation in cutting regimes is essential, it is also important to replicate 'treatments' to provide some statistical power in the experimental design and allow for valid contrasts. Data on the intensity of disturbance and the extent of vegetation retention on logged sites have to be carefully recorded. This is because activities such as site preparation, the retention of biological legacies and regeneration methods can significantly influence the composition of forests following logging (Hazell and Gustafsson, 1999).

Information on species or ecosystem processes gathered from monitoring studies could be used to assess the efficacy of logging prescriptions (e.g. the extent of canopy retention needed across the landscape, the width of wildlife corridors, or the number and spacing of trees retained within cutover sites). Prescriptions and strategies can be modified (based on the results of the adaptive management program) and added to an expanded logging and monitoring study.

Adaptive management in a political context

Adaptive management of natural resources for conservation requires consideration of uncertainty in ecology, resource economics, sociology and politics. Ludwig *et al.* (1993) criticised the Brundland Report (WCED, 1987) for making numerous references to sustainability without specifying how it might be achieved. Ludwig *et al.* (1993) outlined some principles for effective management:

1. Manage human motivations and responses because human greed and short-sightedness underlie most difficulties in resource management. Economic, social and political climates change rapidly and unpredictably and create as much uncertainty as changing natural conditions (Salwasser, 1993).
2. Act before scientific consensus is achieved because consensus is difficult to attain, even in cases where the resource has collapsed. Understanding, but not complete explanation, is needed to form policies and make decisions (Holling, 1993).
3. Rely on scientists to recognise problems but not to remedy them, because the judgement of scientists is coloured by their training and by political pressure, and they are not trained to find social and political solutions (Salwasser, 1993).
4. Distrust claims of sustainability, because past resource exploitation has seldom been sustainable (Rosenberg *et al.*, 1993; Underwood, 1995; Rosenberg, 2003).

5. Manage explicitly for uncertainty by considering a variety of different strategies, favouring actions that are informative, reversible and robust to uncertainty. Experiment with the system, monitor the results and modify management as knowledge grows (i.e. employ adaptive management; Beese *et al.*, 2003; Bunnell *et al.*, 2003).

Rosenberg *et al.* (1993) and Raff (2003) suggest that the solution is to define property rights such that duties and responsibilities are included. Countries including Germany, Australia, New Zealand, Canada, Iceland and the USA have recognised this problem and, in the case of fisheries management, granted individual quotas in some fisheries.

Adaptive management in the real world

Most wild populations for which an assessment of current resource status is available are over-utilised in the USA and Europe (Rosenberg, 2003). Australia has similar problems with its fish stocks (see Chapter 8). Thus, despite the existence of ecological understanding and management principles to achieve sustainable management, in most instances, prevailing political or social conditions make sustainable or adaptive management difficult.

Although adaptive management is potentially valuable, it remains untested (Dovers and Lindenmayer, 1997; Simberloff, 1998) and there are few successful applications (but see Taylor *et al.*, 1998; Innes *et al.*, 1999; Beese *et al.*, 2003; Bunnell *et al.*, 2003). This may be due to the relatively recent emergence of the concept. Adaptive management is a central element in the 'forest plan' adopted for Federal lands in north-west USA (Forest Ecosystem Management Team, 1993; Tuchmann *et al.*, 1996). The plan includes adaptive management areas, that is, places (totalling more than 500 000 hectares) where adaptive approaches are emphasised. Research and monitoring are central elements of the plan (e.g. Haynes and Perez, 2001). Failure to conduct adequate monitoring programs can be used to challenge the legal validity of the plan.

The British Columbia Coastal Forest Project of Weyerhaeuser Company is developing an extraordinary adaptive management program for forest resource management in connection with a shift from clear-felling to variable retention harvesting (Bunnell *et al.*, 2003). This program includes extensive monitoring, research, large-scale experimentation, and major interactions with scientists and stakeholders.

Altered cutting methods in the adaptive management project on forest owned by Weyerhaeuser Company on Vancouver Island, British Columbia. (Photo by Jerry Franklin.)

An emerging operational strategy uses detailed models of biological processes to challenge the performance of management options. Punt and Smith (1999) described the management system for a stock of Gemfish, which commenced as an open access fishery in southern Australia in the 1960s. The bulk of the historical catch came from bottom trawls of the upper continental shelf, at depths from 300 to 500 metres. After the populations declined substantially during the 1970s and 1980s, quota management was introduced in 1988, and trawl catch was banned from 1993 to 1996. Following relatively strong recruitment into the population, trawling was permitted again. However, effort was limited to the extent that there was a greater than 50% chance that biomass (measured as 5+ males and 6+ females) exceeded a target stock (set to be equal to 40% of the biomass in 1979). To resolve differences in attitude towards management, Punt and Smith (1999) used models to explore a range of harvest strategies and procedures, using different assumptions about the ecology of the recruitment of the species. One management procedure used best estimates of model parameters and another used estimates based on quantiles of the distributions to generate conservative judgments. One harvest strategy aimed at maximum sustainable yield, and the other aimed at providing at least 50% certainty of having at least 40% of the 1979 population.

The results demonstrated trade-offs between the expected catch and stock depletion. Performance (in terms of yield and depletion) was sensitive to assumptions about the underlying biology of recruitment. The results of the simulations showed that the economic benefits of monitoring greatly exceeded the costs of running the surveys. The information had substantial value in providing assurances that management objectives had been achieved and in informing the development of future management strategies. The results were embedded in monitoring strategies and harvesting recommendations, which were updated in subsequent years.

Punt and Smith (1999) involved a management committee composed of fishery managers (regulators), representatives from the catching and processing industries, government and independent scientists, an economist and a 'conservation member'. Smith *et al.* (1999) argued that even the most sophisticated risk assessment strategies fail if they cannot accommodate effective stakeholder involvement in all stages of the assessment. They recommended stakeholder involvement in stock assessment, setting research priorities, enforcement and decision-making, none of which are the traditional domains of non-expert stakeholders. Harwood and Stokes (2003) urged more general adoption of this system.

19.3 Ecosystem management

The discussions in this chapter so far view management of renewable resources from the rather myopic context of single species or single utilitarian values. Sensible resource management must consider the context of the resource and the impacts of management on the ecosystem (Clark and Minta, 1994). The Ecological Society of America published a committee report on the scientific basis for ecosystem management. The committee considered ecosystem management to be (after Christensen *et al.*, 1996, p. 665):

> *management driven by explicit goals, executed by policies, protocols, and practices, and made adaptable by monitoring and research based on our best understanding of the ecological interactions and processes necessary to sustain ecosystem composition, structure and function.*

Many of the principles underpinning ecosystem management are like those that aim to achieve sustainability (see Christensen *et al.*, 1996, for a detailed discussion). In particular, the approach highlights the need to maintain ecosystem functions to allow such systems to meet human demands, while also acknowledging that there is a finite limit to the capacity to provide such resources in perpetuity (Christensen *et al.*, 1996). Some of these ecosystem processes include hydrological fluxes, biological productivity, decomposition, and maintenance of biological diversity. On this basis, Grumbine (1994) argued that there are five core components to the conservation of biodiversity within a framework of ecosystem management. These are:

- maintain viable populations of species
- maintain representative ecosystems
- maintain key ecological processes (including natural disturbance regimes)
- maintain the evolutionary potential of species
- ensure that human uses and needs can be accommodated.

It is challenging to move from general rhetoric to on-ground actions (Smith, 1997). Christensen *et al.* (1996) considered that it would require defining sustainable goals and objectives, and identifying appropriate temporal and spatial scales for management, including the implementation of long-term planning and monitoring.

Ecosystem management has been recommended for some areas of the USA, for example the Greater Yellowstone area (which encompasses Yellowstone National Park and the surrounding environs; Clark and Minta, 1994). However, although the concept of ecosystem management, like adaptive management, remains an appealing goal, it (also like adaptive management) remains largely untested (Dovers and Lindenmayer, 1997).

Resource management involves adapting to changing human values and social priorities. Resource sustainability cannot be divorced from the sustainability of human economies, natural communities and ecosystems (Ludwig *et al.*, 1993; Salwasser, 1993; Bunnell and Johnson, 1998). As outlined earlier in this chapter, sustainability is a moving target, not only because ecosystems change over time and have long recovery times (Hilborn and Ludwig, 1993; Bunnell *et al.*, 2003), but also because of the changing economic, social and political climate in which decisions are made (Salwasser, 1993; Lindenmayer and Franklin, 2003).

19.4 Policy and science in conservation biology

Much of this book has focused on the scientific principles and technical methods that underpin conservation biology. Whatever the merit for a given conservation action (and the strength of the science that supports it), the final planning decision almost always has social, economic and political dimensions (Ludwig *et al.*, 2001; Dovers, 2002). Although conservation science must contribute to decisions, decision-making will ultimately balance other (often competing) interests (Ludwig *et al.*, 2001). This is true of virtually all Australian case studies of environmental conflicts (e.g. see Toyne, 1994; Dovers and Wild River, 2003); examples include the proposed culling of populations of Koalas on Kangaroo Island (see Chapter 1), setting quotas on harvested fish populations (see Chapter 8), establishing reserve systems within wood production forests (see Chapter 16), and clearing native vegetation on private agricultural and grazing lands (see Chapter 9).

The development of the forest reserve system in Australia involves competing interests, and the outcomes are determined largely by political and social realities. A major initiative in Australian forest environments is to establish a comprehensive, adequate and representative (CAR) reserve system (see Chapter 16; Mendel and Kirkpatrick, 2002). Current world practice for protected areas is that reservation targets should be a minimum of 10% of each biome (McNeely, 1992). In the case of Australian forest reservation policy, the Commonwealth Government's proposed criterion is that areas set aside should cover 'a total of 15% of the pre-1750 distribution of each forest ecosystem' (JANIS, 1996; Dickson *et al.*, 1997; Commonwealth of Australia, 2001a). However, the importance of other factors in decision-making is highlighted by statements in the JANIS report (1996):

> *15% of the pre-European distribution is seen as a desirable objective...where socio-economic impacts are not acceptable...a lower level of reservation (e.g. 10%) may prove adequate.*

Notably, neither the 10% nor the 15% reservation targets are based on scientific principles or empirical evidence. Rather, they were chosen because they are likely to be politically acceptable. For example, the International Union for the Conservation of Nature proposed a level of 10% reservation of each biome on a global basis, which was based on a recommendation resulting from a workshop on national parks and protected areas held by the International Union for the Conservation of Nature in 1992. The 10% global figure was agreed to by the majority of participants at the workshop as a pragmatic political response – a figure that would be politically tolerable in most countries. The figure has no specific scientific basis (SAG, 1995; Lindenmayer and Franklin, 2002).

As we noted in Chapters 4 and 16, simply capturing a sample of a particular type of forest does not guarantee that the biodiversity associated with that forest type will persist. Any scientific discussion about the amount of land that should be dedicated to conservation reserves is likely to be inconclusive, simply because so many factors affect the answer (e.g. the activities on adjacent land, the resilience of the species, off-reserve conservation, and the state of knowledge about ecological processes). However, there is sufficient scientific evidence to suggest that, in isolation, 10% or even 15% is not sufficient to guarantee the persistence of all (or even most) species and ecological processes (reviewed by Soulé and Sanjayan, 1998; Lindenmayer and Franklin, 2002). For instance, the International Union for the Conservation of Nature, in recommending 10% reservation of each world biome, recognised that the persistence of biodiversity in protected areas depends on the condition and management of surrounding land (SAG, 1995).

In other cases, it is obvious that even when the lessons from conservation biology are clear, better management practices depend on policy and legislation as much as on 'good' science. For example, rates of clearing of native vegetation in Queensland exceed those of many developing countries, despite knowledge of the impacts associated with the clearing (Nix, 1994; Smith, 1994; Fensham, 1996; State of the Environment, 2001; Morton *et al.*, 2002; Cogger *et al.*, 2003; see Chapter 9). Efforts to reverse these trends will require legislation, changes to government policies (Fensham, 1996; Morton *et al.*, 2002), and economic incentive schemes for farmers and graziers (Whitten *et al.*, 2002).

19.5 Conclusions

When we wrote *Conservation Biology for the Australian Environment* in 1995–1996, the single most important biological threat in Australia was land clearing for agriculture, grazing and urban development. Nearly a decade later this is still true. The failure to control this problem remains a major policy failure in modern Australia (Australian and New Zealand Environment and Conservation Council, 2001). It is our belief that land clearing is the largest single mistake that our generation is making and, in the future, Australians will regret this action. The land clearing is taking place irrespective of environmental initiatives such as the Natural Heritage Trust (Morton *et al.*, 2002). We are doing irreparable damage to the environment, the consequences of which will be felt by future generations.

The most obvious strategy that would have the most long-lasting and beneficial consequences for Australia's biodiversity would be to bring land clearance to an immediate halt (Morton *et al.*, 2002). There are enough problems associated with how to best manage land that is already cleared and how to control other problems deriving from that cleared land (e.g. secondary salinity and the contribution to greenhouse gas emissions from vegetation clearance).

Given the political and social reality of continued environmental degradation, the methods described in Part III of this book have ready application in problems of land clearance, the management of fragmented populations, and setting priorities for protecting land. There is a need for a companion text to this one that describes the social, legal and political processes that affect vegetation clearance for urban and agricultural development (e.g. Dovers and Wild River, 2003), and more attention needs to be focused on mechanisms to halt and reverse the damage to the Australian environment.

Land clearing near Mundubbera in central Queensland.

There is no tradition of sustainable management in urban planning or development. The concept of sustainable management in agriculture is limited to stocking rates and water supply on individual farms, and is more closely related to maximum sustainable yield than to ecologically sustainable management. Thus, the two areas that have the greatest potential for adverse impacts on the Australian environment are the most poorly regulated and have the weakest traditions of sustainability.

Most assessments of impact on the Australian environment conclude that threatening processes should be managed (e.g. Recher and Lim, 1990; Saunders *et al.*, 1998; State of the Environment, 2001a). If changes in the conservation status of species or communities trigger conservation action, then most conservation action will be remedial. Proactive management would make more efficient use of scarce conservation and environmental management resources. Yet this is still not part of State or Federal Government approaches to resource management (Figure 19.7). Figure 19.7 is

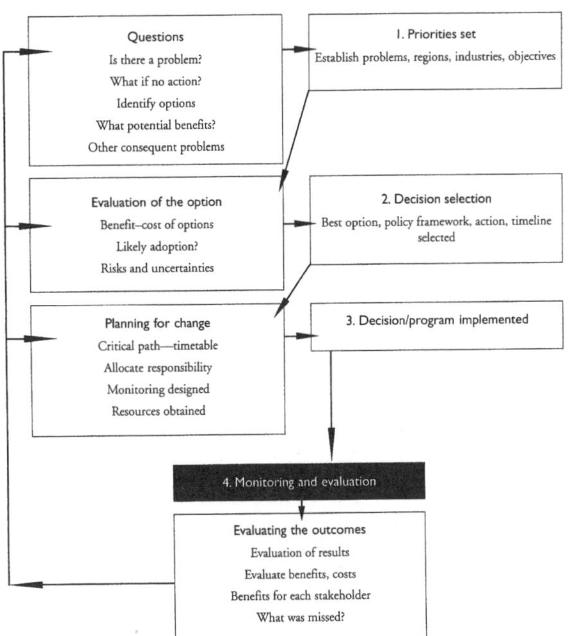

Figure 19.7. Phases in natural resource management decision-making as proposed under the National Land and Water Resources Audit. (Redrawn from the Commonwealth of Australia, 2002a.)

Table 19.1. Area of land in various land tenure categories in Australia in 1993 (×1000 square kilometres; from data in Australian Bureau of Statistics, 2001).

	New South Wales	Victoria	Queensland	South Australia	Western Australia	Tasmania	Northern Territory	Australian Capital Territory	Total
Public land	85.7	72.3	118.0	217.6	1095.0	40.6	137.2	1.5	1767.9
Aboriginal and Torres Strait Islander land	1.5	–	42.2	189.6	325.5	–	536.0	–	1094.8
Private land	714.4	155.3	1567.0	576.8	1105.0	27.2	673.0	0.9	4819.6
Total	801.6	227.6	1727.2	984.0	2525.5	67.8	1346.2	2.4	7682.3

from the National Land and Water Resources Audit (Commonwealth of Australia, 2002a) and the flowchart begins with: 'Is there a problem?'. Clearly, new perspectives on natural resource management decision-making are needed.

A unique feature of the development of conservation prescriptions in Western Europe has been the implementation of special conservation regulations for private land in the public interest (Raff, 2003), and there is also increasing recognition of this issue in North America (Hilty and Merenlender, 2003). There is a need to further develop private land conservation and sustainable management in Australia (Figgis, 1999) and build upon early initiatives (e.g. the BushTender Scheme in Victoria; see also Whitten *et al.*, 2002). The need for this is underscored by the fact that the amount of private land in Australia is about three times greater than the amount of public land (Australian Bureau of Statistics, 2001) (Table 19.1). Further discussion of this topic is beyond the scope of this book. To paraphrase Ferguson (1996), sustainable management requires the clear allocation of property rights and responsibilities, the management of supply through regional management plans, enforceable codes of practice, management of the demand-pressures on publicly owned resources, and public participation in management. This approach acknowledges the roles of the public as consumers, owners of production, and custodians of the future resource.

The Australian environment is a 'human artifact' (Langton, 1998) – the product of tens of thousands of years of management. Indigenous knowledge is not a subset of Western conservation knowledge, and Aboriginal land management exists wherever there are living Aboriginal customary systems (see Hill, 2003). Sustainable development is essential for

Box 19.5

Why resource managers have problems in implementing conservation science

There is always a lag time in the implementation of new research, if the research is implemented at all – indeed many resource and conservation managers use their experience rather than being guided by research findings (Pullin *et al.*, 2004). There are many reasons for this:

- Often management plans are poorly developed (if they exist at all) and there have been few systematic processes for reviewing them (Pullin, 2002; Fazey *et al.*, 2005b).
- Most conservation biologists are poor communicators. They operate under what Hamel and Prahalad (1989) call the 'strategy of hope' – they hope their work will be useful for management professionals.
- Many resource managers lack the scientific training to interpret field data and translate it into practice.
- Most conservation theory remains poorly developed and is still hotly debated by conservation biologists.
- Finally, many conservation biologists have a limited understanding of the complications and social, economic and other constraints associated with real-world natural resource management. This is illustrated by the complexities associated with diary farming shown in Figure 19.8.

There is a need to identify new ways to better communicate research results and increase the speed with which they are adopted. Conservation biologists also need to communicate more clearly their visions about ways to better integrate conservation and production goals (Saunders, 1996; Fazey *et al.*, 2005b).

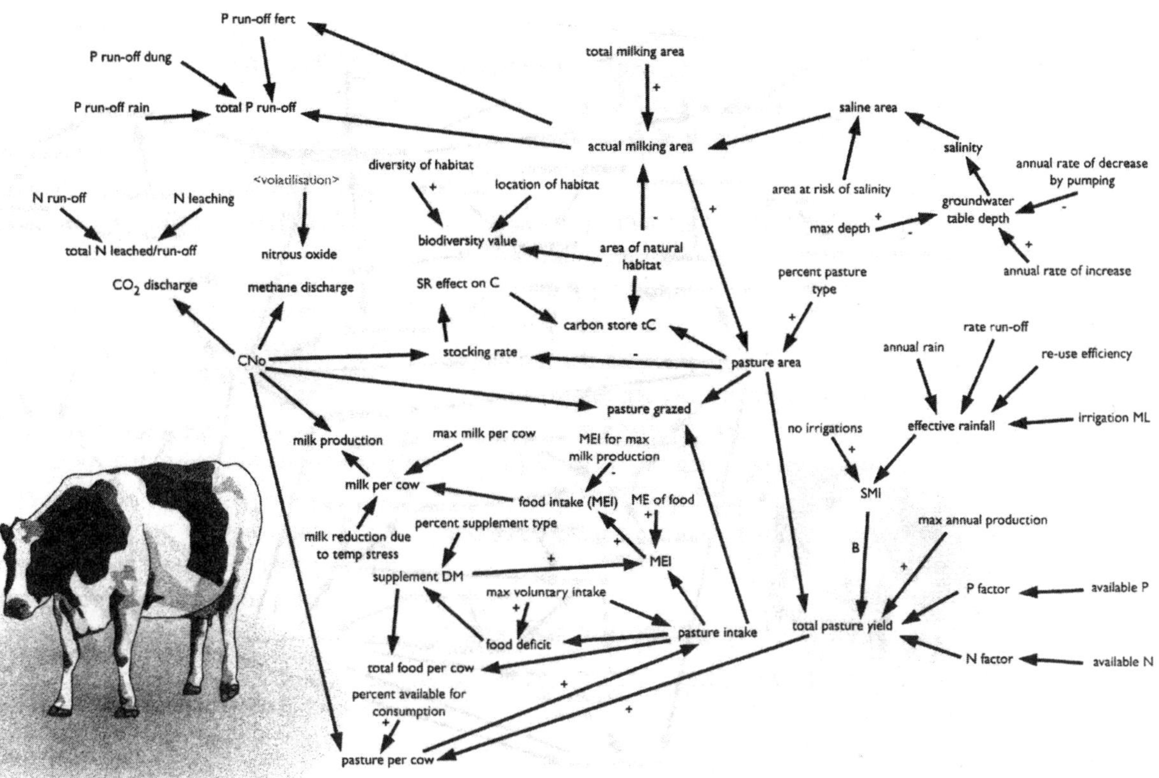

Figure 19.8. The complexities of management decision-making faced by a dairy farmer (from Commonwealth of Australia, 2002a).

permanent residents. Aboriginal people in northern Australia, for instance, have a preference for small-scale economic activities such as crocodile egg harvesting and bioprospecting. Such activities are governed traditionally by monitoring that ensures the sustainability of species and their habitats, and they are suitable companions for traditional hunting and gathering, law and culture (Langton, 1998). However, there are no grounds to assume that Aboriginal land owners do not want development of their land; as Langton (1998) states: '*Sustainable use of natural resources presents options for evolving Aboriginal approaches to the stewardship of their estates*'. Traditional and Western paradigms for sustainable management have much in common and will be successful if they work in partnership, with mutual respect.

Ewel (1993) pointed out that one of the wasteful tragedies in the biophysical sciences is the rift (real or perceived) that separates some biologists and conservationists from some agriculturalists, foresters, engineers and other land-use managers. In fact, all these people have a common goal of sustainable land use. However, definitions of sustainability differ, leading to misunderstanding and differing objectives (Lindenmayer and Franklin, 2003). Without economically, socially and ecologically sustainable land use systems, it is unlikely that the conservation of natural ecosystems can be sustained politically, unless the social landscape in Australia changes to embrace ecocentric perspectives on the value of biodiversity (see Chapter 1). Rosenberg *et al.* (1993), Rosenberg (2003) concluded that the sustainable use of natural resources is attainable if scientific advice based on biological, social and economic considerations is part of the development of policies for renewable resource use. Australia has an extraordinary body of natural, physical and social scientists. It now also has, courtesy of State of the Environment reporting and the National Land and Water Resources Audit, some excellent environmental data. Therefore, sustainable management of natural resources in Australia *should* be an achievable goal, particularly if guided by adaptive management principles. Indeed, as noted in the State of the Environment Report (1996): '*Australia has a better opportunity than perhaps any other nation to protect its environment and use its natural and heritage resources wisely*'. We agree with this comment, but to grasp this opportunity will require continued revision of scientific and bureaucratic practices and a change in public attitudes (Recher, 2002).

19.6 Practical considerations

Technical and scientific analysis should be subordinated to analysing and substantiating the conceptual models of stakeholders – those people who will carry the burden of the outcomes of decisions. Models such as those described in earlier chapters should support the thinking of those affected by a decision, and provide a platform to resolve arbitrary differences of opinion. In this context, they become part of an adaptive management framework that will be successful only if supported by monitoring capable of detecting changes that matter, and if they are part of an iterative process in which natural resource management decisions are revised routinely following experimentation and research.

19.7 Further reading

The journal *Ecological Applications* published a series of articles on sustainability, ecological research and resource management in 1993 (volume 3, issue 4). Christensen *et al.* (1996) reported on the scientific basis of ecosystem management. Dale *et al.* (2000) outline a series of ecological principles and guidelines for the ecologically sustainable management of land. Many books, including those by Yencken and Wilkinson (2000), Hamilton (2001), Dovers (2002), Goldie *et al.* (2004), and Grafton *et al.* (2004), discuss an array of issues on sustainability in an Australian context. Palmer *et al.* (2005) present an informative discussion on the role of ecological science and its application in transitions to sustainability.

Ferguson (1996), Bunnell and Johnson (1998) and Lindenmayer and Franklin (2002, 2003) outline conditions for the development of practices for sustainable forest use. Ward *et al.* (2001) and Rosenberg (2003) outline some of the changes needed for sustainable fisheries management. Figgis (1999) is a detailed discussion of issues associated with conservation on private land in Australia.

Taylor *et al.* (1998), Bunnell *et al.* (2003) and Berger *et al.* (2004) provide some examples of adaptive management in a real world management context. Dovers and Mobbs (1997) discuss adaptive management in a social context. Dovers and Wild River (2003) contains a series of excellent essays on the policy domain as it relates to an array of sustainability issues in Australian natural resource management. Toyne (1994), Mercer (1995) and Dovers (2002) describe some of the political, policy and social dimensions of resource conflicts in Australia. Ludwig *et al.* (2001) present an interesting discussion on the intersection of ecology, conservation and public policy.

Appendix I: Taxonomic names

Common name	Taxonomic name
Abalone	*Haliotus* spp.
Acorn Barnacle	*Elminius modestus*
African Soap Berry	*Phytolacca dodecandra*
Agile Antechinus	*Antechinus agilis*
albatross	family Diomedeidae
algae	kingdom Protista
Alligator Weed	*Alternanthera philoxeroides*
Alpine Ash	*Eucalyptus delegatensis*
Alpine Tree Frog	*Litoria verreauxii alpina*
Altona Skipper Butterfly	*Hesperilla flavescens flavescens*
American Chestnut	*Castanea dentata*
American Eel	*Anguila rostrata*
amphibians	division Amphibia
annelids	phylum Annelida
Antilopine Wallaroo	*Macropus antilopinus*
ants	order Hymenoptera
Apple	*Malus* spp.
Apple Box	*Eucalyptus bridgesdiana*
Arabian Oryx	*Oryx leucoryx*
arthropods	phylum Arthropoda
Asian Mussel	*Musculista senhousia*
Ass	*Equus hemionus/Equus africanus*
Athel Pine	*Tamarix aphylla*
Atherton Antechinus	*Antechinus godmani*
Atlantic Cod	*Gadus morhua*
Atlantic Salmon	*Salmo salar*
Australian Freshwater Mussel	*Velesunio ambiguu*
Australian Grayling	*Prototroctes maraena*
Australian Magpie	*Gymnorhina tibicen*
Australian Smelt	*Retropinna semoni*
Bald Eagle	*Haliaeetus leucocephalus*
Banana	*Musa* × *paradisiaca*
Banded Hare-wallaby	*Lagostrophus fasciatus*
bandicoots	family Peramelidae
Banksia	*Banksia* spp.
Banksia Bee	*Hylaeus alcyoneus*
Banteng	*Bos javaanicus*
Barley	*Hordeum* spp.
Barn Owl	*Tyto alba*
Beautiful Firetail	*Emblema bella*
bee-eaters	family Meropidae
bees, wasps	order Hymenoptera
beetles	order Coleoptera
Bell Miner	*Manorina melanophrys*
Bennett's Wallaby	*Macropus rufogriseus*
Bilby	*Macrotis lagotis*
birds of paradise	family Paradisaeidae
Bison	*Bison bison*
Bitou Bush	*Chrysanthemoides monilifera* ssp. *rotundata*
Bitou Tip Moth	*Comostolopsis germana*
Black Bear	*Ursus americanus*
Black Box	*Eucalyptus largiflorens*
Black Gin	*Kingia australis*
Black Kite	*Milvus milvans*
Black Noddy	*Anous minutus*
Black Peppermint	*Eucalyåptus amygdalina*
Black Rat	*Rattus rattus*
Black Swan	*Cygnus atratus*
Blackberry	*Rubus* spp.
Blackberry Leaf Rust	*Phragmideum violaceum*
Blackbird	*Turdus merula*
Black-flanked Rock Wallaby	*Petrogale lateralis*
Black-footed Ferret	*Mustela nigripes*
Black-footed Rock Wallaby	*Petrogale lateralis*
Black-shouldered Kite	*Elanus notatus*
Black-striped Mussel	*Mytilopsis sallei*
Blakely's Red Gum	*Eucalyptus blakelyi*
Blue Grass	*Dichanthium* spp.
Blue Gum	*Eucalyptus globulus*
Blue Whale	*Balaenoptera musculus*
Blue-breasted Fairy Wren	*Malurus pulcherrimus*
Blunt Wattle	*Acacia aprica*
Bog Onion	*Owenia cepiodora*
Boneseed	*Chrysanthemoides monilifera*
Boobook Owl	*Ninox novaseelandiae*
Bottlebrush	*Callistemon* spp.
Bourke's Parrot	*Neopsephotus bourkii*
Box Mistletoe	*Amyema miquelli*
Bridal Creeper	*Asparagus asparagoides*
Bridled Goby	*Arenigobius bifrenatus*
Bridled Nailtail Wallaby	*Onychoglea fraenata*
Brigalow	*Acacia harpophylla*
Brittle Gum	*Eucalyptus mannifera*
Broad-faced Potoroo	*Potorous platyops*
Broad-headed Snake	*Holocephalus bungaroides*
Broad-toothed Rat	*Mastacomys fuscus*

Brook's Striped Skink	*Ctenotus brooksi iridis*
Broombush	*Melaleuca* spp.
brown algae	phylum Phaeophyta
Brown Antechinus	*Antechinus stuartii*
Brown Barrel	*Eucalyptus fastigata*
Brown Bear	*Ursus arctos*
Brown Hare	*Lepus capensis*
Brown Quail	*Coturnix ypsilophora/australis*
Brown Rat	*Rattus norvegicus*
Brown Thornbill	*Ancanthiza pusilla*
Brown Tree Frog	*Litoria ewingii*
Brown Tree Snake	*Boiga irregularis*
Brown Treecreeper	*Climacteris picumnus*
Brown-headed Honeyeater	*Melithreptus brevirostris*
Brush Bronzewing	*Phaps elegans*
Brush-tailed Bettong	*Bettongia penicillata*
Brush-tailed Rock Wallaby	*Petrogale pencilliata*
Budgerigar	*Melopsittacus undulates*
Buffalo	*Bubalus bubalis*
Buff-rumped Thornbill	*Acanthiza reguloides*
bugs	order Hemiptera
Buloke	*Allocasuarina luehmannii*
Bumblebee	*Bombus terrestris*
Burrowing Bettong	*Bettongia lesueur*
Bush Rat	*Rattus fuscipes*
Bush Thick-knee	*Burhinus grallarius*
butterflies, moths	order Lepidoptera
Buttongrass	*Gymnoschoenus sphaerocephalus*
Buxton Gum	*Eucalyptus crenulata*
Bynoe's Gecko	*Heteronotia bynoeii*
Cabbage	*Brassica oleracea*
Cabomba	*Cabomba caroliniana*
Caerulean butterfly	*Eutrichomyias rowleyi*
Cairns Rainbowfish	*Cairnsichthys rhombosomoides*
California Mouse	*Peromyscus californicus*
California Spiny Lobster	*Panuluris interruptus*
California Tiger Salamander	*Ambystoma californiense*
Californian Condor	*Gymnogyps californianus*
Camel	*Camelus dromedarius*
Candle Bark	*Eucalyptus rubida*
Cane Toad	*Bufo marinaris*
Canola	*Brassica napus*
Cape Weed	*Arctotheca calendula*
Capercaillie Grouse	*Tetrao urogallus*
Capricorn Silvereye	*Zosterops lateralis chloronotus (gouldi)*
Caribou	*Rangifer tarandus*
Carnaby's Cockatoo	*Calyptorhynchus funereus latirostris*
Carpentarian Rock Rat	*Zyzomys palatalis*
Carpet Beetle	*Anthrenocerus australis*
Carpet Python	*Morelia spilota variegata*
Cassava	*Manihot esculenta*
Cattle	*Bos taurus*
Chestnut-breasted Mannikin	*Lonchura castaneothorax*
Chilean Needle Grass	*Nassella neesiana*
Chiming Wedgebill	*Psophodes occidentalis*
Chital Deer	*Axis axis*
Cholera	*Vibrio cholerae*
Chuditch	*Dasyurus geoffroii*
Cicadabird	*Coracina tenuirostris*
Cinnamon Fungus	*Phytophthora cinnamomi*
clams	class Bivalvia
Climbing Galaxias	*Galaxias brevipinnis*
Coffee	*Coffea arabica/robusta*
Colonial Spider	*Badumma socialis*
Common Blossom Bat	*Syconycteris australis*
Common Brushtail Possum	*Trichosurus vulpecula*
Common Carp	*Cyprinus carpio*
Common Correa	*Correa reflexa*
Common Eastern Froglet	*Crinia signifera*
Common Mynah	*Acridotheres tristis*
Common Ringtail Possum	*Pseudocheirus peregrinus*
Common Starling	*Sturnus vulgaris*
Common Tree Snake	*Denrelaphis punctulata*
Common Wombat	*Vombatus ursinus*
conifers	division Coniferophyta
Cootamundra Wattle	*Acacia baileyana*
Cord-grasses	*Spartina* spp.
Corn	*Zea mays*
Corrigin Grevillea	*Grevillea scapigera*
Corroboree Frog	*Pseudophyme corroboree*
Cox's Gudgeon	*Gobiomophus coxii*
crabs	class Decapoda
Crazy Ant	*Anoplolepis gracilipes*
Crescent Honeyeater	*Phylidornis pyrrhoptera*
Crested Shrike Tit	*Falcunculus frantatus*
Crimson Rosella	*Platycercus elegans*
crows	family Corvidae
cuckoos	family Cuculidae
cuckoo-shrikes	family Campephagida
Cup Gum	*Eucalyptus cosmophylla*
Cycad	*Macrozamia* spp.
Cypress Pine	*Callitris* spp.
deer	family Cervidae
Delicate Skink	*Lampropholis delicata*
Diamond Firetail	*Stagonopleura guttata*
Diamond Python	*Morelia spilota spilota*
Dibbler	*Parantechinus apicalis*
Dingo	*Canis familiaris dingo*
Dinoflagellates	*Gymnodinium* and *Alexandrium* spp.

Dog	*Canis familiaris*
Domestic Cattle	*Bos indicus/Bos taurus*
Donkey	*Equus asinus*
Douglas Fir	*Pseudotsuga menziesii*
Drooping She-oak	*Allocasuarina verticillata*
ducks	family Anatidae
Dugong	*Dugong dugon*
Dunnarts	*Sminthopsis* spp.
Dusky Hopping mouse	*Notomys fuscus*
Dwarf Tree Frog	*Litoria fallax*
East Coast Tuna	*Tunnus maccoyi*
Eastern Barred Bandicoot	*Perameles gunnii*
Eastern Bristlebird	*Dasyornis brachypterus*
Eastern Grey Kangaroo	*Macropus giganteus*
Eastern Pygmy Possum	*Cercartetus nanus*
Eastern Quoll	*Dasyurus viverrinus*
Eastern Rosella	*Platycercus eximus*
Eastern Screech-Owl	*Otus asio*
Eastern Spinebill	*Acanthorhynchus tenuirostris*
Eastern Whipbird	*Psophodes olivaceus*
Eastern Yellow Robin	*Eopsaltria australis*
Echidna	*Tachyglossus aculeatus*
Edith's Checkerspot Butterfly	*Euphydras editha*
elephant birds	general term for a wide array of large extinct bird taxa on Madagascar
Eltham Copper Butterfly	*Paralucia pyrodiscus lucida*
Emu	*Dromaius novaehollandiae*
Endod	*Phytolacca dodecandra*
Eurasian (Ethiopian) Wolf	*Canis lupus simensis*
Euro	*Macropus robustus*
European Carp	*Cyprinus carpio*
European Clam	*Corbula gibba*
European Shore Crab	*Carcinus maenas*
Eyebright	*Euphrasia collina* aff. ssp. *diemenica*
Fallow Deer	*Dama dama*
Fan-tailed Cuckoo	*Cacomantis pyrrhopanus*
Feathertail Glider	*Acrobates pygmaeus*
Feral Cat	*Felis catus*
ferns	division Pterophyta
Ferret	*Mustela furio*
Flame Robin	*Petroica phoenicea*
Flathead	*Pseudaphritis urvillii*
Flatweed	*Hypochoeris glabra*, *Hypochoeris radicata*
Flay-headed Galaxias	*Galaxias rostratus*
flies	order Diptera
Florida Panther	*Felis concolor coryi*
Florida Scrub Jay	*Aphelocoma coerulescens*
flycatchers	family Muscicapidae
Freshwater Catfish	*Tandanus tandanus*
Freshwater Hardyhead	*Craterocephalus stercusmuscarum*
Fruit Fly	*Drosophila* spp.
fungi	kingdom Fungi
Galah	*Cactua roseicapilla*
Gang-gang Cockatoo	*Callocephalon fimbriatum*
Gemfish	*Rexea solandri*
Giant Gippsland Earthworm	*Megascoloides australis*
Giant Muntjac Deer	*Megamuntiacus vaguangenis*
Giant Sensitive Plant	*Mimosa pigra*
Giant Tortoise	*Geochelone nigra vandenburghii*
Gilbert's Potoroo	*Potorous tridactylus gilberti*
Gland Flower	*Adenanthos terminals*
Glanville Fritillary Butterfly	*Melitaea cinixia*
Glossy Black Cockatoo	*Calyptorhyncus lathami*
Glossy Grass Skink	*Leiolopisma rawlinsoni*
goannas	family Varanidae
Goat	*Capra hiraus*
gobies	family Gobidae
Golden Bandicoot	*Isoodon auratus*
Golden Wattle	*Acacia pycnantha*
Golden Whistler	*Pachycephala pectoralis*
Golden-shouldered Parrot	*Psephotus chrysopterygius*
Golden-tipped Bat	*Phoniscus papuensis*
Goldfish	*Carassius auratus*
Gorse	*Ulex europaeus*
grass trees	family Xanthorrhoeaceae
Grasshopper	*Caledia captiva*
Grassland Earless Dragon	*Tympanocryptis lineata pinguicolla*
Great Australian Bight demersalan	general term to describe array of bottom dwelling fish species from the Great Australian Bight
Greater Bilby	*Macrotis lagostis*
Greater Glider	*Petauroides volans*
Greater Pipistrelle	*Falsistrellus tasmaniensis*
Greater Stick-nest Rat	*Leporillus conditor*
Green and Gold Bell Frog	*Litoria aurea*
Green Sawfish	*Pritis zijsron*
Green Swamp Frog	*Litoria reniformis*
Green Turtle	*Chelonia mydas*
Grey Back Cane Beetle	*Dermolepida albohirtum*
Grey Box	*Eucalyptus mouccana*
Grey Butcherbird	*Cracticus torquatus*
Grey Honeyeater	*Conopophila whitei*

Grey Shrike Thrush	*Colluricincla harmonica*
Grey Teal	*Anas gracilis*
Grey Wolf	*Canis lupus*
Grey-headed Flying Fox	*Pteropus poliocephalus*
Grizzly Bear	*Ursus arctos*
Ground Boa	*Bolyeria multocarinata*
Ground Parrot	*Pezoporus wallicus*
Growling Grass Frog	*Litoria raniformis*
Gungurru	*Eucalyptus caesia*
Hardhead Catfish	*Arius felis*
Hastings River Mouse	*Pseudomys oralis*
Hazel Pomaderris	*Pomaderris aspera*
Heath Banskia	*Banksia ericifolia*
Heath Rat	*Pseudomys shortridgeii*
Helmeted Honeyeater	*Lichenostomus melanops cassidix*
Hidden Beard Heath	*Leucopogon obtectus*
Hog Deer	*Axis porcinus*
Honey Possum	*Tarsipes rostratus*
Honeybee	*Apis mellifera*
honeyeaters	family Meliphagidae
Hoop Pine	*Araucaria cunninghamii*
Horse	*Equus caballus*
House Mouse	*Mus musculus*
Humpback Whale	*Megaptera novaeangliae*
Huon Pine	*Dacrydium franklinii*
Hymenachne	*Hymenachne amplexicaulus*
Iberian Lynx	*Lynx pardinus*
Ibex	*Capra ibex*
Ice minus	*Pseudomonas syringae*
Ice Plant	*Mesembryanthemum crrstallinum*
Indian Palm Squirrel	*Funambulus pennati*
Jack Mackerel	*Trachurus declivis*
Jack Pine	*Pinus banksiana*
Japanese Sea Bass	*Lateolabrax japoninicus*
Jarrah	*Eucalyptus marginata*
Javanese Banteng	*Bos javanicus*
jewel beetles	family Bupressidae
Johnson's Mallee	*Eucalyptus johnsoniana*
Kangaroo Island Emu	*Dromaius baudinianus*
Kentucky Warbler	*Geothlypis formosa*
Killer Whale	*Orcinus arca*
King Billy Pine	*Arthrotaxis selaginoides*
King Parrot	*Alisterus scapularis*
kingfishers	family Alcedinidae
Kirtland's Warbler	*Dendroica kirtlandii*
Kiwi	*Apteryx australis*
Koala	*Phascolarctos cinereus*
Komodo Dragon	*Varanus komodoensis*
lace coral	phylum Bryozoa
Lake Eacham Rainbowfish	*Melanotaenia eachamensis*
Lantana	*Lantana camara*
Large Quaking-grass	*Briza maxima*
Laterite Mallee	*Eucalyptus lateritica*
Laughing Kookaburra	*Dacelo novaeguineae*
Leadbeater's Possum	*Gymnobelideus leadbeateri*
Leaden Flycatcher	*Myiagra rubecula*
leaf-hoppers	order Homoptera
Leatherback Turtle	*Dermochelys coriacea*
leeches	subclass Hirudinea
Lesser Stick-nest Rat	*Leporillus apicalis*
Lewin's Honeyeater	*Meliphaga lewinii*
Little Pygmy Possum	*Cercartetus lepidus*
Little Raven	*Corvus mellori*
Little Sumba Hawk Owl	*Ninox sumbaensis*
Little Wattlebird	*Anthochaera chrysoptera*
liverworts	class Hepaticae
Long-billed Corella	*Cacatua tenuirostris*
Long-footed Potoroo	*Potorous longipes*
Long-nosed Bandicoot	*Perameles nasuta*
Long-nosed Potoroo	*Potorous tridactylus*
Long-tailed Dunnart	*Sminthopsis longicaudata*
Lord Howe Boobook	*Nixox novaeseelandiae albaria*
Lord Howe Fantail	*Rhipidura cervina*
Lord Howe Gerygone	*Gerygone insularis*
Lord Howe Island Guineafowl	*Notornis alba*
Lord Howe Island Parakeet	*Cyanoramphus novaezelandiae*
Lord Howe Island Pigeon	*Columba vitensis godmanae*
Lord Howe Island Woodhen	*Tricholimnas slyvestris*
Lord Howe Starling	*Aplonis fuscus bullianus*
lorikeets	subfamily Loriinae
Lyrebird	*Menura novaehollandiae*
Magpie	*Gymnorhina tibicen*
Magpie Goose	*Anseranas semipalmata*
Mahogany	*Swietenia macrophylla*
Mahogany Glider	*Petaurus gracilis*
Maize	*Zea mays*
Mala	*Lagorchestes hirsutus* ssp. *bernieri* and ssp. *dorrae*
Mallee Broombrush	*Templetonia egena*
Mallee Emu-wren	*Stripiturus mallee*
Malleefowl	*Leipoa ocellata*
Manna Gum	*Eucalyptus viminalis*
Marri	*Eucalyptus (Corymbia) calophylla*
Masked Owl	*Tyto novaehollandiae*
Matchstick Banksia	*Banksia cuneata*
Mediterranean Fanworm	*Sabella spallanzanii*
Mesquite	*Prosoois* spp.

Messmate	*Eucalyptus obliqua*
Mimosa	*Mimosa pigra*
mistletoe	family Loranthaceae
Mistletoe Bird	*Dicaeum hirundinacaeum*
Mitchell Grass	*Astrebla* spp.
Mitchell's Hopping Mouse	*Notomys mitchelli*
mites	order Acarina
Moa	*Dinornis* spp.
molluscs	phylum Mollusca
Monkey Puzzle	*Araucaria araucana*
Mosquitofish	*Gambusia holbrooki*
moss	class Musci
Mount Taylor Mallee	*Eucalyptus remota*
Mountain Ash	*Eucalyptus regnans*
Mountain Brushtail Possum	*Trichosurus cunninghamii*
Mountain Duck	*Tadorna tadornoides*
Mountain Galaxias	*Galaxias oldidus*
Mountain Grey Gum	*Eucalyptus cypellocarpa*
Mountain Pygmy Possum	*Burramys parvus*
Mountain Swamp Gum	*Eucalyptus camphora*
Mulga	*Acacia aneura*
Murray Cod	*Maccullochella peelii*
Muttonbirds	*Puffinus* spp.
Myrtle Beech	*Nothofagus cunninghamii*
Myrtle Elbow Orchid	*Arthrochilus huntianus nothogagicola*
Narrow-leaved Ironbark	*Eucalyptus crebra*
Native Cats	*Dasyurus* spp.
nematodes (round worms)	phylum Nematoda
Nepean River Gum	*Eucalyptus benthamii*
New Holland Honeyeaters	*Phylidonyris novaehollandiae*
New Holland Mouse	*Pseudomys novaehollandiae*
New Zealand Screwshell	*Maoricolpus roseus*
Night Parrot	*Geopsittacus occidentalis*
nightjar	family Caprimulgidae
Noisy Miner	*Monorina melanocephala*
Noisy Scrub-bird	*Antrichornis clamosus*
Northern Bettong	*Bettongia tropica*
Northern Hairy-nosed Wombat	*Lasiorhinus krefftii*
Northern Pacific Starfish	*Asterias amurensis*
Northern prawn fishery	general term for a range of prawn species harvested in northern Australian waters
Northern Quoll	*Dasyurus hallacatus*
Norwegian Lobster	*Nephrops norvegicus*
Numbat	*Myrmecobius fasciantus*
octopus	class Cephalopoda
Olive	*Olea europea*
Olive Whistler	*Pachycephala olivacea*
Olive-backed Oriole	*Oriolus sagittatus*
Orange Roughy	*Hoplostethus atlanticus*
Orange-bellied Frog	*Geocrinia vitellina*
Orange-bellied Parrot	*Neophema chryogaster*
Orangutan	*Ponga pygmaeus*
Ostrich	*Struthio camelus*
Owlet Nightjar	*Aegotheles cristatus*
Pacific Oyster	*Crassostrea gigas*
Paperbark	*Melaleuca* spp.
Parkinsonia	*Parkinsonia aculeata*
Parma Wallaby	*Macropus parma*
Parthenium Weed	*Parthenium hysterophorus*
Passenger Pigeon	*Ectopistes migratorius*
Patagonian Toothfish	*Dissostichus eleginoides*
Paterson's Curse	*Echium plantagineum*
pea	family Fabaceae
Peaceful Dove	*Geopelia placida*
Pelican (Australian)	*Pelecanus conspicillatus*
pelicans	family Pelicanidae
Pencil Pine	*Arthrotaxis cupressoides*
Pied Butcherbird	*Cracticus nigrogularis*
Pied Currawong	*Strepera graculina*
Pig	*Sus scrofa*
pigeons, doves	family Columbidae
Pine	*Pinus* spp.
Pineapple	*Ananus* spp.
Pink Robin	*Petroica rodinogaster*
Pitch Pine	*Pinus rigida*
Plains Wanderer	*Pedionomus torquatus*
Platypus	*Ornithoryhncus anatinus*
Polar Bear	*Ursus maritimus*
Pomaderris	*Pomaderris* spp.
Pond Apple	*Annona glabra*
Port Orford Cedar	*Chamaecyparis lawsoniana*
Potato	*Solanum tuberosum*
potoroos	family Potoroidae
Pouched Lamprey	*Geotria australis*
Powerful Owl	*Ninox strenua*
Prawn fishery	general term for a range of harvested prawn species
Prickly Acacia	*Acacia nilotica* ssp.*indica*
Prickly Pear	*Opuntia inermis*
Princess Parrot	*Polytelis alexandrae*
Puerto Rican Parrot	*Amazona vittata*
Purdie's Donkey Orchid	*Diuris purdiei*
Pyralid Moth	*Cactoblastis cactorum*
Quokka	*Setonix brachyurus*
Quolls	*Dasyurus* spp.
Rabbit	*Oryctolagus cuniculus*
Radiata Pine	*Pinus radiata*
Rainbow Bee-eater	*Merops ornatus*
Rainbow Lorikeet	*Trichoglossus haematodus*

red algae	phylum Rhodophyta
Red Cedar	*Toona australis*
Red Deer	*Cervus elphas*
Red Fin	*Perca fluviatilis*
Red Fox	*Vulpes vulpes*
Red Imported Fire Ant	*Solenopsis invicta*
Red Kangaroo	*Macropus rufus*
Red Land Crab	*Gecaroidia natalis*
Red Stringybark	*Eucalyptus macrorhynca*
Red-bellied Black Snake	*Pseudechis porphyriacus*
Red-browed Finch	*Aegintha temporalis*
Red-browed Treecreeper	*Climacteris erythrops*
Red-fronted Parakeet	*Cyanoramphus novaezelandiae subflavescens*
Red-lored Whistler	*Pachycephala rufogularis*
Red-necked Wallaby	*Macropus rufogriseus*
Red-tailed Black Cockatoo	*Calyptorhynchus magnificus*
Red-tailed Hawk	*Buteo jamaicensis socorroensis*
Red-tailed Phascogale	*Phascogale calura*
Redwing Blackbird	*Agelaius phoeniceus*
Regent Honeyeater	*Xanthomyza phrygia*
Regent Parrot	*Polytelis anthopeplus*
Ribbon Gum	*Eucalyptus viminalis*
Rice	*Oryza* spp.
Risdon Peppermint	*Eucalyptus risdonii*
River Red Gum	*Eucalyptus camaldulensis*
Robust White Eye	*Zosterops strenua*
Rock Isotome	*Isotoma petraea*
Rock Squirrel	*Thomomys bottae*
rodents	order Rodentia
Roman-nosed Goby	*Awaous crassilabrus*
Roseate Frog	*Geocrinia rosea*
Round-leafed Honeysuckle	*Lambertia orbifolia*
Rubber Vine	*Cryptostegia grandiflora*
Rufous Bristlebird	*Dasyornis broadbenti*
Rufous Hare Wallaby	*Lagorchestes hirsutus* ssp. *bernierii* and ssp. *dorrae*
Rufous Scrub Bird	*Atrichornis rufescens*
Rufous Treecreeper	*Climacteris rufa*
Rufous Whistler	*Pachycephala rufiventris*
Rusa	*Cervus timorensis*
Sacred Kingfisher	*Todiramphus sanctus*
Sallow Wattle	*Acacia longifolia*
Salmon Gum	*Eucalyptus saomonophloia*
Salt Paperbark	*Melaleuca halmaturorum*
Saltwater Crocodile	*Crocodylus porosus*
Salvinia	*Salvinia molesta*
Sambar	*Cervus unicolor*
Sandhill Dunnart	*Sminthopsis laevis*
Satin Bowerbird	*Ptilinorhynchus violaceus*

Satin Flycatcher	*Myiagra cyanoleuca*
Scallops	*Pecten* spp.
Scarlet Bottlebrush	*Callistemon rugulosus*
Scarlet Oak	*Quercus coccinea*
Scarlet Robin	*Petroica multicolor*
sea-lions	subfamily Otariinae
seals	family Otariidae
Sea-otter	*Enhydra lutris*
sea-squirts	phylum Ascidia
seastars	class Asteroidea
Serrated Tussock	*Nassella trichotoma*
Sheep	*Ovis aries*
shelf demersal fishery	general term used to describe a range of shelf bottom dwelling marine fish species
Sheoak	*Casaurina* spp./ *Allocasaurina* spp.
She-oak Skink	*Tiliqua casuarinae*
Shevolski's Horse	*Equus callabus przewalski*
Shining Gum	*Eucalyptus nitens*
Short-eared Possum	*Trichosurus caninus*
Short-tailed Shearwater	*Puffinus tenuirostris*
Shrimps	*Mysidacea*
Silver Top	*Eucalyptus sieberi*
Silver Top Gum	*Eucalyptus nitens*
Skeleton Weed	*Chondrilla juncea*
slaters	order Isopoda
Small-leaved Gum	*Eucalyptus parvifolia*
Smoky Mouse	*Pseudomys fumeus*
Smooth Oreo	*Pseudocyttus maculatus*
snails/slugs	class Gastropoda
Snow Gum	*Eucalyptus pauciflora*
Soala	*Pseudoryx nghetinhensis*
Sockeye Salmon	*Oncorhynchus nerka*
Sooty Owl	*Tyto tenebricosa*
Sorghum	*Sorghum bicolor*
south east trawl fishery	general term to describe an array of deep level fish species in the south-east of Australia
Southern Barred Frog	*Mixophyes balbus*
Southern Beech	*Nothofagus* spp.
Southern Blue Gum	*Eucalyptus globulus*
Southern Blue-fin Tuna	*Thunnus maccoyii*
Southern Brown Bandicoot	*Isoodon obesulus*
Southern Cassowary	*Casuarius casuarius*
Southern Emu Wren	*Stipiturus malachurus*
Southern Pigmy Perch	*Nannoperca australis*
Southern Rock Lobster	*Jasus edwardsii*
Southern Sassafras	*Artherospermum moschatum*
Southern Scrub-robin	*Drymodes brunneopygia*

southern shark fishery	general term used to describe a range of shark species harvested in southern Australia
Soyabean	*Glycine max*
Spanish Heath	*Erica lusitanica*
Spencer's Skink	*Pseudomoia spenceri*
spiders	order Araneae
Spiky Oreo	*Neocyttus rhomboidalis*
Spinifex-dominated communities	*Triodia* spp./ *Symplectrodia* spp./ *Monodia*
Spiny Dogfish	*Squalus* spp.
Spiny-cheeked Honeyeater	*Acanthagenys rufogularis*
Splendid Fairy Wren	*Malurus splendens*
Spotted Gum	*Eucalyptus maculata*
Spotted Handfish	*Branchionichthys hirsutus*
Spotted Pardolote	*Pardalotus punctatus*
Spotted Tree Frog	*Litoria spenceri*
Spotted-tailed Quoll	*Dasyurus maculatus*
Squirrel Glider	*Petaurus norfolcensis*
Striated Thornbill	*Ancanthiza lineata*
Striped Legless Lizard	*Delma impar*
Stubble Quail	*Coturnix pectoralis*
Sugar Cane	*Saccharum officinarum*
Sugar Glider	*Petaurus breviceps*
Sulphur-crested Cockatoo	*Cacatua galerita*
Sumatran Rhino	*Dicerorhinus sumatrensis*
Superb Fairy Wrens	*Malaurus cyaneus*
Superb Parrot	*Polytelis swainsonii*
Swamp Antechinus	*Antechinus minimus*
Swamp Gum (New South Wales)	*Eucalyptus camphora*
Swamp Gum (Tasmania)	*Eucalyptus regnans*
Swamp Gum (Victoria)	*Eucalyptus ovata*
Swamp Mahogany	*Eucalyptus robusta*
Swamp Rat	*Rattus lutreolus*
Swamp Skink	*Eulamprus leuraensis*
Swamp Tea-tree	*Melaleuca irybana*
Swamp Wallaby	*Wallabia bicolor*
Sweet Bursaria	*Bursaria spinosa* var. *spinosa*
Swift Parrot	*Lathamus discolor*
Sydney Rock Oyster	*Saccostrea commercialis*/ *Saccostrea glomerata*
Tall Astelia	*Astelia australiana*
Tall Wheat-grass	*Lophopyrum elongatum*
Tammar Wallaby	*Macropus eugenii*
Tanaids	*Tanaidacea*
Tasman Booby	*Sula tasmani*
Tasmanian Bettong	*Bettongia gaimardi*
Tasmanian Blue Gum	*Eucalyptus globulus*
Tasmanian Devil	*Sarcophilus harrisii*
Tasmanian Native Hen	*Gallinula mortierii*
Tasmanian Pademelon	*Thylogale billardierii*
Tasmanian Tiger/Tasmania Wolf	*Thylacinus cynocephalus*
Tawny Frogmouth	*Podargus strigoides*
Teak	*Tectona grandis*
teals	subfamily Anatinae
Tea-tree	*Leptospermum* spp.
termites	order Isoptera
Three-awned Spear Grass	*Aristida* spp.
Thylacine	*Thylacinus cynocephalus*
Tiger Quoll	*Dasyurus maculatus*
Tiger Salamander	*Ambystoma tigrinum*
Tiger Snake	*Notechis scutatus*
Tomato	*Lycopersicon esculentum*
Toowoomba Canary-grass	*Phalaris aquatica*
Torres Strait prawn	general term used to describe a range of prawn species harvested in the Torres Strait
Toxoplasmosis	*Toxoplasma gondii*
trapdoor spiders	suborder Mygalomorphae
Tree Ferns	*Dicksonia antarctica* and *Cyathea australis*
Tree Goanna	*Varanus varius*
Trigger Plants	*Stylidium coroniforme*
Trout Cod	*Maccullochella macquariensis*
Tuan	*Phascogale tapoatafa*
Tuatara	*Sphenodon* spp.
Varied Sitella	*Daphoenositta chrysoptera*
velvet worms	phylum Onychophora
Vinous-tinted Blackbird	*Turdus xanthopus vinitinctus*
Virginia Opossum	*Didelphis virginiana*
Wadbilliga Ash	*Eucalyptus paliformis*
Wakame	*Undaria pinnatifida*
wallabies	family Macropodidae
Wallaroo	*Macropus robustus*
Water Buffalo	*Bubalus bubalis*
Wattles	*Acacia* spp.
Wedge-tailed Eagle	*Aquila audax*
Welcome Swallow	*Hirundo neoxena*
Western Grey Kangaroo	*Macropus fuliginosus*
Western Rock Lobster	*Panulirus cyngus*
Western Swamp Tortoise	*Pseudomydura umbrina*
Western Whipbird	*Psophodes nigrogularis*
whales	order Cetacea
Wheat	*Triticum aestivum*
Whiptail Wallaby	*Macropus parryi*
Whistlers	*Pachycephala* spp.
White Box	*Eucalyptus albens*
White Cypress Pine	*Callatris glaucophylla*
White Gallinule	*Notornis alba*

White-backed Swallow	*Cheramoeca leucosternum*
White-bellied Frog	*Geocrinia alba*
White-browed Scrub Wren	*Sericornis frontalis*
White-footed Mouse	*Peromyscus leucopus novaeboracensis*
White-naped Honeyeater	*Melipthreptus brevirostris*
White-plumed Honeyeater	*Lichenostomus penicillatus*
White-tailed Deer	*Odocoileus virginianus*
White-throated Pigeon	*Columba vitiensis godmanae*
White-throated Treecreeper	*Cormobates leucophaeus*
Whiting	*Haletta* spp.
Whooping Crane	*Grus americana*
Willow	*Salix* spp.
Wollemi Pine	*Wollemia nobilis*
Wood Duck	*Chenonetta jubuta*
woodpeckers	order Piciformes
worms	class Polychaeta
Woylie	*Bettongia peniicilliata ogilbyi*
Yabbie	*Cherax destructor*
Yam	*Dioscorea* spp.
Yellow Box	*Eucalyptus mellidora*
Yellow Thornbill	*Acanthiza nana*
Yellow-bellied Glider (northern form)	*Petaurus australis reginae*
Yellow-faced Honeyeater	*Lichenostomus chrysops*
Yellowfin Goby	*Acanthogobius flavimanus*
Yellow-tailed Black Cockatoo	*Calyptorhynchus funereus*
Yellow-tufted Honeyeater	*Lichenostomus melanops*
Zebu Cattle	*Bos indicus*

Appendix II: Glossary

The following references were used to construct the glossary: Specht (1970), Luke and McArthur (1977), Holmes (1979), Collin (1988), Borowski and Borwein (1989), Begon *et al.* (1990, 1996), Fowler and Cohen (1990), Costermans (1994, 2000), Knox *et al.* (1994), Meffe and Carroll (1994), Thain and Hickman (1994), Collins (1995), and Department of Natural Resources and Environment (1996).

abiotic Pertaining to physical and inorganic components of the environment; non-living.

abundance The density of individuals within a local area. The terms rare and common (or abundant) describe extremes of abundance.

accumulation curves Plots of the cumulative number of species versus sampling or collection effort.

accumulator species Species that concentrate pollutants in their body tissues or in the environment immediately around them.

acid rain Rain with a very low pH (<5.6), resulting from emissions to the atmosphere of pollutants (oxides of sulfur and nitrogen).

adaptive management Management practices that accommodate uncertain future events.

aerial photo interpretation Mapping and measurement of features (e.g. vegetation type or tree height) using aerial photos; typically these are taken sequentially from an aircraft so that they can be viewed stereo-scopically, resulting in a three-dimensional image.

aestivate To go into a state of dormancy during the summer or extended dry periods.

agrestal weeds Weeds of agricultural land.

algal bloom Enrichment of a water body with nutrients resulting in sudden growths (blooms) of planktonic algae that may produce toxins. See also **eutrophication.**

algorithm A logical arithmetical or computational step-by-step procedure, which if correctly applied ensures the solution to a problem.

alien species See **exotic species.**

allele One of two or more forms of a gene located in the same position on homologous chromosomes.

allelopathy Inhibition of one plant by chemicals produced by another plant.

allozyme One of several possible forms of an enzyme that is the product of a particular allele at a particular gene locus.

alpha (α) diversity The diversity of a community within a local area, composed of the number of species and their relative abundances.

alternative hypothesis Any hypothesis that does not conform with a given null hypothesis.

altitudinal migration Migration by animals or birds to lower or higher elevations depending on the season.

amino acid An organic molecule, which is a structural component of proteins.

analysis of variance Any of a number of techniques for resolving the observed variance between sets of data into components, especially to determine whether the difference between samples can be explained as random variation within the same underlying population.

angiosperm The division that is commonly referred to as flowering plants; strictly, those seed-bearing plants that develop their seeds from ovules within a closed cavity (the ovary).

annelid Soft-bodied, metamerically segmented, coelomate worm.

annual A plant species with a life cycle that takes approximately 12 months or less to complete (cf. **perennial**).

anther In a flower, that part of the stamen that houses developing male reproductive cells (pollen).

anthropocentric Any human-orientated perspective of the environment; regarding humans as central.

arable Land being or capable of being tilled for the production of crops.

arboreal Living in or among trees.

area of occupancy The area within the extent of occurrence that is occupied by a taxon, excluding cases of vagrancy. It includes the smallest area essential at any stage to the survival of a taxon (e.g. nesting and feeding sites).

arithmetic mean See **true mean.**

arthropod Phylum of metamerically segmented animals with jointed legs and an exoskeleton (e.g. insects and spiders).

asexual reproduction Reproduction that does not involve fusion of gametes – offspring are clones of the parent organism.

assessment endpoints Formal expressions of the environmental values to be protected, which provide a means by which management goals can be identified, measured, audited and evaluated.

assignment test A method for estimating inter- population dispersal rates in which the individuals are assigned to a source population based on their 'assignment indices', which are defined as the expected frequencies of their genotypes in those populations. Individuals whose genotypes are more likely in populations other than the one they are found in are said to be 'misassigned' and may be interpreted as immigrants.

asymptotic dynamics of a population Very long-term dynamics; in an undisturbed ecosystem, approximating dynamics in a climax community.

Australasian region Biogeographic region including the Australian mainland and islands on the continental shelf, such as Tasmania and New Guinea.

autecology The biological relations between a single species and its environment; ecology of an individual species.

autocorrelation The condition occurring when successive items in a series are correlated; that is, they are not independent. For example, if samples close together experience relatively similar environmental conditions or are within the range of movement of a group of animals, then the appearance of a species in one sample will not be independent of the appearance of the same species at an adjacent site.

axiomatic Self evident; a logical system consisting of a set of statements that are stipulated to be true.

background rate of extinction The average rate of loss of species globally in geological time, outside the periods of mass extinction.

ballast water Water used to stabilise a vessel, especially one that is not carrying cargo.

base See **logarithm.**

base-pairs Two bases, one purine and one pyrimidine, which pair together in the genetic code of DNA and RNA.

benthic Organisms living at the bottom of a sea, lake or other aquatic ecosystem.

bequest values (of preserved wilderness areas) Based on the concern that large undisturbed landscapes should be available for future generations.

Bernoulli trial An experiment in which a single action, for example flipping a coin, is repeated in an identical manner over and over. The possible results of the action are classified as 'success' or 'failure'. The binomial formula is used to find probabilities for Bernoulli trials.

beta (β) diversity Represents the changes in species composition that occur along an ecological gradient within a community.

binary data Data with two options (e.g. present or absent).

binomial coefficient Any of the numerical factors that multiply the successive terms in a binomial expansion (an expression of the form $(x + a)^n$).

binomial distribution Distribution that is appropriate when observations can be classified into one of two possible categories (e.g. male or female); a statistical distribution giving the probability of obtaining a specified number of successes in a specified number of Bernoulli trials.

binomial naming A system devised by Linnaeus whereby the name of each kind of organism (each species) consists of two words: a genus name and a specific epithet.

bioassay organisms Used in laboratory conditions to determine the activity or effect of concentrations of pollutants by testing the effect of toxins on an organism and comparing this with the activity of an agreed standard.

biocentric ethic A view that argues for the value of all individuals. It suggests that all wild plants and animals are worthy of moral consideration and that humans have an ethical obligation towards them.

biocide A chemical (e.g. a pesticide) capable of killing living organisms.

bioclimatic modelling An approach that can be used to predict the broad potential limits of the distribution of a species. The first step in bioclimatic modelling is to assemble information on the known distribution of a particular organism. Then a summary of the climatic conditions that characterise the locations where the target species has been recorded is generated. Maps of matched areas, or homoclimes, that highlight a species' predicted potential distribution are the final result.

bioclimatic The effects of climatic conditions such as temperature, rainfall and radiation regimes on the distribution of living organisms.

biodiversity surrogates Measures such as species, assemblages, vegetation communities or climatic domains that are thought to indicate the distribution and abundance of unmeasured species, assemblages or other elements of biodiversity.

biodiversity Number, relative abundance, and genetic diversity of organisms from all habitats, including terrestrial, marine and other aquatic systems, and the ecological complexes and processes of which they are a part. This includes diversity within species, between species and of ecosystems.

bioeconomics The study of the interaction of economic and biological systems.

biogeographic province Regions of the world identified by their dominant animals and plants.

biogeography The study and interpretation of the geographical distribution of animals and plants.

bioindicator species A species that responds to environmental changes.

biological control The use of a predatory or parasitic organism (often the pest's natural enemy) to control a targeted pest species.

biological diversity See **biodiversity.**

biological legacies The biological material left after a natural disturbance such as a fire (e.g. seeds and charred trees and logs).

biological resources Consist of natural and semi-natural ecosystems and the genetic resources of these systems.

biological species Groups of interbreeding (or potentially interbreeding) natural populations that are reproductively isolated from other such groups.

biomass The living or dry weight per unit area of organisms in a particular area.

biome Ecological communities with the same structure, delineated by climate, and usually defined by some dominant vegetative pattern.

Biosphere Reserves Reserves designated for a range of reasons, including research, monitoring, training, demonstration, and conservation.

biosphere The part of the earth and its atmosphere inhabited by living organisms.

biotic Pertaining to living organisms, and usually applied to the biological aspects of an organism's environment (i.e. the influences of other organisms).

boreal A term meaning 'northern' that is often used in reference to the forests of northern Canada, northern Scandinavia and northern Russia.

bottleneck A decrease to a small population size, which results in a substantial decrease in genetic variability, usually over several generations.

Brigalow Refers to both the species *Acacia harpophylla* and to the vegetation communities it dominates.

buffer A protective margin of vegetation abutting a stream, wetland, rainforest or any other significant ecosystem, which protects it from potentially detrimental disturbances in the surrounding area.

by-catch Those fish captured during fishing operations targeted at other species.

captive breeding Breeding of animals in captivity.

carrying capacity The maximum number of individuals that a given environment can support indefinitely, usually determined by the organism's resource requirements.

catastrophe (with reference to risk assessment) Large, sudden, widespread change in the environment that causes extensive and relatively rapid change in vital rates such as fecundity and survival; such changes are usually precipitated by extreme environmental conditions such as floods, fires and droughts.

census A complete count of all individuals in a population.

centinelan extinction The loss of a species before it is known to (or described by) science.

central tendency The tendency of the values of a random variable to cluster around the mean, median and mode.

CFCs See **chlorofluorocarbons.**

chenopod Any flowering plant of the family Chenopodiaceae (e.g. many halophytes of arid and coastal salt-affected land).

chlorofluorocarbons (CFCs) Compounds used as aerosol propellants and in refrigerators, air conditioners and freezers. Chlorine in CFCs reacts with and breaks down atmospheric ozone.

chloroplast An organelle containing light-absorbing pigments; has an important role in photosynthesis.

chromosome analysis Study of the chromosomes of an organism to determine its ancestry.

chromosome Any of the microscopic rod-shaped structures that appear in a cell nucleus during cell division, consisting of nucleoprotein arranged into units (genes) that are responsible for the transmission of hereditary characteristics.

cladogram A graphical representation of the evolutionary relationships among taxa.

classification table A table comparing predictions and observations, and showing the number of true positive, true negative, false positive and false negative predictions.

clear-felling Method of harvesting in which an entire stand is removed in one felling with regeneration obtained artificially, by natural seeding from adjacent stands, or from trees cut in the clearing operation.

climax community Community of organisms, with composition more or less stable and in equilibrium with existing natural environmental conditions; the presumed end point of a successional sequence; a community that has reached a steady state.

clone An organism or group of organisms of identical genotype that are descended from a common ancestor by asexual reproduction.

cloning Production of a set of genetically identical individuals or cells, or production of a set of identical DNA molecules from a single starting molecule.

closed forest Forest that has a tallest stratum of trees 10–30 metres tall and dense foliage cover (a closed canopy).

coefficient of variation A measure of the relative variation of a distribution independent of the units of measurement; the standard deviation divided by the mean.

co-extinction The loss of a species that occurs when the species on which it is strongly dependent becomes extinct. An example is the loss of a species-specific parasite that goes extinct with the extinction of its host.

co-linearity Lying in the same straight line. In statistics, referring to variables that are highly correlated.

colour banding A field study method, used particularly for birds, in which coloured bands are attached to the legs to facilitate the recognition of individual animals.

community All living things in a particular area. In an ecological sense, an assemblage of interdependent populations of different species (plants and animals) interacting with one another, and living in a particular area.

competition An interaction between two or more organisms or species, in which, for each, the birth and/or growth rates are depressed and/or the death rates increased by the other organism or species because they both strive for the same resource, which is in limited supply.

complementarity (of reserves) The degree to which a given area contributes previously unrepresented features (e.g. species or environmental classes) to a reserve system.

comprehensiveness (of reserves) The degree to which a system of reserves samples the full range of biodiversity, including community, species and genetic variation.

confidence intervals An interval of values bounded by confidence limits derived by sampling, within which the true value of a population parameter is stated to lie with specified probability. Confidence intervals provide an estimate of the reliability of a prediction.

configuration The spatial arrangement of the patches or units in a landscape.

congener Belonging to the same genus.

conifer Any gymnosperm tree or shrub typically bearing cones and evergreen leaves.

connectivity (of reserves) Refers to the ease with which organisms move between particular landscape elements; the number of connections between patches, relative to the maximum number of potential connections.

conspecific Belonging to the same species.

consumptive use The value of the products of nature that do not pass through a market.

Convention on Wetlands of International Importance (Also known as the Ramsar Convention.) An international agreement signed in Ramsar (Iran) in 1971 to enhance the protection of wetland habitats.

cooperative breeders Those species in which more than two individuals provide care in rearing their young.

correlation coefficient A statistic measuring the degree of correlation between two variables. The closer the correlation coefficient is to 1 or −1, the greater the correlation; if the association between two variables is random, the correlation coefficient is equal to zero.

correlation The extent of correspondence between the ordering of two variables. Correlation is positive when two variables move in the same direction and negative or inverse when they move in opposite directions.

corridor (for wildlife) A strip of habitat that facilitates fauna movement between otherwise isolated patches of habitat.

cost-benefit analyses Examination, in economic terms, of the advantages and disadvantages of a particular course of action.

cpDNA (**chloroplast DNA**) DNA found in plants that encodes chloroplast ribosomal RNA, transfer RNA, and part of RuBP carboxylase. Many mutations affecting the chloroplast are transmitted maternally.

critical weight range species Taxa that weigh between 0.035 and 5.5 kilograms.

critically endangered A category of risk that species are assigned to if they are in the highest risk category (species most likely to become extinct).

cross-validation In regression, holding back some data and then using those data to test how well the model predicts new observations.

crown cover The percentage of the surface of a site above which there is the crowns of trees or shrubs. See also **projective foliage cover**.

cryostorage techniques Storage of seeds or other plant material in liquid nitrogen.

cryptic species Distinct species that show few or no morphological differences.

culling Reducing the size of a herd, flock or population of a species by killing a proportion of its members.

current account That part of the balance of payments (the difference over a given time between total payments to foreign nations and total receipts from foreign nations) composed of the balance of trade (difference in value between total export and total imports of goods) and the invisible balance (the difference in value between total exports of services plus payment of property incomes from abroad and total imports of services plus payment abroad of property incomes).

custodial responsibility The need for current generations of humans to manage natural resources to ensure the maintenance of benefits of the environment and biological resources for future generations.

decision tree A graphical representation of decision pathways, usually within a computer program, that specifies the actions to be taken when certain conditions arise.

declining-population paradigm A model that deals largely with the deterministic influences that influence population persistence (e.g. ongoing habitat loss). See also **small population paradigm.**

deep ecology An environmental philosophy that believes in the inherent rights of nature, and a human existence that does minimal damage to natural systems.

defoliation The loss of leaves from a plant.

deforestation Clearing of trees or depletion of crown cover to less than (arbitrarily) 10%.

deleterious allele A form of a gene occupying a specific locus (position) on a chromosome that has a detrimental effect on an individual's fitness.

deme Any group of inter-breeding individuals that possess a suite of clearly definable genetic or morphological characteristics.

demersal Living or occurring in deep water on or near the bottom of a sea or lake.

demographic stochasticity (with particular reference to risk assessment) The variation in the average chances of survivorship and reproduction that occur because of the demographic structure of populations made up of a finite, integer number of individuals; the random sampling of distributions for variables that must logically take a discrete integer value.

demographic uncertainty See **demographic variation.**

demographic variation The variation in the chances of survival and reproduction that arise because a population is small, discrete and structured; a consequence of sampling a finite number of integers; a synonym of demographic stochasticity.

dendritic Having a branching form.

dependent variable See **response variable.**

detector species Species that exhibit changes in such things as behaviour, mortality patterns and age-class structure in response to environmental changes.

deterministic model A model in which there is no representation of variability.

dieback Death of trees or shrubs characterised by death of the young shoots, which later spreads to the larger branches; in rural areas, the more or less concurrent deaths of large numbers of remaining trees in agricultural land, or in forests affected by disease.

diploid Having two sets of chromosomes.

discounting An economic term describing a process in which future monetary values are translated into present-day monetary values.

disequilibrium Gene frequencies in a population that deviate substantially from those expected in a large randomly mating population without migration, mutation or selection, sometimes maintained by selection or linkage (see **linkage disequilibrium**).

dispersal The movement of individuals among spatially separate patches of habitat, including all immigration and emigration events.

DNA (deoxyribonucleic acid) The self-replicating molecule of nucleic acid that forms the genetic hereditary material of all cells, some organelles, and many viruses; a major component of chromosomes and the sole component of plasmids.

DNA fingerprinting Technique that makes use of the hypervariable repetitive DNA sequences of genes to identify the relatives of individuals.

DNA sequencing Determination of a DNA sequence, either by the chemical cleavage method or by controlled interruption of enzymatic replication. Various automatic methods and machines based on these methods are available.

dominant The plant species in the upper stratum of a formation, which has the greatest foliage cover.

dryland salinisation The percolation of salt near to or on the soil surface following the removal of native vegetation; water is discharged from the ground water system and moves more or less vertically upwards through salt-laden sublayers of the soil, taking the salt with it.

ecocentric ethic A point of view whereby the object of concern is the biotic community as a whole.

ecological extinction The situation where given species still occurs in a community, but the numbers of the taxon are so depleted that it no longer interacts significantly with other species.

ecological footprint A method used to estimate the impacts of humans based on the amount of land required to provide the energy and natural resources to support a person.

ecological redundancy The degree to which a given species can be substituted by different species to provide particular ecosystem functions.

ecologically redundant species Species for which their loss appears to have no flow-on negative impacts on the function of an ecosystem and/or does not result in other changes in community composition.

ecology The study of the relationships between organisms and groups of organisms and their biotic and abiotic environments.

ecosystem services The ecological services to human societies provided by ecosystems, for example provision of clean water and air, and pollination of crop plants.

ecosystem An ecological community together with the physical environment with which its members interact.

ecotone The boundary line or transitional area in between two ecological communities. See also **edge effect.**

ecotourism Tourist activity specifically for environmental values.

ecotype A subset or population of individuals within a species that are adapted to different environmental conditions from other subsets or populations, and are therefore different in structure or physiology.

edaphic Environmental conditions determined by physical, chemical and biological characteristics of the soil or substratum.

edge effect All changes at an ecosystem boundary and within adjacent ecosystems; an increase in species diversity in places where two habitats meet; the negative influence of a disturbed habitat edge on the interior conditions of a habitat, or on species that use the interior habitat.

effect size The level of impact that is required to be detected by a study, given that there is an effect. The magnitude of change that we wish to detect confidently, if there is a change.

effective population size The number of individuals in a hypothetical randomly mating population that would have the same rate of loss of genetic variation as the population under consideration.

EIS See **environmental impact statement.**

electrophoresis A process used to separate molecules (proteins or DNA fragments) to identify genotypes in populations. Samples are placed on a gel through which an electric current is passed. Molecules migrate through the gel depending on their charge, weight and configuration. The gel is treated (termed blotting) so that the patterns formed by the migrating molecules can be visualised.

embryo A young developing organism.

empirical Derived from or relating to experiment and observation rather than theory.

endangered A classification of risk intermediate between critical and vulnerable.

endemic Any species that is naturally highly localised or with a restricted geographic distribution; species native to a specified region.

Environment Protection and Biodiversity Conservation Act (1999) An Australian Federal Government act introduced in July 2000 under which actions likely to have a significant impact on matters of national environmental significance are prohibited unless prior approval is granted by the Federal Minister for Environment and Heritage.

environmental audit Monitoring or testing of the predictions of an environmental impact statement.

environmental domains The classification of an area into zones differentiated by different climatic regimes, slope, and soils. Environmental domains enclose a continuous range of physical environmental parameters that are expected to be important in determining the distributions of species.

environmental flows Water released into managed river systems with the objective of contributing to ecosystem health, maintaining ecosystem processes and enhancing the viability of riparian and in-stream flora and fauna.

environmental impact statement (EIS) (or environmental effects statement) Procedure that involves preparation of a document describing the existing environment, predictions of the kind, risks and magnitude of the likely environmental effects of a proposal, recommendations to mitigate the effects, recommendations on whether and how the proposal should proceed, and the involvement of government officers and public interest groups in assessing the accuracy and completeness of the document.

environmental uncertainty See **environmental variation.**

environmental variation The unpredictable temporal variability in the environment. See also **environmental stochasticity.**

environmental weeds Those plant species that invade natural vegetation.

enzyme Biological catalyst, usually a protein, which increases the rate of a reaction.

EPBC Act See *Environment Protection and Biodiversity Conservation Act (1999).*

ephemeral Transitory, short-lived.

epicormic bud See **epicormic growth.**

epicormic growth A shoot or branch growing from a dormant bud on the trunk of a tree, usually as a result of damage to the tree.

epigeal Living or growing on or close to the surface of the ground.

estuary The widening of a channel of a river where it enters the sea, with a mixture of fresh and salt water.

ESU See **evolutionary significant unit.**

Ethiopian region Biogeographic region comprising Africa south of the Sahara.

eukaryote Organism made up of cells with a nucleus and other membrane-bound organelles (e.g. protozoans, algae, fungi, plants and animals), as opposed to **prokaryotes** (bacteria and blue-green algae), which do not.

eusocial (of insects) Species of insects in which members of a colony cooperate in the care of the young and exhibit reproductive divisions of labour, with more or less sterile individuals working on behalf of others involved in reproduction. Most colonies have an overlap of at least two generations able to contribute to colony activities.

eutrophication Rapid increase in the nutrient status of a body of water, which can be natural or occur as a by-product of human activity (e.g. run-off of fertilisers), and is often accompanied by massive growth of dominant species (see **algal bloom**). Excessive production stimulates respiration, increasing dissolved oxygen demand and leading to anaerobic conditions.

evapotranspiration The return of water vapour to the atmosphere by evaporation from land and water surfaces and by the transpiration of vegetation.

event trees Logic trees that link possible outcomes following an initiating event, constructed around a series of dichotomies (yes/no). Each node is an event (or a decision). The focus is on a primary (initiating) event and consequences are traced forwards from it. Branches can show possibilities or specify the probabilities of alternative outcomes.

evolutionarily significant unit (ESU) A population that is reproductively isolated from other populations of the same species and which has a unique evolutionary history and genetic value.

exogenous Having an external origin; developing or originating outside an organism or species.

exoskeleton Skeleton covering the outside of the body, or located in the skin.

exotic species (Also termed alien or introduced species.) In Australia, usually those species introduced accidentally or deliberately since European settlement.

expert judgements Judgements and opinions by relevant people with specialist knowledge, often used when empirical and theoretical information is scarce; estimates of parameters and ecological relationships that are undocumented, qualitative and largely subjective.

explanatory variable A variable in a mathematical equation whose value determines the dependent variable; the variable that defines a set of distinct experimental conditions, or that an experimenter deliberately manipulates in order to observe its relationship with some other quantity.

exploiter species Species that are abundant in polluted areas because competing species have failed to survive, or because they make use of the polluted environment or the pollutants themselves.

exponential growth Population growth in which the size of the population grows exponentially.

extensive land-use zone A term based on a land classification in Australia where land-use practices have largely involved vegetation utilization (rather than clearing; e.g. grazing). See also **intensive land-use zone.**

extent of occurrence The area containing all known observations of a species.

extinct An extinct species is one that no longer has any living individuals.

extinction debt The time lag between the extinction of a species and habitat loss or some other kind of human disturbance. This translates to a difference between the current number of species in an area and the expected, lower number of species that will occur at equilibrium.

extirpation Localised extinction.

factorial The product of all the positive integers from one up to and including a given integer.

family Any of the taxonomic groups into which an order is divided and which contains one or more genera with a number of features in common.

fault trees Logic trees that show the necessary conditions to initiate a specified event. Fault trees express 'failure logic' and the contributing causes (and, optionally, associated probabilities for an unwanted event).

faunal relaxation The loss of species from a patch of habitat following patch isolation.

fecundity Fertility, fruitfulness; probability of giving birth; the number of eggs, or seeds, or generally offspring in the first stage of the life cycle, produced by an individual.

feldmark A highly specialised and restricted vegetation community found in extreme alpine areas (e.g. the main range of Kosciuszko National Park of southern New South Wales) that is characterised by either very wind-exposed ridges or late summer-lying snowdrift on leeward slopes.

feral Wild, or escaped from cultivation or domestication and reverted to the wild state; often implies that the species is exotic.

field gene banks Areas of land in which collections of growing plants have been assembled in populations containing as wide a sample of genetic variation as possible, ensuring that material is conserved for breeding, reintroduction and research.

fire intensity The amount of heat released at a point on a fire edge.

fire regime The sequence of fires in an area. Differences in fire regimes are characterised by differences in fire intensity, fire type, fire frequency and the season in which a fire occurs.

flagship species A charismatic species that attracts public attention, financial resources or additional conservation efforts to promote the protection of an area (e.g. a nature reserve or national park) or the protection of a suite of other associated (less charismatic) taxa.

flexibility (of reserves) Allows different areas to be substituted in a reserve system if they contribute the same values to the reserve system.

floristics The branch of botany concerned with the types, numbers and distribution of plant species in a particular area.

focal species approach A method used to guide landscape restoration efforts by identifying the species most at risk from a given threatening process. Remediation programs are then focused on that species under the assumption that other less threatened species will also be conserved by the remediation action.

forest type maps Developed for most of the productive forests of Australia, these are based on the distribution of commercial tree species, together with their growth stage and condition.

founders The few individuals of a species who establish a new population.

fragmentation A process of landscape alteration in which natural areas are subdivided into smaller patches.

frequency distribution A set of values of a variable divided into classes of associated rates of occurrence (frequencies).

frugivore Fruit-eating animal.

fuel reduction burn See **prescribed fire.**

fully fished A descriptive term for a fishery in which fishing pressure is close to optimum.

functional extinction See **ecological extinction.**

functional groups Suites of taxa that are classified based on an understanding of the ecology of the species attributed to each functional group (e.g. hot climate specialists, opportunists, and large solitary foragers).

fundamental niche That region of the environment within which a species could persist indefinitely.

fynbos The South African term for low shrubby heath that is typical of nutrient-depleted sands in Mediterranean climates; analogous to Western Australian kwongan and Californian chaparral.

gamete Mature male and female germ cells that fuse to form the zygote; a haploid reproductive cell that can undergo fertilisation.

gametophyte Haploid stage of a plant life cycle that produces gametes.

gamma (γ) diversity The turnover of species that occurs between spatially separate sites, within a more or less homogeneous vegetation type or habitat; measures the rate at which the composition of a community changes as a function of geographical distance, rather than ecological distance.

GDP See **gross domestic product.**

gene flow The uni- or bi-directional exchange of genes through cross-fertilisation within or between populations due to the migration of individuals and subsequent successful reproduction, which results in the spread of genes among populations.

gene locus See **locus.**

gene pool The sum total of genes (i.e. the alleles present and their frequencies) in a sexually reproducing population.

generalised linear model A generalised procedure for estimating linear relationships among variables, which is composed of three components. The linear predictor is the sum of the effects of the independent variables. The error function is related to the type of data being examined. The link function relates the linear predictor to the predictor value and it transforms the mean to produce a linear model; that is, a linear combination of explanatory variables determined by the error distribution.

genetic algorithm A computer-based procedure that mimics the processes of genetic evolution to evolve rules relating dependent variables to explanatory variables.

genetic differentiation The increasing variation between populations within a species, usually resulting from geographic isolation and selection or genetic drift.

genetic distance A measure of the degree of difference of the genotypes between individuals in a population, between populations in a species, or between species in a higher taxon.

genetic diversity The heritable variation that exists between individuals within populations, between populations within species, and between species. This genetic diversity may be represented by different sequences of base-pairs on strands of DNA, by different chromosome numbers and arrangements between populations or between species, or by the different phenotypic traits exhibited by individuals within populations.

genetic drift Gene frequency changes in small isolated breeding populations due to random sampling of genes between generations.

genetic load The average number of alleles that reduce fitness per individual in a population.

genetic neighbourhoods Genotypic arrays within a species that are non-randomly distributed in space; the average distance dispersed by reproductive adults between their birth and the birth of their offspring; the size of the area within which individuals mate at random.

genetic pollution In a particular species, the spread of alleles derived from occasional hybridization with closely related species.

genetic stochasticity In population viability analysis, the unpredictable variation in population vital rates (births and deaths) that arises from the different genetic composition of individuals in a species.

genetic swamping Process by which the genes of one species spread through the genome of another species.

genetic variation Represented by the presence of alternative alleles at a single locus, or by differences among individuals in quantitative traits that are determined by many genes.

genetically modified organism An organism that contains a gene or genes from an unrelated organism that could not be transferred to the species by traditional breeding or reproductive methods.

genome The total genetic material within a cell or individual.

genotype The entire genetic constitution of an organism, or the genetic composition at a specific gene locus or set of loci, or a group of organisms with the same genetic constitution.

genotype-environment interaction Variation in response to genotype that depends on the environment in which the individual occurs.

geocoding The process of assigning latitude, longitude and elevation values for location records.

geographic range/extent of occurrence The spatial limits of the localities at which a species has been recorded in a region of interest.

geometric mean The average value of a set of *n* integers, terms or quantities, expressed as the *n*th root of their product.

geomorphology Branch of geology concerned with the structure, origin and development of the topographical features of the earth's crust.

germ plasm The part of a germ cell (germ cells are the cells that give rise to the gametes) that contains the hereditary material (chromosomes and genes).

germinate To sprout, grow, develop or form new tissue.

glaciation To cover or become covered with glaciers or masses of ice.

global extinction The loss of all members of a species from all populations of that species.

global warming (or the greenhouse effect) The accumulation of gases (carbon dioxide, methane, nitrous oxide and chlorofluorocarbons) in the atmosphere, which increases the reflective capacity of the atmosphere, thus retaining heat on earth.

Gondwanaland Supercontinent that existed in the past and consisted of all southern land masses.

goodness of fit The extent to which observed sample values of a variable agree with the values derived from predictions.

grafting A process by which a relatively small part of one organism is transplanted either onto another part of the same organism or onto a different organism or a part of it.

graminoid Pertaining to grass; grass-like.

granivore A species that feeds on seeds or grain.

grassland Land such as prairie on which grass predominates.

greenhouse effect See **global warming.**

gross domestic product (GDP) The total value of all goods and services produced domestically by a nation during a year. It is equal to gross national product minus net investment incomes from foreign nations.

ground water Underground water that is held in the soil and in pervious rocks; the source of fresh water beneath the earth's surface. Usually refers to only that part of the subsurface water in the zone of saturation. The top of this saturated zone is termed the watertable.

guild A group of species that exploit the same class of environmental resources in a similar way and between which competition can be expected.

habitat evaluation procedure (HEP) A process in which attributes of a species' habitat are assigned weights reflecting their relative importance.

habitat specificity A term describing the degree of generalisation or specialisation of a species' habitat requirements.

habitat suitability indices (HSI) Predictive tools to evaluate the quality of habitat for wildlife management; a framework that uses (usually subjective) descriptions of the biology and life history of a species to identify regions of high-quality habitat (see also **habitat evaluation procedure**).

habitat variegation model A conceptual model of landscape cover in which four landscape alteration states are recognised: intact, variegated, fragmented and relictual.

habitat The subset of physical environmental factors that permit an organism to survive and reproduce. Implicitly these factors are associated with a geographic location.

haploid A cell or an individual possessing only one set of chromosomes (e.g. an egg or sperm, or the worker class in ant and bee species).

harmonic mean A kind of average of a set of n numbers in which the reciprocals of the numbers in the set are added, the sum is divided by n, and the reciprocal of the result is taken.

heath Vegetation composed mainly of shrubs, usually with few or no trees, and often on nutrient-deficient sands.

heavy metals Metals such as lead, cadmium and zinc, with isotopes that have relatively high atomic weights and which, if present in soil or water, inhibit plant growth and are toxic to marine and terrestrial vertebrates and invertebrates.

HEP See **habitat evaluation procedure.**

herbfield Vegetation dominated by prostrate annual plants or perennial plants whose aerial parts do not persist throughout the year.

heritable variation The proportion of variation of a given trait that can be attributed to genetic factors.

heterogeneity Refers to something composed of unrelated parts, or elements not of the same type.

heterosis (Also known as hybrid vigour.) The enhanced vigour and improved performance in fitness-related traits that is observed in individuals with relatively many heterozygous loci (as a result of outbreeding).

heterozygous The situation in which an individual has two different alleles at a given gene locus.

heuristic algorithms Algorithms based on past experience.

heuristic Using or obtained by reasoning from past experience, often since no precise causal mechanism is known.

holotype The type specimen used in the taxonomic description of a species.

homoclimes Areas that experience similar climatic conditions.

homozygous The situation in which an individual has two of the same alleles at a given gene locus.

host-specificity Physiological or behavioural restrictions of parasites that limit them to a few individuals, a single species, or a few related species.

hotspots Areas that are species rich for a particular taxon.

HSI See **habitat suitability indices.**

hummock grasslands Areas where grasses of the genera *Triodia* and *Plectrachne* are dominant. Soil builds up around each clump of grass, forming a hummock; the ground in between hummocks is usually bare.

hybrid vigour See **heterosis.**

hybrid zone A geographic area between two populations or species in which hybrids between the parent populations/species proliferate - often a result of the breakdown of geographic or reproductive isolation.

hybridisation Breeding resulting in offspring between two different varieties or species.

hypogeal Occurring or living below the surface of the ground.

IBRA See **Interim Biogeographic Regionalisation for Australia.**

improved pastures Pastures that have been improved in terms of nutritional value by fertilising, planting legumes, or planting higher producing species of grasses or other pasture species.

***in vitro* storage** The storage of species in laboratory conditions.

in vitro Refers to biological processes occurring experimentally in isolation from a species' environment; usually refers to laboratory conditions.

inbreeding coefficient (*F*) The probability that the pair of alleles carried by the male and female gametes are identical and derived from a common ancestor, or a measure of the reduction in heterozygosity that results from inbreeding.

inbreeding depression A loss of vigour among offspring that occurs when closely related individuals mate; results from elevated levels of homozygosity of deleterious recessive alleles and from a low average level of heterozygosity.

inbreeding A tendency for breeding between individuals who are more closely related than are randomly selected individuals.

incertitude Uncertainty that arises from a lack of knowledge, which can be reduced by acquiring more data and better understanding.

independent variable See **explanatory variable.**

indicator species Species that are highly sensitive to a particular land-use activity and whose presence is also likely to indicate habitat quality for a range of other species, communities or ecosystem processes.

indigenous An organism originating or occurring naturally in a particular locality or habitat; not introduced.

influence diagrams Diagrams that show cause and effect relationships in a system, reflecting mental models and conceptual compartments used by people to simplify and understand complex dynamics.

intensive land-use zone A term for a land classification in Australia where land-use practices have largely involved vegetation clearing (e.g. cropping, forestry plantations, urban development). See also **extensive land-use zone.**

intergenerational responsibility The notion that the current generation holds biodiversity values in trust for future generations, so that they retain the option to use them or preserve them.

Interim Biogeographic Regionalisation for Australia (IBRA) A biogeographic classification of the Australian terrestrial landmass based on the key physical processes (e.g. climate) that drive ecological processes and which, in turn, are thought to influence biological productivity and biodiversity.

intragenerational responsibility The notion that a human population in a particular region holds biodiversity values in trust for people in other regions, so that others retain the option to contribute to decisions to use them or preserve them.

intrinsic co-adaptation The correlated variation and adaptation of sets of genes involved in maintaining the fitness of organisms; interactions between loci in co-adapted gene complexes.

intrinsic values (biocentric and ecocentric values) Ethical positions that place value on species and communities, independent of people.

introduced species See **exotic species.**

introduction The accidental or deliberate movement of an organism to an area previously unoccupied by that organism.

introgression Hybridisation between two species in which the genes of one species gradually diffuse into the gene pool of another.

intron Intervening non-coding sequence present in the coding region of a eukaryotic gene that is transcribed but excised before translation. There are usually about 10 introns per gene.

invertebrate Any animal lacking a backbone.

irreplaceability (in reserve selection) The extent to which complete representativeness can be achieved if a given area is not included in a suite of protected areas; the potential contribution a given area makes to a reservation goal.

irrigation salinity Salinisation caused by salt stored in the soil or ground water being mobilised by irrigated water moving through the soil profile. In addition, water applied through irrigation may raise the level of the watertable, which may then either evaporate, leaving the salts at, or close to, the soil surface, or move through the soil profile and enter a watercourse. These processes leave salt at the soil surface and increase the salinity of watercourses.

island biogeography See **theory of island biogeography.**

island model See **theory of island biogeography.**

isoenzymes See **isozymes.**

isotherm A line on a map that joins places that have the same mean temperature.

isozymes (Also known as isoenzymes.) Variant of a given enzyme coded by a single locus, separable by methods such as electrophoresis.

iterative Refers to the repeated application of a mathematical procedure, where each step is applied to the output of the preceding step.

keystone species Those species which, if removed, precipitate important changes in ecosystem structure and function.

land degradation The process whereby land deteriorates because of either the natural agents of wind, water, gravity and temperature, or the actions of humans. Land degradation is characterised by a reduction in soil depth and/or water quality.

landscape context A term describing the conditions and characteristics of the landscape surrounding a site.

landscape continuum model See **habitat variegation model.**

landscape contour model A conceptual model of landscape cover in which empirical measures of habitat suitability are represented spatially on a map by contour lines to delineate the potential distribution of a species.

landscape indices Spatial metrics or measures of patterns of landscape characteristics.

larva A juvenile free-living form of many animals that develops into a different adult form by metamorphosis.

latitudinal migration Annual migration from north to south, in response to changing seasons.

least-squares regression A method for determining the best value of an unknown quantity by relating one or more sets of measurements by seeking the solution for which the sum of the squares of the differences between the observed and the theoretical values is least.

lekking system An animal mating system in which there is a communal mating ground.

Levins structure A kind of metapopulation characterised by all of the patches being (approximately) the same size. Localised extinctions in the patches are recovered by recolonisation resulting from migration from other patches in the system.

lichen Small plants formed by the symbiotic association of a fungus and an alga, and which occur as crusty patches or bushy growths on tree trunks, bare ground, rocks etc.

lignotuber A woody swelling at the base of some species of plants (especially mallee-form eucalypts), which is found underground or partly underground. Lignotubers contain reserves that assist a plant to recover if its aerial parts are destroyed.

linear algebra The branch of mathematics concerned with variables that are not raised to any power, consisting only of constant multiples of the variables (e.g. linear equations, matrices, determinants, vector spaces).

linear programming The study of optimisation problems that can be solved by seeking the maximum or minimum values of a linear function of non-negative variables subject to constraints expressed as linear equalities or inequalities.

linear regression A regression that is linear in the unknown parameters, whatever the order of the known parameters.

linkage disequilibrium The condition in which alleles associate non-randomly because they are close together on a chromosome, or because the combination is under selective pressure.

liverworts Any of a class of bryophytes growing in wet places and resembling green seaweeds or leafy mosses.

local extinction The loss of all individuals in a population inhabiting a relatively isolated patch of habitat. See also **extirpation.**

locus The position of a gene on a chromosome. At any particular gene locus there are two alleles, or copies of the gene inherited from the two parents.

log scale A measurement scale, distances along which are proportional to the logarithms of the marked indices.

logarithm The power to which a base must be raised to yield a given number.

logic trees Diagrams displaying cause and effect relationships in a system. See also **fault tree, event tree, decision tree.**

logistic equation An equation describing an s-shaped (sigmoidal) relationship between two variables.

logistic regression A regression that works by developing a prediction of the probability of an event (e.g. the presence or absence of a species) and fitting an s-shaped curve to the series of observations such that a prediction based on that curve is most likely to be correct.

macroinvertebrates Relatively large invertebrates in a community, within an arbitrary size range (e.g. 2–20 millimetres body width in soil invertebrates).

macropod Marsupials that are characterised by powerful hindlimbs and long hindfeet, usually having a fast hopping gait.

mainland–island structure A kind of metapopulation that has two major features: a large 'mainland' area in which populations are secure and very rarely (if ever) suffer extinction, and an array of small patches in which extinctions are common events. The mainland provides a source to recolonise the patches following localised extinctions.

mallee Small (generally less than 6 metres tall) eucalypt trees with multiple stems arising from a single root stock or lignotuber from which the tree can regenerate following drought or fire.

management goals The broad socially-mandated objectives of a management action or practice.

management indicator species Species that reflect either the impacts of a disturbance regime or active management efforts to mitigate the effects of such a disturbance regime.

management units Demographically independent (genetic) populations with distinctive allele frequencies.

mangroves Any of various tropical evergreen trees or shrubs that have stilt like intertwining aerial roots and form dense thickets along coasts, usually on muddy substrates.

mariculture A type of fish farming where sea fish or shellfish are grown in farms in seawater.

marker A gene or DNA sequence with a known location in the genome and known effect, which is used to track the inheritance or behaviour of other genes.
mass effects Cases where a species occurs in places other than its primary habitat because of a flux of its propagules from adjacent areas of primary habitat.
mass extinctions Substantial biodiversity losses that are global in extent, taxonomically broad, and rapid relative to the average duration of the taxa involved.
matrix The dominant or most extensive patch type in a landscape, which has the most extensive cover, and which exerts a major influence on landscape dynamics. The term is also often used to describe the heavily modified areas (e.g. agricultural fields) within which remnants of native vegetation remain in fragmented landscapes.
maximum sustainable yield The largest amount of a renewable natural resource that can be taken, while ensuring the indefinite availability of that resource. Harvest is at the population size representing the maximum rate of recruitment into the population, based on a logistic growth curve.
mean Average.
measurement endpoints (and test endpoints) Indices of the environment that are the goal of field measurements or laboratory tests.
mechanical disturbance Refers to disturbance of soil that is caused by machinery or the action of water, as opposed to chemical disturbance.
mechanistic Of or reflecting the simple causal interactions among the components of a system.
median time to extinction The time by which half the species within a taxon have become extinct.
median The middle number, or average of the two middle numbers, in an ordered sequence of numbers.
megadiverse nations The seventeen countries that together harbour 60–70% of the earth's species.
megafauna The largest arbitrary size categorisation of animals in a community (e.g. those with body width larger than 20 millimetres in soil invertebrates, or relatively large mammal species). The term is also used to describe the large vertebrates that lived during the Pleistocene and became extinct during that time as a consequence of hunting, climate change or both.
Mendelian genetics The view that in eukaryote genomes, alleles separate into different nuclei during meiosis, after which any member of a pair of alleles has equal probability of finding itself in the nucleus with either of the members of any other pair (if the loci are unlinked). Ratios of characters among offspring are predictable, given knowledge of the parental genotypes.
meristem A plant tissue responsible for growth whose cells divide and differentiate to form the tissues and organs of the plant.
meta-analysis An analytical approach where the results of many similar studies are assessed to identify general patterns among them.
metapopulation A set of populations (or sub-populations) of the same species, usually more or less isolated from one another in discrete patches of habitat, which interact via individuals moving (dispersing) between local populations.
microclimate The atmospheric conditions affecting an individual, a small group of organisms, or a localised area of the environment, especially when they differ from the climate of the rest of the area; local climatic conditions dependent on vegetation, soil conditions and small-scale topography.
microinvertebrates The smallest arbitrary size categorisation of invertebrates in a community.
microsatellites Tandem repeats of a simple DNA sequence on a chromosome; smaller and more random than minisatellites, they provide a tool for genetic mapping and kinship studies.
migration The regular annual movement of animals between different habitats, each of which is occupied for specific parts of the year; movement of individuals or whole populations from one region to another.
minimum convex polygon The smallest polygon containing all sites in which no internal angle exceeds 180 degrees.
minimum expected population The smallest size to which a population is most likely to fall in a given period (e.g. 10 years).
minimum viable population The smallest isolated population that has a reasonable probability of surviving for some time into the future given foreseeable demographic, genetic and environmental stochasticities, and natural catastrophes.
minisatellites Tandem repeats of a simple DNA sequence on a chromosome, providing a useful genetic marker in DNA fingerprinting.
mitochondria Organelles of cells containing circular DNA molecules, RNA and small ribosomes; site of cellular respiration.
mode The value that has the highest frequency in a set of values.
molluscs Phylum of bilaterally symmetrical soft unsegmented invertebrates. Molluscs are largely aquatic and often have a shell.
monitoring Sampling and analysis designed to ascertain the extent of compliance with a predetermined standard or the degree of deviation from an expected norm.
monoculture Refers to a large area covered by a single plant species (or for crops, a single variety).

monogamous Having only one mate.
monotonic Unvarying, or changing in a single direction.
monotypic Having only one species in the genus.
montane Of or inhabiting mountainous areas below the subalps; a region that is generally above 900 metres above sea level in Australia, and receives some snow each year.
Monte Carlo simulation Usually, a computer-based stochastic simulation model representing the dynamics and uncertainties in a system, designed to estimate the probabilities of specific outcomes.
morphological species concept Concept whereby species are defined by a set of morphological characters that are shared by members of that species, and are distinguished from other species by morphological discontinuities.
morphology The form and structure of an organism.
mtDNA (mitochondrial DNA) DNA of the mitochondria that evolves rapidly, with a mutation rate up to 10 times that of nuclear DNA.
mudflat A tract of low muddy land that is covered at high tide and exposed at low tide.
multidimensional niche The range of tolerances of a species to abiotic factors that affect its survival and reproduction.
multiple regression A regression function that gives the expected value of a random variable, x, or any function of x, given that an event, B, is known to have occurred, in terms of more than one other random variable.
multivariate Involving a number of distinct, though not necessarily independent, random variables.
mutation A spontaneous change in the genotype of an organism at the gene, chromosome, or genomic level, often occurring during DNA replication. Mutations usually refer to alterations to new allelic forms, and so are the ultimate origin of genetic variation.
mutational meltdown Phenomenon that occurs when populations are reduced and breeding pressure leads to the selection of mates with deleterious recessive genes who otherwise would not have mated. Deleterious genes are passed on to the next generation, leading to even fewer mate choices, further increasing the odds that individuals with relatively harmful mutations will breed.
mutualistic Interspecific relationship in which both organisms benefit without sustaining injury; frequently a relationship of complete dependence.
narrow-sense heritability The proportion of the variation in a character that is genetically influenced and can be modified by natural selection.
natal territory The territory where an individual organism was born.
national park An area where the primary objective of land management is nature conservation and/or human recreation.
naturalised Refers to exotic species that are able to persist and subsequently reproduce in the wild.
Nearctic region Biogeographical region consisting of Greenland and North America south to mid-Mexico.
negative exponential A function and a statistical distribution that is useful for describing processes with maxima near the origin, and with a long right-hand tail.
neighbourhood The area within which individuals mate at random; the average distance dispersed by reproductive adults between their birth and the birth of their offspring, or the average dispersal distance of seed or pollen.
Neotropical region Biogeographical region of the world including South America and lower Central America.
nested subset theory An approach examining changes in the composition of biotic assemblages on islands and fragments of habitat in which species-poor small islands (or habitat fragments) support depauperate subsets of species found on species-rich larger islands. The ordered addition of species with increasing island or fragment size is termed 'nestedness'.
net primary productivity The total energy accumulated by plants during photosynthesis (total photosynthetic production minus respiration).
niche The total range of conditions within which individuals of a species can survive, grow and reproduce.
nitrogen fixation Conversion of gaseous atmospheric nitrogen (N_2) by certain bacteria (and a little by lightning, ultraviolet radiation and combustion) to inorganic or organic compounds (ammonia, nitrites and nitrates).
non-equilibrial dynamics Describes a situation in which populations fluctuate because environmental stochasticity disrupts the equilibrium.
non-parametric statistics Statistics that study data measured on a nominal or an ordinal scale.
non-probabilistic uncertainty methods Mathematical and logical methods that calculate quantities and bounds without relying on probabilities. Examples include fuzzy numbers and interval arithmetic.
non-vascular plant Plant without a vascular system (i.e. algae, fungi and bryophytes).
normal distribution A continuous symmetrical distribution of a random variable with its mean, median and mode equal (the histogram verges towards a smooth, symmetrical, bell-shaped curve).
normal forest A forest in which each age class is represented by a fixed area, so that the forest produces an approximately constant supply of wood products.
noxious weed See **weed**.
nucleotides Molecules incorporated into nucleic acids; components of RNA and DNA.
null hypothesis The hypothesis that the treatment has no effect; that is, there is no difference between the various groups in the experiment – the observed pattern of data and an expected pattern are effectively the same, differing only by chance.

nutrient Particular nourishing substance required by an organism, which must be obtained from its environment; includes organic compounds (carbohydrates, amino acids and fats), vitamins and minerals.
obligate seeding plant A plant species that relies exclusively on seed to reproduce; that is, a species with no means of vegetative reproduction.
old growth forest Forests that are both little disturbed and ecologically mature. Old growth forests differ significantly from young forests in features such as the size and shape of the understorey, the dominant tree species, and ecological processes (such as rates and paths of energy and nutrient flows, water cycling, tree population dynamics, and the decay and nutrient dynamics of coarse woody debris and leaf litter).
oligotrophic Lakes, rivers and soils that are low in nutrients.
open forest Forest in which the tallest stratum is composed of trees with 30-70% projective foliage cover.
open shrubland Vegetation in which the tallest stratum is composed of shrubs and the projective foliage cover of the shrubs is less than 10%.
open woodland Vegetation in which the tallest stratum is composed of trees less than 30 metres tall and the projective foliage cover is less than 10%.
open-access management Management systems for natural resources in which there is no restriction on who can take or harvest a resource.
opportunistic Describes a species that is capable of exploiting spasmodically occurring environments.
optimum catch See **maximum sustainable yield.**
option demand The desire by non-users to retain the option to use a resource at some time in the future.
ordinate The vertical or '*y*' coordinate of a point in a two-dimensional graphical system.
ordination An arrangement or order; a mathematical system for categorising communities on a graph so that those that are most similar in species composition appear closest together.
organismal diversity The diversity of organisms at any given taxonomic level (e.g. genera, families).
Oriental region Biogeographical region consisting of south eastern Asia from India to Borneo, Java and the Philippines.
orthodox seeds Seeds that can be stored for long periods at sub-zero temperatures if their moisture contents are reduced below 10%.
outbreeding depression Reduced fitness of progeny from inter-population crosses resulting from the breakdown of co-adapted gene complexes.
outbreeding Breeding between individuals that have been isolated from one another in more or less separate gene pools.
over-fished A descriptive term for a fishery in which few fish are left or for which there are too many fish being taken for the fishery to be sustainable.
ovule Structure found in seed plants that develops into a seed following fertilisation.
Palaearctic region Biogeographic region consisting of Europe, North Africa and Asia south to the Himalayas and the Red Sea.
panmictic Characterised by, or resulting from, random matings; completely mixed.
parameter An arbitrary constant that determines the specific form of a mathematical expression; a characteristic constant in a statistical population, for example its variance or mean; any constant or limiting factor; a variable that is not being considered and that may for present purposes be regarded as a constant.
parametric analysis Method of analysis by which data are measured on a scale of measurement in which the differences between the values can be quantified so that arithmetic operations are applicable to them.
parasite An organism that lives and feeds on or in another organism (the host).
parthenogenetic species Species that use a method of reproduction whereby eggs develop without fertilisation. Parthenogenesis has been observed in some plants, invertebrates and even vertebrates (e.g. geckoes and snakes).
passerines An order of birds characterised by a perching habit.
patch A term used in landscape ecology to describe a relatively homogenous non-linear area that differs markedly in vegetation structure and composition from its surroundings (or the matrix).
pathogen An organism capable of causing disease.
patristic distances The branch lengths of a cladogram.
PCR See **polymerase chain reaction.**
perennial Descriptive of a plant that lives for at least several years, and usually flowers each year (as opposed to **annual** plants, which only live for 1 year).
phenotype The physical expression of a trait of an organism that may be due to genetics, environment, or an interaction of the two.
phenotypic variation The variation between individuals in the same population.
phyletic Of, or relating to, the evolutionary development of organisms.
phylogenetic differentiation Variation between groups or species that results from evolutionary processes, reflecting common ancestry.
phylogenetic species concept The concept whereby species are the basis of evolutionary lineages; a hypothesis of the true genealogical relationships among species based on the concept of shared, derived characteristics.
phylum Taxonomic category particular to animals, which includes one or more classes and is included within a kingdom.

phytogeographic Relating to the science concerned with the geographical distribution of plants (phytogeography).

plant exudates Material discharged through pores or incisions in plants (e.g. resins).

Poisson distribution The number of (relatively rare) events occurring randomly in a fixed time at an average rate of δ. In a Poisson distribution, the mean is equal to the variance.

Poisson regression model A form of generalised linear model in which the underlying distribution is the Poisson distribution. This model is often appropriate for use when the data are skewed and discontinuous, as is usually the case with counts (the response variable is a frequency count of the number of 'incidents').

pollution indicator species Species that reflect the effects of pollution on biota.

polyandrous Relating to female animals that mate with more than one male in a breeding season; flowers that have a large number of stamens.

polygamous Relating to male animals that have more than one mate during a breeding season; plants that have male, female and hermaphrodite flowers on the same plant, or on separate plants of the same species.

polymerase chain reaction (PCR) A chain reaction that creates molecules that are an exact replica of a smaller molecule and are directly obtainable from it.

polymorphic Describes a situation where a given gene locus has two or more alleles within a population.

population bottleneck See **bottleneck**.

population viability analysis An analysis of a population's chances of decline or extinction within a specified time frame.

population A group of conspecific individuals, commonly forming a breeding unit within which the exchange of genetic material is more or less unrestricted; a group sharing a particular habitat at a particular time.

precautionary principle Argues for the cautious use of biodiversity and the environment, and that the consequences of a development action need to be demonstrated to be negligible.

precision (of a prediction) The ratio of the actual to the predicted magnitude of impact, with the smaller of the two as the numerator.

predation The consumption of one organism by another, where the consumed organism is alive when the consumer first attacks it.

prescribed fire Fires that are deliberately lit with the aim of reducing the amount of available fuel (i.e. live and dead vegetative matter) in forests or grasslands.

presumed extinct The status of species that have not been definitely located in nature for some time, despite thorough searching.

primary production The total organic material synthesised in a given time by the autotrophs (organisms independent of outside sources for organic food materials) of an ecosystem; the rate at which biomass is produced per unit area by plants.

primary salinity Refers to soils that are believed to have been saline for thousands of years as the result of natural landscape processes (e.g. saltpans and claypans).

primer In PCR, a primer is a synthetic DNA sequence that complements the flanking regions of a piece of DNA, and binds to the DNA before replication commences, to ensure that replication begins at the right places.

principle of inter-generational equity See **custodial responsibility**.

probability distribution A function that constitutes a measure of probability.

probability A measure of the relative frequency or likelihood of occurrence, and therefore of the degree of confidence with which we can predict that an event will occur. Values are derived from a theoretical distribution or from observations.

probe A known, labelled fragment of DNA or RNA used to detect and identify corresponding sequences in nucleic acids by hybridising with them.

productive use values Values for biological resources that are commercially harvested (i.e. directly measurable economic values).

projective foliage cover The percentage of the surface of a site above which there is foliage or branches.

prokaryote Organisms (typically bacteria or blue-green algae) with cells that lack a membrane-bound nucleus (cf. **eukaryote**).

propagate To reproduce, or cause to reproduce.

provenance The location or source of genetic material (a term usually employed for botanical samples).

pseudoextinction The transformation of species in the fossil record by mutation, natural selection and speciation that gives the appearance of a loss or gain of species.

***P*-value** The probability that a given test statistic takes either the observed value or one that is less likely under the null hypothesis. If the *P*-value is fixed in advance, this is the significance level of the test.

quadrat Delineated area (rectangular or square) of vegetation chosen for sampling or study.

quartile One of three values of a variable dividing its distribution into four groups with equal frequencies (the 25th, 50th and 75th percentiles).

quasi-extinction The risk of decline within some specified time frame to some small population threshold (a lower bound that may be unacceptable for conservation, management, economic or aesthetic purposes).

radio-tracking A field study method for animals in which radio transmitters are attached externally or internally to enable the spatial location of an individual to be determined.

rainforest Dense (closed canopy), broad-leaved forest vegetation with a more or less continuous tree canopy of variable height, and with a characteristic diversity of species and life forms.

Ramsar Convention See **Convention on Wetlands of International Importance.**

random numbers A sequence of numbers with the property that no member can be predicted from the preceding elements.

random sample A sample devised to avoid any interference from any property of the elements selected. In a random sample, each object in the population has an equal probability of being sampled, so the sample can be taken to be representative of the population.

range The spatial distribution of a species. The terms restricted (or local) and widespread describe extremes of spatial distribution.

rare Generally refers to species that have low abundance, limited distribution, or both.

rarefaction A method for estimating the mean and variance of the number of species in two regions when the sampling effort differs.

rarity A combined measure of the prevalence of a species, reflecting its abundance, range or spatial distribution, and its occurrence in different kinds of habitats.

ratchet effect A term used in natural resource management in which direct and indirect subsidies mean that economic investment is maintained in good years and disinvestment does not occur in poor years. The effect can lead to subsidised industries that over-harvest their resources.

ratio estimation A method of estimating abundance that assumes that a fixed number of species exist in a taxon for every one of the species in another taxon.

ratites Large flightless birds.

rDNA DNA that codes for ribosomal RNA. Ribosomes are the site of protein synthesis in cells and RNA takes coded information from DNA and translates it into specific enzymes and proteins.

realised niche The niche that is available to a species after biotic factors such as competition, predation and facilitation affect its potential and abundance.

recalcitrant seeds Seeds that cannot survive low temperatures or dehydration.

recessive (of phenotypic traits) Refers to traits that are only expressed when the genes determining them are homozygous.

recolonisation A metapopulation process in which individuals re-establish in a habitat patch where previous populations have become extinct.

recombination The process in sexual reproduction in which DNA is exchanged during meiosis between homologous chromosomes.

recovery indicator species Species that reflect the extent of recovery of an ecosystem following disturbance.

recruitment The number of new individuals in a population between two census periods; usually the sum of fecundity of all reproductive age classes.

re-establishment reintroduction Situation where captive-bred individuals are used to re-establish an extinct wild population.

regeneration The renewal of habitat by natural or artificial means.

regression analysis The analysis or measure of the mathematical association between a dependent variable and one or more independent variables, usually formulated as an equation in which the independent variables have parametric coefficients, which may enable future values of the dependent variables to be predicted.

regrowth A young, usually even-aged, forest stand that has regenerated after harvesting or fire.

reintroduction The release into natural habitat of captive-bred animals, garden-raised plants, or wild-caught animals or plants.

relative abundance The number of individual specimens of an organism seen over a certain period of time in a certain place.

relative richness The number of species per unit area, or per unit of sampling effort.

remnant vegetation An isolated patch of original vegetation in a landscape that has been altered by human activity such as agriculture.

representativeness The extent to which a reserve system contains examples of all the natural features of a region.

rescue effect The recolonisation of habitat patches by immigrants; the maintenance of a population by migration from other, partially isolated, populations.

response variable The dependent variable; that is, a variable whose value is determined by that taken by the independent variable.

resprouting plant A plant capable of regeneration after it has lost its aerial parts (e.g. to fire or grazing).

restocking reintroduction Situation where a declining population is supplemented with captive-bred stock.

restriction endonucleases Endonucleases produced by micro-organisms that recognize short sequences of DNA, cutting the double helix at a particular point. Different endonucleases recognize different DNA sequences, providing a means for DNA cloning and gene mapping.

restriction fragment polymorphisms (RFLPs) Variation in the length of fragments of DNA, which is generated by the presence or absence of restriction sites, and which is used to distinguish different individuals. RFLPs are detected by separating fragments using electrophoresis.

RFLPs See **restriction fragment polymorphisms.**

riparian Of, on, or inhabiting the bank of a river.

risk assessment The process of estimating the probability of an adverse event, or the measurement of the probability of an event and the magnitude of its consequences.

rotation length The planned number of years between the regeneration of a forest stand and its final harvesting for wood products.

ruderal weeds Weeds of wasteland or disturbed places.

run-off That portion of rainfall that runs into streams as surface water rather than being absorbed by the soil.

saline (of soil) Refers to soil that contains such a quantity of salt that plant growth is significantly reduced.

salinisation The process whereby the concentration of total dissolved solids in water and soil is increased due to natural or human-induced processes. See also **primary salinity, secondary salinity, salinity, dryland salinity** and **irrigation salinity.**

salt marsh In estuaries and tidal marsh areas, salt marshes are broad flat areas that extend to the extreme limit of tidal influence and support both flowering plants and algae.

savanna Open grasslands with scattered broad-leaved deciduous and evergreen trees.

scansorial Formed or adapted for climbing; habitually climbing.

scarcity A statistical measure of the frequency of a species in samples, summarising its abundance, range and specificity of habitat requirements.

sclerophyll Referring to plants characterised by rigid, often small leaves and small internodes. Sclerophyll plants are able to survive low soil nutrients, water stress and fire.

scrub Vegetation consisting of stunted trees, bushes and other plants growing in an arid area.

secondary salinity Refers to areas of previously productive land that are now salt-affected to the point where plant growth is inhibited. Usually results from irrigation or excessive clearing.

sedge A grass-like plant that grows on wet ground and has rhizomes, triangular stems, and minute flowers in spikelets.

seepage salting Any salinisation process induced by the action of water flowing slowly through the soil (e.g. irrigation salinity).

selective tree harvesting Removal of mature trees, either as single scattered individuals or in small groups at relatively short intervals, repeated indefinitely, by means of which the continuous establishment of regeneration is encouraged and an uneven-aged stand is maintained.

selfing Self-pollination or self-fertilisation.

senescing Advancing in age; the complex of aging processes that eventually lead to death.

sensitivity analysis Manipulation of parameters, dependencies and other assumptions in a model, to measure the degree to which model output and decisions depend on uncertainties in these factors.

sensitivity of statistical tests The response or change induced in the output of a statistical test by variations in parameters; usually measured as the proportional change in model predictions that result from a small change in each parameter; used to identify parameters that are most important in determining the model predictions.

sentinel species Highly susceptible species that provide an early warning of a pollutant.

service value The ecological functions and processes on which consumptive and productive values of ecosystems depend (e.g. photosynthetic fixation, and soil production and protection).

shifting mosaic A temporal pattern of occupancy of local populations in a metapopulation that changes over time as patches become extinct and are recolonised.

shrub A woody perennial plant, smaller than a tree, with several major branches arising from near the base of the main stem.

shrubland Vegetation community whose tallest stratum is composed of shrubs and that has only sparse projective foliage cover.

sigmoidal Relating to an s-shaped response curve in which there is an initial acceleration phase followed by a subsequent deceleration phase leading to a plateau.

silviculture The theory and practice of managing forest establishment, composition and growth to achieve specified objectives.

sink population A population in which local mortality exceeds local reproductive success.

site-type indicator Species that are thought to indicate certain types of environmental conditions (e.g. particular rock or soil types).

skewed distribution A distribution that does not have equal probabilities above and below the mean.

SLOSS (single large or several small) The question of whether one large reserve is better than several smaller ones with the same total area.

small-population paradigm A model that deals largely with the stochastic influences (e.g. demographic and genetic stochasticity) that influence small populations. See also **declining-population paradigm.**

smelter An industrial plant for extracting metals from ores by heating.

sodicity The amount of exchangeable sodium cations in water or soil.

soil acidification The reduction in soil pH by human activities to such an extent that plant growth is inhibited. Acid soils are those with a pH value of less than 6.5. Soils can be acidified by human-induced processes such as acid rain, removal of organic matter (for example by cropping) and therefore of nutrients, and through-leaching by irrigation. Few plants are able to grow or reproduce if the pH falls below 4.0–4.5. High acidity affects plants directly by damaging protoplasm (if pH falls below 3) or affecting metabolism, and indirectly (if pH falls below 6) by increasing the availability of toxins (e.g. heavy metals), and by inhibiting some important processes (e.g. nitrogen fixation).

soil structure Describes the extent to which soil particles are cemented together by humus, inorganic salts, and mucilage to form clumps, and the size and shape of these clumps.

source population A population that because of its size and the quality of its habitat produces a net number of emigrants.

spatial autocorrelation A statistical measure describing cases when the probability of occurrence of a species at two sites is dependent on the distance between those sites. A measure of the degree to which environmental conditions at two locations co-vary.

spatial contagion index Index designed to examine the spatial extent and adjacency of landscape units or patches.

spatial correlation Measures the amount of association between measurements of a variable made at points a given distance apart. See also **autocorrelation**.

spatial variation (with reference to risk assessment) The variation in environmental conditions between spatially separate patches of habitat; the different conditions experienced by each of several
populations in a metapopulation.

spawn To produce or deposit eggs.

speciation Any of the processes by which new species form.

species (abundance) curve A term describing the abundance and frequency of species in samples. These curves are typically right-skewed because many assemblages are dominated by many individuals of the same species, with a large number of other species being uncommon or rare and represented by only a few individuals.

species diversity Variability (richness and abundance) of biota in an area; an index of community diversity that takes into account both species richness and the relative abundance of species.

species richness (sometimes called species diversity) The total number of species within a community or a defined area.

species-area curves Curves illustrating the relationship between number of species and area or amount of habitat.

specificity (of habitat) The range of ecological conditions, both physical and biotic, within which a species can survive. The terms generalised (or wide) and specialised (or narrow) describe extremes of habitat specificity.

spoil heaps Heaps of waste material produced by an excavation.

stags Large old dead trees that remain standing in a forest.

standard deviation A measure of dispersion of a frequency distribution, obtained by extracting the square root of the mean of the squared deviations of the observed values from their mean.

State forest In Australia, State forests comprise publicly owned land that is managed for the conservation of flora and fauna; for the protection of water catchments and water quality; for the provision of timber and other forest products on a sustainable basis; for the protection of landscape, archaeological and historical values; and to provide recreational and educational opportunities.

statistical power The likelihood of correctly rejecting the null hypothesis. Statistical power depends on the sample size used in the test, the effect size to be detected, and the variability inherent in the data.

stepwise regression A procedure for building multiple regression models in which alternative models are explored, and terms included or excluded depending on their explanatory power.

stigma The terminal cells of a pistil of a flower that receive and recognise pollen grains during
interactions that can lead to fertilisation.

stochastic perturbation An unpredictable change in a parameter, environmental variable or ecosystem process.

stochastic Any random process; a process governed by probabilistic laws.

stomata Specialised pores in the epidermis of leaves and stems that allow uptake of carbon dioxide from the atmosphere for photosynthesis and through which transpiration occurs.

stratified A sample that is not drawn at random from the whole population, but is drawn separately from a number of disjoint levels of the population in order to ensure a more
representative sample.

stratum A layer in a plant community or formation produced by the occurrence of an aggregation of plants of similar growth habit at approximately the same level, or a sampling unit in a stratified random sample.

studbook A record of all the captive individuals of a given species that is often used in managing captive breeding stock.

subculturing To transfer a small portion of, for example, bacterial culture, onto a new medium.

subpopulation A subgroup in a population; geographically isolated groups of a population.

subspecies A taxonomic term referring to a group of individuals within a species that have certain distinguishing characteristics and that form a breeding group.

subtropical The region lying between the tropics and the temperate lands.

succession Process of replacement over time of one ecological community by another.

surrogate An entity (e.g. a plant community, soil type, or climate domain) that is thought to act as an indicator of another ecological characteristic (e.g. the occurrence of a particular species or community), or an aggregate measure (e.g. overall species diversity).

sustainability Refers to management practices that attempt to meet economic objectives in ways that do not degrade the environment; meeting the needs of the present without compromising the ability of future generations to meet their own needs; using, conserving and enhancing resources so that ecological processes in ecosystems are maintained.

symbiotic Describes a situation where two organisms live together in a close relationship that is beneficial to at least one of them.

synonym Redundant names used to describe a species, which result from a taxon being described more than once, usually by different people.

synonymy The character of being synonymous, equivalence; in taxonomy, a specific epithet ascribed to a species that has previously been formally recognised.

systematic pressure In population viability analysis, a process that drives a species into decline and eventual extinction, more or less deterministically and continuously.

systematics The study of systems and taxonomy, including identification, the practice of classification, and nomenclature.

t-distribution A continuous distribution obtained from a ratio of a normal and a chi-square variable, approximating the normal distribution for a small population.

tailings dams Dams designed to retain wastes left over after processes such as ore crushing on a mine site.

taxon The organisms comprising a particular taxonomic entity (e.g. class, family, genus or species).

taxonomy Methods and principles used for the classification of organisms.

temperate Having a climate intermediate between tropical and polar; moderate to mild in temperature.

test endpoints Measures of environmental impact, usually of a chemical or toxicant, made on specific ecological variables such as the growth rate of plants or the survival of juvenile fish, usually under laboratory conditions.

tetrapod Vertebrate with four limbs with separate fingers and toes.

theory of island biogeography A paradigm used to explain species-area relationships for islands, based on island size and island isolation from mainland sources of colonists. The balance between extinction and colonisation is a function of island size and isolation, resulting in an equilibrium number of species.

thermal pollution Entrainment of relatively warm air or water into the environment, usually from industrial or urban sources, with detrimental consequences for biota or ecological processes.

thinning The removal of part of a forest stand or crop, with the aim of increasing the growth rate or health of retained plants.

threatening process A process that detrimentally impacts on, or could detrimentally impact on, the survival, abundance, distribution, or potential for evolutionary development of a native species or ecological community.

threshold suitability values Values below which a species simply cannot exist and above which an increase in a given measure will make little or no difference in habitat suitability.

tolerance The ability of a species to withstand a range of conditions.

trans-equatorial migration Movement in which animals cross the equator on a seasonal basis to breed or feed.

transgene Any gene introduced into a plant or animal using the techniques of genetic engineering.

transgenic variety A variety that contains a gene or genes from an unrelated organism (e.g. a bacterial gene in a plant) that could not be transferred to the species using traditional breeding techniques; an organism containing a gene that has been artificially transferred from a member of another species.

translocation The transfer of plants and animals from one part of their range to another.

transpiration Loss of water from a plant by evaporation through stomata in leaves.

trophic level Position of an organism in the food chain as assessed by the number of energy transfer steps to reach that level.

tropical rainforest The richest biome in terms of both plant and animal diversity. Neither water nor low temperatures are limiting for photosynthesis in tropical rainforest. Trees are evergreen with leathery leaves, the herbaceous layer is poorly developed because of low light, there are many vines and epiphytes, and there is little litter because nutrient cycling rates are high.

true mean (arithmetic mean) Calculated by dividing the sum of a set of observations by the number of observations.

turbidity Level of cloudiness of a suspension.

two-tailed test A statistical test concerned with the hypothesis that an observed value of a test statistic differs significantly from a given value, where an error in either direction is relevant.

type I error Error made when a true null hypothesis is rejected, to conclude incorrectly that there is a difference among the groups examined. In the context of environmental impact assessment, a type I error would involve the study leading to a conclusion that an impact has occurred when in fact it has not.

type II error Error made when a false null hypothesis is accepted, to conclude incorrectly that there is no difference among the groups examined. A type II error would involve an environmental impact assessment leading to a conclusion that an impact has not occurred when in fact it has.

ubiquitous Having or seeming to have the ability to be everywhere at once.

ubiquity (with reference to sampling) The frequency of species in samples (i.e. the proportion of samples in which a species is found).

umbrella species A species whose protection would automatically also provide protection for other species or even an entire community.

uncertainty Quantitative and qualitative unpredictability around judgements, predictions, estimates or decisions.

under-fished A descriptive term for a fishery that can sustain higher catches than those being presently taken.

understorey A general term for the plants (usually shrubs) of a community occurring at levels lower than the top stratum (usually trees).

ungulate Hoofed animal, usually herd-forming and adapted for grazing.

utilitarian view A philosophy on valuing biological resources from the perspective of producing products of use and value for humans.

vagrant species Species that are found within an area but are not permanent members of the assemblage of species at that point, or species that do not breed and do not have self-sustaining populations within the area of interest.

validation A test of a predictive model involving a comparison of predictions with new data.

variability Quantitative unpredictability in natural processes, or difference among measurements.

variables Characteristics of a population that differ from individual to individual, which can have numbers or values assigned to them so they can be measured. Variables can be continuous (e.g. values for daily temperature), or they can be categorical, where values are assigned to discrete categories (sometimes called factors; e.g. classes of rock types).

variance A measure of the dispersion of a random variable, obtained by taking the square of the standard deviation.

vascular plant Plant that possesses conducting tissue for the transmission or circulation of fluid (i.e. ferns, gymnosperms and angiosperms).

vegetation mapping Mapping involving the interpretation of aerial photographs, stratification of the landscape into relatively uniform units, ground sampling for vegetation structure and floristic composition (usually for vascular plants), classification of ground samples, description of map units, and, finally, detailed mapping of each of the vegetation types.

vertebrate Any animal of a subphylum characterised by a bony skeleton and a well-developed brain.

volcanism The processes that collectively result in the formation of volcanoes and their products.

vulnerable A classification of risk less than endangered, but usually suggesting that the species will require active management action if it is to persist.

watertable The boundary between the unsaturated and saturated soil.

weed A plant that is considered undesirable by humans because it threatens the persistence of native plants or competes with other plants that are used for food or fibre, or to graze domestic livestock.

wetlands Areas of marsh, fen, peatland or water, whether natural or artificial, permanent or temporary, with water that is static or flowing, fresh, brackish or salt, including areas of marine waters, the depth of which at low tide does not exceed 6 metres.

Wheat belt An area in which Wheat is the chief agricultural product.

wilderness A natural, uncultivated, uninhabited region with non-consumptive values such as aesthetic and spiritual aspects, or scientific, educational, recreational or water resource uses.

wildfire An unplanned fire.

windrows A row or line of heaped waste wood, usually stacked prior to burning in harvested forests or on land cleared for agriculture.

woodland Land that is mostly covered with woods or dense growths of trees and shrubs.

World Heritage Areas Reserve areas designated by the Convention Concerning the Protection of the World Cultural and Natural Heritage as areas of 'outstanding universal value'.

zygote Diploid cells resulting from the fusion of male and female gametes.

Bibliography

Abbott, R.J. (1992). Plant invasions, interspecific hybridisation and the evolution of new plant taxa. *Trends in Ecology and Evolution,* **7**, 401–405.

Abbott, R.J. (1994). Ecological risks of transgenic crops. *Trends in Ecology and Evolution,* **9**, 280–282.

Abensperg-Traun, M. and Smith, G.T. (2000). How small is too small for small animals? Four terrestrial arthropod species in different-sized remnants in agricultural Western Australia. *Biodiversity and Conservation,* **8**, 709–726.

Aberg, J., Swenson, J.E. and Angelstam, P. (1995). The effect of matrix on the occurrence of Hazel Grouse (*Bonasa bonasia*) in isolated habitat fragments. *Oecologia,* **103**, 265–269.

Acaba, Z., Jones, H., Preece, R., Rish, S., Ross, D. and Daly, H. (2000). *The Effects of Large Reservoirs on Water Temperature in Three NSW Rivers Based on the Analysis of Historical Data.* Centre for Natural Resources, NSW Department of Land and Water Conservation.

Accad, A., Neldner, V.J., Wilson, B.A. and Niehus, R.E. (2001). Remnant vegetation in Queensland: analysis of pre-clearing, remnant 1997–1999 regional ecosystem information. *Queenland Herbarium and Environmental Protection Agency, Brisbane.*

Adamson, D.A. and Fox, M.D. (1982). Change in Australian vegetation since European settlement. pp. 109–146. In: *A History of Australasian Vegetation.* Smith, J.M.B. (Ed). McGraw-Hill, Sydney.

AFFA (Australian Agriculture, Fisheries and Forestry) (2003a). *At a glance 2003.* Australian Agriculture, Fisheries and Forestry, Commonwealth of Australia, Canberra.

AFFA (Australian Agriculture, Fisheries and Forestry) (2003b). *Fisheries Status Report 2000–2001 – Overview.* Australian Agriculture, Fisheries and Forestry, Commonwealth of Australia, Canberra.

Agee, J.K. (1999). Fire effects on landscape fragmentation in interior west forests. pp. 43–60. In: *Forest Fragmentation. Wildlife Management Implications.* Rochelle, J.A., Lehmann, L.A. and Wisniewski, J. (Eds). Brill, Leiden, Germany.

Agee, J.K. and Huff, M.H. (1987). Fuel succession in a Western Hemlock/Douglas Fir forest. *Canadian Journal of Forest Research,* **17**, 697–704.

Agius, J. (2001). Biodiversity credits: creating missing markets for biodiversity. *Environmental Planning and Law Journal,* **18**, 481–484.

Aguirre, A.A., Ostfeld, R.S., Tabor, G.M., House, C. and Pearl, M.C. (Eds). (2002). *Conservation Medicine: Ecological Health in Practice.* Oxford University Press, Oxford.

Akçakaya, H.R. (1990). A method for simulating demographic stochasticity. *Ecological Modelling,* **54**, 133–136.

Akçakaya, H.R. (1992). The scientific figure processor. *Quarterly Review of Biology,* **67**, 99–100.

Akçakaya, H.R. (2002). Estimating the variance of survival rates and fecundities. *Animal Conservation,* **5**, 333–336.

Akçakaya, H.R. and Burgman, M.A. (1995). PVA in theory and practice. *Conservation Biology,* **9**, 705–707.

Akçakaya, H.R. and Ferson, S. (1992). *RAMAS/space.* Applied Biomathematics, Setauket, New York.

Akçakaya, H.R., Burgman, M.A. and Ginzburg, L. (1997). *Applied population ecology: principles and computer exercises with RAMAS/EcoLab 1.0.* Applied Biomathematics, Setauket, New York.

Akçakaya, H.R., Ferson, S., Burgman, M.A., Keith, D.A., Mace, G.M. and Todd, C.R. (2000). Making consistent IUCN classifications under uncertainty. *Conservation Biology* **14**, 1001–1013.

Akçakaya, H.R., McCarthy, M.A. and Pearce, J.L. (1995). Linking landscape data with population viability analysis – management options for the Helmeted

Honeyeater (*Lichenostomus melanops cassidix*). *Biological Conservation,* **73**, 169–176.

Aldrich, P.R. and Hamrick, J.L. (1998). Reproductive dominance of pasture trees in a fragmented tropical forest mosaic. *Science,* **281**, 103–105.

Alexandra, J. (1995). Bush bashing. *Habitat Australia,* **23**, 19–26.

Alford, R.A. and Richards, S.J. (1999). Global amphibian declines: a problem in applied ecology. *Annual Reviews of Ecology and Systematics,* **30**, 133–165.

Alford, R.A., Dixon, P.M. and Pechmann, J.H. (2001). Global amphibian population declines. *Nature,* **412**, 499–500.

Allan, B.F., Keesing, F. and Ostfeld, R.S. (2003). Effect of forest fragmentation on Lyme disease risk. *Conservation Biology,* **17**, 267–274.

Allan, G.E. and Southgate, R. (2002). Fire regimes in the spinifex landscapes of Australia. pp. 145–176. In: *Flammable Australia. The Fire Regimes and Biodiversity of a Continent.* Bradstock, R.A., Williams, J.E. and Gill, A.M. (Eds). Cambridge University Press, Melbourne.

Allee, W.C., Emerson, A.E., Park, O., Park, T. and Schmidt, K.P. (1949). *Principles of Animal Ecology.* Saunders, Philadelphia.

Ambrose, G.J. (1982). An ecological and behavioural study of vertebrates using hollows in Eucalypt branches. PhD Thesis, La Trobe University, Bundoora, Melbourne.

Amor, R.L. and Richardson, R.G. (1980). The biology of Australian weeds. 2. *Rubus fruticosus* L. agg. *J. Aust. Inst. Agr. Sci.,* **1980**, 87–97.

Amos, B. and Hoelzel, A.R. (1992). Application of molecular genetic techniques to the conservation of small populations. *Biological Conservation,* **61**, 133–144.

Amos, N., Kirkpatrick, J.B. and Giese, M. (1993). *Conservation of Biodiversity, Ecological Integrity and Ecologically Sustainable Development.* Australian Conservation Foundation and World Wide Fund for Nature, Victoria.

Anagnostakis, S.A. (1972). Biological control of chestnut blight. Science, **215**, 446–71.

ANCA (1994). (Australian Nature Conservation Agency) *Australian Nature Conservation Agency: annual report.* Australian Government Publishing Service, Canberra. 1993–94.

ANCA (1995a). *Overview of background information for kangaroo management in Australia.* Population Assessment Unit, The Australian Nature Conservation Agency, Canberra. Unpublished Report. October 1995.

ANCA (1995b). Summary report. *Proceedings of the Conference Workshop on Encounters with Whales 95,* Hervey Bay, Queensland, 26–30 July 1995. Australian Nature Conservation Agency, Canberra.

Andelman, S.J. and Fagan, W.F. (2000). Umbrellas and flagships: efficient conservation surrogates or expensive mistakes? *Proceedings of the National Academy of Science,* **97**, 5954–5959.

Andelman, S.J. and Willig, M.R. (2003). Present patterns and future prospects for biodiversity in the Western Hemisphere. *Ecology Letters,* **6**, 818–824.

Andersen, A.A., Hoffman, B.D. and Somes, J. (2003). Ants as indicators of minesite restoration: community recovery at one of eight rehabilitation sites in central Queensland. *Ecological Management and Restoration (Supplement),* **4**, S12–19.

Andersen, A.N. (1990). The use of ant communities to evaluate change in Australian terrestrial ecosystems: a review and recipe. *Proceedings of the Ecological Society of Australia,* **16**, 317–357.

Andersen, A.N. (1991). *The ants of southern Australia: A guide to the Bassian fauna.* CSIRO, Canberra.

Andersen, A.N. (1993). Ants as indicators of restoration success at a uranium mine in tropical Australia. *Restoration Ecology,* **12**, 156–167.

Andersen, A.N. and Majer, J.D. (2004). Ants show the way down under: invertebrates as bioindicators in land management. *Frontiers in Ecology and Environment,* **2**, 291–298.

Andersen, A.N., Cook, G.D. and Williams, R.J. (Eds) (2003). *Fire in tropical savannas. The Kapalga Experiment.* Springer, New York.

Andersen, A.N., Cook, G.D., Corbett, L.K., Douglas, M.M., Eager, R.W., Russell-Smith, J., Setterfield, S.A., Williams, R.J. and Woinarski, J.C. (2005). Fire frequency and biodiversity conservation in Australian tropical savannas: implications from the Kapalga fire experiment. *Austral Ecology,* **30**, 155–167.

Andersen, L.E., Granger, C.W., Reis, E., Weinhold, D. and Wunder, S. (2002). *The dynamics of deforestation and economic growth in the Brazilian Amazon.* Cambridge University Press, Cambridge.

Anderson, J.L. (2001). Stone-age minds at work on 21st century science. *Conservation Biology in Practice,* **2**, 20–27.

Anderson, S. (1994). Area and endemism. *The Quarterly Review of Biology,* **69**, 451–471.

Andow, D.A. and Ives, A.R. (2002). Monitoring and adaptive resistance management. *Ecological Applications,* **12**, 1378–1390.

Andreassen, H.P., Halle, S. and Ims, R. (1996). Optimal width of movement corridors for root voles: not too narrow and not too wide. *Journal of Applied Ecology,* **33**, 63–70.

Andrén, H. (1992). Corvid density and nest predation in relation to forest fragmentation: a landscape perspective. *Ecology*, **73**, 794–804.

Andrén, H. (1994). Effects of habitat fragmentation on birds and mammals in landscapes with different proportions of suitable habitat: a review. *Oikos*, **71**, 355–366.

Andrén, H. (1996). Population responses to habitat fragmentation: statistical power and the random sample hypothesis. *Oikos*, **76**, 235–242.

Andrén, H. (1997). Habitat fragmentation and changes in biodiversity. *Ecological Bulletin*, **46**, 171–181.

Andrén, H. (1999). Habitat fragmentation, the random sample hypothesis and critical thresholds. *Oikos*, **84**, 306–308.

Andrén, H. and Angelstam, P. (1988). Elevated predation rates as an edge effect in habitat islands: experimental evidence. *Ecology*, **6**, 544–547.

Andrén, H. and Delin, A. (1994). Habitat selection in the Eurasian red squirrel, *Sciurus vulgaris*, in relation to forest fragmentation. *Oikos*, **70**, 43–48.

Andrén, H., Delin, A. and Seiler, A. (1997). Population response to landscape changes depends on specialization to different landscape elements. *Oikos*, **80**, 193–196.

Andrew, D. (1993). Birds. Our natural bio-indicators. *Trees and Natural Resources*, **December 1993**, 27–29.

Andrew, N., Rodgerson, L. and York, A. (2000). Frequent fuel-reduction burning: the role of logs and associated leaf litter in the conservation of ant biodiversity. *Austral Ecology*, **25**, 99–107.

Andrewartha, H.G. and Birch, L.C. (1954). *The distribution and abundance of animals*. University of Chicago Press, Chicago.

Angelstam, P. (1996). The ghost of forest past – natural disturbance regimes as a basis for reconstruction for biologically diverse forests in Europe. pp. 287–337. In: *Conservation of faunal diversity in forested landscapes*. DeGraaf, R.M. and Miller, R.I. (Eds). Chapman and Hall, London.

Angelstam, P. and B. Pettersson. (1997). Principles of Swedish forestry biodiversity management. *Ecological Bulletin*, **46**, 91–203.

Angelstam, P., Boutin, S., Schmiegelow, F., Villard, M., Drapeau, P., Host, G., Innes, J., Isachenko, G., Kuuluvainen, T., Monkonnen, M., Niemi, G., Roberge, J., Spence, J. and Stone, D. (2004). Targets for boreal forest biodiversity conservation – a rationale for macroecological research and internal co-operation. *Ecological Bulletins* (in press).

Angermeier, P.I. (1995). Ecological attributes of extinction-prone species: loss of freshwater fishes of Virginia. *Conservation Biology*, **9**, 143–158.

Angulo, E. (2001). When DNA research menaces diversity. *Nature*, **410**, 739.

Angulo, E. and Cooke, B. (2002). First synthesize new viruses then regulate their release? The case of the wild rabbit. *Molecular Ecology*, **11**, 2703–2709.

Animal and Plant Control Commission (1999). *Olive Policy*. Animal and Plant Control Commission, Adelaide.

Anonymous (2001). *Paddock trees. Who'll miss them when they are gone?* Pamphlet produced by NSW National Parks and Wildlife Service, NSW Department of Land and Water Conservation, and Greening Australia.

Anonymous (a) (1995). *National Forest Conservation Reserves, Commonwealth proposed criteria – a position paper*. Australian Government Publishing Service, Canberra. July, 1995.

Anonymous (b) (1995). *Biodiversity: the UK Steering Group report*. Volume 2. Action Plans. HMSO, London.

ANPWS (1989). *Whale watching guidelines*. Australian National Parks and Wildlife Service, Canberra.

Arborvitae (1995). Arborvitae. *IUCN/WWF Forest Conservation Newsletter*, **1** (September 1995), p. 5.

Armbruster, P. and Lande, R. (1993). A population viability analysis for African Elephant (*Luxodonta africana*): How big should reserves be? *Conservation Biology*, **7**, 602–610.

Armesto, J.J., Rozzi, R., Smith-Ramirez, C. and Arroyo, M.T. (1998). Conservation targets in South American temperate forests. *Science*, **282**, 1271–1272.

Armstrong, D.P., Soderquist, T. and Southgate, R. (1995). Designing experimental reintroductions as experiments. pp. 27–29. In: *Reintroduction biology of Australasian fauna*. Serena, M. (Ed). Surrey Beatty, Chipping Norton, Australia.

Arnold, G.W., Steven, D.E. and Weeldenburg, J.R. (1993). Influences of remnant size, spacing pattern and connectivity on population boundaries and demography in *Euros Macropus robustus* living in a fragmented landscape. *Biological Conservation*, **64**, 219–230.

Arnold, R.J. and Midgley, S.J. (1995). Conserving biodiversity – the work of the Australian Tree Seed Centre. *Commonwealth Forestry Review*, **74**, 121–128.

Aronson, R.B. and Precht, W.F. (1995). Landscape pattern of coral reef diversity: A test of the intermediate disturbance hypothesis. *Journal of Experimental and Marine Biology and Ecology*, **192**, 1–14.

Arrhenius, O. (1921). Species and area. *Journal of Ecology*, **9**, 95–99.

Arthington, A.H. (1998). Comparative evaluation of environmental flow assessment techniques: review of holistic methodologies. *LWRRDC Occasional Paper,* **27/98.** Canberra.

Arthington, A.H. and Pusey, B.L. (2003). Flow restoration and protection in Australian rivers. *River Research and Applications,* **19,** 377–395.

Ås, S. (1999). Invasion of matrix species in small habitat patches. *Conservation Ecology,* **3.** [online] URL: http://www.ecologyandsociety.org/vol3/iss1/art1/

Ashton, D.H. (1975). The root and shoot of *Eucalyptus regnans* F. Muell. *Australian Journal of Botany,* **23,** 867–887.

Ashton, D.H. (1976). The development of even-aged stands of *Eucalyptus regnans* F. Muell. in central Victoria. *Australian Journal of Botany,* **24,** 397–414.

Ashton, D.H. (1981a). The ecology of the boundary between *Eucalyptus regnans* F. Muell. and *E. obliqua* L.Herit in Victoria. *Proceedings of the Ecological Society of Australia,* **11,** 75–94.

Ashton, D.H. (1981b). Fire in tall open forests (wet sclerophyll forests). In: *Fire and the Australian Biota.* Gill, A.M., Groves R.H. and Noble, I.R. (Eds). Australian Academy of Science, Canberra.

Ashton, D.H., (1986). Ecology of bryophytic communities in mature *Eucalyptus regnans* F. Muell. forest at Wallaby Creek, Victoria. *Australian Journal of Botany,* **34,** 107–129.

Ashton, D.H. and Martin, D.G. (1996). Regeneration in a pole-stage forest of *Eucalyptus regnans* subjected to different fire intensities in 1982. *Australian Journal of Botany,* **44,** 395–410.

Ashton, D.H. and Sandiford, E.M. (1988). Natural hybridisation between *Eucalyptus regnans* F. Muell and *E. macrorhynca* F.Muell. in the Cathedral Range, Victoria. *Australian Journal of Botany,* **36,** 1–22.

Aslin, H. (1991). Towards a national biological diversity strategy for Australia. *Proceedings of the Conference on Conservation Biology in Australia and Oceania,* University of Queensland, Brisbane, 30 Sept.–4 Oct. 1991.

Attiwill, P., Florence, R., Hurditch, W.E. and Hurditch, W.J. (1994). *The burning continent. Forest ecosystems and fire management in Australia.* Institute of Public Affairs, Melbourne, September 1994.

Attiwill, P.M. (1994a). The disturbance of forest ecosystems: The ecological basis for conservative management. *Forest Ecology and Management,* **63,** 247–300.

Attiwill, P.M. (1994b). Ecological disturbance and the conservative management of eucalypt forests in Australia. *Forest Ecology and Management,* **63,** 301–346.

AUSLIG (1990). *Atlas of Australian resources – vegetation.* Commonwealth Government Printer, Canberra.

Austin, J.J. and Parkin, D.T. (1995). Female-specific restriction fragments revealed by DNA fingerprinting and implications for extra-pair fertilisations in the Short-tailed Shearwater (*Puffinus tenuirostris*: Procellariiformes: Procellariidae). *Australian Journal of Zoology,* **43,** 443–447.

Austin, M.P. (1985). Continuum concept, ordination methods and niche theory. *Annual Review of Ecology and Systematics,* **16,** 39–61.

Austin, M.P. (1999). A silent clash of paradigms: some inconsistencies in community ecology. *Oikos,* **86,** 170–178.

Austin, M.P. and Heyligers, P.C. (1989). Vegetation survey design for conservation: Gradsect sampling of forests in north-eastern New South Wales. *Biological Conservation,* **50,** 13–32.

Austin, M.P. and Margules, C.R. (1986). Assessing representativeness. pp. 45–67. In: *Wildlife conservation evaluation.* Usher, M.B. (editor). Chapman and Hall, London.

Austin, M.P., Cawsey, E.M., Baker, B.L., Yialeloglou, M.M., Grice, D.J. and Briggs, S.V. (2000). *Predicted vegetation cover in the Central Lachlan Region.* Report and Appendices. Project conducted under the Bushcare program of the Natural Heritage Trust. CSIRO. Natural Heritage Trust. September 2000.

Austin, M.P., Meyers, J.A. and Doherty, M.D. (1994a). *Data capability, sub-project 3. Modelling of landscape patterns and processes using biological data.* Division of Wildlife and Ecology, CSIRO, Canberra.

Austin, M.P., Nicholls, A.O. and Margules, C.R. (1990). Measurement of the realised qualitative niche: Environmental niches of five *Eucalyptus* species. *Ecological Monographs,* **60,** 161–177.

Austin, M.P., Nicholls, A.O., Doherty, M.D. and Meyers, J.A. (1994b). Determining species response functions to an environmental gradient by means of ß-function. *Journal of Vegetation Science,* **5,** 215–228.

Australian and New Zealand Environment and Conservation Council (2001). *Review of the national strategy for the conservation of Australia's biology diversity.* Commonwealth of Australia, June, 2001.

Australian Bureau of Statistics (ABS) (1992). *Australia's environment. Issues and facts.* Commonwealth of Australia, Canberra. *ABS Catalogue* No. 4140.0. 354.

Australian Bureau of Statistics (1997). *Fish Account.* Australian Bureau of Statistics, Canberra.

Australian Bureau of Statistics (1999a). *Themes. Environment, land and soil, agriculture*. Australian Bureau of Statistics, Canberra. [online] URL: http://www.abs.gov.au/websitedbs/c311215.nsf/0/7C9D820AA44A645ACA2569E70011E273?Open

Australian Bureau of Statistics (1999b). *Environmental Issues – People's Views and Practices, March 1998*. Australian Bureau of Statistics, Canberra.

Australian Bureau of Statistics (2000a). *Australia now, a statistical profile, tourism, international inbound tourism*. Australian Bureau of Statistics, Canberra. [online] URL: http://www.abs.gov.au/Ausstats/abs@.nsf/94713ad445ff1425ca25682000192af2/0a41c3d685211114ca256f7200832f02!OpenDocument

Australian Bureau of Statistics (2000b). *Australia now, a statistical profile, tourism, domestic tourism*. Australian Bureau of Statistics, Canberra. [online] URL: http://www.abs.gov.au/Ausstats/abs@.nsf/94713ad445ff1425ca25682000192af2/b546486cdefc4bb2ca256f7200832f36!OpenDocument

Australian Bureau of Statistics (2001). *Australia's environment: Issues and Trends*. Australian Bureau of Statistics, Canberra.

Australian Bureau of Statistics (2003). *Measuring Australia's economy. Section 5. Production and industry tourism*. [online] URL: http://www.abs.gov.au/Ausstats/abs@.nsf/2.7.1?OpenView

Australian Conservation Foundation (2001). *Clearing the common wealth. Land clearing in commonwealth countries*. Australian Conservation Foundation, Melbourne.

Australian Government Publishing Service (1992). *Intergovernmental Agreement on the Environment (IGAE)*. Commonwealth of Australia, Canberra. May 1992.

Australian Government Survey Organisation (1999). *Australia's identified mineral resources*. Australian Government Survey Organisation, Canberra.

Australian Greenhouse Office (2000). *National Greenhouse Strategy, 2000 Progress Report – Implementing the Strategic Framework for Advancing Australia's Greenhouse Response*. Australian Greenhouse Office, Canberra.

Australian Greenhouse Office (2002). *National Greenhouse Gas Inventory 2000. Australia's net greenhouse gas emissions*. Australian Greenhouse Office, Canberra. [online] URL: http://www.greenhouse.gov.au/inventory/

Australian Greenhouse Office (2005). National Greenhouse Gas Inventory 2003. Australian Greenhouse Office, Canberra.

Australian Office of the Gene Technology Regulator (2003). *Office of the Gene Technology Regulator website*. Department of Health and Ageing, Canberra. [online] URL: http://www.ogtr.gov.au/

Avise, J.C. (1989). A role for molecular genetics in the recognition and conservation of endangered species. *Trends in Ecology and Evolution*, **4**, 279–281.

Avise, J.C. and Aguardro, C.F. (1982). A comparative study of genetic distances in the vertebrates. In: *Evolutionary biology*. Hecht, M.K., Wallace, B. and Prance, G.T. (Eds). Vol. 15, Plenum Press, New York.

Avise, J.C. and Hamrick, J.L. (Eds) (1996). *Conservation genetics*. Chapman and Hall, New York.

Avise, J.C., Ball, R.M. and Arnold, J. (1988). Current versus historical population sizes in vertebrate species with high gene flows: A comparison based on mitochondrial DNA lineages and inbreeding theory for nutral mutations. *Molecular Biology and Evolution*, **5**, 331–344.

Awimbo, J.A., Norton, D.A. and Overmars, F.B. (1996). An evaluation of representativeness for nature conservation, Hokitika Ecological District, New Zealand. *Biological Conservation*, **75**, 177–186.

Backhouse, D. and Burgess, L.W. (2002). Climatic analysis of the distribution of *Fusarium graminearum*, *F. pseudograminearum* and *F. culmorum* on cereals in Australia. *Australasian Plant Pathology*, **31**, 321–327.

Backhouse, G.N. (1987). Management of remnant habitat for the conservation of the Helmeted Honeyeater *Lichenostomus melanops cassidix*. pp. 287–294. In: *Nature Conservation: The Role of Remnants of Native Vegetation*. Saunders, D.A., Arnold, G.W., Burbidge A.A. and Hopkins, A.J. (Eds). Surrey Beatty, Chipping Norton, Sydney.

Backhouse, G.N., Clark, T.W. and Reading, R. (1995). Reintroductions for recovery of the Eastern Barred Bandicoot *Parameles gunnii* in Victoria, Australia. pp. 209–218. In: *Reintroduction biology of Australasian fauna*. Serena, M. (Ed). Surrey Beatty, Chipping Norton, Australia.

Baillie, S.R., Sutherland, W.J., Freeman, S.N. Gregory, R.D. and Paradis, E. (2000). Consequences of large-scale processes for the conservation of birds. *Journal of Applied Ecology*, **37** (Supplement 1), 88–102.

Baker, J. (1997). The decline, response to fire, status and management of the Eastern Bristlebird. *Pacific Conservation Biology*, **3**, 235–243.

Baker, J. (1998). Ecotones and fire and the conservation of the Eastern Bristlebird. PhD thesis, University of Wollongong, Wollongong.

Baker, J. and Whelan, R. (1994). Ground parrots and fire at Barren Grounds, New South Wales: a long-term study and an assessment of management implications. *Emu*, **94**, 300–304.

Baker, J., French, K. and Whelan, R.J. (2002). The edge effect and ecotonal species: bird communities across a natural edge in southeastern Australia. *Ecology*, **83**, 3048–3059.

Baker, J.R. (2000). The Eastern Bristlebird: cover dependent and fire sensitive. *Emu*, **100**, 286–298.

Baker, R. (2003). Yanyuwa classical burning regimes, indigenous science and cross-cultural communication. pp. 198–204. In: *Australia Burning. Fire Ecology, Policy and Management Issues*. Cary, G., Lindenmayer, D.B. and Dovers, S. (Eds). CSIRO Publishing, Melbourne.

Baker, W.L. (1992). The landscape ecology of large disturbances in the design and management of nature reserves. *Landscape Ecology*, **7**, 181–194.

Ball, S., Lindenmayer, D.B. and Possingham, H.P. (2003). The predictive accuracy of viability analysis: a test using data from two small mammal species in a fragmented landscape. *Biodiversity and Conservation*, **12**, 2393–2413.

Balmford, A. (1996). Extinction filters and current resilience: the significance of past selection pressures for conservation biology. *Trends in Ecology and Evolution*, **11**, 193–196.

Bangs, E.E. and Fritts, S.H. (1996). Reintroducing the Gray Wolf to central Idaho and Yellowstone National Park. *Wildlife Society Bulletin*, **24**, 402–413.

Banister, K. (1992). Fishes. pp. 116–135. In: *Global diversity: status of the earth's living resources*. Centre, W.C.M. (Ed). Chapman and Hall, London.

Banks, J.C. (1982). The use of dendrochronology in the interpretation of the dynamics of the Snow Gum forest. PhD thesis, Department of Forestry, The Australian National University, Canberra.

Banks, J.C. (2002). Wollemi Pine: tree find of the 20th century. pp. 85–89. In: *Australia's Ever-changing forests. V. Proceedings of the Fifth National Conference on Australian Forest History*. Dargarvel, J., Gaughwin, D. and Libbis, B. (Eds). Centre for Resource and Environmental Studies, The Australian National University, Canberra.

Banks, J.C.G. (1997). Tree ages and ageing in yellow box. pp. 35–47. In: *Australia's Ever-changing Forests III*. Dargavel, J. (Ed). Australian National University, Canberra.

Banks, S.C., Hoyle, S.D., Horsup, A., Sunnicks, P., Witton, A. and Taylor, A.C. (2003). Demographic monitoring of an entire species by genetic analysis of non-invasively collected material. *Animal Conservation*, **6**, 101–108.

Baptista, L.F. and Gaunt, S.L.L. (1997). Bioacoustics as a tool in conservation studies. pp. 212–213. In: *Behavioral Approaches to Conservation in the Wild*. Clemmons, J.R. and Buchholz, R. (Eds). Cambridge University Press, Cambridge.

Barbour, M.S. and Litvaitis, J.A. (1993). Niche dimensions of New England cottontails in relation to habitat patch size. *Oecologia*, **95**, 321–327.

Barbour, R.C., Potts, B.M. and Vaillancourt, R.E. (2003). Gene flow between introduced and native Eucalyptus species: exotic hybrids are establishing in the wild. *Australian Journal of Botany*, **51**, 429–439.

Barbour, R.C., Potts, B.M., Vaillancourt, R.E., Tibbits, W.N. and Wiltshire, W.E. (2000). Hybridisation between plantation and native eucalypts in Tasmania. pp. 395–399. In: *Hybrid Breeding and Genetics of Forest Trees*. Proceedings of QFRI/CRC-SPF Symposium 9–14 April 2000. Dungey, H.S., Dieters, M.J. and Nikles, D.G. (Eds). Department of Primary Industries, Brisbane, Queensland.

Barclay, S.D., Ash, J.E. and Rowell, D.M. (2000). Environmental factors influencing the presence and abundance of a log-dwelling invertebrate *Euperipatodies rowelli* (Onychophora: Peripatopsidae). *Journal of Zoology*, **250**, 425–436.

Barclay, S.D., Rowell, D.M. and Ash, J.E. (1999). Phermonally mediated colonization patterns in Onychophora. *Journal of Zoology*, **250**, 437–446.

Bardgett, R.D. and Cook, R. (1998). Functional aspects of soil animal diversity in agricultural grasslands. *Applied Soil Ecology*, **10**, 263–276.

Barnes, R.L. (2000). Why the American sobean association supports transgenic soybeans. *Pest Management Science*, **56**, 580–583.

Barnett, J. (1994). Fire protection and national parks. *Park Watch*, **March 1994**, 14–15.

Barnett, J.L., How, R.A. and Humphreys, W.F. (1978). The use of habitat components by small mammals in eastern Australia. *Australian Journal of Ecology*, **3**, 277–285.

Barratt, D., Garvey, J. and Chesson, J. (2001). *Implementing selected indicators of marine disturbance in parts of the AEEZ*. Environment Australia, Canberra.

Barrett, G. and Davidson, I. (2000). Community monitoring of woodland habitats – the birds on farms survey. pp. 382–399. In: *Temperate Eucalypt Woodlands in Australia: Biology, Conservation, Management and Restoration*. Hobbs, R.J. and Yates, C.J. (Eds). Surrey Beatty and Sons, Chipping Norton.

Barrett, G.W., Ford, H.A. and Recher. H.F. (1994). Conservation of woodland birds in a fragmented rural landscape. *Pacific Conservation Biology,* **1**, 245–256.

Barrett, G.W., Silcocks, A., Barry, S., Cunningham, R. and Poulter, R. (2003). *The new atlas of Australian brids.* Birds Australia, Melbourne.

Barson, M., Randall, L. and Boardas, V. (2000). Agricultural land cover change project. Bureau of Rural Sciences, Canberra.

Barthell, J.F., Randall, J.M., Thorp, R.W. and Wenner, A.M. (2001). Promotion of seed set in Yellow Star-Thistle by Honeybees: Evidence of an invasive mutualism. *Ecological Applications*, **11**, 1870–1883.

Barton, J.L. and Davies, P.E. (1993). Buffer strips and streamwater contamination by atrazine and pyrethenoids aerially applied to *Eucalyptus nitens* plantations. *Australian Forestry,* **56**, 201–210.

Bascompte, J. and Rodriguez, M.A. (2000). Self-disturbance as a source of spatiotemporal heterogeneity: the case of the Tallgrass Prairie. *Journal of Theoretical Biology*, **204**, 153–164.

Bascompte, J., Melian, C.J. and Sala, E. (2005). Interaction strength combinations and the overfishing of a marine food web. *Proceedings of the National Academy of Sciences*, **102**, 5443–5447.

Basset, Y. and Kitching, R. (1991). Species number, species abundance and body length of arboreal arthropods associated with an Australian rainforest tree. *Ecological Entomology,* **16**, 391–402.

Bates, D. (1938). *The Passing of the Aborigines. A Lifetime Spent Among the Natives of Australia.* Heinemann, Melbourne.

Batley, G.E., Fuhua, C., Brockbank, C.I. and Clegg, K.J. (1989). Accumulation of Tributyltin by the Sydney Rock Oyster, *Saccostrea commercialis. Australian Journal of Marine and Freshwater Research,* **40**, 49–57.

Bauhaus, J. (1999). Silvicultural practices in Australian native forests – an introduction. *Australian Forestry,* **62**, 217–222.

Baum, J.K. and Myers, R.A. (2004). Shifting baselines and the decline of pelagic sharks in the Gulf of Mexico. *Ecology Letters*, **7**, 135–145.

Baur, A. and Baur, B. (1992). Effect of corridor width on animal dispersal: a simulation study. *Global Ecology and Biogeography Letters,* **2**, 52–6.

Baur, B. and Schmid, B. (1996). Spatial and temporal patterns of genetic diversity within species. pp. 169–201. In: *Biodiversity.* Gaston, K.J. (Ed). Blackwell, Oxford.

Baxter, C. and Henderson, R. (2000). A literature summary of the Princess Parrot *Polytelis alexandrae* and a suspected recent breeding event in South Australia. *South Australian Ornithologist*, **33**, 93–108.

Baxter, C.V., Frissell, C.A. and Hauer. F.R. (1999). Geomorphology, logging roads, and the distribution of Bull Trout (*Salvelinus confluentus*) spawning in a forested river basin: implications for management and conservation. *Transactions of the North American Fisheries Society,* **128**, 854–867.

Baxter, M. (2002). *Shorebirds of the Yellow Sea. Importance, threats and conservation status. Wetlands International.* Global Series 6, International Wader Studies 12, Canberra.

Bayer, M. and Porter, W.F. (1988). Evaluation of the guild approach to habitat assessment of forest-dwelling birds. *Environmental Management,* **12**, 797–801.

Bayley, P.B. (1995). Understanding large river-floodplain ecosystems. BioScience **45**: 153–158.

Bayne, E.M. and Hobson, K.A. (1997). Comparing the effects of landscape fragmentation by forestry and agriculture on predation of artificial nests. *Conservation Biology,* **11**, 1418–1429.

Bayne, E.M. and Hobson, K.A. (1998). The effects of habitat fragmentation by forestry and agriculture on the abundance of small mammals in the southern boreal mixed-wood forest. *Canadian Journal of Zoology,* **76**, 62–69.

Beattie, A. and Ehrlich, P.R. (2001). *Wild Solutions.* Melbourne University Press, Melbourne.

Beattie, A.J. (1994). Invertebrates as economic resources. *Memoirs of the Queensland Museum,* **36**, 7–11.

Beattie, A.J. and Oliver, I. (1994). Taxonomic minimalism. *Trends in Ecology and Evolution,* **9**, 488–490.

Beaumont, L.J. and Hughes, L. (2002). Potential changes in the distributions of latitudinally restricted Australian butterfly species in response to climate change. *Global Change Biology*, **8**, 954–971.

Beckage, B. and Stout, J.I. (2000). Effects of repeated burning on species richness in a Florida pine savanna: A test of the intermediate disturbance hypothesis. *Journal of Vegetation Science*, **11**, 113–122.

Bedward, M., Pressey, R.L. and Keith, D.A. (1992). A new approach for selecting fully representative reserve networks: Addressing efficiency, reserve design and land suitability with iterative analysis. *Biological Conservation,* **62**, 115–125.

Beese, W.J., Dunswoth, B.G., Zielke, K. and Bancroft, B. (2003). Maintainaing attributes of old growth forests in coastal B.C. through variable retention. *The Forestry Chronicle*, **79**, 570–578.

Begg, C.B. and Leung, D.H.Y. (2000). On the use of surrogate end points in randomized trials. *J.R. Statist Soc. A.*, **163**, 15–28.

Begon, M., Harper, J.L. and Townsend, C.R. (1990). *Ecology: Individuals, Populations and Communities.* 2nd edition. Blackwell, Oxford.

Begon, M., Harper, J.L. and Townsend, C.R. (1996). *Ecology.* 3rd edition. Blackwell Science, Oxford.

Beheregaray, L.B., Ciofi, C., Geist, D., Gibbs, J.P., Caccone, A. and Powell, J.R. (2003). Genes record a prehistoric volcano eruption in the Galapagos. *Science,* **302,** 75.

Beier, P. (1993). Determining minimum habitat areas and habitat corridors for cougars. *Conservation Biology,* **7,** 94–108.

Beier, P. and Noss, R. (1998). Do habitat corridors provide connectivity. *Conservation Biology,* **12,** 1241–1252.

Beissenger, S.R. and McCullough, D.R. (2002). *Population Viability Analysis.* University of Chicago Press, Chicago.

Beissenger, S.R. and Westphal, M.I. (1998). One the use of demographic models of population viability in endangered species management. *Journal of Wildlife Management,* **62,** 821–841.

Bekessy, S.A., Ennos, R.A., Burgman, M.A., Newton, A.C. and Ades, P.K. (2003). Neutral DNA markers fail to detect genetic divergence in an ecologically important trait. *Biological Conservation,* **110,** 267–275.

Belbin, L. (1992). Comparing two sets of community data: A method for testing reserve adequacy. *Australian Journal of Ecology,* **17,** 255–262.

Bell, A. (1987). Alien dune plants reshape our beaches. *Ecos,* **54,** 3–6.

Bell, P.R.F. and Elemtri, I. (1995). Ecological indicators of large-scale eutrophication in the Great Barrier Reef Lagoon. *Ambio,* **24,** 208–215.

Bellinger, R.G., Ravlin, F.W. and McManus, M.L. (1989). Forest edge effects and their influence on the Gypsy Moth (Lepidoptera: Lymantriidae) egg mass distribution. *Environmental Entomology,* **18,** 840–843.

Belthoff, J.R. and G. Ritchison. (1989). Natal dispersal of Eastern Screech-Owls. *The Condor,* **91,** 254–265.

Bender, D.J., Contreras, T.A. and Fahrig, L. (1998). Habitat loss and population decline: a meta-analysis of the patch size effect. *Ecology* **79**:517–529.

Bender, D.J., Tischendorf, L. and Fahrig, L. (2003). Evaluation of patch isolation metrics for predicting animal movement in binary landscapes. *Landscape Ecology,* **18,** 17–39.

Bengtsson, J., Ahnstrom, J. and Weibull, A-C. (2005). The effects of organic agriculture on biodiversity and abundance: a meta-analysis. *Journal of Applied Ecology,* **42,** 261–269.

Bennett, A.F. (1990a). *Habitat corridors. Their role in wildlife management and conservation.* Department of Conservation and Environment, Melbourne.

Bennett, A.F. (1990b). Land use, forest fragmentation and the mammalian fauna at Naringal, south-western Victoria. *Australian Wildlife Research,* **17,** 325–347.

Bennett, A.F. (1991). Roads, roadsides and wildlife conservation: A review. pp. 99–117. In: *Nature conservation 2: The role of corridors.* Saunders, D.A. and Hobbs, R.J. (Eds). Surrey Beatty, Sydney.

Bennett, A.F. (1998). *Linkages in the landscape. The role of corridors and connectivity in wildlife conservation.* IUCN, Gland, Switzerland.

Bennett, A.F. (2003). Habitat fragmentation. pp. 440–445. In: *Ecology: An Australian Perspective.* Attiwill, P. and Wilson, B. (Eds). Oxford University Press, Melbourne.

Bennett, A.F. and Ford,. L.A. (1997). Land use, habitat change and the conservation of birds in fragmented rural environments: a landscape perspective from the Northern Plains, Victoria, Australia. *Pacifiic Conservation Biology,* **3,** 244–261.

Bennett, A.F., Kimber, S. and Ryan, P. (2000). *Revegetation and wildlife. A guide to enhancing revegetated habitats for wildlife conservation in rural environments.* Bushcare Research Report 2/00. Bushcare National Research and Development Program Research Report.

Bennett, E.L. (2000). Timber certification: where is the voice of the biologist? *Conservation Biology,* **14,** 921–923.

Bennett, J. and Ann, E. (2004). The economic value of biodiversity. A summary report on the National workshop. Report to Department of Environment and Heritage and Land and Water Australia, Canberra.

Bennett, S., Brereton, R., Mansergh, I., Berwick, S., Sandiford, K. and Wellington, C. (1991). *The potential effect of the enhanced greenhouse climate change on selected Victorian fauna.* Department of Conservation and Environment, Arthur Rylah Institute for Environmental Research, Melbourne. Technical Report Series No. 123.

Benson, D.H. and Howell, J. (1990). Sydney's vegetation 1788–1988: Utilisation, degradation and rehabilitation. pp. 115–127. In: *Proceedings of the Ecological Society of Australia.* Vol. 16. Saunders, D.A., Hopkins, A.J.M. and How, R.A. (Eds). Surrey Beatty, Chipping Norton.

Benson, J. (1996). Threatened by discovery: research and management of the Wollemi Pine *Wollemia nobilis* Jones, Hill and Allen. pp. 105–109. In: *Back from the Brink: Refining the Species Recovery Process.* Stephens, S. and Maxwell, S. (Eds). Surrey Beatty and Sons, Sydney.

Benson, J.S. (1991). The effect of 200 years of European settlement on the vegetation and flora of New South Wales. *Cunninghamia,* **2,** 343–370.

Benson, J.S. (1999). Setting the scene. The native vegetation of New South Wales. *Native Vegetation* Advisory Council, Sydney.

Berg, A., Ehnstrom, B., Gustaffson, L., Hallingback, T., Jonsell, M. and Weslien, J. (1994). Threatened plant, animal and fungus species in Swedish forests: distribution and habitat associations. *Conservation Biology,* **8**, 718–731.

Berg, A., Nilsson, S.G. and Bostrom, U. (1992). Predation on artificial wader nests on large and small bogs along a south-north gradient. *Ornis Scandinavia*, **23**, 13–16.

Berger, J. (1990). Persistence of different-sized populations: An empirical assessment of rapid extinctions in Bighorn Sheep. *Conservation Biology,* **4**, 91–98.

Berger, J., Jeitner, C., Clark, K. and Niles, L.J. (2004). The effect of human activities on migrant shorebirds: successful adaptive management. *Environmental Conservation*, **31**, 283–288.

Bergeron, Y., Harvey, B., Leduc, A. and Gauthier, S. (1999). Forest management guidelines based on natural disturbance dynamics: stand- and forest-level considerations. *Forestry Chronicle,* **75**, 49–54.

Bergeron, Y., Richard, P.J., Carcailler, C., Gauthier, S., Flannigan, M. and Prarie, Y.T. (1998). Variability in fire frequency and forest composition in Canada's south-eastern boreal forest: A challenge for sustainable forest management. *Conservation Ecology*, **2**. [online] URL: http://www.ecologyandsociety.org/vol2/iss2/art6/

Berglund, H. and Jonsson, B.G. (2005). Verifying an extinction debt among lichens and fungi in northern Swedish boreal forests. *Conservation Biology*, **19**, 338–348.

Bergstrom, J.C. (1990). Concepts and measures of the economic value of environmental quality: A review. *Journal of Environmental Management,* **31**, 215–228.

Best, L.B., Bergin, T.M. and Freemark, K.E. (2001). Influence of landscape composition on bird use of rowcrop fields. *Journal of Wildlife Management*, **65**, 442–449.

Bezkorowajny, P.G., Gordon, A.M. and McBride, R.A. (1993). The effect of cattle foot traffic on soil compaction in a silvo-pastoral system. *Agroforestry Systems*, **21**, 1–10.

Bierregaard, R.O. and Stouffer, P.C. (1997). Understorey birds and dynamic habitat mosaics in Amazonian rainforests. pp. 138–153. In: *Tropical Forest Remnants. Ecology, Management and Conservation.* Laurance, W.F. and Bierregaard R.O. (Eds). The University of Chicago Press, Chicago.

Billington, H.L. (1991). Effect of population size on genetic variation in a dioecious conifer. *Conservation Biology,* **5**, 115–119.

Birckhead, J., De Lacy, T. and Smith, L. (1992). *Aboriginal involvement in parks and protected areas.* Australian Institute of Aboriginal and Torres Strait Islander Studies, Report Series. Birckhead, J., De Lacy T. and Smith. L. (Eds). Aboriginal Studies Press, Canberra.

BirdLife International (2000). *Threatened birds of the world.* BirdLife International. Lynx Editions, London.

Birds Australia (2000). Birds on Farms. Ecological management for agricultural sustainability. *Supplement to Wingspan*, **10**, December 2000.

Birds Australia (2005). Bird Atlas. Map Viewer. http://www2.abc.net.au/birds/mapviewer.html

Bissonette, J.A. and Storch, I. (2002). Fragmentation: Is the message clear? *Conservation Ecology*, **6**. [online] URL: http://www.ecologyandsociety.org/vol6/iss2/art14/

Blab, J. and Vogel H. (1996). *Amphibien und Reptilien erkennen und schuetzen: Alle mitteleuropaeischen Arten.* BLV Verlagsgesellschaft, Munich.

Blackburn, T.M. and Gaston, K.J. (1995). What determines the probability of discovering a species?: A study of South American oscine passerine birds. *Journal of Biogeography,* **22**, 7–14.

Blakers, M., Davies, S.J. and Reilly, P.N. (1984). *The atlas of Australian birds.* Royal Australasian Ornithithological Union. Melbourne University Press, Melbourne.

Blamey, R. and Hatch, D. (1998). Profiles and motivations of nature-based tourists visiting Australia. *Bureau of Tourism Research Occasional Paper,* No. **25**.

Blanks, P. (1996). The predictive value of Ecological Vegetation Classes for the distribution of bryophytes. Honours thesis. School of Botany, University of Melbourne.

Blaustein, A.R. (1994). Chicken Little or Nero's fiddle? A perspective on declining amphibian populations. *Herpetologica*, **50**, 85–97.

Blaustein, A.R. and Johnson, P.T. (2002). The complexity of deformed amphibians. *Frontiers in Ecology and Environment*, **1**, 87–94.

Blaustein, A.R. and Johnson, P.T.J. (2003). Explaining frog deformities. *Scientific American*, **288**, 60–65.

Blaustein, A.R. and Kiesecker, J.M. (2002). Complexity in conservation: lessons from the global decline of amphibian populations. *Ecology Letters*, **5**, 597–608.

Blaustein, A.R., Root, T.L., Kiesecker, J.M., Belden, L.K., Olson, D.H. and Green, D.M. (2003). Amphibian breeding and climate change: Reply to Corn. *Conservation Biology*, **17**, 626–627.

Block, W.M. and Brennan, L.A. (1993). The habitat concept in ornithology. Theory and applications. pp.

35–91. In: *Current ornithology.* Vol. 11. Power D.M. (Ed). Plenium Press, New York.

Block, W.M., Brennan, L.A. and Gutierrez, R.J. (1986). The use of guilds and guild-indicator species for assessing habitat suitability. pp. 109–113. In: *Wildlife 2000: Modelling Habitat Relationships of Terrestrial Vertebrates.* Verner, J., Morrison M.L. and Ralph C.J. (Eds). University of Wisconsin Press, Madison.

Blouin, M.S. and Connor, E.F. (1985). Is there a best shape for nature reserves? *Biological Conservation,* **32**, 277–288.

Boback, S.M. (2003). Body size evolution in snakes: evidence from island populations. *Copeia,* **2003**, 81–94.

Bock, W.J. (1992). The species concept in theory and practice. *Zoological Science,* **9**, 697–712.

Boekel, C. (1980). Birds of Victoria River Downs Station and of Yarralin, Northern Territory. Part 1. *Australian Bird Watcher,* **8**, 171–193.

Bohnsack, J.A. (1998). Application of marine reserves to reef fisheries management. *Australian Journal of Ecology,* **23**, 296–304.

Boland, C.R. (2004). Introduced cane taods Bufo Marinus are active nest predators and competitors of rainbow bee-eaters *Merops ornatus*: observational and experimental evidence. *Biological Conservation,* **120**, 53–62.

Boland, D.J., Brooker, M.I. and Turnbull, J.W. (1980). *Eucalyptus seed.* CSIRO Division of Forest Research, Canberra.

Boland, D.J., Brooker, M.I., Chippendale, G.M., Hall, N., Hyland, B.P., Johnston, R.D., Kleinig, D.A. and Turner, J.D. (1984). *Forest Trees of Australia.* CSIRO, Melbourne. 687 pp.

Bolton, B.L. and Latz, P.K. (1978). The Western hare-Wallaby *Lagorchestes hirsutus* (Gould) (Macropodictae) in the Tanami Desert. *Australian Wildlife Research,* **5**, 285–293.

Bomford, M. and Sinclair, R. (2002). Australian research on bird pests: impact, management and future directions. *Emu,* **102**, 29–45.

Bonham, K.J., Mesibov, R. and Bashford, R. (2002). Diversity and abundance of some ground-dwelling invertebrates in plantation vs. native forests in Tasmania, Australia. *Forest Ecology and Management,* **158**, 237–247.

Bonner, F.T. (1990). Storage of seeds: Potential and limitations for germplasm conservation. *Forest Ecology and Management,* **35**, 35–43.

Boone, R.B. and Hunter, M.L. (1996). Using diffusion models to simulate the effects of land use on Grizzly Bear dispersal in the Rocky Mountains. *Landscape Ecology,* **11**, 51–64.

Booth, D.E. (1991). Estimating pre-logging old-growth in the Pacific Northwest. *Journal of Forestry,* **89**, 25–29.

Booth, D.E. (1992). The economics and ethics of old-growth forests. *Environmental Ethics,* **14**, 43–62.

Booth, R. (1994). Medicine and husbandry: Dasyurids, possums and bats. pp. 423–441. In: *Proceedings of the T.G. Hungerford Refresher Course for Veterinarians on Wildlife.* Proceedings 233. Westerns Plains Zoo, Dubbo, 19–23 September 1994. Post Graduate Committee, University of Sydney.

Booth, T.H. (1985). A new method for assessing species selection. *Commonwealth Forestry Review,* **64**, 241–250.

Booth, T.H. and Pyror, L.D. (1991). Climatic requirements of some commercially important Eucalypt species. *Forest Ecology and Management,* **43**, 47–60.

Booth, T.H., Nix, H.A., Hutchinson, M.F. and Busby, J.R. (1987). Grid matching: A new method for homoclime analysis. *Agricultural and Forest Meteorology,* **39**, 241–255.

Booth, T.H., Nix, H.A., Hutchinson, M.F. and Javanovic, T. (1988). Niche analysis and tree species distribution. *Forest Ecology and Management,* **23**, 47–59.

Booth, T.H., Stein, J.A., Nix, H.A. and Hutchinson, M.F. (1989). Mapping regions climatically suitable for particular species: An example using Africa. *Forest Ecology and Management,* **28**, 19–31.

Borowski, E.J. and Borwein, J.M. (1989). *Collins dictionary of mathematics.* 2nd edition. HarperCollins.

Botkin, D B., Woodby, D.A. and Nisbet, R.A. (1991). Kirtland's Warbler Habitats: A Possible Early Indicator of Climatic Warming. *Biological Conservation,* **56**, 63–78.

Botkin, D.B. (1990). *Discordant Harmonies: A New Ecology for the 21st Century*. Oxford University Press, N.Y.

Botkin, D.B. and Keller, E.A. (1995). *Environmental Science: Earth as a Living Planet.* First Edition. John Wiley, New York.

Bowman, D.J., Walsh, A. and Prior, L.D. (2004). Landscape analysis of Aboriginal fire management in Central Arnhem Land, north Australia. *Journal of Biogeography,* **31**, 207–233.

Bowman, D.M. (1998). Tansley Review No. 101. The impact of Aboriginal landscape burning on the Australian biota. *New Phytologist,* **140**, 385–410.

Bowman, D.M. (1999). *Australian rainforests. Islands of green in a land of fire.* Cambridge University Press, Melbourne.

Bowman, D.M. (2003). Bushfires: a Darwinian perspective. In: *Australia burning: Fire Ecology, Policy and Management Issues.* Cary, G., Lindenmayer, D.B. and Dovers, S. (Eds). CSIRO Publishing, Melbourne.

Bowman, D.M. and Woinsarski, J.C. (1994). Biogeography of Australian monsoon rainforest mammals: Implications for the conservation of rain-forest mammals. *Pacific Conservation Biology,* **1**, 98–106.

Bowne, D.R., Peles, J.D. and Barrett, G.W. (1999). Effects of landscape spatial structure on movement patterns of the hispid cotton rat (*Sigmodon hispidus*). *Landscape Ecology*, **14**, 53–65.

Boyce, M.S. (1992). Population viability analysis. *Annual Review of Ecology and Systematics,* **23**, 481–506.

Boyden, S. and Dovers, S. (1992). Natural-resource consumption and its environmental impacts in the western world: Impacts of increasing per capita consumption. *Ambio,* **21**, 63–69.

Braby, M.F., Crosby, D.F. and Vaughan, P.J. (1992). Distribution and range reduction in Victoria of the Eltham Copper Butterfly *Paralucia pyrodiscus lucida* Crosby. *Victorian Naturalist,* **109**, 154–167.

Braby, M.F., Van Praagh, B.D. and New, T.R. (1999). The dull copper, Paralucia pyrodiscus (Lycanidae). pp. 247–260. In: *Biology of Australian butterflies.* Kitching, R.L., Scheermeyer, E., Jones R.E. and Pierce, N. (Eds). CSIRO Publishing, Melbourne.

Bradfield, P.J., Schulz, C.E. and Stone, M.J. (1996). Regulatory approaches to environmental management. pp. 46–73. In: *Environmental management in the Australian minerals and energy industries.* Mulligan D.R. (Ed). University of New South Wales Press and Australian Minerals and Energy Environment Foundation, Sydney.

Bradshaw, F.J. (1992). Quantifying edge effect and patch size for multiple-use silviculture – a discussion paper. *Forest Ecology and Management,* **48**, 249–264.

Bradstock, R.A. and Cohn, J. (2002). Fire regimes and biodiversity in semi-arid mallee ecosystems. pp. 238–258. In: *Flammable Australia. The fire regimes and biodiversity of a continent.* Bradstock, R.A., Williams, J.E. and Gill, A.M. (Eds). Cambridge University Press, Melbourne.

Bradstock, R.A., Bedward, M., Scott, J. and Keith, D.A. (1997). Simulation of the effect of temporal and spatial variation in fire regimes on the population viability of a *Banksia* spp. *Conservation Biology*, **10**, 776–784.

Bradstock, R.A., Williams, J.E. and Gill, A.M. (Eds) (2002). *Flammable Australia. The fire regimes and biodiversity of a continent.* Cambridge University Press, Melbourne.

Braithwaite, L.W. (1984). The identification of conservation areas for possums and gliders in the Eden woodpulp concession district. pp. 501–508. In: *Possums and Gliders.* Smith, A.P. and Hume, I.D. (Eds). Surrey Beatty and Sons, Sydney, Australia.

Braithwaite, L.W., Belbin, L., Ive, J. and Austin, M.P. (1993). Land use allocation and biological conservation in the Batemans Bay forests of New South Wales. *Australian Forestry,* **56**, 4–21.

Braithwaite, L.W., Turner, J. and Kelly, J. (1984). Studies on the arboreal marsupial fauna of Eucalypt forests being harvested for woodpulp at Eden, NSW. III: Relationships between faunal densities, Eucalypt occurrence and foliage nutrients and soil parent materials. *Australian Wildlife Research,* **11**, 11–48.

Braithwaite, R.W. (1990). A new savanna fire experiment. *Bulletin of the Ecological Society of Australia,* **20**, 47–48.

Braithwaite, R.W. (1995). Biodiversity and fire in the savanna landscape. pp. 121–140. In: *Biodiversity and Savanna Ecosystem Processes.* Solbrig, O.T., Medina, E. and Silva, J. (Eds). Springer-Verlag, Berlin.

Braithwaite, R.W. and Griffiths, A.D. (1994). Demographic variation and range contraction in the Northern Quoll (*Dasyurus hallacatus*) (Marsupialia: Dasyuridae). *Wildlife Research,* **21**, 203–214.

Brand, D. (1997). Forest management in New South Wales, Australia. *The Forestry Chronicle,* **73**, 578–585.

Bredesn, B., Haugan, R., Anderaa, R., Lindbald, I., Økland, B. and Rosok, Ø. (1997). Wood-inhabiting fungi as indicators of ecological continuity within Spruce forests of southeastern Norway. *Blyttia* (Oslo), **54**, 131–140 (in Norwegian with an English summary).

Brereton, R. (1997). *Management prescriptions for the Swift Parrot in production forests.* Report to the Tasmanian Regional Forest Agreement and Heritage Technical Comittee. Tasmanian Parks and Wildlife Service. June 1997.

Brereton, R., Bennett, S. and Mansergh, I. (1995). Enhanced greenhouse climate change and its potential effect on selected fauna of south-eastern Australia: A trend analysis. *Biological Conservation,* **72**, 339–354.

Brewer, C. (2001). Cultivating conservation literacy: "trickle-down" is not enough. *Conservation Biology*, **15**, 1203–1205.

Bridgeman, H.A., Warner, R. and Dodson, J. (1995). *Urban biophysical environment.* Oxford, Melbourne.

Bridgewater, P.B., Walton, D.W., Busby, J.R. and Reville, B.J. (1992). Theory and practice in framing a national system for conservation in Australia. pp. 3–16. In: *Biodiversity – broadening the debate.* Australian National Parks and Wildlife Service, Canberra.

Briese, D.T. (2000). Classical biological control. pp. 139–160. In: *Australian weed management systems.* Sindel, B. (Ed). R.G. and F.J. Richardson, Melbourne.

Briggs, J.D. and Leigh, J.H. (1988). *Rare or threatened Australian plants.* Australian National Parks and Wildlife Service, Canberra. Special Publication No. 14.

Briggs, J.D. and Leigh, J.H. (1996). *Rare or threatened Australian plants.* CSIRO, Melbourne.

Briggs, S. and Taws, N. (2003). Impacts of salinity on biodiversity – clear understanding or muddy confusion? *Australian Journal of Botany*, **51**, 609–617.

Bright, P.W. (1993). Habitat fragmentation – problems and predictions for British mammals. *Mammal Review*, **23**, 101–111.

Brinkman, R.P. (1990a). Advanced forest planning in the Otways. *Australian Forestry,* **53**, 290–294.

Brinkman, R.P. (1990b). *Wood, water and wildlife yields of the Otways forests.* Centre for Resource and Environmental Studies, The Australian National University, Canberra. Otway Project Working Paper 1990/1.

Briskie, J.V. and Mackintosh, M. (2004). Hatching failure increases with severity of population bottlenecks in birds. *Proceedings of the National Academy of Sciences*, **101**, 558–561.

Brock, M. (1981). The ecology of halophytes in the south–east of Australia. *Hydrobiologica*, **81**, 23–32.

Brook, B.W. (2000). Pessimistic and optimistic bias in population viability analysis. *Conservation Biology*, **14**, 564–566.

Brook, B.W. and Bowman, D.M. (2004). The uncertain blitzkrieg of Pleistocene megafauna. *Journal of Biogeography*, **31**, 517–523.

Brook, B.W. and Kikkawa, J. (1998). Examining threats faced by island birds: a population viability analysis on the Capricorn Silvereye using long-term data. *Journal of Applied Ecology*, **35**, 491–503.

Brook, B.W., Burgman, M.A., Akçakaya, H.R., O'Grady, J.J. and Frankham, R. (2002a). Critiques of PVA ask the wrong questions: throwing out the heuristic baby with the numerical bath water. *Conservation Biology*, **16**, 262–263.

Brook, B.W., Griffiths, A.D. and Puckey, H.L. (2002b). Modelling strategies for the management of the critically endangered Carpentarian Rock-Rat (*Zyzomys palatalis*) of northern Australia. *Journal of Environmental Management*, **65**, 355–368.

Brook, B.W., O'Grady, J.J., Burgman, M.A., Akáakaya, H.R. and Frankham, R. (2000). Predictive accuracy of population viability analysis in conservation biology. *Nature*, **404**, 385–387.

Brooker, L. (2002). The application of focal species knowledge to landscape design in agricultural lands. *Landscape and Urban Planning*, **60**, 185–210.

Brooker, L. and Brooker, M. (2002). Dispersal and population dynamics of the Blue-breasted Fairy-wren, *Malurus pulcherrimus*, in fragmented habitat in the Western Australian wheatbelt. *Wildlife Research*, **29**, 225–233.

Brooker, L.C. and Brooker, M.G. (1994). A model for the effects of fire and fragmentation on the viability of the Splendid Fairy-Wren. *Pacific Conservation Biology,* **1**, 344–358.

Brooker, M. and Brooker, L. (2001). Breeding biology, reproductive success and survival of blue-breasted fairy-wrens in fragmented habitat in the Western Australian wheatbelt. *Wildlife Research*, **28**, 205–214.

Brookes, A.J. (1999). The essence of SNPs. *Gene,* **234**, 177–186.

Brooks, D.R., Mayden, R.L. and McLennan, D.A. (1992). Phylogeny and diversity: Conserving our evolutionary legacy. *Trends in Ecology and Evolution,* **7**, 55–59.

Brooks, T.M., Pimm, S.L. and Oyugi, J.O. (1999). Time lag between deforestation and bird extinction in tropical forest fragments. *Conservation Biology,* **13**, 1140–1150.

Broome, L. and Mansergh, I. (1995). *The Mountain Pygmy Possum and the Australian Alps.* University of New South Wales Press, Sydney.

Brothers, N.P. (1982). Feral cat control on Tasman Island. *Australian Ranger Bulletin*, **2**, 9.

Brothers, T.S. and Spingarn, A. (1992). Forest fragmentation and alien plant invasion of central Indiana old-growth forests. *Conservation Biology,* **6**, 91–100.

Brown, A.H. and Brubaker, C.L. (2000). Genetics and the conservation and use of Australian wild relatives of crops. *Australian Journal of Botany*, **48**, 297–303.

Brown, A.H.D. and Weir, B.S. (1983). Measuring genetic variability in plant populations. pp. 219–239. In: *Isozymes in plant genetics and breeding.* Tansley, S.D. and Orton, T.J. (Eds). Elsevier, Amsterdam.

Brown, G.W. and Nelson, J.L. (1992). *Habitat utilisation by heliothermic reptiles of different successional stages of Eucalyptus regnans (Mountain Ash) forest in the Central Highlands, Victoria.* Department of Conservation and Natural Resources, Victoria. SSP Report No. 17.

Brown, J.C. (1883). *French Forest Ordinance of 1669.* Oliver and Boyd, Edinburgh, United Kingdom.

Brown, J.H. (1984). On the relationship between distribution and abundance. *The American Naturalist,* **124**, 255–279.

Brown, J.H. and Kodric-Brown. A. (1977). Turnover rates in insular biogeography: effect of immigration on extinction. *Ecology*, **58**, 445–449.

Brown, K.A. and Gurevitch, J. (2004). Long-term impacts of logging on forest diversity in Madagascar.

Proceedings of the National Academy of Sciences, **101**, 6045–6049.

Brown Jr, K.S. and Hutchings, R.W. (1997). Disturbance, fragmentation, and the dynamics of diversity in Amazonian forest butterflies. pp. 91–110. In: *Tropical forest remnants: ecology, management and conservation of fragmented communities.* Laurance, W.R. and Bierregaard, Jr, R.O. (Eds). University of Chicago Press, Chicago.

Brown, M.J. (1988). *Distribution and conservation of King Billy Pine.* Forestry Commission of Tasmania, Hobart, Tasmania.

Brown, M.J. and J. Hickey. (1990). Tasmanian forest – genes or wilderness. *Search,* **21**, 86–87.

Brown, P.B., Holdsworth, M.C. and Rounsevell, D.E. (1995). Captive breeding and release as a means of increasing the Orange-bellied Parrot population in the wild. pp. 135–141. In: *Reintroduction biology of Australasian fauna.* Serena, M. (Ed). Surrey Beatty, Chipping Norton, Australia.

Brown, R.S., Caputi, N. and Barker, E. (1995). A preliminary assessment of increases in fishing power on stock assessment and fishing effort expended in the Western Rock Lobster (*Panulirus Cygnus*) fishery. *Crustaceana,* **68**, 226– 405.

Brown, V.B., Davies, S.A. and Synnot, R.N. (1990). Long-term monitoring of the effects of treated sewage effluent on the intertidal macroalgal community near Cape Schank, Victoria. *Botanical Marina,* **33**, 85–98.

Brummitt, N. and Lughadha, E.N. (2003). Biodiversity: where's hot and where's not. *Conservation Biology,* **17**, 1442–1448.

Brunckhorst, D. (1999). *Models to integrate sustainable conservation and resource use – bioregional reserves beyond Bookmark.* Nature Conservtaion Council Annual Conference on Integrated Natural Resource management. University of Sydney, March 1999.

Brunner, H. and Coman, B.J. (1972). Food habits of the feral house cat in Victoria. *Journal of Wildlife Management,* **36**, 849–853.

Brunton, R. (1994). *The precautionary principle: the greatest risk of all.* Environmental Policy Unit, Institute of Public Affairs Ltd, Victoria. Environmental Backgrounder No. 20.

Brussard, P. (1984). Geographical patterns and environmental gradients: The central marginal model in Drosophila revisited. *Annual Review of Ecology and Systematics,* **15**, 25–64.

Brussard, P. and Ehrlich, P. (1992). The challenges of conservation biology. *Ecological Applications,* **2**, 1–2.

Brussard, P.F. (1994). Why do we want to conserve biodiversity anyway? *Society for Conservation Biology Newsletter,* **1**(1).

BTR (1992) (Bureau of Tourism Research). *Australian tourism forecasts: international visitor arrivals, 1991.* Bureau of Tourism Research, Canberra.

Buckley, R. (1989). *Precision in environmental impact prediction. First national environmental audit, Australia.* Centre for Resource and Environmental Studies, The Australian National University, Canberra.

Buckley, R. (1991a). Auditing the precision and accuracy of environmental impact predictions in Australia. *Environmental Monitoring and Assessment,* **18**, 1–24.

Buckley, R. (1991b). *Sustainable development in the Australian mining and petroleum industries: Case studies and general issues.* Griffith Environmental Monographs I. Griffith University, Gold Coast Campus.

Bui, E. (2000). Risk assessment in the face of controversy: tree clearing and salinization in North Queensland. *Environmental Management,* **26**, 447–456.

Bulinski, J. (1999). A survey of mammalian browsing damage in Tasmanian eucalypt plantations. *Australian Forestry,* **62**, 59–65.

Bullock, S., Summerell, B.A. and Gunn, L.V. (2000). Pathogens of the Wollemi Pine, *Wollemia nobilis. Australasian Plant Pathology,* **29**, 211–214.

Bunnell, F. (1998). Evading paralysis by complexity when establishing operational goals for biodiversity. *Journal of Sustainable Forestry,* **7**, 145–164.

Bunnell, F. (1999a). Foreword. Let's kill a panchreston. Giving fragmentation a meaning. pp. vii–xiii. In: *Forest wildlife and fragmentation. Management implications.* Rochelle, J., Lehmann L.A. and Wisniewski J. (Eds). Brill, Leiden, Germany.

Bunnell, F. (1999b). What habitat is an island? pp. 1–31. In: *Forest wildlife and fragmentation. Management implications.* Rochelle, J., Lehmann, L.A. and Wisniewski, J. (Eds). Brill, Leiden, Germany.

Bunnell, F. and Johnson, J.F. (Eds) (1998). *The living dance. Policy and practices for biodiversity in managed forests.* UBC press, Vancouver.

Bunnell, F., Dunsworth, G., Huggard, D. and Kremsater, L. (2003). *Learning to sustain biological diversity on Weyerhauser's coastal tenure.* Weyerhauser Company, Vancouver, British Columbia.

Burbidge, A. (2003). Princess Parrot. *Nature Australia,* **Summer 2003**, 24–25.

Burbidge, A. and Brown, A. (1993). *Setting the priorities for the conservation of Western Australia's threatened plants and animals.* Department of Conservation and Land Management. WATSCU Discussion Paper No. 3.

Burbidge, A.A. and Kuchling, G. (1994). *Western Swamp Tortoise Recovery Plan.* Wildlife Management Program No. 11. Department of Conservation and Land Management, Perth.

Burbidge, A.A. and McKenzie, N.L. (1989). Patterns in modern decline of Western Australia's vertebrate fauna: causes and conservation implications. *Biological Conservation,* **50**, 143–198.

Burbidge, A.A., Johnson, K.A., Fuller, P.J. and Southgate, R.L. (1988). Aboriginal knowledge of the mammals of the central deserts of Australia. *Australian Wildlife Research*, **15**, 9–39.

Burbidge, N.T. (1960). The phytogeography of the Australian region. *Australian Journal of Botany,* **8**, 75–209.

Burdon J.J. and Chilvers, G.A. (1994). Demographic changes and the development of competition in a native eucalypt forest invaded by exotic pines. *Oecologia,* **97**, 419–423.

Bureau of Meteorology. (2002). Atmospheric indicators for the State of the Environment Report 2001. In: *Technical Report 74.* Bureau of Meteorology, Melbourne.

Bureau of Rural Sciences (1998). *Australia's The State of the Forests Report.* Bureau of Rural Sciences, Canberra.

Bureau of Rural Sciences (2003a). Australia's forests at a glance. Bureau of Rural Sciences, Department of Agriculture, Fisheries and Forestry, Canberra.

Bureau of Rural Sciences (2003b). Australia's State of the Forests Report 2003. Bureau of Rural Sciences, Department of Agriculture, Fisheries and Forestry, Canberra.

Bureau of Rural Sciences (2004). *Australia's Forests at a Glance.* Australian Government Publishing Service, Canberra.

Burgess, G.K. and Vanderbyl, T.L. (1996). Habitat Analysis Method for determining environmental flow requirements. In: *Water and the Environment.* Proceedings of the 23rd Hydrological and Water Resources Symposium, Hobart. Institution of Engineers, ACT.

Burgman, M.A. (1989). The habitat volumes of scarce and ubiquitous plants: a test of the hypothesis of environmental control. *American Naturalist,* **133**, 228–239.

Burgman, M.A. (1996). Characterisation and delineation of Eucalypt old-growth forest in Australia: A review. *Forest Ecology and Management,* **83**, 149–161.

Burgman, M.A. (2000). Population viability analysis for bird conservation: prediction, heuristics, monitoring and psychology. *Emu*, **100**, 347–353.

Burgman, M.A. (2002). Are listed threatened plant species actually at risk? *Australian Journal of Botany,* **50**, 1–13.

Burgman, M.A. (2004). Evaluating methods for assessing extinction risk. *Acta Oecologica-International Journal of Ecology*, **26**, 65–66.

Burgman, M.A. (2005). *Risks and decisions for conservation and environmental management.* Cambridge University Press, Cambridge.

Burgman, M.A. Akçakaya, H.R. and Loew, S.S. (1988). The use of extinction models for species conservation. *Biological Conservation,* **43**, 9–25.

Burgman, M.A. and Ferguson, I.S. (1995). *Rainforest in Victoria: a review of the scientific basis of current and proposed protection measures.* Report to Victorian Department of Conservation and Natural Resources. Forest Services Technical Reports 95–4.

Burgman, M.A. and Fox, J. (2003). Bias in species range estimates from minimum convex polygons: implications for conservation and options for improved planning. *Animal Conservation,* **6**, 19–28.

Burgman, M.A. and Gerard, V.A. (1990). A stage-structured, stochastic population model for the Giant Kelp *Macrocystis pyrifera. Marine Biology,* **105**, 15–23.

Burgman, M.A. and Lamont, B.B. (1992). A stochastic model for the viability of *Banksia cuneata* populations: Environmental, demographic and genetic effects. *Journal of Applied Ecology,* **29**, 719–727.

Burgman, M.A. and Lindenmayer, D.B.L. (1998). *Conservation Biology for the Australian Environment.* Surrey Beatty, Chipping Norton, N.S.W.

Burgman, M.A., Breininger, D.R., Duncan, B.W. and Ferson, S. (2001). Setting reliability bounds on habitat suitability indices. *Ecological Applications,* **11**, 70–78.

Burgman, M.A., Ferson, S. and Akçakaya, H.R. (1993). *Risk assessment in conservation biology.* Chapman and Hall, London.

Burgman, M.A., Ferson, S. and Lindenmayer, D.B. (1995a). The effect of the initial age-class distribution on extinction lists: implications for the reintroduction of Leadbeater's Possum. pp. 15–19. In: *Reintroduction biology of Australian and New Zealand fauna.* Serena, M. (Ed). Surrey Beatty, Chipping Norton.

Burgman, M.A., Grimson, R.C. and Ferson, S. (1995b). Inferring threat from scientific collections. *Conservation Biology,* **9(4)**, 923–928.

Burgman, M.A., Lindenmayer, D.B. and Elith, J. (2005). Managing landscapes for conservation under uncertainty. (*Ecology*) (in press).

Burgman, M., Maslin, B., Andrewartha, D., Keatley, M., Boek, C. and McCarthy, M. (2000). Detecting trends in sighting data: power and an application to Western Australian Acacia species. pp. 7–26. In: *Quantitative*

Methods for Conservation Biology. Ferson, S. and Burgman, M.A. (Eds). Springer-Verlag, New York.

Burke, D.M. and Nol, E. (1998). Influence of food abundance, nest-site habitat and forest fragmentation on breeding Ovenbirds. *Auk,* **115**, 96–104.

Burke, V.J. (Ed). (2000). *Gap analysis for landscape conservation.* Special Issue. Landscape Ecology Volume **15**.

Burkey, T.V. (1989). Extinction in nature reserves: the effect of fragmentation and the importance of migration between reserve fragments. *Oikos*, **55**, 75–81.

Burnett, S. (1992). Effects of a rainforest road on movements of small mammals: Mechanisms and implications. *Wildlife Research,* **19**, 95–104.

Burnham, K.P. and Anderson, D.R. (1998). *Model selection and inference: a practical information-theoretic approach.* Springer-Verlag, New York.

Burns, K., Walker, D. and Hansard, A. (1999). *Forest plantations on cleared agricultural land in Australia: A regional and economic analysis.* Research Report No. 99/11. Australian Bureau of Agricultural and Resource Economics, Canberra.

Burris, R.K. and Canter, L.W. (1997). Cumulative impacts are not properly addressed in environmental assessments. *Environmental Impact Assessment Review*, **67**, 5–18.

Burrough, P.A., Brown, L. and Morris, E.C. (1977). Variations in vegetation and soil patterns across the Hawkesbury Sandstone Plateau from Barren Grounds to Fitzroy Falls, New South Wales. *Australian Journal of Ecology*, **2**, 137–159.

Burrows, N.D. and Christensen, P.E. (1990). A survey of aboriginal fire patterns in the Western Desert of Australia. pp. 297–305. In: *Fire and the Environment: ecological and cultural perspectives.* Nodvin, S.C. and Waldorp, T.A. (Eds). USDA Forest Service, General Technical Report SE-69. Southeastern Forest Experimental Station: Asheville, North Carolina.

Burton, B., Kinrade, P., Amos, N., Giese, M. and Krockenberger, M. (1994). *Mining and ecologically sustainable development: A discussion paper.* Australian Conservation Foundation, Melbourne.

Busby, J.R. (1986). A biogeoclimatic analysis of *Nothofagus cunninghamii* Hook Oesrt in south-eastern Australia. *Australian Journal of Ecology,* **11**, 1–7.

Busby, J.R. (1988). Potential implications of climate change on Australia's flora and fauna. pp. 387–398. In: *Greenhouse: Planning for climate change.* Pearman, G.I. (Ed). E.J. Brill, New York.

Bush, G.L. (2001). Process of speciation. pp. 371–381. In: *Encyclopedia of biodiversity*. Levin, S.A. (Ed). Academic Press, San Diego.

Byram, G.M. (1959). Combustion of forest fuels. pp. 61–89. In: *Forest Fire Control and Use*. Davies, K.P. (Ed). McGraw-Hill, New York.

Cabeza, M. (2003). Habitat loss and connectivity of reserve networks in probability approaches to reserve design. *Ecology Letters*, **6**, 665–672.

Cabeza, M. and Moilanen, A. (2001). *Reserve design and the persistence of biodiversity.* Trends in Ecology and Evolution, **16**, 242–248.

Cain, N. (1962). Companies and squatting in the Western Division of New South Wales, 1896–1905. In: *The simple fleece: studies in the Australian wool industry.* Barnard, A. (Ed). Melbourne University Press, Melbourne.

Calaby, J.H. (1960). Australia's threatened animals. *Oryx,* **5**, 381–386.

Caldwallader, P.L. (1986). Flow regulation in the Murray River system and its effect on the native fish fauna. pp. 115–133. In: *Stream protection. The management of rivers for instream uses.* Campbell, I.C. (Ed). Water Studies Institute, Chisholm Institute of Technology, Melbourne.

Cale, P. (1999). The spatial dynamics of the White-browed Babbler in a fragmented agricultural landscape. PhD thesis, University of New England, Armidale, N.S.W.

Cale, P.G. and Hobbs, R.J. (1994). Landscape heterogeneity indices: Problems of scale and applicability, with particular reference to animal habitat description. *Pacific Conservation Biology,* **1**, 183–193.

Calhoun, A. (1999). Forested wetlands. pp. 300–331. In: Managing biodiversity in forest ecosystems. Hunter, III M. (Ed). Cambridge University Press, Cambridge.

Callicott, J.B. and Mumford, K. (1997). Ecological sustainability as a conservation concept. *Conservation Biology*, **11**, 32–40.

CALM (1994). *Setting priorities for the conservation of Western Australia's threatened flora and fauna.* Department of Conservation and Land Management, Perth. Policy Statement No. 50.

Calver, M. and Dell, J. (1998). Conservation status of mammals and birds in south-western Australian forests. I. Is there any evidence of direct links between forestry practices and species decline and extinction? *Pacific Conservation Biology*, **4**, 296–314.

Cameron, M. and Soderquist, T. (2002). Ecological triage. *Nature Australia*, **27** (7), 4.

Campbell, A.J. (Ed) (1999). *Declines and disappearances of Australian frogs.* Environment Australia, Canberra.

Campbell, A.J. and Tanton, M.T. (1981). Effects of fire on the invertebrate fauna of soil and litter of a eucalypt forest. pp. 215–241. In: *Fire and the Australian Biota.* Gill, A.M., Groves R.H. and Noble, I.R. (Eds). Australian Academy of Science, Canberra.

Campbell, L.M. (2002). Science and sustainable use: views of marine turtle conservation experts. *Ecological Applications,* **12**, 1229–1246.

Campbell, R.G. (1984). The eucalypt forests. pp. 1–12. In: *Silvicultural and environmental aspects of harvesting some major commercial eucalypt forests in Victoria: A review.* Campbell, R.G., Chesterfield, E.A., Craig, F.G., Fagg, P.C., Farrell, P.W., Featherstone, G.R., Flinn, D.W., Hopmans, P., Kellas, J.D., Leitch, C.J., Loyn, R.H., Macfarlane, M.A., Pederick, L.A., Squire, R.O., Stewart, H.T. and Suckling G.C. (Eds). Forests Commission Victoria, Division of Education and Research. Forests Commission Victoria, Melbourne, Australia.

Caputi, N., Brown, R.S. and Chubb, C.F. (1995a). Regional prediction of the Western Rock Lobster, *Panulirus cygnus*, commercial catch in Western Australia. *Crustaceana,* **68**, 245–256.

Caputi, N., Chubb, C.F. and Brown, R.S. (1995b). Relationship between spawning stock, environment, recruitment and fishing effort for the Western Rock Lobster, *Panulirus cygnus*, fishery in Western Australia. *Crustaceana,* **68**, 214–226.

Cardale, J.C. (1993). Hymenoptera: Apoidea. In: *Zoological catalogue of Australia.* Volume 10. Huston, W.W. and Maynard, G.V. (Eds). Australian Government Printing Service, Canberra.

Cardillo, M. and Bromham, L. (2001). Body size and risk of extinction in Australian mammals. *Conservation Biology,* **15**, 1435–1440.

Cardinal, B.R. and Christidis, L. (2000). Mitchondrial DNA and morphology reveal three geographically distinct lineages of the Large Bentwing Bat (*Miniopterus schreibersii*) in Australia. *Australian Journal of Zoology,* **48**, 1–19.

Carlson, A. and Edenhamn, P. (2000). Extinction dynamics and the regional persistence of a tree frog metapopulation. *Proceedings of the Royal Society of London Series B,* **267**, 1311–1313.

Caro, T. (2004). Preliminary assessment of the flagship species concept at a small scale. *Animal Conservation,* **7**, 63–70.

Caro, T. and O'Doherty, G. (1999). On the use of surrogate species in conservation biology. *Conservation Biology,* **13**, 805–814.

Caro, T., Engilis, A., Fitzherbert, E. and Gardner, T. (2004). Preliminary assessment of the flagship species concept at a small scale. *Animal Conservation,* **7**, 63–70.

Caro, T.M. (2001). Species richness and abundance of small mammals inside and outside an African national park. *Biological Conservation,* **98**, 251–257.

Carolin, R. and Clarke, P. (1991). *Beach plants of south eastern Australia.* Sainty and Associates, Sydney.

Carpenter, G., Gillison, A.N. and Winter, J. (1993). DOMAIN: A flexible modelling procedure for mapping potential distributions of plants and animals. *Biodiversity and Conservation,* **2**, 667–680.

Carpenter, S., Chisholm, S.W., Krebs, C.J., Schindler, C.J. and Wright, R.F. (1995). Ecosystems experiments. *Science,* **269**, 324–327.

Carr, G.W. (1993). Exotic flora of Victoria and its impact on indigenous biota. pp. 256–297. In: *Flora of Victoria.* Volume 1. Foreman, D.B. and Walsh, N.G. (Eds). Inkata Press, Melbourne.

Carr, G.W., Yugovic, J.V. and Robinson, K.E. (1992). *Environmental weed invasions in Victoria: conservation and management implications.* Department of Conservation and Environment and Ecological Horticulture, Victoria.

Carrick, R. (1963). Ecological significance of territory in the Australian Magpie, *Gymnorhina tibicen. Proceedings of the International Ornithological Congress,* **13**, 740–753.

Carroll, C., Noss, R.F. and Paquet, P.C. (2001). Carnivores as focal species for conservation planning in the Rocky Mountain region. *Ecological Applications,* **11**, 961–980.

Carruthers, R. I. (2004). Biological control of invasive species, a personal perspective. *Conservation Biology,* **18**, 54–57.

Carson, H.L. (1991). Increased genetic variance after a population bottleneck. *Trends in Ecology and Evolution,* **5**, 228–230.

Carson, R. (1962). *The silent spring.* Houghton Mifflin, New York.

Carstairs, J.L. (1976). Population dynamics and movement of *Rattus villosissimus* (Waite) during the 1966–1969 plague of Brunette Downs, Northern Territory. *Australian Wildlife Research,* **3**, 1–10.

Cary, G. (2002). Importance of a changing climate for fire regimes in Australia. pp. 26–48 In: *Flammable Australia.* Bradstock, R.A., Williams, J.E. and Gill, A.M. (Eds). (2002). The fire regimes and biodiversity of a continent. Cambridge University Press, Melbourne.

Cary, G. and Bradstock, R. (2003). Sensitivity of fire regimes to management. pp. 65–81. In: *Australia Burning: Fire Ecology, Policy and Management Issues.* Cary, G., Lindenmayer, D. and Dovers S. (Eds). CSIRO Publishing, Melbourne.

Cary, G., Lindenmayer, D.B. and Dovers, S. (Eds) (2003). *Australia Burning. Fire Ecology, Policy and Management Issues.* CSIRO Publishing, Melbourne.

Catchpole, W.R. (2002). Fire properties and burn patterns in heterogeneous landscapes. pp. 49–75. In: *Flammable Australia: The Fire Regimes and Biodiversity of a Continent.* Bradstock, R.A., Williams, J.E. and Gill, M.A. (Eds). Cambridge University Press, Cambridge.

Catling, P.C. and Burt, R.J. (1995). Studies of the ground-dwelling mammals of the eucalypt forests in south-eastern New South Wales – the effect of habitat variables on distribution and abundance. *Wildlife Research,* **22**, 271–88.

Catling, P.C., Burt, R.J. and Forrester, R.I. (2000). Models of the distribution and abundance of ground-dwelling mammals in the eucalypt forests of north-eastern New South Wales in relation to habitat variables. *Wildlife Research*, **27**, 639–54.

Catling, P.C., Coops, N.C. and Burt, R.J. (2001). The distribution and abundance of ground-dwelling mammals in relation to time since wildlife and vegetation structure in south-eastern Australia. *Wildlife Research*, **28**, 555–564.

Catling, P.C., Hertog, A., Burt, R.J., Wombey, J.C. and Forrester, R.I. (1999). The short-term effect of cane toads (*Bufo marinus*) on native fauna in the Gulf country of the Northern Territory. *Wildlife Research*, **26**, 161–185.

Catterall, C.P., Green, R.J. and Jones, D.N. (1991). Habitat use by birds across a forest-suburb interface in Brisbane: implications for corridors. pp. 247–58. In: *Nature Conservation 2: The Role of Corridors.* Saunders, D.A. and Hobbs R.J. (Eds). Surrey Beatty and Sons, Chipping Norton, NSW.

Caughley, G. (1987). Ecological relationships. pp. 159–187. In: *Kangaroo: Their ecology and management in the sheep rangelands of Australia.* Caughley, G., Shepherd N. and Short, J. (Eds). Cambridge University Press, Cambridge.

Caughley, G. (1994). Directions in conservation biology. *Journal of Animal Ecology,* **63**, 215–244.

Caughley, G.C. (1978). *Analysis of vertebrate populations.* John Wiley, Sydney.

Caughley, G.C. and Gunn, A. (1996). *Conservation biology in theory and practice.* Blackwell Science, Cambridge, Massachusetts.

Caughley, J., Bayliss, P. and Giles, J. (1982). Trends in kangaroo numbers in western New South Wales and their relation to rainfall. *Australian Wildlife Research,* **11**, 415–422.

Cauley, H.A., Peters, C.M., Donovan, R.Z. and O'Connor, J.M. (2001). Forest Stewardship Council Forest Certification. *Conservation Biology,* **15**, 311–312.

CCST (1994). (Coordination Committee on Science and Technology) *Access to Australia's biological resources.* Office of the Chief Scientist, Department of the Prime Minister and Cabinet. Australian Government Publishing Service, Canberra.

Celebrezze, T. and Paton, D.C. (2004). Do introduced honeybees (*Apis mellifera*, Hymenoptera) provide full pollination to bird-adapted Australian plants with small flowers? An experimental study of *Brachyloma erocoides* (Epacridaceae). *Austral Ecology*, **29**, 129–136.

Chalson, J. and Keith, D.A. (1995). *RAVAS: A risk assessment scheme for vascular plants: pilot application to the flora of New South Wales.* Commonwealth Endangered Species Program, New South Wales National Parks and Wildlife Service, Sydney. Project No. 450.

Chambers, L.E., Hughes, L. and Weston, M.A. (2005). Climate change and its impact on Australia's avifauna. *Emu*, **105**, 1–20.

Chape, S., Blyth, S., Fish, L., Fox, P. and Spalding, M. (2003). *United Nations List of Protected Areas.* IUCN Publications Services Unit, Cambridge, UK.

Chapman, A.D. and Busby, J.R. (1994). Linking plant species information to continental biodiversity inventory, climate modeling and environmental monitoring. pp. 179–195. In: *Mapping the diversity of nature.* Miller, R.I. (Ed). Chapman and Hall, London.

Chartres, C.J., Helyar, K.R., Fitzpatrick, R.W. and Williams, J. (1992). Land degradation as a result of European settlement of Australia and its influence on soil properties. pp. 3–33. In: *Australia's renewable resources: Sustainability and global change.* Gifford, R.M. and Barson, M.M. (Eds). Bureau or Rural Resources and CSIRO Division of Plant Industry. Australian Government Publishing Service, Canberra.

Chase, M.K., Kristan, W.B., Lynam, A.J., Price, M.V. and Rotenberry, J.T. (2000). Single species as indicators of species richness and composition in California coastal Sage Scrub birds and small mammals. *Conservation Biology*, **14**, 474–487.

Chattaway, M.M. (1958). The regenerative powers of certain eucalypts. *Victorian Naturalist*, **75**, 45–46.

Chee, Y. E. (2004). An ecological perspective on the valuation of ecosystem services. Biological Conservation, **120**, 549–565.

Chen, J. (1991). Edge effects: microclimatic pattern and biological responses in old-growth Douglas Fir forests. PhD thesis, University of Washington, Seattle.

Chen, J., Franklin, J.F. and Spies, T.A. (1990). Microclimatic pattern and basic biological responses at the clearcut edges of old-growth Douglas Fir stands. *Northwest Environmental Journal*, **6**, 424–5.

Chen, J., Franklin, J.F. and Spies, T. (1991). Vegetation responses to edge environments in old-growth Douglas-fir forest. *Ecological Applications*, **2**, 387–396.

Chepko-Sade, B.D. and Halpin, Z.T. (1987). *Mammalian dispersal patterns*. University of Chicago, Chicago.

Chesson, P. (2001). Metapopulations. pp. 161–176. In: *Encyclopedia of biodiversity*. Levin, S.A. (Ed). Academic Press, San Diego.

Chichilnisky, G. and Heal, G. (1998). Economic returns from the biosphere. *Nature*, **391**, 629–630.

Chilvers, B.L., Lawler, I.R., Macknight, F., Marsh, H., Noad, M. and Paterson, R. (2005). Moreton Bay, Queensland, Australia: an example of the co-existence of significant marine mammal populations and large-scale coastal development. *Biological Conservation*, **122**, 559–571.

Chindarsi, K.A. (1997). The logging of Australian native forests: A critique. *The Australian Quarterly*, **69**, 86–104.

Chiras, D.D. (1994). Environmental science: action for a sustainable future. Fourth Edition. Benjamin/Cummings, Redwood City, California.

Chisholm, A. (1999). Land resource and the idea of carrying capacity. Business Council of Australia, Melbourne.

Chisholm, R. and Burgman, M.A. (2004). The unified neutral theory of biodiversity and biogeography: a comment. *Ecology* (in press).

Chou, Y.E. (2004). An ecological perspective on the valuation of ecosystem services. *Biological Conservation*, **120**, 549–565.

Chou, Y.H., Minnich, R.A. and Dezzani, R.J. (1993). Do fire sizes differ between southern California and Baja California. *Forest Science*, **39**, 835–844.

Christensen, N.L., Bartuska, A.M., Brown, J.H., Carpenter, S., D'Antonio, C., Francis, R., Franklin, J.F., MacMahon, J.A., Noss, R., Parsons, D.J., Peterson, C.H., Turner, M.G. and Woodmansee, R.G. (1996). The report of the Ecological Society of America on the scientific basis for ecosystem management. *Ecological Applications*, **6**, 665–691.

Christensen, P. and Abbott, I. (1989). Impact of fire in the eucalypt forest ecosystem of southern Western Australia: a critical review. *Australian Forestry*, **52**, 103–121.

Christensen, P. and Burrows, N. (1995). Project desert dreaming: Experimental reintroduction of mammals to the Gibson desert, Western Australia. pp. 199–207. In: *Reintroduction biology of Australasian fauna*. Serena, M. (Ed). Surrey Beatty, Chipping Norton, Australia.

Christensen, P., Recher, H. and Hoare, J. (1981). Responses of Open forest to Fire Regimes. pp. 367–394. In: *Fire and the Australian Biota*. Gill, A.M., Groves R.H. and Noble I.R. (Eds). Australian Academy of Science, Canberra.

Christensen, P.E.S. (1980). The biology of *Bettongia penicillata* Gray, 1837, and *Macropus eugenii* (Desmarest, 1817) in relation to fire. *Forests Department of Western Australia Bulletin*, **91**.

Christensen-Szalanski, J. and Bushyhead, J. (1981). Physicians use of probabilistic information in a real clinical setting. *Journal of Experimental Psychology: Human Perception and Performance*, **7**, 928–935.

Christie, E.K. (1993). Ecosystem change and land degradation. pp. 307–342. In: *Land degradation processes in Australia*. McTainsh, G. and Boughton, W.C. (Eds). Longman, Cheshire.

Churchill, S. (1998). *Australian Bats*. Reed New Holland, Sydney.

Claridge, A. (2002). Use of bioclimatic analysis to direct survey effort for the Long-footed Potoroo (*Potorous longipes*), a rare forest-dwelling rat-kangaroo. *Wildlife Research*, **29**, 193–202.

Claridge, A.W. and Lindenmayer, D.B. (1994). The need for a more sophisticated approach toward wildlife corridor design in the multiple-use forests of south-eastern Australia: The case for mammals. *Pacific Conservation Biology*, **1**, 301–307.

Claridge, A.W. and May, T.W. (1994). Mycophagy among Australian mammals. *Australian Journal of Ecology*, **19**, 251–275.

Claridge, A.W., Barry, S.C., Cork, S.J. and Trappe, J.M. (2000a). Diversity and habitat relationships of hypogeous fungi. II. Factors influencing the occurrence and number of taxa. *Biodiversity and Conservation*, **9**, 175–199.

Claridge, A.W., Cork, S.J. and Trappe, J.M. (2000b). Diversity and habitat relationships of hypogeous fungi. I. Study design, sampling techniques and general survey results. *Biodiversity and Conservation*, **9**, 151–173.

Clark, C.W. (1976). *Mathematical bioeconomics: the optimal management of renewable resources*. Wiley, New York.

Clark, C.W. (1984). Strategies for multispecies management: objectives and constraints. pp. 303–312. In: *Dahlem Konferenzen*. May, R.M. (Ed). Springer-Verlag, Berlin.

Clark, R.G. and Shutler, D. (1999). Avian habitat selection: pattern from process in nest-site use by ducks? *Ecology*, **80**, 272–287.

Clark, T.W. (2002). *The policy process. A practical guide for natural resource professionals.* Yale University Press, New Haven, Connecticut.

Clark, T.W. and Minta, S.C. (1994). *Greater Yellowstone's future. Prospects for ecosystem science, management and policy.* Homestead Publishing, Moose, Wyoming.

Clark, T.W. and Seebeck, J.H. (Eds) (1990). *Management and conservation of small populations* . Chicago Zoological Society, Brookfield, Illinois.

Clark, T.W. and Zaunbrecker, D. (1987). The Greater Yellowstone ecosystem: the ecosystem concept in natural resource policy and management. *Renewable Resources Journal,* **5**, 8–16.

Clark, T.W., Warneke, R.M. and George, G.G. (1990). Management and conservation of small populations. pp. 1–18. In: *Management and conservation of small populations.* Clark, T.W. and Seebeck, J.H. (Eds). Proceedings of a conference Melbourne, September 26–27 1989.

Clarke, K.R. (1993). Non-parametric multivariate analyses of changes in community structure. *Journal of Ecology,* **18**, 117–143.

Clarke, M.F. (1995). Co-operative breeding in Australasian birds: A review of hypotheses and evidence. *Corella,* **19**, 73–90.

Clarke, P.J. (2002). Habitat insularity and fire response traits: evidence from a sclerophyll archipelago. *Oecologia*, **132**, 582–591.

Clarke, R.H., Oliver, D.L., Boulton, R.L., Cassey, P. and Clarke, M.F. (2003). Assessing programs for monitoring threatened species – a tale of three honeyeaters (Meliphagidae). *Wildlife Research*, **30**, 427–435.

Clarkson, E.N. (1979). *Invertebrate palaeontology and evolution.* George Allen and Unwin, London.

Clarkson, R.W. and Childs, M.R. (2000). Temperature effects of hypolimnial-release from dams on early life stages of Colorado River Basin Big-River Fishes. *Copeia,* **2000**, 402–412.

Clemen, R.T. and Winkler, R.L. (1999). Combining probability distributions from experts in risk analysis. *Risk Analysis,* **19**, 187–203.

Clinchy, M., Taylor, A.C., Zanette, L.Y., Krebs, C.J. and Jarman, P.J. (2004). Body size, age and paternity in Common Brushtail Possums (*Trichosurus vulpecula*). *Molecular Ecology*, **13**, 195–202.

Clode, D. and Burgman, M.A. (Eds). (1997). *Joint old growth forest project: Summary report.* NSW National Parks and Wildlife Service and State Forests of NSW, Sydney. 272 pp.

Coates, D.J. (1992). Genetic consequences of a bottleneck and spatial genetic structure in the triggerplant *Stylidium coroniforme* (Stylidaceae). *Heredity,* **69**, 512–520.

Coates, D.J. (2000). Defining conservation units in a rich and fragmented flora: implications for the management of genetic resources and evolutionary processes in south-west Australian plants. *Australian Journal of Botany,* **48**, 329–339.

Coates, D.J. and Hamley, V.L. (1999). Genetic divergence and the mating system in the endangered and geographically restricted species *Lambertia orbifolia* Gardner (Proteaceae). *Heredity,* **83**, 418–427.

Coates, D.J. and Hnatiuk, R.J. (1990). Systematic and evolutionary inferences from isozyme studies in the genus *Eremaea* (Myrtaceae). *Australian Systematic Botany,* **3**, 59–74.

Cochran, T. (2003). Managing threatened species and communities on Bruny Island. Threatened Species Unit, Department of Primary Industries, Water and Environment, Hobart.

Cochrane, G.R. (1963). Vegetation studies in forest-fire areas of the Mount Lofty Ranges, South Australia. *Ecology*, **44**, 41–52.

Cockburn, A.C. (1991). *An Introduction to Evolutionary Ecology*. Blackwell Scientific, Boston.

Cockburn, A.C. (1992). Habitat heterogeneity and dispersal: environmental and genetic patchiness. pp. 65–69. In: *Animal Dispersal. Small mammals as a model.* Stenseth, N. and Lidicker W.Z. (Eds). Chapman and Hall, London.

Cocks, D. (1996). *People, Policy: Australia's Population Choices.* UNSW Press, Syndey.

Cody, M.L. (1986). Diversity, rarity and conservation in Mediterranean climate regions. In: *Conservation Biology: The Science of Scarcity and Diversity.* Soulě, M. (Ed.) Sinaur Associates, Sunderland, Massachusetts.

Cogger, H. (2000). *Reptiles and Amphibians of Australia.* Reed New Holland, Sydney.

Cogger, H., Ford, H., Johnson, C., Holman, J. and Butler, D. (2003). *Impacts of land clearing on Australian wildlife in Queensland.* Report to WWF Australia. 48 pp.

Cogger, H.G. (1994). *Reptiles and Amphibians of Australia.* Reed, Sydney.

Cogger, H.G. (1995). *Reptiles and Amphibians of Australia.* Revised edition. Reed, Sydney.

Cohn, J.P. (1993). The flight of the Condor. *BioScience,* **43**, 206–209.

Cole, E.W. (1947). *Cole's Funny Picture Book.* 73rd Edition. Cole Publications, Melbourne.

Cole, F.R., Reeder, D.M. and Wilson, D.E. (1994). A synposis of the distribution patterns and the conservation of mammal species. *Journal of Mammalogy*, **75**, 266–276.

Collett, D.A. (1991). *Modelling Binary Data*. Chapman and Hall, London, England.

Collett, N. (2003). Short and long-term effects of prescribed fires in autumn and spring on surface-active arthropods in dry sclerophyll eucalypt forests of Victoria. *Forest Ecology and Management*, **182**, 117–138.

Collier, U. (1998). The environmental dimension of deregulation. In: *Deregulation in the European Union: Environmental Perspectives*. Collier, U. (Ed). Routledge, London.

Collin, P.H. (1988). *Dictionary of Ecology and the Environment*. Peter Collin Publishing, Great Britain.

Collins (1995). *Collins English Dictionary*. HarperCollins, Sydney.

Collins, S.L. (1992). Fire frequency and community heterogeneity in tallgrass prarie vegetation. *Ecology*, **73**, 2001–2006.

Collins, S.L., Glenn, S.M. and Gibson, D.J. (1995). Experimental analysis of intermediate disturbance and initial floristic composition: decoupling cause and effect. *Ecology*, **76**, 486–492.

Colwell, R.K. and Coddington, J.A. (1995). Estimating terrestrial biodiversity through extrapolation. pp. 5–12. In: *Biodiversity: Measurement and Estimation*. Hawksworth, D.L. (Ed). Chapman and Hall, London.

Commonwealth Department of Tourism (1992). *Tourism, Australia's Passport to Growth: A National Tourism Strategy*. Commonwealth Department of Tourism, Canberra.

Commonwealth of Australia (1992). *National Forest Policy Statement*. Australian Government Publishing Service, Canberra.

Commonwealth of Australia (1997). *Australia's First Approximation Report for the Montreal Process*. Commonwealth of Australia, Canberra, Australia. June 1997.

Commonwealth of Australia (1998). *A Framework of Regional (Sub-national) Level Criteria and Indiators of Sustainable Forest Management in Australia*. Commonwealth of Australia, Canberra, Australia.

Commonwealth of Australia (1999). *International Forest Conservation: Protected Areas and Beyond*. A discussion paper for the intergovernmental forum on forests. Commonwealth of Australia, Canberra, Australia. March 1999.

Commonwealth of Australia (2000). *Our Vital Resources – National Action Plan for Salinity and Water Quality in Australia*. Environment Australia, Canberra.

Commonwealth of Australia (2001a). *Australia's Forests – the Path to Sustainability*. Commonwealth of Australia, Canberra.

Commonwealth of Australia (2001b). *National Land and Water Audit 2001 – Australian Dryland Salinity Assessment 2000*. Commonwealth of Australia, Canberra.

Commonwealth of Australia (2001c). *National Land and Water Audit 2001 – Australian Native Vegetation Assessment 2001*. Commonwealth of Australia, Canberra.

Commonwealth of Australia (2001d). *National Land and Water Audit 2001 – Australian Agriculture Assessment 2001*. Commonwealth of Australia, Canberra.

Commonwealth of Australia (2001e). *National Land and Water Audit 2001 – Rangeland – tracking changes*. Commonwealth of Australia, Canberra.

Commonwealth of Australia (2002a). *Australia's natural resources 1997–2002 and beyond*. National Land and Water Resources Audit. Commonwealth of Australia, Canberra.

Commonwealth of Australia (2002b). *National Land and Water Audit 2001 – Australian Catchment, River and Estuary Assessment 2002*. Commonwealth of Australia, Canberra.

Commonwealth of Australia and Department of Natural Resources and Environment (1997). *Comprehensive Regional Assessment – Biodiversity. Central Highlands of Victoria*. The Commonwealth of Australia and Department of Natural Resources and Environment, Canberra.

Connell, J.H. (1978). Diversity in tropical rainforests and coral reefs. *Science*, **199**, 1302–1310.

Conner, R.N. (1988). Wildlife populations: minimally viable or ecologically functional? *Wildlife Society Bulletin*, **16**, 80–84.

Connor, E.F. and McCoy, E.D. (1979). The statistics and biology of the species-area relationship. *American Naturalist*, **113**, 791–833.

Conradt, L., Bodsworth, E.J., Roper, T.J. and Thomas, C.D. (2000). Non-random dispersal in the butterfly *Manioal jurtina*: Implications for metapopulation models. *Proceedings of the Royal Society of London Series B*, **267**, 1505–1510.

Conservation International (2000). *Megadiversity Data Tables*. Washington, DC.

Conway, W.G. (1986). The practical difficulties and financial implications of endangered species breeding programmes. *International Zoo Yearbook*, **24**(25), 210–219.

Cooke, B., Henzell, R. and Murphy, E. (2004). International issues in the use of GMOs for the management of mammal populations. *Biocontrol News and Information*, **25**, 37N–41N.

Cooke, B.D. (1987). The effects of rabbit grazing on regeneration of sheoaks, *Allocasuarina verticillata* and saltwater ti-trees, *Melaleuca halmaturorum*, in the Coorong National Park, South Australia. *Australian Journal of Ecology*, **13**, 11–20.

Cooke, R.M. (1991). *Experts in Uncertainty: Opinion and Subjective Probability in Science.* Oxford University Press, Oxford.

Cooper, C.B. and Walters, J.R. (2002). Experimental evidence of disrupted dispersal causing decline of an Australian passerine in fragmented habitat. *Conservation Biology*, **16**, 471–478.

Cooper, C.B., Walters, J.R. and Ford, H. (2002) Effects of remnant size and connectivity on the response of Brown Treecreepers to habitat fragmentation. *Emu*, **102**, 249–256.

Copley, P. (1988). *The Stick-nest Rats of Australia: A Final Report to World Wildlife Fund.* Department of Environment and Planning, Adelaide, Australia.

Copley, P. (1995). Translocations of native vertebrates in South Australia: A review. pp. 35–42. In: *Reintroduction Biology of Australasian Fauna.* Serena, M. (Ed). Surrey Beatty, Chipping Norton, Australia.

Corbett, L. (2001). *The Dingo in Australia and Asia.* Australian Natural History. JB Books, Adelaide.

Cork, S.J., Margules, C.R. and Braithwaite, L.W. (1990). Implications of Koala nutrition and the ecology of other arboreal marsupials in south-eastern N.S.W. for the conservation management of Koalas. pp. 48–57. In: *Proceedings of the Koala Summit held at the University of Sydney 7–8 November 1988.* Lunney, D., Urquhart, C. and Read, P. (Eds). National Parks and Wildlife Service, NSW, Sydney.

Corn, P.S. (2003). Amphibian breeding and climate change: importance of snow in the mountains. *Conservation Biology*, **17**, 622–625.

Costanza, R., d'Arge, R., de Groot, R., Farber, S., Grasso, M., Hannon, B., Limburg, K., Nacem, S., O'Neill, R.V., Paruelo, J., Raskin, R.G., Sutton, P. and van den Belt, C. (1997). The value of the world's ecosystem services and natural capital. *Nature*, **387**, 253–260.

Costermans, L. (1994). *Native Trees and Shrubs of South-eastern Australia.* 2nd edition. Rigby, Sydney.

Costin, A.B., Gray, M., Totterdell, C.J. and Wimbush, D.J. (1979). *Koscuisko Alpine Flora.* CSIRO, Melbourne.

Cotton, P.A. (2003). Avian migration phenology and global climate change. *Proceedings of the National Academy of Science*, **100**, 12219–12222.

Coulson, T., Mace, G.M., Hudson, E. and Possingham, H. (2001). The use and abuse of Population Viability Analysis. *Trends in Ecological Evolution*, **16**, 219–221.

Courtenay, J., Start, T. and Sinclair, E. (1996). An update on the status and ecology of Gilbert's Potoroo. *Australian Mammal Society Newsletter* November 1996, 11.

Cowan, P. (1990). Family Phalangeridae. pp. 67–98. In: *The Handbook of New Zealand mammals.* King, C.M. (Ed). Oxford University Press, Auckland.

Cracraft, J. (1973). Phylogeny and evolution of ratite birds. *Ibis*, **116**, 494–521.

Cracraft, J. (1991). Patterns of diversification within continental biotas: Hierarchical congruence among the areas of endemism of Australian vertebrates. *Australian Systematic Botany*, **4**, 211–227.

Cracraft, J. (1992). The species of the birds-of-paradise (Paradisaeidae): applying the phylogenetic species concept to a complex pattern of diversification. *Cladistics*, **8**, 1–43.

Craig, J.L., Saunders, D.A. and Mitchell, N. (Eds) (2000). *Conservation in Production Environments: Managing the Matrix.* Surrey Beatty and Sons, Chipping Norton, Australia.

Cranston, P.S. (1990). Biomonitoring and invertebrate taxonomy. *Environmental Monitoring and Assessment*, **14**, 265–273.

Cranston, P.S. and Trueman, J.H. (1997). "Indicator" taxa in invertebrate biodiversity assessment. *Memoirs of the Museum of Victoria*, **56**, 267–274.

Crawley, M.J. (1993). *GLIM for Ecologists.* Blackwell Scientific Publications, Oxford.

Crawley, M.J., Brown, S.L., Hails, R.S., Kohn, D.D. and Rees, M. (2001). Biotechnology: transgenic crops in natural habitats. *Nature*, **409**, 682–683.

Crisp, M.D., Laffan, S., Linder, H.P. and Monro, A. (2001). Endemism in the Australian flora. *Journal of Biogeography*, **28**, 183–198.

Crockford, H. and Richardson, D.P. (1998a). Litterfall, litter and associated chemistry in a dry eucalypt forest and a pine plantation. 1. Litter and litterfall. *Hydrological Processes*, **12**, 365–384.

Crockford, H. and Richardson, D.P. (1998b). Litterfall, litter and associated chemistry in a dry eucalypt forest and a pine plantation. 2. Nutrient cycling by litter, throughfall and streamflow. *Hydrological Processes*, **12**, 365–400.

Crockford, H. and Richardson, D.P. (2002). Decomposition of litter in a dry eucalypt forest and a

Pinus radiata plantation in south-eastern Australia. *Hydrological Processes,* **16**, 3317–3327.

Croizat, L.C. (1960). *Principia Botanica: or, Beginnings of Botany.* Weldon and Wesley. Codicator, Hitchin, England.

Crome, F.H. (1976). Some observations on the biology of the Cassowary in northern Queensland. *Emu,* **76**, 8–14.

Crome, F.H. (1990). Vertebrates and succession. pp. 53–64. In: *Australian Tropical Rainforests: Science, Values, Meaning.* Webb, L.J. and Kikkawa, J. (Eds). CSIRO, Melbourne.

Crome, F.H. (1994). Tropical rainforest fragmentation: some conceptual and methological issues. pp. 61–76. In: *Conservation Biology in Australia and Oceania.* Moritz, C. and Kikkawa, J. (Eds). Surrey Beatty, Chipping Norton.

Crome, F.H. and Bentrupperbaumer, J. (1993). Special people, a special animal and a special vision: The first steps to restoring a fragmented tropical landscape. pp. 267–279. In: *Nature Conservation 3: Reconstruction of Fragmented Ecosystems.* Saunders, D.A., Hobbs, R.J. and Ehrlich, P.R. (Eds). Surrey Beatty, Chipping Norton.

Crome, F.H. and Moore, L.A. (1990). The Southern Cassowary (*Casuarius casuarius*) in north Queensland. *Australian Wildlife Research,* **17**, 369–385.

Cronk, Q.C. and Fuller, J.L. (1995). *Plant Invaders.* Chapman and Hall, London.

Cropper, S.C. (1993). *Management of Endangered Plants.* CSIRO Publishing, Melbourne.

Crossland, M.R. (2001). Ability of predatory native Australian fishes to learn to avoid toxic larvae of the introduced *Bufo marinus. Journal of Fish Biology,* **59**, 319–329.

Crossland, M.R. and Alford, R.A. (1998). Evaluation of the toxicity of eggs, hatchlings, and tadpoles of the introduced toad *Bufo marinus* (Anura: Bufonidae) to native aquatic predators. *Australian Journal of Ecology,* **23**, 129–137.

Crossman, N.D. (1999). The impact of European olive (*Olea europaea* L.) on Grey Box (*Eucalyptus microcarpa* Maiden) woodland in South Australia. Thesis submitted for Honours Degree in Environmental Management. School of Geography, Population and Environmental Management, Flinders University, South Australia.

Crow, J.F. and Dennison, C. (1988). Inbreeding and variance effective population numbers. *Evolution,* **42**, 482–495.

Crow, J.F. and Kimura, M. (1970). *An Introduction to Population Genetics Theory.* Burgess, Minneapolis, Minnesota.

Crowley, G.M. and Garnett, S.T. (1998). Vegetation change in the grasslands and grassy woodlands of east-central Cape York Peninsula. *Pacific Conservation Biology,* **4**, 132–148.

Crozier, R.H. (1990). From population genetics to phylogeny: uses and limits of mitochondrial DNA. *Australian Systematic Botany,* **3**, 111–124.

Crozier, R.H. (1992). Genetic diversity and the agony of choice. *Biological Conservation,* **61**, 11–15.

Crozier, R.H. and Kusmierski, R.M. (1994). Genetic distances and the setting of conservation priorities. pp. 227–237. In: *Conservation Genetics.* Loeschcke, V., Tomiuk, J. and Jain, S.K. (Eds). Birkhauser Verlag, Basel.

Crozier, R.H., Agapow, P. and Pedersen, K. (1999). Toward complete biodiversity assessment: an evaluation of the subterranean bacterial communities in the Oklo region of the sole surviving nuclear reactor. *FEMS Microbiology Ecology,* **28**, 325–334.

Crumpacker, D.W., Hodge, S.W., Friedley, D.F. and Gregg, W.P. (1988). A preliminary assessment of the status of major terrestrial and wetland ecosystems on federal and Indian lands in the United States. *Conservation Biology,* **2**, 103–115.

Csuti, B., Polasky, S., Williams, P.H., Pressey, R.L., Camm, J.D., Kershaw, M., Kiester, A.R., Downs, B., Hamilton, R., Huso, M. and Sahr, K. (1997). A comparison of reserve selection algorithms using data on terrestrial vertebrates in Oregon. *Biological Conservation,* **80**, 83–97.

Cubbage, F.W., Dvorak, W.S., Abt, R.C. and Pacheco, G. (1996). World timber supply and prospects: Models, projections, plantations and implications. Central America and Mexico Coniferous (CAMCORE) Annual Meeting, Bali, Indonesia.

Culver, D.C., Master, L.S., Christman, M.C. and Hobbs, H.H. (2000). Obligate cave fauna of the 48 contiguous United States. *Conservation Biology,* **14**, 386–401.

Cunningham, M. and Moritz, C. (1998). Genetic effects of forest fragmentation on a rainforest restricted lizard (Scincidae, *Gnypetoscincus queenslandiae*). *Biological Conservation,* **83**, 19–30.

Cunningham, R.B. and Welsh, A.H. (1996). *Coral Sea Reserves: Monitoring seabird populations 1992–1996. Survey design and analysis.* (1996) Report to The Australian Nature Conservation Agency, Canberra, Australia.

Cunningham, R.B., Chambers, R. and Donnelly, C. (1995). *Coral Sea Reserves: North East Herald Islet Seabird Survey – Statistical Analysis of July 1994 Survey.* Unpublished report for The Australian Nature Conservation Agency, Canberra, Australia.

Cunningham, R.B., Lindenmayer, D.B., Nix, H.A. and Lindenmayer, B.D. (1999). Quantifying observer heterogeneity in bird counts. *Australian Journal of Ecology,* **24**, 270–277.

Cunningham, S.A. (2000). Depressed pollination in habitat fragments causes low fruit set. *Proceedings of the Royal Society of London. Series B*, **267**, 1149–1152.

Cunningham, S.A., Floyd, R.B. and Weir, T.A. (2005). Do *Eucalyptus* plantations host an insect community similar to remnant *Eucalyptus* forest? *Austral Ecology*, **30**, 103–117.

Cutler, A. (1991). Nested faunas and extinction in fragmented habitats. *Conservation Biology,* **5**, 496–505.

Daily, G.C. (1997a). Introduction: what are ecosystem services? pp. 1–10. In: *Nature's Services: Societal Dependence on Natural Ecosystems.* Daily, G.C. (Ed). Island Press, Washington, DC.

Daily, G.C. (Ed) (1997b). *Nature's Services: Societal Dependence on Natural Ecosystems.* Island Press, Washington, DC.

Daily, G.C. (1999). Developing a scientific basis for managing Earth's life support systems. *Conservation Ecology*, **3**. [online] URL: http://www.ecologyandsociety.org/vol3/iss2/art14/

Daily, G.C. (2000). Management objectives for the protection of ecosystem services. *Environmental Science and Policy,* **3**, 333–339.

Daily, G.C. and Walker, B.H. (2000). Seeking the great transition. *Nature*, **403**, 243–245.

Dale, M., Kershaw, P., Kikkawa, J., Parsons, P. and Webb, L. (1980). Resolution by the Ecological Society of Australia on Australian rainforest conservation. *Bulletin of the Ecological Society of Australia,* **10**, 6–7.

Dale, V.H., Brwon, S., Haeuber, R.A., Hobbs, N.T., Huntly, N., Naiman, R.J., Riebsame, W.E., Turner, M.G. and Valone, T.J. (2000). Ecological principles and guidelines for managing the use of land. *Ecological Applications*, **10**, 639–670.

Dangerfield, M., Pik, A., Britton, D., Holmes, A., Gikllings, M., Olivwer, I., Briscoe, D. and Beattie, A. (2003). Patterns of invertebrate biodiversity across a natural edge. *Austral Ecology*, **28**, 227–236.

Daniels, M.J. and Corbett, L. (2003). Redefining introgressed protected mammalss: when is a wildcat a wild cat and a dingo a wild dog? *Wildlife Research*, **30**, 213–218.

Danks, A. (1995). Noisy Scrub-bird translocations: 1983–1992. pp. 129–134. In: *Reintroduction Biology of Australasian Fauna.* Serena, M. (Ed). Surrey Beatty, Chipping Norton, Australia.

Dargarvel, J. (1995). *Fashioning Australia's Forests.* Oxford University Press, Melbourne, Australia.

Darveau, M., Beauchesne, P., Belanger, L., Hout, J. and Larue, P. (1995). Riparian forest strips as habitat for breeding birds in boreal forest. *Journal of Wildlife Management,* **59**, 67–78.

Daszak, P., Berger, L., Cunningham, A.A., Hyatt, A.D., Green, D.E. and Speare, R. (2000). Emerging infectious diseases and amphibian population declines. *Emerging Infectious Diseases*, **5**, 735–748. [online] URL: http:// www.cdc.gov/ncidod/eid/vol5no6/daszak.htm

Date, E.M., Ford, H.A. and Recher, H.F. (1991). Frugivorous pigeons, stepping stones and weeds in northern NSW. pp. 241–245. In: *Nature Conservation 2: The Role of Corridors.* Saunders, D.A. and Hobbs, R.J. (Eds). Surrey Beatty, Chipping Norton, NSW.

Date, E.M., Recher, H.F., Ford, H.A. and Stewart, D.A. (1996). The conservation and ecology of rainforest pigeons in northern New South Wales. *Pacific Conservation Biology,* **2**, 299–308.

Davey, C. (1997). Observations on the Superb Parrot within the Canberra District. *Canberra Bird Notes*, **22**, 1–14.

Davey, S.M. (1984). Habitat preferences of arboreal marsupials within a coastal forest in southern New South Wales. pp. 509–516. In: *Possums and Gliders.* Smith, A.P. and Hume, I.D. (Eds). Australian Mammal Society and Surrey Beatty, Sydney.

Davey, S.M. (1989a). The environmental relationships of arboreal marsupials in a Eucalypt forest: a basis for Australian wildlife management. PhD thesis, The Australian National University, Canberra.

Davey, S.M. (1989b). Thoughts towards a forest wildlife management strategy. *Australian Forestry,* **52**, 56–67.

Davidson, S. (2004). A bold blueprint. *Ecos*, **118**, 13–16.

Davie, J. (1997). Is biodiversity really the link between conservation of ecologically sustainable management? A reflection on paradigm and practice. *Pacific Conservation Biology,* **3**, 83–90.

Davies, K.F. (1993). The effects of habitat fragmentation on carabid beetles (Coleoptera: Carabidae) at Wog Wog, south-eastern NSW. Honours thesis, The Australian National University, Canberra.

Davies, K.F. and Margules, C.R. (1998). Effects of habitat fragmentation on carabid beetles: experimental evidence. *Journal of Animal Ecology*, **67**, 460–471.

Davies, K.F., Gascon, C. and Margules, C.R. (2001b). Habitat fragmentation. pp. 81–97. In: *Conservation Biology. Research Priorities for the Next Decade.* Soulé, M.E. and Orions, G.H. (Eds). Island Press, Washington DC.

Davies, K.F., Margules, C.R. and Lawrence, J.F. (2000). Which traits of species predict population declines in experimental forest fragments? *Ecology,* **81**, 1450–1461.

Davies, K.F., Melbourne, B.A. and Margules, C.R. (2001a). Effects of within- and between-patch processes on community dynamics in a fragmentation experiment. *Ecology*, **82**, 1830–1846.

Davies, R. (1995). *Threatened Plant Species Management in National Parks and Wildlife Act Reserves in South Australia.* Botanic Gardens of Adelaide and State Herbarium, Adelaide.

Davies, R. (1996). *Threatened Plant Species on Roadsides: Kangaroo Island, South Australia.* Department of Environment and Natural Resources, Adelaide.

Davis, J. (2003). Indigenous land management. pp. 219–223. In: *Australia Burning. Fire Ecology, Policy and Management Issues.* Cary, G., Lindenmayer, D.B. and Dovers, S. (Eds). CSIRO Publishing, Melbourne.

Davis, L.S., Johnson, K.N., Bettinger, P.S. and Howard, T.E. (2001). *Forest Management to Sustain Ecological, Economic, and Social Values.* 4th edition. McGraw-Hill, New York.

Dawson, T.J. (2000). *Kangaroos, Biology of the Largest Marsupials.* UNSW Press, Sydney, Australia.

de Forges, B.R., Koslow, J.A. and Poore, G.C.B. (2000). Diversity and endemism of the benthic seamount fauna in the southwest Pacific. *Nature*, **405**, 944–947.

De Marco, P. and Coelho, F. (2004). Services performed by the ecosystem: forest remnants influence agricultural culture's pollination and production. *Biodiversity and Conservation*, **13**, 1245–1255.

Debinski, D.M. and Holt, R.D. (2000). A survey and overview of habitat fragmentation experiments. *Conservation Biology,* **14**, 342–355.

Delin, A.E. and Andrén, H. (1999). Effects of habitat fragmentation on Eurasian Red Squirrel (*Sciurus vulgaris*) in a forest landscape. *Landscape Ecology,* **14**, 67–72.

Delong, S.C. (1996). Defining biodiversity. *Wildlife Society Bulletin*, **24**, 738–749.

Delong, S.C. and Kessler, W.B. (2000). Ecological characteristics of mature forest remnants left by wildfire. *Forest Ecology and Management,* **131**, 93–106.

den Boer, P.J. (1968). Spreading of risk and stabilization of animal numbers. *Acta Biotheoretica,* **18**, 165–194.

den Boer, P.J. (1981). On the survival of populations in a heterogeneous and variable environment. *Oecologia,* **50**, 39–53.

Dennis, B., Brown, B.E., Stage, A.R., Burkhart, H.E. and Clark, S. (1985). Problems of modelling growth and yield of renewable resources. *The American Statistician,* **39**, 374–383.

Denny, C.M. and Babcock, R.C. (2004). Do partial marine reserves protect reef fish assemblages? *Biological Conservation*, **116**, 119–129.

Department of Conservation and Environment (1991). *Annual Report 1990–1991.* Department of Conservation and Environment, Melbourne.

Department of Conservation and Land Management (2000). *Salinity Action Plan. Biological survey of the agricultural zone: Status Report.* June 2000. Department of Conservation and Land Management, Perth.

Department of Conservation Forests and Lands (1989). *Code of Practice. Code of Forest Practices for Timber Production.* Revision No. 1, May 1989 edition. Department of Conservation, Forests and Lands, Melbourne.

Department of Environment and Heritage (2003). *Australia's World Heritage – location map.* Department of Environment and Heritage, Canberra. [online] URL: http:// www.deh.gov.au/heritage/worldheritage/index.html

Department of Environment and Heritage (2005a). *Key Threatening Processes.* Department of Environment and Heritage, Canberra. http://www.deh.gov.au/cgi-bin/sprat/ publicgetkeythreats.pl

Department of Environment and Heritage (2005b). *Threatened species and threatened ecological communities.* Department of Environment and Heritage, Canberra. [online] URL: http:// www.deh.gov.au/biodiversity/threatened/ index.html

Department of Environment and Heritage (2005c). *Commercial kangaroo harvest quotas – 2005.* Department of Environment and Heritage, Canberra. [online] URL: http:// www.deh.gov.au/biodiversity/trade-use/ publications/kangaroo/quotas-background-2005.html

Department of Environment and Heritage (2005d). *Parks and reserves. National reserves system.* Department of Environment and Heritage, Canberra. [online] URL: http:// www.deh.gov.au/parks/nrs/capad/2002/national/ nat-type02.html

Department of Environment and Heritage (2005e). *The convention on wetlands.* Department of Environment and Heritage , Canberra. http:// www.deh.gov.au/water/wetlands/publications/ index.html

Department of Environment and Heritage (2005f). *Commonwealth marine protected areas.*

Department of Environment and Heritage, Canberra. [online] URL: http://www.deh.gov.au/coasts/mpa/nrmpsa/index.html

Department of Indigenous Affairs (2002). *Fact Sheet.* Department of Indigenous Affairs, Perth January 2002. [online] URL: http://www.dia.wa.gov.au

Department of Infrastructure, Planning and Environment (2002). *Threatened species list of the Northern Territory*. Department of Infrastructure, Planning and Environment, Darwin [online] URL: http://www.ipe.nt.gov.au/news/2002/10/threatened/classification_summary.html

Department of Natural Resources and Environment. (1996). *Code of Forest Practice for Timber Production.* Revision No. 2. Department of Natural Resources and Environment, Melbourne.

Department of Natural Resources and Environment (2002). BushTender Trial – Gippsland. Department of Natural Resources and Environment, Melbourne.

Department of Primary Industries, Water and Environment (2002). *Threatened species.* Department of Primary Industries, Water and Environment. [online] URL: http://www.dpiwe.tas.gov.au/inter.nsf/ThemeNodes/RLIG-53KUPV?open

Department of Sustainability and Environment (2003). *Advisory List of Threatened Vertebrate Fauna in Victoria – 2003.* Department of Sustainability and Environment, Melbourne.

Department of the Environment and Heritage (2002). *Annual Report of the Supervising Scientist 2000–2001.* Commonwealth of Australia, Canberra.

Department of the Environment Sports and Territories (1995). *Native Vegetation Clearance, Habitat Loss and Biodiversity Decline.* Department of the Environment Sports and Territories, Canberra. Biodiversity Series Paper No. 6.

Department of the Environment Sports and Territories (1996). *National strategy for the Conservation of Australia's Biological Diversity.* Australian Government Publishing Service, Canberra, Australia.

Desrochers, A. and Hannon, S.J. (1997). Gap crossing decisions by forest songbirds during the post-fledging period. *Conservation Biology,* **11**, 1204–1210.

Devall, B. and Sessions, G. (1985). *Deep Ecology.* G.M. Smith, Salt Lake City, Utah.

Deville, A. and Harding, R. (1997). *Applying the Precautionary Principle.* Federation Press, Sydney.

Deweerdt, S. (2003). Lost and found. *Conservation in Practice,* **4**, 32–38.

Dial, R.J. (1995). Species-area curves and Koopowitz *et al.*'s simulation of stochastic extinctions. *Conservation Biology,* **9**, 960–961.

Diamond, J. (1986). Overview: laboratory experiments, field experiments and natural experiments. pp. 3–22. In: *Community Ecology*. Diamond, J. and Case, T.J. (Eds). Harper and Row, New York.

Diamond, J. (1989). Overview of recent extinctions. pp. 37–41. In: *Conservation for the Twenty-first Century.* Western, D. and Pearl, M. (Eds). Oxford University Press, New York.

Diamond, J.M. (1973). Distributional ecology of New Guinea birds. *Science,* **179**, 759–769.

Diamond, J.M. (1975). The island dilemma: Lessons of modern biogeographic studies for the design of natural preserves. *Biological Conservation,* **7**, 129–146.

Diamond, J.M. (1984). 'Normal' extinctions of isolated populations. pp. 191–246. In: *Extinctions.* Nitecki, M.H. (Ed). University of Chicago Press, Chicago.

Diamond, J.M. (1985). How many unknown species are yet to be discovered? *Nature,* **315**, 538–539.

Diamond, J.M. (1987). Extant unless proven extinct? Or, extinct unless proven extant. *Conservation Biology,* **1**, 77–79.

Diamond, J.M. and May, R.M. (1976). Island biogeography and the design of nature reserves. pp. 163–186. In: *Theoretical Ecology: Principles and Applications.* May, R.M. (Ed). W.B. Saunders, Philadelphia.

Diamond, J.M., Bishop, K.D. and van Balen, S. (1987). Bird survival in an isolated Javan woodlot: island or mirror. *Conservation Biology,* **2**, 132–42.

DiCatstri, F. and Younes, T. (1989). Ecosystem function of biological diversity. *Biology International,* **22**, 1–18.

Dickman, C.R. (1993). Raiders of the last ark: Cats in island Australia. *Australian Natural History,* **24**, 44–52.

Dickman, C.R., Parnaby, H.E., Crowther, M.S. and King, D.H. (1998). *Antechinus agilis* (Marsupialia: Dasyuridae), a new species from the *A.stuartii* complex in south-eastern Australia. *Australian Journal of Zoology,* **46**, 1–26.

Dickman, C.R., Pressey, R.L., Lim, L. and Parnaby, H.E. (1993). Mammals of particular conservation concern in the Western Division of New South Wales. *Biological Conservation,* **65**, 219–248.

Dickson, D.L. (1993). The Red-throated Loon as an indicator of environmental quality. Canadian Wildlife Service. Occasional paper No. 73.

Dickson, R., Aldred, T. and Baird, I. (1997). National forest conservation reserves: recent developments. pp. 359–368. In: *National Parks and Protected Areas:*

Selection, Delimitation and Management. Pigram, J.J. and Sundell, R. (Eds). Centre for Water Policy Research, University of New England, Armidale, Australia.

Didham, R.K. (1997). The influence of edge effects and forest fragmentation on leaf litter invertebrates in central Amazonia. pp. 55–70. In: *Tropical Forest Remnants: Ecology, Management and Conservation of Fragmented Communities.* Laurance, W.F. and Bierregaard Jr, R.O. (Eds). University of Chicago Press, Chicago, Illinois, USA.

Dieckmann, U., O'Hara, B. and Weisser, W. (1999). The evolutionary ecology of dispersal. *Trends in Ecology and Evolution*, **14**, 88–90.

Digby, P.G. and Kempton, R.A. (1987). *Multivariate analysis of ecological communities.* Chapman and Hall, London.

Disney, H.J. de S. and Stokes, A. (1976). Birds in pine and native forests. *Emu,* **76**, 133–138.

Dixon, K.W. (1994). Towards integrated conservation of Australian endangered plants – the Western Australian model. *Biodiversity and Conservation,* **3**, 148–159.

Doak, D. (1989). Spotted Owls and old growth logging in the Pacific Northwest. *Conservation Biology,* **3**, 389–396.

Doak, D. and Mills, L.S. (1994). A useful role for theory in conservation. *Ecology*, **75**, 615–626.

Dodd, C.K. and Siegel, R.A. (1991). Relocation, repatriation, and translocation of amphibians and reptiles: Are they conservation strategies that work? *Herpetologica,* **47**, 336–350.

Dodson, J.J., Gibson, R.J., Cunjak, R.A., Friedland, K.D., de Lainiz, C.G., Gross, M.R., Newbury, R., Nielsen, J.L., Power, M.E. and Roy, S. (1998). Elements in the development of conservation plans for Atlantic Salmon (*Salmo salar*). *Canadian Journal of Fisheries and Aquatic Sciences*, **55** (Supplement 1), 312–323.

Dooley, J.L. and Bowers, M.A. (1998). Demographic responses to habitat fragmentation: experimental tests at the landscape and patch scale. *Ecology,* **79**, 969–980.

Dorges, B. and Huecke, J. (1995). One-humped Camel. pp. 718–721. In: *The Mammals of Australia.* Strahen, R. (Ed). New Holland Publishers, Sydney.

Doughty, R.W. (2001). *The Eucalyptus.* John Hopkins University Press, Baltimore.

Douglas, J.W., Gooley, G.J. and Ingram, B.A. (1994). *Trout Cod (Maccullochella Macquariensis) resource handbook and research and recovery plan.* DCNR, Melbourne.

Dovers, S. (1994). Sustainability and 'pragmatic' environmental history: a note from Australia. *Environmental History* Review, **18**, 21–36.

Dovers, S. (2002). Sustainability: Reviewing Australia's progress, 1992–2002. *International Journal of Environmental Studies*, **59**, 559–571.

Dovers, S. and Lindenmayer, D.B. (1997). Managing the environment: Rhetoric, policy and reality. *Australian Journal of Public Administration,* **56**, 65–80.

Dovers, S. and Mobbs, C. (1997). An alluring prospect? Ecology and the requirements of adaptive management. pp. 39–52 In: *Frontiers in Ecology: Building the Links.* Klomp, N. and Lunt, I. (Eds). London: Elsevier.

Dovers, S. and Norton, T. (1994). Population, environment and sustainability: Reconstructing the debate. *Sustainable Development,* **2**, 1–7.

Dovers, S. and Wild River, S. (Eds) (2003). *Managing Australia's environment.* Sydney: Federation Press.

Dovers, S.R. (1990). *Sustainable development: theory and implementation in Australia.* Commonwealth Department of Arts, Sport, the Environment, Tourism and Territories and Australian National University, Canberra.

Dow, D.D. (1978). Reproductive behaviour of the Noisy Miner, a communally breeding honeyeater. *Living Bird,* **16**, 163–185.

Dow, D.D. (1979). The influence of nests on the social behaviour of males in *Manorina melanocephala*, a communally breeding honeyeater. *Emu,* **79**, 71–83.

Dow, D.D. and Whitmore, M.J. (1990). Noisy Miners: Variations on the theme of communality. pp. 559–592. In: *Cooperative Breeding in Birds.* Stacey, P.B. and Koenig W.D. (Eds). Cambridge University Press, Cambridge.

Dowling, T.E., DeMarais, B.D., Minckley, W.L., Douglas, M.E. and Marsh, P.C. (1992). Use of genetic characters in conservation biology. *Conservation Biology,* **6**, 7–8.

Driml, S. and Common, M. (1995). Economic and financial benefits of tourism in major protected areas. *Australian Journal of Environmental Management,* **2**, 19–29.

Driscoll, D., Milkovits, G. and Freudenberger, D. (2000). *Impact and use of firewood in Australia.* CSIRO Sustainable Ecosystems Report, Canberra. November 2000.

Drury, W.H. (1974). Rare species. *Biological Conservation,* **6**, 162–169.

Duffy, J.E. (2003). Biodiversity loss, trophic skew and ecosystem functioning. *Ecology Letters,* **6**, 680–687.

Dugteren, A. (1999). Conserving the future of the Great Artesian Basin. *Geo,* **20**, 1999.

Duncan, R.P., Bomford, M., Forysyth, D.M. and Conibear, L. (2001). High predictability in introduction outcomes and the geographic range size of introduced Australian birds: a role for climate. *Journal of Animal Ecology*, **70**, 621–632.

Dunstan, C.E. and Fox, B.J. (1996). The effects of fragmentation and disturbance of rainforest on ground-dwelling mammals on the Robertson Plateau, New South Wales, Australia. *Journal of Biogeography*, **23**, 187–201.

Dyne, G. (1991). *Attributes of old growth forest in Australia.* Bureau of Rural Resources, Department of Primary Industries and Energy, Canberra. Working Paper No. WP/4/92.

Dyne, G.R. and Walton, D.W. (1987). Exploitations and introductions. In: *Fauna of Australia.* Vol. 1a. Dyne, G.R. and Walton, D.W. (Eds). Australian Government Publishing Service, Canberra. 1999). *The Environmental Law Handbook: planning and land use in New South Wales.* 3rd edn. Redfern Legal Centre Publishing, Redfern, NSW.

East, R. (1981). Species-area curves and populations of large mammals in African savannah reserves. *Biological Conservation*, **21**, 111–126.

Eberhardt, L.L. and Thomas, J.M. (1991). Designing environmental field studies. *Ecological Monographs*, **61**, 53–74.

Ecological Society of Australia (2003). *Position Statement. Environmental Impact Assessment.* http://life.csu.edu.au/esa/esaPSeia.html

Eddy, D. (2002). *Managing native grassland. A guide to management for conservation, production and landscape protection.* WWF Australia, Sydney.

Edgar, G.J. (2001). *Australian Marine Habitats in Temperate Waters.* Reed New Holland Publishers (Australia), Sydney.

Edgar, G.J. and Samson, C.R. (2004). Catastrophic decline in mollusc diversity in eastern Tasmania and its concurrence with shelfish fisheries. *Conservation Biology*, **18**, 1579–1588.

Edgecombe, J. (1999). *A history of Flinders Island.* Second Edition. Published by the author. Whitemark, Flinders Island.

Edwards, G.P., Saalfelf, K. and Clifford, B. (2004). Population trend of feral camels in the Northern Territory, Australia. *Wildlife Research*, **31**, 509–517.

Edwards, M.E. (1986). Disturbance history of four Snowdonian woodlands and their relation to Atlantic bryophyte distributions. *Biological Conservation*, **37**, 301–320.

Edwards, T.C., Deshler, E.T., Foster, D. and Moisen G.G. (1996). Adequacy of wildlife habitat relation models for estimating spatial distributions of terrestrial vertebrates. *Conservation Biology*, **10**, 263–270.

Egler, F.E. (1954). Vegetation science concepts. I: Initial floristic composition – a factor in old field vegetation development. *Vegetation*, **4**, 412–417.

Ehrenfield, D.W. (1988). Why put a value on biodiversity? pp. 212–216. In: *Biodiversity.* Wilson, E.O. and Peters, F.M. (Eds). National Academy Press, Washington.

Ehrlich, P.R. (1990). Habitats in crisis: Why we should care about the loss of species. *Forest Ecology and Management*, **35**, 5–11.

Ehrlich, P.R. (1992). Population biology of Checkerspot Butterflies and the preservation of global diversity. *Oikos*, **63**, 6–12.

Ehrlich, P.R. (1997). *A world of wounds: ecologists and the human dilemma.* Ecology Institute, Oldendorf, Germany.

Ehrlich, P.R. and Daily, G.C. (1993). Science and the management of natural resources. *Ecological Applications*, **3**, 558–560.

Ehrlich, P.R. and Ehrlich, A.H. (1990). *The population explosion.* Simon and Schuster, New York.

Eichel, H. (2000). Statement by Mr Hans Eichel German Minister of Finance to the International Monetary and Financial Committee Meeting Prague, September 24, 2000 (http://www.imf.org/external/am/2000/imfc/eng/deu.htm).

Eldredge, L.G. and Miller, S.E. (1995). How many species are there in Hawaii? *Bishop Museum Occasional Paper*, **41**, 1–18.

Eldridge, K., Davidson, J., Harwood, C. and van Wyk, G. (1994). *Eucalypt domestication and breeding.* Clarendon Press, Oxford.

Eldridge, K.G. and Griffin, A.R. (1989). Selfing effects in *Eucalyptus regnans. Silvae Genetica*, **32**, 216–221.

Eldridge, M.D., King, J., Loupis, A.K., Spencer, P.B., Taylor, A.C., Pope, L. and Hall, G. (1999). Unprecedented low levels of genetic variation and inbreeding depression in an island population of the Black-footed Rock-wallaby. *Conservation Biology*, **13**, 531–541.

Elith, J. (2000). Quantitative methods for modeling species habitat: comparative performance and an application to Australian plants. pp. 39–58. In: *Quantitative Methods in Conservation Biology.* Ferson, S. and Burgman, M. (Eds). Springer, Berlin.

Elith, J. (2002). Quantitative methods for modelling species habitat: comparative performance and an application to Australian plants. pp. 39–58. In: *Quantitative Methods for Conservation Biology.* Ferson, S. and Burgman, M. (Eds). Springer-Verlag, New York, USA.

Elith, J. and Burgman, M.A. (2003). Habitat models for PVA. pp. 203–235. In: *Population Viability in Plants: Conservation, Management and Modelling of Rare Plants*. Brigham, C.A. and Schwartz, M.W. (Eds). Springer-Verlag, New York, USA.

Elliott, C.P., Yates, C.J., Ladd, P.G. and Coates, D.J. (2002). Morphometric, genetic and ecological studies clarify the conservation status of a rare Acacia in Western Australia. *Australian Journal of Botany*, **50**, 63–73.

Ellison, G.T., Taylor, P.J., Nix, H.A., Bronner, G.N. and McMahon, J.P. (1993). Climatic adaptation of body size among pouched mice (*Saccostomus campestris*: Cricetidae) in the southern African subregion. *Global Ecology and Biogeography Letters*, **3**, 41–47.

Ellner, S.P. and Fieberg, J. (2003). Using PVA for management despite uncertainty: effects of habitat, hatcheries, and harvest on Salmon. *Ecology*, **84**, 1359–1369.

Ellner, S.P., Fieberg, J., Ludwig, D. and Wilcox, C. (2002). Precision of Population Viability Analysis. *Conservation Biology*, **16**, 258–261.

Ellsworth, J.W. and McComb, B.C. (2003). Potential effects of Passenger Pigeon flocks on the structure and composition of presettlement forests of eastern North America. *Conservation Biology*, **17**, 1548–1558.

Elton, C.S. (1927). *Animal Ecology*. Sidgwick and Jackson, London.

Engeman, R.M. and Linnell, M.A. (1998). Trapping strategies for deterring the spread of Brown Tree Snakes from Guam. *Pacific Conservation Biology*, **4**, 348–353.

Engeman, R.M., Linnell, M.A., Aguon, P., Manibusan, A., Sayama, S. and Techaira, A (1999). Implications of Brown Tree Snake captures from fences. *Wildlife Research*, **26**, 111–116.

Enoksson, B., Angelstam, P. and Larsson, K. (1995). Deciduous forest and resident birds: the problem of fragmentation within a coniferous forest landscape. *Landscape Ecology*, **10**, 267–275.

Environment Australia (2000). *Revision of the Interim Biogeographic Regionalisation of Australia (IBRA) and the Development of Version 5.1. – Summary Report*. Department of Environment and Heritage, Canberra.

Environmental Protection Agency (1996). *Best practice environmental management in mining. Environmental auditing*. Commonwealth of Australia, Canberra.

Environmental Resources Management Australia (2001). *Environmental Impact Statement. Wood processing and metallurgical carbon facility for Australian Silicon Operations Pty Ltd*. Environmental Resources Management Australia November 2001.

Er, K. (1997). Effects of eucalypt dieback on bird species diversity in remnants of native woodland. *Corella*, **21**, 69–76.

Erwin, D.H. (1998). The end and the beginning: recoveries from mass extinctions. *Trends in Ecology and Evolution*, **13**, 344–349.

Erwin, T.L. (1982). Tropical forests: Their richness in coleoptera and other species. *Coleopterists Bulletin*, **36**, 74–75.

Erwin, T.L. (1991). How many species are there? Revisited. *Conservation Biology*, **5**, 1–4.

ESDWG (1991a). (Ecologically Sustainable Developement Working Group Report) *Final report – forest use*. Australian Government Publishing Service, Canberra.

ESDWG (1991b). (Ecologically Sustainable Developement Working Group Report) *Final report – tourism*. Australian Government Publishing Service, Canberra.

Esseen, P. (1994). Tree mortality patterns after experimental fragmentation of an old-growth conifer forest. *Biological Conservation*, **68**, 19–28.

ESSS (1995). (Endangered Species Scientific Subcommittee) *Listing ecological communities under the Endangered Species Protection Act*. Australian Nature Conservation Agency, Canberra. Public Discussion Paper, 1992.

Estades, C.F. (2001). The effect of breeding-habitat patch size on bird population density. *Landscape Ecology*, **16**, 161–173.

Estades, C.F. and Temple, S.A. (1999). Deciduous-forest bird communities in a fragmented landscape dominated by exotic pine plantations. *Ecological Applications*, **9**, 573–585.

Estes, J.A. (1998). Concerns about rehabilitation of oiled wildlife. *Conservation Biology*, **12**, 1156–1157.

Estes, J.A., Duggins, D.O. and Rathbun, G.B. (1989). The ecology of extinctions in kelp forest communities. *Conservation Biology*, **3**, 252–264.

Estes, J.A., Tinker, M.T., Williams, T.M. and Doak, D.F. (1998). Killer Whale predation on Sea Otters linking oceanic and nearshore ecosystems. *Science*, **282**, 473–476.

Euskirchen, E.S., Chen, J.Q. and Bi, R.C. (2001). Effects of edges on plant communities in a managed landscape in northern Wisconsin. *Forest Ecology and Management*, **148**, 93–108.

Everett, R.A. (2000). Patterns and pathways of biological invasions. *Trends in Ecology and Evolution*, **15**, 177–178.

Ewel, J.J. (1993). The power of biology in the sustainable land use equation. *Biotropica*, **25**, 250–251.

Fagan, W.F., Meir, E., Prendergast, J., Folarin, A. and Karieva, P. (2001). Characterizing population vulnerability for 758 species. *Ecology Letters*, **4**, 132–138.

Fagan, W.F., Unmack, P.J., Burgess, C. and Minckley, W.L. (2002). Rarity, fragmentation and extinction risk in desert fishes. *Ecology*, **83**, 3250–3256.

Fahrig, L. (1992). Relative importance of spatial and temporal scales in a patchy environment. *Theoretical Population Biology*, **41**, 300–314.

Fahrig, L. (1997). Relative effects of habitat fragmentation and habitat loss on population extinction. *Journal of Wildlife Management*, **61**, 603–610.

Fahrig, L. (1998). When does fragmentation of breeding habitat affect population survival? *Ecological Modelling*, **105**, 273–292.

Fahrig, L. (1999). Forest loss and fragmentation: Which has the greater effect on persistence of forest-dwelling animals? pp. 87–95. In: *Forest Fragmentation. Wildlife Management Implications*. Rochelle, J.A., Lehmann, L.A. and Wisniewski, J. (Eds). Brill, Leiden, Germany.

Fahrig, L. (2003). Effects of habitat fragmentation on biodiversity. *Annual Review of Ecology, Evolution and Systematics*, **34**, 487–515.

Fahrig, L. and Merriam, G. (1994). Conservation of fragmented populations. *Conservation Biology*, **8**, 50–59.

Fairweather, P.G. (1991). Statistical power and design requirements for environmental monitoring. *Australian Journal of Marine Freshwater Research*, **42**, 555–67.

Faith, D.P. (1992). Conservation evaluation and phylogenetic diversity. *Biological Conservation*, **61**, 1–10.

Faith, D.P. (1994). Genetic diversity and taxonomic priorities for conservation. *Biological Conservation*, **68**, 69–74.

Faith, D.P. and Walker, P.A. (1993). *DIVERSITY: a software package for sampling phylogenetic and environmental diversity*. User's Guide. 1.0. CSIRO Division of Wildlife and Ecology, Canberra.

Faith, D.P. and Walker, P.A. (1996). How do indicator groups provide information about the relative biodiversity of different sets of areas?: On hotspots, complementarity and pattern-based approaches. *Biodiversity Letters*, **3**, 18–25 .

Falconer, D.S. (1989). *Introduction to quantitative genetics*. 3rd edition. Longman, London.

Falk, D.A. (1990). Endangered forest resources in the U.S.: Integrated strategies for conservation of rare species and genetic diversity. *Forest Ecology and Management*, **35**, 91–107.

FAO (1990). *Interim report on forest resources assessment 1990 project*. Food and Agricultural Organisation, Rome, Italy. Committee on Forestry Tenth Session.

FAO (1995). *Harmonisation of criteria and indicators for sustainable forest management*. Food and Agricultural Organisation, Rome, Italy. Report on the FAO/ITTO expert consulation. February, 1995.

FAO (2000). On definition of forest and forest change. *FRA Working Paper No. 33*. Food and Agriculture Organization of the United Nations, Rome. (www.fao.org/forestry/fo/fra/index.jsp).

FAO (2001). Global *Forest Resource Assessment 2000*. Food and Agriculture Organization of the United Nations, Rome. (http://www.fao.org/forestry/fo/fra/index.jsp).

FAO (2003a). *Indigenous knowledge and biodiversity*. Food and Agriculture Organization of the United Nations, Rome, Italy. [online] URL: http://www.fao.org/DOCREP/004/V1430E/V1430E03.htm

FAO (2003b). *Harvesting nature's diversity*. Food and Agriculture Organization of the United Nations, Rome, Italy. [online] URL: http://www.fao.org/DOCREP/004/V1430E/V1430E06.htm

FAO Fisheries Department (1994). *The precautionary approach to fisheries with reference to straddling fish stocks and highly migratory fish stocks*. Food and Agricultural Organisation, Rome. Fisheries Circular No. 871.

Farrell, T.P. and Kratzing, D.C. (1996). Environmental effects. pp. 14–45. In: *Environmental Management in the Australian Minerals and Energy Industries*. Mulligan, D.R. (Ed). UNSW Press, Sydney, and Australian Minerals and Energy Environment Foundation.

Farrier, D. and Tucker, L. (2000). Wise use of wetlands under the Ramsar Convention: A challenge for meaningful interpretation of international law. *Journal of Environmental Law*, **12**, 21–42.

Farrier, D., Lyster, R. and Pearson, L. (1999). *Environmental Law Handbook: Planning and Land Use in New South Wales*. Redfern Legal Centre Publishing, Redfern, NSW.

Fazey, I., Fischer, J. and Lindenmayer, D.B. (2005a). What do conservation biologists publish? *Biological Conservation*, **124**, 63–73.

Fazey, I., Salisbury, J., Lindenmayer, D.B., Douglas, R. and Maindonald, J. (2005b). Can methods applied in medicine be used to summarize and disseminate conservation research? *Environmental Conservation*, **31**, 190–198.

Fensham, R.J. (1996). Land clearance and conservation of inland dry rainforest in North Queensland, Australia. *Biological Conservation* **75**, 289–298.

Fensham, R.J. and Fairfax, R. (1996). The grassy balds on the Bunya Mountains, south-eastern Queensland:

floristics and conservation issues. *Cunninghamia*, **4**, 511–530.

Fensham, R.J., Fairfax, R.J., Butler, D.W. and Bowman, D.M. (2003). Effects of fire and drought in a tropical eucalypt savanna colonized by rainforest. *Journal of Biogeography*, **30**, 1405–1414.

Ferguson, I.S. (1996). *Sustainable forest management.* Oxford University Press, Melbourne, Australia.

Ferreras, P. (2001). Landscape structure and asymm-etrical inter-patch connectivity in a metapopulation of the endangered Iberian Lynx. *Biological Conservation*, **100**, 125–136.

Ferrier, S. (1985). Habitat requirements of a rare species, the rufous scrub bird. pp. 241–248. In: *Birds of the Eucalypt Forests and Woodlands: Ecology, Conservation and Management.* Keast, A., Recher, H.F., Ford H. and Saunders, D. (Eds). Surrey Beatty, Sydney, Australia.

Ferrier, S. (1991). Computer-based spatial extension of forest-fauna survey data: Current issues, problems and directions. pp. 221–227. In: *Conservation of Australia's Forest Fauna.* Royal Zoological Society of NSW, Sydney.

Fieberg, J. and Ellner, S.P. (2001). Stochastic matrix models for conservation and management: a comparative review of methods. *Ecology Letters* , **4**, 244–266.

Fiedler, P.L. and Jain, S.K. (1990). *Conservation biology: the theory and practice of nature conservation.* Chapman and Hall, London.

Figgis, P. (1999). Conservation on private lands: the Australian experience. IUCN, Gland, Switzerland.

Fisher, A. and Goldney, D.C. (1998). Native forest fragments as critical habitat in a softwood forest landscape. *Australian Forestry*, **61,** 287–295.

Fischer, J. and Lindenmayer, D.B. (2000). A review of relocation as a conservation management tool. *Biological Conservation*, **96**, 1–11.

Fischer, J. and Lindenmayer, D.B. (2002a). Treating the nestedness temperature calculator as a black box can lead to false conclusions. *Oikos*, **99**, 193–199.

Fischer, J. and Lindenmayer, D.B. (2002b). The conservation value of paddock trees for birds in a variegated landscape in southern New South Wales. 2. Paddock trees as stepping stones. *Biodiversity and Conservation*, **11**, 833–849.

Fischer, J. and Lindenmayer, D.B. (2004). Beyond fragmentation – a contour-based landscape model. (*Conservation Biology*) (in press).

Fischer, J., Fazey, I., Briese, R. and Lindenmayer, D.B. (2005). Making the matrix matter: challenges in Australian grazing landscapes. *Biodiversity and Conservation*, **14**, 561–578.

Fischer, J., Lindenmayer, D.B. and Fazey, I. (2004). Appreciating ecological complexity: Habitat contours as a conceptual landscape model. *Conservation Biology*, **18**, 1245–1253.

Fischer, J., Lindenmayer, D.B., Nix, H.A. and Stein, J. (2001). The bioclimatic domains of the Mountain Brushtail Possum. *Journal of Biogeography*, **28**, 293–304.

Fischhoff, B. (1995). Risk perception and communication unplugged: twenty years of progress. *Risk Analysis*, **15**, 137–145.

Fischhoff, B., Slovic, P. and Lichtenstein, S. (1982). Lay foibles and expert fables in judgements about risk. *The American Statistician*, **36**, 240–255.

Fisher, R.A. (1930). *The genetical theory of natural selection.* Clarendon Press, London.

Fisheries Research and Development Corporation (2000). *Investing for tomorrow's fish: the FRDC's Research and Development Plan 2000 to 2005.* Fisheries Research and Development Corporation, Canberra.

Fitzgerald, S. (1989). *International wildlife trade: whose business is it?* World Wide Fund for Nature, USA, Washington.

Fitzgibbon, C.D. (1997). Small mammals in farm woodlands: the effects of habitat, isolation and surrounding land-use patterns. *Journal of Applied Ecology*, **34**, 530–539.

Flannery, T.F. (1994). *The future eaters.* Reed Books, Sydney.

Flather, C.H. and Sauer, J.H. (1996). Using landscape ecology to test hypotheses about large-scale abundance patterns in migratory birds. *Ecology*, **77**, 28–35.

Fleishman, E., Murphy, D.D. and Brussard, P.F. (2000). A new method for selection of umbrella species for conservation planning. *Ecological Applications*, **10**, 569–579.

Fletcher, M., Southwell, C.J., Sheppard, N.W., Caughley, G.C., Grice, D., Grigg, G.C. and Geard, L.A. (1990). Kangaroo population trends in the Australian rangelands, 1980–87. *Search*, **21**, 28–29.

Fletcher, R.J. (2005). Multiple edge effects and their implications in fragmented landscapes. *Journal of Applied Ecology*, **74**, 342–352.

Florence, R.G. (1994). The ecological basis for forest fire management in New South Wales. pp. 15–33. In: *The Burning Continent. Forest Ecosystems and Fire Management in Australia.* Attiwill, P., Florence, R., Hurditch, W.E. and Hurditch, W.J. (Eds). Institute of Public Affairs, Melbourne, September 1994.

Floyd, A.G. (1990). *Australian rainforests in New South Wales.* Vol. 1. Surrey Beatty, Sydney.

Foley, J.A., Costa, M.H., Delire, C., Ramankutty, N. and Snyder, P. (2003). Green surprise? How terrestrial ecosystems could affect earth's climate. *Frontiers of Ecology and Environment*, **1**, 38–44.

Foran, B. and Poldy, F. (2002). *Future dilemmas. Options to 2050 for Australia's population, technology, resources and environment.* Working paper 02/01. CSIRO Sustainable Ecosystems, Canberra.

Forbes, S.H. and Boyd, D.K. (1997). Genetic structure and migration in native and reintroduced Rocky Mountain wolf populations. *Conservation Biology*, **11**, 1126–1234.

Ford, H.A. (1979). Birds. pp. 103–114. In: *Natural History of Kangaroo Island.* Tyler, M.J., Twidale, C.R. and Ling, J.K. (Eds). Royal Society of South Australia, Adelaide.

Ford, H.A. (1989). *Ecology of Birds: An Australian Perspective.* Surrey Beatty and Sons Pty Limited, Australia

Foreman, D., Davis, J., Johns, D., Noss, R. and Soulé, M. (1992). The Wildlands Project Mission Statement. *Wild Earth* (Special Issue): 3–4.

Forest Ecosystem Management Assessment Team (1993). *Forest ecosystem management: an ecological, economic, and social assessment.* Various pagination. USDA Forest Service, Portland, Oregon.

Forest Practices Board (1998). *Threatened fauna manual for production forests in Tasmania.* Forest Practices Board, Hobart, Tasmania.

Forest Stewardship Council (1996). *FSC Document No. 1.2. Principles and crieria for forest management.* Revised March 1996.

Forestry Tasmania (1999). *Annual Report.* Forestry Tasmania, Hobart, Tasmania.

Forestry Tasmania (2004). *Towards new silviculture in Tasmania's old growth forests. 2. Sustaining wood yields.* Forestry Tasmania Draft Report. February 2004. Forestry Tasmania, Hobart.

Forman, R.T. (1964). Growth under controlled conditions to explain the hierarchical distributions of a moss, *Tetraphis pellucida. Ecological Monographs,* **34**, 1–25.

Forman, R.T. (1995). *Land mosaics. The ecology of landscapes and regions.* Cambridge University Press, New York.

Forman, R.T. and Godron, M. (1986). *Landscape ecology.* Wiley and Sons, New York.

Forman, R.T., Sperling, D., Bissonette, J.A., Clevenger, A.P., Cutshall, C.D., Dale, V.H., Fahrig, L., France, R., Goldman, C.R., Heanue, K., Jones, J.A., Swanson, F.J., Turrentine, T. and Winter, T.C. (Eds) (2002). *Road ecology.* Science and solutions. Island press, Washington.

Forney, K.A. and Gilpin, M.E. (1989). Spatial structure and population extinction: A study with *Drosophila* flies. *Conservation Biology,* **3**, 45–51.

Forshaw, J.M. (2002). Australian parrots. Third Edition. Avi-Trader Publishing, Brisbane.

Foster, D.R., Aber, J.B., Melillo, J.M., Bowden, R.D. and Bazzaz, F.A. (1997). Forest response to disturbance and anthropogenic stress. *BioScience,* **47**, 437–445.

Foster, D.R., Knight, D.H. and. Franklin, J.F. (1998). Landscape patterns and legacies resulting from large infrequent forest disturbances. *Ecosystems,* **1**, 497–510.

Fowler, J. and Cohen, L. (1990). *Practical statistics for field biology.* Wiley, London.

Fox, A.M. (1978). The '72 fire of Nadgee Nature Reserve. *Parks and Wildlife,* **2**, 5–24.

Fox, B.J. (1982). Fire and mammalian secondary succession in an Australian coastal heath. *Ecology,* **63**, 1332–1341.

Fox, B.J. and Fox, M.D. (2000). Factors determining mammal species richness on habitat islands and isolates: habitat diversity, disturbance, species interactions and guild assembly rules. *Global Ecology and Biogeography,* **9**, 19–37.

Fox, B.J. and McKay, G.M. (1981). Small-mammal responses to pyric successional changes in eucalypt forest. *Australian Journal of Ecology,* **6**, 29–41.

Fox, L.R. and Morrow, P.A. (1983). Estimates of damage by herbivorous insects on *Eucalyptus* trees. *Australian Journal of Ecology,* **8**, 139–147.

Fox, M.D. (1990). Interactions of native and introduced species in new habitats. pp. 141–147. In: *Australian Ecosystems: 200 years of Utilisation, Degradation and Reconstruction.* Proceedings of the Ecological Society of Australia. Vol. 16. Saunders, D.A., Hopkins, A.J. and Majer, J.D. (Eds). Surrey Beatty, Chipping Norton.

Fox, M.D. and Fox, B.J. (1986). The susceptibility of natural communities to invasion. pp. 57–66. In: *Ecology of Biological Invasions; An Australian Perspective.* Groves R.H. and Burdon, J.J. (Eds). Australian Academy of Science, Canberra.

Frankham, R. (1995a). Conservation genetics. *Annual Review of Genetics,* **29**, 305–327.

Frankham, R. (1995b). Inbreeding and extinction: A threshold effect. *Conservation Biology,* **9(4)**, 792–799.

Frankham, R. (1995c). Effective population size/adult population size ratios in wildlife: a review. *Genetics Research,* **66**, 95–107.

Frankham, R. (1996). Relationship of genetic variation to population size in wildlife. *Conservation Biology,* **10**, 1500–1508.

Frankham, R., Briscoe, J.D. and Ballou, D.A. (2002). Introduction to conservation genetics. Cambridge University Press, Cambridge.

Frankham, R., Hemmer, H., Ryder, O.A., Cothran, E.G., Soulé, M.E., Murray, N.D. and Snyder, M. (1986). Selection in captive populations. *Zoo Biology*, **5**, 127–138.

Franklin, I.R. (1980). Evolutionary change in small populations. pp. 135–149. In: *Conservation Biology: An Evolutionary-Ecological Perspective.* Soulé, M. and Wilcox, B. (Eds). Sinauer, Massachusetts.

Franklin, J.F. (1988). Structural and functional diversity in temperate forests. pp. 166–175. In: *Biodiversity.* Wilson, E.O. and Peters, F.M. (Eds). National Academy Press, Washington.

Franklin, J.F. (1993a). Preserving biodiversity: Species, ecosystems, or landscapes? *Ecological Applications*, **3**, 202–205.

Franklin, J.F. (1993b). Lessons from old-growth. *Journal of Forestry* **December 1993**, 11–13.

Franklin, J.F. (1993c). Scientific basis for new perspectives in forests and streams. pp. 25–72. In: *Watershed Management Balancing Sustainability and Environmental Change.* Naiman, R.J. (Ed). Springer-Verlag, New York.

Franklin, J.F. and Forman, R.T. (1987). Creating landscape patterns by forest cutting: ecological consequences and principles. *Landscape Ecology*, **1**, 5–18.

Franklin, J.F. and MacMahon, J.A. (2000). Messages from a mountain. *Science*, **288**, 1183–1185.

Franklin, J.F., Berg, D.E., Thornburgh, D.A. and Tappeiner, J.C. (1997). Alternative silvicultural approaches to timber harvest: variable retention harvest systems. pp. 111–139. In: *Creating a Forestry for the 21st Century.* Kohm, K.A. and Franklin, J.F. (Eds). Island Press, Covelo, California.

Franklin, J.F., Cromack, K.J., Denison, W., McKee, A., Maser, C., Sedell, J., Swanson, F. and Juday, G. (1981). *Ecological attributes of old-growth Douglas-fir forests.* Pacific Northwest Forest and Range Experimental Station, USDA Forest Service, Portland, Oregon. USDA Forest Service General Technical Report PNW-118.

Franklin, J.F., Harmon, M.E. and Swanson, F.J. (1999). *Complementary roles of research and monitoring: lessons from the U.S. LTER Program and Tierra del Fuego.* Paper presented to the Symposium, Toward a unified framework for inventorying and monitoring forest ecosystem resources. Guadalajara, Mexico, November 1998.

Franklin, J.F., Lindenmayer, D.B., MacMahon., J.A., McKee, A., Magnusson, J., Perry, D.A., Waide, R. and Foster, D.R. (2000). Threads of continuity: ecosystem disturbances, biological legacies and ecosystem recovery. *Conservation Biology in Practice*, **1**, 8–16.

Franklin, J.F., MacMahon, J.A., Swanson, F.J. and Sedell. J.R. (1985). Ecosystem responses to the eruption of Mount St. Helens. *National Geographic Research*, **Spring 1985**, 198–216.

Franklin, J.F., Spies, T.A., van Pelt, R., Carey, A., Thornburgh, D., Berg, D.R., Lindenmayer, D.B., Harmon, M., Keeton, W. and Shaw D.C. (2002). Disturbances and the structural development of natural forest ecosystems with some implications for silviculture. *Forest Ecology and Management*, **155**, 399–423.

Fraser, F., Lawson, V., Morrison, S., Christophersen, P., MacGregg, S. and Rawlinson, M. (2003). Fire management experiment for the declining Partridge Pigeon, Kakadu National Park. *Ecological Management and Restoration*, **4**, 94–102.

Fraser, G.S. and Stutchbury, B.J. (2004). Area-sensitive forest birds move extensively among forest patches. *Biological Conservation*, **118**, 377–387.

Freedman, B. (1989). *Environmental ecology.* Academic Press, New York.

Freeland, W.J. (1986). Invasion north: successful conquest by the Cane Toad. *Australian Natural History*, **22**, 69–73.

Freudenberger, D. (2001). *Bush for birds: biodiversity enhancement guidelines for the Saltshaker Project, Borowa, NSW.* CSIRO Wildlife and Ecology, Greening Australia, A.C.T. and SE NSW Inc.

Friend, G.R. (2003). Gilbert's Potoroo. *Nature Australia*, **27(9)**, 22–23.

Fries, C., Carlsson, M., Dahlin, B., Lamas, T. and Sallnas, O. (1998). A review of conceptual landscape planning models for multiobjective forestry in Sweden. *Canadian Journal of Forest Research-Revue (Canadienne De Recherche Forestiere)*, **28**, 159–167.

Frith, H.G. and Davies, S.J.J.F. (1961). Ecology of the Magpie Goose, *Aseranas semipalmata Latham* (Anstidae). *CSIRO Wildlife*, **6**, 91–141.

Frith, H.J. (1990). Migrants and nomads. pp. 31. In: *Reader's Digest Complete Book of Australian Birds.* Reader's Digest, Sydney.

Fritz, R. (1979). Consequences of insular population structure: distribution and extinction of spruce grouse populations. *Oecologia*, **42**, 57–65.

Frumhoff, P.C. (1995). Conserving wildlife in tropical forests managed for timber. *BźoScience*, **45**, 456–464.

Funch, P. and Kristensen, R.M. (1995). Cycliophora is a new phylum with affinities to Entoprocta and Ectoprocta. *Nature*, **378**, 711–714.

Furusawa, T., Pahari, K., Umezaki, M. and Ohtsuka, R. (2004). Impacts of selective logging on New Georgia

Island, Solomon Islands evaluated using very-high-resolution satellite (IKONOS) data. *Environmental Conservation*, **31**, 349–355.

Ganley, R.J., Brunsfeld, S.J. and Newcombe, G. (2004). A community of unknown, endophytic fungi in western white pine. *Proceedings of the National Academy of Sciences of the United States Of America*, **101**, 10107–10112.

Gardener, C.J., McIvor, J.G. and Williams, J. (1990). Dry tropical rangelands: Solving one problem and creating another. pp. 279–286. In: *Australian Ecosystems: 200 years of Utilisation, Degradation and Reconstruction.* Proceedings of the Ecological Society of Australia. Vol. 16. Saunders, D.A., Hopkins, A.J.M. and How, R.A. (Eds). Surrey Beatty, Chipping-Norton.

Gardner, R.H., Milne, B.T., Turner, M.G. and O'Neill R.V. (1987). Neutral models for analysis of broad-scale landscape patterns. *Landscape Ecology*, **1**, 19–28.

Garnett, S.T. (1992). *The Action Plan for Australian Birds.* Australian National Parks and Wildlife Service, Canberra.

Garnett, S.T. and Crowley, G.M. (1995). *Golden-shouldered Parrot: Options for management.* Report to Queensland Department of Environment and Heritage, Cairns, Queensland.

Garnett, S.T. and Crowley, G.M. (1997). Golden-shouldered Parrot of Cape York Peninsula: the importance of cups of tea to effective conservation. pp. 201–205. In: *Conservation Outside Nature Reserves.* Hale P. and Lamb, D. (Eds). University of Queensland, Brisbane.

Garnett, S.T. and Crowley, G.M. (2000). *The Action Plan for Australian birds.* Natural Heritage Trust, Canberra.

Garnett, S.T. and Loyn, R.H. (1992). *Loss of hollow-bearing trees from Victorian native forests as a threatening process.* Action statement. Draft 1. Department of Conservation and Natural Resources, Victoria.

Garrod, G. and Willis, K.G. (1999). *Economic valuation of the environment.* Edward Elgar, Cheltenham.

Gascon, C. (1993). Breeding-habitat use by five Amazonian frogs at forest edge. *Biodiversity and Conservation*, **2**, 438–444.

Gascon, C. and Lovejoy, T.E. (1998). Ecological impacts of forest fragmentation in central Amazonia. *Zoology-Analysis of Complex Systems*, **101**, 273–280.

Gascon, C., Lovejoy, T., Bieeregaard, R.O., Malcolm, J.R., Stouffer, P.C., Vasconcelos, H.L., Laurance, W.F., Zimmerman B., Tocher, M. and Borges S. (1999). Matrix habitat and species richness in tropical forest remnants. *Biological Conservation*, **91**, 223–229.

Gaston, K.J. (1991). The magnitude of global insect species richness. *Conservation Biology*, **5**, 283–296.

Gaston, K.J. (1994). *Rarity.* Chapman and Hall, London.

Gaston, K.J. (1996). Species richness: Measure and measurement. pp. 77–113. In: *Biodiversity.* Gaston, K.J. (Ed). Blackwell, Oxford.

Gaston, K.J. and Spicer, J.I. (1998). *Biodiversity: An introduction.* Blackwell Science, Oxford.

Gaston, K.J. and Spicer, J.I. (2004). *Biodiversity: An introduction.* Second Edition. Blackwell Publishing, Oxford.

Gaston, K.J. and Williams, P.H. (1993). Mapping the world's species – the higher taxon approach. *Biodiversity Letters*, **1**, 2–8.

Gaston, K.J., New, T.R. and Samways, M.J. (Eds) (1993). *Perspectives in Insect Conservation.* Intercept Ltd, Andover, England.

Gates, J.E. and Gysel, L.W. (1978). Avian nest dispersion and fledging success in field-forest ecotones. *Ecology*, **59**, 871–83.

Gégout, J. and Krizova, E. (2003). Comparison of indicator values of forest understorey plant species in Western Carpathians (Slovakia) and Vosges Mountains (France). *Forest Ecology and Management*, **182**, 1–11.

George, R.J. (1994). Rising water tables – the hidden threat to remnant native vegetation. *Biolinks*, 3–5.

George, R.J., McFarlane, D.J. and Speed, R.J. (1996). The consequences of a changing hydrologic environment for native vegetation in southwestern Australia. pp. 9–22. In: *Nature Conservation 4: The Role of Networks.* Saunders, D.A., Hobbs, R.J. and Ehrlich, P.R. (Eds). Surrey Beatty, Chipping Norton.

Gerard, P.W. (1995). *Agricultural practices, farm policy and the conservation of biological diversity.* USDI Biological Sciences Report. No. 4. Washington, D.C.

Gerrodette, T. (1987). A power analysis for detecting trends. *Ecology*, **68**, 1364–1372.

Ghassemi, F., Jakeman, A.J. and Nix, H.A. (1995). *Salinisation of land and water resources.* University of New South Wales Press, Sydney.

Ghazoul, J. (2001). Barriers to biodiversity conservation in forest certification. *Conservation Biology*, **15**, 315–317.

Ghilarov, A. (1996). What does "biodiversity" mean – scientific problem of convenient myth? *Trends in Ecology and Evolution*, **11**, 304–306.

Gibbons, P. and Boak, M. (2002). The value of paddock trees for regional conservation in an agricultural landscape. *Ecological Management and Restoration*, **3**, 205–210.

Gibbons, P. and Lindenmayer, D.B. (1996). A review of issues associated with the retention of trees with hollows in wood production forests. *Forest Ecology and Management,* **83**, 245–279.

Gibbons, P. and Lindenmayer, D.B. (2002). *Tree hollows and wildlife conservation in Australia.* CSIRO Publishing, Melbourne.

Gibbs, J.P. (2001). Demography versus habitat fragmentation as determinants of genetic variation in wild populations. *Biological Conservation,* **100**, 15–20.

Gibbs, J.P. and Faaborg, J. (1990). Estimating the viability of Ovenbird and Kentucky Warbler populations in forest fragments. *Conservation Biology,* **4**, 193–196.

Gibson, D., Johnson, K.A., Langford, D.G., Cole, J.R., Clarke, D.E. and Community, W. (1995). The Rufous Hare-Wallaby *Lagorchestes hirsutus*: A history of experimental reintroduction in the Tanami Desert, Northern Territory. pp. 171–176. In: *Reintroduction Biology of Australasian Fauna.* Serena, M. (Ed). Surrey Beatty, Chipping Norton, Australia.

Gibson, D., Lundie-Jenkins, G., Langford, D.G., Cole, J.R., Clarke, D.E. and Johnson, K.A. (1993). Predation by feral cats, *Felis catus*, on the Rufous Hare-Wallaby, *Lagorchestes hirsutus*, in the Tanami Desert. *Australian Mammalogy,* **17**, 103–107.

Gibson, L.A., Wilson, B.A., Cahill, D.M. and Hill, J. (2004). Spatial prediction of Rufous Bristlebird habitat in a coastal heathland: a GIS-based approach. *Journal of Applied Ecology,* **41**, 213–223.

Gibson, L.M. and Young, M.D. (1987). *Kangaroos counting the cost. The economic effects of kangaroos and kangaroo culling on agricultural production.* CSIRO Division of Wildlife and Rangelands Research, Deniliquin, NSW. Report to Australian National Parks and Wildlife Service. CSIRO Project Report No. 4.

Giese, M. (1996). Effects of human activity on Adelie Penguin, Pygoscelis adeliae breeding success. *Biological Conservation,* **75**, 157–164.

Gilbert, F., Gonzalex, A. and Evens-Freke, I. (1998). Corridors maintain species richness in the fragmented landscapes of a microsystem. *Proceedings of the Royal Society of London Series B,* **265**, 577–582.

Gilbert, L.E. (1980). The equilibrium theory of island biogeography: Fact or fiction? *Journal of Biogeography,* **7**, 209–235.

Gilfedder, L. and Kirkpatrick, J.B. (1999). Factors influencing the integrity of remnant bushland in subhumid Tasmania. *Biological Conservation,* **84**, 89–96.

Gill, A.M. (1975). Fire and the Australian flora: A review. *Australian Forestry,* **38**, 4–25.

Gill, A.M. (1977). Management of fire-prone vegetation for plant species conservation in Australia. *Search,* **8**, 20–26.

Gill, A.M. (1981). Past settlement fire history in Victorian landscapes. pp. 77–98. In: *Fire and the Australian Biota.* Gill, A.M., Groves R.H. and Noble I.R. (Eds). Australian Academy of Science, Canberra.

Gill, A.M. (1999). Biodiversity and bushfires: an Australia-wide perspective on plant-species changes after a fire event. pp. 9–53. In: *Australia's Biodiversity – Responses to Fire.* Gill, A.M, Woinarski, J. and York, A. (Eds). Environment Australia Biodiversity Technical Paper, **1**, 1–266.

Gill, A.M. (2000). *Fire-pulses in the Heart of Australia. Fire Regimes and Fire Management in Central Australia.* Report to Environment Australia, Canberra, August 2000.

Gill, A.M. (2001). Economically destructive fires and biodiversity conservation: an Australian perspective. *Conservation Biology,* **15**, 1558–1560.

Gill, A.M. and Bradstock, R. (1995). Extinction of biota by fires. pp. 309–322. In: *Conserving Biodiversity: Threats and solutions.* Bradstock, R., Auld, T.D., Keith, D.A., Kingsford, R.T., Lunney, D. and Sivertsen, D.P. (Eds). Surrey Beatty and Sons, Sydney.

Gill, A.M. and Bradstock, R. (2003). Fire regimes and biodiversity: a set of postulates. pp. 15–25. In: *Australia Burning: Fire Ecology, Policy and Management Issues.* Cary, G., Lindenmayer, D.B. and Dovers, S. (Eds). CSIRO Publishing, Melbourne.

Gill, A.M. and Catling. P.C. (2002). Fire regimes and biodiversity of forested landscapes. pp. 351–369. In: *Flammable Australia. The Fire Regimes and Biodiversity of a Continent.* Bradstock, R.A., Williams, J.E. and Gill, A.M. (Eds). Cambridge University Press, Melbourne.

Gill, A.M. and McCarthy, M.A. (1998). Intervals between prescribed fires in Australia: what intrinsic variation should apply. *Biological Conservation,* **85**, 161–169.

Gill, A.M. and Moore, P.H. (1994). Some ecological research perspectives on the disastrous Sydney fires of January 1994. pp. 63–72. In: *Proceedings of the 2nd International Conference on Forest Fire Research.* Coimbra, Portugal.

Gill, A.M., Groves, R.H. and Noble, I.R. (Eds). (1981). *Fire and the Australian biota.* Australian Academy of Science, Canberra.

Gill, A.M., Woinarski, J.C.Z. and York, A. (1999). *Australia's biodiversity – responses to fire.* Biodiversity Technical paper No. 1. Environment Australia, Canberra.

Gill, F.B. (1995). *Ornithology.* Second edition. W.H. Freeman and Company, New York.

Gill, P. and Burke, C. (1999). *Whale watching in Australian and New Zealand waters.* New Holland, Sydney.

Gillespie, G.R. (1995). Tadpoles and trout don't mix. *Victorian Frog Group Newsletter*, **1**, 14–15.

Gilligan, D.M., Briscoe, D.A. and Frankham, R. (2005). Comparative losses of quantitative and molecular genetic variation in finite populations of *Drosophila melanogaster. Genetical Research*, **85**, 47–55.

Gilligan, D.M., Woodworth, L.M., Montgomery, M.E., Briscoe, D.A. and Frankham, R. (1997). Is mutation accumulation a threat to the survival of endangered populations? *Conservation Biology,* **11**, 1235–1241.

Gilmore, A.M. (1985). The influence of vegetation structure on the density of insectivorous birds. pp. 21–31. In: *Birds of Eucalypt Forests and Woodlands: Ecology, Conservation, Management.* Keast, A., Recher, H.F., Ford, H. and Saunders, D. (Eds). Surrey Beatty, Chipping Norton.

Gilmore, A.M. (1990). Plantation forestry: Conservation impacts on terrestrial vertebrate fauna. pp. 377–388. In: *Prospects for Australian Plantations.* Dargavel J. and Semple, N. (Eds). Centre for Resource and Environmental Studies, Australian National University, Canberra.

Gilna, B., Lindenmayer, D.B. and Viggers, K.L. (2005). New conservation tools should not be new conservation threats – biocontrol of introduced brushtail possums in New Zealand and the conservation of Australian possum fauna (*Conservation Biology*) (in press).

Gilpin, M.E. (1987). Spatial structure and population vulnerability. pp. 126–139. In: *Viable Populations for Conservation.* Soulé, M.E. (Ed). Cambridge University Press, New York.

Gilpin, M.E. and Soulé, M.E. (1986). Minimum viable populations: Processes of species extinctions. pp. 19–34. In: *Conservation biology: The science of scarcity and diversity.* Soulé, M.E. (Ed). Sinauer Associates, Sunderland, Massachusetts.

Ginzburg, L.V., Slobodkin, L.B., Johnson, K. and Bindman, A.G. (1982). Quasi-extinction probabilities as a measure of impact on population growth. *Risk Analysis,* **2**, 171–181.

Gippel, C.J., Finlayson, B.L. and O'Neill, I.C. (1996). Disturbance and hydraulic significance of large woody debris in a lowland Australian river. *Hydrobiology,* **318**, 179–194.

Given, D.R. and Norton, D.A. (1993). A multi variate approach to assessing threat and for priority setting in threatened species conservation. *Biological Conservation,* **64**, 57–66.

Gjerdrum, C., Vallee, A., Cassay St. Clair, C., Bertram, D., Ryder, J. and Blackburn, G. (2003). Tufted puffin reproduction reveals ocean climate variability. *Proceedings of the National Academy of Sciences*, **100**, 9377–9382.

Gladstone, W. (2002). The potential value of indicator groups in the selection of marine reserves. *Biological Conservation*, **104**, 211–220.

Glanznig, A. (1995). *Native vegetation clearance, habitat loss and biodiversity decline: an overview of recent native vegetation clearance in Australia and its implications for biodiversity.* Biodiversity Unit, Department of the Environment, Sport and Territories, Canberra. Biodiversity Series Paper No. 6.

Godfree, R.C., Young, A.G., Londsdale, W.M., Woods, M.J. and Burdon, J.J. (2004). Ecological risk assessment of transgenic pasture plants: a community gradient modeling approach. *Ecology Letters*, **7**, 1077–1089.

Godsell, J. (1983). Eastern Quoll. pp. 20–21. In: *Complete Book of Australian Mammals.* Strahan, R. (Ed). Angus and Robertson, Sydney.

Goklany, I.M. (2001). *The precautionary principle: a critical appraisal of environmental risk assessment.* CATO Institute, Washington D.C.

Goldammer, J.G. and Price, C. (1998). Potential impacts of climate change on fire regimes in the tropics based on MAGICC and GISS GCM-derived lightning model. *Climate Change*, **39**, 273–296.

Goldberg, T.L., Grant, E.C., Inendino, K.R., Kassler, T.W., Claussen, J.E. and Phillip, D.P. (2005). Increased infectious disease susceptibility resulting from outbreeding depression. *Conservation Biology*, **19**, 455–462.

Goldie, J., Douglas, R. and Furnass, B. (Eds). 2004). *In Search of Sustainability*. CSIRO Publishing, Melbourne.

Goldingay, R.G. and Kavanagh, R.P. (1991). The Yellow-bellied Glider: A review of its ecology and management considerations. pp. 365–75. In: *Conservation of Australia's Forest Fauna.* Lunney, D. (Ed). Surrey Beatty, Sydney.

Goldingay, R.L. and Possingham, H.P. (1995). Area requirements for viable populations of the Australian gliding marsupial *Petaurus australis. Biological Conservation,* **73**, 161–167.

Goldney, D. and Wakefield, S. (1997). *Save the Bush Toolkit.* Charles Sturt University, Bathurst.

Goldsborough, C.L., Hochuli, D.F. and Shine, R. (2003). Invertebrate biodiversity under hot rocks: habitat use by the fauna of sandstone outcrops in the Sydney region. *Biological Conservation*, **109**, 85–93.

Goldsmith, B. (Ed) (1991). *Monitoring for Conservation and Ecology.* Chapman and Hall, London.

Goldstein, D.B., Linares, A.R. and Feldman, M.W. (1995). An evolution of genetic distances for use with microsatellite loci. *Genetics,* **139,** 463–471.

Goodman, D. (1987a). Considerations of stochastic demography in the design and management of biological reserves. *Natural Resources Modelling,* **1,** 205–34.

Goodman, D. (1987b). The demography of chance extinction. pp. 11–34. In: *Viable Populations for Conservation.* Soulé, M.E. (Ed). Cambridge University Press, Cambridge.

Gordon, D.M. and Fitzgibbon, F. (1999). The distribution of enteric bacteria from Australian mammals: host and geographical effects. *Microbiology,* **145,** 2663–2671.

Gordon, G. (1984). Fauna of the Brigalow belt. pp. 61–70. In: *The Brigalow Belt of Australia.* Symposium held at the John Kindler Memorial Theatre, Queensland Institute of Technology, 23 October 1982. Bailey, A. (Ed). The Royal Society of Queensland, Brisbane.

Gordon, G., Brown, A.S. and Pulsford, T. (1988). A Koala (*Phascolarctos cinereus Goldfuss*) population crash during drought and heatwave conditions in south-western Queensland. *Australian Journal of Ecology,* **13,** 451–461.

Goss, K.F. (2003). Environmental flows, river salinity and biodiversity conservation: managing trade-offs in the Murray-Darling Basin. *Australian Journal of Botany,* **51,** 619–625.

Gould, J. (1870). *Mammals of Australia.* The Author, London.

Gove, J.H., Patil, G.P., Swindel, B.F. and Taillie, C. (1994). Ecological diversity and forest management. pp. 409–462. In: *Handbook of Statistics.* Vol. 12. Patil, G.P. and Rao, C.R. (Eds). Elsevier, Amsterdam.

Government of Queensland. (2005). State-wide landcover and trees study. www.nrm.qld.gov.au/slats/index.html

Government of Western Australia (1996). *Guidelines for the Preparation of an Annual Environmental Report.* Department of Minerals and Energy, Western Australia.

Graetz, R.D., Wilson, M.A. and Campbell, S.K. (1995). *Landcover disturbance over the Australian continent: a contemporary assessment.* Department of the Environment Sports and Territories, Canberra. Biodiversity Series, Paper No. 7. pp. 1–86.

Grafton, R.G., Kompas, T. and Lindenmayer, D.B. (2005). Marine reserves and ecological uncertainty. *Bulletin of Mathematical Biology* (in press).

Grafton, R.Q. and Pezzey, J. (2005). Economics of the environment. pp. 40–56. In: *Understanding the Environment. Bridging the Disciplinary Divides.* Grafton, R.Q., Robin, L. and Wasson, R.J. (Eds). UNSW Press, Sydney.

Grafton, R.Q. Robin, L. and Wasson, R.J. (Eds). (2004). Understanding the environment. *Bridging the Disciplinary Divides.* UNSW Press, Sydney.

Gray, J.S. (1990a). Letters: Professor Gray replies. *Marine Pollution Bulletin,* **21(12),** 599–600.

Gray, J.S. (1990b). Statistics and the precautionary principle. *Marine Pollution Bulletin,* **21(4),** 174–176.

Gray, P.A., Cameron, D. and Kirkham, I. (1996). Wildlife habitat evaluation in forested ecosystems: some examples from Canada and the United States. pp. 407–533. In: *Conservation of Faunal Diversity in Forested Landscapes.* Degraaf, R.M. and Miller, R.I., (Eds). Chapman and Hall, New York.

Greathead, D.J. (1983). The multi-million dollar weevil that pollinates oil palms. *Antenna,* **7,** 105–107.

Green, K. and Osborne, W. (1994). *Wildlife of the Australian Snow-country.* Reed Books, Sydney.

Green, R.H. (1993). Application of repeated measures designs in environmental impact and monitoring studies. *Australian Journal of Ecology,* **18,** 81–98.

Greenslade, P. (1996). Fuel reduction burning: Is it causing the extinction of Australia's rare invertebrates? *National Parks Journal,* **April 1996,** 18–19.

Greenslade, P.J.M. and Yeatman, E.M. (1980). Ants as indicators of habitat in three conservation parks in South Australia. *South Australian Naturalist,* **55,** 20–26.

Greenstein, B.J., Curran, H.A. and Pandolfi, J.M. (1998). Shifting ecological baselines and the demise of Acropora cervicornis in the western North Atlantic and Caribbean province: a Pleistocene perspective. *Coral Reefs,* **17,** 249–261.

Gregory, S.V. (1997). Riparian management in the 21st century. pp. 69–85. In: *Creating a Forestry for the 21st Century.* Kohm, K.A. and Franklin, J.F. (Eds). Island Press, Washington D.C.

Grey, M.J., Clarke, M.F. and Loyn, R.H. (1997). Initial changes in the avian communities of remnant eucalypt woodlands following a reduction in the abundance of noisy miners, *Manorina melanocephala. Wildlife Research,* **24,** 631–648.

Grey, M.J., Clarke, M.F. and Loyn, R.H. (1998). Influence of the Noisy Miner *Manorina melanocephala* on avian diversity and abundance in remnant Grey Box woodland. *Wildlife Research,* **24,** 631–648.

Griffin, G.F. and Friedel, M.H. (1985). Discontinuous change in central Australia: Some implications of major ecological events for land management. *Journal of Arid Environments*, **9**, 63–80.

Griffith, B., Michael Scott, J., Carpenter, J.W. and Reed, C. (1989). Translocation as a species conservation tool: Status and strategy. *Science*, **245**, 477–480.

Grigg, G. (1982). Roo harvesting. Are kangaroos really under threat ? *Australian Natural History*, **21**, 123–127.

Grigg, G. (1988). Kangaroo harvesting and the conservation of the sheep rangelands. *Australian Zoologist*, **24**, 124–128.

Grigg, G. (1991). Kangaroos, land care and animal welfare: A proposal for change. *Bulletin of the Ecological Society of Australia*, **21**, 30–35.

Grigg, G. (1995). Kangaroo harvesting for conservation of rangelands, kangaroos ... and graziers. pp. 161–165. In: *Conservation Through the Sustainable Use of Wildlife*. Grigg, G.C., Hale, P.T. and Lunney, D. (Eds). Centre for Conservation Biology, University of Queensland, Brisbane.

Grigg, G. (2000). Cane toads vs native frogs. *Nature Australia*, **Winter 2000**, 32–41.

Grigg, G., Hale, P. and Lunney, D. (Eds) (1995). *Conservation Through the Sustainable Use of Wildlife*. Centre for Conservation Biology, The University of Queensland, Brisbane.

Grimson, R.C., Aldrich, T.E. and Drane, J.W. (1992). Clustering in sparse data and an analysis of Rhabdomyosarcoma incidence. *Statistics in Medicine*, **11**, 761–768.

Groombridge, B. (Ed) (1992). *IUCN Red List of Threatened Animals*. Gland, Switzerland.

Groombridge, B. (Ed) (1994). *Biodiversity Data Sourcebook*. World Conservation Monitoring Centre, Biodiversity Series No 1. World Conservation Press, Cambridge.

Groombridge, B. and Jenkins, M.D. (2002). *World Atlas of Biodiversity. Earth's Living Resources in the 21st Century*. UNEP-WCMC. University of California Press, Berkeley.

Gross, C.L. and Mackay, D. (1998). Honeybees reduce fitness in the pioneer shrub *Melastoma affine* (Melastomataceae). *Biological Conservation*, **86**, 169–178.

Grove, S.J. (2001). Developing appropriate mechanisms for sustaining mature timber habitat in managed natural forest stands. *International Forestry Review*, **3**, 272–283.

Grove, S.J., Meggs, J.F. and Goodwin, A. (2002). *A Review of Biodiversity Conservation Issues Relating to Coarse Woody Debris Management in the Wet Eucalypt Production Forests of Tasmania*. A report to Forestry Tasmania. August 2002.

Grover, D.R. and Slater, P.J. (1994). Conservation value to birds of remnants of Melaleuca forest in suburban Brisbane. *Wildlife Research*, **21**, 433–444.

Groves, R.H. (1997). *Recent Incursions of Weeds to Australia 1971–1995*. CRC for Weed Management Systems Technical Series No. 3.

Grumbine, R.E. (1990). Viable populations, reserve design, and Federal lands management: a critique. *Conservation Biology*, **4**, 127–134.

Grumbine, R.E. (1994). What is ecosystem management? *Conservation Biology*, **8**, 27–38.

Guiler, E. (1961). The former distribution and decline of the Thylacine. *Australian Journal of Science*, **23**, 207–210.

Guiler, E. (1983). Tasmanian Devil. pp. 27–28. In: *Complete book of Australian Mammals*. Strahan, R. (Ed). Collins Angus and Robertson Publishers, Sydney.

Guiler, E. (1985). *Thylacine: The Tragedy of the Tasmanian Tiger*. Oxford University Press, Melbourne.

Guisan, A. and Zimmerman, N.E. (2000). Predictive habitat distribution models in ecology. *Ecological Modelling*, **135**, 147–186.

Gullan, P.J. and Cranston, P.S. (1994). *The Insects. An Outline of Entomology*. Chapman and Hall, Melbourne.

Gullan, P.K. and A.C. Robinson. (1980). Vegetation and small mammals of a Victorian forest. *Australian Mammalogy*, **3**, 87–96.

Gulland, J.A. (1971). The effect of exploitation on the numbers of marine animals. pp. 450–468. In: *Dynamics of Numbers in Populations*. den Boer, P.J. and Gradwell, G.R. (Eds). Proceedings of the Advanced Study Institute, Oosterbeek, Netherlands, 7–18 September, 1970. Centre for Agricultural Publishing and Documentation, Wageningen.

Gurd, D.B. and Nudds, T.D. (1999). Insular biogeography of mammals in Canadian National Parks: a re-analysis. *Journal of Biogeography*, **26**, 973–982.

Gurd, D.B., Nudds, T.D. and Rivard, D.H. (2001). Conservation of mammals in eastern North American reserevs: How small is too small? *Conservation Biology*, **15**, 1355–1363.

Gustafsson, L., de Jong, J. and Noren, M. (1999). Evaluation of Swedish woodland key habitats using red-listed bryophytes and lichens. *Biodiversity and Conservation*, **8**, 1101–1114.

Gustafsson, E.J. and Gardner, R.H. (1996). The effect of landscape heterogeneity on the probability of patch colonisation. *Ecology*, **77**, 94–107.

Gustafsson, E.J. and Parker, G.R. (1992). Relationships between landcover proportion and indices of landscape spatial pattern. *Landscape Ecology*, **7**, 101–110.

Haddad, N.M. and Baum, K.A. (1999). An experimental test of corridor effects on butterfly densities. *Ecological Applications*, **9**, 623–633.

Haila, Y. (1999). Islands. pp. 234–264. In: *Maintaining Biodiversity in Forest Ecosystems.* Hunter, M.L. (Ed). Cambridge University Press, Cambridge.

Haila, Y. (2002). A conceptual genealogy of fragmentation research from island biogeography to landscape ecology. *Ecological Applications*, **12**, 321–334.

Hails, R.S. (2000). Genetically modified plants – the debate continues. *Trends in Ecology and Evolution*, **15**, 14–18.

Haines-Young, R. and Chopping, M. (1996). Quantifying landscape structure: a review of landscape indices and their application to forested environments. *Progress in Physical Geography*, **20**, 418–445.

Hairston, N.G. (1989). *Ecological Experiments: Purpose, Design, and Execution.* Cambridge University Press, Cambidge, New York.

Hale, P. (1995). Conservation through sustainable use of wildlife. Conference report. *Pacific Conservation Biology*, **2**, 158–160.

Hale, P. and Lamb, D. (Eds) (1997). *Conservation Outside Reserves.* Centre for Conservation Biology, University of Queensland, Brisbane, Australia.

Hall, A.V. (1993). Setting conservation priorities for threatened species: A joint grouping and sequencing method. *South African Journal of Botany*, **59**, 581–591.

Hall, C.M. (1988). The 'worthless lands hypothesis' and Australia's national parks and reserves. pp. 441–59. In: *Australia's Ever Changing Forests.* Frawley, K.J. and Semple, N.M. (Eds). Australian Defence Force Academy, Canberra.

Hall, L. and Richards, G. (2000). *Flying Foxes. Fruit and Blossum Bats of Australia.* Australian Natural History Series. UNSW Press, Sydney.

Hall, L.A. (1987). Transplantation of sensitive plants as mitigation for environmental impacts. pp. 413–420. In: *Conservation and Management of Rare and Endangered Plants.* Elias, T.S. (Ed). California Native Plant Society, Sacremento.

Hall, L.S., Krausman, P.A. and Morrison, M.L. (1997). The habitat concept and a plea for the use of standard terminology. *Wildlife Society Bulletin*, **25**, 173–182.

Hall, N. and Chubb, C. (2001). The status of the Western Rock Lobster, *Panulirus cygnus*, fishery and the effectiveness of management controls in increasing egg production of the stock. *Marine and Freshwater Research*, **52**, 1657–1667.

Hall, N.G. (1994). New developments in modelling the Rock Lobster fishery of Western Australia. *Agricultural Systems and Information Technology*, **6**, 25–28.

Halse, S.A., Ruprecht, J.K. and Pinder, M. (2003). Salinisation and prospects for biodiversity in rivers and wetlands of south-west Western Australia. *Australian Journal of Botany*, **51**, 673–688.

Hamel, G. and Prahalad, C.K. (1989). Strategic intent. *Harvard Business Review*, **89**, 63–76.

Hamer, A.J., Lane, S.J. and Mahony, M.J. (2002). The role of introduced mosquitofish (*Gambusia holbrooki*) in excluding the native green and gold bell frog (*Litoria aurea*) from original habitats in south-eastern Australia. *Oecologia*, **132**, 445–452.

Hamilton, C. (2001). *Running from the Storm.* UNSW Press, Sydney.

Hamilton, C., Schlegelmilch, K., Hoerner, A. and Milne, J. (2000). *Environmental Tax Reform: Using The Tax System to Protect the Environment and Promote Employment.* Tela series, Australian Conservation Foundation.

Hammond, H. (1991). *Seeing the Forest Among the Trees.* Polestar Press, Vancouver, BC.

Hammond, P. (1992). Species inventory. pp. 17–39. In: *Global Diversity: Status of the Earth's Living Resources.* World Conservation Monitoring Centre (Ed). Chapman and Hall, London.

Hammond, P.M. (1995). Practical approaches to the estimation of the extent of biodiversity in speciose groups. pp. 119–136. In: *Biodiversity: Measurement and Estimation.* Hawksworth, D.L. (Ed). Chapman and Hall, London.

Hampe, A. and Petit, R. (2005). Conserving biodiversity under climate change: the rear edge matters. *Ecology Letters*, **8**, 461–467.

Hamrick, J.L. and Godt, M.J. (1989). Allozyme diversity in plant species. pp. 43–63. In: *Plant Population Genetics, Breeding and Genetic Resources.* Brown, A.H.D., Clegg, M.T., Kahler, A.L. and Weir, B.S. (Eds). Sinauer, Sunderland, Massachusetts.

Hamrick, J.L. and Godt, M.J. (1996). Conservation genetics of endemic plant species. pp. 281–304. In: *Conservation Genetics.* Avise, J.C. and Hamrick, J.L. (Eds). Chapman and Hall, New York.

Hancock, P. (1993). *Green and Gold. Sustaining Mineral Wealth, Australians and their Environment.* Centre for Resource & Environmental Studies, The Australian National University, Canberra.

Hannah, D.S., Smith, G.C. and Agnew, G. (1998). Reptile and amphibian composition in prescribed burnt dry sclerophyll forest, southern Queensland. *Australian Forestry*, **61**, 34–39.

Hannah, L., Midgley, G.F., Lovejoy, T., Bond, W.J., Bush, M., Lovett, J.C., Scott, D. and Woodward, F.I. (2002).

Conservation of biodiversity in a changing climate. *Conservation Biology*, **16**, 264–268.

Hannon, S.J. and Cotterill, S.E. (1998). Nest predation in Aspen woodlots in an agricultural area in Alberta: the enemy from within. *The Auk*, **115**, 16–25.

Hansen, A. and Rotella, J. (1999). Abiotic factors. pp. 161–209. In: *Managing Biodiversity in Forest Ecosystems*. Hunter III, M. (Ed). Cambridge University Press, Cambridge.

Hanski, I. (1994a). Patch occupancy dynamics in fragmented landscapes. *Trends in Evolution and Ecology*, **9**, 131–134.

Hanski, I. (1994b). A practical model of metapopulation dynamics. *Journal of Animal Ecology*, **63**, 151–62.

Hanski, I. (1997). Metapopulation dynamics: from concepts and observations to predictive models. pp. 69–91. In: *Metapopulation Biology: Ecology, Genetics and Evolution*. Hanksi, I. and Gilpin, M.E. (Eds). Academic Press, San Diego.

Hanski, I. (1999a). *Metapopulation Ecology*. Oxford University Press, Oxford.

Hanski, I. (1999b). Habitat connectivity, habitat continuity, and metapopulations in dynamic landscapes. *Oiko,s* **87**, 209–219.

Hanski, I. and Gilpin, M. (1991). Metapopulation dynamics: Brief history and conceptual domain. *Biological Journal of the Linnean Society*, **42**, 3–16.

Hanski, I. and Gyllenberg, M. (1993). Two general metapopulation models and the core-satellite hypothesis. *The American Naturalist*, **142(1)**, 17–41.

Hanski, I. and Simberloff, D. (1997). The metapopulation approach, its history, conceptual domain, and application to conservation. pp. 5–26. In: *Metapopulation Biology: Ecology, Genetic and Evolution*. Hanski, I. and Gilpin, M.E. (Eds). Academic Press, San Diego.

Hanski, I. and Thomas, C.D. (1994). Metapopulation dynamics and conservation: A spatially explicit model applied to butterflies. *Biological Conservation*, **68**, 167–180.

Hanski, I.K., Fenske, T.J. and Niemi, G.J. (1996). Lack of edge effect in nesting success of breeding birds in managed forest landscapes. *The Auk*, **113**, 578–585.

Hansson, L. (1983). Bird numbers across edges between mature conifer and clearcuts in central Sweden. *Ornis Scandinavia*, **14**, 97–103.

Hansson, L. (1991). Dispersal and connectivity in metapopulations. *Biological Journal of the Linnean Society*, **42**, 89–103.

Hansson, L. (1998). Nestedness as a conservation tool: plants and birds of oak-hazel woodland in Sweden. *Ecology Letters*, **1**, 142–145.

Harcourt, A.H. and Parks, S.A. (2003). Threatened primates experience high human densities: adding an index of threat to the IUCN Red List criteria. *Biological Conservation*, **109**, 137–149.

Harcourt, A.H., Parks, S.A. and Woodroffe, R. (2001). Human density as an influence on species/area relationships: double jeopardy for small African reserves? *Biodiversity and Conservation*, **10**, 1011–1026.

Harcourt, C. (1992). Tropical moist forests. pp. 256–275. In: *Global Diversity: Status of the Earth's Living Resources*. World Conservation Monitoring Centre (Ed). Chapman and Hall, London.

Hardin, G. (1993). Living within limits: ecology, economics and population taboos. Oxford University Press, New York.

Harley, K.L.S. and Forno, I.W. (1992). *Biological Control of Weeds: A Handbook for Practitioners and Students* . Inkata Press, Melbourne.

Harmelin-Vivien, M.L. (1989). Reef fish community structure: an Indo-Pacific comparison. pp. 21–60. In: *Vertebrates in Complex Tropical Systems*. Harmelin-Vivien, M.L. and Bourlière, F. (Eds). Springer Verlag, New York.

Harmon, M.E., Franklin, J.F., Swanson, F.J., Sollins, P., Gregory, S.V., Lattin, J.D., Anderson, N.H., Cline, S.P., Aumen, N.G., Sedell, J.R., Lienkaemper, G.W., Cromack, K.J. and Cummins, K.W. (1986). Ecology of coarse woody debris in temperate ecosystems. *Advances in Ecological Research*, **15**, 133–302.

Harper, J.L. (1981). The meanings of rarity. pp. 189–203. In: *The Biological Aspects of Rare Plant Conservation*. Synge, H. (Ed). Wiley, Sydney.

Harper, J.L. (1982): After description. pp. 11–26. In: *The Plant Community as a Working Mechanism*. Newman, E. (Ed). Blackwell Scientific Publishing, Oxford.

Harper, J.L. and Hawksworth, D.L. (1995). Preface. pp. 5–12. In: *Biodiversity: Measurement and Estimation*. Hawksworth, D.L. (Ed). Chapman and Hall, London.

Harrington, G.N. and Sanderson, K.D. (1994). Recent contraction of wet sclerophyll forest in the wet tropics of Queensland due to invasion by rainforest. *Pacific Conservation Biology*, **1**, 3319–3327.

Harris, J.H. (1995). The use of fish in ecological assessments. *Australian Journal of Ecology*, **20**, 65–80.

Harris, L.D. (1984) *The Fragmented Forest: Island Biogeography Theory and the Preservation of Biotic Diversity*. University of Chicago Press, Chicago.

Harrison, S. (1991). Local extinction in a metapopulation context: an empirical evaluation. *Biological Journal of the Linnean Society*, **42**, 73–88.

Harrison, S. and Bruna, E. (2000). Habitat fragmentation and large-scale conservation: What do we know for sure? *Ecography*, **22**, 225–232.

Harrison, S. and Taylor, A.D. (1997). Empirical evidence for metapopulation dynamics. pp. 27–42. In: *Metapopulation Biology: Ecology, Genetics and Evolution.* Hanski, I. and Gilpin, M.E. (Eds). Academic Press, San Diego.

Harrison, S., Maron, J. and Huxel, G. (2000). Regional turnover and fluctuation of five plants confined to serpentine seeps. *Conservation Biology*, **14**, 769–779.

Harrison, S., Murphy, D.D. and Ehrlich, P. (1988). Distribution of the Bay Checkerspot butterfly, *Euphydras editha bayensis*: evidence for a metapopulation model. *American Naturalist*, **132**, 360–382.

Hart, B.T., Bailey, P., Edwards, R., Hortle, K., James, K., McMahon, A., Meredith, C. and Swalding, K. (1991). A review of salt sensitivity of the Australian freshwater biota. *Hydrobiologica*, **210**, 105–144.

Hartley, S. and Kunin, W.E. (2003). Scale dependency of rarity, extinction risk and conservation priority. *Conservation Biology*, **17**, 1559–1570.

Harwood, J. and Stokes, K. (2003). Coping with uncertainty: lessons from fisheries. *Trends in Ecology and Evolution*, **18**, 617–622.

Has, A.H. and Dietsch, T.V. (2004). Linking shade coffee certification to biodiversity conservation: Butterflies and birds in Chippas, Mexico. *Ecological Applications*, **14**, 642–654.

Hastie, T.J. and Tibshirani, R.J. (1990). *Generalised additive models.* Chapman and Hall, London.

Hastings, A. (1993). Complex interactions between dispersal and dynamics: lessons from coupled logistic equations. *Ecology*, **74**, 1362–1372.

Hastings, A. and Harrison, S. (1994). Metapopulation dynamics and genetics. *Annual Review of Ecology and Systematics*, **25**, 167–188.

Hatch, S.A. (2003). Statistical power for detecting trends with applications to seabird monitoring. *Biological Conservation*, **111**, 317–329.

Hatchwell, P. (1989). Opening Pandora's box: the risks of releasing genetically engineered organisms. *The Ecologist*, **19**, 130–136.

Hawksworth, D.L. (1992). Microorganisms. pp. 47–54. In: *Global Diversity: Status of the Earth's Living Resources.* W.C.M. Centre (Ed). Chapman and Hall, London.

Haynes, R.W. and Perez, G.P. (Eds). (2001). *Northwest Forest Plan Research Synthesis.* USDA General Technical Report PNW-GTR-498.

Hayward, M.W., Tores de, P.J., Dillon, M.J. and Fox, B.J. (2003). Local population structure of a naturally occurring metapopulation of the Quokka (*Setonix brachyurus* Macropodidae: Marsupialia). *Biological Conservation*, **110**, 343–355.

Hazell, D., Cunningham, R.B., Lindenmayer, D.B. and Osborne, W. (2001). Use of farm dams as frog habitat in an Australian agricultural landscape: factors affecting species richness and distribution. *Biological Conservation*, **102**, 155–169.

Hazell, P. and Gustafsson, L. (1999). Retention of trees at final harvest – evaluation of a conservation technique using ephiphytic bryophyte and lichen transplants. *Biological Conservation*, **90**, 133–142.

Head, I.M., Saunders, J.R. and Pickup, R.W. (1998). Microbial evolution, diversity and ecology: A decade of ribosomal RNA analysis of uncultivated organisms. *Microbial Ecology*, **35**, 1–21.

Heinsohn, R., Lacy, R.C., Lindenmayer, D.B., Marsh, H., Kwan, D. and Lawler, I.R. (2004). Unsustainable harvest of dugongs in Torres Strait and Cape York (Australia) waters: two case studies using population viability analysis. *Animal Conservation*, **7**, 417–425.

Heinze, D. and Williams, L. (1998). The discovery of the Mountain Pygmy possum, *Burramys parvus* on Mount Buller, Victoria. *Victorian Naturalist*, **115**, 132–134.

Hellawell, J.M. (1991). Development of a rationale for monitoring. pp. 1–14. In: *Monitoring for Conservation and Ecology.* Goldsmith, F.B. (Ed). Chapman and Hall, London.

Helsinki Process (1995). European criteria and indicators for sustainable forest management. *Proceedings of the Second Ministerial Conference on protection of Forests in Europe.* Antayla January, 1995.

Henderson, N. and Sutherland, W.J. (1996). Two truths about discounting and their environmental consequences. *Trends in Ecology and Evolution*, **11**, 527–528.

Hendricks, M. (2004). Deoxygenation purges ballast water of invasives – and its cheap. *Conservation in Practice*, **5**(**3**), 38–39.

Henrickson, D.A. and Brooks, J.E. (1991). Transplanting short-lived fishes in North American deserts: Review, assessment and recommendations. pp. 283–298. In: *Battle Against Extinction: Native Fish Management in the American West.* Minckley, W.L. and Deacon, J.E. (Eds). University of Arizona Press, Tucson.

Henzell, R. (2002). Rabbits and possums in the GMO potboiler. *Biocontrol News and Information*, **23**, 89–96.

Hewitt, C.L., Campbell, M.L., Thresher, R.E. and Martin, R.B. (Eds) (1999). *Marine Biological Invasions of Port Phillip Bay, Victoria.* CRIMP, Melbourne.

Heywood, V.H. and Stuart, S.N. (1992). Species extinctions in tropical forests. pp. 91–117. In: *Tropical*

Deforestation and Species Extinction. Whitmore, T.C. and Sayer, J.A. (Eds). Chapman and Hall, London.

Higgins, P.J. (2001). *Handbook of Australian, New Zealand and Antarctic Birds. Volume 4. Parrots to Dollarbirds.* Oxford University Press, Melbourne.

Hilborn, R. (1987). Living with uncertainty in resource management. *North American Journal of Fisheries Management,* 7, 1–5.

Hilborn, R. (1992). Can fisheries agencies learn from experience? *Fisheries,* **17**, 6–14.

Hilborn, R. and Ludwig, D. (1993). The limits of applied ecological research. *Ecological Applications,* **3**, 550–552.

Hilborn, R., Quinn, T.P., Schindler, D.E. and Rogers, D.E. (2003). Biocomplexity and fisheries sustainability. *Proceedings of the National Academy of Science,* **100**, 6564–6568.

Hill, C.J. (1995). Linear strips of rain forest vegetation as potential dispersal corridors for rain forest insects. *Conservation Biology,* **9**, 1559–1566.

Hill, C.J., Gillison, A. and Jones, R.E. (1992). The spatial distribution of rainforest butterflies at three sites in North Queensland. *Australian Journal of Tropical Ecology,* **8**, 37–46.

Hill, D.G., Fitzsimons, P.F. and Thomas, W.H. (1985). *Coastal erosion control guidelines.* Land Protection Service, Department of Conservation, Forests and Lands, Victoria.

Hill, J.K., Thomas, C.D. and Lewis, O.T. (1996). Effects of habitat patch size and isolation on dispersal by *Hesperia comma* butterflies: implications for metapopulation structure. *Journal of Animal Ecology,* **65**, 725–735.

Hill, J.K., Thomas, C.D. and Lewis, O.T. (1999). Flight morphology in fragmented populations of a rare British butterfly, *Hesperia comma. Biological Conservation,* **87**, 277–283.

Hill, M.O. (1973). Diversity and evenness: A unifying notion and its consequences. *Ecology,* **54**, 427–431.

Hill, R. (2003). Frameworks to support indigenous managers: the key to fire futures. pp. 175–186. In: *Australia Burning. Fire Ecology, Policy and Management Issues.* Cary, G., Lindenmayer, D.B. and Dovers, S. (Eds). CSIRO Publishing, Melbourne.

Hill, R.S., Read, J. and Busby, J.R. (1988). The temperature-dependence of photosynthesis of some Australian temperate rainforest trees and its biogeographic significance. *Journal of Biogeography,* **15**, 431–449.

Hillis, D.M. and Moritz, C. (1996). Molecular systematics: context and controversies. pp. 1–13. In: *Molecular Systematics.* Second Edition. Hillis, D.M., Moritz, C. and Mable, B.K. (Eds). Sinauer Associates: Sunderland.

Hilty, J. and Merenlender, A. (2000). Faunal indicator taxa selection for monitoring ecosystem health. *Biological Conservation,* **92**, 185–197.

Hilty, J. and Merenlender, A.M. (2003). Studying biodiversity on private lands. *Conservation Biology,* **17**, 132–137.

Hinds, L.A., Hardy, C.M., Lawson, M.A. and Singleton, G.R. (2003). Developments in fertility control for pest animal management. In: *Rats, mice and people: rodent biology and management.* Hinds, L.A., Krebs, C.R. and Spratt, D.M. (Eds). ACIAR, Canberra.

Hingston, A.B. and 31 others (2002). Extent of invasion of Tasmanian native vegetation by the exotic Bumblebee *Bombus terrestris* (Apoidea: Apidae). *Austral Ecology,* **27**, 162–172.

Hinsley, S.A., Bellamy, P.E., Newton, I. and Sparks, T.H. (1995). Habitat and landscape factors influencing the presence of individual breeding bird species in woodland fragments. *Journal of Avian Biology,* **26**, 94–104.

Hinsley, S.A., Rothery, P., Bellamy, P.E. (1999). Influence of woodland area on breeding success in great tits *Parus major* and blue tits *Parus caeruleus. Journal of Avian Biology,* **30**, 271–281.

Hiura, T. (1995). Gap formation and species diversity in Japanese beech forests: a test of the intermediate disturbance hypothesis on a geographic scale. *Oecologia,* **104**, 265–271.

Hnatiuk, R. (1990). *Census of Australian vascular plants.* Australian Government Publishing Service, Canberra.

Hobbs, R. and Huenneke, L.F. (1992). Disturbance, diversity, and invasion: implications for conservation. *Conservation Biology,* **6**, 324–337.

Hobbs, R.H. and Mooney, H.A. (1998). Broadening the extinction debate: population deletions and additions in California and Western Australia. *Conservation Biology,* **12**, 271–283.

Hobbs, R.J. (1992). The role of corridors in conservation: Solution or bandwagon? *Trends in Ecology and Evolution,* **7**, 389–392.

Hobbs, R.J. (2001). Synergisms among habitat fragmentation, livestock grazing, and biotic invasions in southwestern Australia. *Conservation Biology,* **15**, 1522–1528.

Hobbs, R.J. and Atkins, L. (1990). Fire-related dynamics of a *Banksia* woodland in south-western Australia. *Australian Journal of Botany,* **38**, 97–110.

Hobbs, R.J. and Atkins, L. (1991). Interactions between annuals and woody perennials in a Western Australian wheatbelt reserve. *Journal of Vegetation Science,* **2**, 643–654.

Hobbs, R.J. and Hopkins, A.J.M. (1990). From frontier to fragments: European impact on Australia's vegetation.

pp. 93–114. In: *Australian Ecosystems: 200 years of Utilisation, Degradation and Reconstruction.* Proceedings of the Ecological Society of Australia. Vol. 16. Saunders, D.A., Hopkins, A.J.M and How, R.A (Eds). Surrey Beatty, Chipping-Norton.

Hobbs, R.J. and Yates, C.J. (Eds) (2000). *Temperate eucalypt woodlands in Australia: Biology, conservation, management and restoration.* Surrey Beatty and Sons, Chipping Norton.

Hobbs, R.J. and Yates, C.J. (2003). Impacts of ecosystem fragmentation on plant populations: generalising the idiosyncratic. *Australian Journal of Botany*, **51**, 471–488.

Hobbs, R.J., Catling, P., Wombey, J.C., Clayton, M., Atkins, L. and Reid, A. (2004). Faunal use of Bluegum (*Eucalyptus globulus*) plantations in southwestern Australia. *Agroforestry Systems* (in press).

Hobbs, R.J., Floyd, R., Cunningham, S., Catling, P. and Ive, J. (2002). *Hardwood plantations: quantifying conservation and environmental service benefits.* Report to the Joint Agroforestry Porgram. Rural Industries Research and Development Corporation, Canberra.

Hobbs, R.J., Saunders, D.A. and Arnold, G.W. (1993). Integrated Landscape Ecology- A Western-Australian Perspective. *Biological Conservation*, **64**, 231–238.

Hobson, K.A. and Schieck, J. (1999). Changes in bird communities in boreal mixedwood forest: Harvest and wildfire effects over 30 years. *Ecological Applications*, **9**, 849–863.

Hodder, K.H. and Bullock, J.M. (1997). Translocations of native species in the UK: implications for diversity. *Journal of Applied Ecology*, **34**, 547–565.

Hoddle, M.S. (2004). Restoring balance: Using exotic species to control invasive exotic species. *Conservation Biology*, **18**, 38–49.

Hodgkinson, K.C. and Harrington, G.N. (1985). The case for prescribed burning to control shrubs in eastern semi-arid woodland. *Australian Rangeland Journal*, **7**, 64–74.

Hoffman, B.D. and Andersen, A.N. (2003). Responses of ants to disturbance in Australia, with particular reference to functional groups. *Austral Ecology*, **28**, 444–464.

Hoffman, E.O. and Hammonds, J.S. (1994). Propagation of uncertainty in risk assessments: the need to distinguish between uncertainty due to lack of knowledge and uncertainty due to variability. *Risk Analysis*, **14**, 707–712.

Hogbin, P.M., Peakall, R. and Sydes, M.A. (2000). Achieving practical outcomes from genetic studies of rare Australian plants. *Australian Journal of Botany*, **48**, 375–382.

Holdren, J.P. (1991). Population and the energy problem. *Population and Environment*, **12**, 231–255.

Holdsworth, A.R. and Uhl, C. (1997). Fire in Amazonian selectively logged rain forest and the potential for fire reduction. *Ecological Applications*, **7**, 713–725.

Holling, C.S. (1973). Resilience and stability of ecological systems. *Annual Review of Ecology and Systematics*, **4**, 1–23.

Holling, C.S. (1993). Investing in research for sustainability. *Ecological Applications*, **3**, 552–555.

Holling, C.S. (Ed) (1978). *Adaptive environmental assessment and management.* International Series on Applied Systems Analysis 3, International Institute for applied systems analysis. John Wiley, Toronto.

Holling, C.S. and Meffe, M. (1996). Command and control and the pathology of natural resource management. *Conservation Biology*, **10**, 328–337.

Holmes, S. (1979). *Henderson's Dictionary of Biological Terms.* 9th edition. Longman, London.

Holt, S.J. (1987). Categorisation of threats to and status of wild populations. pp. 19–30. In: *The Road to Extinction.* Fitter, R. and Fitter, M. (Eds). IUCN, Gland, Switzerland.

Holtkamp, R.H. (1996). Integrated control of *Chrysanthemoides monilifera* in New South Wales. pp. 511–515. In: *Proceedings of the Eleventh Australian Weeds Conference.* Sheperd, R.C.H. (Ed). Melbourne Weed Science Society of Victoria, Frankston.

Holyoak, M. (2000). Habitat subdivision causes changes in food web structure. *Ecology Letters*, **3**, 509–515.

Honnay, O., Hermy, M. and Coppin, P. (1999). Effects of area, age and diversity of forest patches in Belgium on plant species richness, and implications for conservation and reaforestation. *Biological Conservation*, **87**, 73–84.

Hooker, S.K., Whitehead, H. and Gowans, S. (1999). Marine protected area design and the spatial and temporal distribution of cetaceans in a submarine canyon. *Conservation Biology*, **13**, 592–602.

Hooper, D.U., Bignell, D.E., Brown, V.K., Brussaard, L., Dangerfield, J.M., Wall, D.H., Wardle, D.H., Coleman, D.C., Giller, K.E., Lavelle, P., Van der Putten, W.H., De Ruiter, P.C., Rusek, J., Silver, W.L., Tiedje, J.M. and Wolters, V. (2000). Interactions between aboveground and belowground biodiversity in terrestrial ecosystems: patterns, mechanisms, and feedbacks. *BioScience*, **50**, 1049–1061.

Hooper, D.U., Chapin, F.S. and 13 others. (2005). Effects of biodiversity on ecosystem functioning: a consensus of current knowledge. *Ecological Monographs*, **75**, 3–35.

Hopper, S.D. (1997). An Australian perspective on plant conservation biology in practice. pp. 255–278. In: *Conservation biology for the coming decade.* Fiedler, P.and Karieva, P. (Eds). Chapman and Hall, New York.

Hopper, S.D. (2000). Floristics of Autralian granitoid inselberg vegetation. pp. 391–407. In: *Inselbergs.* Ecological Studies. Volume 146. Porembski, S. and Barthlott, W. (Eds). Spriner–Verlag, Berlin.

Hopper, S.D. and Coates, D.J. (1990). Conservation of genetic resources in Australia's flora and fauna. *Proceedings of the Ecological Society of Australia,* **16**, 567–577.

Hopper, S.D. and Wardell-Johnson, G. (2004). *Eucalyptus virginea* and *E. relicta* (Myrtaceae), two new rare forest trees from south-western Australia allied to *E. lanepoolei*, and a new phantom hybrid. *Nuytsia*, **15**, 227–240.

Horowitz, H. (1995). An environmental critique of some freshwater captive breeding and reintroduction programmes in Australia. pp. 75–80. In: *Reintroduction Biology of Australasian Fauna.* Serena, M. (Ed). Surrey Beatty, Chipping Norton.

Horowitz, P. and Calver, M. (1998). Credible science? Evaluating the Regional Forest Agreement Process in Western Australia. *Australian Journal of Environmental Management,* **5**, 213–225.

Hose, G. (2003). Colonial Spider. *Nature Australia,* **27(10)**, 22–23.

Houlahan, J.E., Findlay, C.S., Meyer, A.H., Kuzmin, S.L. and Schmidt, B.R. (2001). Global amphibian population declines Response to Alford *et al. Nature*, **412**, 499–500.

How, R.A., Barnett, J.L., Bradley, A.J., Humphreys, W.F. and Martin, R. (1984). The population biology of *Pseudocheirus peregrinus* in a *Leptospermum laevigatum* thicket. pp. 261–268. In: *Possums and Gliders.* Smith, A.P. and Hume, I.D. (Eds). Surrey Beatty and Sons, Sydney.

Howard, T.M. (1973). Studies in the ecology of *Nothofagus cunninghamii* Oerst. I. Natural regeneration on the Mt. Donna Buang massif, Victoria. *Australian Journal of Botany,* **21**, 67–78.

Howden, S.M. (1985). The potential distribution of Bitou Bush in Australia. *Proceedings of the Conference on Chrysanthemoides monilifera*, Port Macquarie, Sydney, 1984. National Parks and Wildlife Service and NSW Department of Agriculture.

Howe, R.G., Davies, G.J. and Mosca, V. (1991). The demographic significance of "sink" populations. *Biological Conservation,* **57**, 239–255.

Howe, R.W. (1984). Local dynamics of bird assemblages in small forest habitat islands in Australia and North America. *Ecology,* **65**, 1585–1601.

Howie, H. (2002). Bookmark biosphere. *Environment South Australia*, **9**. [online] URL: http://www.ccsa.asn.au/esa/bookmark.htm

Howitt, A.W. (1890). The eucalypts of Gippsland. *Transactions of the Royal Society of Victoria*, **22**, 81–120.

Hoyt, E. (1995). Whale watching: a global overview of the industry's rapid growth and some implications for Australia. Summary report. *Proceedings of the Encounters with Whales 95, Conference Workshop*, Hervey Bay, Queensland, 26–30 July 1995. Australian Nature Conservation Agency, Canberra. [online] URL:

http://www.deh.gov.au/biodiversity/threatened/nominations/index.html

http://www.dse.vic.gov.au/dse/index.htm

http://www.ea.gov.au/parks/index.html

http://www.ea.gov.au/water/wetlands/ramsar/index.html

http://www.fisheries.nsw.gov.au/threatened_species/threatened_species2/conservationhttp://www.npws.nsw.gov.au/news/tscdets/index/html

http://www.ruralfutures.une.edu.au/text/tpublications/tconfpapers/tconfpapers.htm

Hubbell, S.P. (2001). *The Unified Neutral Theory of Biodiversity and Biogeography.* Princeton University Press, Princeton.

Hughes, A.R. and Stachowicz, J.J. (2004). Genetic diversity enhances the resistance of a seagrass ecosystem to disturbance. *Proceedings of the National Academy of Sciences*, **101**, 8998–9002.

Hughes, G. and Madden, K.V. (2003). Evaluating predictive models with application in regulatory policy for invasive weeds. *Agricultural Systems*, **76**, 755–774.

Hughes, J., Goudkamp, K., Hurwood, D., Hancock, M. and Bunn, S. (2003). Translocation causes extinction of a local population of the freshwater shrimp Paratya australiensis. *Conservation Biology*, **17**, 1007–1012.

Hughes. L. (2000). Biological consequences of global warming: is the signal already apparent? *Trends in Ecology and Evolution*, **15**, 56–61.

Hughes, L. (2003). Climate change and Australia: Trends, projections and impacts. *Austral Ecology*, **28**, 423–443.

Hughes, L., Cawsey, E.M. and Westoby, M. (1996). Climate range sizes of *Eucalyptus* species in relation to future climate change. *Global Ecology and Biogeography*, **5**, 23–29.

Hughes, R.G. (2004). Climate change and loss of saltmarshes: consequences for birds. *Ibis*, **146**, 21–28.

Hugo, G. (1999). *Demographic Trends Influencing Housing Needs and Demands in Australia.* AHURI Workshop on innovation in housing. Melbourne, April, 1999.

Hume, I. (1999). *Marsupial Nutrition.* Cambridge University Press, Cambridge.

Humphries, S.E., Groves, R.H. and Mitchell, D.S. (1991). *Plant Invasions–The Incidence of Environmental Weeds in Australia.* Part One. Plant invasions of Australian ecosystems: a status review and management directions. Australian National Parks and Wildlife Service, Canberra.

Hunter, M.L. (1990). *Wildlife, Forests and Forestry. Principles for Managing Forests for Biological Diversity.* Prentice Hall, Englewood Cliffs.

Hunter, M.L. (1991). Coping with ignorance: The coarse filter strategy for maintaining biodiversity. pp. 266–281. In: *Balancing on the Edge of Extinction.* Kohm, K. (Ed). Island Press, Washington, DC.

Hunter, M.L. (1993). Natural fire regimes as spatial models for managing boreal forests. *Biological Conservation,* **65**, 115–120.

Hunter, M.L. (1994). *Fundamentals of Conservation Biology.* Blackwell, Cambridge, Melbourne.

Hunter, M.L. (1996). Benchmarking for managing natural ecosystems: are human activities natural? *Conservation Biology,* **10**, 695–697.

Hunter, M.L. (2002). *Fundamentals of conservation biology.* Blackwell Science, Melbourne.

Hunter, M.L. and Yonzon, P. (1993). Altitudinal distributions of birds, mammals, people, forests, and parks in Nepal. *Conservation Biology,* 7, 420–423.

Hurditch, W.E. and Hurditch, W.J. (1994). The politics of bushfires. pp. 37–51. In: *The Burning Continent. Forest Ecosystems and Fire Management in Australia.* Attiwill, P., Florence, R, Hurditch, W.E. and Hurditch, W.J. (Eds). Institute of Public Affairs, Melbourne.

Hurlbert, S.H. (1972). The nonconcept of species diversity: A critique and alternative parameters. *Ecology,* **52**, 577–586.

Hurlbert, S.H. (1984). Pseduoreplcation and the design of ecological field experiments. *Ecological Monographs,* **54**, 187–211.

Huston, M.A. (1997). Hidden treatments in ecological experiments: re-evaluating the ecosystem function of biodiversity. *Oecologia,* **110**, 449–460.

Hutchings, H.R. (1991). Dimamica de tres comunidades de Papilionoidea (Insecta: Lepidoptera) en fragmentos de floresta na Amazonia central. MSc thesis, INPA/Fundaçâo Universidade de Amazons, Manuas, Brazil.

Hutchinson, G.E. (1958). Concluding remarks. *Cold Spring Harbor Symposium on Quantitative Biology,* **22**, 415–427.

Hutchinson, G.E. (1959). Homage to Santa Rosalia; or why are there so many different kinds of animals? *American Naturalist,* **93**, 145–159.

Hutchinson, M.F. (1989). A new technique for gridding elevation data and streamline data with automatic removal of spurious pits. *Journal of Hydrology,* **106**, 211–232.

Hutchinson, M.F. (1995). Stochastic space-time weather models from ground-based data. *Agricultural and Forest Meteorology,* **73**, 237–264.

Hutchinson, M.F. (1999). ANUSPLIN Version 4.0. http://cres.anu.edu.au/software/anusplin.html

Hutchinson, M.F. and Dowling, T.I. (1989). A continental assessment of hydrological applications of a new grid-based digital elevation model of Australia. *Hydrological Processes,* **5**, 45–58.

Hutha, E., Jokimaki, J. and Rahko, P. (1998). Distribution and reproductive success of the pied flycatcher *Fidecula hypoleuca* in relation to forest patch size and vegetation characteristics: the effects of scale. *Ibis,* **140**, 214–222.

Hutto, R.L. (1995). Composition of bird communities following stand replacing fires in northern Rocky Mountain (USA) conifer forests. *Conservation Biology,* **10**, 1041–1058.

Hutton, I. (1986). *Lord Howe Island. Discovering Australia's World Heritage.* Conservation Press, Melbourne.

Hutton, I. (1990). *Birds of Lord Howe Island.* Lithocraft, Melbourne.

IIED (1995). Citizens action to lighten Britain's ecological footprint. International Institute for Environment and Development, London.

Ims, R.A., Rolstad, J. and Wegge, P. (1993). Predicting space use responses to habitat fragmentation: Can voles *Microtus oeconomus* serve as an experimental model system (EMS) for Capercaillie Grouse *Tetrao urogallus* in boreal forest? *Biological Conservation,* **63**, 261–268.

Incoll, R.D. (1995). Landscape ecology of glider populations in old-growth forest patches. BSc Hons thesis. Department of Zoology, University of Melbourne, Melbourne, Australia.

Incoll, R.D., Loyn, R.H., Ward, S.J., Cunningham, R.B. and Donnelly, C.F. (2001). The occurrence of gliding possums in old-growth patches of Mountain Ash (*Eucalyptus regnans*) in the Central Highlands of Victoria. *Biological Conservation,* **98**, 77–88.

Inger, R.F. and Colwell, R.K. (1977). Organisation of contiguous communities of amphibians and reptiles in Thailand. *Ecological Monographs,* **47**, 229–253.

Ingham, D.S. and Samways, M.J. (1996). Application of fragmentation and variegation models to epigaeic invertebrates in South Africa. *Conservation Biology,* **10**, 1353–1358.

Inglis, G. and Underwood, A.J. (1992). Comments on some designs proposed for experiments on the biological importance of corridors. *Conservation Biology*, **6**, 581–586.

Inions, G., Tanton, M.T. and Davey, S.M. (1989). Effects of fire on the availability of hollows in trees used by the Common Brushtail Possum, *Trichosurus vulpecula* Kerr 1792, and the Ringtail Possum, *Pseudocheirus peregrinus* Boddaerts 1785. *Australian Widlife Research*, **16**, 449–458.

Innes, J., Hay, R., Flux, I., Bradfield, P., Speed, H. and Jansen, P. (1999). Successful recovery of North Island Kokako *Callaeus cinerea wilsoni* populations by adaptive management. *Biological Conservation*, **87**, 201–214.

Intergovernmental Panel on Climate Change (2001a). *Climate Change 2001: The Scientific Basis*. Intergovernmental Panel on Climate Change. Cambridge University Press, Port Chester, New York.

Intergovernmental Panel on Climate Change. (2001b). *Climate Change 2001: Impacts, Adaptation and Vulnerability*. Intergovernmental Panel on Climate Change. Cambridge University Press, Port Chester, New York.

International Road Federation (1997). *World Road Statistics*. 1997 Edition. International Road Federation, Geneva, Switzerland.

IRC (2002). *Western Rock Lobster Ecological Risk Assessment*. IRC Environment, Project JOO–207. Report to the Western Australian Department of Fisheries, Perth.

ITTO (1990). *Guidelines for the Sustainable Management of Natural Tropical Forests*. International Tropical Timber Organisation, Yokohama. Technical Series No. 5.

ITTO (1992). *Criteria for the Measurement of Sustainable Tropical Forest Management*. International Tropical Timber Organisation, Yokohama. Policy Development Series No. 3.

ITTO (1993). *Guidelines on Biodiversity Conservation of Production Tropical Forests*. International Tropical Timber Organisation, Yokohama. Policy Development Series No. 5.

IUCN (1980). *World Conservation Strategy*. International Union for the Conservation of Nature, Gland, Switzerland.

IUCN (1984). Categories and criteria for protected areas. pp. 47–53. In: *National Parks, Conservation and Development. The Role of Protected Areas in Sustaining Society*. McNeely, J.A. and Miller, K.R. (Eds). Smithsonian Institution Press, Washington.

IUCN (1987). *The IUCN Position Statement on Translocation of Living Organisms: Introductions, Re-introductions and Restocking*. International Union for the Conservation of Nature, Gland, Switzerland.

IUCN (1994). *IUCN Red List Categories*. As approved by the 40th Meeting of the IUCN Council. Prepared by the International Union for the Conservation of Nature, Species Survival Commission, Gland, Switzerland.

IUCN (1997). *IUCN Red List of Threatened Plants*. UNEP World Conservation Monitoring Centre, International Union for the Conservation of Nature, Gland, Switzerland. [online] URL: http://www.wcmc.org.uk/species/plants/geographic_table.htm

IUCN (1998). *Guidelines for Re-introductions*. IUCN/SSC-Reintroduction Specialist Group, Gland, Switzerland.

IUCN (2001). *Red List Categories*. Version 3.1. Gland, Switzerland

IUCN (2003). *The IUCN Red List of Threatened vertebrates*. International Union for the Conservation of Nature, Gland, Switzerland. [online] URL: http://www.redlist.org/

Ive, J.R. and Lambeck, R.J. (1997). *Practical Integration of Farm Forestry and Biodiversity. Milestone 2 Report – Review of Scientific Evidence*. Report to Rural Industries Research and Development Corporation, Canberra. August 1997.

Jablonski, D. (1995). Extinctions in the fossil record. pp. 25–44. In: *Extinction Rates*. Lawton, J.H. and May, R.M. (Eds). Oxford University Press, Oxford.

Jackson, J.B.C. and 18 others (2001). Historical overfishing and the recent collapse of coastal ecosystems. *Science*, **293**, 629–636.

Jackson, S. (1999). Preliminary predictions of the impacts of habitat area and catastrophes on the viability of Mahogany Glider *Petaurus gracilis* populations. *Pacific Conservation Biology*, **5**, 56–62.

Jackson, S. (2000). The Mahogany Glider (*Petaurus gracilis*): A review of its ecology and conservation. pp. 87–98. In: *Biology of Gliding Mammals*. Goldingay, R.L. and Scheibe, J.S. (Eds). Filander Verlag, Fürth, Germany.

Jackson, S. (2003). *Australian Mammals: Biology and Captive Management*. CSIRO Publishing, Melbourne.

Jackson, S. and Claridge, A. (1999). Climate modelling of the distribution of the Mahogany Glider *Petaurus gracilis* and Squirrel Glider *Petaurus norfolcensis*, with implications for their evolutionary history. *Australian Journal of Zoology*, **47**, 47–57.

Jackson, W.D. (1968). Fire, air, water and earth – an elemental ecology of Tasmania. *Proceedings of the Ecological Society of Australia*, **3**, 9–16.

Jacobs, M.R. (1955). *Growth habits of the Eucalypts.* Forestry & Timber Bureau, Department of the Interior, Canberra.

James, C. (2002). *Global status of commercialised transgenic crops: 2002.* ISAAA Briefs No. 27. International Service for the Acquisition of Agri-biotech Applications, Ithaca, New York.

James, C., Landsberg, J. and Morton, S. (1999). Provision of watering points in Australian arid zone: a review of effects on the biota. *Journal of Arid Environments*, **41**, 87–121.

James, S.H. (1982). The relevance of genetic systems in *Isotoma petraea* to conservation practice. pp. 63–71. In: *Species at risk: Research in Australia.* Groves R.H. and Ride W.D.L. (Eds). Australian Academy of Science, Canberra.

JANIS (1996). (Joint Australian and New Zealand National Forest Policy Statement Implementation Sub-committee) *National Forest Conservation Reserves: Commonwealth Proposed Criteria.* Commonwealth of Australia, Canberra.

JANIS (1997). *Nationally Agreed Criteria for the Establishment of a Comprehensive, Adequate and Representative Reserve System for Forests in Australia.* A Report by the Joint ANZECC/MCFFA National Forest Policy Statement Implementation Sub-committee. ANZECC, Canberra.

Jansen, A. and Robertson, A.I. (2001). Relationships between livestock management and the ecological condition of riparian habitats along an Australian floodplain river. *Journal of Applied Ecology*, **38**, 63–75.

Jansson, G. and Angelstam, P. (1999). Threshold levels of habitat composition for the presence of the Long-tailed Tit (*Aegithalos caudatus*) in a boreal landscape. *Landscape Ecology,* **14**, 283–290.

Janzen, D.H. (1983). No park is an island: Increase in interference from outside as park size decreases. *Oikos* **41**, 402–410.

Jarvis, P.G. and McNaughton, K.G. (1986). Stomatal control of transpiration: scaling up from leaf to region. *Advances in Ecological Research,* **15**, 1–49.

Jeans, D.N. (1972). *An Historical Geography of New South Wales to 1901.* Reed, Sydney.

Jellinek, S., Driscoll, D.A. and Kirkpatrick, J.B. (2004). Environmental and vegetation variables have a greater influence than habitat fragmentation in structuring lizard communities in remnant urban bushland. *Austral Ecology*, **29**, 294–304.

Jenkins, M. (1992). Species extinction. pp. 192–233. In: *Global Diversity: Status of the Earth's Living Resources.* World Conservation Monitoring Centre (Ed). Chapman and Hall, London.

Jenkins, M.B. and Smith, E.T. (1999). *The business of sustainable forestry. Strategies for an industry in transition.* Island Press, Covelo, California.

Jensen, D.C. (1984). *A Conspiracy Called Conservation.* Veriats Publishing Company, Perth.

Jha, D.K., Sharma, G.D. and Mishra, R.R. (1992). Ecology of soil microflora and mycorrhizal symbionts in degraded forests at two altitudes. *Biology and Fertility of Soils,* **12**, 272–278.

Ji., W., Clout, M. and Sarre, S.D. (2000). Responses of male brushtail possums to sterile females: implications for biological control. *Journal of Applied Ecology*, **37**, 926–934.

Jimenez, J.A., Hughes, K.A., Alaks, G., Graham, L. and Lacy, R.C. (1994). An experimental study of inbreeding depression in a natural habitat. *Science,* **265**, 271–273.

Johnson, A.S., Hale, P.E., Ford, W.M., Wentworth, J.M., French, J.R., Anderson, O.F. and Pullen G.B. (1995). White-tailed deer in relation to successional stage, overstorey type and management of southern Appalachian forests. *American Midland Naturalist,* **133**, 18–35.

Johnson, C.N. and Wroe, S. (2003). Causes of extinction during the Holocene of mainland Australia: arrival of the Dingo, or human impact? *The Holocene*, **13**, 941–948.

Johnson, D.H. (1981). *The Use and Misuse Of Statistics in Wildlife Habitat Studies.* USDA Forest Service. Technical Report No. RM–87 pp. 11–19.

Johnson, K. (1988). Rare and endangered. Rufous Hare-Wallaby. *Australian Natural History*, **22**, 406–407.

Johnson, K.A., Burbidge, A.A. and McKenzie, N.L. (1989). Australian Macropodidea: Status, causes of decline and future research and management. pp. 641–657. In: *Kangaroos, Wallabies and Rat-kangaroos.* Grigg, G., Jarman, P. and Hume, I. (Eds). Surrey Beatty and Sons, Chipping Norton.

Johnson, M.L. and Gaines, M.S. (1985). Selective basis for emigration of the Prairie Vole, *Microtus ochrogaster*: an open field experiment. *Journal of Animal Ecology*, **54**, 399–410.

Johnson, M.L. and Gaines, M.S. (1990). Evolution of dispersal: theoretical models and empirical tests using birds and mammals. *Annual Review of Ecology and Systematics*, **21**, 449–480.

Johnson, P. (1994). Environmental ambassadors or global canaries? *Park Watch*, **March**, 4–7.

Johnson, P.M., Nolan, B.J. and Schaper, D.N. (2003). Introduction of the Proserpine Rock-Wallaby *Petrogale persephone* from mainland Queensland to nearby Hayman island. *Australian Mammalogy*, **25**, 61–71.

Johnson, R.W. (1984). Flora and vegetation of the Brigalow belt. pp. 41–60. In: *The Brigalow belt of Australia.* Symposium held at the John Kindler Memorial Theatre, Queensland Institute of Technology, 23 October 1982. Bailey, A. (Ed). The Royal Society of Queensland, Brisbane.

Johnson, S. (1992). Protected areas. pp. 447–478. In: *Global Diversity: Status of the Earth's Living Resources.* World Conservation Monitoring Centre (Ed). Chapman and Hall, London.

Johnston, F.H., Kavanagh, A., Bowman, D.M. and Scott, R. (2002). Exposure to bushfire smoke and asthma: an ecological study. *Australian Medical Journal*, **176**, 535–538.

Joint SCC/SCFA National Taskforce (2000). *Joint SCC/SCFA National Taskforce on the Prevention and Management of Marine Pest Incursions.* Final Report, 2000.

Jones, D. (1986). Exotic birds: Selected examples. pp. 92–107. In: *The Ecology of Exotic Animals and Plants: Some Australian Case Histories.* Kitching, R.L. (Ed). Wiley, Brisbane.

Jones, D. and Finn, P.G. (1999). Translocation of aggressive Australian magpies: a preliminary assessment of a potential management action. *Wildlife Research*, **26**, 271–279.

Jones, E. (1977). Ecology of the feral cat, *Felis catus* (L.), (Carnivora: Felidae) on Macquarie Island. *Australian Wildlife Research*, **4**, 249–262.

Jones, M.M. (1991). *Marine organisms transported in ballast water. A review of the Australian scientific position.* Australian Government Publishing Service. Commonwealth of Australia, Canberra. Australian Bureau of Rural Resources. Bulletin No. 7.

Jones, R. (1969). Fire-stick farming. *Australian Natural History*, **17** (7), 224–228.

Jones, R. and Crome, F.H. (1990). The biological web – plant-animal interactions in the rainforest. pp. 74–87. In: *Australian Tropical Rainforest – Science, Values, Meaning.* Webb, L.J. and Kikkawa, J. (Eds). CSIRO, Australia.

Jones, R.N. and Pittock, B. (1997). Assessing the impacts of climate change: The challenge for ecology. pp. 311–322 In: *Frontiers of Ecology.* Klomp N. and Lunt, I. (Eds). Elsiever Science, Oxford.

Jones, W.G., Hill, K.D. and Allen, J.M. (1995). *Wollemia nobilis*, a new living Australian genus and species in the Araucariaceae. *Telopea*, **6**, 173–176.

Josefson, A.B. (1990). Letters: Statistics and the precautionary principle. *Marine Pollution Bulletin*, **21**(12), 598.

Journet, A.R.P. (1981). Insect herbivory on the Australian woodland eucalypt, *Eucalyptus blakelyi* M. *Australian Journal of Ecology*, **6**, 135–138.

Kadmon, R., Farber, O. and Danin, A. (2004). Effect of roadside bias on the accuracy of predictive maps produced by bioclimatic models. *Ecological Applications*, **14**, 401–413.

Kahneman, D. and Tversky, A. (1979). Prospect theory: an analysis of decision under risk. *Econometrica*, **47**, 263–291.

Kailola, P.J., Williams, M.J., Stewart, P.C., Reichelt, R.E., McNee, A. and Grieve, C. (1993). *Australian Fisheries Resources.* Bureau of Resource Sciences, Commonwealth of Australia, Canberra.

Kanowski, J. (2001). Effects of elevated CO_2 on the foliar chemistry of seedlings of two rainforest trees from north-east Australia: implications for folivorous marsupials. *Austral Ecology*, **26**, 165–172.

Kanowski, J., Catterall, C.P. and Wardell-Johnson, G.W. (2005). Consequences of broadscale timber plantations for biodiversity in cleared forest landscapes of tropical and subtropical Australia. *Forest Ecology and Management*, **208**, 359–372.

Kanowski, P. (1997). *Plantation Forestry in the 21st Century. Special paper: Afforestation and Plantation Forestry.* XI World Forestry Congress, 13–22 October 1997.

Kapos, V. (1989). Effects of isolation on the water status of forest patches in Brazilian Amazonia. *Journal of Tropical Ecology*, **5**, 173–185.

Karr, J.R. (1991). Avian survival rates and the extinction process on Barro Coarado Island, Panama. *Conservation Biology*, **4**, 391–397.

Kati, V., Devillers, P., Dufrene, M., Legakis, A., Vokou, D. and Lebrun, P. (2004). Hotspots, complementarity or representativeness? Designing optimal small-scale reserves for biodiversity conservation. *Biological Conservation*, **120**, 471–480.

Kauffman, J.B. and Uhl, C. (1990). Interactions of anthropegenic activities, fire, and rainforests in the Amazon Basin. pp. 117–134. In: *Fire in the Tropical Biota.* Goldammer, J.G. (Ed). Springer-Verlag, New York.

Kavanagh, R.P. (1988). The impact of predation by the powerful owl *Ninox strenua* on a population of the Greater Glider *Petauroides volans. Australian Journal of Ecology*, **13**, 445–450.

Kavanagh, R.P. (1991). The target species approach to wildlife management: Gliders and owls in the forests of southeastern New South Wales. pp. 377–383. In: *Conservation of Australia's Forest*

Fauna. Lunney, D. (Ed). Royal Zoological Society of NSW, Mosman.

Kavanagh, R.P. (1997). Ecology and management of large forest owls in south-eastern Australia. PhD thesis, University of Sydney, Sydney.

Kavanagh, R.P. and Lambert, M.J. (1990). Food selection by the Greater Glider, *Petauroides volans*: Is foliar nitrogen a determinant of habitat quality? *Australian Wildlife Research*, **17**, 285–299.

Keals, N. and Majer, J.D. (1991). The conservation of ant communities along the Wubin-Perenjori corridor. pp. 387–393. In: *Nature Conservation 2: The Role of Corridors*. Saunders, D.A. and Hobbs, R.J. (Eds). Surrey Beatty and Sons, Chipping Norton.

Kearney, M., Moussalli, A., Strasburg, J., Lindenmayer, D. and Moritz, C. (2003). Geographic parthenogenesis in the Australian arid zone: I. A climatic analysis of the *Heteronotia binoei* complex (Gekkonidae). *Evolutionary Ecology Research*, **5**, 953–976.

Kearney, R. (1990). Resource assessment and the joys of always being wrong! *Australian Fisheries*, **July 1990**, 35–38.

Kearney, R.E. (1995). Biodiversity and fisheries management: The implications of extracting maximum yields from interactive ecosystems. pp. 300–305. In: *Conserving Biodiversity: Threats and Solutions*. Bradstock, R.A., Auld, T.A., Keith, D.A., Kingsford, R.T., Lunney, D. and Sivertsen, D.P. (Eds). Surrey Beatty, Chipping Norton.

Keast, A. (1968). Seasonal movements in the Australian honeyeaters (Meliphagidae) and their ecological significance. *Emu*, **67**, 159–210.

Keast, A. (Ed) (1981). *Ecological Biogeography of Australia*. Junk, The Hague.

Keating, K.A., Quinn, K A., Ivie, M.A. and Ivie, L.L. (1998). Estimating the effectiveness of further sampling in species inventories. *Ecological Applications*, **8**, 1239–1249.

Keating, P. (1992). *Australia's Environment – A National Asset: State on the Environment*. Australian Government Publishing Service, Canberra.

Keenan, R. and Ryan, P. (2004). *Old Growth Forests in Australia*. Department of Agriculture, Fisheries and Foerstry.

Keith, D., McCaw, W.L. and Whelan, R.J. (2002b). Fire regimes in Australian heathlands and their effects on plants. pp. 199–237. In: *Flammable Australia. The Fire Regimes and Biodiversity of a Continent*. Bradstock, R., Williams, J. and Gill, A.M. (Eds). Cambridge University Press, Cambridge.

Keith, D., Williams, J. and Woinarski, J. (2002a). Fire management and biodiversity conservation: key approaches and principles. pp. 401–425. In: *Flammable Australia. The Fire Regimes and Biodiversity of a Continent*. Bradstock, R., Williams, J. and Gill, A.M. (Eds). Cambridge University Press, Cambridge.

Keith, D.A. and Burgman, M.A. (2004). The Lazarus effect: can the dynamics of extinct species lists tell us anything about the status of biodiversity? *Biological Conservation*, **117**, 41–48.

Kellas, J.D. and Hateley, A.J.M. (1991). Management of dry sclerophyll forests in Victoria. I. The low elevation mixed forests. pp. 142–162. In: *Forest Management in Australia*. McKinnell, F.H. and Fox, J.E.D. (Eds). Surrey Beatty, Chipping Norton.

Kelleher, G., Bleakley, C. and Wells, S. (Eds) (1995). *Introduction. A Global Representative System of Marine Protected Areas*. Volume IV. Great Barrier Reef Marine Park Authority, World Bank and The World Conservation Union. World Bank, Environment Department, Washington, DC.

Kelly, A. (Ed) (1998). *Old-Growth Forest in South-east Queensland. Old Growth Forest Assessment Project*. Department of Natural Resources, Indooropilly, Brisbane.

Kelly, C.L., Pickering, C.M. and Buckley, R.C. (2003). Impacts of tourism on threatened plant taxa and communities in Australia. *Ecological Management and Restoration*, **4**, 37–44.

Kelly, E.J. and Campbell, K. (2000). Separating variability and uncertainty in environmental risk assessment – making choices. *Human and Ecological Risk Assessment*, **6**, 1–13.

Kelly, J.R. and Harwell, M.A. (1990). Indicators of ecosystem recovery. *Environmental Management*, **14**, 527–545.

Kelly, P.A. and Rotenberry, J.T. (1993). Buffer zones and ecological reserves in California: replacing guesswork with science. pp. 85–92. In: *Interface Between Ecology and Land Development in California*. Keeley, J.E. (Ed). Southern California Academy of Sciences, Los Angeles.

Kennedy, M. (1992). *Australian Marsupials and Monotremes: An Action Plan for their Conservation*. IUCN, Gland.

Keogh, J.S. (1999). Evolutionary implications of hemipenal morphology in the terrestrial Australian elapid snakes. *Zoological Journal of the Linnean Society*, **125**, 239–278.

Keogh, J.S. (2000). Snake penises. *Nature Australia*, **Winter 2000**, 42–49.

Keough, M.J. and Mapstone, B.D. (1995). *Protocols for Designing Marine Ecological Monitoring Programs Associated with BEK Mills*. National Pulp Mills Research Program, CSIRO, Canberra. Technical Report No. 11.

Kerle, A. (2001). *Possums. The Brushtails, Ringtails and Greater Glider.* UNSW Press, Sydney.

Kerr, J.T. (1997). Species richness, endemism, and the choice of areas for conservation. *Conservation Biology,* **11**, 1094–1100.

Kerr, J.T., Sugar, A. and Packer, L. (2000). Indicator taxa, rapid biodiversity assessment, and nestedness in an endangered ecosystem. *Conservation Biology,* **14**, 1726–1734.

Kershaw, A.P., Clark, J.S., Gill, A.M. and D'Costa, D.M. (2002). A history of fire in Australia. pp. 3–25. In: *Flammable Australia. The Fire Regimes and Biodiversity of a Continent.* Bradstock, R., Williams, J. and Gill, A.M. (Eds). Cambridge University Press, Cambridge.

Kevan, P.G. (1999). Pollinators as bioindicators of the state of the environment: species, activity, diversity. *Agriculture Ecosystems and Environment,* **74**, 373–393.

Khan, M.L., Menon, S. and Bawa, K.S. (1997). Effectiveness of the protected area network in biodiversity conservation: a case study of Meghalaya state. *Biodiversity and Conservation,* **6**, 853–868.

King, A.P. (1963). *The Influences of Colonization on the Forests and the Prevalence of Bushfires in Australia.* CSIRO Division of Physical Chemistry, Melbourne, Australia.

King, F.W. (1988). Animal rights: a growing moral dilemma. *Animal Kingdom,* **91**, 33–35.

King, G.C. (1985). The effect of fire on small mammal fauna and their resources in two forest types. Thesis for Master of Natural Resources, University of New England, Armidale.

Kingsford, R.T. (2000). Ecological impacts of dams, water diversions and river management on floodplain wetlands in Australia. *Austral Ecology,* **25**, 109–130.

Kinnear, J.E., Sumner, N.R. and Onus, M.L. (2002). The Red Fox in Australia – an exotic predator turned biocontrol agent. *Biological Conservation,* **108**, 335–359.

Kinrade, P. (1995). Australia's greenhouse obligations and options. *Habitat Australia,* **February**, 1995.

Kirby, M.X. (2004). Fishing down the coast: Historical expansion and collapse of oyster fisheries along continental margins. *Proceedings of the National Academy of Sciences,* **101**, 13096–13099.

Kirkpatrick, J. (1994). *A Continent Transformed. Human Impact on the Natural Vegetation of Australia.* Oxford University Press, Melbourne.

Kirkpatrick, J., Massey, J.S. and Parsons, R.F. (1974). Natural history of Curtis Island, Bass Strait. 2. Soils and vegetation with notes on Rodondo Island. *Papers and Proceedings of the Royal Society of Tasmania,* **107**, 131–144.

Kirkpatrick, J.B. (1983). An iterative method for establishing priorities for the selection of nature reserves: An example from Tasmania. *Biological Conservation,* **25**, 127–134.

Kirkpatrick, J.B. (1998). Nature conservation and the Regional Forest Agreement prcess. *Australian Journal of Environmental Management,* **5**, 31–37.

Kirkpatrick, J.B. and Brown, M.J. (1991). *Reservation Analysis of Tasmanian forests.* Resource Assessment Commission, Canberra.

Kirkpatrick, J.B. and Brown, M.J. (1994). A comparison of direct and environmental domain approaches to planning reservation of forest higher-plant communities and species in Tasmania. *Conservation Biology,* **8**, 217–224.

Kirkpatrick, J.B. and Fowler, M. (1998). Locating likely glacial forest refugia in Tasmania using palynological and ecological information to test alternative climatic models. Biological Conservation, **85**, 171–182.

Kirkpatrick, J.B. and Gilfedder, L. (1995). Maintaining integrity compared with maintaining rare and threatened taxa in remnant bushland in subhumid Tasmania. *Biological Conservation,* **74**, 1–8.

Kirkpatrick, J.B. and Gilfedder, L. (1998). Conserving weedy natives: two Tasmanian endangered herbs in the Brassicaeae. *Australian Journal of Ecology,* **23**, 466–473.

Kirkpatrick, J.B., Brown, M.J. and Moscal, A. (1980). *Threatened Plants of the Tasmanian Central East Coast.* Tasmanian Conservation Trust, Hobart.

Kirkpatrick, J.B., McDougall, K. and Hyde, M. (1995). *Australia's Most Threatened Ecosystems, the Southeastern Lowland Native Grasslands.* Surrey Beatty and Sons, Chipping Norton, and World Wide Fund for Nature Australia.

Kirkpatrick, J.F., Turner, J.W., Liu, I.K., Frayer-Hosken, R. and Rutberg, A.T. (1997). Case studies in wildlife immunocontraception: wild and feral equids and white-tailed deer. *Reproduction, Fertility and Development,* **9**, 105–110.

Kitching, R. (1971). A simple simulation model of dispersal of animals among units of discrete habitat. *Oecologia,* **7**, 95–116.

Klein, B.C. (1989). Effects of forest fragmentation on dung and carrion bettle communities in central Amazonia. *Ecology,* **70**, 1715–1725.

Kleinman, D.G. (1989). Reintroduction of captive mammals for conservation. *BioScience,* **39**, 152–161.

Klenner, W. and Kroeker, D.W. (1990). Denning behaviour of Black Bears, *Ursus americanus*, in western Manitoba. *Canadian Field-Naturalist,* **104**, 540–544.

Klinka, K., Krajina, V.J., Ceska, A. and Scagel, A.M. (1989). *Indicator Plants of Coastal British Columbia.* UBC Press, Vancouver, Canada.

Kloppenburg, J.J. and Kleinman, D.L. (1987). The plant germplasm controversy: analyzing empirically the distribution of the world plant genetic resources. *BioScience,* **37**, 190/198–207.

Knaapen, J.P., Bottom, M. and Harms, B. (1992). Estimating habitat isolation in landscape planning. *Landscape and Urban Planning,* **23**, 1–16.

Knight, E.H. and Fox, B.J. (2000). Does habitat structure mediate the effects of forest fragmentation and human-induced disturbance on the abundance of *Antechinus stuartii*? *Australian Journal of Zoology,* **48**, 577–595.

Knopf, F.L. (1992) Faunal mixing, faunal integrity, and the biopolitical template for diversity conservation. *Proceedings of the 57th North American Wildlife and Natural Resource Conference,* pp. 330–342.

Knox, B., Ladiges, P. and Evans, B. (Eds) (1994). *Biology.* McGraw-Hill, Sydney.

Koehn, J.D. (1993). Fish need trees. *Victorian Naturalist,* **110**, 255–257.

Koehn, J.D., Doeg, T.J., Harrington, D.J. and Milledge, G.A. (1996). Dartmouth Dam: effects on downstream aquatic fauna. pp. 49–56. In: *Riverine Environment Research Forum, Proceedings of the Inaugural Riverine Environment Research Forum of MDBC, National Resource Management Strategy Funded Projects.* October 4–6, 1995, Attwood Victoria. Banens, R.J. and Lehane, R. (Eds). Murray-Darling Basin Commission, Canberra.

Koenig, W.D. (1998). Spatial autocorrelation in California land birds. *Conservation Biology,* **12**, 612–619.

Kohlmann, B., Nix, H. and Shaw, D.D. (1988). Environmental predictions and distributional limits of chromosomal taxa in the Australian grasshopper *Caledia captiva* (F.). *Oecologia,* **75**, 483–493.

Kolm, N. and Berglund, A. (2003). Wild populations of a reef fish suffer from the "nondestructive aquarium trade fishery. *Conservation Biology,* **17**, 910–914.

Konstant, W.R. and Mittermeier, R.A. (1982). Introduction, reintroduction and translocation of neotropical primates: past experiences and future possibilities. *International Zoo Yearbook,* **22**, 69–77.

Koopowitz, H., Thornhill, A.D. and Anderson, M. (1994). A general stochastic model for the prediction of biodiversity losses based on habitat conservation. *Conservation Biology,* **8**, 425–438.

Korpilahti, E. and Kuuluvainen, T. (Eds) (2002). Disturbance dynamics in boreal forests: defining the ecological basis of restoration and management of biodiversity. *Silva Fennica,* **36**, 1–447.

Koslow, J.A. and Gowlett-Holmes, K. (1998). *The Seamount Flora off Southern Tasmania: Benthic Communities, their Conservation and Impacts of Trawling.* Environment Australia, Canberra.

Kotliar, N.B. and Wiens, J.A. (1990). Multiple scales of patchiness and patch structure – a hierarchical framework for the study of heterogeneity. *Oikos,* **59**,253–260.

Krebs, C. (1999). Current paradigms of rodent population dynamics – what are we missing? pp. 33–48. In: *Ecologically-based Management of Rodent Pests.* Singleton, G.R., Hinds, L.A., Leirs, H. and Zhang, Z. (Eds). Australian Centre for International Agricultural Research, Canberra.

Krebs, C.J. (1985). *Ecology: The Experimental Analysis of Distribution and Abundance.* 3rd edition. Harper and Row, New York.

Kremen, C. (1992). Assessing the indicator properties of species assemblages for natural areas monitoring. *Ecological Applications,* **2**, 203–217.

Kremen, C. (1994). Biological inventory using target taxa: a case study of the butterflies of Madagascar. *Ecological Applications,* **4**, 407–422.

Kremen, C. (2005). Managing ecosystem services: what do we need to know about their ecology? *Ecology Letters,* **8**, 468–479.

Kremen, C., Raymond, I. and Lance, K. (1998). An interdisciplinary tool for monitoring conservation impacts in Madagascar. *Conservation Biology,* **12**, 549–563.

Kremen, C., Razafimahatratra, V., Guillery, R.P., Rakotomalala, J., Weiss, A. and Ratsisompatrarivo, J. (1999). Designing the Masoala National Park in Madagascar based on biological and socioeconomic data. *Conservation Biology,* **13**, 1055–1068.

Kremsater, L. and Bunnell, F.L. (1999). Edge effects: theory, evidence and implications to management of western North American forests. pp. 117–153. In: *Forest Wildlife and Fragmentation. Management Implications.* Rochelle, J., Lehmann, L.A. and Wisniewski, J. (Eds). Brill, Leiden, Germany.

Kriticos, D.J. (2001). Use of simulation models to explore the effects of climate change on exotic woody weeds in Australia. PhD thesis, University of Queensland, Brisbane.

Krohne, D.T. (1997). Dynamics of metapopulations of small mammals. *Journal of Mammalogy,* **78**, 1014–1026.

Kruess, A. and Tscharntke, T. (2002). Contrasting responses of plant and insect diversity to variation in grazing intensity. *Biological Conservation,* **106**, 293–302.

Kuchling, G., Dejose, J.P., Burbidge, A.A. and Bradshaw, S.D. (1992). Beyond captive breeding: the Western Swamp Tortoise *Pseudemydura umbrina* recovery programme. *International Zoo YearBook,* **31**, 37–41.

Kurki, S., Nikula, A., Helle, P. and Linden, H. (2000). Landscape fragmentation and forest composition effects on grouse breeding success in boreal forests. *Ecology,* **81**, 1985–1997.

Kurlansky, M. (1999). *Cod. A biography of the fish that changed the world.* Random House, London.

Lacy, R.C. (1987). Loss of genetic diversity from managed populations: Interacting effects of drift, mutation, immigration, selection and population subdivision. *Conservation Biology*, **1**, 143–158.

Lacy, R.C. (1993a). Impacts of inbreeding in natural and captive populations of vertebrates: Implications for conservation. *Perspectives in Biology and Medicine,* **36**, 480–496.

Lacy, R.C. (1993b). Vortex: a computer simulation model for population viability analysis. *Wildlife Research,* **20**, 45–65.

Lacy, R.C. (2000). Structure of the VORTEX simulation model for population viability analysis. *Ecological Bulletins,* **48**, 191–203.

Lacy, R.C. and Clark, T.W. (1990). Population viability assessment of Eastern Barred Bandicoot. pp. 131–146. In: *Management and Conservation of Small Populations.* Clark, T.W. and Seebeck, J.H. (Eds). Chicago Zoological Society, Chicago.

Lacy, R.C. and Lindenmayer, D.B. (1995). Using Population Viability Analysis (PVA) to explore the impacts of population sub-division on the Mountain Brushtail Possum, *Trichosurus caninus* Ogilby (Phalangeridae: Marsupialia) in south-eastern Australia. II. Changes in genetic variability in sub-divided populations. *Biological Conservation,* **73**, 131–142.

Lacy, R.C., Flesness, N.R. and Seal, U.S. (1989). *Puerto Rican Parrot Population Viability Analysis.* Captive Breeding Specialist Group, Species Survival Commission, I.U.C.N., Apple Valley: Minnesota. Report to the U.S. Fish and Wildlife Service.

Lahaye, W.S., Gutierrez, R.J. and Akcakaya, H.R. (1994). Spotted Owl metapopulation dynamics in southern California. *Journal of Animal Ecology*, **63**, 775–785.

Lahti, D.C. (2001). The 'edge effect on nest predation' hypothesis after twenty years. *Biological Conservation,* **99**, 365–374.

Lajeunesse, M.J. and Forbes, M.R. (2003). Variable reporting and quantitative reviews: a comparison of three meta-analytical techniques. *Ecology Letters*, **6**, 448–454.

Lake, P.S. (1986). *Ecology of the Yabby Cherax destructor Clark (Crustacea: Decapoda: Parastcidae) and its Potential as a Sentinel Animal For Mercury and Lead Pollution.* Australian Government Publishing Service, Canberra.

Lambeck, R.J. (1997). Focal species: A multi-species umbrella for nature conservation. *Conservation Biology,* **11**, 849–856.

Lambeck, R.J. (1999). Landscape planning for biodiversity conservation in agricultural regions. A case study from the wheatbelt of Western Australia. *Biodiversity Technical Paper No.* **2**. 1–96. Environment Australia, Canberra, Australia.

Lamberson, R.H., Noon, B.R., Voss, C. and McKelvey, R. (1994). Reserve design for territorial species: the effects of patch size and spacing on the viability of the Northern Spotted Owl. *Conservation Biology,* **8**, 185–195.

Lambert, J. and Elix, J. (undated). *Grassy White Box Woodlands.* Information kit. Community Solutions, Sydney.

Lambshead, P.J. (1993). Recent developments on marine benthic biodiversity research. *Oceanis,* **19**, 5–24.

Lambshead, P.J.D. and Boucher, G. (2003). Marine nematode deep-sea biodiversity – hyperdiverse or hype? *Journal of Biogeography*, **30**, 475–485.

Lamm, D.W. and Calaby, J.H. (1970). Seasonal variation of bird populations along the Murrumbidgee in the Australian Capital Territory. *Emu,* **50**, 114–122.

Lampo, M. and De Leo, G.A. (1998). The invasion ecology of the toad *Bufo marinus* from South America to Australia. *Ecological Applications*, **8**, 388–396.

Land and Water Resources Research and Development Corporation (1999). *Cost of Algal Blooms.* Land and Water Resources Research and Development Corporation Occasional Paper 26/99. Land and Water Resources Research and Development Corporation, Canberra.

Land Protection (2001). *Giant Sensitive Plant (Mimosa diplotricha*). NRM Facts, Pest series. Queensland Department of Natural Resources and Mines, Brisbane.

Lande, R. (1988). Genetics and demography in biological conservation. *Science,* **241**, 1455–1460.

Lande, R. (1993). Risks of population extinction from demographic and environmental stochasticity and random catastrophes. *The American Naturalist,* **142**, 911–927.

Landers, J.L., Hamilton, R.J., Johnson, A.S. and Marchington, R.L. (1979). Food and habitat of Black Bears in southeastern Carolina. *Journal of Wildlife Management,* **43**, 143–153.

Landres, P.B. (1983). Use of the guild concept in environmental impact assessment. *Environmental Management,* **7**, 393–398.

Landres, P.B. and McMahon, J.A. (1980). Guilds and community organisation: analysis of an oak woodland avifauna in Sonora, Mexico. *Auk,* **97**, 351–365.

Landres, P.B., Verner, J. and Thomas, J.W. (1988). Ecological uses of vertebrate indicator species: A critique. *Conservation Biology,* **2(4)**, 316–328.

Landsberg, J. and Wylie, F.R. (1983). Water stress, leaf nutrients and defoliation: a model of dieback of rural Eucalypts. *Australian Journal of Ecology,* **8**, 27–41.

Landsberg, J. and Wylie, F.R. (1988). Dieback of rural trees in Australia. *GeoJournal,* **17**, 231–237.

Landsberg, J. and Wylie, R. (1991). A review of rural dieback in Australia. pp. 3–11. In: *Growback '91.* Offor, T. and Watson, R.J. (Eds). Growback Publications, Melbourne.

Landsberg, J., James, C., Morton, S., Hobbs, T.J., Stol, J., Drew, A. and Tongway, H. (1997). *The Effects of Artificial Sources of Water on Rangeland Biodiversity.* A report to the Department of Environment, Sport and Territories, Canberra.

Landsberg, J., James, C., Morton, S., Muller, W.J. and Stol, J. (2001). Abundance and composition of plant species along grazing gradients in Australian rangelands. *Journal of Applied Ecology,* **40**, 1008–1024.

Langdon, J.S. (1990). Disease risks of fish introductions and translocations. *Bureau of Rural Resources Proceedings,* **8**, 98–107.

Langford, D. (1999). *The Mala (*Lagorchestes hirsutus*) recovery plan.* Parks and Wildlife Commission of the Northern Territory, Darwin [online] URL: http://www.deh.gov.au/biodiversity/threatened/publications/recovery/mala/index.html

Langton, M. (1998). *Burning questions: emerging environmental issues for indigenous peoples in northern Australia.* Centre for Indigenous Natural and Cultural Resource Management, Northern Territory University, Darwin.

Larkin, P.A. (1977). An epitaph for the concept of maximum sustained yield. *Transactions of the American Fisheries Society,* **106**, 1–11.

Larsen, T.B. (1994). Butterfly biodiversity and conservation in the Afrotropical region. pp. 290–303. In: *Ecology and Conservation of Butterflies.* Pullin, A.S. (Ed). Chapman and Hall, London.

Larson, D.W., Matthes, U. and Kelly, P.E. (2000). *Cliff Ecology: Pattern and Process in Cliff Ecosystems.* Cambridge University Press, Cambridge.

Lassau, S.A. and Hochuli, D.F. (2005). Wasp community responses to habitat complexity in Sydney sandstone forests. *Austral Ecology,* **30**, 179–187.

Lasseau, D. (2003). Effects of tour boats on the behavior of bottlenose dolphins: Using Markov chains to model anthropogenic impacts. *Conservation Biology,* **17**, 1785–1793.

Last, P.R. and Stevens, J.D. (1994). *Sharks and Rays of Australia.* CSIRO Division of Fisheries, Hobart, Australia.

Laurance, W.F. (1991a). Ecological correlates of extinction proneness in Australian tropical rain forest mammals. *Conservation Biology,* **5**, 79–89.

Laurance, W.F. (1991b). Edge effects in tropical rainforest fragments: application of a model for the design of nature reserves. *Biological Conservation,* **57**, 205–219.

Laurance, W.F. (1996). Catastrophic declines of Australian rainforest frogs: is unusual weather responsible? *Biological Conservation,* **77**, 203–212.

Laurance, W.F. (1997a). A distributional survey and habitat model for the endangered Northern Bettong *Bettongia tropica* in tropical north Queensland. *Biological Conservation,* **82**, 47–60.

Laurance, W.F. (1997b). Responses of mammals to rainforest fragmentation in tropical Queensland: a review and synthesis. *Wildlife Research,* **24**, 603–612.

Laurance, W.F. and Bierregaard, R.O. (Eds) (1997). *Tropical Forest Remnants. Ecology, Management and Conservation of Fragmented Communities.* The University of Chicago Press, Chicago.

Laurance, W.F. and Yensen, E. (1991). Predicting the impacts of edge effects in fragmented habitats. *Biological Conservation,* **55**, 77–92.

Laurance, W.F., Bierregaard, R.O., Gascon, C., Didham, R.K., Smith, A.P., Lynam, A.J., Viana, V.M., Lovejoy, T.E., Sieving, K.E., Sites, J.W., Andersen, M., Tocher, M.D., Kramer, E.A., Restrepo, C. and Moritz, C. (1997). Tropical forest fragmentation: synthesis of a diverse and dynamic discipline. pp. 502–525. In: *Tropical Forest Remnants. Ecology, Management and Conservation of Fragmented Communities.* Laurance, W.F. and Bierregaard, R.O. (Eds). The University of Chicago Press, Chicago.

Lavoral, S., Stafford Smith, M. and Reid, N. (1999). Spread of mistletoes (*Amyema preissii*) in fragmented Australian woodlands: a simulation study. *Landscape Ecology,* **14**, 147–160.

Law, B. and Dickman, C.R. (1998). The use of habitat mosaics by terrestrial vertebrate fauna: implications for conservation and management. *Biodiversity and Conservation,* **7**, 323–333.

Law, B.S., Anderson, J. and Chidel, M. (1999). Bat communities in a fragmented forest landscape on the south-west slopes of New South Wales, Australia. *Biological Conservation,* **88**, 333–345.

Law, S.B. (1994). Climatic limitation of the southern distribution of the common blossom bat *Synconycteris australis* in New South Wales. *Australian Journal of Ecology,* **19**, 366–374.

Lawrence, J. and Taylor, D. (1990). Letters: Statistics and the precautionary principle. *Marine Pollution Bulletin,* **21(12)**, 598–599.

Lawton, J. (1991). Are species useful? *Oikos,* **62**, 3–4.

Lawton, J. (1997). The science and non-science of conservation biology. *Oikos,* **79**, 3–5

Lawton, J.H. (2002). Conservation biology: Where next? *Society for Conservation Biology Newsletter,* **9(4)**, 1–2.

Lawton, J.H. and Gaston, K.J. (2001). Indicator species. pp. 437–450. In: *Encylopedia of Biodiversity.* Volume 3. Levin, S. (Ed). Academic Press, San Diego.

Lawton, J.H., Nee, S., Letcher, A.J. and Harvey, P. (1984). Animal distribution patterns and processes. pp. 41–58. In: *Large Scale Ecology and Conservation Biology.* Edwards, P.J., May, R.M. and Webbs, N.R. (Eds). Blackwell Scientific Publications, Oxford.

Laymon, S.A., Salwasser, H. and Barrett, R.H. (1985). *Habitat Suitability Index Models: Spotted Owl.* United States Fish and Wildlife Service. Biological Report No. 82 1–14.

LCC (1988). (Land Conservation Council) *Background Report on Remnant Vegetation.* Department of Planning and Urban Growth, Melbourne.

Lee, K.E. (1996). Biodiversity of soil organisms: community concepts and ecosystem function. *Biodiversity and Conservation,* **5**, 133–134.

Lees, B.G. (1994). Decision trees, artificial neural networks and genetic algorithms for classification of remotely sensed and ancilliary data. *Proceedings of the Seventh Australasian Remote Sensing Conference,* Melbourne.

Legendre, L. and Legendre, P. (1998). *Numerical Ecology.* 2nd English Edition. Elsevier, Amsterdam.

Lehmkuhl, J.F. (1984). Determining size and dispersion of minimum viable populations for land management planning and species conservation. *Environmental Management,* **8**, 167–176.

Leigh, E.G.J. (1981). The average lifetime of a population in a varying environment. *Journal of Theoretical Biology,* **90**, 213–239.

Leigh, J., Boden, R. and Briggs, J. (1984). *Extinct and Endangered Plants of Australia.* MacMillan, Melbourne.

Leigh, J.H. and Briggs, J.D. (1992). *Threatened Australian Plants: Overview and Case Studies.* Australian National Parks and Wildlife Service, Canberra.

Leigh, J.H. and Holgate, M.D. (1979). The responses of the understorey of forests and woodlands of the Southern Tablelands to grazing and burning. *Australian Journal of Ecology,* **4**, 25–45.

Lenghaus, C., Obendorf, D.L. and Wright, G.H. (1990). Veterinary aspects of *Perameles gunnii* biology with special reference to conservation. pp. 89–102. In: *Management and Conservation of Small Populations.* Clark, T.W. and Seebeck, J.H. (Eds). Chicago Zoological Society, Brookfield, Illinois.

Leopold, A. (1933). *Game Management.* Charles Scribners, New York.

Leopold, A. (1949). *A Sand County Almanac.* Oxford University Press, New York.

Leopold, A. (1953). *Round River.* Oxford University Press, New York.

Lepczyk, C.A., Mertig, A.G. and Liu, J. (2003). Landowners and cat predation across rural-urban landscapes. *Biological Conservation,* **115**, 191–201.

Lesslie, R. and Maslen, M. (1995). *National Wilderness Inventory Handbook.* Second Edition. Australian Heritage Commission. Australian Government Publishing Service, Canberra, Australia.

Letnic, M. (2003). The effects of experimental patch burning and rainfall on small mammals in the Simpson Desert, Queensland. *Wildlife Research,* **30**, 547–563.

Leung, K.P., Dickman, C.R. and Moore, L.A. (1993). Genetic variation in fragmented populations of an Australian rainforest rodent, *Melomys cervinipes. Pacific Conservation Biology,* **1**, 58–65.

Leuschner, W.A. (1990). *Forest Regulation, Harvest Scheduling and Planning Schedules.* Wiley, New York.

Lever, C. (2001). *The Cane Toad: The History and Ecology of a Successful Colonist.* Westbury Academic and Scientific Publishing, London.

Levin, S.A. (Ed) (2001). *Encyclopedia of Biodiversity.* Volumes 1–5. Academic Press, Sydney.

Levins, R. (1969). The effect of random variation of different types on population growth. pp. 1061–1065. In: *Proceedings of the National Academy of Science, USA.* Vol. 62.

Levins, R.A. (1970). Extinction. *Lecture Notes in Mathematics and Life Sciences,* **2**, 75–107.

Levinton, J.S. (2001). Rates of extinction. pp. 715–729. In: *Encyclopedia of Biodiversity.* Volume 2. Levin, S.A. (Ed). Academic Press, San Diego.

Levinton, J.S. and Ginzburg, L. (1984). Repeatability of taxon longevity in successive foraminifera radiations and a theory of random appearance and extinction. *Proceedings of the National Academy of Sciences,* **81**, 5478–5481.

Lewis, R. (1996). Unravelling the weave of spider silk. *BioScience*, **46**, 638.

Liddle, L. (2003). Fire in a jointly managed landscape: fire at Uluru Kata-Tjuta National Park. pp. 187–197. In: *Australia Burning. Fire Ecology, Policy and Management Issues.* Cary, G., Lindenmayer, D. and Dovers, S. (Eds). CSIRO Publishing, Melbourne.

Lindborg, R. and Eriksson, O. (2004). Historical landscape connectivity affects present plant species diversity. *Ecology*, **85**, 1840–1845.

Lindburg, D.G. (1992). Are wildlife reintroductions worth the cost? *Zoo Biology*, **11**, 1–2.

Lindenmayer, D.B. (1994). Wildlife corridors and the mitigation of logging impacts on forest fauna in south-eastern Australia. *Wildlife Research*, **21**, 323–40.

Lindenmayer, D.B. (1995). Disturbance, forest wildlife conservation and a conservative basis for forest management in the Mountain Ash forests of Victoria. *Forest Ecology and Management*, **74**, 223–231.

Lindenmayer, D.B. (1996). *Wildlife and woodchips: Leadbeater's Possum as a testcase of sustainable forestry.* University of New South Wales Press, Sydney.

Lindenmayer, D.B. (1997). Differences in the biology and ecology of arboreal marsupials in southeastern Australian forests and some implications for conservation. *Journal of Mammalogy*, **78**, 1117–1127.

Lindenmayer, D.B. (1998). The design of wildlife corridors in wood production forests. NSW National Parks and Wildlife Service, Occasional Paper Series. *Forest Issues Paper*, **No. 4** : 1–41.

Lindenmayer, D.B. (2000). Factors at multiple scales affecting distribution patterns and its implications for animal conservation – Leadbeater's Possum as a case study. *Biodiversity and Conservation*, **9**, 15–35.

Lindenmayer, D.B. (2002a). The Greater Glider as a model to examine key issues in Australian forest ecology and management. pp. 46–58. In: *Perspectives on Wildlife Research. Celebrating 50 years of CSIRO Wildlife and Ecology.* D.A. Saunders, D. Spratt and M. van Wensveen (Eds). Surrey Beatty and Sons, Chipping Norton.

Lindenmayer, D.B. (2002b). *Gliders of Australia. A Natural History*. UNSW Press, Sydney.

Lindenmayer, D.B. and Cunningham, R.B. (1996). A habitat-based microscale forest classification system for zoning wood production areas to conserve a rare species threatened by logging operations in south-eastern Australia. *Environmental Monitoring and Assessment*, **39**, 543–557.

Lindenmayer, D.B. and Cunningham, R.B. (1997). Patterns of co-occurrence among arboreal marsupials in the forests of central Victoria. *Australian Journal of Ecology*, **22**, 340–346.

Lindenmayer, D.B. and Fischer, J.F. (2003). The focal species approach: sound science or social hook. *Landscape and Urban Planning*, **62**, 149–158.

Lindenmayer, D.B. and Franklin, J.F. (1996). *The Importance of Stand Structure for the Conservation of Wildlife in Logged Forests: A Case Study from Victoria.* CRES Working Paper No. 96/1.

Lindenmayer, D.B. and Franklin, J.F. (1997). Managing stand structure as part of ecologically sustainable forest management in Australian Mountain Ash Forests. *Conservation Biology*, **11**, 1053–1068.

Lindenmayer, D.B. and Franklin, J.F. (2002). *Conserving Forest Biodiversity: A Comprehensive Multiscaled Approach*. Island Press, Washington.

Lindenmayer, D.B. and Franklin, J.F. (Eds) (2003). *Towards Forest Sustainability.* CSIRO Publishing, Melbourne.

Lindenmayer, D.B. and Gibbons, P. (1998). Timber harvesting and extinction. *Institute of Foresters of Australia Newsletter*, **June 1998**, 35–38.

Lindenmayer, D.B. and Gibbons, P. (2004). On charcoal, increased logging intensity and a flawed EIS process. In *Forest Fauna II.* Lunney, D. (Ed). (in press).

Lindenmayer, D.B. and Hobbs, R.J. (2004). Biodiversity conservation in plantation forests – a review with special reference to Australia. *Biological Conservation*, **119**, 151–168.

Lindenmayer, D.B. and Lacy, R.C. (1993). Using a computer simulation package for PVA to model the dynamics of sub-divided meta-populations: an example using hypothetical meta-populations of the Mountain Brushtail Possum. pp. 615–620. In: *Proceedings of the International Congress on Modelling and Simulation.* McAleer, M. and Jakeman, A. (Eds). UniPrint, Western Australia.

Lindenmayer, D.B. and Lacy, R.C. (1995). Metapopulation viability of Leadbeater's Possum, *Gymnobelideus leadbeateri*, in fragmented old growth ash forests. *Ecological Applications*, **5**, 164–182.

Lindenmayer, D.B. and Lacy, R.C. (2002). Small mammals, patches and PVA models: a field test of model predictive ability. *Biological Conservation*, **103**, 247–265.

Lindenmayer, D.B. and McCarthy, M.A. (2001). The spatial distribution of non-native plant invaders in a pine-eucalypt mosaic in south-eastern Australia. *Biological Conservation*, **102**, 77–87.

Lindenmayer, D.B. and McCarthy, M.A. (2002). Congruence between natural and human forest disturbance – an Australian perspective. *Forest Ecology and Management*, **155**, 319–335.

Lindenmayer, D.B. and Possingham, H.P. (1994). *The Risk of Extinction: Ranking Management Options for Leadbeater's Possum.* Centre for Resource and Environmental Studies, The Australian National University and The Australian Nature Conservation Agency, Canberra. 204 pp.

Lindenmayer, D.B. and Possingham, H.P. (1995). Modeling the impacts of wildfire on metapopulation behaviour of the Australian arboreal marsupial, Leadbeaters Possum, *Gymnobelideus leadbeateri. Forest Ecology and Management,* **74**, 197–222.

Lindenmayer, D.B. and Possingham, H.P. (1996). Modeling the relationships between habitat connectivity, corridor design and wildlife conservation within intensively logged wood production forests of south-eastern Australia. *Landscape Ecology,* **11**, 79–105.

Lindenmayer, D.B. and Viggers, K.L. (1994). An extension of the range limits of the Long-nosed Potoroo (*Potorous tridactylus*). *Memoirs of the Queensland Museum,* **335**, 180.

Lindenmayer, D.B., Ball, I., Possingham, H.P., McCarthy, M.A. and Pope, M.L. (2001a). A landscape-scale test of the predictive ability of a meta-population model in an Australian fragmented forest ecosystem. *Journal of Applied Ecology,* **38**, 36–48.

Lindenmayer, D.B., Burgman, M.A., Akcakaya, H.R., Lacy, R.C. and Possingham, H.P. (1995a). A review of the generic computer programs ALEX, RAMAS/space and VORTEX for modelling the viability of metapopulations. *Ecological Modelling,* **82**, 161–174.

Lindenmayer, D.B., Claridge, A.W., Gilmore, A.M., Michael, D. and Lindenmayer, B.D. (2002b). The ecological role of logs in Australian forest and the potential impacts of harvesting intensification on log-using biota. *Pacific Conservation Biology,* **8**, 121–140.

Lindenmayer, D.B., Claridge, A.W., Hazell, D., Michael, D.R., Crane, M., MacGregor, C.I. and Cunningham, R.B. (2003a). *Wildlife on Farms. How to Conserve Native Wildlife.* CSIRO Publishing, Melbourne.

Lindenmayer, D.B., Crane, M., Michael, D., MacGregor, C. and Cunningham, R.B. (2005a). *Woodlands. A Disappearing Landscape.* CSIRO Publishing, Melbourne.

Lindenmayer, D.B., Cunningham, R.B. and Donnelly, C.F. (1993a). The conservation of arboreal marsupials in the montane ash forests of the Central Highlands of Victoria, south-east Australia. IV. The distribution and abundance of arboreal marsupials in retained linear strips (wildlife corridors) in timber production forests. *Biological Conservation,* **66**, 207–221.

Lindenmayer, D.B., Cunningham, R.B. and Donnelly, C.F. (1994). The conservation of arboreal marsupials in the montane ash forests of the Central Highlands of Victoria, south-east Australia. VI. Tests of the performance of models of nest tree and habitat requirements of arboreal marsupials. *Biological Conservation,* **70**, 143–147.

Lindenmayer, D.B., Cunningham, R.B. and Donnelly, C.F. (1997). Tree decline and collapse in Australian forests: implications for arboreal marsupials. *Ecological Applications,* **7**, 625–641.

Lindenmayer, D.B., Cunningham, R.B. and Fischer, J. (2005b). Vegetation cover thresholds and species responses. *Biological Conservation,* **124**, 311–316.

Lindenmayer, D.B., Cunningham, R.B. and McCarthy, M.A. (1999c). The conservation of arboreal marsupials in the montane ash forests of the Central Highlands of Victoria, south-eastern Australia. VIII. Landscape analysis of the occurrence of arboreal marsupials in the montane ash forests. *Biological Conservation,* **89**, 83–92.

Lindenmayer, D.B., Cunningham, R.B., Donnelly, C.F. and Franklin, J.F. (2000b). Structural features of Australian old growth montane ash forests. *Forest Ecology and Management,* **134**, 189–204.

Lindenmayer, D.B., Cunningham, R.B., Donnelly, C.F. and Lesslie, R. (2002a). On the use of landscape indices as ecological indicators in fragmented forests. *Forest Ecology and Management,* **159**, 203–216.

Lindenmayer, D.B., Cunningham, R.B., Donnelly, C.F., Nix, H.A. and Lindenmayer, B.D (2002c). The distribution of birds in a novel landscape context. *Ecological Monographs,* **72**, 1–18.

Lindenmayer, D.B., Cunningham, R.B., Donnelly, C.F., Tanton, M.T. and Nix, H.A. (1993b). The abundance and development of cavities in montane ash-type Eucalypt trees in the montane forests of the Central Highlands of Victoria, south-eastern Australia. *Forest Ecology and Management,* **60**, 77–104.

Lindenmayer, D.B., Cunningham, R.B., MacGregor, C. and Incoll, R.D. (2003b). A long-term monitoring study of the population dynamics of arboreal marsupials in the Central Highlands of Victoria. *Biological Conservation,* **110**, 161–167.

Lindenmayer, D.B., Cunningham, R.B., MacGregor, C., Tribolet, C.R. and Donnelly. (2001c). A prospective longitudinal study of landscape matrix effects on woodland remnants: experimental design and baseline data for mammals, reptiles and nocturnal birds. *Biological Conservation,* **101**, 157–169.

Lindenmayer, D.B., Cunningham, R.B., Pope, M. and Donnelly, C.F. (1999d). The response of arboreal marsupials to landscape context: A large-scale fragmentation study. *Ecological Applications,* **9**, 594–611.

Lindenmayer, D.B., Cunningham, R.B., Tanton, M.T. and Smith, A.P. (1990). The conservation of arboreal marsupials in the montane ash forests of the Central Highlands of Victoria, south-east Australia: II. The loss of trees with hollows and its implications for the conservation of Leadbeater's Possum *Gymnobelideus leadbeateri* McCoy (Marsupialia: Petauridae). *Biological Conservation,* **54**, 133–145.

Lindenmayer, D.B., Cunningham, R.B., Tanton, M.T., Nix, H.A. and Smith, A.P. (1991a). The conservation of arboreal marsupials in the montane ash forests of the Central Highlands of Victoria, south-east Australia. III. The habitat requirements of Leadbeater's Possum, *Gymnobelideus leadbeateri* McCoy and models of the diversity and abundance of arboreal marsupials. *Biological Conservation,* **56**, 295–315.

Lindenmayer, D.B., Dubach, J. and Viggers, K.L. (2002d). Geographic dimorphism in the Mountain Brushtail Possum – the case for a new species. *Australian Journal of Zoology*, **50**, 369–393.

Lindenmayer D.B., Franklin, J. and Foster, D. (2004b). Salvage harvesting fire-damaged wet eucalypt forests – some ecological perspectives. *Australian Forestry* (in press).

Lindenmayer, D.B., Hobbs, R.J. and Salt, D. (2004a). Biodiversity conservation in plantation forests – a review with special reference to Australia. *Biological Conservation* (in press).

Lindenmayer, D.B., Incoll, R.D., Cunningham, R.B. and Donnelly, C.F. (1999a). Attributes of logs in the Mountain Ash forests in south-eastern Australia. *Forest Ecology and Management,* **123**, 195–203.

Lindenmayer, D.B., Lacy, R.C. and Pope, M.L. (2000d). Testing a simulation model for Population Viability Analysis. *Ecological Applications*, **10**, 580–597.

Lindenmayer, D.B., Lacy, R.C., Thomas, V.C. and Clark, T.W. (1993c). Predictions of the impacts of changes in population size and of environmental variability on Leadbeater's Possum, *Gymnobelideus leadbeateri* McCoy (Marsupialia: Petauridae) using population viability analysis: An application of the computer program VORTEX. *Wildlife Research,* **20**, 68–87.

Lindenmayer, D.B., Lacy, R.C., Tyndale-Biscoe, H., Taylor, A., Viggers, K. and Pope, M.L. (1999f). Integrating demographic and genetic studies of the Greater Glider (*Petauroides volans*) at Tumut, south-eastern Australia: Setting hypotheses for future testing. *Pacific Conservation Biology,* 5, 2–8.

Lindenmayer, D.B., Mackey, B. and Nix, H.A. (1996). Climatic analyses of the distribution of four commercially-important wood production Eucalypt trees from south-eastern Australia. *Australian Forestry,* **59**, 11–26.

Lindenmayer, D.B., Mackey, B., Mullins, I., McCarthy, M.A., Gill, A.M., Cunningham, R.B. and Donnelly, C.F. (1999b). Stand structure within forest types – are there environmental determinants? *Forest Ecology and Management,* **123**, 55–63.

Lindenmayer, D.B., Mackey, B.G., Cunningham, R.B., Donnelly, C.F., Mullen, I.C., McCarthy, M.A. and Gill, A.M. (2000a). Statistical and Environmental Modelling of Myrtle Beech (*Nothofagus cunninghamii*) in southern Australia. *Journal of Biogeography,* **27**, 1001–1009.

Lindenmayer, D.B., Manning, A., Smith, P.L, McCarthy, M., Possingham, H.P., Fischer, J. and Oliver, I. (2002e). The focal species approach and landscape restoration: A critique. *Conservation Biology,* **16**, 338–345.

Lindenmayer, D.B., Margules, C.R. and Botkin, D. (2000c). Indicators of forest sustainability biodiversity: the selection of forest indicator species. *Conservation Biology*, **14**, 941–950.

Lindenmayer, D.B., McCarthy, M.A. and Pope, M.L. (1999e). A test of Hanski's simple model for metapopulation model. *Oikos,* **84**, 99–109.

Lindenmayer, D.B., McCarthy, M.A., Possingham, H.P. and Legge, S. (2001b). A simple landscape-scale test of a spatially explicit population mode: patch occupancy in fragmented south-eastern Australian forests. *Oikos,* **92**, 445–458.

Lindenmayer, D.B., McIntyre, S. and Fischer, J. (2003d). Birds in eucalypt and pine forests: landscape alteration and its implications for research models of faunal habitat use. *Biological Conservation*, **110**, 45–53.

Lindenmayer, D.B., Nix, H.A., McMahon, J.P., Hutchinson, M.F. and Tanton, M.T. (1991b). The conservation of Leadbeater's Possum *Gymnobelideus leadbeateri* (McCoy): A case study of the use of bioclimatic modelling. *Journal of Biogeography,* **18**, 371–383.

Lindenmayer D.B., Norton, T.W. and Tanton, M.T. (1991c). Differences between the effects of wildfire and clearfelling in montane ash forests of Victoria and its implications for fauna dependent on tree hollows. *Australian Forestry*, **53**, 61–68.

Lindenmayer, D.B., Pope, M.L. and Cunningham, R.B. (1999g). Roads and nest predation: an experimental study in a modified forest ecosystem. *Emu,* **99**, 148–152

Lindenmayer, D.B., Possingham, H.P., Lacy, R.C., McCarthy, M.A. and Pope, M.L. (2003c). How accurate are population models? Lessons from landscape-scale population tests in a fragmented system. *Ecology Letters*, **6**, 41–47.

Lindenmayer, D.B., Ritman, K.R., Cunningham, R.B., Smith, J. and Howarth, D. (1995b). Predicting the spatial distribution of the Greater Glider, *Petauroides volans* Kerr in a timber production forest block in south-eastern Australia. *Wildlife Research,* **22**, 445–456.

Linington, S.H. and Pritchard, H.W. (2002). Gene banks. pp. 901–912. In: *Encyclopedia of biodiversity*. Volume 53. Levin, S.A. (Ed). Academic Press, San Diego.

Linklater, W.L. (2003). Science and management in a conservation crisis: a case study with Rhinoceros. *Conservation Biology*, **17**, 968–975.

Linnell, J.D.C., Swenson, J.E. and Andersen, R. (2000). Conservation of biodiversity in Scandinavian boreal forests: large carnivores as flagships, umbrellas, indicators or keystones? *Biodiversity and Conservation*, **9**, 857–868.

Loch, A.D. and Floyd, R.B. (2001). Insect pests of Tasmanian blue gum, *Eucalyptus globules globules,* in south-western Australia: History, current perspectives and future prospects. *Austral Ecology,* **26**, 458–466.

Logan, J.A., Regniere, J. and Powell, J.A. (2003). Assessing the impacts of global warming on forest pests. *Frontiers in Ecology*, **1**, 130–137.

Loman, J. and von Schantz, T. (1991). Birds in a farmland – more species in small than in large habitat islands. *Conservation Biology,* **5**, 176–188.

Lomborg, B. (2000). *The skeptical environmentalist.* Cambridge University Press, Cambridge.

Lomolino, M.V. and Channell, R. (1995). Splendid isolation: patterns of geographic range collapse in endangered mammals. *Journal of Mammalogy,* **76**, 335–347.

Long, J. (1981). *Introduced Birds of the World.* David and Charles, London.

Long, M. (1999). Biological inventory of the Mt. Loft Ranges, South Australia. Department of Environment, Heritage and Aboriginal Affairs, Adelaide.

Longmore, R. (Ed.) (1986). *Atlas of Elapid Snakes of Australia.* Australian Government Publishing Service, Canberra.

Lonsdale, W.M. (1992). The impact of weeds in national parks. pp. 145–149. In: *Proceedings of the First International Weed Control Congress.* Melbourne.

Lonsdale, W.M. (1994). Inviting trouble: introduced pasture species in northern Australia. *Australian Journal of Ecology,* **19**, 345–354.

Lonsdale, W.M. (1999). Global patterns of plant invasions and the concept of invisibility. *Ecology*, **80**, 1522–1536.

Loppi, S., Putorti, E., Signorini, C., Fommei, S., Pirintsos, S.A. and de Dominicis, V. (1998). A retrospective study using epiphytic lichens as biomonitors of air quality: 1980 and 1996 (Tuscany, central Italy). *Acta Oecologica*, **19**, 405–408.

Losos, E., Hayes, J., Phillips, A., Wilcove, D. and Alkire, C. (1995). Taxpayer-subsidized resource extraction harms species. *BioScience,* **45**, 446–455.

Love, A. (1985). Distribution of the Bitou Bush along the NSW coast. *Proceedings of the Conference on Chrysanthemoides monilifera,* Port Macquarie, Sydney, 1984. National Parks and Wildlife Service and NSW Department of Agriculture.

Lovei, G. (2001). Modern examples of extinctions. pp. 731–743. In: *Encyclopedia of Biodiversity*. Volume 2. Levin, S.A. (Ed). Academic Press, San Diego.

Lovejoy, T.E. (1980). *A Projection of Species Extinctions. Volume 2.* Washington DC.

Lovejoy, T.E., Bierregaard, R.O.J., Rylands, A.B., Malcolm, J.R., Quintela, C.E., Harper, L.H., Brown, K.S.J., Powell, A.H., Powell, G.V.N., Schubart, H.O.R. and Hays, M.B. (1986). Edge and other effects of isolation on Amazon forest fragments. pp. 258–285. In: *Conservation Biology: The Science of Scarcity and Diversity.* Soulé, M.E. (Ed). Sinauer, Sunderland, Massachusetts.

Lovett, S. and Price, P. (Eds) (1999). *Riparian Land Management Technical Guidelines.* Volume 1: Principles of sound management. Land and Water Resources Research and Development Corporation, Canberra, Australia.

Low, T. (1988). *Wild Food Plants of Australia.* Angus and Robertson, Sydney.

Low, T. (1999). *Feral Future.* Penguin, Australia.

Low, T. (2002). Ant wars. *Nature Australia,* **Spring 2002**, 56–63.

Lowman, M.D. and Heatwole, H. (1992). Spatial and temporal variability in defoliation of Australian eucalypts. *Ecology*, **73**, 129–142.

Loyn, R.H. (1985). Bird populations in successional forests of Mountain Ash *Eucalyptus regnans* in Central Victoria. *Emu,* **85**, 213–230.

Loyn, R.H. (1987). Effects of patch area and habitat on bird abundances, species numbers and tree health in fragmented Victorian forests. pp. 65–77. In: *Nature Conservation: The Role of Remnants of Native Vegetation.* Saunders, D.A., Arnold, G., Burbidge, A. and Hopkins, A. (Eds). Surrey Beatty, Chipping Norton.

Loyn, R.H. (1993). Effects of an extensive wildfire on birds in far eastern Victoria. *Pacific Conservation Biology*, **3**, 221–234.

Lu, H., Gallant, J., Prosser, I.P., Moran, C. and Priestly, G. (2001). *Prediction of sheet and rill erosion over the Australian continent, incorporating monthly soil loss contribution.* Department of Environment and Heritage, Canberra.

Lubchenco, J. and 16 others (1991). The sustainable biosphere initiative: an ecological research agenda. *Ecology*, 72, 371–412.

Luck, G. (2002a). The habitat requirements of the Rufous Treecreeper (*Climacteris rufa*). 1. Preferential habitat use demonstrated at multiple spatial scales. *Biological Conservation*, **105**, 383–394.

Luck, G. (2002b). The habitat requirements of the Rufous Treecreeper (*Climacteris rufa*). 1. Validating predictive habitat models. *Biological Conservation*, **105**, 395–403.

Luck, G., Ricketts, T.H., Daily, G. and Imhoff, M. (2004). Alleviating spatial conflict between people and biodiversity. *Proceedings of the National Academy of Sciences*, **101**, 182–186.

Luck, G.W., Possingham, H.P. and Paton, D.C. (1999a). Bird responses at inherent and induced edges in the Murray Mallee, South Australia. I. Differences in abundance and diversity. *Emu*, **99**, 157–169.

Luck, G.W., Possingham, H.P. and Paton, D.C. (1999b). Bird responses at inherent and induced edges in the Murray Mallee, South Australia. 2. Nest predation as an edge effect. *Emu*, **99**, 170–175.

Ludwig, D. (1993). Environmental sustainability: magic, science, and religion in natural resource management. *Ecological Applications*, **3**, 555–558.

Ludwig, D. (1999). Is it meaningful to estimate a probability of extinction? *Ecology*, **80**, 298–310.

Ludwig, D., Hilborn, R. and Walters, C. (1993). Uncertainty, resource exploitation, and conservation: Lessons from history. *Science*, **260**, 17–36.

Ludwig, D., Mangel, M. and Haddad, B. (2001). Ecology, conservation and public policy. *Annual Reviews of Ecology and Systematics*, **32**, 481–517.

Luke, R.H. and McArthur, A.G. (1977). *Bushfires in Australia*. Australian Government Publishing Service, Canberra.

Lumsden, L.F. (1993). Bats. Nature's nocturnal insect controllers. *Trees and Natural Resources*, **December 1993**, 8–11.

Lumsden, L.F., Bennett, A.F. and Silins, J.E. (2002). Location of roosts of the lesser long-eared bat *Nyctophilus geoffroyi* and Gould's wattled bat *Chalinolobus gouldii* in a fragmented landscape in south-eastern Australia. *Biological Conservation*, **106**, 237–249.

Lumsden, L.F., Bennett, A.F., Silins, J. and Krasna, S. (1994). *Fauna in a Remnant Vegetation-farmland Mosaic, Movement, Roosts and Foraging Ecology of Bats.* A report to the Australian Nature Conservation Agency 'Save the Bush Program'. Flora and Fauna Branch, Department of Conservation and Natural Resources, Melbourne.

Lunney, D. and Leary, T. (1988). The impact on native mammals of land-use changes of exotic species in the Bega district, New South Wales. *Australian Journal of Ecology*, **13**, 67–92.

Lunney, D., Curtin, A., Ayers, D., Cogger, H.G. and Dickman, C.R. (1996). An ecological approach to identifying endangered fauna of New South Wales. *Pacific Conservation Biology*, **2**, 212–231.

Lunt, I., Coates, F. and Spooner, P. (2005). Grassland indicator species predict flowering of endangered Gaping Leek-orchid (Prasophyllum correctum D.L. Jones). *Ecological Restoration and Management*, **6**, 69–71.

Lutter, S. (1990). Letters: Statistics and the precautionary principle. *Marine Pollution Bulletin*, **21(11)**, 547–548.

Lutz, W. (1999). Population, number of households and global warming. *Population Network Newsletter*, **27**.

Lutze, M.T., Campbell, R.G. and Fagg, P.C. (1999). Development of silviculture in the native State forests of Victoria. *Australian Forestry*, **62**, 236–244.

LWRRDC (1997). *Terms of reference; remnant native vegetation in Australia*. Land and Water Resources Research and Development Corporation with Environment Australia, Canberra.

Lyell, C. (1832). *Principles of Geology*. Vol. 2. Murray, London.

Lyles, A. (2002). Zoos and zoological parks. pp. 901–912. In: *Encyclopedia of Biodiversity*. Volume 5. Levin, S.A. (Ed). Academic Press, San Diego.

Lynch, M. (1996). A quantitative-genetic perspective on conservation issues. pp. 471–501. In: *Conservation Genetics*. Avise, J.C. and Hamrick, J.L (Eds). Chapman and Hall, New York.

Lynch, M. and Ritland, K. (1999). Estimation of pairwise relatedness with molecular markers. *Genetics*, **152**, 1753–1766.

Lynch, M., Conery, J. and Burger, R. (1995). Mutation accumulation and the extinction of small populations. *American Naturalist*, **146**, 489–518.

Lyons, M.T., Brooks, R.R. and Craig, D.C. (1974). The influence of soil composition of the Coolac Serpentinite belt in New South Wales. *Journal and Proceedings, Royal Society of New South Wales*, **107**, 67–75.

McAlpine, C.A., Mott, J.J., Grigg, G.C. and Sharma, P. (1999). The influence of landscape structure on kangaroo abundance in disturbed semi-arid woodland. *The Rangeland Journal*, **21**, 104–34.

McAlpine, C.M., Lindenmayer, D.B., Eyre, T. and Phinn, S. (2002). Landscape surrogates for conserving Australia's forest fauna: Synthesis of Montreal Process Case Studies. *Pacific Conservation Biology*, **8**, 108–120.

McArdle, B.H. (1990). When are rare species not there? *Oikos*, **57**, 276–277.

MacArthur, A.G. (1968). The fire resistance of Eucalyptus. *Proceedings Ecological Society Australia*, **3**, 83–90.

McArthur, C. (2000). Balancing browsing damage management and fauna conservation in plantation forestry. *Tasforests*, **12**, 167–169.

MacArthur, R.H. (1972). *Geographical ecology: Patterns in the distribution of species*. Princeton University Press, Princeton, New Jersey.

MacArthur, R.H. and Wilson, E.O. (1963). An equilibrium theory of insular zoogeography. *Evolution*, **17**, 373–387.

MacArthur, R.H. and Wilson, E.O. (1967). *The Theory of Island Biogeography*. Princeton University Press, Princeton.

McCallum, H., Kikkawa, J. and Catterall, C. (2000). Density dependence in an island population of silvereyes. *Ecology Letters*, **3**, 95–100.

McCallum, H., Timmers, P. and Hoyle, S. (1995). Modelling the impact of predation on reintroductions of Bridled Nailtail Wallabies. *Wildlife Research*, **22**, 163–171.

McCarthy, M. (1996a). Report on the eighth meeting of the Montreal Process. *Bulletin of the Ecological Society of Australia*, **26**, 123–124.

McCarthy, M., Lindenmayer, D.B. and Dreschler, M. (1997). Extinction debts and the risks faced by abundant species. *Conservation Biology*, **11**, 221–226.

McCarthy, M.A. (1996b). Extinction dynamics of the Helmeted Honeyeater: Effects of demography, stochasticity, inbreeding, and social structure. *Ecological Modelling*, **85**, 151–163.

McCarthy, M.A. (1998). Identifying declining and threatened species with museum data. *Biological Conservation*, **83**, 9–17.

McCarthy, M.A. and Broome, L.S. (2000). A method for validating stochastic models of population viability: a case study of the mountain pygmy-possum (*Burramys parvus*). *Journal of Animal Ecology*, **69**, 599–607.

McCarthy, M.A. and Burgman, M.A. (1995). Coping with uncertainty in forest wildlife planning. *Forest Ecology and Management*, **74**, 23–36.

McCarthy, M.A. and Lindenmayer, D.B. (1998). Multi-aged mountain ash forest, wildlife conservation and timber harvesting. *Forest Ecology and Management*, **104**, 43–56.

McCarthy, M.A. and Lindenmayer, D.B. (1999). Incorporating metapopulation dynamics of Greater Gliders into reserve design in disturbed landscapes. *Ecology*, **80**, 651–667.

McCarthy, M.A. and Lindenmayer, D.B. (2000). Spatially correlated extinction in a metapopulation model of Leadbeater's Possum. *Biodiversity and Conservation*, **9**, 47–63.

McCarthy, M.A. and Thompson, C. (2001). Expected minimum population size as a measure of threat. *Animal Conservation*, **4**, 351–355.

McCarthy, M.A., Andelman, S.J. and Possingham, H.P. (2003). Reliability of relative predictions in population viability analysis. *Conservation Biology*, **17**, 982–989.

McCarthy, M.A., Franklin, D.C. and Burgman, M.A. (1994). The importance of demographic uncertainty: an example from the helmeted honeyeater. *Biological Conservation*, **67**, 135–142.

McCarthy, M.A., Gill, A.M. and Lindenmayer, D.B. (1999a). Fire regimes in mountain ash forest: evidence from forest age structure, extinction models and wildlife habitat. *Forest Ecology and Management*, **124**, 193–203.

McCarthy, M.A., Lindenmayer, D.B. and Possingham, H.P. (2000). Australian Treecrepers and landscape fragmentation: a test of a spatially-explicit PVA model. *Ecological Applications*, **10**, 1722–1731.

McCarthy, M.A., Lindenmayer, D.B. and Possingham, H.P. (2001b). Assessing spatial PVA models of arboreal marsupials using significance tests and Bayesian statistics. *Biological Conservation*, **98**, 191–200.

McCarthy, M.A., Menkhorst, P.W., Quin, B.R., Smales, I.J. and Burgman, M.A. (2004). Assessing options for establishing a new population of the Helmeted Honeyeater *Lichenostomus melanops cassidix*. In: *Species Conservation and Management*. Akcakaya, H.R., Burgman, M.A., Kindvall, O., Wood, C.C., Sjorgen-Gulve, P., Hatfield, J. and McCarthy, M.A. (Eds). Oxford University Press, New York.

McCarthy, M.A., Possingham, H.P. and Gill, A.M. (2001c). Using stochastic dynamic programming to determine optimal fire management for *Banksia ornata*. *Journal of Applied Ecology*, **38**, 585–592.

McCarthy, M.A., Possingham, H.P., Day, J.R. and Tyre, A.J. (2001a). Testing the accuracy of population viability analysis. *Conservation Biology*, **73**, 143–150.

McCarthy, M.A., Webster, A., Loyn, R. and Lowe, K.W. (1999b). Uncertainty in assessing the viability of the Powerful Owl *Ninox strenua* in Victoria, Australia. *Pacific Conservation Biology*, **5**, 144–154.

McCarty, J.P. (2001). Ecological consequences of recent climate change. *Conservation Biology*, **15**, 320–331.

MacCleery, D.W. (1996). *American Forests. A History of Resiliency and Recovery*. Forest History Society Issues Series, Durham, North Carolina.

McCollin, D. (1998). Forest edges and habitat selection in birds: a functional approach. *Ecography*, **21**, 247–260.

McComb, W.C., McGarigal, K., Fraser, J.D. and Davis, W.H. (1991). Planning for basin-level cumulative effects in the Appalachian coal fields. pp. 138–151. In: *Wildlife Habitats in Managed Landscapes.* Rodiek, J.E. and Bolen, E.C. (Eds). Island Press, California.

McCoy, E.D. and Mushinsky, H.R. (1999). Habitat fragmentation and the abundances of vertebrates in the Florida Scrub. *Ecology,* **80**, 2526–2538.

McCullagh, P. and Nelder, J.A. (1988). *Generalized Linear Models.* 2nd edition. Chapman and Hall, New York.

Macdonald, L.H. and Smart, A. (1993). Beyond the guidelines: practical lessons for monitoring. *Environmental Monitoring and Assessment,* **26**, 203–218.

MacDonald, M.A. and Kirkpatrick, J.B. (2003). Explaining bird species composition and richness in eucalypt-dominated remnants in subhumid Tasmania. *Journal of Biogeography,* **30**, 1415–1426.

McDougall, K. and Kirkpatrick, J.B. (Eds) (1994). *Conservation of Lowland Native Grasslands in South-eastern Australia.* World Wide Fund for Nature, Sydney.

McDougall, K. and Wright, G.T. (2004). The impact of trampling on feldmark vegetation in Kosciuszko National Park, Australia. *Australian Journal of Botany,* **52**, 315–320.

Mace, G.M. and Lande, R. (1991). Assessing extinction threats: towards a re-evaluation of IUCN threatened species categories. *Conservation Biology,* **5**, 148–157.

Macfarlane, M.A. and Seebeck, J.H. (1991). *Draft Management Strategies for the Conservation of Leadbeater's Possum* Gymnobelideus leadbeateri *in Victoria.* Technical Report Series No. 111. Arthur Rylah Institute for Environmental Research, Department of Conservation and Environment, Melbourne.

Macfarlane, M.A., Smith, J. and Lowe, K. (1998). *Leadbeater's Possum Recovery Plan.* 1998–2002. Department of Natural Resources and Environment, Melbourne.

McGarical, K. and McComb, W.C. (1995). Relationships between landscape structure and breeding birds in the Oregon Coast Range. *Ecological Monographs,* **65**, 235–260.

McGarigal, K. and Cushman, S.A. (2002). Comparative evaluation of experimental approaches to the study of fragmentation studies. *Ecological Applications,* **12**, 335–345.

McGeoch, M.A. (1998). The selection, testing and application of terrestrial insects as bioindicators. *Biological Reviews,* **73**, 181–201.

McGill, B.J. (2003). Does Mother Nature really prefer rare species or are log-left-skewed SADs a sampling artefact? *Ecology Letters,* **6**, 766–773.

McGrady-Steed, J., Harris, P.M. and Morin, P.J. (1997). Biodiversity regulates ecosystem predictability. *Nature,* **390**, 162–165.

McIntosh, B.A., Sedell, J.R., Thurlow, R.F., Clarke, S.E. and Chandler, G.L. (2000). Historical changes in pool habitats in the Columbia River Basin. *Ecological Applications,* **10**, 1478–1496.

McIntyre, S. (1992). Risks associated with the setting of conservation priorities from rare plant species lists. *Biological Conservation,* **8**, 521–531.

McIntyre, S. (1994). Integrating agriculture and land use and management for conservation of a native grassland flora in a variegated landscape. *Pacific Conservation Biology,* **1**, 236–244.

McIntyre, S. and Barrett, G.W. (1992). Habitat variegation, an alternative to fragmentation. *Conservation Biology,* **6** 146–147.

McIntyre, S. and Hobbs, R. (1999). A framework for conceptualizing human effects on landscapes and its relevance to management and research models. *Conservation Biology,* **13**, 1282–1292.

McIntyre, S. and Lavorel, S. (1994a). How environmental and disturbance factors influence species composition in temperate Australian grasslands. *Journal of Vegetation Science,* **5**, 373–384.

McIntyre, S. and Lavorel, S. (1994b). Predicting richness of native, rare and exotic plants in response to habitat and disturbance variables across a variegated landscape. *Conservation Biology,* **8**, 521–531.

McIntyre, S. and Martin, T.G. (2002). Managing intensive and extensive land uses to conserve grassland plants in sub-tropical eucalypt woodlands. *Biological Conservation,* **107**,241–252.

McIntyre, S., Barrett, G.W. and Ford, H.A. (1996). Communities and ecosystems. pp. 154–170. In: *Conservation Biology*. Spellerberg, I.F. (Ed). Longman, Harlow, England.

McIntyre, S., McIvor, J.G. and Heard, K.M. (Eds) (2002). *Managing and Conserving Grassy Woodlands.* CSIRO Publishing, Melbourne.

McIntyre, S., McIvor, J.G. and MacLeod, N.D. (2000). Principles for sustainable grazing in eucalypt woodlands: landscape-scale indicators and the search for thresholds. pp. 92–100. In: *Management for Sustainable Ecosystems.* Hale, P., Moloney, D. and Sattler, P. (Eds). Centre for Conservation Biology, The University of Queensland, Brisbane.

McIver, J.D. and Starr, L. (2001). A literature review on the environmental effects of postfire logging. *Western Journal of Applied Forestry,* **16**, 159–168.

McKay, G.M. (1983). The Greater Glider. pp. 134–135. In: *Complete book of Australian mammals.* Strahan, R. (Ed). Angus and Robertson, Sydney.

McKay, R.J. (1984). Introductions of exotic fishes in Australia. In: *Distribution, Biology and Management of Exotic Fishes.* Courtenay, W.R. and Strauffer jnr, J.R. (Eds). John Hopkins University Press, Sydney.

McKee, J.K., Sciulli, B., Fooce, C.D. and Waite, T.A. (2003). Forecasting global biodiversity threats associated with human population growth. *Biological Conservation,* **115**, 161–164.

McKenney, D.W., Sims, R.A., Soulé, M.E., Mackey, B.G. and Campbell, K.L. (1994). Towards a set of biodiversity indicators for Canadian forests. *Proceedings of the Forest Biodiversity Indicators Workshop*, Sault Ste. Marie, Canada Natural Resources Canada, Canadian Forest Service.

McKenny, H.J. and Kirkpatrick, J.B. (1999). The role of fallen logs in the regeneration of tree species in Tasmanian mixed forest. *Australian Journal of Botany,* **47**, 745–753.

McKenzie, N.L. and Belbin, L. (1991). Kimberley rainforest communities: reserve recommendations and management considerations. pp. 453–480. In: *Kimberely Rainforests.* McKenzie, N.L., Johnson, R.B. and Kendrick, P.G. (Eds). Surrey Beatty and Sons, Chipping Norton.

McKenzie, N.L., Burbidge, A. and Rolfe, J.K. (2003). Effect of salinity on small, ground-dwelling animals in the Western Australian wheatbelt. *Australian Journal of Botany*, **51**, 725–740.

Mackey, B.G. (1993). A spatial analysis of the environmental relations of rainforest structural types. *Journal of Biogeography,* **20**, 303–336.

Mackey, B.G. (1994). Predicting the potential distribution of rainforest structural types. *Journal of Vegetation Science,* **5**, 43–54.

Mackey, B.G. and Lindenmayer, D.B. (2001). A hierarchical framework for linking surveys for forest wildlife at various spatial scales. *Journal of Biogeography,* **28**, 1147–1166.

Mackey, B.G. and Sims, R.A. (1993). A climatic analysis of selected boreal tree species and potential responses to global climate change. *World Resources Review,* **5**, 469–487.

Mackey, B.G., Lesslie, R.G., Lindenmayer, D.B. and Nix, H.A. (1998). The role of wilderness and wild rivers in nature conservation. *Pacific Conservation Biology,* **4**, 182–185.

Mackey, B.G., Lesslie, R.G., Lindenmayer, D.B., Incoll, R.D. and Nix, H.A. (1999). *The role of wilderness and wild rivers in nature conservation.* Major report to Environment Australia.

Mackey, B.G., Lindenmayer, D.B., Gill, A.M., McCarthy, M.A. and Lindesay, J. (2002). *Wildlife, Fire and Future Climate. A Forest Ecosystem Analysis.* CSIRO Publishing, Melbourne.

Mackey, B.G., Nix, H.A. and Hitchcock, P. (2001). *The Natural Heritage Significance of Cape York Peninsula.* Queensland Environmental Protection Agency, Brisbane.

Mackey, B.G., Nix, H.A., Hutchinson, M.F., MacMahon, J.P. and Fleming, P.M. (1988). Assessing representativeness of places for conservation reserves and heritage listing. *Environmental Management,* **12**, 501–504.

Mackey, B.G., Nix, H.A., Stein, J.A. and Cork, S.E. (1989). Assessing the representativeness of the Wet Tropics of Queensland World Heritage Property. *Biological Conservation,* **50**, 279–303.

Mackey, R. (2003). *The Atlas of Endangered Species.* Earthscan Publications, London.

MacKinnon, K. and MacKinnon, J. (1991). Habitat protection and re-introduction programmes. pp. 173–198. In: *Beyond Captive Breeding: Reintroducing Endangered Species Through Captive Breeding.* Symposia of the Zoological Society of London 62. Gipps, J.H.W. (Ed). Clarendon Press, Oxford.

McKinnon, L.J. (1993). A significant record of the endangered Trout Cod *Maccullochella macquariensis* (Pisces: Percichthyidae) made during fish surveys of the Barmah Forest, Victoria. *Victorian Naturalist,* **110**, 186–190.

Mackowski, C. (1986). Distribution, habitat and status of the Yellow-bellied Glider, *Petaurus australis* Shaw (Marsupialia: Petauridae) in northeastern New South Wales. *Australian Mammalogy,* **9**, 141–144.

Mackowski, C.M. (1987). Wildlife hollows and timber management in Blackbutt forest. Master of Natural Resources, University of New England, Armidale.

McLean, I.G., Lundie-Jenkins, G. and Jarman, P. (1995). Training captive Rufous Hare-wallabies to recognise predators. pp. 177–182. In: *Reintroduction Biology of Australasian Fauna.* Serena, M. (Ed). Surrey Beatty, Chipping Norton.

McLean, I.G., Lundie-Jenkins, G. and Jarman, P.J. (1996). Teaching an endangered mammal to recognise predators. *Biological Conservation,* **75**, 51–62.

MacLeod, C.D., Bannon, S.M., Pierce, G.J., Schweder, C., Learmonth, J.A., Herman, J.S. and Reid, R.J. (2005). Climate change and the cetacean community of north-west Scotland. *Biological Conservation,* **124**, 477–483.

McMahon, T.A., Finlayson, B.L., Haines, A.T. and Srikanthan, R. (1992a). *Global runoff – Continental Comparisons of Annual Flows and Peak Discharges.* Catena Verlag, Gremlingen, Germany.

McMahon, T.A., Gan, K.C. and Finlayson, B.L. (1992b). Anthropogenic changes to the hydrologic cycle in Australia. pp. 35–66. In: *Australia's Renewable Resources: Sustainability and Global Change.* Gifford, R.M. and Barson, M.M. (Eds). Bureau or Rural Resources and CSIRO Division of Plant Industry.

McMillan, M.A., Nekola, J.C. and Larson, D.W. (2003). Effects of rock climbing on the land snail community of the Niagara Escarpment in southern Ontario, Canada. *Conservation Biology*, **17**, 616–621.

Mac Nally, R. (1994a). Habitat-specific guild structure of forest birds in south-eastern Australia: a regional scale perspective. *Journal of Animal Ecology,* **63**, 988–1001.

Mac Nally, R. (1994b). A winter's tale: Among-year variation in bird community structure in a southeastern Australian forest. *Australian Journal of Ecology*, **21**, 280–291.

Mac Nally, R. (1997). Monitoring forest bird communities for impact assessment: the influence of sampling intensity and spatial scale. *Biological Conservation,* **82**, 355–367.

Mac Nally, R. and Bennett A.F. (1997). Species-specific predictions of the impact of habitat fragmentation: local extinction of birds in the box-ironbark forests of central Victoria, Australia. *Biological Conservation*, **82**, 147–155.

Mac Nally, R. and Fleishman, E. (2002). Using "indicator" species to model species richness: model development and predictions. *Ecological Applications*, **12**, 79–92.

Mac Nally, R. and Horrocks, G. (2000). Landscape-scale conservation of an endangered migrant: The Swift Parrot (*Lathamus discolor*) in its winter range. *Biological Conservation,* **92**, 335–343.

Mac Nally, R., Bennett, A.F. and Horrocks, G. (2000). Forecasting the impacts of habitat fragmentation. Evaluation of species-specific predictions of the impact of habitat fragmentation on birds in the box-ironbark forests of central Victoria, Australia. *Biological Conservation,* **95**, 7–29.

Mac Nally, R., Bennett, A.F., Brown, G.W., Lumsden, L., Yen, A., Hinkley, S., Lillywhite, P. and Ward, D. (2002). How well do ecosystem-based planning units represent different components of biodiversity? *Ecological Applications*, **12**, 900–912.

Mac Nally, R.C. (1989). The relationship between habitat breadth, habitat position, and abundance in forest and woodland birds along a continental gradient. *Oikos,* **54**, 44–54.

McNeeley, J.A., Neville, L.E. and Rejmanek, M. (2003). When is eradication a sound investment? *Conservation in Practice*, **4**, 30–41.

McNeely, J.A. (1988). *Economics and biological diversity: developing and using economic incentives to conserve biological resources.* International Union for the Conservation of Nature and Natural Resources, Gland, Switzerland.

McNeely, J.A. (1994a). Lessons for the past: Forests and biodiversity. *Biodiversity and Conservation,* **3**, 3–20.

McNeely, J.A. (1994b). Protected areas for the 21st century: working to provide benefits to society. *Biodiversity and Conservation*, **3**, 390–405.

McNeely, J.A. (Ed) (1992). Caracas Action Plan. *Parks for Life.* Report of the IVth World Congress on National Parks and protected areas. IUCN, Gland, Switzerland.

McNeely, J.A., Miller, K.R., Reid, W.V., Mittermeier, R.A. and Werner, T.B. (1990). *Conserving the World's Biological Diversity.* International Union for the Conservation of Nature and Natural Resources, Gland, Switzerland.

McPhail, M., Hill, K., Partridge, A., Truswell, E. and Foster, C. (1995). Wollemi Pine – old pollen records for a newly discovered species. *Geology Today*, **March–April 1995**, 48–49.

McPhee, D.P., Leadbitter, D. and Skilleter, G.A. (2002). Swallowing the bait: is recreational fishing in Australia ecologically sustainable? *Pacific Conservation Biology*, **8**, 40–51.

McPhee, M.E. (2003). Generations in captivity increases behavioral variance: considerations for captive breeding and reintroduction programs. *Biological Conservation*, **115**, 71–77.

McShea, W.J. and Rappole, J.H. (2000). Managing the abundance and diversity of breeding bird populations through manipulation of deer populations. *Conservation Biology,* **14**, 1161–1170.

McTainsh, G. and Boughton, W.C. (1993). Land degradation in Australia – an introduction. pp. 1–18. In: *Land Degradation Processes in Australia.* McTainsh, G. and Boughton, W.C. (Eds). Longman, Cheshire.

Madden, B., Hayes, G. and Duggan, K. (2000). *National investment in rural landscapes.* An investment scenario for NFF and ACF with assistance from LWRDDC. Virtual Consulting Group and Griffin nrm Pty Ltd.

Maddox, J. (2000). Positioning the goal posts. *Nature,* **403**, 139.

Madsen, T., Shine, R., Olsson, M. and Wittzell, H. (1999). Restoration of an inbred adder population. *Nature,* **402**, 34–35.

Maes, D. and van Dyck, H. (2005). Habitat quality and biodiversity indicator performances of a threatened butterfly versus a multispecies group for wet heathlands in Belgium. *Biological Conservation*, **123**, 177–187.

Maestas, J.D., Knight, R.L. and Gilger, W.C. (2003). Biodiversity across a rural land-use gradient. *Conservation Biology*, **17**, 1425–1434.

Magin, C.D., Johnson, T.H., Groombridge, B., Jenkins, M. and Smith, H. (1994). Species extinctions, endangerment and captive breeding. pp. 3–31. In: *Creative Conservation*. Olney, P., Mace, G. and Feistner, A. (Eds). Chapman and Hall, London.

Maguire, L.A., Lacy, R.C., Begg, R.J. and Clark, T.W. (1990). An analysis of alternative strategies for recovering the Eastern Barred bandicoot. pp. 147–164. In: *Management and Conservation of Small Populations.* Clark, T.W. and Seebeck, J.H. (Eds). Chicago Zoological Society, Chicago

Magurran, A.E. (2003). *Measuring biological diversity.* Blackwell, London.

Maher, W. and Norris, R.H. (1990). Water quality assessment programs in Australia: Deciding what to measure and how and where to use bioindicators. *Environmental Monitoring and Assessment,* **14**, 115–130.

Main, A.R. (1984). Rare species: Problems of conservation. *Search,* **15**, 93–97.

Main, B.Y. (1999). Biological anachronisms among trapdoor spiders reflect Australia's environmental changes since the Mesozoic. pp. 236–245. In: *The Other 99%: The Conservation and Biodiversity of Invertebrates.* Ponder, W. and Lunney, D. (Eds). Transactions of the Royal Society of New South Wales and Surrey Beatty and Sons, Chipping Norton.

Main, G. (2000). *Gunderbooka: a Stone Country Story.* Resource Policy and Management, ACT.

Majer, J.D. (1978). Preliminary survey of the epigaeic invertebrate fauna with particular reference to ants in areas of different landuse at Dwellingup, Western Australia. *Forest Ecology and Management,* **1**, 321–334.

Majer, J.D. (1985). Recolonisation by ants of rehabilitated mineral sand mines on North Stradbroke Island, Queensland, with particular reference to seed removal. *Australian Journal of Ecology,* **10**, 31–48.

Majer, J.D. (1987) Invertebrates as indicators for management. pp. 353–354. In: *Nature Conservation: The Role of Remnants of Vegetation.* Saunders, D.A., Arnold, G.W., Burbidge, A.A. and Hopkins, A.J. (Eds). Surrey Beatty, Chipping Norton.

Majer, J.D., Day, J.E., Kabbay, E.D. and Perriman, W.S. (1984). Recolonisation by ants in bauxite mines rehabilitated by a number of different methods. *Journal of Applied Ecology,* **21**, 355–375.

Majer, J.D., Recher, H.F. and Postle, A.C. (1994). Comparison of arthropod species richness in eastern and western Australian canopies: a contribution to the species number debate. *Memoirs of the Queensland Museum,* **36**, 121–131.

Major, R., Christie, F.J., Gowing, G., Cassis, G. and Reid, C.A. (2003). The effect of habitat configuration on arboreal insects in fragmented woodlands of south-eastern Australia. *Biological Conservation,* **113**, 35–48.

Malcolm, C.V. (1971). *Plant Collection for Pasture Improvement in Saline and Arid Environments.* Technical Bulletin No. 6. Western Australia Department of Agriculture.

Male, B. (1996). Listing process. *On the Brink,* **8 (June)**, 12–13.

Mallet, J. (1996). The genetics of biological diversity: from variety to species. pp. 13–53. In: *Biodiversity.* Gaston, K.J. (Ed). Blackwell, Oxford.

Mallick, S.A., Hocking, G.J. and Driessen, M.M. (1997). Habitat requirements of the Eastern Barred Bandicoot, *Parameles gunnii*, on agricultural land in Tasmania. *Wildlife Research,* **24**, 237–243.

Mangel, M. and 41 others. (1996). Principles for the conservation of wild living resources. *Ecological Applications,* **6**, 338–362.

Manly, B.F.J. (1991). *Randomization and Monte Carlo Tests in Biology.* Chapman and Hall, London.

Manning, A.D. (2004). A multi-scale study of the Superb Parrot (*Polytelis swainsonii*): implications for landscape-scale ecological restoration. PhD thesis, The Australian National University, Canberra.

Manning, A.D., Lindenmayer, D.B. and Nix, H.A. (2004). Continua and umwelt: alternative ways of viewing landscapes. *Oikos,* **104**, 621–628.

Manning, R. (1997). The honey bee debate: a critique of scientific studies of honey bees *Apis mellifera* and their alleged impact on Australian wildlife. *The Victorian Naturalist* **114**, 13–22.

Mansergh, I. and Bennett, S. (1989). 'Greenhouse' and wildlife management in Victoria. *Victorian Naturalist,* **106**, 243–251.

Mapstone, B.D. (1995). Scaleable decision rules for Environmental Impact Studies: effect size, Type I and Type II errors. *Ecological Applications,* **5(2)**, 401–410.

Marcot, B.G. (1997). Biodiversity of old forests of the west: a lesson from our elders. pp. 87–105. In: *Creating a Forestry for the 21st Century.* Kohm, K.A. and Franklin, J.F. (Eds). Island Press, Washington, DC.

Margules, C. and Usher, M.B. (1981). Criteria used in assessing wildlife conservation potential: A review. *Biological Conservation,* **21**, 79–109.

Margules, C.R. (1992). The Wog Wog habitat fragmentation experiment. *Environmental Conservation,* **19**, 316–325.

Margules, C.R. and Austin, M.P. (1991). *Nature Conservation: Cost Effective Biological Surveys and Data Analysis.* Margules, C.R. and Austin, M.P. (Eds). CSIRO, Canberra.

Margules, C.R. and Austin, M.P. (1994). Biological models for monitoring species decline: The construction and use of data bases. *Philosophical Transactions of the Royal Society of London, B* **344**, 69–75.

Margules, C.R. and Nicholls, A.O. (1987). Assessing the conservation value of remnant 'islands': Mallee patches on the western Eyre Peninsula, South Australia. pp. 89–102. In: *Nature Conservation: The Role of Remnants of Native Vegetation.* Saunders, D.A., Arnold, G.W., Burbudge, A.A. and Hopkins, A.J.M. (Eds). Surrey Beatty, Chipping Norton.

Margules, C.R. and Pressey, R.L. (2000). Systematic conservation planning. *Nature,* **405**, 243–253.

Margules, C.R. and Stein, J.L. (1989). Patterns of distributions of species and the selection of nature reserves: an example from *Eucalyptus* forests in south-eastern New South Wales. *Biological Conservation,* **50**, 219–238.

Margules, C.R., Davies, K.F., Meyers, J.A. and Milkovits, G.A. (1995a). The responses of some selected arthropods and the frog *Crinia signifera* to habitat fragmentation. pp. 94–103. In: *Conserving Biodiversity: Threats and Solutions.* Bradstock, R.A., Auld, T.A., Keith, D.A., Kingsford, R.T., Lunney, D. and Sivertsen, D.P. (Eds). Surrey Beatty, Chipping Norton.

Margules, C.R., Higgs, A.J. and Rafe, R.W. (1982). Modern biogeographic theory: are there any lessons for nature reserve design? *Biological Conservation,* **24**, 115–128.

Margules, C.R., Milkovits, G.A. and Smith, G.T. (1994a). Contrasting effects of habitat fragmentation on the scorpion *Cercophonius squama* and an amphipod. *Ecology,* **75**, 2033–2042.

Margules, C.R., Nicholls, A.O. and Pressey, R.L. (1988). Selecting networks of reserves to maximise biological diversity. *Biological Conservation,* **43**, 663–676.

Margules, C.R., Nicholls, A.O. and Usher, M.B. (1994b). Apparent species turnover, probability of extinction and the selection of nature reserves: a case study of the Ingleborough limestone pavements. *Conservation Biology,* **8**, 398–409.

Margules, C.R., Redhead, T.D., Faith, D.P. and Hutchinson, M.F. (1995b). *BioRap. Guidelines for Using the BioRap Methodology and Tools.* CSIRO, Canberra.

Marine Stewardship Council (2003). *MSC online.* Marine Stewardship Council, London. [online] URL: http://www.msc.org

Marks, G.C. and Smith, I.W. (1991). *The Cinnamon Fungus in Victorian Forests.* Lands and Forests Bulletin No. 31. Department of Conservation and Environment, Victoria.

Maron, M. and Lill, A. (2005). The influence of livestock grazing and weed invasion on habitat use by birds in grassy woodland remnants. *Biological Conservation,* **124**, 439–450.

Marsh, H., Harris, N.M. and Lawler, I.R. (1997). The sustainability of the indigenous dugong fishery in Torres Strait, Australia/Papua New Guinea. *Conservation Biology,* **11**, 1375–1386.

Martin, P.S. and Klein, R.G. (1984). *Quaternary Extinctions: A Prehistoric Evolution.* Arizona University Press, Tucson.

Martin, R. and Handasyde, K. (1999). *The Koala. Natural History, Conservation and Management.* UNSW Press, Sydney.

Martin, T.E. and Karr, J.R. (1986). Patch utilisation by migrating birds: resource orientated? *Ornis Scandinavia,* **17**, 165–74.

Martin, T.G. and Possingham, H.P. (2005). Predicting the impact of livestock grazing on birds using foraging height data. *Journal of Applied Ecology*, **42**, 400–408.

Martin, W.K. (1996). The current and potential distribution of the Common Myna *Acridotheres tristis* in Australia. *Emu,* **96**, 166–173.

Marvier, M. and van Acker, R.C. (2005). Can crop transgenes be kept on a leash? *Frontiers in Ecology and Environment,* **3**, 99–106.

Marzluff, J.M. and Restani, M. (1999). The effect of forest fragmentation on avian nest predation. pp. 155–169. In: *Forest Wildlife and Fragmentation. Management Implications.* Rochelle, J., Lehmann, L.A. and Wisniewski, J. (Eds). Brill, Leiden, Germany.

Matarczyk, J.A., Willis, A.J., Vranjic, J.A. and Ash, J.E. (2002). Herbicides, weeds and endangered species: management of Bitou Bush (*Chrysanthemoides monilfera ssp. rotundata*) with glyphosate and impacts on the endangered shrub, *Pimelia spicata. Biological Conservation,* **108**, 133–141.

Mather, A.S. (1990). *Global Forest Resources.* Belhaven Press, London.

Matlack, G.R. (1993). Microenvironment variation within and among forest edge sites in the eastern United States. *Biological Conservation,* **66**, 185–194.

Matlack, G.R. and Litvaitis, J.A. (1999). Forest edges. pp. 210–233. In: *Managing Biodiversity in Forest Ecosystems.* Hunter III, M. (Ed). Cambridge University Press, Cambridge.

Maunder, M. (1992). Plant reintroduction: An overview. *Biodiversity and Conservation,* **1**, 51–61.

Maurer, B.A. and Heywood, S.G. (1993). Geographic range fragmentation and abundance in neotropical migratory birds. *Conservation Biology*, **7**, 501–509.

Mawson, P.R. and Long, J.L. (1994). Size and age parameters of nest trees used by four species of parrot and one species of cockatoo in south-west Australia. *Emu*, **94**, 149–155.

Maxwell, S., Burbidge, A.A. and Morris, K. (1996). *The 1996 Action Plan for Australian Marsupials and Monotremes.* Australian Nature Conservation Agency, Canberra.

May, R.M. (1973). *Stability and Complexity in Model Ecosystems.* Princeton University Press, Princeton, New Jersey.

May, R.M. (1976). *Theoretical Ecology.* Blackwell, Oxford.

May, R.M. (1990a). How many species? *Philosophical Transactions of the Royal Society, B* **30**, 293–304.

May, R.M. (1990b). Taxonomy as destiny. *Nature*, **347**, 129–130.

May, R.M. (1994). Resource management – the economics of extinction. *Nature*, **372**, 42–43.

May, R.M., Lawton, J.H. and Stork, N.E. (1995). Assessing extinction rates. pp. 1–24. In: *Extinction Rates.* Lawton, J.H. and May, R.M. (Eds). Oxford University Press, Oxford.

May, S. and Norton, T.W. (1996). Influence of fragmentation and disturbance on the potential impact of feral predators on native fauna in Australian forest ecosystems. *Wildlife Research*, **23**, 387–400.

May, S.A. (2001). Aspects of the ecology of the cat, dog and fox in the south-east forests of NSW. PhD thesis, The Australian National University, Canberra.

May, S.A. and Norton, T.W. (1996). Influence of fragmentation and disturbance on the potential impact of feral predators on native fauna in Australian forest ecosystems. *Wildlife Research*, **23**, 387–400.

May, T. and Simpson, J.A. (1997). Fungal diversity and ecology in eucalypt systems. pp. 246–277. In: *Eucalypt Ecology: Individuals to Ecosystems.* Williams, J.E. and Woinarski, J.C. (Eds). Cambridge University Press, Melbourne.

Maynard-Smith, J. (1989). *Evolutionary Genetics.* Oxford University Press, Oxford.

Mayr, E. (1942). *Systematics and the Origin of Species, from the Viewpoint of a zoologist.* Columbia University Press, New York.

Mayr, E. (1965). Avifauna: Turnover on islands. *Science*, **150**, 1587–1588.

Meffe, G.K. and Carroll, C.R. (1994). *Principles of Conservation Biology.* Sinauer, Sunderland, Massachusetts.

Meffe, G.K., Carroll, C.R. and contributors (1997). *Principles of Conservation Biology.* Sinauer Associates, Inc. Sunderland, Massachusetts.

Meilleur, A., Bouchard, A. and Bergeron, Y. (1992). The use of understorey species as indicators of landform ecosystem type in heavily disturbed forest: An evaluation of the Haut-Saint-Laurent, Quebec. *Vegetatio*, **102**, 13–32.

Melbourne Zoo (2003). *World's Rarest Insect Breeds.*Melboune Zoo, Melbourne. [online] URL: http://www.zoo.org.au/

Mendel, L.C. and Kirkpatrick, J.B. (2002). Historical progress of biodiversity conservation in the protected-area system of Tasmania, Australia. *Conservation Biology*, **16**, 1520–1529.

Menkhorst, P. (1995a). Eastern Quoll. pp. 53–54. In: *Mammals of Victoria. Distribution, Ecology and Conservation.* Menkhorst P. (Ed). Oxford University Press, Melbourne.

Menkhorst, P. and Knight, F. (2001). *A Field Guide to the Mammals of Australia.* Oxford University Press, Melbourne.

Menkhorst, P. and Middleton, D. (1991). *The Helmeted Honeyeater Recovery Plan: 1989–1993.* Department of Conservation and Environment, Victoria.

Menkhorst, P., Smales, I. and Quin, B. (1997). *Helmeted Honeyeater Recovery Plan 1999–2003.* Victorian Department of Natural Resources and Environment, Melbourne, May 1997.

Menkhorst, P.W. (Ed) (1995b). *Mammals of Victoria.* Oxford University Press and Department of Conservation and Natural Resources, Melbourne.

Menkhorst, P.W., Loyn, R.H. and Brown, P.B. (1990). Management of the Orange-bellied Parrot. pp. 239–252. In: *Management and Conservation of Small Populations.* Clark, T.W. and Seebeck, J.H. (Eds). Chicago Zoological Society, Chicago.

Menkhorst, P.W., Weavers, B.W. and Alexander, J.S.A. (1988). Distribution, habitat and conservation status of the Squirrel Glider, *Petaurus norfolcensis* (Petauridae: Marsupialia). *Australian Wildlife Research*, **15**, 59–71.

Mercer, D. (1995). *A Question of Balance: Natural Resource Conflict Issues in Australia.* 2nd edition. The Federation Press, Sydney.

Meredith, C.W. (1984). Possums or poles? – The effects of silvicultural management on the possums of Chiltern State Park, north-eastern Victoria. pp. 517–525. In: *Possums and Gliders.* Smith, A.P. and Hume, I.D. (Eds). Surrey Beatty, Chipping Norton.

Meredith, C.W., Gilmore, A.M. and Isles, A.C. (1984). The Ground Parrot (*Pezoporus wallicus* Kerr) in south-eastern Australia: a fire-adapted species? *Australian Journal of Ecology*, **9**, 367–380.

Merriam, G. and Saunders, D.A. (1993). Corridors in restoration of fragmented landscapes. pp. 71–87. In: *Nature Conservation 3: Reconstruction of Fragmented Ecosystems.* Saunders, D.A., Hobbs, R.J. and Ehrlich, P.R. (Eds). Surrey Beatty, Chipping Norton.

Merrilees, D. (1968). Man the destroyer: late Quaternary changes in the Australian marsupial fauna. *Proceedings of the Royal Society of Western Australia,* **51**, 1–24.

Merritt, B. and Wallis, R. (2004). Are wide revegetated strips better for birds and frogs than narrow ones? *The Victorian Naturalist*, **121**, 288–292.

Mesibov, R. (1990). Velvet worms: A special case of invertebrate fauna conservation. *Tasforests,* **2**, 53–56.

Mesquita, R.C., Delamonica, P. and Laurance, W.F. (1999). Effect of surrounding vegetation on edge-related tree mortality in Amazonian forest fragments. *Biological Conservation,* **91**, 129–134.

Metzeling, L., Doeg, T. and O'Conner, W. (1995). The impact of salinisation and sedimentation on aquatic biota. pp. 126–136. In: *Conserving Biodiversity: Threats and Solutions.* Bradstock, R.A., Auld, T.A., Keith, D.A., Kingsford, R.T., Lunney, D. and Sivertsen, D.P. (Eds). Surrey Beatty, Chipping Norton.

Metzger, J.P. (1997). Relationships between landscape structure and tree species diversity in tropical forests of south–east Brazil. *Landscape and Urban Planning,* 37, 29–35.

Michener, C.D. (1965). A classification of the bees of the Australian and South Pacific regions. *Bulletin of the American Museum of Natural History*, **130**, 1–362.

Michener, W.K. and Brunt, J.W. (2000). *Ecological Data: Design, Management and Processing.* Blackwell Science, Oxford.

Middleton, J. and Merriam, G. (1981). Woodland mice in a farmland mosaic. *Journal of Applied Ecology*, **18**, 703–710.

Milledge, D.R., Palmer, C.L. and Nelson, J.L. (1991). 'Barometers of change': The distribution of large owls and gliders in Mountain Ash forests of the Victorian Central Highlands and their potential as management indicators. pp. 53–65. In: *Conservation of Australia's Forest Fauna.* Lunney, D. (Ed). Royal Zoological Society of NSW, Mosman.

Millennium Ecosystem Assessment. (2005). Millennium Ecosystem Assessment Synthesis Report. http://www.millenniumassessment.org/

Miller, B. and Mullette, K.J. (1985). Rehabilitation of an endangered Australian bird: the Lord Howe Island Woodhen *Tricholimnas sylivestris* (Sclater). *Biological Conservation,* **34**, 55–95.

Miller, B., Conway, W., Reading, R., Wemmer, C., Wildt, D., Kleiman, D., Monfort, S., Rabinowitz, A., Armstrong, B. and Hutchins, M. (2004). Evaluating the conservation mission of zoos, aquariums, botanical gardens, and natural history museums. *Conservation Biology*, **18**, 86–93.

Miller, D.R., Lin, J.D. and Lu, Z.N. (1991). Some effects of surrounding forest canopy architecture on the wind field in small clearings. *Forest Ecology and Management,* **45**, 79–91.

Miller, G.H., Magee, J.W., Johnson, B.J., Fogel, M.L., Spooner, N.A., McCulloch, M.T. and Ayliffe, L.K. (1999). Pleistocene extinction of *Genyornis newtonii*: human impact on Australian megafauna. *Science*, **283**, 205–208.

Miller, P.S. (1994). Is inbreeding depression more severe in a stressful environment? *Zoo Biology,* **13**, 195–208.

Miller, S.G., Bratton, S.P. and Hadidian, J. (1992). Impacts of white-tailed deer on endangered and threatened vascular plants. *Natural Areas Journal,* **12**, 67–74.

Millington, P.J. and Walker, K.F. (1983). Australian Freshwater Mussel *Velesunio ambiguus* (Phillipi) as a biological indicator for zinc, iron and manganese. *Australian Journal of Marine and Freshwater Research,* **34**, 873–892.

Mills, L.S. and Allendorf, F.W. (1996). The one-migrant-per-generation rule in conservation and management. *Conservation* Biology, **10**, 1509–1518.

Mills, L.S., Soulé, M.E. and Doak, D.F. (1993). The keystone species concept in ecology and conservation. *BioScience*, **43**,219–224.

Millsap, B.A., Gore, J.A., Runde, D.E. and Cerulean, S.I. (1990). Setting priorities for the conservation of fish and wildlife species in Florida. *Wildlife Monographs,* **111**, 1–57.

Milne, B.T., Turner, M.G., Wiens, J.A. and Johnson, A.R. (1992). Interactions between fractal geometry of landscapes and allometric herbivory. *Theoretical Population Biology,* **41**, 337–353.

Minerals Council of Australia (2002). *Sustainable development report 2002.* Minerals Council of Australia, Canberra. [online] URL: http://www.minerals.org.au/environment

Minister for the Environment and Natural Resources (1996). *The kangaroo conservation and management program in South Australia.* Kangaroo Management Review Task Group, Department of Environment and Natural Resources.

Minta, S.C. and Clark, T.W. (1989). Habitat suitability analysis of potential translocation sites for black-footed ferrets in northcentral Montana. pp. 29–45. In: *The Prarie Dog Ecosystem.* Clark, T.W., Hinckley, D. and Rich, T. (Eds). Montana Bureau Land Management Wildlife, Billings, Montana. Technical Bulletin No. 2, MT5907.

Mirande, C., Lacy, R.C. and Seal, U.S. (1991). *Whooping Crane population viability analysis and species survival plan.* Captive Breeding Specialist Group, Species Survival Commission, I.U.C.N., Apple Valley: Minnesota.

MIRCEN-Stockholm (2003). *Networking activities.* MIRCEN, Stockholm. [online] URL: http://www.bdt.fat.org.br/bin21/ws92

Mitchell, T.L. (1848). *Journal of an Expedition into the Interior of Tropical Australia in Search of a Route to the Gulf of Carpentaria.* Longman, Green and Longmans, London. (Reprinted in 1969 by Greenwood, USA).

Mittermeier, R.A., Gill, P. and Mittermeier, C.G. (1997). *Megadiversity. Earth's Biologically Richest Nations.* Cemex, Mexico.

Mittermeier, R.A., Mittermeier, C.G., Brooks, T.M., Pilgrim, J.D., Konstant, W.R., da Fonseca, G.A.B. and Kormos, C. (2003). Wilderness and biodiversity conservation. *Proceedings of the National Academy of Sciences*, **100**, 10309–10313.

Mittermeier, R.A., Myers, N., Thomsen, J.G., da Fonseca, G.A. and Oliveri, S. (1998). Biodiversity hotspots and major tropical wilderness areas: approaches to setting conservation priorities. *Conservation Biology*, **12**, 516–520.

Mladenoff, D.J., White, J., Crow, T.R. and Pastor, J. (1994). Applying principles of landscape design and management to integrate old-growth forest enhancement and commodity use. *Conservation Biology*, **8**, 752–762.

Mladenoff, D.J., White, J., Pastor, J. and Crow, T.R. (1993). Comparing spatial pattern in unaltered old-growth and disturbed forest landscapes. *Ecological Applications*, **3**, 294–306.

Mlot, C. (1989). Blueprint for conserving plant diversity. *BioScience*, **39**, 364–369.

Moilanen, A. and Cabeza, M. (2002). Single-species dynamic site selection. *Ecological Applications*, **12**, 913–926.

Molino, J. and Sabatier, D. (2001). Tree diversity in tropical rain forests: A validation of the intermediate disturbance hypothesis. *Science*, **294**, 1702–1704.

Molloy, J. and Davis, A. (1992). *Setting Priorities for Conservation of New Zealand's Threatened Plants and Animals.* Department of Conservation, Te Papa Atawhai, Wellington.

Monge-Najera, J. (1995). Phylogeny, biogeography and reproductive trends in the Onychophora. In: Onychophora: past and present. Walker, M.H. and Norman, D.B. (Eds). *Zoological Journal of the Linnean Society*, **114(1)**, 21–60.

Monkkonen, M. and Reunanen, P. (1999). On critical thresholds in landscape connectivity: a management perspective. *Oikos*, **84**, 302–305.

Montague, T.L. (Ed) (2000). *The Brushtail Possum. Biology, Impact and Management of an Introduced Marsupial.* Manaaki Whenua Press, Lincoln, New Zealand.

Montreal Process (1995). Criteria and indicators for the conservation and sustainable management of boreal and temperate forest, Santiago. *Annex to the Santiago Declaration: Sixth meeting of the working group.* February, 1995.

Mooney, N. (2002). Native pets: an unnatural disaster. *Nature Australia*, **Summer 2001–2002**, 84.

Moore, M.K. (1977). *Factors contributing to blowdown in streamside leave strips on Vancouver Island.* Land Management Report **3**, 1–34. British Columbia Forest Service, Vancouver.

Moore, N.W. (1962). The heaths of Dorset and their conservation. *Journal of Ecology*, **50**, 369–391.

Moran, D. (1994). Contingent valuation and biodiversity: measuring the user surplus of Kenyan protected areas. *Biodiversity and Conservation*, **3**, 663–684.

Moran, G.F. and Hopper, S.D. (1983). Genetic diversity and insular population structure of the rare granite rock species, *Eucalyptus caesia Benth. Australian Journal of Botany*, **31**, 161–172.

Moran, G.F. and Hopper, S.D. (1987). Conservation of the genetic resources of rare and widespread Eucalypts in remnant vegetation. pp. 151–162. In: *Nature Conservation: The Role of Remnants of Native Vegetation.* Saunders, D.A., Arnold, G.W., Burbidge, A.A. and Hopkins, A.J.M. (Eds). Surrey Beatty, Chipping Norton.

Morell, M.K., Peakall, R., Appels, R., Preston, L.R. and Lloyd, H.L. (1995). DNA profiling techniques for plant variety and identification. *Australian Journal of Experimental Agriculture*, **35**, 807–819.

Moreton Bay Waterways and Catchments Partnership (2003). Annual Report 2002–2003. Moreton Bay Waterways and Catchments Partnership, Brisbane.

Moreton Bay Waterways and Catchments Partnership (2004). Report Card 2004 for the waterways of south east Queensland. Moreton Bay Waterways and Catchments Partnership, Brisbane. October 2004.

Morgan, G. (2001). Landscape health in Australia. A rapid assessment of the relative condition of Australia's bioregions and subregions. Environment Australia and National Land and Water Resources Audit, Canberra.

Morgan, M.G. and Henrion, M. (1990) *Uncertainty: A Guide to Dealing with Uncertainty in Quantitative Risk and Policy Analysis*. Cambridge University Press, Cambridge.

Moritz, C. (1994a). Applications of mitochondrial DNA analysis in conservation: A critical review. *Molecular Ecology*, **3**, 401–411.

Moritz, C. (1994b). Defining 'Evolutionarily Significant Units' for conservation. *Trends in Ecology and Evolution*, **9**, 373–375.

Moritz, C. (1999). Conservation units and translocations: strategies for conserving evolutionary processes. *Hereditas*, **130**, 217–228.

Moritz, C., Lavery, S. and Slade, R. (1995). Using allele frequency and phylogeny to define units for conservation and management. *American Fisheries Society Symposium*, **17**, 249–262.

Morley, S., Grant, C., Hobbs, R. and Cramer, V. (2004). Long-term impact of prescribed burning on the nutrient status and fuel loads of rehabilitated bauxite mines in Western Australia. *Forest Ecology and Management*, **190**, 227–239.

Moro, D. (2003). Translocation of captive-bred dibblers *Parantechinus apicalis* (Marsupialia: Dasyuridae) to Escape Island, Western Australia. *Biological Conservation*, **111**, 305–315.

Morris, S.C. (1995). A new phylum from the lobster's lips. *Nature*, **378**, 661–662.

Morrison, M.L., Marcot, B.G. and Mannan, R.W. (1992). *Wildlife habitat relationships: concepts and applications*. University of Wisconsin Press, Madision.

Morton, S. (1990). The impact of European settlement on the vertebrate animals of arid Australia – a conceptual model. pp. 201–213. In: *Australian Ecosystems: 200 years of Utilisation, Degradation and Reconstruction*. Proceedings of the Ecological Society of Australia. Vol. 16. Saunders, D.A., Hopkins, A.J.M. and How, R.A. (Eds). Surrey Beatty, Chipping-Norton.

Morton, S., Bourne, G., Cristofani, P., Cullen, P., Possingham, H. and Young, M. (2002). *Sustaining our natural ecosystems and biodiversity: an independent report to the Prime Minister's Science, Engineering and Innovation Council.* CSIRO Sustainable Ecosystems and Environment Australia, Canberra.

Moseby, K.E. and Bice, J.K. (2004). A trial re-introduction of the greater Stick-nest Rat (*Leporillus conditor*) in arid South Australia. *Ecological Restoration and Management*, **5**, 118–124.

Moss, A.J., Rayment, G.E., Reilly, N. and Best, E.K. (1992). *A preliminary assessment of sediment and nutrient exports from Queensland coastal catchments*. Report of the Queensland Department of Environment and Heritage and Department of Primary Industries, No. 33.

Mueck, S.G. and Peacock, R. (1992). *Impacts of intensive timber harvesting on the forests of East Gippsland, Victoria*. VSP Technical Report No. **15**. Department of Conservation and Environment, Melbourne, Australia.

Mueck, S.G., Ough, K. and Banks, J.C.G. (1996). How old are Wet Forest understoreys? *Australian Journal of Ecology*, **21**, 345–348.

Mulder, R.A., Dunn, P.O., Cockburn, A., Lazenby-Cohen, K. and Howell, M. (1994). Helpers liberate female Fairy-wrens form constraints on extra-pair mate choice. *Proceedings of the Royal Society of London (Series B)*, **255**, 223–229.

Mullen, I. (1995). Environmental processes and landscape pattern. Rainforest on the south coast of New South Wales. BSc (Hons) thesis, Department of Geography, The Australian National University.

Mullner, A., Linsenmair, K.E. and Wikelski, M. (2004). Exposure to ecotourism reduces survival and affects stress response in Hoatzin chicks (*Opisthocomus hoazin*). *Biological Conservation*, **118**, 549–558.

Mungomery, R.W. (1936). A survey of the feeding habitats of the Giant Toad (*Bufous marinus* L.), and notes on its progress since its introduction into Queensland. *Proceedings of the Annual Conference of the Queensland Society of Sugar Cane Technologists*, 63–74.

Munton, P. (1987). Concepts of threat and status of wild populations. pp. 72–95. In: *The Road to Extinction*. Fitter, R. and Fitter, M. (Eds). IUCN, Gland, Switzerland.

Murphy, D.D. (1989). Conservation and confusion: wrong species, wrong scale, wrong conclusion. *Conservation Biology*, **3**, 82–84.

Murphy, D.D. and Noon, B.R. (1992). Integrating scientific methods with habitat conservation planning: reserve design for Northern Spotted Owls. *Ecological Applications*, **2**, 3–17.

Murphy, D.M., Freas, K.E. and Weiss, S.T. (1990). An environment-metapopulation approach to population viability for a threatened invertebrate. *Conservation Biology*, **4**, 41–51.

Murray Darling Basin Commission (1999). *Salinity audit of the Murray Darling Basin*. Murray Darling Basin Commission, Canberra.

Murray, A.J. and Poore, R.N. (2004). Potential impact of aerial baiting for wild dogs on a population of Spotted-tailed Quolls. *Wildlife Research*, **31**, 639–644.

Murray, B.D. and Lepschi, B.J. (2004). Are locally rare species abundant elsewhere in their geographical range? *Austral Ecology*, **29**, 287–293.

Murray, B.R., Thrall, P.H., Gill, A.M. and Nicotra, A.B. (2002). How plant life-history and ecological traits relate to species rarity and commonness at varying spatial scales. *Austral Ecology*, **27**, 291–310.

Murtaugh, P.A. (2002). Journal quality, effect size and publication bias in meta-analysis. *Ecology*, **83**, 1162–1166.

Myers, J., Simberloff, D.A., Kuris, A.M. and Carey, J.R. (2000). Eradication revisited: dealing with exotic species. *Trends in Ecology and Evolution*, **15**, 17–20.

Myers, K. (1986). Introduced vertebrates in Australia, with emphasis on the mammals. pp. 120–136. In: *Ecology of Biological Invasions: An Australian Perspective*. Groves, R.H. and Burdon, J.J. (Eds). Australian Academy of Science, Canberra.

Myers, N. (1979). *The Sinking Ark: A New Look at the Problem of Disappearing Species.* Pergamon Press, Oxford.

Myers, N. (1988). Threatened biotas: 'Hot spots' in tropical forests. *The Environmentalist*, **8**, 187–208.

Myers, N. (1990a). *The Wild Supermarket: The Importance of Biodiversity to Food Security*. World Wide Fund for Nature, Gland, Switzerland.

Myers, N. (1990b). The biodiversity challenge: expanded hot-spots analysis. *The Environmentalist*, **10**, 243–256.

Myers, N. (1993). The question of linkages in environment and development. *BioScience*, **43**, 302–310.

Myers, N. and Mittermeier, R.A. (2003). Impacts and acceptance of the hotspots strategy: response to Ovadia and to Brummitt and Lughadha. *Conservation Biology*, **17**, 1449–1450.

Myers, R.A., Hutchings, J.A. and Barrowman, N.J. (1997). Why do fish stocks collapse? The example of Cod in Atlantic Canada. *Ecological Applications*, **7**, 91–106.

Myroniuk, P. (Ed) (1995). *International Studbook for Leadbeater's Possum*, Gymnobelideus leadbeateri. Royal Melbourne Zoo, Melbourne.

Nabhan, G.P. and Buchmann, S.L. (1997). Services provided by pollinators. pp. 133–150. In: *Nature's services: Societal Dependence on Natural Ecosystems.* Daily, G.C. (Ed). Island Press, Washington, DC.

Nachman, G. (1981). A mathematical model of the functional relationships between density and spatial distribution of a population. *Journal of Animal Ecology*, **50**, 453–460.

Naeem, S. (1998). Species redundancy and ecosystem reliability. *Conservation Biology*, **12**, 39–45.

Naeem, S. (2002). Ecosystem consequences of biodiversity loss: the evolution of a paradigm. *Ecology*, **83**, 1537–1552.

Naeem, S. and Wright, J.P. (2003). Disentangling biodiversity effects on ecosystem functioning: deriving solutions to a seemingly insurmountable problem. *Ecology Letters*, **6**, 567–579.

Naeem, S., Chain, F.S., Costanza, R., Ehrlich, P., Golley, F., Hoper, D.U., Kawton, J.H., O'Neill, R.V., Mooney, H.A., Sala, O.E., Symstad, A.J. and Tilman, D. (1999). Biodiversity and ecosystem functioning: maintaining natural life support systems. *Issues in Ecology*, **4**, Fall 1999.

Naeem, S., Thompson, L.J., Lawler, S.P., Lawton, J.H. and Woodfin, R.M. (1994). Declining biodiversity can alter the performance of ecosystems. *Nature*, **368**, 734–737.

National Forest Inventory. (2003). *Australia's State of the Forests Report 2003*. Bureau of Rural Sciences, Canberra. (http://www.affa.gov.au/nfi).

National Population Council, Population Issues Committee (1991). *Population Issues and Australia's Future: Environment, Economy and Society, Final Report.* Australian Government Publishing Service, Canberra.

Nattrass, N. and Vanderwoude, C. (2001). A preliminary investigation of the ecological effects of Red Imported Fire Ants (*Solenopsis invicta*) in Brisbane, Australia. *Ecological Management and Restoration*, **2**, 220–223.

Nei, M. (1972). Genetic distance between populations. *American Naturalist*, **106**, 283–292.

Nei, M. and Li Wi, H. (1979). Mathematical model for studying genetic variation in terms of restriction endonucleases. *Proceedings of the Natural Academy of Science, USA*, **76**, 5269–5273.

Nell, J.A. (1993). Farming the Sydney Rock Oyster (*Saccostrea commercialis*) in Australia. *Reviews Fisheries Science*, **1**, 97–120.

Nelson, G. (1989). Species and taxa: Systematics and evolution. pp. 60–81. In: *Speciation and its Consequences.* Otto, D. and Endler, J.A. (Eds). Sinauer, Sunderland, Massachusetts.

Nelson, G.J. and Platnick, N. (1981). *Systematics and Biogeography.* Columbia University, New York.

Nelson, J.G. (1987). National parks and protected areas, national conservation strategies and sustainable development. *Geoforum*, **18**, 291–319.

Nelson, J.G. (1991). Beyond parks and protected areas: from public and private stewardship to landscape planning and management. *Environments*, **21**, 23–34.

Nelson, J.S. (1984). *Fishes of the World.* Wiley, New York.

Nelson, M.E. (1993). Natal dispersal and gene flow in white-tailed deer in northeastern Minnesota. *Journal of Mammalogy*, **74**, 316–22.

Neubert, M.G. (2003). Marine reserves and optimal harvesting. *Ecology Letters*, **6**, 833–849.

Neumann, F.G. (1979). Insect pest management in Australian Radiata Pine plantations. *Australian Forestry*, **42**, 30–38.

New, T.R. (1991). *Butterfly Conservation.* Oxford University Press, Oxford.

New, T.R. (1995). Onychophora in invertebrate conservation: priorities, practice and prospects. In: Onychophora: past and present. Walker, M.H. and Norman, D.B. (Eds). *Zoological Journal of the Linnean Society,* **114(1)**, 77–89.

New, T.R. (2000). *Conservation Biology. An Introduction for Southern Australia.* Oxford University Press, Melbourne.

Newman, D. and Pilson, D. (1997). Increased probability of extinction due to decreased effective population size: experimental populations of *Clarkia pulchella*. *Evolution*, **51**, 354–362.

Newman, D.M.R. (1983). One-humped camel. pp. 497–498. In: *Complete Book of Australian Mammals.* Strahan, R. (Ed). Angus and Robertson, Sydney.

Newman, J.A., Gibson, D.J., Parsons, A.J. and Thornley, J.H. (2003). How predictable are aphid responses to elevated CO_2? *Journal of Animal Ecology*, **72**, 556–566.

Newmark, W.D. (1985). Legal and biotic boundaries of western North American national parks: a problem of congruence. *Biological Conservation,* **33**, 197–208.

Newmark, W.D. (1987). A land-bridge island perspective on mammalian extinctions in western North American parks. *Nature*, **325**, 430–432.

Newsome, A. (1966). The influence of food on breeding in the Red Kangaroo in central Australia. *CSIRO Wildlife Research*, **11**, 187–196.

Newsome, A. and Noble, I.R. (1986). Ecological and physiological characters of invading species. pp. 1–20. In: *Ecology of Biological Invasions.* Groves, R.H. and Burdon, J.J. (Eds). Australian Academy of Science, Canberra.

Newton, G. (1999). Australia's fisheries – a finite resource. *Australian Marine Science Association Bulletin,* **146**, 20–22.

Newton, I. (1998). *Population Limitation in Birds.* Academic Press, London.

Nicholls, A.O. (1989). How to make biological surveys go further with Generalised Linear Models. *Biological Conservation,* **50**, 51–75.

Nicholls, A.O. (1991). Examples of the use of generalised linear models in analysis of survey data for conservation evaluation. pp. 54–63. In: *Nature Conservation: Cost Effective Biological Surveys and Data Analysis.* Margules, C.R. and Austin, M.P. (Eds). CSIRO Australia, Melbourne.

Nicholls, A.O. (1994). Variation in mosaic diversity in the forests of coastal northern New South Wales. *Pacific Conservation Biology,* **1**, 177–182.

Nicholls, A.O. and Margules, C.R. (1991). The design of studies to demonstrate the biological importance of corridors. pp. 49–61. In: *Nature Conservation 2: The Role of Corridors.* Saunders, D.A. and Hobbs, R.J. (Eds). Surrey Beatty, Chipping Norton.

Nicholls, A.O. and Margules, C.R. (1993). An upgraded reserve selection algorithm. *Biological Conservation,* **64**, 165–169.

Niemelä, J., Langor, D. and Spence, J.R. (1993). Effects of clear-cut harvesting on boreal ground-beetle assemblages (Coleoptera: Carabidae) in western Canada. *Conservation Biology,* **7**, 551–561.

Niemi, G.J., Hanowski, J.M., Lima, A.R., Nicholls, T. and Weiland, N. (1997). A critical analysis on the use of indicator species in management. *Journal of Wildlife Management*, **61**, 1240–1252.

Niemien, M. and Hanski, I. (1998). Metapopulations of moths on islands: a test of two contrasting models. *Journal of Animal Ecology,* **67**, 149–160.

Niesenbaum, R.A. and Lewis, T. (2003). Ghettoization in conservation biology: how interdisciplinary is our teaching? *Conservation Biology*, **17**, 6–10.

Niewenhuis, A. (1987). The effect of fire frequency on the sclerophyll vegetation of the West Head, New South Wales. *Australian Journal of Ecology*, **12**, 373–385.

Nix, H.A. (1978). Determinants of environmental tolerance limits in plants. pp. 195–206. In: *Biology and quarternary environments.* Walker, D. and Guppy, J.C. (Eds). Australian Academy of Science, Canberra.

Nix, H.A. (1986). A biogeographic analysis of the Australian elapid snakes. pp. 4–15. In: *Atlas of Elapid Snakes.* Australian Flora and Fauna Series Number 7. Longmore R. (Ed). Australian Government Publishing Service, Canberra.

Nix, H.A. (1994). The Brigalow. pp. 198–233. In: *Australian environmental history.* Dovers, S. (Ed). Oxford University Press, Melbourne.

Nix, H.A. (1997). Management of parks and reserves for the conservation of biological diversity. pp. 11–36. In: *National Parks and Protected Areas: Selection, Delimitation and Management.* Pigram, J.J. and Sundell, R.C. (Eds). Centre for Water Policy Research, University of New England, Armidale, Australia.

Nix, H.A. and Gillison, A.N. (1985). Towards an operational framework for habitat and wildlife management. pp. 39–55. In: *Wildlife Management in the Forests and Forestry-controlled Lands in the Tropics of the Southern Hemisphere.* Kikkawa, J. (Ed). Queensland University Press, Brisbane.

Nix, H.A. and Kalma, J.D. (1972). Climate as a dominant control in the biogeography of northern Australia and New Guinea. pp. 61–91. In: *Bridge and Barrier: The Natural Climate of Torres Strait.* Walker, D. (Ed). Australian National University Press, Canberra.

Nix, H.A. and Switzer, M.A. (1991). Rainforest animals. Atlas of vertebrates endemic to Australia's wet tropics. *Kowari,* **1**, 1–112.

Nix, H.A., Faith, D.P., Hutchinson, M.F., Margules, C.R., West, J., Alison, A., Kesteven, J.L., Natera, G., Slater, W., Stein, J.L. and Walker, P. (2000). *The BIORAP toolbox. A national study of biodiversity assessment and planning for Papua New Guinea.* Consultancy Report to the World Bank. Centre for Resource and Environmental Studies, The Australian National University, Canberra.

Nix, H.A., Stein, J.A. and Stein, J.L. (1992). *Developing an environmental geographic information system for Tasmania. An application to assessing the potential for hardwood plantation forestry.* Consultant's Report to the Department of Primary Industries and Energy, March 1992.

Nixon, K.C. and Wheeler, Q.D. (1990). An amplification of the phylogenetic species concept. *Cladistics,* **6**, 211–223.

Noble, I.R. and Slatyer, R.O. (1980). The use of vital attributes to predict successional changes in plant communities subject to recurrent disturbances. *Vegetatio,* **43**, 5–21.

Noble, J.C., MacCould, N. and Griffin, G.F. (1997). The rehabilitation of landscape function in rangelands. pp. 107–120. In: *Landscape Ecology: Function and Management.* Ludwig, J.A., Tongway, D.G., Freudenberger, D., Noble, J.C. and Hodgkinson, K.C. (Eds). CSIRO Publishing, Melbourne.

Noble, W.S. (1977). *Ordeal by fire. The week a state burned up.* Hawthorn Press, Melbourne.

Norman, J., Olsen, P. and Christidis, L. (1998). Molecular genetics confirms taxonomic affinities of the endangered Norfolk Island Boobook Owl *Ninox novaeseelandiae undulata. Biological Conservation,* **86**, 33–36.

Norris, R.H., Liston, P., Davies, N., Coysh, J., Dyer, F., Linke, S., Prosser, I. and Young, W. (2001). *Snapshot of the Murray-Darling River Basin Condition.* CSIRO Land and Water Report to the Murray-Darling Basin Commission.

Norton, B.G. (1987). *Why Preserve Natural Variety?* Princeton University Press, Princeton, New Jersey.

Norton, B.G. (1988). Commodity, amenity and morality: the limits of quantification in valuing biodiversity. pp. 200–205. In: *Biodiversity.* Wilson, E.O. and Peters, F.M. (Eds). National Academy Press, Washington DC.

Norton, D.A. (1996). Monitoring biodiversity in New Zealand's terrestrial ecosystems. pp. 19–41. In: *Papers from a Seminar Series on Biodiversity.* McFadgen, B. and Simpson, P. (Compilers). Department of Conservation, Wellington, New Zealand.

Norton, D.A. (1999). Forest Reserves. pp. 525–555. In: *Managing Biodiversity in Forest Ecosystems.* Hunter III, M. (Ed). Cambridge University Press, Cambridge.

Norton, D.A., Hobbs, J.J. and Atkins, L. (1995). Fragmentation, disturbance, and plant distribution – mistletoes in woodland remnants in the Western Australian wheatbelt. *Conservation Biology,* **9**, 426–438.

Norton, T. and Possingham, H. (1992). *Position statement on the conservation values of old growth forests.* Bureau of Rural Resources, Department of Primary Industries and Energy, Canberra. Working Paper No. WP/4/92, 72 pp.

Norton, T.W. and Nix, H.A. (1991). Application of biological modelling and G.I.S. to identify regional wildlife corridors. pp. 19–26. In: *Nature Conservation 2: The Role of Corridors.* Saunders, D.A. and Hobbs, R.J. (Eds). Surrey Beatty, Chipping Norton.

Noss, R.F. (1987). Corridors in real landscapes: a reply to Simberloff and Cox. *Conservation Biology,* **1**, 159–164.

Noss, R.F. (1990). Indicators for monitoring biodiversity: A hierachical approach. *Conservation Biology,* **4**, 355–364.

Noss, R.F. (1991). Landscape connectivity: different functions at different scales. pp. 27–39. In: *Landscape Linkages and Biodiversity.* Hudson, W.E. (Ed). Island Press, Covelo.

Noss, R.F. (1999). Assessing and monitoring forest biodiversity: a suggested framework for indicators. *Forest Ecology and Management,* **115**, 135–146.

Noss, R.F. and Cooperrider, A.Y. (1994). *Saving Nature's Legacy. Protecting and Restoring Biodiversity.* Island Press, Covelo, California.

Noss, R.F. and Harris, L.D. (1986). Nodes, networks, and MUMs: preserving diversity at all scales. *Environmental Management,* **10**, 299–309.

Noss, R.F. and Murphy, D.D. (1995). Endangered species left homeless in sweet home. *Conservation Biology,* **9**, 229–231.

Novacek, M.J. and Cleland, E.E. (2001). The current biodiversity extinction event: Scenarios for mitigation and recovery. *Proceedings of the National Academy of Sciences,* **98**, 5466–5470.

Nowell, K., Wei-Lien, C. and Chia-Jai, P. (1992). *The Horns of Dilemma: The Market for Rhino Horn in Taiwan.* Traffic International, Cambridge.

NSW Fisheries (2003). *Threatened Species.* NSW Fisheries.

NSW Forestry Commission (undated). *Notes on the Silviculture of Major NSW Forest Types.* Forestry Commission, Sydney.

NSW National Parks and Wildlife Service (2002). *White Box – Yellow Box – Blakely's Red Gum (Box-Gum) Woodland Fact-sheet.* New South Wales Parks and Wildlife Service, Sydney.

NSW National Parks and Wildlife Service (2003a). *Threatened Species Conservation Act (1995).* Final determinations of the Scientific Committee.

NSW National Parks and Wildlife Service (2003b). *Gundabooka National Park: Draft Plan of Management.* NSW National Parks and Wildlife Service, Bourke, New South Wales.

Oakleaf, B., Luce, B., Thorne, E.T. and Williams, B. (1993). Black-footed ferret reintroduction in Shirley Basin. *1992 Completion Report.* Wyoming Game and Fish Department, Cheyene, Wyoming.

Obendorf, D.L. and Munday, B.L. (1983). Toxoplasmosis in wild Tasmanian wallabies. *Australian Veterinary Journal,* **60**, 62.

O'Brien, P. (1990). Managing Australian wildlife. *Search,* **21**, 24–27.

O'Brien, S.J. and Mayr, E. (1991). Bureaucratic mischief: recognising endangered species and subspecies. *Science,* **251**, 1187–1188.

OCS (Office of the Chief Scientist) (1992). *Scientific aspects of major environmental issues: biodiversity.* Department of Prime Minister and Cabinet. Australian Government Publishing Service, Canberra.

O'Dowd, D.J., Green, P.T. and Lake, P.S. (2003). Invasional "meltdown" on an oceanic island. *Ecology Letters,* **6**, 812–818.

O'Dwyer, T.W., Buttemer, W.A. and Priddel, D.M. (2000). Inadvertent translocation of amphibians in the shipment of agricultural produce into New South Wales: its extent and conservation implications. *Pacific Conservation Biology,* **6**, 40–45.

Offord, C.A., Porter, C.L., Meagher, P.F. and Errington, G. (1999). Sexual reproduction and early plant growth of Wollemi Pine (*Wollemia nobilis*), a rare and threatened conifer. *Annals of Botany,* **84**, 1–9.

Ogden, J. (1981). Dendrochronological studies and the determination of tree ages in the Australian tropics. *Journal of Biogeography,* **8**, 405–420.

Ogle, C.C. (1987). The incidence and conservation of animal and plant species in remnants of native vegetation in New Zealand. pp. 79–87. In: *Nature Conservation: The Role of Remnants of Vegetation.* Saunders, D.A., Arnold, G.W., Burbidge, A.A. and Hopkins, A.J. (Eds). Surrey Beatty, Chipping Norton.

Ogutu, J.O. and Owen-Smith, N. (2003). ENSO, rainfall and temperature influences on extreme population declines among African savanna ungulates. *Ecology Letters,* **6**, 412–419.

Økland, B. (1996). Unlogged forests: important sites for preserving the diversity of mycetophilids (Diptera : Sciaroidea). *Biological Conservation,* **76**, 297–310.

Oksanen, L. (2001). Logic of experiments in ecology: Is pseudoreplication a pseudoissue? *Oikos,* **94**, 27–38.

Old, K.M., Kylie, G.A. and Ohmart, C.P. (1981). *Eucalypt dieback in forests and woodlands.* CSIRO, Melbourne.

Oldroyd, B.P., Lawler, S.H. and Crozier, R.H. (1994). Do feral Honey Bees (*Apis mellifera*) and Regent Parrots (*Polytelis anthopeplus*) compete for nest sites? *Australian Journal of Ecology,* **19**, 444–450.

Oliver, C.D. and Larson, B.C. (1996). *Forest Stand Dynamics.* John Wiley and Sons, New York.

Oliver, I., Beattie, A.J. and York, A. (1998). Spatial fidelity of plant, vertebrate, and invertebrate assemblages in multiple-use forest in eastern Australia. *Conservation Biology,* **12**, 822–835.

Ollier, C. (1998). Applying the precautionary principle. *The Australian Geologist,* **106**, 28–29.

Olney, P.J.S. and Ellis, P. (Eds) (1991). *International Zoo Yearbook.* Vol. 30, Zoological Society of London, London.

Olsen, J., Wink, M., Sauer-Gürth, H. and Trost, S. (2002). A new *Ninox* owl from Sumba, Indonesia. *Emu,* **102**, 223–231.

Olsen, P. (1995). *Australian Birds of Prey.* UNSW Press, Sydney.

Olsen, P. and Weston, M.A. (Compliers). (2005). Fire and birds: fire management and biodiversity. *Wingspan* (Supplement), **15** (in press).

Olsen, P., Fuller, P. and Marples, T.G. (1993). Pesticide-related eggshell thinning in Australian raptors. *Emu,* **93**, 1–11.

Olsen, P., Weston, M., Cunningham, R. and Silcocks, A. (2003). The state of Australia's birds 2003. *Supplement to Wingspan,* **13 (4)**, I–XXIII.

Olsen, P.D. (1996). Re-establishment of an endangered subspecies: the Norfolk Island Boobook Owl *Ninox novaeseelandiae undulata. Bird Conservation International,* **6**, 63–80.

O'Neill, R.V., Krummel, J.R., Gardener, R.H., Sugihara, G., DeAngelis, D.L., Milne, B.T., Turner, M.G., Zysmunt, B., Christensen, S.W., Dale, V.H. and Graham, R.L. (1988). Indices of landscape pattern. *Landscape Ecology,* **1**, 153–162.

Oram, R.N. and Freebairn, R.D. (1984). Genetic improvement of drought survival ability in *Phalaris aquatica* L. *Australian Journal of Experimental Agriculture and Animal Husbandry*, **24**, 403–409.

Oredsson, A. (1997). Threatened species not necessarily rare, rare species not necessarily threatened. *Environmental Conservation*, **24**, 207–209.

Orr, A.G. (1994). Inbreeding depression in Australian butterflies: some implications for conservation. *Memoirs of the Queensland Museum*, **36**, 179–184.

Orr, D.W. (2002). Four challenges of sustainability. *Conservation Biology*, **16**, 1457–1460.

Oryx (1990). Navy bans TBT. *Oryx*, **24**, 75.

Osborne, W. and Green, K. (1992). Seasonal changes in the composition, abundance and foraging behaviour of birds in the Snowy Mountains. *Emu*, **92**, 93–105.

Osborne, W. and Norman, J.A. (1991). Conservation genetics of Corroboree Frogs, *Pseudophyrne corroboree* Moore (Anura: Myobatrachidae): Population subdivision and genetic divergence. *Australian Journal of Zoology*, **39**, 285–297.

Osborne, W.S. (1989). Distribution, relative abundance and conservation status of Corroboree frogs, *Pseudophyrne corroboree* Moore (Anura: Myobatrachidae). *Australian Wildlife Research*, **16**, 537–547.

Osborne, W.S., Zentilis, R.A. and Lau, M. (1996). Geographical variation in Corroboree Frogs, *Pseudophyrne corroboree* Moore (Anura: Myobatrachidae): a reappraisal supports recognition of *P. pengilleyi* Wells and Wellington. *Australian Journal of Zoology*, **44**, 569–587.

O'Shaughnessy, P. and Jayasuriya, J. (1991). pp. 341–363. In: *Managing the Ash-type Forest for Water Production in Victoria. Forest Management in Australia*. McKinnell, F.H., Hopkins, E.R. and Fox, J.E.D. (Eds). Surrey Beatty and Sons, Chipping Norton.

Ostrowski, S., Bedin, E., Lenain, D.M. and Abuzinada, A.H. (1998). Ten years of Arabian Oryx conservation breeding in Saudi Arabia: achievements and regional perspectives. *Oryx*, **32**, 209–222.

Ough, K. (2001). Regeneration of wet forest flora a decade after clearfelling or wildfire – is there a difference? *Australian Journal of Botany*, **49**, 645–664.

Ough, K. and Ross, J. (1992). Floristics, fire and clearfelling in wet forests of the Central Highlands of Victoria. Department of Conservation and Environment, Melbourne. *Silvicultural Systems Project Technical Report No. 11.*

Ovadia, O. (2003). Ranking hotspots of varying size: a lesson from the nonlinearity of the species-area relationship. *Conservation Biology*, **17**, 1440–1441.

Paddle, R.N. (1996). Mueller's magpies and marsupial wolves: a window into "what might have been". *Victorian Naturalist*, **113**, 215–218.

Paddle, R.N. (2000). *The Last Tasmanian Tiger*. Cambridge University Press, Melbourne.

Padilla, D.K. and Williams, S.L. (2004). Beyond ballast water: aquarium and ornamental trades as sources of invasive species in aquatic ecosystems. *Frontiers in Ecology and Environment*, **2**, 131–138.

Paine, R.T., Tegner, M.J. and Johnson, E.A. (1998). Compounded perturbations yield ecological surprises. *Ecosystems*, **1**, 535–545.

Paini, D.R. (2004). Impact of the introduced honey bee (*Apis mellifera*) (Hymenoptera: Apidae) on native bees: A review. *Austral Ecology*, **29**, 399–407.

Paini, D.R. and Roberts, J.D. (2005). Commercial honey bees (*Apis mellifera*) reduce the fecundity of an Australian native bee (*Hylaeus alcyoneus*). *Biological Conservation*, **123**, 103–112.

Pallin, N. (2000). Ku-ring-gai flying fox reserve. *Ecological Management and Restoration*, **1**, 10–20.

Palmer, C. and Woinarski, J.C. (1999). Seasonal roosts and foraging movements of the Black Flying Fox (*Pteropus alecto*) in the Northern Territory: resource tracking in a landscape mosaic. *Wildlife Research*, **26**, 823–838.

Palmer, M.A., Bernhardt, E.S. and 18 others. (2005). Ecological science and sustainability for the 21st century. *Frontiers in Ecology and Environment*, **3**, 4–11.

Palomares, F., Delibes, M., Ferreras, P., Fedriani, J.M., Calazada, J. and Revilla, E. (2000). Iberian lynx in a fragmented landscape: predispersal, dispersal, and postdispersal habitats. *Conservation Biology*, **14**, 809–818.

Pamilo, P. and Crozier, R.H. (1997). Population ecology of social insect conservation. *Memoirs of the Museum of Victoria*, **56**, 411–419.

Panetta, F.D. (1993). A system of assessing proposed plant introductions for weed potential. *Plant Protection Quarterly*, **8**, 10–14.

Panetta, F.D. and Dodd, J. (1987). Bioclimatic prediction of the potential distribution of skeleton weed *Chrondrilla juncea* L. in Western Australia. *Journal of Australian Institute of Agricultural Science*, **53**, 11–16.

Pardon, L.G., Brook, B.W., Griffiths, A.D. and Braithwaite, R.W. (2003). Determinants of survival for the Northern Brown Bandicoot under a landscape fire experiment. *Journal of Animal Ecology*, **72**, 106–115.

Park, G.N. (1975). Nutrient dynamics and secondary ecosystem management. PhD thesis, Department of Forestry, Australian National University, Canberra.

Parker, I.M. (1997). Pollinator limitation of *Cytisus scoparius* (Scotch Broom), an invasive exotic shrub. *Ecology*, **78**, 1457–1470.

Parker, M. and Mac Nally, R. (2002). Habitat loss and the habitat fragmentation threshold: an experimental evaluation of impacts on richness and total abundances using grassland invertebrates. *Biological Conservation*, **105**, 217–229.

Parks Victoria (2002). *Victoria's System of Marine National Parks and Sanctuaries. Draft Management Strategy for Public Comment*. October 2002. Parks Victoria, Melbourne.

Parks, S.A. and Harcourt, A.H. (2002). Reserve size, local human density, and mammalian extinction in U.S. protected areas. *Conservation Biology*, **16**, 800–808.

Parmesan, C. (1996). Climate and species' range. *Nature* 383, 765–766.

Parris, K.M. and Hazell, D.L. (2005). Biotic effects of climate change in urban environments: The case of the Grey-headed Flying Fox (*Pteropus poliocephalus*) in Melbourne, Australia. *Biological Conservation*, **124**, 267–276.

Parry, B.B. (1997). Abiotic edge effects in wet sclerophyll forest in the central highlands of Victoria. MSc thesis, School of Botany, University of Melbourne.

Parsons, P. (1991). Biodiversity conservation under global climatic change: the insect Drosophila as a biological indicator ? *Global Ecology and Biogeography Letters*, **1**, 77–83.

Parsons, W.T. and Cuthbertson, E.G. (1992). *Noxious weeds of Australia.* Inkata Press, Melbourne.

Pärt, T. and Söderstrom, B. (1999). Conservation value of semi-natural pastures in Sweden: contrasting botanical and avian measures. *Conservation Biology*, **13**, 755–765.

Pastoral Review (1945). A drought lesson: settlement of light rainfall country. *The Pastoral Review and Graziers Recod*, **March 16**.

Patkeau, D. (1999). Using genetics to identify intraspecific conservation units: a critique of current methods. *Conservation Biology*, **13**, 1507–1509.

Patkeau, D., Calvert, W., Stirling, I. and Strobeck, C. (1995). Microsatellite analysis of population structure in Canadian polar bears. *Molecular Ecology*, **4**, 347–354.

Paton, D. (1991). Loss of wildlife to domestic cats. pp. 64–69. In: *The Impacts of Cats on Native Wildlife.* Potter, C. (Ed). The Australian National Parks and Wildlife Service.

Paton, D.C. (1985). Impact of introduced honeybees on the abundance and behaviour of honeyeaters and their potential impact on seed production by *Callistemon macropunctatus. Australian Pollination Ecology Society Newsletter*, **10**.

Paton, D.C. (1993). Honeybees in the Australian environment. *BioScience*, **43**, 95–103.

Paton, D.C. (1997). Honeybees *Apis mellifera* and the disruption of plant-pollinator systems in Australia. *Victorian Naturalist*, **114**, 23–29.

Paton, D.C. (1999). *Impact of Commercial Honeybees on Flora and Fauna.* Report 99/15 to Rural Industries Research and Development Corporation, Canberra.

Paton, D.C. (2000). Disruption of plant-pollination systems in southern Australia. *Conservation Biology*, **14**, 1232–1234.

Paton, D.C., Prescott, A.M., Davies, R. and Heard, L.M. (2000). The distribution, status and threats to temperate woodlands in South Australia. pp. 57–85. In: *Temperate Eucalypt Woodlands in Australia: Biology, Conservation, Management and Restoration.* Hobbs, R.J. and Yates, C.J. (Eds). Surrey Beatty and Sons, Chipping Norton.

Paton, P.W. (1994). The effect of edge on avian nest success: how strong is the evidence? *Conservation Biology*, **8**, 17–26.

Patterson, B.D. (1987). The principle of nested subsets and its implications for biological conservation. *Conservation Biology*, **1**, 247–293.

Patterson, B.D. and Atmar, W. (1986). Nested subsets and the structure of insular mammalian faunas and archipelagos. *Biological Journal of the Linnean Society*, **28**, 65–82.

Patton, D.R. (1974). Patting increases deer and edges of a Pine Forest in Arizona. *Journal of Forestry*, **December 1974**, 764–766.

Pauly, D. and MacLean, J. (2003). *In a Perfect Ocean: The State of Fisheries and Ecosystems in the North Atlantic.* Island Press, Washington, DC.

Pavey, C. (1995). Food of the Powerful Owl, *Ninox strenua*, in suburban Brisbane, Queensland. *Emu*, **95**, 231–232.

Pavlik, B.M., Nickrent, D.L. and Howlad, A. (1993). The recovery of an endangered plant. Creating a new population of *Amsinckia grandiflora. Conservation Biology*, **7**, 510–526.

Peakall, R. and Sydes, M.A. (1996). Defining priorities for achieving practical outcomes from the genetic studies of rare plants. pp. 119–129. In: *Back from the Brink: Refining the Threatened Species Recovery Process.* Stephens, S. and Maxwell, S. (Eds). Surrey Beatty, Chipping Norton.

Peakall, R., Ebert, D., Scott, L., Meagher, P. and Offord, C. (2003). Comparative genetic study confirms exceptionally low genetic variation in the ancient and endangered relictual conifer, *Wollemia nobilis* (Araucarieaceae). *Molecular Ecology*, **12**, 2331–2343.

Peakall, R., Jones, L., Bower, C. and Mackey, B. (2002). Bioclimatic assessment of the geographic and climatic limits to hybridisation in a sexually deceptive orchid system. *Australian Journal of Botany*, **50**, 21–30.

Peakall, R., Ruibal, M. and Lindenmayer, D.B. (2003). Spatial autocorrelation analysis offers new insights into gene flow in the Australian Bush Rat, (*Rattus fuscipes*) biology. *Evolution*, **57**, 1182–1195.

Peakall, R., Smouse, P.E. and Hoff, D.R. (1994). Evolutionary implications of allozyme and RAPD variation in diploid populations of dioecious Buffalo Grass *Buchloe dactyloides. Molecuar Ecology,* **4**, 135–147.

Pearce, J. and Ferrier, S. (2000). Evaluating the predictive performance of habitat models developed using logistic regression. *Ecological Modelling*, **133**, 225–245.

Pearce, J. and Lindenmayer, D.B. (1998). The use of BIOCLIM to enhance reintroduction biology – a case study of the endangered Helmeted Honeyeater in south-eastern Australia. *Restoration Ecology*, **6**, 238–243.

Pearce, J. and Minchin, P. (2001). Vegetation of the Yellingbo Nature Conservation Reserve and its relationship to the distribution of the Helmeted Honeyeater, Bell Miner and White-eared Honeyeater. *Wildlife Research*, **28**, 41–52.

Pearce, J. and Venier, L. (2005). Small mammals as biodindicators of sustainable forest management. *Forest Ecology and Management*, **208**, 153–175.

Pearce, J.L., Burgman, M.A. and Franklin, D.C. (1994). Habitat selection by Helmeted Honeyeaters. *Wildlife Research,* **21**, 53–63.

Pearman, G.I. (Ed) (1988). *Greenhouse: Planning for Climate Change.* CSIRO, Melbourne.

Pearson, H. (2003). *Lost forest fuels malaria.* News@Nature [online] URL: http:// www. nature.com/nsu/031124/031124–12.html

Pearson, R.G. and Dawson, T. (2003). Predicting the impacts of climate change on the distribution of species: are bioclimate envelope models useful? *Global Ecology and Biogeography*, **12**, 361–371.

Pearson, S.M., Turner, M.G., Gardner, R.H. and O'Neill, R.V. (1996). An organism perspective of habitat fragmentation. pp. 77–95. In: *Biodiversity in Managed Landscapes: Theory and Practice.* Szaro, R.C. and Johnston, D.W. (Eds). Oxford University Press, New York.

Pech, R., Hood, G.M., McIlroy, J. and Saunders, G. (1997). Can foxes be controlled by reducing their fertility? *Reproduction, Fertility and Development*, **9**, 41–51.

Pechmann, J.H., Scott, D.I., Semlitch, R.D., Caldwell, J.P., Vitt, L.J. and Gibbons, J.W. (1991). Declining amphibian populations: the problem of separating human impacts from natural fluctuations. *Science,* **253**, 892–895.

Peck, A.J. (1993). Salinity. pp. 234–270. In: *Land degradation processes in Australia.* McTainsh, G. and Boughton, W.C. (Eds). Longman, Cheshire.

Peek, M.S., Leffler, A.J., Flint, S.D. and Ryel, R.J. (2003). How much variance can be explained by ecologists and evolutionary biologists? Additional perspectives. *Oecologia*, **137**, 161–170.

Pender, P.J., Willing, R.S. and Cann, B. (1992). NPF by-catch a valuable resource? *Australian Fisheries,* **51**, 30–31.

Perry, D.A. (1994). *Forest Ecosystems.* Johns Hopkins Press, Baltimore.

Petchey, O.L., McPhearson, P.T., Casey, T.M. and Morin, P.J. (1999). Environmental warming alters food-web structure and ecosystem function. *Nature*, **402**, 69–72.

Peterman, R.M. and M'Gonigle, M. (1992). Statistical power analysis and the precautionary principle. *Marine Pollution Bulletin,* **24(5)**, 231–234.

Peters, R.L. and Darling, J.D. (1985). The greenhouse effect and nature reserves. *BioScience,* **35**, 707–717.

Peters, R.L. and Lovejoy, T.E. (Eds) (1992). *Global warming and biological diversity.* Yale University Press, New Haven.

Pettigrew, J.D. (1993). A burst of feral cats in the Diamantina – a lesson for the management of pest species? pp. 25–32. In: *Proceedings of the Cat Management Workshop.* Siepen, G. and Owens, C. (Eds). Queensland Department of Environment and Heritage, Brisbane.

Pfefferkorn, H.W. (2004). The complexity of mass extinction. *Proceedings of the National Academy of Sciences,* **101**, 12779–12780.

Pharo, E.J., Beattie, A.J. and Binns, D. (1999). Vascular plant diversity as a surrogate for bryophyte and lichen diversity. *Conservation Biology*, **13**, 282–292.

Pharo, E.J., Beattie, A.J. and Pressey, R.L. (2000). Effectiveness of using vascular plants to select reserves for bryophytes and lichens. *Biological Conservation*, **96**, 371–378.

Pheloung, P.C., Williams, P.A. and Halloy, S.R. (1999). A weed risk assessment model for use as a biosecurity tool evaluating plant introductions. *Journal of Environmental Management*, **57**, 239–251.

Philip, M.S. (1994). *Measuring Trees and Forests.* CAB International, Wallingford, UK.

Phillips, B.L. and Shine, R. (2004). Adapting to an invasive species: toxic cane toads induce morphological change in Australian snakes. *Proceedings of the National Academy of Sciences*, **101**, 17150–17155.

Phillips, B.L., Brown, G.P. and Shine, R. (2003). Assessing the potential impact of Cane Toads on Australian snakes. *Conservation Biology,* **17**, 1738–1747.

Philpott, S. and Dietsch, T. (2003). Coffee and conservation: A global context and the value of farmer involvement. *Conservation Biology,* **117**, 1844–1846.

Pickett, S.T. and Thompson, J.H. (1978). Patch dynamics and the design of nature reserves. *Biological Conservation,* **13**, 27–37.

Pietsch, R.S. (1995). The fate of Common Brushtail Possums translocated to sclerophyll forest. pp. 239–246. In: *Reintroduction Biology of Australasian Fauna.* Serena, M. (Ed). Surrey Beatty, Chipping Norton, Australia.

Pigram, J.J. (1986). *Issues in the Management of Australia's Water Resources.* Longman Cheshire, Melbourne.

Pimentel, D., Lach, L, Zuniga, R. and Morrison, D. (2000). Environmental and economic costs of nonindigenous species introductions in the United States. *BioScience,* **50**, 53–65.

Pimentel, D., Stachow, U., Takacs, D.A., Brubaker, H.W., Dumas, A.R., Meaney, J.J., O'Neill, J.A., Onsi, D.E. and Corzilius, D.B. (1992). Conserving bio-logical diversity in agricultural/forestry systems. *BioScience,* **42**, 354–362.

Pimentel, D., Wilson, C., McCullum, C., Huang, R., Dwen, P., Flack, J., Tran, Q., Saltman, T. and Cliff, B. (1997). Economic and environmental benefits of biodiversity. *BioScience,* **47**, 747–757.

Pimm, S.L. (1993). Life on an intermittent edge. *Trends in Evolution and Ecology,* **8**, 45–46.

Pimm, S.L. (1995). Dead reckoning – getting hard numbers on extinction rates in paradise. *Sciences-New York,* **35**, 15–17.

Pimm, S.L. (2001). *The World According to Pimm.* MacGraw Hill, New York.

Pimm, S.L., Jones, H.L. and Diamond, J. (1988). On the risk of extinction. *American Naturalist,* **132**, 757–785.

Pimm, S.L., Russell, G.J., Gittleman, J.L. and Brooks, T.M. (1995). The future of biodiversity. *Science,* **269**, 347–350.

Pinnegar, J.K., Polunin, N.V., Francour, P., Badalamenti, F., Chemello, R., Harmelin-Vivien, M., Hereu, B., Milazzo, M., Zabala, M., D'Anna, G. and Pipitone, C. (2000). Trophic cascades in benthic marine ecosystems: lessons from fisheries and protected area management. *Environmental Conservation,* **27**, 179–200.

Piper, S., Catterall, C.P. and Olsen, M. (2002). Does adjacent land use affect predation of artificial shrub-nests near eucalypt forest edges? *Wildlife Research,* **29**, 127–133.

Piper, S.D. and Catterall, C.P. (2003). A particular case and a general pattern: hyperaggressive behaviour by one species mediates avifaunal decreases in fragmented Australian forests. *Oikos,* **101**, 602–614.

Pither, J. and Taylor, P.D. (1998). An experimental assessment of landscape connectivity. *Oikos,* **83**, 166–174.

Pittock, A.B. and Nix, H.A. (1986). The effect of changing climate on Australian biomass production. *Climate Change,* **8**, 243–255.

Pizzey, G. and Knight, F. (1997). *Field Guide to the Birds of Australia.* Angus and Robertson, Sydney.

Platt, S. (1995). *Dieback Lessons: Learning How to Manage Sustainably.* Department of Conservation and Natural Resources, Melbourne. Land for Wildlife Note No. 34. August 1995.

Plenet, S. (1995). Freshwater amphipods as biomonitors of metal pollution in surface and interstitial systems. *Freshwater Biology,* **33**, 127–137.

Plous, S. (1993). *The Psychology of Judgment and Decision Making.* McGraw-Hill, New York.

Podger, F.D. and Brown, M.J. (1989). Vegetation damage caused by *Phytophthora cinnamomi* on disturbed sites in temperate rainforest in western Tasmania. *Australian Journal of Botany,* **37**, 443–480.

Podger, F.D., Mummery, D.C., Palzer, C.R. and Brown, M.J. (1990). Bioclimatic analysis of the distribution of damage to native plants in Tasmania by *Phytophthora cinnamomi. Australian Journal of Ecology,* **15**, 281–289.

Pogonoski, J. (2002). Green Sawfish. *Nature Australia,* **Spring 2002**, 26–27.

Pogonoski, J., Pollard, D.A. and Paxton, J.R. (2002). *Conservation overview and action plan for Australian threatened and potentially threatened marine and estuarine fishes.* Report to Environment Australia. February, 2002. [online] URL: http://www.ea.gov.au/biodiversity/threatened/action/index.html

Poldmaa, T., Montgomerie, R. and Boag, P. (1995). Mating system of the cooperatively breeding Noisy Minor *Manorina melanocephala*, as revealed by DNA profiling. *Behavioural Ecology and Sociobiology,* **37**, 137–143.

Pope, M.L., Lindenmayer, D.B. and Cunningham, R.B. (2004). Patch use by the greater glider (*Petauroides volans*) in a fragmented forest system. 1. Home range size and movements. *Wildlife Research,* **31**, 559–568.

Pope, S.E., Fahrig, L. and Merriam, H.G. (2000). Landscape complementation and metapopulation effects on Leopard Frog populations. *Ecology,* **81**, 2498–2508.

Pople, A. and Grigg, G. (undated). *Overview of background information for kangaroo harvesting.* Report to Environment Australia.

Pople, T. and Grigg, G. (1994). Commercial use of wildlife for conservation. pp. 363–366. In: *Conservation Biology in Australia and Oceania.* Moritz, C. and Kikkawa, J. (Eds). Surrey Beatty and Sons, Sydney.

Population Action International (2000). *Nature's Place. Human Population and the Future of Biological Diversity.* Population Action International, Washington, DC.

Possingham, H., Nadolny, C., Catterall, C. and Trail, B. (1995). Position statement on vegetation clearance. *Bulletin of the Ecological Society of Australia,* **25**, 3–5.

Possingham, H.P. (1993). The impact of elevated CO_2 on biodiversity: a mechanistic population-dynamic perspective. *Australian Journal of Botany,* **41**, 11–21.

Possingham, H.P. (2001). The business of biodiversity: applying decision theory principles to nature conservation. *Tela Series* **No. 9**. The Australian Conservation Foundation, Melbourne.

Possingham, H.P. and Davies, I. (1995). ALEX: A model for the viability analysis of spatially structured populations. *Biological Conservation,* **73**, 143–150.

Possingham, H.P. and Gepp, B. (1993). The application of population viability analysis to assess management options for the Southern Brown Bandicoot (*Isoodon obesulus*) in the south-east of south Australia. pp. 633–638. In: *Proceedings of the International Congress on Modelling and Simulation.* December 1993. Jakeman, A.J. and McAleer, M. (Eds). UniPrint, Perth, Western Australia.

Possingham, H.P. and Noble, I.R. (1991). *An Evaluation of Population Viability Analysis for Assessing the Risk of Extinction.* Research consultancy for the Resource Assessment Commission, Forest and Timber Inquiry. Canberra, Australia.

Possingham, H.P., Andelman, S.J., Burgman, M.A., Medellin, R.A., Master, L.L. and Keith, D.A. (2002). Limits to the use of threatened species lists. *Trends in Ecology and Evolution,* **17**, 503–507.

Possingham, H.P. Lindenmayer, D.B. and Norton, T.W. (1993). A framework for improved threatened species management using Population Viability Analysis. *Pacific Conservation Biology,* **1**, 39–45.

Possingham, H.P., Lindenmayer, D.B. and McCarthy, M.A. (2001). Population Viability Analysis. *Encyclopedia of Biodiversity,* Volume 4, 831–843.

Possingham, H.P., Lindenmayer, D.B., Norton, T.W. and Davies, I. (1994). Metapopulation viability of the Greater Glider in a wood production forest. *Biological Conservation,* **70**, 265–276.

Potts, B.M. and Wiltshire, R.J.E. (1997). Eucalypt Genetics and Genecology. pp. 56–91. In: *Eucalypt Ecology: Individuals to Ecosystems.* Williams J.E. and Woinarski J.C.Z. (Eds). Cambridge University Press, Cambridge UK.

Potts, B.M., Barbour, R.C. and Hingston, A.B. (2001). *The risk of genetic pollution from farm forestry using eucalypt species and hybrids.* A report for RIRDC/Land and Water Australia/FWPRDC Joint Venture Agroforestry Program. February 2001.

Potts, B.M., Barbour, R.C., Hingston, A.B. and Vaillancourt, R.E. (2003). Genetic pollution of native eucalypt gene pools. *Australian Journal of Botany,* **51**, 1–25.

Potts, W. (2000). Recovery plan for threatened lowland Tasmanian lowland *Euphrasia* species. Department of Primary Industries, Water and Environment, Hobart.

Pouliquen-Young, O. (1997). Evolution of the system of protected areas in Western Australia. *Environmental Conservation,* **24**, 168–181.

Pounds, J.A. (2001). Climate and amphibian declines. *Nature,* **410**, 639–640.

Pounds, J.A., Fogden, M.L. and Campbell, J.H. (1999). Biological response to climate change on a tropical mountain. *Nature,* **398**, 611–615.

Powell, A.H. and Powell, G.V. (1987). Population dynamics of male euglossine bees in Amazonian forest fragments. *Biotropica,* **19**, 176–179.

Powell, G.V.N. and Bjork, R. (1995). Implications of intratropical migration on reserve design – a case-study using *Pharomachrus mocinno. Conservation Biology,* **9**, 354–362.

Powell, J.M. (1993). *The Emergence of Bio-regionalism in the Murray-Darling Basin.* Murray-Darling Basin Commission, Canberra.

Pracy, L.T. (1962). *Introduction and Liberation of the Opossum (*Trichosurus vulpecula*) into New Zealand.* New Zealand Forest Service. Information Series No. 45.

Prendergast, H.D. and Hattersley, P.W. (1985). Distribution and cytology of Australian Neurachne and its allies (Poaceae), a group containing C3, C4 and C3-C4-intermediate species. *Australian Journal of Botany,* **33**, 317–336.

Prendergast, J.R. and Eversham, B.C. (1997). Species richness covariance in higher taxa: empiricial tests of the biodiversity indicator concept. *Ecography,* **20**, 210–216.

Prendergast, J.R., Quinn, R.M. and Lawton, J.H. (1999). The gaps between theory and practice in selecting nature reserves. *Conservation Biology,* **13**, 484–492.

Prendergast, J.R., Quinn, R.M., Lawton, J.H., Eversham, B.C. and Gibbons, D.W. (1993). Rare species, the coincidence of diversity hotspots and conservation strategies. *Nature,* **365**, 335–337.

Pressey, R. and Cowling, R.M. (2001). Reserve selection algorithms and the real world. *Conservation Biology,* **15**, 275–277.

Pressey, R.L. (1994a). Ad hoc reservations: forward or backward steps in developing representative reserve systems. *Conservation Biology,* **8**, 662–668.

Pressey, R.L. (1994b). Land classifications are necessary for conservation planning but what do they tell us about fauna. pp. 31–41. In: *Future of the Fauna of Western New South Wales.* Lunney, D., Hand, S., Redd, P. and Butcher, D. (Eds). Royal Society of NSW, Mosman, Sydney.

Pressey, R.L. (1995). Conservation reserves in NSW. Crown jewels or leftovers? *Search,* **26(2)**, 47–51.

Pressey, R.L. (1997). Algorithms, politics and timber: an example of the role of science in a public, political negiotation process over new conservation areas in production forests. pp. 73–87. In: *Ecology for Everyone: Communicating Ecology to Scientists, the Public and the Politicans.* Wills, R.T., Hobbs, R.J. and Fox, M.D. (Eds). Surrey Beatty and Sons, Sydney.

Pressey, R.L. and Logan, V.S. (1994). Level of geographical subdivision and its effects on assessments of reserve coverage: A review of regional studies. *Conservation Biology,* **8**, 1037–1046.

Pressey, R.L. and Nicholls, A.O. (1989). Application of a numerical algorithm to the selection of reserves in semi-arid New South Wales. *Biological Conservation,* **50**, 263–278.

Pressey, R.L. and Tully, S.L. (1994). The cost of adhoc reservation: A case study in Western New South Wales. *Australian Journal of Ecology,* **19**, 375–384.

Pressey, R.L., Ferrier, S., Hager, T.C., Woods, C.A., Tully, S.L. and Weinman, K.M. (1996). How well protected are the forests of north-eastern New South Wales? – Analyses of forest environments in relation to formal protection measures, land tenure, and vulnerability to clearing. *Forest Ecology and Management,* **85**, 311–333.

Pressey, R.L., Hager, T.C., Ryan, K.M., Wall, S., Ferrier, S. and Creaser, P.M. (2000). Using abiotic data for conservation assessments over extensive regions: Quantitative methods applied across New South Wales, Australia. *Biological Conservation,* **96**, 55–82.

Pressey, R.L., Humphries, C.J., Margules, C.R., Vane-Wright, R.I. and Williams, P.H. (1993). Beyond opportunism: key principles for systematic reserve selection. *Trends in Evolution and Ecology,* **8**, 124–128.

Pressey, R.L., Johnson, I.R. and Wilson, P.D. (1994). Shades of irreplaceability: towards a measure of the contribution of sites to a reservation goal. *Biodiversity and Conservation,* **3**, 242–262.

Pressey, R.L., Possingham, H.P. and Day, J.R. (1997). Effectiveness of alternative heuristic algorithms for identifying minimum requirements for conservation reserves. *Biological Conservation,* **80**, 207–219.

Preston, F.W. (1962). The canonical distribution of commonness and rarity. *Ecology,* **43**, 185–215, 410–432.

Price, O., Woinarski, J.C., Liddle, D.L. and Russell-Smith, J. (1995). Patterns of species composition and reserve design for a fragmented estate: monsoon rainforests in the Northern Territory. *Biological Conservation,* **74**, 9–19.

Price, O.F., Woinarski, J.C.Z. and Robinson, D. (1999). Very large requirements for frugivorous birds in monsoon rainforests of the Northern Territory, Australia. *Biological Conservation,* **91**, 169–180.

Priddel, D. and Wheeler, R. (1994). Mortality of captive-raised Malleefowl, *Leipoa ocellata,* released into a Mallee remnant within the wheat-belt of New South Wales. *Wildlife Research,* **21**, 543–552.

Priddel, D. and Wheeler, R. (2004). An experimental translocation of Brush-tailed Bettongs (*Bettongia penicillata*) to western New South Wales. *Wildlife Research,* **31**, 421–432.

Primack, R. (2001). Causes of extinction. pp. 697–713. In: *Encyclopedia of Biodiversity.* Volume 2. Levin, S.A. (Ed). Academic Press, San Diego.

Primack, R. (2002). *The Principles of Conservation Biology.* Sinauer, Sunderland, Massachusetts.

Prober, S. and Thiele, K.R. (1995). Conservation of the Grassy White Box Woodlands: Relative contributions of size and disturbance to floristic composition and diversity of remnants. *Australian Journal of Botany,* **43**, 349–366.

Prober, S., Tompkins, C., Moran, G.F. and Bell, J.C. (1990). The conservation genetics of *Eucalyptus paliformis* L. Johnson et Blaxell and *E. parviflora* Cambage, two rare species from south-eastern Australia. *Australian Journal of Botany,* **38**, 79–95.

Prober, S.M. and Austin, M.P. (1990). Habitat peculiarity as a cause of rarity in *Eucalyptus paliformis. Australian Journal of Ecology,* **16**, 189–205.

Prober, S.M. and Brown, A.H. (1994). Conservation of grassy white box woodlands: population genetics and fragmentation of *Eucalyptus albens. Conservation Biology,* **8**, 1003–1013.

Probst, J. (1991). What about human populations? *Journal of Forestry,* **89**, 5.

Pruett-Jones, S.G. and Lewis, M.J. (1990). Sex ratio and habitat limitation promote delayed dispersal in Superb Fairy-Wrens. *Nature, London,* **348**, 541–542.

Pulliam, H.R. and Haddad, N.M. (1994). Human population growth and the carrying capacity concept. *Bulletin of the Ecological Society of America*, **75**, 141–157.

Pulliam, H.R., Dunning, J.B. and Liu, J. (1992). Population dynamics in complex landscapes: a case study. *Ecological Applications*, **2**, 165–177.

Pullin, A.S. (2002). *Conservation biology*. Cambridge University Press, Cambridge.

Pullin, A.S. and Knight, T.M. (2001). Effectiveness in conservation practice: Pointers from medicine and public health. *Conservation Biology*, **15**, 50–54.

Pullin, A.S., Knight, T.M., Stone, D.A. and Charman, K. (2004). Do conservation managers use scientific evidence to support their decision-making? *Biological Conservation*, **119**, 245–252.

Punt, A.E. and Smith, A.D.M. (1999). Harvest strategy evaluation for the eastern stock of gemfish (*Rexea solandri*). *ICES Journal of Marine Science*, **56**, 860–875.

Putz, F.E. and Romero, C. (2001). Biologists and timber certification. *Conservation Biology*, **15**, 313–314.

Putz, F.E., Redford, K.H., Robinson, J.G., Fimbel, R. and Bate, G.M. (2000). *Biodiversity Conservation in the Context of Tropical Forest Management.* World Bank Environment Department Papers. Paper No. 75. Biodiversity Series – Impact studies. The World Bank, Washington D.C., U.S.A. September, 2000.

Pyke, G. (1990). Apiarists vs scientists: a bittersweet case. *Australian Natural History*, **23**, 386–392.

Pyke, G. (1995). Fauna impact statements: a review of processes and standards. *Australian Zoologist*, **30**, 93–110.

Pyke, G. and Balzer, L. (1985). *The Effects of the Introduced Honey Bee (*Apis mellifera*) on Australian Native Bees.* N.S.W. National Parks and Wildlife Service Report. Occasional Papers No. 7.

Pyke, G., Saillard, R. and Smith, R. (1995). Abundance of eastern bristlebirds in relation to habitat and fire history. *Emu*, **95**, 106–110.

Queensland Department of Natural Resources (1997). *Salinity Management Handbook.* Queensland Department of Natural Resources, Brisbane.

Quinn, G.P. and Keough, M.J. (2002). *Experimental Design and Data Analysis for Biologists.* Cambridge University Press.

Quinn, H. and Quinn, H. (1993). Estimated number of snake species that can be managed by species survival plans in North America. *Zoo Biology*, **12**, 243–255.

Quinn, J.F. and Hastings, A. (1987). Extinction in subdivided habitats. *Conservation Biology*, **1**, 198–208.

Quinn, J.F., Wolin, C.L. and Judge, M.L. (1989). An experimental analysis of patch size, habitat subdivision, and extinction in a marine intertidal snail. *Conservation Biology*, **3**, 242–251.

Rabinowitz, A. (1995). Helping a species go extinct: The Sumatran Rhino in Borneo. *Conservation Biology*, **9**, 482–488.

Rabinowitz, D., Cairns, S. and Dillon, T. (1986). Seven forms of rarity and their frequency in the flora of the British Isles. pp. 184–204. In: *Conservation Biology: the Science of Scarcity and Diversity.* Soulé, M.E. (Ed). Sinaur Associates, Sunderland.

RAC (1991). (Resource Assessment Commission) *Forest and Timber Inquiry, Draft Report.* Volume 1. Australian Government Publishing Service, Canberra.

RAC (1992a). (Resource Assessment Commission) *Forest and Timber Inquiry, Final Report.* Overview. Australian Government Publishing Service, Canberra.

RAC (1992b). (Resource Assessment Commission) *Forest and Timber Inquiry, Final Report.* Volume 1. Australian Government Publishing Service, Canberra.

RAC (1992c). (Resource Assessment Commission) *Forest and Timber Inquiry.* Volume 2. Australian Government Printing Service, Canberra.

RAC (1993). (Resource Assessment Commission) *Coastal Zone Inquiry. Final report.* Overview. Australian Government Publishing Service, Canberra.

Raff, M. (2003). *Private Property and Environmental Responsibility: A Comparative Study of German Real Property Law.* Aspen Law and Business.

Ralls, K., Ballou, J.D. and Templeton, A.R. (1988). Estimates of lethal equivalents and the cost of inbreeding in mammals. *Conservation Biology*, **2**, 185–193.

Ramsey, D. (2005). Population dynamics of brushtail possums subject to fertility control. *Journal of Applied Ecology*, **42**, 348–360.

Rand, G.M. and Newman, J.R. (1998). The applicability of habitat evaluation methodologies in ecological risk assessment. *Human and Ecological Risk Assessment*, **4**, 905–929.

Rapaport, E.H. (1993). The process of plant colonisation in small settlements and large cities. pp. 190–207. In: *Humans as Components of Ecosystems.* McDonnell, M.J. and Pickett, S.T.A. (Eds). Springer-Verlag, New York.

Raup, D.M. (1986). Biological extinction in earth history. *Science*, **231**, 1528–1533.

Raup, D.M. (1991). Extinction: bad genes or bad luck. *New Scientist*, **14 September**, 38–41.

Raven, P. (1985). Disappearing species: a global tragedy. *The Futurist*, **19**, 8–14.

Raven, P.H. (1987). The scope of the plant conservation problem world wide. pp. 19–29. In: *Botanic Gardens and the World Conservation Strategy.* Bramwell, D., Hamann, O., Heywood, V. and Synge, H. (Eds). Academic Press, London.

Raven, P.H. (1988a). Biological resources and global stability. pp. 3–27. In: *Coadaptation in Biotic Communities.* Kawano, S., Connell, J.H. and Hidaka, T. (Eds). University of Tokyo Press, Tokyo.

Raven, P.H. (1988b). Our diminishing tropical forests. pp. 119–122. In: *Biodiversity.* Wilson, E.O. and Peter, F.M. (Eds). National Academy Press, Washington DC.

Rawling, J. (1996). Managing bushland remnants in the urban environment. pp. 348–353. In: *Proceedings of the Eleventh Australian Weeds Conference.* Sheperd, R.C.H. (Ed). Melbourne Weed Science Society of Victoria, Frankston.

Rawlinson, P.A. (1988). Kangaroo conservation and kangaroo harvesting: intrinsic value versus instrumental value of wildlife. *Australian Zoologist,* **24**, 129–137.

Rawlinson, P.A. (1991). Taxonomy and distribution of the Australian Tiger Snakes (*Notechis*) and Copperheads (*Austrelaps*) (Serpentes, Elapidae). *Proceedings of the Royal Society of Victoria,* **103**, 125–135.

Ray, S., Robinson, D. and Werren, G. (1983). *Pines Versus Native Forests: An Annotated Bibliography.* Native Forests Action Council, Melbourne.

Read, D.G. and Fox, B.J. (1991). Assessing the habitat of the Parma Wallaby, *Macropus parma* (Marsupialia: Macropodidae). *Wildlife Research,* **18**, 469–478.

Read, J. and Brown, M. (1996). Ecology of Australian *Nothofagus* forests. pp. 131–181. In: *The Ecology and Biogeography of Nothofagus forests.* Veblen, T.T., Hill, R.S. and Read, J. (Eds). Yale University Press, New Haven.

Read, J.L., Reid, N. and Venables, W.N. (2000). Which birds are useful bioindicators of mining and grazing impacts in arid South Australia? *Environmental Management,* **26**, 215–232.

Reader, R. (1988). Using the guild concept in the assessment of tree harvesting effect on understorey herbs: a cautionary note. *Environmental Management,* **12**, 803–808.

Reader's Digest (Ed) (1990). *Reader's Digest Complete Book of Australian Birds.* Reader's Digest, Sydney.

Reading, R.P., Clark, T.W., Seebeck, J.H. and Pearce, J. (1996). Habitat suitability index model for the Eastern Barred Bandicoot, *Parameles gunnii. Wildlife Research,* **23**, 221–235.

Recher, H.F. (1996). Conservation and management of eucalypt forest vertebrates. pp. 339–388. In: *Conservation of Faunal Diversity in Forested Landscapes.* DeGraff, R. and Miller, I. (Eds). Chapman and Hall, London.

Recher, H.F. (2002). The past, future and present of biodiversity conservation in Australia. *Pacific Conservation Biology,* **8**, 8–11.

Recher, H.F. and Lim, L. (1990) A review of current ideas of the extinction, conservation and management of Australia's terrestrial vertebrate fauna. *Proc. Ecol. Soc. Aust.,* **16**, 287–301.

Recher, H.F., Allen, D. and Gowing, G. (1985). The impact of wildfire on birds in an intensively logged forest. pp. 283–290. In: *Birds of Eucalypt Forests and Woodlands: Ecology, Conservation and Management.* Surrey Beatty and Sons, Chipping Norton.

Recher, H.F., Lunney, D. and Posamentier, H. (1975). A grand natural experiment – the Nadgee wildfire. *Australian Natural History,* **18**, 150–163.

Recher, H.F., Rohan-Jones, W. and Smith, P. (1980). *Effects of the Eden Woodchip Industry on Terrestrial Vertebrates with Recommendations for Management.* Forest Commission of New South Wales. Research Note No. 42 pp. 1–85.

Recher, H.F., Shields, J., Kavanagh, R. and Webb, G. (1987). Retaining remnant mature forest for nature conservation at New South Wales. pp. 177–194. In: *Nature Conservation: The Role of Remnants of Native Vegetation.* Saunders, D.A., Arnold, G.W., Burbridge, A.A. and Hopkins, A.J. (Eds). Surrey Beatty and Sons, Chipping Norton.

Redford, K.H. (1992). The empty forest. *BioScience,* **42**, 412–422.

Redford, K. and da Fonseca, G. (1986). The role of gallery forests in the zoogeography of the Cerrado's non-volant mammalian fauna. *Biotropica,* **18**, 126–135.

Redford, K. and Taber, A. (2000). Writing the wrongs: developing a safe-fail culture in conservation. *Conservation Biology,* **14**, 1567–1568.

Redford, K., Coppollilo, P., Sanderson, E.W., da Fonseca, G., Dinerstein, E., Groves, C., Mace, G., Maginnis, S., Mittermeier, R., Noss, R., Olson, D., Robinson, J., Vedder, A. and Wright, M. (2002). Mapping the conservation landscape. *Conservation Biology,* **17**, 116–131.

Reed, D.H. and Bryant, E.H. (2000). Experimental tests of minimum viable population size. *Animal Conservation,* **3**, 7–14.

Reed, D.H. and Frankham, R. (2003). Correlation between fitness and genetic diversity. *Conservation Biology,* **17**, 230–237.

Reed, J.M. and Blaustein, A.R. (1995). Assessment of 'nondeclining' amphibian populations using power analysis. *Conservation Biology,* **9**, 1299–1300.

Reed, J.M., Mills, L.S., Dunning, J.B., Menges, E.S., McKelvey, K.S., Frye, R., Beissenger, S.R., Antett, M.

and Miller, P. (2002). Emerging issues in Populatin Viability Analysis. *Conservation Biology*, **16**, 7–19.

Regan, H.M., Akçakaya, H.R., Ferson, S., Root, K.V., Carroll, S. and Ginzburg, L.R. (2003). Treatments of uncertainty and variability in ecological risk assessment of single-species populations. *Human and Ecological Risk Assessment*, **9**, 889–906.

Regan, H.M., Colyvan, M. and Burgman, M.A. (2002). A taxonomy and treatment of uncertainty for ecology and conservation biology. *Ecological Applications*, **12**, 618–628.

Reid, J. and Fleming, M. (1992). The conservation status of birds in arid Australia. *Rangelands Journal*, **14**, 65–91.

Reid, M.A., Tibby, J.C., Penny, D. and Gell, P.A. (1995). The use of diatoms to assess past and present water quality. *Australian Journal of Ecology*, **20**, 57–64.

Reid, N. and Landsberg, J. (2000). Tree decline in agricultural landscapes. pp. 127–166. In: *Temperate Eucalypt Woodlands in Australia: Biology, Conservation, Management and Restoration*. Hobbs, R.J. and Yates, C.J. (Eds). Surrey Beatty and Sons, Chipping Norton.

Reid, W.V. (1992). How many species will there be? pp. 55–73. In: *Tropical Deforestation and Species Extinction.* Whitmore, T.C. and Sayer, J.A. (Eds). Chapman and Hall, London.

Reiter, N., Weste, G. and Guest, D. (2004). The risk of exctinction resulting from disease caused by *Phyophthora cinnamomi* to endangered, vulnerable or rare plant species endemic to the Grampians, western Victoria. *Australian Journal of Botany*, **52**, 425–433.

Renjifo, L.M. (2001). Effect of natural and anthropogenic landscape matrices on the abundance of sub-Andean bird species. *Ecological Applications*, **11**, 14–31.

Resource Planning and Development Commission (2002). *Inquiry on the progress with implementation of the Tasmanian Regional Forest Agreement (1997).* Background Report. Resource Planning and Development Commission, Hobart.

Reville, B.J., Tranter, J.D. and Yorkston, H.D. (1990). Impact of forest clearing on the endangered seabird *Sula abbotti*. *Biological Conservation*, **51**, 23–38.

Rhoades, C.C., Brosi, S.L., Dattilo, A.J. and Vincelli, P. (2003). Effect of soil compaction and moisture on incidence of phytophthora root rot on American chestnut (*Castanea dentata*) seedlings. *Forest ecology and Management*, **184**, 47–54.

Rice, B. and Westoby, M. (1983). Plant species richness at the 0.1 scale in Australian vegetation compared to other continents. *Vegetatio*, **52**, 129–140.

Rice, K.J., Matzner, S.L., Byer, W. and Brown, J.R. (2004). Patterns of tree dieback in Queensland, Australia: the importance of drought stress and the role of resistance to cavitation. *Oecologia*, **139**, 190–198.

Rich, A.C., Dobkin, D.S. and Niles, L.J. (1994). Defining forest fragmentation by corridor width: the influence of narrow forest-dividing corridors on forest-nesting birds in southern New Jersey. *Conservation Biology*, **8**, 1109–1121.

Richards, B.N., Bridges, R.G., Curtin, R.A., Nix, H.A., Shepherd, K.R. and Turner, J. (1990). *Biological Conservation of the South-east Forests.* Australian Government Publishing Service, Canberra. Report of the Joint Scientific Committee.

Richards, G.C. (1991). The conservation of forest bats in Australia: do we really know the problems and solutions? pp. 81–90. In: *Conservation of Australia's Forest Fauna.* Lunney, D. (Ed). Royal Zoological Society of NSW, Sydney.

Richards, J.D. and Short, J. (2003). Reintroduction and establishment of the Western Barred bandicoot *Perameles bougainville* (Marsupialia: Peramelidae) at Shark Bay, Western Australia. *Biological Conservation*, **109**, 181–195.

Richards, J.D., Copley, P. and Morris, K. (2001). The Wopilkara's return. *Nature Australia*, **26**, 52–61

Richards, R.J., Applegate, R.J. and Ritchie, A.I. (1996). The Rum Jungle rehabilitation project. pp. 530–553. In: *Environmental Management in the Australian Minerals and Energy Industries.* Mulligan, D.R. (Ed). University of New South Wales Press, Sydney, and Australian Minerals and Energy Environment Foundation.

Richards, S.A., Possingham, H.P. and Tizard, J. (1999). Optimal fire management for maintaining community diversity. *Ecological Applications*, **9**, 880–892.

Richards, S.J., McDonald, K.R. and Alford, R.A. (1993). Declines in populations of Australia's endemic tropical frogs. *Pacific Conservation Biology*, **1**, 66–77.

Richardson, D.M., Williams, P.A. and Hobbs, R.J. (1994). Pine invasions in the southern hemisphere: determinants of spread and invadability. *Journal of Biogeography*, **21**, 511–527.

Richardson, K. (1992). Assessing the impacts of roads on small mammal movement in forest habitats of Victoria. BSc (Hons) thesis, Deakin University, Melbourne.

Richardson, K.C., Webb, G.J. and Manolis, S.C. (2002). *Crocodiles: Inside out. A Guide to Crocodilians and their Functional Morphology*. Surrey Beatty and Sons: Chipping Norton.

Ricketts, T.H. (2001). The matrix matters: effective isolation in fragmented landscpes. *American Naturalist*, **158**, 87–99.

Ricketts, T.H., Daily, G.C., Ehrlich, P.R. and Michener, C.D. (2004). Economic value of tropical forest to coffee production. *Proceedings of the National Academy of Sciences*, **101**, 12579–12582.

Ride, W.D.L. (1975). Towards an integrated system: A study of selection and acquisition of National Parks and reserves in Australia. In: *A National System of Ecological Reserves in Australia*. Australian Academy of Science. Report No. 19.

Riley, S.P., Shaffer, H.B., Voss, S.R. and Fitzpatrick, B.M. (2003). Hybridization between a rare, native tiger salamander (*Ambystoma californiense*) and its introduced congener. *Ecological Applications*, **13**, 1263–1275.

Ringold, P.L., Alegria, J., Czaplewksi, R.L., Mulder, B.S., Tolle, T. and Burnett, K. (1996). Adaptive monitoring design for ecosystem management. *Ecological Applications*, **6**, 745–747.

Risbey, D.A., Calver, M.C., Short, J., Bradley, J.S. and Wright, I.W. (2000). The impacts of cats and foxes on the small vertebrate fauna of Herrison Prong, Western Australia. *Wildlife Research*, **27**, 223–235.

Roberge, J-M. and Angelstam, P. (2004). Usefulness of the umbrella species concept as a conservation tool. *Conservation Biology*, **18**, 76–85.

Roberts, K.A. (1991). Field monitoring: Confessions of an addict. pp. 179–212. In: *Monitoring for Conservation and Ecology*. Goldsmith, F.B. (Ed). Chapman and Hall, London.

Roberts, W.B. (1973). Air movements within a plantation and an open area and their effects on fire behaviour. *Australian Forest Research*, **4**, 41–47.

Robertson, G. (1986). The mortality of kangaroos in drought. *Australian Wildlife Research*, **13**, 349–354.

Robertson, G.G. (2000). Effect of line sink rate on Albatross mortality in the Patagonian toothfish long-line fishery. *CCAMLR Science*, **7**, 133–150.

Robertson, P. (1999). *Terrick Terrick National park habitat assessment for five species of threatened vertebrates.* Report of Parks Victoria, Melbourne.

Robertson, P. and Cooper, P. (1997). Recovery plan for the Grassland Earless Dragon (*Tympanocryptis lineata pinguicolla*). Report to Environment Australia, Canberra.

Robinson, D. (1993). Threatened birds in Victoria: their distributions, ecology and future. *Victorian Naturalist*, **108**, 67–77.

Robinson, G.R., Holt, R.D., Gaines, M.S., Hamburg, S.P., Johnson, M.L., Fitch, H.S. and Martinko, E.A. (1992). Diverse and contrasting effects of habitat fragmentation. *Science*, **257**, 524–526.

Robinson, S.K. (1998). Another threat posed by forest fragmentation: Reduced food supply. *Auk*, **115**, 1–3.

Robinson, S.K., Thompson, E.R., Donovan, T.M., Whitehead, D.R. and Faaborg, J. (1995). Regional forest fragmentation and the nesting success of migratory birds. *Science*, **267**, 1987–1990.

Robinson, W.D. (1999). Long-term changes in the avifauna of Barro Colorado Island, Panama, a tropical forest isolate. *Conservation Biology*, **13**, 85–97.

Robley, A.J., Short, J. and Bradley, S. (2002). Do European Rabbits (*Oryctolagus cuniculus*) influence the population ecology of the Burrowing Bettong (*Bettongia lesuer*)? *Wildlife Research*, **29**, 423–429.

Rock Lobster Industry Advisory Committee (2002). *Management arrangements for the 2002/03 season and measures proposed for 2003/04.* Rock Lobster Industry Advisory Committee, Perth.

Rodenhouse, N.L. and Best, L.B. (1994). Foraging patterns of vesper sparrows (*Pooecetes gramineus*) breeding in cropland. *American Midland Naturalist*, **131**, 196–206.

Rodrigues, A.S., Gregory, R.D. and Gaston, K.J. (2000). Robustness of reserve selection procedures under temporal species turnover. *Proceedings of the Royal Society of London Series B*, **267**, 49–55.

Rogers, A.L., Anderson, G.W., Biddiscombe, E.F., Arkell, P., Glencross, R., Nicholas, D.A. and Paterson, J.G. (1979). *Perennial Pasture Grasses in South Western Australia. I. Preliminary Evaluation of Species.* Western Australian Department of Agriculture. Technical Bulletin No. 45.

Rogers, C.S. (1993). Hurricanes and coral reefs: the intermediate disturbance hypothesis revisited. *Coral Reefs*, **12**, 127–137.

Rogers, K. (1997). Operationalising ecology under a new paradigm: An African perspective. pp. 60–77. In: *The Ecological Basis for Conservation: Heterogeneity, Ecosystems, and Biodiversity*. Pickett, S.T., Ostfeld, R.S., Shachak, M. and Likens, G.E. (Eds). Chapman and Hall, New York.

Rojas, M. (1992). The species problem and conservation: What are we protecting? *Conservation Biology*, **6**, 170–178.

Rolls, E.C. (1969). *They All Ran Wild.* Angus and Robertson, Sydney.

Rolstad, J. and Wegge, P. (1987). Distribution and size of capercaillie leks in relation to old forest fragmentation. *Oecologia*, **72**, 389–394.

Rolstad, J., Gjerdde, I., Gundersen, V.S. and Saetersal, M. (2002). Use of indicator species to assess forest continuity: a critique. *Conservation Biology*, **16**, 253–257.

Root, R.B. (1967). The niche exploitation pattern of the Blue-gray Gnatcatcher. *Ecological Monographs,* **37**, 317–350.

Rose, A. (1995). The effects of rainfall on the abundance and species richness of small vertebrates in the Stirling ranges National Park. *Western Australian Naturalist,* **20**, 53–59.

Rosenberg, A.A. (2003). Managing to the margins: the overexploitation of fisheries. *Frontiers in Ecology and Environment,* **1**, 102–106.

Rosenberg, A.A., Fogarty, M.J., Sissenwine, M.P., Beddington, J.R. and Shepherd, J.G. (1993). Achieving sustainable use of renewable resources. *Science,* **262**, 828–829.

Rosenberg, D.K., Noon, B.R. and Meslow, E.C. (1997). Biological corridors: form, function and efficacy. *BioScience,* **47**, 677–687.

Rosenberg, D.M. and Resh, V.H. (1993). *Freshwater Biomonitoring and Benthic Macroinvertebrates.* Chapman and Hall, New York.

Rosenberg, D.M., Danks, H.V. and Lehmkuhl, D.M. (1986). Importance of insects in environmental impact assessment. *Environmental Management,* **10**, 773–783.

Rosenberg, M.S., Adams, D.C. and Gurewich, J. (2000). *Metawin. Statistical software for meta-analysis.* Version 2.0. Sinauer Associates, Sunderland, Massachusetts.

Rosenzweig, M.L. (1995). *Species diversity in space and time.* Cambridge University Press, Cambridge.

Rossetto, M., Weaver, P.K. and Dixon, K.W. (1995). Use of RAPD analysis in devising conservation strategies for the rare and endangered *Grevillea scapigera* (Proteaceae). *Molecular Ecology,* **4**, 321–329.

Rounsevell, D.E. (1988). Thylacine. pp. 82–83. In: *Complete Book of Australian Mammals.* Strahan, R. (Ed). Collins Angus and Robertson Publishers, Sydney.

Rounsevell, D.E. and Smith, S.J. (1982). Recent alleged sightings of the Thylacine (Marsupialia: Thylacinidae) in Tasmania. pp. 233–236. In: *Carnivorous Marsupials.* Royal Zoological Society of NSW, Sydney.

Rouphael, A.B. and Inglis, G.J. (1997). Impacts of recreational SCUBA diving at sites with different reef topography. *Biological Conservation,* **82**, 329–336.

Rowell, D.M., Higgins, A.V., Briscoe, D.A. and Tait, N.N. (1995). The use of chromosomal data in the systematics of the viviparous onychophorans from Australia (Onychophora: Peripatopsidae). In: Onychophora: past and present. Walker, M.H. and Norman, D.B. (Eds). *Zoological Journal of the Linnean Society,* **114(1)**, 139–153.

Rowley, I. (1965). The life history of the Superb Blue Wren. *Emu,* **64**, 251–297.

Rowley, I. (1990). *The Galah. Behavioural Ecology of Galahs.* Surrey Beatty, Chipping Norton.

Royal Botanic Gardens (2003). *Millenium seed bank project.* Royal Botanic Gardens, Kew. [online] URL: http://www.rbgkew.org.uk/msbp/internat/index.html

Royal Commission (1931). *Report of the Royal Commission on the Development of North Queensland (Land Settlement and Forestry).* Government Printer, Brisbane.

Rübsamen, K., Hume, I.D., Foley, W.J. and Rübsamen, U. (1984). Implications of the large surface area to body mass ratio on the heat balance of the greater glider (*Petauroides volans*: Marsupialia). *Journal of Comparative Physiology,* **154**, 105–111.

Ruckelshaus, M.H., Levin, P., Johnson, J.B. and Kareiva, P.M. (2002). The Pacific Salmon wars: what science brings to the challenge of recovering species. *Annual Review of Ecology and Systematics,* **33**, 665–706.

Rudnicky, T.C. and Hunter, M.L. (1993). Avian nest predation in clearcuts, forests, and edges in a forest-dominated landscape. *Journal of Wildlife Management,* **57**, 358–364.

Ruesink, J.L., Parker, I.M., Groom, M.J. and Kareiva, P.M. (1995). Reducing the risks of nonindigenous species introductions. *BioScience,* **45**, 465–477.

Rülcker, C., Angelstam, P. and Rosenberg, P. (1994). Natural forest-fire dynamics can guide conservation and silviculture in boreal forests. *SkogForsk,* **2. 1994**, 1–4.

Rushton, S.P., Ormerod, S.J and Kerby, G. (2004). New paradigms for modelling species distributions? *Journal of Applied Ecology,* **41**, 193–200.

Russ, G.R. and Alcala, A.C. (2004). Marine reserves: long-term protection is required for full recovery of predatory fish populations. *Oecologia,* **138**, 622–627.

Russell, B.G., Smith, B. and Augee, M.L. (2003). Changes to a population of common ringtail possums (*Pseudocheirus peregrinus*) after bushfire. *Wildlife Research,* **30**, 389–396.

Russell-Smith, J. and Bowman, D.M. (1992). Conservation of monsoon rainforest isolates in the Northern Territory, Australia. *Biological Conservation,* **59**, 51–63.

Rutberg, A.T., Naugle, R.E., Thiele, L.A. and Liu, I.K. (2004). Effects of immunocontraception on a suburban population of White-tailed Deer *Odocoileus virginainus. Biological Conservation,* **116**, 243–250.

Ryan, P. (2000). The use of revegetated areas by vertebrate fauna in Australia: A review. pp. 318–335.

In: *Temperate Eucalypt Woodlands in Australia: Biology, Conservation, Management and Restoration*. Hobbs, R.J. and Yates, C.J. (Eds). Surrey Beatty and Sons, Chipping Norton.

Saab, V. (1999). Importance of spatial scale to habitat use by breeding birds in riparian forests: a hierarchical approach. *Ecological Applications,* **9**, 135–151.

Saab, V.J. and Dudley, J. (1998). Responses of cavity-nesting birds to stand-replacement fire and salvage logging in Ponderosa Pine/Douglas-Fir forests of southwestern Idaho. Research paper RMRS-RP-11. Ogden, Utah. USDA, Forest Service, Rocky Mountain Research Station.

Saari, L., Aberg, J. and Swenson, J.E. (1998). Factors influencing the dynamics of occurrence of the Hazel Grouse in a fine-grained managed landscape. *Conservation Biology,* **12**, 586–592.

Saccheri, I., Kuussaari, M., Kankare, M., Vikman, P., Fortelius, W. and Hanski, I. (1998). Inbreeding and extinction in a butterfly metapopulation. *Nature,* **392**, 491–494.

SAG (1995). (Scientific Advisory Group) *National Forest Conservation Reserves, commonwealth proposed criteria: a position paper.* Australian Government Publishing Service, Canberra.

Salinity Planning Working Group (1992). *Regional Salinity Impact (Draft).* Salinity Bureau, Department of Premier and Cabinet, Melbourne. 42 pp.

Salt, D., Lindenmayer, D.B. and Hobbs, R.J. (2004). *Trees and biodiversity. A guide for Australian farm forestr.* Rural Industries Research and Development Corporation, Canberra, Australia.

Salwasser, H. (1990). Conserving biological diversity: A perspective on scope and approaches. *Forest Ecology and Management,* **35**, 79–90.

Salwasser, H. (1993). Sustainability needs more than better science. *Ecological Applications,* **3**, 587–589.

Sampson, J.F., Hopper, S. and James, S.H. (1995). The mating system and genetic diversity in the arid zone mallee, *Eucalyptus rameliana. Australian Journal of Botany*, **43**, 461–474.

Santiago Declaration (1995). *The Montreal Process. Criteria and Indicators for the Conservation and Sustainable Management of Temperate and Boreal Forests.* Canadian Forest Service, Hull, Quebec.

Sargent, R.A., Kilgo, J.C., Chapman, B.R. and Miller, K.V. (1998). Predation of artificial nests in hardwood fragments enclosed by pine and agricultural habitats. *Journal of Wildlife Management,* **62**, 1438–1442.

Sarre, S. (1995). *Mitochondrial-DNA variation among populations of Oedura-reticulata* (Gekkonidae) in remnant vegetation – implications for metapopulation structure and population decline. *Molecular Ecology*, **4**, 395–405.

Sarre, S., Smith, G.T. and Meyers, J.A. (1995). Persistence of two species of gecko (*Oedura reticulata* and *Gehyra variegata*) in remnant habitat. *Biological Conservation,* **71**, 25–33.

Sattler, P.S. and Webster, R.J. (1984). The conservation status of Brigalow (*Acacia harpophylla*) communities in Queensland. pp. 149–160. In: *Conference on The Brigalow Belt of Australia.* Proceedings of a symposium held at the John Kindler Memorial Theatre, Queensland Institute of Technology, 23 October 1982. The Royal Society of Queensland, Brisbane.

Saunders, D., Beattie, A., Eliott, S., Fox, M., Hill, B., Pressey, B., Veal, D., Venning, J., Maliel, M. and Zammit, C. (1996). Biodiversity. pp. 4–1 to 4–59. In: *State of the Environment Australia 1996.* State of the Environment Advisory Council (Ed). CSIRO, Melbourne.

Saunders, D.A. (1994). The effects of habitat reduction and fragmentation on the mammals and birds of the Western Australian central wheat belt. pp. 99–105. In: *The Future of the Fauna of Western New South Wales.* Lunney, D., Hand, S. and Butcher D. (Eds). Royal Zoological Society of New South Wales, Sydney.

Saunders, D.A. (1996). Does our lack of vision threaten the viability of the reconstruction of disturbed ecosystems? *Pacific Conservation Biology,* **2**, 321–326.

Saunders, D.A. and de Rebeira, C.P. (1991). Values of corridors to avian populations in a fragmented landscape. pp. 221–240. In: *Nature Conservation 2: The Role of Corridors.* Saunders, D.A. and Hobbs, R.J. (Eds). Surrey Beatty, Chipping Norton.

Saunders, D.A. and Hobbs, R.J. (Eds) (1991). *Nature Conservation 2: The Role of Corridors.* Surrey Beatty, Chipping Norton.

Saunders, D.A. and Ingram, J. (1987). Factors affecting survival of breeding populations of Carnaby's cockatoo *Calyptorhyncus funerus latirostris* in remnants of native vegetation. pp. 249–258. In: *Nature Conservation: The Role of Remnants of Native Vegetation.* Saunders, D.A., Arnold, G.W., Burbudge, A.A. and Hopkins, A.J.M. (Eds). Surrey Beatty, Chipping Norton.

Saunders, D.A. and Ingram, J. (1995). *Birds of Southwestern Australia.* Surrey Beatty, Chipping Norton.

Saunders, D.A., Arnold, G.W., Burbridge, A.A. and Hopkins, A.J. (Eds) (1987). *Nature Conservation: The*

Role of Remnants of Native Vegetation. Surrey Beatty and Sons, Chipping Norton.

Saunders, D.A., Hobbs, R.J. and Ehrlich, P. (Eds) (1993). *Nature Conservation 3: Reconstruction of Fragmented Ecosystems.* Surrey Beatty, Chipping Norton.

Saunders, D.A., Hobbs, R.J. and Margules, C.R. (1991). Biological consequences of ecosystem fragmentation: a review. *Conservation Biology,* **5(1)**, 18–32.

Saunders, D.A., Margules, C.R. and Hill, B. (1998). *Environmental Indicators for National State of the Environment Reporting – Biodiversity.* Environment Australia, Canberra.

Saunders, D.A., Rowley, I. and Smith, G.T. (1985). The effects of clearing for agriculture on the distribution of cockatoos in the southwest of Western Australia. pp. 309–321. In: *Birds of Eucalypt Forests and Woodlands: Ecology, Conservation and Management.* Keast, A., Recher, H.F., Ford, H. and Saunders, D. (Eds). Surrey Beatty, Chipping Norton.

Saurez, A.V., Bolger, D.T. and Case, T.J. (1998). Effects of fragmentation and invasion on native ant communities. *Ecology,* **79**, 2041–2056.

Savidge, J.A. (1987). Extinction of an island forest avifauna by an introduced snake. *Ecology,* **68**, 660–668.

Savolainen, P., Leitner, T., Wilton, A.N., Matisoo-Smith, E. and Lundeberg, J. (2004). A detailed picture of the origin of the Australian Dingo, obtained from the study of mitochondrial DNA. *Proceedings of the National Academy of Sciences,* **101**, 12387–12390.

Sax, J.L. (1980). *Mountains without Handrails: Reflections on the National Parks.* University of Michigan Press, Ann Arbor.

SCA (1991). (Standing Committee on Agriculture) *Guidelines for the Control of Exotic Vertebrate Animals.* Working Group on Vertebrate Pests, CSIRO, Canberra. SCA Technical Report Series No. 37.

Schall, R. (1991). Estimation of generalised linear models with random effects. *Biometrika,* **78**, 719–727.

Schaller, G. and Rabinowitz, A. (1995). The Soala or Spindle-horned Bovid Pseudoryx nghetinhensis in Laos. *Oryx,* **29**, 107–114.

Schaller, G. and Vrba, E.S. (1996). Description of the Giant Muntjac (*Megamuntiacus vaguangenis*) in Laos. *Journal of Mammalogy,* **77**, 675–683.

Scharenberg, W. (1991). Cormorants (*Phalacrocorax carbo sinensis*) as bioindicators for polychlorinated biphenyls. *Archives of Environmental Contamination and Toxicology,* **21**, 536–540.

Schindler, D.E., Rogers, D.E., Scheuerell, M.D. and Abrey, C.A. (2005). Effects of changing climate on zooplankton and juvenile Sockeye Salmonin southwestern Alaska. *Ecology,* **86**, 198–209.

Schmiegelow, F.K. and Monkkonen, M. (2002). Habitat loss and fragmentation in dynamic landscapes: Avian perspectives from the boreal forest. *Ecological Applications,* **12**, 375–389.

Schmiegelow, F.K., Machtans, C.S. and Hannon, S.J. (1997). Are boreal birds resilient to forest fragmentation? An experimental study of short-term community responses. *Ecology,* **78**, 1914–1932.

Schmitz, H. and Bleckmann, H. (1998). The photomeachnic infrad receptor for the detection of forest fires in the beetle *Melanophilia acuminata. Journal of Comparative Physiology A,* **182**, 647–611.

Schodde, R. and Mason, I.J. (1999). *The Directory of Australian birds.* CSIRO Publishing, Melbourne.

Schuler, A. (1998). Sustainability and biodiversity – forest historical notes on two main concerns of environmental utilisation. pp. 353–360. In: *Assessment of Biodiversity for Improved Forest Planning.* Bachmann, P., Köhl, P. and Päivenen, R. (Eds). Kluwer, Dordrecht.

Schultz, C.B. (1998). Dispersal behavior and its implications for reserve design in a rare Oregon butterfly. *Conservation Biology,* **12**, 284–292.

Schur, B. (1990). W.A.'s biggest nature conservation problem: Land clearing in the south west. *Land and Water Research News,* **5**, 6 – 9.

Schwaner, T.D. (1985). Population structure of the Black Tiger Snakes, Notechis ater niger, on offshore islands of South Australia. pp. 35–46. In: *Biology of Australasian Frogs and Reptiles.* Grigg, G., Shine, R. and Ehmann, H. (Eds). Royal Society of New South Wales, Sydney.

Schwaner, T.D. and Sarre, S.D. (1988). Body size of Tiger Snakes in South Australia, with particular reference to *Notechis ater serventyi* (Elapidae) on Chappell Island. *Journal of Herpetology,* **22**, 24–33.

Schwartz, M.W. (1999). Choosing the appropriate scale of reserves for conservation. *Annual Review of Ecology and Systematics,* **30**, 83–108.

Schwartz, M.K. and Mills, L.S. (2005). Gene flow after inbreeding leads to higher survival in deer mice. *Biological Conservation,* **123**, 413–420.

Schwartz, M.W. and van Mantgem, P.J. (1997). The value of small preserves in chronically fragmented landscapes. pp. 379–394. In: *Conservation in Highly Fragmented Landscapes.* Schwartz, M.W. (Ed). Chapman and Hall, New York.

Schwilk, D.W., Keeley, J.E. and Bond, W.J. (1997). The intermediate disturbance hypothesis does not explain fire and diversity pattern in fynbos. *Plant Ecology*, **132**, 77–84.

Scientific Advisory Group (1995). *National Forest Conservation Reserves, Commonwealth Proposed Criteria: A Position Paper.* Australian Government Publishing Service, Canberra, Australia.

Scott, D., Malcolm, J. and Lemieux, C. (2002). Climate change and modelled biome representation in Canada's national park system: implications for system planning and park mandates. *Global Ecology and Biogeography*, **11**, 475–484.

Scott, J.K. and Adair, R.J. (1991). The commencement of biological control of Bitou Bush and Boneseed. Plant invasions – the incidence of environmental weeds in Australia. Part Two. pp. 161–163. In: *Proceedings of the Ninth Australian Weeds Conference.* Adelaide, August 1990. Humphries, S.E., Groves, R.H. and Mitchell, D.S. (Eds). Australian National Parks and Wildlife Service, Canberra.

Scott, J.M. (1999). Vulnerability of forested ecosystems in the Pacific Northwest to loss of area. pp. 33–42. In: *Forest Wildlife and Fragmentation. Management Implications.* Rochelle, J., Lehmann, L.A. and Wisniewski, J. (Eds). Brill, Leiden, Germany.

Scott, J.M., Abbitt, R.J. and Groves, C.R. (2001). What are we protecting? The United States conservation portfolio. *Conservation Biology in Practice,* **2**, 18–19.

Scott, J.M., Davis, F., Csuti, B., Noss, R., Butterfield, B., Groves, C., Anderson, H., Caicco, S., Derchia, F., Edwards, T.C., Ulliman, J. and Wright, R.G. (1993). Gap analysis – a geographic approach to protection of biological diversity. *Wildlife Monographs*, **123**, 1–41.

Scott, P., Burton, J.A. and Fitter, R. (1987). Red Data books: The historical background. pp. 1–6. In: *The Road to Extinction* . Fitter, R. and Fitter, M. (Eds). IUCN, Gland, Switzerland.

Scotts, D.J. (1991). Old-growth forests: their ecological characteristics and value to forest-dependent fauna of south-east Australia. pp. 147–159. In: *Conservation of Australia's Forest Fauna.* Lunney, D. (Ed). Royal Zoological Society of New South Wales, Mosman.

Scotts, D.J. (1994). Sustaining sensitive wildlife within temperate forest landscapes: regional systems of retained habitat as a planning framework. pp. 85–106. In: *Ecology and Sustainability of Southern Temperate Ecosystems.* Norton, T. and Dovers S. (Eds). CSIRO, Melbourne.

Scougall, S.A., Majer, J.D. and Hobbs, R.J. (1993). Edge effects in grazed and ungrazed Western Australia wheatbelt remnants in relation to ecosystem reconstruction. pp. 163–178. In: *Nature Conservation 3: Reconstruction of Fragmented Ecosystems.* Saunders, D.A., Hobbs, R.J. and Ehrlich, P.R. (Eds). Surrey Beatty and Sons, Chipping Norton.

Scully, L. (2003). Relationships between vegetation clearance and the introduction of legislation in the Nyngan region, central western NSW. *Ecological Management and Restoration*, **4**, 150–153.

Seagle, S.W. and Shugart, H.H. (1985). Faunal richness and turnover on dynamics landscapes: a simulation study. *Journal of Biogeography,* **15**, 759–774.

Searle, S.R., Casella, G. and McCulloch, C.E. (1992). *Variance components analysis.* Wiley, New York.

Seebeck, J.H. (1977). Mammals of the Melbourne metropolitan area. *Victorian Naturalist,* **94**, 165–171.

Seebeck, J.H., Bennett, A.F. and Dufty, A.C. (1990). Status, distribution and biogeography of the Eastern Barred Bandicoot, *Perameles gunnii*, in Victoria. pp. 253–274. In: *The Conservation of Small Populations.* Clark, T. and Seebeck, J.H. (Eds). Chicago Zoological Board, Chicago.

Semlitsch, R.D. and Bodie, J.R. (1998). Are small, isolated wetlands expendable? *Conservation Biology,* **12**, 1129–1133.

Semlitsch, R.D. and Bodie, J.R. (2003). Biological criteria for buffer zones around wetlands and riparian habitat for amphibians and reptiles. *Conservation Biology*, **17**, 1219–1228.

Semmens, T.D., Turner, E. and Buttermore, R. (1993). *Bombus terrestris* (L.) (Hymenoptera: Apidae) now established in Tasmania. *Journal of Australian Entomological Society*, **32**, 346.

Sepkowski, J.J. (1993). Ten years in the library: new data confirm paleontological patterns. *Paleobiology,* **19**, 43–5.

Serena, M. (Ed) (1995). *Reintroduction Biology of Australian and New Zealand Fauna.* Surrey Beatty, Chipping Norton.

Severinghaus, W.D. (1981). Guild theory development as a mechanism for assessing environmental impact. *Environmental Management,* **51**, 230–237.

Shafer, C.L. (1990). *Nature Reserves: Island Theory and Conservation Practice.* Smithsonian Institution Press, Washington.

Shaffer, M.L. (1981). Minimum population sizes for species conservation. *Bioscience,* **31**, 131–134.

Shaffer, M.L. (1990). Population viability analysis. *Conservation Biology,* **4**, 39–40.

Shaffer, M.L. and Samson, F.B. (1985). Population size and extinction: a note on determining critical population sizes. *American Naturalist,* **125**, 144–152.

Shaner, D.L. (1996). Introduction of transgenic crops and their potential as future weeds. pp. 581–585.

In: *Proceedings of the Eleventh Australian Weeds Conference*. Sheperd, R.C.H. (Ed). Melbourne Weed Science Society of Victoria, Frankston.

Shaw, D., Greenleaf, J. and Berg, D. (1993). Monitoring new forestry. *Environmental Monitoring and Assessment*, **26**, 187–193.

Shaw, J. (1979). The introduction of captive hand-reared Ringtail Possums (*Pseudocheirus peregrinus*) into the Coranderk Reserve, Healesville. *Bulletin of the Australian Mammal Society*, **5**, 27–28.

Shea, K., Roxburgh, S.H., Rauschert, E.S.J. (2004). Moving from pattern to process: coexistence mechanisms under intermediate disturbance regimes. *Ecology Letters*, **7**, 491–508.

Shea, S.R., Abbott, I., Armstrong, J.A. and McNamara, K.J. (1997). Sustainable conservation: a new integrated approach to nature conservation in Australia. pp. 39–48. In: *Conservation Outside Nature Reserves*. Hale, P. and Lamb, D. (Eds). University of Queensland, Brisbane, Australia.

Shearer, B.L., Crane, C.E. and Cochrane, A. (2004). Quantification of the susceptibility of the native flora of the South-West Botanical Province, Western Australia, to *Phytophthora cinnamoni*. *Australian Journal of Botany*, **52**, 435–443.

Sheppard, C. (1995). Propagation of endangered species in US institutions: How much space is there? *Zoo Biology*, **14**, 197–210.

Sherwin, W.B. and Brown, P.R. (1990). Problems in the estimation of the effective population size of the Eastern Barred Bandicoot *Perameles gunnii* at Hamilton, Victoria. pp. 367–374. In: *Bandicoots and Bilbies*. Seebeck, J.H., Brown, P.R., Wallis, R.L. and Kemper, C.M. (Eds). Surrey Beatty, Sydney.

Sherwin, W.B. and Murray, N.D. (1990). Population and conservation genetics of marsupials. *Australian Journal of Zoology*, **37**, 161–180.

Sherwin, W.B., Murray, N.D., Graves, J.A.M. and Brown, P.R. (1989). Minimum research on the conservation genetics of wild populations. pp. 211–220. In: *The Conservation of Threatened Species and their Habitats*. Hicks, M. and Eiser, P. (Eds). Australian Committee for IUCN, Canberra. Occasional Paper Number No. 2.

Sherwin, W.B., Murray, N.D., Graves, J.A.M. and Brown, P.R. (1991). Measurement of genetic variation in endangered populations: Bandicoots (Marsupialia: Peramelidae) as an example. *Conservation Biology*, **5**, 103–108.

Sherwin, W.B., Timms, P., Wilcken, J. and Houlden, B. (2000). Analysis and conservation implications of Koala genetics. *Conservation Biology*, **14**, 639–649.

Shiel, D. and Burslem, F.R. (2003). Disturbing hypotheses in tropical forests. *Trends in Ecology and Evolution*, **18**, 18–26.

Shields, J. and Kavanagh, R. (1985). *Wildlife Research and Management in the Forestry Commission of NSW. A Review*. Forestry Commission of NSW, Sydney. Technical Paper No. 32.

Shields, J., York, A. and Binns, D. (1992). *Flora and Fauna Survey, Mt. Royal Management Area, Newcastle Region*. Forestry Commission of N.S.W, Sydney. Forest Resources Series No. 16.

Shine, R. (1987). Ecological comparisons of island and mainland populations of Australian Tigersnakes (Notechis: Elapidae). *Herpetologica*, **43**, 233–240.

Shine, R. and Fitzgerald, M. (1989). Conservation and reproduction of an endangered species: the Broad-headed Snake, *Hoplocephalus bungaroides* (Elapidae). *Australian Zoologist*, **25**, 65–66.

Shine, R., Webb, J.K., Fitzgerald, M. and Sumner, J. (1998). The impact of bush-rock removal on an endangered snake species, *Hoplocephalus bungaroides* (Serpentes: Elapidae). *Wildlife Research*, **25**, 285–295.

Shmida, A. and Ellner, S. (1984). Coexistence of plant species with similar niches. *Vegetatio*, **58**, 29–55.

Short, H.L. (1984). *Habitat Suitability Index Models: the Arizona Guild and Layers of Habitat Models*. U.S. Fish and Wildlife Service. FWS/OBS-8210.70. 37 pp.

Short, H.L. and Burnham, K.P. (1982). Techniques for structuring wildlife guilds to evaluate impacts on wildlife communities. *Wildlife*, **244**.

Short, J. and Smith, A.P. (1994). Mammal decline and recovery in Australia. *Journal of Mammalogy*, **75**, 288–297.

Short, J. and Turner, B. (1994). A test of the vegetation mosaic hypothesis: An hypothesis to explain the decline and extinction of Australian mammals. *Conservation Biology*, **8**, 439–449.

Short, J., Bradshaw, S.D., Giles, J., Prince, R.I. and Wilson, G.R. (1992). Reintroductions of macropods (Marsupialia: Macropodoidea) in Australia – a review. *Biological Conservation*, **62**, 189–204.

Shrader-Frechette, K.S. and McCoy, E.D. (1994). Biodiversity, biological uncertainty, and setting conservation priorities. *Biology and Philosophy*, **9**, 167–195.

Sieving, K.E., Willson, M.F. and de Santo, T.L. (2000). Defining corridor functions for endemic birds in fragmented south-temperate rainforest. *Conservation Biology*, **14**, 1120–1132.

Sigg, D.P., Goldizen, A.W. and Pople, A.R. (2005). The importance of mating system in translocation programs: reproductive success of released male

Bridled Nailtail Wallabies. *Biological Conservation*, **123**, 289–300.

Silsbee, D.G. and Peterson, D.L. (1993). Planning for implementation of long-term resource monitoring programs. *Environmental Monitoring and Assessment*, **26**, 177–185.

Simandl, J. (1992). The distribution of Pine Sawfly cocoons (Diprionidae) in Scots pine stands in relation to stand edge and tree base. *Forest Ecology and Management*, **54**, 193–203.

Simberloff, D. (1986). Are we on the verge of a mass extinction in tropical rainforests? pp. 165–180. In: *Dynamics of Extinction*. Elliott, D.K. (Ed). Wiley, New York.

Simberloff, D. (1992). Do species-area curves predict extinction in fragmented forest? pp. 75–89. In: *Tropical Deforestation and Species Extinction*. Whitmore, T.C. and Sayer, J.A. (Eds). Chapman and Hall, London.

Simberloff, D. (1995). Why do introduced species appear to devastate islands more than mainland species? *Pacific Science*, **49**, 87–97.

Simberloff, D. and Dayan, T. (1991). The guild concept and the structure of ecological communities. *Annual Review of Ecology and Systematics*, **22**, 115–143.

Simberloff, D. and Martin, J.L. (1991). Nestedness of insular avifaunas: simple summary statistics masking species patterns. *Ornis Fennica*, **68**, 178–192.

Simberloff, D., Parker, I.M. and Windle, P.N. (2005). Introduced species policy, management and future research needs. *Frontiers in Ecology and Environment*, **3**, 12–20.

Simberloff, D.A. (1988). The contribution of population and community biology to conservation science. *Annual Review of Ecology and Systematics*, **19**, 473–511.

Simberloff, D.A. (1998). Flagships, umbrellas, and keystones: is single-species management passe in the landscape era. *Biological Conservation*, **83**, 247–257.

Simberloff, D.A. (1999). The role of science in the preservation of forest biodiversity. *Forest Ecology and Management*, **115**, 101–111.

Simberloff, D.A., Farr, J.A., Cox, J. and Mehlman, D.W. (1992). Movement corridors: conservation bargains or poor investments? *Conservation Biology*, **6**, 493–504.

Simberloff, D.S. and Cox, J. (1987). Consequences and costs of conservation corridors. *Conservation Biology*, **1**, 63–71.

Simon, J. and Kahn, H. (1984). *The Resourceful Earth: A Response to Global 2000*. Blackwell, Oxford.

Sims, R.A., Corns, I.G.W. and Klinka, K. (Eds) (1995). *Global to Local: Ecological Land Classification*. Kluwer Academic Publishers, London.

Sinclair, E.A., Costello, B., Courtenay, J.M., Crandall, K.A. (2002). Detecting a genetic bottleneck in Gilbert's Potoroo (*Potorous gilbertii*) (Marsupialia: Potoroidae), inferred from microsatellite and mitochondrial DNA sequence data. *Conservation Genetics*, **3**, 191–196.

Sinclair, E.A., Danks, A. and Wayne, A.F. (1996). Discovery of Gilbert's Potoroo, *Potorous tridactlylus* in Western Australia. *Australian Mammalogy*, **19**, 69–72.

Singer, P. (2000). Darwinian dominion: Animal welfare and human interests . *British Journal for the Philosophy of Science*, **51**, 495–498.

Singh, S., Smyth, A.K. and Blomberg, S.P. (2002). Effect of a control burn on lizards and their structural environment in a eucalypt open-forest. *Wildlife Research*, **29**, 447–454.

Singleton, G., Hind, L., Leirs, H. and Zhang, Z. (Eds) (1999). *Ecologically-based rodent management*. Australian Centre for International Agricultural Research, Canberra.

Sivertsen, D. (1994). The native vegetation crisis in the wheat belt of NSW. *Search*, **25**, 5–8.

Sjörberg, K. and Ericson, L. (1992). Forested and open wetlands. pp. 326–351. In: *Ecological Principles of Nature Conservation*. Hansson, L. (Ed). Elsevier, London.

Skelly, D.K., Yurewicz, K.L., Werner, E.E. and Relyea, R.A. (2003). Estimating decline and distributional change in amphibians. *Conservation Biology*, **17**, 744–751.

Skidmore, A.K., Brinkhof, W., Delaney, J. and Turner, B.J. (1994). Using neural networks to analyse spatial data. pp. 235–246. In: *Proceedings of the Seventh Australasian Remote Sensing Conference*. Melbourne.

Skinner, J.L., Gilliam, E. and Rohlin, C.-J. (1998). The demographic future of the Moreton region. pp. 245–265. In: *Moreton Bay and catchment*. Tibbetts, R., Hall, N. and Dennison, W.C. (Eds). School of Marine Science, University of Queensland.

Slatkin, M. (1985). Rare alleles as indicators of gene flow. *Evolution*, **39**, 53–65.

Slatkin, M. (1987). Gene flow and the geographic structure of natural populations. *Science*, **236**, 787–792.

Slatkin, M. (1995). A measure of population subdivision based on microsatellite allele frequencies. *Genetics*, **139(1)**, 457–462.

Slatkin, M. and Barton, N.H. (1989). A comparison of three indirect methods for estimating average levels of gene flow. *Evolution*, **43**, 1349–1368.

Slatkin, M. and Maddison, W.P. (1990). Detecting isolation by distance using phylogenies of genes. *Genetics*, **126**, 249–260.

Slee, B. (2001). Resolving production-environment conflicts: the case of the Regional Forest Agreement process in Australia. *Forest Policy and Economics*, **3**, 17–30.

Slovic, P. (1999). Trust, emotion, sex, politics, and science: surveying the risk-assessment battlefield. *Risk Analysis*, **19**, 689–701.

Slovic, P., Monahan, J. and MacGregor, D.G. (2000). Violence risk assessment and risk communication: the effects of using actual cases, providing instruction, and employing probability versus frequency formats. *Law and Human Behavior*, **24**, 271–296.

Smith, A. and Marsh, H. (1990). Management of traditional hunting of dugongs (*Dugong dugon* (Muller, 1776)) in the northern Great Barrier Reef, Australia. *Environmental Management*, **14**, 47–55.

Smith, A.D.M., Sainsbury, K.J. and Stevens, R.A. (1999). Implementing effective fisheries-management systems – management strategy evaluation and the Australian partnership approach. *ICES Journal of Marine Science*, **56**, 967–979.

Smith, A.P. (1982). Diet and feeding strategies of the marsupial Sugar Glider in temperate Australia. *Journal of Animal Ecology*, **51**, 149–166.

Smith, A.P. (1984a). Diet of Leadbeater's possum *Gymnobelideus leadbeateri* (Marsupialia). *Australian Wildlife Research*, **11**, 265–273.

Smith, A.P. (1984b). Demographic consequences of reproduction, dispersal and social interaction in a population of Leadbeater's Possum (*Gymnobelideus leadbeateri*). pp. 359–73. In: *Possums and Gliders.* Smith, A.P. and Hume, I.D. (Eds). Surrey Beatty and Sons, Sydney.

Smith, A.P. (1997). Ecosystem management in Australia. pp. 12–31. In: *Saving our Natural Heritage? The Role of Science in Managing Australia's ecosystems.* Copeland, C. and Lewis, D. (Eds). Halstead Press, Rushcutters Bay, Sydney.

Smith, A.P. and Quin, D.G. (1996). Patterns and causes of extinction and decline in Australian conilurine rodents. *Biological Conservation*, **77**, 243–267.

Smith, A.P. and Winter, J.W. (1984). A key and field guide to the Australian possums, gliders and Koala. pp. 579–594. In: *Possums and Gliders.* Smith, A.P. and Hume, I.D. (Eds). Surrey Beatty, Sydney.

Smith, A.P., Horning, N. and Moore, D. (1997). Regional biodiversity planning and lemur conservation with GIS in western Madagascar. *Conservation Biology*, **11**, 498–512.

Smith, A.P., Moore, D.M. and Andrews, S.P. (1992). *Proposed Forestry Operations in the Glen Innes Forest Management Area. Fauna Impact Statement.* Supplement to the Environmental Impact Statement. Report for the Forestry Commission of New South Wales by Austeco Pty. Ltd.

Smith, A.P., Nagy, K.A., Fleming, M.R. and Green, B. (1982). Energy requirements and turnover in free-living Leadbeater's Possums, *Gymnobelideus leadbeateri* (Marsupialia: Petauridae). *Australian Journal of Zoology*, **30**, 737–749.

Smith, A.P., Wellham, G.S. and Green, S.W. (1989). Seasonal foraging activity and microhabitat selection by echidnas (*Tachyglossus aculeatus*) on the New England Tablelands. *Australian Journal of Ecology*, **14**, 457–466.

Smith, A.T. (1980). Temporal changes in insular populations of the Pika (*Ochotona principes*). *Ecology*, **61**, 8–13.

Smith, A.T. and Peacock, M.M. (1990). Conspecific attraction and the determination of metapopulation colonization rates. *Conservation Biology*, **4**, 320–323.

Smith, C.S., Lonsdale, W.M. and Fortune, J. (1999). When to ignore advice: invasion predictions and decision theory. *Biological Invasions*, **1**, 89–96.

Smith, D. (1994). *Saving a Continent. Towards a Sustainable Future.* UNSW Press, Sydney.

Smith, D. (1998). *Water in Australia. Resources and Management.* Oxford University Press, Melbourne.

Smith, D.M., Shields, P.G. and Danahar, T.J. (1994). *An assessment of the extent of clearing in south-central Queensland.* Department of Primary Industries, Queensland. Land Resources Bulletin No. QV94001.

Smith, D.S. and Hellmund, P.C. (Eds) (1993). *Ecology of Greenways: Design and Function of Linear Conservation Areas.* University of Minnesota Press, Minneapolis.

Smith, F.D.M., May, R.M., Pellew, R., Johnson, T.H. and Walter, K.R. (1993). How much do we know about the current extinction rate? *Trends in Ecology and Evolution*, **8**, 375–378.

Smith, G.C. and Agnew, G. (2002). The value of 'bat boxes' for attracting hollow-dependent fauna to farm forestry plantations in southeast Queensland. *Ecological Management and Restoration*, **3**, 37–46.

Smith, J.N. and Hellman, J.J. (2002). Population persistence in fragmented landscapes. *Trends in Ecology and Evolution*, **17**, 397–399.

Smith, P.A. (1994). Autocorrelation in logistic regression modeling of species' distribution. *Global Ecology and Biogeography Letters*, **4**, 47–61.

Smith, P.J., Pressey, R.L. and Smith, J.E. (1994). Birds of particular conservation concern in the Western Division of New South Wales. *Biological Conservation*, **69**, 315–338.

Smith, R.B. and Woodgate, P. (1985). Appraisal of fire damage for timber salvage by remote sensing in mountain ash forests. *Australian Forestry,* **48**, 252–263.

Smith, R.J. (2000). An investigation into the relationships between anthropogenic forest disturbance patterns, population viability and landscape indices. MSc thesis. Institute of Land and Food Resources, The University of Melbourne, Australia.

Smith, S.J. (1980). *The Tasmanian Tiger – 1980.* Tasmanian National Parks and Wildlife Service. Wildlife Division Technical Report No. 91/1.

Snow, A.A., Andow, D.A., Gepts, P., Hallerman, E.M., Power, A., Tiedje, J.M. and Wolfenbarger, L.L. (2005). Genetically engineered organisms and the environment: current status and recommendations. *Ecological Applications*, **15**, 377–404.

Soberon, J.M. and Llorente, J.B. (1993). The use of species accumulation functions for the prediction of species richness. *Conservation Biology,* **7**, 480–488.

Sokal, R.R. and Oden, N.L. (1978). Spatial autocorrelation in biology. I. Methodology. *Biological Journal of the Linnean Society,* **10**, 199–228.

Sokal, R.R. and Rohlf, F.J. (1995). *Biometry.* Freeman, San Francisco.

Solow, A.R. (1993a). Inferring extinction from sighting data. *Ecology,* **74**, 962–964.

Solow, A.R. (1993b). A simple test for change in community structure. *Journal of Animal Ecology,* **62**, 191–193.

Solow, A.R. and Roberts, D.L. (2003). A nonparametric test for extinction based on a sighting record. *Ecology,* **84**, 1329–1332.

Soulé, M., Gilpin, M., Conway, W. and Foose, T. (1986). The millennium ark: How long a voyage, how many staterooms, how many passengers? *Zoo Biology,* **5**, 101–113.

Soulé, M.E. (1980). Thresholds of survival: Maintaining fitness and evolutionary potential. pp. 151–169. In: *Conservation Biology: An Evolutionary-ecological Perspective.* Soulé, M.E. and Wilcox, B. (Eds). Sinauer, Massachusetts.

Soulé, M.E. (1983). What do we really know about extinction? pp. 111–124. In: *Genetics and Conservation: A Reference for Managing Wild Animal and Plant Populations.* Schonewald-Cox, C.M., Chambers, S.M., MacBryde, B. and Thomas, L. (Eds). Benjamin/Cummings, Meulo Park, California.

Soulé, M.E. (1985). What is conservation biology? *BioScience*, **35**, 727–734.

Soulé, M.E. (Ed) (1987). *Viable Populations for Conservation.* Cambridge University Press, Cambridge.

Soulé, M.E. (1991). Conservation: Tactics for a constant crisis. *Science*, **253**, 744–750.

Soulé, M.E. and Kohm, K.A. (1989). *Research Priorities for Conservation Biology.* Island Press, Covelo.

Soulé, M.E. and Sanjayan, M.A. (1998). Conservation targets do they help. *Science*, **279**, 2060–2061.

Soulé, M.E. and Simberloff, D. (1986). What do genetics and ecology tell us about the design of nature reserves? *Biological Conservation,* **35**, 19–40.

Southgate, R. and Possingham, H.P. (1995). Modelling the reintroduction of the Greater Bilby *Macrotis lagotis* using the metapopulation model analysis of the likelihood of extinction (ALEX). *Biological Conservation,* **73**, 151–160.

Specht, R.L. (1970). Vegetation. pp. 44–67. In: *The Australian Environment.* 4th edition. Leeper, G.W. (Ed). CSIRO and Melbourne University Press, Melbourne.

Specht, R.L. (1981). Major vegetation formations in Australia. In: *Ecological biogeography of Australia.* Keats, A. (Ed). W. Junk, The Hague.

Specht, R.L. and Specht, A. (1999). *Australian Plant Communities. Dynamics of Structure, Growth and Biodiversity.* Oxford University Press, Melbourne.

Specht, R.L., Specht, A., Whelan, M.B. and Hegarty, E.E. (1995). *Conservation Atlas of Plant Communities in Australia.* Centre for Coastal Management and Southern Cross University Press, Lismore.

Spellerberg, I.F. (1994). *Monitoring Ecological Change.* 2nd edition. Cambridge University Press, Cambridge.

Spencer, J. (1991). Indications of antiquity: some observations on the nature of plants associated with ancient woodland. *British Wildlife*, **2**, 90–102.

Spennemann, D.H. and Allen, L.R. (2000). The avian dispersal of olives *Olea europeae*: Implications for Australia. *Emu*, **100**, 264–273.

Spies, T.A. and Franklin, J.F. (1996). The diversity and maintenance of old-growth forests. pp. 296–314. In: *Biodiversity in Managed Landscapes: Theory and Practice.* Szaro, R.C. and Johnson, D.W. (Eds). Oxford University Press, New York.

Spies, T.A. and Turner, M.G. (1999). Dynamic forest mosaics. pp. 95–160. In: *Managing Biodiversity in Forest Ecosystems.* Hunter III, M. (Ed). Cambridge University Press, Cambridge.

Spooner, P., Lunt, I. and Robinson, W. (2002). Is fencing enough? The short-term effects of stock exclusion in remnant grassy woodlands in southern NSW. *Ecological Management and Restoration*, **3**, 117–126.

Springer, A.M., Estes, J.A., van Vliet, G.B., Williams, T.M., Doak, D.F., Danner, E.M., Forney, K.A. and Pfister, B. (2003). Sequential megafaunal collapse in the North Pacific Ocean: an ongoing legacy of industrial whaling?

Proceedings of the National Academy of Sciences, **100**, 1223–12228.

Squire, R.O. (1987). *Silvicultural Systems for Victoria's Commercially Important Mountain Eucalypt Forests: Project Brief.* Public Lands and Forest Division, Melbourne.

Srivastava, D.S. and Lawton, J.H. (1998). Why more productive sites have more species: An experimental test of theory using tree-hole communities. *The American Naturalist*, **152**, 510–529.

SSCAVA (*Report of the Senate Select Committee on Agricultural and Veterinary chemicals in Australia*). Australian Government Publishing Service, Canberra. 1990.

St. John, B. (1997). Risk assessment and Koala management in South Australia. *Australian Biologist*, **10**, 47–56.

Stacey, P.B. and Taper, M. (1992). Environmental variation and the persistence of small populations. *Ecological Applications*, **2**, 18–29.

Stachowicz, J.J., Terwin, J.R., Whitlatch, R.B. and Osman, R.W. (2002). Linking climate change and biological invasions: ocean warming facilities nonindigenous species invasions. *Proceedings of the National Academy of Sciences*, **99**, 15497–15550.

Stafford-Smith, M., Morton, S. and Ash, A. (2000). Towards sustainable pastoralism in Australia's rangelands. *Australian Journal of Environmental Management*, **7**, 190–203.

Stallard, R.F. (2001). Possible environmental factors underlying amphibian decline in eastern Puerto Rico: analysis of U.S. Government data archives. *Conservation Biology*, **15**, 943–953.

Stanbury, P.J. (1987). The discovery of the Australian fauna and the establishment of collections. pp. 202–226. In: *Fauna of Australia.* Volume 1A. General Articles. Dyne, G.R. and Walton, D.W. (Eds). Australian Government Publishing Service, Canberra.

Stapp, P., Antolin, M.F. and Ball, M. (2004). Patterns of extinction in Prarie Dog metapopulations: plague outbreaks follow El Niño events. *Frontiers in Ecology and Environment*, **2**, 235–240.

Starfield, A.M. and Bleloch, A.L. (1992). *Building models for conservation and wildlife management.* Burgess International Group, Edina.

Starfield, A.M., Roth, J.D. and Ralls, K. (1995). 'Mobbing' in Hawaiian Monk Seals (*Monachus schauinslani*): the value of simulation modeling in the absence of apparently crucial data. *Conservation Biology*, **9**, 166–174.

Start, T., Burbidge, A., Sinclair, E. and Wayne, A. (1995). Gilbert's Potoroo. *Landscope*, **1995**, 29–34.

State Forests of New South Wales Northern Region (1996). *Fuel Management Plan Casino District, Northern Region State Forests of New South Wales.* Murwillumbah Management Area EIS Supporting Document No. 2. Forestry Commission of New South Wales.

State Forests of NSW (1994). *Eden Forest Management Area Plan, proposed forestry operations, Environmental Impact Statement.* Sydney.

State Forests of NSW (2000). *Annual Report 2001/02.* State Forests of NSW, Pennant Hills, NSW.

State Government of Victoria (1988). *Flora and Fauna Guarantee Act.* State Government of Victoria, Melbourne. (http://www.dpi.vic.gov.au/dse/nrenpa.nsf/FID/-0488335CD48EC1424A2567C10006BF6D?OpenDocument).

State Government of Victoria (2003). *Report of the Inquiry into the 2002–2003 Bushfires in Victoria.* State Government of Victoria, Melbourne. [online] URL: http://www.dpc.vic.gov.au

State of the Environment Advisory Council (1996). *State of the environment Australia 1996.* CSIRO, Melbourne. Report for Commonwealth Minister for the Environment.

State of the Environment Report (2001a). *Biodiversity.* Commonwealth of Australia, Canberra.

State of the Environment Report (2001b). *Oceans.* Commonwealth of Australia, Canberra.

State of the Environment Report (2001c). *Atmosphere.* Commonwealth of Australia, Canberra.

State of the Environment Report (2001d). *Land.* Commonwealth of Australia, Canberra.

State of the Environment Report (2001e). *Human settlements.* Commonwealth of Australia, Canberra.

State of the Environment Report (2001f). *Inland Waters.* Commonwealth of Australia, Canberra.

Steinbauer, M.J., Yonow, T., Reid, I.A. and Cant, R. (2002). Ecological biogeography of species of *Gelonus*, *Acantholybas* and *Amorbus* in Australia. *Austral Ecology*, **27**, 1–25.

Steinke, E. and Walton, C. (1999). Weed risk assessment of plant imports to Australia: Policy and process. *Australian Journal of Environmental Assessment*, **6**, 157–163.

Stenseth, N. and Lidicker, W. (Eds) (1992). *Animal Dispersal.* Chapman and Hall, London.

Stirzaker, R., Vertessey, R. and Sarre, A. (Eds) (2002). *Trees, Water and Salt. An Australian Guide to Using Trees for Healthy Catchments and Productive Farms.* Joint Venture Agroforestry Program, Canberra.

Stocker, G.C. and Unwin, G.L. (1986). Fire and the functioning of tropical plant communities. pp. 91–103. In: *Tropical Plant Communities*. Clifford, H.T. and Specht, R.L. (Eds). University of Queensland, Brisbane.

Stockwell, D.R.B., Davey, S.M., Davis, J.R. and Nobel, I.R. (1990). Using decision trees to predict Greater Glider density. *AI Applications*, **4**, 33–43.

Stodart, E. and Parer, I. (1988). Colonisation of Australia by the rabbit *Oryctolagus cuniculus* (L.). *CSIRO Wildlife and Ecology*. Project Report No. 6.

Stoneham, G., Crowe, M., Platt, S., Chaudhri, V., Soligo, J. and Strappazzon, L. (2000). *Mechanisms for biodiversity conservation on private land*. Department of Natural Resources and Environment, Melbourne.

Stork, N. (1992). Measuring global biodiversity and its decline. pp. 41–68. In: *Biodiversity II: Understanding and Protecting our Biological Resources*. Reaka-Kudla, M.L., Wilson, D.E. and Wilson, E.O. (Eds). Joseph Henry Press, Washington DC.

Strahan, R. (Ed) (1995). *Complete Book of Australian Mammals*. Collins Angus and Robertson Publishers, Sydney.

Strauss, S.Y. (2001). Benefits and risks of biotic exchange between *Eucalyptus* plantations and native Australian forests. *Austral Ecology*, **26**, 447–457.

Struhsaker, T.T., Struhsaker, P.J. and Siex, K.S. (2005). Conserving Africa's rain forests: problems in protected areas and possible solutions. *Biological Conservation*, **123**, 45–54.

Suckling, G.C. (1982). Value of reserved habitat for mammal conservation in plantations. *Australian Forestry*, **45**, 19–27.

Suckling, G.C. and Goldstraw, P. (1989). Progress of Sugar Glider, *Petaurus breviceps*, establishment at Tower Hill State Game Reserve, Victoria. *Victorian Naturalist*, **106**, 179–83.

Suckling, G.C. and Heislers, A. (1978). Populations of four small mammals in radiata pine plantations and eucalypt forests of north-eastern Victoria. *Australian Wildlife Research*, **5**, 305–315.

Suckling, G.C., Backen, E., Heislers, A. and Neumann, F.G. (1976). *The flora and fauna of Radiata Pine plantations in north-eastern Victoria*. Forest Commission of Victoria Bulletin No. **24**. Forest Commission of Victoria, Melbourne, Australia.

Sugden, E.A. and Pyke, G.H. (1991). Effects of Honeybees on colonies of *Exoneura asimillima*, an Australian native bee. *Australian Journal of Ecology*, **16**, 171–181.

Sugihara, G. (1980). Minimal community structure; an explanation of species abundance patterns. *American Naturalist*, **116**, 770–787.

Summerville, K.S., Ritter, L.M. and Crist, T.O. (2004). Forest moth taxa as indicators of lepidopteran richness and habitat disturbance: a preliminary assessment. *Biological Conservation*, **116**, 9–18.

Sumner, J. and Dickman, C.R. (1998). Distribution and identity of species in the *Antechinus stuartii-A. flavipes* group (Marsupialia: Dasyuridae) in south-eastern Australia. *Australian Journal of Zoology*, **46**, 27–41.

Sundquist, F. (1993). Should we put them all back? *International Wildlife*, **22**, 35–40.

Sunnucks, P. (2000). Efficient genetic markers for population biology. *TREE*, **15**, 199–203.

Surridge, A.K., Timmins, R.J., Hewitt, G.M. and Bell, D.J. (1999). Striped rabbits in Southeast Asia. *Nature*, **400**, 726.

Suter, G. (1993). *Environmental Risk Assessment*. Lewis, Baton Roca.

Suter, W., Graf, R.F. and Hess, R. (2002). Capercaillie (*Tetrao urogallus*) and avian biodiversity: testing the umbrella-species concept. *Conservation Biology*, **16**, 778–788.

Sutherland, W. (1996). *Ecological Census Techniques. A Handbook*. Cambridge University Press, Cambridge.

Sutherland, W.J. (2000). *The Conservation Handbook. Research, Management and Policy*. Blackwell Science, Oxford.

Swanson, F.J., Kratz, T.J., Caine, N. and Woodmansee, R.G. (1988). Landform effects on ecosystem patterns and processes. *BioScience*, **38**, 92–98.

Swenson, J.E., Alt, K.L. and Eng, R.L. (1986). Ecology of Bald Eagles in the Greater Yellowstone ecosystem. *Wildlife Monographs*, **95**, 1–46.

Swets J.A., Dawes, R.M. and Monahan, J. (2000). Better decisions through science. *Scientific American*, **October 2000**, 82–87.

Swift Parrot Recovery Team (2000). *Draft Swift Parrot Recovery Plan 2001–2005*. Parks and Wildlife Service, Hobart, Tasmania, Australia.

Sydes, M. (1994). Orchids: Indicators of management success. *Victorian Naturalist*, **111**, 213–217.

Sydes, M. (1995). Is the concept of provenance relevant to biodiversity conservation? A genetic viewpoint. pp. 15–28. In: *Linking Provenance and Biodiversity Conservation*. Greening Australia, Victoria.

Szaro, R.C. (1986). Guild management: An evaluation of avian guilds as a predictive tool. *Environmental Management*, **10**, 681–688.

Szaro, R.C. and Jakle, M.D. (1985). Avian use of a desert riparian island and its adjacent scrub habitat. *The Condor*, **87**, 511–519.

Tait, C.J., Daniels, C.B. and Hill, R.S. (2005). Changes in species assemblages within the Adelaide metropolitan area, Australia, 1836–2002. *Ecology*, **15**, 346–359.

Tait, N.N., Briscoe, D.A. and Rowell, D.M. (1995). Onycophora – ancient and modern radiations. *Association of Australasian Paleontologists,* Memoir **18**, 21–30.

Takahata, N. and Slatkin, M. (1990). Genealogy of neutral genes in two partially isolated populations. *Theoretical Population Biology,* **38**, 331–350.

Talbot, L.M. (1993). *Principles for Living Resource Conservation.* Draft preliminary report on consultations. The Marine Mammal Commission, Washington, D.C. September, 1993.

Tang, S.M. and Gustafson, E.J. (1997). Perception of scale in forest management planning: Challenges and implications. *Landscape and Urban Planning,* **39**, 1–9.

Taylor, B., Kremsater, L. and Ellis, R. (1998). *Adaptive Management of Forests in British Columbia.* British Columbia Ministry of Forests – Forest Practices Branch. Victoria, British Columbia.

Taylor, B.L. (1995). The reliability of using population viability analysis for risk classification of species. *Conservation Biology,* **9**, 551–558.

Taylor, B.L. and Gerrodette, T. (1993). The uses of statistical power in conservation biology: the Vaquita and Northern Spotted Owl. *Conservation Biology,* **7**, 489–500.

Taylor, C.R., Caldwell, S.L. and Rowntree, V.J. (1972). Running up and down hills: some consequences of size. *Science,* **178**, 1096–1097.

Taylor, P.D. and Merriam, G. (1995). Habitat fragmentation and parasitism of a forest damselfly. *Landscape Ecology,* **11**, 181–189.

Taylor, P.D., Fahrig, L., Henein, K. and Merriam, G. (1993). Connectivity is a vital element of landscape structure. *Oikos,* **68(3)**, 571–573.

Taylor, R. (1979). How the Macquarie Island Parakeet became extinct. *New Zealand Journal of Ecology,* **2**, 42–45.

Taylor, R., Duckworth, P., Johns, T. and Warren, B. (1997). Succession in bird assemblages over a seven-year period in regrowth dry sclerophyll forest in south-east Tasmania. *Emu,* **97**, 220–230.

Taylor, R.J. (1990). Occurrence of log-dwelling invertebrates in regeneration and old-growth wet sclerophyll forest in southern Tasmania. *Papers and Proceedings of the Royal Society of Tasmania,* **124**, 27–34.

Taylor, R.J., Bryant, S.L., Pemberton, D. and Norton, T.W. (1985). Mammals of the Upper Henty River region, Western Tasmania. *Papers and Proceedings of the Royal Society of Tasmania,* **119**, 7–15.

Tellez-Valdes, O. and Davila-Aranda, P. (2003). Protected areas and climate change: A case study of the cacti in the Tehuacan-Cuicatlan Biospehere Reserve, Mexico. *Conservation Biology,* **17**, 846–853.

Temple, S.A. (1986). The problem of avian extinctions. *Current Ornithology,* **3**, 453–485.

Temple, S.A. and Cary, J.R. (1988). Modelling dynamics of habitat interior bird populations in fragmented landscapes. *Conservation Biology,* **2**, 340–7.

Temple, S.A. and Wiens, J.A. (1989). Bird populations and environmental changes: Can birds be bio-indicators? *American Birds,* **43**, 260–270.

Templeton, A. (1986). Coadaptation and outbreeding depression. pp. 105–116. In: *Conservation Biology. The science of scarcity and diversity.* Soulé, M.E. (Ed). Sinauer Associates, Sunderland.

Templeton, A.R. (1991). Genetics and conservation biology. pp. 15–29. In: *Species Conservation: A Population-biological Approach.* Seitz, A. and Loeschcke, V. (Eds). Birkhauser, Basel.

Terborgh, J. (1974). Preservation of natural diversity. The problem of extinction prone species. *BioScience,* **24**, 715–722.

Terborgh, J. (1986). Keystone plant resources in the tropical forest. pp. 330–344. In: *Conservation Biology: the Science of Scarcity and Diversity.* Soulé, M.E. (Ed). Sinauer, Sunderland.

Terborgh, J. (1989). *Where Have all the Birds Gone?* Princeton University Press, Princeton, New Jersey.

Terborgh, J. (1992). Maintenance of diversity in tropical forests. *Biotropica,* **24**, 283–292.

Thackway, R. and Creswell, I. (Eds) (1995). *An Interim Biogeographic Regionalisation for Australia: A Framework for Establishing the National System of Reserves.* Version 4.0. Australian Nature Conservation Agency, Canberra.

Thackway, R. and Creswell, I. (1997). A bioregional framework for planning the national system of protected areas in Australia. *Natural Areas Journal,* **17**, 241–247.

Thackway, R. and Olsson, K. (1999). Public/private partnerships and protected areas: selected Australian case studies. *Landscape and Urban Planning,* **44**, 87–97.

Thain, M. and Hickman, M. (1994). *The Penguin Dictionary of Biology.* 9th edition. Penguin, London.

Tharme, R.E. (2003). A global assessment on environmental flow assessment: emerging trends in the development and application of environmental flow methodologies for rivers. *River Research and Applications,* **19**, 397–441.

The Australian Forestry Standard (2003). *The Australian Forestry Standard.* The Australian Forestry Standard Project Office, Canberra, Australia.

The Nature Conservancy (2003). *The Nature Conservancy's Invasive Species Initiative.* The Nature Conservancy, Arlington, Virginia, USA.

Thiollay, J.M. (1992). Influence of selective logging on bird species-diversity in a Guianan rain-forest. *Conservation Biology,* **6**, 47–63.

Thomas, C.D. (1990). What do real population dynamics tell us about minimum viable population sizes. *Conservation Biology,* **4**, 324–327.

Thomas, C.D., Cameron, A., Green, R.E., Bakkenes, M., Beaumont, L.J., Collingham, Y.C., Erasmus, B.F., Siqueiria, M., Grainer, A., Hannah, L., Hughes, L., Huntley, B., Jaarsveld, A., Midgley, G.F., Miles, L., Heurta-Ortega, M.A., Townsend Peterson, A., Philips, O.L. and Williams, S.E. (2004). Extinction risk from climate change. *Nature,* **427**, 145–148.

Thomas, C.D., Thomas, J.A. and Warren, M.S. (1992). Distributions of occupied and vacant habitats in fragmented landscapes. *Oecologica,* **92**, 563–567.

Thomas, I. (1996). *Environmental Impact Assessment in Australia: Theory and Practice.* The Federation Press, Sydney.

Thomas, J.A. and Morris, M.G. (1995). Rates and patterns of extinction among British invertebrates. pp. 111–130. In: *Extinction Rates.* Lawton, J.H. and May, R.M. (Eds). Oxford University Press, Oxford.

Thomas, J.W. (Ed) (1979). Wildlife habitats in managed forests the Blue Mountains of Oregon and Washington. USDA Agricultural Handbook 553. U.S. Government Printing Office, Washington, DC.

Thompson, I.D. and Angelstam, P. (1999). Special species. pp. 434–459. In: *Maintaining Biodiversity in Forest Ecosystems.* Hunter, M.L. (Ed). Cambridge University Press, Cambridge.

Thompson, M.B., Medlin, G., Hutchinson, R. and West, N. (1989). Short-term effects of fuel reduction burning on populations of small terrestrial mammals. *Australian Wildlife Research,* **16**, 117–129.

Thoms, M. and Sheldon, F. (2000). Water resource development and hydrological change in a large dryland river: The Barwon-Darling River, Australia. *Journal of Hydrology,* **228**, 10–21.

Thoms, M., Suter, P., Roberts, J., Koehn, J., Jones, G., Hillman, T. and Close, A. (2000). *Report of the River Murray Scientific Panel on Environmental Flows. River Murray – Dartmouth to Wellington and the Lower Darling River.* Murray Darling Basin Commission, Canberra.

Threatened Species Network (2003). *Western Swamp Tortoise.* Fact Sheet. World Wide Fund for Nature, Perth.

Thrush, S.F., Hewitt, J.E., Cunnings, V.J., Green, M.O., Funnell, G.A. and Wilkinson, M.R. (2000). The generality of field experiments: interactions between local and braod-scale processes. *Ecology,* **81**, 399–415.

Tickle, P., Hafner, S., Lesslie, R., Lindenmayer, D.B., McAlpine, C., Mackey, B., Norman, P. and Phinn, S. (1998). *Scoping Study: Final report.* Montreal Indicator 1.1e. Fragmentation of forest types – identification of research priorities. A study prepared for the Forest and Wood Products Research and Development Corporation. Canberra, Australia.

Tidemann, C.R. (2003). Displacement of a flying-fox camp using sound. *Ecological Restoration and Management,* **4**, 224–226.

Tilghman, N.G. (1989). Impacts of White-tailed Deer on forest regeneration in northwestern Pennsylvania. *Journal of Wildlife Management,* **53**, 524–532.

Tilman, D., May, R.M., Lehman, C.L. and Nowak, M.A. (1994). Habitat destruction and the extinction debt. *Nature,* **371**, 65–66.

Timbal, B. (2004). Southwest Australia past and future rainfall trends. *Climate Research,* **26**, 233–249.

Tisdell, C., Wilson, C. and Nantha, H.S. (2005). Policies for saving a rare Australian glider: economics and ecology. *Biological Conservation,* **123**, 237–248.

Tissot, B.N. and Hallacher, L.E. (2003). Effects of aquarium collectors on coral reef fishes in Kona, Hawaii. *Conservation Biology,* **17**, 1759–1768.

Todd, C.R., Jenkins, S. and Bearlin, A.R. (2002). Lessons about extinction and translocation: models for Eastern Barred Bandicoots (*Parameles gunnii*) at Woodland Historic Park, Victoria, Australia. *Biological Conservation,* **106**, 211–223.

Todd, C.R., Ryan, T., Nicol, S.J. and Bearlin, A.R. (2005). The impact of cold-water releases on the critical period of post-spawning survival and its implications for Murray Cod (*Maccullochella peelii peelii*): a case study of the Mitta Mitta River in south-eastern Australia. *River Research and Applications* (in press).

Torsvik, V., Goksoyr, J. and Daae, F. (1990). High diversity in DNA of soil bacteria. *Applied and Environmental Microbiology,* **56**, 782–787.

Toyne, P. (1994). *The reluctant nation. Environment, law and politics in Australia.* ABC Books, Sydney.

Traill, B.J. (1993). Forestry, birds, mammals, and management in Box and Ironbark forests. *Victorian Naturalist,* **110**, 11–14.

Tratado de Cooperacion Amazonica (1995). *Tarapoto proposal on criteria and indicators for sustainability of*

Amazonian forests. Amazonian Cooperation Treaty, Peru. February, 1995.

Troll, C. (1939). Luftbildplan and okologische Bodenforschung. Zeitschraft der Gesellschraft fur Erdkunde Zu Berlin.

Trzcinski, M.K., Fahrig, L. and Merriam, G. (1999). Independent effects of forest cover and fragmentation on the distribution of forest breeding birds. *Ecological Applications,* **9**, 586–593.

Tscharntke, T. (1992). Fragmentation of Phragmites habitats, minimum viable population size, habitat suitability, and local extinction of moths, midges, flies, aphids and birds. *Conservation Biology,* **6**, 530–536.

Tubelius, D.P., Lindenmayer, D.B., Saunders, D.A., Cowling, A. and Nix, H.A. (2004). Landscape supplementation provided by an exotic matrix: implications for bird conservation and forest management in a softwood plantation system in south-eastern Australia (*Oikos*) (in review).

Tuchmann, E.T., Connaughton, K.P., Freedman, L.E. and Moriwaki, C.B. (1996). The Northwest Forest Plan. A report to the President and Congress. USDA Office of Forestry and Economic Assistance, Portland, OR.

Tuck, G.N., Polacheck, T. and Bulman, C.M. (2003). Spatio-temporal trends of longline fishing effort in the Southern Ocean and implications for seabird catch. *Biological Conservation,* **114**, 1–27.

Tuckey, W. (2000). http://www.affa.gov.au/ministers/tuckey/releases/00/00_73tu.html

Turner, I.M. (1996). Species loss in fragments of tropical rain forest: a review of the evidence. *Journal of Applied Ecology,* **33**, 200–209.

Turner, M.G. (1989). Landscape ecology: the effect of pattern on process. *Annual Review of Ecology and Systematics,* **20**, 171–197.

Turner, M.G., Baker, W.L., Peterson, C.J. and Peet, R.K. (1998). Factors influencing succession: lessons from large, infrequent natural disturbances. *Ecosystems,* **1**, 511–523.

Turner, M.G., Romme, W.H. and Tinker, D.B. (2003). Surprises and lessons form the 1988 Yellowstone fires. *Frontiers in Ecology and Environment,* **1**, 351–358.

Twigg, L.E., Loew, T.J., Martin, G.R., Wheeler, A.G., Gray, G.S., Griffin, S.L., O'Reilly, C.M., Robinson, D.J. and Huback, P.H. (2000). Effects of surgically imposed sterility on free-ranging rabbit populations. *Journal of Applied Ecology,* **37**, 16–39.

Twyford, K.L., Humphrey, P.G., Nunn, R.P. and Willoughby, L. (2000). Eradication of Feral Cats (*Felis catus*) from Gabo Island, south-east Victoria. *Ecological Management and Restoration,* **1**, 42–49.

Tyndale-Biscoe, C. and Calaby, J.H. (1975). Eucalypt forests as refuges for wildlife. *Australian Forestry,* **38(2)**, 117–33.

Tyndale-Biscoe, C.H. (1995). Vermin and viruses. Risks and benefits of viral-vectored immunosterilisation. *Search,* **26**, 239–244.

Tyndale-Biscoe, C.H. (1997). A fresh approach to quarantine. *Search,* **28**, 54–58.

Tyndale-Biscoe, C.H. and Smith, R.F. (1969). Studies of the marsupial glider, *Schoinobates volans* (Kerr). III. Response to habitat destruction. *Journal of Animal Ecology,* **38**, 651–659.

Tyre, D., Possingham, H.P. and Lindenmayer, D.B. (2001). Inferring process from pattern: can territory occupancy provide information about life history paramters? *Ecological Applications,* **11**, 1051–1061.

Uhl, C. and Kauffman, J.B. (1990). Deforestation, fire susceptibility, and potential tree response to fire in the eastern Amazon (Brazil). *Ecology,* **71**, 437–449.

UN (1982). *The world charter for nature.* Annex 2. General Assembly of the United Nations.

UN (1998). *World population projections to 2150.* UN Population Division, New York.

UNCED (1992). *Agenda 21, Chapter 11: Combating deforestation.* United Nations Conference on Environment and Development, UNO, New York.

Underwood, A.J. (1993). The mechanics of spatially replicated sampling programmes to detect environmental impacts in a variable world. *Australian Journal of Ecology,* **18**, 99–116.

Underwood, A.J. (1995). Ecological research and environmental management. *Ecological Applications,* **5**, 232–247.

UNEP (1999). *Global Environmental Outlook 2000.* United Nations Environment Programme, Nairobi, Kenya.

UNESCO (2003). *The World Heritage List.* http://whc.unesco.org/nwhc/pages/doc/mainf3.htm

United States Fish and Wildlife Service (1980). *Habitat Evaluation Procedures.* U.S. Government Printing Office, Washington, DC, Ecological Services Manual No. 102.

United States Fish and Wildlife Service (1981). *Standard for the Development of Habitat Suitability Index Models.* U.S. Government Printing Office, Washington, DC, Ecological Services Manual No. 103.

Urban, D.L. (2000). Using model analysis to design monitoring programs for landscape management and impact assessment. *Ecological Applications,* **10**, 1820–1832.

USDA (2003). *National Plant Germplasm System.* United States Department of Agriculture, Agricultural Research Service, National Plant Germplasm System. [online] URL: http://www.ars-grin.gov/npgs/

Usher, M.B. (1986). Wildlife conservation evaluation: attributes, criteria and values. pp. 1–45. In: *Wildlife Conservation Evaluation.* Usher, M.B. (Ed). Chapman and Hall, London.

Usher, M.B. (1991). Scientific requirements of a monitoring program. pp. 15–32. In: *Monitoring for Conservation and Ecology.* Goldsmith, F.B. (Ed). Chapman and Hall, London.

Van der Meer, P.J., Dignan, P. and Savaneh, A.G. (1999). Effects of gap size on seedling establishment, growth and survival at three years in mountain ash (*Eucalyptus regnans* F. Muell.) forest in Victoria, Australia. *Forest Ecology and Management,* **117**, 33–42.

van Dam, R.A., Walden, D.J. and Begg, G.W. (2002). A preliminary risk assessment of cane toads in Kakadu National Park. *Scientist Report 164.* Supervising Scientist, Darwin, Northern Territory.

Van Dugteren, A. (1999). Conserving the future of the Great Artesian Basin. *Geo,* **20**, 22–37.

Van Dyck, S. (1993). The taxonomy and distribution of *Petaurus gracilis* (Marsupialia: Petauridae), with notes on its ecology and conservation status. *Memoirs of the Queensland Museum,* **33**, 77–122.

Van Horne, B. (1983). Density as a misleading indicator of habitat quality. *Journal of Wildlife Management,* **47**, 893–901.

Van Horne, B. and Wiens, J.A. (1991). *Forest bird habitat suitability models and the development of general habitat models.* U.S. Fish and Wildlife Service, Washington, DC, Fisheries and Wildlife Research Report No. 8. pp. 1–31.

van Jaarsveld, A.S., Ferguson, J.H. and Bredenkamp, G.J. (1998). The Groenvaly grassland fragmentation experiment: design and initiation. *Agriculture, Ecosystems and Environment,* **68**, 139–150.

Van Nieuwstadt, M.G., Shiel, D. and Kartawinata, K. (2001). The ecological consequences of logging in the burned forests of east Kalimantan, Indonesia. *Conservation Biology,* **15**, 1183–1186.

van Valen, L.M. (1985). A theory of origination and extinction. *Evolutionary Theory,* **7**, 133–142.

Vane-Wright, R.I. (1992a). Species concepts. pp. 13–16. In: *Global diversity: status of the earth's living resources.* World Conservation Monitoring Centre (Ed). Chapman and Hall, London.

Vane-Wright, R.I. (1992b). Systematics and diversity. pp. 7–12. In: *Global Diversity: Status of the Earth's Living Resources.* World Conservation Monitoring Centre (Ed). Chapman and Hall, London.

Vane-Wright, R.I., Humphries, C.J. and Williams, P.H. (1991). What to protect? – Systematics and the agony of choice. *Biological Conservation,* **55**, 235–254.

Vaughan, P.J. (1987). *The Eltham Copper Butterfly Draft Management Plan.* Department of Conservation, Forests and Lands, Victoria. Arthur Rylah Institute Technical Report No. 657.

Vaughan, P.J. (1988). *Management Plan for the Eltham Copper Butterfly (*Paralucia pyrodiscus lucida Crosby*).* Department of Conservation, Forests, and Lands, Victoria. Technical Report Series No. 79. 54 pp.

Vaughton, G. (1996). Pollination disruption by European honeybees in the Australian bird-pollinated shrub *Grivellea barklyeana* (Proteaceae). *Plant Systematics and Evolution,* **200**, 89–100.

Vavilov, N.I. (1949). The origin, variation, immunity and breeding of cultivated plants. (Translated by K.S. Chester). *Chronica Botanica,* **13**, 1–6.

Verner, J. (1984). The guild concept applied to management of bird populations. *Environmental Management* **8**, 1–14.

Veron, J.E.N. (2000). *Corals of the World.* Volumes 1–3. Australian Institute of Marine Science, Townsville.

Viana, V.M., Ervin, J., Donovan, R.Z., Elliott, C. and Gholz, H. (1996). *Certification of Forest Products.* Island Press, Covelo.

Vietch, C.R. (1995). Habitat repair: A necessary prerequisite to translocation of threatened birds. pp. 97–104. In: *Reintroduction Biology of Australasian Fauna.* Serena, M. (Ed). Surrey Beatty, Chipping Norton.

Viggers, K.L. and Lindenmayer, D.B. (2005). A review of the biology of the Short-eared Possum *Trichosurus caninus* and the Mountain Brushtail Possum Trichosurus cunninghamii. pp. 490–505. In: The biology of Australian possums and gliders. Goldingay, R.L. and Jackson, S.M. (Eds). Surrey Beatty and Sons, Chipping Norton.

Viggers, K.L. and Hearn, J. (2005). The kangaroo conundrum: home range studies and implications for land management. *Journal of Applied Ecology,* **42**, 99–107.

Viggers, K.L. and Lindenmayer, D.B. (2002). Problems with keeping native Australian mammals as companion animals. pp. 130–151. In: *A Zoological Revolution: Using Native Fauna to Assist in its own Survival.* Royal Zoological Society of NSW, Mosman.

Viggers, K.L. and Spratt, D.M. (1995). The parasites recorded from *Trichosurus* species (Marsupialia, Phalangeridae). *Wildlife Research,* **22**, 311–332.

Viggers, K.L., Lindenmayer, D.B. and Spratt, D.M. (1993). The role and importance of disease in reintroduction biology. *Wildlife Research,* **20**, 687–698.

Villard, M.A. (1998). On forest-interior species, edge avoidance, area sensitivity, and dogmas in avian conservation. *The Auk,* **115,** 801–805.

Villard, M.A. (2002). Habitat fragmentation: major conservation issue or intellectual attractor? *Ecological Applications,* **15,** 319–320.

Villard, M.A., Trzcinski, M.K. and Merriam, G. (1999). Fragmentation effects on forest birds: relative influence of woodland cover and configuration on landscape occupancy. *Conservation Biology,* **13,** 774–783.

Vincent, P.J. and Haworth, J.M. (1983). Poisson regression models of species abundance. *Journal of Biogeography,* **10,** 153–160.

Virkkala, R., Rajasarrkka, A., Vaisanen, R.A., Vickholm, M. and Virolainen, E. (1994). The significance of protected areas for the land birds of southern Finland. *Conservation Biology,* **8,** 532–544.

Virolainen, K.M., Virola, T., Suhonen, J., Kuitunen, M., Lammin, A. and Siikamaki, P. (1999). Selecting networks of nature reserves: methods do affect the long-term outcome. *Proceedings of the Royal Society of London Series B,* **266,** 1141–1146.

Visscher, H., Looy, C.V., Collinson, M.E., Brinkhuis, H., van-Konijnenburg-Cittert, J.H., Kurschner, W.M. and Sephton, M.A. (2004). Environmental mutagenesis during the end-Permian ecological crisis. *Proceedings of the National Academy of Sciences,* **101,** 12952–12956.

Vitousek, P.M. (1994). Beyond global warming: Ecology and global change. *Ecology,* **75,** 1861–1876.

Vitousek, P.M., Ehrlich, P.R., Ehrlich, A.H. and Matson, P.M. (1986). Human appropriation of the products of photosynthesis. *BioScience,* **36,** 368–373.

Vogler, A.P. and DeSalle, R. (1994). Diagnosing units of conservation management. *Conservation Biology,* **8,** 354–363.

Von Uexkhüll, J. (1926). *Theoretical Biology.* Kegan Paul, Trench, Trubner and Co. Ltd, London.

Wace, N. (1977). Assessment of dispersal of plant species – the car borne flora in Canberra. *Proceedings of the Ecological Society of Australia,* **10,** 166–186.

Wackernagel, M. and Rees, W.E. (1996). *Our Ecological Footprint: Reducing Human Impact on the Earth.* New Society Publishers, Garbiola Island, British Columbia.

Wackernagel, M., Onisto, L., Linares, A., Falfan, I., Garcia, J., Guerrero, A. and Guerrero, G. (1997). *Ecological Footprints of Nations. How Much Nature do they Use? – How Much Nature do they Have?* Centre for Sustainability Studies. Earth Council, Costa Rica.

Waits, L.P., Talbot, S.L., Ward, R.H. and Shields, G.F. (1998). Mitochondrial DNA phylogeography of the North American Brown Bear and implications for conservation. *Conservation Biology,* **12,** 408–417.

Walker, B. (1998). The art and science of wildlife management. *Wildlife Research,* **25,** 1–9.

Walker, B.H. (1992). Biodiversity and ecological redundancy. *Conservation Biology,* **6,** 18–23.

Walker, J., Bullen, F. and Williams, B.G. (1993). Ecohydrological changes in the Murray-Darling Basin. I. The number of trees cleared over two centuries. *Journal of Applied Ecology,* **30,** 265–273.

Walker, K.F. (1981). *Ecology of freshwater mussels in the River Murray.* Australian Government Publishing Service, Canberra. Australian Water Resources Council Technical Paper No. 63.

Wall, J. (2000). Fuelwood in Australia: impacts and opportunities. pp. 372–381. In: *Temperate Eucalypt Woodlands in Australia: Biology, Conservation, Management and Restoration.* Hobbs, R.J. and Yates, C.J. (Eds). Surrey Beatty and Sons, Chipping Norton.

Wallace, A.R. (1863). On the physical geography of the Malay Archipelago. *Journal of the Royal Geographical Society, London,* **33,** 217–234.

Wallace, I.F., Lindner, R.K. and Dole, D.D. (1997). Evaluating stock and catchability trends: annual catch average per unit effort is an inadequate indicator of stock and catchability trends in fisheries. *Marine Policy,* **22,** 45–55.

Wallace, K.J., Beecham, B.C. and Bone, B.H. (2003). *Managing natural biodiversity in the Western Australian wheatbelt.* Department of Conservation and Land Management. November 2003.

Wallis, A., Stokes, D., Wescott, G. and McGee, T. (1997). Certification and labelling as a new tool for sustainable forest management. *Australian Journal of Environmental Management,* **4,** 224–238.

Walsh, B. (1992). *Feral camels. Distribution and population characteristics.* Summary of talk given to the Camel Industry Steering Committee. March 1992.

Walsh, F.J. (1990). An ecological study of traditional Aboriginal use of 'country': Martu in the Great and Little Sandy Deserts, Western Australia. pp. 23–37. In: *Australian Ecosystems: 200 years of Utilisation, Degradation and Reconstruction. Proceedings of the Ecological Society of Australia.* Vol. 16. Saunders, D.A., Hopkins, A.J.M. and How, R.A. (Eds). Surrey Beatty, Chipping Norton.

Walsh, R.G., Bjonback, R.D., Aiken, R.A. and Rosenthal, D.H. (1990). Estimating the public benefits of protecting forest quality. *Journal of Environmental Management,* **30,** 175–189.

Walter, H.S. (1990). Small viable population: The Red-tailed Hawk of Socorro Island. *Conservation Biology*, **4**, 441–443.

Walters, C. (1997). Adaptive policy design: Thinking at large spatial scales. pp. 386–394. In: *Wildlife and Landscape Ecology. Effects of Pattern and Scale.* Bissonette, J. (Ed). Springer-Verlag, New York.

Walters, C.J. (1986). *Adaptive Management of Renewable Resources.* Macmillan Publishing Company, New York.

Walters, C.J. and Holling, C.S. (1990). Large scale management experiments and learning by doing. *Ecology*, **71**, 2060–2068.

Walters, J.R., Ford, H.A. and Cooper, C.B. (1999). The ecological basis of sensitivity of brown treecreepers to habitat fragmentation: a preliminary assessment. *Biological Conservation*, **90**, 13–20.

Walton, D. (1997). *Appeal to Expert Opinion: Arguments from Authority.* Pennsylvania State University Press, Pennsylvania.

Walton, D.W., Busby, J.R. and Woodside, D.P. (1992). Recorded and predicted distribution of the golden-tipped bat *Phoniscus papuensis* (Dobson 1878) in Australia. *Australian Zoology*, **28**, 52–54.

Ward, D.J., Lamont, B.B. and Burrows, C.L. (2001). Grasstrees reveal contrasting fire regimes in eucalypt forest before and after European settlement of south-western Australia. *Forest Ecology and Management*, **150**, 323–329.

Ward, T.J., Heinemann, D. and Evans, N. (2001). *The role of marine reserves as fisheries management tools: A review of concepts, evidence and international experience.* Bureau of Rural Sciences, Canberra.

Wardell-Johnson, C. and Nichols, O. (1991). Forest wildlife and habitat management in southwestern Australia: knowledge, research and direction. pp. 161–192. In: *Conservation of Australia's Forest Fauna.* Lunney, D. (Ed). Surrey Beatty and Sons, Chipping Norton.

Wardell-Johnson, G. and Horowitz P. (1996). Conserving biodiversity and the recognition of heterogeneity in ancient landscapes: a case study from south-western Australia. *Forest Ecology and Management*, **85**, 219–238.

Wardell-Johnson, G. and Roberts, J.D. (1993). Biogeographic barriers in a subdued landscape: The distribution of the *Geocrinia rosea* (*Anura Myobatrachidae*) complex in south-western Australia. *Journal of Biogeography*, **20(1)**, 95–108.

Warnken, J. and Buckley, R. (1998). Scientific quality of tourism environmental impact assessment. *Journal of Applied Ecology*, **35**, 1–8.

Warwick, R.M. (1993). Environmental impact studies on marine communities: pragmatical considerations. *Australian Journal of Ecology*, **18**, 63–80.

Wasson, R.J., White, I., Mackey, B. and Fleming, M. (1999). *The Jabiluka Project. Environmental Issues that Threaten Kakadu National Park.* Submission to UNESCO World Heritage Committee Delegation to Australia.

Watson, D.M. (2001). Mistletoe – A keystone resource in forests and woodlands worldwide. *Annual Review of Ecology and Systematics*, **32**, 219–249.

Watson, D.M. (2002). Effects of mistletoe on diversity: a case-study from southern New South Wales. *Emu*, **102**, 275–281.

Watson, D.M., Croft, D.B. and Crozier, R.H. (1992). Paternity exclusion and dominance in captive Red-necked Wallabies, *Macropus rufogriseus* (Marsupialia: Macropodidae). *Australian Mammalogy*, **15**, 31–36.

Watson, J., Freudenberger, D. and Paull, D. (2001). An assessment of the focal-species approach for conserving birds in variegated landscapes in southeastern Australia. *Conservation Biology*, **15**, 1364–1373.

Watts, D. (1987). *Tasmanian Mammals, A Field Guide.* Tasmanian Conservation Trust, Hobart.

Wayne, P.K. and Motin, P.A. (2004). Conservation genetics and the new molecular age. *Frontiers in Ecology and the Environment*, **2**, 89–97.

WCED (World Commission on Environment and Development (1987).) *Our Common Future.* Oxford University Press, Oxford.

WCMC (World Conservation Monitoring Centre) (1992). *Global Diversity: Status of the Earth's Living Resources.* Chapman and Hall, London.

WCMC (World Conservation Monitoring Centre) (1998). *WCMC Protected Areas Database.* [online] URL: http://www.wcmc.org.uk/protected_areas/

Weaver, J.C. (1995). Indicator species and the scale of observation. *Conservation Biology*, **9**, 939–942.

Webb, G.J. (2002). Conservation and sustainable use of wildlife – an evolving concept. *Pacific Conservation Biology*, **8**, 12–26.

Webster, R. (1988). *A survey of breeding, distribution and habitat requirements of the Superb Parrot.* Australian Parks and Wildlife Service Report Series No. 12. Australian Parks and Wildlife Service, Canberra.

Webster, R. and Ahern, L. (1992). *Management for Conservation of the Superb Parrot (*Polytelis swainsonii*) in New South Wales and Victoria.* Department of Conservation and Natural Resources.

Weed Management Society of South Australia (2003). *Olive Position Paper.* Weed Management Society of South Australia Inc., Adelaide.

Weeds Australia (2000). *Noxious weed list*. Weeds Australia. [online] URL: http://www.weeds.org.au/noxious.htm

Wegner, J. (1994). *Ecological landscape variables for monitoring and management of forest biodiversity in Canada*. GM Group, Ecological Land Management, Manotick, Canada. Report to Canadian Ministry of Natural Resources.

Wegner, J.F. and Merriam, G. (1979). Movements by birds and small mammals between a wood and adjoining farmland habitats. *Journal of Applied Ecology*, **16**, 349–357.

Weisberg, S. (1980). *Applied Linear Regression*. John Wiley and Sons, New York.

Weishampel, J.F., Shugart, H.H. and Westman, W.E. (1997). Phenetic variation in insular populations of a rainforest centipede. pp. 111–123. In: *Tropical Forest Remnants. Ecology, Mangement and Conservation*. Laurance, W.F. and Bierregaard, R.O. (Eds). The University of Chicago Press, Chicago.

Weiss, P.W. (1986). The biology of Australian weeds, *Chrysanthemoides monilifera*. *Journal of the Australian Institute of Agricultural Science*, **52**, 127–134.

Weller, M.W. (1995). Use of two waterbird guilds as evaluation tools for Kissimmee River restoration. *Restoration Ecology*, **3**, 211–224.

Welsh, A., Cunningham, R.B. and Lindenmayer, D.B. (1996). Modelling the abundance of rare species – statistical models for counts with extra zeros. *Ecological Modelling*, **88**, 297–308.

Welsh, A.H., Cunningham, R.B. and Chambers, R.L. (2000). Methodology for estimating the abundance of rare animals: seabird nesting on North East Herald Cay. *Biometrics*, **56**, 22–30.

Welsh, C.J. and Healy, W.M. (1993). Effects of even-aged timber management on bird species diversity and composition in northern hardwoods of New Hampshire. *Wildlife Society Bulletin*, **21**, 143–154.

Wescott, G. (2001). Integrated coastal zone management in the Australian states. pp. 133–147. In: *Integrated Oceans Management: Issues in Implementing Australia's Oceans Policy*. Haward, M. (Ed). Cooperative Research Centre for Antarctica and the Southern Ocean, Hobart. Research Report No. 26.

West, P.W. (1979). Estimation of height, bark thickness and plot volume in regrowth eucalypt forest. *Australian Forest Research*, **9**, 295–308.

Weste, G. and Ashton, D.H. (1994). Regeneration and survival of indigenous and dry sclerophyll species in the Brisbane Ranges, Victoria, after a *Phytophthora cinnamomi* epidemic. *Australian Journal of Botany*, **42**, 239–253.

Weste, G. and Marks, G.C. (1987). The biology of *Phytophthora cinnamomi* in Australasian forests. *Annual Review of Phytopathology*, **25**, 207–229.

Weste, G. and Vithanage, K. (1978). Effect of *Phytophthora cinnamomi* on microbial populations associated with the roots of forest flora. *Australian Journal of Botany*, **25**, 153–167.

Western, D. and Gichohi, A. (1993). Segregation effects and the impoverishment of savanna parks: the case for ecosystem viability analysis. *African Journal of Ecology*, **31**, 269–281.

Westoby, J. (1989). *Introduction to World Forestry*. Blackwell, Oxford.

Westoby, M. (1988). Comparing Australian ecosystems to those elsewhere. *BioScience*, **38**, 549–556.

Whelan, R., Rodgerson, L., Dickman, C.R. and Sutherland, E.F. (2002). Critical life cycles of plants and animals: developing a process-based understanding of population changes in fire-prone landscapes. pp. 94–124. In: *Flammable Australia. The Fire Regimes and Biodiversity of a Continent*. Bradstock, R.A., Williams, J.E. and Gill, A.M. (Eds). Cambridge University Press, Melbourne.

Whelan, R.J. (1994). Conserving nature's ways. *Today's Life Science*, **June 1994**, 20–25.

Whelan, R.J. (1995). *The Ecology of Fire*. Cambridge Studies in Ecology. Cambridge University Press, Cambridge.

Whelan, R.J. (2002). Managing fire regimes for conservation and property: an Australian response. *Conservation Biology*, **16**, 1659–1661.

White, I. (2001). *Safeguarding Environmental Conditions for Oyster Cultivation in NSW*. Healthy Rivers Commission of New South Wales. Occasional Paper. OCP 104. August 2001.

White, P.S. and Pickett, S.T.A. (1985). Natural disturbance and patch dynamics: An introduction. pp. 3–13. In: *The Ecology of Natural Disturbance and Patch Dynamics*. Pickett, S.T.A. and White, P.S. (Eds). Academic Press, Orlando, U.S.A.

Whitehead, P. and Dawson, T. (2000). Let them eat grass. *Nature Australia*, **Autumn 2000**, 47–55.

Whitely, G.P. (1974). From First Fleet to El Torito. *Australian Natural History*, **18**, 38–44.

Whitely, P.L. (1989). Taxoplasmosis transmitted by feral cats may affect the distribution and abundance of marsupials. pp. 10. In: *Proceedings of the Second Conference of the Australasian Wildlife Management Society (Abstracts)*, Melbourne, 6–8 December 1989.

Whittaker, R.H. (1954a). The ecology of serpentine soils. I. Introduction. *Ecology*, **35**, 258–259.

Whittaker, R.H. (1954b). The ecology of serpentine soils. IV. The vegetational response to serpentine soils. *Ecology,* **35**, 275–288.

Whittaker, R.H. (1960). Vegetation of the Siskiyou Mountains, Oregon and California. *Ecological Monographs,* **30**, 279–338.

Whittaker, R.H., Levin, S.A. and Root, R.B. (1973). Niche, habitat and ecotope. *American Naturalist,* **107**, 321–328.

Whitten, S., Bennett, J., Moss, W., Handley, M. and Phillips, W. (2002). *Incentive Measures for Conserving Freshwater Ecosystems.* Environment Australia, Canberra.

Wiens, J. (1989). *The Ecology of Bird Communities.* Cambridge University Press, Cambridge, New York.

Wiens, J. (1994). Habitat fragmentation: island vs landscape perspectives on bird conservation. *Ibis,* **137**, S97–S104.

Wiens, J. (1995). Landscape mosaics and ecological theory. pp. 1–26. In: *Landscape Mosaics and Ecological Processes.* Hansson, L., Fahrig, L. and Merriam, G. (Eds). Chapman and Hall, London.

Wiens, J. (1997a). Wildlife in patchy environments: metapopulations, mosaics and management. pp. 53–84. In: *Metapopulations and Wildlife Conservation.* McCullogh, D.R. (Ed). Island Press, Covelo.

Wiens, J. (1997b). Metapopulation dynamics and landscape ecology. pp. 43–68. In: *Metapopulation Biology – Ecology, Genetics and Evolution.* Hanski, I. and Gilpin, M. (Eds). Academic Press, San Diego.

Wiens, J. (1999). The science and practice of landscape ecology. pp. 37–383. In: *Landscape Ecological Analysis. Issues and Applications.* Klopatek, J.M. and Gardner, R.H. (Eds). Springer, New York.

Wiens, J.A., Schooley, R.L. and Weekes, R.D. (1997). Patchy landscapes and animal movements: do beetles percolate? *Oikos,* **78**, 257–264.

Wilcove, D.S. (1985). Nest predation in forest tracts and the decline of migratory songbirds. *Ecology,* **66**, 1211–1214.

Wilcove, D.S. (1989). Protecting biodiversity in multiple-use lands: lessons from the US Forest Service. *Trends of Evolution and Ecology,* **4**, 385–388.

Wilcove, D.S., McLellen, C.H. and Dobson, A.P. (1986). Habitat fragmentation in the temperate zone. pp. 237–256. In: *Conservation Biology: the Science of Scarcity and Diversity.* Soulé, M.E. (Ed). Sinauer, Sunderland.

Wilcove, D.S., Rothstein, D., Dubow, J., Phillips, A. and Losos, E. (1998). Quantifying threats to imperilled species in the United States. *Bioscience,* **48**, 607–615.

Wilcox, B.A. and Murphy, D.D. (1984). Conservation strategy: the effects of fragmentation on extinction. *American Naturalist,* **12**, 879–887.

Wiles, G.J., Bart, J., Beck, R.E. and Aguon, C.F. (2003). Impacts of the Brown Tree Snake: patterns of decline and species persistence in Guam's avifauna. *Conservation Biology,* **17**, 1350–1360.

Wilkinson, M.J., Elliott, L.J., Allainguillaume, J., Shaw, M.W., Norris, C., Welters, R., Alexander, M., Sweet, J. and Mason, D.C. (2003). Hybridization between *Brassica napus* and *B. rapa* on a national scale in the United Kingdom. *Science,* **302**, 457–459.

Williams, C.B. (1944). Some applications of the logarithmic series and the index of diversity to ecological problems. *Journal of Ecology,* **32**, 1–44.

Williams, J. (1991). Biogeographic patterns of three sub-alpine Eucalypts in south-east Australia with special reference to *Eucalyptus pauciflora* Sieb. ex Spreng. *Journal of Biogeography,* **8**, 223–230.

Williams, J., Bui, E.N., Gardner, E.A., Littleboy, M. and Probert, M.E. (1997). Tree clearing and dryland salinity hazard in the upper Burdekin catchment of north Queensland. *Australian Journal of Soil Research,* **35**, 785–801.

Williams, J.C., Revelle, C.S. and Levin, S.A. (2004). Using mathematical optimisation models to design reserves. *Frontiers in Ecology and Environment,* **2**, 98–105.

Williams, J.E. and Gill, A.M. (1995). The impact of fire regimes on native forests in eastern New South Wales. *Environmental Heritage Monograph Serices* **No. 2**, 1–68. New South Wales National Parks and Wildlife Service, Sydney.

Williams, J.E. and West, C.J. (2000). Environmental weeds in Australia and New Zealand: issues and approaches to management. *Austral Ecology,* **25**, 425–444.

Williams, K., Parer, I., Coman, B., Burley, J. and Braysher, M. (1995). *Managing Vertebrate Pest Rabbits.* Bureau of Resource Science, CSIRO Division of Wildlife and Ecology, Australian Government Publishing Service, Canberra.

Williams, M.R., Abbott, I., Liddelow, G.L., Vellios, C., Wheeler, I.B. and Mellican, A.E. (2001). Recovery of bird populations after clearfelling of tall open eucalypt forest in Western Australia. *Journal of Applied Ecology,* **38**, 910–920.

Williams, N.S.G., McDonnell, M.J. and Seager, E.J. (2004). Factors influencing the loss of an endangered ecosystem in an urbanising landscape: a case study of native grasslands from Melbourne, Australia. *Landscape and Urban Planning* (in press).

Williams, P., Gibbons, D., Margules, C., Rebelo, A., Humphries, C. and Pressey, R. (1996). A comparison of richness hotspots, rarity hotspots, and complementary areas for conserving diversity of British birds. *Conservation Biology*, **10**, 155–174.

Williams, P.H. and Gaston, K. (1994). Measuring more of biodiversity: can higher taxon richness predict wholesale species richness? *Biological Conservation*, **67**, 211–217.

Williams, P.H. and Humphries, C.J. (1993). Biodiversity, taxonomic relatedness and endemism in conservation. In: *Systematics and Conservation Evaluation.* Forey, P.L., Humphries, C.J. and Vane-Wright, R.I. (Eds). Oxford University Press, Oxford.

Williams, P.H., Humphreys, C.J. and Vane-Wright, R.I. (1991). Measuring biodiversity: taxonomic relatedness for conservation priorities. *Australian Systematic Botany*, **4**, 665–679.

Williams, R.J., Griffiths, A.D. and Allan, G.E. (2002). Fire regimes and biodiversity in the savannas of northern Australia. pp. 281–304 In: *Flammable Australia. The Fire Regimes and Biodiversity of a Continent.* Bradstock, R.A., Williams, J.E. and Gill, A.M. (Eds). Cambridge University Press, Melbourne.

Williams, S.E. and Hero, J.M. (2001). Multiple determinants of Australian tropical frog biodiversity. *Biological Conservation*, **98**,1–10.

Williams, W.D. (1987). Salinisation of rivers and streams: An important environmental hazard. *Ambio*, **16**, 180–185.

Williamson, D., Russ, G. and Ayling, A. (2004). No-take marine reserves increase abundance and biomass of reef fish on inshore fringing reefs of the Great barrier Reef. *Environmental Conservation*, **31**, 149–159.

Williamson, I. (1999). Competition between the larvae of the introduced Cane Toad *Bufo marinus* (Anura: Bufonidae) and native anurans from the Darling Downs area of southern Queensland. *Australian Journal of Ecology*, **24**, 636–643.

Wilson, A.M. and Lindenmayer, D.B. (1996). *The role of wildlife corridors in the conservation of biodiversity in multi-use landscapes.* Centre for Resource & Environmental Studies, Greening Australia and The Australian Nature Conservation Agency.

Wilson, B.A. and Clark, T.W. (1995). The Victorian Flora and Fauna Guarantee Act 1988: A five-year review of its implementation. pp. 87–103. In: *People and Nature Conservation. Perspectives on Private Land Use and Endangered Species Recovery.* Bennett, A., Backhouse, G. and Clark, T. (Eds). Surrey Beatty, Chipping Norton.

Wilson, B.A., Laidlaw, W.S. and Newell, G.R. (2002). The impact of *Phytophthora cinnamomi* on mammals in south-eastern Australia. *Australian Mammal Society Newsletter*, **October 2002**, 29.

Wilson, B.A., Lewis, A. and Aberton, J. (2003). Spatial model for predicting the presence of Cinnamon Fungus (*Phytophthora cinnamomi*) in sclerophyll vegetation communities in south-eastern Australia. *Austral Ecology*, **28**, 108–115.

Wilson, E.O. (1988). The current state of biological diversity. pp. 3–18. In: *Biodiversity.* Wilson, E.O. and Peter, F.M. (Eds). National Academy Press, Washington DC.

Wilson, E.O. (1989). Threats to biodiversity. *Scientific American*, **September 1990**, 108–116.

Wilson, E.O. (1992). *The Diversity of Life.* Harvard University Press, Cambridge, Massachusetts.

Wilson, G., Dexter, N., O'Brien, P. and Bomford, M. (1992) *Pest Animals in Australia. A Survey of Introduced Wild Mammals.* Bureau of Rural Resources and Kangaroo Press, Kenthurst.

Wilson, J.B. (1990). Mechanisms of species coexistence: twelve explanations for Hutchinson's "Paradox of the Plankton": evidence from New Zealand plant communities. *New Zealand Journal of Ecology*, **13**, 17–42.

Wilson, J.B. (1994). The "intermediate disturbance hypothesis" of species coexistence is based on patch dynamics. *New Zealand Journal of Ecology*, **18**, 176–181.

Wilson, M.H., Kepler, C.B., Synder, N., Derrickosn, S.R., Dein, F.J., Wiley, J.W., Wunderle, J.M., Lugo, A.E., Graham, D.L. and Toone, W.D. (1994). Puerto Rican Parrots and the potential limitations of the metapopulation approach to species conservation. *Conservation Biology*, **8**, 114–123.

Wilson, S.M. (2000). *The Costs of Dryland Salinity to Non-agricultural Stakeholders in Selected Victorian and New South Wales Catchments – Interim Report.* Part 1. Report to the Murray Darling Basin Commission, Canberra.

Winter, J.W. (1979). The status of Australian Phalangeridae, Petauridae, Burramyidae, Tarsipedidae and the Koala. pp. 45–59. In: *Proceedings of the Centennial Symposium on The status of Endangered Wildlife.* Tyler, M.J. (Ed). Adelaide Royal Zoological Society, South Australia.

Winter, W.H., Cameron, A.G., Reid, R., Stockwell, T.G. and Page, M.C. (1985). Improved pasture plants. In: *Agro-research for the Semi-arid Tropics: North-west Australia.* Muchow, R.C. (Ed). University of Queensland Press, Brisbane.

Wipf, S., Rixen, C., Fischer, M., Schmid, B. and Stoeckili, V. (2005). Effects of ski piste preparation on alpine vegetation. *Journal of Applied Ecology*, **42**, 306–316.

Wirgin, I.I. and Waldman, J.R. (1994). What DNA can do for you. *Fisheries*, **19**, 16–27.

With, K.A. (1997). The application of neutral landscape models in conservation biology. *Conservation Biology*, **11**, 1069–1080.

With, K.A. (1999). Is landscape connectivity necessary and sufficient for wildlife management? pp. 97–115. In: *Forest Wildlife and Fragmentation. Management Implications*. Rochelle, J., Lehmann, L.A. and Wisniewski, J. (Eds). Brill, Leiden, Germany.

With, K.A. and Crist, T.O. (1995). Critical thresholds in species' responses to landscape structure. *Ecology*, **76**, 2446–2459.

With, K.A. and King, A.W. (1999). Extinction thresholds for species in fractal landscapes. *Conservation Biology*, **13**, 314–326.

Witham, T.G., Morrow, P.A. and Potto, B.M. (1991). Conservation of hybrid plants. *Science*, **254**, 779–780.

Withler, F. (1982). Transplanting Pacific Salmon. *Canadian Technical Report on Fisheries and Aquatic Science*, **1079**, 1–27.

Witting, L. and Loeschcke, V. (1995). The optimization of biodiversity conservation. *Biological Conservation*, **71**, 205–207.

Woinarski, J.C. and Recher, H.F. (1997). Impact and response: a review of the effects of fire on the Australian avifauna. *Pacific Conservation Biology*, **3**, 183–205.

Woinarski, J.C. and Sykes, B.J. (1983). Decline and extinction of the Helmeted Honeyeater at Cardinia Creek. *Biological Conservation*, **27**, 7–21.

Woinarski, J.C., Whitehead, P.J., Bowman, D. and Russell-Smith, J. (1992). Conservation of mobile species in a variable environment: the problem of reserve design in the Northern Territory, Australia. *Global Ecology and Biogeography Letters*, **2**, 1–10.

Woinarski, J.C.Z. (1989). The vertebrate fauna of broombush *Melaleuca uncinata* vegetation in north-western Victoria, with reference to effects of broombush harvesting. *Australian Wildlife Research*, **16**, 217–238.

Woinarski, J.C.Z. (1992). Biogeography and conservation of reptiles, mammals and birds across north-western Australia: An inventory and base for planning an ecological reserve system. *Wildlife Research*, **19**, 665–691.

Woinarski, J.C.Z. (1999). Fire and Australian birds: A review. pp. 55–112. In: *Australia's Biodiversity – Responses to Fire*. Biodiversity Technical Paper No. 1. Gill, A.M., Woinarski, J.C.Z. and York, A. (Eds). Environment Australia, Canberra.

Woinarski, J.C.Z. and Tidemann, S.C. (1991). The bird fauna of a deciduous woodland in the wet-dry tropics of Northern Australia. *Wildlife Research*, **18**, 479–500.

Woinarski, J.C.Z., Price, O. and Faith, D.P. (1996). Application of a taxon priority system for conservation planning by selecting areas which are most distinct from environments already reserved. *Biological Conservation*, **76**, 147–159.

Woinarski, J.C.Z., Rilser, J. and Kean, L. (2004). Response of vegetation and vertebrate fauna to 23 years of fire exclusion ina tropical *Eucalyptus* open forest, Northern Territory, Australia. *Austral Ecology*, **29**, 156–176.

Woinarski, J.M. and Cullen, J.C. (1984). Distribution of invertebrates on foliage in forests in south-eastern Australia. *Australian Journal of Ecology*, **9**, 207–232.

Wojcik, D.P., Allen, C.R., Brenner, R.J., Forys, E.A., Jouvenaz, D.P. and Lutz, R.S. (2001). Red Imported Fire Ants: impact on biodiversity. *American Entomologist*, **47**, 16–21.

Wolf, C.M., Griffith, B., Reed, C. and Temple, S.A. (1996). Avian and mammalian translocations: update and reanalysis of 1987 survey data. *Conservation Biology*, **10**, 1142–1154.

Wolfenbarger, D.O. (1946). Dispersion of small organisms. *American Midland Naturalist*, **35**, 1–152.

Wolff, J.O., Schauber, E.M. and Edge, W.D. (1997). Effects of habitat loss and fragmentation in the behaviour and demography of Gray-tailed Voles. *Conservation Biology*, **11**, 945–956.

Wolseley, P.A. and Aguirre-Hudson, B. (1991). Lichens as indicators of environmental change in the tropical forests of Thailand. *Global Ecology and Biogeography Letters*, **1**, 170–175.

Wood, G.W. (1991). Owl conservation strategy flawed. *Journal of Forestry*, **89**, 39–41.

Wood, M. and Allison, B. (2000). *National Plantation Inventory, Tabular Report*. Bureau of Rural Sciences, Canberra.

Wood, M.S., Stephens, N.C., Allison, B.K. and Howell, C.I. (2001). *Plantations of Australia – A Report from the National Plantation Inventory and the National Farm Forest Inventory.* National Forest Inventory, Bureau of Rural Sciences, Canberra.

Wood, P.J. and Burley, J. (1983). Ex situ conservation stands. *Silvicultura*, **8**, 158–160.

Woodgate, P.W., Peel, B.D., Coram, J.E., Farrell, S.J., Ritman, K.T. and Lewis, A. (1996). Old-growth forest studies in Victoria, Australia. Concepts and principles. *Forest Ecology and Management*, **85**, 79–84.

Woodgate, P.W., Peel, W.D., Ritman, K.T., Coram, J.E., Brady, A., Rule, A.J. and Banks, J.C.G. (1994). *A Study of the Old-growth Forests of East Gippsland.* Department of Conservation and Natural Resources, Melbourne.

Woods, L. (1984). *Land Degradation in Australia.* Australian Government Publishing Service, Canberra.

Woodward, F.I. (1987). *Climate and Plant Distribution.* Cambridge University Press, Cambridge.

Woodwell, G.M. and Rebuck, A.L. (1967). Effects of chronic gamma radiation on the structure and diversity of an Oak-Pine forest. *Ecological Monographs,* **37**, 53–69.

World Commission on Forests and Sustainable Development (1999). *Our Forests Our Future.* Report of the World Commission on Forests and Sustainable Development. Cambridge University Press, Cambridge, England.

World Resources Institute (1992). *World Resources 1992–1993.* Oxford University Press, New York.

World Resources Institute (1997). *The Last Frontier Forests: Ecosystems and Economies on the Edge.* Bryant, D., Nielsen, D. and Tangley, L. (Eds). World Resources Institute, New York.

World Resources Institute (1999). *World Energy Consumption.* World Resources Institute, New York.

World Wide Fund for Nature (2002). *Bukit Barisan Seletan. Sumatran Rhinos on the Edge.* World Wide Fund for Nature Report, 2002.

Worm, B., Lotze, H.K. and Myers, R.A. (2003). Predator diversity hotspots in the blue ocean. *Proceedings of the National Academy of Science*, **100**, 9884–9888.

Worthington-Wilmer, J.M., Melzer, A., Carrick, F. and Moritz, C. (1993). Low genetic diversity and inbreeding depression in Queensland Koalas. *Wildlife Research,* **20**, 177–188.

Wright, D.H. (1991). Correlations between incidence and abundance are expected by chance. *Journal of Biogeography,* **18**, 463–466.

Wright, D.H., Patterson, B.D., Mikkelson, G.M., Cutler, A. and Atmar, W. (1998). A comparative analysis of nested subset patterns of species composition. *Oecologia*, **113**, 1–20.

Wright, R.A.D. and Nunn, R.M. (1996). Regulatory approaches to environmental management. pp. 645–654. In: *Environmental Management in the Australian Minerals and Energy Industries.* Mulligan, D.R. (Ed). University of New South Wales Press and Energy Environment Foundation, Sydney.

Wylie, F.R. and Landsberg, J. (1987). The impact of tree decline on remnant woodlots on farms. pp. 331–332. In: *Nature Conservation: The Role of Remnants of Vegetation.* Saunders, D.A., Arnold, G.W., Burbidge, A.A. and Hopkins, A.J. (Eds). Surrey Beatty, Chipping Norton.

Yaffee, S.L. (1994). *The Wisdom of the Spotted Owl. Policy Lessons for a New Century.* Island Press, Washington, DC.

Yahner, R.H. (1983). Seasonal dynamics, habitat relationships and management of avifauna in farmstead shelterbelts. *Journal of Wildlife Management,* **47**, 85–104.

Yahner, R.H. (1988). Changes in wildlife communities near edges. *Conservation Biology,* **2**, 333–339.

Yamada, K., Ansari, M., Harrington, R., Morgan, D. and Burgman, M. (2004). Sindh Ibex in Khirthar National Park, Pakistan. In: *Species Conservation and Management: Case Studies Using RAMAS GIS.* Akçakaya, H.R., Burgman, M., Kindvall, O., Wood, C.C., Sjogren-Gulve, P., Hatfield, J. and McCarthy, M. (Eds). Oxford University Press, New York.

Yanai, R.D. (1991). Soil solution phosphorus dynamics in a whole-tree-harvested northern hardwood forest. *Soil Science Society of America Journal,* **55**, 1746–1752.

Yates, C.J., Hobbs, R.J. and True, D.T. (2000). The distribution and status of woodlands in Western Australia. pp. 86–106. In: *Temperate Woodlands in Australia: Biology, Conservation, Management and Restoration.* Hobbs, R.J. and Yates, C. (Eds). Surrey Beatty and Sons, Chipping Norton.

Yen, A. and Butcher, R. (1997). *An Overview of the Conservation of Non-marine Invertebrates in Australia.* Environment Australia, Canberra.

Yen, A., Hinckley, S. and Lillywhite, P. (1999). *Bugs in the System. Wildlife in Box-Ironbark Forests. Linking Research and Biodiversity Management.* Department of Natural Resources and Environment, Bushcare and Land and Water Resources Research and Development Corporation.

Yencken, D. and Wilkinson, D. (2000). *Resetting the Compass: Australia's Journey Towards Sustainability.* CSIRO Publishing, Melbourne.

York, A. (1999). Long-term effects of repeated prescribed burning on forest invertebrates: Management implications for the conservation of biodiversity. pp. 181 –259. In: *Australia's Biodiversity – Responses to Fire. Plants, Birds and Invertebrates.* Gill, A.M., Woinarski, J.C. and York, A. (Eds). Environment Australia Biodiversity Technical Paper **1**, 1–266.

York, A. (2000). Long-term effects of frequent low-intensity burning on ant communities in coastal blackbutt forests of southeastern Australia. *Austral Ecology*, **25**, 83–98.

Young, A., Boyle, T. and Brown, A. (1996). The population genetic consequences of habitat fragmentation for plants. *Trends in Ecology and Evolution,* **11**, 413–418.

Young, M. and Wilson, A. (1995). When will it pay to farm kangaroos? pp. 106–109. In: *Conservation Through the Sustainable Use of Wildlife.* Grigg, G.C., Hale, P.T. and Lunney, D. (Eds). Centre for Conservation Biology, University of Queensland, Brisbane.

Zacharias, M.A. and Roff, J.C. (2001). Use of focal species in marine conservation and management: a review and critique. *Aquatic Conservation: Marine and Freshwater Ecosystems,* **11**, 59–76.

Zallar, S.H. (1980). *Soil Stabilisation and Revegetation Manual.* 3rd edition. Soil Conservation Authority, Victoria.

Zammit, C. (1988). Dynamics of resprouting in the lignotuberous shrub *Banksia oblongifolia. Australian Journal of Ecology,* **13**, 311–320.

Zanette, L., Doyle, P. and Tremont, S.M. (2000). Food shortage in small fragments: Evidence from an area- sensitive passerine. *Ecology,* **81**, 1654–1666.

Zaniewski, A.E., Lehmann, A. and Overton, J. (2002). Predicting species spatial distributions using presence-only data: a case study of New Zealand ferns. *Ecological Modelling,* **157**, 261–280.

Zann, L. and Kailola, P. (Eds) (1995). *State of the Marine Environment Report for Australia: Technical Annex 1 – The Marine Environment.* Department of Environment, Sport and Territories, Canberra.

Zeckhauser, R.J. and Viscusi, W.K. (1990). Risk within reason. *Science,* **248**, 559–564.

Zedler, J.B. (2003). Wetlands at your service: reducing impacts of agriculture at the watershed level. *Frontiers of Ecology and Environment,* **1**, 65–72.

Zimmerman, B.L. and Bierregaard, R.O. (1986). Relevance of equilibrium theory of island biogeography and species-area relations to conservation with a case study from Amazonia. *Journal of Biogeography,* **13**, 133–143.

Zobel, D.B., Roth, L.F. and Hawk G.M. (1985). *Ecology, Pathology, and Management of Port-Orford-cedar.* USDA Forest Service General Technical Report GTR-PNW-184, 161pp.

Zuidema, P.A., Sayer, J. and Dijkman, W. (1996). Forest fragmentation and biodiversity: the case for intermediate-sized reserves. *Environmental Conservation,* **2**, 290–297.

Zweig, M. and Campbell, G. (1983). Receiver-operating characteristic (ROC) polts: a fundamental evaluation tool in clinical medicine. *Clinical Chemistry,* **39**, 561–577.

Index